VERTEBRATES OF THE UNITED STATES

VERTEBRATES OF THE UNITED STATES
Second Edition

W. Frank Blair, *Professor of Zoology, University of Texas at Austin*
Albert P. Blair, *Professor of Zoology, University of Tulsa*
Pierce Brodkorb, *Professor of Zoology, University of Florida*
Fred R. Cagle, *Professor of Zoology and Vice-president, Tulane University*
George A. Moore, *Professor Emeritus of Zoology, Oklahoma State University*

McGraw-Hill Book Company
New York   St. Louis   San Francisco   Toronto   London   Sydney

VERTEBRATES OF THE UNITED STATES

ISBN 07-005591-2

678910 VHVH 76543

Our main purpose in revising this manual of vertebrates is to bring the taxonomic accounts up to date as nearly as possible by including the results of monographic revisions and the descriptions of new species that have occurred since publication of the first edition. Another purpose is, of course, to correct any errors in the first edition that have come to our attention. It is our hope and belief that the taxonomic keys are now better after several years of testing. The title of the book remains unchanged even though there are now fifty states, of which two, Alaska and Hawaii, are not covered by the treatments in this book. There seems no strong biological merit in including such disjunct geographic areas as are represented by these two states just because of their political inclusion in the United States.

We have continued to be relatively conservative with respect to changing long-established names where these changes, as proposed by various authors, represent no increase in the biological knowledge of the taxa but rest only on legalistic interpretations of the rules of zoological nomenclature.

The amount of change from the first edition in each section of this book reflects in general the amount of taxonomic activity with respect to the class or classes of vertebrates treated in that section. Not unexpectedly, the amount of change is greatest in Part Two, because the bony fishes remain the most poorly known group of vertebrates in the United States and hence are undergoing the greatest amount of change in their classification. With respect to this group, some decisions regarding taxa presently under investigation were necessary and may not agree with the thinking of some leading taxonomists. Furthermore, a policy of status quo in the mergence or separation of some taxa as effected without published reasons or without general acceptance has been maintained. Retention of the genus *Thoburnia* is based on advice of ichthyologists presently working with the catostomids. *Noturus* is retained as monotypic and distinct from *Schilbeodes*. The pygmy sunfishes are accorded family status because in practically all characteristics they show scant relationship to the centrarchids. On the other hand, the genus *Siphateles* is merged with *Gila* because of evidence presented by Bailey and Uyeno. Similarly, *Clinostomus* is merged with *Richardsonius*. V. D. Vladykov prefers to reinstate the genus *Cristivomer*, presently included in *Salvelinus*, but this change lacks general acceptance, at present. Euryhaline species of fishes are excluded because of limitations of space. The scientific and common names follow almost exclusively the second edition of Special Publication no. 2, 1960, American Fisheries Society. However, for the considerable number of species described since 1960, the common names utilized by their authors have been used without consultation with the Committee on Common Names.

Changes in the treatment of the amphibians (Part Three) mostly reflect improvements in the classification as the result of the increasing use of modern experimental methods to refine the knowledge of relationships. The use of these methods has resulted in the discovery of cryptic species and, on the other hand, in the merging of nominal species for which the stage of speciation is now clearly demonstrated. In Part Four, the principal change involves the inclusion of the marine turtles, which were not treated in the first edition. In Part Five, an extensive bibliography has been added. In Part Six, the changes have been relatively minor except as they reflect increased knowledge of distributions.

W. Frank Blair

Albert P. Blair

Pierce Brodkorb

Fred R. Cagle

George A. Moore

Many more persons than can be individually acknowledged made either direct or indirect contributions to this revision. We are especially indebted to several colleagues as follows:

Part Two (Fishes). Advice and counsel were sought from ichthyologists with interests in several troublesome groups. Reeve M. Bailey loaned specimens of *Cottus*, offered helpful suggestions, and sent reprints of very recent papers concerning cottids and percids. Robert R. Miller was most helpful with the western minnows, particularly the genus *Gila*. Manuscripts of important papers in press were generously supplied by Edward C. Raney. Mr. Chu-fa Tsai took data from cottids in Cornell collections to aid in the preparation of the key to the species of *Cottus*. Leslie W. Knapp presented a copy of his doctoral dissertation concerning the description of *Percina lenticula* and a most careful study of *Etheostoma caeruleum* and its subspecies. Mr. Morgan Sisk assisted in the refinement of several keys by using them to determine specimens from Georgia supplied by Walter Whitworth. R. H. Gibbs gave many helpful suggestions concerning the genus *Notropis*. Assistance in delineating the ranges of southeastern species of *Hybopsis* was given by Rudolph Miller. Donn E. Rosen was most helpful with advice concerning the families Amblyopsidae and Aphredoderidae and the phylogenetic sequence of several orders. Dr. Rosen also promptly supplied important revisionary papers as soon as they became available.

Thanks are especially due Bruce Collette, who supplied characters and even prepared keys to separate darter species that have received little or no attention since their original descriptions were published many years ago. Dr. Collette also supplied his then unpublished manuscript concerned with percid taxa. D. E. McAllister and V. D. Vladykov, Canadian ichthyologists, supplied much helpful information. Milton R. Curd helped with the keys.

Part Three (Amphibians). Advice on nomenclatural problems was given or gifts of specimens were made by: Kraig E. Adler, W. Frank Blair, Ronald A. Brandon, Anthony J. Gaudin, Coleman J. Goin, Richard Highton, William F. Pyburn, Francis L. Rose, and Robert C. Stebbins. Miss Eloise Janssen typed the manuscript.

Part Four (Reptiles). Helpful suggestions were made by Donald Tinkle, Robert G. Webb, Harold Dundee, and James Dobie.

Part Five (Birds). Thanks are due Robert W. McFarlane for his many suggestions. All new drawings are by Miss Linda Free and Miss Linda Wilson.

Part Six (Mammals). Thanks are due Michael Smith for information about the variation of *Peromyscus* taxa in Florida. The two new drawings are by William F. Martin.

CONTENTS

Part One

INTRODUCTION

W. F. Blair

VERTEBRATES possess a notochord and pharyngeal gills or pouches at some stage of development, and they have a dorsal, hollow, fluid-filled nerve cord. These characters are shared with some small marine animals, which, with the vertebrates, are classified in the phylum Chordata. These relatives of the vertebrates include 3 subphyla, which are often referred to collectively as protochordates:

Subphylum Hemichordata. Acorn worms. Marine, burrowing animals, with superficially worm-like body form. Body divided into proboscis, collar, and trunk. Dorsal, hollow nerve cord in collar. Diverticulum at base of proboscis thought to represent notochord. Gill slits in pharyngeal region.

Subphylum Cephalochordata. Lancelets. Small, free-swimming, translucent, marine animals, with fishlike body form. A well-developed notochord extending length of animal. Well-developed, dorsal, hollow central nervous system. Numerous gill slits. Muscular, digestive, and circulatory systems are simple prototypes of these systems in vertebrates.

Subphylum Urochordata. Tunicates. Small, marine, free-floating or sessile as adults. Some with free-swimming larval stage in which there is a well-developed notochord, a dorsal, hollow central nervous system, and rudimentary brain. Notochord and part of nervous system degenerate in adult.

The vertebrates are placed in the separate subphylum, Vertebrata. They differ from the protochordates in having a vertebral column of cartilage or bone, which supplements or replaces the notochord as the main support of the long axis of the body. They also have a cranium, or braincase, of cartilage or bone, or both, for support and protection of the brain and major special sense organs. No living protochordate may be seriously regarded as ancestral to the vertebrates, but the common ancestry is evident. The structure of lancelets is suggestive of a stage in the evolution of the ancestors of the vertebrates. These animals, along with the other protochordates, however, seem to have reached an impasse in evolution. The vertebrates, on the other hand, have progressed to a dominant place in the living world in part because of the advantages conferred by their light, internal, continually growing skeleton.

Vertebrate History

The earliest remains of vertebrates are known from Ordovician rocks, indicating that this group has been in existence for more than 400 million years. These earliest remains are mostly remnants of bony armor such as has been found in the jawless ostracoderms of Silurian and Devonian age. These earliest known vertebrates apparently lived in fresh water. The small group of living jawless fishes (Cyclostomata) appear to be a highly specialized (mostly parasitic) remnant of this primitive vertebrate stock. Much modification of body form, of organ systems, and of life habits has occurred during the long history of the vertebrates. Outstanding developments in the progressive evolution of the vertebrates have been (1) appearance of paired appendages, (2) appearance of jaws, (3) movement to land, and (4) development of homoiothermy.

PAIRED APPENDAGES. Unpaired fins were present in the ostracoderms as accessory organs of locomotion. Such structures have persisted through the fishlike vertebrates, and analogous structures have developed in terrestrial vertebrates which have reinvaded the aquatic environment. Paired fins were

Fig. 1-1. Acorn worm, a representative of the subphylum Hemichordata. (After Bateson, from Neal and Rand, *Chordate Anatomy,* The Blakiston Division, McGraw-Hill Book Company.)

Proboscis
Collar
Mouth   Gills
Balanoglossus

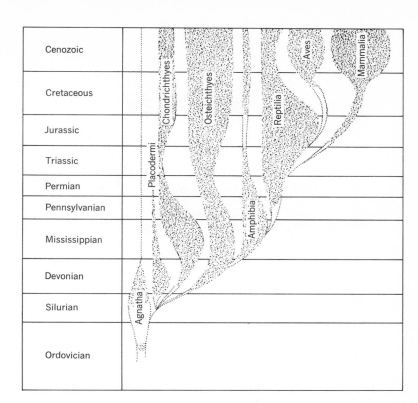

Fig. 1-2. Time-space relationships of the vertebrate classes. Thickness of the various branches gives a rough indication of comparative abundance. (After Romer, *Vertebrate Paleontology*, University of Chicago Press.)

developed as stabilizers. The beginnings are seen in the ostracoderms, which developed paired flaps, spines, or folds. In the class Placodermi, which flourished in the Silurian and Devonian and persisted into the Permian, various combinations of paired fins were present. Some placoderm fossils show evidence of as many as 7 pairs of fins. In the cartilaginous fishes (class Chondrichthyes) and bony fishes (class Osteichthyes), which apparently branched from placoderm ancestors in the Devonian or earlier, there is an anterior (pectoral) and a posterior (pelvic) pair except where the fins have been secondarily lost. While useful to fishes as stabilizers in the water, the paired appendages were vitally important as locomotor organs when the move came to land life.

JAWS. The development of jaws came early in vertebrate history, and the possession of such structures conferred important food-getting advantages on the aquatic vertebrates. More importantly, this development was an essential forerunner of the migration to land. The jawless ostracoderms were probably limited to the ingestion of small organisms and dead organic material which could be sucked into the pharynx and strained out by the gill mechanism. The jaws of placoderms and higher vertebrates permitted the grasping, holding, and ingestion, and in some cases the crushing or chopping up, of food organisms of enormously wider variety than was available to the jawless ostracoderms.

MOVE TO LAND. The movement of vertebrates to land life may have been a by-product of adaptations to permit the aquatic ancestors of the land vertebrates to survive in water. It has been theorized that the modification of the paired fins of crossopterygian fishes into tetrapod appendages came as an adaptation to permit overland movement of the primitive amphibians from a drying pool to the nearest pool still containing water. The change to land life began in the Devonian, which was apparently a time of periodic drought. The crossopterygian fishes possessed simple lungs, but only those which were able to move overland participated in the migration to land. The gradual change

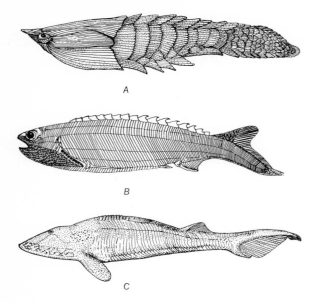

Fig. 1-3. Representative ostracoderms (jawless fishes). (*A* and *C* after Heintz, *B* after Kiaer; from Romer, *The Vertebrate Body,* W. B. Saunders Company.)

to land life could have resulted from selectional pressures when the primitive amphibians tended to linger longer and longer on land as they moved from pool to pool.

The migration of vertebrates to land life posed many adaptational problems. The new and previ-

Fig. 1-4. Representative placoderms (primitive jawed fishes). (From Romer, *The Vertebrate Body,* W. B. Saunders Company.)

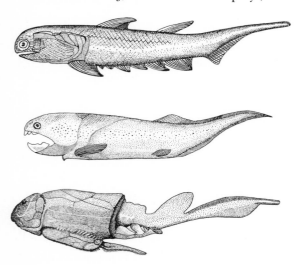

ously unexploited environment had many advantages. There was a plentiful and constant supply of oxygen. There was a solid substratum on which to move, and air was a much less dense medium in which to live and move about than had been water. The new environment also had formidable disadvantages. There was the constant danger of drying up, of losing a fatal amount of water from the body protoplasm. There were great ranges of temperature to be adapted to or avoided. The change to land life, consequently, resulted either directly or indirectly in marked changes in virtually every organ system of the vertebrate body.

The amphibians have not become well adapted to life on land. The thin stratum corneum of their epidermis and their numerous skin glands render them susceptible to water loss through the skin. Even the most land-adapted must seek out moist environments to replenish and to reduce their water loss, and many have remained closely associated with the aquatic environment. Their unshelled eggs must be deposited in water or in moist soil on land.

The reptiles, derived from amphibian ancestors in the Carboniferous, were the first vertebrates which were really well adapted for land life. The most significant advance was in the type of egg, which was now adapted to survive on land. The amniote egg of these and higher vertebrates is protected by a hard or tough porous shell. The embryo is surrounded by a series of membranes, of which the amnion provides the embryo with its own minute aquatic environment, the allantois serves as respiratory and excretory tissue, and the chorion encloses and protects the whole lot of embryonic structures. The newly emerged young are similar in general to the adult and are ready to meet the problems of survival in the terrestrial environment. The adult reptile is much better adapted than the adult amphibian for land life. Loss of most of the skin glands and thickening and cornification of the epidermis render reptiles highly resistant to water loss through the skin. Many excrete their metabolic wastes as crystalline uric acid rather than in urine as an adaptation that conserves body water.

The birds and mammals are offshoots of reptilian stocks and, like the reptiles, are well adapted to resist desiccation on land. The birds apparently branched from archosaurian reptiles in the Jurassic or possibly earlier. The mammals appear to have branched from synapsid reptiles in the Triassic. These groups have gone beyond the reptiles in adaptation to the terrestrial environment, as they have also adapted to control their body temperatures in the variable temperature environment on land.

HOMOIOTHERMY. Birds and mammals have acquired the ability to control their body temperatures within narrow limits and so maintain internal temperatures that are independent of the external environment. Among the bony fishes, it has been suggested that the tunas, which inhabit warm seas, may maintain constant body temperatures, but homoiothermy is in general an adaptation to the variable temperatures of the terrestrial environment.

It has been theorized that the giant Mesozoic reptiles were essentially homoiothermal in the sense that their great body mass in relation to their heat-radiating surfaces resulted in fairly stable internal temperatures. Unlike true homoiotherms, however, the dinosaurs did not possess effective insulative coverings. The great bulk of these reptiles was disadvantageous in that progressive warming of climate might have exceeded their tolerance of high temperatures, or progressive cooling might have exceeded their tolerance of low temperatures. The reptiles which have persisted to the present are mostly small, or at most, moderate-sized animals.

Fig. 1-5. Eryops, an extinct labyrinthodont amphibian. (After Gregory, *Evolution Emerging,* The Macmillan Company.)

Their internal temperatures remain close to the external temperatures, although they are capable of some control through behavioral mechanisms, and their activities are largely dictated by the external temperature environment.

The acquisition of homoiothermy has rendered birds and mammals relatively free of dependence on the external temperature. The constant maintenance of body temperatures at optimal temperatures for metabolic processes permits more sustained activity than is possible for poikilotherms. The higher intelligence of homoiotherms, culminating in the reasoning forebrain of man, is largely made possible by this constant maintenance of high metabolic activity. The ability to maintain body temperatures above that of the external environment has enabled the birds and mammals to spread into parts of the world which are closed to terrestrial poikilotherms because of sustained cold temperatures.

Heat conservation is effected by integumental structures, feathers in birds, hair in mammals, which trap an insulative coat of air next to the body surface. Mammals which have reinvaded the aquatic environment have tended to lose the hair and to replace it with an insulative layer of fat (blubber) beneath the skin, as in the whales. The principal mechanisms to cool the body below that of the environment are heat loss through a highly developed skin vascular system and through expired air. The presence of a well-developed skin vascular system is the chief evidence that tunas may be essentially homoiothermal fishes. The evaporation of water from the moist skin of amphibians may result in the cooling of the body below the temperature of the environment. Some mammals use this method by evaporating water which reaches the surface through sweat glands.

Vertebrate Characters

BODY FORM.   Most fishes are fusiform (torpedo-shaped), which permits the body to pass through the dense medium of water with a mimimum of resistance. The anteriorly tapering head passes gradually into the trunk with no constriction or neck, and the trunk narrows gradually into the caudal region. The greatest diameter is near the middle of the body. Various modifications have occurred. Many bottom-living fishes have the body flattened dorsoventrally; other fishes have the body laterally compressed. Some fishes have the body greatly elongated (anguilliform) and flexible. Larval amphibians have the fusiform body of generalized fishes. Adult aquatic salamanders do not differ far from it, and some are anguilliform. Mammals which have reinvaded the aquatic environment (e.g., whales) tend to be fusiform.

The change to land life and to locomotion in air, which offers relatively little resistance to movement, brought major changes in body form. The head became readily movable on the constricted and more or less elongated neck. Streamlining was lost, and the body became relatively compacted. The caudal region became progressively constricted in diameter, but it usually remained as a balancing organ. The bipedal method of locomotion appeared as a specialization in ancient reptiles, in birds, and in some lines of mammals, and brought changes in body form. Saltatorial (jumping) locomotion reached high development in modern anuran amphibians and brought additional foreshortening of the body, great development of the posterior appendages, and loss of the tail. Saltatorial mammals, such as the kangaroos and kangaroo rats, retained the tail as a well-developed balancer. Elongation comparable to that of the anguilliform fishes and reduction or loss of the limbs appeared in some amphibian and reptilian lines as an adaptation for burrowing.

Flying locomotion has appeared in each of the truly land-adapted classes of vertebrates. The flying reptiles became extinct, but this has become the principal method of locomotion in most birds and in bats among the mammals. The body of volant vertebrates tends to be foreshortened and relatively rigid, although the neck is long in birds. The contour feathers of birds are so arranged as to streamline these fast-flying vertebrates.

APPENDAGES.   The fins of fishes are typically thin webs of membranous tissue with an inner support of hardened tissue. These membranous structures serve to propel and stabilize the fish in its aquatic environment. With the move to land the unpaired fins were lost, and the paired fins became modified into limbs to support and move the vertebrate on land. In most living fishes the fin skeleton consists of a series of parallel rods of supporting tissue. The mostly extinct crossopterygian fishes, however, possessed a dichotomously branched fin skeleton that could be modified into the limb of a tetrapod by losing some of its elements, and muscular tissue extending into the base of the fin. The earliest known amphibians possessed a limb skeleton intermediate between a crossopterygian fin and the limb skeleton of the more advanced tetrapods.

The tetrapod limb differs from the fish fin in its definite segmentation into proximal, intermediate, and terminal parts, with often highly developed joints between the segments. The limb contains large amounts of muscular tissue, since its principal function is to support and move the body on a solid substratum. The posterior limbs are usually more highly developed than the anterior pair, as they support a larger part of the body weight. In generalized amphibians and reptiles the upper segment of the limb extends more or less laterally from the body, the feet are placed to the side of the body, and the

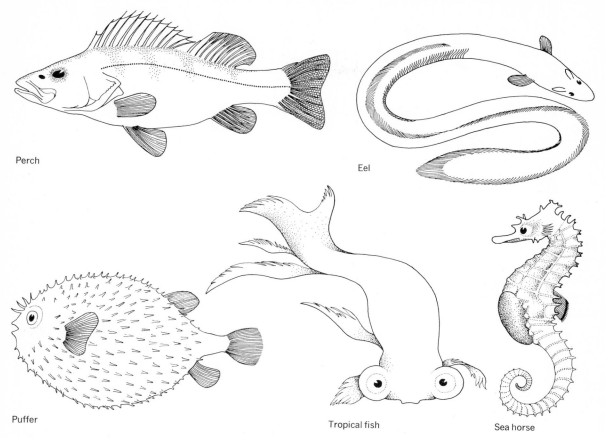

Perch

Eel

Puffer

Tropical fish

Sea horse

Fig. 1-6. **Teleost fishes, showing diversity of body form. (After Kent,** *Comparative Anatomy of the Vertebrates,* **The Blakiston Division, McGraw-Hill Book Company.)**

weight must be borne through the bent limb. In birds and mammals the legs have moved directly under the body, thus reducing the size of the limb required to support an equal amount of body weight. The basic number of digits is 5, but various reductions from this number have occurred in all classes of tetrapods. Enormous modifications of the limbs have occurred in the tetrapods in exploitation of the many ecological opportunities available on land. Drastic reduction in the number of digits tends to be associated with the running type of locomotion, as in various ancient reptiles, in ostriches among living birds, and in horses among living mammals.

The modification of the anterior pair of append-

ages into a wing (probably through intermediate stages as a gliding organ) permitted the development of flight. The ancient flying reptiles (pterosaurs) utilized an elongated finger to support the flight membrane. In birds, the hand elements are reduced and fused to serve as a strong base for attachment of the feathers which form the flight surface. Bats use several elongated fingers to support the wing membrane.

The return of various lines of tetrapods to the aquatic environment has resulted in modification of the tetrapod limbs into finlike structures, without, however, the loss of the internal tetrapod structure. This is seen in various lines of extinct reptiles and in sea turtles among living ones, in birds such

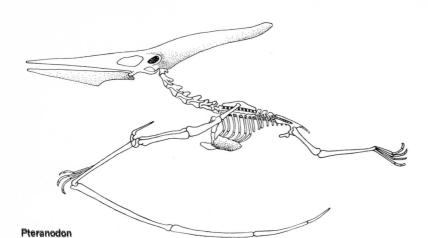

Pteranodon

Fig. 1-7. **Flying reptile (pterosaur) from Upper Cretaceous of Kansas. These had a wingspread of up to 22 ft. (After Eaton, from Gregory,** *Evolution Emerging,* **The Macmillan Company.)**

as the penguins, and in mammals such as the whales, seals, and manatees.

INTEGUMENT. The skin of vertebrates is an outer, protective layer which is readily separable from the underlying tissues. It has an inner layer, the dermis, and an outer layer, the epidermis. Major changes, particularly in the epidermis, have occurred as vertebrates adapted to life in water and later adapted to the new life on land.

The entire epidermis of fishes is made up of living cells. Numerous glands in the epidermis secrete a coating which retards passage of water through the skin, resists entrance of foreign materials, and reduces friction as the fish moves through the water. The protective function of the skin is augmented by scales of one kind or another in the great majority of fishes. Bony armor plates were present in the earliest known fishes (ostracoderms), and they were well developed in the extinct placoderms. Cartilaginous fishes have scales of a complex (placoid) type, and these became the teeth of these and higher vertebrates. Ancient bony fishes generally had an external armor of heavy interlocking cosmoid or ganoid scales; these are represented in living fishes by the ganoid scales of the garfishes. Most living bony fishes have very thin, bony scales which have evolved through reduction and simplification of the ancestral ganoid scale.

The move to land brought a subdivision of the epidermis into an inner layer of living cells (stratum germativum) and an outer layer of dead cornified cells (stratum corneum). The stratum corneum is thin in amphibians and relatively thick in the xeric-adapted reptiles, birds, and mammals, where it serves to retard water loss through the skin. The land vertebrates have developed various integumental structures as an adaptation to land life. Scales are found in modern amphibians only in the tropical, burrowing coecelians, in which they are vestigial and embedded, but many ancient amphibians were well covered with scales. The epidermal scales of reptiles serve in part to reduce water loss through the skin; in some cases (snakes) they aid in locomotion, and in others they may serve to protect the owner from aggression. The feathers of birds are modified reptilian scales which provide an insulative and contouring cover for the body and form the flight surfaces of the wings and tail fan.

Some land vertebrates have the epidermal scales underlain by bony plates to form a body armor. The turtles have been successful with this type of integumental structure through the span of time that saw the rise and virtual extinction of the giant reptiles. Armadillos among mammals have a similar body armor.

Cornified epidermal tissue (keratin) has been

variously modified into adaptive structures in the land vertebrates. The tips of the digits are protected by this material in the shape of claws, nails, or hooves. The horny beaks of various extinct reptiles, of living turtles, and of birds have the same origin. Many reptiles have spines of keratin, and this material enters into the formation of various kinds of horns which have appeared in mammals. The food strainers of baleen whales are similarly of keratin.

The colors of vertebrates are due to pigments in the skin and in its appendages. Colors are important in concealing animals from predators, in identifying them to others of their own and related kinds, and in controlling heat absorption. Most of the color of poikilotherms is produced by living color cells or chromatophores. Some of these animals are able to produce rapid and striking changes in color through rearrangement of the pigment granules in the cells and through changes in the relative position of the various kinds of chromatophores. Much of the color of birds and mammals is produced by pigments in the nonliving structures of the feathers or hair, so rapid color change is not possible. Various homoiotherms show seasonal color change, however, by molting the feathers or hair.

SKELETON. The central element of the skeleton is the vertebral column, which is made up of individual vertebrae. Each vertebra consists of a main element, the centrum, and various processes. The vertebral column of the fishlike vertebrates is characterized by its flexibility. The centra are typically concave at both ends (amphiocoelous). 2 regions of the vertebral column are differentiated in fishes. Neural arches project dorsally to protect the nerve cord. In cartilaginous fishes complete, closed hemal arches project ventrally in the caudal region, but these are open in the trunk region. In land vertebrates, the adjacent centra usually articulate by a ball-and-socket mechanism. When the concave face is anterior the vertebrae are procoelous; when it is posterior they are opisthocoelous. In some mammals the faces of the centra are flat (acoelous), and in the neck of birds the faces of the centra are typically saddle-shaped (heterocoelous). In the tetrapods, accessory articulations (pre- and postzygapophyses and sometimes others) add strength to the column. The vertebral column of tetrapods is differentiated into a neck or cervical region, trunk, sacral region,

and tail or caudal region. In some reptiles and in birds and mammals the trunk is divided into a rib-bearing thoracic region and a ribless lumbar region. 2 or more sacral vertebrae are often fused in tetrapods for better support of body weight through the attached pelvic girdle. This is carried to an extreme in birds with fusion of lumbars, sacrals, and some caudals.

The skull supports and protects the brain and the major special sense organs. This is a cartilaginous structure (chondrocranium) in jawless and cartilaginous fishes. In bony fishes and all higher vertebrates, bones of dermal origin invest the chondrocranium and tend to progressively obscure it. The vertebrate jaw is originally formed of elements of the first gill arch and is braced by elements of the second. In bony fishes and higher vertebrates dermal plates invest the old cartilage jaw and eventually replace it.

Teeth are topographically associated with the skull, although they are derived embryologically from integument and functionally are a part of the digestive system. The original function of teeth was probably simple grasping and holding of food organisms. Such teeth are simple and conical in shape and usually numerous. In fishes the teeth may be located on various bones of the palate and even on the tongue in addition to those along the margin of the jaw. Cutting or crushing teeth have developed in most vertebrate lines from cartilaginous fishes to mammals. The teeth of modern amphibians and reptiles are of the conical type; a few palatal teeth are retained in some members of both groups. The teeth of mammals are restricted to the margins of the jaw and are typically (but not always) differentiated into nipping teeth (incisors), killing teeth (canines), and grinding teeth (premolars and molars). Teeth have been lost completely by some representatives of most vertebrate lines. In turtles and birds the teeth have been replaced by a horny beak.

The limb girdles of the cartilaginous fishes are simple U-shaped or V-shaped structures for attachment of the fin bases. In the bony fishes dermal elements have been added to the anterior (pectoral) girdle. In the tetrapods the pelvic girdle becomes a triradiate structure, which attaches to the sacrum and which articulates with the femur at the juncture of its 3 main elements.

MUSCULATURE. The greatest bulk of the musculature of fishes is made up by chevron-shaped masses of muscles (myomeres) arranged segmentally along the long axis of the body. Coordinated contractions in the axial musculature provide the main locomotor mechanism of fishes. In the change to land life, the axial musculature decreases in bulk as the locomotor function is taken over by the appendages and their musculature. The original segmentation becomes obscured as the musculature of the limbs and limb girdles spreads out over the axial muscles. The appendicular musculature reaches enormous development, and the axial musculature is proportionately reduced, in flying vertebrates such as birds and bats.

BREATHING. In the gill breathing of fishes, dissolved oxygen is taken from the water as it bathes the gills in the pharyngeal region. The gills are highly vascularized tissues through which gas transfer is readily accomplished. In fishes, the gill tissues are arranged in platelike (lamellar) structures in the pharynx. In jawless fishes and in most cartilaginous fishes, each gill has its separate opening (gill slit) to the exterior. In chimaeras among the latter group and in bony fishes, there is a single opening, protected by a valvelike flap, the operculum. Larval amphibians and those adults which retain the gills have the gills located on the outside of the gill arch so that the gills are external. Such gills are in more danger of damage than internal ones, but they have the advantage that the oxygen-bearing water in contact with them can be changed more readily.

In the lung breathing of land vertebrates, oxygen from a mixture of gases (air) passes through moist, respiratory membranes deep within the body in the lungs. The moist skin of amphibians permits a considerable amount of integumental breathing, and land-living members of one large family of salamanders (Plethodontidae) utilize no other method as adults. Structures homologous to the lungs of land vertebrates first appear in bony fishes, and some of these (lungfishes, crossopterygians, garfishes, bowfin) use lung breathing as a supplement to gill breathing. In most living bony fishes, these structures (air sacs) serve as hydrostatic organs or may be lost.

CIRCULATION. In fishes, a simple tubular heart pumps blood anteriorly, where it passes through the aortic arches and through the capillaries of the gill tissues before being distributed throughout the body. With this diagrammatically simple system of circulation, the blood is oxygenated on each circuit through the body.

The change to lung breathing involved major changes in circulation, mainly to provide a separate circuit to the lungs. The heart became progressively divided into a right side which receives oxygen-depleted blood from the general circulation and pumps it to the lungs and a left side which receives oxygen-rich blood from the lungs and pumps it into the body circulation. This separation of the heart began in the bony fishes (lungfishes) and became complete in birds and mammals.

In the venous system, the postcardinal veins of fishes were progressively replaced by the postcava as the principal vessels returning blood from the body to the heart. A hepatic-portal system, because of its function in food absorption, is present throughout the vertebrates. A renal-portal system appeared in the jawed fishes as an apparent adaptation to ensure some blood going through the kidney on each circuit. This system became reduced in the lung-breathing vertebrates and disappeared in birds and mammals possibly because of the greater efficiency of higher blood pressures possible in lung breathers.

DIGESTION. Vertebrates, like other animals, get most of their food by eating parts of a green plant or by eating another animal which has, in turn, obtained its food from the same source. In fishes, the food is taken in along with the respiratory medium of water. Food and water are separated in the pharynx, with gill rakers usually functioning to strain out the food as the water passes out through the gill slits. In land vertebrates, many mucous glands are present in the mouth to lubricate the food, which is ingested in air as a medium.

The digestive tube is variously modified in the vertebrates, mostly in relation to kinds of food utilized and to problems of food absorption. The short esophagus of the fishes becomes elongated as the land vertebrates develop a neck and as the digestive organs move posteriorly with the development of lungs. The stomach is a conservative structure throughout the vertebrates; it does become highly specialized, however, in birds, where it serves a grinding function in the absence of teeth. The intestine is generally longer in herbivorous than in carnivorous vertebrates. Internally, the intestine is variously modified to slow the passage of food materials and to increase its absorptive surface. A typhlosole serves this function in jawless fishes, and a spiral valve serves in cartilaginous fishes and in primitive bony fishes. Pyloric caeca serve the same function in most bony fishes. 1 or 2 colic caeca may be present in reptiles, birds, and mammals. Villi serve the same function in mammals.

EXCRETION. Nitrogenous wastes and excess salts are mostly removed through the kidney in which the functional unit is the renal corpuscle. Excretion serves to maintain the proper concentrations of salts and other dissolved materials in the body fluids. Fresh-water fishes live in water which has lower salt concentrations than their own body fluids; they have large renal corpuscles and use water freely in excreting metabolic wastes. Marine fishes, on the other hand, live in water in which the salt concentrations are higher than in their own body fluids and are in danger of losing water to their environment. Bony fishes have solved this problem by reduction in size of the renal corpuscle and by excretion of salts through the gills. Cartilaginous fishes have solved it by retaining nitrogenous wastes in the body fluids in the form of urea, thereby raising the total osmotic pressure without increasing the salt concentration.

Land vertebrates also are faced with the problem of water conservation when they excrete metabolic wastes. The renal corpuscle is relatively small, and, in addition, there is much resorption of water in the kidney tubule. Many reptiles and most birds excrete crystalline uric acid; mammals excrete a solution of urea, although the solution may be very concentrated in desert inhabitants.

COORDINATION. The brain, as the most important center of nervous coordination, undergoes great changes in the course of vertebrate evolution. Various special sense organs are developed for coordination of the animal with its external environment.

The relative development of regions of the brain in the lower vertebrates is mostly related to which sense organs are most used in food getting. The forebrain (telencephalon) of jawless and cartilaginous fishes is highly developed in these groups, which find food mainly by olfactory stimuli. The midbrain (mesencephalon) is highly developed in most bony fishes and in birds in connection with the use of sight in food getting and in flying by visual reference in the case of the birds. From slight beginnings in the early tetrapods to culmination in the mammals, the cerebral hemispheres of the forebrain (formerly olfactory in function only) become increasingly important association centers of the brain. The function of most other centers of the brain is diverted to a new tissue (neopallium) which develops here in mammals. The reasoning center of the human brain is located here.

The eye is a remarkably constant structure through vertebrate history. It tends to be lost in those vertebrates which have adopted cave or subterranean life, where light is wanting. Unpaired eyes are present in jawless fishes and have persisted in some reptiles. A well-developed parietal eye is present in the "living fossil," *Sphenodon*, among the reptiles, and a vestigial one may be seen as a light spot beneath a median head scale in many kinds of lizards.

The ear is an organ of equilibrium in fishes, and in the land vertebrates it has developed the accessory function of sound reception. The lower aquatic vertebrates have systems of hair cells (neuromasts) located widely over the body surface and responsive to vibrations in the water, but for obvious reasons these have been lost in the land vertebrates.

The olfactory sense occurs throughout the vertebrates. The olfactory region is a blind sac in all fishes except the myxinoids and the choanichthyans, which gave rise to the land vertebrates. In all other vertebrates the olfactory region is connected to the oral cavity.

REPRODUCTION. Reproduction is almost universally bisexual in the vertebrates. Its chief advantage is the maintenance of genetic variability on which natural selection can operate to keep the population adapted to its environment. A few of the jawless fishes are hermaphroditic. A few genera of bony fishes and of lizards are known to contain parthenogenetic species.

The trend in reproductive systems in the vertebrates has been toward a reduction in the number of zygotes which must be produced. This has been accomplished principally through (1) internal as opposed to external fertilization, (2) viviparity as opposed to oviparity, and (3) parental care of the young.

Most fishes and many amphibians practice external fertilization. The eggs are discharged into the water, and the sperm are discharged over the eggs or in the general vicinity. Many eggs must be discharged to ensure that some are fertilized, and the zygotes are exposed immediately to the vicissitudes of existence. Internal fertilization increases the chances of fertilization and consequently reduces the number of eggs that must be produced. Internal fertilization has appeared in various groups of bony fishes, and it is universal in living cartilaginous fishes. It occurs in some amphibians, and it is universal in amniotes. Retention of the developing zygotes within the reproductive tract of the mother has the advantage of protecting them at a stage when they are helpless to escape predators or unfavorable environmental conditions. Most of the fishes which practice internal fertilization are also viviparous. They retain the zygotes until they are ready to emerge as free-swimming individuals. Among the land vertebrates, some reptiles are oviparous and some are viviparous. All birds are oviparous. The monotreme mammals are oviparous; all other mammals are viviparous. A logical consequence of the retention of the zygote by the mother is the attachment of the zygote to the wall of the reproductive tract so that it may receive nourishment from the mother and utilize her excretory and respiratory systems. Primitive attachments are developed in various viviparous groups, including fishes. In the eutherian mammals a highly efficient connection (placenta) is developed.

Some kinds of fishes protect the nest (redd) until the young have hatched, and some kinds even carry the zygotes about in the mouth or in a pouch until they hatch. Most amphibians and reptiles, with a few exceptions in each group, show no care for the nest or the young. Birds and mammals both show highly developed parental care. In both cases, the parents feed the young during their early helpless stages of development. In birds, parental care seems related to the fact that the young birds are mostly helpless until they have learned to fly. In the case of mammals, nourishment is furnished by the mammary glands of the mother.

Classification

The basic unit of classification is the species. Classification involves the recognition of species and the placing of species in higher categories in order to show phylogenetic relationships. The species is a reality in sexually reproducing animals. It is a population of animals which has reached the stage of evolutionary divergence at which the population does not normally exchange hereditary materials with other, related populations. By modern concepts, different species are actually or potentially prevented from interbreeding by the existence of reproductive isolation mechanisms. In practical taxonomy, it is often necessary to infer interbreeding or lack of it from morphological evidence. The higher taxonomic categories are less objective than are species.

RECOGNITION OF SPECIES. A positive decision as to whether or not a population represents a separate species is sometimes very difficult. This is so principally because of the present inadequate knowledge of species. Most of our knowledge is based on the morphology in small samples of individuals collected from species populations. Comparatively little is known about species as the dynamic, living populations which they are. Even where the latter type of information is available, however, the decision may have to be arbitrary, because evolution is a continuous process. 2 populations may stand in such relationship to one another that they are literally on the border line of separation into different species. In such cases, only by the most critical study of the relationships of such populations in nature and by experiment can the status be established. Studies of this kind in the vertebrates have been too few to be of much aid to the taxonomist.

In usual practice, the decision as to whether or not a population should be called a species is made on the morphological evidence from specimens in hand and on ecological and geographical evidence. The kinds of situations that may be encountered in making a decision are shown graphically in Fig. 1-8. These situations are as follows:

POLYTYPIC SPECIES. In this situation, different kinds of animals replace one another geographically, but the kinds are connected by intervening populations in which the characters are intermediate.

These morphologic intermediates (intergrades) are evidence that interbreeding takes place between the populations showing morphologic extremes. This is the most common kind of species in vertebrates. It is often referred to as a polytypic species to distinguish it from those species in which there is no conspicuous geographic variation in morphologic characters. No matter how different the extremes may be morphologically, they belong to the same species so long as they are connected by geographically intermediate, crossbreeding populations.

OVERLAPPED POLYTYPIC SPECIES. In widely distributed, polytypic species it sometimes happens that end populations of a long chain of geographic races occur in the same region and exist there without interbreeding. The 2 kinds can exist together in the region of overlap if they have developed different ecological preferences so that they are not in contact and in competition with one another. We might say that the 2 kinds are behaving as species in the region of overlap, although they are connected by a chain of races outside of it. It is consistent with the biological concept of the species to treat such a case as a single, polytypic species, for each population in the overlap zone potentially has access to the hereditary variability of the other through interconnecting races outside of the overlap zone.

SECONDARY INTERBREEDING. One of the chief sources of new species seems to be the fragmentation of former ranges by such things as major climatic changes. The geographically isolated fragments no longer have access to the same pool of hereditary variability, and they may differentiate under such isolation, through mutation and selection, to populations that will not interbreed if the opportunity to do so again occurs. When they will no longer interbreed in nature they are reproductively isolated.

If once geographically isolated populations extend their ranges so that they again are in contact before reproductive isolation mechanisms have evolved, they may again interbreed freely where they are in contact. Such secondary interbreeding is often marked by great morphological variability in the zone of interbreeding; the whole range of characters of the parent populations may be present there along with intermediates. To treat such

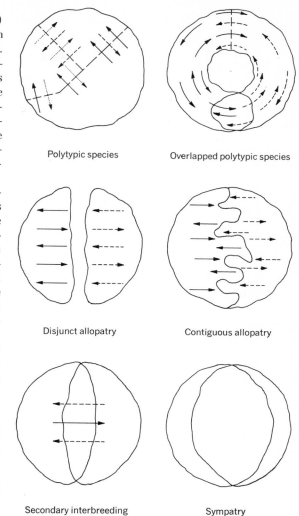

Polytypic species

Overlapped polytypic species

Disjunct allopatry

Contiguous allopatry

Secondary interbreeding

Sympatry

Fig. 1-8. Diagrams showing relationships of species and speciating populations. Arrows show potential gene flow pertinent to interpopulation relationships.

secondarily interbreeding populations as a single species is consistent with the biological concept of the species. The 2 once geographically separate populations have reestablished their access to a common hereditary pool.

Not all cases of secondary interbreeding are as easily treated taxonomically as the one discussed above. Previously isolated populations may interbreed only to a limited extent after they have reestablished contact. Such limited hybridization produces morphological intergrades, but to take the mere fact of existence of intermediates as evidence that the parent populations represent 1 species could be biologically incorrect. Dobzhansky has theorized that reproductive isolation could arise and be reinforced under just such conditions of limited hybridization if the hybrids were at a disadvantage in competition with the parental types. If this theory is correct, selection would operate consequently against those individuals that tended to cross-mate, and it would favor those that preferred to mate within their own type. Under such conditions, hybridization would be serving to separate, rather than fuse, the 2 populations. A correct taxonomic decision in cases of this sort requires more information about the breeding systems of the populations involved than is available for more than a few vertebrates. It seems apparent that, with rigorous selection against hybrids, limited interbreeding could occur without transfer of characters from one population to the other. It seems best to treat populations that show such limited, secondary interbreeding as distinct species. However, it is true that most workers of 20 or more years ago, and some who are still active, would lump the 2 populations as a single species if they found so much as 1 intermediate individual.

CONTIGUOUS ALLOPATRY. The ranges of 2 kinds may meet and interdigitate, but not overlap, without interbreeding of the 2 kinds. Absence of morphological intermediates in collections is the usual evidence available. Different ecological requirements, such as adaptations for clayey and sandy soils, respectively, may account for interdigitation where the respective environments interdigitate. On the other hand, the 2 populations may fail to overlap because they are so similar in ecological adaptations that competition keeps them from occupying the same areas. In both cases the allopatric populations are treated as separate species.

DISJUNCT ALLOPATRY. The most difficult cases to treat taxonomically are those in which 2 more or less morphologically differentiated populations are separated by a wide gap in which neither occurs. That such cases are numerous is not surprising when it is recalled that geographic separation of populations is an important method by which speciation is initiated. Since such populations are prevented from interbreeding by a geographic barrier, the natural situation gives no clue to whether or not they have developed reproductive isolation mechanisms. A fairly reliable decision as to the status of such populations may be reached if representatives of the two are brought into the laboratory and tested for reproductive isolation in mating-choice experiments. Thorough knowledge of the biology of the respective populations in nature may reveal attributes, such as differences in time of breeding, which would act as isolation mechanisms if geographic isolation were to break down. Little information from either of these sources is available for disjunct, allopatric populations of vertebrates. The degree of morphological difference has been the most frequently used criterion in such cases, but this makes for considerable subjectivity in the classification. In general, the taxonomic "splitters" tend to treat disjunct allopatrics as separate species if there is a moderate amount of morphological differentiation. Taxonomic "lumpers" tend to treat them as subspecies of the same species unless there is very considerable differentiation. Either system is sure to produce a goodly number of wrong guesses. The policy of lumping tends to obscure the important evolutionary fact that certain populations have undergone at least the early stages toward speciation. The policy of splitting accords specific distinction to some populations which have not reached that stage in evolution. For morphologically differentiated allopatric populations for which no evidence bearing on reproductive isolation is available, it might be best to use the term allopatric species. This serves to distinguish such populations from species which have proved their distinctness in nature, and it distinguishes them from the interbreeding geographic races of polytypic species.

SYMPATRY. When 2 populations exist in the same region, with either a broad or narrow zone of overlap, and do not interbreed, they are demonstrating that they are "good" species. By their coexistence as separate breeding systems, such populations show that they have developed effective isolation mechanisms. Occasional hybrids between related sympatric species provide no reason for treating these other than as distinct species.

SPECIES IN FISHES. By their discontinuous distribution in separate lakes and separate stream systems, fresh-water fishes present problems of their own. Morphologically rather similar populations may occur in separate stream systems, where they are completely isolated from one another. The classification, therefore, has been built almost entirely on morphological evidence. It is not unusual to have species and even subspecies with discontinuous distributions in separate stream systems. Only by extensive experimentation with living fishes can it be determined whether or not this morphological classification is a biologically correct one.

HIGHER CATEGORIES. Species are arranged in higher categories in order to achieve an orderly classification and in order to show phylogenetic relationships. To the pre-Darwinian classifier of animals, a genus or family or other category was simply a group of similar kinds of animals without any clear concept of evolutionary relationship. To the post-Darwinian worker, the scheme of classification presents a picture of relationship by descent. The branching (dichotomous) system of classification shows, at least crudely, the evolutionary splits which have occurred in past history of a group. Each higher category would ideally correspond to a dichotomy. As animals have been studied in more and more detail, the number of apparent dichotomies that can be recognized from

morphological evidence has greatly increased. As they have increased it has been necessary to invent more and more categories to supplement the genus, family, order, class, and phylum of the early classifications. Table 1-1 illustrates the system of classification and includes most categories used in vertebrate classification. For the example used, a crude estimate (largely guess) of the time that has elapsed since the dichotomy is given.

Nomenclature

INTERNATIONAL RULES. Nomenclature refers to names. It is important to the zoologist only in that names are symbols for the things with which he works. The naming of animals is governed by a complex set of rules embodied in the International Rules of Zoological Nomenclature. These rules are administered by the International Commission on Zoological Nomenclature, which derives its authority from the International Congresses of Zoology. The International Commission is vested with the power to interpret, amend, or suspend provisions of the International Rules. Most of the work of the Commission is reflected in a long list of Opinions of the International Commission. The Commission has no way of enforcing its decisions,

TABLE 1-1. *System of Classification*

Category and example		Crude estimate of time elapsed since dichotomy
Phylum	Chordata	Cambrian—500 million years ???
Subphylum	Vertebrata	Ordovician—450 million years
Class	Mammalia	Early Triassic—175 million years
Subclass	Theria	Late Triassic—more than 155 million years
Infraclass	Eutheria	Jurassic—130 million years
Cohort	Glires	Late Cretaceous—80 million years
Order	Rodentia	Paleocene—60 million years
Suborder	Myomorpha	Early Eocene—50 million years
Superfamily	Muroidea	Middle Eocene—45 million years
Family	Cricetidae	Late Eocene—35 million years
Subfamily	Cricetinae	Oligocene—25 million years
Tribe	Hesperomyini	Miocene—15 million years
Genus	*Peromyscus*	Pliocene—less than 10 million years
Subgenus	*Peromyscus*	Early Pleistocene—less than 1 million years
Species	*polionotus*	Late Pleistocene—10,000–100,000 years

and the burden of holding names to conformity with the International Rules falls on the individual worker and on editors of scientific publications.

Only a few basic principles of the International Rules can be mentioned here, and for further information reference is made to the *Bulletin of Zoological Nomenclature,* which is an official publication of the International Commission, or to discussions such as that in *Methods and Principles of Systematic Zoology* by Mayr, Linsley, and Usinger.

The one purpose of the International Rules is to ensure a stable and uniform system of nomenclature. Stated at its simplest, the basic principle of the International Rules is that an animal can have only 1 name, and no 2 animals can have the same name. In application, the International Rules have fallen far short of establishing a stable and uniform nomenclature. It is ironical that provisions of the International Rules specifically designed to stabilize the nomenclature have proved the greatest source of instability.

CHANGES OF NAMES.   The names applied to animals have changed, and are changing, so much so that it would be ridiculous to pretend that we have a uniform nomenclature. The name of an animal may be changed for a variety of reasons. Some reasons are biological ones, and this source of change can never be eliminated. For example, 2 kinds of animals are named as species on the basis of incomplete information. As additional information about these populations of animals is accumulated, it becomes apparent that the 2 kinds of animals which have been named are merely geographic variants of the same species. Since no species can have 2 names, one name becomes a synonym of the other. By application of the law of priority, the first-proposed name takes precedence as the senior synonym and the other becomes a junior synonym. The oldest name applies even though it has been used for a relatively obscure kind, while the junior synonym has been extensively used for a common and widely distributed one. Recent examples of this kind of name change are the application of the name *Cnemidophorus sacki* instead of *C. gularis* to a common U.S. racerunner and *Didelphis marsupialis* instead of *D. virginiana* to the common opossum. A genus may be split because it is found to be an artificial assemblage, and some of its species must, therefore, be given a new

generic name. Conversely, 2 named genera may best be lumped in a single genus, with one, the latest proposed, generic name becoming a junior synonym of the other. A species may have been erroneously referred to a genus and therefore must be transferred to another. While any change from a commonly used name is inconvenient and annoying, there can be no serious objection to these changes, as they result from increased biological information.

Other changes result from strict application of the International Rules and have a purely legalistic, rather than biological, basis. A major source of such change is the discovery of an earlier name than the one which has been in general use. Under strict application of the law of priority a familiar name, which has been used for many years, may be replaced by one which has gone unnoticed in some obscure publication. However, the latest revision of the Rules carried the provision that a senior synonym that has not been used in the primary zoological literature for more than 50 years is to be treated as a *nomen oblitum,* which must not be used unless the Commission so directs.

Another recurrent source of change is the discovery that the name has been misapplied, i.e., the name has been in use for a species or genus entirely different from the species or genus to which the name was originally applied. Most of the early taxonomic work was inadequate by modern standards, and it has been easy for such mistaken application of names to occur. It is possible to conserve misapplied names which have had long usage by appeal to the Commission.

A fairly uncommon source of change in vertebrate nomenclature today is the discovery that 2 things have been given the same name. No 2 genera of animals may have the same name; no 2 species in the same genus may have the same name. A name which has been applied to 2 or more things is a homonym. It is apparent that a homonym may be created by original error in overlooking prior use (primary homonym) or by lumping 2 genera each of which contains a species with the same specific name (secondary homonym). In each case, the oldest name takes precedence as the senior homonym and another name must be found to replace the later (junior) homonym.

Many ambiguities exist in the International Rules, and different interpretations have been made by different workers. General usage has been followed where possible in this book, but in some instances an arbitrary decision has had to be made between different proposed usages of names.

Distribution

The terrestrial and fresh-water vertebrates on a given continent are ones which have either originated there or have been able to cross from other continents. Very few terrestrial or fresh-water vertebrates have been able to cross from one continent to another unless there has been a land connection between the continents even if they were volant types which might fly over the ocean barrier. Each continental land mass, therefore, has its own distinctive faunal assemblage. Differences and similarities between continental faunas are directly related to past and present opportunities for exchange of species between the continents.

North America has, or has had in the geologic past, connections to 2 other major continental masses. A connection to the Eurasian land mass through the Bering Straits has existed at times in the geological past; many groups of vertebrates crossed in both directions in Pleistocene time and earlier. North America has been connected to South America by the Central American land bridge since the middle Pliocene, and interchanges in both directions have taken place between these continents. As a result of these past and present interchanges, various genera, and even species, are shared between Asia and North America and between South America and North America.

The distribution of animals on the continent is influenced primarily by their adaptations to various features of the physical environment and to vegetational types, which, in turn, are a reflection of the physical environment. The basic pattern of distribution of physical factors shows a rough correspondence from one continent to another, with the pattern on the Southern Hemisphere continents a mirror image of the pattern on the Northern continents. The major variables are temperature and moisture. There is a general decrease in temperature and an increase in seasonality (i.e., contrast between summer and winter) going poleward from the Equator. The general north-south zonation is interrupted by mountain masses, which project up into increasingly colder air. The ascent of a high mountain in the middle latitudes roughly duplicates the south-north temperature gradient. Proximity of land to relatively warm ocean currents also disrupts the latitudinal gradient, for shores bathed by such currents are relatively warmer than the interior of the continent at the same latitude. Moisture available on land is mostly that picked up from the oceans by winds and carried inland, where it is precipitated. Since warm air can carry more moisture than cold, those shores bathed by warm ocean currents tend to have warm, moist climates; those bathed by cold currents tend to have dry climates. Continental interiors tend to be dry because in-blowing winds tend to lose much of their moisture before reaching the interior.

In general, the east coasts of the continents are bathed by warm currents poleward to about 40° latitude, and the adjacent land has a warm, moist climate. The west coasts of the continents poleward of about 50° latitude are bathed by relatively warm currents and have moist climates. The dry climates extend from the west coasts of the continents, between about 20 and 40° latitude, which are bathed by cold currents, into the interior and bend poleward as they reach into the interior. In North America, the Sierra Nevada chain forces the air masses upward, causing them to give up much of the moisture that might otherwise reach the dry interior. On the other hand, moist winds off the Gulf of Mexico cause the boundary between the dry lands and the eastern, humid climates to be displaced westward of the area where it theoretically would come.

The basic pattern of climatic distribution in North America has probably existed without major modifications since Miocene time when a milder climate existed. The greatest changes have been the southward shifting of cold climates with the advance of continental glaciers into the northern United States and the concurrent increase in moisture in regions which are arid today; the cold climates shifted northward again in interglacial time. This process was repeated several times during the Pleistocene, and the poleward withdrawal of cold faunas is still continuing today.

The North American vertebrate fauna is roughly divisible into 2 major elements, as pointed out

Fig. 1-9. Principal vegetation types of North America.
(From Transeau, Sampson, and Tiffany,
Textbook of Botany, Harper & Brothers.)

many years ago by C. Hart Merriam. One is a Boreal fauna, which has numerous affinities with Eurasia. This fauna contains virtually no amphibians or reptiles because of the limitations imposed on these animals by persistent cold. The species of this fauna are widely distributed in Alaska and Canada, and many reach southward into the northern United States. Paleontological evidence shows that this fauna moved far southward during the periods of Pleistocene glaciation and returned northward in interglacial periods. The last retreat northward followed the retreat of the last (Wisconsin) glaciers. Remnants were left, however, on higher elevations of the 3 great mountain chains (Cascade–Sierra Nevada, Rocky Mountain, and Appalachian) and on other, isolated peaks. The Boreal fauna, therefore, tends to be distributed transcontinentally in Canada today and to finger southward along the main mountain chains.

The other major faunal element is a southern or Sonoran element. This includes many species of birds and mammals and most North American amphibians and reptiles. This faunal group is distributed over much of the United States and was displaced southward for the most part by Pleistocene glaciation. The present distributions have been reached by northward spread from southern "refuges" as the ice and the cold climates retreated northward. This element actually consists of 2 groups. One is adapted to the relatively warm moist climate and deciduous forest vegetation of the eastern United States. The other is adapted to life in the relatively arid grasslands and deserts of central and southwestern North America. Some ecologically tolerant species fit concisely in neither group but occur widely in both moist and arid environments. The former ranges of various southern species were fractured as these species retreated southward in times of Pleistocene glaciation. Peninsular Florida and the Southwest were probably the chief refuges of these southern groups. The northward spread from these refuges has mostly followed a few general patterns. From the Florida center, species have spread northward along the Atlantic Coastal Plain and westward on the Gulf Coastal Plain, and others have spread widely through the eastern United States. From the southwestern center, some species have spread widely through the United States. Others have followed the Gulf Coastal Plain eastward and subsequently spread widely through the eastern forests. Others have spread northward through the central grasslands or western deserts, or they have moved northward in both.

Some warmth-adapted species have spread into the southern United States from Central America and northern South America. These are found today mostly along the Mexican boundary from southern Texas to southern California, although a few have even reached Florida.

KEY TO CLASSES OF VERTEBRATES

1. Jawless   **Agnatha**
   With jaws derived from modified gill arch, or from dermal bone, or from both   **2**
2. Internal gills present in adult; auricle undivided   **3**
   Gills absent in adult, or if present, external; auricle divided by longitudinal septum   **4**
3. Gills with separate openings through pharynx and without cover of dermal bone; placoid scales present   **Chondrichthyes**
   Gills with a single cover (operculum) of dermal bone; scales ganoid, cosmoid, cycloid, ctenoid or absent, never placoid   **Osteichthyes**
4. Tetrapods with metamorphosis; skin naked (in U.S. groups); ventricle undivided   **Amphibia**
   Tetrapods without metamorphosis; skin with scales or other appendages; ventricle more or less completely divided by longitudinal septum   **5**
5. Poikilothermal; left and right elements of 4th aortic arch present; skin without feathers or hair   **Reptilia**
   Homoiothermal; either left or right element of 4th aortic arch absent; skin with feathers or hair   **6**
6. Feathers present; right element only of 4th aortic arch present   **Aves**
   Hair present, at least embryologically; left element only of 4th aortic arch present   **Mammalia**

REFERENCES

Colbert, E. H. 1955. *Evolution of the Vertebrates*, John Wiley & Sons, Inc., New York.

Darlington, P. J., Jr. 1957. *Zoogeography: The Geographical Distribution of Animals*, John Wiley & Sons, Inc., New York.

Gregory, W. K. 1951. *Evolution Emerging*, 2 vols., The Macmillan Company, New York.

Haurwitz, B., and J. M. Austin. 1944. *Climatology,* McGraw-Hill Book Company, New York.

Mayr, E. 1963. *Animal Species and Evolution,* Harvard University Press, Cambridge, Mass.

——, E. G. Linsley, and R. L. Usinger. 1953. *Methods and Principles of Systematic Zoology,* McGraw-Hill Book Company, New York.

Orr, R. T. 1966. *Vertebrate Biology,* 2d ed., W. B. Saunders Company, Philadelphia.

Romer, A. S. 1966. *Vertebrate Paleontology,* 2d ed., University of Chicago Press, Chicago.

——. 1962. *The Vertebrate Body,* 3d ed., W. B. Saunders Company, Philadelphia.

——. 1959. *The Vertebrate Story,* University of Chicago Press, Chicago.

Simpson, G. G. 1931. *Principles of Animal Taxonomy,* Columbia University Press, New York.

Young, J. Z. 1962. *The Life of Vertebrates,* 2d ed., Oxford University Press, Fair Lawn, N.J.

THOSE aquatic vertebrates which bear fins, either paired or unpaired, have long been known as Pisces. Some authorities prefer to use the term to distinguish between the lower vertebrates (superclass Pisces) and the higher vertebrates (superclass Tetrapoda). Other writers use the term as a series under the superclass Gnathostomata, the jawed vertebrates. The more primitive superclass Agnatha, by virtue of fundamental differences such as the absence of jaws and paired fins, is set apart from the fishes taxonomically, but still treated by ichthyologists. For the classification of fishes see Table 2-1.

Students seriously interested in ichthyological taxonomy should consult the new and provisional classification of living teleostean fishes proposed by P. H. Greenwood, D. E. Rosen, S. H. Weitzman, and G. S. Myers (1966).

The oldest known vertebrate remains, the ostracoderms, are of an agnathous type which probably originated in Ordovician times. The most ancient fishlike vertebrates represented by an abundance of fossils are found in Silurian rocks. Most of these agnathous animals became extinct during Devonian times, but they are believed to be represented by the modern lampreys (class Petromyzones) and hagfishes (class Myxini). The Myxini are strictly marine and therefore outside the limits of this treatise.

The most primitive jawed vertebrates are the placoderms; they flourished during Devonian times and probably were derived from ancestral types similar to the progenitors of the Petromyzones. Most of these odd fishes became extinct during the Devonian, but the class Acanthodii persisted until early Permian times.

The class Elasmobranchii (Chondrichthyes), the sharks and their relatives, dates back to Devonian times, and although some became extinct during Permian or early Triassic times, the class has been successful in maintaining numbers through to the present. These cartilaginous fishes with jaws, paired fins, and placoid scales were probably derived from placodermlike ancestors which are believed to have been also the progenitors of the class Holocephali (regarded as a subclass by some writers), commonly known as the chimaeras or ratfishes. The elasmobranchs have no gill cover, whereas the holocephalans do have one. Few elasmobranchs

and no holocephalans enter fresh water. Since this section treats only fresh-water forms, the elasmobranchs are omitted. Some primarily marine teleosts which occasionally enter fresh water are not included in this book.

The lungfishes, class Dipnoi, are not represented by living species in the United States although some Devonian fossils are known.

The class Teleostomi (Osteichthyes) are sometimes known as the bony fishes although in some the skeleton is not wholly bony. The teleosts probably originated from some Devonian stock which also produced the elasmobranchs. The most primitive recent teleosts are the coelacanths which, until 1938, were believed to have become extinct in late Cretaceous times. The highly publicized *Latimeria* from marine waters off Africa is probably the most important ichthyological discovery of recent years. It is quite possible that other "living fossils" will be discovered.

The Petromyzones are represented by few species, but the Teleostomi are present in a great profusion of highly adaptive animals that have invaded practically every possible aquatic niche from subterranean waters to desert pools and high-mountain brooks and lakes. We are therefore primarily concerned with the teleosts in this section.

CLASS PETROMYZONES

Order Petromyzoniformes. Family Petromyzonidae. Lampreys. These recent fishlike vertebrates, closely allied to the Cephalaspides (Upper Silurian to Upper Devonian), are worldwide in distribution and are characterized by the absence of jaws and paired fins. The Myxini (hagfishes) are placed with the lampreys in the class Agnatha by some writers. The lampreys have a single median nostril, and the gills, unlike those in true fishes, are not covered by an operculum but are externally represented by 7 gill openings which form a row behind each eye. Lampreys are either parasitic or nonparasitic, and therefore the 2 types exhibit somewhat different adult structure and habits. All of our lampreys ascend streams to spawn. A small depression is excavated in the sand or gravel. Many individuals may use the same redd (nest) in which the eggs and sperm are deposited. After spawning

TABLE 2-1. *Classification of Fishes* [from Berg (1947) with Minor Modification]
[Families of uncertain position (*inc. sedis.*) are included.]

		Families Extinct	Families Living
Subphylum Acrania			
Class Amphioxi	Order Amphioxiformes	0	1
Subphylum Craniata			
Superclass Agnatha			
†Class Cephalaspides . . .	†Order Cephalaspidiformes	5	0
	†Order Tremataspidiformes	2	0
	†Subclass? Birkeniae		
	†Order Birkeniiformes	5	0
	†Order Lasaniiformes	1	0
	†Order Endeiolepiformes	1	0
Class Petromyzones . . .	Order Petromyzoniformes	0	1
†Class Pteraspides	†Order Astraspiformes	1	0
	†Order Psammosteiformes	3	0
	†Order Pteraspiformes	1	0
	†Order Cyathaspiformes	10	0
	†Order Amphiaspiformes	1	0
	†Subclass? Coelolepides		
	†Order Coelolepiformes	1	0
	†Order Phlebolepiformes	2	0
†Class? Palaeospondyli . . .	†Order Palaeospondyliformes	1	0
Class Myxini	Order Myxiniformes	0	3
Superclass Gnathostomata			
Series Pisces			
†Class Pterichthyes	†Order Remigolepiformes	1	0
	†Order Asterolepiformes	4	0
†Class Coccostei	†Subclass Euarthrodira		
	†Order Arctolepiformes	3	0
	†Order Coccosteiformes	12	0
	†Order Mylostomiformes	1	0
	†Order Ptyctodontiformes	1	0
	†Subclass Phyllolepida		
	†Order Phyllolepiformes	1	0
	†Subclass Macropetalichthys		
	†Order Macropetalichthyiformes	1	0
	†Order Stensioelliformes	1	0
	†Order Gemuendiniformes	1	0
	†Order Jagoriniformes	1	0
†Class Acanthodii	†Order Climatiiformes	3	0
	†Order Mesacanthiformes	1	0
	†Order Ischnacanthiformes	1	0
	†Order Gyracanthiformes	1	0
	†Order Diplacanthiformes	1	0
	†Order Cheiracanthiformes	1	0
	†Order Acanthodiformes	1	0

† Fossil forms.

TABLE 2-1. *Classification of Fishes* (*continued*)

		Families Extinct	Living
Class Elasmobranchii . .	†Subclass Xenacanthi		
	†Order Xenacanthiformes	1	0
	†Subclass Cladoselachii		
	†Order Cladoselachiformes	2	0
	†Order Cladodontiformes	2	0
	Subclass Selachii		
	Superorder Selachoidea (sharks)		
	Order Heterodontiformes	4	1
	Order Hexanchiformes	0	2
	Order Lamniformes	0	6
	Order Squaliformes	1	3
	Superorder Batoidea (rays)		
	Order Rajiformes	1	8
	Order Torpediniformes	0	1
Class Holocephali . . .	†Subclass Chondrenchelyes		
	†Order Chondrenchelyiformes	1	0
	Subclass Chimerae		
	Order Chimaeriformes	9	3
Class Dipnoi	†Superorder Dipteri		
	†Order Dipteriformes	2	0
	†Order Phaneropleuriformes	3	0
	†Order Uronemiformes	2	0
	†Order Ctenodontiformes	1	0
	Superorder Ceratodi		
	Order Ceratodiformes	0	1
	Order Lepidosireniformes	0	2
	†Order Rhynchodipteriformes	1	0
Class Teleostomi	Subclass Crossopterygii		
	†Superorder A. Osteolepides		
	†Order Osteolepiformes	3	0
	†Order Holoptychiiformes	1	0
	†Order Rhizodontiformes	3	0
	Superorder B. Coelacanthi		
	Order Coelacanthiformes	3	1
	Subclass Actinopterygii		
	Group A		
	Order Polypteriformes	0	1
	Group B		
	†Order Tarrasiiformes	1	0
	†Order Palaeonisciformes	8	0
	†Order Gymnonisciformes	1	0
	†Order Phanerorhynchiformes	1	0
	†Order Dorypteriformes	1	0
	†Order Bobasatraniiformes	1	0
	†Order Redfieldiiformes	2	0

† Fossil forms.

TABLE 2-1. *Classification of Fishes* (*continued*)

	Families	
	Extinct	Living
†Order Perleidiformes	3	0
†Order Ospiiformes	3	0
†Order Pholidopleuriformes	1	0
†Order Saurichthyiformes	1	0
Order Acipenseriformes	1	2
Order Amiiformes	5	1
†Order Aspidorhynchiformes	1	0
†Order Pycnodontiformes	3	0
†Order Pachycormiformes	2	0
Order Lepisosteiformes	0	1
†Order Pholidophoriformes	3	0
Order Clupeiformes	12	44
Order Bathyclupeiformes	0	1
Order Galaxiiformes	0	1
Order Scopeliformes	3	9
Order Ateleopiformes	0	1
Order Giganturiformes	0	1
Order Saccopharyngiformes	0	3
Order Mormyriformes	0	2
Order Cypriniformes	0	43
Order Anguilliformes	4	24
Order Halosauriformes	0	1
Order Notocanthiformes	0	2
Order Beloniformes	0	4
Order Gadiformes	0	4
Order Macruriformes	0	2
Order Gasterosteiformes	1	3
Order Syngnathiformes	0	6
Order Lampridiformes	0	6
Order Cyprinodontiformes	0	7
Order Phallostethiformes	0	2
Order Percopsiformes	0	2
Order Stephanoberyciformes	0	2
Order Beryciformes	2	13
Order Zeiformes	0	3
Order Mugiliformes	0	3
Order Polynemiformes	0	1
Order Ophiocephaliformes	0	1
Order Symbranchiformes	0	3
Order Perciformes	4	160
Order Dactylopteriformes	0	1
Order Thunniformes	0	1
Order Pleuronectiformes	1	5
Order Icosteiformes	0	1
Order Chaudhuriiformes	0	1
Order Mastacembeliformes	0	1
Order Echeneiformes	1	1

† Fossil forms.

TABLE 2-1. *Classification of Fishes* (*continued*)

	Families	
	Extinct	Living
Order Tetrodontiformes	2	7
Order Gobiesociformes	0	1
Order Batrachoidiformes	0	1
Order Lophiiformes	0	16
Order Pegasiformes	0	1

the adults die. The ammocoetes (larvae) form a U-shaped burrow in the stream bottom where they obtain food by straining microorganisms and organic detritus from the water. Some differences in structure and habits of parasitic and nonparasitic species follow.

Parasitic lampreys, in comparison with nonparasitic forms, have a shorter larval life and a longer adult life, during which they feed on fishes by attaching the disclike mouth. Within the disc are usually many strong, sharp, horny teeth which, with the rasping tongue, are used to break the skin of the prey and cause the body fluids to flow (Fig. 2-1).

Fig. 2-2. Sucking disc of a lamprey (*Lampetra richardsoni*). (After Vladykov and Follett, 1965, with modification.) *AR*, anterior row of teeth; *C*, circumoral teeth; *EO*, esophageal opening; *F*, oral fimbriae; *IL*, infraoral lamina; *LL*, longitudinal lingual lamina; *LLO*, lateral-line organ; *LR*, lateral row of teeth; *MR*, marginal row of teeth; *SL*, supraoral lamina; *TL*, transverse lingual lamina. NOTE: the labels *AR*, *C*, and *LR* lead to lines transecting teeth included in the respective counts; included in the count of circumorbital teeth (*C*) are those of the opposite side, not transected by the line in Fig. 2-1. The figure shows 18 circumoral teeth.

Fig. 2-1. Schematic drawing of the disc of *Petromyzon marinus*. Abbreviations are as in Fig. 2-2. (Modified after Vladykov and Mukerji, 1961.)

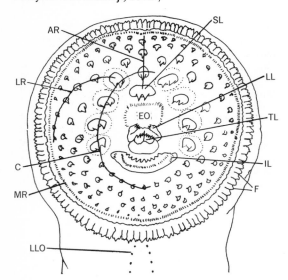

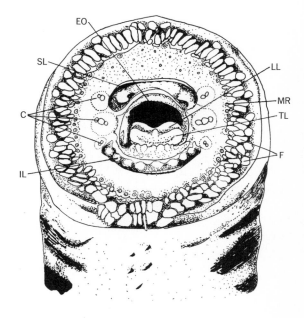

Even though the prey may not be killed at once, death may be brought on by fungi which can live in the open sore.

Nonparasitic lampreys live a longer larval life and a shorter adult life. The ammocoetes just prior to metamorphosis are about as large as adults. After the larvae transform, the gut degenerates and fewer teeth, than in parasitic species, develop in the buccal disc (Fig. 2-2).

One order, the Petromyzoniformes, containing a single family, the Petromyzonidae, is represented in the United States by 4 genera containing 13 species. Some ichthyologists combine *Entosphenus* with *Lampetra*.

Ammocoetes (Fig. 2-3) are very difficult to identify and within a single genus are often incapable of specific diagnosis.

Characters

The distinguishing characters of the Petromyzonidae are so radically different from those of the true fishes that some definitions are prerequisite to an understanding of the following diagnoses and keys.

MYOMERES. The number of muscular impressions or segments which lie between the last gill opening and the anus. The myomeres are often difficult to count since in preserved specimens they are hidden under the thick skin which may be covered by mucus. It may be necessary to scrape the skin toward the tail in order to count the myomeres. Sometimes false creases appear between the myocommata (creases between the myomeres); if these are present the normal width of a myomere must be determined, after which dividers can be used to estimate the number present, keeping in mind that

Fig. 2-3. Ventral views of the heads of larvae and an adult of the American brook lamprey (*Entosphenus lamottei*). (From Vladykov, 1949.) *A*, an ammocoete; *B*, the same ammocoete during metamorphosis; *C*, adult female.

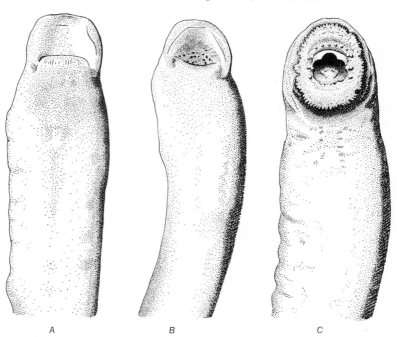

A   B   C

the myomeres diminish in width near the anus. Ammocoetes and adults of the same species may have slightly different numbers.

SUPRAORAL LAMINA. A band or patch of tough connective tissue immediately anterior to the oral opening. This structure usually bears blunt or sharp cornified cusps. Care must be taken to count as a cusp a small one on either or both sides of and at the base of a large cusp. A cusp divided at its tip is counted as 2.

INFRAORAL LAMINA. A crescentic ridge, posterior to the oral opening, bearing cusps that may be difficult to count in specimens not completely transformed. A jet of compressed air is of great assistance in making this and other counts. Care must be exercised to count all cusps.

CIRCUMORAL TEETH. Included in this count are the innermost disc teeth (not including a small tooth sometimes developed at the edge of the esophageal opening) lateral to the mouth, those connecting the anteriormost laterals of each side just in advance of the supraoral lamina, and the similar posterior arc of teeth paralleling the infraoral lamina. The circumoral teeth may be unicuspid. In *Entosphenus* and *Lampetra* these teeth, sometimes called laterals, are arranged in a row of 3 or 4 unicuspid, bicuspid, or tricuspid teeth on each side of the oral opening.

TEETH IN ANTERIOR ROW. Those teeth which form the apexes of the V-shaped rows of the anterior field and including the 1 or 2 teeth in advance of the first V. All teeth of the anterior row are median and are best seen near the edge of the disc by employing an air jet.

TEETH IN LATERAL ROWS. The lateral rows usually counted are those rows that originate in the circumoral tooth of each side that lies opposite the supraoral lamina. Since these rows are deflected backward near the disc edge, care must be taken to see all the teeth which become progressively smaller outward and backward. In *Ichthyomyzon*, the lateral deflected ends of these rows comprise the marginals. In *Entosphenus* and *Lampetra* the lateral rows are absent, the space being occupied by patches of teeth known as lateral teeth which are not in radiating rows.

TRANSVERSE LINGUAL LAMINA. This structure lies on the floor of the esophageal opening and is usually visible just in front of the infraoral lamina.

It is diversified in form and degree of cornification and is described as being linear or bilobed. A row of denticles, of varying degrees of development, forms a fringe on its posterior edge.

TOTAL LENGTH. Distance from anteriormost end to tip of tail. Standard length, so widely used in measuring fishes, is not a practical measurement for lamprey description.

LENGTH OF DISC. The longitudinal diameter between the outer bases of the marginal fimbriae.

LONGITUDINAL LINGUAL LAMINA. This structure lies on the floor of the esophageal opening behind the transverse lingual lamina. Its denticles may be so tiny as to render counts impossible.

KEY TO GENERA OF ADULT PETROMYZONIDAE

1. Dorsal fin not completely divided, but often emarginate   *Ichthyomyzon*
   Dorsal fin completely divided into 2 parts or deeply notched   **2**
2. Disc teeth large and numerous; lateral teeth bicuspid. Size large. Myomeres about 70. Parasitic   *Petromyzon*
   Disc teeth inconspicuous and few; lateral teeth sometimes multicuspid. Size smaller. Varying numbers of myomeres. Parasitic or not   **3**
3. Posterior field of disc with visible (sometimes minute) teeth other than marginals   *Entosphenus*
   Posterior field of disc lacking teeth other than tiny marginals   *Lampetra*

Genus *Ichthyomyzon* Girard. Small lampreys (usually not longer than 380 mm) with dorsal fin continuous with caudal, separated only by a broad notch. Anterior lingual tooth grooved; circumoral teeth unicuspid or bicuspid; transverse lingual lamina linear to bilobed. Myomeres between last gill opening and anus 47–62. Both parasitic and free-living forms are represented. 7 species restricted to the fresh waters of the United States and southern Canada from the Mississippi Valley eastward.

KEY TO SPECIES OF *ICHTHYOMYZON*

1. Gut functional until a length of 175 mm is reached. Gonad and secondary sexual characters inconspicuous in examples under 175 mm. Disc contained

7.7–16.9 times in total length. Teeth well developed even on posterior field of disc. Parasitic **2**
Gut nonfunctional in adults. Small lampreys, less than 175 mm. Ova enlarged before metamorphosis. Disc length contained 15.4–27.7 times in total length. Teeth degenerate, at least on posterior field of disc. Nonparasitic **4**
2. Circumoral teeth unicuspid; supraoral cusps 2 (1–4);[1] 3 (2–4) teeth in anterior row; lateral teeth 6 or 7 (4–9). Disc usually more than 12.5 in total length. Myomeres 49–52 (48–57). Transverse lingual lamina moderately to strongly bilobed
I. unicuspis
Circumoral teeth 6–8 (1–11), in part bicuspid; supraoral teeth 2 or 3, anterior teeth 5 (rarely 3); lateral teeth 8 or 9 (6–11). Disc more than 12.5 in total length. Myomeres 51–58 (49–62) **3**
3. Transverse lingual lamina linear or weakly bilobed. Myomeres 51–54 (49–58). Disc contained fewer than 14.3 times in total length. Bicuspid circumorals 6–8 (rarely all unicuspid or as many as 13)
I. castaneus
Transverse lingual lamina moderately to strongly bilobed. Myomeres 56–58 (53–62). Disc contained more than 13.3 times in total length. Bicuspid circumorals usually 8 *I. bdellium*
4. Circumoral teeth unicuspid; supraoral cusps 2 (rarely 1); anterior teeth 2 (1–3); visible lateral teeth 3–5 (2–6). Disc contained 24.4 (20.4–27.7) times in total length. Myomeres 50–52 (47–56). Transverse lingual lamina moderately to strongly bilobed
I. fossor
Circumoral teeth in part bicuspid; supraoral cusps 3 (2–4); anterior teeth 4 (3–5); lateral teeth 7–8 (5–9). Disc contained 15.4–26.3 times in total length. Myomeres 52–60 (51–61) **5**
5. Transverse lingual lamina linear or weakly bilobed, not denticulate. Myomeres 51–54. Disc contained 17.2–26.3 times in total length. Bicuspid circumorals 8 (2–8) *I. gagei*
Transverse lingual lamina moderately to strongly bilobed. Myomeres 55–61. Disc contained 15.4–23.3 times in total length. Bicuspid circumorals 7–11 (rarely 7) **6**
6. Height of 1st dorsal fin greater than distance from anterior margin of eye to nostril *I. greeleyi*
Height of 1st dorsal fin less than distance from anterior margin of eye to nostril *I. hubbsi*

Ichthyomyzon unicuspis Hubbs and Trautman.
Silver lamprey. Myomeres 47–55. Supraoral cusps

1–4 (usually 2); infraoral cusps 5–11, well developed and strongly cornified; circumoral teeth 15–25; teeth of anterior row 2–4 (usually 3); teeth in lateral rows 5–8 (usually 6–7); bicuspid circumorals 0–2 (rarely 1 or 2). Transverse lingual lamina moderately to strongly bilobed. Maximum length about 380 mm. From Minnesota eastward through Wisconsin, Illinois, Indiana, Michigan, Ohio, to New York and southward in the Mississippi Valley to Nebraska and Missouri. Parasitic.

Ichthyomyzon fossor Reighard and Cummins.
Northern brook lamprey. Myomeres 47–56 (usually 50–52). Supraoral cusps 1–2 (seldom 1); infraoral cusps 6–11; circumorals 15–25; teeth in anterior row 1–3 (usually 2); teeth in lateral rows 2–6 (usually 4 or 5); bicuspid circumorals 0. Transverse lingual lamina strongly bilobed, weak, and little cornified; denticulations none or weak and numerous. Maximum size about 150 mm. Wisconsin, Michigan, Ohio, and New York. Nonparasitic.

Ichthyomyzon castaneus Girard. Chestnut lamprey. Myomeres 49–56 (usually 51–54). Supraoral cusps 2–3; infraoral cusps 6–11; circumoral teeth 17–25; teeth in anterior row 3–5; teeth in lateral rows 6–11 (usually 8 or 9); bicuspid circumorals 1–10 (usually 6–8). Transverse lingual lamina linear to weakly bilobed. Maximum length slightly over 300 mm. From Minnesota and North Dakota southward to Texas. From Tennessee and Georgia north to Ohio, Michigan, Illinois, and Wisconsin. Parasitic.

Ichthyomyzon gagei Hubbs and Trautman. Southern brook lamprey. Myomeres 51–54 (usually 52). Supraoral cusps 2–3 (usually 3); infraoral cusps 7–10; circumoral teeth 20–22; teeth in anterior row 3–4 (usually 4); teeth in lateral rows 6–9 (usually 7 or 8); bicuspid circumorals 2–8. Transverse lingual lamina linear to weakly bilobed, degenerate, without denticulations, and little or no cornification. Maximum size about 140 mm. From the Ochlockonee River system and tributaries of Choctawhatchee Bay in Florida and the Chattahoochee River system in Georgia and Alabama westward to the Neches, San Jacinto, and Trinity Rivers in Texas. Northward through the Red River system to the Arkansas River system in Oklahoma. Nonparasitic.

Ichthyomyzon bdellium (Jordan). Ohio lamprey. Myomeres 53–62 (usually 56–58). Supraoral

---

[1] Such figures, not in parentheses, indicate the usual or average count or measurement. Figures in parentheses indicate extremes of variation.

cusps 2 or 3; infraoral cusps 7–10 (usually 8); circumoral teeth 19–22 (usually 21); teeth in anterior row 4 or 5 (usually 4); teeth in lateral rows 7–10 (usually 8); bicuspid circumorals 5–10 (usually 8). Transverse lingual lamina moderately to strongly bilobed. Maximum length about 260 mm. Confined to the Ohio River and its tributaries. Parasitic.

Ichthyomyzon greeleyi Hubbs and Trautman. Allegheny brook lamprey. Myomeres 55–61 (usually 57–59). Supraoral cusps 2–4; infraoral cusps 7–12 (usually 8–10); circumoral teeth 19–24 (usually 21); teeth in anterior row 3–5 (usually 4); teeth in lateral rows 5–9 (usually 7); bicuspid circumorals 7–11 (usually 8). Transverse lingual lamina moderately to strongly bilobed. Maximum length about 163 mm. Upper Ohio River system. Ohio, Pennsylvania, West Virginia, and Kentucky. Nonparasitic.

Ichthyomyzon hubbsi Raney. Tennessee brook lamprey. Small (103–150 mm). Nonparasitic; related to *I. greeleyi*. Body deep. First dorsal and anal fins low. Myomeres 57 (55–59). Buccal disc small, dentition weak; supraoral cusps usually 2; infraoral cusps 7–11; circumoral teeth 20–27; teeth in anterior row 3–4; lateral teeth 5–7; bicuspid circumorals 5–10; teeth slightly, if at all, cornified. Transverse lingual lamina moderately bilobed, with or without minute denticulations. Dorsal lateral-line (L.l.) organs dark. Mountain tributaries of the Tennessee River in North Carolina and Georgia.

Genus Petromyzon Linnaeus. Large (over 900 mm). Supraoral cusps 2 or 3, with sharp points close together, not in crescentlike plate; anterior lingual tooth with a median depression. Buccal disc large, with numerous teeth in rows radiating from oral opening. Dorsal fins separate, the second confluent with caudal. 1 species.

Petromyzon marinus Linnaeus. Sea lamprey. Circumoral teeth bicuspid, others simple, 4–7 in each row. Lips fringed. Adults in spring with a fleshy ridge before dorsal fin. About 70 myomeres. Atlantic coasts of Europe and North America, recently established in all the Great Lakes. Ascending streams in the spring to spawn.

Genus Entosphenus Gill. Small to large lampreys with supraoral lamina crescent-shaped and bearing 2 or 3 cusps; anterior lingual tooth wedge-shaped and finely serrate; teeth on all fields of disc, but not in radiating rows. Dorsal fins separate. Parasitic or not.

KEY TO SPECIES OF *ENTOSPHENUS*

1. Supraoral cusps 2. Adults under 200 mm. Atlantic drainage   **E. lamottei**
   Supraoral cusps 3. Adults 450 mm or more. Pacific drainage   **E. tridentatus**

Entosphenus tridentatus (Gairdner). Pacific lamprey. Teeth on all fields of disc; a row of small marginals; lateral teeth usually multicuspid; teeth on posterior field a row parallel to marginal series; supraoral lamina crescentic, bearing 3 cusps, the middle one smaller and all connected by bridges. Myomeres 57–74. Unalaska to Santa Ana River in southern California. Parasitic; anadromous.

Entosphenus lamottei (LeSueur). American brook lamprey. Similar to *E. tridentatus* but much smaller. Supraoral cusps 2; lateral bicuspid teeth 3; all teeth blunt and weak. Intestine in adult nonfunctional. Myomeres 63–70. Eastern and northern United States from Minnesota to New Hampshire south to Tennessee and Missouri; on the Atlantic Coast from Connecticut to Maryland. Nonparasitic.

Genus Lampetra Gray. No teeth other than the marginals on the posterior field of disc; 3 enlarged lateral (circumoral) elevations on each side of the esophageal opening, usually bearing unicuspid, bicuspid, or tricuspid teeth; supraoral lamina, without a median cusp, a transverse band with usually a blunt cusp at each end; infraoral lamina with a row of more or less reduced cusps; transverse lingual lamina with a few blunt cusps or crenulations; cusps on the longitudinal lingual lamina indistinct to well developed. Dorsal fin completely or incompletely divided into anterior and posterior portions. Eurasia and North America.

KEY TO SPECIES OF *LAMPETRA*

1. Visible portion of supraoral lamina consisting of 2 separated lobes, with or without blunt cusps; lateral (circumoral) cusps poorly developed. Eastern United States   **L. aepyptera**
   Supraoral lamina a transverse band, with a cusp at

each end; lateral (circumoral) cusps well developed, enlarged. Pacific Coast 2

2. Middle lateral (circumoral) tooth on each side tricuspid; longitudinal lingual cusps 19–22

L. ayresi

Middle lateral (circumoral) tooth on each side usually bicuspid; longitudinal lingual cusps small, uncountable (see Fig. 2-2, but note the unusual tricuspid, middle lateral tooth) *L. richardsoni*

Lampetra ayresi (Günther). River lamprey. Formerly confused with *L. fluviatilis*, a European brook lamprey (see Vladykov and Follett, 1958). Myomeres 65–70 (rarely 60 or 71). Middle tooth of lateral (circumoral) rows tricuspid, the other 2 bicuspid; transverse lingual cusps about 14, the middle one enlarged; infraoral cusps 7–10, the end ones often bicuspid; longitudinal lingual cusps 19–22. Back and sides brownish gray, venter whitish; caudal fin with a dark blotch. Anadromous, parasitic. From the Skeena River, British Columbia, to San Francisco Bay, California.

Lampetra richardsoni Vladykov and Follett (Fig. 2-2). Western brook lamprey. Similar to *L. planeri* (Bloch) of Europe (see Vladykov and Follett, 1965). Myomeres 57–67. Middle tooth of lateral (circumoral) rows usually bicuspid, the other two unicuspid or bicuspid; transverse lingual cusps small, the middle one enlarged; infraoral cusps 7–10; longitudinal lingual cusps small, impossible to count. Back and sides grayish, venter whitish; caudal fin, heavily pigmented. Nonparasitic. From British Columbia, Washington, Oregon, and possibly Alaska.

Lampetra aepyptera (Abbott). Ohio brook lamprey. Myomeres 51–60. Lateral (circumoral) cusps poorly developed, sometimes represented by small elevations without cusps; transverse and longitudinal lingual lamina toothless; infraoral cusps blunt, 5–7. From the Wabash Basin in Indiana, upper Ohio, along the Atlantic drainage from Maryland to North Carolina, the Alabama River system of Alabama and Georgia, and the Tombigbee River of Mississippi.

CLASS TELEOSTOMI

The teleosts or true fishes all have well-developed jaws. Paired fins, at least pectorals, are present in all fresh-water species of the United States. All fins except the adipose are supported by rays which may be branched or unbranched and segmented or not segmented. The notochord is persistent or more or less replaced by vertebral centra. 3 semicircular canals are regularly present in the inner ear. Unlike the Petromyzones, teleosts have only 1 pair of gill slits. The gills are covered by an operculum usually composed of 4 plates, the opercle, preopercle, subopercle, and interopercle. Each gill bears 2 rows of filaments or branchial rays. An air bladder is typically present and is connected or not connected with the pharynx by a duct. In some teleosts, such as the bowfins, gars, and mudminnows, the air bladder is abundantly supplied with blood and is used as a lung. Usually there is no cloaca. The teleosts first appeared in the Lower Devonian period and are at present worldwide in distribution. The 37 families in U.S. fresh water contain about 154 genera and 634 species. A considerable number of species have not been described.

Methods of Counting

The great variation in methods employed in the past for counting fin rays and scales has been a source of trouble, and in order to avoid repetition of these difficulties, ichthyologists for the most part have adopted the methods outlined by Hubbs and Lagler (1947). The advanced student who is obligated to search out facts from the literature must bear in mind that differences between his own and those meristic data of earlier ichthyologists may be due to differences in method rather than differences of a genetic nature.

The accompanying figures (Figs. 2-4 and 2-5) will help to clarify the standard methods briefly discussed here.

The dorsal, anal, and paired fins of teleosts may have spines, soft rays, or both spines and soft rays. Spines are never branched or segmented, whereas soft rays are segmented and often branched. If a species has a spinous dorsal of 11 rays and a distinctly separate soft dorsal of 12 rays, the fin-ray formula is written XI–12 to indicate that there is a space between the spines and soft rays. If the spinous and soft portions are united, a comma replaces the dash and the formula is XI,12. Standard

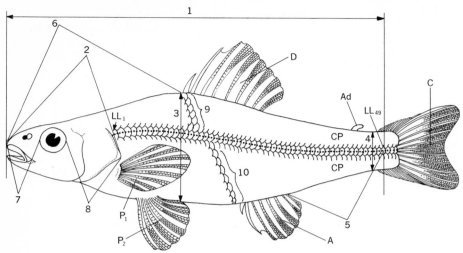

Fig. 2-4. Topography of a fish showing location of structures, regions used in identification, and methods of measurement. (After Bailey, 1951.) *A*, anal fin; *Ad*, adipose fin; *C*, caudal fin; *CP*, caudal peduncle; *D*, dorsal fin; *LL*, lateral line; *P₁*, pectoral fin; *P₂*, pelvic fin. 1, standard length; 2, head length (to tip of membrane); 3, body depth; 4, least depth of caudal peduncle; 5, length of caudal peduncle; 6, predorsal length; 7, snout length; 8, postorbital length of head; 9, scales above lateral line; 10, scales below lateral line.

abbreviations for fins are: D., dorsal; C., caudal; A., anal; P₁, pectoral; P₂, pelvic. These abbreviations are used in the species descriptions in this text.

In some teleost groups rudimentary rays in dorsal and anal fins grade from tiny ones to well-developed rays making it necessary to count all of them, but if the rudiments are few and distinctly different in length from the developed rays, only the developed rays are counted. Thus in catfishes, salmonids, and pikes all rays are counted, but in minnows and suckers only the developed or principal rays in the dorsal and anal fins are counted. The last ray is often double and is counted as 1 ray.

Unless otherwise stated only the principal rays of a caudal fin are counted. Sometimes the caudal rays, except one at the upper and lower edges, are branched. In such fishes 2 is added to the number of branched rays.

All rays in pectoral and pelvic fins are counted.

Scales in the lateral line are counted from the first scale in contact with the shoulder girdle to the one which is over the end of the hypural plate or structural caudal base (Fig. 2-6). In some fishes there is no lateral line, and a count of the rows of scales crossing the mid-sides is taken. In salmonids the lateral-line scale count is often very different from a count of scales made in the second row above the lateral line. The count in the second row above the lateral line gives a more reliable and useful separation of species in these fishes.

The scales above the lateral line, unless otherwise indicated, are counted from the dorsal origin diagonally downward and backward, to but not including, the lateral-line scale.

The scales below the lateral line are counted from the anal origin diagonally upward and forward to, but not including, the lateral-line scale. In species descriptions and keys the words "lateral line" are abbreviated L.l.

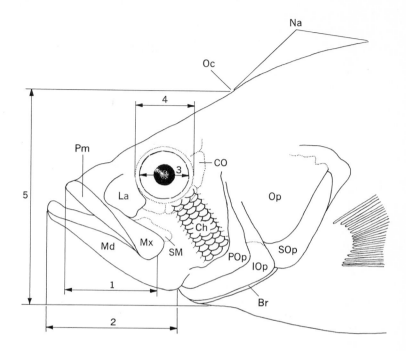

Fig. 2-5. Head of fish showing structures, regions used in identification, and measurement methods. (After Bailey, 1951.) *Br*, branchiostegal ray; *Ch*, cheek; *CO*, circumorbital; *IOp*, interopercle; *La*, lacrimal (or preorbital); *Md*, mandible; *Mx*, maxilla; *Na*, nape; *Oc*, occiput; *Op*, opercle; *Pm*, premaxilla; *POp*, preopercle; *SM*, supramaxilla; *SOp*, subopercle. 1, length of upper jaw; 2, length of mandible; 3, diameter of eye; 4, diameter of orbit; 5, depth of head.

Scales around the caudal peduncle are counted around the slenderest portion, in zigzag fashion.

In counting cheek scales the rows crossed by a line from the orbit to the angle of the preopercle are included.

Predorsal scales may be counted in 2 ways: as the number of scales impinging on the middorsal line from the dorsal origin to the occiput or as the number of scale rows crossing the middorsal line. The first count is usually the greater.

Sometimes scale rows around the body in front of the dorsal fin are counted. It is best to proceed in a zigzag manner from about a scale row or 2 in front of the dorsal origin and progress around the body to end just short of the midline where the count began. In recording this count, it is well to keep separate the scale number above and below the lateral line, since it may happen that 2 species with the same total number of scales around the body will have a significantly different distribution with respect to the lateral line.

The method of counting pharyngeal teeth is given below in connection with the family Cyprinidae.

Methods of Measuring

Total length is the distance from the anteriormost projection of the head to the end of the longest caudal ray, with lobes squeezed together.

Standard length, most useful in taxonomic work, is the distance from the snout tip to the end of the hypural plate or structural base of the caudal fin. In fishes having a heterocercal tail, total length is used.

Body depth is the greatest depth, not including fleshy or scaly structures at fin bases.

Caudal-peduncle depth is the least depth of the caudal peduncle.

Caudal-peduncle length is the oblique distance between the posterior end of the anal-fin base to the hidden base of the middle caudal ray.

Predorsal length is the distance between the dorsal origin and the snout tip.

Fin-base length is the distance between the bases of the first and last fin ray.

Dorsal- or anal-fin height is taken from the origin to the end of the anterior lobe.

Length of longest ray is the distance from the structural base of the ray to its tip.

Length of paired fins is measured from that end of the fin base under the best-developed rays. This end is the uppermost, outermost, or anterior-most one, and the measurement extends to the tip of the longest ray, including a filament, if present.

Spine and soft-ray length is taken, in the case of spines, from the structural base of the spine to its bony tip and, in the case of soft rays, from the structural base to the tip of the ray, including any filamentous structure.

Head length is measured from the snout tip to the end of the opercular membrane.

Head width is the greatest distance between the opercles when they are in normal (resting) position.

Snout length is the distance from the midanterior snout tip to the anterior bony orbital rim.

Postorbital head length is taken from the posterior bony orbital rim to the end of the opercular membrane.

Interorbital width is either the least bony or the least fleshy width, between the orbits, as specified. Gentle pressure of the dividers will serve to take the bony width.

Orbital length is the greatest length, often oblique, between the free orbital rims.

Eye length is the greatest distance across the cornea.

Upper-jaw length is the distance from the anterior premaxilla tip to the posterior end of the maxilla.

Mandible length is the greatest distance between the tip of the lower jaw to its hinder end.

Gape width is the greatest transverse distance across the mouth opening.

KEY TO ORDERS OF TELEOSTOMI

1. Caudal fin heterocercal (Fig. 2-6) **2**
   Caudal fin not heterocercal, but usually homocercal (Fig. 2-6) **4**
2. Caudal fin strongly heterocercal, the lower lobe well developed. Mouth inferior, the jaws almost or entirely toothless. Endoskeleton largely cartilaginous
   Acipenseriformes
   Caudal fin abbreviate-heterocercal (Fig. 2-6) and

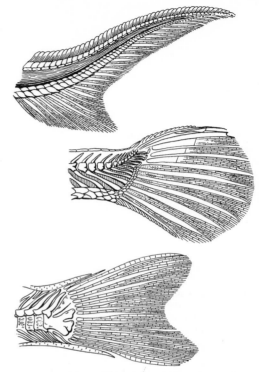

Fig. 2-6. Types of caudal fins in fishes. (Forbes and Richardson, 1920.) Heterocercal fin of sturgeon (upper); abbreviate heterocercal fin of gar (middle); homocercal fin of pikeperch (lower).

rounded behind. Mouth terminal and bearing strong teeth. Endoskeleton largely bony **3**
3. Scales ganoid (Fig. 2-7). No gular plate. Dorsal short, its origin behind anal origin. Snout elongated into a beak **Lepisosteiformes**
   Scales cycloid (Fig. 2-8). Gular plate present. Dorsal long, its origin in front of pelvic insertion. Snout blunt **Amiiformes**

Fig. 2-7. Ganoid scale of *Lepisosteus osseus*.

5 mm

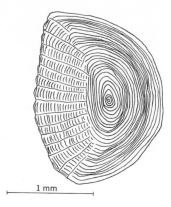

Fig. 2-8. Cycloid scale of *Gambusia affinis*.

1 mm

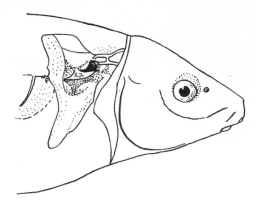

Fig. 2-10. Lateral view of the Weberian apparatus in *Moxostoma erythrurum*. (From Krumholz, 1943.)

4. Skull asymmetrical, both eyes on same side of head. The eyed side of the body with most pigment **Pleuronectiformes**
 Skull symmetrical, eyes (if present) on opposite sides. Pigmentation of the 2 sides similar **5**
5. Dorsal, caudal, and anal fins continuous **Anguilliformes**
 Dorsal, caudal, and anal fins separate **6**
6. Body completely covered with dermal plates. Snout elongated, tubular. Anal fin rudimentary or absent. Males with a large subcaudal brood pouch **Syngnathiformes**
 Body scaled or naked, never with dermal plates. Snout never tubelike. Anal fin present. Males without a brood pouch **7**
7. Anterior vertebrae modified (Fig. 2-10). Head naked; body naked or scaly **Cypriniformes**
 Anterior vertebrae not greatly unlike those behind. Head scaly, except in the Clupeiformes, Gasterosteiformes, Percopsiformes (in part), and some Perciformes; body scaly or with few to many bony plates (if naked, free dorsal spines are present) **8**

8. Spines wholly absent (if anus is far forward, not near anal fin origin, take couplet 14) **9**
 Spines present in at least some of the fins (spines may be flexible or stiff, but not segmented) **12**
9. Barbels absent **10**
 A single median chin barbel present (Fig. 2-11) **Gadiformes**
10. Pectoral fins placed high on the sides, above the axis of the body **Beloniformes**
 Pectoral fins placed near or much below the body axis **11**
11. Maxillae not forming border of upper jaw. Jaw teeth conical, in villiform bands or incisorlike with 1, 2, or 3 points (Fig. 2-82) **Cyprinodontiformes**
 Maxillae form part of the border of upper jaw. Jaw teeth, if present, conical and never in villiform bands or incisorlike **Clupeiformes**
12. Dorsal fin preceded by free spines, not connected by membranes (Fig. 2-12) **Gasterosteiformes**
 Dorsal spines connected by membranes, to each other or to soft dorsal **13**
13. Pelvic fins abdominal or subabdominal in position (Fig. 2-13). Pectoral fins inserted high on the sides, usually above body axis. Scales cycloid **Mugiliformes**

Fig. 2-9. Ctenoid scale of *Percina caprodes*. (Redrawn from Storer and Usinger, 1965.)

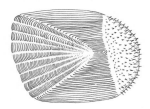

Fig. 2-11. *Lota lota*. Note the single chin barbel.

Fig. 2-12. Eucalia inconstans. (Forbes and Richardson, 1920.)

Pelvic fins thoracic (Fig. 2-13) in position or nearly so (abdominal in Percopsidae, which have an adipose dorsal fin). Pectoral fins usually inserted below body axis. Scales ctenoid (Fig. 2-9) **14**

14. Adipose fin present, or, if absent, the anus is far forward in region of isthmus **Percopsiformes**
    Adipose fin absent, and anus in usual position just in front of anal-fin origin **Perciformes**

Order Acipenseriformes. Sturgeons and paddle-fishes. Snout elongated and mouth ventral. Body with 5 rows of bony plates or largely naked. 2–4 barbels. Tail heterocercal. Premaxillae fused with maxillae. Preopercle absent or rudimentary, interopercle absent. Large anadromous and fresh-water fishes of the Northern Hemisphere. Upper Cretaceous. 2 families and 3 genera in the United States.

KEY TO FAMILIES OF ACIPENSERIFORMES

1. Body with tiny embedded scales, some large ones on upper caudal lobe **Polyodontidae**
   Body with 5 rows of bony plates, 1 dorsal, 2 lateral, and 2 ventral **Acipenseridae**

Family Polyodontidae. The paddlefishes. Snout long, depressed, and spatulate, often wider at its tip than at base. 2 minute barbels on lower posterior surface of snout. Gill cover greatly produced backward. Gill rakers very long, slender and numerous. Spiracles present. Numerous teeth in young, disappearing with age. Eyes tiny. Skeleton largely of cartilage. Heterocercal tail. Scales tiny, embedded, some larger ones on C. fin base. 2 living representatives, *Polyodon* of the United States and *Psephurus gladius* of China. Upper Cretaceous.

Genus *Polyodon* Lacépède.

Polyodon spathula (**Walbaum**). Paddlefish. Formerly abundant in larger streams. Now needing protection since sexual maturity is not reached early in life. Spawning over gravel bars has been observed in the Osage River, Missouri. Great Lakes and Mississippi Valley.

Family Acipenseridae. The sturgeons. Body with 5 rows of bony plates, 1 dorsal, 2 lateral, and 2 ventral. 4 barbels on lower surface of snout. Jaws toothless in adult. Mouth protractile. Branchiostegals absent. Gill rakers few. D. behind P_2. First P_1 ray a spine. Circumpolar in distribution. Upper

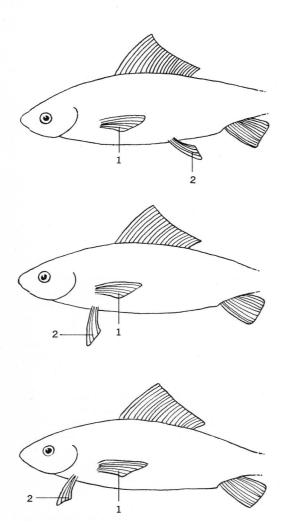

Fig. 2-13. Position of pelvic fins: abdominal (**upper**); thoracic (**middle**); jugular (**bottom**).
1, pectoral fin; 2, pelvic fin. (Brauer, 1909.)

Cretaceous. 2 genera and 7 species in the United States.

KEY TO GENERA OF ACIPENSERIDAE

1. Caudal peduncle incompletely armored. No caudal filament. Spiracles and pseudobranchiae present
Acipenser
   Caudal peduncle completely armored. A filament on upper lobe of caudal. Spiracles and pseudobranchiae absent *Scaphirhynchus*

Genus *Acipenser* Linnaeus. Snout subconical. A small spiracle above the eye. C. peduncle without plates. Pseudobranchiae present. 4 barbels form a transverse row on the lower surface of snout.

KEY TO SPECIES OF *ACIPENSER*

1. Atlantic Coast   2
   Pacific Coast   4
2. Gill rakers on lower arm of 1st gill arch fewer than 15   *A. oxyrhynchus*
   Gill rakers on lower arm of 1st gill arch more than 14 (rarely 14)   3
3. Gill rakers on lower arm of 1st gill arch 14–19. Last dorsal plate more than ½ size of one before it. Anal fin with about 37 rays, its origin below middle of dorsal base (Fig. 2-14)   *A. fulvescens*

Fig. 2-14. Acipenser fulvescens. (Jordan and Evermann, 1900.)

Gill rakers on lower arm of 1st gill arch 17–27. Last dorsal plate less than ½ size of one before it. Anal fin with about 22 rays and entirely below dorsal fin   *A. brevirostris*
4. Plates between pelvics and anal small, in 2 rows of 4–8. Dorsal rays 44–88. Anal rays 28–30
    A. transmontanus
   Plates between pelvics and anal large, in 1 or 2 rows of 1–4 each. Dorsal rays 33–35. Anal rays 22–28
    A. medirostris

Acipenser transmontanus **Richardson.** White sturgeon. Largest sturgeon of the United States, reaching a length of nearly 390 cm and a weight of more than 1200 lb. Color dark grayish, without stripes. Snout sharp in young, blunter in adult. Barbels nearer tip of snout than mouth. Gill rakers about 26. Dorsal plates 11 or 12, laterals 36–50 (usually 44), ventrals 10–12. D. 45 (44–48); A. 28–30. From Alaska to Monterey, ascending large rivers in spring as far east as the Kootenai in Montana.

Acipenser medirostris **Ayres.** Green sturgeon. Color olive green, an olive stripe on midventral line, and one on each side above the ventral plates. Snout as in *A. transmontanus*. Barbels about midway between snout tip and mouth. Gill rakers about 17. Dorsal plates 10 (9–11); laterals 26–30; ventrals 9 (7–10). D. 33–40; A. 22–28. Pacific Coast from the Fraser River south to Point Vicente, ascending rivers.

Acipenser oxyrhynchus **Mitchill.** Atlantic sturgeon. Color olive gray, paler below. Snout rather sharp, nearly as long as head, growing blunter in adults. Barbels short, midway between snout tip and mouth. Gill rakers small, slender, pointed, and fewer than 25 on 1st arch. Dorsal plates 7–14; laterals 24–35; ventral plates 10 (8–12); preanal plates in double series (4–8); postdorsal plates in 2–4 pairs. D. 38–46; A. 27 (26–28). Atlantic Coast of North America from Quebec south to Gulf and tributary waters, ascending rivers. Critical study may prove this sturgeon to be conspecific with *A. sturio* of Europe.

Acipenser brevirostris **LeSueur.** Shortnose sturgeon. Color dusky, paler below. Snout very short, blunt, about ¼ head. Gill rakers on lower arm of 1st gill arch 14–19. Barbels short. Dorsal plates 11 (8–11); laterals 32 (22–33); ventral plates 9 (6–9). D. about 41; A. about 22. Cape Cod to Florida, more abundant in the South than in the North.

Acipenser fulvescens **(Rafinesque).** Lake sturgeon. Color dark olive above, sides paler or reddish, frequently spotted with blackish. Snout slender in young, but very blunt in adults. Gill rakers short, blunt, and numerous (27–39 on 1st arch). Dorsal plates 13 (9–17); laterals 35 (29–42); ventral plates 9 (7–12), deciduous in old individuals; preanal plates 1–4; postdorsal plates 1–3 in single series.

D. 38 (36–40); A. 28 (25–30). Great Lakes, Lake Champlain to Alabama, Missouri, and Nebraska.

Genus *Scaphirhynchus* Heckel. Snout broad, depressed, and shovel-shaped. Spiracles absent. C. peduncle long, slender, depressed, and completely covered with bony plates, the rows confluent below D. D. lobe of C. with a long filament. Mississippi and Rio Grande Basins.

KEY TO SPECIES OF *SCAPHIRHYNCHUS*

1. Belly with small dermal plates (except in young). Inner barbel more than ⅔ length of outer. Dorsal fin rays 30–36. Anal fin rays 18–23. Eye longer than anterior nostril (in young and half-grown) and equal to it in adults   *S. platorynchus*
   Belly largely naked. Inner barbel less than ⅔ length of outer. Dorsal fin rays 37–43. Anal fin rays 24–28. Eye about equal to (young and half-grown) or less than (adults) anterior nostril   *S. albus*

Scaphirhynchus platorynchus (**Rafinesque**). Shovelnose sturgeon. Color buff or drab above and light below. Small, seldom weighing as much as 8 lb, most specimens less than half that weight. Mississippi Valley south to northern Alabama and Texas. 1 record from the Rio Grande in New Mexico.

Scaphirhynchus albus (**Forbes and Richardson**). Pallid sturgeon. Similar to S. *platorynchus,* but belly naked, color much paler, inner barbels shorter, eye smaller, snout longer and sharper, and reaching a larger size (almost 70 lb). Mississippi Valley in Illinois, the Dakotas, Montana, Kansas, and Iowa, south to Louisiana. Apparently rare.

Order Amiiformes. Bowfin. C. abbreviate-heterocercal. Premaxilla nonprotractile. Lower jaws broad, a large bony gular plate between the rami. Entire subconical head covered with bony plates. Jaws with sharp conical teeth; a band of rasplike teeth on hinder part of lower jaw. Vomer, palatines, and pterygoids toothed. Anterior nostril tubular. No pseudobranchiae. Gill rakers very short. L.l. present. D. long and low. Vertebrae double-concave. Air bladder functioning as a lung, richly supplied with blood and connected with the pharynx. No closed oviduct. No pyloric caeca. Upper Permian. Eastern North America. 1 family; 1 species in the United States.

Family Amiidae. Bowfin.

Genus *Amia* Linnaeus.

Amia calva **Linnaeus.** Bowfin (Fig. 2-15). Blackish olive with greenish reticulations. Male with a black ocellus edged with orange at C. base. D. 48; A. 11. Carnivorous, reaching about 90 cm in length. Lakes and sluggish waters of North America from Minnesota, Quebec, and Vermont southward west of the Appalachians to Florida and Texas; north on the Atlantic Coast to the Susquehanna River. Introduced in Connecticut.

Order Lepisosteiformes. Gars. C. abbreviate-heterocercal. Ganoid scales. Nasal openings at end of greatly elongated snout. Lacrimal divided into a series of several toothed bones, forming a large portion of upper-jaw margin. Maxilla, at least in young, a small toothed bone at angle of mouth and behind the lacrimal series. The complex lower jaw articulated with skull in front of eye. Vomer paired. No gular plate; branchiostegals 3. Body greatly elongated. D. placed well back toward C. 1 family, living in fresh waters of North America east of the Rocky Mountains south to Guatemala, Cuba, and the Lake Nicaragua Basin; on Pacific slope of

Fig. 2-15. Amia calva.

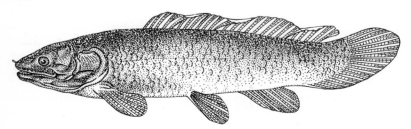

middle America from southern Mexico to Honduras.

Family Lepisosteidae. Gars. 1 genus.
Genus *Lepisosteus* Lacépède.

KEY TO SPECIES OF *LEPISOSTEUS*

1. Palatine bones of adult with a row of enlarged teeth.
   Gill rakers of 1st arch 59–66   **L. spatula**
   Palatine bones of adult lacking enlarged teeth. Gill
   rakers of 1st arch 14–33   2
2. Least snout width contained 13–25.5 times in its
   length in specimens more than 50 mm long
       L. osseus
   Least snout width contained 4.5–11 times in its length
   in specimens more than 50 mm long   3
3. L.l. scales 60–63 (extremes 59–65); predorsal scales
   52–53 (extremes 50–55). Body and head without
   dark spots   **L. platostomus**
   L.l. scales 54–58 (53–59); predorsal scales 47–50
   (45–54), rarely more than 51. Body and head with
   dark spots   4
4. Adults with bony plates on isthmus   **L. oculatus**
   Adults without bony plates on isthmus
       L. platyrhincus

Lepisosteus oculatus **Winchell.** Spotted gar (Fig. 2-16). Similar to *L. platyrhincus*, but with bony plates on isthmus; snout and lower jaw longer and narrower. Body much heavier than that of *L. platostomus*. Anterior part of body and head with large, round, dark spots. L.l. scales about 56; predorsal scales 45–54; scale rows between P_2 insertion and D. origin 27–32. From the Great Lakes through the Mississippi Valley to the Gulf of Mexico and from western Florida to central Texas.

Lepisosteus osseus (**Linnaeus**). Longnose gar (Fig. 2-16). Large young to adults easily recognized by the long slender snout. Vertical fins and posterior part of body usually with black blotches. L.l. scales 57–63; predorsal scales 47–55; scale rows between P_2 insertion and D. origin 31–35. From Quebec to Florida, except the eastern part of New England states; from the Great Lakes to northern Mexico. Abundant in the Mississippi Valley.

Lepisosteus platyrhincus **DeKay.** Florida spotted gar. Similar to *L. oculatus* but lacking bony plates on isthmus and having longer and broader snout and lower jaw. L.l. scales 54–59; predorsal

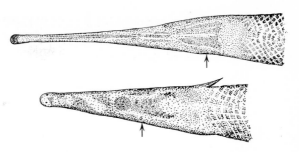

Fig. 2-16. Dorsal views of heads of 2 gars to show differences in proportions and coloration. *Lepisosteus osseus* (upper) and *Lepisosteus oculatus* (lower). Arrows indicate end of beak at anterior orbital rim. (From Hubbs and Lagler, 1947.)

scales 47–51; scale rows between P_2 insertion and D. origin 30–33. From the lowlands of Georgia south to southern Florida.

Lepisosteus platostomus **Rafinesque.** Shortnose gar. Recognized by the short snout, large teeth of upper jaw in single series, unspotted body (often with streaks between scale rows), and its especially slender, but short, body. Larger muddy rivers of Mississippi River drainage.

Lepisosteus spatula **Lacépède.** Alligator gar (Fig. 2-17). Adults readily identified by enlarged teeth on palatine bones. Snout short, broad, and stout, L.l. scales 58–62; predorsal scales about 52; scale rows between P_2 insertion and D. origin 34–38. Maximum known size 9 ft 8.5 in. and weight 302 lb. Mississippi River and lower parts of major tributaries from the Ohio and Missouri Rivers to the Gulf of Mexico; also in brackish and salt waters from Choctawhatchee Bay, Florida to northern Mexico.

Order Clupeiformes.[1] Salmons, chars, trouts, whitefishes, pikes, mooneyes, herrings, shad, tarpons, and anchovies. C. homocercal. Scales cycloid. Weberian apparatus absent. Intermuscular bones present. Upper jaws usually bordered by premaxillae and maxillae. P_2 abdominal. An artificial as-

[1] Greenwood et al. (1966) proposed recognition of order Elopiformes and suborder Elopoidei for the Elopidae and Megalopidae; order Clupeiformes and suborder Clupeoidei for the Clupeidae and Engraulidae; order Salmoniformes with suborders Salmonoidei, for the Salmonidae and Osmeridae, and Esocoidei for the Esocidae and Umbridae; and the relegation of the suborder Notopteroidei to the order Osteoglossiformes (bony tongues).

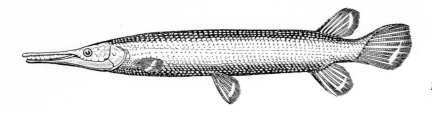

Fig. 2-17. Lepisosteus spatula.
(Jordan and Evermann, 1900.)

semblage that may, on basis of future studies, be divided into several orders. Middle Triassic. Worldwide distribution.

KEY TO SUBORDERS OF CLUPEIFORMES

1. Adipose fin present   Salmonoidei
   Adipose fin absent   2
2. Head at least partially scaled   Esocoidei
   Head naked   3
3. L.l. present. No gular plate   Notopteroidei
   L.l. absent, or, if present, there is a gular plate
        Clupeoidei

Suborder Clupeoidei. Adipose fin absent. Oviducts present. Lower Cretaceous. Widely distributed in temperate to tropical seas; some in fresh water.

KEY TO FAMILIES OF CLUPEOIDEI

1. L.l. and gular plate both present   Elopidae
   L.l. and gular plate both absent   2
2. Maxillae not reaching beyond middle of eye
        Clupeidae
   Maxillae reaching beyond middle of eye
        Engraulidae

Family Elopidae. Tarpons and tenpounders. Gular plate present; no angular; 2 supramaxillae; branchiostegals numerous; upper jaw bordered by premaxillae and maxillae. Air bladder not connected with ear. Pseudobranchiae present or absent. Widely distributed in tropical and subtropical seas; 2 genera and 3 species enter fresh water in the United States.

KEY TO GENERA OF ELOPIDAE

1. Pseudobranchiae absent. Scales large, about 42 in L.l. Anal fin larger than dorsal; last dorsal ray filamentous   *Megalops*

Pseudobranchiae present. Scales small, about 120 in L.l. Anal fin smaller than dorsal; last dorsal ray not filamentous   *Elops*

Genus *Megalops* Valenciennes. Filament of D. longer than head. Mouth large, the maxilla extending beyond eye. Color silvery, the back darker. Jaws, vomer, palatines, tongue, sphenoid, and pterygoid bones toothed. L.l. almost straight, with tubes radiating on scales. Gill rakers long and slender. D. short (12 rays), its origin behind pelvics; A. long (20 rays), its last ray elongated and its base with a scaly sheath. 1 species in the United States.

Megalops atlantica **Valenciennes.** Tarpon. The largest herringlike fish, maximum weight 300 lb. An important sport fish. Nova Scotia to Brazil, common along the Gulf Coast. Fresh and salt water.

Genus *Elops* Linnaeus. D. long, 20 rays, and depressible into a scaly sheath, last D. ray not filamentous; A. smaller, 13 rays, and similar to D. L.l. straight, its tubes simple. Gular plate 3 or 4 times as long as broad. Fins and back blackish green, otherwise silvery. 2 species.

Elops saurus **Linnaeus.** Ladyfish. Maximum size 90 cm. Tropical and subtropical seas from Cape Cod to Recife, Brazil, but not common north of North Carolina. Enters the mouths of rivers.

Elops affinis **Regan.** Machete. Known as far up the Colorado River as Laguna Dam, Arizona, and California; also in nearby Salton Sea, southeastern California.

Family Clupeidae. Herrings and shad. Deciduous cycloid scales. Dentition weak or absent in adult. Air bladder connected with ear. 1 or 2 supramaxillae; L.l. short, on only 2–5 scales. Belly weakly or strongly keeled along midline. Temperate and tropical seas, 2 genera and 9 species in fresh waters of the United States.

KEY TO GENERA OF CLUPEIDAE

1. Mouth terminal, jaws equal or the upper included. Posterior dorsal fin ray not elongated (Fig. 2-18)
Alosa

Mouth subterminal or inferior; jaws not equal, the lower included. Posterior dorsal fin ray greatly elongated *Dorosoma*

Genus *Alosa* Linck. Silvery fishes with thin, cycloid, deciduous scales. Jaws about equal or with lower projecting, upper jaw shallowly or deeply notched. Teeth feebly developed or absent on jaws, strongly developed or absent on tongue. Adipose eyelid variously developed. D. short, nearly midway between snout and C. base. 7 species, mostly anadromous.

KEY TO SPECIES OF *ALOSA*

1. Peritoneum black *A. aestivalis*
Peritoneum silvery or pale 2
2. Premaxillae meeting at a midanterior point in an acute angle to form a deep notch receiving lower jaw 3
Premaxillae meeting in an obtuse angle to form a shallow notch 6
3. Tongue and jaws toothless 4
Tongue with a small patch or a single row of teeth; jaws with weak teeth, lost in adult 5
4. Gill rakers long and slender, about 60 on lower arm of 1st arch *A. sapidissima*
Gill rakers shorter, about 40 on lower arm of 1st arch *A. alabamae*
5. Tongue with a small patch of teeth. A black spot back of opercle *A. pseudoharengus*
Tongue with single row of teeth. No black spot back of opercle *A. ohiensis*
6. Tongue teeth sometimes present; lower-jaw teeth present only in young. Sides with faint longitudinal stripes *A. mediocris*
Tongue teeth always present; lower-jaw teeth present in young and usually in adults. Sides plain *A. chrysochloris*

Alosa chrysochloris (Rafinesque). Skipjack herring. Head slender and pointed, the upper profile straight. Lower jaw strongly projecting, upper jaw notched. Teeth present on premaxillae, tongue, and usually on dentaries. Eye large, with conspicuous adipose lid. C. peduncle slender. Gill rakers about 23 below angle on 1st arch. Blue above, sides sil-

very with golden iridescence; no dark spots behind opercles. Peritoneum pale. Resident in larger streams of Mississippi Valley and along Gulf Coast from Galveston, Texas, to the Escambia River, Florida. Not valued as food.

Alosa mediocris (Mitchill). Hickory shad. Head rather long, profile straight. Lower jaw projecting, upper jaw notched. Teeth sometimes present on tongue and, in young, on lower jaw. Eye relatively small. Bluish silvery, with faint longitudinal stripes and 5 or 6 spots in a row back of opercle. Peritoneum pale. Atlantic Coast from Bay of Fundy to Florida, ascending streams to spawn (according to Elser, 1950, but not Jordan and Evermann, 1896).

Alosa pseudoharengus (Wilson). Alewife (Fig. 2-18, upper). Head short, profile steep. Jaws nearly equal, upper notched. Teeth in a small patch on tongue. Eye large, longer than snout. Lower C. lobe longer than upper. Bluish above, sides silvery; indistinct dark stripes along scale rows; a single spot behind opercle. Peritoneum pale. Native from Newfoundland to Florida, and Lake Ontario. Entered by way of the Welland Canal and migrated to Lakes Huron, Michigan, and Superior. Landlocked in New England lakes, but usually anadromous. Large numbers taken by commercial fishermen.

Alosa aestivalis (Mitchill). Blueback herring. Similar to *A. pseudoharengus*, but with a black peritoneum, more elongate body, smaller eyes, and darker color. Not a highly valued food fish. Atlantic Coast from Nova Scotia to northern Florida. Anadromous.

Alosa sapidissima (Wilson). American shad (Fig. 2-18, lower). Jaws about equal, the lower fitting into the notch of the upper. Cheeks much deeper than long. Preopercle, with short lower arm, joining mandible at a point behind eye. Gill rakers long and slender, about 60 below angle on 1st arch. Bluish above, sides silvery; 1 to several spots in a row behind opercles. Peritoneum white. Atlantic Coast from Newfoundland to the St. Johns River, Florida. Introduced on the Pacific Coast where it occurs from Monterey northward. An excellent anadromous food fish.

Alosa alabamae Jordan and Evermann. Alabama shad. Similar to *A. sapidissima*, but smaller and with fewer and shorter gill rakers. Black War-

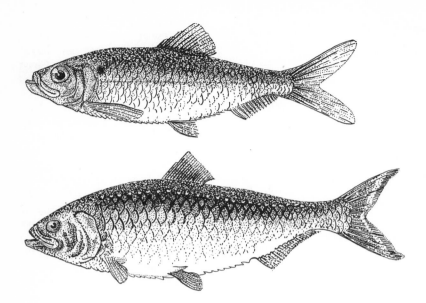

Fig. 2-18. Alosa pseudo-harengus (upper) and *Alosa sapidissima* (lower).

rior River, Alabama, and Escambia River, Florida.

Alosa ohiensis Evermann.   Ohio shad. Jaws almost equal, the upper with an acute notch. Teeth in a single median row on tongue, present on lower jaw only in young. Gill rakers 30–40 below angle of 1st arch. Mandible with dark pigment along most of its length. Ohio River valley and in the Mississippi Valley proper from Iowa to Oklahoma and Tennessee. Formerly more abundant than at present. No recent Iowa records, and only 2 specimens have been taken in Oklahoma. A resident stream fish not valued as food.

Genus Dorosoma Rafinesque.   Body short, deep, and compressed. Scales deciduous, thin, cycloid, 40–83 in lateral series. Belly sharply keeled with sharp-pointed scales (16–18, usually 17–18, + 10–12 scutes), forming a sawlike edge. Head naked. Mouth small to moderate, not reaching middle of eye, inferior, subterminal or terminal, oblique (lower jaw included or the jaw subequal), toothless in adult; young have row of fine teeth on upper jaw. Premaxilla not protractile. Maxilla with two supramaxillary bones. Snout short and rounded. Eye with adipose lid. D. about midway between tip of snout and C. base, last ray filamentous in adult, undeveloped in young. D. rays 9–15; A. rays long and shallow with 22–38 rays, P_1 rays 12–17, P_2 rays 8; C. rays 19 (17 branched). Pseudo-branchiae large, stomach thick-walled, gizzardlike, the intestine long and much convoluted, with numerous pyloric caeca. Vertebrae (including urostyle) 40–51. 2 species in the United States.

KEY TO SPECIES OF *DOROSOMA*

1. Anal fin speckled with melanophores, its rays 17–25. Young, less than 30 mm, with many melanophores along anal base   *D. petenense*
   Anal fin immaculate, its rays 29–35. Young, less than 30 mm, with few melanophores along anal base
                                               D. cepedianum

Dorosoma cepedianum (LeSueur).   Gizzard shad (Fig. 2-19). D. 12; A. 25–36 (usually 29–34). Silvery with bluish back; a round dark spot, about size of eye, behind operculum, disappearing with age. Maximum length about 450 mm. From the Dakotas, and Nebraska to St. Lawrence River near Quebec and Ohio Valley in western Pennsylvania; southward through Mississippi Valley to the Gulf and northeastern Mexico; north on Atlantic Coast to central New Jersey (rarely to New York) and southeastern Pennsylvania. The gizzard shad entered the Great Lakes by way of canals or as a migrant during postglacial times. It is uncertain and impossible to decide which of

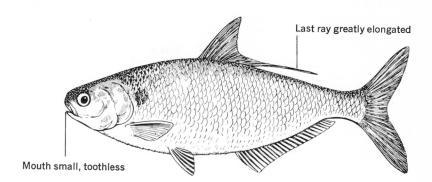

Last ray greatly elongated

Mouth small, toothless

Fig. 2-19. Dorosoma cepedianum. (Drawings by **Wallace Hughes**, in *Know Your Oklahoma Fishes*, by H. C. Ward.)

these 2 possibilities is correct. An extremely bony species not valued as food, but important as a forage fish, when young.

Dorosoma petenense (**Günther**). Threadfin shad. D. 14–15; A. 17–25. Scales in lateral series 40–43. Color silvery, with a postopercular spot smaller than eye. Size smaller than *D. cepedianum.* Gulf Coast from Florida to Texas, northward in Mississippi Valley to Tennessee and southern Arkansas and Oklahoma and southward to British Honduras. Eastward to West Virginia (by planting). Recently introduced as a forage fish in California and Arizona.

Family Engraulidae. Anchovies. Body elongate and compressed. Scales deciduous. Mouth very large, the maxilla extending behind eye. Piglike snout projects over mouth. Teeth usually very small. No angular, L.l., or adipose fin. Air bladder connected with ear. 1 species enters fresh waters of the United States.

Genus Anchoa Jordan and Evermann. With characters of the family. Belly not serrated. Vertebrae about 40. Gill membranes narrowly connected. Wide gill openings. No P_1 filaments. A. origin behind D. Lower jaw included; maxilla not extending behind operculum. Teeth very small. Gill rakers 20–40 below angle of 1st arch.

Anchoa mitchilli (**Valenciennes**). Bay anchovy. Depth more than $\frac{1}{5}$ standard length. D. origin nearer C. base than snout tip; D. 14; A. about 26. Both jaws toothed. Eye very large. Pale, with a narrow silvery lateral band. Atlantic and Gulf Coasts from Maine to Texas and southward to Yucatán, Mexico. Entering streams.

Suborder Salmonoidei. Salmons, chars, gray-

lings, whitefishes, and smelts. Cycloid scales. Head naked. Adipose fin present (in all U.S. species). Oviducts absent or incomplete. No lower intermuscular bones. Large (Chinook salmon, 100 lb) to small (less than 1 lb in some smelts). Lower Eocene. Northern seas and fresh waters of Northern Hemisphere.

KEY TO FAMILIES OF SALMONOIDEI

1. Last 3 vertebrae not turned upward. Entopterygoid toothed (Fig. 2-20)   **Osmeridae**
   Last 3 vertebrae turned upward. Entopterygoid toothless   **Salmonidae**

Fig. 2-20. Osmerus eperlanus.

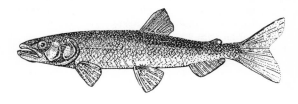

Family Salmonidae. Salmons, trouts, chars, graylings, and whitefishes. Body elongate. Mouth small to large, the maxilla often reaching beyond eye and forming the lateral margin of upper jaw; supplementary maxilla present. Jaw teeth weak or moderately to strongly developed. Pseudobranchiae present. Gill membranes free from isthmus. C. forked or somewhat truncate; P_2 abdominal. L.l.

present. Pyloric caeca numerous. Northern Hemisphere. Important food and game fishes. The nominal families Thymallidae and Coregonidae are included in the Salmonidae (see Norden, 1961).

KEY TO GENERA OF SALMONIDAE

1. Dorsal fin with fewer than 17 rays. Scales rounded, large or small   **2**
   Dorsal fin with 17 or more rays. Embedded edges of scales indented (Fig. 2-21)   ***Thymallus***

Fig. 2-21. Thymallus arcticus.
(Jordan and Evermann, 1900.)

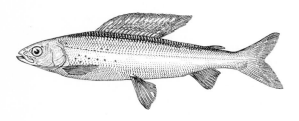

2. Jaw hinge behind orbit. Maxillae with teeth   **3**
   Jaw hinge not behind orbit. Maxillae without teeth
                                                                                 5
3. Fontanels absent in adult. Palatines widely separated from prevomer. Branchiostegals 13–19. Anal fin with 13 or more rays   ***Oncorhynchus***
   Fontanels present in adult. Palatines and prevomer narrowly separated. Branchiostegals 12 or fewer. Anal fin with 13 (rarely) or fewer rays   **4**
4. Color pattern of light markings on dark background. No teeth on prevomer shaft (Fig. 2-22)
                                                                        Salvelinus

Fig. 2-22. Salvelinus namaycush.
(Jordan and Evermann, 1909.)

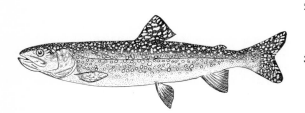

Color pattern of dark markings on a lighter background. Teeth on prevomer shaft   ***Salmo***
5. Nasal flap single (Fig. 2-25A). Juveniles with parr marks. Supraethmoid elongate   ***Prosopium***
   2 nasal flaps (Fig. 2-25B). No parr marks. Supraethmoid short (Fig. 2-23)   ***Coregonus***

Fig. 2-23. Coregonus clupeaformis.
(Jordan and Evermann, 1900.)

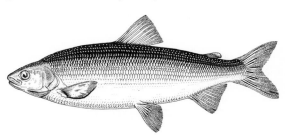

Genus *Oncorhynchus* Suckley. Jaws with moderate teeth, greatly enlarged in breeding male. Vomer long, narrow, and flat with teeth on both the head and shaft; teeth on tongue in a marginal series on each side; no basibranchial teeth; vomerine and tongue teeth often lost in adults. Both upper and lower jaws hooked in old males. North Pacific, ascending streams of Asia and North America. 5 species in the United States.

KEY TO SPECIES OF *ONCORHYNCHUS*

1. Sea-run fish without spots; fine stippling, red or calico marks may be present (old landlocked *O. nerka* may be spotted, but with 30 or more gill rakers). Parr marks, in young, small and irregular above L.l.
                                                                                 2

   Sea-run fish with black spots on dorsum and sometimes on dorsal and caudal fins. Parr marks, in young, absent or, if present, long and vertical, extending well below L.l.   **3**
2. Gill rakers 30 or more. Pyloric caeca 95 or fewer
                                                                           O. nerka
   Gill rakers fewer than 30. Pyloric caeca 140 or more
                                                                            O. keta
3. Teeth in adults set in white gum line. Pyloric caeca 45–85. Lower lobe of caudal fin and upper half of dorsal fin usually unspotted. Parr marks, in young, vertical and usually narrower than interspaces. 1st

rays of anal fin white and elongate, reaching well past the end of the fin base. First dorsal rays sometimes elongate (Fig. 2-24) *O. kisutch*

Fig. 2-24. Oncorhynchus kisutch: A, adult male (drawing by Neil H. Douglas, after Carl and Clements, 1953); B, young, 65 mm standard length. Free-hand sketch.

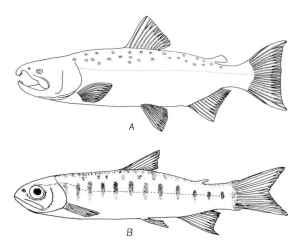

Teeth in adults set in black gum line. Pyloric caeca more than 140. Both lobes of caudal fin typically heavily spotted. Parr marks, in young, usually wider than interspaces or absent. 1st rays of dorsal and anal fins not greatly elongate **4**
4. L.l. scales 131–151, 140–153 in 1st row above L.l. Gill rakers 20–28. Dorsal and caudal fins usually spotted. No predorsal hump in breeding males. Parr marks, in young, wider than interspaces
O. tschawytscha

Fig. 2-25. Nostril flaps of coregonids: A, the single flap in Prosopium cylindraceum; B, the double flap in Coregonus artedii. (After Hubbs and Lagler, 1949.)

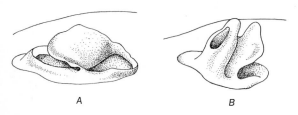

L.l. scales 150–198, 170–229 in 1st row above L.l. Gill rakers 26–34. Very large spots on dorsum and caudal fin. Predorsal hump present in breeding males. Young without parr marks *O. gorbuscha*

Oncorhynchus gorbuscha (**Walbaum**). Pink salmon. Recognized by its small scales and oblong C. spots. Northern Japan to Alaska and south to La Jolla, California. Accidentally introduced into Lake Superior; 2 specimens taken in 1959.
Oncorhynchus keta (**Walbaum**). Chum salmon. Fins mostly blackish. Breeding males with brick-red sides, often barred or mottled. Gill rakers 19–26, short and stout. From Kamchatka to Alaska and south to San Francisco, California.
Oncorhynchus tschawytscha (**Walbaum**). Chinook salmon. Small, distinct, irregular black spots on back and both C. lobes. From Northern China to Alaska and south to the Ventura River, southern California. Introduced but not established in Maine.
Oncorhynchus kisutch (**Walbaum**). Coho salman (Fig. 2-24). Small distinct black spots on back; C. spots, if present, on upper lobe only. From Japan to Alaska and south to Monterey Bay, California. Now established in Maine and Montana.
Oncorhynchus nerka (**Walbaum**). Sockeye salmon. No distinct black spots on back and C. Rakers on 1st arch 30–40, long and slender. Japan to Alaska and south to Klamath River, California. Introduced into several states, including Maine, Montana, Idaho, and Connecticut.
Genus *Salmo* Linnaeus. Mouth large; jaws, palatines, and tongue toothed; vomer head with a few teeth; vomer shaft with 1 or 2 rows of deciduous teeth. Scales large or small (about 110–200 above L.l.). D. and A. with 8–12 rays each; C. forked to nearly truncate (especially in old adults). Rivers and lakes of North America, Europe, and Asia, widely introduced elsewhere. A most difficult genus to which no entirely satisfactory key seems possible. In the past many forms of trout belonging to 2 distinct series, the rainbows and cutthroats, have been variously treated as species by some and as subspecies by other authors. Some noted ichthyologists, for example, the late J. O. Snyder, regarded any attempt to write artificial keys to the trouts as futile. The nomenclature followed here is that of Shapovalov, Dill, and Cordone (1959).

KEY TO SPECIES OF *SALMO*

1. Basibranchial teeth present (minute teeth in midline behind large ones on tip of tongue; sometimes obsolete in old adults). Reddish (yellow-orange to red-orange) dash on each side of throat along inner border of dentary bone (cutthroat mark)   *S. clarki*
   No basibranchial teeth. No cutthroat mark or, at most, a restricted pale yellow or reddish dash   **2**
2. Dorsum emerald green (may be black in preservation), sides silvery, and venter dead white. Restricted to Pyramid and Winnemucca Lakes, Nevada   *S. smaragdus*
   Color not as above   **3**
3. Black or brown spots larger and more diffuse, scarcely developed on caudal fin. Orange or reddish spots usually rather strongly developed, often ringed with bluish   **4**
   Black or brown spots smaller and sharper (quite restricted in some of the golden trouts), typically well developed on caudal fin. No orange or red spots   **5**
4. Scales in row extending downward and forward from rear of adipose fin to, but not including, L.l. row, 11–13, rarely 10 or 14. Adipose fin smaller, olive in young. Teeth on shaft of vomer weak or absent except in young   *S. salar*
   Scales from adipose to L.l. 13–16, rarely 12, 17, or 18. Adipose fin larger, orange in young. Teeth on shaft of vomer well developed   *S. trutta*
5. Parr marks typically disappearing rather early in life. Scales (2 rows above lateral line) about 115–165, usually fewer than 150. Upper jaw (except in old males) shorter, 1.9–2.2, usually 2.0–2.1 times in head length   *S. gairdneri*
   Parr marks usually retained throughout life. Scales variable, about 130 to more than 200. Upper jaw longer, about 1.7–1.9 times in head length   **6**
6. Ventral surface of body rich golden to reddish orange in life (the golden trouts). Scales in lateral series about 165–200. Spots on dorsal and caudal fins large, widely spaced   *S. aguabonita*
   Ventral surface of body white, pinkish, or golden yellow (in largest adults) in life. Scales in lateral series about 130–150. Spots tiny and numerous on dorsal and caudal fins   *S. gilae*

Salmo salar **Linnaeus.** Atlantic salmon. Color, in fresh waters, dull silvery and steel blue; scattered round or X-shaped black spots. D., C., and P_1 dusky; P_2 and A. white. Males on spawning run may have orange or red spots. North Atlantic Ocean, running up rivers to spawn from southern Greenland and Labrador to the Delaware River, now only to Maine. Landlocked in cold lakes of eastern Canada and New England.

Salmo trutta **Linnaeus.** Brown trout. Color golden brown; dark-brown or black spots on body and D., adipose, and C.; body spots, above L.l. sometimes posteriorly edged with orange; those below L.l. outlined with pink or red. Originally from Europe. Widely planted in the United States.

Salmo clarki **Richardson.** Cutthroat trout. Variously spotted with black, the spots few to many and restricted or well distributed on body; a pink or red cutthroat mark present; operculum and underparts may be pink or red. Coastal streams from Alaska to northern California eastward through the Intermountain region to the Upper Missouri, Arkansas, Platte, Colorado, and Rio Grande drainages; reportedly southward into southern Chihuahua, Mexico. Widespread introduction of rainbow and brown trout has resulted in much replacement of cutthroat.

Salmo gairdneri **Richardson.** Rainbow trout (fresh-water) or steelhead trout (sea-run). Sides with pinkish-blue to red longitudinal band; back and D., adipose, and C. with black spots, cutthroat mark absent. Alaska to the Rio Santo Domingo in Baja, California. Widely introduced elsewhere.

Salmo aguabonita **Jordan.** South Fork golden trout. Brilliantly colored; sides with bright carmine stripe, below L.l. golden yellow grading to greenish white; belly orange-yellow; cheeks and opercles bright red. D. with orange tip; P_2 and A. white-tipped. Dorsum above L.l. with black spots. Kern River, California. Now established elsewhere by planting.

Salmo smaragdus **Snyder.** Emerald trout. This species is distinguished from others of the genus in having an emerald-green dorsum, silvery sides, and white venter. Pyramid and Winnemucca Lakes, Nevada, now probably extinct.

Salmo gilae **Miller.** Gila trout. Recognized by numerous tiny black spots above L.l. Restricted to the headwaters of the Gila River system in New Mexico, where now almost extinct.

Genus Salvelinus Richardson. Mouth large or small. Teeth on jaws, palatines, and tongue. Vomer boat-shaped, its shaft depressed or with a raised crest which may or may not bear teeth. Background dark with light spots of gray or red. Scales very small (154–254, usually more than 180). C. truncate to forked. Streams and lakes of Northern Eurasia and North America. About 5 species in the United States.

KEY TO SPECIES OF *SALVELINUS*

1. Pyloric caeca 95–170. Caudal fin deeply forked. No bright colors   *S. namaycush*
   Pyloric caeca fewer than 65. Caudal fin little forked. Red or orange spots on sides   **2**
2. Mouth smaller, maxilla reaching to middle of eye. Dorsal rays 9. Anal rays usually 8   *S. aureolus*
   Mouth large, maxilla reaching to or beyond posterior margin of eye. Dorsal rays 10 or 11. Anal rays 9 or 10   **3**
3. Back with conspicuous wormlike dark markings   *S. fontinalis*
   Back without wormlike markings   **4**
4. Red spots on sides about size of eye, those on back paler and smaller. Mouth very large; maxilla reaching well beyond eye. Dorsal rays usually 11   *S. malma*
   Red spots on sides smaller than pupil. Mouth smaller; maxilla reaching only posterior margin of eye. Dorsal rays usually 10   *S. oquassa*

Salvelinus fontinalis (**Mitchill**). Brook trout. Olive green with wormlike markings on back and D. Sides with many small greenish spots, some red with blue borders. From Saskatchewan east to northern Labrador and southward in the Appalachians to the headwaters of the Savannah River of South Carolina and Georgia. Successfully introduced in northwestern United States.

Salvelinus aureolus **Bean.** Sunapee trout. Color brown above; sides silver gray with small orange spots above and below L.l.; C. grayish; A. and P_2 orange with white anterior and outer edges, respectively; not mottled. Lakes of New Hampshire, Vermont, and Maine.

Salvelinus oquassa (**Girard**). Blueback trout. Color dark blue, small red spots and indistinct bars on sides; A. and P_2 variegated and light-tipped. Lakes of Maine, principally the Rangeley Lakes.

Salvelinus malma (**Walbaum**). Dolly Varden trout. Color olivaceous with many round orange or red spots as large as pupil; paired fins and A. with anterior edges white. Japan and Korea to Alaska,

southward to the headwaters of the Sacramento River, California, and eastward to Montana.

Salvelinus namaycush (**Walbaum**). Lake trout (Fig. 2-22). Color green, gray, or brown; many light spots on body and fins; no bright colors. Great Lakes region and the lakes of New England north to the Arctic Circle and south on the West Coast to the headwaters of the Fraser and Columbia Rivers. Known also in the upper Missouri Basin in Montana. Introduced where waters are suitable.

Genus *Prosopium* Milner. Body slender. Head short. No jaw teeth. 1 flap between nostrils (Fig. 2-25A). 5 species in northern and northwestern United States, but see *P. coulteri* on p. 48.

KEY TO SPECIES OF *PROSOPIUM*

1. Gill rakers 17 (extremes 15–20)   *P. cylindraceum*
   Gill rakers 19 or more   **2**
2. Snout long and sharp; lower jaw projecting. Gill rakers 37–45   *P. gemmiferum*
   Snout short and blunt; upper jaw longer than lower. Gill rakers 19–26   **3**
3. L.l. scales 67–78 (usually 69–74). Scales around body, in front of dorsal fin 38–43   *P. abyssicola*
   L.l. scales 74–94 (usually 80–90). Scales around body 44 or more   **4**
4. Length of upper jaw 2.1–2.14 times in dorsal base. Anal base 1.8–2.1 in head. Dorsal and anal rays 12 or 13   *P. williamsoni*
   Length of upper jaw 1.3–1.9 times in dorsal base. Anal base 2.4–3.5 times in head. Dorsal rays 10–12; anal rays 9–11   *P. spilonotus*

Prosopium cylindraceum (**Pallas**). Round whitefish. L.l. scales about 80–100. D. 11–14; A. 10–13. Maxilla not reaching eye. Gill rakers 15–20 on first arch (Fig. 2-26A). Color uniform bronze to sepia brown tinged with green above, silvery below. Siberia and northern North America, south to Connecticut. In all Great Lakes except Erie.

Prosopium gemmiferum (**Snyder**). Bonneville cisco. The long, sharply pointed snout, prominently projecting lower jaw, and numerous gill rakers (37–45) readily identify this species, formerly assigned to the genera *Coregonus* and *Leucichthys*. Known only from Bear Lake in Idaho and Utah.

Prosopium williamsoni (**Girard**). Mountain whitefish (Fig. 2-27). L.l. scales usually 70–90. D. and A. 11–13. Maxilla reaching almost to an-

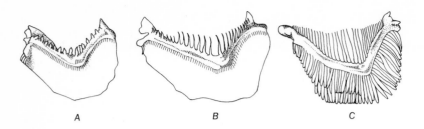

Fig. 2-26. Anterior gill arches of coegonids: A, *Prosopium cylindraceum;* B, *Coregonus clupeaformis;* C, *Coregonus nigripinnis.* (After Hubbs and Lagler, 1947, after Koelz.)

terior margin of eye. Gill rakers 17–25 on 1st arch. Back with few large spots; fins tipped with black. Streams and lakes of the Pacific Coast from Fraser River and Jasper Park south to the interior basins of Lakes Lahontan and Bonneville. Also in head-waters of the Saskatchewan and Missouri Rivers.

Prosopium coulteri (**Eigenmann and Eigenmann**). Pygmy whitefish. L.l. scales 58–63. No spots on back. D. 10–12; A. 9–11. Maxilla reaching eye in large specimens. $P_1$7. Height of D. 7, and head length 4.5–5 in standard length. Snout blunt and decurved. Alaska south to the headwaters of the Columbia River. Recently taken in Lake Superior.

Prosopium spilonotus (**Snyder**). Bonneville whitefish. L.l. scales 75–90. D. and A. rays 11. Maxilla reaching eye. Eye diameter less than 2 in postorbital length. Small numerous spots on back. Bear Lake in Utah and Idaho.

Prosopium abyssicola (**Snyder**). Bear Lake whitefish. L.l. scales 68–69. D. 11; A. 10. Maxilla not reaching eye. Color silvery. Restricted to the deep water of Bear Lake, Utah, and Idaho.

Genus *Coregonus* Linnaeus. At present most authorities include the nominal genus *Leucichthys* Dybowski in *Coregonus*. It is admittedly impossible to produce a satisfactory key to the whitefishes (see

Hubbs and Lagler, 1947). However, the following key, adapted from Hubbs and Lagler, may help when used with a knowledge of the geographic source of the specimens being studied. See the distributional list.

KEY TO SPECIES OF *COREGONUS*

1. Premaxillae wider than long, anterior edge of upper jaw directed downward and backward. Gill rakers on 1st arch fewer than 32   *C. clupeaformis*
   Premaxillae longer than wide, anterior edge of upper jaw directed forward and slightly downward. Gill rakers usually more than 31   2
2. Body deepest anteriad to its middle   3
   Body deepest near its middle   5
3. Gill rakers usually fewer than 33   *C. johannae*
   Gill rakers usually more than 33   4
4. Body small, compressed. Mandible thin, with a knob at its tip   *C. kiyi*
   Body large, thick. Mandible thick, no knob   *C. nigripinnis*
5. Gill rakers usually 34–52   6
   Gill rakers usually 51–59   *C. hubbsi*
6. Gill rakers usually 43–52 (39–43 in 1 subspecies, *greeleyi*)   *C. artedii*
   Gill rakers usually 30–43 (40–47 in some deepwater races of *hoyi*)   7

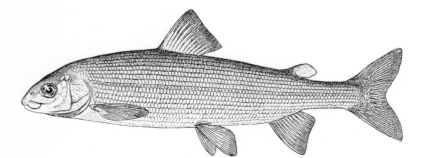

Fig. 2-27. *Prosopium williamsoni.* (From Dymond, 1932.)

7. Small species. Mandible thin, a knob at its tip
                                     C. hoyi
    Larger species. Mandible thick, usually without a knob   **8**
8. Body little compressed. Lower jaw blackish, included   *C. reighardi*
    Body usually more compressed. Lower jaw not blackish, equal or projecting   **9**
9. Jaws usually equal   *C. zenithicus*
    Upper jaw included   **10**
10. L.l. scales usually 71–77   *C. bartletti*
    L.l. scales usually 78–85   *C. alpenae*

Coregonus artedii LeSueur. Cisco or lake herring. An important food fish widely distributed in the streams and lakes of northern North America, south to Mississippi drainage of Wisconsin, Great Lakes region, and Hudson River system in New York. Westward in Canada, probably to the Yukon River Basin.

Coregonus hubbsi (Koelz). Ives Lake cisco. Deep water of Ives Lake, Huron Mountains, Michigan.

Coregonus bartletti (Koelz). Siskiwit Lake cisco. Siskiwit Lake, Isle Royale.

Coregonus reighardi (Koelz). Shortnose chub. Lakes Superior, Michigan, Ontario, and Nipigon (Canada). Most common in shallower water, but found between the extreme depths of 6–90 fathoms.

Coregonus zenithicus (Jordan and Evermann). Shortjaw chub. Lakes Superior, Michigan, Ontario, and some Canadian lakes (Nipigon to the Northwest Territories). Found in water from 11 to 100 fathoms in depth, but usually in less than 30 fathoms.

Coregonus alpenae (Koelz). Longjaw chub. Lakes Michigan, Huron, and Erie 3–100 fathoms.

Coregonus hoyi (Gill). Bloater. From Lakes Superior, Michigan, Huron, and Ontario, northward in Canada.

Coregonus johannae (Wagner). Deepwater cisco. Deeper water of Lakes Michigan and Huron (16–100 fathoms).

Coregonus kiyi (Koelz). Lakes Superior, Michigan, and Ontario (30–100 fathoms).

Coregonus nigripinnis (Gill). Blackfin cisco. Lakes Superior, Michigan, Huron, and Ontario, northward to Hudson Bay drainage.

Coregonus clupeaformis (Mitchill). Lake whitefish (Fig. 2-23*B*). In all the Great Lakes and in many inland lakes of New England and Canada; entering the upper portions of the Fraser River in British Columbia and in the North to Victoria Island. Recently found in Cheesman Reservoir, Colorado. Introduced elsewhere. An important food species.

Genus *Thymallus* Cuvier. The grayling. Similar to the Salmonidae, but with large flaglike D. of more than 15 soft rays, the anterior half unbranched. Teeth slender, few, and absent from tongue. C. deeply forked. Cold waters of Northern Hemisphere, 1 species in the United States.

Thymallus arcticus (Pallas). Arctic grayling (Fig. 2-21). Body compressed and elongate, purplish gray with small black spots. D. 22–26 total rays, with red and dusky streaks and greenish and rose-colored spots. A. of 11–16 rays. Maxilla reaches a point below middle of eye. Jaws subequal. Gill rakers slender, 11 below the angle. L.l. scales 82–100. Formerly abundant in streams of northern Michigan where now extinct; from the Yukon River Basin, Alaska, to upper Missouri River system, Montana. Widely planted in cold waters. A choice game fish.

Family Osmeridae. The smelts. Slender and small; similar to salmonids, but with stomach usually a blind sac and few or no pyloric caeca. Scales rather small and cycloid. Terminal mouth of varying sizes. Maxilla forms upper-jaw margin. Teeth various, sharp; vomer shaftless. Gill membranes separate and free from isthmus; branchiostegals 6–10. D. short, nearly median; C. forked; L.l. present. No oviducts. Almost wholly confined to the Northern Hemisphere. Very fine food fishes.

KEY TO GENERA OF OSMERIDAE

1. Gill rakers on upper half of arch 4–6. Opercle and subopercle with strong concentric striae. Pyloric caeca 8–11   *Thaleichthys*
    Gill rakers on upper half of arch 8–14. No opercular striae. Pyloric caeca 0–8   **2**
2. Tongue teeth minute, filliform. Maxilla not reaching posterior orbital margin   *Hypomesus*
    Tongue teeth of canine type. Maxilla reaching beyond orbital margin (mouth closed)   **3**
3. Tiny, pointed vomerine teeth arranged in an arc. Gill rakers on lower half of arch 24–32   *Spirinchus*
    1 or 2 large canines on either side of vomer. Gill rakers on lower half of arch 15–24   *Osmerus*

Genus *Spirinchus* Jordan and Evermann. Vomerine teeth uniserial, forming an arc and not canine; palatine teeth small. Concentric striae, if present, weak on subopercle and opercle. Male sometimes with a midlateral ridge and anal shelf. Gill rakers on 1st arch 32–47. Pyloric caeca 4–8. A. 14–19; P_1 10–12. Midlateral scales 54–65; L.l. incomplete. Adipose base $\frac{2}{3}$ of or equal to orbital diameter. Orbit $\frac{2}{3}$ or less of caudal peduncle depth.

KEY TO SPECIES OF *SPIRINCHUS*

1. Pectoral fin long, 84% or more of distance to pelvic insertion. Midlateral scales usually 55–62. Premaxillae forming an angle of 68–90° with the head dorsum profile. Longest anal ray 1.4–2.2 in head
   S. *thaleichthys*
   Pectoral fin short, 84% or less of distance to pelvic insertion. Midlateral scales usually 62–66. Premaxillae forming an angle of 54–65° with head dorsum profile. Longest anal ray 2.2–3.1 in head
   S. *starksi*

***Spirinchus thaleichthys* (Ayers).** Longfin smelt. Recognized by the long A. rays, 1.4–2.2 in head, midlateral scales 55–62, P_1 almost reaching or passing P_2 insertion, and the presence or absence of 3 or fewer subopercular, longitudinal striae. From the west coast of North America, Hinchinbrook Island, Alaska to San Francisco Bay, California. Anadromous. The nominal species S. *dilatus* Schultz and Chapman is included here because of demonstrated clinal variation.

***Spirinchus starksi* (Fisk).** Night smelt. This species and S. *thaleichthys* are sympatric, S. *starksi* differing from the latter in having short A. rays (2.5–3.1 in head), midlateral scales 62–65, and P_1 not reaching P_2 insertion. 3 or fewer longitudinal striae on subopercle.

Genus *Osmerus* Linnaeus. A monotypic genus with 1 large canine tooth on either side of vomer, palatine teeth enlarged anteriad, no striae on opercle or subopercle, a midlateral ridge but no anal shelf, gill rakers 25–37 on 1st arch, pyloric caeca 3–8. A. 11–16, P_1 11–14, midlateral scales 58–72, L.l. incomplete.

***Osmerus eperlanus* (Linnaeus).** Rainbow smelt. Anadromous or landlocked in the Pacific, Arctic, and Atlantic Oceans and their drainages. 2 subspecies are recognized. O. e. *mordax* (Mitchill) is widely distributed in northern North America and Asia, replaced by O. e. *eperlanus* in northern Europe. In the United States *mordax* occurs in the Great Lakes region and on the New England Coast as far south as the Delaware River, Pennsylvania. An excellent food fish.

Genus *Thaleichthys* Girard. A monotypic genus with two canine teeth on vomer, anterior palatine teeth larger, concentric striae on opercle and subopercle, a midlateral ridge but no anal shelf, gill rakers on 1st arch 17–23, pyloric caeca 8–11, A. 18–23, P_1 10–12, L.l. complete with 70–78 scales and orbital diameter less than $\frac{2}{3}$ C. peduncle depth.

***Thaleichthys pacificus* (Richardson).** Eulachon. Eastern Pacific from the Pribilof Islands south to the Klamath River, California. This is the species sometimes called "candlefish." The dried oily fish supplied with a wick can be used as a candle.

Genus *Hypomesus* Gill. Vomerine and palatine teeth similar to those of *Spirinchus*. Opercular and subopercular striae, midlateral ridge, and anal shelf absent. Gill rakers 26–36. Pyloric caeca 0–7. A. 11–17. P_1 10–17. Midlateral scales 51–73; L.l. incomplete. 3 species, 1 in United States.

***Hypomesus transpacificus* McAllister.** Wakasagi. Restricted to the lower parts of the Sacramento and San Joaquin Rivers, California. Referred in the literature to *H. olidus*.

Suborder Esocoidei. Pikes, pickerels, muskellunges, and mudminnows. Air bladder connected by a duct to alimentary canal. P_2 abdominal; P_1 placed low; no adipose fin. No spines. Scales cycloid, head scaly. Premaxillae (nonprotractile) and maxillae form edges of upper jaws. No maxillary teeth. Lower Eocene. Fresh waters of the Northern Hemisphere.

KEY TO FAMILIES OF ESOCOIDEI

1. Snout short and rounded. Dorsal origin only slightly behind pelvic insertion, much less than halfway to caudal base. No L.l. Caudal fin rounded
   Umbridae
   Snout resembling a duck's bill. Dorsal origin about midway between caudal base and pelvic insertion. L.l. complete. Caudal fin forked   **Esocidae**

Family Umbridae. Mudminnows. Body oblong. C. rounded; P_2 6 or 7, slightly farther forward than

D.; A. smaller than D. No L.l. Branchiostegals 4–7. Air bladder functioning as a lung. 2 genera in the United States, 1 in Washington and 1 in north central and eastern states.

KEY TO GENERA OF UMBRIDAE

1. Scales fewer than 47. Pectoral rays fewer than 18. A dark bar at caudal base (Fig. 2-28)   *Umbra*
   Scales more than 47. Pectoral rays 18–25. No dark basicaudal bar (Fig. 2-29)   *Novumbra*

Fig. 2-28. Umbra pygmaea. (Jordan and Evermann, 1900.)

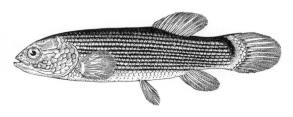

Fig. 2-29. Novumbra hubbsi. (From Schultz, 1936.)

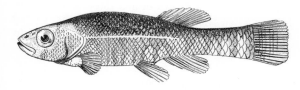

Genus *Umbra* Walbaum. Small (about 100 mm). Represented in north central and eastern United States by 2 species and in central Europe by 1.

KEY TO SPECIES OF *UMBRA*

1. Body with about 14 vertical bars; sometimes faint longitudinal streaks in breeding individuals   *U. limi*
   Body without vertical bars; about 12 longitudinal streaks   *U. pygmaea*

***Umbra limi* (Kirtland).** Central mudminnow. Small (maximum size about 100 mm). Color dull

olive green with about 14, often indistinct, bars and a dark basicaudal bar. Inhabitant of sluggish waters over bottom of soft mud used as a refuge from enemies. Very hardy, used for bait. Great Lakes region and south to northwestern Tennessee and northeastern Arkansas, westward to eastern Kansas.

***Umbra pygmaea* (DeKay).** Eastern mudminnow (Fig. 2-28). Similar to *U. limi*, but with stripes instead of bars. Basicaudal bar very distinct. Lower jaw dark in contrast to that of *U. limi*. Lowland coastal streams from New York to Florida.

Genus *Novumbra* Schultz. No basicaudal bar as in *Umbra*. More than 47 scale rows. A. base longer; P_1 fins with more rays than in *Umbra* (see key). In some respects intermediate between *Umbra* and *Dallia* Bean, the Alaskan and Siberian blackfish. 1 species locally in western Washington.

***Novumbra hubbsi* Schultz.** Western mudminnow (Fig. 2-29). Formerly considered rare, this beautiful species, according to J. W. Meldrim, is known to occur not only in the Chehalis and Deschutes River drainages, but from Coal Creek of the upper Chehalis to Cook Creek in the Quinault Basin. A comprehensive study is being made to clarify its taxonomic status.

Family Esocidae. Pickerels, pikes, and muskellunges. Holarctic, containing a single genus, *Esox*. Beak ducklike; teeth sharp. Head partly scaled. Body cylindrical, slightly compressed. Scales cycloid, with deeply scalloped anterior edges. Lower jaw articulating with skull behind eye. C. forked. Branchiostegals 11–20.

Genus *Esox* Linnaeus. With the characters of the family. Circumpolar in distribution.

KEY TO SPECIES OF *ESOX*

1. Cheeks and opercles completely scaled   **2**
   Cheeks partly or fully scaled. Opercles with lower half naked   **3**
2. Adult with chainlike reticulations on sides; young with irregular light bars confluent with light middorsal stripe. Subocular bar vertical. Snout length 7 or less in total length   ***E. niger***
   Adult with irregular vertical bars on sides; young with light lateral stripe with little or no connection with light middorsal stripe. Subocular bar inclined downward and backward. Snout 8.7 or more in total length   ***E. americanus***

3. Cheeks and opercles each with upper half scaled.
    Mandibular pores 6–9. Branchiostegals 17–19. L.l.
    scales about 150   *E. masquinongy*
    Cheeks wholly scaled, but lower half of opercles
    naked. Mandibular pores 5. Branchiostegals 14–16.
    L.l. scales about 123 (Fig. 2-30)   *E. lucius*

Fig. 2-30. Esox lucius.

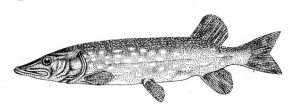

Esox americanus Gmelin. Redfin pickerel. Cheeks and opercles fully scaled. Small species widely distributed from Nebraska through southern Michigan and Ontario to New Hampshire and south on Atlantic Coast to Gulf drainage. Throughout the Mississippi lowlands from eastern Oklahoma and Texas eastward. Probably introduced in eastern Washington where locally abundant. *E. vermiculatus* regarded as subspecies of *E. americanus.*

Esox niger LeSueur. Chain pickerel. Cheeks and opercles fully scaled. Differing from above species in having dark reticulations resembling chains and in reaching a greater length (about 600 mm). New Brunswick, the St. Lawrence River, and Lake Ontario southward east of Appalachian Mountains to Florida. Lower Mississippi Valley as far north as southern Missouri and in the Tennessee River system of Alabama. West to Arkansas and eastern Texas. Introduced in the Lake Erie drainage of New York.

Esox lucius Linnaeus. Northern pike (Fig. 2-30). Cheeks fully scaled and opercles scaled on upper half. Large species sometimes more than 130 cm long and weighing as much as 46 lb. Canada and northern United States; northern Eurasia and North America. Northern New England through Great Lakes region to eastern Nebraska and south to Missouri. Introduced in Montana and Connecticut. This and the next are highly prized game species.

Esox masquinongy Mitchill. Muskellunge. Cheeks and opercles scaled on upper half only. Largest of the pikes reaching 140 cm and weighing over 60 lb. Headwaters of the Mississippi River system in northern United States and Canada, south to South Carolina and Tennessee.

Suborder Notopteroidei. Mooneyes. Silvery herringlike fishes with subopercle present. Scales cycloid. Air bladder connected with ear. Head naked, with short blunt snout and oblique mouth; premaxillae not protractile; maxillae small and slender, with no supramaxillae, and articulated with the posterior ends of premaxillae. Teeth present on jaws, tongue, vomer, palatines, sphenoid, and ectopterygoids. Eye large, with small adipose lid. Nostrils large, set close together and separated by a flap. Gill membranes not connected, free from isthmus. Branchiostegals 8–10. Gill rakers few, short, and thick. Pseudobranchiae obsolete. L.l. present. C. deeply forked. A single pyloric caecum. Oviducts absent. 1 family, 1 genus, and 2 species in the United States. Others in tropical Africa, East Indies, and Indo-Malayan archipelago.

Family Hiodontidae. The mooneyes. With characters of the suborder.

Genus Hiodon LeSueur. See characters of suborder above. Visual cells unlike those of any other fishes known from the fresh waters of the United States; no cones; rods bound in bundles; a bright guanine tapetum.

KEY TO SPECIES OF *HIODON*

1. Dorsal fin of 11 or 12 rays, its origin before anal and
    its base about ½ anal base. Iris silvery (Fig. 2-31)
                                            H. tergisus

Fig. 2-31. Hiodon tergisus.
(Jordan and Evermann, 1900.)

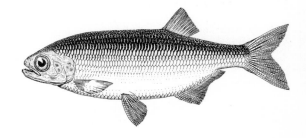

Dorsal fin of 9 or 10 rays, its origin behind anal and its base about ⅓ anal base. Iris golden

H. alosoides

Hiodon alosoides (**Rafinesque**). Goldeye. D. 9 or 10, its origin behind A. Belly keeled between P₂ and A. Absent from Great Lakes Basin; abundant in Hudson Bay drainage of Manitoba; Mississippi Valley as far west as the Yellowstone and as far south as Mississippi and Alabama. Primarily a lake inhabitant, but ascends streams to spawn over gravel beds in spring and early summer. In Manitoba an important food fish and when better known southward may become commercially valuable there; many new reservoirs with large populations.

Hiodon tergisus **LeSueur**. Mooneye (Fig. 2-31). D. 11 or 12, its origin before A. Belly usually not keeled. More widely distributed in the north than *H. alosoides*. From Hudson Bay tributaries and the lower Great Lakes south to Oklahoma, Missouri, and Tennessee in Mississippi Valley and in the Tombigbee River. Not a good food fish, although utilized along with the goldeye in Manitoba.

Order Cypriniformes. Suckers, minnows, catfishes, and characins. Anterior vertebrae modified to form the peculiar Weberian apparatus (Fig. 2-10) which serves as a hearing aid connecting the air bladder with the ear. Air bladder connected by a duct to the alimentary canal, except in a few forms having air bladder reduced and enclosed by a bony capsule. P₂ abdominal in position, sometimes absent. Spines present or absent. An adipose fin present in most characins and in the catfishes. Head always, and the body sometimes, scaleless.

The largest order of fresh-water fishes, with perhaps 4500 species, some native to all continents except Australia. Upper Cretaceous. 3 suborders and 4 families in the United States.

KEY TO SUBORDERS OF CYPRINIFORMES

1. Adipose fin present   2
   Adipose fin absent   **Cyprinoidei**
2. Barbels present (Fig. 2-80). Body naked   **Siluroidei**
   Barbels absent. Body scaled   **Characoidei**

Suborder Characoidei. Characins. Lower pharyngeals rarely falciform. Jaws with or without teeth. Pseudobranchiae absent. Adipose fin present or absent. Scales cycloid.

Family Characidae. The characins. Texas to South America. Africa.

Genus *Astyanax* Baird and Girard. Texas to South America.

Astyanax mexicanus (**Philippi**). Banded tetra (Fig. 2-32). Pugnacious little fish with adipose fin; strong and sharp jaw teeth; conspicuous black lateral band intensified on C. peduncle and extending to end of C.; lateral band overlaid by a broad silvery band. Premaxillae with 2 rows of large sharp teeth. D. 10 or 11; A. 21–23. Nueces and Rio Grande systems in Texas and New Mexico, southward into South America. Introduced and established in coastal streams of Texas as far east as the Colorado River.

Suborder Cyprinoidei. Minnows and suckers. Largest group of Cypriniformes. Characterized by complete absence of jaw teeth. Falciform lower

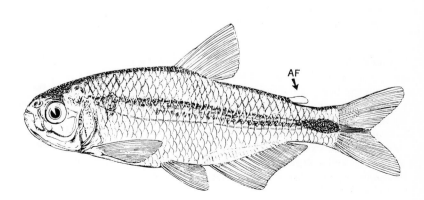

Fig. 2-32. Astyanax mexicanus (Philippi). (From R. R. Miller, 1952.) *AF*, adipose fin.

pharyngeals with 1–3 rows of teeth. Upper and lower intermuscular bones present, but no upper ribs.

KEY TO FAMILIES OF CYPRINOIDEI

1. Barbels 5 or 6 on each side. Largest otolith in the
   utriculus   **Cobitidae**
   Barbels absent or if present, never more than 2 on
   each side. Largest otolith in the lagena   **2**
2. Pharyngeal teeth in a single comblike series (sometimes molarlike, but always with more than 16 teeth) (Fig. 2-33). Mouth typically inferior, with

Fig. 2-33. Pharyngeal arch of Moxostoma erythrurum showing comblike teeth. (From Hubbs and Lagler, 1947.)

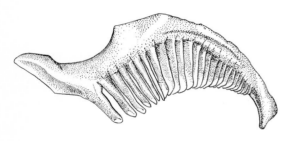

thick, fleshy papillose or plicate lips. Fins without spines, the dorsal usually with 10 or more rays; caudal fin with 16 branched rays   **Catostomidae**
Pharyngeal teeth in 1–3 rows, but principal row with not more than 8 teeth. Mouth variable in position, but with lips typically thin. Fins usually spineless, but sometimes with 1 or 2 spines in dorsal or 1 spine in dorsal and anal. Dorsal rays usually fewer than 10. Caudal typically with 17 branched rays   **Cyprinidae**

Family Cyprinidae. The minnows. Highly variable. Upper jaw formed by premaxillae only; lower pharyngeal bones falciform (Fig. 2-40) with 1–3 rows of teeth; never fewer than 3 or more than 7 teeth in the main row. Head naked; body scaly (except *Meda* and *Plagopterus*). Gill membranes always united with isthmus as in suckers (Fig. 2-64). Branchiostegals 3. 4 gills always present. Pseudobranchiae absent in some genera. Spines absent; the 1 or 2 rudimentary fin rays at D. and A.

origins are not regarded as spines since these are never stiff or sharp-pointed (hardened rays in the exotic *Carassius* and *Cyprinus* and our native *Meda, Lepidomeda,* and *Plagopterus* are called spines) (Fig. 2-34). P_2 abdominal. Herbivorous and carnivorous habits are correlated with long gut and usually blunt heavy teeth in the former and a short S-shaped gut and slender hooked teeth in the latter. Although a black peritoneum is frequently found in herbivorous minnows and a silvery peritoneum is of common occurrence in the carnivores, color of peritoneum is not sharply correlated with food habits. Sexual dimorphism often conspicuous, the breeding males being brightly colored and having nuptial tubercles on the head, body, and along fin rays. Color pattern and arrangement, extent of development, and numbers of nuptial tubercles often peculiar to a species or genus.

It is not possible to arrange the minnows in a phylogenetic sequence acceptable to all ichthyologists. However, in this treatment the Old World genera are regarded as the more primitive.

The student who wishes to become familiar with the many species of this great family should spend some time practicing the removal of pharyngeal arches beginning with some large and common species and after mastery of this is attained trying smaller and smaller specimens. The specimen may be held in one hand so that the thumbnail is holding the operculum and the 4 gill arches forward out of the way. The other hand, holding a sharp needle, a narrow lancet, or dental hooked knife, is used to cut the muscle free from the posterior edge of the pharyngeal arch so that it can be tipped forward. After the arch has been severed from the skull and also loosened at its lower end, a pair of fine forceps may be used to remove it. When studying pharyngeal arches of small species it is best to effect final cleaning under magnification in order to avoid breaking the teeth. After the arch·is cleaned it should be carefully examined to determine the number of teeth that have been lost naturally or by operational accident. The number of teeth actually present often is not the full complement since fishes often replace lost teeth. To avoid confusion the beginning worker must learn to recognize replacement teeth.

After a tentative identification is attained one should consult the annotated list.

KEY TO GENERA OF CYPRINIDAE

1. Caudal fin not forked   *Tinca*
   Caudal fin forked   **2**
2. Spines present in dorsal fin (Fig. 2-34)   **3**
   Spines absent, sometimes 1 or 2 rudimentary spine-like rays fastened to or separated from next ray by a membrane (Fig. 2-35)   **7**
3. A strong serrated spine (Fig. 2-34) at origin of dorsal and anal fins. Dorsal fin with more than 15 rays. Inner pelvic ray not attached to body by a membrane   **4**
   Dorsal fin with 2 spines, the anterior one grooved to receive the second. Inner pelvic ray attached to body by a membrane   **5**
4. 2 barbels on each upper jaw (Fig. 2-36). L.l. scales 35–38 (sometimes few to no scales). Gill rakers on 1st arch 21–27. 3 rows of pharyngeal teeth (Fig. 2-40D)   *Cyprinus*

No barbels. L.l. scales 26–29. Gill rakers on 1st arch 37–43. Pharyngeal teeth in 1 row   *Carassius*
5. Scales absent. Teeth 2,5–5,2 or 2,5–4,2   **6**
   Scales present, but small and difficult to see. Teeth 2,4–4,2 or 2,5–4,2   *Lepidomeda*

Fig. 2-35. Anterior dorsal rays in *Pimephales* and *Notropis. A*, male, *B*, female, of *P. notatus* in which the second unbranched ray is thickened (less so in *B*) and separated from the third or first principal ray; *C, Notropis cornutus* in which the second unbranched ray is slender and closely adherent to the third ray. The first ray is so small that it is overlooked without dissection. (After Bailey, 1951.)

Fig. 2-34. Spines of minnows. *A*, dorsal spine of *Cyprinus carpio; B*, cross section of *A* showing 2 serrae with the groove between; *C*, dorsal spines of *Meda fulgida; D*, cross section of *C* showing how the second spine fits into the first.

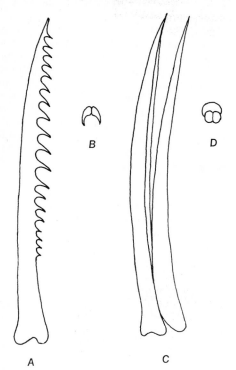

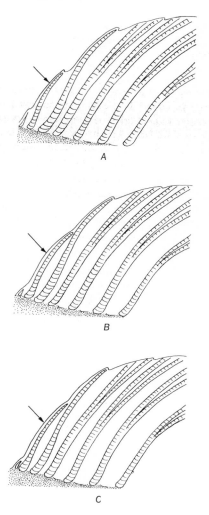

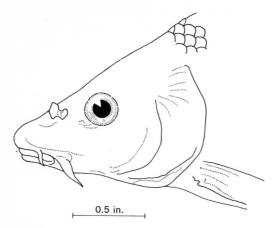

Fig. 2-36. Head of *Cyprinus carpio* showing 2 left maxillary barbels. (After a drawing by Claud Ward.)

6. A single barbel at the end of maxilla (Fig. 2-37)
   Plagopterus
   No barbels   *Meda*
7. Abdomen behind pelvic fins compressed as a keel
   (Fig. 2-38)   8
   Abdomen rounded, without a keel   9
8. Pharyngeal teeth 5–5   *Notemigonus*
   Pharyngeal teeth 2 or 3, 4 or 5–5, 3 *Scardinius*
9. Female with an ovipositor often over 2 times as
   long as longest anal ray   *Rhodeus*
   Female lacking an ovipositor   10

Fig. 2-37. Head of *Hybopsis biguttata* showing the single terminal maxillary barbel. (After a drawing by Claud Ward.)

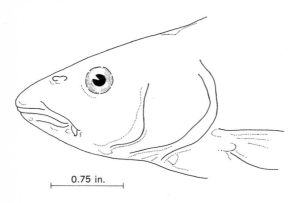

0.75 in.

10. 1st ray of dorsal fin, in adult males, blunt, short,
    and separated from 2d by a membrane
    Pimephales
    1st ray of dorsal fin slender and closely joined to
    2d, there being no membrane between them. If
    a membrane is present, spinelike ray is slender
    at its tip   11
11. Intestine long, with coils and loops, its total length
    often several times length of body   12
    Intestine short, with 1 anterior loop (sometimes 1
    or 2 loops lying across stomach), stomach and
    intestine forming a flattened S   21
12. Air bladder encircled many times by intestine
    Campostoma
    Air bladder wholly above alimentary canal   13
13. Pharyngeal teeth 5–4, 6–5, or 6–6 (4–4 in *Agosia*,
    a barbeled minnow)   14
    Pharyngeal teeth 4–4 (barbels absent)   20
14. Pharyngeal teeth 6–6, 6–5, or 5–5. Scales very small,
    more than 100 in the lateral line   *Orthodon*
    Pharyngeal teeth 4–4, 5–4, or 5–5. Scales larger,
    fewer than 100 in the L.l.   15
15. Lower jaw or both jaws with sharp-edged horny
    sheaths   16
    Jaws without horny plates   17
16. Horny plate on lower jaw only (not a conspicuous
    character in young). Scales (difficult to see)
    more than 80 in L.l. Gill rakers about 16
    Acrocheilus
    Horny plates present on both jaws. Scales 70–80 in
    L.l. Gill rakers 9–11 (Fig. 2-39)   *Eremichthys*
17. Dorsal rays 10–13. Anal rays 10–15. No conspicu-
    ous lateral bands. L.l. complete   *Lavinia*
    Dorsal and anal rays 7 or 8. Lateral bands present
    or not. L.l. complete or incomplete   18
18. Maxilla with a small barbel. Anal fin much elon-
    gated in adult female   *Agosia*
    Maxilla without a barbel. Anal fin not notably
    elongated   19

Fig. 2-38. Side view of the belly of *Notemigonus crysoleucas* showing the fleshy keel. *P₂*, pelvic fin; *K*, keel; *A*, anus; *AO*, anal origin.

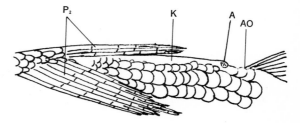

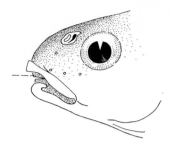

Fig. 2-39. Side view of anterior portion of the head of *Erimichthys acros* showing the horny sheaths on upper and lower jaws. (After Hubbs and Miller, 1948.)

19. Scales very small, but not deeply embedded. L.l. incomplete, almost or quite lacking. 2 lateral bands (in different degrees of completeness), the upper narrower than the lower one. No caudal spot   *Chrosomus*

Fig. 2-40. Pharyngeal teeth of minnows: *A*, hooked teeth in *Notropis venustus* (note 1 tooth in outer row); *B*, 1 of the 4 teeth in *Dionda nubila* is hooked; *C*, no hooks in *Hybognathus nuchalis*; *D*, molar teeth of *Cyprinus carpio.*

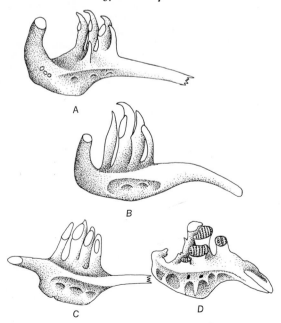

Scales small (70–80) and deeply embedded, skin with a leathery texture. L.l. complete. No lateral band, but a strong, round, caudal spot

                         Moapa

20. Pharyngeal teeth with distinct hooks (at least on 1 tooth) (Fig. 2-40*B*). Sides with a conspicuous lateral band   *Dionda*

Pharyngeal teeth without well-developed hooks. Lateral band poorly developed or absent (Fig. 2-40*C*)

                   Hybognathus

21. Some or all of the pharyngeal teeth in main row blunt and molarlike, with no or poorly developed hooks (Fig. 2-40*D*)   **22**

Pharyngeal teeth in main row slender and with well-developed hooks   **23**

Fig. 2-41. Ventral views of the mouths in 3 minnow genera. *A, Exoglossum maxillingua* with dentary bones parallel and devoid of fleshy tissue; *B, Parexoglossum laurae* with dentary bones parallel, but lips less modified than in *A; C, Phenacobius mirabilis* with dentary bones divergent and lips less modified than in *A* or *B.* (From Hubbs and Lagler, 1947.)

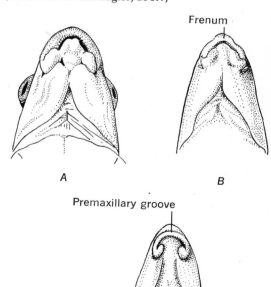

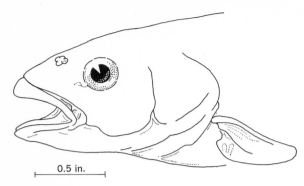

Fig. 2-42. Head of *Semotilus atromaculatus* showing position of the leaflike barbel. (After a drawing by Claud Ward.)

22. Pharyngeal teeth 2,5–4,2, with 2 or 3 molar teeth in main row. Upper jaw not protractile (Fig. 2-98*B*). No barbels   ***Mylopharodon***

    Pharyngeal teeth 1,5–5,1 the main teeth hooked in young and becoming molarlike in adults. Upper jaw protractile (Fig. 2-98*A*). Maxilla with a barbel   ***Mylocheilus***

23. Lower jaw of unusual form (Fig. 2-41), with lower lip thickened posteriorly and forming a conspicuous lobe, or if there is no lobe, lip extends no farther than about ⅔ of distance to anterior tip   **24**

    Lower lip normally formed, with no posterior thickening, or if wider posteriad, lip extends on to tip of lower jaw   **26**

24. Lower lip thickened only at inner posterior angle of mouth, forming a round fleshy lobe. Premaxillae protractile. Dentary bones not fused, except at symphysis. Pharyngeal teeth in 1 row (4–4)   ***Phenacobius***

Fig. 2-43. Position of dorsal fin in A, *Notropis zonatus* with dorsal fin over the pelvics; B, *Notropis umbratilis* with dorsal fin behind pelvics.

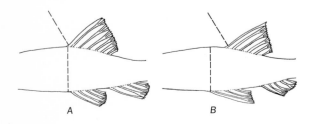

A   B

    Lower lip restricted to posterior ½ or ⅔ of lower jaw. Premaxillae not protractile. Dentary bones fused their full length or at least part way. Pharyngeal teeth in 2 rows, lesser row sometimes missing on 1 side   **25**

25. Barbels absent   ***Exoglossum***

    Barbels present   ***Parexoglossum***

26. Pharyngeal teeth in main row 5–5 or 5–4 (if in a single row, 5–5 or 5–4, rarely 4–4)   **27**

    Pharyngeal teeth in main row 4–4   **37**

27. Barbels present   **28**

    Barbels absent   **29**

28. Barbel not terminal, but set below the maxilla and some distance forward from its posterior end (Fig. 2-42)   ***Semotilus***

    Barbel terminal at posterior end of maxilla   ***Pogonichthys***

29. Pharyngeal teeth in a single row, 5–5 or 5–4 (rarely 4–4 in *Gila*)   **30**

    Pharyngeal teeth in a double row, 2,5–5,2; 2,5–4,2; or 1,5–4,2   **32**

30. Mouth very small and extremely oblique, the maxilla reaching only about halfway to anterior orbital margin. L.l. scales fewer than 40   ***Opsopoeodus***

    Mouth larger and nearly horizontal to moderately oblique, maxilla reaching nearly or quite to anterior orbital margin. L.l. scales more than 40   **31**

31. Dorsal origin behind pelvic insertion (Fig. 2-43*B*). Gill rakers 8–10, rarely 11   ***Hesperoleucus***

    Dorsal origin over (Fig. 2-43*A*) or before pelvic insertion. Gill rakers 9–40, rarely fewer than 11   ***Gila*** (in part)

32. L.l. incomplete to wholly absent   **33**

    L.l. complete or nearly so   **35**

33. L.l. absent in most specimens. Teeth 2,5–4,2   ***Iotichthys***

    L.l. with at least 14 pores   **34**

34. Scales tiny, more than 70 rows along body   ***Chrosomus*** (in part)

    Scales larger, fewer than 70 rows along body   ***Hemitremia***

35. Body pikelike. Snout long and pointed. Lower limb of pharyngeal arch very long and slender (Fig. 2-44). Upper end of preopercle nearer end of opercle than eye   ***Ptychocheilus***

    Body not pikelike. Snout shorter. Lower limb of pharyngeal arch short and stout. Upper end of preopercle nearer eye than end of opercle   **36**

36. Scales not crowded before dorsal fin. Breeding male with red on sides   ***Richardsonius***

    Scales, if present, crowded before dorsal fin. Breeding males (except *Gila ditaenia*, a southern species) without red colors   ***Gila***

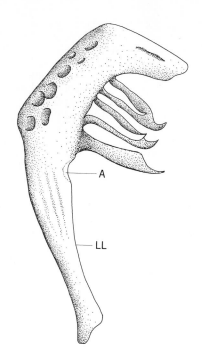

Fig. 2-44. Pharyngeal arch of *Ptychocheilus oregonensis*. *A*, position of lost tooth; *LL*, lower limb of arch.

37. Maxilla with 1 or 2 barbels (frequently absent in desert-spring populations of *Rhinichthys*) **38**

Maxilla without barbels **39**

38. Gill rakers usually 7–9. Predorsal scales usually numerous (29–44) and often embedded, making them difficult to count. Teeth 2,4–4,2 or 1,4–4,1 .... *Rhinichthys*

Gill rakers usually 5 or 6. Predorsal scales usually less numerous, not embedded, easy to count. Teeth 1,4–4,1; 1,4–4,2; 1,4–4,0; or 4–4 .... *Hybopsis*

39. L.l. system of head with greatly enlarged and cavernous tubes easily visible to unaided eye .... *Ericymba*

L.l. system of head normal and not readily visible to unaided eye **40**

40. Scales relatively large, fewer than 60 in L.l. Pharyngeal teeth variable, in 1 or 2 rows: 4–4; 1 or 2,4–4,2 or 1 .... *Notropis*

Scales relatively small, more than 60 in L.l. Pharyngeal teeth in 2 rows: 1,4–4,1 or 2,4–4,2 (teeth may be missing in either lesser row) .... *Tiaroga*

Genus *Cyprinus* Linnaeus. D. long; a strong serrated spine (Fig. 2-34*A*, *B*) at origin of both D. and A.; 2 pairs of barbels (Fig. 2-36). Native to Asia, introduced in many countries.

***Cyprinus carpio* Linnaeus.** Carp. Well established throughout the United States.

Genus *Carassius* Nilsson. D. long, as in the carp; D. and A. with serrated spines. Barbels absent. Asiatic, widely introduced and now extensively propagated as a bait fish.

***Carassius auratus* (Linnaeus).** Goldfish. Not as successful in wild state as carp, but usually common where established.

Genus *Tinca* Cuvier. Recognized by following characters: Body chunky. All fins rounded, C. not forked. 1 pair of maxillary barbels. Scales very small, 90–115 in L.l. Teeth 5–4 (4–4 or 5–5). European, sporadically planted with little success in the United States.

***Tinca tinca* (Linnaeus).** Tench. Best established in California, Colorado, Washington, and Idaho.

Genus *Rhodeus* Agassiz. Scales moderate; L.l. incomplete. D. 9–12, branched rays extending from P_2 to beyond A. origin; A. 10–12. Mouth subinferior, small; lower jaw without labial fold; no barbels. Gill rakers short. Pseudobranchiae present. Teeth 5–5, compressed, nonserrate. Breeding males with tubercles on snout; females with a long ovipositor. An Old World genus.

***Rhodeus sericeus* (Pallas).** Bitterling. D. 10; A. 10. L.l. with 3 or 4 pores; scales in lateral series 33–35, around body at D. origin 26 or 27 and around C. peduncle 13 or 14. Sides with lateral band beginning in a thin line below D., broadening posteriad. Introduced in the United States about 1925, established in Saw Mill River, Westchester County, New York.

Genus *Scardinius* Heckel. D. far behind P_2. Mouth terminal, oblique; jaws equal. Pharyngeal teeth in 2 rows. Abdomen behind P_2 compressed as a keel. An Old World genus.

***Scardinius erythrophthalmus* (Linnaeus).** Rudd. D. 11–12; A. 13–15; P_2 9–10. L.l. scales 39–42. Pharyngeal teeth 2 or 3, 4 or 5–5, 3, serrated. Eye red and body brassy yellow. Lakes and sluggish rivers from the British Isles to Asia Minor. Introduced in the United States. Now believed to be established in Roeliff Jansen Kill, tributary of the Hudson River, Columbia County, New York, from whence it was first reported by Greeley (1936), *fide* Dr. C. W. Greene, Chief, Bureau of Fish, New York Conservation Department.

Genus *Notemigonus* Rafinesque. Another unique, monotypic genus characterized by a fleshy keel (Fig. 2-38) on midventral line of abdomen between P_2 and anus. A. relatively long, its rays varying from 11 to 13 in the North and West and 14 to 17 in the South and East. Mouth very oblique. Teeth 5–5.

Notemigonus crysoleucas (**Mitchill**). Golden shiner. Southern Canada east of the Rockies, to the Gulf states. Introduced in western United States.

Genus *Orthodon* Girard. Large (300–400 mm) western minnow, peculiar in having teeth in 1 row 6–6, 6–5, or 5–5. Scales very small, over 100 in the decurved L.l. D. 9; A. 8.

Orthodon microlepidotus (**Ayres**). Sacramento blackfish. Kern River and Sacramento Basin; tributaries of Monterey and San Francisco Bays, California.

Genus *Gila* Baird and Girard. Extraordinarily variable, with biserial or uniserial teeth (usually 2,5–4,2, but 5–5, 5–4, or 4–4 in some species) and no barbels. As currently treated, *Gila* includes the nominal genera *Snyderichthys* and *Siphateles*. Scales relatively large or small (48–96 in L.l.), uniformly distributed on body or virtually absent (in certain Colorado River populations) and with radii developed only on the apical, apical and lateral fields, or all fields. D. placed over or somewhat behind P_2 insertion. Body relatively short and stout or long and slender, usually normally formed, but in 1 species of the Colorado River a prominent predorsal hump developed. A difficult genus now known to contain 13 U.S. species in the Great Basin, Pacific slope, and the Rio Grande.

KEY TO SPECIES OF *GILA*

1. Pharyngeal teeth in 1 row   **2**
   Pharyngeal teeth in 2 rows   **3**
2. Pharyngeal arch slender, the lower limb particularly long and narrow throughout. Gill rakers usually 21–27 (18–29). Mouth not notably oblique   *G. mohavensis*
   Pharyngeal arch various, slender to moderate (if slender, there are 20–40 gill rakers). Gill rakers 9–

40. Mouth notably oblique in lacustrine types, less oblique to nearly horizontal in others   *G. bicolor*
3. Pharyngeal teeth in main row 4–4   *G. copei*
   Pharyngeal teeth in main row 5–4   **4**
4. Dorsal origin almost or directly over insertion of pelvics (Fig. 2-43A). Scales with apical radii only or with very weak radii on other fields   **5**
   Dorsal origin behind pelvic insertion (Fig. 2-43B). Scales with radii on all fields or at least on apical and lateral fields   **7**
5. Scales small, shield-shaped at base, 60–70 in L.l.; teeth 2,5–5,2   *G. coerulea*
   Scales large, no basal shield, fewer than 65 in L.l.; teeth 2,5–4,2   **6**
6. Dorsal rays typically 8. Caudal peduncle very deep, about as deep as head depth   *G. crassicauda*
   Dorsal rays typically 9. Caudal peduncle slenderer, only about $\frac{1}{2}$ to $\frac{2}{3}$ as deep as head depth   *G. atraria*
7. Scales with strong radii on all fields, 48–75 in L.l.   **8**
   Scales with radii on apical and lateral (rarely basal) fields, 56–96 in L.l.   **10**
8. 63–75 in L.l. (Fig. 2-45)   *G. ditaenia*
   Scales 48–62 in L.l.   **9**

Fig. 2-45. *Gila ditaenia.* (From R. R. Miller, 1945.)

9. L.l. complete and decurved   *G. purpurea*
   L.l. incomplete (sometimes complete, but often poorly developed posteriad) and little decurved   *G. orcutti*
10. Nape with a conspicuous hump. Body often almost naked   *G. cypha*
    Napa without a hump. Body more fully scaled   **11**
11. L.l. scales 75–95. Colorado River system   *G. robusta*
    L.l. scales 51–78. Rio Grande and Pecos Basins   **12**
12. Dorsal rays 8. Gill rakers 6–10. L.l. scales 51–67   *G. pandora*
    Dorsal rays 9. Gill rakers 9–14. L.l. scales 67–78   *G. nigrescens*

Gila robusta **Baird and Girard.** Bonytail. D. usually 8 (8–11); A. 7–9. Body usually more completely scaled than in *G. cypha*. C. peduncle deep to quite slender. D. origin behind P₂. Colorado River system in Nevada, Utah, Wyoming, Colorado, California, Arizona, and New Mexico.

Gila cypha **Miller.** Humpback chub. Probably close to *G. robusta*, but D. usually 9; A. usually 10. Body almost naked, but variable. Nape with a conspicuous hump with an angular profile abruptly declivous to occiput. Distance from A. origin to C. base projected forward from A. origin falls between preopercular margin and end of opercle. Snout fleshy anteriad. Eye small, but variable. Interorbital broad and nearly flat, 0.9–1.1 in snout and 2.7–3.0 in head. Head depth, over middle of pupil 3.1–3.9 in its length. A bizarre species confined to the torrential canyons of the Colorado River, recently in danger of extinction by large-scale poisoning.

Gila nigrescens **(Girard).** Rio Grande chub. Radii on apical and lateral fields or on all fields. L.l. scales 53–68. D. origin behind P₂. Basin of the Rio Grande in Colorado, New Mexico, and Texas.

Gila pandora **(Cope).** Pecos chub. Similar to *G. nigrescens*, but with larger scales, fewer gill rakers, and usually 8 rather than 9 D. rays. The Pecos Basin in New Mexico and Texas (Davis Mountains).

Gila copei **(Jordan and Gilbert).** Leatherside chub. Formerly assigned to the genus *Snyderichthys*. Bonneville Basin in Utah, Wyoming, and Idaho; upper Snake River in Idaho and Wyoming and the Little Wood River drainage of western Idaho. Recently taken in northeastern Nevada.

Gila crassicauda **(Baird and Girard).** Thicktail chub. D. usually 8, its origin over P₂ insertion. L.l. scales fewer than 65; radii on apical field of scales only or weakly developed on other fields. Teeth 2,5–4,2. Formerly abundant in lower Sacramento and San Joaquin Rivers, lower courses of tributaries of San Francisco Bay, and also in Clear Lake, California.

Gila atraria **(Girard).** Utah chub. D. usually 9, its origin over P₂ insertion. L.l. scales 45–65; radii on apical field of scales only or weakly developed on other fields. C. peduncle rather deep. Teeth 2,5–4,2. Widespread in the Lake Bonneville Basin and in the upper Snake River in eastern Nevada, Utah, Idaho, and Wyoming; introduced as a bait fish in parts of Nevada, Utah, Wyoming, and

Montana (established as far down Madison River as mouth).

Gila orcutti **(Eigenmann and Eigenmann).** Arroyo chub. Radii strongly developed on all fields. D. origin behind P₂ insertion. L.l. incomplete, not decurved. Coastal streams of southern California from Santa Ynez (introduced) to San Luis Rey; introduced in Mohave River, San Bernardino County, California.

Gila purpurea **(Girard).** Yaqui chub. D. origin behind P₂ insertion. L.l. scales 52–60; radii strong on all fields of scales. L.l. complete, decurved. Teeth 2,5–4,2. In the United States only in San Bernardino Creek, tributary of the Yaqui River in southeastern Arizona.

Gila coerulea **(Girard).** Blue chub. D. origin over P₂ insertion. L.l. scales 60–70; radii on apical field only or weakly developed on other fields. Teeth 2,5–5,2. Abundant in Klamath River system of southern Oregon and northern California. Before the merger of *Siphateles* with *Gila*, this species was *G. bicolor*.

Gila ditaenia **Miller.** Sonora chub (Fig. 2-45). D. origin behind P₂ insertion. L.l. scales 63–75; radii strongly developed on all fields of scales. Teeth 2,5–4,2. Upper tributaries of Rio de la Concepcion, a flood tributary to the Gulf of California; in the United States only from Sycamore Canyon, in extreme southern Arizona, tributary to Rio Altar, Mexico.

Gila bicolor **(Girard).** Tui chub. This species was formerly known as *Siphateles bicolor*; several subspecies are recognized. The nomenclatural changes involved are explained by Bailey and Uyeno (1964). Widely distributed in the Columbia Basin (possibly introduced), Klamath and Sacramento–San Joaquin systems, and also some interior basins in California, Oregon, and Nevada. Teeth 5–5, 5–4, or 4–4.

Gila mohavensis **Snyder.** Mohave chub. This is another member of the nominal genus *Siphateles* believed, by R. R. Miller, to represent descendents of inhabitants of Pleistocene lakes of the Mohave system. In 1939 and 1940, Dr. Miller transferred a stock from a spring in the Mohave drainage to Sentenac Canyon in San Diego County, California. It appeared to be practically extinct in the Mohave. According to Dr. Miller there is another, but undescribed, species in the Death Valley system.

Genus *Lepidomeda* **Cope.** Body completely scaled except where paired fins may be pressed against the body in some species. 2 D. spines, the 2d stronger and longer than 1st; neither spine pungent. P_2 attached to body by a membrane, the lower branched rays with soft tips extending to margin of fin. Segments of P_1 rays in mature males little dilated; those on front margin of ventral branches smooth-edged. Head and belly rounded; head depth about or more than $\frac{2}{3}$ its length. Pharyngeal teeth in 2 rows. Peritoneum silvery with brown dots. Intestine with 1 S loop. No barbels.

KEY TO SPECIES OF *LEPIDOMEDA*

1. Anal rays 9 (often 8 in *L. albivallis*). Teeth in main row 5–4 (rarely 4–4). Scales fewer than 90 (except in *L. altivelis*)   **2**

     Anal rays 8 (rarely 9). Teeth in main row 4–4. Scales usually more than 90   *L. vittata*

2. Mouth less oblique, a line from upper end of premaxilla to middle of caudal peduncle is below middle of pupil. Snout more rounded. Dorsal spines weaker. Pigment on shoulder girdle usually extending forward beyond scapular bar   **3**

     Mouth more oblique, a line from upper end of premaxilla to middle of caudal peduncle is above middle of pupil. Snout sharper. Dorsal spines stronger. Almost no pigment on shoulder girdle in front of scapular bar   *L. altivelis*

3. Melanophores confined to upper half of opercle and upper portion of upper limb of preopercle; lower half of outer face of shoulder girdle in adults lacking pigment in front of vertical from pectoral insertion. Size smaller, rarely more than 80 mm standard length. Pharyngeal arch and teeth smaller and more delicate; the whole arch flatter; anterior angle usually sharp and conspicuous   *L. mollispinis*

     Melanophores typically extending across opercle and subopercle and to preopercular angle; lower half of outer face of shoulder girdle in adults with pigment in front of vertical from pectoral insertion. Size larger, 80–103 mm standard length. Pharyngeal arch and teeth massive; the whole arch thicker; anterior angle usually less conspicuous, more evenly rounded   *L. albivallis*

Lepidomeda **mollispinis** **Miller and Hubbs.** Middle Colorado spinedace. Recognized by having 5–4 teeth in the main row, rather weak, soft-tipped

2d D. spine, 9 A. rays, typically fewer than 90 L.l. scales, the depressed D. less than head, sides mostly silvery, and melanophores as indicated in key above. Virgin River and its tributaries in Utah, Arizona, and Nevada. Also Big Spring near Panaca, Nevada, where now apparently extinct.

Lepidomeda albivallis **Miller and Hubbs.** White River spinedace. Similar to *L. mollispinis*, but with melanophores extending well below level of L.l. Found in White River and its tributary springs in eastern Nevada.

Lepidomeda altivelis **Miller and Hubbs.** Pahranagat spinedace. Similar to *L. mollispinis*, but mouth more oblique, the D. high and sharp, head more compressed, mandible longer than in *L. albivallis*, and finer scales. Known only from Pahranagat Valley, Nevada.

Lepidomeda vittata **Cope.** Little Colorado spinedace. This species differs from its relatives in having the dental formula 1 or 2,4–4,2 or 1, 8 A. rays, and usually more than 90 L.l. scales. Restricted to the upper parts of the Little Colorado system, eastern Arizona. Through man's activities this species is believed to be extirpated from its original range except in Clear Creek.

Genus *Meda* **Girard.** D. II,7 (rarely 8); A. 9; pharyngeal teeth usually 1,4–4,1 (rarely 5–4 in main row). No barbels. Body naked except for a few anterior scales along the L.l. 1st D. spine stronger than 2d and almost as long, always large, hardened, and sharp. P_2 attached to body by membrane, the lower branches spiny and sharp but not reaching margin. Segments of P_1 rays in mature males greatly dilated. Head and belly flattened; head depth much less than $\frac{2}{3}$ its length.

Meda fulgida **Girard.** Spikedace. Gila River system in New Mexico and Arizona.

Genus *Plagopterus* **Cope.** D. II,8 or 9; A. 10. Similar to *Meda* but with a few scales only at the base of tiny dermal ridges on anterior dorsum, the presence of well-developed barbels, longer fins, and pharyngeal teeth in main row usually 5–4 (rarely 4–4).

Plagopterus argentissimus **Cope.** Woundfin. Virgin River system and formerly the Gila River system in Arizona, Nevada, and Utah.

Genus *Richardsonius* **Girard.** Redsides and daces. The nominal genus *Clinostomus* is included here. Similar to *Gila*, but with larger scales, not crowded before D. Breeding males with bright-red sides, unlike *Gila* (except *G. ditaenia,* a southern species). D. 8–11; A. 8–22. Pharyngeal teeth typically 2,5–4,2. L.l. complete and decurved. Barbels always absent. Intestine short, only 1 anterior loop.

KEY TO SPECIES OF *RICHARDSONIUS*

1. Width of gape more than ½ length of lower jaw   2
   Width of gape less than ½ length of lower jaw   3
2. Anal rays 10–22 (usually 11–18, rarely 9)
                                             R. balteatus
   Anal rays 8–10 (usually 9)   *R. egregius*
3. Scales finer, 62–70 in L.l.   *R. elongatus*
   Scales coarser, fewer than 65 in L.l.   *R. funduloides*

Richardsonius balteatus (**Richardson**). Redside shiner. D. usually 9; A. 10–22 (usually 11–18, rarely 9). L.l. scales 55–63. Teeth 2,5–4,2. Skeena and Fraser Rivers of British Columbia south through the Columbia River system to the Bonneville Basin in Idaho, Montana, Washington, Oregon, Nevada, and Utah; in the Colorado River system as a baitfish introduction.

Richardsonius egregius (**Girard**). Lahontan redside. D. usually 8; A. 9. L.l. scales 52–57. Teeth 2,5–4,2. Streams and lakes of the Lahontan Basin of Nevada, California, southern Idaho, and western Utah.

Richardsonius elongatus (**Kirtland**). Redside dace. D. usually 8; A. 9. L.l. complete, with 65–70 scales; scales with radii only on apical field. Teeth 2,5–4,2. Sides bright red in adult. Upper Mississippi River system from Minnesota, Iowa, and Wisconsin east to New York and south to Kentucky.

Richardsonius funduloides (**Girard**). Rosy-side dace. Similar to *R. elongatus,* but with coarser scales; L.l. scales 48–53. Sides not so bright red. East coast of the United States from New York to Georgia and west of the mountains in tributaries of the Ohio and Tennessee Rivers.

Genus *Ptychocheilus* Agassiz. Includes largest members of U.S. cyprinidae (*P. lucius* as long as 150 cm and weighing as much as 80 lb). Body slender and pikelike, snout long and pointed (Fig. 2-46), with premaxillae protractile. Scales small (73–95 in the complete and decurved L.l.). D. 8–10 and placed well back; A. 8–9. Gill rakers very short; pseudobranchs present; teeth 2,5–4,2 and

without grinding surfaces. Lower limb of pharyngeal arch very long and slender, about 2 times length of upper. Intestine short, as usual in carnivorous fishes.

KEY TO SPECIES OF *PTYCHOCHEILUS*

1. Anal rays 9   2
   Anal rays 8   3
2. Distance from posterior orbital margin to snout tip less than or equal to postorbital head length (except in young less than 100 mm). L.l. scales 79–80. Restricted to the Colorado River system   *P. lucius*
   Distance from posterior orbital margin to snout tip greater than postorbital head length. L.l. scales 67–75. From coastal streams of Oregon and Washington (Fig. 2-46)   *P. oregonensis*

Fig. 2-46. **Head of *Ptychocheilus oregonensis*.**

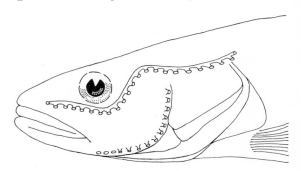

3. Dorsal rays 9. Scales above L.l. 19–24. Found in Oregon   *P. umpquae*
   Dorsal rays 8. Scales above L.l. 13–15. Found in California   *P. grandis*

Ptychocheilus grandis (**Ayres**). Sacramento squawfish. Similar to *P. oregonensis* but with larger scales and D. 8 rather than 9. Sacramento River system and tributaries of San Francisco Bay, Monterey Bay, and the Russian River, California.

Ptychocheilus oregonensis (**Richardson**). Northern squawfish. Large species 60–120 cm in length. Scales intermediate in size between those of *P. umpquae* and *P. grandis*. Columbia River system north to the Skeena River Basin in British Columbia. Also in the disconnected basin of Malheur Lake, eastern Oregon.

Ptychocheilus umpquae **Snyder.** Umpqua squawfish. A fine-scaled species of large size. Common in Siuslaw and Umpqua Rivers, and adjacent streams and lakes, Oregon.

Ptychocheilus lucius **Girard.** Colorado squawfish. Largest of North American minnows, as large as 150 cm long and weighing 80 lb; no recent reports of specimens larger than 30 lb. Formerly abundant, but now rare in the lower Colorado. Colorado River system in California, Arizona, Utah, New Mexico, Colorado, and Wyoming. A valuable food fish.

Genus *Moapa* Hubbs and Miller. Monotypic. Teeth 5–4, the upper ones moderately hooked and lower ones with obsolescent hooks and a stumpy form. No barbels; premaxillary frenum hidden. Lower jaw included; maxilla reaching nearly to or beyond anterior orbital rim. 1st gill slit restricted, being about as long as eye. Gill rakers 5–9. No pseudobranchs. Intestine short, but slightly coiled on itself anteriad; peritoneum blackish. Scales tiny (70–80 in the complete and slightly decurved L.l.), deeply embedded and with radii on all fields; skin of leathery texture. D. 8 (sometimes 7), its origin over or slightly behind P_2 insertion. A. 7 or 8.

Moapa coriacea **Hubbs and Miller.** Moapa dace (Fig. 2-47). Isolated in Warm Springs, Clark County, Nevada. Colder water below type locality seems an effective barrier to this very interesting relict.

Genus *Semotilus* Rafinesque. Barbel unique (Fig. 2-42), sometimes absent, partially or wholly concealed (in young often seen with difficulty) in groove below maxilla, some distance forward from its posterior tip. Body thickset; head large and mouth terminal. Premaxillae protractile. Gut short. Teeth 2,5–4,2 (rarely with teeth in 3 rows). L.l. complete. D. 7 or 8. A. usually 8. 3 species east of Rocky Mountains.

KEY TO SPECIES OF *SEMOTILUS*

1. Dorsal origin directly over pelvic insertion (Fig. 2-43A). Scales large, about 45 in L.l.   **S. corporalis**
   Dorsal origin behind pelvic insertion. Scales smaller 50–75   **2**
2. Black spot at dorsal origin. Mouth large, upper jaw reaching front of eye. Sides without irregular and conspicuous black blotches   **S. atromaculatus**
   No black spot at dorsal origin. Mouth smaller, not reaching eye. Sides with irregular and conspicuous black blotches   **S. margarita**

Semotilus corporalis (**Mitchill**). Fallfish. D. 8; A. 8. D. directly over P_2 insertion. Black spot at D. origin absent. Scales large, not crowded anteriad. Maximum size about 450 mm. Inhabiting clear streams and lakes from the James Bay region, northern St. Lawrence tributaries, eastern Lake Ontario drainage, and southward east of the Appalachians to Virginia.

Semotilus atromaculatus (**Mitchill**). Creek chub (Fig. 2-42). D. 7; A. 8. D. origin behind P_2 insertion. Black spot at D. origin. Scales crowded anteriad. Length about 250 mm. An inhabitant of small clear streams from Montana through the Red River of the North to Gaspé Peninsula, Canada. Southward on both sides of the Appalachians to the Gulf states. Southwestward through Kansas, Ozark region, and the Arkansas and Pecos River systems in New Mexico.

Semotilus margarita (**Cope**). Pearl dace. D. 8; A. 9. D. origin slightly behind P_2 insertion. Dark spots irregularly placed on sides and back. Length about 75 mm. A fish of cool streams and lakes over large part of Canada, south of the tundra and through northern portion of Great Lakes region and west to Montana. South on Atlantic Coast to Pennsylvania and Virginia.

Genus *Pogonichthys* Girard. Monotypic. Re-

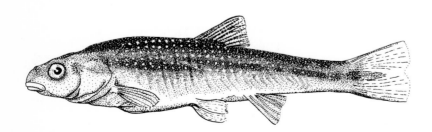

Fig. 2-47. Moapa coriacea. (From Hubbs and Miller, 1948.)

markable for unequal development of C. lobes, a unique feature among North American Cyprinidae; upper lobe much longer than lower in adults, but not in young. D. 9; A. 8. Scales about 66 in decurved L.l. Teeth 2,5–5,2 or 2,5–4,2.

Pogonichthys macrolepidotus (**Ayres**). Splittail. Sacramento and San Joaquin Rivers, California. Entering brackish water.

Genus *Mylocheilus* **Agassiz.** Similar to *Mylopharodon,* but smaller (about 300 mm). Premaxillae protractile; maxilla with terminal barbel, rudimentary in young; mouth smaller, maxilla not reaching orbit. Teeth (1,5–5,1) in main row hooked in young, larger ones becoming molarlike with increasing age.

Mylocheilus caurinus (**Richardson**). Peamouth. Columbia River system and north to the Fraser River, Recorded from Montana, Idaho, Oregon, Washington, and British Columbia. Known to enter the sea.

Genus *Mylopharodon* **Ayres.** Large, about 60 cm; body long and slender; snout long and pointed; mouth rather large, maxilla reaching orbit; premaxillae not protractile. D. origin slightly behind P_2 insertion. Teeth 2,5–4,2, the anterior 2 or 3 teeth in main row heavy, molarlike, and without hooks, posterior teeth tending to become slender and hooked. Scales small 70–80 in L.l. (complete and anteriorly decurved) and 50 before D. origin. D. and A. 8. Gill rakers short, 10–14 (including all rudiments) on 1st arch.

Mylopharodon conocephalus (**Baird and Girard**). Hardhead. Sacramento and San Joaquin River systems and Russian River, California.

Genus *Hemitremia* **Cope.** Monotypic. Scales large, about 43 in lateral series; L.l. very short (about 14 pores). Black lateral band terminating in a small C. spot. Teeth 2,5–4,2. Mouth small and oblique. In the spring, males are bright scarlet.

Hemitremia flammea (**Jordan and Gilbert**). Flame chub. Tennessee River system in Georgia, Tennessee, and Alabama.

Genus *Iotichthys* **Jordan and Evermann.** Monotypic. Size very small (about 35 mm). Body short, chunky, and compressed. Lateral band poorly developed, largely limited to sides of C. peduncle. Middorsal stripe, well developed in adult, weak in young. About 33 scale rows cross the body; L.l. absent; scales in front of D. in about 33 rows. D.

8 or 9 (usually 8). A. 7 or 8 (usually 8). Mouth strongly oblique, the gape almost reaching anterior orbital rim. Eye 3.5 in head. D. origin slightly behind P_2 insertion. Pharyngeal teeth 2,5–4,2, slightly hooked. Intestine short. Peritoneum silvery.

Iotichthys phlegethontis (**Cope**). Least chub. Bonneville Basin of Utah.

Genus *Chrosomus* **Rafinesque.** Small (about 60–100 mm), with very small scales. Intestine long or short; peritoneum black. L.l. short or absent. Teeth 5–5 or 5–4 (2,5–4 or 5,2 in *C. neogaeus*). 1 or 2 dark lateral bands; brilliantly colored with red or yellow in spring.

KEY TO SPECIES OF *CHROSOMUS*

1. Lateral band single and dusky. Intestine usually short, with a single main loop   *C. neogaeus*
   2 black lateral bands. Intestine long, with more than a single main loop   **2**
2. Lateral bands not horizontal, lower one extending downward and backward from snout to anal base and upper beginning above anus and ending at caudal case   *C. oreas*
   Lateral bands horizontal, lower one beginning on snout and extending to caudal base, upper beginning above operculum, extending parallel to lower and ending on caudal peduncle   **3**
3. Distance from tip of snout to back of eye distinctly longer than rest of head. Mouth little oblique   *C. erythrogaster*
   Distance from tip of snout to back of eye equal or little longer than rest of head. Mouth strongly oblique   *C. eos*

Chrosomus oreas **Cope.** Mountain redbelly dace. D. and A. 8. Upper lateral band beginning above anus, extending to C. base, and ending in a spot; lower band beginning on snout and ending above A. base. Appalachian Mountain region of West Virginia, Virginia, and North Carolina.

Chrosomus eos **Cope.** Northern redbelly dace. Similar to *erythrogaster,* but the 2 lateral bands uniting on C. peduncle. L.l. less distinct. Mouth strongly oblique. Rather widely distributed in the North from British Columbia and Hudson Bay drainage south in the West to Colorado and in the East to New York and Pennsylvania. In all Great Lakes drainages except that of Erie in Ohio.

Chrosomus erythrogaster **Rafinesque.** Southern

redbelly dace. D. 7; A. 8. 2 lateral bands, the upper extending from upper angle of gill cleft to C. base, and the broader lower band from snout to C. base. L.l. developed about halfway from gill cleft to C. base. Mouth little oblique. Southern Minnesota, Wisconsin, southern Michigan, the Ohio River system of Pennsylvania and West Virginia, and southward, on both sides of the Mississippi, to Mississippi, northern Alabama and Oklahoma. Living in cool, clear, gravelly creeks and in the South in cooler springs. An isolated population in the Arkansas River drainage of New Mexico.

Chrosomus neogaeus (**Cope**). Finescale dace. Contrasted with other *Chrosomus* in having gut usually short; teeth in 2 rows (2,5–4 or 5,2); and in having a single dark lateral band, with a pale streak above and abruptly white below it. From northern Canada south to the Black Hills of South Dakota, and the Sand Hills of Nebraska; Great Lakes region to the Hudson and St. Lawrence Rivers. An inhabitant of cool, sluggish creeks, ponds, and lakes.

Genus *Eremichthys* Hubbs and Miller. Monotypic. Differing most strikingly from *Acrocheilus* in having horny sheaths on both upper and lower jaws. Fins low and rounded, C. with shallow emargination. L.l. nearly complete. Teeth 5–4.

Eremichthys acros **Hubbs and Miller.** Desert dace. From type locality only, in desert of southwestern Humboldt County, Nevada, where the springs and stream of Soldier Meadows are completely isolated from other waters.

Fig. 2-48. Head of a nuptial male *Opsopoeodus emiliae* showing the extremely oblique mouth and nuptial tubercles on the snout and chin.

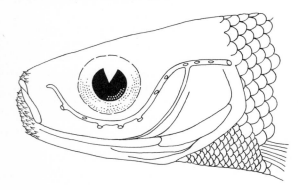

Genus *Opsopoeodus* Hay. Monotypic. Differing from other small minnows of the Great Lakes and Mississippi drainages in having D. 9; A. 8. Mouth very small and oblique (Fig. 2-48), less so in specimens from the Gulf Coast. Pharyngeal teeth 5–5, slender, strongly hooked, and serrated. Silvery, with a lateral band extending along sides and not terminating in a C. spot. Anterior rays of D. blackened in adults. C. peduncle long and slender. Length about 55 mm.

Opsopoeodus emiliae **Hay.** Pugnose minnow (Fig. 2-48). Lowland streams of Great Lakes region; throughout the Mississippi Valley and Gulf Coast from Texas to Florida, Georgia, and South Carolina.

Genus *Hesperoleucus* Snyder. D. origin behind P_2 insertion. Teeth 5–4 with narrow but well-developed grinding surfaces. Head relatively short, lower jaw nearly horizontal. D. 8 or 9; A. 7 or 8. Scales rather large 47–61 in L.l. Coastal streams of California from Monterey Bay north through the San Joaquin and Sacramento systems to the Navarro River, Mendocino County. 4 species currently recognized, but may be only subspecifically distinct; if so, proper name will be *H. symmetricus*. Size to about 140 mm.

KEY TO SPECIES OF *HESPEROLEUCUS*

1. Scales cuplike, with radii present on all fields
   H. mitrulus
   Scales not cuplike, radii poorly developed on lateral fields and always absent on basal field **2**
2. Gill rakers on 1st arch 9 or 10. Radii on lateral fields of scales absent or rarely present (usually present in specimens from tributaries of Monterey Bay)
   H. symmetricus
   Gill rakers on 1st arch 8. Radii on lateral fields of scales present, but weak and often irregular **3**
3. Scales above L.l. 11–13, those below L.l. 7. Anal rays usually 8 *H. navarroensis*
   Scales above L.l. 14–16, those below L.l. 8. Anal rays usually 7 *H. parvipinnis*

Hesperoleucus symmetricus (**Baird and Girard**). California roach. D. 8 or 9; A. 7 or 8. L.l. scales 47–63; scales before D. 28–33; scales above L.l. 12–14, those below 6–7. Gill rakers on 1st arch 9–11. Tributaries of Monterey and San Francisco

Bays: Pajaro, San Joaquin, Sacramento, Suisun, San Pablo, and Russian Rivers; also Clear Lake, California.

Hesperoleucus navarroensis **Snyder.** Navarro roach. D. 8; A. 8 (rarely 7 or 9). L.l. scales 51–59; scales before D. about 30; scales above L.l. 11–13, those below 7. Gill rakers on 1st arch 8, short and stumpy. Navarro River in northwestern California.

Hesperoleucus parvipinnis **Snyder.** Shortfin roach. D. 8; A. 7 (rarely 6, occasionally 8). L.l. scales 54–59; scales before D. about 30; scales above L.l. 14–16, those below 8. Gill rakers on 1st arch 8, short and stumpy. Gualala River Basin in Napa County, California.

Hesperoleucus mitrulus **Snyder.** Northern roach. D. 8; A. 7. L.l. scales 48–61; scales before D. 28–38; scales above L.l. 12–15, those below 8. Gill rakers on 1st arch 9, very small. Northern tributaries of Goose Lake in Oregon and in Pit River, tributary of the Sacramento River, in California.

Genus *Lavinia* **Girard.** Closely resembling the golden shiner (*Notemigonus*), but lacking the distinctive fleshy abdominal keel. D. 10–13; A. 10–15. Teeth 5–5 or rarely 5–4. Gill rakers about 17–26. Size about 300 mm.

Lavinia exilicauda **Baird and Girard.** Hitch. Sacramento River Basin and tributaries of San Francisco Bay; Pajaro-Salinas Basin and Clear Lake.

Genus *Acrocheilus* **Agassiz.** Monotypic. Prominent horny sheath on lower jaw. Intestine long and much convoluted. Scales small (about 85 in L.l.). Gill rakers about 13–17. Teeth 5–4 or 5–5. Length to about 300 mm.

Acrocheilus alutaceus **Agassiz and Pickering.** Chiselmouth. Columbia River system (below Snake River Falls) and Malheur Lake drainage in Oregon, Washington, Idaho, and Nevada.

Genus *Hybopsis* **Agassiz.** Barbels (*H. hypsinota, H. harperi,* and, in the South, *H. amblops* often lack them) usually at terminus of maxilla (Figs. 2-37 and 2-49). In *H. plumbea* the barbel is not terminal but placed on the anterior surface of maxilla anteriad to its posterior end. Eye variable in size, small to large. Mouth usually horizontal, rather small, and with lower jaw included, but in some species rather large, oblique, and nearly terminal (Fig. 2-37). Body usually quite slender and

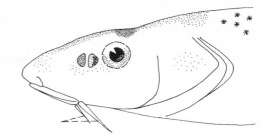

Fig. 2-49. Head of *Hybopsis aestivalis* showing 2 barbels of left side.

little compressed. Teeth vary from 4–4 to 2,4–4,2, usually with poorly developed grinding surfaces. Color pattern variable, some species silvery and without bright colors or a lateral band, others with a lateral band or conspicuous spots on back and sides; fins bright-colored in breeding individuals of some species. Intestine usually short but looped across the stomach in *H. leptocephala* and *H. bellica.* Size small (50 mm) to large (300 mm). Studies in progress, at the time of this writing, are expected to result in description of some new species, *fide* R. E. Jenkins, Cornell University. Habitat varying from large muddy plains rivers to clear gravelly mountain brooks.

KEY TO SPECIES OF *HYBOPSIS*

1. Body deep and compressed. Caudal peduncle very slender, its depth contained about 11 times in standard length. Restricted to Willamette and Umpqua Rivers in Oregon   **H. crameri**
   Body more cylindrical, little compressed. Species east of Rocky Mountains   2
2. Pharyngeal teeth 2,4–4,2   3
   Pharyngeal teeth 1,4–4,1 (tooth in outer row of either side may be lacking) or 4–4   4
3. Barbels terminal (Fig. 2-37), placed at posterior end or very near end of maxilla   **H. gracilis**
   Barbels not terminal, usually placed on anterior side of maxilla anteriad to end of maxilla   **H. plumbea**
4. Mouth rather large, almost horizontal to somewhat oblique and subterminal   5
   Mouth small (if large, lower lips thick and papillose) and horizontal or nearly so, often overhung by fleshy snout   8
5. Pharyngeal teeth 1,4–4,1 or 1,4–4,0. Caudal spot distinct, large, and round. Eye large. Barbel small.

A red spot behind eye in adult. Caudal fin red in young. Nuptial tubercles about 45 and located from internasal area to occiput. No occipital or interorbital swellings (Fig. 2-37) **H. biguttata**
Pharyngeal teeth 4–4. Caudal spot small and round or diffuse and indistinct. Eye small. Barbel long. No red spot behind eye; caudal fin of young, amber. Nuptial males with about 7–40 tubercles located from internasal area to interorbital area, with only a few approaching the occiput (none in *H. bellica*). Occipital and interorbital swellings present **6**

6. Breeding males with few (7–10) large tubercles restricted to internasal and interorbital areas. Interorbital swellings in breeding males very prominent **H. bellica**
Breeding males with many (about 40) smaller tubercles, some below nostrils or on occiput as well as on internasal and interorbital areas. Interorbital swellings less prominent **7**

7. Snout contained about 8 times in standard length and somewhat pointed. Distance from dorsal origin to hypural equal to distance from dorsal origin to nostril. Caudal spot indistinct. Nuptial tubercles primarily on internasal area and on sides and front of snout, not approaching occiput **H. micropogon**
Snout about 9 in standard length and blunter. Distance from dorsal origin to hypural equal to distance from dorsal origin to eye. Caudal spot small, round, and distinct. Nuptial tubercles, on internasal and interorbital areas, approach occipital region and sometimes sides and front of snout **H. leptocephala**

8. Pharyngeal teeth 1,4–4,1 or 1,4–4,0 (occasionally 2 teeth in lesser row on one side) **9**
Pharyngeal teeth 4–4 **13**

9. Mouth large, lower lip thick and papillose on its inner edge. 2 conspicuous black spots, one at origin and another at end of dorsal base **H. labrosa**
Mouth small, inferior or subinferior, lower lip thin and not papillose. No black spots at dorsal base **10**

10. No lateral band. Lower caudal lobe blackened near its white edge. Scales about 42 in L.l. **H. storeriana**
A dark, dusky or plumbeous lateral band. Lower caudal lobe not peculiarly marked. L.l. scales 35–40 **11**

11. Breeding males with violet-colored body and bright-red fins, barbels often absent **H. hypsinota**
Breeding males with or without bright colors, but never as above **12**

12. Never with red colors. Scales about 38 in L.l. and about 16 before dorsal fin. No caudal spot **H. amblops**
Breeding male with a red snout and fins largely red. Scales about 36 in L.l. and 13 before dorsal fin. A caudal spot present **H. rubrifrons**

13. L.l. scales fewer than 39 **14**
L.l. scales 39–62 (rarely 38) **15**

14. Body marked with scattered giant melanophores (not readily visible in specimens from muddy water). No caudal spot. Frontal area of head dorsum dark. Mouth strictly ventral, often overhung by a fleshy snout. With 2–4 maxillary barbels **H. aestivalis**
Body with many small melanophores on dorsum. Caudal spot present. Frontal area of head dorsum with a median light streak. Mouth slightly oblique. Never more than 2 barbels (sometimes absent on one or both sides) **H. harperi**

15. Anal rays 9. Scales of dorsum with longitudinal keels **H. gelida**
Anal rays 7 or 8. Scales without keels **16**

16. Lower lobe of caudal fin with black pigment above its white edge **H. meeki**
Lower lobe of caudal fin uniformly pigmented **17**

17. L.l. scales 52–62. Dorsal fin with black pigment in posterior interradial membranes. No dark blotches or speckles on body. Gill membranes narrowly connected with isthmus; width between lower ends of gill openings contained 6.5–16.0 times in head **H. monacha**
L.l. scales fewer than 52. Dorsal fin lacking black pigment. Body with black blotches, X-shaped markings or, posteriad, with V-shaped markings on each myomere. Gill membranes more broadly connected, isthmus width contained fewer than 6 times in head **18**

18. Dorsum with irregular X-shaped marks or, lacking these, the posterior part of the body with V-shaped markings on each myomere; never with oval or rectangular blotches **19**
Color pattern with a lateral row of oval or rectangular blotches and a middorsal row of spots; No X- or V-shaped markings **20**

19. Dorsum with irregular X-shaped marks **H. x-punctata**
Body, posteriad, with V's on each myomere **H. cahni**

20. Sides with a row of oval blotches. Eye diameter usually contained 3.5–4.0 times in head. Width of isthmus between lower ends of gill openings contained 3.6–4.6 (extremes 3.2–5.5) times in head **H. dissimilis**

Sides with a row of rectangular blotches. Eye diameter usually contained 2.9–3.5 times in head. Isthmus width contained 2.7–3.6 (extremes, 2.6–4.1) times in head   *H. insignis*

Hybopsis gracilis (**Richardson**). Flathead chub. Head dorsoventrally flattened. D. and A. 8, rarely 9. Size large, about 300 mm. Abundant in the Mississippi River and its principal tributaries as far south as the South Canadian River in Oklahoma and New Mexico and in the West, north to Colorado, Wyoming, Montana, Texas, and, in Canada, to the Mackenzie Delta and Saskatchewan Basin east to Lake Winnepeg.

Hybopsis plumbea (**Agassiz**). Lake chub. Barbels not quite terminal. Widespread through Canada and northern United States from northern New England west to Montana, Wyoming, and Colorado; also headwaters of the Fraser River in British Columbia, the Mackenzie Delta, northwest territories, and south to Lake Pend d'Oreille, Idaho. *H. greeni* (Jordan) is included here.

Hybopsis biguttata (**Kirtland**). Hornyhead chub. A postorbital red spot; C. spot black, distinct. An inhabitant of clear gravelly streams from New York and Ohio westward to Colorado and Wyoming and south to Tennessee, Arkansas, and Oklahoma.

Hybopsis micropogon (**Cope**). River chub. Similar to *H. biguttata,* but lacking postorbital red spot; C. spot indistinct. In clear creeks and rivers from the James River system in Virginia north to the Lake Ontario drainage, west to southern Michigan, and south to Tennessee and Alabama.

Hybopsis leptocephala (**Girard**). Bluehead chub. Similar to *H. micropogon,* but snout shorter and blunter. C. spot distinct. Atlantic slope from the York River, Virginia, south to the Savannah River system in South Carolina.

Hybopsis bellica (**Girard**). Alabama chub. A near relative of *H. leptocephala,* but with fewer nuptial tubercles and more prominent interorbital swellings in breeding males. From the Savannah River in Georgia westward through Alabama and Mississippi.

Hybopsis labrosa (**Cope**). Thicklip chub. Mouth large, lips thick and papillose. Santee River Basin of North and South Carolina where it is common.

Hybopsis amblops (**Rafinesque**). Bigeye chub. Scales large, 12–15 predorsal rows and 20–24 rows around body. Mouth horizontal; maxillary barbel terminal. In sandy or gravelly streams from western Lake Ontario (southern drainage), west to southern Michigan and south to Oklahoma, Arkansas, the Gulf states, and Georgia.

Hybopsis rubrifrons (**Jordan**). Rosyface chub. Similar to *H. amblops,* but with snout and fins red in breeding males. Altamaha River system in Georgia, also recorded from South Carolina. Specimens have been taken from the Chattahoochee and Savannah Rivers.

Hybopsis hypsinota (**Cope**). Highback chub. A. 8; fins bright red in breeding males. Santee Basin of North and South Carloina.

Hybopsis crameri **Snyder**. Oregon chub. Body deep and compressed; C. peduncle slender. Willamette and Umpqua Rivers, Oregon.

Hybopsis storeriana (**Kirtland**). Silver chub. Size large (250 mm). Lower C. lobe with black streak near a white border. Widely distributed in large silty rivers and in lakes from Ontario westward through the Red River of the North to Wyoming; southward through Ohio to the Red River of the South, and west of the Appalachians to the Alabama, Pascagoula, and Pearl Rivers.

Hybopsis harperi (**Fowler**). Redeye chub. Sides with a distinct dark band, terminating in a small, inconspicuous basicaudal spot; a light streak above the lateral band. A median light streak on frontal area. A. 8 (7–9). In springs, spring-fed creeks, sinks, and caves of northern Florida and adjacent parts of Georgia and Alabama.

Hybopsis monacha (**Cope**). Spotfin chub. Scales small, 52–62 in L.l. C. spot conspicuous and black. Tennessee River system in Tennessee, Virginia, North Carolina, Georgia, and Alabama.

Hybopsis cahni **Hubbs and Crowe**. Slender chub. Body and fins devoid of conspicuous blotches, but posterior part of body with a median shading produced by a series of dark V's, one on each myomere. A. 7 (rarely 6). Clinch and Powell Rivers of the upper Tennessee system in Tennessee.

Hybopsis dissimilis (**Kirtland**). Streamline chub. Sides with a row of oval blotches; midline of dorsum with a row of blotches. A. 7 (rarely 6). Ohio and Tennessee River systems from Indiana to New York and southward to Tennessee and North Carolina. Also in the Ozark upland of southern Missouri and northern Arkansas.

Hybopsis x-punctata Hubbs and Crowe. Gravel chub. Sides with numerous irregularly scattered X-shaped specklings; no middorsal row of spots. A. 7. Ohio River Basin (not in the Cumberland and Tennessee River systems) from Illinois to Pennsylvania and south to Kentucky; Thames River in Ontario and possibly in the Lake Erie drainage of Ohio. In the Mississippi drainage basin from Wisconsin and Minnesota to the Ozark upland in Kansas, Missouri, Oklahoma, and Arkansas.

Hybopsis insignis Hubbs and Crowe. Blotched chub. Sides with a row of rectangular spots; midline of back with a row of spots. A. 7. In the Cumberland Basin in Kentucky and Tennessee and the Tennessee River system in Tennessee, Virginia, North Carolina, Georgia, and northern Alabama.

Hybopsis meeki Jordan and Evermann. Sickle-fin chub. Eye small, often partially covered by skin. In large silty rivers, the Mississippi and Missouri. Records for Illinois, Iowa, Kansas, Missouri, Nebraska, and South Dakota.

Hybopsis aestivalis (**Girard**). Speckled chub (Fig. 2-49). Body conspicuously speckled with black. Size small, about 60 mm. In main rivers and larger tributaries of the Mississippi and Rio Grande and in the Gulf Coast rivers from Texas to western Florida.

Hybopsis gelida (**Girard**). Sturgeon chub. Scales keeled. Size small, about 50 mm. From main channels of upper portion of the Mississippi River system from the Mississippi westward to Montana and Wyoming and south as far as southern Illinois, Missouri, and Kansas.

Genus *Rhinichthys* Agassiz. Currently including *Apocope* Cope. Difficult to define because of the variable array of included forms. Teeth always 2-rowed, but with either 1 or 2 teeth in outer rows and always 4–4 in main row; outer teeth occasionally lacking; main row rarely 5–4. A. almost invariably 7; D. 7–9. Scale radii consistently developed on all fields except in *R. falcatus;* L.l. scales about 35–90. D. origin invariably behind P_2 insertion. A capacity to develop barbels (sometimes absent). Horny sheath often on lower jaw. Intestine very short, with only 1 loop as in *Notropis.* Premaxillae protractile or nonprotractile. 5 species.

KEY TO SPECIES OF *RHINICHTHYS*

1. Scale radii present on all fields   2
   Scale radii not present on basal field (dorsal 9: anal 7)   *R. falcatus*
2. Head narrow and slender, with fleshy snout overhanging an inferior mouth. Premaxillae never protractile, frenum broad, always present   3
   Head broader and shorter, snout less fleshy; mouth not strictly inferior, sometimes oblique. Premaxillae protractile or nonprotractile with frenum variously developed   4
3. Dorsal fin of 9 rays   *R. evermanni*
   Dorsal fin of 8 rays   *R. cataractae*
4. Barbels conspicuous. Teeth 1,4–4,1. From Dakotas and Nebraska eastward   *R. atratulus*
   Barbels less conspicuous or absent. Teeth 2,4–4,2, occasionally 1,4–4,1. West slope of Rocky Mountains westward   *R. osculus*

Rhinichthys atratulus (**Hermann**). Blacknose dace. D. 8; A. 7. Scales in L.l. about 53; before D. crowded and embedded, about 31; scale radii present on all fields. Mouth terminal, not overhung by snout. Many scales solid black; lateral band indistinct or absent. C. spot poorly developed. Premaxillae not protractile, frenum broad. Teeth 1,4–4,1. From the Dakotas eastward through the Great Lakes region to the Atlantic Coast and southward on both slopes of the Appalachians to Georgia, Alabama, and Mississippi.

Rhinichthys osculus (**Girard**). Speckled dace. D. 8–9; A, 6–8. Scales: in L.l. about 55–75; before D. 29–44, sometimes crowded and embedded; scale radii present on all fields. Premaxillae protractile or nonprotractile, frenum present or absent and sometimes hidden. Scales with poorly defined blotches; lateral band absent; C. spot poorly developed or absent. Teeth 2,4–4,2. Widely distributed from western slopes of the Rocky Mountains to the West Coast, from the Columbia River system south to the Colorado River system in California, Arizona, New Mexico, and Sonora, Mexico. A highly variable species in need of exhaustive study.

Rhinichthys falcatus (**Eigenmann and Eigenmann**). Leopard dace. D. 9; A. 7. L.l. scales 52–57. Radii absent on basal field. Premaxillae protractile. Mottled or blotched on upper parts. Columbia River Basin east of the Cascades.

Rhinichthys evermanni **Snyder**. Umpqua dace.

Close relative of *R. cataractae,* but with 9 dorsal rays. Umpqua River in Oregon.

Rhinichthys cataractae (**Valenciennes**). Longnose dace. D. 7 or 8; A, 7. Scales: in L.l. 58–68; before D. about 31; around C. peduncle about 28; radii present on all fields. Premaxillae not protractile. Snout long, prominently pointed; mouth inferior. Mottled with black, the back often almost black; no lateral band. Teeth 2,4–4,2. Widely distributed from coast to coast in the North, south to North Carolina and Iowa, and in the West to northern Mexico. Recently reported from northeastern Nevada.

Genus *Agosia* **Girard.** Monotypic. Mouth small, subinferior, lower jaw included, frenum hidden; small maxillary barbel. Eye rather small, about 5 in head. Scales small (73–95 rows across L.l., about 45 in front of D. and 18 above the complete L.l.), with radii on all fields. D. 8, its origin nearer C. base than snout tip; A. 7 (rarely 6), elongate in adult female. Teeth 4–4, hooked. Gill rakers rudimentary, 5–9. Intestine rather long; peritoneum jet black. Middorsal stripe broad, blackish; spot in front of D. base roughly circular, cream-colored; lateral band intensified in a conspicuous black C. spot.

Agosia chrysogaster **Girard.** Longfin dace. Lower Colorado River system from Arizona and New Mexico southward into northern Mexico.

Genus *Tiaroga* **Girard.** Monotypic. Pharyngeal teeth 1 or 2,4–4,2 or 1. Scales small (about 65–70 in the complete L.l.), with radii on all fields; breast, belly, and most of back naked. Premaxillae with broad frenum. Isthmus very broad, gill openings much restricted, with lower angle of gill cleft opening just anterior to P_1 base. Mouth very small and oblique. D. 8, its origin slightly behind P_2 insertion; A. 7. Body olivaceous, with dusky specks above; black spot at C. base. Gila River system of Arizona, New Mexico, and northern Mexico.

Tiaroga cobitis **Girard.** Loach minnow. A brightly colored species from riffle situations in the Gila River Basin of Arizona, New Mexico, and northern Sonora, Mexico. Rare and threatened with extinction.

Genus *Phenacobius* **Cope.** Closely allied to *Parexoglossum* and *Exoglossum*. Body slender and little compressed. Mouth inferior, lower lip mesially thin and expanded at angle of mouth, forming

a fleshy lobe (Fig. 2-41C). Dentary bones separate except at symphysis. No barbels. Upper jaw protractile. Teeth 4–4. Scales small. L.l. complete. D. origin in front of P_2. Intestine short; peritoneum white.

KEY TO SPECIES OF *PHENACOBIUS*

1. L.l. scales 42–53   2
   L.l. scales 56–63   3
2. No distinct basicaudal spot. Body depth at dorsal origin contained about 6 times in standard length
              P. teretulus
   A distinct round caudal spot. Body depth at dorsal origin contained about 4 or 5 times in standard length (Fig. 2-41C)   *P. mirabilis*
3. Body very slender, its depth at dorsal origin contained more than 6 times in standard length. Caudal spot distinct   *P. uranops*
   Body less slender, its depth at dorsal origin contained fewer than 6 times in standard length. Basicaudal spot vague and dusky   4
4. Pattern with black and silvery colors; lateral band pigment not extending along myosepta. Caudal peduncle scales 15–19. Pelvic fins not reaching anus   *P. catostomus*
   Pattern of brown colors; lateral band pigment extending along myosepta. Caudal peduncle scales 20. Pelvic fins reaching to or beyond anus
              P. crassilabrum

Phenacobius teretulus **Cope.** Kanawha minnow. Scales relatively coarse. No C. spot. Kanawha River in West Virginia, Virginia, and North Carolina.

Phenacobius mirabilis (**Girard**). Suckermouth minnow (Fig. 2-41C). Scales relatively coarse. C. spot distinct. Widely distributed in the Mississippi Valley; also in the drainage of western Lake Erie.

Phenacobius uranops **Cope.** Stargazing minnow. Scales fine. C. spot distinct. Tennessee River system in Alabama, Kentucky, Tennessee, and Virginia.

Phenacobius catostomus **Jordan.** Riffle minnow. Scales fine. C. spot indistinct. Upper C. lobe larger than lower. Alabama and Black Warrior River systems.

Phenacobius crassilabrum **Minckley and Craddock.** Fatlips minnow. D. 8; A. 7. L.l. scales 56–68 (usually 60); C. peduncle scales 20. Lips greatly expanded, the distance across them almost equal to snout length. Upper Tennessee River system in Tennessee and North Carolina.

Genus *Parexoglossum* Hubbs. Closely allied to *Exoglossum*, but differing in the possession of barbels, absence of fleshy lobes at mandibular bases, dentary bones less swollen posteriad, the absence of radii on lateral fields of scales, and the nonlobular lip. Teeth 1,4–4,1.

***Parexoglossum laurae* Hubbs.** Tonguetied minnow (Fig. 2-41*B*). Virginia, West Virginia, North Carolina, Pennsylvania, Ohio, and New York. The nominal species *P. hubbsi* Trautman from the Mad and Stillwater Rivers in Ohio is included in *P. laurae* on advice of Dr. Trautman.

Genus *Exoglossum* Rafinesque. Monotypic. Mandible peculiarly curved inward, producing a strongly trilobed outline; dentary bones united their full length; lower lip with a broad fleshy lobe on each side. Upper jaw not protractile, the lip thick and somewhat plicate. Teeth 1,4–4,1. L.l. complete. D. a little behind the P_2; A. 7. Alimentary canal short; peritoneum white.

***Exoglossum maxillingua* (LeSueur).** Cutlips minnow (Fig. 2-41*A*). From eastern end of the Lake Ontario watershed, the St. Lawrence River system, and the Lake Champlain drainage southward east of the mountains to the Ronanoke River system in Virginia.

Genus *Notropis* Rafinesque. This genus contains more species than any other genus of American minnows, therefore is most difficult to characterize. Teeth 4–4; 1,4–4,1 or 2,4–4,2 (teeth sometimes missing in the outer row of either side); rarely 5 teeth in the inner row. Scales relatively large and often deciduous; in counting scales it is often necessary to count scale pockets. D. 8, with slender rudiments closely joined to 1st developed ray; D. origin directly over P_2 insertion or in varying distances behind it; A. short (6 or 7) or long (12 or 13). Barbels never present,[1] but mouth highly variable, being horizontal to oblique and, if horizontal, highly reminiscent of some species of *Hybopsis*. Eye size variable, being small or large.

Peritoneum spotless silvery, silvery with scattered melanophores, or inky black. Body color often silvery, with or without conspicuous lateral bands and C. spots. Some species brilliantly colored, usually with bright yellows or reds, others with irridescent greens and blues. Intestine regularly short, with only 1 anterior loop, paralleling the stomach in its course forward and then extending directly to anus. Most species small, but some as long as 150 or 200 mm.

Except for 3 species in Mexico, *Notropis* was originally confined to North America east of the Rocky Mountains. Representatives of *N. formosus* are native to the Pacific drainage of the Southwest. Some species introduced on the West Coast of the United States.

While using the key to the genus *Notropis* one must not be content with one count of A. rays since the A. rays vary in number and only the modal number is given in the key. Care should be exercised to examine the pharyngeal arch to ascertain the presence of a broken tooth or an old socket where a tooth once was. Superficially, a species with the dental formula 2,4–4,2 may appear to be 1,4–4,1. If, in a collection of specimens of a species, counts of both 2,4–4,2 and 1,4–4,1 are obtained, follow the former in the key. Some species may have either of these dental formulas and therefore complicate the problem of key construction.

Some species of *Notropis* awaiting description have no available names and are not treated in this manual. If it is suspected that such a species is at hand, the specimens should be sent to an ichthyologist who will be able to give an answer to the problem or to refer the specimens to someone who can.

Sometimes exceptions are inserted parenthetically in the key. These exceptions must be eliminated or accepted on other grounds. For example, in couplet 2 the fish being keyed may be *N. zonistius,* and therefore, before going on to couplet 53, try 3. If the A. rays are 10, take 4, and if a C. spot is present, go on to 5, etc. If the fish fails to satisfy the key requirements, it is not *N. zonistius*.

KEY TO SPECIES OF *NOTROPIS*

1. Pharyngeal teeth in 2 rows on each side; 2,4–4,2 or 1,4–4,1 with occasionally a tooth missing on either side **2**

---

[1] The confusing relationship of *Hybopsis amblops winchelli* with *Notropis amnis* was called to my attention by Carl L. Hubbs, who indicates a strong possibility that *H. amblops* should be regarded as a monotypic species and that *N. amnis* should become *N. winchelli* consisting of 3 subspecies: *N. winchelli winchelli* (often barbeled and therefore exceptional), *N. winchelli amnis,* and *N. winchelli pinnosus.*

Pharyngeal teeth in one row; 4–4   **75**
2. Usually 2 teeth in the outer row (1 in *N. leuciodus* and sometimes in *N. zonistius*)   **3**
     Usually 1 tooth in the outer row or 0 in some, e.g., *N. jemezanus*   **53**
3. Mode of anal-ray counts 10 or 11 (sometimes 9 in *N. ardens* and *N. altipinnis* and as many as 13 in *N. umbratilis* and *N. fumeus*)   **4**
     Mode of anal-ray count 7, 8, or 9   **28**
4. Caudal base with an evident spot (with melanophores in contracted state the spot appears as a cluster of fine dots) which may be large or small and clearly set off from a lateral stripe or an intensification (sometimes diffuse) of the lateral band. In some species caudal spot may show black streaks extending onto caudal rays. No black pigment at dorsal origin   **5**
     Caudal base without a spot or a very vague one (*N. altipinnis* young have a caudal spot). With or without a spot at dorsal origin   **8**
5. Dorsal fin devoid of black markings, membranes and rays pale   *N. stilbius*
     Dorsal fin with black markings or with numerous melanophores on membranes   **6**
6. Pigment of dorsal fins as scattered melanophores on membranes, except distal portions of anteriormost ones   *N. euryzonus*
     Pigment of dorsal fins concentrated as a band   **7**
7. Anal rays usually 11. Dark lateral band broadest below dorsal origin. L.l. scales usually 36 or 37   *N. hypselopterus*
     Anal rays usually 10. Lateral band narrow and best developed posteriad. L.l. scales about 40   *N. zonistius*
8. An area of intensified black pigment (absent in *N. lirus*, a slender species with tiny predorsal scales and the lateral band extended onto snout) on the back at dorsal origin and often on bases of first dorsal rays   **9**
     No intensification of black pigment at dorsal origin (a weak spot sometimes in *N. fumeus*)   **12**
9. Scale rows crossing midline in front of dorsal fin fewer than 27   *N. matutinus*
     Scale rows crossing midline in 27 or more rows   **10**
10. Snout conspicuously blackened by continuation of a black lateral band which is continued forward through the eye and onto the snout   *N. lirus*
     Snout not blackened, lateral band poorly developed anteriad   **11**
11. Anal rays 9–11. Body width contained less than 1.8 times in depth. Breeding males with nuptial tubercles on head dorsum, but not on cheeks; mandibles with 1 row of tubercles   *N. ardens*

Anal rays 10–13. Body width contained more than 1.8 times in depth. Breeding males with nuptial tubercles on head dorsum and cheeks; mandibles with 2 rows of tubercles   *N. umbratilis*
12. A black blotch on upper portion of dorsal fin, intense in adult males but only dusky in females and immature specimens   *N. roseipinnis*
     No black blotch on dorsal fin (except that fin may be margined with black)   **13**
13. Melanophores on body large, few, and irregularly scattered. Pharyngeal teeth, 2,4–4,2 or 1,4–4,1   *N. perpallidus*
     Melanophores on body, small, numerous, and forming a regular pattern. Pharyngeal teeth almost always 2,4–4,2   **14**
14. Width of the eye much less than snout length   **15**
     Width of the eye equal to or greater than snout length (except sometimes in breeding males)   **16**
15. Snout blunt and declivitous. Caudal fin very long, its length greater than head length   *N. simus*
     Snout sharp, dorsal contour of the head a straight line. Caudal fin length about equal to head length   *N. oxyrhynchus*
16. A pair of black crescents between nostrils   *N. photogenis*
     No black crescents between nostrils (except sometimes in *N. ariommus*)   **17**
17. Species restricted to Atlantic Coast north of Georgia   **18**
     Species of Great Lakes, Gulf Coast, Mississippi, and Rio Grande drainages. (Note: *N. rubellus*, included here, also occurs on Atlantic Coast)   **21**
18. Dorsal origin nearer base of caudal fin than front margin of eye. Height of dorsal fin about $\frac{1}{2}$ distance from dorsal origin to occiput   *N. amoenus*
     Dorsal origin nearer front of eye than base of caudal fin or about midway between these points. Height of dorsal fin more than $\frac{1}{2}$ distance from dorsal origin to occiput   **19**
19. Snout marked by a dark preorbital blotch, extending onto anterior half of lips and bordered above by a light streak passing through nostrils and around snout tip   *N. altipinnis*
     Region in front of snout darkened by thickset color cells and lacking the light line   **20**
20. Nuptial tubercles granular, small, and present in both sexes and even immature. L.l. scales usually 38–41; predorsal scales usually 19–23   *N. semperasper*
     Nuptial tubercles granular but larger and only on adult males. L.l. scales 34–38; predorsal scales usually 15–18   *N. scepticus*

21. Species restricted to Gulf Coast east of Mississippi River 22
    Species not so restricted 23
22. Scales in front of dorsal fin more crowded, 26–31 rows crossing midline. Dorsal, pelvic, and anal fins margined with black *N. bellus*
    Scales in front of dorsal fin larger, 15–18 rows crossing midline. Dorsal, pelvic, and anal fins lacking a dark margin *N. signipinnis*
23. Distance from dorsal origin to end of hypural equal to or greater than distance from dorsal origin to nostrils. L.l. pores with a very black dash above and below them *N. ariommus*
    Distance from dorsal origin to end of hypural less than distance from dorsal origin to nostrils. L.l. pores with or without black dashes above and below them 24
24. Middorsal stripe rather broad and prominent before and behind the dorsal fin. L.l. pores often marked by melanophores at least anteriorly 25
    Middorsal stripe consisting of fine lines or rows of dots before dorsal fin and faintly or well developed behind. L.l. pores usually not marked by melanophores 26
25. Lateral band diffuse, but with about same intensity throughout its length. Chin with some black pigment *N. percobromus*
    Lateral band intense black on the caudal peduncle and diminishing in intensity forward. Chin without black pigment, except on lips. *N. rubellus*
26. Eye about equal to snout. Pigment along anal base and undersurface of caudal peduncle absent in adults *N. jemezanus*
    Eye width greater than snout length. Pigment along anal base and undersurface of caudal peduncle present, poorly to well developed 27
27. Middorsal stripe distinct behind dorsal fin. Lateral band poorly developed anteriorly, but most distinct and broadest on caudal peduncle. Pigment along anal base and undersurface of caudal peduncle poorly developed. Predorsal scale rows 17–20 *N. atherinoides*
    Middorsal stripe behind dorsal fin faintly developed or only anteriad. Lateral band posteriorly prominent, wider and more diffuse forward. Pigment along anal base and undersurface of caudal peduncle rather prominent. Predorsal scale rows 23–28 *N. fumeus*
28. Modal number of anal rays 9 29
    Modal number of anal rays 7 or 8 35
29. Sides of adults (young also in some specimens) marked with conspicuous irregularly placed black cross blotches or with some scales (conspicu-

ously higher than long) darkened with many small melanophores 30
    Sides not so conspicuously marked 31
30. Sides with irregularly placed black cross blotches. Breeding males flushed with pink and with all fins red. Both sexes with nuptial tubercles curved forward *N. cerasinus*
    Sides of adults with irregularly blackened scales which are much higher than long. Breeding males with lower fins, sides, and underparts rosy. Nuptial tubercles erect, present only on males (Fig. 2-50) *N. cornutus*

Fig. 2-50. Notropis cornutus (nuptial male). (Forbes and Richardson, 1920.)

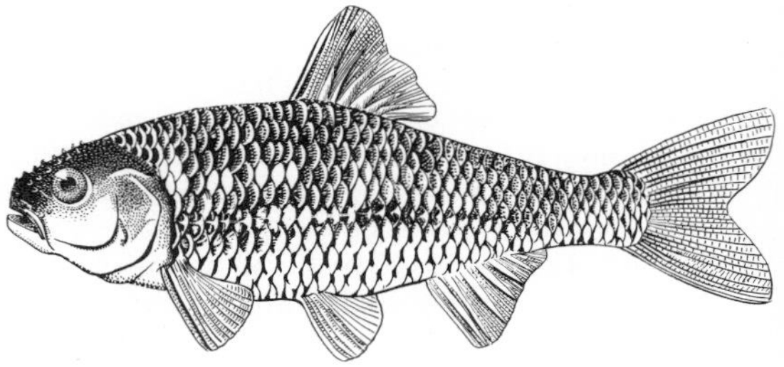

31. L.l. scales fewer than 40 32
    L.l. scales more than 40 33
32. Lateral band silvery, with little or no black pigment. Dorsal origin over or very little behind insertion of pelvic fins (Fig. 2-43A) *N. shumardi*
    A dark lateral band present. Dorsal origin usually behind pelvic insertion (Fig. 2-43B). Southern Texas from the Colorado River west to New Mexico and south into Mexico *N. amabilis*
33. A broad lateral band and a narrower one above it, disappearing posteriad. In life the bands are masked by silvery pigment. Restricted to the Ozark uplands 34
    Lateral band represented by a faint lateral streak. Uplands from Kentucky to Georgia *N. coccogenis*
34. Adults with a conspicuous, black scapular bar behind operculum, less evident but present in young *N. zonatus*
    Black scapular bar absent *N. pilsbryi*
35. Modal number of anal rays 8 (often 9 in *N. shumardi*) 36
    Modal number of anal rays 7 46

36. A dark or dusky caudal spot usually present (sometimes as a termination of lateral band)   **37**

No caudal spot although lateral band may extend over hypural region onto caudal fin   **43**

37. Caudal spot distinct and wedge-shaped. Restricted to Ozark region (Fig. 2-51)   *N. greenei*

Fig. 2-51. Caudal fin of *Notropis greenei* showing the wedge-shaped caudal spot.

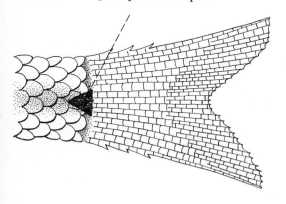

Caudal spot not wedge-shaped, but round or merely an intensification of lateral band   **38**

38. Dorsal origin before, over, or only slightly behind 'pelvic insertion (if mouth lining is pigmented take couplet 42)   **39**

Dorsal origin distinctly behind pelvic insertion   **41**

39. Coloration silvery, very pale, and no bright colors. Caudal spot conspicuous, round, and black in young and becoming dusky to absent in adults. Mouth nearly horizontal   *N. hudsonius*

Coloration less silvery, breeding males with bright red or green colors. Dark caudal spot usually present. Mouth oblique   **40**

40. L.l. pores boldly outlined with elongated, dashlike melanophores for full length of L.l. Light stripe above lateral band from below dorsal fin to caudal fin. Breeding males with red snout and dorsal base   *N. leuciodus*

L.l. irregularly marked with rounded melanophores. In preserved specimens a light stripe above lateral band from snout to caudal fin. Breeding males with scarlet bars on dorsal, caudal, and anal fins. A narrow scarlet band from upper edge of opercle to caudal fin and below this a silvery band   *N. chrosomus*

41. L.l. scales about 39. Caudal spot tiny, round, and often inconspicuous. Breeding males with top of head and middorsal stripe green, and sides of head, belly, lateral band, dorsal and caudal bases red   *N. chlorocephalus*

L.l. scales 33–36. Caudal spot larger, somewhat rectangular or V-shaped. Breeding males with red band on dorsal and anal or with no bright colors   **42**

42. Mouth lining with much black pigment; intense black pigment around anus, anal base, and under caudal peduncle; lateral band intense, continuous around snout. Scales about 35   *N. chalybaeus*

Mouth lining with little or no black pigment; little or no pigment about anus, anal base, and under caudal peduncle; lateral band dusky and covered with silvery pigment (except in old museum specimens), not continuous around snout. Scales about 36   *N. chiliticus*

43. Lateral band silvery, very little black pigment. L.l. scales 33–35. Lower jaw definitely included.   *N. shumardi*

Lateral band black. L.l. scales 38–40. Lower jaw equal to or included in upper   **44**

44. Eye width less than snout length. Breast naked. Lateral band on caudal peduncle as wide as pupil   *N. lutipinnis*

Eye width equal to or greater than snout length. Breast scaly, at least a few scales between pectoral bases. Lateral band on caudal peduncle not as wide as pupil   **45**

45. Sides with few or no melanophores below L.l. Lateral band less intense   *N. scabriceps*

Sides with melanophores outlining first row of scales below L.l. Lateral band very black   *N. rubricroceus*

46. A dark lateral band present and well developed   **47**

Dark lateral band poorly developed, at least anteriad   **51**

47. Mouth inferior. Anterior L.l. scales elevated. Predorsal dark streak absent or obsolescent   *N. asperifrons*

Mouth terminal. Anterior L.l. scales little or not elevated. Predorsal dark streak well developed   **48**

48. Middorsal stripe dark, continuous, and encircling dorsal-fin base, extending posteriad to procurrent caudal rays. Upper jaw much longer than eye. Dark lateral band uniform in intensity throughout, its lower margin even-edged   *N. baileyi*

Middorsal stripe poorly developed or absent posteriad, not encircling dorsal base. Upper jaw equal to or shorter than eye. Lateral band broken into scattered melanophores anteriad   **49**

49. Scales below L.l. without pigment. Lower jaw included, snout slightly overhanging mouth
N. petersoni

 Scales below L.l. with pigment. Jaws equal, nearly so, or the upper included **50**

50. No light stripe above lateral band; no pigment paralleling last anal rays **N. xaenocephalus**

 A light stripe above lateral band; last anal rays paralleled by rows of melanophores
N. texanus

51. Upper lips greatly swollen posteriad (Fig. 2-52). Teeth uniformly 2,4–4,2 **N. potteri**

 Upper lips not markedly swollen posteriad (Fig. 2-53). Teeth 1 or 2,4–4,2 or 1 **52**

52. Black patches on body at bases of first 4 or 5 dorsal and anal rays **N. hypsilepis**

 Black patches absent at bases of dorsal and anal fins **N. blennius**

Fig. 2-52. Head of *Notropis potteri* **showing the thickened upper lip.**

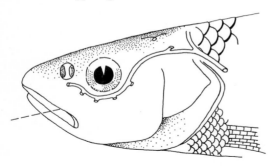

Fig. 2-53. Head of *Notropis blennius* **with upper lip not thickened.**

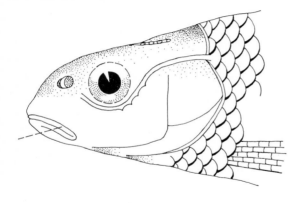

53. Posterior interradial membranes of dorsal fin not blackened (Fig. 2-54B). Exposed areas of scales often not much higher than wide, not presenting a diamond-shaped appearance **69**

 Posterior interradial membranes of dorsal fin with concentrations of melanophores (Fig. 2-54A). Exposed areas of scales higher than wide, usually (see couplets 57 and 65) presenting a diamond-shaped appearance (subgenus *Cyprinella*, in part) **54**

Fig. 2-54. Dorsal fins of *A*, *Notropis whipplei* **with blackened interradial membranes;** *B*, *Notropis lutrensis* **without blackened interradial membranes.**

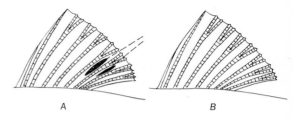

A   B

54. Anal rays 10 or 11 **55**
 Anal rays 8 or 9 **56**

55. Scales below L.l. on caudal peduncle, 7 (count from one side to the other, not including L.l. scales). Caudal fin not black-tipped
N. xaenurus

 Scales below L.l. on caudal peduncle, 5. Caudal fin black-tipped **N. pyrrhomelas**

56. Anal rays usually 8 **57**

 Anal rays usually 9 (often 8 in *N. venustus* in Chattahoochee system and regularly so in the Red River drainage, where a conspicuous round caudal spot is present) **65**

57. Scales diamond-shaped **59**
 Scales not diamond-shaped **58**

58. Scales in front of dorsal fin, 11 (counted from L.l. over back to the opposite side, not including L.l. scales); scales below L.l. on caudal peduncle, 7. Lateral band present but less prominent
N. callistius

 Scales in front of dorsal fin, 13 (counted as above); scales below L.l. on caudal peduncle, 5. Prominent dark lateral band present **N. leedsi**

59. Predorsal scales above L.l. usually 11 **60**
 Predorsal scales above L.l. usually 13 or more **61**

60. Scales counted from below L.l., below dorsal origin, to L.l. of other side, 9. Body depth usually

21 or 22% of standard length. Lateral band extending backward from opercle **N. caeruleus**
Scales below L.l., below dorsal origin, to L.l. of other side, 11. Body depth 25–28% of standard length. Lateral band, if present, extending no farther forward than dorsal origin (Santee River system. See notes for *N. analostanus*, couplet 68.) **N. chloristius**

61. Caudal spot black and conspicuous (round or elongate) **N. venustus**
    Caudal spot absent or poorly developed, not in contrast to lateral band **62**

62. Pigment in dorsal membranes most dense posteriad. Teeth 1,4–4,1 **64**
    Pigment in dorsal membranes absent or, if present, more conspicuous anteriad. Teeth variable 4–4 to 1,4–4,1 **63**

63. A tooth present in lesser row, at least on one side. A black line extending downward from eye to middle of upper jaw. A narrow caudal spot, darker than lateral band, usually present **N. callitaenia**
    Teeth 4–4. No black line downward from eye to upper jaw. Caudal spot absent or obscure **N. callisema**

64. A prominent blue or black lateral band. Breeding male with tubercles in rows on back and lower caudal peduncle. Mouth inferior, nearly horizontal **N. niveus**
    Lateral band, if present, diffuse, wide. Breeding male with scattered tubercles. Mouth terminal or subterminal, oblique **N. spilopterus**

65. Scales above L.l. not diamond-shaped, their posterior edges rounded. A prominent lateral band and caudal spot usually merging but sometimes separate. Breeding males with tubercles in rows on head dorsum, nape, and lower caudal peduncle. Dorsal fin in breeding males bright orange **N. trichroistius**
    Scales above L.l. diamond-shaped. No lateral band or caudal spot. Breeding tubercles scattered. Dorsal fin in breeding males not orange, but red or without bright colors **66**

66. Bases of caudal rays lacking melanophores, forming either a dorsal and ventral light area or a complete white bar. Dorsal fin red in breeding males and tubercles continuous on head dorsum and snout; each mandible with 2 rows of tubercles **67**
    Caudal base with the usual pigment, no white patches or bar. Dorsal fin without bright colors, and tubercles on snout separated by a naked area from head dorsum; each mandible usually with 1 row of tubercles **68**

67. L.l. scales usually 39–41. Body depth usually 20–23% of standard length. Snout conical or subconical **N. galacturus**
    L.l. scales usually 36–38. Body depth usually 24–27% of standard length. Snout blunt. **N. camurus**

68. Pectoral rays usually 13 or 14 **N. analostanus**
    Pectoral rays usually 15 **N. whipplei**

69. Anal rays 9 to 11. Dark lateral band extending around snout, through eye, and terminating in a caudal spot streaking backward on caudal fin. From the Neuse River to western Florida **N. cummingsae**
    Anal rays 7 or 8. Lateral band present or absent; caudal spot poorly developed to absent. Great Lakes region, Mississippi Valley, and Gulf of Mexico tributaries **70**

70. Posterior one-third of upper jaw concealed beneath suborbital **N. amnis**
    Upper jaw normally formed and lacking a conspicuous extension behind angle of mouth **71**

71. Mouth horizontal or nearly so **72**
    Mouth oblique **73**

72. Lateral band well developed, more conspicuous posteriad. Eye about equal to snout. Body depth about 4 in standard length **N. dorsalis**
    Lateral band poorly developed. Eye much shorter than snout. Body depth about 5 in standard length **N. longirostris**

73. Lateral band a zigzag line; no caudal spot **N. heterodon**
    Lateral band with straighter borders; caudal spot poorly to well developed **74**

74. Anal rays 8 (7 to 9). L.l. scales about 36 **N. boops**
    Anal rays 7. L.l. scales about 32 **N. braytoni**

75. Anal rays 9 (sometimes 10 in *N. ortenburgeri* and 8 in *N. spectrunculus* and *N. welaka*) **76**
    Anal rays 7 or 8 **79**

76. Lateral band black, from snout through eye to caudal base **77**
    Lateral band poorly developed, diffuse anteriad and intensified posteriad **78**

77. Mouth very oblique, chin curved anteriad. Caudal spot absent or not conspicuous. Dorsal origin over pelvic insertion **N. ortenburgeri**
    Mouth slightly oblique. Caudal spot present. Dorsal origin behind pelvic insertion **N. welaka**

78. Border of upper jaw forming an arc posteroventrad; snout fleshy, overhanging mouth. Interorbital distance equal to or less than eye length. Caudal peduncle slender, its width 3 or more in its length **N. spectrunculus**
    Border of upper jaw straight; snout not fleshy, not overhanging mouth. Interorbital distance greater

than eye length. Caudal peduncle depth less
than 3 in its length   **N. lutrensis**

79. Anal rays typically 8 (sometimes 7 in *N. anogenus*)
    80

Anal rays typically 7 (8 in *N. proserpinus* and
often in *N. chihuahua;* 8 or 9 in *N. lepidus*)   **87**

80. L.l. scales with black crescents, their concavities
posteriad   **N. heterolepis**

L.l. scales without black crescents, but often with
a black dot on each side of L.l. pores   **81**

81. Anterior L.l. scales much higher than long (Fig.
2-55A). (If anterior dorsal rays are basally black
and body depth 5 or more times in standard
length, take 86)   **82**

Anterior L.l. scales not noticeably elevated (Fig.
2-55B)   **83**

**Fig. 2-55. Lateral-line scales of A, *Notropis buchanani;*
B, *Notropis stramineus.***

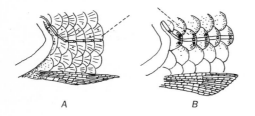

A   B

82. Infraorbital canals complete   **N. volucellus**

Infraorbital canals absent or poorly developed
     N. buchanani

83. L.l. scales about 43. Range restricted to southwestern New Mexico and southeastern Arizona,
south into Mexico   **N. formosus**

L.l. scales fewer than 40   **84**

84. Lateral band distinctly black, extending forward on
head and encroaching on chin. Mouth very
oblique, almost vertical   **N. anogenus**

Lateral band poorly developed, not extending forward on head. Mouth slightly to moderately
oblique or nearly horizontal   **85**

85. Caudal spot large, round, and conspicuous with a
small black triangle at upper and another at
lower edge of caudal base (Fig. 2-56)
     N. maculatus

Caudal spot small and often indistinct, or absent.
No other markings at caudal base   **86**

86. Body depth contained fewer than 5 times in standard length. Tips of pectoral fins approximating or
extending behind pelvic insertion. Mouth larger,

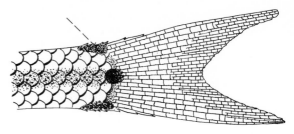

Fig. 2-56. Caudal fin of *Notropis maculatus.*

more terminal, more oblique, and maxilla reaching the vertical from anterior orbital margin
     N. girardi

Body depth contained more than 5 times in standard length. Tips of pectoral fins falling far short
of pelvic insertion. Mouth smaller, overhung by
the snout, less oblique, and maxilla reaching vertical from posterior margin of posterior nasal
opening   **N. ozarcanus**

87. Upper parts conspicuously marked with irregularly
placed large black spots, some of which are as
large as pupil   **N. chihuahua**

Upper parts without such conspicuous spots   **88**

88. Fishes of Atlantic drainage   **89**

Fishes of Mississippi drainage and Gulf Coast
streams from Alabama to Texas   **91**

89. L.l. typically incomplete. Dorsal contour at nape
slightly concave (sometimes nearly straight).
Eye diameter greater than snout length. Lateral-
band extension on snout confined to upper lip
and extreme edge of rostral fold (Fig. 2-57).
Lateral band on body very intense and wider
than pupil   **N. bifrenatus**

L.l. complete or nearly so. Dorsal contour straight
or convex. Eye diameter equal to or less than
snout length. Lateral band on snout restricted
to preorbital spot (Fig. 2-58) or very broad on

Fig. 2-57. Head of *Notropis bifrenatus.*
(After Hubbs and Raney, 1947.)

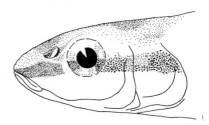

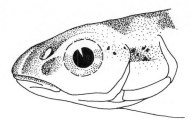

Fig. 2-58. Head of *Notropis procne.*
(After Hubbs and Raney, 1947.)

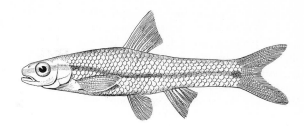

Fig. 2-60. Notropis atrocaudalis.
(Jordan and Evermann, 1900.)

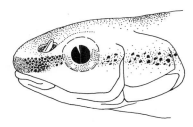

Fig. 2-59. Head of *Notropis alborus.*
(After Hubbs and Raney, 1947.)

origin directly over pelvic insertion; interradial membranes of dorsal fin immaculate
N. sabinae
Anterior dorsal rays forming a blunt and rounded tip and its origin distinctly behind pelvic insertion; posterior interradial membranes of dorsal fin with melanophores   *N. lepidus*
95. Intermandibular area not black   **96**
Intermandibular area black from lower lip to isthmus   *N. proserpinus*
96. A pair of internarial crescents present similar to those in *N. photogenis.* Eyes more superiorly placed, appearing to look upward. Snout more pointed   *N. uranoscopus*
No internarial crescents. Eyes directed laterally. Snout more rounded, blunt   *N. stramineus*
97. Breast and nape often naked   **98**
Breast and nape fully scaled   *N. topeka*
98. Mouth larger, width of gape greater than orbit, and the maxilla extends beyond the vertical from the anterior orbital margin   *N. bairdi*
Mouth smaller, width of gape less than orbit, and the maxilla does not reach the vertical from the anterior orbital margin   *N. buccula*

snout tip (Fig. 2-59). Lateral band on body narrower than pupil or diffuse   **90**
90. Dark band completely ringing snout tip. A light band on snout above the dark band. Caudal spot conspicuous and joined with lateral band. Pigment on lips confined to concealed area about symphysis of upper lip   *N. alborus*
Dark band on snout confined to a preorbital spot. No light band above black and extending around snout. Caudal spot much smaller than pupil and not connected with lateral band. Pigment developed on entire upper lip   *N. procne*
91. Lateral band conspicuous and joined with a distinct caudal spot as wide as the band (Fig. 2-60)   *N. atrocaudalis*
Lateral band conspicuous, but not joined with a small caudal spot, or the band poorly developed (at least anteriorly), indistinct, or absent   **92**
92. Mouth horizontal or nearly so   **93**
Mouth oblique   **97**
93. Lateral band anteriorly absent and poorly developed posteriad   **94**
Lateral band, though not conspicuous, developed from operculum to caudal base, represented anteriad by scattered dots in some specimens   **95**
94. Anterior dorsal rays forming a sharp point; dorsal

Notropis atherinoides **Rafinesque.** Emerald shiner. D. origin well behind P_2 insertion; A. 11 (10–11). L.l. decurved. Widely distributed from southern Canada and the Great Lakes region through the Mississippi Valley to the Gulf Coast in Alabama.

Notropis percobromus **(Cope).** Plains shiner. Regarded by some as conspecific with *N. atherinoides.* Similar to *N. atherinoides,* but eye smaller, body deeper, and scales less sharply outlined with melanophores. Licking River in Kentucky and the Mississippi Valley from Iowa south to Arkansas and Oklahoma.

Notropis jemezanus (**Cope**). Rio Grande shiner. Aspect of *N. percobromus,* but with more distinct lateral band. L.l. not marked with melanophores. Rio Grande drainage in Texas, New Mexico, and Mexico.

Notropis oxyrhynchus **Hubbs and Bonham.** Sharpnose shiner. Snout very long, sharp-pointed. A. 10 (9–11). C. spot absent. Colorado and Brazos River systems in Texas.

Notropis photogenis (**Cope**). Silver shiner. 2 dark crescents between nostrils, a unique character. From the Wabash River system of Indiana, the western end of Lake Erie drainage east to the Allegheny River of New York and Pennsylvania south to the Little Tennessee in North Carolina.

Notropis rubellus (**Agassiz**). Rosyface shiner. D. origin well behind P_2 insertion. Eye about equal to snout length. L.l. distinctly outlined with 2 rows of dots. Widely distributed on the Atlantic Coast from New York south to Virginia; Great Lakes drainage and south to the Tennessee River system and Arkansas and eastern Kansas and Oklahoma. Recently taken in North Dakota.

Notropis amoenus (**Abbott**). Comely shiner. D. origin well behind P_2 insertion; eye width equal to or longer than snout. C. spot faint, plumbeous, or absent. Lateral band obscure. Clear streams of the eastern slope of the Alleghenies from New York south to North Carolina.

Notropis scepticus (**Jordan and Gilbert**). Sandbar shiner. Eye longer than snout. Mouth moderate, oblique. A. 10 (9–12). Vertebral stripe indistinct. Lower jaw slightly included. Atlantic Coast streams of North Carolina and South Carolina.

Notropis semperasper **Gilbert.** Roughhead shiner. Teeth 2,4–4,2. A. 10 or 11 (usually 10). L.l. scales 37–43 (usually 38–41); predorsal scales 20–23; circumferential scales 26–29. A close relative of *N. scepticus,* but with granular cephalic nuptial tubercles in both sexes; smaller scales; anteriorly, the upper margin of lateral stripe about halfway between L.l. and middorsal line; no light area on C. peduncle above lateral stripe; and the dorsal stripe behind D. complete and broader. Upper James River system in Virginia.

Notropis roseipinnis **Hay.** Cherryfin shiner. Eye equal to snout. D. origin well behind P_2 insertion, black spot on upper margin. Upper jaw included. Gulf Coast, Louisiana to western Florida.

Notropis amabilis (**Girard**). Texas shiner. D. origin well behind P_2 insertion, last ray less than $\frac{1}{2}$ longest ray. Scales on dorsum conspicuously black-bordered. From southcentral Texas (Colorado River) west to New Mexico and south into Mexico.

Notropis leuciodus (**Cope**). Tennessee shiner. D. origin only a little behind P_2 insertion; A. 8. C. spot very distinct, a continuation of lateral band. L.l. nearly straight. Tennessee River system in Virginia, North Carolina, Georgia, Tennessee, and Kentucky.

Notropis stilbius (**Jordan**). Silverstripe shiner. Eye longer than snout. Mouth large, oblique. A. 10. Lateral band broad, silvery, with many dark dots. C. spot evident. Gulf Coast streams from Mississippi, Alabama, and Georgia.

Notropis lirus (**Jordan**). Mountain shiner. Eye longer than snout. Black spot at D. origin, as in *N. umbratilis.* Lateral band passing through eye onto snout. The Alabama and Tennessee River systems in Virginia, Tennessee, Georgia, and Alabama.

Notropis matutinus (**Cope**). Pinewoods shiner. Eye equal to snout. Predorsal black spot small, distal half of D. red; snout and chin red in breeding males. Streams of the Atlantic Coast in Virginia and North Carolina.

Notropis ardens (**Cope**). Rosefin shiner. Similar to *N. umbratilis,* but body less compressed. A. 9–11. Sides sometimes marked with dark bars. Ohio and Tennessee Basins; Shenandoah River to Neuse Basin, Atlantic Coast.

Notropis umbratilis (**Girard**). Redfin shiner. Predorsal scales tiny, crowded. Black spot at D. origin. A. 10–13. Adults often with dark chevrons anteriad on upper sides. Widely distributed in the Mississippi River system from Pennsylvania west to Minnesota and south to Mississippi, Louisiana, and Texas.

Notropis bellus (**Hay**). Pretty shiner. D., P_2, and A. margined with black; no black spot at D. origin. A. 10–11. Dark chevrons as in *N. umbratilis.* Gulf Coast streams of Mississippi and Alabama.

Notropis fumeus **Evermann.** Ribbon shiner. Aspect of *N. umbratilis,* but lacking black spot at D. origin. Predorsal scales not crowded. A. 11 (9–12). From the Tennessee River system westward through southern Missouri, Arkansas, and eastern Oklahoma.

Notropis zonistius (**Jordan**). Bandfin shiner.

D. with jet-black bar; basicaudal spot round, nearly as large as eye. C. dull red, pale at base, milky at tip. Chattahoochee River system in Georgia and Alabama and Florida, and upper Savannah River system in Georgia.

Notropis coccogenis (**Cope**). Warpaint shiner. D. with black bar; basicaudal spot absent; black bar back of operculum. Tennessee uplands from Virginia and Kentucky south to Georgia and Alabama.

Notropis cornutus (**Mitchill**). Common shiner. Teeth 2,4–4,2. A. 9; D. origin over or slightly behind P_2 insertion. Eye large. Mouth terminal, oblique. Peritoneum black. No lateral band, but sides with occasional scales overlaid with black. Gilbert (1961a) has indicated that *N. cornutus* and *N. chrysocephalus* may occur as distinct species or occur sympatrically and behave as subspecies, since they readily hybridize or intergrade. *N. cornutus* differs from *N. chrysocephalus* in having higher scale counts before the D. fin and anterodorsal region, a dorsolateral streak on each side of the middorsal stripe, and pigment confined to the rami of the lower jaw in contrast to the lower scale count, presence of a patch of chin pigment, and a zigzag line on each side of the middorsal stripe in *N. chrysocephalus*. The *cornutus* [1] species group, including *cerasinus*, enjoys a wide range from the Rocky Mountains through the Great Lakes region to the Atlantic Coast and throughout most of the Mississippi Valley.

Notropis cerasinus (**Cope**). Crescent shiner. Aspect of *N. cornutus*, but sides with dark cross blotches; fins red. Found only in mountain brooks of the New and Roanoke River systems in Virginia and North Carolina.

Notropis zonatus (**Agassiz**). Bleeding shiner. Teeth 2,4–4,2. A. 9. Sides with a broad dark band and a narrower one above; a black scapular bar behind the opercle. As in *N. cornutus,* the eye is large, the mouth terminal and oblique. Breeding males with bright-red fins and underparts; the dorsum is intensely black and in life obscures the lateral bands. The bleeding shiner and *N. pilsbryi* behave as species. Northern and eastern Ozark Plateau in the Osage, Current, Black, and St. Fran-

cis Rivers of southcentral Missouri and northeastern Arkansas.

Notropis pilsbryi **Fowler.** Duskystripe shiner. Similar to *N. zonatus* but lacking the scapular bar. Southern and southwestern slope of the Ozarks in the White and Arkansas Basins of northcentral and northwestern Arkansas, northeastern Oklahoma, and southwestern Missouri. Some isolated records for the Red River in Oklahoma may represent bait-bucket transfers, especially since more recent collections have not included the species.

Notropis shumardi (**Girard**). Silverband shiner. Teeth 2,4–4,2 (rarely 5 in main row and 3 in lesser row). A. 8 or 9; D. over or slightly behind P_2 insertion. Sides with silvery band. For nomenclatural reasons Gilbert and Bailey (1962), in merging the nominal species *N. illecebrosus* and *N. brazosensis,* reinstated the name, *Notropis shumardi,* for this shiner. Restricted to large silty rivers: the Mississippi-Missouri Basin and major tributaries, Arkansas and Red Rivers; the Brazos, Colorado, San Bernard, and Lavaca Bay drainages in Texas; and the Alabama and Tombigbee River systems in Alabama.

Notropis boops **Gilbert.** Bigeye shiner. Eye very large (2.3–3.0 in head). D. origin over P_2 insertion. Peritoneum usually jet black. A. 8. From Indiana and Ohio south to Tennessee and west to Kansas and eastern Oklahoma in both the Arkansas and Red Rivers.

Notropis heterodon (**Cope**). Blackchin shiner. Lateral band with zigzag lines extending through eye and onto chin and premaxillae. A. 8. North Dakota to Ontario, Quebec, New York, and Pennsylvania. A northern species, not found south of Iowa and Indiana.

Notropis ariommus (**Cope**). Popeye shiner. Eye very large, much longer than snout. D. origin over or slightly behind P_2 insertion. A. 9–11. From southern Indiana south to the Tennessee River and west to Missouri and Arkansas.

Notropis scabriceps (**Cope**). New River shiner. Sides with little or no pigment below L.l. Eye large (about 3.0 in head). A. 8. Found only above the falls on the Kanawha River in West Virginia, Virginia, and North Carolina.

Notropis ortenburgeri **Hubbs.** Kiamichi shiner. Mouth very oblique. Lateral band extending through eye to jaws and chin, but not on snout; C.

Gilbert (1961a) regarded *N. albeolus* as a subspecies of *N. cornutus,* but see Gilbert (1964), *Bull. Fla. State Mus.,* 8(2): 5–194. *N. chrysocephalus* includes *N. c. chrysocephalus* and *N. isolepis.*

spot absent. A. 9 or 10. Osage Hills region in Oklahoma, Poteau River system of eastern Oklahoma, and the Ouachita Mountains of southwestern Arkansas and southeastern Oklahoma, south into eastern Texas.

Notropis perpallidus **Hubbs and Black.** Colorless shiner. Pigment on dorsum in no regular pattern, large stellate melanophores; no lateral band; no C. spot. A. 10 (9–11). Little River system of southwestern Arkansas and southeastern Oklahoma.

Notropis simus (**Cope**). Bluntnose shiner. D. origin well behind P_2 insertion. Mouth large, oblique. Color silvery; no C. spot. A. usually 9. Rio Grande system in Texas and New Mexico.

Notropis xaenurus (**Jordan**). Altamaha shiner. D., A., and C. bright crimson; C. with a faint spot and without a dark margin. A. usually 10. Apparently restricted to the Altamaha River system in Georgia.

Notropis pyrrhomelas (**Cope**). Fieryblack shiner. Color dark steel blue; belly abruptly milk white; head reddish; snout tip, lower jaw, and iris scarlet. A. usually 10. Pee Dee and Santee River Basins of North and South Carolina.

Notropis camurus (**Jordan and Meek**). Bluntface shiner. D. origin over P_2 insertion. Posterior interradial membranes of D. black; C. base milky white. A. 9. Arkansas, Osage, and White Rivers in Oklahoma, Kansas, Missouri, and Arkansas, and to the lower Mississippi and upper Pearl Rivers in Mississippi.

Notropis callistius (**Jordan**). Alabama shiner. Similar to *N. venustus,* but with smaller C. spot and no black in posterior interradial membranes of D. A. 8. Tombigbee and Alabama River Basins in Alabama and Georgia.

Notropis niveus (**Cope**). Whitefin shiner. D. with large dusky blotch on last rays. Mouth slightly oblique. Fins white-tipped, anal all white; faint C. spot. From the Tar to the Savannah, in coastal streams.

Notropis galacturus (**Cope**). Whitetail shiner. D. with posterior interradial membranes black. Mouth nearly horizontal, lower jaw included. C. base creamy yellow; breeding males with all fins tipped with white. A. 8. From the eastern slope of the Ozarks in Missouri and Arkansas through the Tennessee Valley to Virginia, South Carolina, and Mississippi.

Notropis caeruleus (**Jordan**). Blue shiner. D. origin behind P_2 insertion, posterior D. rays faintly darkened. D., A., and C. tipped with satin white; fins otherwise yellow. Mouth small, overhung by snout. A. 8. Found only in the Alabama River system.

Notropis callisema (**Jordan**). Ocmulgee shiner. D. origin nearly over P_2 insertion, nearer snout than C. base; D. with a black spot; longest D. ray longer than head in male. A. 8. Altamaha Basin in Georgia.

Notropis callitaenia **Bailey and Gibbs.** Bluestripe shiner. Teeth 1,4–4,1 (sometimes 0,4–4,1 or 1,4–4,0). A. 8 (rarely 7). D. origin slightly behind P_2 insertion and nearer snout tip than C. base. L.l. (complete) scales usually 38 or 39; predorsal scales not crowded. Breast and throat scaled. Mouth slightly inferior, lower jaw included. Coloration: generally pale with a lateral band, deep metallic blue in life and lead-colored to black in preserved specimens, extending from slightly anteriad to D. origin to C. base; basicaudal spot in contact with lateral band; a dark middorsal stripe rather broad before D. and narrower on C. peduncle. Apalachicola River system in Alabama, Georgia, and Florida and the Escambia River drainage in Florida.

Notropis leedsi **Fowler.** Ohoopee shiner. Teeth 4–4. A. 7. Scales: about 36 in L.l.; predorsal scales 14; breast naked. D. origin behind P_2 insertion. Color pale blue-gray to lighter below; a gray lateral band 1 scale wide. D. fin with upper anterior one-half gray to blackish gray; posterior interradial membranes without pigment. Nuptial tubercles on head dorsum, in 1 or 2 rows in front of D. origin, and on lower portion of C. peduncle. Mouth rather large, low, and a little oblique; lower jaw included. Eye diameter contained 1.5 times in snout length. From the Savannah River southwestward to western Florida.

Notropis analostanus (**Girard**). Satinfin shiner. D. with posterior interradial membranes dark, last D. ray more than $\frac{1}{2}$ length of longest ray. Lateral band widest posteriad; no C. spot. *N. chloristius* included here as a subspecies. A. 9. From New York to South Carolina east of the mountains.

Notropis whipplei (**Girard**). Steelcolor shiner. D. with posterior interradial membranes blackened, anterior membranes of D. with pigment in young. Lateral band on C. peduncle wide and above the midline. A. 9. In the Mississippi Valley from Indiana, Ohio, and New York south to Alabama and west to Oklahoma and Texas.

Notropis spilopterus (**Cope**). Spotfin shiner. Similar to *N. whipplei*, but no pigment on anterior membranes of D. in young. Lateral band on C. peduncle narrow and little above midline. A. 8. From North Dakota to New York in the North and south to Alabama and Oklahoma. In the Great Lakes (except Superior) and Mississippi drainages. On the Atlantic slope, south of New England to the Potomac River.

Notropis trichroistius (**Jordan and Gilbert**). Tricolor shiner. D. vermilion, with milk-white tip, horizontal dusky band at base; D. origin well behind P_2 insertion. Lateral band very black, expanded into a large C. spot. A. 9. Alabama River system in Georgia and Alabama.

Notropis lutrensis (**Baird and Girard**). Red shiner. D. origin about over P_2 insertion; D. usually without black pigment. Breeding males with purple shoulder crescent. Lower fins red. A. 9. L.l. scales about 35. Although recorded from Kentucky and Mississippi, this species ranges principally west of the Mississippi River from Minnesota and Wyoming south to Mexico; now established, after bait introduction, in the lower Colorado River, California and Arizona.

Notropis lepidus (**Girard**). Plateau shiner. Teeth 4–4. A. 8 or 9. Scales: predorsal about 15; in L.l. 31–36 (usually 34). D. origin nearer C. base than snout tip and behind P_2 insertion. Mouth subinferior, lower jaw included, slightly oblique. D. fin not falcate, its last ray about $\frac{1}{2}$ the longest, and the posterior membranes with melanophores. Lateral band weak anteriad, stronger posteriad; no C. spot. Males with a dark bar, broadest above, behind opercle and with nuptial tubercles larger at occiput than at snout tip. Head springs in the southern part of the Edwards Plateau, Texas, in the Nueces, Frio, Medina, and Guadalupe River systems.

Notropis formosus (**Girard**). Beautiful shiner. Similar to *N. lutrensis,* but A. usually 8. Scales smaller, about 43 in L.l. Southwestern New Mexico and northern Mexico. *N. mearnsi* Snyder from Pacific drainage in southeastern Arizona and Mexico is included here.

Notropis venustus (**Girard**). Blacktail shiner.

Basicaudal spot large, round, and black. Posterior interradial membranes of D. blackened. A. 7 or 8. From Illinois south through Tennessee, Arkansas, Oklahoma, and Texas; eastward through the Gulf states to Florida.

Notropis hypselopterus (**Günther**). Sailfin shiner. D. origin behind P_2 insertion; D. with conspicuous horizontal black blotch. Distinct C. spot smaller than eye. A. usually 11. From the Gulf Coast streams of South Carolina, Georgia, Florida, and Alabama. *N. stonei* Fowler, included here, is from the Santee River in South Carolina to the Satilla River in Georgia; type locality, Pocataligo River in South Carolina, but this may be in error since recent collections yielded no specimens from the Pocataligo.

Notropis euryzonus **Suttkus**. Broadstripe shiner. Teeth 2,4–4,2. A. 10 (often 9 or 11). Scales: from D. origin to opercle, 17–22; in L.l. 35–40. Body very deep and compressed. D. origin nearer C. base than snout tip and behind P_2 insertion. Mouth terminal, oblique; upper jaw longer than eye. L.l. complete, much decurved. D. and A. fins of male greatly elevated, the anterior rays, when depressed, extending beyond posterior rays. Lateral band very wide, covering about $\frac{1}{3}$ the area of the sides; a chevron or lunate basicaudal spot. Interradial membranes of D. and A. fins evenly dotted with melanophores, except for the anterior distal portions of the 1st 2 or 3 membranes of D. From the lower tributaries of the Chattahoochee River in Alabama and Georgia.

Notropis signipinnis **Bailey and Suttkus**. Flagfin shiner (Fig. 2-61). Allied to *N. hypselopterus* but with mandibular nuptial tubercles numerous (22–32), forming a saw-edged comb. Scale rows around body, 22–26. Predorsal streak faint. A. 10 or 11 (9–13). Abundant in the Gulf Coast streams from western Florida to Louisiana.

Notropis rubricroceus (**Cope**). Saffron shiner. D. origin behind P_2 insertion. Mouth large, oblique; jaws equal. Head pale red; all fins red. A. 8 (7–9). Tennessee River system in Virginia, North Carolina, and Tennessee. Introduced in Linville River, North Carolina.

Notropis chlorocephalus (**Cope**). Greenhead shiner. Breeding males with dark-red venter; D. and C. fins red at base; sides of head and lateral band red. Eye large, about 3 in head. Mouth large,

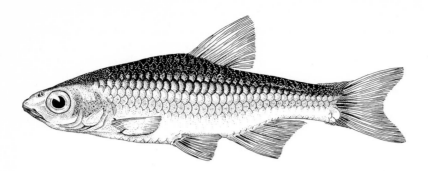

Fig. 2-61. Notropis signipinnis. (From Bailey and Suttkus, 1952.)

oblique. A. 8. Santee River system from North and South Carolina and south to Georgia.

Notropis chrosomus (**Jordan**). Rainbow shiner. D. and C. fins with a scarlet bar; C. spot distinct; narrow scarlet band from upper edge of opercle to C. base, below this a silvery band. Lower jaw included. A. 8 (sometimes 7 or 9). Tennessee River system in Georgia, and Alabama.

Notropis lutipinnis (**Jordan and Brayton**). Yellowfin shiner. Lateral band intense black. Males with whole body red, head and fins golden yellow. Breast naked. Lower jaw included, blackened at tip. A. 8 (rarely 7 or 9). Small clear streams of the pine woods in North and South Carolina and Georgia. Atlantic Coastal drainage.

Notropis baileyi **Suttkus and Raney**. Rough shiner. Allied to *N. chrosomus* and *N. lutipinnis,* but with A. 7 (rarely 6, sometimes 8). Middorsal streak dark, continuous to C. base; lateral band intense throughout. Mouth terminal, jaws about equal. Pascagoula River and Mobile Bay drainages in Mississippi and Alabama, in small wooded streams with sand and gravel bottoms.

Notropis altipinnis (**Cope**). Highfin shiner. Lower jaw blackish; lateral band broad, conspicuous; preorbital bar connects with black lips and bordered above by pigmentless area; young with black C. spot. L.l. incomplete. A. 10 (8–12). Streams of the Atlantic Coast from the Chowan River system in Virginia south to the Santee River system in South Carolina.

Notropis cummingsae **Myers**. Dusky shiner. Lateral band intense, extending around snout, through eye and ending in a black C. spot. A. 9–11. From the Neuse River in North Carolina to northern and western Florida and eastern Alabama.

Notropis chalybaeus (**Cope**). Ironcolor shiner.

Inside of mouth heavily pigmented. A. rays usually 8. Widely distributed on the coastal lowlands from New Jersey to eastern Texas and north in the Mississippi lowlands to Iowa and Indiana.

Notropis chiliticus (**Cope**). Redlip shiner. D. with red band. Eye large, longer than snout; lips and snout red; lateral band silvery. A. 8. Upland streams, Pee Dee River, North Carolina; Dan River, Roanoke tributary.

Notropis texanus (**Girard**). Weed shiner. Last 3 or 4 A. rays conspicuously darkened. A dark oblique scapular bar. A. rays usually 7. From southwestern Michigan and Wisconsin through eastern Iowa. South in the Mississippi Valley to central Texas and east to the Apalachicola drainage in Georgia and Florida.

Notropis xaenocephalus (**Jordan**). Coosa shiner. Lower edge of lateral band not sharply delimited. Breast well scaled. C. spot large and circular. Predorsal stripe narrow and single. Last A. rays not darkened. Upland tributaries of the Alabama and Chattahoochee Rivers in Alabama and Georgia.

Notropis hypsilepis **Suttkus and Raney**. Highscale shiner. Related to *N. petersoni, N. xaenocephalus,* and *N. texanus,* but with black areas on the body at bases of first 4 or 5 D. and A. rays. Anterior L.l. scales highly elevated as in *N. volucellus, N. buchanani,* and *N. heterolepis.* Apalachicola River system in Georgia and Alabama; an inhabitant of sandy-bottomed streams near confluence with main river.

Notropis petersoni **Fowler**. Coastal shiner. Closely allied with *N. roseus* and *N. xaenocephalus.* Lower edge of lateral band sharply delimited. A. rays usually 7. Coastal streams from North Carolina to Florida.

Notropis asperifrons **Suttkus and Raney.** Burhead shiner. Similar to *N. hypsilepis,* but snout more sharply pointed. Head subtriangular rather than subquadrate. Mouth more oblique. Color darker, lateral band well developed anteriad. C. spot large, quadrate rather than wedge-shaped. Area behind anus dark rather than light. Scales around body in 19–21 rows. Preoperculomandibular pores 10. Drainages of Mobile Bay, Black Warrior and Alabama Rivers, Alabama and Georgia.

Notropis hudsonius **(Clinton).** Spottail shiner. Mouth moderate, nearly horizontal; jaws nearly equal. Lateral band pale, silvery, sometimes dusky. C. spot usually present. A. 8. From Canada south through the Great Lakes region and as far south as Georgia in the East and Kansas in the West.

Notropis greenei **Hubbs and Ortenburger.** Wedgespot shiner. C. spot wedge-shaped, unique. A. 8. Ozark uplands in Missouri, Arkansas, and Oklahoma.

Notropis blennius **(Girard).** River shiner. D. short, blunt, its origin over P_2 insertion and its longest ray contained more than 2 times in predorsal length. Mouth horizontal or nearly so. No lateral band or C. spot. A. 7. This species occurs from Alberta and Manitoba, Canada, through the Red River of the North and the Mississippi Basin where it is known from Wyoming to Pennsylvania in the North and southward to Alabama and eastern Texas.

Notropis potteri **Hubbs and Bonham.** Chub shiner. Aspect of *N. blennius,* but with upper lip widest posteriad. Lateral band weakly developed anteriad; a diffuse C. spot. Colorado and Brazos Rivers in Texas and the Red River between Texas and Oklahoma.

Notropis dorsalis **(Agassiz).** Bigmouth shiner. D. without pigment. Mouth large, horizontal. Lateral band absent. A. 8. Wyoming and Colorado to New York and Pennsylvania. Some rivers tributary to the Great Lakes in Wisconsin, Michigan, Ohio, and New York, and in the Mississippi River system south to West Virginia, Missouri, and Kansas.

Notropis longirostris **(Hay).** Longnose shiner. D. origin over P_2 insertion. Eye shorter than snout. Mouth large, horizontal; lower jaw included. Lateral band obsolete. A. 7. In Gulf Coast streams from Louisiana to Florida.

Notropis amnis **Hubbs and Greene.** Pallid

shiner. Mouth small, horizontal; upper jaw unique, its posterior one-third extending behind angle of mouth beneath suborbital. A. 8. Mississippi River system from Wisconsin south to Louisiana and Texas. In the Ohio River system in Indiana and Kentucky and some Gulf Coast streams of central Texas.

Notropis ⸲trocaudalis **Evermann.** Blackspot shiner (Fig. 2-60). Lateral band about width of pupil, extending around snout and on upper lip, terminating in a C. spot. Eye 1.0–1.3 in snout. Mouth moderately oblique. A. 7. Eastern Texas, western Louisiana, and southeastern Oklahoma.

Notropis bifrenatus **(Cope).** Bridled shiner (Fig. 2-57). Lateral band much wider than pupil, intensely black and ringing snout on upper lip and edge of rostral fold; C. spot separate from lateral band. Eye 0.7–0.9 in snout. Mouth strongly oblique. A. 7. Atlantic Coast from Maine to the Potomac River, Virginia; westward through Lakes Champlain and Ontario drainages.

Notropis anogenus **Forbes.** Pugnose shiner. Mouth very small, extremely oblique. Lateral band black, distinct, extending through eye to cross preorbital area, lateral portion of upper lip, and to cover the lower lip and chin; postdorsal stripe narrow and inconspicuous; postanal streak absent or faint; a light stripe above the lateral band. Peritoneum black. A. 8 (sometimes 7). Pharyngeal teeth 4–4, with serrate cutting edges. From eastern North Dakota through the Great Lakes region to Ontario and New York, as far south as northern Iowa. Rare.

Notropis procne **(Cope).** Swallowtail shiner (Fig. 2-58). Lateral band narrow, diffuse, but with strong markings about L.l. pores; C. spot small, separate from lateral band; a preorbital blotch; pigment on upper lip, but weak on chin. A. 7. From a southern tributary of Lake Ontario southward on the Atlantic slope to the James River system, Virginia.

Notropis alborus **Hubbs and Raney.** Whitemouth shiner (Fig. 2-59). Lateral band strong, black, narrower than pupil, ringing snout and terminating in a C. spot; no pigment on lips. Mouth nearly horizontal. A. 7. Roanoke, Cape Fear, and Pee Dee Rivers of Virginia and North Carolina.

Notropis spectrunculus **(Cope).** Mirror shiner. D. and A. bases black; all fins orange; C. spot conspicuous, black. A. 9. Headwaters of the Tennessee River in Virginia, Tennessee, North Carolina, and Georgia.

Notropis welaka **Evermann and Kendall.** Bluenose shiner. Teeth 4–4. D. 8; A. 8 or 9. L.l. scales about 35; L.l. incomplete on only 6–10 scales. Mouth somewhat oblique, maxilla scarcely reaching the large eye. D. origin behind P_2 insertion, nearer C. base than snout tip. A broad black band extending from snout through eye and ending in a rather distinct C. spot. From the St. Johns River in Florida to the Pearl River Basin in Mississippi.

Notropis volucellus **(Cope).** Mimic shiner. L.l. scales highly elevated. Infraorbital canals complete. Widely distributed from southeastern Canada and northern United States south, in the Mississippi River system, to the Gulf and westward to the Guadalupe River. Also on the east coast south of the James and north of the Cape Fear River.

Notropis buchanani **Meek.** Ghost shiner. Similar to *N. volucellus,* but with infraorbital canals absent or poorly developed. In the Mississippi River system from Iowa to Ohio and southward through Virginia, West Virginia, Tennessee, and Alabama; west of the Mississippi through Kansas, Missouri, Arkansas, Oklahoma, western Louisiana, and Texas to Mexico.

Notropis maculatus **(Hay).** Taillight shiner. C. spot conspicuous, large, and round; small black spot above and below C. spot. From Missouri southward in the lowlands through Arkansas and southeastern Oklahoma to eastern Texas and Mississippi; on the Atlantic Coast from North Carolina to Florida.

Notropis heterolepis **Eigenmann and Eigenmann.** Blacknose shiner. Pigment on L.l. scales concentrated to form black crescents. A. 8. Lateral band extending through eye and around snout above premaxillae, no chin pigment. Widely distributed from the Hudson Bay drainage to the New England states, westward to Iowa, and southward in the Mississippi Valley to Kansas, Missouri, and Tennessee.

Notropis ozarcanus **Meek.** Ozark shiner. Similar to *N. spectrunculus,* but slenderer. Lateral band of dark dots, faint; a small C. spot. A. 8. White River system in Arkansas and Missouri.

Notropis girardi **Hubbs and Ortenburger.** Arkansas River shiner. D. origin nearer snout tip than C. base. Mouth rather large, slightly oblique. C. spot small, often indistinct; vertebral stripe distinct. A. 8. Arkansas River and its larger silty tributaries in Kansas, Oklahoma, Texas, and Arkansas.

Notropis bairdi **Hubbs and Ortenburger.** Red River shiner. Similar to *N. girardi,* but nape and breast usually naked. A. 7. Red River, above Lake Texoma, in Oklahoma and Texas.

Notropis buccula **Cross.** Smalleye shiner. Similar to *N. bairdi* but having a smaller mouth, longer snout, and lesser head depth. Teeth 4–4; A. 7; P_1 14–16; nape and breast usually naked; L.l. scales 33–37. Brazos River, Texas.

Notropis sabinae **Jordan and Gilbert.** Sabine shiner. Teeth 4–4. A. 7 (8). Scales: predorsal, 12; in L.l. 32–33; 13 around C. peduncle. D. origin midway between snout tip and C. base or nearer C. base, directly over P_2 insertion; last D. ray about $\frac{1}{2}$ longest ray; anterior D. rays form a sharp tip. Mouth inferior, horizontal; upper jaw length about equal to snout. Lateral band faint on C. peduncle, practically absent anteriad; median dorsal and postanal stripes poorly developed; dots on each side of L.l. or a vertical row above and below tubes; scales of dorsum lacking melanophores in central areas; interradial membranes of all fins clear. Peritoneum silvery. White River in Missouri and Arkansas, Sabine River on Texas-Louisiana boundary, and Neches River, Texas.

Notropis stramineus **(Cope).** Sand shiner. D. origin over P_2 insertion. Mouth nearly horizontal. Lateral band weak; C. spot indistinct; vertebral stripe present, but interrupted in D. base. From the Rocky Mountains to the Appalachians and from the Great Lakes to Mexico, but apparently absent on the Gulf Coast east of Mississippi. A common fish in sandy streams of small to moderate size.

Notropis uranoscopus **Suttkus.** Skygazer shiner. Teeth 4–4. A. 7; P_1 14–16. Scales: predorsal rows 14, not crowded; in L.l. 35 or 36. This slender species is similar to *N. stramineus,* but differs in having 2 black crescents between the nostrils, eyes that seem to look upward and a more pointed snout. Known only from the Cahaba River in Alabama.

Notropis proserpinus **(Girard).** Proserpine shiner. Mouth subinferior, nearly horizontal, lateral band of dark metallic points, not silvery. A. 8. Rio Grande region.

Notropis topeka **Gilbert.** Topeka shiner. Eye shorter than snout. Mouth small, terminal, oblique. Lateral band dusky; small C. spot; sides and lower fins red or orange. A. 7. Breeding males with coarse sharp nuptial tubercles. From Minnesota and South Dakota south to Iowa, Nebraska, Kansas, and Missouri.

Notropis braytoni **Jordan and Evermann.** Tamaulipas shiner. D. origin nearer snout tip than C. base; D. rays extend to same point posteriad when depressed. Mouth large, oblique; lower jaw included. Lateral band brownish, with silver; C. spot small. A. 7. Rio Grande River system of Texas and Mexico.

Notropis chihuahua **Woolman.** Chihuahua shiner. Eye slightly longer than snout. Mouth a little oblique. Lateral band weak; C. spot present; dorsum with many large dark spots as in *Hybopsis aestivalis*. A. 7. In the Rio Grande drainage of western Texas and Mexico.

Genus Ericymba Cope. Monotypic. Closely related to *Notropis*, but differing most strikingly in the remarkable development of cephalic L.l. canals, the canals being cavernous with broad channels easily visible with unaided eye. Lips thin, mouth small and almost horizontal. No barbels. Pharyngeal teeth, 1,4–4,1 or 1,4–4,0. D. and A. rays 8. Intestine short, as in *Notropis;* peritoneum silvery. Scales in L.l. 31–36; about 15 scale rows before D.; breast naked. D. origin directly over P_2 insertion.

Ericymba buccata **Cope.** Silverjaw minnow. From southeastern Missouri to southeastern Michigan and through Ohio to western Pennsylvania; southward to northern Florida and Mississippi. On the Atlantic Coast restricted to the Potomac River system, above Great Falls.

Genus Hybognathus Agassiz. Resembling *Notropis*, but with a long intestine thrown into several coils. Teeth (4–4), hooked to very slight extent or, more often, not at all. Peritoneum very black. Mouth horizontal or nearly so and crescent-shaped; jaws normally formed, the lower with a small hard protuberance in front. Suborbitals broad, $\frac{1}{2}$ as wide as cheek. Scales large, never markedly crowded before D.; D. over or in front of P_2 insertion. A. 7 or 8. Sexual dimorphism very slight. No bright colors in breeding season. 4 species, difficult of separation.

KEY TO SPECIES OF *HYBOGNATHUS*

1. Dorsal fin rounded, not falcate. 1st developed dorsal ray shorter than 2d or 3d. Scale radii about 20 in adult. Color brassy   **H. hankinsoni**
   Dorsal fin falcate. 1st developed dorsal ray equal to or longer than 2d or 3d. Scale radii 10–15 in adult. Silvery fishes   **2**
2. Scale borders more heavily pigmented, producing a diamond pattern and crosshatched appearance. Snout broadly rounded, not easily seen from below; mouth more oblique. Distance from anal origin to caudal base 30–34% of standard length. Intestine shorter   **H. hayi**
   Scale borders less conspicuous, the pigment more evenly distributed over the scales (rounded distally). Snout more pointed and projecting beyond the more horizontal mouth. Distance from anal origin to caudal base 17–32% of standard length. Intestine longer   **3**
3. Posterior process of the basioccipital bone so narrow that the large, inferior pharyngeal retractor muscles attached to it present a V-shaped appearance. Eye smaller; width of head greater than distance from snout tip to posterior orbital rim
       H. placitus
   Posterior process of basioccipital bone broadened posteriad so that the inferior pharyngeal retractor muscles have their origins well separated, not presenting a V-shaped appearance. Eye larger; width of head equal to or less than distance from snout tip to posterior orbital rim   **H. nuchalis**

Hybognathus nuchalis **Agassiz.** Silvery minnow. Widely distributed, in most states east of the Rocky Mountains and from the watershed of Lake Ontario to the Gulf. Not known in the drainages of any of the Great Lakes except Ontario.

Hybognathus placitus **Girard.** Plains minnow. From Montana, Wyoming, and North Dakota to Texas and New Mexico.

Hybognathus hayi **Jordan.** Cypress minnow. Lowland species in Indiana, Illinois, Missouri, Arkansas, northeast Texas, Mississippi, Alabama, Tennessee, and western Florida.

Hybognathus hankinsoni **Hubbs.** Brassy minnow. In the drainage basins of all the Great Lakes, except Lake Erie in Ohio, and from Montana and Manitoba eastward to Quebec and New York. In the West as far south as Colorado, Kansas, and Missouri. Introduced into British Columbia.

Genus Dionda Girard. Closely allied to *Hu-*

bognathus, differing in having a U-shaped mouth; pharyngeal teeth 4–4, with tendency to be hooked (Fig. 2-40*B*); narrow suborbitals less than $\frac{1}{2}$ the cheek width and a distinct dusky lateral band. Fins of breeding males usually red. 3 species in the United States, others in Mexico.

KEY TO SPECIES OF *DIONDA*

1. Mouth small, posterior end of maxilla reaching perpendicular line through nostrils   2
   Mouth larger, posterior end of maxilla to orbital rim
                                                       D. nubila
2. Caudal spot rounded. Dark edges of dorsal scale pockets lacking or poorly developed
                                                     D. episcopa
   Caudal spot wedge-shaped. Dark margins of dorsal scale pockets conspicuous   *D. diaboli*

Dionda episcopa **Girard.** Roundnose minnow. Texas, New Mexico, and Mexico.
Dionda diaboli **Hubbs and Brown.** Devils River minnow. Headwaters of Devils River and other tributaries of the Rio Grande, Texas.
Dionda nubila (**Forbes**). Ozark minnow. From the Ozark region north to Minnesota and Wisconsin.
Genus *Pimephales* Rafinesque. In many respects similar to *Notropis,* but with a blunt rudimentary, spinelike ray attached by a membrane to first developed D. ray (Fig. 2-35). Intestine variable in length, being short (as in *Notropis*) or quite long. In *P. notatus* the intestine is usually long, but often quite as short as in *Notropis.* Gut regularly long in *P. promelas,* but in subgenus *Ceratichthys* (3 species), regularly short. 1 nominal species, *P. callarchus* (Hubbs and Black), is represented only by the holotype and is here regarded as an aberrant specimen of *P. tenellus.*
Body chubby to very slender, never compressed. D. origin over P_2 insertion; A. short, usually 7. Peritoneum black to silvery. Mouth horizontal or nearly so, overhung by fleshy snout or quite terminal. Jaws in breeding males of some species, cornified and broadened to form a parrotlike beak. Lateral band dark, usually present, and ends in an often conspicuous C. spot. Anterior D. rays regularly with a dark spot in *P. vigilax* and *P. tenellus,* and adult male *P. notatus.* Pharyngeal teeth always 4–4, hooked or not.

KEY TO SPECIES OF *PIMEPHALES*

1. Body stout, its greatest depth contained about 3.2–4.0 times in standard length. Caudal spot often indistinct or absent. Breeding males with 3 rows (sometimes a 4th row between nostrils) of tubercles on snout. Nuptial chin tubercles present or not   *P. promelas*
   Body slender, its greatest depth contained about 3.9–5.1 times in standard length. Caudal spot distinct. Breeding males with 1–3 rows of nuptial tubercles on snout. Chin tubercles absent   2
2. Peritoneum black or at least with some black pigment. Upper lip not mesially wider, but overhung by fleshy snout. Mouth more ventral. Intestine variable in length, but usually with at least 1 transverse loop across anterior end of stomach
                                                      P. notatus
   Peritoneum silvery. Upper lip widest at its mid-anterior point. Mouth more terminal. Intestine without a transverse loop, only 1 longitudinal loop paralleling stomach   3
3. Body slenderer, its greatest depth contained 4.5–5.1 times in standard length. Crosshatching more distinct. 3 rows of nuptial tubercles on snout of breeding males. No melanophores in the crotches formed by bifurcation of dorsal rays (Fig. 2-62*B*)
                                                      P. tenellus

Fig. 2-62. **Dorsal fins in *Pimephales*: A, *Pimephales vigilax* with concentrations of melanophores in the crotches of the rays; B, *Pimephales tenellus* without such pigment.**

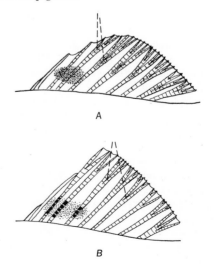

Body stouter, its greatest depth contained 3.9–4.8 in standard length. Crosshatching less distinct. Melanophores present in crotches of dorsal rays (Fig. 2-62A)   **P. vigilax**

***Pimephales tenellus* (Girard).** Slim minnow. From northern Oklahoma to Arkansas, Kansas, and Missouri. Largely in the Ozark region.

***Pimephales vigilax* (Baird and Girard).** Bullhead minnow. In the Mississippi River system from Minnesota to Texas, Louisiana, and Mississippi. In the East from Pennsylvania to Georgia and Alabama, west of the mountains.

***Pimephales notatus* (Rafinesque).** Bluntnose minnow. Widely distributed from southern Canada, throughout the Great Lakes region, to southern Quebec and Lake Champlain and southward on the Atlantic slope from New York to the James River system of Virginia. West of the Appalachians from the Red River of the North to the Gulf of Mexico, east of Texas to the Escambia River in Alabama and Florida. Apparently absent from the western portions of the Great Plains.

***Pimephales promelas* Rafinesque.** Fathead minnow. More widely distributed than *P. notatus,* ranging throughout the Great Plains region of Canada and the United States as well as much of the regions east of the Great Plains from the southern drainage of Hudson Bay and the Maritime Provinces of Canada southward through the Ohio and Cumberland systems to the Tennessee River Basin. Apparently absent on the Atlantic slope and the Gulf states east of the Mississippi River, but present as far west as New Mexico and Chihuahua, Mexico, in the South. Now established in California.

Genus *Campostoma* Agassiz. Readily distinguished by the remarkably long intestine wound many times around the air bladder. Mouth strictly ventral, lower lip not extending to anterior edge of lower jaw (Fig. 2-63). Premaxillae protractile; no barbels. Pharyngeal teeth usually 4–4, but sometimes a tooth developed in outer row. Peritoneum black. D. 8, its origin almost over P_2 insertion; A. 7. Sexual dimorphism well developed, males having large nuptial tubercles over much of body, but best developed on head and dorsum. Breeding males with orange-colored patch on D., but otherwise without bright colors. 2 species.

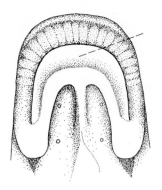

Fig. 2-63. Ventral view of the head of *Campostoma anomalum* showing the lower jaw projecting before the lower lip.

KEY TO SPECIES OF *CAMPOSTOMA*

1. L.l. scales 41–58. A very wide range in the United States   ***C. anomalum***
   L.l. scales about 62–76. Found only in the Rio Grande Basin in Texas   ***C. ornatum***

***Campostoma anomalum* (Rafinesque).** Stoneroller. Range includes most of the United States east of the Rocky Mountains and from the Great Lakes to the Gulf of Mexico, including northeastern Mexico.

***Campostoma ornatum* Girard.** Mexican stoneroller. Primarily Mexican, included here only because of specimens from creeks in Brewster County, Texas, in the Big Bend region of the Rio Grande, and 1 record (as *C. pricei*) from Rucker Canyon, southeastern Arizona.

Family Catostomidae. Suckers. Body variable in shape, varying from slender and cylindrical to deep and compressed. Mouth typically inferior and protractile, but rather oblique and thin in a few species. Barbels never present; lips usually fleshy, plicate or papillose. Both premaxillae and maxillae forming margin of upper jaw. No teeth except those borne on falciform pharyngeals (Fig. 2-33). Branchiostegals 3. Gill membranes always united to isthmus, usually very broadly so that gill openings are markedly restricted (Fig. 2-64). Pseudobranchiae present. L.l. usually present. Spines never present. D. 9–50; A. short (usually 7). Alimentary canal long with a simple stomach and no

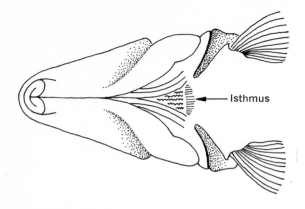

Fig. 2-64. Ventral view of head of *Catostomus commersoni* showing gill membranes united to isthmus. (After Hubbs and Lagler, 1947.)

Fig. 2-66. Lips of *Catostomus clarki*. Note the papillose nature, the deep notches at the outer angles of the mouth, and the hardened sheaths on the inner edges of the lips. (After Hubbs, Hubbs, and Johnson, 1943.)

pyloric caeca. Air bladder with 2 or 3 chambers, often greatly reduced in *Thoburnia*. Central and North America, northeastern Siberia, and Yangtse Kiang. 12 genera comprising 63 species in the United States.

KEY TO GENERA OF CATOSTOMIDAE

1. Dorsal fin with more than 20 rays   2
   Dorsal fin with fewer than 20 rays   4
2. Eye posteriorly placed. Body long and slender. Head abruptly slenderer than body. Scales small, in more than 50 rows. Lips papillose. Posterior fontanel closed   *Cycleptus*
   Eye anteriorly placed. Body deep. Head not abruptly slenderized. Scales large, in 40 or fewer rows. Lips smooth or weakly plicate. Posterior fontanel well developed (Fig. 2-65)   3
3. Anterior fontanel small or closed (Fig. 2-65B). Distance from eye to lower posterior angle of pre-

opercle about ¾ distance from eye to upper corner of gill cleft. Subopercle somewhat semicircular   *Ictiobus*
     Anterior fontanel open (Fig. 2-65A). Distance from eye to lower posterior angle of preopercle about equal to distance from eye to upper corner of gill cleft. Subopercle subtriangular   *Carpiodes*
4. Nape region abruptly elevated in a compressed ridge   *Xyrauchen*
   Nape region not abruptly elevated or markedly compressed   5
5. Air bladder often greatly reduced in adult, with 2 chambers in young   *Thoburnia*
   Air bladder of 2 or 3 chambers at all ages   6
6. Mouth oblique, about 45° when closed. L.l. present. Lips thin and almost smooth   *Chasmistes*
   Mouth inferior or if oblique there is no L.l. Lips usually thick, papillose (Fig. 2-66) or plicate (Fig. 2-67)   7
7. Air bladder of 3 chambers   8
   Air bladder of 2 chambers   9
8. Premaxillae protractile. Halves of lower lip widely joined together on the midline (Figs. 2-67 and 2-68)   *Moxostoma*

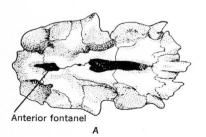

Anterior fontanel

A

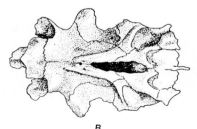

B

Fig. 2-65. Dorsal views of skulls of *A, Carpiodes carpio*, with anterior fontanel well developed; *B, Ictiobus bubalus*, without anterior fontanel. (From Hubbs and Langler, 1947.)

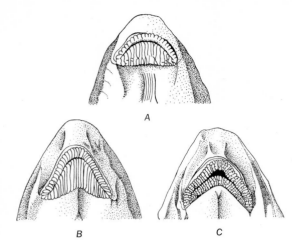

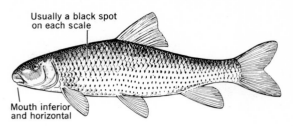

Usually a black spot
on each scale

Mouth inferior
and horizontal

Fig. 2-69. *Minytrema melanops.*
(**Drawing by Wallace Hughes in Ward, no date.**)

Fig. 2-67. Ventral views of mouths of *Moxostoma* showing plicate lips. *A, Moxostoma macrolepidotum,* posterior margin of lower lip straight with plicae somewhat broken by cross creases near angle of mouth; *B, Moxostoma erythrurum,* posterior margin of lower lip an angle, with unbroken plicae; *C, Moxostoma anisurum,* posterior margin of lower lip an angle, the plicae broken into papillalike elements. (After Hubbs and Lagler, 1947.)

Premaxillae not protractile. Halves of lower lip separated   *Lagochila*

9. L.l. poorly developed or absent in adult   **10**
    L.l. well developed in adult   **11**

10. L.l. poorly developed. Mouth horizontal, inferior. Each scale with a black spot (Fig. 2-69)
                     Minytrema

    L.l. absent. Mouth oblique. No black spot on each scale, but body with 2 black bands (young) which may be present with or replaced by bars (adult) (Fig. 2-74)   *Erimyzon*

Fig. 2-68. *Moxostoma macrolepidotum.*
(**From Trautman and Martin, 1951.**)

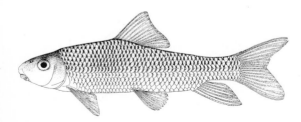

11. Head convex above, the orbital rim not elevated. Eye near middle of head. L.l. scales small, more than 50   *Catostomus*
    Head concave above, the orbital rims elevated. Eye behind middle of head. L.l. scales larger, fewer than 50   *Hypentelium*

Genus *Cycleptus* Rafinesque. Monotypic. Readily recognized by its long slender body; very long D. (about 30), the short and slender head and the inferior mouth with lips covered by coarse papillae.

Cycleptus elongatus (**LeSueur**). Blue sucker. Large streams and artificial impoundments from the Pearl and Mississippi Rivers to the Rio Grande.

Genus *Ictiobus* Rafinesque. D. 25–30. Body deep and short. Head heavy; anterior fontanel (Fig. 2-65B) much reduced or lacking; subopercle almost semicircular. Large fishes of the Great Lakes and Mississippi drainage systems. South into Mexico.

KEY TO SPECIES OF *ICTIOBUS*

1. Mouth large and oblique (Fig. 2-70). Lower pharyngeal arch thin and cancellate, teeth weak. Gill rakers nearly 100 as seen from posterior face. Upper jaw about as long as snout   *I. cyprinellus*
    Mouth smaller and almost or quite horizontal. Lower pharyngeal arch heavy, teeth strong. Gill rakers fewer than 60 as seen from posterior face. Upper jaw shorter than snout   **2**

2. Body depth about 2.5. Head width more than 5. Distance from posterior end of maxilla to front of mandible less than length of eye, except in large fish   *I. bubalus*
    Body depth about 3. Head width less than 5. Distance from posterior end of maxilla to front of

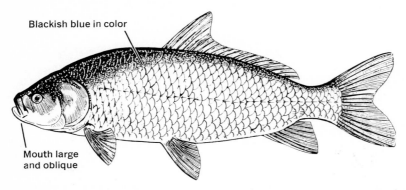

Fig. 2-70. Ictiobus cyprinellus. (Drawing by Wallace Hughes in Ward, no date.)

mandible greater than length of eye, except in young   *I. niger*

Ictiobus cyprinellus (**Valenciennes**). Bigmouth buffalo (Fig. 2-70). Differing most markedly from other buffalofish in having a very large oblique mouth with anterior end of upper jaw near or above level of lower margin of orbit. Fine food fish. From Saskatchewan and the Red River of the North, south through the Mississippi Valley to Alabama, Louisiana, and Texas. Absent from the Great Lakes except western Lake Erie. An inhabitant of large rivers and especially oxbow lakes and sloughs. Well established in some Arizona impoundments, and in California.

Ictiobus niger (**Rafinesque**). Black buffalo. Back low-arched; darker coloration; broad head and large mouth. From Nebraska through the Great Lakes region to Lake Erie south through the Mississippi Valley to the Gulf Coast and Coahuila, Mexico. An inhabitant of larger rivers, oxbow lakes, and sloughs. In artificial impoundments of the Great Plains region *I. niger* and *I. bubalus* evidently interbreed, rendering positive identification difficult or impossible.

Ictiobus bubalus (**Rafinesque**). Smallmouth buffalo. Back highly arched and compressed; light coloration, narrow head, and small mouth. Western portion of the Hudson Bay drainage in Canada to the Ohio River and southward in the Mississippi Valley to the Gulf states and northeastern Mexico. The habitat is similar to that of *niger*.

Genus *Carpiodes* Rafinesque. Similar to *Ictiobus*, but with the mouth strictly ventral, anterior fontanel (Fig. 2-65A) always open, silvery coloration, and a subtriangular suboperculum. Not ordinarily large as the buffaloes, although when isolated in ponds and lakes sometimes outweighs those of same species from river habitats. Generally discarded, but excellent food fishes of fine texture and flavor.

Fig. 2-71. Ventral view of mouth of Carpiodes carpio showing knob at tip of mandible.

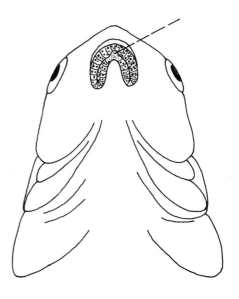

KEY TO SPECIES OF *CARPIODES*

1. Anterior rays of dorsal fin greatly elongated (Fig. 2-73), often equal to or greater than length of dorsal base **2**

   Anterior rays of dorsal fin not greatly elongated, seldom longer than ½ length of dorsal base **3**
2. A small knob at tip of mandible (Fig. 2-71). L.l. scales 35 or fewer. Dorsal rays 27 or fewer
   C. velifer

   No knob at tip of mandible (Fig. 2-72). L.l. scales 36 or more. Dorsal rays 27 or more (Fig. 2-73)
   C. cyprinus

Fig. 2-72. Ventral view of mouth of *Carpiodes cyprinus* showing a lack of mandibular knob.

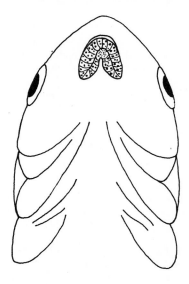

Fig. 2-73. *Carpiodes cyprinus.*
(Jordan and Evermann, 1900.)

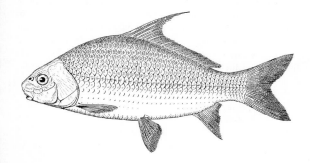

3. A small knob at tip of mandible (Fig. 2-71). L.l. scales usually 35 (33–37). Dorsal rays 23–27. Distance from anterior nostril to tip of snout less than length of eye *C. carpio*

   No knob at tip of mandible (Fig. 2-72). L.l. scales 37–40. Dorsal rays 25–30. Distance from anterior nostril to tip of snout greater than length of eye *C. forbesi*

Carpiodes forbesi **Hubbs.** Plains carpsucker. Separated from *C. cyprinus* by the longer, more posterior D., lacking excessively elongated anterior rays; L.l. scales more numerous; form more slender. Distribution apparently discontinuous; records from Colorado, Wyoming, Nebraska, Iowa, Tennessee, Missouri, northern Arkansas, and central Texas.

Carpiodes cyprinus (**LeSueur**). Quillback carpsucker (Fig. 2-73). Similar to *C. forbesi*, but differing as indicated above. Both *C. forbesi* and *C. cyprinus* differ from other carpsuckers in lacking a nipplelike knob at mandibular tip. Southern Canada and the Great Lakes (except Superior) and rare in Kansas and Oklahoma. On the Atlantic Coast from the St. Lawrence River system to western Florida. Cook (1959) lists it for Mississippi.

Carpiodes carpio (**Rafinesque**). River carpsucker. Most closely related to *C. velifer;* recognized by its mandibular knob and low D. In the North from Wyoming to Pennsylvania and southward throughout the Mississippi Valley in large silty rivers, southward into northeastern Mexico.

Carpiodes velifer (**Rafinesque**). Highfin carpsucker. Easily recognized (except the young) since the long filamentous D. rays are shared only by *C. cyprinus,* among its close relatives, which lacks the mandibular knob. Absent from the Great Lakes region, but in the clearer rivers and lakes of the Mississippi Valley from Illinois, Indiana, and Ohio south to Mississippi, Alabama, and western Florida. West of the Mississippi from Nebraska to Oklahoma.

Genus *Moxostoma* Rafinesque. Air bladder of 3 chambers. Mouth inferior. Lips plicate, papillose, or semipapillose. D. short, of fewer than 20 rays. Pharyngeal teeth (Fig. 2-33) in single row, varying from a slender and compressed form in a comblike series to a stout, cylindrical molar shape. Body never much compressed except in *M. hubbsi.* Posterior fontanel always open. Jaws without cartilaginous sheaths. L.l. well developed and usually straight, but sometimes curved anteriad. **A.** short and high. Spring spawners, usually using smaller streams of clear water. Gill openings restricted, and therefore *Moxostoma* is apparently unable to withstand silty-water conditions.

A difficult genus, as yet imperfectly known. Since some species are known only from the types and original descriptions or are scantily represented in museum collections, much work is needed to determine their true status. Although 19 species are listed, available descriptions of *M. coregonus,* *M. collapsum,* and *M. lachrymale* are so imperfect that these species could not be included in the key.

KEY TO SPECIES OF *MOXOSTOMA*

1. Scales around caudal peduncle 16   **2**
Scales around caudal peduncle 12, rarely 13   **9**
2. Dorsal rays 13 or more. Dorsal fin margin convex. Length of dorsal base contained fewer than 2 times in distance between dorsal origin and occiput   **3**
Dorsal rays 12 or fewer, rarely 13. Dorsal margin falcate or S-shaped. Dorsal base contained more than 2 times (except in *M. ariommum* and *M. robustum*)   **4**
3. Pharyngeal teeth greatly enlarged, bases of 7 largest ones more than half length of arch. Middle of pupil nearer snout tip than posterior opercular margin. Nostril above posterior margin of upper lip. Scales below L.l., immediately before pelvics, 19   *M. hubbsi*
Pharyngeal teeth not enlarged, bases of 7 largest ones less than half length of arch. Middle of pupil nearer posterior opercular margin than snout tip. Nostril well behind posterior margin of upper lip. Scales below L.l., immediately before pelvics, 17   *M. valenciennesi*
4. Lower lip flattened and flaring, papillae large and well separated. Upper lip flattened and folded in, its inner surface with many papillae. Eye diameter greater than least distance from orbit to opercular edge. Anterior edge of eye above or slightly behind posterior edge of lower lip. Nostril far in front of hind margin of lower lip. Occipital line straight. Last dorsal ray contained fewer than 1½ times in dorsal base   *M. ariommum*
Lips plicate or semipapillose. Eye diameter less than least distance from orbit to opercular edge. Anterior edge of eye well behind posterior edge of lower lip. Nostril behind posterior edge of

lower lip. Occipital line curved forward. Last dorsal ray shorter, contained more than 1½ times in dorsal base   **5**
5. Tips of anterior dorsal rays black; caudal lobes black-tipped, except in very young. L.l. scales 43 or fewer. Lips plicate. Eye length greater than ⅔ the outside width of mouth in adults   *M. cervinum*
Tips of dorsal and caudal rays not black, dorsal sometimes with dusky edges. L.l. scales 44 or more. Lips plicate or semipapillose. Eye length less than ⅔ the outside width of mouth in adults   **6**
6. Head depth at occiput contained 3 or more times in predorsal length. Middle L.l. scales much smaller than eye. Eye diameter contained more than 5 times in head length. Body depth at dorsal origin less than head length   **7**
Head depth at occiput contained fewer than 3 times in predorsal length. Middle L.l. scales equal to or larger than eye. Eye diameter contained fewer than 5 times in head. Depth of body at dorsal origin greater (equal in young) than head length   **8**
7. Head deeper than wide. Lips plicate. Lowermost caudal ray light colored. Dorsal rays usually 12   *M. lachneri*
Head wider than deep. Lips semipapillose or, if plicate, the plicae divided posteriad. Lowermost caudal ray dusky. Dorsal rays usually 10 or 11   *M. rupiscartes*
8. Anterior base of pelvic fin nearer anal origin than anterior base of pectoral. Preoperculomandibular canal connected to infraorbital canal on both sides. Pectoral rays 34 or more. Anal fin membranes light   *M. robustum*
Anterior base of pelvic fin nearer anterior base of pectoral than anal origin. Preoperculomandibular canal not connected with infraorbital. Pectoral rays 33 or fewer. Anal fin membranes dark   *M. congestum*
9. Lips papillose (Fig. 2-66)   *M. papillosum*
Lips plicate (Fig. 2-67A, B), but plicae sometimes broken by cross furrows   **10**
10. Pharyngeal teeth few and enlarged, little or not at all compressed   *M. carinatum*
Pharyngeal teeth numerous and compressed   **11**
11. Caudal fin with a black longitudinal streak through lower lobe which is whiter on its margin   *M. poecilurum*
Caudal fin with black pigment more evenly distributed   **12**
12. Body slender. Least depth of caudal peduncle less than ⅔ its length. Snout usually longer than post-

orbital head length. Eye usually less than ½ snout in large young and less than ⅔ snout in small adults. Scales usually 44–47 (extremes 42–49). Pelvic rays 10 (8–11). (Halves of lower lip meeting at a rather obtuse angle, almost straight in adult.) (Fig. 2-67*B*)   *M. duquesnei*
Body chunkier. Least depth of caudal peduncle usually more than ⅔ its length. Snout usually shorter than postorbital head length. Eye usually more than ½ snout (except in *M. valenciennesi*) in young and more than ⅔ snout in small adults. Scales usually 39–45 (extremes 38–47). Pelvic rays 9 (7–10), 10 in *M. breviceps*   **13**

13. Halves of lower lip meeting at a rather sharp angle (Fig. 2-67*C*) with mouth closed. Head 3.7–4.4 in adult length, 3.3–3.7 in young. Dorsal fin rounded in front and nearly straight-edged. Region from occiput to dorsal origin not strongly arched   **14**
Halves of lower lip meeting in a straight line (obtuse angle in young) (Fig. 2-67*A*). Head 4.3–5.4 in adult length, 3.5–3.8 in young. Dorsal fin rather sharply pointed in front, with falcate edge. Region from occiput to dorsal origin more arched   **15**
14. Plicae of lips with transverse creases forming papil-lalike elements (Fig. 2-67*C*). Dorsal rays 15 or 16 (14–17). Depressed-dorsal fin length more than ⅔ distance from dorsal origin to snout tip. Dorsal base about equal to distance from dorsal origin to occiput   *M. anisurum*
Plicae of lips lacking transverse creases, except at angle of mouth (Fig. 2-67*A*). Dorsal rays 11–15 (usually fewer than 15). Depressed-dorsal fin length less than ⅔ distance from dorsal origin to snout tip. Dorsal base less than distance from dorsal origin to occiput   *M. erythrurum*
15. Pelvic rays usually 10. Dorsal fin strongly falcate, longest ray, when depressed, extending to or beyond end of last ray. Upper lobe of caudal longer and narrower than lower lobe. Lips plicate, never with upper lip swollen into a pea-shaped knob   *M. breviceps*
Pelvic rays usually 9 (8–10). Dorsal fin weakly falcate, longest ray when depressed not extending to end of last ray. Upper caudal lobe not narrower and longer than lower lobe. Lips predominantly plicate, but upper sometimes swollen at its tip to form a pealike knob or lower lip broken to form papillae near angle of mouth
    M. macrolepidotum

Moxostoma lachneri **Robins and Raney.** Greater jumprock. D. 12 (rarely 11 or 13), its edge S-shaped. C. peduncle scales 16. L.l. scales 45–46 (rarely 44 or 47). Edges of P_1 and P_2 rounded. Lips plicate, lower meeting in obtuse angle (about 105°). Body with prominent dark and light stripes. C. bicolored, the white lower ray contrasting with darker rays immediately above. Leading edge of D. and C. black. Shoulder with an oblique dark bar. Maximum length 15 in. Chattahoochee-Apalachicola River system in Georgia and Florida.

Moxostoma rupiscartes **Jordan and Jenkins.** Striped jumprock. D. modally 10 or 11, its edge slightly concave or straight. C. peduncle scales 16. L.l. scales 43–50. P_1 moderately pointed, P_2 rounded. Lips plicate or semipapillose, the plicae of lower lip branching posteriad; lower lips meeting in obtuse angle (about 140°). Body with prominent dark and light stripes, young with 4 or 5 lateral blotches forming complete or incomplete saddles. C. yellow in young. Leading edge of D. dusky. Shoulder bar poorly to well developed. Maximum length slightly more than 9 in. In the piedmont and mountain portions of the Santee, Savannah, Altamaha, and Chattahoochee systems in South Carolina and Georgia.

Moxostoma cervinum **(Cope).** Black jumprock. D. 11 (rarely 10 or 12), slightly falcate in front and convex near rear; anterior tip of D. pointed. C. peduncle scales 16. L.l. scales 42–44. P_1 and P_2 rather pointed. C. small, with shallow notch, its upper lobe more pointed than the definitely rounded lower lobe. Lips plicate, the lower meeting at an obtuse angle and broadly joined. Body, especially in young, with prominent dark and light stripes; young with dark lateral blotches which form saddles across back. C. tipped with black. D. and, in some, A. with black tips. Leading edge of D. and upper lobe of C. light, except for dusky base and black tips. Maximum length 6½ in. Abundant in Upper Piedmont and mountains of major river systems from the James River to the Neuse River in Virginia and North Carolina.

Moxostoma ariommum **Robins and Raney.** Bigeye jumprock. D. 11 (rarely 10 or 12), its free edge slightly concave to straight; posterior rays very long. C. peduncle scales 16. L.l. scales 44–46 (rarely 47). P_1 and P_2 rounded on free edges, but pointed posteriad. C. short, not deeply forked; upper lobe longer and more pointed than lower lobe. Lips flattened, papillose; the upper folded inward; lower lip deeply incised, meeting at an obtuse angle. Body sharply bicolored, the dark color of dorsum extending downward to first scale row below L.l. Light and dark stripes poorly seen in preserved fish. Leading edges of all fins white. C. dusky, with small dark dashes on scaly base. Maximum size slightly over $5\frac{1}{2}$ in. Limited to the Roanoke River, Virginia.

Moxostoma congestum (**Baird and Girard**). Gray redhorse. Similar to *M. macrolepidotum* but lacking red or orange in fins. Head short. Mouth rather small. D. low, anterior ray reaching middle of last ray when depressed. New Mexico and Texas, southward into northeastern Mexico.

Moxostoma robustum (**Cope**). Robust redhorse. D. 12 (rarely 11 or 13), its tip pointed and its edge slightly falcate. C. peduncle scales 16. L.l. scales 44–46 (extremes 43–49). P_1 pointed, P_2 less so. C. with a shallow rounded notch, upper lobe more pointed than lower. Lips broadly joined, meeting in almost a straight line in adults; plicae of lower lips branched, occasionally broken; upper lip definitely wider at its center. Body not bicolored, but with light and dark streaks, especially in young. Adults with each scale on sides bearing a dark basal bar. Unlike *M. congestum*, venter devoid of streaks. Young with 4 lateral blotches and dorsal saddles. D. membranes dark, the rays light. Maximum size in museum specimens 11 in. Coastal streams and lakes from the Yadkin River system in North Carolina to the Altamaha River in Georgia.

Moxostoma hubbsi **Legendre.** Copper redhorse. Similar in dentition to *M. carinatum* but with deeper body and shorter head. Apparently restricted to Lake Ontario and the St. Lawrence River.

Moxostoma valenciennesi **Jordan.** Greater redhorse. Largest member of genus, growing to 16 lb. In all drainages of the Great Lakes and sparingly in the Ohio River system. South to Virginia and Kentucky. Red River, North Dakota.

Moxostoma carinatum (**Cope**). River redhorse. Pharyngeal teeth few, molarlike. Body rather thick and heavy. Mouth large, lips thick, plicate. Iowa east to Michigan, Ohio, and Pennsylvania, south to Georgia, western Florida, and Alabama, and west to Oklahoma and Louisiana, where common in clearer rivers. A good food fish.

Moxostoma papillosum (**Cope**). Suckermouth redhorse. Virginia to the Carolinas and Georgia in the Atlantic drainage.

Moxostoma anisurum (**Rafinesque**). Silver redhorse. Lips relatively thin with plicae broken into papillalike elements. Both sexes with nuptial tubercles on A. and C. Head short. Eye rather large, midway in head. Upper lip thin, lower strongly V-shaped. Lower fins white or pale red; D. dark at margin; C. olive. No dark spots on scales. Scales 40–42 (38–44). From Canada to Alabama west of the Appalachian Mountains and east of the Mississippi River. North Dakota, Iowa, and Missouri. An inhabitant of lakes and large rivers.

Moxostoma macrolepidotum (**LeSueur**). Pealip redhorse (Fig. 2-68). Easily recognized by the straight or nearly straight posterior margin of lower lip; plicae of lower lip broken by cross creases near angles of mouth. A thickening at tip of upper lip in specimens from the Ozark region. Central Canada, Great Lakes region, and St. Lawrence drainage south to tributaries of Delaware and Chesapeake Bays. In the Mississippi Basin from Montana to Ohio and south to Oklahoma and Colorado. Lakes and rivers.

Moxostoma coregonus (**Cope**). Carolina redhorse. Lips plicate. D. 14. Lower lip narrow, infolded, V-shaped, with crease on midline forming an acute angle. Body compressed, back elevated. Muzzle projecting beyond very small mouth. A poorly defined species reported from Virginia to North Carolina.

Moxostoma breviceps (**Cope**). Shorthead redhorse. D. strongly falcate; upper lobe of C. narrower and longer than lower lobe. D. and C. deep red. Eastern tributaries of the Mississippi River from Ohio to Alabama, eastward to New York, Virginia, and West Virginia.

Moxostoma collapsum (**Cope**). V-lip redhorse. Lips plicate. D. 15. L.l. scales 42. Lower lip V-shaped. Eye in middle of head. Edge of D. straight, its first ray as long as D. base. C. lobes subequal. No dark spots on scale bases. D. and C. membranes blackish, other fins orange and plain. A little-known species from the Roanoke, Neuse, Catawba, and Yadkin Rivers.

Moxostoma duquesnei (**LeSueur**). Black redhorse. Body and C. peduncle very slender. Eye small, lobes of lower lip meeting at an obtuse angle. Scales fine. Mississippi and Great Lakes drainages, western Florida. Chiefly in mountainous regions and cool clear streams of moderate size. An excellent food fish.

Moxostoma erythrurum (**Rafinesque**). Golden redhorse. Eyes large. Scales coarse. Fins red or orange. One of the smaller redhorses in the drainages of lakes Michigan and Huron, the Mississippi River system south to Arkansas and Oklahoma, and in Gulf tributaries of Georgia, Alabama, and Mississippi. A good food fish.

Moxostoma lachrymale (**Cope**). Neuse redhorse. Cranium with oblique superopercular region and vertex with a ridge on each side as in *M. macrolepidotum*. Snout projecting, mouth large, lips thick. Eye small, contained 2 times in interorbital space. Back gently arched. D. 12 or 13. Lower fins white. Body color olivaceous, scales pale or smoky at base. A poorly defined species of the Neuse River, North Carolina.

Moxostoma poecilurum Jordan. Blacktail redhorse. A small species recognized by a black-margined lower C. lobe. In coastal streams from eastern Texas to Florida and Georgia.

Genus *Lagochila* Jordan and Brayton. Monotypic. Similar in most respects to *Moxostoma* but differing in having the lower lip reduced and divided into 2 separate and weakly papillose lobes. Lobes of lower lip separated from upper lips by a deep furrow at angles of mouth.

Lagochila lacera Jordan and Brayton. Harelip sucker. Formerly present in clear streams of the Mississippi Valley from the Maumee in Ohio to the Tennessee River in Georgia. Jordan and Evermann found it common only in the Ozark uplands (White River). Since it has not been reported for many years, *Lagochila* is now probably extinct.

Genus *Minytrema* Jordan. Monotypic. Similar to *Moxostoma*, but air bladder with 2 instead of 3 chambers. Lower lip plicate, narrow, and deeply incised, V-shaped. L.l. imperfectly developed, absent in young. D. about 12; P_2 9. Dusky above, lighter beneath with a coppery luster. Each scale with a dark spot in adults, young with spots indistinct or wanting.

Minytrema melanops (**Rafinesque**). Spotted sucker (Fig. 2-69). Mississippi Valley from Minnesota to Pennsylvania and southward. From the San Jacinto River in Texas eastward to Florida and north on the Atlantic Coast to North Carolina.

Abundant in smaller streams and, at least in the South, in impounded waters.

Genus *Erimyzon* Jordan. Small suckers (as long as 340 mm standard length) with mouth rather oblique for a sucker; lips plicate; no cartilaginous sheath on lower jaw, the lips incised, V-shaped. Body oblong and compressed, relatively deep and short. Sexual dimorphism rather strong, the males with tubercles on snout and on the swollen and usually emarginate A. D. 10–12, short and high. L.l. absent. Air bladder of 2 chambers. 3 species. Fishes of sluggish streams, backwaters, lakes, and ponds.

KEY TO SPECIES OF *ERIMYZON*

1. Anal fin of male not bilobed, but longer than head. Dorsal fin in half-grown rather sharply pointed. 4th tubercle (just before upper half of eye) of breeding males about as strong as other 3

    E. tenuis

    Anal fin of male bilobed, little or no longer than head in either sex. Dorsal fin in half-grown and adult quite rounded. 4th tubercle of breeding males absent or reduced in size **2**

2. Scales 34–38. Developed gill rakers on lower limb of 1st arch 6–9, the longest about ⅔ length of raker-bearing portion of the limb. Head 3.25–3.8 in standard length. Young with caudal fin usually reddish; large young and yearlings usually in barred-striped color phase (Fig. 2-74)

    E. sucetta

    Scales 39–45. Developed gill rakers on lower limb of 1st arch 7–10, the longest about ½ length of raker-bearing portion of limb. Head 3.45–4.2 in standard length. Young with amber-colored caudal fin;

Fig. 2-74. Erimyzon sucetta.
(Jordan and Evermann, 1900.)

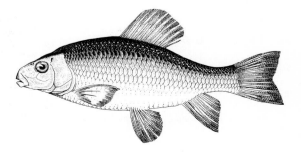

large young and yearlings usually in striped-color phase   *E. oblongus*

Erimyzon tenuis (**Agassiz**). Sharpfin chubsucker. D. in half-grown rather sharply pointed; A. of male not bilobed. Nuptial tubercles 4, of about equal length. Striking golden color, unlike other chubsuckers. Gulf Coast from Louisiana to Florida.

Erimyzon sucetta (**Lacépède**). Lake chubsucker (Fig. 2-74). D. not pointed but rounded; A. of male bilobed; nuptial tubercles unequal, the 4th absent or smaller than other 3. Color brownish; C. of young reddish. Widely distributed from the Great Lakes region and through the Mississippi Valley to Alabama and eastern Texas, westward to Missouri and Arkansas. The Atlantic Coastal Plain from New York to Florida.

Erimyzon oblongus (**Mitchill**). Creek chubsucker. Similar to *E. sucetta*, but with smaller scales, shorter gill rakers and head. C. of young amber. From Minnesota east to New Brunswick and Nova Scotia, southward on the Atlantic Coastal Plain to South Carolina. In the Mississippi River system south to Georgia and Florida and through the Gulf states to Texas, Oklahoma, Arkansas, Missouri, and Iowa.

Genus *Chasmistes* Jordan. Differing from the usual sucker pattern of ventral mouth and thick lips in having a large, terminal, and oblique mouth with relatively thin lips that are nearly smooth. Lower lips well separated at midline. Tip of lower jaw about on level of eye when mouth is closed. Air bladder in 2 parts. Fontanel open. D. 10–12; A. 7 or 8. Scales anteriorly crowded, about 60–80 in L.l. Large fishes known from the Great Basin in Utah and Nevada and from Klamath Lakes, Oregon. Conspicuous only during the brief spawning period, retiring to deep water the remainder of the year.

KEY TO SPECIES OF *CHASMISTES*

1. Scales small, 70–80 in L.l.   *C. brevirostris*
   Scales larger, 58–65 in L.l.   **2**
2. About 9–11 scales above the L.l. and 8 or 9 below it
   C. liorus
   About 13 or 14 scales above the L.l. and 10–12 below
   C. cujus

Chasmistes liorus **Jordan.** June sucker. Similar to *C. cujus*, but with 9–11 instead of 13 or 14 scales above the L.l. and 8 or 9 instead of 10–12 below. Formerly abundant in Utah Lake, Utah, but now believed to be extinct.

Chasmistes cujus **Cope.** Cui-ui. See *C. liorus* for description. Pyramid Lake and lower Truckee River, Nevada.

Chasmistes brevirostris **Cope.** Shortnose sucker. Characterized by its smaller scales, 70–80 in L.l. Klamath Lake, Oregon, and northern California. *C. stomias* and *C. copei* are regarded as synonyms of *C. brevirostris*.

Genus *Catostomus* LeSueur. D. 7–15. Air bladder of 2 parts, L.l. complete, with 50–122 scales. Fontanel closed or open. Mouth large, inferior, except in *C. luxatus* and *C. fecundus*. Lips thick, papillose, and always protractile; lower lips shallowly to deeply incised behind, with or without notches (Fig. 2-66) at angles of upper and lower lips. Body long and slender. 21 species are recognized here, including the nominal genus *Pantosteus* Cope, reduced to a subgenus by Smith (1966).

KEY TO SPECIES OF *CATOSTOMUS*

1. Notch at angle of mouth inconspicuous or absent; lower lip usually deeply incised, sometimes with only 1 or 2 rows of papillae crossing midline. Intestine usually with fewer than 6 loops. Posterior fontanel usually open. Pleural ribs (Weberian apparatus) with angle of divergence less than 90°   **2**
   Notch at angle of mouth usually well developed; lower lip shallowly incised, with more than 3 rows of papillae (2 or 3 in *C. plebeius*) crossing midline. Intestine usually with more than 6 loops. Posterior fontanel usually closed. Pleural ribs (Weberian apparatus) with angle of divergence greater than 90°   **16**
2. Mouth subterminal to terminal, lower jaw oblique; lips weakly papillose. Gill rakers very short and triangular, each with a broad base   *C. luxatus*
   Mouth inferior, lower jaw typically horizontal; lips papillose. Gill rakers longer, slender, and not triangular   **3**
3. Scales coarse, 54–79 in L.l.   **4**
   Scales fine, 80–115 (as low as 73 in *C. microps*) in L.l.   **12**

4. Species of Great Basin of Utah and Nevada and upper Snake River, Idaho, and Wyoming **5**

  Species not in Great Basin or upper Snake River **6**

5. Mouth large, oblique; lips moderately papillose. Scales above L.l. usually 8. Utah Lake, Utah

  C. fecundus

  Mouth inferior; lips fleshy, thickly papillose. Scales above L.l. usually 12. Bonneville Basin of Utah and Nevada and upper Snake River, Idaho and Wyoming *C. ardens*

6. L.l. scales usually fewer than 60 (54–67). A spot present on each scale. Restricted to Gila River system in Arizona *C. insignis*

  L.l. scales usually more than 60. Scale spots absent. Not found in Gila River system **7**

7. Dorsal rays 10 or 11. Restricted to Warner and Klamath Lakes Basins **8**

  Dorsal rays 12–15 (as few as 11 in *C. bernardini*, which is restricted to Yaqui River system in Arizona and Sonora, Mexico) **9**

8. Lower lip not deeply incised, 2 or more rows of papillae cross the midline *C. warnerensis*

  Lower lip deeply incised, not more than 1 row of papillae across the midline of the lower lip

  C. snyderi

9. Scales above L.l. usually 10. From Mississippi River system from Continental Divide eastward through Great Lakes region to Atlantic Coast

  C. commersoni

  Scales above L.l. 11 or more. Not in Mississippi drainage, but on Pacific slope **10**

10. Caudal peduncle abruptly slenderized, its depth contained about 2 times in its length. Upper lip with 6–8 rows of papillae *C. macrocheilus*

  Caudal peduncle deeper, its depth contained fewer than 2 times in its length. Upper lip with 5 or 6 rows of papillae **11**

11. Predorsal scales 25–37. Head conical and small. Body depth in length 4.6–4.8 *C. occidentalis*

  Predorsal scales 27–29. Head larger. Body depth in length 4.0–4.2 (restricted in the United States to the headwaters of Yaqui River system)

  C. bernardini

12. Lower lip deeply incised, with 1, 2, or no rows of papillae crossing the midline **13**

  Lower lip less deeply incised, with 2 or more rows of papillae crossing the midline *C. rimiculus*

13. From the Pit River drainage of California

  C. microps

  From the Colorado, Lahontan Basin, or elsewhere

  14

14. Upper lip very broad, with 5 or 6 rows of papillae, and its width, on the midline, greater than half length of eye. Dorsal fin with 11–15 rays. Caudal

peduncle conspicuously slender. Colorado River system *C. latipinnis*

  Upper lip narrow, with 2–4 rows of papillae. Dorsal fin with 9–12 rays. Caudal peduncle deeper **15**

15. Body slender. Head long and conical. Widely distributed from upper Columbia and Missouri Basins eastward through the Great Lakes region, but not south of latitude 40° *C. catostomus*

  Body stouter. Head short, not so conical. Limited to Lahontan Basin of Nevada and California

  C. tahoensis

16. Dorsal rays 9 (8–10). Predorsal scales usually 40–50 (32–55). Gill rakers on outer row (arch 1) fewer than 27 and 37 (inner row). Lower lip deeply incised, 2 or 3 rows of papillae crossing midline *C. plebeius*

  Dorsal rays 10–12. Predorsal scales and gill rakers variable. Lower lip usually with more than 3 rows of papillae crossing midline (except in *C. columbianus*, which has more than 55 predorsal scales)

  17

17. Anterior papillae of lower lip arranged in an arc not parallel to the lower-lip margin; anterolateral corners of lower lip almost or quite bare. Caudal interradial membranes almost without pigment. Axillary process of pelvic fins well developed

  C. platyrhynchus

  Anterior papillae of lower lip in a row parallel to lower-lip margin; anterolateral corners of lower lip with papillae. Caudal interradial membranes with pigment. Axillary process of pelvic fins poorly developed or absent **18**

18. Predorsal scale rows fewer than 43 **21**

  Predorsal scale rows 43 or more **19**

19. Notches at outer angles of mouth weak or absent; 2 or 3 rows of papillae crossing midline of lower lip. Gill rakers with rosette clusters of spines. Dorsal rays 10–13 *C. columbianus*

  Notches at outer angles of mouth well developed (Fig. 2-66); more than 3 rows of papillae crossing the midline of lower lip. Gill rakers usually with spines in double row. Dorsal rays 10 or 11, rarely 12 **20**

20. Predorsal scales usually more than 50 (44–75). Caudal peduncle slender, its least depth more than 12 in standard length, except in some specimens from the Great Basin and Snake River (in which there are more than 45 gill rakers in the inner row of arch 1) and some specimens from the Little Colorado River, with lips similar to those of *C. plebeius* *C. discobolus*

  Predorsal scales fewer than 47 (as many as 52 in specimens from the Virgin River, Utah. Such specimens have fewer than 46 gill rakers on inner row of arch 1). Least depth of caudal peduncle less than 12 in standard length

C. clarki (in part)

21. Gill rakers usually 30–43 (27–43) in outer row and usually 40–59 (36–59) in inner row of arch 1. Predorsal scales 13–52   *C. clarki* (in part)
    Gill rakers 21–28 in outer row and 27–36 in inner row of arch 1. Predorsal scales 27–41

C. santaanae

Catostomus commersoni (Lacépède). White sucker. Mouth large; lips strongly papillose; papillae in 2–6 rows. Scales crowded anteriad, larger on sides than below. The common sucker of North America east of the Rocky Mountains as far north as the Mackenzie River and the Hudson Bay drainage and south on both sides of the Appalachian Mountains to Georgia, Arkansas, and northeastern Oklahoma.

Catostomus ardens **Jordan and Gilbert.** Utah sucker. Upper lip wide and pendant, papillae coarse, irregular, and in 4–8 rows; lower lip deeply incised. Locally abundant in the Lake Bonneville Basin and the upper Snake River, Wyoming, Idaho, Utah, and Nevada.

Catostomus macrocheilus **Girard.** Largescale sucker. Color dark; a dusky lateral band; venter pale. Lips as in *C. ardens;* papillae of upper lip in 6–8 rows. Pacific Coast drainage from the Skeena and Fraser Rivers of British Columbia south to the Sixes River in Oregon and eastward to Montana.

Catostomus occidentalis **Ayres.** Sacramento sucker. Head small. Mouth rather small, upper lip with 5 or 6 rows of small papillae; lips rather thin. Sacramento River Basin, streams tributary to San Francisco Bay, coastal streams from Monterey Bay drainage north to the Russian River. Also in the Bear, Eel, and Mad Rivers in northwestern California.

Catostomus snyderi **Gilbert.** Klamath largescale sucker. Lower lip deeply incised; not more than 1 row of papillae crossing midline. Scales coarse (69–77 in L.l.). Restricted to the Klamath River of Oregon and California.

Catostomus warnerensis **Snyder.** Warner sucker. Lower lip not deeply incised; 2 or more rows of papillae crossing midline. Scales coarse (73–79 in L.l.). Found only in Warner Lakes Basin in southeastern Oregon.

Catostomus insignis **Baird and Girard.** Sonora sucker. Spots (often obscure) along each scale row. Scales larger above than below, 54–67 in L.l. Limited to the Gila River system of Arizona, New Mexico, and Mexico.

Catostomus bernardini **Girard.** Yaqui sucker. Similar to *C. occidentalis,* but with larger scales and head and slenderer body. Known from the Yaqui River system of Arizona and Sonora, Mexico. In the United States only in San Bernardino Creek, southeastern Arizona, where rare or possibly extirpated. Possibly conspecific with *C. insignis.*

Catostomus catostomus (**Forster**). Longnose sucker. Body slender; head long, conical; scales small (95–115 in L.l.), crowded anteriad. Widely distributed; Asia and nothern North America from the Atlantic to the Pacific drainages and south to Washington and Colorado in the West, the Great Lakes drainage, and in the East to Ohio and New York.

Catostomus microps **Rutter.** Modoc sucker. Lower lip deeply incised, 1 row of papillae crossing midline. Scales small (80–93 in L.l.), little crowding anteriad. Known only from Rush Creek, Modoc County, California, tributary to Pit River of the Sacramento Basin.

Catostomus tahoensis **Gill and Jordan.** Tahoe sucker. Lower lip deeply incised, 1 row of papillae crossing midline. D. 10–12. C. peduncle rather deep. Originally confined to the basin of ancient pluvial Lake Lahontan in Nevada and California, introduced into the Sacramento River system.

Catostomus rimiculus **Gilbert and Snyder.** Klamath smallscale sucker. Lower lip not deeply incised, 2 or more rows crossing midline. D. 10–12. C. peduncle rather deep. Found only in the Rogue and Klamath Rivers in Oregon and California.

Catostomus latipinnis **Baird and Girard.** Flannelmouth sucker. Lower lip divided to its base. D. usually 11, sometimes 12 or 13. In swift portions of the Colorado River system of Arizona, California, Colorado, Nevada, Utah, and Wyoming.

Catostomus fecundus **Cope and Yarrow.** Webug sucker. Utah Lake, Utah. Lower lip broadly divided and oblique, at an angle of about 30° from the horizontal line. Status uncertain.

Catostomus luxatus (**Cope**). Lost River sucker. Mouth terminal, premaxillae projecting to form a hump on top of snout. Assigned to the genus *Deltistes* Seale by some writers. Confined to the drainage of the Klamath Lakes, Oregon and California.

Catostomus plebeius **Baird and Girard.** Rio Grande mountain-sucker. Rio Grande Basin in Colorado and New Mexico; Rio Mimbres in New Mexico and southward in several river basins of Mexico as far as the Rio Piaxtla and Rio Mezquital.

Catostomus santaanae (**Snyder**). Santa Ana sucker. Southern California from the Santa Clara to the Santa Ana River. A variable species known to hybridize with other species of *Catostomus*.

Catostomus platyrhynchus (**Cope**). Mountain sucker. Widely distributed from southwestern Canada to California, Nevada, Utah, and Colorado and east to western South Dakota and northwestern Nebraska. *C. platyrhynchus* is known to hybridize with *C. tahoensis, C. commersoni,* and *C. catostomus.*

Catostomus clarki **Baird and Girard.** Gila mountain-sucker. Found in tributaries of the Colorado River, below the Grand Canyon in Nevada and Arizona; the Gila River Basin of Arizona and Sonora. Hybrids are known between this species and *C. insignis* and *C. latipinnis*.

Catostomus discobolus **Cope.** Green sucker. Found in the Colorado River Basin above the Grand Canyon in Wyoming, Utah, Colorado, New Mexico, and Arizona; in the upper Snake River Basin of Idaho and Wyoming; and in the Bear and Weber River systems of the Bonneville Basin in Idaho, Wyoming, and Utah. Hybridization of *C. discobolus* with *C. commersoni, C. latipinnis,* and *C. platyrhynchus* has been reported.

Catostomus columbianus (**Eigenmann and Eigenmann**). Bridgelip sucker. Known from the Fraser River, British Columbia, and the Columbia River Basin below the great falls of the Snake River in Idaho, Nevada, Oregon, Washington, and British Columbia. This species is known to hybridize with *C. macrocheilus*.

Genus *Xyrauchen* Eigenmann and Kirsch. Monotypic. Closely allied with *Catostomus*. Easily recognized by an odd compressed and elevated hump, on the back between occiput and D. origin, produced by the high development of interneural bones.

Xyrauchen texanus (**Abbott**). Humpback sucker. D. rays 13–16. L.l. scales 68–87. Gill rakers 36–50. The specific name of this odd species is a misnomer since it is unknown in Texas. Colorado

and Gila River systems in Arizona, Colorado, California, Nevada, Utah, and Wyoming. Commercially utilized in certain Arizona impoundments.

Genus *Hypentelium* Rafinesque. Resembling *Catostomus* in many respects, but differing markedly in having a very broad head which is distinctly and transversely concave above. D. 10 or 11. Found only in clear and cool streams.

KEY TO SPECIES OF *HYPENTELIUM*

1. Dorsal usually 10. Sides above L.l. with dark longitudinal stripes. Lower fins red, dorsal basally orange. Snout reddish and lips orange
H. etowanum
   Dorsal usually 11. Sides with or without longitudinal stripes. Colors dull. Lower fins light orange or brown, snout and lips dusky or light, not orange 2
2. Posterior fontanel reduced to a slit. Air bladder thin and shorter. L.l. scales 41 (38–44). Pectoral fin rays about 31 (both fins). Dark saddle before dorsal fin obsolescent, not crossing midline of back. Light streaks developed along scale rows on back. A dwarf species (max. 6 in.) *H. roanokense*
   Posterior fontanel large, rectangular. Normal air bladder. L.l. scales 46 (44–50). Pectoral rays about 34 (both fins). Saddle before dorsal well developed, crossing midline of back. No light streaks on scale rows of dorsum. Larger (13.4 in.) *H. nigricans*

Hypentelium etowanum (**Jordan**). Alabama hog sucker. D. 10. Lower fins red. Alabama River system of Alabama and Georgia.

Hypentelium nigricans (**LeSueur**). Northern hog sucker. D. 10–12 (most often 11); P_1 (both sides) 30–38 (mean 34). L.l. scales 44–50 (mean 46). Lower fins light orange or brown. Posterior fontanel large, rectangular. Eastern United States and southern Canada from the Great Lakes to the Gulf of Mexico. Lake Ontario and southwestward to eastern Oklahoma.

Hypentelium roanokense **Raney and Lachner.** Roanoke hog sucker. Similar to *H. nigricans*, but P_1 (both sides) 28–33 (mean 31). L.l. scales 38–44 (mean 41). Posterior fontanel reduced to a slit. Upper Roanoke River system in Virginia.

Genus *Thoburnia* Jordan and Snyder. Closely allied with *Hypentelium*, but differing in having an air bladder often greatly reduced. Head smaller than in the hog sucker and somewhat like that of

Moxostoma, with which it has been confused. Snout quite rounded or somewhat pointed. Lips papillose or partly plicate. Small fishes of mountain streams of Virginia, West Virginia, Kentucky, and Tennessee. 3 species.

KEY TO SPECIES OF *THOBURNIA*

1. Dorsal fin with a large black blotch on distal portion of anterior rays. Dorsum and sides boldly striped with black. Peritoneum pale, with scattered melanophores   *T. atripinne*
   Dorsal fin without black blotch. Dorsum and sides weakly striped with black. Peritoneum black   **2**
2. Head rounded. Lower lips entirely covered with papillae (almost circular on posterior half). A vertical line from posterior edge of lower lip passes through eye. Lower lips more truncate, the halves meeting in a shallow notch. Greatest outside width of lips contained 1 or more times in postocular head length and 2.3 (2.2–2.4) in standard length   *T. hamiltoni*
   Head bluntly pointed. Lower lips anteriorly plicate, plicae broken into oblong ridges posteriad and becoming papillose on posterior edge. A vertical line from posterior edge of lower lips passes in front of eye. Halves of lower lip meeting in a deep notch. Greatest outside width of lips less than 1 in postocular head length and 2.6 (2.4–2.8) in standard length   *T. rhothoeca*

Thoburnia rhothoeca (Thoburn). Torrent sucker. Similar to *T. hamiltoni,* but head more pointed; mouth smaller, less inferior, lower lip not entirely papillose, but less fleshy and truncate; pigment absent on snout above upper lip; light streaks absent from anterior dorsum; eye larger. James, upper Kanawha, and Potomac River systems in Virginia.

Thoburnia hamiltoni Raney and Lachner. Rustyside sucker. D. 10; P_1 16–16; P_2 9–9; C. 18. L.l. scales 43–50 (mean 46); scales above L.l. 5–7 (usually 6), below 6–7 (usually 6); scales around C. peduncle 16–18 (most often 16). Roanoke River in Virginia.

Thoburnia atripinne (Bailey). Blackfin sucker. D. 9; A. 7; P_1 14–16 (each side); P_2 8 or 9 (each side). L.l. scales 46–50 (mean 47); scales above L.l. 6 or 7, below 7–10; scales around C. peduncle 15 or 16. Barren River drainage of the Green River in Kentucky and Tennessee.

Family Cobitidae. Loaches. This Old World family differs from the Cyprinidae in having the largest otolith in the utriculus rather than the lagena. The orbitosphenoid bone is always present and in contact with the mesethmoid. Pharyngeal teeth in one series. No pseudobranchiae. Europe, Asia, and Northern Africa.

Genus *Misgurnus* Lacépède. Body elongate, compressed. No suborbital spine. 10 to 12 barbels, 4 on mandibles. D. opposite P_2; C. rounded. Air bladder enclosed in a bony capsule. Europe and Asia.

Misgurnus anguillicaudatus (Cantor). Oriental weatherfish. D. 9; A. 7–8; P_2 6–7. Barbels 10, 4 on mandible. D. origin midway between occiput and C. base. P_2 equal to or shorter than head. Usually a small black spot at upper portion of C. base. Established in Genesee and Oakland Counties, Michigan. In 1939 a shipment of weatherfish was received from Kobe, Japan. Presumedly some of the original stock escaped and were successfully acclimated (E. E. Schultz, 1960).

Suborder Siluroidei.[1] Catfishes. Maxillae rudimentary, serving to support a barbel on each side. Symplectic and parietals absent. Bony plates in some species, but those in the United States naked, or usually with plates covered by skin. The Ariidae (sea catfishes) differ from the Ictaluridae (the only other siluroid family in the United States) in the absence of nasal barbels and in having bony plates on head dorsum and at D. origin or near D. origin only. *Bagre marinus* (Mitchill), the gafftopsail, with 2 chin barbels, and *Galeichthys felis* (Linnaeus), the sea catfish, with 4 chin barbels, are not treated here because they are marine, entering fresh water to limited extent.

Family Ictaluridae. North American freshwater catfishes. Barbels 8: 2 nasal, 2 maxillary, and 4 on chin. Skin always naked. D. and P_1 always armed with strong spines, with or without serrae on anterior or posterior edges or both. Teeth (absent in *Trogloglanis*) in villiform bands, present only on premaxillae and dentaries. Adipose fin always present. Lower pharyngeals separate. Air bladder large. 6 genera containing 24 species in the United States.

---

[1] Greenwood et al. (1966) proposed recognition of full ordinal rank, Siluriformes, for the catfishes.

KEY TO GENERA OF ICTALURIDAE

1. Eyes absent. Body without pigment   **2**
   Eyes present. Body pigmented   **3**
2. Jaw teeth well developed. Jaws strong. Lower jaw normal and slightly shorter than upper. Mouth not inverted   ***Satan***
   Jaw teeth absent. Jaws paper-thin. Lower jaw much shortened and turned into mouth   ***Trogloglanis***
3. Adipose fin free (Fig. 2-75), at its posterior end, from back   **4**
   Adipose fin adnate (Fig. 2-76) to back   **5**
4. Premaxillary band of teeth with lateral backward processes (Fig. 2-77C), forming a broad U with its open end backward   ***Pylodictis***
   Premaxillary band of teeth in a nearly straight band (Fig. 2-77A), not forming a U, lacking backward processes   ***Ictalurus***
5. Premaxillary band of teeth with lateral backward processes (Fig. 2-77D)   ***Noturus***
   Premaxillary band of teeth without backward processes (Fig. 2-77B)   ***Schilbeodes***

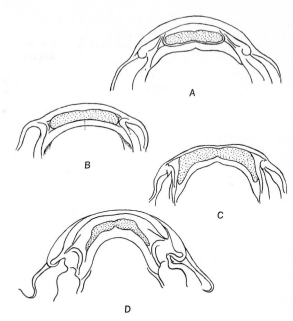

Fig. 2-77. **Premaxillary teeth of A,** *Ictalurus melas;* **B,** *Schilbeodes gyrinus;* **C,** *Pylodictis olivaris;* **D,** *Noturus flavus.* (Forbes and Richardson, 1920.)

Fig. 2-75. **Tail of** *Ictalurus furcatus.* (Forbes and Richardson, 1920.)

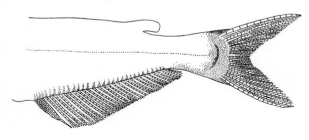

Genus *Ictalurus* **Rafinesque.** The channel catfishes and bullheads. Highly variable, with either forked or truncate tails and with or without a bony ridge from the supraoccipital process to D. A. 16–36. Size range from less than 1 lb to 150 lb. Good food fishes. 9 species in the United States.

KEY TO SPECIES OF *ICTALURUS*

1. Caudal fin usually deeply forked (Figs. 2-75 and 2-78). Supraoccipital process extending far back and making a bony connection with or nearly with dorsal spine   **2**
   Caudal fin not deeply forked. Supraoccipital process not extending so far back and not usually making a bony connection with dorsal spine   **6**
2. Anal rays 18–22. Bony ridge between head and dorsal fin not quite complete. Tail lobes not pointed, the upper longer   ***I. catus***
   Anal rays 23–36. Bony ridge between head and dorsal fin complete or nearly so. Tail lobes sharply pointed, about equal in length   **3**

Fig. 2-76. **Tail of** *Schilbeodes exilis* with adnate adipose fin.

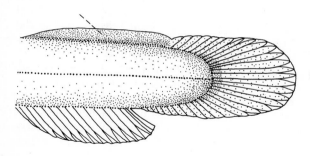

3. Anal base very long, anal rays 30–36. Tips of anal rays lie in almost a straight line (Fig. 2-75) (pectoral spines with simple, bifid, or trifid serrae, not recurved, none on anterior edge)   *I. furcatus*
   Anal base shorter, anal rays fewer than 30. Tips of anal rays forming an arc (Fig. 2-78)   **4**

Fig. 2-78. Tail of *Ictalurus punctatus*. (Forbes and Richardson, 1920.)

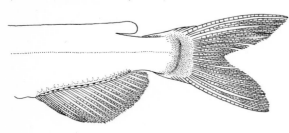

4. Anal rays 24–29. Caudal fin more deeply forked, distance from tip of shortest caudal ray to tip of longest contained fewer than 6 times in standard length (pectoral spines with simple, recurved serrae, none on anterior edge)   *I. punctatus*
   Anal rays 23–25. Caudal fin less deeply forked, distance from tip of shortest caudal ray to tip of longest contained more than 6 times in standard length   **5**
5. Anal base longer, contained fewer than 4 times in standard length   *I. lupus*
   Anal base shorter, contained 4 or more times in standard length   *I. pricei*
6. Anal rays 16–18. Restricted to Atlantic coastal waters from North Carolina to Florida   *I. platycephalus*
   Anal rays 17–27. More widely distributed   **7**
7. Chin barbels without melanophores, always white. Anal fin low anteriorly so that anterior rays are very little longer than those near posterior end; posterior anal-ray tips reach to or almost to caudal base   *I. natalis*
   Chin barbels with melanophores usually black, but may appear to be white if fish is taken in muddy water. Anal rays near anterior end of fin much longer than those near posterior end   **8**
8. Body never mottled. Serrae on posterior edge of pectoral spines weak or absent. Fin membranes conspicuously blackened. A whitish bar at caudal base   *I. melas*
   Body mottled. Strong serrae on posterior edge of pectoral spines. Fin membranes not conspicuously blackened. Caudal base without whitish bar   *I. nebulosus*

Ictalurus punctatus (**Rafinesque**).   Channel catfish. Body slender. Color silvery, often with small round spots; old examples darker, chunkier, and spotless. C. deeply forked. Original range from the Great Plains region of Canada, the Hudson Bay drainage, and St. Lawrence through the Great Lakes and the Mississippi Valley. Successfully planted in drainages of both the Atlantic and Pacific Coasts, now almost anywhere in lakes and rivers of the United States.

Ictalurus lupus (**Girard**).   Headwater catfish. Similar to *I. punctatus*, but A. rays fewer; C. less deeply forked. Rivers of New Mexico and Mexico.

Ictalurus pricei (**Rutter**).   Yaqui catfish. Close to *I. lupus*, A. base shorter. In the United States only from San Bernardino Creek in southeast Arizona. Introduced into Monkey Spring, northeast of Nogales, Arizona. Recent concerted efforts by B. A. Branson have failed to yield specimens at either locality.

Ictalurus catus (**Linnaeus**).   White catfish. Olive blue above, silvery below; no spots, but sometimes mottled. C. weakly forked, upper lobe longer than lower. On the Atlantic Coast from New York to Florida. Introduced on the Pacific Coast and in Nevada.

Ictalurus furcatus (**LeSueur**).   Blue catfish. Eye smaller than in *I. punctatus*. Back elevated at D. origin. A. margin straight; C. deeply forked. Originally only in the Mississippi Valley west to Mexico, principally in large rivers, but often succeeding in large impoundments. This is the largest member of the genus, maximum weight 150 lb.

Ictalurus platycephalus (**Girard**).   Flat bullhead. Body very slender. C. slightly emarginate. Lower jaw included. From North Carolina to Florida in the coastal streams.

Ictalurus natalis (**LeSueur**).   Yellow bullhead. Body chunky. C. not emarginate. Lower jaw included. Most of A. rays about equal in length. Chin barbels white. From the Great Lakes region to the Gulf Coast and from the Great Plains to the Atlantic Coast. Introduced on the west coast.

Ictalurus melas (**Rafinesque**).   Black bullhead. Body usually slenderer than in *I. natalis*. C. very slightly emarginate. Jaws about equal. Anterior A. rays longest. Chin barbels black. A light bar at C. base. Fins with dark membranes. From southeastern Canada through the Great Lakes region southward throughout the Mississippi Valley to the Gulf and westward to the Rocky Mountains. Introduced in Connecticut and on the Pacific Coast; recent records from Oregon, Idaho, and Arizona.

Ictalurus nebulosus (**LeSueur**). Brown bullhead. Similar to *I. melas,* but fins more unicolored, lacking dark membranes; C. base without light bar. From southern Canada and North Dakota south to Arkansas and southeastern Oklahoma; on the east coast from Maine to Florida. Introduced on the west coast.

Genus *Noturus* **Rafinesque.** The stonecat differs from other catfishes having an adnate adipose fin, in having backward lateral processes on the premaxillary band of teeth. Head broad and flat; back behind D. keeled, and the adipose fin deeply notched. D. spine short; P_1 spines retrorsely serrated in front and smooth or slightly roughened behind. Coloration yellowish brown with yellow-edged fins. Sometimes reaching a length of 325 mm. 1 species.

Noturus flavus **Rafinesque.** Stonecat. From Montana, Wyoming, and Manitoba eastward to Quebec and Ontario. Southward west of the Appalachians and east of the Rocky Mountains to Oklahoma and Tennessee.

Genus *Pylodictis* **Rafinesque.** Monotypic. Backward processes at each end of premaxillary band of teeth. Adipose fin not adnate. Head greatly flattened dorsoventrally. Eyes very small. Mouth very large, anterior, and transverse, lower jaw projecting. D. origin anterior to P_2 insertion. C. rounded, its upper portion white in small young. A. 12–15. Color yellowish with dark mottlings. A large catfish often reaching 75 lb; specimens of nearly 100 lb are sometimes taken. A good food fish.

Pylodictis olivaris (**Rafinesque**). Flathead catfish. From South Dakota to Pennsylvania and southward in the Mississippi Valley to the Gulf Coastal Plain and in the Rio Grande to northeastern Mexico.

Genus *Satan* **Hubbs and Bailey.** Monotypic, subterranean. Blind as in *Trogloglanis,* but with well-developed teeth on strong jaws; mouth transverse. No skin pigment.

Satan eurystomus **Hubbs and Bailey.** Widemouth blindcat. From artesian wells at or near San Antonio, Texas.

Genus *Trogloglanis* **Eigenmann.** Monotypic, subterranean. Body devoid of pigment. Mouth toothless; jaws paper-thin, the lower curiously curved upward and into mouth.

Trogloglanis pattersoni **Eigenmann.** Toothless blindcat. From artesian wells at or near San Antonio, Texas.

Genus *Schilbeodes* **Bleeker.** Madtoms. Large genus of small catfishes imperfectly known at present and combined by some authors with *Noturus.* Several undescribed species not mentioned here have been discovered and await description. Small catfishes (maximum size about 75–125 mm) without backward processes on premaxillary band of teeth (Fig. 2-77*B*). Adipose fin adnate (Fig. 2-76), usually joined by a membrane to C. rudimentary rays and deeply or shallowly notched. Mouth transverse, lower lip usually included (equal in *S. gyrinus*). D. I,7, its origin anterior to P_2 insertion. A. 12–23. P_1 with a sharp serrated or smooth spine (Fig. 2-79), with a poison gland at base. L.l. complete.

KEY TO SPECIES OF *SCHILBEODES*

1. Pectoral spines without serrae on posterior edges (Fig. 2-79*F*) **2**

   Pectoral spines with serrae (weak in *S. nocturnus,* Fig. 2-79*E*) on posterior edges **3**

2. Pectoral spines devoid of serrae on anterior edges and grooved behind. Sides with a narrow black lateral streak. Jaws equal (Fig. 2-80) ***S. gyrinus***

   Pectoral spines with anterior serrae and grooved behind. Color yellowish, somewhat mottled. Lower jaw included ***S. leptacanthus***

3. Anal fin long, with 20 or more rays (including rudiments) ***S. funebris***

   Anal fin shorter, with fewer than 20 rays **4**

4. Adipose fin continuous with caudal, notch very poorly developed. Pectoral serrae limited to a few near tip of spine on anterior edge and a few weak ones near base on posterior edge (Fig. 2-79*E*). Color dark, with many fine punctulations. At least the sides of belly with black or brownish dots. Fins white-edged ***S. nocturnus***

   Adipose fin not continuous with caudal, a well-developed notch separating them. Pectoral serrae covering a large portion of posterior edge and usually anterior edge as well. Color pattern various **5**

5. Pectoral spines never with retrorse serrae on posterior edges, serrae shorter, weaker, erect, and sometimes bifurcate (Fig. 2-79*D*) **6**

 Pectoral spines with retrorse serrae on posterior edges, serrae longer, stronger, and sometimes longer than width of spine (Fig. 2-79*A–C*) **8**

6. Adipose fin very low, completely separated from caudal by a notch. Pectoral spine very short, its length contained 4 or 5 times in head. Caudal fin slightly emarginate. Base of dorsal and lower caudal lobe black ***S. gilberti***

Fig. 2-79. Pectoral spines of *Schilbeodes*: A, *Schilbeodes eleutherus*; B, *Schilbeodes miurus*; C, *Schilbeodes hildebrand*; D, *Schilbeodes exilis*; E, *Schilbeodes nocturnus*; F, *Schilbeodes gyrinus*. (After drawings by R. J. Ellis; A, B, and C after Bailey and Taylor, 1950.) NOTE: The upper edge in each figure represents the posterior edge.

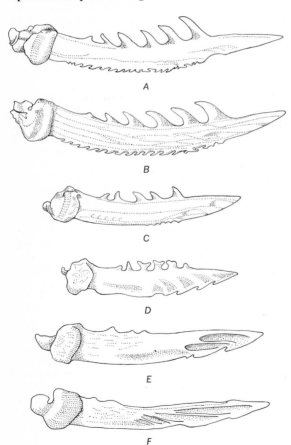

Adipose fin moderate, notch not completely separating adipose from caudal fin. Pectoral spine long, its length contained fewer than 4 times in head. Caudal fin rounded. Dorsal base and lower caudal lobe not black, fins margined with dusky or black **7**

7. Body very slender. Head narrow and very flat. Fins usually not margined with black, but sometimes dusky. Jaws nearly equal. Mississippi Valley ***S. exilis***

 Body rather elongate. Head broad and flat. Fins usually black-edged. Lower jaw included. Atlantic Coast and Kanawha River above the falls ***S. insignis***

8. Restricted to eastern North Carolina. (Similar to *S. miurus*, but with much larger spines) ***S. furiosus***

 Species of the Mississippi Valley and the Pearl, Tombigbee, and Cahaba Rivers in Mississippi and Alabama **9**

9. Anterior pectoral serrae weak and few, usually 6 or fewer. Posterior pectoral serrae strong, but their length less than width of spine. Pectoral rays usually I,9. Pelvic rays 8 ***S. hildebrandi***

 Anterior pectoral serrae strong and numerous (14–33). Posterior pectoral serrae very strong, their length equal to or greater than width of spine. Pectoral rays usually I,8. Pelvic rays 9 **10**

10. Dark blotch on adipose fin not reaching to margin; anterior rays of dorsal fin without a dark blotch ***S. eleutherus***

 Dark blotch on adipose fin reaching to margin; anterior rays of dorsal fin with a dark blotch weakly to strongly developed **11**

11. Total caudal rays 57–63. Anterior 4 or 5 rays and membranes of dorsal fin with a dark blotch ***S. miurus***

 Total caudal rays 45–52. Anterior 1 or 2 rays and membranes of dorsal fin covered by pigment ***S. munitus***

 Schilbeodes gyrinus (Mitchill). Tadpole madtom (Fig. 2-80). P_1 spines grooved behind, smooth. Black lateral streak. Jaws equal. From the Red River system of North Dakota eastward through the drainage systems of Lakes Michigan, Huron, and Ontario and southward on both sides of the Appalachians to the Gulf states including Texas. West of the Mississippi as far west as Lake Altus, southwestern Oklahoma, Kansas, and northeastern Texas. Recent records from Snake River system in Oregon and Idaho.

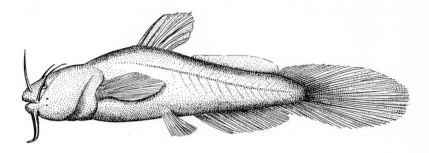

Fig. 2-80. Schilbeodes gyrinus. (Forbes and Richardson, 1920.)

Schilbeodes nocturnus (**Jordan and Gilbert**). Freckled madtom. P_1 spines with a few short serrae on anterior edges and few to none on posterior edges. Belly, at least on sides, with black dots. From the Mississippi Valley in Indiana and Illinois southward to Alabama, Mississippi, and Texas. In the Arkansas and Red River systems west to about the 97th meridian in Kansas and Oklahoma.

Schilbeodes funebris (**Jordan and Swain**). Black madtom. P_1 spines roughened anteriorly, with 3 or 4 weak serrae on posterior edges. A. long, with 20 or more rays. Gulf Coast from western Florida to Louisiana.

Schilbeodes leptacanthus (**Jordan**). Speckled madtom. P_1 spines with anterior serrae and a posterior groove. Lower jaw included. From the Florida parishes, eastern Louisiana to Florida, and north to South Carolina.

Schilbeodes exilis (**Nelson**). Slender madtom. Posterior P_1 serrae not retrorse, erect, sometimes bifurcate; P_1 spine contained fewer than 4 times in head. Fins without black margins, sometimes dusky. Jaws nearly equal. Abundant in the Ozark region of Missouri, northwestern Arkansas, eastern Oklahoma, and southeastern Kansas; northward to central Iowa, southern Minnesota, southeastern Wisconsin and Michigan; east of the Mississippi River to West Virginia and as far south as northern Alabama. Records from the Great Lakes Basin of Wisconsin and Michigan doubtful (Hubbs and Lagler, 1947).

Schilbeodes insignis (**Richardson**). Margined madtom. Similar to *S. exilis,* but with black-margined fins and lower jaw definitely included. From New Hampshire and the southern tributaries of Lake Ontario southward on the Atlantic Coast to northern Georgia.

Schilbeodes gilberti (**Jordan and Evermann**). Orangefin madtom. P_1 spine contained 4 or 5 times in head. Adipose fin separate from the emarginate C. Roanoke River system in Virginia.

Schilbeodes hildebrandi **Bailey and Taylor.** Least madtom (Fig. 2-81). Anterior serrae of P_1 spine weak, 6 or fewer; posterior serrae less, in length, than width of spine. P_1 usually I,9; P_2 usually 8. From a tributary of the Homochitto River, southwestern Mississippi.

Schilbeodes eleutherus (**Jordan**). Mountain madtom. Similar to *S. hildebrandi,* but P_1 spines stronger with more numerous, longer serrae. P_1 usually I,8; P_2 usually 9. In tributaries of western Lake Erie, the Ohio River system including the Tennessee River system, and Little River of southeastern Oklahoma.

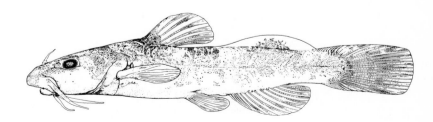

Fig. 2-81. Schilbeodes hildebrandi. (From Bailey and Taylor, 1950.)

Schilbeodes miurus (Jordan). Brindled madtom. Recognized by the black adipose blotch, extending to fin margin, and the conspicuous black blotch on first 4 or 5 D. rays and membranes. From Illinois, Indiana, and the drainage basins of Lakes Erie, St. Clair, and Ontario; southward in the Mississippi Valley to eastern Oklahoma and Mississippi.

Schilbeodes furiosus (Jordan and Meek). Carolina madtom. Similar to *S. miurus*, but with much larger spines. Eastern North Carolina in the Neuse, Tar, and Little Rivers.

Schilbeodes · munitus (Suttkus and Taylor). Frecklebelly madtom. Similar to *S. miurus*, but the anterior D. rays bearing less pigment, fewer C. rays, and stronger but fewer anterior P_1 serrae. 1 or 2 basal serrae on posterior edge of P_1 spine curved distally, the others retrorse as in *miurus*. Common in the Pearl River in Mississippi and known also from the Tombigbee River in Mississippi and the Cahaba River in Alabama. Maximum size 53 mm.

Order Anguilliformes.[1] Body very long and slender, compressed posteriad. Scales cycloid, embedded, and in oblique rows. L.l. well developed. Teeth small and nearly equal, set in bands on jaws and vomer. Tongue free at tip. Upper jaw included. D. far back, continuous with C. and A.; P_1 well developed; P_2 absent. Anus close in front of A. Catadromous. Tropical, warm and temperate seas. Upper Cretaceous. 1 family, 1 genus, and 1 species entering fresh waters of the United States.

Family Anguillidae. With characters of the order.

Genus *Anguilla* Shaw. The eels.

Anguilla rostrata (LeSueur). American eel. Eastern North America from the southern portion of the Great Plains to the Atlantic Coast and from southern Greenland and Labrador to northern South America. In the Mississippi Valley as far north as Minnesota and South Dakota.

Order Beloniformes.[2] Physoclistic fishes without spines. P_2 abdominal, with 6 rays; P_1 insertion

very high; C. with 13 branched rays. Lower pharyngeals fused. Upper jaw bordered by premaxillae. Scales cycloid. L.l. very low on sides. Branchiostegals 9–15. Intestine straight. Eocene. In warm and temperate seas. 1 family, 1 genus, and 1 species entering fresh waters of the United States.

Family Belonidae. With characters of the order.

Genus *Strongylura* Van Hasselt. Needlefishes.

Strongylura marina (Walbaum). Atlantic needlefish. Ascending rivers from Maine to Texas.

Order Cyprinodontiformes. Physoclistic. P_2, if present, typically of not more than 7 (as many as 9) rays, abdominal. No spines. D. single. Maxillae not entering gape. Lateral scales without L.l. tubes. P_1 high on sides, their bases lateral and vertical. Upper and lower ribs present, but no intermuscular bones. Eyes well developed. Represented on all continents, except Australia, many in salt water. The Cyprinodontiformes, as here treated, include 2 families, 11 genera, and 53 species. Lower Oligocene.

KEY TO FAMILIES OF CYPRINODONTIFORMES

1. Anal fin in male not modified as an intromittent organ, 3d anal ray branched. Oviparous
   Cyprinodontidae
   Anal fin in male with its anterior rays modified into an intromittent organ, 3d anal ray unbranched. Viviparous **Poeciliidae**

Family Cyprinodontidae. Scales cycloid. Mouth small, terminal, and protractile, premaxillae forming edge of upper jaw. Gill membranes conjoined, free from isthmus. Head scaly. Size small, usually less than 175 mm. 10 genera and 42 species.

KEY TO GENERA OF CYPRINODONTIDAE

1. Jaw teeth conical (Fig. 2-82A) 2
   Jaw teeth compressed, bi- or tricuspid (Fig. 2-82B, C) 7
2. Female with a membranous oviducal pouch covering the base of anal fin, at least anteriorly 3
   Female lacking an oviducal pouch or with a pair of enlarged scales at genital opening 4
3. Mandibular pores absent. Jaw teeth usually uniserial,

[1] The eels are placed close to the Elopiformes by Greenwood et al. (1966).
[2] Greenwood et al. (1966) proposed the order Atheriniformes to accommodate the suborders Exocoetoidei, including the Belonidae; Cyprinodontoidei, including the Cyprinodontidae and Poeciliidae; and Atherinoidei for the Atherinidae. The family Mugilidae, placed here with the Atherinidae in the order Mugiliformes, is placed in the suborder Mugiloidei under the Perciformes.

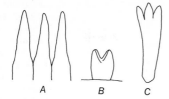

Fig. 2-82. Jaw teeth of cyprinodontids: *A*, conical teeth of *Lucania goodei*; *B*, bifid tooth of *Crenichthys nevadae* (after Hubbs, 1932); *C*, trifid tooth of *Cyprinodon rubrofluviatilis*.

but sometimes irregular, with a few strong inner teeth or many weak teeth   *Lucania*
Mandibular pores present, usually 4 or 5. Jaw teeth multiserial   *Fundulus*
4. Female with a pair of enlarged scales near genital opening. Branchiostegal rays 3   *Leptolucania*
Female lacking enlarged scales near genital opening   **5**
5. Pelvic fins absent. Pharyngeal teeth molariform
                                                                    Empetrichthys
Pelvic fins present. Pharyngeal teeth not molariform
                                                                    6
6. A large ocellus (light and dark spots in young) at upper caudal base. Body slender, depth about 5.5 in standard length   *Rivulus*
No ocellus, but sides with vertical bars. Body depth about 3 in standard length   *Adinia*
7. Pelvic fins absent. Jaw teeth bicuspid *Crenichthys*
Pelvic fins present. Jaw teeth tricuspid   **8**
8. Anal rays 9. Mandibular pores 3   *Floridichthys*
Anal rays 10 or more. Mandibular pores 2 or more **9**
9. Dorsal rays 14–16; anal rays 11–13   *Jordanella*
Dorsal rays 9–11; anal rays 10–11   *Cyprinodon*

Genus Adinia Girard. Killifish. Body and C. peduncle very deep. D. origin before A. Snout sharp, conical. 1 species.

Adinia xenica (**Jordan and Gilbert**). Diamond killifish. Gulf Coast from Florida to Texas.

Genus Empetrichthys Gilbert. Killifishes. Jaw teeth conical, biserial or weakly triserial, a few outer teeth enlarged. Lower pharyngeals variable in degree of enlargement, in close contact but not united and with conical or molariform teeth. Branchiostegals 6, the 1st one broad anteriad. Pelvic girdle and fins absent. Epiotic process absent or rudimentary.

KEY TO SPECIES OF *EMPETRICHTHYS*

1. Scales in lateral series 29 or 30. Snout markedly sloped downward so that mouth lies below median frontal plane   *E. merriami*
Scales in lateral series 29–33 (usually 30–32). Downward slope of snout less, so that mouth lies about on median frontal plane   *E. latos*

Empetrichthys merriami **Gilbert.** Ash Meadows killifish. Found only in Ash Meadows of the Amargosa River Basin on boundary between California and Nevada.

Empetrichthys latos **Miller.** Pahrump killifish. Pahrump Valley, Nevada.

Genus Crenichthys Hubbs. Killifishes. A near relative of *Empetrichthys,* but outer row of jaw teeth strongly bicuspid; inner row or rows irregular, conical. Lower pharyngeals not enlarged; teeth conical. Branchiostegals 5, anterior one with a long, narrow, ʻanterior projection. Pelvic girdle and fins absent. Epiotic process absent or rudimentary.

KEY TO SPECIES OF *CRENICHTHYS*

1. Lateral dark spots in 2 series   *C. baileyi*
Lateral dark spots in a single series   *C. nevadae*

Crenichthys baileyi (**Gilbert**). White River killifish. Basin of Pleistocene White River, eastern Nevada.

Crenichthys nevadae **Hubbs.** Railroad Valley killifish. Railroad Valley, Nye County, Nevada.

Genus Fundulus Lacépède. Killifishes. Body never much elevated or compressed anteriad, sometimes distinctly compressed posteriad. Dentary bones firmly united. Gill openings usually not restricted above, but in some species the gill slit reaches only to or slightly above the upper end of the P_1 insertion. P_2 always present in abdominal position. Intestine short, except in *F. kansae* and *F. zebrinus.* An oviducal pouch almost always present in adult females; in some the pouch covers a large portion of the anterior A. rays, and rarely it is obsolescent. 24 species inhabit salt, brackish, and fresh waters of southeastern Canada, the United States, and Mexico; in the United States and Canada from southeastern Canada along the coast to Texas, west to New Mexico, and northward through the Mississippi Valley to the Red River of the North; west of the Continental Divide, only in coastal waters of southern and Baja California. The genus is also represented in Bermuda, northern Cuba, and some coastal areas of Mexico.

KEY TO SPECIES OF *FUNDULUS*

1. A bold, black band extending from snout to caudal base, sometimes with barlike extensions above and below it, especially in males   **2**
    No single, black band from snout to caudal base, but with pattern of vertical bars, several longitudinal streaks or broken, dashed lines, scattered spots or plain   **3**
2. Spots above lateral band discrete, black, and conspicuous   ***F. olivaceus***
    Spots above lateral band diffuse, dusky or brownish, and inconspicuous   ***F. notatus***
3. A conspicuous, black blotch (teardrop) below eye   ***F. notti***
    Teardrop absent, but sometimes a less-discrete aggregation of pigment present   **4**
4. Vertical bars absent in both sexes (pale ones may be present in *F. rathbuni* young)   **5**
    Vertical bars present in one or both sexes; if absent in one, the sides in the other sex are marked with longitudinal streaks that may be interrupted to form series of dashes; the bars may be strong, weak, or restricted posteriad   **10**
5. Sides plain, without streaks or bars   ***F. sciadicus***
    Sides with random spots or longitudinal streaks or dashes following scale rows   **6**
6. Upper end of gill slit opposite or only slightly above upper end of pectoral base, the distance from pectoral base to angle of gill slit less than $\frac{1}{2}$ length of pectoral base   **7**
    Upper end of gill slit farther above pectoral base, the distance from pectoral base to angle of gill slit more than $\frac{1}{2}$ length of pectoral base   **8**
7. Dorsal and caudal fins of male black-edged; random, broken, or irregular longitudinal rows of orange or dark flecks on sides. Scales around caudal peduncle 19–23. Anal rays usually 14–15   ***F. stellifer***
    Dorsal fin of male without black edge; caudal fin of male white-edged, with a submarginal black band. Orange or dark flecks in more orderly arrangement, following scale rows. Caudal peduncle scales 16–20. Anal rays usually 16 or 17   ***F. catenatus***
8. A fine line extending from angle of gape to below eye; a thicker line on opercle, sometimes not evident in old specimens; males plain, with few or no spots on body; female with scattered spots on sides, seldom with vague bars or lines   ***F. rathbuni***
    Fine line from angle of gape to eye absent; no opercular line; males and females with spots or lines; no vertical bars   **9**
9. Sides with 12–30 round spots, sometimes in two rows along mid-sides and occasionally merged to form short, indistinct vertical bars. Dorsal rays 8 or 9   ***F. jenkinsi***
    Sides with interrupted white (males) or black (females) streaks. Dorsal 10 or 11   ***F. albolineatus***
10. Upper end of gill slit opposite or only slightly above upper end of pectoral base, this distance less than $\frac{1}{2}$ length of pectoral base   **11**
    Upper end of gill slit farther above pectoral base, more than $\frac{1}{2}$ length of pectoral base   **13**
11. Male vertically barred, with an ocellus at hinder end of dorsal fin; female without bars or an ocellus. Dorsal rays 8, inserted over anal fin in female and behind in male. Anal rays 10. Scale rows crossing mid-sides 31 or 32   ***F. luciae***
    Both sexes barred, no ocelli. Dorsal rays 14–17, inserted before anal fin; anal rays 13–16. Scale rows crossing mid-sides 41–64   **12**
12. Scale rows crossing mid-sides 41–49   ***F. zebrinus***
    Scale rows crossing mid-sides 52–64   ***F. kansae***
13. Color pattern of sides variable, but female with one to several irregularly placed dark stripes; vertical bars in either sex, narrow and less than $\frac{2}{3}$ width of interspaces   ***F. majalis***
    Sides never with irregular horizontal dark stripes; vertical bars variable in width and present in one or both sexes   **14**
14. Females with a posterior ocellus or dark blotches on dorsal fin. Genital pouch of female very small (highest on 1st anal ray) or lacking   **15**
    Females without ocellus or blotches on dorsal fin. Genital pouch of female well developed or highest on other than 1st ray of anal fin   **16**
15. Female with many dark spots, some as large as pupil or confluent as longitudinal lines. Dorsal rays usually 10 (9–11). Gulf Coast, Alabama to Texas   ***F. pulvereus***
    Female with many vertical bars, no spots. Dorsal rays usually 11 (10–12). Coastal streams from Chesapeake Bay to Alabama   ***F. confluentus***
16. Vertical bars present in both sexes, sometimes diffuse and variable   **17**
    Vertical bars in males only   **20**
17. Scale rows crossing mid-sides 40 or more   **18**
    Scale rows crossing mid-sides fewer than 40   **21**

18. Scale rows 40–49 (35–55). Dorsal rays 10–14; anal rays 9–13   *F. diaphanus*
    Scale rows more than 48. Dorsal rays 13–20; anal rays 10–15   **19**
19. Scales around caudal peduncle 19–25. Pectoral rays 15–18. Restricted to Lake Waccamaw, North Carolina   *F. waccamensis*
    Scales around caudal peduncle 24–29. Pectoral rays 18–21. Restricted to Florida   *F. seminolis*
20. Male with about 20 vertical bars; female with an indistinct lateral band. Coastal waters of southern California   *F. parvipinnis*
    Male with 6–10 vertical bars; females with small, scattered "pearl" spots, not evident when long preserved in alcohol   *F. chrysotus*
21. Dorsal rays 12–13 (11–14). Scales crossing midsides usually 30–32   *F. similis*
    Dorsal rays usually 12 or fewer. Scales 29–30 or 31–35   **22**
22. Total mandibular pores 6. Dorsal rays 7–9; anal rays 10   *F. cingulatus*
    Total mandibular pores 8 or 10. Dorsal rays 10–12; anal rays 10–12   **23**
23. Total mandibular pores 8. Longest anal ray 1.2–1.9 in head   *F. heteroclitus*
    Total mandibular pores 10. Longest anal ray 1.9–2.6 in head   *F. grandis*

***Fundulus parvipinnis* Girard.** California killifish. Male with about 20 short black bars on midsides; female without bars, but with obscure dusky lateral band. D. 13 or 14; A. about 11. Morro Bay, California, south to Magdalena Bay, Baja California. Occasionally in fresh water.

***Fundulus heteroclitus* (Linnaeus).** Mummichog. Sides with alternating dark and silvery bars, less numerous in males than in females. D. usually 11; A. 10–12. Abundant along the coast from Maine to northeastern Florida in salt, brackish, and fresh waters.

***Fundulus grandis* Baird and Girard.** Gulf killifish. Similar to *F. heteroclitus* but with 10 instead of 8 mandibular L.l. pores. Gulf Coast from northeastern Mexico to Florida, entering fresh water in southern portion of range.

***Fundulus majalis* (Walbaum).** Striped killifish. A silvery species; sides of females with variable color pattern of 1 to several broken, longitudinal, dark stripes, these sometimes bent downward anteriad as vertical bars; often with vertical bars on the C. peduncle. Male with vertical bars and often a posteriorly placed dark spot on D. D. usually 14 or 15 (12–16). Lateral scale rows 33–35. In salt, brackish, and sometimes fresh waters from New Hampshire to northeastern Florida.

***Fundulus similis* (Baird and Girard).** Longnose killifish. With a silvery sheen as in *F. majalis*. Males and females with 10–15 narrow, vertical bars; males with large diffuse spot above P_1 base behind opercle; a small black spot at upper C. base. Young males with a posterior spot in D. D. usually 12–13 (11–14). Lateral scales 30–32. Along the Atlantic and Gulf Coasts from northeastern Florida nearly to Tampico, Mexico; usually in salt and brackish waters.

***Fundulus confluentus* Goode and Bean.** Marsh killifish. Color yellowish gray, with a longitudinal streak along each row of scales; about 14 vertical dark bars on sides. Females with a D. ocellus. D. 10–12 and A. 10. From Chesapeake Bay, Maryland, southward and around the Florida coast to Big Lake, Gulf Shores, Baldwin County, Alabama, usually in brackish water.

***Fundulus pulvereus* (Evermann).** Bayou killifish. Closely related to *F. confluentus*, but females with many small dark dots on fins and back and 10–12 larger brown spots, sometimes in 2 rows on sides; upper spots sometimes fused to form oblong blotches; males barred. From Bayou Minette at Old Spanish Fort, Baldwin County, Alabama, west to Corpus Christi, Texas. Usually in brackish water.

***Fundulus diaphanus* (LeSueur).** Banded killifish (Fig. 2-83). About 20 silvery vertical bars, narrower than dark interspaces; in female the interspaces pale, sometimes black-spotted. D. 10–15; A. 9–13. Scales in lateral series 40–52 (35–55). Widely distributed from eastern Canada south to South Carolina and from the Yellowstone River in Montana through the Great Lakes Basin. An inhabitant of lakes, quiet rivers, and estuaries. Two subspecies.

***Fundulus waccamensis* Hubbs and Raney.** Waccamaw killifish. Similar to *F. diaphanus*, but with slenderer body and 54–64 scales in lateral series. Lake Waccamaw, North Carolina.

***Fundulus seminolis* Girard.** Seminole killifish. Sides with longitudinal rows of spots. Old females and all young with 12–14 faint dark bars. D. 13–20; A. 10–15. Rivers and swamps of central and northern Florida.

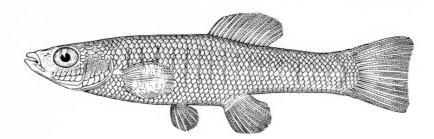

Fig. 2-83. Fundulus diaphanus.
(Jordan and Evermann, 1900.)

Fundulus catenatus (**Storer**). Northern studfish. Color bluish or greenish, with an orange spot on each scale in male, female with smaller brown spots. D. about 14; A. 15 or 16. Upland tributaries of the Tennessee, Cumberland, and Green Rivers in Kentucky, Tennessee, Virginia, and Alabama; clear streams of the Ozark region in Kansas, Missouri, Oklahoma, and Arkansas; Red River tributaries in southwestern Arkansas; the Homochitto River in Mississippi; and the upper part of the East Fork of White River, Indiana.

Fundulus stellifer (**Jordan**). Southern studfish. Color livid blue above, silvery below; body and cheeks with large orange spots not in rows or in middle of scales. Apparently restricted to the Alabama River and tributaries in Alabama and Georgia.

Fundulus albolineatus **Gilbert**. Whiteline topminnow. Male blackish brown, sides with plumbeous whitish streaks along rows of scales, a black vertebral stripe; D., A., and C. black-edged; female with black lines on scales, C. as in male. D. and A. 10 or 11. Lateral scales 42. Tennessee River system in Alabama and Tennessee.

Fundulus rathbuni **Jordan and Meek**. Speckled killifish. Male plain, with few if any spots; females with irregular spots on sides; fins usually without spots; a fine line from angle of gape to beneath eye and a wider stripe on opercle, distinct in young, but diffuse in adults; scales with dark borders. D. 11–14; A. 10–12; P_1 15–18. Lateral scales 33–38. Piedmont region of North Carolina in the Roanoke, Cape Fear, Neuse, and Pee Dee River systems.

Fundulus zebrinus **Jordan and Gilbert**. Rio Grande killifish. With 14–18 vertical dusky bars, narrower in female. D. 14–15; A. 13–14. Scales in lateral series 41–49. From the Brazos, Colorado, and Pecos drainages of Texas and New Mexico.

Fundulus kansae **Garman**. Plains killifish. Similar to *F. zebrinus*, but with scales in lateral series 52–64. Throughout the Great Plains region from South Dakota to Texas and New Mexico (Arkansas River Basin) and east to Missouri.

Fundulus luciae (**Baird**). Spotfin killifish. Male with 10–12 vertical bars, ending abruptly short of the middorsal and midventral lines; female plain. D. 8, with pale edge and, in male, an ocellated black spot on posterior portion; A. 10; lower fins yellowish. Coastal waters from Long Island to North Carolina.

Fundulus jenkinsi (**Evermann**). Saltmarsh topminnow. Scales edged with numerous fine dots presenting a crosshatched appearance; large spots in 2 irregular rows above body axis; latter spots sometimes form indistinct vertical bars. D. 8 or 9; A. 12 (11–13). From Mississippi, Louisiana, and Texas, east to the Escambia River in Florida. In salt, brackish, and fresh water.

Fundulus cingulatus **Valenciennes**. Banded topminnow. Color olivaceous, scales edged with dusky; usually 12 or more faint bars. D. 7–9; A. 10; fins blood red; belly orange. In fresh waters from Georgia to Florida and Alabama.

Fundulus chrysotus **Holbrook**. Golden topminnow. Male with fewer than 12 narrow bars; female plain, with pearl-colored dots; an elongate dark area above P_1 base; median fins with small black dots, paired fins plain. D. 8; A. 10. From South Carolina to eastern Texas, north to Missouri and Tennessee.

Fundulus sciadicus **Cope**. Plains topminnow. Color uniform olivaceous, no spots or lines. D. 10–12; A. 12–14. From Wyoming, South Dakota, and Iowa south to northeastern Oklahoma.

Fundulus notti (**Agassiz**). Starhead topminnow (Fig. 2-84). This species has been divided into 3 subspecies, *F. n. dispar*, *F. n. notti*, and *F. n.*

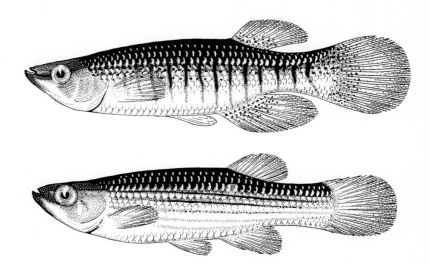

Fig. 2-84. Fundulus notti.
(Male above.) (Forbes and
Richardson, 1920.)

lineolatus (Brown, 1958), but Rivas (1966), without considering *F. n. dispar*, gave full species rank to *F. notti* and *F. lineolatus*. Populations in the Poteau and Little River Basins of southeastern Oklahoma lack conspicuous sexual dimorphism and were not studied by either Brown or Rivas. Until a more comprehensive study of these forms has been made, it seems best to retain them as 1 species. *F. notti*, including the 3 forms, is easily separated from all other *Fundulus* by the absence of a single lateral band and the presence of a dark subocular bar, "teardrop." Southeastwardly the females have about 6 narrow longitudinal black or brown stripes and males may have 9 dark vertical bars. The following distributions of the subspecies are as given by Brown: *F. n. lineolatus*, southeastern Virginia along the Coastal Plain to central peninsular Florida and west along the Coastal Plain to the Ochlockonee River in Georgia and western Florida; *F. n. notti*, from New River in western Florida west along the Gulf Coast to Louisiana and eastern Texas; *F. n. dispar*, Mississippi Valley from Iowa, southern Wisconsin, and the Lake Michigan drainage of Michigan and Indiana south to northeastern Arkansas and western Tennessee.

Fundulus notatus (Rafinesque). Blackstripe topminnow. A conspicuous black lateral band, with vertical streaks extending from it above and below in males; spots on dorsum indistinct, diffuse; predorsal stripe often present. D. 9 (8–12); A. 11–13. From Iowa, southeastern Wisconsin, and southern Michigan south to Tennessee. Along the Gulf Coast from Alabama to central Texas and north to Missouri and Kansas.

Fundulus olivaceus (Storer). Blackspotted topminnow. Similar to *F. notatus*, but spots on dorsum distinct, round; predorsal stripe sometimes weak in young, absent in adult. From Florida to northeastern Texas and northward in the Mississippi Valley to Kentucky, Illinois, Missouri, and Oklahoma.

Genus *Lucania* Girard. Body oblong and compressed. Mouth short and very oblique, lower jaw projecting. Jaw teeth conical and in a single series. Scales very large (25–29 in lateral series). Gill openings not restricted. Very small (37–50 mm) fishes of the coastal swamps, entering rivers.

KEY TO SPECIES OF *LUCANIA*

1. Lateral band from muzzle to caudal base conspicuous, ending in a basicaudal spot, most noticeable in young. Dorsal, anal, and pectoral fins of males black-edged; dorsal and anal fins basally black in males. Lateral scale rows 28 or 29   *L. goodei*

   Lateral band inconspicuous; no caudal spot. Anal and pectoral fins but not dorsal fin of male black-edged; dorsal and anal not conspicuously black at their bases, but male with anterior dorsal rays blackened. Lateral scale rows 25–26   *L. parva*

Lucania parva (**Baird**).   Rainwater killifish. D. 9–14 (usually 11); A. 8–13 (usually 9 or 10). Scales in lateral series 23–29 (usually 27). A middorsal stripe often developed before and behind D.; stripe on lower edge of C. peduncle; cheeks pigmented, but not forming a conspicuous subocular bar; pigment on mandible extends farther downward than in *L. goodei;* fins of females without conspicuous markings. See key above for other characters. Mostly in salt and brackish waters from Massachusetts to Florida and along the Gulf Coast to the lower Río Pánuco, along the Rio Grande and the Pecos Rivers in Texas and New Mexico. Recently found, unaccountably, in western United States: San Francisco Bay area, California; Yaquina Bay area, Oregon; Blue Lake and Timpie Springs, Utah; and Irvine Lake, southern California.

Lucania goodei Jordan.   Bluefin killifish. D. and A. 9. No middorsal stripe, but C. peduncle with a midventral stripe. Cheeks immaculate. Scales outlined with dark pigment, especially the enlarged lateral series. See key for other description. From Georgia and Florida.

Genus *Leptolucania* Myers.   Ocellated killifish. Similar to *Lucania,* but body slender and less compressed. D. 6–7; A. 9–10.

Leptolucania ommata (**Jordan**).   Pygmy killifish. From fresh-water swamps of Georgia, southeastern Alabama, and Florida.

Genus *Cyprinodon* Lacépède.   The pupfishes. Body typically short and stout. Small, reaching a length of about 75 mm. Teeth tricuspid, incisorlike, a single series in each jaw. Scales large, 20–34 (usually 25 or 26) from opercular angle to C. base. Preorbital region below nostrils scaled (except in *C. rubrofluviatilis, C. salinus, C. diabolis,* and some subspecies of *C. nevadensis*). Free edges of preorbital scales directed upward. Prepelvic scales 15–30 or more (usually 20–25), absent in *C. rubrofluviatilis.* D. 8–13, typically 10 or 11. A. 8–12, usually 9–11; P_1 12–18; C. 13–22. Intestine much longer than body and much convoluted.

KEY TO SPECIES OF *CYPRINODON*

1. Species west of the Continental Divide   2
   Species east of the Continental Divide   6

2. Circuli of scales with spinelike projections. Interspaces between circuli without conspicuous reticulations   *C. macularius*
   Circuli of scales without spinelike projections. Interspaces between circuli with conspicuous reticulations   3
3. Pelvic rays usually 6–6, occasionally lacking on one or both sides   4
   Pelvic rays usually 7–7 or 0–0   5
4. Predorsal scales typically 25–27 (22–30) and scales around body 37–42 (33–46). Body slender
                                                 C. salinus
   Predorsal scales typically 17–19 (15–24), those around body 22–26 (18–32). Body consistently deeper   *C. nevadensis*
5. Pelvic fins absent   *C. diabolis*
   Pelvic fins present   *C. radiosus*
6. Body slender, back not greatly arched. No crossbars
                                                 C. elegans
   Body deep, back notably arched. Crossbars present
                                                          7
7. Abdomen naked   *C. rubrofluviatilis*
   Abdomen well scaled   8
8. Humeral scale not notably enlarged. Pectoral fin shorter and broader, its length contained more than 2.0 times in predorsal length in mature males. Gill rakers fewer than 20   *C. bovinus*
   Humeral scale conspicuously enlarged. Pectoral fin long and narrow, its length contained fewer than 2.0 times in predorsal length in mature males. Gill rakers typicaly 21–26   9
9. Preorbital well scaled up to or beyond level of uppermost preorbital pore. Body deeper, its depth in adults contained fewer than 3 times in standard length   *C. variegatus*
   Preorbital squamation reduced, typically no scales above level of middle of pupil. Body slenderer, its depth in adults contained more than 3 times in standard length   *C. hubbsi*

Cyprinodon variegatus **Lacépède.**   Sheepshead pupfish. No lateral band, but vertical bars posteriad; D. 11, with spots or blotches; A. 10, with black dots; C. with black tip. In fresh, brackish, and salt waters along the Atlantic Coast, from Massachusetts to Yucatán, Mexico, and in the West Indies to northern South America. Common.

Cyprinodon macularius **Baird and Girard.**   Desert pupfish. Recognized by the spinelike projections from scale circuli and absence of reticulations between circuli. From the lower Colorado and the Gila River Basins, the Salton Sea, southern Arizona to southeastern California, northeastern Baja California, and the Sonoyta River of northern Sonora, Mexico.

Cyprinodon nevadensis **Eigenmann and Eigenmann.** Nevada pupfish. P_2 6–6, occasionally absent on one or both sides. Scales large: predorsal 15–24, usually 17–19; around the deep body 18–32, usually 22–26. Desert springs and streams in Amargosa River Basin, from Ash Meadows, Nye County, Nevada, to Saratoga Springs and Amargosa River in the southeastern arm of Death Valley, San Bernardino County, California.

Cyprinodon diabolis **Wales.** Devils Hole pupfish. P_2 absent. Devil's Hole, in Ash Meadows, Nye County, Nevada.

Cyprinodon salinus **Miller.** Salt Creek pupfish. Similar to *C. nevadensis*, but with smaller scales: predorsal 22–30, usually 25–27; around body 33–46, usually 37–42. Salt Creek in Death Valley, California.

Cyprinodon radiosus **Miller.** Owens River pupfish. P_2 present, the rays 7–7. Interspaces between scale circuli with conspicuous reticulations. Owens Valley, California, where formerly abundant and now possibly extinct. The introduction of predatory exotic fishes and the removal of water from the valley are thought to be the most important factors in depleting the numbers of this species.

Cyprinodon rubrofluviatilis **Fowler.** Red River pupfish. Unique in having a naked breast. Upper reaches of the Brazos and Red Rivers, Texas and Oklahoma. Locally very abundant.

Cyprinodon elegans **Baird and Girard.** Comanche Springs pupfish. Body slender. Crossbars absent on sides. Comanche Springs and Phantom Lake in Texas; and many springs tributary to the Pecos River.

Cyprinodon bovinus **Baird and Girard.** Leon Springs pupfish. Leon Springs near Fort Stockton, Texas; now extinct.

Cyprinodon hubbsi **Carr.** Lake Eustis pupfish. No scales on preorbital region above level of pupil. Lake Eustis, Florida.

Genus *Floridichthys* **Hubbs.** Killifish. D. 11–13; A. 9; P_1 18–20; C. 16–18. Lateral scales about 24. Mandibular pores 3 on each side. C. without terminal band. Supraorbital canal not interrupted. D. long, relatively low anteriad, 1st ray not enlarged, posterior rays of adult male much elongated. Sides silvery, adults with brassy or orange spots ringed with blue; faint, irregular vertical bars or blotches; no suborbital bar. Monotypic.

Floridichthys carpio **(Günther).** Goldspotted killifish. From the Florida Keys north to about 29° N. lat. Coastal parts of the Yucatán Peninsula. Occasionally taken in fresh water.

Genus *Jordanella* **Goode and Bean.** Flagfish. Monotypic. Body short, deep, and compressed. Mouth small and protractile, upper jaw included. No mandibular pores. Tricuspid teeth in single series. Branchiostegals 5. Gill openings restricted above. D. 16–18 rays, its origin in front of A. origin, the 1st ray thick and spinelike; A, 11–13; P_1 14–16; C. 17–21. Scales in lateral series 25–27. Intestine about 3 times as long as body. Dark suborbital bar present.

Jordanella floridae **Goode and Bean.** Flagfish. From the streams and swamps of Florida.

Genus *Rivulus* **Poey.** Small fishes with elongate body, moderate scales, small mouth with protractile upper jaw; conical teeth in several rows, the outer row of teeth recurved and some enlarged. C. peduncle with a posterodorsal ocellus. Anterior nostrils tubular, extending forward beyond the snout tip.

Rivulus marmoratus **Poey.** Rivulus. D. usually 8 (7); A. 10 (11); P_1 13 (12–14); P_2 6; branched C. 10–16. Scales: lateral series 49 (47–51); longitudinal series 12 (10–13); predorsal 37 (34–39). Teeth conical in several rows; about 14 teeth in outer row, with 3 on each side enlarged and inner rows smaller and irregular, except for 3 larger ones on each side; lower jaw with 10 large teeth in outer row and other rows similar to upper-jaw teeth, but without enlarged ones. Color in life deep maroon, in alcohol, pale reddish brown; venter pale, sharply demarked; C. base with a dorsally placed ocellus encircled by cream color; a dark round humeral spot; D., A., and C. mottled with cream and brown; P_1 and P_2 plain, except at P_2 bases. Lower jaw protrusible. Eyes far forward; anterior nostril tubular, extending beyond the small upper lip. P_2 undeveloped in young, appearing at about 9.6 mm. Indian River, Florida; also Cuba. In salt, brackish, and fresh waters.

Family Poeciliidae. Livebearers. For a detailed definition of this family see Rosen and Bailey (1963), who use many osteological characters of both sexes, but emphasize those of males. The poeciliids differ from the cyprinodontids most conspicuously in having the A. of males modified for placing sperm bundles in the female (Fig. 2-85). The male A. is placed farther forward, and the 3d A. ray in both sexes is unbranched. Adult females usually contain embryos in various stages of development. The 1st few P_1 rays, in some species, are modified in males; the P_2 may also be modified in males or small and poorly developed. Gonads fused as a median saclike or tubular structure with an oviduct or sperm duct leading to the genital opening near the A. origin. Three subfamilies, 21 genera, and 138 species are recognized; in the United States, there are 5 genera and 12 species, largely in the South and Southeast. One other family, Embiotocidae, in the United States gives birth to young.

KEY TO GENERA OF POECILIIDAE

1. Premaxillae and dentaries much elongated as a beak
   Belonesox

   Premaxillae and dentaries not forming a conspicuous beak 2
2. Teeth in villiform bands. Lateral band absent 3
   Teeth in single series or inner teeth weak to obsolete. Lateral band or streak present 4
3. Dorsal rays 12–16; dorsal origin over or in front of anal origin. Intestine long, with many convolutions
   Poecilia

   Dorsal rays 8–10; dorsal origin well behind anal origin. Intestine short, with few convolutions
   Gambusia
4. Dorsal, anal, and caudal fins each with a black spot near its base. Lateral band crossed by 6–9 vertical bars
   Heterandria
   Dorsal, anal, and caudal fins without spots. A lateral streak, but no vertical bars
   Poeciliopsis

Genus *Poecilia* Bloch and Schneider. Mollies and guppies. The name *Poecilia*, following Rosen and Bailey (1963), replaces *Mollienesia* and several generic names not pertinent to the fish fauna of the United States, although some (e.g., *Lebistes*) are widely used by aquarists.

Gonopodium bilaterally symmetrical, with a large fleshy palp arising from ventral surface of ray 3; ray 3 with or without spines near its tip; ray 4*a* rarely with serrae near its tip; ray 4*p* usually with 10 or more serrae; ray 5 often with a

retrorse claw. P_2 highly modified, its specializations involving rays 1, 2, and 3 (rarely 4 and 5); ray 2 longest, with dentate, serrate, or crenulate leading edge and with many fused basal segments. Coastal Plain of southeastern United States, Atlantic and Pacific coastal waters of Mexico, throughout Central America, Cuba, Haiti, and southward along the Atlantic Coast of South America to the Rio de la Plata. About 32 species, 2 in the United States.

KEY TO SPECIES OF *POECILIA*

1. Dorsal rays 10–12. Dorsal origin over anal origin
   P. formosa

   Dorsal rays 13–16. Dorsal origin in front of anal origin
   P. latipinna

Poecilia formosa (**Girard**). Amazon molly. A few males presumed to be of this species have been found, but the Amazon molly regularly uses males of other species. Regardless of the male parent, the offspring are females with characters of the mother. Northeastern Mexico and southern Texas as far north as San Marcos.

Poecilia latipinna (**LeSueur**). Sailfin molly. From South Carolina to Texas and Mexico, in coastal swamps. Established in tributaries of the Salton Sea, California.

Genus *Gambusia* Poey. The mosquitofish and gambusias. A taxonomically difficult group in which

Fig. 2-85. Gonopodial tip of *Gambusia heterochir*. (Modified from Hubbs, 1957.)
In the text, the numbers 3, 4*a*, 4*p*, 5*a*, and 5*p* refer to anal rays 3, 4, and 5. Rays 4 and 5 are branched to form anterior *a* and posterior *p* portions.

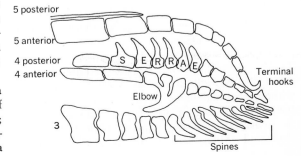

the principal characters are found in the gono-podium of the male (see Fig. 2-85 and legend).

Small fishes (seldom longer than 50 mm); males much smaller than females. Scales large. Gill openings not restricted. D. and A. low and small. Teeth sharply conical, in villiform bands, with a single arc following jaw outline. Over 30 species in all, but only 7 known in the United States.

KEY TO SPECIES OF *GAMBUSIA*

1. Gonopodial spines of ray 3 short, inconspicuous, and thick, the longest one shorter than combined length of its segment and the segment distal to it. Most segments distal to elbow of 4a anteriorly coalesced
   G. affinis

   Gonopodial spines of ray 3 elongate, conspicuous, and slender, the longest one about equal to or longer than combined length of its segment and the segment distal to it. Few or none of segments distal to elbow of ray 4a anteriorly coalesced **2**

2. Longest spine of ray 3 (including inner process) longer than apical portion of 4a distal to elbow; ray 4a not or rarely reaching beyond tip of ray 4p; the 2 or 3 segments proximal to elbow of 4a more or less abruptly narrower than the 2 or 3 segments proximal to them; anterior marginal outline formed by spine tips of ray 3 rather straight or slightly concave or convex; ray 3 with 6 or 7 spine-bearing segments; usually 2–5 segments distal to elbow of 4a **3**

   Longest spine of ray 3 (including inner process) equal to or shorter than apical part of 4a distal to elbow; ray 4a reaches past tip of 4p; the 2 or 3 segments proximal to elbow of 4a not abruptly narrower than the 2 or 3 segments proximal to them; anterior marginal outline formed by spine tips of ray 3 slightly to strongly convex; ray 3 with 7–15 spine-bearing segments; usually 5–8 segments distal to elbow of 4a **5**

3. 1st and/or 2d proximal enlarged spines of ray 3 with recurved hook; terminal hooks of 4p and 5a angular at tip. Dorsal rays 7 (8). Postanal streak prominent; a median row of spots on caudal fin and some at its base; a median row of spots on dorsal fin. Distance from dorsal origin to caudal fin equal to distance from dorsal origin to 1st predorsal scale in females. Caudal peduncle depth contained 1.5–1.8 times in caudal length **G. geiseri**
   Ray 3 lacking recurved hooks; terminal hooks of rays 4p and 5a rounded at tip. Dorsal rays 8 or 9 (7 or 10). Postanal streak weaker than pigment on scale pockets. No prominent spots on caudal fin; no spots or a subbasal row on dorsal fin. Distance from dorsal origin to caudal fin equal to distance from dorsal origin to 2d or 3d predorsal scale; caudal peduncle depth contained 0.8–1.4 times in caudal length **4**

4. Elbow on 4a of 2–4 segments. Dark markings on scale edges below midline more prominent than those above; no chin bar; no black spot around anus **G. senilis**
   Elbow on 4a of one (rarely 2) segments. Dark markings more frequent above midline; chin bar present (sometimes weak); a dark spot around anus, at least in females **G. gaigei**

5. Elbow of 4a falcate, antrorse, opposite serrae of 4p. Ray 5p reaching to serrae of 4p **6**
   Elbow of 4a not falcate, but distal to serrae of 4p. Ray 5p not reaching serrae of 4p **G. punctata**

6. Lateral stripe thin, but distinct; caudal with a dark margin; predorsal streak strong. Elbow of 4a shorter than longest modified spine of ray 3; distal serra of 4p opposite elbow; distal segments of ray 5 meet axis of gonopodium at an angle of 45° or less; spines of ray 3 without hooks **G. nobilis**
   Lateral stripe indistinct; caudal without dark margin; predorsal streak absent. Elbow of 4a longer than longest spines of ray 3 and often overlaps the unmodified segments of ray 3; distal serra of 4p distal to elbow; distal segments of ray 5 meet axis of gonopodium at an angle of more than 45°; at least one spine of ray 3 with recurved hook
   G. heterochir

Gambusia affinis (Baird and Girard). Mosquitofish. Originally in central United States from southern Illinois to Alabama and southern Texas; on the Atlantic Coast from New Jersey to Florida. Now more widely distributed by planting.

Gambusia geiseri Hubbs and Hubbs. Large-spring gambusia. Found in the Rio Colorado and the San Marcos, Guadalupe, and Pecos Rivers in Texas.

Gambusia senilis Girard. Blotched gambusia. Known in the United States only from Devils River, Texas; also in Chihuahua and Durango, Mexico.

Gambusia gaigei Hubbs. Big Bend gambusia. Known only from one spring in Big Bend National Park, Texas, where it has been threatened with extinction. In an effort to save stock of the species Dr. Clark Hubbs has propagated the fish in aquaria at Austin.

Gambusia nobilis (**Baird and Girard**). Pecos gambusia. Pecos River system in western Texas and eastern New Mexico.

Gambusia heterochir **Hubbs.** Clear Creek gambusia. From head springs of Clear Creek, Menard County, Texas.

Gambusia punctata **Poey.** Spotted gambusia. Known from extreme southern Florida and Cuba.

Genus *Belonesox* Kner. Pike killifishes. Similar to *Gambusia*, but with the premaxillae and dentaries greatly elongated, with the inner teeth much longer than the outer. Scales minute and numerous. This fish reaches the largest size in the family. 1 species in fresh and brackish waters of the Atlantic lowlands in southern Mexico and northern Central America.

Belonesox belizanus **Kner.** Pike killifish. Characters of the genus. Introduced into southern Florida, where now common.

Genus *Heterandria* Agassiz. Least killifishes. Gonopodium bilaterally symmetrical, without fleshy appendages; rays 3 and 5 reaching to tip of membrane; 5–10 rudimentary distal spines on ray 3; tip of ray 4*a* with 1 to 3 or more segments often fused to form a hook; 7 to 10 or more serrae near tip of ray 4*p*; and tip of ray 5, segmented or not, curves downward and forward to terminal portion of ray 4. Jaw teeth in single series. A lateral band from snout tip ends in a basicaudal spot. 1 species in the United States and another on the Atlantic slope from Río Tamesí, Mexico to Nicaragua.

Heterandria formosa **Agassiz.** Least killifish. From South Carolina to Florida and west to New Orleans in the coastal swamps.

Genus *Poeciliopsis* Regan. Topminnows. Marginal jaw teeth recurved, conical, and distributed evenly; or compressed or hairlike, in clusters of 5 or 6. Gonopodium asymmetrical to the left, rays 3, 4, and 5 folded or twisted to form a trough; ray 3 with some fused segments; ray 4*p* sometimes with paired symmetrical serrae or unicuspid serrae on one and bicuspid serrae on the other branch; rays 7 and 8 close together near their middles or tips. Pacific drainage from southwestern United States to Colombia and on the Atlantic side in southeastern Mexico, Guatemala, and Honduras. 15 species, 1 in the United States.

Poeciliopsis occidentalis (**Baird and Girard**). Gila topminnow. From the Gila Basin in Arizona

and Mexico through coastal streams in Sonora, Mexico.

Order Gasterosteiformes. Physoclistic fishes with 2–15 free spines (Fig. 2-12) before D.; P_2 (I,1 or 2) subabdominal, not far behind P_1 and well behind gill openings. Pelvic bones not articulated with cleithra. 2d infraorbital extending across cheek to preopercle. Mouth bordered above by premaxillae only. Body elongate, with slender C. peduncle, naked or with bony plates. Teeth sharp, in jaws only. Branchiostegals 3. Opercles unarmed. A. with 1 spine. Tertiary. Small fishes of fresh, brackish, and salty waters of the Northern Hemisphere. 1 family, 4 genera, and 4 species in fresh waters of the United States.

Family Gasterosteidae. The sticklebacks.

KEY TO GENERA OF GASTEROSTEIDAE

1. Pelvic bones not joined, each extending backward as a strong process under skin. Gill membranes joined to isthmus. Dorsal spines usually 4 *Apeltes*
   Pelvic bones joined, forming a median plate on belly behind pelvic fins. Gill membranes connected or not connected across isthmus **2**
2. Caudal peduncle without a lateral keel. Body wholly naked. Dorsal spines usually 5 *Eucalia*
   Caudal peduncle with a lateral keel of bony plates. Body mailed, at least posteriorly, except in some naked fresh-water forms. Dorsal spines typically 3 or 7–11 **3**
3. Dorsal spines 7–11 (usually 9 or 10), without serrae, but their points divergent. Body anteriorly naked. Pelvic spines shorter, not serrate *Pungitius*
   Dorsal spines 3 (0–4), serrate and not divergent. Body usually with plates over much of its length, but wholly naked in some Californian populations. Pelvic spines longer and serrate *Gasterosteus*

Genus *Apeltes* DeKay. Body naked. D. spines usually 4. Pelvic bones not joined. Gill membranes joined to isthmus.

Apeltes quadracus (**Mitchill**). Fourspine stickleback. From Labrador to Maryland. Abundant in salt water, but sometimes entering fresh waters.

Genus *Eucalia* Jordan. Body naked. D. spines usually 5. Pelvic bones united. Gill membranes joined to form a fold across the isthmus.

Eucalia inconstans (**Kirtland**). Brook stickleback (Fig. 2-12). From Canada and in the United States from Montana to Maine and south to the Ohio River system in Ohio, Pennsylvania, and Indiana and the Missouri River system in Kansas. Always found in fresh water.

Genus *Pungitius* Costa. Body naked except for some small plates along the bases of the median fins. D. spines 9–11. P_2 bones small and joined together. Gill membranes joined across isthmus.

Pungitius pungitius (**Linnaeus**). Ninespine stickleback. Circumpolar in distribution, southward to central Europe, northern China; Great Lakes (except Erie); on the east coast as far north as Newfoundland and southward to New Jersey. Recently found in North Dakota. In fresh and salt waters.

Genus *Gasterosteus* Linnaeus. Sides of body usually with bony plates. D. spines usually 3. P_2 bones joined. Gill membranes free, not united across isthmus.

Gasterosteus aculeatus **Linnaeus.** Threespine stickleback. Northern Hemisphere south to northern Africa and China, southern Japan, and Baja California, and Hudson and Chesapeake Bays. In both fresh and salt waters. Extremely variable. *G. wheatlandi* Putnam, the twospine stickleback, occasionally enters fresh water; it differs from *G. aculeatus* in having 2 D. spines, a deeper body, no keel on C. peduncle, and fewer soft D. rays.

Order Syngnathiformes. Physoclistic. D., A., and P_1 rays unbranched. P_2 and C. rays partly branched. First D., if present, spinous. P_2, if present, abdominal or nearly so in position. Snout tube-shaped and bordered by premaxillae or both premaxillae and maxillae. Branchiostegal rays 1–5. The first 3–6 vertebrae united to one another. Tropical to temperate seas. Lower Eocene. 1 family, 1 genus, and 1 species in U.S. fresh waters.

Family Syngnathidae. The pipefishes.

Genus *Syngnathus* Linnaeus.

Syngnathus scovelli (**Evermann and Kendall**). Gulf pipefish. From Florida to Texas and Mexico in salt, brackish, and fresh water.

Order Percopsiformes. Trout-perches, pirate perches, and cavefishes. This order, containing only 3 families, 5 genera, and 8 species, is currently subject to considerable study and speculation. Although the cavefishes, Amblyopsidae, have been placed previously in the order Cyprinodontiformes,

they appear to be closely related to the genus *Aphredoderus* (pirate perch), a relative of the genus *Percopsis* (including the nominal genus *Columbia*). In some respects the 3 families are strikingly different: spines are present or absent (cavefishes), an adipose fin is present only in *Percopsis,* scales are ctenoid or, in cavefishes, cycloid, the anus is in the usual position only in *Percopsis* (2 species), and only the pirate perch has scales on the head. Eyes are well developed, small, or rudimentary. However, in some fundamental characters there is close agreement. In all species the cutaneous sense organs, occurring in rows, bands, or clusters, are highly developed and are easily seen under some magnification. In the trout-perches and pirate perches the cavernous, cephalic L.l. canals are greatly enlarged. Also important in tracing their relationships are the similarities of their bones and muscles. One of the osteological characters is easily seen without dissection; the premaxilla narrows to a point distally, in contrast to its broad trapezoidal shape in the killifishes. Palatine teeth present; operculum spiny; gill rakers complex, in clusters. For many other similarities, see Rosen, 1962, and Gosline, 1963. See also order Cyprinodontiformes. Eocene. Small fishes restricted to fresh waters of North America.

KEY TO FAMILIES OF PERCOPSIFORMES

1. Adipose fin present. Anus in "normal" position, near anal fin　　　　**Percopsidae**

   Adipose fin absent. Anus, in adult, in jugular position
   　　　　　　　　　　　　　　　　　　　2

2. Spines present in dorsal, anal, and pelvic fins. Cephalic L.l. canals greatly enlarged; sensory papillae in rows, bands, or clusters, mainly on the head. Eyes well developed　　　　**Aphredoderidae**

   Spines absent. Cephalic L.l. canals not conspicuous; sensory papillae arranged in vertical or horizontal rows on the head, sides, and, in some species, on the caudal fin. Eyes small or degenerate
   　　　　　　　　　　　　　Amblyopsidae

Family Amblyopsidae. Cave, spring, and swamp fishes with eyes greatly reduced or rudimentary and buried in flesh. D. far back over A.; C. rounded, not forked; P_2 small or absent. Body little compressed; head elongate, depressed; mouth large, lower jaw projecting. Head naked; body scales cycloid. Anus far forward, in jugular position. Gill membranes joined to isthmus; branchiostegals usually 6. Since at present there exists a lack of agreement concerning affinities of several taxa (see Rosen, 1962, and Gosline, 1963) and since further studies are in progress, the family Amblyopsidae is tentatively placed in the order Percopsiformes. Nearctic fishes of central and eastern United States; 3 genera and 5 species.

KEY TO GENERA OF AMBLYOPSIDAE

1. Postcleithrum present.[1] Sensory papillae absent or in 1 row on upper and lower half of caudal fin **2**
   Postcleithrum absent. Sensory papillae in 2 or 3 rows on upper and lower half of caudal fin *Amblyopsis*
2. Eyes present *Chologaster*
   Eyes absent *Typhlichthys*

Genus *Chologaster* Agassiz. Eyes well developed, but covered by skin that is more or less transparent in life, translucent when preserved. Body always pigmented, dark stripes on lower sides present or absent. C. rays 9–16.

KEY TO SPECIES OF *CHOLOGASTER*

1. Branched caudal rays 9–11. Lower sides with dark stripes *C. cornuta*
   Branched caudal rays 12–16 (rarely 11). Lower sides without dark stripes *C. agassizi*

***Chologaster cornuta* Agassiz.** Swampfish. Atlantic Coastal Plain from Virginia to central Georgia. Common in swamps and streams.

***Chologaster agassizi* Putnam.** Spring cavefish. Central Tennessee, southern and western Kentucky, to southwestern Illinois. Living in springs, caves, and subterranean streams.

Genus *Typhlichthys* Girard. Eyes absent. Body pigment usually absent but developing on examples living or kept in light, occasional specimens with minute pigment specks, not in stripes. 1 species.

***Typhlichthys subterraneus* Girard.** Southern cavefish. Middle Tennessee and northern Alabama;

[1] To determine presence or absence of postcleithrum, make an incision in skin of axil and probe gently with a needle. The bone is readily visible, in large specimens, without magnification.

in and along the eastern side of Dripping Springs escarpment of Kentucky; Ozark region of Missouri and northeastern Oklahoma.

Genus *Amblyopsis* DeKay. Eyes wholly absent or degenerate. Body pigment absent.

KEY TO SPECIES OF *AMBLYOPSIS*

1. Pelvic fins absent *A. rosae*
   Pelvic fins present, small *A. spelaea*

***Amblyopsis rosae* (Eigenmann).** Ozark cavefish. Southwestern Missouri in Jasper, Greene, Newton, Stone, and Barry Counties; northwestern Arkansas.

***Amblyopsis spelaea* DeKay.** Northern cavefish. Mammoth Cave region northward in Kentucky and caves of unglaciated south central Indiana.

Family Percopsidae. The trout-perches. Body rather slender. Scales ctenoid; head naked. Mouth small, with very small villiform teeth, except on vomer and palatines. Premaxillae nonprotractile, forming margin of upper jaw. Gill rakers short. Cephalic L.l. canals very large. Stomach siphonal, intestine issuing from stomach near esophagus. 2 genera found in northern United States and Canada.

Genus *Percopsis* Agassiz. Trout-perches. D. II,9; A. I or II,6–8. Scales 44–50.

KEY TO SPECIES OF *PERCOPSIS*

1. Dorsal fin with 2 weak spines; anal with 1 slender spine. Scales most strongly ctenoid on caudal peduncle. Preopercle with or without weak serrae. L.l. complete *P. omiscomaycus*
   Dorsal and anal fins with 2 strong spines each. Scales most strongly ctenoid anteriad. Preopercle with a few strong spines. L.l. incomplete
   P. transmontanus

***Percopsis omiscomaycus* (Walbaum).** Trout-perch. C. peduncle long and slender. Color olivaceous, with a silver stripe on sides and obscure spots on dorsum. Canada from the Yukon and Alberta to Hudson Bay and Quebec. On the Atlantic Coast south to the Potomac River and in the Mississippi Valley south to West Virginia, Kentucky, Missouri, and Kansas.

Percopsis transmontanus (**Eigenmann and Eigenmann**). Sand roller. Body deeper and more compressed than in *P. omiscomaycus*. Back and sides with rows of oblong spots. Columbia River Basin in Oregon, Washington, and western Idaho (Snake River).

Family Aphredoderidae. The pirate perches. Body oblong and elevated anteriad. Scales ctenoid. C. peduncle rather deep and thick. Sides of head scaly. Mouth rather large and somewhat oblique. Teeth in villiform bands on jaws, vomer, palatines, and pterygoids. Premaxillae nonprotractile, forming margin of gape. Gill rakers very short. Cephalic L.l. canals large, but smaller than in *Percopsis*.

Genus *Aphredoderus* LeSueur. Pirate perches. D. III,11, or IV,10; A. II,6. P_2 7–7. Color, depending on clarity of water, very dark to pinkish. Anus in young in normal position, but shifting forward as age advances. 1 species.

Aphredoderus sayanus (**Gilliams**). Pirate perch (Fig. 2-86). From southeastern Minnesota to the southern tributaries of Lake Ontario southward to the Gulf States. On the Atlantic slope from New York to Florida.

Order Gadiformes. Physoclistic fishes without spines. Cycloid scales. P_2 in jugular position. C. isocercal. No pseudobranchiae or intermuscular bones, but upper ribs present. Worldwide distribution, mostly marine. Paleocene. 1 family, 1 genus, and 1 species in the United States.

Family Gadidae. The codfishes and burbots. Characters of the order. Mostly marine.

Genus *Lota* Cuvier. The burbots. Unique in having a single chin barbel.

Lota lota (**Linnaeus**). Burbot (Fig. 2-11). Fresh waters of Europe, Siberia, and northern North America. Eastern and central United States

to the Columbia River and north to arctic waters. South as far as Missouri and Kansas.

Order Mugiliformes.[1] Similar to the Perciformes, but with P_2 abdominal. Scales cycloid or ctenoid. Teeth not in deep sockets. L.l. absent or rudimentary. P_1 placed high. Upper jaw protractile. Gill membranes free from isthmus. Pseudobranchiae present. Gill rakers slender. Lower Eocene. Warm and temperate seas, some entering fresh waters. A few species live wholly in fresh water.

KEY TO FAMILIES OF MUGILIFORMES

1. Anal spine single. Dorsal spines slender and flexible
   Atherinidae
   Anal spines 2 or 3. Dorsal spines heavy and stiff
   Mugilidae

Family Atherinidae. Silversides. Body long (fresh-water species about 100 mm, some marine forms much larger) and slender, covered with cycloid scales. Teeth small and in villiform bands. Preopercles not serrated. 2 well-separated D. fins, the 1st of slender weak spines and the 2d of soft rays. A. with 1 weak spine. P_2 with 1 small spine and 5 soft rays. Marine and fresh-water fishes. 2 genera.

KEY TO GENERA OF *ATHERINIDAE*

1. Premaxillae viewed from above appear triangular (Fig. 2-87A), forming a pointed beak. Scales small,

[1] Greenwood et al. (1966) proposed the order Atheriniformes to accommodate the suborders Exocoetoidei, including the Belonidae; Cyprinodontoidei, including the Cyprinodontidae and Poeciliidae; and Atherinoidei for the Atherinidae. The family Mugilidae, placed here with the Atherinidae in the order Mugiliformes, is placed in the suborder Mugiloidei under the Perciformes.

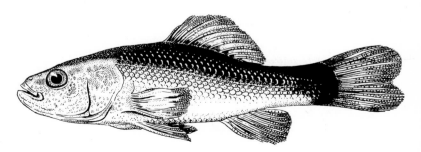

Fig. 2-86. Aphredoderus sayanus. (Forbes and Richardson, 1920.)

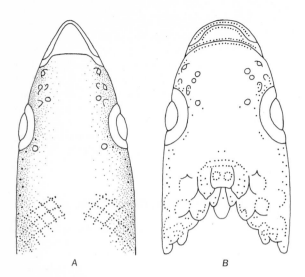

Fig. 2-87. Dorsal views of atherine snouts to show the shape of premaxillae in *A, Labidesthes sicculus; B, Menidia audens.* (After drawings by Kenneth Stewart.)

more than 50 in lateral series. Predorsal scales crowded, more than 23   **Labidesthes**
Premaxillae viewed from above crescent-shaped (Fig. 2-87*B*), not forming a pointed beak. Scales larger, not more than 37–50 in lateral series. Predorsal scales not crowded, 14–22   *Menidia*

Genus *Labidesthes* Cope. Brook silversides.
Labidesthes sicculus (**Cope**). Brook silverside. Widely distributed from Minnesota eastward through Wisconsin, Lower Peninsula of Michigan, Ontario to the St. Lawrence, and south to the Gulf Coast. In the Mississippi Valley as far west as central Oklahoma.
Genus *Menidia* Jordan and Gilbert. Silversides.

KEY TO SPECIES OF *MENIDIA*

1. Predorsal scales about 14–16. Body depth 17–21% of standard length   **M. beryllina**
   Predorsal scales about 19–22. Body depth 13–16% of standard length   **2**
2. Vertebrae 38–42. Scales about 39–44. Body depth about 16% of standard length   **M. audens**
   Vertebrae 42–45. Scales about 44–50. Body depth about 13% of standard length   **M. extensa**

Menidia beryllina (**Cope**). Tidewater silverside. In salt, brackish, and coastwise fresh waters from Massachusetts to northern Veracruz.
Menidia audens **Hay.** Mississippi silverside. Fresh waters of the lower Mississippi Valley as far north as southern Arkansas and Oklahoma. In Texas only from the Red River drainage.
Menidia extensa **Hubbs and Raney.** Waccamaw silverside. Lake Waccamaw, North Carolina.
Family Mugilidae. The mullets. Body oblong and somewhat compressed. Cycloid or ctenoid scales. Jaws with weak teeth or none. Branchiostegals 5 or 6. Gill rakers long and slender. Pseudobranchiae large. Spinous D. of 4 stiff spines, well separated from soft D. A. with 2 or 3 graduated spines. P_2 with 1 spine and 5 soft rays. 2 genera enter our fresh waters.

KEY TO GENERA OF MUGILIDAE

1. Lower jaw rounded, not angular, without a symphyseal knob. No adipose eyelid. Scales with rather prominent ctenii   **Agonostomus**
   Lower jaw angular, with a symphyseal ridgelike prominence. Adipose eyelid present, except in young. Scales cycloid to weakly ctenoid   **Mugil**

Genus *Agonostomus* Bennett. Mountain mullet. Aspect of *Mugil* but lacking a mandibular symphyseal knob and adipose eyelids. Scales with numerous fine ctenii making scale borders conspicuous in slightly dried alcohol-preserved specimens. D., A., and C. fins with bright yellow. The distally broad lacrimal bone strongly serrate. Mouth terminal, lips rather thick. Primarily marine, but entering fresh water along the Gulf Coast.
Agonostomus monticola (**Bancroft**). Mountain mullet. D. IV–9; A. II,10; C. 14. Predorsal scales 18–23; scales in lateral series 38–44; circumferential scales 25–29; scales around C. peduncle 20–23. D. and A. fins basally scaled. Known from fresh water in the lower Mississippi Valley and the Atlantic coastal streams of Florida.
Genus *Mugil* Linneaus. The mullets. Similar to *Agonostomus*, but with a mandibular symphyseal knob and well-developed adipose eyelids. Lacrimal bone distally narrow and serrate. Mouth subterminal, lips thin. Marine fishes, entering streams on both the Atlantic and Pacific Coasts.

KEY TO SPECIES OF *MUGIL*

1. Soft dorsal and anal fins almost naked. Sides with dark longitudinal streaks in adult. Anal II,9 (young) to III,8. Scales 38–42   **M. cephalus**
   Soft dorsal and anal fins scaly. Sides without dark streaks. Anal usually II,10 (young) to III,9. Scales 33–39   **M. curema**

Mugil cephalus **Linnaeus.**   Striped mullet. Widespread along the southern shores of Europe, north shores of Africa, east coast of America from Cape Cod to Brazil, west coast from Monterey to Chile, and Haiti, Hawaii, and Japan.

Mugil curema **Valenciennes.**   White mullet. Common on the Atlantic Coast from Cape Cod to Brazil and on the Pacific side from Magdalena Bay to Chile. Recorded from fresh waters of Texas.

Order Perciformes.   Basses, sunfishes, perches, sculpins, etc. Physoclistic spiny-rayed fishes. D. single or double, but always the anterior fin or portion spiny. A. with varying number of spines (absent in Cottidae) and soft rays, but always single. P_2 with not more than 6 rays and usually in thoracic position. Pelvic bones usually attached to cleithra. C. with not more than 17 principal rays. Scales typically ctenoid (Fig. 2-9). L.l. usually present. Gill rakers usually stout and armed with teeth. Jaw teeth and those of vomer, palatines, hyoid, and pterygoids, when present, pointed and arranged in bands. Gill membranes free from or joined to and sometimes connected across the isthmus. Pseudobranchiae usually present, but sometimes concealed. Intestine usually short, long in herbivorous forms.

A very large order containing fishes of highly diversified form and making characterization very difficult. The above characters, it is hoped, will give a general picture of the group. Upper Cretaceous. Worldwide in fresh and salt waters.

KEY TO SUBORDERS OF PERCIFORMES

1. 2d suborbital without bony connection with preopercle   2
   2d suborbital with bony connection   **Cottoidei**
2. Dorsal spines stiff and sharp. Pelvic fins always separate, never as a sucking disc   **Percoidei**
   Dorsal spines flexible, not sharp to the touch. Pelvic fins separate or not, often used as a sucking disc (Fig. 2-88)   **Gobioidei**

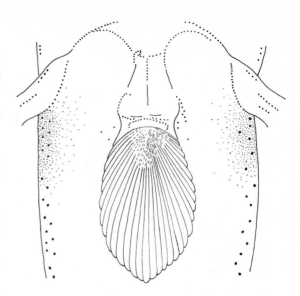

Fig. 2-88. Ventral view of the belly of a goby (*Gillichthys mirabilis*, marine) to show the union of the pelvic fins as a sucking disc. (After a drawing by Kenneth Stewart.)

Suborder Percoidei.   Spiny-rayed fishes with P_2 thoracic or jugular, with 1 spine, and not used as a sucking disc. Pelvic bones attached directly to cleithra.

KEY TO FAMILIES OF PERCOIDEI

1. Anal spines 3 or more   2
   Anal spines 1 or 2   8
2. Pseudobranchiae (Fig. 2-89) well developed, exposed. Anal spines 3   3
   Pseudobranchiae, if present, small and concealed by a membrane. Anal spines 3 or more   6
3. Lower pharyngeals joined. Anal fin long, with more than 15 soft rays. Viviparous   **Embiotocidae**
   Lower pharyngeals separate. Anal fin short, with fewer than 15 soft rays. Oviparous   4
4. L.l. extending to end of caudal fin. Dorsal and anal fins with a basal scaly sheath   **Centropomidae**
   L.l. not extending to end of caudal fin. Dorsal and anal fins without a basal scaly sheath   5
5. Maxilla not sheathed by preorbital when mouth is closed. Pelvic axillary scale small or absent. Jaws without incisorlike teeth   **Serranidae**

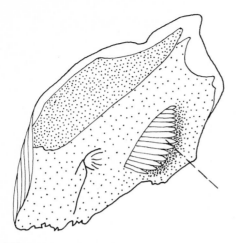

Fig. 2-89. Inner surface of the opercle of *Roccus chrysops* showing the well-developed pseudobranchiae.

Maxilla sheathed for most of its length when mouth is closed. Pelvic axillary scale well developed. jaws with incisorlike teeth   **Sparidae**
6. L.l. absent on sides; infraorbital bones represented only by the lacrimal; no mandibular canal. No occular sulcus   **Elassomatidae**
   L.l. present, sometimes interrupted or incomplete; infraorbital bones, in addition to lacrimal, present; mandibular canal present. Ocular sulcus present   **7**
7. 2 pairs of nasal openings. Lower pharyngeals not fused. L.l. pores in a single series. Vomer with teeth   **Centrarchidae**
   1 pair of nasal openings. Lower pharyngeals fused. L.l. in 2 series, the anterior portion higher than the posterior. Vomer toothless   **Cichlidae**
8. L.l. not extending far onto caudal fin. 2d anal spine (when present) not long and stout. Head bones not conspicuously cavernous. Lower pharyngeals separate, slender, and with sharp teeth   **Percidae**
   L.l. extending well onto caudal fin. 2d anal spine very long and stout. Head bones conspicuously cavernous. Lower pharyngeals broad, heavy, fused, and with blunt molar teeth   **Sciaenidae**

Family Centropomidae. Snooks. Body elongate and covered with ctenoid scales. L.l. complete and extending onto C. Mouth large, lower jaw projecting. D. and A. basally enclosed in a scaly sheath; C. forked; P_2 large, I,5, and placed well behind P_1, with a scaly basal process; P_1 rather narrow, the upper rays longest. Atlantic, Indian, and Pacific Oceans; 1 species enters our fresh waters.

Genus *Centropomus* Lacépède.
Centropomus undecimalis (**Bloch**). Snook. From Florida to Texas, southward to Brazil, and in the West Indies. A fine food and game fish reaching a weight of about 50 lb and often ascending rivers.

Family Serranidae. Sea basses. Body oblong and somewhat compressed. Scales usually ctenoid. Mouth moderately large; premaxillae protractile. Conical teeth in bands on jaws, vomer, hyoid, and palatines. Gill rakers variable in length, but usually stiff and armed with teeth. Pseudobranchiae large and conspicuous. Gill membranes separate and free from isthmus. Branchiostegals normally 7. Cheeks and opercles always scaly. Preopercle serrate. Nostrils double. L.l. single. P_2 with 1 spine and 5 soft rays, in thoracic position. Intestine short. Many marine species. Widely distributed in tropical and warm seas; 4 species in our fresh waters.

Genus *Roccus* Mitchill.

KEY TO SPECIES OF *ROCCUS*

1. Teeth in 1 or 2 patches on tongue base. Dorsal fins separate. Anal spines graduated, 2d much shorter than 3d. Lower jaw projecting. Sides marked with bold stripes that are typically continuous   **2**
   Teeth absent on tongue base but present on sides of tongue. Dorsal fins joined by a membrane. Anal spines not graduated, 2d almost if not equal in length to 3d. Sides marked with pale stripes or bold discontinuous ones   **3**
2. Hyoid teeth in 2 parallel patches. 2d anal spine contained about 5 times in head   ***R. saxatilis***
   Hyoid teeth in a single patch. 2d anal spine contained about 3 times in head   ***R. chrysops***
3. Longest dorsal spine about $\frac{1}{2}$ head length. Sides marked with faint streaks. Dorsal fins well connected   ***R. americanus***
   Longest dorsal spine more than $\frac{1}{2}$ head. Sides marked with bold and interrupted lines. Dorsal fin slightly connected   ***R. mississippiensis***

Roccus saxatilis (**Walbaum**). Striped bass. On the Atlantic Coast from New Brunswick to Florida. and also recorded from Mississippi. Introduced on the west coast. An excellent anadromous food and game fish. Sometimes reaching a size of over 70 lb.
Roccus chrysops (**Rafinesque**). White bass. From Minnesota, Wisconsin, and Michigan south

to the Gulf states, Florida, Alabama, Mississippi, and Texas. A food and game species of smaller size, strictly a fresh-water inhabitant.

Roccus americanus (Gmelin). White perch. Eastern Canada, Lake Ontario, and the St. Lawrence River to South Carolina. A good pan fish, ascending rivers to spawn, sometimes landlocked in ponds and lakes.

Roccus mississippiensis (Jordan and Eigenmann). Yellow bass. Minnesota, Wisconsin, and Indiana south to Alabama, Louisiana, and eastern Texas. West of the Mississippi River as far west as eastern Oklahoma. Introduced elsewhere. Less attractive, as a pan fish, than the other serranids because of smaller size.

Family Centrarchidae. The sunfishes. Body oblong or nearly circular and compressed. Mouth terminal. Premaxillae protractile. Preopercle serrate. Maxillae not lying under preorbitals when mouth is closed. Gill membranes separate, not conjoined or united with isthmus. Pseudobranchiae small and concealed. Branchiostegals 6 or 7. Lower pharyngeals separate. L.l. present. D. single, sometimes deeply emarginate. Intestine short. Originally North America, east of Rocky Mountains, except *Archoplites* of the west coast.

KEY TO GENERA OF CENTRARCHIDAE

1. Anal spines typically 3. Branchiostegals 5   **2**
   Anal spines rarely fewer than 5 and sometimes more. Branchiostegals 7 (6 in *Ambloplites*)   **6**
2. Scales 55–81. Body elongate and less compressed. Precaudal vertebrae 14 or 15   *Micropterus*
   Scales 26–54. Body deep and more compressed. Precaudal vertebrae usually 12   **3**
3. Opercle rounded behind (sometimes emarginate in young). Scales 31–54. Caudal fin emarginate. Lacrimal with 4 pores   **4**
   Opercle emarginate behind. Scales 26–35. Caudal fin convex. Lacrimal with 3 pores   *Enneacanthus*
4. Teeth on tongue, ectopterygoids and entopterygoids. Supramaxilla longer than width of maxilla (Fig. 2-90)   *Chaenobryttus*
   No teeth on tongue or pterygoids. Supramaxilla shorter than maxillary width   *Lepomis*
5. Gill rakers on 1st arch fewer than 15. Preopercle entire or weakly serrate   **6**
   Gill rakers on 1st arch 25 or more. Preopercle finely serrate   **7**

Fig. 2-90. Chaenobryttus gulosus. Lake Texoma, June 1965.

6. Scales ctenoid. Caudal fin emarginate   *Ambloplites*
   Scales cycloid. Caudal fin convex   *Acantharchus*
7. Dorsal base about 2 times length of anal base   *Archoplites*
   Dorsal base about equal to anal base   **8**
8. Dorsal spines 5–8   *Pomoxis*
   Dorsal spines 11–13   *Centrarchus*

Genus *Micropterus* Lacépède. The basses. Body elongate; mouth large, lower jaw prominent. D. shallowly or deeply emarginate. Largest members of sunfish family; perhaps most important fresh-water game fishes of America.

KEY TO SPECIES OF *MICROPTERUS*

1. Dorsal fin deeply emarginate, next to last spine less than 0.5 of the longest. Maxilla extends distinctly behind eye in adults (Fig. 2-91*B*). Membranes of soft dorsal and anal fins almost always without scales. Pyloric caeca typically branched near their bases. Sides of young and small adults with a dark, rather even-edged lateral band   *M. salmoides*
   Dorsal fin shallowly emarginate, next to last spine more than 0.5 of the longest. Maxilla extends to below pupil and sometimes to hinder margin of orbit (Fig. 2-91*A*). Membranes of soft dorsal and anal with small scales near their bases. Pyloric caeca almost never branched. Sides never marked with an even-edged lateral band; lateral band, if present, always broken into blotches   **2**
2. Scale rows above L.l. 12 or 13, those below 20–23; scales around caudal peduncle usually 29–31. Sides of young to half-grown with vertical bars or broad blotches with light centers   *M. dolomieui*
   Scale rows above L.l. 7–10, those below 14–19; scales around caudal peduncle 22–31. Sides with a broken lateral band or vertical bars and blotches

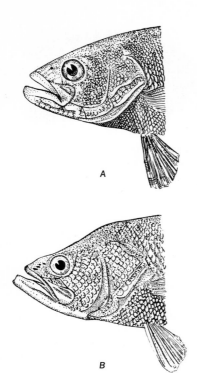

Fig. 2-91. Bass heads: *A, Micropterus dolomieui; B, Micropterus salmoides.*
(Jordan and Evermann, 1900.)

without light centers (except on caudal peduncle of *M. coosae*)   3

3. Sides with vertical bars (not light-centered) thickened near midlateral line, in young; older fish with horizontal rows of spots on sides of belly to well above mid-sides   *M. treculi*

Sides with a broken lateral band of more or less confluent blotches or with vertical bars. Ventrolateral rows of spots imperfectly to well developed, but never occurring above mid-sides   4

4. A broken lateral band of often confluent blotches present at all ages, but less distinct in very large specimens. Lateral blotches not distinctly vertically elongated. Scale rows around caudal peduncle 24 or 25. Basicaudal spot and subterminal caudal band very prominent   *M. punctulatus*

Color pattern of vertical bars or vertically elongated blotches. Scale rows around caudal peduncle usually 26–30. Basicaudal spot well developed or not; subterminal caudal band poorly developed or absent   5

5. Basicaudal spot large and prominent; ventrolateral streaks poorly developed; subterminal caudal band faintly present   *M. notius*

Basicaudal spot not prominent; ventrolateral streaks well developed; subterminal caudal band absent   *M. coosae*

Micropterus dolomieui **Lacépède.** Smallmouth bass. Emargination of D. shallow, shortest D. spine longer than $\frac{1}{2}$ the longest; D. soft rays 13–15; P_1 rays 16–18; A. soft rays 11 (10–12). L.l. scales 69–77; scales above L.l. 12–13, below L.l. 20–23; cheek scale rows 14–18. No lateral band. Originally from northern Minnesota to Lake Nipissing and Quebec south to northern Alabama and west to eastern Kansas and Oklahoma. A game species of high reputation, widely distributed by man.

Micropterus notius **Bailey and Hubbs.** Suwannee bass. Similar to *M. dolomieui*, but D. soft rays 12; P_1 soft rays 15–16; A. soft rays 10. L.l. scales 59–63; scales above L.l. usually 8 (8–9), below L.l. 16–19; cheek scale rows 10–15. Lateral band imperfect. From type locality only, Ichtucknee Springs, a tributary of Santa Fe River, northern Florida.

Micropterus punctulatus (**Rafinesque**). Spotted bass. Emargination of D. shallow, shortest D. spine 0.6 of longest; D. soft rays 12 (11–14); P_1 rays 15 (often 16); A. soft rays 10 (9–11). L.l. scales 64 (55–71); scales above L.l. 8 (7–9), below L.l. 15–16 (14–18); cheek scale rows 14–15 (11–18). Sides with confluent dark blotches. From the Ohio River system of Illinois, Indiana, and Ohio south to the Gulf states and northward through Texas to Oklahoma and Kansas.

Micropterus treculi (**Vaillant and Bocourt**). Guadalupe bass. Similar to *M. punctulatus*, but sides with vertical bars (without light centers) changing to longitudinal lines in adults. Basicaudal spot prominent in young to half-grown, poorly developed in adult. From the Colorado, San Marcos, and Guadalupe Rivers and western tributaries of the Brazos River in Texas.

Micropterus coosae **Hubbs and Bailey.** Redeye bass. Similar to *M. treculi*, but vertical bars on sides with light centers. L.l. scales 67–72 (63–77). In southeastern streams, from the Alabama to the Savannah Rivers in Alabama and Georgia. Recently taken in Florida.

Micropterus salmoides (**Lacépède**). Large-mouth bass. Emargination of D. deep, shortest D. spine less than ½ longest. Lateral band, in young, sharply defined on C. peduncle. Originally from southeastern Canada throughout the Great Lakes region southward through the Mississippi Valley to Mexico and Florida. On the Atlantic Coast as far north as Virginia. One of the most important game fishes, widely introduced.

Genus *Chaenobryttus* Gill. Warmouth sunfish. Similar to *Lepomis,* but with teeth on tongue, ectopterygoids, and entopterygoids. Supramaxilla longer than width of maxilla. 1 species.

Chaenobryttus gulosus (**Cuvier**). Warmouth (Fig. 2-90). From the Mississippi River system, in Kansas, Iowa, and southern Wisconsin to southern Michigan, Lake Erie, western Pennsylvania, and south through the Mississippi Valley. On the Atlantic Coast from New York southward to Florida and the Rio Grande.

Genus *Lepomis* Rafinesque. Sunfishes. Largest sunfish genus containing those with a deep compressed body, 3 anal spines, no hyoid or pterygoid teeth, and opercles not emarginate (except sometimes in young). Some grow large enough to be good pan fishes, others too small.

KEY TO SPECIES OF *LEPOMIS*

1. Opercle stiff to its bony margin, not fimbriate on its posterior edge (Fig. 2-92A)   **2**
 Opercle extended backward as a thin, flexible flap, usually fimbriate posteriad (Fig. 2-92B)   **6**
2. Supramaxilla (Fig. 2-93) well developed, its length about ⅔ the greatest maxillary width. Gill rakers

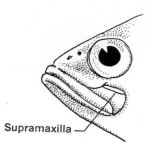

Supramaxilla —

Fig. 2-93. Head of *Chaenobryttus gulosus* illustrating the well-developed supramaxilla. (From Hubbs and Lagler, 1947.)

long (Fig. 2-94A, B), when depressed reaching base of 2d or 3d raker below (longer in young)   **3**
Supramaxillary length about ⅓ the maxillary width. Gill rakers short (Fig. 2-94C, D), when depressed

Fig. 2-94. Sunfish heads with opercles cut away to show lengths of gill rakers: A, *Lepomis macrochirus;* B, *Lepomis cyanellus;* C, *Lepomis gibbosus;* D, *Lepomis megalotis.* (From Hubbs and Lagler, 1947.)

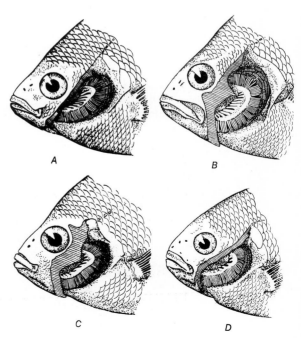

A

B

C

D

Fig. 2-92. Opercular bones in *Lepomis:* **A, stiff opercle of *L. cyanellus;* B, flexible and fimbriate opercle of *L. megalotis.* Stippled portion represents the opercle.**

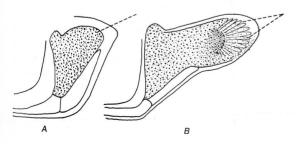

A

B

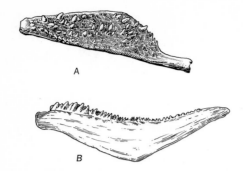

Fig. 2-95. Lower left pharyngeal bone of *Lepomis macrochirus: A*, upper aspect; *B*, lateral aspect. (Forbes and Richardson, 1920.)

usually not reaching base of 2d raker below (except in young) **4**
3. L.l. complete. L.l. scales 43–53; those above L.l. 8–10, and below 16–19 *L. cyanellus*
   L.l. incomplete. L.l. scales 31–40; those above L.l. 5–7, and below 12–14 *L. symmetricus*
4. Lower pharyngeals elongate, outer margins straight as in Fig. 2-95A, teeth rather sharp. Pectoral fins short and rounded. Palatine teeth variable, well developed to absent *L. punctatus*
   Lower pharyngeals broad and heavy, outer margins sigmoid, teeth blunt and paved (Fig. 2-96). Pectorals long and pointed. Palatine teeth typically absent **5**

Fig. 2-96. Lower left pharyngeal bone of *Lepomis gibbosus: A*, upper aspect; *B*, lateral aspect. (Forbes and Richardson, 1920.)

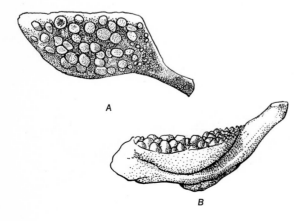

5. Opercle with a small red semicircular spot near its margin. Pectoral fins about 3.0–3.3 in adults *L. gibbosus*
   Opercle broadly margined with scarlet. Pectoral fins about 2.7–3.0 in adults *L. microlophus*
6. Gill rakers, when depressed, not reaching past 1st raker below (except in young). Pectoral fin short and rounded **7**
   Gill rakers, when depressed, reaching base of 2d raker below (3d in young). Pectoral fins moderate to long, pointed **9**
7. Palatine teeth present. Opercle long and narrow, its margin deeply fimbriate in adult. Opercular membrane dark to its margin. Gill rakers short, but not mere knobs *L. auritus*
   Palatine teeth absent. Opercle shorter and broader, its margin shallowly fringed in adult. Opercular membrane not dark to its margin, but bordered by greenish, red, or white. Gill rakers reduced to mere knobs (short even in young) (Fig. 2-94D) **8**
8. Pectoral rays usually 12. Cheek scales in 4 rows. Opercular membranes with a greenish margin *L. marginatus*
   Pectoral rays usually 13–15. Usually 5 or more rows of cheek scales. Opercular membranes with a red or white border *L. megalotis*
9. Palatine teeth present. Cephalic L.l. system well developed, interorbital canals wider than space between them. Opercle extending little into its membranous flap, its margin not fringed. Anal fin III,7–9 *L. humilis*
   Palatine teeth absent. Cephalic L.l. system with narrow tubes, much narrower than space between them. Opercle fimbriate and extending almost to its membranous margin. Anal fin III,10–12 *L. macrochirus*

Lepomis cyanellus **Rafinesque.** Green sunfish. Body not so deep as in other *Lepomis,* but thicker. Mouth large. Opercle stiff to its margin, not fimbriate. Gill rakers long. L.l. complete, its scales 43–53; scales 8–10 above and 16–19 below L.l. Widely distributed from Colorado through North Dakota and the Great Lakes region to southern Ontario and New York. Southward through the Mississippi Valley to Georgia, Alabama, and on to northeastern Mexico and New Mexico. Introduced in many places; recently taken in Florida. A good pan fish.

Lepomis symmetricus **Forbes.** Bantam sunfish. Similar, in some respects, to *L. cyanellus* but with L.l. incomplete; scales 31–40 in L.l. and 5–7 above, 12–14 below L.l. In the Mississippi Valley from Illinois and Tennessee through Arkansas and southeastern Oklahoma to Mississippi, Louisiana, and eastern Texas.

Lepomis punctatus (Valenciennes). Spotted sunfish. Gill rakers short, when depressed, not reaching 2d raker below (except in young). Lower pharyngeals elongate, their outer margins straight. Body sometimes with red spots. From southern Indiana southward in the Mississippi Basin to the Gulf States. On the east coast as far north as North Carolina and in the West to central Texas, southeastern Oklahoma, Arkansas, and Missouri.

Lepomis gibbosus (Linnaeus). Pumpkinseed. Gill rakers as in *L. punctatus.* Lower pharyngeals broad and heavy, their teeth blunt, paved. P_1 long, pointed, 3.0–3.6 in standard length. Palatine teeth absent. Opercle with a red spot. Southern Canada south on the Atlantic Coast to South Carolina. In the Mississippi Valley as far south as Missouri and as far west as Colorado and Wyoming. A good pan fish.

Lepomis microlophus (Günther). Redear sunfish. Similar to *L. gibbosus,* but opercle broadly margined with scarlet; P_1 longer, contained 2.6–3.0 in standard length. From the Mississippi River Basin in Indiana and Missouri south to Alabama, Florida, Louisiana, and Texas. Introduced in Oklahoma and California, probably in many other states. A good food fish.

Lepomis auritus (Linnaeus). Redbreast sunfish. Opercular flap long and narrow, black to its margin. Palatine teeth present. Belly orange-red. Posterior D. rays without a black blotch. On the east coast from New Brunswick to Florida and thence to Texas, where introduced. North in the Mississippi Valley to southern Oklahoma where recently introduced. An excellent pan fish.

Lepomis marginatus (Holbrook). Dollar sunfish. Aspect of *L. megalotis,* but P_1 usually 12. Cheek scales in 4 rows. Opercular flap with greenish margin. However, the coloration is variable. Specimens from Beaumont, Texas, have a light-bordered earflap, silver spots on the black portion of flap, and the belly bright red. Lower Mississippi Valley from Tennessee, Arkansas, and southeastern Oklahoma to Texas and Florida and north on the east coast to South Carolina. Much too small to be of food value.

Lepomis megalotis (Rafinesque). Longear sunfish. Body deep and compressed. Mouth very small. Palatine teeth absent. P_1 short and rounded, its rays 13–15. Cheek scales in 5–7 rows. Opercular flap with red or white border, the flap variable in length, sometimes broader posteriad. Nape often gibbous. From Minnesota east to Ontario, Ohio, and western Pennsylvania. Southward through the Mississippi Basin to the Gulf states and Mexico; as far north on the east coast as South Carolina.

Lepomis humilis (Girard). Orangespotted sunfish. Body rather deep and compressed. Mouth quite large; palatine teeth present. Gill rakers long. P_1 moderately long, pointed. L.l. canals in head unique, very large. Opercular flap long, white-edged. Males with orange spots on sides. North Dakota east to western Ohio and south through western Kentucky, Tennessee, Alabama, and Louisiana. Throughout the Great Plains region from the Dakotas to Texas. Recently taken in Florida. Too small to be useful for food.

Lepomis macrochirus Rafinesque. Bluegill. Similar in some respects to *L. humilis,* but differing most conspicuously as follows: palatine teeth absent; L.l. canals in head narrow; A. soft rays 10–12 instead of 7–9. Widespread in the Great Lakes region to the St. Lawrence drainage, throughout the Mississippi Valley, and along the Gulf Coast from Mexico to Florida, and thence north on the Atlantic Coast to Virginia. Introduced in Connecticut and elsewhere. A favorite fish for fly casting.

Genus *Enneacanthus* Gill. Sunfishes. Small and short, maximum length about 100 mm. Body

Fig. 2-97. Emarginate opercle of *Enneacanthus obesus.* Stippled area represents the opercle.

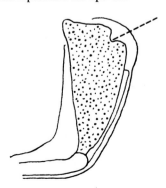

deep and compressed. Mouth small, supramaxilla small or well developed; vomerine teeth present, palatine teeth present or absent. Opercle emarginate (Fig. 2-97). Gill rakers short. D. spines 9 or 10, A. spines 3. Very small, pretty fishes.

KEY TO SPECIES OF *ENNEACANTHUS*

1. Palatine teeth present. Dorsal fin not emarginate   2
 Palatine teeth absent. Dorsal fin emarginate
                            E. chaetodon
2. Body marked with broad vertical dark bands. Opercular spot larger than pupil. Caudal peduncle with 19–22 rows of scales   *E. obesus*
 Body marked with longitudinal light stripes. Opercular spot about ⅔ diameter of pupil. Caudal peduncle with usually 16–18 (15–19) scales
                            E. gloriosus

Enneacanthus obesus (**Girard**). **Banded sunfish.** Southeastern New Hampshire to Florida, in coastwise waters.
 Enneacanthus gloriosus (**Holbrook**). **Blue-spotted sunfish.** Southeastern New York to Florida. In clear sluggish streams.
 Enneacanthus chaetodon (**Baird**). **Black-banded sunfish.** In coastwise streams from New Jersey to Florida.
 Genus *Ambloplites* **Rafinesque.** The rock-basses. Body oblong, somewhat elevated and compressed. Mouth large; maxilla broad, with a well-developed supramaxilla; lower jaw projecting; teeth on vomer, palatines, tongue, ectopterygoids, and entopterygoids. Pharyngeal teeth sharp. Branchiostegals 6. Opercle emarginate; preopercle serrate. L.l. complete. D. with 10 or 11 and A. usually with 6 spines. C. emarginate. Game fishes.

KEY TO SPECIES OF *AMBLOPLITES*

1. Cheeks incompletely scaled (occasionally naked). Rows of scales above L.l. 10–12   *A. cavifrons*
 Cheeks completely scaled. Scale rows above L.l. 7–9
                            A. rupestris

Ambloplites cavifrons **Cope.** Roanoke bass. From the Roanoke River and possibly the James River in Virginia.
 Ambloplites rupestris (**Rafinesque**). Rockbass.

From Manitoba and North Dakota through southern Ontario and the Great Lakes to Quebec; St. Lawrence River system, and Lake Champlain southward to the Tennessee River system in Alabama and Georgia, the Catawba River, Choctawhatchee River in Florida, and the Escambia River in Alabama. Southward west of the Mississippi River through Arkansas and eastern Oklahoma to Mississippi and Louisiana. It is believed that the rockbass reached the Atlantic drainage through the canals, since it is now common there.
 Genus *Pomoxis* **Rafinesque.** The crappies. Body deep and much compressed. Mouth large and oblique, lower jaw prominent; maxilla broad, with a well-developed supramaxilla; teeth on vomer, palatines, tongue, ectopterygoids, and entopterygoids. Lower pharyngeals narrow, with sharp teeth. Gill rakers long, slender, and numerous. Opercle emarginate; preopercle finely serrate. A. larger than large D. C. emarginate; preopercle finely serrate. Branchiostegals 7. L.l. complete. Game fishes of considerable importance.

KEY TO SPECIES OF *POMOXIS*

1. Dorsal spines 7 or 8. Dorsal base equal to or greater than distance from dorsal origin to posterior orbital rim   *P. nigromaculatus*
 Dorsal spines usually 6. Dorsal base less than distance from dorsal origin to posterior orbital rim
                            P. annularis

Pomoxis nigromaculatus (**LeSueur**). **Black crappie.** Color pattern of irregular dark blotches. Southern Canada from Manitoba and Lake of the Woods to Quebec, through the Great Lakes and Lake Champlain. In the Mississippi Valley from eastern North Dakota and Nebraska to western New York and Pennsylvania and south to eastern Texas and Florida. Northward on the Atlantic Coast to Virginia. Introduced elsewhere.
 Pomoxis annularis **Rafinesque.** **White crappie.** Color pattern of dark blotches arranged in vertical bars. From Nebraska and the Mississippi River system in Minnesota eastward to Lake Erie tributaries in Ontario and the western portion of the Lake Ontario drainage. South in the Mississippi Valley to the Gulf states and north on the Atlantic Coast to North Carolina. Widely introduced elsewhere.

Genus *Acantharchus* Gill. Monotypic. Shape oblong, not much compressed. Scales cycloid, an unusual feature in the family. Mouth rather small; maxilla broad, with supramaxilla well developed; lower jaw projecting; teeth on vomer, palatines, pterygoids, and tongue. Pharyngeal teeth sharp. Gill rakers few and long. Opercle emarginate; preopercle entire. L.l. complete. C. rounded.

Acantharchus pomotis (**Baird**). Mud sunfish. New York to Florida along the coastal plain. Small (150 mm), not valued for food.

Genus *Centrarchus* Cuvier. Monotypic. Body very deep, short, and compressed. Mouth moderate, lower jaw projecting; teeth on vomer, tongue, palatines, and pterygoids. Opercle emarginate. Long gill rakers with fine teeth. D. and A. very high, with about 12 and 8 spines, respectively. Scales large, weakly ctenoid. L.l. complete. A pretty fish, too small to be used as food.

Centrarchus macropterus (**Lacépède**). Flier. Virginia to eastern Texas and north to Illinois and Indiana.

Genus *Archoplites* Gill. Monotypic. Body oblong, compressed. Mouth large, oblique; maxilla broad, supramaxilla well developed; vomer, palatines, tongue, and pterygoids toothed, the tongue teeth in 2 patches. Gill rakers long and numerous. Branchiostegals 7. Opercle emarginate, other members of opercular series as well as preorbitals and suborbitals serrate. L.l. canals in the dentaries and preopercle large. C. emarginate. Scales strongly ctenoid.

Archoplites interruptus (**Girard**). Sacramento perch. The only native centrarchid west of the Rocky Mountains. Utah, Nevada, and California; introduced, 1961, in Nebraska; originally restricted to the Sacramento River system, San Francisco Bay tributaries, and the Russian River, Pajaro-Salinas drainage, and Clear Lake. An excellent food fish, reported to be decreasing in numbers because of competition imposed by introduced species.

Family Elassomatidae. The pygmy sunfishes. These tiny fishes have been considered as members of the family Centrarchidae, but actually have so little in common with the sunfishes that it seems desirable to return to the status recognized by Jordan. In structure and behavior the pygmy sunfishes bear almost no resemblance to any centrarchid, although Jordan regarded them as distant relatives of the crappies. The single genus, *Elassoma*, may be recognized by the following characters not shared by the centrarchids: no suborbitals, except the lacrimal; no medial extrascapular; L.l. segment of posttemporal separate from pectoral girdle; no mandibular or angular canal; all L.l. bones as open grooves; reduced number of hypurals; no lateralis canals; superficial neuromasts numerous and structurally different from those of sunfishes; eyes with many structural features in contrast to those of centrarchids, the absence of the ocular sulcus and presence of a spectacle, large and less-numerous rods and, therefore, a narrow outer nuclear layer in the retina and more numerous ganglion cells; crepuscular, more sedentary habits; and different breeding habits. These characters and others make it difficult to consider the pygmy sunfishes as even a subfamily of the Centrarchidae.

Genus *Elassoma* Jordan. Pygmy sunfishes. D. III–V,8–13; A. III,4–8; C. 10–13, rounded or truncate; P. 13–17, short and round; P_2 I,5, rather long and pointed. Scales without ctenii, 27–40 in lateral series and present on cheeks, opercles, and top of head, except in *E. okefenokee*. Jaws with strong, conical teeth; vomer with few, weak teeth; lower pharyngeals narrow, with sharp teeth. Gill membranes broadly united, free from isthmus. Mouth small and terminal, lower jaw protecting. Inhabitants of lowland springs, sluggish streams, and swamps; 3 species known, but probably more await description.

KEY TO SPECIES OF *ELASSOMA*

1. Scales in lateral series 38–45. 1 or 2 dark blotches below dorsal origin *E. zonatum*
   Scales in lateral series fewer than 38. Color pattern of vertical bars or light-centered scales, giving a streaked appearance **2**
2. Dorsal rays 9 (8–10); anal rays 5 (4 or 5); sum of soft dorsal, soft anal, and pectoral (both sides) rays 40–44. Snout in head 4.2–4.4 *E. evergladei.*
   Dorsal rays 11 (10–13); anal rays 7 (6–8); sum of soft dorsal, soft anal, and pectoral (both sides) rays 48–53. Snout in head 3.4–3.9 *E. okefenokee*

Elassoma zonatum **Jordan.** Banded pygmy sunfish. Sides often marked with dark vertical bars and 1 or 2 dark spots on sides below **D.** origin. In the lowlands from Illinois to eastern Texas, east through the Gulf states to Florida, and north to North Carolina.

Elassoma evergladei **Jordan.** Everglades pygmy sunfish. Sides without conspicuous dark bars or spots, but with faint longitudinal pale streaks. From Florida to South Carolina.

Elassoma okefenokee **Böhlke.** Okefenokee pygmy sunfish. Sides in females, conspicuously barred with brown; D. of male dark throughout. Southern Georgia to central Florida.

Family Percidae. The perches, pikeperches, and darters. Percoid fishes similar to the Serranidae,

having the spinous D. well developed; P_2 thoracic, with 1 spine and 5 branched rays. 2 nostrils. 5 to 8 branchiostegal rays. 4 gills, a slit behind the 4th. Gill membranes free from isthmus. Villiform or cardiform teeth on upper and lower jaws. Subocular shelf absent as in Centrarchidae and Cichlidae. 1 or 2 A. spines. 1 or no auxiliary interneural bone. Vertebrae 32–50. According to Collette (1963), 2 subfamilies are recognized: Luciopercinae (*Stizostedion*) and Percinae (*Perca, Percina, Ammocrypta,* and *Etheostoma*). Fresh-water fishes of the Northern Hemisphere, except North America west of the Rocky Mountains (*Etheostoma pottsi,* west of divide in Mexico).

KEY TO GENERA OF PERCIDAE

1. Preopercle strongly serrate. Mouth large, the maxilla extending at least to below middle of eye. Posterior upper border of maxilla not concealed under preorbital. Branchiostegals 7 (rarely 8). Genital papilla absent   **2**

    Preopercle entire or weakly serrate. Mouth small, maxilla not usually extending to below middle of eye, never farther. Maxilla, posteriorly, slipping under preorbital. Branchiostegals usually 6 (sometimes 5 or 7). Genital papilla present   **3**

2. Canine teeth absent (Fig. 2-99*B*). L.l. extending very little, if at all, past hypural onto caudal fin.

Fig. 2-98. Anterior views of percid heads; *A,* protractile premaxillae in *Percina copelandi,* note absence of frenum; *B,* nonprotractile premaxillae in *Percina phoxocephala,* note the frenum. (After drawings by Kenneth Stewart.)

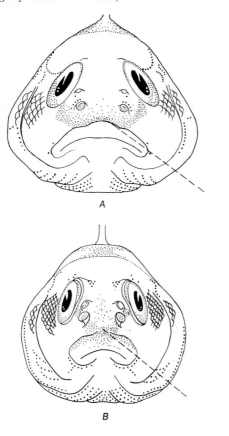

Fig. 2-99. Lower jaw bones of *A, Stizostedion vitreum* with canine teeth; *B, Perca flavescens* without canine teeth. (From Hubbs and Lagler, 1947.)

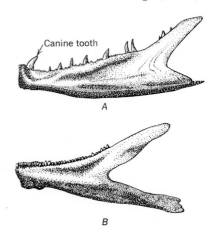

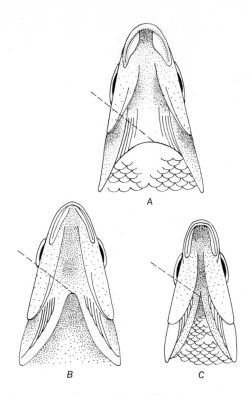

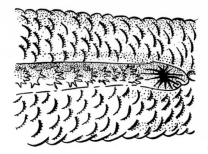

Fig. 2-101. Hinder portion of the belly of *Percina caprodes* with a row of enlarged scales. (From Hubbs and Lagler, 1947.)

Fig. 2-100. Ventral views of percid heads: A, *Etheostoma zonale*, gill membranes broadly connected; B, *Etheostoma caeruleum*, gill membranes slightly connected; C, *Etheostoma chlorosomum*, gill membranes not connected.

4. Body not extremely long and slender, its depth contained fewer than 7 times in standard length. Flesh, in life, opaque. Body well scaled, except on nape, breast, cheeks, and belly of some species. 1 or 2 anal spines   *Etheostoma*
   Body extremely long and slender (Fig. 2-102), its depth contained 7–9 times in standard length. Flesh, in life, hyaline. Scales usually absent on belly, sometimes on back and sides. A single anal spine   5
5. Premaxillae not protractile. Body scutulation complete (except on belly). Dorsal spines about 13–15   *Crystallaria*
   Premaxillae protractile. Body scutulation incomplete (at times complete). Dorsal spines 7–11   *Ammocrypta*

Genus *Stizostedion* Rafinesque. Pikeperches. Body elongate; head somewhat conical; mouth quite large, teeth in villiform bands containing caninelike teeth on jaws and palatines. Gill rakers slender and strong. Gill membranes (Fig. 2-100) separate. Preopercle serrate. D. fins well separated. A. with 2 slender spines. P_2 fins separated by a space about equal to length of base of 1 pelvic fin.

Pelvic fins inserted close together. Body with bold, dark vertical bars   *Perca*
   Canine teeth present (Fig. 2-99A). L.l. extending well onto caudal fin where there are also 1 above and below the main line. Pelvic fins far apart, space between them equaling either base. Body with bars or saddles often indistinct   *Stizostedion*
3. Midline of belly with a row of enlarged and spiny scales (Fig. 2-101) that are sometimes shed, leaving a naked strip (the modified scales are sometimes inconspicuous, and, if absent, anterior belly is naked and there is a rather narrow band of scales just in front of anus). Premaxillary frenum (Fig. 2-98) usually present. L.l. complete   *Percina*
   Midline of belly covered with ordinary scales or naked only anteriad. Premaxillary frenum present or absent. L.l. often incomplete   4

Fig. 2-102. *Ammocrypta pellucida.* (Forbes and Richardson, 1920.)

Scales ctenoid, small. L.l. complete, extending onto C., with supplementary line above and below the main one on C. Branchiostegals 7. Pseudobranchiae well developed. 2 species in the United States and 3 in Europe.

KEY TO SPECIES OF *STIZOSTEDION*

1. Dorsal fins with rows of round, black spots (not evident in young). No prominent black blotch on posterior end of spinous dorsal. Cheeks usually closely scaled. Soft dorsal rays 17–20

    S. canadense

    Dorsal fins with indistinct dusky mottlings. Posterior end of spinous dorsal with a large black blotch. Cheeks usually sparsely scaled. Rays of soft dorsal 19–22   *S. vitreum*

Stizostedion canadense (**Smith**). Sauger. Easily separated from S. *vitreum* by the absence of the large black blotch on posterior portion of spinous D. From the Red and Assiniboine Rivers, Hudson Bay region, and New Brunswick southward through the Great Lakes region to the Tennessee River in Alabama. Found also in Montana, Arkansas, eastern Kansas, Oklahoma, Louisiana, and Mississippi. As far north as Wyoming in Mississippi drainage. Recently taken in North Dakota. An important food and game fish.

Stizostedion vitreum (**Mitchill**). Walleye (Fig. 2-103) and blue pike. Posterior portion of spinous D. with a conspicuous black spot. In Canada from Great Slave Lake, the Saskatchewan River, Hudson Bay region to Labrador, and south, west of the Appalachians, throughout the Great Lakes region, to the Gulf Coast in Alabama and west of the Mississippi River to northern Arkansas, Kansas,

Nebraska, and North Dakota. Introduced in Montana and Connecticut. An important food and game fish, widely introduced in areas it did not originally occupy.

Genus *Perca* Linnaeus. The perches. Readily distinguished from *Stizostedion* by the deeper, more compressed body, marked on sides by bold, dark vertical bars. Canine teeth absent. A food fish of considerable importance. 1 species in the United States.

Perca flavescens (**Mitchill**). Yellow perch. From Lesser Slave Lake, Hudson Bay region, and southeastern Canada southward through the Great Lakes region to South Dakota, northern Iowa, Illinois, Indiana, Ohio, and Pennsylvania; in the eastern states south to Florida. Although successfully introduced as far south as southern Oklahoma, the fish there are stunted. Also introduced in some western states.

Genus *Percina* Haldeman. The logperches and blackside darters. Body elongate, slightly compressed. Scales small and ctenoid, those on midline of belly enlarged or absent, leaving a naked strip, sometimes crossed by a preanal scaly bridge. L.l. complete or nearly so. The head somewhat depressed and otherwise conical. Mouth variable, being large and terminal to ventral (overhung by fleshy snout in most *P. rex* and *P. caprodes*). Teeth on vomer and palatines. Gill membranes separate to broadly connected. Premaxillae usually not protractile. D. fins well separated, spinous fin larger and with 10–17 spines. A. spines typically 2. P_2 separated by a space about equal to length of a P_2 base. Size varying from about 63–124 mm. Included here are the nominal genera *Hadropterus*, *Hypohomus* (in part), *Imostoma*, and *Cottogaster*. 23 species.

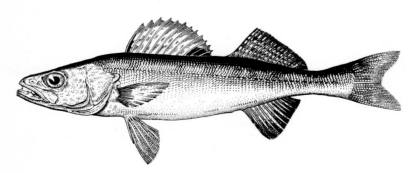

Fig. 2-103. Stizostedion vitreum. (Jordan and Evermann, 1900.)

KEY TO SPECIES OF *PERCINA*

1. Snout usually conical and fleshy, protruding over a ventral mouth. Interorbital broader and flatter **2**

   Snout not conical and fleshy, mouth terminal. Interorbital area narrower and less flat **3**

2. Depth of body contained fewer than 5 times in standard length **P. rex**

   Depth of body contained 5 or more times in standard length **P. caprodes**

3. Anal fin of adult males extremely long, posterior rays extending, in large specimens, beyond posterior end of depressed soft dorsal **4**

   Anal fin shorter, its posterior rays extending very slightly, if at all, behind posterior end of depressed soft dorsal **5**

4. Spinous dorsal with a conspicuous black spot on its posterior rays and a smaller one on anterior end (Fig. 2-104). Sides marked with indistinct blotches **P. shumardi**

Fig. 2-104. Percina shumardi.
(Forbes and Richardson, 1920.)

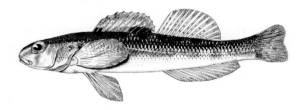

Spinous dorsal without conspicuous blotches, or with basal blotches on all interradial membranes. Sides with about 9 conspicuous blotches and the back with 4 dark saddles **P. uranidea**

5. Premaxillary frenum absent (sometimes a narrow and rarely a broad one) (Fig. 2-98A). Back with tessellated markings. **P. copelandi**

   Premaxillary frenum always present (Fig. 2-98B). Color pattern not of tessellated markings **6**

6. Caudal base with a vertical row of 3 spots (lower 2 sometimes coalesced). Preopercle usually serrate. Gill membranes moderately conjoined (Fig. 2-100B) **7**

   Caudal base with a single black spot or no spot at all. Preopercle entire. Gill membranes most often separate **10**

7. Dark blotches on base of pectoral fin and near anterior end of soft dorsal base. L.l. scales 77–90 **P. lenticula**

Pectoral fin and soft dorsal not as above. L.l. scales fewer than 76 **8**

8. L.l. scales usually 46–64 **P. nigrofasciata**

   L.l. scales usually 64–72 (59–74) **9**

9. Dark dorsal saddles present, the predorsal one separated from dorsolateral stripes by light area. **P. sciera**

   Dark dorsal saddles absent, with little or no middorsal pigment **P. aurolineata**

10. Midventral scales on the belly enlarged between pelvic fins only, not modified posteriad **11**

    Midventral scales modified to a greater or lesser degree, sometimes small, indistinct, or absent **12**

11. L.l. scales small and numerous, about 90–100. Scales above L.l. about 14 **P. aurantiaca**

    L.l. scales larger, 64–70. Scales above L.l. about 7 **P. cymatotaenia**

12. Color pattern of 7–10 vertical bars which cross the back to join those of opposite side or tend to coalesce to form a lateral band **13**

    Color pattern of round, horizontally oblong, or vertically elongated lateral blotches, with dorsal spots or blotches alternating, more or less, with lateral ones **15**

13. Scales above L.l. fewer than 7, in L.l. fewer than 50, and below L.l. usually fewer than 10. Sides crossed by 10–11 bars which tend to coalesce in forming a lateral band **P. crassa**

    Scales above L.l. more than 6, in L.l. more than 50, and below L.l. more than 9 (10–14). Sides, in half-grown to adults, crossed by 7–10 distinct well-separated bars which are continuous over the back **14**

14. Subocular bar distinct and inclined downward and forward **P. evides**

    Subocular bar absent **P. palmaris**

15. L.l. scales usually fewer than 66. Species east of the Appalachian Mountains **16**

    L.l. scales usually more than 65 (sometimes as low as 62 in *P. maculata*). Species west of the Appalachians **17**

16. A strong black median stripe on head behind chin **P. peltata**

    No black median stripe on head behind chin **P. notogramma**

17. Color pattern of round, squarish, or oblong lateral blotches (Fig. 2-105). Breast naked, except usually a few enlarged scales between pelvics. Snout not notably long and pointed. Spinous dorsal without submarginal orange band **18**

    Color pattern of vertically elongated lateral blotches, at least anteriorly (except *P. squamata* which has a scaly breast). Snout long and pointed (rather blunt in specimens from Chikaskia River in Oklahoma and Kansas). Spinous dorsal with a submarginal orange band **20**

18. L.l. scales 62–89. Cheeks and opercles scaly. No distinct black spot at soft dorsal origin **19**
    L.l. scales 88–90. Cheeks and opercles largely naked. A distinct black spot at soft dorsal origin
    P. macrocephala

19. L.l. scales 62–77; scales above L.l. 7–10, those below 9–15; scales around caudal peduncle fewer than 32. Color pattern of oblong lateral blotches
    P. maculata

Fig. 2-105. *Percina maculata.*
(Jordan and Evermann, 1900.)

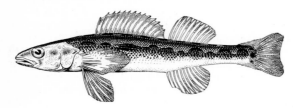

L.l. scales 85–89; scales above L.l. 9–11, those below 16–18; scales around caudal peduncle 32. Color pattern of round dark spots in a lateral row, with smaller spots above *P. pantherina*

20. Breast with small scales in addition to enlarged ones between pelvics. Anal rays II,9 or 10
    P. squamata
    Breast naked except for enlarged scales between pelvics and at anterior end of pelvic girdle. Anal rays II,8 or 9 **21**

21. Width of snout less than ⅔ its length. Snout length 3.0–3.6 in head length *P. nasuta*
    Width of snout greater than ⅔ its length. Snout length 3.3–4.0 in head length **22**

22. Dorsal spines usually 11 or 12. L.l. scales 59–79. Scales above L.l. about 12. Eye length 2.5–3.2 in distance from union of gill membranes to tip of mandible *P. phoxocephala*
    Dorsal spines usually 13 or 14. L.l. scales 73–80. Scales above L.l. 9 or 10. Eye length 3.3–4.0 in distance from union of gill membranes to tip of mandible *P. oxyrhyncha*

Percina aurantiaca (**Cope**). Yellow darter. Premaxillary frenum present. L.l. scales about 100; cheeks, opercles, and nape scaly. Sides with confluent black blotches. North Fork of the Holston,

Clinch, Watauga, and French Broad Rivers of the upper Tennessee River system in North Carolina, Virginia, and Tennessee.

Percina cymatotaenia (**Gilbert and Meek**). Bluestripe darter. Premaxillary frenum present. L.l. scales 64–70; cheeks, opercles, breast, and nape scaly. Sides with a wavy lateral band bordered above and below with light color. Southern Missouri and northern Arkansas.

Percina macrocephala (**Cope**). Longhead darter. Premaxillary frenum present. L.l. scales 88–90; scales above L.l. about 11, below 16; cheeks and opercles naked; caducous scales large. Color pattern of 9 black squarish lateral blotches, confluent. Ohio River system from New York and Pennsylvania south and westward to Virginia, West Virginia, Ohio, Kentucky, and Tennessee.

Percina maculata (**Girard**). Blackside darter (Fig. 2-105). Premaxillary frenum present. L.l. scales 62–77; scales above L.l. 7–10, below 9–15; cheeks and opercles scaly; nape naked or scaly; breast naked; scales around C. peduncle 22–30. Color pattern of 7 large confluent blotches, longer than deep; a wavy irregular line above lateral band; subocular bar present. From southern Manitoba and Ontario southward, west of the Appalachians, to Alabama, Arkansas, eastern Oklahoma, and northeastern Texas.

Percina pantherina (**Moore and Reeves**). Leopard darter. Similar to *P. maculata*, but with strikingly different color pattern of round spots. L.l. scales 85–89; scales above L.l. 9–11, below 16–18, and around C. peduncle 32. Little River system in southeastern Oklahoma and southwestern Arkansas.

Percina notogramma (**Raney and Hubbs**). Stripeback darter. L.l. scales 50–65; scales above L.l. 6–8, below 8–12; scales around C. peduncle 18–24; cheeks partly scaled, opercles scaly, nape naked or partly scaled. Color pattern of 6–8 lateral blotches connected by a lateral stripe, longitudinal nuchal blotch or blotches present; subocular bar present; a C. spot at level of lateral stripe. Coastal streams of Maryland, Virginia, and West Virginia.

Percina peltata (**Stauffer**). Shield darter. L.l. scales 52–56; scales above L.l. 6, below 9; cheeks naked or scaly; opercles with few scales; nape and breast naked. Sides with 6 squarish blotches, back with dark saddles; a wavy line as in *P. maculata*.

Coastal streams from New York to North Carolina.

Percina sciera (Swain). Dusky darter. Similar to *P. nigrofasciata,* but L.l. scales 68–71; breast naked, except between P$_2$; caducous scales little enlarged. Gill membranes broadly connected. Lateral blotches tending to form crossbars. Mississippi River system from Ohio and Indiana south to Mississippi, Louisiana, and central Texas.

Percina aurolineata Suttkus and Ramsey. Goldline darter. A close relative of *P. sciera,* from which it differs mainly in lacking dorsal saddles; the middorsal stripe almost devoid of black pigment. L.l. scales 59–74 (usually 64–72); soft D. rays 10–13 (usually 10 or 11); C. peduncle depth less than $\frac{1}{3}$ its length. Breeding color bright amber or russet; a dorsolateral, continuous or interrupted, dark stripe. Alabama River system, Alabama and Georgia.

Percina lenticula Richards and Knapp. Freckled darter. Similar to *P. sciera* but unique, in its group, in having dark blotches on pectoral base and near the anterior base of soft D. Belly unusually fully scaled, with 2 rows of slightly enlarged, modified scales along midline in males. Nape and opercle with partially embedded scales; cheek scales almost fully embedded. Breast naked or scaly, at least 1 modified scale between P$_2$ fins. From the Cahaba and Etowah Rivers of the Alabama River system in Alabama and Georgia.

Percina nigrofasciata (Agassiz). Blackbanded darter. L.l. scales about 58; breast naked or scaled; caducous scales somewhat enlarged. Gill membranes slightly connected. Dorsum olive with black markings; sides with narrow diamond-shaped bars. Gulf states from southeastern Louisiana to Florida and north to South Carolina.

Percina phoxocephala (Nelson). Slenderhead darter. L.l. scales 59–79; scales above L.l. about 12, below 16; cheeks, opercles, and nape scaly; breast naked. Color pattern of numerous lateral blotches, subocular bar absent. D. XI or XII–12 to 14; A. II,8 or 9. Minnesota to western Pennsylvania and south to Tennessee, Arkansas, Kansas, and Oklahoma.

Percina nasuta (Bailey). Longnose darter. L.l. scales 73–83; scales above L.l. 11–12, below 15–16; scales around C. peduncle 27–29; cheeks, opercles, and nape scaly; breast naked. Color pattern of 10–14 vertically elongated spots, subocular bar absent. D. XIII–12 or 13; A. II,8 or 9. Southern

Missouri, northern Arkansas, and eastern Oklahoma.

Percina oxyrhyncha (Hubbs and Raney). Sharpnose darter. L.l. scales 73–80; scales above L.l., below soft D., 9–10, below L.l. 14–16; cheeks, opercles, and nape scaly; breast naked. Color pattern of 10–12 lateral blotches, subocular bar absent. D. XIII or XIV–13 or 14; A. II,8 or 9. Mountain streams in the Kanawha River drainage in Virginia and West Virginia.

Percina squamata (Gilbert and Swain). Olive darter. L.l. scales about 82; scales above L.l. about 10, below 15–18; cheeks, opercles, nape, and breast scaly. Color pattern of 10 broad dusky saddles and 10 dark, broadly connected lateral blotches; subocular bar absent. D. usually XIV–13; A. II,9 or 10. Tennessee River system in Tennessee and North Carolina.

Percina caprodes (Rafinesque). Logperch. Similar to *P. rex,* but slenderer, the body depth contained 5 or more times in standard length. From the Churchill River system in Saskatchewan and the Hudson Bay drainage south through the Great Lakes to Gulf states and the Rio Grande. Inadvertently planted in California.

Percina rex Jordan and Evermann. Roanoke logperch. *P. rex* and *P. caprodes* are recognized by the usually piglike snout and the narrow vertical bars on sides. *P. rex* largest of darters, about 124 mm, possibly conspecific with *P. caprodes.* Roanoke River in Virginia.

Percina palmaris (Bailey). Bronze darter. L.l. scales 59–73; cheeks and nape scaly; breast naked, except between P$_2$; opercles incompletely scaled. Color pattern similar to *P. evides,* but with 8–10 saddle bars. Etowah and Tallapoosa River systems in Georgia and Alabama.

Percina evides (Jordan and Copeland). Gilt darter. L.l. scales 52–67; cheeks, nape, and breast naked; caducous scales little enlarged. Color pattern of 7 broad barlike saddles extending across back and sides; spinous D. with a posterior black spot; subocular bar present. Maumee River in Ohio, the Ohio River system from Illinois to New York, and south to the Tennessee River system; west of the Mississippi River from Minnesota and Wisconsin to Missouri and Arkansas.

Percina crassa (Jordan and Brayton). Piedmont darter. L.l. scales about 48; cheeks, nape, and breast naked; opercles with few or no scales; caducous scales moderately enlarged. Sides with 10 or 11 vertical bars tending to form a lateral band; subocular bar present. From the Roanoke River in Virginia south to, and including, the Santee River system in South Carolina.

Percina shumardi (**Girard**). River darter (Fig. 2-104). Premaxillary frenum present or absent. Caducous scales absent, belly naked anteriad; breast naked; cheeks (rarely naked), opercles and nape scaly. Sides with diffuse blotches. A. very long in adult males, extending beyond end of soft D. From southern Manitoba, western Ontario, the Mississippi River drainage of Wisconsin, southern Lake Michigan, eastern Michigan, and Lake Erie in Ohio, southward to the Tennessee River system in Alabama, and westward to North Dakota, southeastern Kansas, and eastern Oklahoma to some Gulf coastal streams.

Percina uranidea (**Jordan and Gilbert**). Stargazing darter. Premaxillary frenum usually absent. Caducous scales weakly ctenoid; cheeks and breast mostly naked; opercles and nape scaly. Back with 4 black crossbands; sides with about 9 conspicuous blotches. From the Escambia to the Mississippi River system in Alabama, Mississippi, Louisiana, and Arkansas north to Missouri and Indiana.

Percina copelandi (**Jordan**). Channel darter. Premaxillary frenum usually absent, but present or absent in northern part of range. Cheeks naked or scaly; breast naked; opercles and nape with a few scales. Back tessellated; sides with dark lateral blotches. From the Lower Peninsula of Michigan and from Lake Erie to the upper St. Lawrence and Lake Champlain tributaries southward west of the Appalachians to the Alabama River system in Alabama and the Red River system in southern Oklahoma.

Genus *Crystallaria* **Jordan and Gilbert.** Body long, slender, hyaline (in life). Scales small, rough; cheeks and opercles scaly. L.l. complete. Premaxillae not protractile. Vomerine teeth present. Gill membranes slightly connected. Branchiostegals 6. Vertical fins high and long. A. with 1 spine. Maximum length about 100 mm. 1 species. This genus is currently relegated by some writers to the synonymy of *Ammocrypta,* but M. R. Curd (personal communication) finds no evidence of close

relationship based on retinal study; he expects to publish his findings.

Crystallaria asprella (**Jordan**). Crystal darter. From southern Minnesota and Wisconsin to Ohio and south to Tennessee, Arkansas, Oklahoma, Mississippi, and Louisiana.

Genus *Ammocrypta* **Jordan.** Sand darters. Similar in some respects to *Crystallaria,* but with protractile premaxillae, imperfect scutulation, a very different color pattern composed of blotches instead of crossbands, and smaller size (about 75 mm). 4 species.

KEY TO SPECIES OF *AMMOCRYPTA*

1. Body scaled, at least with 1 row of scales above L.l. and 2 or more rows below. Opercular spine present and flattened or with sharp-pointed backward projection   **2**
   Body, except L.l. scales, naked anteriad; scales present on caudal peduncle. Opercular spine obsolete   ***A. beani***
2. Nape scaly, at least with a few scales near occiput. Opercle with a flattened more triangular spine   **3**
   Nape wholly naked (sometimes a few scales on midline). Opercle with a sharp pinlike spine   ***A. clara***
3. Soft dorsal without a dusky bar at its base. Outer row of teeth in upper jaw scarcely enlarged (Fig. 2-102)   ***A. pellucida***
   Soft dorsal with a dusky bar at its base. Outer row of teeth in upper jaw moderately enlarged   ***A. vivax***

Ammocrypta vivax **Hay.** Scaly sand darter. See above key for characters. Eastern Texas to Alabama and north to Missouri and Oklahoma.

Ammocrypta pellucida (**Baird**). Eastern sand darter (Fig. 2-102). See above key for characters. From Quebec and the Lake Champlain drainage in Vermont south and west to southern Ontario, southeastern Michigan, West Virginia, Indiana, and Kentucky.

Ammocrypta clara **Jordan and Meek.** Western sand darter. Characterized in above key. Indiana, Wisconsin, and Minnesota southward through Iowa and Missouri to Arkansas (White River and Saline River in Warren County), Oklahoma (Red River), and eastern Texas.

Ammocrypta beani **Jordan.** Naked sand darter. Characterized above. Coastal streams from Louisiana to western Florida.

Genus *Etheostoma* Rafinesque. Darters. This largest genus of darters as here treated includes the following nominal genera: *Hypohomus* (in part), *Ioa, Poecilichthys, Catonotus, Ulocentra, Boleosoma, Doration, Hololepis, Villora, Microperca,* and *Psychromaster,* since these groups tend to grade from one to another. In a group so large it is to be expected that critical characters at the generic level are difficult to select. The L.l. is usually incomplete, although several species contain individuals with a complete or nearly complete L.l. On the other extreme there is a total absence of the L.l. Most species have a premaxillary frenum, but in some the frenum is wanting. There are no specialized scales on the midline of the belly, which is usually well scaled. The A. has 1 or 2 spines and is always smaller than the soft D. Many species are brilliantly colored. The list is so long, and the variability so great, that the task of preparing a workable key is indeed a precarious one.

KEY TO SPECIES OF *ETHEOSTOMA*

1. Preoperculomandibular canal with 9 or more pores (Fig. 2-106). L.l. complete or incomplete, the pored scales rarely fewer than 10 (if fewer, the preopercle is usually serrate); scales 36–75 (rarely fewer than 40). Vertebrae 33–43 (usually 36 or more) **2**
   Preoperculomandibular canal with 6–8 pores. L.l.

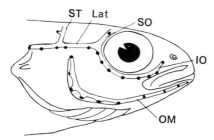

Fig. 2-106. Head of *Etheostoma edwini* indicating the terminology of the lateral-line system. *ST,* supratemporal canal; *Lat,* lateral canal; *SO,* supraorbital canal; *IO,* infraorbital canal (complete with 8 pores); *OM,* preoperculomandibular canal with 10 pores. (From Hubbs and Cannon, 1935.)

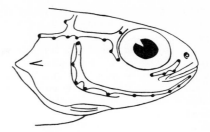

Fig. 2-107. Head of *Etheostoma zoniferum* with infraorbital canal interrupted. (From Hubbs and Cannon, 1935.)

absent or with no more than 7 pores. Preopercle entire. Scales 34–37. Vertebrae 32–36 **79**

2. L.l. (complete or incomplete) almost or quite straight, the least depth between L.l. and first dorsal base contained 4 or fewer than 4 times in projection of depth below L.l. (Regardless of this character, specimens from western Florida, with rugate female anal and genital openings, should be considered under couplet 71) **3**
   L.l. incomplete and arched upward anteriorly, the least depth between canal and first dorsal base contained 4.5 or more times in projection of depth below L.l. **71**

3. Anus encircled by many fleshy villi. Body translucent in life *E. vitreum*
   Anus not encircled by villi. Body opaque **4**

4. Anal spines (1 or 2) thin and flexible, the 1st not appreciably thicker than 2d and usually less than ½ length of 3d anal ray (spines included) **5**
   Anal spines usually 2, the 1st heavy and stiff, appreciably thicker than 2d ray and usually more than ½ length of 3d (*E. tuscumbia,* with head and breast scaly, has 1 anal spine) **12**

5. Anal spines typically 1 (*E. nigrum* usually with 2 spines in Georgia and the Carolinas) **6**
   Anal spines usually 2 **8**

6. L.l. nearly complete, extending farther posteriad than middle of soft dorsal. Dorsal fins, though separate, are placed close together **7**
   L.l. incomplete, extending only to about middle of spinous dorsal which is separated from soft dorsal by a much wider space *E. chlorosomum*

7. Infraorbital canal usually complete; preoperculomandibular pores 10 or 11. Dorsal soft rays usually 12–14; pectoral rays 12–13 *E. olmstedi*
   Infraorbital canal usually incomplete; preoperculomandibular pores 9. Dorsal soft rays usually 10–12; pectoral rays 11–12 *E. nigrum*

8. Pectoral fin longer than head. Anterior portion of belly naked or with only a few scales. L.l. complete **9**

Pectoral fin shorter than head. Anterior portion of belly fully scaled. L.l. incomplete **11**

9. L.l. scales more than 50. Gill membranes scarcely connected, the distance from union of gill membranes to mandible tip contained 1.2 times in head length *E. perlongum*

L.l. scales fewer than 50. Gill membranes broadly connected, the distance from union of gill membranes to mandible tip contained 1.5–1.8 in head length **10**

10. Snout more pointed, its profile forming a more gentle slope. Lateral blotches and dorsal saddles (except for 1 dark blotch at spinous dorsal origin) indistinct *E. podostemone*

Snout profile almost vertical. 7 lateral blotches and 6 dorsal saddles distinct *E. longimanum*

11. Premaxillary frenum absent (occasionally a narrow one present), a groove continuous across snout above premaxillae. Dorsal rays X to XII–10 to 12. Dorsal blotches 6 or 7 and broad on the midline *E. stigmaeum*

Premaxillary frenum present, groove interrupted on midline so that skin of snout is continuous with that of upper lip. Dorsal rays XII or XIII–11 or 12. Dorsal blotches 5–7, shaped like an hourglass with narrowest part on the midline *E. jessiae*

12. Anal spines 2 **15**

Anal spine single (rarely 2) **13**

13. Scales in lateral series 58–66. Dorsal X to XII–9 or 10 *E. australe*

Scales in lateral series 43–52. Dorsal IX to XI–11 to 13 **14**

14. Head dorsum, cheek, prepectoral area, and breast well scaled. Supratemporal canal widely interrupted *E. tuscumbia*

Median part of head dorsum and lower part of cheek naked; prepectoral area with few or no scales; breast naked. Supratemporal canal complete *E. trisella*

15. Pelvic fins far apart, space between them about ¾ of each fin base. Gill membranes rather broadly connected, forming a gentle curve (an acute angle in *E. sellare*). Pectoral fin long and expansive. Snout typically blunt (considerably produced in *E. sellare* and some allies of *E. variatum* which are marked above by 4–6 dark saddles) **16**

Pelvic fins close together, space between them about ⅔ of a pelvic base or less. Gill membranes usually separate or moderately conjoined to form an acute angle (broadly joined only in *E. juliae*

and *E. flabellare*). Pectoral fin shorter and smaller. Snout typically more or less sharp and produced, not steeply declivous **34**

16. Premaxillary frenum typically well developed (frenum absent in *E. blennioides* and narrow or occasionally with a groove in some other species). Branchiostegals usually 6 (often 5 in *E. zonale*). Palatine most often toothed. Dorsal blotches 4–7 **17**

Premaxillary frenum weak or absent. Branchiostegals typically 5. Palatines toothless. Dorsal blotches 7–9 **31**

17. Dorsal blotches 4 or 5 **18**

Dorsal blotches 6 or 7 (sometimes 5 in *E. kanawhae*) **23**

18. Gill membranes forming a V-shaped border. Belly naked *E. sellare*

Gill membranes more broadly connected, not forming a V-shaped border. Belly scaled **19**

19. Gill membranes moderately to broadly connected, forming an angle of 50–90°. Scales 47–73. Least bony interorbital width 2.2–3.8 in eye. Highest dorsal spine 2.2–2.8 in head. Dorsal saddles not bordered posteriad by a lighter color; lateral blotches not fused into a zigzag line **20**

Gill membranes very broadly connected, forming an angle of 89–110°. Scales 42–45. Least bony interorbital width 1.8–2.0 in eye. Highest dorsal spine 1.9–2.0 in head. Dorsal saddles bordered posteriad by a lighter color; lateral blotches fused into a zigzag line *E. blennius*

20. Dorsal blotches 5 *E. osburni*

Dorsal blotches 4 **21**

21. Breast naked (except near pelvic fins). Scales 50–58 *E. variatum*

Breast more or less scaly. Scales 47–73 **22**

22. Scales 47–51. Dorsal spines 10–12 *E. tetrazonum*

Scales 57–73. Dorsal spines 13 (12–14) *E. euzonum*

23. Dorsal blotches 6 (sometimes 5 in *E. kanawhae*) or if 7, breast scaly in *E. zonale* **24**

Dorsal blotches 7, breast naked *E. blennioides*

24. Breast scaly **25**

Breast naked, except for a few embedded scales in some specimens **26**

25. Cheeks and opercles naked. Sides with or without poorly defined blotches, but not with vertical bars. Dorsal spines 12 or 13. Restricted to Kanawha River above the falls *E. kanawhae*

Cheeks and opercles scaly. Sides with vertical bars some of which extend around belly to those of opposite side. Dorsal spines 10 or 11 (rarely 12). Not so restricted in range (Fig. 2-108) *E. zonale* (in part)

26. Cheeks and opercles naked   **27**
    Cheeks naked or scaly and opercles scaly (some-
    times only a few scales)   **29**
27. Head length contained 1.3 times in pectoral fin
    length. Dorsal rays XI or XII–11 to 13. Snout pro-
    file a straight line (45°) from level of eyes to
    snout tip   *E. swannanoa*
    Head length contained 1.1–1.2 times in pectoral fin
    length. Dorsal rays X or XI–11 or 12. Snout pro-
    file a gentle curve from eye level to snout tip   **28**
28. Belly naked anteriorly. Color pattern of irregular
    blotches on sides. Postcleithrum neither enlarged
    nor blackened   *E. inscriptum*
    Belly fully scaled. Color pattern of broad vertical
    bars on sides. Postcleithrum enlarged and black-
    ened   *E. thalassinum*
29. Cheeks and opercles scaly. Sides with vertical bars
    which extend around belly to connect with those
    of opposite side (not always evident in specimens
    from Gulf states)   *E. zonale* (**in part**)
    Cheeks naked and opercles with a few scales. Sides
    irregularly blotched   **30**
30. Breast heavily spotted; one pair of large basicaudal
    spots   *E. histrio*
    Breast with little or no spotting; two pairs of small
    basicaudal spots (known only from the Alabama
    Tombigbee River systems in Mississippi, Ala-
    bama, and Georgia)   *E. rupestre*
31. Dorsal blotches 9. L.l. scales 53–59. Snout strongly
    declivous   *E. atripinne*
    Dorsal blotches 8. L.l. scales fewer than 53. Snout
    less declivous to moderately pointed   **32**
32. Premaxillary frenum absent. Suborbital bar straight
    in female   *E. duryi*
    Premaxillary frenum present, but weak. Suborbital
    bar in female curved posteriorly   **33**
33. Posterior 5 lateral blotches as bands, almost con-
    necting on caudal peduncle. Snout moderately
    pointed   *E. coosae*
    Lateral blotches tending to fuse into lateral band,
    their height about equal to width. Snub-nosed
         E. simoterum
34. Supratemporal canal usually complete (rarely inter-
    rupted at midline, but typically so in *E. parvi-
    pinne, E. exile, E. cragini, E. pallididorsum, E.
    punctulatum,* and *E. ditrema,* all of which have
    complete infraorbital canals, and *E. pottsi, E.
    grahami,* and *E. nuchale* with incomplete infra-
    orbitals). Dorsal spines 8–13, never with terminal
    knobs or swellings   **35**
    Supratemporal canal with a wide median interrup-
    tion (often complete in *E. flabellare*), the pores
    typically 2–2. (Infraorbital canals incomplete.)

Dorsal spines 6–10, often (especially in breeding
males, sometimes in females) with thickened
fleshy tips (not known to develop in some species)
     67
35. Palatine teeth absent. Snout produced, its length
    about ⅓ head. Soft dorsal fin greatly elevated (at
    least in adult males), the highest ray almost equal
    to or much greater in length than fin base (in-
    terradial membranes with lengthwise reddish
    streaks). Cheek fully scaled   *E. cinereum*
    Palatine teeth present. Snout about ¼ head. Soft dor-
    sal fin not excessively elevated, the highest ray
    shorter than fin base. Cheek usually naked, but
    with scales in a few species   **36**
36. Spinous dorsal without orange or red submarginal
    band (a crimson band at margin in *E. jordani* of
    the Alabama system). L.l. complete (except in
    E. tippecanoe of the Ohio drainage) or with only
    1–4 unpored scales. Breast of adult males blue or
    green   **37**
    Spinous dorsal with an orange or red marginal or
    submarginal band, at least in adult males (often
    not evident in *E. parvipinne,* which exhibits a
    pale line along L.l.). L.l. usually more or less
    incomplete (usually complete in *E. mariae* and *E.
    fricksium,* species of the Atlantic slope of Georgia
    and the Carolinas, and in *E. nianguae* and *E.
    juliae*). Breast not blue or green (except in *E.
    caeruleum*)   **44**
37. Belly scaled, except on midline in *E. rubrum.* L.l.
    complete, sometimes 3 or 4 unpored posterior
    scales; 44–67 pored scales. Maximum length 46
    mm standard length   **38**
    Belly naked anteriad and on midline. L.l. incom-
    plete, with 19–34 pored scales and not reaching
    beyond middle of soft dorsal. Maximum length
    34 mm standard length   *E. tippecanoe*
38. Longitudinal dark streaks along edges of scale rows.
    Nape naked. Scales 45–67. Dorsal spines 10–13,
    the fin lacking a red margin   **39**
    Longitudinal dark streaks absent. Nape with a few
    scales posteriad. Scales 44–53. Dorsal spines 9–11,
    the fin with a narrow red margin in adult male
         E. jordani
39. Soft dorsal and anal fins usually not margined with
    black (except narrowly in *E. maculatum* of the
    Cumberland and Tennessee Rivers), without light
    submarginal band and without red or orange.
    Head 3.2–3.6 in body; head depth 1.55–1.8 in its
    length. Breast and posteriad onto belly, blue or
    green. Snout moderately to very sharp   **40**
    Soft dorsal and anal fins (at least in adult males)
    with black margin; a submarginal white, red, or
    orange band. Head 2.1–3.8 in body; head depth

1.3–1.6 in its length. Blue or green of breast not extending onto belly. Snout moderately sharp or blunt **41**

40. Opercle scaly. Gill membranes not joined. Snout moderately sharp; head contour forming an angle of 158–167°, above the eye, with the posterior profile. Adult males with many bright red spots. Spinous dorsal fin sometimes with red or orange spots. Caudal fin basally red or orange. Dark subocular bar distinct. Orbit longer than snout **E. maculatum**

Opercle naked. Gill membranes narrowly joined. Snout very sharp; head contour forming an angle of 171–176° with the posterior profile. No red or orange spots. Subocular bar faint or obsolete. Orbit shorter than snout **E. acuticeps**

41. Cheek with single dark spot behind eye; suborbital bar well developed, obsolete, or absent, not with horizontal patches. Sides with few to many light red or orange spots. Bar in front of pectoral base diffuse or absent **42**

Cheek with 5–7 dark spots or dashes; suborbital bar, if present, represented by two pigmented areas. Sides of males with horizontal red streaks. A strong bar in front of pectoral base **E. rufilineatum**

42. Modal number of dorsal rays (total count) 22 and of anal soft rays 7 (often 8 in *E. rubrum*). Female with numerous dark spots on soft dorsal, caudal, and anal fins and prominent red or light spots on body. Dark edge on vertical fins of male **43**

Modal number of dorsal rays (total count) 24 or 25 and of anal soft rays 8. Female without dark spots on soft dorsal, caudal, and anal rays and lacking prominent light spots on body. Dark edge on first dorsal fin of male absent or present only on posterior half **E. camurum**

43. Area behind eye naked or with 1 or 2 small scales. Eye equal to or shorter than snout. Red spots on sides absent or few in either sex. No red band in dorsal fin. L.l. scales usually 53–57 **E. moorei**

Area behind eye with 6–9 large exposed scales. Eye longer than snout. Red spots on sides present, especially in females. Dorsal fin with red band. L.l. scales usually 50–53 **E. rubrum**

44. Anal fin long, 10–12 (rarely 9) soft rays. Head long, slender, and pointed. Breast naked and head nearly so **45**

Anal fin shorter, rarely more than 9 soft rays. Head shorter, stouter, and often quite blunt. Breast and head scutulation variable, but usually scales developed on 1 or more areas **46**

45. Scales 73-82 along body and 24–27 around caudal

peduncle; L.l. almost complete, usually 0–3 scales unpored; upper cheek with a few embedded scales. 2 discrete, jet-black spots on caudal fin base **E. nianguae**

Scales 52–69 along body and 18–24 around caudal peduncle; L.l. more incomplete, usually 4 or more scales unpored; no cheek scales. Caudal spots usually fused to form a short vertical bar **E. sagitta**

46. Gill membranes broadly to narrowly joined to each other seldom overlapping anteriad **47**

Gill membranes not joined (somewhat connected in *E. grahami* of the Rio Grande drainage and in *E. pallididorsum*, Caddo River, Arkansas) but often overlapping anteriad **58**

47. Breast fully scaled. L.l. scales without dark pigment near the pores, leaving a yellowish white line along sides. Unpored L.l. scales 0–7 **E. parvipinne**

Breast naked or only partly scaled, sometimes a few scattered and embedded scales. L.l. scales rather evenly pigmented (pores lacking pigment in *E. nuchale*). L.l. complete or incomplete **48**

48. A concentraiton of pigment overlying enlarged postcleithrum or supracleithrum, or under the opercular flap as in *E. nuchale* making a conspicuous, though sometimes small, humeral spot **49**

No conspicuous humeral spot, or if one is present, postcleithrum is not greatly enlarged **55**

49. Gill membranes very broadly connected (Fig. 2-100A). 1st dorsal blotch large, broad, and conspicuous, others smaller and less conspicuous. Caudal base with a pair of median spots **E. juliae**

Gill membranes less broadly connected, their edges forming a V. Dorsal blotches sometimes inconspicuous, and when present all about same intensity and size. Caudal base with 3 spots in a vertical row **50**

50. Scales in L.l. series fewer than 45 (35–42) **51**

Scales in L.l. series more than 44 (45–75) **52**

51. Unpored scales in L.l. series fewer than 10; prepectoral area scaly **E. fricksium**

Unpored scales in L.l. series 15 or more, prepectoral area naked **E. nuchale**

52. Dark spot near middle of caudal base subaxial, nearer the lower than the upper spot. Sides of male with red or red-orange bars alternating with green ones; anal fin with little or no red **E. hopkinsi**

Dark spot near middle of caudal base axial, about equidistant between other two. Sides of males with red spots, but not bars (except sometimes on caudal peduncle), or with underparts suffused

with orange; anal fin with conspicuous red band
53

53. Adults with underparts suffused with orange, the sides with dark blotches forming a midlateral row and sometimes poorly defined vertical bars posteriad. No red spots *E. radiosum*
Adults with red spots interspersed in a reticulum of black pigment on sides **54**

54. Scales 45–63 (usually 49–56). Red spots clustered in groups on lower sides. Dorsal rays IX to XII–11 to 14. Anal rays II,6–9 *E. artesiae*
Scales 51–75 (usually 59–71). Red spots more scattered, not clustered in groups on lower sides. Dorsal rays X to XII–13 to 16. Anal rays II,7–9
E. whipplei

55. Opercles usually naked (specimens from San Saba River, Texas, scaled). Pectoral fins shorter than head and contained therein about 1.3 times
E. lepidum
Opercles scaly. Pectorals longer, about equal to head length **56**

56. Adult male coloration very dark, undersurface of head and breast intensely spotted with black. Females and young with 11 or 12 vertically elongated dark blotches. Anal rays II,8–10 *E. mariae*
Adult males not so intensely darkened, underparts not spotted. Color pattern of continuous bands encircling body at least posteriad or with dorsal crossbands and lateral blotches. Anal rays II,7 or 8 (sometimes 9) **57**

57. Color pattern of about 12 continuous bands, encircling body at least posteriad, and about 10 dorsal saddles. Cheeks naked, except for a few scales near eye *E. caeruleum*
Color pattern of 7 dorsal saddles and about 9 lateral blotches. Cheeks scaly *E. luteovinctum*

58. Supratemporal canal widely interrupted, the branch from either side terminating in a pore short of middorsal line **62**
Supratemporal canal complete, continuous across occiput or only narrowly interrupted **59**

59. Infraorbital canal interrupted (Fig. 2-107)
E. spectabile
Infraorbital canal complete (Fig. 2-106) **60**

60. Prepectoral region with ctenoid scales; breast usually scaly *E. ditrema*
Prepectoral region naked or with embedded cycloid scales; breast largely naked **61**

61. L.l. complete or nearly so, usually not more than 4 (but as many as 7) unpored scales. Sides without blotches, but most scales with a central spot
E. swaini
L.l. incomplete, more than 4 unpored scales. Sides with about 9 lateral blotches (sometimes indis-

tinct). Postcleithrum not conspicuously blackened
E. asprigene

62. Infraorbital canal incomplete. Restricted to Rio Grande drainage **63**
Infraorbital canal complete. Not in Rio Grande drainage **64**

63. Cheeks and opercles completely naked *E. pottsi*
Cheeks with scales around eye, opercles completely scaled *E. grahami*

64. Cheeks and opercles scaly *E. exile*
Cheeks and opercles naked or with few scales **65**

65. L.l. with 5–20 pores. Sides variable or with 12 small, dusky blotches. Dorsal blotches often indistinct. Infraorbital bar present in young and adult. Infraorbital pores 6–9 (usually 7 or 8)
66

L.l. with more than 20 pores. Sides with irregular dark, diagonal bars (sometimes absent, leaving a plain appearance). Dorsal blotches distinct: 1 at dorsal origin, 1 between dorsals, 1 about at middle of soft dorsal, and 1 near caudal base. Infraorbital bar absent in young, present in adults. Infraorbital pores 9 *E. punctulatum*

66. Venter behind pelvics fully scaled; prepectoral and prepelvic areas with a few scales; vertical bar at caudal base poorly developed or absent
E. cragini
Venter behind pelvics naked; prepectoral and prepelvic areas naked; prominent vertical bar at caudal base *E. pallididorsum*

67. Opercles scaly (scales sometimes inconspicuous and embedded) *E. squamiceps*
Opercles naked **68**

68. 1st dorsal fin in adults (both sexes in *E. kennicotti*) with knobs at tips of spines **69**
Spinous dorsal fin without conspicuous knobs **70**

69. Gill membranes connected, but not broadly
E. kennicotti
Gill membranes broadly connected *E. flabellare*

70. Scales of back and sides without continuous stripes
E. obeyense
Scales of back and sides with spots forming continuous stripes *E. virgatum*

71. Genital papilla of breeding female a low tube bearing numerous villi **72**
Genital papilla of breeding female lacking villi **73**

72. Unpored scales in L.l. series usually 7–15. Branchiostegal membranes separate. 1st anal spine shorter than 2d *E. edwini*
Unpored scales in L.l. series 0–4. Branchiostegal membranes narrowly conjoined. 1st anal spine longer than 2d *E. okaloosae*

73. Infraorbital canal complete **74**
Infraorbital canal incomplete **75**

74. Preopercle strongly serrate. Infraorbital pores 6
E. serriferum

Preopercle entire. Infraorbital pores 8   *E. gracile*

75. Preoperculomandibular pores 10; interorbital pores absent. Anal spines 2   *E. zoniferum*

Preoperculomandibular pores 9; interorbital pores 0, 1, or 2. Anal spines 1 or 2   76

76. Interorbital pores absent. Breast completely scaly; interorbital area with 0–37 scales   77

Interorbital pores 0, 1, or 2. Breast naked or 80% scaly; interorbital area naked   78

77. Interorbital area with 0–12 scales, usually 0–4. Infraorbital pores usually 2 + 3 (note 2 behind, 3 in front of eye)   *E. fusiforme*

Interorbital with 1–36 scales, usually 5–20. Infraorbital pores usually 1 + 3   *E. barratti* [1]

78. Anal spines 1 or 2. Interorbital pores present
E. saludae

Anal spines 1. Interorbital pores usually absent
E. collis [2]

79. Cheeks and opercles scaly. L.l. with 3–7 pores anteriorly   *E. proeliare*

Cheeks and opercles naked or nearly so. L.l. absent or with only 1 or 2 pores anteriorly   80

80. Anal spines 2   *E. microperca*

Anal spine single   *E. fonticola*

Etheostoma vitreum (**Cope**). Glassy darter. Anus encircled by many fleshy villi. Body translucent in life. A. with 2 (rarely 1) spines. Dorsal blotches 7–9. Piedmont and Coastal Plain from Maryland and Virginia to North Carolina.

Etheostoma perlongum (**Hubbs and Raney**). Waccamaw darter. D. X to XII–15; A. II,9 or 10. L.l. complete, its scales 58–66 (seldom fewer than 60); cheeks, opercles, and nape scaly; breast naked or with a small scaly patch. Infraorbital canal complete. Lake Waccamaw, North Carolina.

Etheostoma podostemone **Jordan and Jenkins.** Riverweed darter. D. IX or X–12 to 14; A. II,7 or 8. L.l. complete, with about 37–39 scales; cheeks, nape, and breast naked; opercles scaly. Infraorbital canal complete. Snout pointed. Color pattern less distinct than in *E. longimanum*. Roanoke River system in Virginia.

Etheostoma longimanum **Jordan.** Longfin darter. Similar to *E. podostemone*, but snout profile

almost vertical. Color pattern distinct, of 7 lateral blotches and 6 dorsal saddles. From the James River system in Virginia and West Virginia. An inhabitant of fast water.

Etheostoma jessiae (**Jordan and Brayton**). Blueside darter. D. XII or XIII–10 to 12. Premaxillae nonprotractile. Dorsal blotches, narrowest on middorsal line, 5–7. Tennessee River system in Georgia and Tennessee.

Etheostoma stigmaeum (**Jordan**). Speckled darter. D. X to XII–10 to 12. Premaxillae usually protractile. Dorsal blotches, broad on middorsal line, 6 or 7. Southern Missouri and the Green and Cumberland River systems in Kentucky and Tennessee, south through eastern Kansas and Oklahoma to Louisiana and through the Gulf Coast to the Choctawhatchee River system in Alabama and western Florida.

Etheostoma nigrum **Rafinesque.** Johnny darter. A. with 1 or 2 (usually 1) spines. Nape naked or scaly near midline; cheeks naked. L.l. nearly complete. D. fins placed close together. From the Great Lakes region and the Maritime Provinces southward on both sides of the Appalachians to Georgia, Alabama, and Mississippi. West of the Mississippi River to eastern Oklahoma and Kansas. Typically a stream fish.

Etheostoma olmstedi **Storer.** Tessellated darter. In addition to characters used in the key above, this species further differs from *E. nigrum* in having 4 or 5 scales above the L.l. instead of 3 or 4, breeding males not conspicuously blackened, P_2 ray tips only slightly enlarged and whitened, mouth subterminal and slightly oblique instead of horizontal and terminal, 9–11 rather than 6–7 X- or W-shaped markings on sides. Massachusetts and southern New Hampshire south along Atlantic Coastal Plain and Piedmont to the St. Johns River, Northeastern Florida. It hybridizes with *E. nigrum* in western New York.

Etheostoma chlorosomum (**Hay**). Bluntnose darter. A. with 1 spine. L.l. incomplete. D. fins far apart. Cheeks scaly. Mississippi Valley from Minnesota southward to Alabama and central Texas. An inhabitant of quiet waters, oxbow lakes, and backwater pools.

Etheostoma variatum **Kirtland.** Variegate darter. D. XI to XIII–12 to 14; A. with 2 stiff spines and 8–10 (commonly 9 or 10) soft rays.

---

[1] Shown to be a subspecies of *E. fusiforme*.
[2] *E. collis* contains 2 subspecies: *E. c. collis* (Hubbs and Cannon) and *E. c. lepidinion* Collette; *Hololepis thermophilus* Hubbs and Cannon is synonymized with *E. f. fusiforme* according to Collette (1962).

L.l. complete or nearly so. Back with 4 blackish saddles; lateral blotches not fused in a zigzag line. L.l. scales 50–58; breast naked, except near P_2. From Indiana to New York and southward, west of the Appalachians, to the Tennessee River system. A riffle inhabitant.

Etheostoma tetrazonum (**Hubbs and Black**). Missouri saddled darter. Similar to *E. variatum,* but breast scaly; L.l. scales 47–51. Niangua and Gasconade Rivers in Missouri.

Etheostoma euzonum (**Hubbs and Black**). Arkansas saddled darter. Similar to *E. variatum,* but breast scaly; L.l. scales 57–73. Southern Missouri and northern Arkansas.

Etheostoma kanawhae (**Raney**). Kanawha darter. D. XII or XIII–12 or 13; A. II,7–9. L.l. complete, with scales 48–57. Cheeks and opercles naked; breast with scales only between P_2 fins; nape scaly. Color pattern: dorsal saddles 6; lateral blotches (9 or 10) poorly defined or absent. Gill membranes broadly connected. Kanawha River system above the falls in Virginia and North Carolina.

Etheostoma osburni (**Hubbs and Trautman**). Finescale saddled darter. Similar to *E. kanawhae,* but with smaller scales (64–71 in L.l.) and gill membranes less broadly connected. Kanawha River system above the falls in Virginia and West Virginia.

Etheostoma blennius **Gilbert and Swain**. Blenny darter. Similar to *E. variatum,* but lateral blotches fused to form a zigzag line. Tennessee River system in Tennessee and Alabama.

Etheostoma sellare (**Radcliffe and Welsh**). Maryland darter. D. X or XI–11 or 12; A. II,8 or 9. L.l. nearly complete. Scales in lateral series about 46–50; opercles and posterior part of head scaly, rest of head naked; cheek, nape, breast, and belly naked. Premaxillae not protractile. Back with 4 black saddles; sides below L.l. with 4 blotches

alternating with the saddles. Rare, represented only by types from Swan Creek near Havre de Grace, Maryland, and one specimen taken in 1962.

Etheostoma thalassinum (**Jordan and Brayton**). Seagreen darter. D. IX to XI–10 to 12; A. II,7–9. Gill membranes broadly connected. L.l. scales 39–48; cheeks, opercles, and breast naked; nape scaly; belly entirely scaled. Color pattern: dorsal saddles 6, extending below L.l.; spinous D. with black spot on first 2 interradial membranes. Santee River Basin of North and South Carolina.

Etheostoma inscriptum (**Jordan and Brayton**). Turquoise darter. D. X or XI–11 or 12; A. II,7 or 8. Similar to *E. thalassinum,* but lacking spot on spinous D. Belly naked anteriad. Savannah and Oconee Rivers, Georgia.

Etheostoma swannanoa **Jordan and Evermann**. Swannanoa darter. D. XI or XII–11 to 13; A. II,7 or 8. L.l. scales 45–57. Snout profile a straight line at about 45°. From the Tennessee River system in Virginia, Tennessee, and North Carolina.

Etheostoma histrio **Jordan and Gilbert**. Harlequin darter. D. about X–13; A. II,7. L.l. scales 48–54; cheeks and breast naked; opercles with a few scales; nape scaly. Dorsal blotches 6 (rarely 7); lateral blotches 7–9, poorly defined. Premaxillary frenum absent or weakly developed. Discontinuous distribution from Indiana (record of 1890) and Green River, Kentucky, through Missouri, Arkansas, southeastern Oklahoma, and eastern Texas to Alabama and western Florida.

Etheostoma zonale (**Cope**). Banded darter (Fig. 2-108). Aspect of *E. histrio,* but premaxillae nonprotractile and sides usually with conspicuous bands. Breast scaly or naked. From the drainage of Lake Michigan eastward to Pennsylvania and Virginia and south to Georgia and the Gulf Coast. Throughout the Mississippi Valley in clear streams from Minnesota to Mississippi and westward to eastern Oklahoma and Kansas.

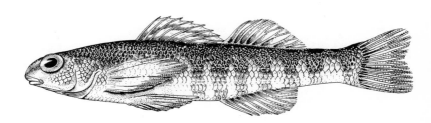

Fig. 2-108. Etheostoma zonale.
(**Forbes and Richardson, 1920.**)

Etheostoma rupestre **Gilbert and Swain.** Rock darter. D. X to XIII, 10 to 13; A. II, 5 to 8. L. 1. scales 47–61; cheeks and breast naked; nape scaly to naked; opercles and belly partially scaled. Dorsal blotches 6; brown-edged, lateral bars 7(6–9); P_1 base with a dark spot. Alabama and Tombigbee River systems in Mississippi, Alabama, and Georgia.

Etheostoma blennioides **Rafinesque.** Greenside darter (Fig. 2-109). D. XII to XIV–12 to 15; A. II,8 or 9. L.l. scales 58–78; cheeks, opercles, and nape scaly; breast naked. Mouth small, inferior; upper jaw concealed in a groove under snout. Coloration: sides with about 8 double bars, each pair forming a Y-shaped figure; males brilliant green in spring; sides sometimes with red spots. Widely distributed in riffle habitats of the Mississippi River system from Illinois to New York and southward to Alabama, Georgia, and North Carolina. West of the Mississippi River, in the Ozark region of Missouri, Arkansas, and eastern Oklahoma. In the Great Lakes drainage: Ontario tributaries of Lake St. Claire; Lake Erie and in the southern tributaries of Lake Ontario.

Etheostoma simoterum **(Cope).** Tennessee snubnose darter. D. X to XII–10 or 11; A. II,7. Scales in complete L.l. 45–52; cheeks, opercles, and nape scaly; breast naked; belly completely scaled. Coloration: dorsal blotches 7 or 8; sides with 9 lateral blotches. Premaxillary frenum narrow. Tennessee River system in Virginia, Tennessee, Georgia, and Alabama.

Etheostoma coosae **(Fowler).** Coosa darter. D. IX to X–10 to 13; A. II,7; P_1 12 or 13; P_2 I,5. Lateralis almost complete, 40–45 pores, 2 or 3 unpored scales. Fins except C. plain; C. with 5 cross bars; back with 8 dark saddles, with indistinct bars on sides; dark-brown streak diagonally from eye to occiput; a short postocular bar; a narrow, vertical subocular streak; and a dark dash diagonally down from eye to snout. Cheeks, opercles, and prepectoral region scaly; venter naked anteriad. Presumed by Fowler to be closely related to *E. squamiceps*. Originally known only from the Coosa River Basin, Cherokee County, Alabama.

Etheostoma duryi **Henshall.** Blackside snubnose darter. D. XI–11 or 12; A. II,6 or 7. L.l. complete, with 40–45 scales; cheeks, opercles, and nape scaly; breast naked. Coloration: dorsal blotches 8; lateral blotches 9, run together to form a band; preorbital and infraorbital bars present, each extending downward and forward; P_1 base without a spot. Tennessee River system in Tennessee and Alabama.

Etheostoma atripinne **(Jordan).** Cumberland snubnose darter. D. XI or XII–10 to 12; A. II,7 or 8. L.l. complete, with about 55 or 56 scales; cheeks, opercles, nape, and belly scaly; breast naked. Coloration: dorsal blotches 9; lateral blotches 9 or 10, vertically elongated below D.; infraorbital bar vertical; preorbital bar extending downward and forward. Premaxillary frenum narrow. Cumberland River system in Tennessee.

Etheostoma tippecanoe **Jordan and Evermann.** Tippecanoe darter. D. XII–10 or 11; A. II,7 or 8. L.l. incomplete, unpored scales about 18–22, total scales in lateral series 45–50; cheeks, nape, and breast naked; opercles scaly; belly naked anteriad or with embedded scales. Coloration: dorsal blotches absent; sides plain, with small black spots, indistinct bars on C. peduncle; breeding males brilliant golden orange. Riffles of some small streams of the Ohio River system in Pennsylvania, Ohio, Indiana, Kentucky, as far west as Green

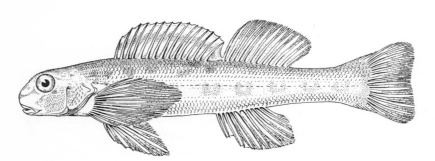

Fig. 2-109. *Etheostoma blennioides.* (Forbes and Richardson, 1920.)

River, and the Cumberland in central Tennessee.

Etheostoma maculatum **Kirtland.** Spotted darter. D. XII to XIV–12 or 13; A. II,7–9. L.l. incomplete with 56–63 scales; cheeks, nape, and breast naked; opercles and belly scaly. Coloration: dorsal blotches absent; sides with longitudinal stripes; infraorbital and preorbital bars present; males with many irregularly placed red spots. Snout sharply pointed. Ohio and Tennessee River systems from New York and Pennsylvania south and westward to Virginia, Kentucky, and Tennessee.

Etheostoma rufilineatum **(Cope).** Redline darter. D. X to XII–11 to 13; A. II,7–9. L.l. complete, with 45–51 scales; cheeks, nape, and breast naked; opercles and belly scaly. Coloration: dorsal blotches 7 or 8, indistinct; sides with logitudinal stripes and vertical rows of 2 or 3 blotches each; C. base with 2 white blotches tending to be separated by a median dark spot or chevron; all fins in female boldly black-spotted; males with soft D., A. and C. black-edged. Tennessee River system in Virginia, Tennessee, North Carolina, Georgia, and Alabama.

Etheostoma camurum **(Cope).** Bluebreast darter. D. XI to XIII–12 or 13; A. II,6–8. L.l. complete, with 50–60 scales; cheeks, nape, and breast naked; opercles and belly scaly. Coloration: dorsal blotches absent; sides with longitudinal stripes and red spots (dull red or olive in females and young). Snout very blunt. Body compressed. From Pennsylvania to Indiana and southward to Tennessee and North Carolina.

Etheostoma acuticeps **Bailey.** Sharphead darter. D. XII–11 or 12; A. II,7 or 8; P_1 13 or 14. L.l. complete or nearly so, with 54–61 scales, 0–4 being unpored; 22–26 scale rows around C. peduncle; head and nape naked. Infraorbital canal complete (8 or 9 pores); preoperculomandibular canal complete (9 pores); supratemporal canal complete (3 pores). Gill membranes narrowly conjoined. Snout very sharp. Coloration lacking red or orange, but fins and lower surface blue-green or pale green; body with 13–14 narrow vertical bars and horizontal dark lines between scale rows; subocular bar faint; fins not dark-edged. A rare and now apparently extinct species taken only from the south fork of the Holston River prior to filling the

South Holston Reservoir near Bristol, Sullivan County, Tennessee.

Etheostoma moorei **Raney and Suttkus.** Moore's darter. Closely related to *E. camurum* and *E. rufilineatum* but differs in having fewer vertebrae (36 instead of 38) and fewer rays in D., A., and P_1 fins. This is the only bluebreast darter (subgenus *Nothonotus*) known to occur west of the Mississippi River. Tributaries of Little Red River of the White River system in Arkansas.

Etheostoma rubrum **Raney and Suttkus.** Bayou darter. Closely related to *E. moorei*, but differs in having large and exposed scales behind the larger eye, red spots on the sides in both sexes and especially in the female, a red band in D., and fewer L.l. scales (usually 50–53 instead of 53–57). Known only from Bayou Pierre and its tributary, Oak Creek, Copiah County, Mississippi.

Etheostoma jordani **Gilbert.** Greenbreast darter. D. X or XI–10 to 12; A. II,7 or 8. L.l. complete, with 42–55 scales; cheeks and breast naked; opercles and belly scaly; nape naked anteriad. Coloration: dorsal blotches 8 or 9; sides without stripes, irregularly blotched; infraorbital bar poorly developed; preorbital bar present. Alabama River system in Georgia and Alabama.

Etheostoma nianguae **Gilbert and Meek.** Niangua darter. C. base with 2 unique, discrete, round spots. Scales small, 74–80 along the sides and 24–26 around C. peduncle; cheek scales present. Lower tributaries of the Osage River in Missouri.

Etheostoma sagitta **(Jordan and Swain).** Arrow darter. Head slender. Branchiostegals 6 (rarely 7). Upper portions of Kentucky and Cumberland River systems in Kentucky.

Etheostoma juliae **Meek.** Yoke darter. Readily recognized by a conspicuous, large black saddle in front of spinous D., extending down to P_1. White and James River systems in Missouri and Arkansas.

Etheostoma whipplei **(Girard).** Redfin darter (Fig. 2-110). D. X to XII–13 to 16; A. II,7–9. L.l. incomplete, scales in lateral series 51–75 (usually 59–71); cheeks, opercles, and breast mostly naked; nape scaly; belly fully scaled. Coloration: sides with dark reticulations and spotted with yellow or red; C. base with 3 dark spots. Arkansas River Basin of eastern Kansas and Oklahoma,

Fig. 2-110. Etheostoma whipplei.

Missouri, and Arkansas; Saline River Basin in Arkansas.

Etheostoma artesiae (**Hay**). Eastern redfin darter. Similar to *E. whipplei,* but D. IX to XII–11 to 14. Scales 45–63 (usually 49–56). Red spots in clusters on lower sides, especially on C. peduncle. From the southern Red River tributaries of eastern Texas southward and east to the Alabama River Basin in Alabama. Very close to *E. whipplei,* possibly only subspecifically distinct.

Etheostoma radiosum (**Hubbs and Black**). Orangebelly darter (Fig. 2-111). D. IX to XII–11 to 17; A. II,6–9. L.l. incomplete, scales in lateral series 47–66. In many ways similar to *E. whipplei,* but never with red spots on sides; lateral blotches 8 or 9, sometimes forming bars posteriad. Tributaries of Red River in southern Oklahoma and Arkansas.

Etheostoma hopkinsi (**Fowler**). Christmas darter. D. X or XI–11 to 13; A. II,7–9; P₁ 12 or 13. L.l. scales 40–49, with 3–13 unpored; nape, opercle, and cheek naked to fully scaled; breast naked or with embedded scales; belly partly scaled. Infraorbital canal complete, usually with 8 pores; supratemporal canal complete; 10 preoperculomandibular pores. Dorsal blotches usually 7 or 8; male with 10 red or red-orange bars alternating with green on sides; D. with a submarginal red or red-orange band; A. with little or no red color; predorsal area with 2 poorly to well-developed dark blotches; subocular bar present; prepectoral dark

Fig. 2-111. Etheostoma radiosum.

spot usually present. Confined to the Altamaha, Ogeechee, and Savannah Rivers, above and below Fall Line, in Georgia and South Carolina.

Etheostoma mariae (**Fowler**). Pinewoods darter. D. IX or X–11 to 13; A. II,8–10. L.l. complete or nearly so, scales in lateral series about 37 or 38; cheeks, opercles, and nape scaly; breast with a few scales. Adult males very dark. Pee Dee River Basin in North Carolina.

Etheostoma fricksium **Hildebrand.** Savannah darter. D. X or XII–11 to 13; A. II,8 or 9. Scales in lateral series about 41 or 42; cheeks and opercles scaly; nape with few scales; breast naked; belly fully scaled. Coloration: indistinct short bars, especially on C. peduncle; basicaudal spots 3, middle one below center; dorsal blotches absent; 5 blotches on upper sides near D. base; infraorbital bar an arc curving downward and backward; preorbital bar present; sides with dark stripes. Tributaries of the Savannah and Combahee Rivers in South Carolina and Georgia.

Etheostoma swaini (**Jordan**). Gulf darter. D. X or XI–10 to 12; A. II,6 or 7. L.l. complete or nearly so, lateral scale rows 36–44; cheeks, opercles, and belly scaly; nape and breast naked (nape scales sometimes present). Dorsal blotches indistinct; sides with rows of dark spots. From the Florida parishes of Louisiana to the Apalachicola River system, in coastal drainage.

Etheostoma ditrema **Ramsey and Suttkus.** Coldwater darter. D. VIII to XII–9 to 12; A. II,6–8. L.l. moderately arched anteriad and incomplete, 41–54 scales in the series, pored scales 19–35 and unpored 13–30. Infraorbital canal complete, usually with 8 pores; commonly there are 2 coronal pores; supratemporal canal usually interrupted. Cheeks, opercles, and posterolateral corners of head dorsum scaly; nape naked to scaly; breast usually scaly but naked anteriad. Gill membranes overlapping anteriad. Breeding male dark brown, with orange belly and lower C. peduncle; female indistinctly mottled dark brown; C. base with a vertical row of 3 dark blotches; spinous D. with submarginal orange band in both sexes. Springs of the Alabama River Basin, Alabama and Georgia, above Fall Line.

Etheostoma nuchale **Howell and Caldwell.** Watercress darter. D. VIII to XI–10 to 12; A. II,6–8. L.l. straight, incomplete 35–42 scales in the series, pored scales 12–24 and unpored 15–27; infraorbital canal usually incomplete with pores 3 + 5; supratemporal canal interrupted. The L.l. scales unpigmented, forming a light line. Cheeks largely naked; opercles scaly; nape, mesially, and breast naked. Gill membranes moderately to narrowly joined. Breeding male with belly bright red-orange ventrolaterally, with a narrow midventral light stripe; 6 indistinct bars posteriorly; 7 irregular dorsal saddles; C. base with a vertical row of 3 indistinct blotches, with 2 orange spots behind them on a blue C. base. Spinous D. margined with blue, followed by red-orange; A. bright blue; female without bright colors, except D. is similar to that of the male. Known only from the type locality, Glen Spring near Bessemer, Alabama.

Etheostoma asprigene (Forbes). Mud darter. Similar to *E. swaini*, but with smaller scales, about 47–55 in lateral series. Sides without rows of spots. A lowland species of the Mississippi Valley from Iowa, Minnesota, and Wisconsin south to the Gulf states and westward as far as the Neches River in Texas.

Etheostoma punctulatum (Agassiz). Stippled darter. D. X or XII–14 or 15; A. II,8 or 9. L.l. incomplete. Lateral scale rows 63–80; cheeks, opercles, and breast largely naked. Nape and belly scaly. Dorsal blotches 4; sides with indistinct or irregular bars. In quiet, cool waters of spring-fed Ozarkian streams in southeastern Kansas, southwestern Missouri, and northeastern Oklahoma.

Etheostoma cragini Gilbert. Arkansas darter. D. VIII to XI (usually IX)–11 to 13 (usually 11 or 12); A. II,5–8 (usually II,7). L.l. incomplete, scales in lateral series 43–55; cheeks and opercles usually naked; breast naked; nape and belly scaly. Coloration: dorsal blotches inconspicuous; sides with 12 dusky blotches. Infraorbital canal complete. Arkansas River drainage in Missouri, Oklahoma, Kansas, and Colorado.

Etheostoma pallididorsum Distler and Metcalf. Paleback darter. A close relative of *E. cragini*. D. VIII to XI–10 to 14; A. II,7–9 (rarely I,7–9). L.l. incomplete, with 8–19 (usually fewer than 13) pores. Scales in lateral series 43–55. Head, breast, and anterior belly naked; nape with embedded scales. Gill membranes slightly connected. Coloration: wide pale stripe on dorsum from occiput to C. base; 1–6 dorsal saddles poorly de-

veloped; isthmus and breast light orange; cheek pale orange, branchiostegals bright orange. Spinous D. black at margin and base, with a submarginal band bright orange. Soft D. and A. with horizontal bars formed by pigment on the rays; C. with similar vertical bars. Infraorbital canal complete, supratemporal widely interrupted. Caddo River system in Arkansas.

Etheostoma parvipinne **Gilbert and Swain.** Goldstripe darter. D. IX to XI (rarely VIII)–9 to 12; A. II,7–9. L.l. complete or nearly so, lateral rows of scales 49–57; cheeks, opercles, nape, breast, and belly scaly. Dorsal blotches indistinct; sides often with a single spot on each scale; L.l. scales without dark pigment. From Red River tributaries in southeastern Oklahoma south through eastern Texas, and east along the Gulf Coast to southwestern Georgia and on the Atlantic Coast in the Altamaha Basin.

Etheostoma caeruleum **Storer.** Rainbow darter. D. IX to XII–12 to 14; A. II,7 or 8. L.l. incomplete, scales in lateral series 37–50; cheeks naked, a few scales near eye; opercles scaly; nape and breast usually naked; belly fully scaled. Coloration: about 10 dorsal saddles; sides with about 12 blue bars with red interspaces, encircling C. peduncle. Infraorbital canal complete. Gill membranes rather broadly connected. From the Lake Ontario drainage of Ontario and New York westward through the Great Lakes drainages of Michigan, Minnesota, and southward in the Ohio and Mississippi Basins to northern Alabama. West of the Mississippi from Iowa to northern Arkansas.

Etheostoma pottsi (Girard). D. X to XII–10 to 12; A. II,6–7. L.l. incomplete, with 17–31 pored scales, 11–27 unpored, and total scales in lateral series 34–45. Cheeks, opercles, nape, and breast naked; belly scaly. Infraorbital canal incomplete, pores 4 + 2; supratemporal canal rather widely interrupted. Tributaries of the Chihuahua River in Mexico. This species and *Etheostoma australe* are included, as suggested by Dr. Bruce Collette, for sake of completeness. Neither species is recorded from the United States.

Etheostoma grahami (Girard). Rio Grande darter. L.l. incomplete; infraorbital canal incomplete; supraorbital canal with a narrow interruption. Cheeks scaly only below and behind eye; opercles fully scaled. Infraorbital bar conspicuous.

Tributaries of the Rio Grande in Texas and Mexico.

Etheostoma spectabile (**Agassiz**). Orange-throat darter. Similar to *E. caeruleum,* but with infraorbital canal interrupted, the C. peduncle not encircled by bands and the gill membranes less broadly connected. From southeastern Michigan and western Ohio westward through Iowa and Nebraska to Colorado and south to Tennessee and Texas.

Etheostoma lepidum (**Baird and Girard**). Greenthroat darter. D. usually IX–10 to 13; A. II,6–8. L.l. incomplete, scales in lateral series 48–55; cheeks and belly fully scaled; opercles and breast naked; nape naked anteriad. Infraorbital canals interrupted. Dorsal blotches 6–8, roundish. Limestone streams of the Edwards Plateau in central Texas.

Etheostoma luteovinctum **Gilbert and Swain.** Redband darter. D. IX or X–13; spinous D. with a posterior black blotch; A. II,7 or 8. L.l. incomplete, scales in lateral series 49–55; cheeks, opercles, and nape scaly; breast naked or partly scaly. Gill membranes narrowly connected. Tributaries of the Tennessee and Cumberland Rivers in Tennessee.

Etheostoma exile (**Girard**). Iowa darter. D. VII to X–10 to 12; II,6–8. L.l. incomplete, scales in lateral series 55–63 (27–42 without pores); cheeks, opercles, nape, and belly scaly; breast naked (sometimes scaly). Dorsal blotches 7 or 8; lateral blotches 10, vertically elongated. Widely distributed in southern Canada from Alberta to Quebec and south to Colorado, Kansas, and Iowa. East of the Mississippi River from Illinois, Indiana, Ohio, and New York.

Etheostoma cinereum **Storer.** Ashy darter. D. XI or XII–11 to 13; A. II,8. L.l. complete, its scales 57–60; cheeks, nape, and breast naked or with few scales; opercles scaly. Dorsal saddles 4; sides above L.l. with 2 or 3 rows of brown spots extending to below soft D.; lateral blotches 11 or 12, oblong. Tennessee River system in Tennessee and Alabama.

Etheostoma squamiceps **Jordan.** Spottail darter. D. VIII to X (usually VIII)–12 to 14 (usually 13); A. II,7 (6–8). L.l. incomplete, scales in lateral series 41–54 (unpored 9–28); cheeks, opercles, nape, and breast scaly; infraorbital canal incomplete. Coloration: mottled, never with a single spot

on each scale. From Indiana south to the Tennessee River system in Alabama.

Etheostoma kennicotti (**Putnam**). Stripetail darter. D. VI to VIII–11 or 12, spinous portion very low, with knobs in both sexes; A. II,7 or 8. L.l. incomplete, lateral scales 41–47 (unpored, 17–33); cheeks, opercles, nape, and breast naked; belly scaly. Dorsal saddles absent; sides finely punctate. From Illinois through Kentucky to the Tennessee River system in Tennessee.

Etheostoma flabellare **Rafinesque.** Fantail darter. Similar to *E. kennicotti,* but spinous D. with knobs in male only; gill membranes more broadly connected. Sides often with longitudinal rows of spots. Dorsal blotches 6; lateral blotches 10 or 11, poorly defined. Widely distributed from Ontario and Quebec, Indiana and the Lower Peninsula of Michigan south to North Carolina. Westward in the Mississippi Valley from Wisconsin and Minnesota southward through Iowa and Kansas to Oklahoma and Alabama.

Etheostoma obeyense **Kirsch.** Barcheek darter. D. VIII or IX–11 to 13; A. II,8 or 9. L.l. incomplete, scales in lateral series 44 to 49 (about 30 without pores); cheeks, opercles, nape, breast, and belly naked. Spinous D. with first 4 or 5 interradial membranes blackened, knobs weakly developed. Cumberland River system in Kentucky and Tennessee.

Etheostoma virgatum (**Jordan**). Striped darter. D. IX or X–12 or 13; A. II,8. L.l. incomplete, with about 12–18 pores; scales in lateral series about 46–52; cheeks, opercles, nape, and breast naked. Coloration: first 4 or 5 interradial membranes of D. blackened; dorsal blotches none; sides with 9 or 10 longitudinal streaks and about 10 lateral blotches. Tributaries of the Cumberland River in Kentucky and Tennessee.

Etheostoma tuscumbia **Gilbert and Swain.** Tuscumbia darter. D. IX to XI–11 to 13; A. I (sometimes II),8. L.l. incomplete, 15–20 pores, not arched upward; scales in lateral series about 48–50; top of head, cheeks, opercles, and breast scaly. Premaxillae not protractile. Coloration: grayish or greenish olive, mottled and speckled with black; back with 6 broad dark saddles; sides with 8–10 elongated blotches along L.l.; 2 black blotches at base. Tennessee River system in Tennessee and Alabama and Mississippi.

Etheostoma trisella **Bailey and Richards.** Tri-spot darter. D. IX–11; A. I,8; P₁ 13; principal C. rays 16. Nape, belly, and opercle scaly; breast naked and prepectoral area almost so; head dorsum naked except for a few scales at lower end of supratemporal canal. Infraorbital canal complete (8 or 9 pores); preoperculomandibular pores 10; supratemporal canal complete (3 pores). 3 dark, dorsal saddles, 1 just before D., 1 between the D. fins, and 1 on the C. peduncle; no humeral blotch. Only the holotype and 1 other specimen are known, and the type locality is now inundated by Weiss Lake near Centre, Alabama. The second specimen is from a tributary of the Conasauga River in Georgia.

Etheostoma australe **Jordan.** Mexican darter. D. X to XII–9 or 10; A. I,7 or 8. Scales 58–66. Color clear olive; about 10 obscure bars on sides, most prominent in males, the lighter areas reddish. Soft D. and C. with transverse bars. Although this species does not occur in the United States, it is included, on request, for sake of completeness. Chihuahua River, Mexico.

Etheostoma edwini **(Hubbs and Cannon).** Redspot darter. D. IX or X–9 to 12; A. II,7. L.l. incomplete, moderately elevated and with 23–32 pores; scales in lateral series 36–41; cheeks, nape, opercles, and breast scaly. Infraorbital and supratemporal canals complete. Coloration: back with 8 or 9 dark saddles; sides with about 9 dark blotches, obscure anteriad; no basicaudal spots. From the Suwannee River Basin in Florida to the Perdido River, Florida and Alabama.

Etheostoma okaloosae **(Fowler).** Okaloosa darter. This darter, formerly confused with *E. swaini,* has been shown to be most closely related to *E. edwini* (Collette and Yerger, 1962). It differs from *E. swaini* in having the genital villi, more highly arched L.l., slenderer C. peduncle, fewer unpored L.l. scales, and some scales on the prepelvic area, in addition to several other characters. See the key above for differences between *E. okaloosae* and *E. edwini.* Confined to streams of western Florida in Okaloosa and Walton Counties.

Etheostoma serriferum **(Hubbs and Cannon).** Sawcheek darter. D. X to XII–12 to 15; A. II,6 to 8. L.l. incomplete, highly elevated, and usually 28–38 pores; scales in lateral series usually 50–58; cheeks, opercles, nape, breast, and top of head scaly. Infraorbital and supratemporal canals com-

plete. Dorsal saddles and blotches absent; lateral blotches large, irregular; C. base with vertical row of 4 spots. Preopercle strongly serrate. Atlantic Coast usually below the Fall Line from the Dismal Swamp in Virginia to the Altamaha River system in Georgia.

Etheostoma gracile **(Girard).** Slough darter. Similar to *E. serriferum,* but D. VIII to X–9 to 13; breast, top of head, and often the nape naked. L.l. with fewer pores (13–27). Coloration: in life the sides are barred with green; males with a submarginal red-orange band in spinous D., which has the anterior interradial membranes blackened; a basicaudal spot and often one above and below it. From Indiana and Illinois southward in the Mississippi Valley through Missouri, Kentucky, Arkansas, eastern Kansas, and Oklahoma to Mississippi and Louisiana. Also in Gulf Coast streams from the Nueces River in Texas to the Tombigbee River Basin in Alabama.

Etheostoma zoniferum **(Hubbs and Cannon).** Backwater darter. D. IX–10 to 12; A. II,5 to 7. L.l. incomplete, elevated anteriad and with 13–19 pores. Lateral scale rows 41–50; breast naked, top of head and usually the nape naked; cheeks and opercles scaly. Coloration: breeding individuals much like *E. gracile;* back with about 9 indistinct saddles. Infraorbital canal incomplete, supratemporal canal complete. Alabama and Tombigbee Basins in Alabama and Mississippi.

Etheostoma saludae **(Hubbs and Cannon).** Saluda darter. D. IX or X–10 or 11; A. I or II,7. L.l. incomplete, elevated anteriad and with 9–24 pores; lateral scale rows 36–50; breast, nape, and top of head naked; belly weakly scaled; cheeks and opercles scaly. Infraorbital and supratemporal canals incomplete. Coloration: sides with about 10 rectangular, dark blotches; back with about 10 dark saddles; C. base with vertical row of 3 or 4 spots. Santee River system in South Carolina.

Etheostoma collis **(Hubbs and Cannon).** Carolina darter. D. VIII or IX–11 or 12; A. II,6 or 7. L.l. incomplete, elevated anteriad, and with 9–24 pores; breast, nape, and top of head naked; cheeks and opercles scaly; scales in lateral series 35–49. Infraorbital canal interrupted; supratemporal canal complete. Coloration: sides with a narrow median dark stripe, broken on C. peduncle into blotches; back without saddles; C. base with vertical row of spots. Roanoke River in Virginia, Neuse River in North Carolina, and the Santee Basin in South Carolina.

Etheostoma barratti (**Holbrook**). Scalyhead darter. D. VIII to XII–9 to 12; A. II,6 or 9. L.l. incomplete, strongly elevated anteriad and with 0–35 pores; scales in lateral series 42–63; cheeks, opercles, nape, breast, and top of head more or less scaly. Infraorbital canal incomplete, supratemporal canal complete (incomplete in young). Coloration: sides with 9–12 rectangular blotches; back with about 9 dark saddles; C. base usually with vertical row of 4 dark spots. Regarded as a subspecies of *E. fusiforme* by Collette (1962). Widely distributed in coastal streams from North Carolina to Florida and westward to the Red River system in Texas and Oklahoma; also Reelfoot Lake in Tennessee.

Etheostoma fusiforme (**Girard**). Swamp darter. D. VIII to XII–9 to 12; II,6–9. L.l. usually incomplete, with 0–25 pored scales; scales in lateral series 40–63; cheeks, nape, and breast scaly; interorbital and parietal scales present or absent. Infraorbital canals interrupted; interorbital pores absent; supratemporal canals complete or incomplete. Coloration: sides with a dark band (usually) or a row of distinct blotches; C. base with a vertical row of 3 or 4 dark spots; suborbital bar present, weak, or absent. Body stout, little compressed to slender and compressed. In streams of the Atlantic Coast from southern New Hampshire and Maine to North Carolina.

Etheostoma proeliare (**Hay**). Cypress darter. D. VIII–11; A. II,6. L.l. almost absent, with 2–7 pores; scales in lateral series about 36; cheeks and opercles scaly. Sides with about 10 dark spots. East of the Mississippi River from Illinois to western Florida and west of the Mississippi from Missouri to eastern Texas.

Etheostoma microperca **Jordan and Gilbert.** Least darter. D. VI to VIII–9 or 10; A. II,6. L.l. absent. Scales in lateral series 34–37; cheeks naked; opercles with a few scales. Sides speckled and with vague bars and zigzag markings. From Ontario to Minnesota and North Dakota and southward to Kentucky, Arkansas, and Oklahoma.

Etheostoma fonticola (**Jordan and Gilbert**). Fountain darter. Similar to *E. microperca,* but usually with only 1 A. spine. Back with about 8 indistinct saddles. Mid-sides with a series of short horizontal lines. San Marcos and Comal Springs, Texas.

Family Sciaenidae. Drums. L.l. system extensively developed, the head canals forming deep channels in cranial bones; canals with few pores; lateral canal continuous to end of C. Head scaly. Pharyngeal teeth molarlike; jaw teeth sharp and conical. D. deeply notched, anterior portion spiny. A. with 1 or 2 spines. Paleocene. Mostly marine, 1 species in fresh waters of the United States.

Genus *Aplodinotus* Rafinesque.

Aplodinotus grunniens **Rafinesque.** Freshwater drum (Fig. 2-112). From Canada southward throughout the Mississippi River system; Mississippi and eastern Mexico to Guatemala (Rio Usumacinta Basin). A good food fish, sometimes reaching a size of 60 lb. Otoliths from even larger

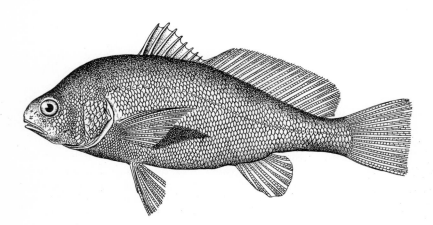

Fig. 2-112. Aplodinotus grunniens. (Jordan and Evermann, 1900.)

drums have been found in Indian mounds and camp sites.

Family Sparidae. Porgies. Distinguished by the unusual dentition; upper and lower jaws with humanlike incisors in front; roof of mouth and sides of lower jaws with enlarged molar teeth. Preopercle not serrate. Body with large smooth scales. Large fishes valued for food. Lower Eocene. Mostly in tropical and subtropical seas. 1 species enters U.S. fresh waters.

Genus *Archosargus* Gill.

Archosargus probatocephalus (**Walbaum**). Sheepshead. Known to occur in the fresh waters of Florida and Texas. Some poorly authenticated records from Oklahoma.

Family Cichlidae. Cichlids. Superficially resembling the sunfishes, but with only 1 nostril on each side and an interrupted L.l. L.l. anteriad much as in sunfishes, but terminating under D. and beginning again farther down on sides. D. single, anterior portion spinous. A. with 3 or more spines. Vomer and palatines toothless. Premaxillae very protractile. Eocene. Texas, Central and South America, West Indies, Africa, and Southern Asia.

KEY TO GENERA OF CICHLIDAE

1. Anal spines usually 5   *Cichlasoma*
   Anal spines 3   *Tilapia*

Genus *Cichlasoma* Swainson.

Cichlasoma cyanoguttatum (**Baird and Girard**). Rio Grande perch. Originally from the Rio Grande southward on the Gulf Coast into Mexico, now found in other streams in Texas.

Genus *Tilapia* Smith. Old World cichlids introduced as aquarium fishes and for use as game-fish forage.

KEY TO SPECIES OF *TILAPIA*

1. Sides plain without conspicuous bars   ***T. heudeloti***
   Sides with vertical bars, sometimes irregular   **2**
2. Sides with 6 or 7 irregular vertical bars. Gill rakers 14–19   ***T. mossambica***
   Sides with 11 or 12 regular bars. Gill rakers 19–28   ***T. nilotica***

Tilapia heudeloti **Dumeril.** African cichlid. D. XV or XVI,10–12; A. III,7–9; P_1 15; L.l. scales 27–30. A dark opercular spot; sides plain and fins not conspicuously barred; chin of male black, lips white.

A native of west Africa from Senegal to Gabon, this species was introduced by tropical fish dealers into Florida and released intentionally or accidentally some time prior to 1958 when the species was first noticed in gill-net catches. During 1962 one commercial dealer sold 3,500 to 4,500 lb. Springer and Finucane (1963) reported the species as well established in Hillsborough and Manatee Counties on the west coast of Florida.

Tilapia mossambica (**Peters**). Mossambique tilapia. D. XV or XVI,10 or 11; A. III,9 or 10; P_1 15; L.l. scales 30–33. Opercle with a dark spot; sides with 6 or 7 irregular vertical bars; external genitalia chalky to yellowish; cheeks yellow-brown to purplish. Gill rakers 14–19 (rarely 20). This species and *T. nilotica*, are being used experimentally in some southern states as game-fish forage. Where winter temperatures are too low, breeding stock must be kept indoors. Because of rapid growth, young *Tilapia* are soon an available food for game species. Established in San Antonio and San Marcos springs in Texas, before 1961.

Tilapia nilotica (**Linnaeus**). Nile bolti. Similar to *T. mossambica*, but with 11 or 12 regular bars on the sides, external genitalia pink, cheeks blue to brilliant blue, and gill rakers 19–28. Not known to be established in the United States except as experimental stock overwintered indoors and planted in ponds as forage for game fishes during the warmer months.

Family Embiotocidae. Surfperches. Scales cycloid, cheeks and opercles scaly. L.l. continuous and complete. Mouth small and terminal; premaxillae protractile; vomerine and palatine teeth absent; maxillae short and lying under the preorbital. Branchiostegals 5 or 6. 2 nostrils on each side. D. single, its anterior portion spinous and with a scaly sheath covering part of its base; A. rather long with 3 spines; C. forked. Viviparous. Middle Miocene. Largely marine, 1 species confined to fresh waters of California.

Genus *Hysterocarpus* Gibbons.

Hysterocarpus traski **Gibbons.** Tule perch. Sacramento River Basin and tributaries of San Francisco Bay, Clear Lake, Russian River, and Pajaro-Salinas Basin.

Suborder Gobioidei. Sleepers and gobies. Spinous D., when present, with 1–8 flexible spines. P_2 thoracic in position and often united as a sucking disc. Parietal bones absent; infraorbitals, if present, not ossified. Air bladder usually absent. Eocene. Shores of warm and temperate seas, some in fresh water.

KEY TO FAMILIES OF GOBIOIDEI

1. Pelvic fins separate   **Eleotridae**
   Pelvic fins united   **Gobiidae**

Family Eleotridae. Sleepers. Gill membranes usually joined to isthmus. L.l. absent. Body scaly. A. spine single or absent; P_1 low on sides; P_2 I,5, thoracic and separate. Although *Eleotris pisonis* (Gmelin) and *E. picta* Kner and Steindachner occasionally enter fresh water, they are essentially salt-water species and are not characterized herein. *E. pisonis* is rarely recorded from fresh water on the Gulf Coast, and 1 specimen of *E. picta* has been taken from the lower Colorado River in California. Another species, *E. amblyopsis,* has been recorded from Florida.

Genus *Dormitator* Gill. Head and body with large ctenoid scales. Vomer toothless. Isthmus broad.

***Dormitator maculatus* (Bloch).** Fat sleeper. Body thickset; head short and broad. Lateral scale rows about 33. D. VII–I,8 or 9; A. I,9 or 10. Dark gray or brown, with lighter spots; large black spot above P_1 base. From North Carolina to Texas and southward through the West Indies and Mexico to Brazil. Often penetrating far inland.

Family Gobiidae. Gobies. Gill membranes broadly joined to isthmus. L.l. absent. P_2 conjoined, thoracic. 1 goby, *Clevelandia ios* (Jordan and Gilbert), enters brackish water from British Columbia to San Diego, California, but not known in fresh water. *Gillichthys mirabilis* Cooper, established in Salton Sea, California, is unable to maintain itself in fresh water.

Genus *Eucyclogobius* Gill. Body short, stout; head naked, depressed, and concave. Scales cycloid.

***Eucyclogobius newberryi* (Girard).** Creek goby. Scales small, irregular, about 60 or 70 in lateral series. Mouth very large, maxilla reaching past orbit. Jaws with caninelike teeth. In creeks and brackish waters near the Pacific Ocean from Mendocino County to San Diego County, California.

Suborder Cottoidei.[1] Sculpins. 2d infraorbital united with preopercle. Parietals fused with extrascapulars. Many species in salt waters; worldwide in both Northern and Southern Hemispheres. Paleocene. 1 family well represented in U.S. fresh waters.

Family Cottidae. Head very large, in proportion to the rest of the body, eyes placed high. Mouth large. Jaw teeth in villiform bands. Vomerine and palatine teeth present or absent. Premaxillae protractile. Gill membranes joined or not joined with isthmus. Body naked except for presence of small prickles in some species (U.S.). L.l. complete or incomplete. D. with a spinous and a soft-rayed portion. No A. spines. P_2 with 1 spine hidden in flesh. Air bladder often absent. 3 genera and 24 species in fresh waters of the United States. The characters used in the following key may break down in some localities, and if so, the list should be consulted for geographic ranges.

Recent works by Robins and Miller (1957), Robins (1961), McAllister and Lindsey (1961), and Bailey and Bond (1963) have indicated need for further study of this difficult family. In my effort to get the most information into the smallest possible space, I may have failed to abstract some useful characters from the above works.

KEY TO GENERA OF COTTIDAE

1. Preopercular spines simple, not branched   2
   Preopercular spines branched, antlerlike
                                                          Leptocottus
2. Gill membranes free from isthmus. 2d preopercular spine conspicuous, directed backward. Dorsal fins widely separated   *Myoxocephalus*
   Gill membranes attached to side of the wide isthmus. Preopercular spines absent or present, the 2d, if present, often concealed by skin and directed downward and backward. Dorsal fins joined, at least at base   *Cottus*

Genus *Leptocottus* Girard. Skin smooth, prickles absent. Suborbital stay narrow. Preopercu-

---

[1] The cottoids are aligned with the largely marine order Scorpaeniformes by Greenwood et al. (1966).

lar spines strong, with 2 or 3 points hooked upward. Gill membranes narrowly connected with isthmus. D. VII–17; A. 17; P_2 I,4. Teeth on jaws, vomer, and palatines. L.l. complete. Coloration: grayish olive above, abruptly lighter below; sides creamy; P_1 creamy yellow; spinous D. dusky with a black blotch, a white band below it, on tips of posterior rays. Soft D., A., and P_1 barred. 1 species.

Leptocottus armatus **Girard.** Staghorn sculpin. Pacific Coast from Alaska to Baja California, occasionally entering fresh water in lower stream courses.

Genus *Myoxocephalus* **Tilesius.** This genus differs from *Cottus* in the form of the preopercle, the type of L.l., and in having the gill membranes free from the isthmus.

Myoxocephalus thompsoni (**Girard**). Deepwater sculpin. Preopercular spines 4, not branched, the uppermost one less than 4.6% of standard length. Nasal, frontal, parietal, and cleithral spines sometimes present. The L.l. usually extends past the end of 2d D. D. VII–X,11–16; A. 11–16; P_1 rays 14–18, P_2 I,3. Deep lakes of northern North America. A closely related species, *M. quadricornis,* occurs in shallow salt and fresh waters of extreme northern North America.

Genus *Cottus* **Linnaeus.** Preopercular spines present or absent. Gill membranes united with isthmus. L.l. complete or incomplete. Prickles present or absent. Teeth on jaws, vomer, and sometimes palatines. A difficult group, with about 21 species in the United States.

In identifying examples of *Cottus,* some inaccuracies may be expected. Not all populations of *Cottus bairdi* and *Cottus carolinae* are easily separable on the basis of characters used here. The accompanying species list should be consulted for geographic ranges.

KEY TO SPECIES OF *COTTUS*

1. Preopercular spine long, sickle-shaped   *C. ricei*
   Preopercular spines, if present, not sickle-shaped   2
2. Pelvic rays never branched. Dorsal fins separate or narrowly joined; the first dorsal fin with an anterior and posterior dark blotch, sometimes joined as a dark band or the blotches absent. Median chin pores 2, rarely 1   3
   Pelvic rays sometimes branched. Dorsal fins usually joined (often separate in *C. pitensis,* sometimes

in *C. gulosus*). 1st dorsal fin with posterior blotch or a dark median band. Chin pores 1 or 2   12
3. Palatine teeth present, well developed   4
   Palatine teeth absent or, if present, in a short, narrow, concealed row   8
4. L.l. complete or incomplete. 2d preopercular spine hooked anteroventrad. Color pattern of bold saddles or crossbands (often replaced by mottlings in *C. bairdi*)   5
   L.l. incomplete. 2d preopercular spine directed posteroventrad. Color pattern plain, without saddles or crossbands   7
5. L.l. usually incomplete (complete, or nearly so, in Columbia and Bonneville Basins). Color pattern of mottlings, sometimes with weak crossbands. 1st dorsal fin with or without anterior and posterior blotches   *C. bairdi*
   L.l. usually complete (incomplete in *C. carolinae* from the Alabama and Kanawha Basins). Color pattern of bold crossbands beneath 2d dorsal fin and at caudal base. Pigment of 1st dorsal fin largely confined to fin rays   6
6. Caudal peduncle depth usually less than orbital length. A northwestern species   *C. rhotheus*
   Caudal peduncle depth typically equal to or greater than orbital length. Central and southeastern United States   *C. carolinae*
7. Prickles numerous and well distributed over body   *C. echinatus*
   Prickles absent on breast and belly   *C. extensus*
8. Palatine teeth in a short, narrow, concealed row   *C. confusus*
   Palatine teeth absent   9
9. Preopercle without spines   *C. leiopomus*
   Preopercle with 1–4 spines   10
10. Anal rays 10–12. L.l. with 18–22 pores   *C. cognatus*
    Anal rays 11–13. L.l. pores more than 23   11
11. L.l. with 24–26 pores. An eastern species, Tennessee and Virginia   *C. baileyi*
    L.l. nearly complete. A western species: Lahontan, Bonneville, and Columbia Basins   *C. beldingi*
12. Preopercle with only one well-developed spine (sometimes absent in *C. princeps*)   13
    Preopercle with 2 or 3 blunt to well-developed spines   17
13. Pectoral rays simple; pelvic rays I,3. L.l. very incomplete   *C. greenei*
    Pectoral rays usually branched; pelvic rays I,4 (if I,3, the rays are sometimes branched). L.l. more complete, with 15 or more pores   15
14. L.l. complete. Dorsal VIII or IX,18 or 19. Posterior nostril tubular, equal to anterior one in length   *C. aleuticus*

L.l. incomplete, with 14–26 pores. Dorsal V–
VIII,17–23. Posterior nostril not so long   **15**
15. Pelvic fin I,3, its rays branched   *C. asperrimus*
    Pelvic fin I,4, its rays simple   **16**
16. Anal rays 13–14. Prickles sometimes absent in
adults, present in young. Preopercle with 1 spine
    C. klamathensis
    Anal rays 16–18. Prickles well developed. Preopercle
with 1 spine, sometimes reduced to a mere knob
or absent   *C. princeps*
17. Pelvic fin I,3, its rays branched   *C. tenuis*
    Pelvic fin I,3 or 4, the rays simple   **18**
18. Palatine teeth present (if absent, as sometimes in *C.
gulosus*, caudal peduncle deep and compressed)
    19
    Palatine teeth absent   **20**
19. Caudal peduncle narrow and rounded. Palatine
teeth present   *C. asper*
    Caudal peduncle deep and compressed. Palatine
teeth present or absent   *C. gulosus*
20. Pelvic fin I,3 or the 4th ray reduced. Preopercular
spines usually 2, short and blunt   *C. marginatus*
    Pelvic fin I,4 (sometimes I,3 in *C. pitensis*). Pre-
opercular spines, usually 2 sharp ones   **21**
21. Pectoral rays simple; dorsal VIII or IX,17 or 18.
L.l. pores 33–37; median chin pore absent
    C. pitensis
    Pectoral rays branched; dorsal VII, 18–20. L.l. pores
20–32. Median chin pore present   *C. perplexus*

Cottus bairdi **Girard.** Mottled sculpin. D. VI–
IX,16–18; P_1 rays simple, 13–15; P_2 I,3 or 4; A. 13
or 14. Preopercular spines usually 3, the 2d
antrorse. L.l. variable, incomplete to complete.
Palatine teeth present. Axillary prickles present
or absent. Color pattern of 3 indistinct saddles at
soft D. or with scattered blotches, often with
anterior and posterior blotches, well separated from
soft D. Widespread over a discontinuous range
from the Potomac Basin across northern United
States and southern Canada to the Columbia River,
Lake Malheur, and Bonneville Basins. The nominal
species *C. bendirei* and *C. hubbsi* are included here.

Cottus baileyi **Robins.** Black sculpin. D.
VII,17; P_1 rays 15, simple; P_2 I,4; A. 15. 1 well-
developed preopercular spine, an accessory 1 above
and 1 or 2 knoblike ones below. Palatine teeth and
prickles absent. L.l. incomplete, 24–26 pores. D.
dark with orange-red border; 2 dark saddles at
soft D.; C. base black. D. fins narrowly joined.
Holston, Watauga Rivers, Tennessee, Virginia.

Cottus cognatus **Richardson.** Slimy sculpin. D.
VII–IX,15–19; P_1 rays 12–15, simple; P_2 I,3 or 4;
A. 10–12. Preopercular spines 1–3. No palatine
teeth. Prickles few, sometimes a sparse axillary
patch. L.l. incomplete, with 18–22 pores. Color pattern of 3 saddles at soft D.; spinous D. in breeding
males dark, with orange edge. North America,
north of and including the Columbia River Basin,
east and south through the Great Lakes, Iowa,
Connecticut, and Virginia.

Cottus beldingi **Eigenmann and Eigenmann.**
Piute sculpin. D. VI–VIII,15–18; P_1 rays simple;
P_2 I,4; A. 11–13. 1 long, slender preopercular spine
with a weak one below. Palatine teeth and prickles
absent. L.l. incomplete to complete. Lahontan,
Bonneville, Columbia, and upper Colorado Basins
of Nevada, Utah, Idaho, Washington, Oregon, and
Colorado. The nominal species *C. tubulatus* and
C. annae are included.

Cottus leiopomus **Gilbert and Evermann.** Wood
River sculpin. D. VII,18 or 19; P_1 rays simple;
P_2 I,4. Palatine teeth, preopercular spines and
prickles absent. L.l. incomplete. Little Wood River,
Idaho.

Cottus echinatus **Bailey and Bond.** Utah Lake
sculpin. D. VII or VIII,16–18; P_1 16–18; P_2 I,4;
A. 13–14. Preopercular spines 3 or 4 sharp ones
plus a blunt knob. Palatine teeth exposed, in 2 or
3 rows. Prickles numerous and distributed all over
body. L.l. incomplete, with 26–29 pores. Color tan
or brownish, with no conspicuous markings. Restricted to Utah Lake, Utah.

Cottus extensus **Bailey and Bond.** Bear Lake
sculpin. D. VII or VIII,16–19; P_1 15–17; P_2 I,3 or
4; A. 13–15. 3 sharp preopercular spines plus a
blunt knob. Palatine teeth exposed, in 2 to 4 rows.
Prickles present, but not on breast and belly. L.l. incomplete, 22–31 pores. No conspicuous color
pattern. Bear Lake, Utah and Idaho.

Cottus confusus **Bailey and Bond.** Shorthead
sculpin. D. VII–IX,16–18; P_1 rays 13–14, simple;
P_2 I,4; A. 12–14. Preopercular spines usually 2.
Palatine teeth concealed in a short, narrow row.
Prickles variable, present mesial to P_1 or absent.
L.l. incomplete, 22–33 pores. Sides with dark
mottlings; D. with anterior and posterior blotches.
Restricted to the drainages of Puget Sound and the
Columbia River.

Cottus asper **Richardson.** Prickly sculpin. D. VIII or IX,19–22; P_1 rays 15–18, simple; P_2 I,4; A. 15–19. Preopercular spines 2 or 3. Palatine teeth present. Prickles well developed on back and sides or mesial to P_1 only. L.l. complete, fewer than 28 pores. D. with a large posterior blotch and a thin orange edge in both sexes. Coastal rivers from Alaska to Ventura River, southern California.

Cottus pitensis **Bailey and Bond.** Pit sculpin. D. VIII or IX,17 or 18; P_1 rays 13–15, simple; P_2 I,3 or 4; A. 13–15. Preopercular spines 2 plus 1 blunt knob. Palatine teeth absent. Axillary prickles only. L.l. complete or nearly so, 33–37 pores. Color pattern of 5 or 6 indistinct saddles; spinous D. with large posterior blotch, sometimes fused with an anterior one to form a broad band; C. peduncle often with an almost complete, dark band. Pit River system in Oregon and California, in the upper Sacramento River system.

Cottus gulosus **(Girard).** Riffle sculpin. D. VII or VIII,16–19; P_1 rays 15–16, some branched; P_2 I,4; A. 14–16. Preopercular spines usually 2 or 3. Palatine teeth present or absent. Axillary prickles present. L.l. complete or incomplete; 22–36 pores. Spinous D. with a large posterior blotch. Sacramento system and coastal streams north of San Francisco Bay to the Noyo River; coastal streams from Coquille River, Oregon, to the Puget Sound drainage.

Cottus perplexus **Gilbert and Evermann.** Reticulate sculpin. D. VII,18–20; P_1 rays 14–16, some branched; P_2 I,4; A. 13–16. Preopercular spines usually 2 or 3. Palatine teeth absent. L.l. complete or incomplete, 20–32 pores. Spinous D. with posterior blotch. Coastal streams: Rogue River, lower Columbia, Willamette, and Crescent Creek, tributaries of Deschutes River.

Cottus marginatus **(Bean).** Margined sculpin. D. soft rays 17–19; P_1 rays branched; P_2 I,3 or with a 4th reduced. Usually 2 short, blunt preopercular spines. Palatine teeth absent. Prickles present. L.l. complete or nearly so. Walla Walla and Umatilla Rivers in Oregon.

Cottus klamathensis **Gilbert.** Marbled sculpin. D. VII or VIII,18 or 19; P_1 rays 14–16, some branched; P_2 I,4. C. rounded. Usually 1 preopercular spine. Palatine teeth absent. Prickles present on young but often absent on adults. L.l. incom-

plete, 14–22 pores. Restricted to the Klamath Basin, Oregon.

Cottus greenei **(Gilbert and Culver).** Shoshone sculpin. D. VI,19; P_1 rays simple; P_2 I,3. One preopercular spine. L.l. very incomplete. Palatine teeth and prickles present. Snake River, Idaho.

Cottus princeps **Gilbert.** Klamath Lake sculpin. D. VI or VII,20–23; P_1 rays 15–16, usually branched; P_2 I,4; A. 16–18. Preopercular spine single or absent. No palatine teeth. Prickles well developed. L.l. incomplete, 15–25 pores; sensory pores on the head remarkably enlarged, the preorbital pores longer than the interspaces. Restricted to Klamath Lake, Oregon.

Cottus tenuis **(Evermann and Meek).** Slender sculpin. D. V–VII,17–19; P_1 rays 13–15, profusely branched; P_2 I,3, its rays branched; A. 15–16. Preopercular spines 2 or 3, upper one long and pointed, the next below antrorse. No palatine teeth. Prickles strong to absent, never on head. L.l. incomplete, reaching to or beyond last D. ray, 25–30 pores. Venter nearly immaculate. Klamath Basin, Oregon.

Cottus asperrimus **Rutter.** Rough sculpin. D. V–VII,17–19; P_1 rays 14–15, some branched; P_2 I,3, some rays branched; A. 14–16. Preopercular spines 1 or 2, the upper one short and blunt, the next one below weak, not antrorse. No palatine teeth. Prickles well developed. L.l. incomplete reaching to 2d and 5th from last D. ray, 21–26 pores. Venter profusely dotted with chromatophores. Apparently restricted to the Pit River, tributary of Sacramento River, California.

Cottus aleuticus **Gilbert.** Coastrange sculpin. D. VIII or IX,18 or 19; P_1 14–16; P_2 I,3 or 4 long almost reaching vent; A. 12–14. 1 preopercular spine with another weak one below. No palatine teeth. Prickles few, if any. L.l. complete. Posterior nostril tubular and as long as the anterior one. Coastal streams from San Luis Obispo County, California, to Alaska.

Cottus rhotheus **(Smith).** Torrent sculpin. D. VII–IX,15–18; P_1 rays 15–16; P_2 I,4; A. 11–13. Preopercular spines 3 or 4. Palatine teeth present. Prickles variable, but usually on dorsum. L.l. usually complete. Color pattern of 2 bars under soft D. slanting anteroventrad; spinous D. without blotches. Puget Sound drainage, Columbia and

Kootenai Rivers, British Columbia, south to Ne-halem River, Oregon.

Cottus carolinae (**Gill**). Banded sculpin. D. VII–IX,16–18; P_1 15–17, with simple rays; P_2 I,4; A. 12–13. Palatine teeth present. Prickles few to absent. L.l. complete 27–34 pores. Bold bars as in *C. rhotheus* under D. fins. Widely distributed in southeastern United States in upland spring-fed streams from eastern Kansas and Oklahoma to West Virginia, north to Indiana and Illinois and south to Mississippi, Alabama, and Georgia.

Cottus ricei (**Nelson**). Spoonhead sculpin. D. VII–X,16–18; P_1 rays 14–16; P_2 I,4; A. 12–13. Preopercular spines 3 (sometimes 2 or 4), the upper one sickle-shaped. No palatine teeth. Prickles variable, may extend over whole body, reduced in number or absent. L.l. usually complete, 33–36 pores. 2 to 4 dark bars under soft D.; spinous D. without blotches. Widespread in Canada and the Great Lakes, but no reliable records from Alaska.

Order Pleuronectiformes. Flatfishes. Skull asymmetrical, with both eyes on one side. Body very flat. P_2 usually with not more than 6 rays. Usually no spines. Mostly marine coastal fishes, but some commonly enter fresh water. Lower Eocene.

KEY TO FAMILIES OF PLEURONECTIFORMES

1. Pelvic rays 6. Preopercular margin free
Pleuronectidae
Pelvic rays 5. Preopercular margin hidden by skin
Achiridae

Family Pleuronectidae. Flounders. Preopercle with free margin. 1 or 2 postcleithra. Ribs present. P_2 6.

Genus *Platichthys* Girard.

Platichthys stellatus (**Pallas**). Starry flounder. Japan to Alaska and south to Santa Barbara County, California. Often enters fresh water, having been taken 75 miles up the Columbia River.

Family Achiridae. Soles. The author follows Berg in uniting the families Achiridae and Soleidae. Preopercular margin hidden. Ribs absent. P_2 5.

Genus *Trinectes* Rafinesque.

Trinectes maculatus (**Bloch and Schneider**). Hogchoker. From Maine to Texas and south to Panama. Commonly taken far up coastal streams.

REFERENCES

General

Bailey, R. M., et al. 1960. A List of Common and Scientific Names of Fishes from the United States and Canada, *Am. Fisheries Soc. Spec. Publ.* 2, pp. 1–102.

Berg, L. S. 1947. *Classification of Fishes Both Recent and Fossil* (English and Russian), J. W. Edwards, Publisher, Inc., Ann Arbor, Mich.

Chute, W. H., et al. 1948. A List of Common and Scientific Names of the Better Known Fishes of the United States and Canada, *Am. Fisheries Soc. Spec. Publ.* 1, pp. 1–45.

Collette, B. B. 1963. The Subfamilies, Tribes and Genera of the Percidae (Teleostei), *Copeia*, **1963**(4): 615–623.

Gosline, W. A. 1963. Considerations Regarding the Relationships of the Percopsiform, Cyprinodontiform, and Gadiform Fishes, *Occasional Papers Museum Zool. Univ. Mich.*, 629, pp. 1–38.

Greenwood, P. H., D. E. Rosen, S. H. Weitzman, and G. S. Myers. 1966. Phyletic studies of teleostean fishes, with a provisional classification of living forms, *Bull. Am. Museum Nat. Hist.*, **131**(4):339–456. Text figs. 1–9; pls. 21–23; charts 1–32.

Gregory, W. K. 1933. Fish Skulls: A Study of the Evolution of Natural Mechanisms, *Trans. Am. Phil. Soc.*, **23**:75–481.

Jordan, D. S. 1929. Manual of the Vertebrate Animals of the Northeastern United States Inclusive of Marine Species, 13th ed., World Book Company, Yonkers, N.Y.

——— and B. W. Evermann. 1896. The Fishes of North and Middle America . . . , *Bull. U.S. Natl. Museum*, **47**(1–4):1–3313.

———, B. W. Evermann, and H. W. Clark. 1930. Check List of the Fishes and Fishlike Vertebrates of North and Middle America North of the Northern Boundary of Venezuela and Colombia, *Rept. U.S. Comm. Fisheries*, **1928**(2):1–670.

Krumholz, L. A. 1943. A Comparative Study of the Weberian Ossicles in North American Ostariophysine Fishes, *Copeia*, **1943**(1):33–40.

LaMonte, F. 1945. *North American Game Fishes*, Doubleday & Company, Inc., New York.

Nelson, E. M. 1948. The Comparative Morphology of the Weberian Apparatus of the Catostomidae and Its Significance in Systematics, *J. Morphol.*, **83**(2):225–245.

Norden, C. R. 1961. Comparative Osteology of Representative Salmonid Fishes, with Particular Reference to the Grayling (*Thymallus arcticus*) and Its Phylogeny, *J. Fisheries Res. Board. Can.*, **18**(5):679–791.

Romer, A. S. 1945. *Vertebrate Paleontolgy*, 2d ed., University of Chicago Press, Chicago.

Rosen, D. E. 1962. Comments on the Relationships of the North American Cave Fishes of the Family Amblyopsidae, *Am. Museum Novitates*, 2109, pp. 1–35.

Schrenkeisen, R. 1938. *Field Book of Fresh-water Fishes of North America North of Mexico*, G. P. Putnam's Sons, New York.

Storer, Tracey I., and Robert L. Usinger. 1965. *General Zoology*, 4th ed., p. 562, fig. 31–8, McGraw-Hill Book Company, New York. Used by permission.

Regional

Alvarez, J. 1950. *Claves para la Determinación de Especies en los Peces de las Aguas Continentales Méxicanas*, Secretaria de Marina, Dirección General de Pesca e Industrias Conexas, Mexico.

Bailey, J. R., and J. A. Oliver. 1939. The Fishes of the Connecticut Watershed, in: Biological Survey of the Connecticut Watershed, *N.H. Fish and Game Dept. Surv. Rept.* 4, pp. 150–189.

Bailey, R. M. 1938. The Fishes of the Merrimack Watershed, in: Biological Survey of the Merrimack Watershed, *N.H. Fish and Game Dept. Surv. Rept.* 3, pp. 149–185.

———. 1951. A Check List of the Fishes of Iowa with Keys for Identification, in: *Iowa Fish and Fishing*, pp. 187–238, Iowa State Conservation Comm.

Bailey, R. M., and M. O. Allum. 1962. Fishes of South Dakota, *Mis. Publ. Museum Zool. Univ. Mich.*, 119, pp. 1–131, 1 pl.

Beckman, W. C. 1952. Guide to the Fishes of Colorado, *Univ. Colo. Museum Leaflet* 11, pp. 1–110.

Bond, C. E. 1961. Keys to Oregon Freshwater Fishes, *Tech. Bull.* 58, Agr. Expt. Sta., Ore. State Univ., Corvallis.

Brauer, A. 1909. *Süsswasserfauna Deutschlands*, vol. 1.

Briggs, John C. 1958. A List of Florida Fishes and Their Distribution, *Bull. Fla. State Museum*, 2(8):223–318.

Carl, G. Clifford, and W. A. Clemens. 1953. The Freshwater Fishes of British Columbia, *B.C. Prov. Museum Handbook* 5, 2d. ed., rev.

Carpenter, R. G., and H. R. Siegler. 1947. *A Sportsman's Guide to the Fresh-water Fishes of New Hampshire*, pp. 1–87, N.H. Fish and Game Comm.

Carr, A. 1937. A Key to the Fresh Water Fishes of Florida, *Proc. Fla. Acad. Sci.*, 1(1936):72–86.

Carr, Archie, and C. J. Goin. 1955. *Guide to the Reptiles, Amphibians and Fresh-water Fishes of Florida*, University of Florida Press, Gainesville, i–x, 1–341.

Churchill, E. P., and W. H. Over. 1933. Fishes of South Dakota, *S.D. Dept. Game and Fish*, 1933:1–87.

Clay, W. M. 1962. A Field Manual of Kentucky Fishes, *Ky. Dept. Fish Wildlife Res.*, pp. VII + 147.

Cook, Fannye A. 1959. Freshwater Fishes in Mississippi, *Miss. Game and Fish Comm.*, pp. 1–239.

Curtis, B. 1949. The Warm-water Game Fishes of California, *Calif. Fish and Game*, 35(4):255–273.

Dymond, J. R. 1932. *The Trout and Other Game Fishes of British Columbia*, pp. 1–51, Dom. Canada Dept. of Fisheries.

Eddy, S., and T. Surber. 1947. *Northern Fishes with Special Reference to the Upper Mississippi Valley*, 2d ed., University of Minnesota Press, Minneapolis.

Elser, H. J. 1950. The Common Fishes of Maryland: How to Tell Them Apart, *Md. Bd. Nat. Res., Dept. Res. and Educ.*, 88, pp. 1–45.

Everhart, W. H. 1950. *Fishes of Maine*, pp. 1–53, Maine Dept. Inland Fish and Game.

Evermann, B. W., and W. C. Kendall. 1894. The Fishes of Texas and the Rio Grande Basin, Considered Chiefly with Reference to Their Geographic Distribution, *Bull. U.S. Fish Comm.*, 1892:55–126.

——— and ———. 1900. Check List of the Fishes of Florida, *Rept. U.S. Fish Comm.*, 1899:37–103.

Forbes, S. A., and R. E. Richardson. 1920. The Fishes of Illinois, 2d ed., *Ill. Nat. Hist. Surv. Circ.* 3.

Fowler, H. W. 1906. The Fishes of New Jersey, *Ann. Rept. N.J. State Museum*, 1905(2):35–477.

———. 1907. A Supplementary Account of the Fishes of New Jersey, *Ann. Rept. N.J. State Museum*, 1906 (3):251–384.

———. 1919. A List of the Fishes of Pennsylvania, *Proc. Biol. Soc. Wash.*, 32:49–73.

———. 1945. A Study of the Fishes of the Southern Piedmont and Coastal Plain, *Monograph Acad. Nat. Sci. Phila.*, 7, pp. 1–408.

Gerking, S. D. 1945. The Distribution of the Fishes of Indiana, *Invest. Ind. Lakes Streams*, 3(1):1–137.

———. 1955. Key to the Fishes of Indiana, *Invest. Ind. Lakes Streams*, 4(2):49–86.

Greeley, J. R. 1927. Fishes of the Genesee Region with Annotated List, in: A Biological Survey of the Genesee River System, *Suppl. 16th Ann. Rept. N.Y. State Conservation Dept.*, 1926, pp. 47–66.

———. 1928. Fishes of the Oswego Watershed with Annotated List, in: A Biological Survey of the Oswego River System, *Suppl. 17th Ann. Rept. N.Y. State Conservation Dept.*, 1927, pp. 84–107.

———. 1929. Fishes of the Erie-Niagara Watershed with Annotated List, in: A Biological Survey of the Erie-Niagara System, *Suppl. 18th Ann. Rept. N.Y. State Conservation Dept.*, 1928, pp. 150–179.

———. 1930. Fishes of the Lake Champlain Watershed with Annotated List, in: A Biological Survey of the Champlain Watershed, *Suppl. 19th Ann. Rept. N.Y. State Conservation Dept.*, 1929, pp. 44–87.

———. 1934. Annotated List of the Fishes Occurring in

the Watershed, in: A Biological Survey of the Raquette Watershed, *Suppl. 23d Ann. Rept. N.Y. State Conservation Dept.*, 1933, pp. 53–108.

———. 1935. Fishes of the Watershed with Annotated List, in: A Biological Survey of the Mohawk-Hudson Watershed, *Suppl. 24th Ann. Rept. N.Y. State Conservation Dept.*, 1934, pp. 63–101.

———. 1936. Fishes of the Area with Annotated List, in: A Biological Survey of the Delaware and Susquehanna Watersheds, *Suppl. 25th Ann. Rept. N.Y. State Conservation Dept.*, 1935, pp. 45–88.

———. 1937. Fishes of the Area with Annotated List, in: A Biological Survey of the Lower Hudson Watershed, *Suppl. 26th Ann. Rept. N.Y. State Conservation Dept.*, 1936, pp. 45–103.

———. 1938. Fishes of the Area with Annotated List, in: A Biological Survey of the Allegheny and Chemung Watersheds, *Suppl. 27th Ann. Rept. N.Y. State Conservation Dept.*, 1937, pp. 48–73.

———. 1939. The Freshwater Fishes of Long Island and Staten Island with Annotated List, in: A Biological Survey of the Fresh Waters of Long Island, *Suppl. 28th Ann. Rept. N.Y. State Conservation Dept.*, 1938, pp. 29–44.

———. 1940. Fishes of the Watershed with Annotated List, in: A Biological Survey of the Lake Ontario Watershed, *Suppl. 29th Ann. Rept. N.Y. State Conservation Dept.*, 1939, pp. 42–81.

——— and S. C. Bishop. 1932. Fishes of the Area with Annotated List, in: A Biological Survey of the Oswegatchie and Black River Systems, *Suppl. 21st Ann. Rept. N.Y. State Conservation Dept.*, 1931, pp. 54–92.

——— and ———. 1933. Fishes of the Upper Hudson Watershed with Annotated List, in: A Biological Survey of the Upper Hudson Watershed, *Suppl. 22d Ann. Rept. N.Y. State Conservation Dept.*, 1932, pp. 64–101.

——— and C. W. Greene. 1931. Fishes of the Area with Annotated List, in: A Biological Survey of the St. Lawrence Watershed, *Suppl. 20th Ann. Rept. N.Y. State Conservation Dept.*, 1930, pp. 44–94.

Greene, C. W. 1935. The Distribution of Wisconsin Fishes, pp. 1–235, Wisconsin Conservation Comm., Madison, Wis.

Hankinson, T. L. 1929. Fishes of North Dakota, *Papers Mich. Acad. Sci., Arts Letters*, **10**(1928):439–460.

Hay, O. P. 1894. The Lampreys and Fishes of Indiana, *19th Ann. Rept. Ind. Dept. Geol. and Nat. Res.*, pp. 147–296.

Henshall, J. A. 1906. A List of the Fishes of Montana with Notes on the Game Fishes, *Bull. Univ. Mont.*, **34**:1–12.

Hinks, D. 1943. *The Fishes of Manitoba*, pp. 1–102, Prov. Manitoba, Dept. Mines and Nat. Res.

Hubbs, C. L., and K. F. Lagler. 1947. Fishes of the Great Lakes Region, *Bull. Cranbrook Inst. Sci.*, **26:** 1–186.

——— and O. L. Wallis. 1948. The Native Fish Fauna of Yosemite National Park and Its Preservation, *Yosemite Nature Notes*, **27**(12):131–144.

Hubbs, Clark. 1954. Corrected Distributional Records for Texas Fresh-water Fishes, *Tex. J. Sci.*, **6**(3):277–291.

Kendall, W. C. 1908. Fauna of New England. 8. List of the Pisces, *Occasional Papers Boston Soc. Nat. Hist.*, 7, pp. 1–152.

———. 1914. The Fishes of New England: The Salmon Family. 1. The Trout or Charrs, *Mem. Boston Soc. Nat. Hist.*, **8**(1):1–103.

———. 1935. The Fishes of New England: The Salmon Family. 2. The Salmons, *Mem. Boston Soc. Nat. Hist.*, **9**(1):1–166.

King, W. 1947. *Important Food and Game Fishes of North Carolina*, pp. 1–54, Dept. Conservation and Development, Div. Game and Inland Fish.

Koster, W. J. 1957. *Guide to the Fishes of New Mexico*, The University of New Mexico Press, Albuquerque, pp. VII–116.

Kuhne, E. R. 1939. *A Guide to the Fishes of Tennessee and the Mid-South*, Tenn. Dept. Conservation, Div. Game and Fish, pp. 1–124.

La Rivers, I. 1952. A Key to Nevada Fishes, *Bull. S. Calif. Acad. Sci.*, **51**(3):86–102.

——— and T. J. Trelease. 1952. An Annotated Check List of the Fishes of Nevada, *Calif. Fish and Game*, **38**(1):113–123.

Legendre, V. 1949. Clef des poissons de pêche sportive et commerciale de la Province de Quebec, **1**:I–XIII, 1–84, Société Canadienne d'Escologie.

———. 1953. The Freshwater Fishes of the Province of Quebec, *9th Rept. Biol. Bur., Prov. Quebec, Game and Fisheries Dept.*, 1951–1952, pp. 190–295.

Michael, E. L. 1906. Catalogue of Michigan Fish, *Bull. Mich., Fish Comm.*, 8, pp. 1–45.

Miller, R. R. 1952. Bait Fishes of the Lower Colorado River from Lake Mead, Nevada, to Yuma, Arizona, with a Key for Their Identification, *Calif. Fish and Game*, **38**(1):7–42.

Moore, G. A. 1952. *A List of the Fishes of Oklahoma*, Oklahoma Game and Fish Dept.

Murphy, G. 1941. A Key to the Fishes of the Sacramento–San Joaquin Basin, *Calif. Fish and Game*, **27**(3):165–171.

Neale, G. 1931. Spiny-rayed Fresh Water Game Fishes of California Inland Waters, *Calif. Fish and Game*, **17**(1):1–16.

O'Donnell, J. D. 1935. Annotated List of the Fishes of Illinois, *Ill. Nat. Hist. Surv. Bull.,* **20**(5):473–500.

Osburn, R. C., E. L. Wickliff, and M. B. Trautman. 1930. A Revised List of the Fishes of Ohio, *Ohio J. Sci.,* **30**:169–176.

Raney, E. C. 1950. Freshwater Fishes, in: *The James River Basin Past, Present and Future,* pp. 151–194, Virginia Academy of Science, Richmond.

Schultz, L. P. 1948. Keys to the Fishes of Washington, Oregon and Closely Adjoining Regions, *Univ. Wash. Publ. Biol.,* **2**(4):103–228.

———. 1941. Fishes of Glacier National Park, Montana, *U.S.D.I. Conservation, Bull.* 22, pp. 1–42.

Shapovalov, L., and W. A. Dill. 1950. A Check List of the Fresh-water and Anadromous Fishes of California, *Calif. Fish and Game,* **36**(4):382–391.

———, ———, and A. J. Cordone. 1959. A Revised Check List of the Freshwater and Anadromous Fishes of California, *Calif. Fish and Game,* **45**(3):159–180.

Sigler, W. F., and R. R. Miller. 1963. *Fishes of Utah,* Utah Dept. of Fish and Game, pp. 1–203.

Simon, J. R. 1939. *Yellowstone Fishes,* The Yellowstone Library and Museum Association.

———. 1946. Wyoming Fishes, *Bull. Wyo. Game Fish Dept.,* **4**:1–129.

——— and F. Simon. 1942. Check List and Keys of the Fishes of Wyoming, *Univ. Wyo. Rev. Publ.,* **6**(4):47–65.

Smith, H. M. 1907. The Fishes of North Carolina, *N.C. Geol. and Econ. Surv.,* 2, pp. 1–445.

Surber, T. 1920. A Preliminary Catalogue of the Fishes and Fish-like Vertebrates of Minnesota. *Appendix Bienn. Rept. State Game and Fish Comm.,* 1920, pp. 1–92.

Tanner, V. M. 1936. A Study of the Fishes of Utah, *Proc. Utah Acad. Sci., Arts Letters,* **13**:155–184.

Trautman, M. B. 1957. *The Fishes of Ohio,* Ohio State University Press, Columbus, pp. i–xvii, 1–683.

Truitt, R. V., B. A. Bean, and H. W. Fowler. 1929. The Fishes of Maryland, *Md. Conservation Dept., Bull.* 3, pp. 1–120.

Vladykov, V. D. 1949. Quebec Lampreys (Petromyzonidae). I. List of Species and Their Economic Importance. Prov. Quebec, Dept. of Fisheries, Contr. 26.

——— and Gerard Beaulieu. 1946. Etudes sur l'Esturgeon (*Acipenser*) de la Province de Quebec. I. Distinction entre deux espèces d'Esturgeon par le nombre de boucliers osseux et de branchiospines, *Naturaliste Canadien,* **73**:143–204. 1951. II. Variation du nombre de branchiospines sur le premier arc branchial, *ibid.,* **78**:129–154.

Ward, H. C. (n.d.) *Know Your Oklahoma Fishes,* Oklahoma Game and Fish Dept.

Weed, A. C. 1925. A Review of the Fishes of the Genus *Signalosa, Field Museum Nat. Hist. Publ.,* 233, **12**(11):137–146.

Specific

Bailey, R. M. 1940. *Hadropterus palmaris,* a New Darter from the Alabama River System, *J. Wash. Acad. Sci.,* **30**(12):524–530.

———. 1941. *Hadropterus nasutus,* a New Darter from Arkansas, *Occasional Papers Museum Zool. Univ. Mich.,* 440, pp. 1–8.

———. 1959. A New Catostomid Fish, *Moxostoma* (*Thoburnia*) *atripinne,* from the Green River Drainage, Kentucky and Tennessee, *Occasional Papers Museum Zool. Univ. Mich.,* 599, pp. 1–19, 2 pls.

———. 1959. *Etheostoma acuticeps,* a New Darter from the Tennessee River System, with Remarks on the Subgenus *Nothonotus, Occasional Papers Museum Zool. Univ. Mich.,* 603, pp. 1–11.

——— and C. E. Bond. 1963. Four New Species of Freshwater Sculpins, Genus *Cottus,* from Western North America, *Occasional Papers Museum Zool. Univ. Mich.,* 634, pp. 1–27.

——— and F. B. Cross. 1954. River Sturgeons of the American Genus *Scaphirhynchus:* Characters, Distribution, and Synonymy, *Papers Mich. Acad. Sci., Arts Letters,* **39**(1953):169–208.

——— and M. F. Dimick. 1949. *Cottus hubbsi,* a New Cottid Fish from the Columbia River System in Washington and Idaho, *Occasional Papers Museum Zool. Univ., Mich.,* 513, pp. 1–18.

——— and R. H. Gibbs. 1956. *Notropis callitaenia,* a New Cyprinid Fish from Alabama, Florida and Georgia, *Occasional Papers Museum Zool. Univ. Mich.,* 576, pp. 1–14, 1 pl.

——— and C. R. Gilbert. 1960. The American Cyprinid Fish *Notropis kanawha* Identified as an Interspecific Hybrid, *Copeia,* **1960**(4):354–357.

——— and C. L. Hubbs. 1949. The Black Basses (*Micropterus*) of Florida, with Description of a New Species, *Occasional Papers Museum Zool. Univ. Mich.,* 516, pp. 1–40.

——— and W. J. Richards. 1963. Status of *Poecilichthys hopkinsi* Fowler and *Etheostoma trisella,* New Species, Percid Fishes from Alabama, Georgia, and South Carolina, *Occasional Papers Museum Zool. Univ. Mich.,* 630, pp. 1–21.

——— and R. D. Suttkus. 1952. *Notropis signipinnis,* a New Cyprinid Fish from Southeastern United States, *Occasional Papers Museum Zool. Univ. Mich.,* 542, pp. 1–15.

——— and W. R. Taylor. 1950. *Schilbeodes hildebrandi,*

a New Ameiurid Catfish from Mississippi, *Copeia,* **1950**(1):31–38.

—— and Teruya Uyeno. 1964. Nomenclature of the Blue Chub and the Tui Chub, Cyprinid Fishes from Western United States, *Copeia,* **1964**(1):238–239.

Berg, L. S. 1931. A Review of the Lampreys of the Northern Hemisphere, *Ann. Museum Zool. Acad. Sci. U.R.S.S.,* **32**:87–116.

Böhlke, J. 1956. A New Pygmy Sunfish from Southern Georgia, *Notulae Naturae Acad. Nat. Sci. Phila.,* 294, pp. 1–11.

Brown, J. L. 1957. A Key to the Species and Subspecies of the Cyprinodont Genus *Fundulus* in the United States and Canada East of the Continental Divide, *J. Wash. Acad. Sci.,* **47**(3):69–77.

——. 1958. Geographic Variation in Southeastern Populations of the Cyprinodont Fish *Fundulus notti* (Agassiz), *Am. Midland Naturalist,* **59**(2):477–488.

Collette, Bruce B. 1962. The Swamp Darters of the Subgenus *Hololepis* (Pisces, Percidae), *Tulane Stud. Zool.,* **9**(4):113–211.

—— and Ralph W. Yerger. 1962. The American Percid Fishes of the Subgenus *Villora, Tulane Stud. Zool.,* **9**(4):213–230.

Cross, F. B. 1953. A New Minnow, *Notropis bairdi buccula,* from the Brazos River, Texas, *Tex. J. Sci.,* **1953**(2):252–259.

Dick, Myvanwy M. 1964. Suborder Esocoidei, in: *Fishes of the Western North Atlantic, Mem. Sears Found. Marine Res.,* **1**(4):550–560.

Distler, D. A., and A. L. Metcalf. 1962. *Etheostoma pallididorsum,* a New Percid Fish from the Caddo River System of Arkansas, *Copeia,* **1962**(3):556–561.

Fingerman, Sue W., and R. D. Suttkus. 1961. Comparison of *Hybognathus hayi* Jordan and *Hybognathus nuchalis* Agassiz, *Copeia,* **1961**(4):462–467.

Fowler, H. W. 1942. Descriptions of Six New Freshwater Fishes (Cyprinidae and Percidae) from the Southeastern United States, *Notulae Naturae, Acad. Nat. Sci. Phila.,* 107, pp. 1–11.

Gibbs, R. H., Jr. 1957. Cyprinid Fishes of the Subgenus *Cyprinella* of *Notropis,* I. Systematic Status of the Subgenus *Cyprinella,* with a Key to the Species Exclusive of the *lutrensis-ornatus* Complex, *Copeia,* **1957**(3):185–195.

——. 1957. Cyprinid Fishes of the Subgenus *Cyprinella* of *Notropis,* II. Distribution and Variation of *Notropis spilopterus,* with the Description of a New Subspecies, *Lloydia,* **20**(3):186–211.

——. 1957. Cyprinid Fishes of the Subgenus *Cyprinella* of *Notropis,* III. Variation and Subspecies of *Notropis venustus* (Girard), *Tulane Stud. Zool.,* **5**(8):175–203.

——. 1961. Cyprinid Fishes of the Subgenus *Cypri-*nella of *Notropis,* IV. The *Notropis galacturus-camurus* Complex, *Am. Midland Naturalist,* **66**(2): 337–354.

——. 1963. Cyprinid Fishes of the Subgenus *Cyprinella* of *Notropis.* The *Notropis whipplei-analostanus-chloristius* Complex, *Copeia,* **1963**(3):511–528.

Gilbert, C. R. 1961a. Hybridization versus Intergradation: An Inquiry into the Relationship of Two Cyprinid Fishes, *Copeia,* **1961**(2):181–192.

——. 1961b. *Notropis semperasper,* a New Cyprinid Fish from the Upper James River System, Virginia, *Copeia,* **1961**(4):450–456.

—— and R. M. Bailey. 1962. Synonymy, Characters, and Distribution of the American Cyprinid Fish, *Notropis shumardi, Copeia,* **1962**(4):807–819.

Gosline, W. A. 1948. Speciation in the Fishes of the Genus *Menidia, Evolution,* **2**(4):306–313.

Harrington, R. W., Jr., and L. R. Rivas. 1958. The Discovery in Florida of the Cyprinodont Fish, *Rivulus marmoratus,* with a Redescription and Ecological Notes, *Copeia,* **1958**(2):125–130.

Hildebrand, S. F. 1943. A Review of the American Anchovies (Family Engraulidae), *Bull. Bingham Oceanographic Collection,* **8**(2):1–165.

Holt, Ramona D. 1960. Comparative Morphometry of the Mountain Whitefish, *Prosopium williamsoni, Copeia,* **1960**(3):192–200.

Howell, W. M., and R. D. Caldwell. 1965. *Etheostoma (Oligocephalus) nuchale,* a New Darter from a Limestone Spring in Alabama, *Tulane Stud. Zool.,* **12**(4):101–108.

Hubbs, C. L. 1930. Materials for a Revision of the Catostomid Fishes of Eastern North America, *Misc. Publ. Museum Zool. Univ. Mich.,* 20, pp. 1–47.

——. 1932. Studies of the Fishes of the Order Cyprinodontes. XII. A New Genus Related to *Empetrichthys, Occasional Papers Museum Zool. Univ. Mich.,* 252, pp. 1–5.

——. 1941. A Systematic Study of Two Carolinian Minnows, *Notropis scepticus* and *Notropis altipinnis, Copeia,* **1941**(3):165–174.

——. 1951. *Notropis amnis,* a New Cyprinid Fish of the Mississippi Fauna, with Two Subspecies, *Occasional Papers Museum Zool. Univ. Mich.,* 530, pp. 1–30. 1 pl.

—— and R. M. Bailey. 1940. A Revision of the Black Basses (*Micropterus* and *Huro*) with Descriptions of Four New Forms, *Misc. Publ. Museum Zool. Univ. Mich.,* 48, pp. 1–51.

—— and J. D. Black. 1940. Percid Fishes Related to *Poccilichthys variatus* with Descriptions of Three New Forms, *Occasional Papers Museum Zool. Univ. Mich.,* 416, pp. 1–30.

—— and ——. 1947. Revision of *Ceratichthys,* a

Genus of American Cyprinid Fishes, *Misc. Publ. Museum Zool. Univ. Mich.,* 66, pp. 1–56.

—— and K. Bonham. 1951. New Cyprinid Fishes of the Genus *Notropis* from Texas, *Tex. J. Sci.,* 3(1): 91–110.

—— and M. D. Cannon. 1935. The Darters of the Genera *Hololepis* and *Villora, Misc. Publ. Museum Zool. Univ. Mich.,* 30, pp. 1–93.

—— and W. R. Crowe. 1956. Preliminary Analysis of the American Cyprinid Fishes, Seven New, Referred to the Genus *Hybopsis,* Subgenus *Erimystax, Occasional Papers Museum Zool. Univ. Mich.,* 578, pp. 1–8.

——, L. C. Hubbs, and R. E. Johnson. 1943. Hybridization in Nature between Species of Catostomid Fishes, *Contr. Lab. Vert. Biol. Univ. Mich.,* 22, pp. 1–76.

—— and R. R. Miller. 1941. Studies of the Fishes of the Order Cyprinodontes. XVII. Genera and Species of the Colorado River System, *Occasional Papers Museum Zool. Univ. Mich.,* 433, pp. 1–9.

—— and ——. 1948. Two New, Relict Genera of Cyprinid Fishes from Nevada, *Occasional Papers Museum Zool. Univ. Mich.,* 507, pp. 1–30, 3 pls.

—— and ——. 1965. Studies of Cyprinodont Fishes, XXII. Variation in *Lucania parva,* Its Establishment in Western United States, and Description of a New Species from an Interior Basin in Coahuila, Mexico, *Misc. Publ. Museum Zool. Univ. Mich.,* 127, pp. 1–104, 3 pls.

—— and E. C. Raney. 1939. *Hadropterus oxyrhynchus,* a new Percid Fish from Virginia and West Virginia, *Occasional Papers Museum Zool. Univ. Mich.,* 396, pp. 1–9.

—— and ——. 1944. Systematic Notes on North American Siluroid Fishes of the Genus *Schilbeodes, Occasional Papers Museum Zool. Univ. Mich.,* 487, pp. 1–37.

—— and ——. 1947. *Notropis alborus,* a New Cyprinid Fish from North Carolina and Virginia, *Occasional Papers Museum Zool. Univ. Mich.,* 498, pp. 1–17.

—— and ——. 1951. Status, Subspecies, and Variations of *Notropis cummingsae,* a Cyprinid Fish of the Southeastern United States, *Occasional Papers Museum Zool. Univ. Mich.,* 535, pp. 1–25.

—— and M. B. Trautman. 1932. *Poecilichthys osburni,* a New Darter from the Upper Kanawha River System in Virginia and West Virginia, *Ohio J. Sci.,* 32(1):31–38.

—— and ——. 1937. A Revision of the Lamprey Genus *Ichthyomyzon, Misc. Publ. Museum Zool. Univ. Mich.* 35, pp. 1–109.

Hubbs, Clark. 1957. *Gambusia heterochir,* A New Poeciliid Fish from Texas, with an Account of Its Hybridization with *G. affinis, Tulane Stud. Zool.,* 5(1):1–16.

—— and V. G. Springer. 1957. A Revision of the *Gambusia nobilis* Species Group, with Descriptions of Three New Species, and Notes on Their Variation, Ecology, and Evolution, *Tex. J. Sci.,* 9(3):279–327.

Koelz, W. 1929. Coregonid Fishes of the Great Lakes. *Bull. U.S. Bur. Fish,* 43(2):297–643.

——. 1931. The Coregonid Fishes of Northeastern North America, *Papers Mich. Acad. Sci., Arts and Letters,* 13(1930):303–432.

Kuehne, R. A., and R. M. Bailey. 1961. Stream Capture and the Distribution of the Percid Fish, *Etheostoma sagitta,* with Geologic and Taxonomic Considerations, *Copeia,* 1961(1):1–8.

Lachner, E. A. 1952. Studies of the Biology of the Cyprinid Fishes of the Chub Genus *Nocomis* of Northeastern United States, *Am. Midland Naturalist,* 48(2):433–466.

McAllister, D. E. 1961. The Origin and Status of the Deepwater Sculpin, *Myoxocephalus thompsonii,* a Nearctic Glacial Relict, *Bull.* 172, *Natl. Museum Can., Contr. Zool.* 1959, pp. 44–65.

——. 1963. A Revision of the Smelt Family, Osmeridae, *Natl. Museum Can., Bull.* 191, pp. i–iv, 1–53.

—— and C. C. Lindsey. 1961. Systematics of the Freshwater Sculpins (*Cottus*) of British Columbia, *Bull.* 172, *Natl. Museum Can., Contr. Zool.,* 1959, pp. 66–89.

Miller, R. R. 1943. The Status of *Cyprinodon macularius* and *Cyprinodon nevadensis,* Two Desert Fishes of Western North America, *Occasional Papers Museum Zool. Univ. Mich.,* 473, 1–25. 7 pls.

——. 1945. A New Cyprinid Fish from Southern Arizona, and Sonora, Mexico, with the Description of a New Subgenus of Gila and a Review of Related Species, *Copeia,* 1945(2):104–110.

——. 1946. *Gila cypha,* a Remarkable New Species of Cyprinid Fish from the Colorado River in Grand Canyon, Arizona, *J. Wash. Acad. Sci.,* 36(12):409–415.

——. 1948. The Cyprinodont Fishes of the Death Valley System of Eastern California and Southwestern Nevada, *Misc. Publ. Museum Zool. Univ. Mich.,* 68, pp. 1–155.

——. 1950. Notes on the Cutthroat and Rainbow Trouts with the Description of a New Species from the Gila River, New Mexico, *Occasional Papers Museum Zool. Univ. Mich.,* 529, pp. 1–42.

——. 1950. A Review of the American Clupeid Fishes of the Genus *Dorosoma, Proc. U.S. Natl. Museum,* 100(3267):387–410.

——. 1956. A New Genus and Species of Cyprinodont

Fish from San Luis Potosi, Mexico, with Remarks on the Subfamily Cyprinodontinae, *Occasional Papers Museum Zool. Univ. Mich.*, 581, pp. 1–17, 2 pls.

———. 1963. Distribution, Variation, and Ecology of *Lepidomeda vitta*, a Rare Cyprinid Fish Endemic to Eastern Arizona, *Copeia*, **1963**(1):1–5.

———. 1963. Synonymy, Characters, and Variations of *Gila crassicauda*, a rare Californian Minnow, with an Account of Its Hybridization with *Lavinia exilicauda*, *Calif. Fish and Game*, 49(1):20–29.

——— and C. L. Hubbs. 1960. The Spiny-rayed Cyprinid Fishes (Plagopterini) of the Colorado River System, *Misc. Publ. Museum Zool. Univ. Mich.*, 115, pp. 1–39, 3 pls.

Minckley, W. L., and J. E. Craddock. 1962. A New Species of *Phenacobius* (Cyprinidae) from the Upper Tennessee River System, *Copeia*, **1962**(2):369–377.

Moore, George A., and J. D. Reeves. 1955. *Hadropterus pantherinus*, a New Percid Fish from Oklahoma and Arkansas, *Copeia*, **1955**(2):89–92.

Morton, W. M., and R. R. Miller. 1954. Systematic Position of the Lake Trout, *Salvelinus namaycush*, *Copeia*, **1954**(2):116–124.

Ramsey, J. S., and R. D. Suttkus. 1965. *Etheostoma ditrema*, a New Darter of the Subgenus *Oligocephalus* (Percidae) from Springs of the Alabama River Basin in Alabama and Georgia, *Tulane Stud. Zool.*, **12**(3): 65–77.

Raney, E. C. 1941. *Poecilichthys kanawhae*, a New Darter from the Upper New River System in North Carolina and Virginia, *Occasional Papers Museum Zool. Univ. Mich.*, 434, pp. 1–16.

———. 1952. The Life History of the Striped Bass, *Roccus saxatilis* (Walbaum), *Bull. Bingham Oceanographic Coll., Peabody Museum Nat. Hist.*, 14(1): 5–97.

——— and C. L. Hubbs. 1948. *Hadropterus notogrammus*, a New Percid Fish from Maryland, Virginia, and West Virginia, *Occasional Papers Museum Zool. Univ. Mich.*, 512, pp. 1–26.

——— and E. A. Lachner. 1947. *Hypentelium roanokense*, a New Catostomid Fish from the Roanoke River in Virginia, *Am. Museum Novitates*, 1333, pp. 1–15.

——— and R. D. Suttkus. 1964. *Etheostoma moorei*, a New Darter of the Subgenus *Nothonotus* from the White River System, Arkansas, *Copeia*, **1964**(1): 130–139.

——— and ———. 1966. *Etheostoma rubrum*, a New Percid Fish of the Subgenus *Nothonotus* from Bayou Pierre, Mississippi, *Tulane Stud. Zool.*, 13(3):95–102.

Richards, William J., and Leslie W. Knapp. 1964. *Percina lenticula*, a New Percid Fish, with a Re-description of the Subgenus *Hadropterus, Copeia*, **1964**(4):690–701.

Rivas, L. R. 1963. Subgenera and Species Groups in the Poeciliid Fish Genus *Gambusia* Poey, *Copeia*, **1963**(2):331–347.

———. 1966. The Taxonomic Status of the Cyprinodontid Fishes *Fundulus notti* and *F. lineolatus, Copeia*, **1966**(2):353–354.

Robins, C. R. 1961. Two New Cottid Fishes from the Fresh Waters of Eastern United States, *Copeia*, **1961**(3):305–315.

——— and R. R. Miller. 1957. Classification, Variation, and Distribution of the Sculpins, Genus *Cottus*, Inhabiting Pacific Slope Waters in California and Southern Oregon, with a Key to the Species, *Calif. Fish and Game*, 42(3):213–233.

——— and E. C. Raney. 1956. Studies of the Catostomid Fishes of the Genus *Moxostoma*, with Descriptions of Two New Species, *Cornell Univ. Mem.*, 343, pp. 1–56.

——— and ———. 1957. The Systematic Status of the Suckers of the Genus *Moxostoma* from Texas, New Mexico and Mexico, *Tulane Stud. Zool.*, 5(12):291–318.

Rosen, D. E., and R. M. Bailey. 1963. The Poeciliid Fishes (Cyprinodontiformes), Their Structure, Zoogeography and Systematics, *Bull. Am. Museum Nat. Hist.*, **126**(1):1–176.

Schlicht, F. G. 1959. First Records of the Mountain Mullet, *Agonostomus monticola* (Bancroft), in Texas, *Tex. J. Sci.*, 11(2):181–182.

Schultz, E. E. 1960. Establishment and Early Dispersal of a Loach, *Misgurnus anguillicaudatus* (Cantor), in Michigan, *Trans. Am. Fisheries Soc.*, 89(4):376–377.

Schultz, L. P. 1929. Description of a New Type of Mud-minnow from Western Washington with Notes on Related Species, *Univ. Wash. Publ. in Fisheries*, 2(6):73–82.

Smith, G. R. 1966. Distribution and Evolution of the North American Catostomid Fishes of the Subgenus *Pantosteus*, Genus *Catostomus, Misc. Publ. Museum Zool. Univ. Mich.*, 129, pp. 1–132, 1 pl.

Springer, V. G., and J. H. Finucane. 1963. The African Cichlid, *Tilapia heudeloti* Dumeril, in the Commercial Fish Catch of Florida, *Trans. Am. Fisheries Soc.*, 92(3):317–318.

Suttkus, R. D. 1955. *Notropis euryzonus*, a New Cyprinid Fish from the Chattahoochee River System of Georgia and Alabama, *Tulane Stud. Zool.*, 3(5):85–100.

———. 1956. First Record of the Mountain Mullet, *Agonostomus monticola* (Bancroft), in Louisiana, *Proc. La. Acad. Sci.*, 19, pp. 43–46.

————. 1958. Status of the Nominal Cyprinid Species *Moniana deliciosa* Girard and *Cyprinella texana* Girard, *Copeia*, **1958**(4):307–318.

————. 1959. *Notropis uranoscopus*, a New Cyprinid Fish from the Alabama River System, *Copeia*, **1959**(1):7–11.

————. 1963. Order *Lepisostei*, in: *Fishes of the Western North Atlantic, Mem. Sears Found. Marine Res.*, **I**(3):61–88.

———— and J. S. Ramsey. 1967. *Percina aurolineata*, a New Percid Fish from the Alabama River System and a Discussion of Ecology, Distribution, and Hybridization of Darters of the Subgenus *Hadropterus*, *Tulane Stud. Zool.*, **13**(4):129–145.

———— and E. C. Raney. 1955. *Notropis baileyi*, a New Cyprinid Fish from the Pascagoula and Mobile Bay Drainages of Mississippi and Alabama, *Tulane Stud. Zool.*, **2**(5):71–86.

———— and ————. 1955. *Notropis hypsilepsis*, a New Cyprinid Fish from the Apalachicola River System of Georgia and Alabama, *Tulane Stud. Zool.*, **2**(7):161–170.

———— and ————. 1955. *Notropis asperifrons*, a New Cyprinid Fish from the Mobile Bay Drainage of Alabama and Georgia, with Studies of Related Species, *Tulane Stud. Zool.*, **3**(1):1–33.

———— and W. R. Taylor. 1965. *Noturus munitus*, a New Species of Madtom, Family Ictaluridae, from Southern United States, *Proc. Biol. Soc. Wash.*, **78**:169–178.

Tarp, F. H. 1952. A Revision of the Family Embiotocidae (the Surfperches), *Calif. Dept. Fish and Game Fish, Bull.* 88, pp. 1–99.

Thomerson, J. E. 1966. A Comparative Biosystematic Study of *Fundulus notatus* and *Fundulus olivaceus* (Pisces: Cyprinodontidae), *Tulane Stud. Zool.*, **13**(1):29–47.

Trautman, M. B. 1931. *Parexoglossum hubbsi*, a New Cyprinid Fish from Western Ohio, *Occasional Papers Museum Zool. Univ. Mich.*, 235, pp. 1–11.

———— and R. G. Martin. 1951. *Moxostoma aureolum pisolabrum*, a New Subspecies of Sucker from the Ozarkian Streams of the Mississippi River System, *Occasional Papers Museum Zool. Univ. Mich.*, 534, pp. 1–10.

Vladykov, V. D. 1963. A Review of Salmonid Genera and Their Broad Geographical Distribution, *Trans. Roy. Soc. Can.*, 1(Ser. 4, Sec. 3):459–504.

———— and W. I. Follett. 1958. Redescription of *Lampetra ayresii* (Günther) of Western North America, a Species of Lamprey (Petromyzontidae) Distinct from *Lampetra fluviatilis* (Linnaeus) of Europe, *J. Fisheries Res. Board Can.*, **15**(1):47–77.

———— and ————. 1965. *Lampetra richardsoni*, a New Nonparasitic Species of Lamprey (Petromyzonidae) from Western North America, *J. Fisheries Res. Board Can.*, **22**(1):139–158.

———— and J .R. Greeley. 1963. Order Acipenseroidei in: *Fishes of the Western North Atlantic, Mem. Sears. Found. Marine Res.*, 1(3):24–60.

———— and G. N. Mukerji. 1961. Order of Succession of Different Types of Infraoral Lamina in Landlocked Sea Lamprey (*Petromyzon marinus*), *J. Fisheries Res. Board Can.*, **18**(6):1125–1143.

Woods, L. P., and R. F. Inger. 1957. The Cave, Spring, and Swamp Fishes of the Family Amblyopsidae of Central and Eastern United States, *Am. Midland Naturalist*, **58**(1):232–256.

Part Three

AMPHIBIANS

A. P. Blair

THE Amphibia may be defined as tetrapods which lack protective embryonic membranes. On a practical level, any modern tetrapod with a naked skin may be referred to the Amphibia (some of the tropical vermiform apodans have minute scales embedded in the skin). Typically there is an aquatic gill-breathing larva which metamorphoses into a lung-breathing terrestrial adult; in the course of adaptive radiation some forms have abandoned the aquatic larval stage while others have dispensed with the land stage. In such larval features as gills and lateral-line organs the amphibians reveal their ancestry. The amphibians were derived from rhipidistian crossopterygian fishes in Devonian time, possibly from more than one stock, perhaps as a solution to the problem of transportation from drying-up pool to more permanent pool. The earliest known amphibians were aquatic forms and perhaps came to land a number of times. Radiation produced a multiplicity of forms concerning whose relationships there is little agreement. The reader is referred to the symposium on "Evolution and Relationships in the Amphibia" in *American Zoologist*, vol. 5, 1965.

A brief synopsis of extinct and living Amphibia follows (alternative arrangements will be found in the symposium referred to above):

Superorder Temnospondyli: pleurocentra paired structures, never forming complete discs; tabular and parietal not in contact; vomers broad, widely separating internal nares; laterosphenoid region of braincase completely walled by bone or cartilage. Manus 4-toed. Devonian to Triassic.

Order Ichthyostegalia: intertemporal absent; a short, rounded snout and a relatively long skull table. Devonian to Pennsylvanian.

Order Rhachitomi: vertebrae rhachitomous, with pleurocentra moderately developed and intercentra usually crescentic and not surrounding notochord; intertemporal present or absent. Carboniferous to Triassic.

Order Trematosauria: intertemporal absent; skull high and narrow, triangular in shape with pointed snout; body long and slender with limbs reduced. Perhaps marine. Triassic.

Order Stereospondyli: pleurocentra reduced, intercentra well developed; skull depressed; intertemporal absent; braincase poorly ossified. Permian and Triassic.

Superorder Lepospondyli: with direct deposition of bone around the embryonic notochord. Carboniferous to present.

Order Aistopoda: limbless, elongate, many ribs (to more than 100). Carboniferous.

Order Nectridia: elongate, limbs absent or reduced; caudal vertebrae with expanded fan-shaped neural and hemal spines. Pennsylvanian and Permian.

Order Trachystomata: eel-shaped; hind limbs absent and forelimbs reduced; tail vertebrae with expanded arches. Cretaceous to present.

Order Microsauria: size small, with body moderately long and slender; limbs relatively short and feeble; manus 3-toed. Centra spool-shaped; skull greatly elongated postorbitally; otic notch absent; skull roof primitive, but tabular absent, and a very large supratemporal present between parietal and squamosal. Scales with characteristic radiate striation. Carboniferous and Permian.

Order Apoda: body form elongate, no limbs, tail abbreviated; small scales embedded in skin in some genera; eyes reduced, sensory tentacles present. No known fossils; about 75 living species.

Order Urodela: body form usually normal, sometimes elongate; girdles largely cartilaginous. Jurassic to present.

Superorder Salientia: vertebrae pleurocentral. Triassic to present.

Order Anura: tail lacking in adult; hind legs adapted for jumping. Jurassic to present.

Superorder Anthracosauria: pleurocentra complete discs enclosing notochord; tabular and parietal in contact; vomers narrow, narrowly separating internal nares; laterosphenoid region of braincase open. Carboniferous and Permian.

Order Embolomeri: neural arches normal; skull moderately elongate in facial region; otic notch slitlike; tabulars horned. Carboniferous and Permian.

Order Seymouriamorpha: dorsal neural arches expanded and swollen; skull relatively short; otic notch usually large and rounded; tabulars without horns. Pennsylvanian and Permian.

The ancestry of modern amphibians is obscure. Some workers have derived the Anura from apsidospondyls (temnospondyls plus anthracosaurs) and the Urodela and Apoda from lepospondyls, while others have sought the ancestry of the Apoda in lepospondyls and that of the Anura and Urodela in temnospondyls. At present there is a growing feeling that the living amphibians represent a single phyletic line and should be included in the subclass Lissamphibia, of equal rank with subclasses Lepospondyli and Labyrinthodontia (Apsidospondyli). Possibly the rhachitomes of the genus *Amphibamus,* which form the basis of Romer's order Eoanura, are ancestral to the modern amphibians. At any rate, the living amphibians constitute a scanty and degenerate remnant of a once dominant and abundant group. Among the more conspicuous degenerative changes may be mentioned flattening of the skull, loss of many skull bones, loss of ossification of girdle elements, loss of ossification of tarsus and carpus, loss of one of the digits of the front foot, loss of scales, and loss of the pineal eye.

Adaptive radiation has produced arboreal, terrestrial, subterranean, and aquatic amphibians. Nevertheless, the physiological restrictions of the class are severe, and moisture constitutes a greater limiting factor with amphibians than with any of the other classes of tetrapods. The greatest numbers of species are found in tropical regions with abundant rainfall. Correlated with the moisture requirement of amphibians is the fact that most forms are nocturnal. One great habitat, the ocean, remains closed to modern amphibians, although some species tolerate brackish water and there is evidence that the Triassic trematosaurs lived in salt water.

As a group the amphibians are of modest economic importance. Toads have long been noted as voracious devourers of insects. The legs of several species of frogs are highly prized for food. Both salamanders and anurans have proved exceptionally fine subjects for research in experimental embryology, endocrinology, and other fields of inquiry. In recent years several species of anurans—the African *Xenopus laevis* and the North American *Rana pipiens, Bufo americanus,* and *B.*

terrestris—have been widely used in this country for human pregnancy tests.

Linnaeus originally included in the class Amphibia both amphibians and reptiles; later he added several groups of fishes. In the Tableau Elémentaire, 1795, Cuvier adopted the Linnean classification, using the French word *reptiles* as a class name for the group. Brogniart, 1799, pointed out the basic differences between the Batrachia (amphibians) and the other reptiles, and Latreille, 1804, proposed to set up the Amphibia as a separate class from the Reptilia. Not for more than 50 years, however, were the amphibians to be generally regarded as constituting a separate class.

Amphibian taxonomy at the species level is complicated by a number of considerations. Mensuration is difficult because hard parts are lacking, meristic features are few, color is lost in preservative, there is much local variation, and many species are so secretive that they are difficult to collect in quantity. In the following keys and descriptions body length refers to the distance from tip of snout to anterior end of anal opening. In the identification of salamanders, determination of the number of costal grooves between front and hind limbs has long been standard procedure; in this book this determination is used less extensively than the number of costal folds between appressed limbs. For the latter determination the front leg is extended backward along the axis of the body while the hind leg is extended forward; the number of *complete* costal folds not overlapped by toes is then counted (Fig. 3-1). In a few instances, clearing with potassium hydroxide and staining with alizarin red are required for determination of critical skeletal structure. Observation of some characters, particularly those that separate the major taxonomic groups, calls for dissection, and the student interested in acquiring a basic understanding of the Amphibia would do well to take time and sacrifice specimens for these basic dissections.

Fig. 3-1. Method of determining number of costal grooves between appressed limbs in salamanders.

TABLE 3-1. *Classification of Amphibia*

	Families		Genera	
	Extinct	Living	Extinct	Living
Superorder Temnospondyli				
Order Ichthyostegalia	4	0	6	0
Order Rhachitomi	15	0	50	0
Order Trematosauria	1	0	10	0
Order Stereospondyli	9	0	38	0
Superorder Lepospondyli				
Order Aistopoda	2	0	3	0
Order Nectridia	3	0	9	0
Order Trachystomata	0	1	2	2
Order Microsauria	6	0	17	0
Order Apoda	0	1	0	17
Order Urodela	2	7	20	49
Superorder Salientia				
Order Anura	3	13	14	100
Superorder Anthracosauria				
Order Embolomeri	4	0	16	0
Order Seymouriamorpha	4	0	6	0

Among internal characters of importance are the presence and shape of various skull bones (Figs. 3-2 and 3-3), the shape of the centra of vertebrae, the nature of the girdles [especially the presence of the ypsiloid cartilage (Fig. 3-4) in salamanders and the character of the pectoral girdle, whether arciferous (Fig. 3-5) or firmisternal (Fig. 3-6), in anurans], the presence of ribs, the shape of the terminal phalanges, the shape of the diapophyses of sacral vertebrae, and the presence of lungs. Many of these characters can be demonstrated equally well by dissection or by clearing and staining.

Superorder Lepospondyli. With direct deposition of bone around the embryonic notochord; contains 3 orders of extinct amphibians and the living orders Trachystomata, Urodela, and Apoda. Living representatives of this superorder in the United States are provided with a tail.

Order Trachystomata. Sirens. Aquatic, with 3 pairs of gills present throughout life. Eellike, with posterior limbs absent. Eyes small, lidless. Maxillaries absent; nasals meeting; upper and lower jaws toothless, covered with horn. Fertilization probably internal. Living forms restricted to eastern North America. Cretaceous of Texas, Cretaceous and Eocene of Wyoming, and Miocene, Pliocene, and Pleistocene of Florida. Many workers strongly feel the sirens should be placed in the order Urodela rather than set aside in a distinct order Trachystomata.

Fig. 3-2. **Dorsal view of salamander skull (plethodontid).**

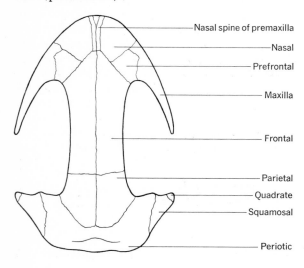

Nasal spine of premaxilla
Nasal
Prefrontal
Maxilla
Frontal
Parietal
Quadrate
Squamosal
Periotic

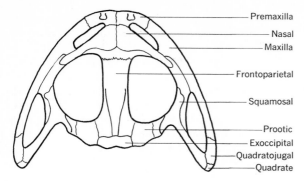

Fig. 3-3. Dorsal view of anuran skull (ranid).

Labels: Premaxilla, Nasal, Maxilla, Frontoparietal, Squamosal, Prootic, Exoccipital, Quadratojugal, Quadrate

Family Sirenidae. Sirens. Characters, distribution, and fossil record as for order.

KEY TO GENERA OF SIRENIDAE

1. 4 toes on each foot   *Siren*
   3 toes on each foot   *Pseudobranchus*

Genus *Siren* Linnaeus. 3 pairs of gill slits. 4 toes. No longitudinal light markings. Size large. Eastern United States into Mexico. Miocene, Pliocene, and Pleistocene of Florida, and Eocene of Wyoming.

KEY TO SPECIES OF *SIREN*

1. Costal grooves 36–39; no small dark spots on dorsum
   S. lacertina
   Costal grooves 31–36; small dark spots on dorsum
   S. intermedia

Siren lacertina **Linnaeus.** Greater siren (Fig. 3-7). Body length to 640 mm. In life light gray

Fig. 3-4. Pelvic girdle of salamander.

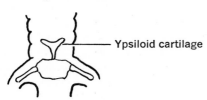

Ypsiloid cartilage

Fig. 3-5. Arciferous pectoral girdle.

above with yellow blotches, sides lighter. In Coastal Plain, District of Columbia to Leon County, Florida, and southern Alabama; peninsular Florida. Pleistocene of Florida.

Siren intermedia **Le Conte.** Lesser siren. Body length to 432 mm. Above dark brown to olive green with many small dark spots. Coastal Plain, North Carolina to northeastern Mexico, south in Florida to Pasco County; up Mississippi River valley to Illinois and Indiana. 3 described subspecies; the Rio Grande population may represent a distinct species.

Genus *Pseudobranchus* Gray. 1 pair of gill slits. 3 toes. Longitudinal light markings. Size small. Southeastern United States. Monotypic. Pliocene and Pleistocene of Florida.

Pseudobranchus striatus (**Le Conte**). Dwarf siren. Body length to 116 mm. Dorsum and sides with longitudinal stripes. Peninsular Florida, Coastal Plain of Georgia, and southeastern South Carolina. 5 described subspecies.

Order Urodela. Salamanders. These amphibians have remained superficially rather unspecialized, retaining the primitive division of the body into head, trunk, and tail. Number of vertebrae not reduced as in the Anura. In most species metamerism very obvious, as seen in costal grooves and

Fig. 3-6. Firmisternal pectoral girdle.

Fig. 3-7. Siren lacertina. **Biven's Arm, 1 mile south of Gainesville, Alachua County, Fla., Jan. 4, 1954. (Coll. Coleman J. Goin. Photo by Isabelle Hunt Conant.)**

folds. Tympanum absent. Eyes with or without lids. Ribs present; pubis cartilaginous (but may be calcified). Lungs present or absent in adults.

External secondary sex differences are not so pronounced in the salamanders as in the anurans. In most salamanders the females are larger than the males, although in *Desmognathus* the converse is true. Differences in body proportions, as longer legs or tails of males (*Plethodon, Ensatina*), or differences in head shape (*Hydromantes, Aneides, Eurycea*), are found. Parts or features of the body may be enlarged in males, as premaxillary teeth (*Plethodon, Batrachoseps, Desmognathus*), hind legs (salamandrids), or the swollen area around the vent (proteids, ambystomatids, salamandrids). Special structures may be present in males, as cirri on the snout (*Eurycea, Typhlotriton*), mental gland beneath the lower jaw (*Plethodon, Aneides*), excrescences on the toes and legs (salamandrids), or papillae lining the vent (plethodontids, ambystomatids). Structures may be lacking in males, as the vomerine teeth in some *Desmognathus*. Skin color and texture may differ in the 2 sexes (salamandrids).

Primitively salamanders breed in water just as anurans do; secondarily many of the plethodontids have adopted terrestrial reproduction. Primitively fertilization is external; secondarily most salamanders have adopted internal fertilization. Most salamanders exhibit a more or less complicated *liebespiel,* with body writhings and rubbing together, tail waving, and frequently amplexus. The male deposits a packet of sperm, the spermatophore, which the female picks up with her cloacal lips. Egg laying may follow almost immediately, or, in some species, may not occur for days or even months. The number of eggs laid by salamanders averages much less than in anurans, many species laying fewer than 100 eggs and few more than 1000. The eggs are covered with 1 or more layers of jelly and may be pigmented or unpigmented.

Salamander larvae have essentially the body form of adults. In general, larval salamanders may be classified as of 3 types: (1) terrestrial, with the abbreviated larval period passed within the egg, with fins absent, and with so-called staghorn gills; (2) pond, with long gills, extensive fins, and balancers; and (3) stream, with reduced gills and fins and without balancers. Metamorphosis does not involve the dramatic changes typical of anurans. Loss of gills and fins, closure of the gill slits, replacement of larval by adult skin, alterations of the eye including change of position and development of lids, development of functional lungs—these are among the more apparent changes. Members of the genera *Necturus, Cryptobranchus,* and *Amphiuma,* like the trachystomes *Siren* and *Pseudobranchus,* remain completely aquatic and undergo but a partial metamorphosis. Among the ambystomatids, salamandrids, and plethodontids larval reproduction (neoteny) is found in certain species, in some populations of species, and in some individuals of populations.

The living urodeles constitute a very small order, and with about 86 species the United States has about a third of the known species. 6 of the 7 known families are represented, with 2 of these not found outside of North America. Only the primitive Asiatic land salamanders, the Hynobiidae, are lacking. In the United States the Great Plains form an extremely effective barrier to east-west dispersal, the range of but 1 species (*Ambystoma tigrinum*) crossing the plains. The eastern half of the United States is abundantly supplied with salamanders, and the Washington-Oregon-California coastal strip represents a second area of concentration. A few relict species occur in the Rocky Mountains. The ambystomatids are well represented in Mexico, and the plethodontids have extended well into South America, far from their center of differentiation in the Appalachian upland.

The oldest fossil salamander, from the Jurassic of Wyoming, is perhaps a proteid. The somewhat scanty paleontological record indicates that the present families have been long established.

An alternative classification of the higher categories of salamanders is given by Regal (1966).

KEY TO SUBORDERS OF URODELA

1. Aquatic; eyes reduced; fully pigmented; body not eellike   2
    Not with above combination of characters   3
2. Adults with gills   **Proteida**
    Adults without gills but with 1 pair of open gill slits   **Cryptobranchoidea**
3. Parasphenoid teeth present (Fig. 3-8)   **Salamandroidea**
    Parasphenoid teeth absent   **Ambystomatoidea**

Suborder Cryptobranchoidea. The most primitive living salamanders. Aquatic, but adults without gills. Large, depressed body. Eyes reduced, lidless. No lacrimals or septomaxillaries; angulars not fused with prearticulars; spines of premaxillaries short, not separating the nasals; vomerine and maxillary teeth in parallel rows; vertebrae amphicoelous. Second epibranchial retained in adult. Fertilization external. Japan, northeastern Asia, and eastern United States. Oligocene, Miocene, and Pliocene of Europe; Miocene of North America.

Family Cryptobranchidae. Characters, range, and fossil record as for suborder. 2 genera, *Megalobatrachus* of China and Japan and the following:

Genus *Cryptobranchus* Leuckart. 1 pair of open gill slits, sometimes closed on one side or the other; 4 epibranchials on hyoid. Eastern United States. Monotypic.

***Cryptobranchus alleganiensis* (Daudin).** Hellbender (Fig. 3-9). Body length to 500 mm. Prominent lateral fleshy folds; digits flattened. Above light brown to dark brown with varied develop-

Fig. 3-8. Parasphenoid teeth of salamandroid.

ment of spotting; below lighter than dorsum, with or without spots. In creeks and rivers, southern New York south to northern Georgia and Alabama, and west to northeastern Mississippi, northern Arkansas, southeastern Kansas, and southeastern Iowa. 2 described subspecies.

Suborder Proteida. Aquatic, with 3 pairs of gills present throughout life. Eyes reduced, lidless. Lungs present. Maxillae absent; palatine and pterygoid teeth not separated; puboischium pointed anteriorly, a separate ypsiloid cartilage absent. Fertilization internal. Eastern United States (*Necturus*) and southern Europe (*Proteus,* a cave dweller). Known from Eocene and Miocene of Europe. Cope, 1889, considered the Proteida an order. Hecht, 1957, concluded that *Necturus* and *Proteus* were not closely related and mentioned the

Fig. 3-9. Cryptobranchus alleganiensis. **Blue River, near White Cloud, Harrison County, Ind., Mar. 28, 1954. (Coll. Sherman Minton. Photo by Isabelle Hunt Conant.)**

possibilities of (1) recognizing the Necturidae and Proteidae as families of the suborder Salamandroidea, or (2) recognizing *Necturus* and *Proteus* as genera in the family Salamandridae. Kezer, Seto, and Pomerat, 1965, compared the chromosomes of *Necturus* and *Proteus* and concluded that the 2 genera represent a single line of evolution.

Family Proteidae. Characters, distribution, and fossil record as for suborder.

Genus *Necturus* Rafinesque. Pigmented permanent larvae. Toes 4–4. Eastern North America. Pleistocene of Florida.

KEY TO SPECIES OF *NECTURUS*

1. Adults becoming sexually mature at 140–150 mm snout-vent length; larvae usually striped
   N. maculosus
   Adults becoming sexually mature at less than 125 mm snout-vent length; larvae never striped   **2**
2. Dorsum without blotches, sometimes a few small punctations   *N. punctatus*
   Dorsum with large black patches   **3**
3. Neuse and Tar River drainage of North Carolina
   N. lewisi
   Gulf drainage   **4**

Fig. 3-10. Necturus maculosus. **West shore of Cayuga Lake, 4 miles north of Ithaca, N.Y., Feb. 15, 1954. (Coll. Douglass Payne. Photo by Isabelle Hunt Conant.)**

4. Body and head flattened; venter immaculate
   N. alabamensis
   Body and head less flattened; venter spotted, at least in larger individuals   *N. beyeri*

Necturus maculosus (**Rafinesque**). Mudpuppy (Fig. 3-10). Body length to 300 mm. Head and body depressed, tail strongly compressed. Dorsal ground color tan, brown, or almost black; venter immaculate or variously spotted. In lakes and streams, eastern half of the United States except for peninsular Florida and Atlantic Coastal Plain as far north as southern North Carolina; into Canada. 3 described subspecies.

Necturus alabamensis **Viosca.** Alabama waterdog. Body length to 150 mm. Head and body flattened. About 100 spots on back and sides as compared with about 200 in *N. beyeri*, with which it is sympatric in parts of Mobile River drainage; venter immaculate. Mobile River drainage east to Ochlockonee River drainage, most abundant above the Fall Line.

Necturus lewisi **Brimley.** Neuse River waterdog. Body length to 175 mm but sexually mature at about 100 mm. Dorsum and venter darker than in *N. beyeri*. In larger rivers and deeper waters of Neuse and Tar River drainage of North Carolina.

Necturus beyeri **Viosca.** Gulf Coast waterdog. Body length to 160 mm. Body subcylindrical in cross section. Venter spotted. Chiefly in sandy, spring-fed streams of Gulf Coast drainage from Angelina River, Texas, to Mobile River, Alabama.

Necturus punctatus (**Gibbes**). Dwarf waterdog. Body length to 135 mm but sexually mature at about 65 mm. Unspotted or very nearly so. In smaller, more sluggish streams, Chowan River system of Virginia south to Altamaha River system of Georgia.

Suborder Ambystomatoidea. Angulars fused with prearticulars; spines of premaxillaries long; vomers short; vertebrae amphicoelous. Second epibranchial absent in adult. Fertilization internal. North America. Paleocene, Eocene, Pliocene, and Pleistocene of North America; Paleocene and Miocene of Europe.

Family Ambystomatidae. Adults normally transforming, sometimes neotenic. Eyes with lids. Vomerine teeth in transverse series, parasphenoid teeth lacking; ypsiloid cartilage present; carpus and tarsus ossified. Tongue large, fleshy, free at sides. Costal grooves usually well defined; gular fold present. Toes 4–5. Range and fossil record as for suborder. 6 living genera, 3 in the United States.

KEY TO GENERA OF AMBYSTOMATIDAE

1. Horizontal diameter of eye equal to or greater than distance from anterior corner of eye to tip of snout *Rhyacotriton*
   Horizontal diameter of eye less than distance from anterior corner of eye to tip of snout   **2**
2. Costal grooves indistinct; dorsum brown with network of black markings; western United States *Dicamptodon*
   Costal grooves distinct; dorsum not brown with network of black markings   *Ambystoma*

Genus *Ambystoma* Tschudi.   Nasals present, lacrimals absent. Costal grooves well defined. Lungs and ypsiloid cartilage normal. Sometimes neotenic. North America; 10 additional species in Mexico. Pliocene and Pleistocene of North America.

KEY TO SPECIES OF *AMBYSTOMA*

1. Sexually mature adults with gills   **2**
   Sexually mature adults without gills   **4**
2. Southeastern United States; body length under 60 mm   *A. talpoideum* (part)
   Western United States; body length over 60 mm   **3**
3. 7–10 gill rakers on anterior face of 3d arch *A. gracile* (part)
   15–24 gill rakers on anterior face of 3d arch (Fig. 3-11)   *A. tigrinum* (part)
4. Plicae of tongue diverging from median furrow (Fig. 3-12)   **5**
   Plicae of tongue not diverging from median furrow   **8**

Fig. 3-11. **Third gill arch of *Ambystoma tigrinum*.**

Fig. 3-12. **Dorsal view of tongue of ambystomatids of *mabeei-annulatum-texanum-cingulatum* group.**

5. Teeth on margin of jaw in a single row *A. mabeei*
   Teeth on margin of jaw in more than 1 row   **6**
6. With yellow-white crossbands   *A. annulatum*
   Without yellow-white crossbands   **7**
7. With strongly reticulated dorsal pattern; North Carolina to Louisiana   *A. cingulatum*
   Dorsum with gray lichenlike splotches, but not strongly reticulated; Ohio to Texas to Alabama *A. texanum*
8. In western half of the United States   **9**
   In eastern half of the United States   **11**
9. With an irregular-edged dorsal band, sometimes broken into spots   *A. macrodactylum*
   No irregular-edged dorsal band   **10**
10. Palmar tubercles present (Fig. 3-13) *A. tigrinum* (part)
    Palmar tubercles absent   *A. gracile* (part)

Fig. 3-13. **Palmar tubercles of *Ambystoma tigrinum*.**

11. Dorsum boldly marked with contrasting black and gray (or white)   *A. opacum*
    Not as above   **12**
12. 10 costal grooves between fore and hind limbs, counting 1 each in axilla and groin; body short and stout   *A. talpoideum* (part)
    More than 10 costal grooves; body not short and stout   **13**
13. Dorsum with yellow or orange spots in 2 rows *A. maculatum*
    Spots, if present, not in 2 rows   **14**
14. Dorsum immaculate in life   *A. jeffersonianum*
    Dorsum with bars, spots, or speckles in life   **15**
15. Palmar tubercles present   *A. tigrinum* (part)
    Palmar tubercles absent   *A. laterale*

Ambystoma cingulatum **Cope.** Flatwoods salamander. Body length to 62 mm; 0–4 intercostal folds between appressed limbs. Dorsum black-brown with gray reticulations; venter dark with light spots or dots. In Coastal Plain, southeastern North Carolina to Mobile Bay area; south in Florida to Marion County. 2 described subspecies, but these not valid according to Martof and Gerhardt (1965).

Ambystoma mabeei **Bishop.** Mabee's salamander. Body length to 60 mm; about 2 intercostal folds between appressed limbs. In life dorsum dark brown with tan flecks; venter light brown with few pale gray flecks. In Coastal Plain, southeastern North Carolina to Charleston County, South Carolina.

Ambystoma annulatum **Cope.** Ringed salamander. Body length to 95 mm; body slender with small head; 3–4 intercostal folds between appressed limbs. Above brown-black with crossbands of yellow (gray in preservative); below slate with or without white dots. Central Arkansas (Mena, Hot Springs) to Sequoyah and Adair Counties, Oklahoma, to Gasconade County, Missouri.

Ambystoma texanum (**Matthes**). Small-mouthed salamander. Body length to 88 mm; 2–3 intercostal folds between appressed limbs; snout short. Above deep brown or slate with light lichen-like markings; venter lighter than back. Ohio to southeastern Iowa, eastern Kansas, and Oklahoma, and eastern half of Texas; east to western Kentucky, Tennessee, and Alabama. Pleistocene of Texas.

Ambystoma talpoideum (**Holbrook**). Mole salamander. Body length to 61 mm; appressed limbs meeting or overlapping; body stout with depressed head. Above dark brown with many small bluish white flecks; below blue-gray with small light flecks. Sometimes neotenic. In Coastal Plain, southeastern South Carolina to eastern Texas and southeastern Oklahoma; not in lower two-thirds of peninsula of Florida; up Mississippi Valley to southern Illinois; Henderson and Transylvania Counties, North Carolina; Polk County, Tennessee; Georgia piedmont.

Ambystoma opacum (**Gravenhorst**). Marbled salamander. Body length to 65 mm; appressed limbs overlapping; body stout. Strongly contrasting gray or white transverse bars on trunk and tail. Northern Florida north to New Hampshire and west to eastern Texas and southeastern Oklahoma; north to southern Illinois, Indiana, and northern Ohio.

Ambystoma macrodactylum **Baird.** Long-toed salamander. Body length to 75 mm; appressed limbs overlapping; body slender. Dorsal markings tan to yellowish or orange, usually forming an irregular-edged stripe; below dark brown with minute light dots. Northern California, Oregon except for southeastern portion, Washington, northern three-fourths of Idaho, and western Montana; into Canada. 5 described subspecies, one of these disjunct in Santa Cruz County, California.

Ambystoma jeffersonianum (**Green**). Jefferson salamander. Body length to 100 mm; appressed limbs meeting or overlapping by as many as 2 intercostal folds. Toes long and slender. Dorsum brownish gray, dark brown, or lead color; belly lighter. Southern Indiana to Virginia, north to Vermont. Hybridizes with *A. laterale;* range largely allopatric to that of *A. laterale*. A triploid female population associated with *A. jeffersonianum* has been called *A. platineum* by Uzzell.

Ambystoma laterale **Hallowell.** Blue-spotted salamander. Body length to 75 mm; appressed limbs not meeting, meeting, or overlapping by 1 intercostal fold. Toes less attenuated than in *A. jeffersonianum*. Dorsum darker than in *A. jeffersonianum*, in life with numerous pale blue to bluish-white flecks. Minnesota to northern Indiana, east to Massachusetts; into Canada. Hybridizes with *A. jeffersonianum;* range largely allopatric to that of *A. jeffersonianum*. A triploid female population associated with *A. laterale* has been called *A. tremblayi* by Uzzell.

Ambystoma maculatum (**Shaw**). Spotted salamander. Body length to 94 mm; appressed limbs meeting or separated by 1 intercostal fold; body stout. Above blue-black with 2 rows of spots, the more anterior of these frequently orange, the rest yellow; below light slate color, eastern United States west to the plains; absent from Minnesota, Iowa except northeastern corner, most of Illinois, northern Missouri, Florida, and southern Georgia; into Canada.

Ambystoma gracile (**Baird**). Northwestern salamander. Body length to 92 mm; appressed limbs overlapping by as many as 6 intercostal folds.

Parotoid glands present; a glandular ridge along upper edge of tail. Above dark brown, below light brown. Sometimes neotenic. Coastal region, southeastern Alaska to Sonoma County, California. 2 described subspecies.

Ambystoma tigrinum (**Green**). Tiger salamander (Fig. 3-14). Body length to 177 mm; appressed limbs overlapping 1–4 intercostal folds; body stout. Tubercles present on soles of fore and hind feet. Above greenish to brown or black, with yellow spots or bars or small dark spots. Frequently neotenic in western half of the United States. Over much of the United States, but distribution erratic; south in Florida at least to central part of state; absent from Appalachian upland, New England, western Washington, Oregon, Nevada, and California except Central Valley; into Canada and Mexico. 9 described subspecies. Pliocene and Pleistocene of Arizona and Pleistocene of Florida.

Genus *Dicamptodon* **Strauch.** Nasals present. Lungs and ypsiloid cartilage normal. Costal grooves poorly defined. Size large. Monotypic.

Dicamptodon ensatus (**Eschscholtz**). Pacific giant salamander. Body length to 175 mm; body stout. Above brown with dark marbling; venter light. Coastal forest, Santa Cruz County, California, to British Columbia; extends farther east, perhaps into Sierra Nevada, in northern California; northern Idaho and northwestern Montana.

Genus *Rhyacotriton* **Dunn.** Size small; snout short, eyes prominent. Nasals absent. Lungs and ypsiloid cartilage reduced. Monotypic.

Rhyacotriton olympicus (**Gaige**). Olympic salamander. Body length to 64 mm; 2–3 intercostal folds between appressed limbs. Above brown, or olive with small dark markings; below orange-yellow or greenish yellow. In or near small streams or seepage areas, coastal forest from Olympic Peninsula to northwestern California. 2 described subspecies.

Suborder Salamandroidea. Vomers extending back as 2 tooth-bearing bones on each side of the parasphenoid bone or split off as 1 or 2 groups of teeth lying on the parasphenoid bone. Fertilization internal. North and South America, northern Africa, Europe, and Asia. Paleocene-Pliocene of Europe; Cretaceous, Paleocene, Miocene, Pliocene, and Pleistocene of North America.

Fig. 3-14. Ambystoma tigrinum. **10 miles south of McRae, Telfair County, Ga., Dec. 22, 1951. (Coll. Bartley Burns. Photo by Isabelle Hunt Conant.)**

KEY TO FAMILIES OF SALAMANDROIDEA

1. Body greatly elongated, eellike; aquatic
   Amphiumidae
   Body not eellike   **2**
2. Nasolabial groove present (Fig. 3-15); costal grooves well developed   **Plethodontidae**
   Nasolabial groove absent; costal grooves poorly developed   **Salamandridae**

Family Amphiumidae. Aquatic; adults with 1 pair of gill slits but no gills. Large; eellike; limbs

Fig. 3-15. **Nasolabial groove of plethodontid.**

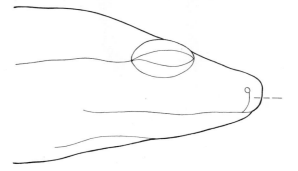

very small, with reduced toes. Eyes reduced, lidless. Maxillary and vomerine teeth in parallel rows; vomers with posterior projections; vertebrae amphicoelous. North America. Paleocene and Pleistocene of North America.

Genus *Amphiuma* Garden. With characters of family.

KEY TO SPECIES OF *AMPHIUMA*

1. Body distinctly bicolored; 3 toes   *A. tridactylum*
    Body not distinctly bicolored; less than 3 toes   **2**
2. Each limb with but a single toe   *A. pholeter*
    Each limb with 2 toes   *A. means*

Amphiuma pholeter **Neill.** 1-toed amphiuma. Body length to 208 mm. Head, eyes, and limbs reduced as compared to *A. means* and *A. tridactylum*. Known only from Levy, Jefferson, and Liberty Counties, Florida.

Amphiuma means **Garden.** 2-toed amphiuma (Fig. 3-16). Body length to 700 mm. 2 toes. Body not distinctly bicolored; no throat patch. In lowland ditches, pools, and swamps of Coastal Plain, Virginia to eastern Louisiana; peninsular Florida. Hybridizes with *A. tridactylum.*

Fig. 3-16. Amphiuma means. Savannah, Chatham County, Ga., Apr. 11, 1954. (Coll. Charles T. Stine. Photo by Isabelle Hunt Conant.)

Amphiuma tridactylum **Cuvier.** 3-toed amphiuma. Body length to 800 mm. 3 toes. Body distinctly bicolored; a dark throat patch. In lowland streams, pools, and swamps of Coastal Plain, extreme western Florida to eastern Texas; up Mississippi River valley to about 25 miles south of Cairo, Illinois. Hybridizes with *A. means*.

Family Salamandridae. Newts. Lungs present. Eyes with lids. Toes 4–5. Tongue small, slightly free at margin. Vomeropalatine teeth in 2 longitudinal rows diverging posteriorly; ypsiloid cartilage present; vertebrae opisthocoelous. Terrestrial stage with relatively rough skin, aquatic adults with relatively smooth skin. North America, Europe, Asia, and North Africa. About 43 living species. Paleocene, Eocene, Oligocene, and Miocene of Europe; Oligocene, Pliocene, and Pleistocene of North America. Some recent workers doubt that salamandrids are closely related to either plethodontids or amphiumids.

KEY TO GENERA OF SALAMANDRIDAE

1. Top of head with 2 longitudinal ridges (Fig. 3-17)   *Notophthalmus*
    Top of head smooth   *Taricha*

Genus *Notophthalmus* Rafinesque. Top of head with 2 longitudinal ridges. Dorsum with spots or stripes. Adults to 55 mm body length. Eastern half of the United States; northeastern Mexico.

KEY TO SPECIES OF *NOTOPHTHALMUS*

1. Spots on anteroventral part of tail distinct, about as large as eye   *N. meridionalis*
    Spots on anteroventral part of tail much smaller than eye or else indistinctly bordered   **2**
2. With dorsolateral stripe (red in life), not heavily bordered with black   *N. perstriatus*
    Without dorsolateral stripe, or, if present, heavily bordered with black   *N. viridescens*

Notophthalmus meridionalis **(Cope).** Black-spotted newt. Body length to 53 mm. Dorsum olive green with black dots, 2 narrow yellow longitudinal lines, and short yellow longitudinal dashes; venter orange with black spots. In lakes and other quiet

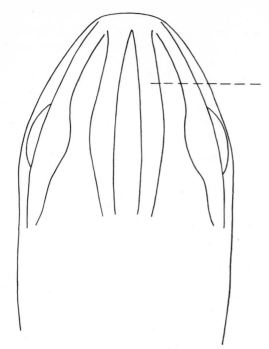

Fig. 3-17. Dorsal view of head of *Notophthalmus* showing longitudinal ridges.

waters of Coastal Plain, Aransas County, Texas, south into northeastern Mexico.

Notophthalmus perstriatus (**Bishop**). Striped newt. Body length to 42 mm. Dorsum dark brown to olive green with 2 dorsolateral red or reddish stripes, and sometimes a middorsal light stripe; venter light yellow, immaculate or with black markings. Southern Georgia and adjacent Florida, south to central Florida.

Notophthalmus viridescens **Rafinesque**. Newt (Fig. 3-18). Body length to 55 mm. Dorsum olive green, yellow-brown, or brown-black, with vari-

Fig. 3-18. Notophthalmus viridescens. 1½ miles northwest of Sellersville, Bucks County, Pa., Mar. 27, 1954. (Coll. Robert G. Hudson. Photo by Isabelle Hunt Conant.)

able number of small black dots and with dorsolateral red stripes, dashes, or rows of spots; venter yellowish, with varying number of small black dots. Subadult terrestrial red-eft stage, with orange-red or red-brown dorsum, may be present. Eastern half of the United States west to plains; into Canada. Sometimes neotenic. 4 described subspecies. Pleistocene of Florida.

Genus *Taricha* Gray. No longitudinal ridges on top of head. Dorsum of uniform color. Adults to 89 mm body length. West of the Rocky Mountains. Lower Pliocene of California.

KEY TO SPECIES OF *TARICHA*

1. Lower eyelid dark-colored   *T. granulosa*
   Lower eyelid light-colored, at least in part   **2**
2. Dark dorsal color sharply demarcated from light venter; iris dark; in life venter tomato red

                        T. rivularis

   Dark dorsal color not sharply demarcated from light venter; iris with pale areas; in life venter yellow or orange   *T. torosa*

Taricha rivularis (**Twitty**). Red-bellied newt. Body length to 75 mm. Dorsum black to brownish; iris dark brown. Dorsal tail fin not so well developed as in *T. granulosa* and *T. torosa.* Larva with balancers reduced or absent. Breeds in rapid, rocky streams. Coastal California north of San Francisco Bay.

Taricha torosa (**Rathke**). California newt. Body length to 89 mm. Above yellowish brown to reddish brown; iris with yellow to orange areas. Breeds in streams and standing waters. Larva with balancers. Coastal California as far north as middle Mendocino County; foothills of the Sierra Nevada. 2 described subspecies.

Taricha granulosa (**Skilton**). Rough-skinned newt. Body length to 89 mm. Above tan to black; iris pale yellow to whitish. Most aquatic of the western newts; breeds in streams and standing waters. Larva with balancers. Pacific coastal region, vicinity of San Francisco Bay north to southeastern Alaska; western slope of Sierra Nevada in northern California; Latah County, Idaho; Thompson Falls, Sanders County, Montana. 4 described subspecies.

Family Plethodontidae. Lungless salamanders.

Nasolabial groove present. Costal grooves well defined. No lungs. Vomerine and parasphenoid teeth present; ypsiloid cartilage reduced or absent; vertebrae amphicoelous or opisthocoelous; terminal phalanges T-shaped except in *Aneides*. Adults neotenic or transforming. Aquatic larval stage present or absent. North America, northern South America, and southern Europe (*Hydromantes* only); a few species of western United States enter northwestern Mexico; approximately 90 species in 7 genera (none represented in the United States) in Central and South America and Mexico. Cretaceous, Miocene, and Pliocene of North America; Miocene of Europe.

KEY TO SUBFAMILIES OF PLETHODONTIDAE

1. Gularis muscle making prominent swellings on each side of posterior throat region   *Desmognathinae*
   Not as above   *Plethodontinae*

Subfamily Desmognathinae. Lower jaw immovable, with opening of mouth effected by elevation of upper jaw and head. Frontal with a ventrolaterally directed process; occipital condyles stalked with convex articulation surfaces. Larva with 4 gill slits. Eastern United States. Cretaceous of North America and Miocene of Europe. Sometimes considered a full family.

KEY TO GENERA OF DESMOGNATHINAE

1. A light line from posterior corner of eye to angle of jaw (Fig. 3-19)   *Desmognathus*
   No light line from eye to angle of jaw   2
2. 3–4 intercostal folds between appressed limbs
        Leurognathus
   About 13 intercostal folds between appressed limbs
        Phaeognathus

Fig. 3-19. Lateral view of head of *Desmognathus* showing light line.

Genus *Desmognathus* Baird. Premaxillary fontanel present; premaxillae fused; no prefrontals; vomerine and parasphenoid teeth not continuous. Internal nares not concealed. Tongue attached in front. Terrestrial or aquatic; a very transitory terrestrial or more prolonged aquatic larval period. Eastern United States.

KEY TO SPECIES OF *DESMOGNATHUS*

1. Tail round in cross section, or depressed   2
   Tail compressed, keeled above   4
2. Larger, to 62 mm body length   *D. ochrophaeus*
   Smaller, to 29 mm body length   3
3. With a distinct middorsal chevron pattern; usually above 5000 feet   *D. wrighti*
   No distinct middorsal chevron pattern; below 5000 feet   *D. aeneus*
4. Venter mottled   *D. fuscus* (part)
   Venter uniformly pigmented   5
5. Venter dark   *D. quadramaculatus*
   Venter light   6
6. Back usually with conspicuous markings; vomerine teeth in adult males   *D. monticola*
   Back usually without conspicuous markings; no vomerine teeth in adult males   *D. fuscus* (part)

Desmognathus fuscus (Rafinesque). Dusky salamander. Body length to 74 mm; 3–6 intercostal folds between appressed limbs. Dorsal color and markings very variable. Essentially terrestrial but frequently in water. In lowlands and uplands, southeastern Oklahoma and eastern Texas to central Florida to Maine to southern Indiana and Illinois; into Canada. 5 described subspecies; Valentine (1963) is of the opinion that *D. f. brimleyorum* of the Ouachita Mountains of Oklahoma and Arkansas and *D. f. auriculatus* of the Coastal Plain from Virginia to Texas should be considered full species.

Desmognathus quadramaculatus (Holbrook). Black-bellied salamander. Body length to 104 mm; 2–3 intercostal folds between appressed limbs. Large individuals black above and below. Tail strongly keeled above. Montane, along streams, northeastern Georgia to southern West Virginia.

Desmognathus monticola Dunn. Seal salamander. Body length to 75 mm; 4–5 intercostal folds between appressed limbs. Dorsum strongly marked, with spots opposite, alternate, or fused; venter very light. Along brooks, southwestern Pennsylvania and north central Kentucky to northern Georgia and northeastern Alabama; Alabama coastal plain, at least in Clarke, Butler, and Conecuh Counties. 2 described subspecies.

Desmognathus ochrophaeus **Cope.** Mountain salamander. Body length to 62 mm; 1–5 intercostal folds between appressed limbs. Dorsum variable, sometimes uniformly dark, frequently with stripe (tan-yellow, gray-red, brown) containing middorsal line of dots or chevrons, sometimes with 2 rows of large spots which may coalesce. Terrestrial. New York to central Alabama, northern Georgia, and western North Carolina; into Canada. According to Martof and Rose, 1963, *D. ocoee* and *D. perlapsus* should be included here.

Desmognathus wrighti **King.** Pygmy salamander. Body length to 29 mm; 1–3 intercostal folds between appressed limbs. Dorsal stripe light tan to reddish, with median row of darker chevrons. Terrestrial, usually above 5000 feet. Western North Carolina and adjacent Tennessee, north to Whitetop Mountain, Virginia.

Desmognathus aeneus **Brown and Bishop.** Cherokee salamander. Body length to 28 mm; 3–5 intercostal folds between appressed limbs. Dorsal stripe reddish bronze, yellow, or tan, with or without dark dots, blotches, or middorsal line; venter with dark blotches or scattered chromatophores. Terrestrial. Western North Carolina to central Alabama. 2 described subspecies.

Genus *Phaeognathus* Highton. Body extremely elongate (22 trunk vertebrae as opposed to 14–16 in other desmognathines); limbs very short (13 intercostal folds between appressed limbs as opposed to 1–6 in other desmognathines). No light line from eye to angle of jaw. Monotypic.

Phaeognathus hubrichti **Highton.** Hubricht's salamander. Body length to 101 mm; appressed limbs separated by 13 intercostal folds. Dorsum and venter dark brown. A fossorial species known only from Butler and Conecuh Counties, Alabama.

Genus *Leurognathus* Moore. Very similar to *Desmognathus* but completely aquatic, with flattened head and small, concealed, widely spaced internal nares. No premaxillary fontanel. Southern Appalachians.

Leurognathus marmoratus **Moore.** Shovel-nosed salamander. Body length to 92 mm; 3–4 in-

tercostal folds between appressed limbs. Above immaculate brown, or brown to black with large light spots in 2 rows; belly mottled, with or without light central area. Tail with sharp dorsal keel. Aquatic. Blue Ridge of western North Carolina; extreme northwestern South Carolina; northeastern Georgia; Blount, Carter, and Sevier Counties, Tennessee; Whitetop Mountain, Virginia. 5 described subspecies.

Subfamily Plethodontinae. Lower jaw movable. Frontal without ventrolaterally directed process; occipital condyles sessile with articulation surfaces plane or concave. Larva with 3 gill slits. Pliocene of North America.

KEY TO GENERA OF PLETHODONTINAE

1. Tail with basal constriction (Fig. 3-20)   2
   Tail without basal constriction   3
2. Venter with small black spots   *Hemidactylium*
   Venter unspotted   *Ensatina*

Fig. 3-20. **Basal constriction of tail of** *Hemidactylium* **and** *Ensatina.*

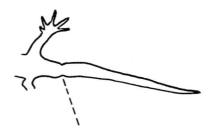

3. 4 toes on hind foot   4
   5 toes on hind foot   5
4. 4–5 intercostal folds between appressed limbs; southeastern United States   *Manculus*
   6–13 intercostal folds between appressed limbs; California and Oregon   *Batrachoseps*
5. Sexually mature adults with gills   6
   Sexually mature adults without gills   11
6. Body pigment lacking or greatly reduced; eyes lacking or greatly reduced   7
   Not as above   8
7. Southern Georgia and northern Florida   *Haideotriton*
   Central Texas   *Eurycea* (part)

8. Size large, to about 75 mm body length; east of Mississippi River   *Gyrinophilus* (part)
   Size smaller, to about 55 mm body length; west of Mississippi River   9
9. Central Texas   *Eurycea* (part)
   Western Ozark area   10
10. Pigmentation reduced; dorsal and lateral markings tend to form longitudinal dashes; 20 rib-bearing vertebrae from skull to sacrum, inclusive
                                   Typhlotriton (part)
    Pigmentation normal; dorsal and lateral markings usually not forming longitudinal dashes; 21 or 22 rib-bearing vertebrae from skull to sacrum, inclusive   *Eurycea* (part)
11. Posterior part of maxilla sharp-edged and without teeth   *Aneides*
    Posterior part of maxilla normal, with teeth   12
12. Tongue free all around   13
    Tongue attached in front   16

Fig. 3-21. Roof of mouth with vomerine and parasphenoid teeth continuous.

Fig. 3-22. Roof of mouth with vomerine and parasphenoid teeth separated.

13. Vomerine and parasphenoid teeth continuous (Fig. 3-21)   14
    Vomerine and parasphenoid teeth not continuous (Fig. 3-22)   15
14. A definite canthus rostralis marked by a light line bordered with black   *Gyrinophilus* (part)
    Not as above   *Pseudotriton*
15. Toes strongly webbed; California   *Hydromantes*
    Toes barely webbed if at all; eastern half of the United States   *Eurycea* (part)

16. Eyes reduced; body virtually pigmentless
                                    Typhlotriton (part)
    Eyes normal; body pigmented   17
17. With alternating dark and light longitudinal lines on sides; semiaquatic   *Stereochilus*
    Without alternating dark and light longitudinal lines on sides; terrestrial   *Plethodon*

Genus *Plethodon* Tschudi. Tongue attached in front. 2 premaxillae; prefrontals present; vomerine and parasphenoid teeth not continuous. No aquatic larval stage. The United States and Canada.

KEY TO SPECIES OF *PLETHODON*

1. East of the 100th meridian   2
   West of the 100th meridian   16
2. Venter light with melanophores forming a mottled or reticulated pattern   3
   Not as above   6
3. 7–9 intercostal folds between appressed limbs   4
   0–4 intercostal folds between appressed limbs   5
4. Adult males with straight premaxillary teeth and small, discrete, round or oval mental gland
                                        P. dorsalis
   Adult males with sickle-shaped premaxillary teeth and large mental gland limited anteriorly by mandible   *P. cinereus*
5. 0–1 intercostal folds between appressed limbs
                                   P. yonahlossee (part)
   2–4 intercostal folds between appressed limbs
                                      P. wehrlei (part)
6. Throat not conspicuously lighter than top of head   7
   Throat conspicuously lighter than top of head   8
7. 0–2 intercostal folds between appressed limbs
                                    P. glutinosus (part)
   5 or more intercostal folds between appressed limbs
                                       P. richmondi
8. West of Mississippi River   9
   East of Mississippi River   11
9. Edwards Plateau region of central Texas
                                    P. glutinosus (part)
   Ouachita uplift of Oklahoma and Arkansas   10
10. North of Ouachita River and west of line south from Mena   *P. ouachitae*
    South of Ouachita River and east of line south from Mena   *P. caddoensis*
11. Web between 3d and 4th digits of hind foot extending halfway between proximal and distal ends of proximal phalanges   *P. wehrlei* (part)
    No web between 3d and 4th digits of hind foot   12
12. Dorsum chestnut color in life; venter usually with

light spots or areas   *P. yonahlossee* (part)
Not as above   **13**
13. Adults to 50 mm body length; 3–4 intercostal folds between appressed limbs   *P. welleri*
    Adults to 100 mm body length; 0–3 intercostal folds between appressed limbs   **14**
14. No white dorsal spots   *P. jordani*
    With white dorsal spots   **15**
15. With chestnut-colored dorsal spots in life   *P. longicrus*
    Not as above   *P. glutinosus* (part)
16. 5th toe of hind foot with only 1 phalanx   **17**
    5th toe of hind foot with 2 phalanges   **18**
17. Jemez Mountains, Sandoval County, New Mexico   *P. neomexicanus*
    Lower Columbia River Gorge of Oregon and Washington   *P. larselli*
18. Parotoid glands present (Fig. 3-23)   *P. vandykei*
    Parotoid glands absent   **19**

Fig. 3-23. Dorsal view of head of *Plethodon vandykei* showing parotoid glands.

19. 18 or 19 trunk vertebrae   **20**
    16 or 17 trunk vertebrae   **21**
20. Venter, sides, and dorsum light brown   *P. stormi*
    Venter, sides, and dorsum dark brown or nearly black   *P. elongatus*
21. 15 costal grooves; dorsal stripe, if present, greenish yellow   *P. dunni*
    16 costal grooves; dorsal stripe, if present, not greenish yellow   *P. vehiculum*

Plethodon vandykei **Van Denburgh.** Van Dyke's salamander. Body length to 60 mm; 2–3 intercostal folds between appressed limbs. Parotoid glands present. A broad dorsal stripe, yellow to tan in color, sometimes poorly defined. Disjunct populations in Willapa Hills, Olympic Mountains, and west central Cascade Mountains of Washington; a disjunct population known from Kootenai, Benewah, and Idaho Counties, Idaho, and Lincoln and Mineral Counties, Montana, is recognized as a distinct subspecies.

Plethodon larselli **Burns.** Larch Mountain salamander. Body length to 55 mm; 1–3 intercostal folds between appressed limbs. A broad dorsal stripe (may be absent), highly variable in color; venter reddish. Lava talus slopes of the Columbia River Gorge (Skamania County, Washington, and Multnomah and Hood River Counties, Oregon).

Plethodon dunni **Bishop.** Dunn's salamander. Body length to 67 mm; 2–4 intercostal folds between appressed limbs. In life a greenish-yellow dorsal stripe (may be absent), bordered with black, containing darker markings, and extending to near tip of tail; sides dark brown with light spots; below slate-colored with pale spots. In wet situations, coastal forest of western Oregon and southwestern Washington.

Plethodon vehiculum **(Cooper).** Western red-backed salamander. Body length to 62 mm; 4–5 intercostal folds between appressed limbs. Dorsal stripe variable in color (may be absent), but not greenish yellow as in *P. dunni;* below mottled. Western Washington and Oregon as far south as Coquille, Coos County; into Canada.

Plethodon stormi **Highton and Brame.** Storm's salamander. Body length to 59 mm; 5–6 intercostal folds between appressed limbs; limbs relatively longer than in *P. elongatus.* Venter, sides, and dorsum light brown; small white spots on head and dorsum. Siskiyou Mountains of Jackson and Josephine Counties, Oregon.

Plethodon elongatus **Van Denburgh.** Del Norte salamander. Body length to 70 mm; 6–8 intercostal folds between appressed limbs. Dorsal stripe pinkish brown, more or less obscured in large adults; slate color below. Coastal forest of central Humboldt County in northwestern California, north to Rogue River in southwestern Oregon.

Plethodon neomexicanus **Stebbins and Riemer.** Jemez Mountains salamander. Body length to 75 mm; 7–8 intercostal folds between appressed limbs. Brown above with pale gold stippling in life. 5th toe reduced with but 1 phalanx. Jemez Mountains, New Mexico, in spruce-fir forest.

Plethodon welleri **Walker.** Weller's salamander. Body length to 50 mm; 3–4 intercostal folds between appressed limbs. Dorsum black with golden blotches which disappear in preservative; venter dark. At high altitudes in Yancey County, North Carolina, and on Whitetop Mountain and

Mt. Rogers, Virginia; in northeastern Tennessee as low as 2300 feet. 2 described subspecies.

Plethodon dorsalis Cope. Zigzag salamander. Body length to 50 mm; 6–7 intercostal folds between appressed limbs. Dorsal stripe absent or present, if present, straight-edged or zigzag; below reticulated. Central Indiana southwest to northeastern Oklahoma, southeast to western North Carolina, and south to northwestern Georgia, northern Alabama, and northeastern Mississippi. 2 described subspecies.

Plethodon cinereus (Green). Red-backed salamander. Body length to 60 mm; 6–9 intercostal folds between appressed limbs. Dorsal stripe absent or present, if present, straight-edged, serrate, or more or less zigzag; below reticulated. Terrestrial. Northeastern United States west to Minnesota, southwest to southeastern Oklahoma, and south to North Carolina and northern Georgia; largely absent from Kentucky and Tennessee; into Canada. 3 described subspecies. The relation of this species to *P. dorsalis* is much in need of investigation.

Plethodon richmondi Netting and Mittleman. Ravine salamander. Body length to 72 mm; 5–10 intercostal folds between appressed limbs. Brown or black above with small white or bronze flecks which disappear in preservative; venter slate gray with scattered white dots. Central Pennsylvania to southeastern Indiana, eastern Kentucky, northeastern Tennessee, and northwestern North Carolina. 3 described subspecies.

Plethodon wehrlei Fowler and Dunn. Wehrle's salamander. Body length to 73 mm; 2–3 intercostal folds between appressed limbs. Above dark brown or black, with or without spots or flecks; sides with light flecks; below light gray or gray mottled. Southwestern New York, western Pennsylvania, southeastern Ohio, West Virginia, to Roanoke County, Virginia.

Plethodon longicrus Adler and Dennis. Bluerock Mountain salamander. Body length to 100 mm; appressed limbs overlap 1 intercostal fold or fail to touch by 1 fold. Dorsum black with white flecking and flecks and blotches of chestnut brown; venter dark gray with white flecks, lighter in pectoral area. Known only from Bluerock Mountain, Rutherford County, North Carolina, at elevation of approximately 1650 feet.

Plethodon yonahlossee Dunn. Yonahlossee salamander. Body length to 91 mm; 0–1 intercostal folds between appressed limbs. Ground color black with a broad dorsal stripe of chestnut brown which disappears quickly in preservative; sides with light blotches. Southern section of the Blue Ridge province north of French Broad River (northeastern Tennessee, northwestern North Carolina, and adjacent Virginia).

Plethodon ouachitae Dunn and Heinze. Rich Mountain salamander. Body length to 60 mm; 0–1 intercostal folds between appressed limbs. Dorsum with white and bronze flecks and with or without chestnut-brown stripe or markings (chestnut-brown color disappears in preservative); pectoral area usually dark; throat light. Ouachita Mountains of Arkansas and Oklahoma north of the Ouachita River and west of line south from Mena.

Plethodon caddoensis Pope and Pope. Caddo Mountain salamander. Body length to 50 mm; 0–1 intercostal folds between appressed limbs. Back and upper sides with profuse pale white spotting; pectoral area usually light; throat light. A slimmer salamander than *P. ouachitae.* Ouachita Mountains of southwestern Arkansas south of the Ouachita River and east of line south from Mena.

Plethodon jordani Blatchley. Jordan's salamander. Body length to 82 mm; 1–3 intercostal folds between appressed limbs. Dorsum with or without light flecks; sides with or without light flecks or markings; venter black to light gray; legs and cheeks with or without red (remains for some time in preservative). Southern Appalachians (northeastern Georgia, northwestern South Carolina, western North Carolina, extreme eastern Tennessee, southwestern Virginia). 7 described subspecies, but none of these recognized by Highton, 1962.

Plethodon glutinosus (Green). Slimy salamander. Body length to 85 mm; 1–3 intercostal folds between appressed limbs. Above black, with or without silver or bronze flecks or spots; lateral markings variable, from few flecks to vertical bars; below slaty. Eastern United States, New York to Indiana, Missouri, eastern Oklahoma, and central Texas; not in lower half of peninsula of Florida. 4 described subspecies, but only 2 of these recognized by Highton, 1962.

Genus Ensatina Gray. Tail with basal constriction. Palmar tubercles present. Tongue attached in front. 2 premaxillae; prefrontals present; vomerine and parasphenoid teeth not continuous. No aquatic larval stage. Western United States, north into Canada.

Ensatina eschscholtzi **Gray.** Ensatina. Body length to 76 mm; appressed limbs meeting or overlapping by as many as 5 intercostal folds. Dorsum brown, unspotted or with black spots, or black with yellow or red spots or blotches. Terrestrial. Coastal mountains, British Columbia to southern California; Sierra Nevada. 7 described subspecies.

Genus *Hemidactylium* Tschudi. Tail constricted at base. Toes 4–4. Tongue attached in front. 2 premaxillae; prefrontals present; vomerine and parasphenoid teeth not continuous. Adults terrestrial, but an aquatic larva with a dorsal fin. East central and northeastern United States, adjacent Canada.

Hemidactylium scutatum **(Schlegel).** 4-toed salamander. Body length to 46 mm; 3–4 intercostal folds between appressed limbs. Venter with black spots. Nova Scotia to Wisconsin and south to the Gulf of Mexico, but apparently the southern records (Missouri, Arkansas, Louisiana, Florida, Georgia, the Carolinas) represent disjunct colonies.

Genus *Aneides* Baird. Posterior end of maxilla sharp-edged and edentulous; premaxillae fused; prefrontals present; vomerine and parasphenoid teeth not continuous; terminal phalanges Y-shaped. Tongue attached in front. No aquatic larval stage. Appalachian Mountains in eastern United States, British Columbia to Baja California in West.

KEY TO SPECIES OF *ANEIDES*

1. Venter and dorsum dark in color   *A. flavipunctatus*
   Venter light, dorsum dark   **2**
2. Dorsum conspicuously mottled in life with yellow-green lichenlike patches; eastern United States
       A. aeneus
   Dorsum not conspicuously mottled in life with yellow-green lichenlike patches; western United States
       3
3. Dorsum with yellow spots   *A. lugubris*
   Dorsum without yellow spots   **4**
4. Tips of digits expanded   *A. ferreus*
   Tips of digits not expanded   *A. hardyi*

Aneides hardyi **(Taylor).** Sacramento Mountain salamander. Body length to 56 mm; 3–4 intercostal folds between appressed limbs. Back with clay to grayish-tan or bronze markings on black or dark brown ground color; venter slate-colored. Toes slightly webbed at base, flattened, and round-tipped. Coniferous forest of Sacramento Mountains, Otero County, New Mexico.

Aneides aeneus **(Cope).** Green salamander. Body length to 60 mm; appressed limbs overlapping by 1–3 intercostal folds. Above black with yellowish green lichenlike patches in life. Toes webbed with expanded tips. Commonly found in crevices in rocks, West Virginia and southern Ohio to northern Alabama, northern Georgia, and northwestern South Carolina.

Aneides ferreus **Cope.** Clouded salamander. Body length to 62 mm; appressed limbs separated by 1 intercostal fold or may overlap by 1 intercostal fold. Above with light mottling on dark background. Tips of digits expanded. Arboreal. Humid coastal forest from Columbia River south to central Mendocino County, California.

Aneides lugubris **(Hallowell).** Arboreal salamander. Body length to 100 mm; appressed limbs separated by 1 intercostal fold or may overlap by 1 intercostal fold. Back light brown to brown with yellow spots. Tips of digits expanded. Arboreal, with tail prehensile. Coastal mountains of California, San Diego County to Humboldt County; Sierra Nevada foothills, Calaveras County to Madera County; islands off California. 2 described subspecies.

Aneides flavipunctatus **(Strauch).** Black salamander. Body length to 80 mm; 3–5 intercostal folds between appressed limbs. Above black to greenish, with or without light spots; below black with scattered gray dots. Tail prehensile. In wet soil or talus, Santa Cruz County to northern Humboldt County in coastal mountains of California; east in Klamath Mountains into Shasta County, California. 2 described subspecies.

Genus *Batrachoseps* Bonaparte. Body elongate. Premaxillae fused; no prefrontals; vomerine and parasphenoid teeth not continuous. Tongue attached in front. Toes 4–4. Terrestrial, with no aquatic larval stage. Western Oregon to Baja California. Lower Pliocene of California. 2 additional species have recently been found in the Kern River Canyon and several disjunct localities of California, and will presumably be described soon.

KEY TO SPECIES OF *BATRACHOSEPS*

1. Venter light, with scattered melanophores
   B. pacificus
   Venter dark, with light spots   2
2. Light spots of venter much smaller than width of intercostal fold   *B. attenuatus*
   Many of light spots of venter as large as width of intercostal fold   *B. wrighti*

Batrachoseps attenuatus (Eschscholtz). California slender salamander. Body length to 62 mm;
8–13 intercostal folds between appressed limbs.
Dorsal ground color brown to black, markings variable (frequently a stripe of variable color); below
dark with many white dots. Coastal region, southwestern Oregon to northern Baja California; Sierra
Nevada foothills, Butte County to Kern County;
Central Valley of California in vicinity of San Francisco Bay. 2 described subspecies. Hybridizes with
B. pacificus.
   Batrachoseps pacificus (Cope). Pacific slender
salamander. Body length to 78 mm; 9–13 intercostal folds between appressed limbs. Above brown
with gold dots and blotches, less obviously striped
than *B. attenuatus;* below pale. Los Angeles, San
Bernardino, Riverside, and Orange Counties, California; Channel Islands of California. 3 described
subspecies. Hybridizes with *B. attenuatus* and
sometimes considered a subspecies of *B. attenuatus.*
   Batrachoseps wrighti (Bishop). Oregon slender
salamander. Body length to 60 mm; 6–7 intercostal
folds between appressed limbs. Above with reddish stripe which is sometimes obscure (tan or
gray); below dark. Multnomah County to Lane
County, Oregon.
   Genus *Stereochilus* Cope. Tongue attached in
front. Premaxillae fused; prefrontals present; vomerine and parasphenoid teeth continuous. Adults
aquatic, at least in the spring; an aquatic larval
stage. Southeastern United States.
   Stereochilus marginatus (Hallowell). Many-
lined salamander. Body length to 52 mm; 8–9 intercostal folds between appressed limbs. Sides with
longitudinal light lines. Top of head with conspicuous pores. In the Coastal Plain, southeastern Virginia to Georgia.
   Genus *Typhlotriton* Stejneger. Adults neotenic
or metamorphosed. Tongue attached in front in
metamorphosed adults. Premaxillae fused; pre-

frontals present. Eyes and pigmentation reduced.
Ozark uplift.

KEY TO SPECIES OF *TYPHLOTRITON*

1. Adults with gills; 5–7 intercostal folds between appressed limbs   *T. nereus*
   Adults without gills; 2–4 intercostal folds between appressed limbs   *T. spelaeus*

Typhlotriton spelaeus Stejneger. Grotto salamander. Body length to 71 mm; 2–4 intercostal
folds between appressed limbs; 19 rib-bearing
vertebrae between skull and sacrum, inclusive.
Body pale flesh color. Eyes reduced, covered by
fused lids. Adults subterranean, larvae in underground streams or spring-fed brooks of Ozark
Plateau of Oklahoma, Arkansas, and Missouri.
   Typhlotriton nereus Bishop. Spring grotto salamander. Body length to 55 mm; 5–7 intercostal
folds between appressed limbs; 20 rib-bearing vertebrae between skull and sacrum, inclusive. Neotenic, in subterranean and open streams of the
Ozark Plateau of Oklahoma, Arkansas, Missouri,
and Kansas. Dowling (1957) doubts the validity
of this species.
   Genus *Haideotriton* Carr. Subterranean permanent larvae with eyes absent or reduced and
pigmentation reduced. Skull simple, larval, mostly
unossified; premaxillae fused; maxillae absent.
Head spatulate. Southern Georgia and adjacent
Florida; the population in Jackson County, Florida,
described by Pylka and Warren (1958) was not
referred to a species.
   Haideotriton wallacei Carr. Georgia blind salamander. Body length 42 mm in the 1 known
specimen; appressed limbs overlapping. Gill rami
very long. Known only from a 200-ft-deep artesian
well at Albany, Georgia.
   Genus *Gyrinophilus* Cope. Tongue free all
around. A single premaxilla in larvae and neotenic
adults, separating into 2 distinct bones in most
metamorphosed adults; nasal processes of premaxilla unfused and separated by a fontanel;
prefrontals present; vomerine and parasphenoid
teeth continuous. An aquatic larval stage. Adults
neotenic or metamorphosing, aquatic or terrestrial.
Montane forms, New England to eastern Tennessee, northern Alabama and Georgia, and north-

western South Carolina. Grobman (1959) was of the opinion that *Gyrinophilus* was not validly distinct from *Pseudotriton,* but Martof and Rose (1962) maintain the validity of the 2 genera.

KEY TO SPECIES OF *GYRINOPHILUS*

1. Eye small, its diameter entering distance from anterior corner of eye to snout tip 4–5 times
   G. palleucus
   Eye larger, its diameter entering distance from anterior corner of eye to snout tip 1½–3½ times
   G. porphyriticus

Gyrinophilus palleucus McCrady. Tennessee cave salamander. Body length to 135 mm; 6–8 intercostal folds between appressed limbs. Eyes reduced. Snout broad and spadelike. In underground streams of Roane, McMinn, Grundy, and Franklin Counties, Tennessee, and Jackson County, Alabama. 3 described subspecies, varying considerably in pigmentation.

Gyrinophilus porphyriticus (Green). Spring salamander. Body length to 136 mm; 5–8 intercostal folds between appressed limbs. Dorsum light yellowish brown clouded or mottled, or reddish with middorsal line of chevrons or few to many small dark spots, or pink with few or no spots but a dorsolateral series of spots; throat immaculate, dotted or heavily reticulated. In and around springs, caves, and brooks, northeastern United States to northern Ohio, south to northern Alabama and Georgia and northwestern South Carolina; into Canada. 4 described subspecies; *G. danielsi* included here.

Genus *Pseudotriton* Tschudi. Tongue free all around. Premaxillae fused; prefrontals present; vomerine and parasphenoid teeth continuous. Adults aquatic or terrestrial; an aquatic larval stage. Eastern United States.

KEY TO SPECIES OF *PSEUDOTRITON*

1. Horizontal diameter of eye 1¼–1½ in snout; no light dark-bordered line from eye to nostril
   P. montanus
   Horizontal diameter of eye 1½–2 in snout; a light dark-bordered line from eye to nostril   **P. ruber**

Fig. 3-24. Pseudotriton ruber. **Elk Neck, Cecil County, Md., summer, 1951. (Coll. George Gifford. Photo by Isabelle Hunt Conant.)**

Pseudotriton ruber (**Latreille**). Red salamander (Fig. 3-24). Body length to 105 mm; 3–7 intercostal folds between appressed limbs. In life dorsum red, dark salmon, or purplish brown, with many black dots which may coalesce; venter flesh color, immaculate or with many black dots. In and around springs, seeps, and brooks, southeastern Louisiana to northwestern Florida, northeast to southern New York; not in Atlantic Coastal Plain. 4 described subspecies.

Pseudotriton montanus **Baird.** Mud salamander. Body length to 117 mm; 5–8 intercostal folds between appressed limbs. In life dorsum red to dark chocolate brown, with well-separated dark spots which are sometimes lost in dark ground color or with yellowish longitudinal dashes; venter flesh color, salmon pink, salmon red, or dusky orange-red, immaculate or dotted or flecked with dark markings. In and around springs, seeps, and brooks, extreme south central Pennsylvania to southern Ohio and eastern half of Kentucky, south to northern half of Florida, thence west to southeastern Louisiana. 4 described subspecies.

Genus *Eurycea* Rafinesque. Tongue free all around in transformed adults. Premaxillae fused; prefrontals present; vomerine and parasphenoid teeth not continuous. Body form usually slender. Adults neotenic or metamorphosing, aquatic or terrestrial; an aquatic larval stage. Eastern United States. *Typhlomolge* included here, in agreement with Mitchell and Reddell (1965).

KEY TO SPECIES OF *EURYCEA*

1. Adults without gills — 2
   Adults with gills — 6
2. 7 or more intercostal folds between appressed limbs
   E. multiplicata (part)
   5 or fewer intercostal folds between appressed limbs
   3
3. With black spots on orange (in life) ground color
   E. lucifuga
   Not with black spots on orange ground color — 4
4. Tail shorter than snout-vent length — *E. aquatica*
   Tail longer than snout-vent length — 5
5. Appressed limbs separated by 2 or fewer intercostal folds — *E. longicauda*
   Appressed limbs separated by 3 or more intercostal folds — *E. bislineata*
6. In Oklahoma, Arkansas, or Missouri — 7
   In central Texas — 8
7. Snout-vent length about $8\frac{1}{2}$ times head width
   E. tynerensis
   Snout-vent length about $6\frac{1}{2}$ times head width
   E. multiplicata (part)
8. Eyes reduced, diameter $\frac{1}{10}$ of head width or less — 9
   Eyes larger, diameter $\frac{1}{8}$ of head width or more — 11
9. Appressed limbs not meeting — *E. troglodytes*
   Appressed limbs meeting or overlapping — 10
10. Appressed limbs overlapping 0–3 intercostal folds
    E. tridentifera
    Appressed limbs overlapping 5–6 intercostal folds
    E. rathbuni
11. Last rib with single articulation — *E. pterophila*
    Last rib with dual articulation — 12
12. Eye diameter not more than $\frac{1}{3}$ interorbital distance
    E. latitans
    Eye diameter $\frac{2}{3}$ of, or equal to, interorbital distance
    13
13. Light yellowish above in a mottled or reticulated pattern — *E. neotenes*
    Uniformly light brown above except for a dorsolateral series of small, separate, yellowish flecks
    E. nana

Eurycea lucifuga (**Rafinesque**). Cave salamander. Body length to 66 mm; appressed limbs meeting or overlapping by as many as 2 intercostal folds. In life dorsum orange-red with black spots; lateral dark markings may form an irregular stripe. In caves and around springs and seeps, Ozark Plateau northeast to southern Indiana and West Virginia, thence south to northern Georgia and northern Alabama.

Eurycea longicauda (**Green**). Long-tailed salamander. Body length to 70 mm; appressed limbs separated by fewer than 3 intercostal folds. A middorsal dark stripe bordered by tan or yellow stripes, or a broad olive, yellow, or orange dorsal stripe containing brown or black spots which may tend to be concentrated in the middorsal line. In damp situations such as mouths of caves, springs, seeps, creek banks, and swampy areas; generally distributed east of Mississippi River south of line from southern Illinois to southeastern New York, but not in peninsular Florida; Ozark Plateau. 4 described subspecies.

Eurycea bislineata (**Green**). Two-lined salamander. Body length to 50 mm; 3–5 intercostal folds between appressed limbs. A broad dorsal stripe (yellow, greenish yellow, orange-yellow, brown) bounded on either side by a narrow black line, containing few to many black dots. Terrestrial or semiaquatic, along brooks and streams, around springs and swamps; Maine to northern Florida to southeastern Louisiana to eastern Illinois; into Canada. 4 described subspecies.

Eurycea aquatica **Rose and Bush**. Water eurycea. Body length to 48 mm; 1–3 intercostal folds between appressed limbs. A brown dorsal stripe bordered with lateral black; dorsal stripe usually with fine black middorsal line; usually a row of lateral light spots. Stockier than *E. bislineata*. Aquatic, in springs and streams 2 miles west of Bessemer, Jefferson County, Alabama.

Eurycea multiplicata (**Cope**). Many-ribbed salamander. Body length to 47 mm; 7–10 intercostal folds between appressed limbs. Dorsal ground color yellowish with overlying brown pigment giving a light brown general impression, or dark gray dorsal stripe with variable darker markings sometimes forming middorsal line; venter pale gray to yellow, with or without melanophores. Sometimes neotenic. In and around springs and brooks of Ouachita uplift of Oklahoma and Arkansas and Ozark uplift of northeastern Oklahoma, northwestern Arkansas, and southwestern Missouri. 2 described subspecies.

Eurycea tynerensis **Moore and Hughes**. Oklahoma salamander. Body length to 33 mm; 8–12 intercostal folds between appressed limbs; 22 rib-bearing vertebrae between skull and sacrum, inclusive. Dorsum with pigment forming irregular reticulum; sides sometimes with light spots; venter immaculate. In creeks and rivers of the Ozark uplift of northeastern Oklahoma, southwestern Missouri, and presumedly northwestern Arkansas.

Eurycea nana **Bishop.** San Marcos salamander. Body length to 28 mm; 6–7 intercostal folds between appressed limbs. A dorsolateral series of 7–9 light spots. Neotenic, known only from head of San Marcos River, San Marcos, Hays County, Texas.

Eurycea pterophila **Burger, Smith, and Potter.** Fern Bank salamander. Body length to 35 mm; 6–7 intercostal folds between appressed limbs. Similar to *E. neotenes* in external appearance. Neotenic. Known only from Fern Bank Spring, 6.3 miles northeast of Wimberley, Hays County, Texas.

Eurycea neotenes **Bishop and Wright.** Bexar County salamander. Body length to 51 mm; 5–7 intercostal folds between appressed limbs. Yellowish above in mottled or reticulated pattern. Neotenic, in cave and surface streams, Val Verde to Williamson County in Edwards Plateau of Texas.

Eurycea latitans **Smith and Potter.** Cascade Cavern salamander. Body length to 53 mm; 3–5 intercostal folds between appressed limbs. Back with reticulations enclosing light spots. Eye small, its diameter $\frac{1}{3}$ to $\frac{1}{4}$ the interorbital distance. Neotenic, in cave and surface streams, Kendall and perhaps Kerr Counties, Texas.

Eurycea troglodytes **Baker.** Valdina Farms neotene. Body length to 40 mm; 0–4 intercostal folds between appressed limbs. Color white to light gray with white specks and indistinct yellow stripes on sides; venter pigmentless and translucent. Known only from Valdina Farms sinkhole, Medina County, Texas.

Eurycea tridentifera **Mitchell and Reddell.** Honey Creek Cave salamander. Body length to 37 mm; appressed limbs overlapping 0–3 intercostal folds; head twice as broad as body; snout depressed; eyes greatly reduced. Depigmented, nearly white. Neotenic, known only from Honey Creek Cave, Comal County, Texas.

Eurycea rathbuni **(Stejneger).** Texas blind salamander. Body length to 76 mm; appressed limbs overlapping 5–6 intercostal folds; limbs very long and slender; snout greatly depressed; eyes reduced to dots. Body nearly white. Neotenic, in subterranean streams of Purgatory Creek system in vicinity of San Marcos, Texas.

Genus *Manculus* Cope. Very similar to *Eurycea* but toes 4–4 and larva with a dorsal fin. Southeastern United States. Monotypic.

Manculus quadridigitatus **(Holbrook).** Dwarf salamander. Body length to 34 mm; 3–5 intercostal folds between appressed limbs. Above frequently with broad stripe, yellowish to bronze in color; venter yellow. Tail long and slender. In moist situations in the Coastal Plain, eastern Texas to North Carolina; not in lower half of peninsula of Florida. 3 described subspecies.

Genus *Hydromantes* Gistel. Tongue free all around. Toes half-webbed. Head and body depressed. 2 premaxillae; no prefrontals; vomerine and parasphenoid teeth not continuous. California and southern Europe.

KEY TO SPECIES OF *HYDROMANTES*

1. Dorsum of uniform color   *H. brunus*
   Dorsum mottled   2
2. Appressed limbs overlapping; Shasta County, California   *H. shastae*
   Appressed limbs not overlapping; south from Alpine County in Sierra Nevada of California   *H. platycephalus*

Hydromantes brunus **Gorman.** Limestone salamander. Body length to 64 mm; appressed limbs meeting or overlapping 1 intercostal fold. Head and body less depressed than in *H. platycephalus*. Dorsal color uniform light to dark brown, greenish in young. Known from 3 closely adjacent localities near Briceburg, Mariposa County, California, at altitudes from 1200 to 2400 ft.

Hydromantes shastae **Gorman and Camp.** Shasta salamander. Body length to 64 mm; appressed limbs overlapping 1–2 intercostal folds. Head and body less depressed than in *H. platycephalus*. Dorsum with chromatophores forming mottled pattern overlying reddish chocolate ground color. Shasta County, California, at altitudes from 1100 to 2500 ft.

Hydromantes platycephalus **(Camp).** Mount Lyell salamander. Body length to 68 mm; 1–2 intercostal folds between appressed limbs. Head and body strongly depressed. Above brown to black with mottling of gray (greenish in young); below gray to black with varying amount of gray blotches. Sierra Nevada from Alpine County to northern Tulare County, California, at altitudes from 5000 to 11,000 ft.

Superorder Salientia. In addition to the living order Anura, some workers include here the extinct order Proanura based on the lower Triassic *Protobatrachus,* the relationship of which is questioned by Hecht, 1962. Vertebral centra reduced or absent, being functionally replaced by a downgrowth of the neural arches; skull bones reduced in number.

Order Anura. Superficially highly modified amphibians without a tail and with the hind legs adapted for jumping, an extra leg segment being added by virtue of elongation of tarsal bones. Tail vertebrae fused, represented by a bone, the urostyle; radius and ulna fused and tibia and fibula fused; pubis cartilaginous; cleithrum present; vertebrae reduced in number, with presacral vertebrae varying from 5 to 9. Tympanum concealed or exposed. The 6 known Jurassic genera of frogs have no primitive characteristics not retained by some living forms.

Anurans exhibit many striking secondary sex characters. Females are usually larger than males. In some species (*Bufo, Ascaphus*) males have longer legs than females. Parts or features of the body may be enlarged in males, as thumbs (*Rana, Ascaphus*), forearms (*Rana, Ascaphus*), tympana (*Rana*), and hind-foot webbing (*Scaphiopus*). Special structures may be present in males, as the intromittent organ of *Ascaphus*, ventral or lateral vocal pouches (*Bufo, Hyla, Scaphiopus, Rana*), and horny excrescences or nuptial pads on fingers and forearms (*Bufo, Hyla, Scaphiopus, Rana*). General skin pigmentation (*Bufo, Scaphiopus*) or that of special areas such as the vocal pouches (*Bufo, Hyla, Scaphiopus, Gastrophryne*) may differ in the 2 sexes. In addition to structural characters there are behavioral characters such as the chirp and sex-warning vibration of male *Bufo* and the sex call of males of most species of anurans.

Most of the anurans of the United States go to the water to breed. Rainfall is a powerful stimulus to breeding activity in many species, although numerous anurans, xeric as well as mesic, breed quite independently of precipitation. In some species large mating aggregations of hundreds or even thousands of individuals may be formed. In most species the male utters a sex call which induces calling and other sexual activity in other males and attracts females. In many species, especially toads (*Bufo*), males clasp indiscriminately any moving object (and sometimes nonmoving) of approximately the right size, releasing if the male sex warning is given, but holding on for hours or even days in the absence of this warning or of oviposition. With normal amplexus the male clasps the female just behind the front limbs (axillary position) or just in front of the hind limbs (inguinal position). Amplexus may last from an hour or so to several days. As the eggs are laid the male sheds sperm over them. The eggs, which are typically pigmented (unpigmented in *Ascaphus*) and covered with jelly, may be laid singly, in small or large masses, in short rods or long strings, or in a surface film. In some of the leptodactylids the eggs are laid on land and are unpigmented.

Typically there is an aquatic larval stage, the tadpole, which has a fishlike body form with compressed tail, lacks legs until well along in development, and lacks true teeth. External gills make their appearance as a transient feature early in the larval period, but throughout most of the larval period internal gills and the skin perform the function of exchange of respiratory gases. Most tadpoles are herbivorous, many are scavengers, and some are carnivorous (*Scaphiopus*) with flesh-tearing mouth parts. Many lead solitary lives, while others (*Bufo, Rana*) form large, compact aggregations of hundreds or thousands of individuals. The larval period lasts from 2 weeks (*Scaphiopus, Bufo*) to 2 years (*Rana*). Metamorphosis, which is controlled by the thyroid gland, involves resorption of the tail, appearance of the front limbs (hidden underneath the operculum during development), shortening of the digestive tract, loss of the gills with closure of the gill slits, development of true teeth (not developed in some groups), modification of the jaws, modification of the position of the eyes, and changes in the skin for a terrestrial existence. In some species sexual maturity is attained in less than a year after metamorphosis, while in others 2 or more years is required.

The United States is not exceptionally rich in number of species of anurans, with about 70 species as compared with a world total of 1500 or more. Of the 13 families into which many workers divide the Anura, 8 are found in the United States. Not one of the families is confined to the United States, however.

Some workers have thought the Anura polyphyletic, or at least diphyletic. Others have maintained that the group is monophyletic with an early bifurcation into the notobatrachid-ascaphid-pipid-rhinophrynid line and into the line that contains pelobatid and all other family groups.

Cope (1889) recognized 3 suborders of anurans: Aglossa, Arcifera, and Firmisternia. Noble (1931) revised and expanded anuran classification to include the following suborders: Amphicoela, Opisthocoela, Anomocoela, Procoela, and Diplasiocoela. Orton (1957) considered the recognition of anuran suborders unnecessary and undesirable, but, basing her opinions on tadpole morphology, divided the families of living anurans into 4 groups. The most primitive group includes the Pipidae and the Rhinophrynidae. The second group contains the Microhylidae, the third group the Ascaphidae and Discoglossidae, and the most advanced group comprises all remaining families.

KEY TO FAMILIES OF ANURA

1. Pupil of eye vertical 2
   Pupil of eye not vertical 4
2. Tongue attached in front 3
   Tongue free in front .... **Rhinophrynidae**
3. With a sharp-edged metatarsal tubercle (Fig. 3-25)
   .... **Pelobatidae**
   No sharp-edged metatarsal tubercle; male with intromittent organ (Fig. 3-31) .... **Ascaphidae**
4. Tympanum absent .... **Microhylidae**
   Tympanum present (Fig. 3-26) 5
5. Tongue bicornuate behind (Fig. 3-27) .... **Ranidae**
   Tongue not bicornuate behind 6

Fig. 3-25. Metatarsal tubercle of pelobatid.

Fig. 3-26. Lateral view of anuran head showing tympanum.

Fig. 3-27. Bicornuate tongue of ranid.

Fig. 3-28. Dorsal view of head of bufonid showing parotoid glands.

Fig. 3-29. Lateral view of toe of hylid.

6. With parotoid glands (Fig. 3-28) .... **Bufonidae**
   Without parotoid glands 7
7. Toes with intercalary bone or cartilage, giving a stepped appearance in lateral view (Fig. 3-29)
   .... **Hylidae**
   Toes without intercalary bone or cartilage
   .... **Leptodactylidae**

Family Rhinophrynidae. Sacral diapophyses somewhat dilated; no teeth; omosternum rudimentary, sternum absent; foot specialized for digging, with the single phalanx of the first toe shovellike and the prehallux covered by a large cornified "spade." No typanum. Pupil vertical. A single living species, but known from the Eocene of Wyoming and the Oligocene of Saskatchewan.

Genus *Rhinophrynus* Dumeril and Bibron. With characters of family. Known from the Oligocene of Saskatchewan.

***Rhinophrynus dorsalis* Dumeril and Bibron.** Burrowing toad. Body length to 65 mm. Skin smooth, pink and brown. Fat and short-legged, head small. A termite feeder, found from the lower Rio Grande Valley of Texas (Starr and Zapata Counties) south into Guatemala.

Family Microhylidae. Sacral diapophyses more or less dilated; premaxillary and maxillary teeth present or absent. Tympanum concealed or exposed. Pupil usually horizontal. A variable and widely distributed family, found in North and South America, Africa, southern and eastern Asia, and the Indo-Australian Archipelago; about 53 genera, 16 in the New World. Miocene and Pleistocene of Florida.

KEY TO GENERA OF MICROHYLIDAE

1. Clavicles absent; without 2 smooth dermal ridges across palate in front of pharynx   *Gastrophryne*
   Clavicles present; with 2 smooth dermal ridges across palate in front of pharynx   *Hypopachus*

Genus *Gastrophryne* Fitzinger. No clavicles or precoracoids; quadratojugal in contact with maxilla. 1 or 2 metatarsal tubercles. Pupil horizontal. Terminal phalanges simple. Digital discs absent. Toe webs reduced or absent. 5 species in North and Central America. Miocene of Florida.

KEY TO SPECIES OF *GASTROPHRYNE*

1. Venter mottled   *G. carolinensis*
   Venter not mottled   2
2. Dorsal surface of thigh with dark crossbar
       G. mazatlanensis
   Dorsal surface of thigh without dark crossbar
       G. olivacea

***Gastrophryne olivacea* (Hallowell).** Great Plains narrow-mouthed toad. Body length to 36 mm. Above olive, sometimes with dark dots. Toes webless. Fossorial; frequently under rocks on dry prairie, especially in tarantula burrows. Breeds only after heavy rainfall. Nebraska to Texas, into Mexico.

***Gastrophryne carolinensis* (Holbrook).** Narrow-mouthed toad (Fig. 3-30). Body length to 38 mm. Above brown, usually with irregular dark markings. Toes webless. Fossorial. Heavy rainfall not necessary for breeding. Maryland to Florida in Coastal Plain, west to eastern Texas and eastern Oklahoma; north to Kentucky, southern Illinois, and southeastern Iowa. Pleistocene of Florida.

***Gastrophryne mazatlanensis* (Taylor).** Sinaloa narrow-mouthed toad. Body length to 31 mm. Above brown, with distinct dark markings; an indistinct black stripe behind eye; venter unmarked. Pima and Santa Cruz Counties, Arizona; range allopatric to that of *G. olivacea*, to which it is closely related and of which it may be a subspecies; into Mexico.

Genus *Hypopachus* Keferstein. Palatines absent; clavicles straight or only slightly curved; omosternum lacking; simple terminal phalanges; no bony ridge behind the choanae. Tongue well developed. 2 smooth dermal ridges across palate in front of pharynx. Tympanum lacking. 1 or 2 metatarsal tubercles. Central America (10 species) and South America (3 species).

***Hypopachus cuneus* Cope.** Sheep frog. Body length to 41 mm. Back brown to olive, with irregular dark markings; a narrow light middorsal line; venter mottled; groin with orange color in life. 2

Fig. 3-30. *Gastrophryne carolinensis.*
Near Micanopy, Alachua County, Fla., Apr. 8, 1953.
(Coll. A. F. Carr, Jr. Photo by Isabelle Hunt Conant.)

prominent metatarsal tubercles. Fossorial. **Southern Texas**, into Mexico. 2 described subspecies, 1 in the United States.

Family Ascaphidae. Vertebrae amphicoelous; pectoral girdle arciferous; 2 pairs of free ossified ribs; 9 presacral vertebrae; sacral diapophyses dilated; vomerine teeth present; lower jaw toothless. Tympanum absent. 2 tail-wagging muscles. New Zealand and western North America.

Genus *Ascaphus* Stejneger. Male with intromittent organ. Western North America.

***Ascaphus truei* Stejneger.** Tailed frog (Fig. 3-31). Body length to 50 mm. 3 prominent palmar tubercles; hind foot fully webbed. Dorsum variable in color, smooth or with tubercles; venter light, variously mottled, or almost black. In cold, rapid, mountain brooks, northwestern California to British Columbia; western Montana and north central Idaho.

Family Pelobatidae. Spadefoot toads. Pectoral girdle arciferous; vertebrae procoelous or amphicoelous; 8 presacral vertebrae; urostyle fused to presacral vertebra; sacral diapophyses dilated; terminal phalanges simple. Widely distributed in the Northern Hemisphere and extending into the Southern Hemisphere in the East Indian region. Oligocene of Mongolia, Miocene of Germany, and Oligocene, Miocene, Pliocene, and Pleistocene of the United States.

Genus *Scaphiopus* Holbrook. Fossorial and nocturnal anurans of toadlike body form. Skin smooth and thin. Pupil of eye vertical. Restricted to North America, the only genus of the family in the New World; some workers recognize 2 genera, *Spea* (*bombifrons, hammondi,* and *intermontanus*) and *Scaphiopus* (*couchi* and *holbrooki*). Miocene of Florida, Miocene-Pliocene contact of Nebraska, Pliocene of Kansas and Nevada, and Pleistocene of Kansas, Texas, and Florida.

KEY TO SPECIES OF *SCAPHIOPUS*

1. Pectoral glands present (Fig. 3-32) **S. holbrooki**
   Pectoral glands absent **2**
2. Interorbital osseous boss present (Fig. 3-33) **S. bombifrons**

Fig. 3-32. Ventral view of pelobatid showing pectoral glands.

Fig. 3-33. Head of *Scaphiopus bombifrons* showing osseous boss.

Fig. 3-34. Hind foot of *Scaphiopus hammondi* showing metatarsal tubercle.

Fig. 3-31. Ascaphus truei. Near the junction of the East and West Forks of the Lostine River, Wallowa County, Ore., July 8, 1955. (Coll. Denzel E. Ferguson. Photo by Isabelle Hunt Conant.)

Fig. 3-35. Hind foot of *Scaphiopus intermontanus* or *S. couchi* showing metatarsal tubercle.

    No interorbital osseous boss (may be dermal thickening in *S. intermontanus*)   3
3. Pigmented edge of metatarsal tubercle rounded, about as wide as long (Fig. 3-34)   **S. hammondi**
    Pigmented edge of metatarsal tubercle elongate, sometimes twice as wide as long (Fig. 3-35)   4
4. Internarial distance about as great as interorbital distance   **S. intermontanus**
    Internarial distance definitely less than interorbital distance   **S. couchi**

Scaphiopus holbrooki (**Harlan**). Eastern spadefoot (Fig. 3-36). Body length to 82 mm. No frontoparietal fontanel. Rounded parotoid glands. Pigmented edge of metatarsal tubercle curved, 3 or more times wider than long. Above olive to brown, usually with 2 or more longitudinal light stripes; throat of male white. Eastern United States as limited by line from Massachusetts to southern Illinois to eastern Oklahoma and Texas; possibly a hiatus in lower Mississippi River valley. 2 described

Fig. 3-36. *Scaphiopus holbrooki.* Sam Houston State Park, Walker County, Tex., Mar. 9, 1952. (Coll. Wilmot Thornton. Photo by Isabelle Hunt Conant.)

subspecies. Hybridizes with *S. couchi.* Pleistocene of Florida.

Scaphiopus couchi **Baird.** Couch's spadefoot. Body length to 80 mm. No frontoparietal fontanel. Pigmented edge of metatarsal tubercle about twice as wide as long. Back variable in color, more or less coarsely vermiculate (males may have less well-defined markings than females); throat of male white. Southwestern Oklahoma to southeastern California; western two-thirds of Texas, into northern Mexico. Hybridizes with *S. holbrooki.*

Scaphiopus intermontanus **Cope.** Great Basin spadefoot. Body length to 63 mm. No frontoparietal fontanel. Dorsum frequently with a pair of light longitudinal stripes; throat discolored in males. Eastern Washington to east central California to northern Arizona to western Colorado to western Montana to southern Idaho.

Scaphiopus hammondi **Baird.** Hammond's spadefoot. Body length to 60 mm. Frontoparietal fontanel present. Tympanum indistinct. Dorsal color variable, olive, green, gray, or brown, with small light dark-edged spots; throat of male discolored. Colorado, eastern half of Arizona south of the Colorado River, New Mexico, western Texas, and western Oklahoma; coastal California as far north as San Francisco area and farther north in Central Valley. 2 described subspecies, 1 in Mexico. Hybridizes with *S. bombifrons.*

Scaphiopus bombifrons **Cope.** Plains spadefoot. Body length to 58 mm. Frontoparietal fontanel present. Pigmented area of metatarsal tubercle rounded, a little wider than long. Above olive gray with darker markings; throat of male discolored. Montana and North Dakota to eastern Kansas and Oklahoma to southern Texas and southeastern Arizona; eastern Wyoming to New Mexico; into Mexico. Pleistocene of Kansas. Hybridizes with *S. hammondi.*

Family Bufonidae. Toads. Vertebrae procoelous; sacral diapophyses dilated; pectoral girdle arciferous; no maxillary teeth. Parotoid glands present. Tympanum present or absent. Bidder's organ (rudimentary ovary) present in both sexes. Worldwide in distribution except for New Guinea, Polynesia, Australia, and Madagascar. Oligocene of Europe and Argentina; Miocene of Colombia, Florida, and Nebraska; Pliocene of Chihuahua, Arizona, Florida, Kansas, and Nebraska; Pleistocene of Asia, Argentina, Yucatán, Arkansas, California, Florida, Kansas, Nevada, Oklahoma, Pennsylvania, Texas, and Virginia. 10 genera, but only *Bufo* in the New World.

Genus *Bufo* Laurenti. Terminal phalanges simple. Nostrils directed laterally. Toes partly webbed. Range and fossil record as for family. About 250 species, of which about 70 are New World forms.

KEY TO SPECIES OF *BUFO*

1. Parotoid glands long as head, broad anteriorly and tapering posteriorly **2**
   Parotoid glands not so long as head, not broad anteriorly and tapering posteriorly **4**
2. Body length of sexually mature adults under 60 mm **3**
   Body length of sexually mature adults over 60 mm; extreme southern Texas and vicinity of Miami, Florida ***B. marinus***
3. Dorsal pattern of black dots or lines on light background ***B. debilis***
   Dorsal pattern of black reticulum on light background ***B. retiformis***
4. Body length of sexually mature adults under 35 mm; a conspicuous vertebral line ***B. quercicus***
   Body length of sexually mature adults over 35 mm **5**
5. Interparotoid distance about width of 1 parotoid gland ***B. canorus***
   Interparotoid distance more than width of 1 parotoid gland **6**
6. Parotoid glands approximately circular or triangular **7**
   Parotoid glands not approximately circular or triangular **8**
7. Parietal cranial crests conspicuously present (Fig. 3-37) ***B. valliceps***
   Parietal cranial crests absent or faint ***B. punctatus***
8. Parietal cranial crest a conspicuous knob (Fig. 3-38) ***B. terrestris***
   Parietal cranial crest not a conspicuous knob **9**
9. Supraorbital cranial crests united into horny boss at level of anterior margin of eyes (Fig. 3-39) ***B. cognatus***
   Supraorbital cranial crests not united into horny boss at level of anterior margin of eyes **10**
10. Thigh skin smooth with 1–3 large glands on dorsal surface; western Arizona and adjacent California ***B. alvarius***
    Thigh skin tuberculate or warty, without 1–3 large glands on dorsal surface **11**
11. Area between supraorbital cranial crests filled in to form a boss (Fig. 3-40) ***B. hemiophrys***
    Area between supraorbital cranial crests not filled in to form a boss **12**
12. Cranial crests present **13**
    Cranial crests absent or but poorly developed **15**
13. Parotoid glands usually reniform (Fig. 3-41); pigment of throat of male largely confined to transverse muscles of posterior part of throat ***B. americanus***
    Parotoid glands usually oval; skin of throat of male pigmented **14**
14. Usually 1 or 2 warts to a dorsal spot; adult females to 120 mm body length; Great Plains and westward ***B. woodhousei***

Fig. 3-38. **Dorsal view of head of *Bufo terrestris* showing knobbed parietal cranial crests.**

Fig. 3-39. **Dorsal view of head of *Bufo cognatus* showing horny boss at junction of supraorbital cranial crests.**

Fig. 3-37. **Dorsal view of head of *Bufo valliceps* showing prominent parietal cranial crests.**

Usually more than 2 warts to a dorsal spot; adult females to 85 mm body length; east of Great Plains   **B. fowleri**

15. Vertebral line present   **B. boreas**
    Vertebral line absent   **16**
16. Inner (large) metatarsal tubercle elongate and sharp-edged   **B. speciosus**
    Inner metatarsal tubercle rounded, not sharp-edged   **B. microscaphus**

Fig. 3-40. Dorsal view of head of *Bufo hemiophrys* showing boss between supraorbital cranial crests.

Fig. 3-41. Dorsal view of head of *Bufo americanus* showing reniform parotoid glands.

Bufo woodhousei Girard. Woodhouse's toad. Body length to 120 mm. Parotoid glands elongate-oval, strongly divergent posteriorly in larger individuals. A prominent vertebral line. In some individuals, especially southwestern ones, the area between the cranial crests may be filled in. Venter usually immaculate or with single pectoral spot, but in southwestern individuals may be stippled. Warts of dorsal surface of hind legs not particularly large and spiny except in individuals from the southwestern part of the range. Call a nasal drone. Hybridizes with *B. fowleri, B. valliceps, B. americanus,* and *B. microscaphus.* Great Plains region generally north to southern Montana and Washington; eastern Oregon, most of Utah, eastern half of Arizona; also in southeastern California and

southern Nevada; into northern Mexico; to 8500 ft in some localities. Pliocene of Arizona and Pleistocene of Kansas and Texas. 3 described subspecies, 1 fossil.

Bufo fowleri Hinckley. Fowler's toad. Body length to 85 mm. Parotoid glands oval. Warts of dorsal surface of hind legs small, not spiny, except in populations of eastern Texas and lower Mississippi River valley (*B. fowleri velatus*). Dorsal color variable, usually olive, sometimes reddish; below usually immaculate or with single pectoral spot, but frequently with pectoral stippling in *B. fowleri velatus.* Call a nasal drone. Hybridizes with *B. terrestris, B. americanus, B. valliceps,* and *B. woodhousei.* Massachusetts to southern Pennsylvania, southern Michigan, southeastern Iowa, eastern Oklahoma and Texas; generally absent from coastal strip, Mississippi to North Carolina; into Canada. Pleistocene of Arkansas, Florida, Pennsylvania, and Virginia. 2 described subspecies.

Bufo terrestris (Bonnaterre). Southern toad. Body length to 107 mm. Parietal cranial crests form prominent knobs. Parotoid glands oval. Dorsal color variable, frequently reddish; below unmarked or with pectoral spots; throat discolored in male. Call a trill. Hybridizes with *B. fowleri.* In Coastal Plain, eastern North Carolina to eastern Louisiana; peninsular Florida. Pleistocene of Florida.

Fig. 3-42. *Bufo americanus.* Comer's Rock, Grayson County, Va., Sept. 20, 1952. (Coll. R. Conant. Photo by Isabelle Hunt Conant.)

Bufo americanus Holbrook.   American toad (Fig. 3-42). Body length to 110 mm. Parotoid glands irregularly reniform. Large spiny warts on dorsal surfaces of hind legs; warts on dorsum usually 1 or 2 to a spot. Dorsal color variable, olive to reddish; venter immaculate or with variable number of spots, sometimes heavily reticulated. Call a trill. Hybridizes with *B. fowleri* and *B. woodhousei.* Eastern United States, Minnesota to northeastern Texas to northwestern South Carolina; into Canada. 4 described subspecies, 3 in the United States; an apparently disjunct population on the southeastern Texas coast, which has been described as *B. houstonensis,* is here considered a subspecies. Pleistocene of Pennsylvania and Virginia.

Bufo hemiophrys Cope.   Dakota toad. Body length to 75 mm. Usually a horny boss on head between eyes; if not present, then supraorbital cranial crests parallel. A pale vertebral line; venter more or less spotted; throat discolored in male. Call a trill. Northwestern Minnesota, northern South Dakota, into Montana and Wyoming; adjacent Canada.

Bufo microscaphus Cope.   Southwestern toad. Body length to 80 mm. Cranial crests poorly developed or absent. Parotoid glands oval. Vertebral stripe absent or faint; dorsal color variable, reddish, pale with dark spots, or otherwise; below unspotted; throat may be slightly discolored posteriorly in males. Call a trill. Hybridizes with *B. woodhousei.* Southwestern Utah, southeastern Nevada, and central Arizona; Catron and Grant Counties, New Mexico; coastal California from San Luis Obispo County southward; into Mexico. 2 described subspecies, ranges disjunct.

Bufo boreas Baird and Girard.   Western toad. Body length to 125 mm. Cranial crests absent. Parotoid glands round, oval, or elongate. A conspicuous light vertebral line; below unspotted to heavily spotted or reticulated; throat not discolored in males. Hind leg with glandular swellings on dorsal surface. A very weak trill in males of at least the southern California subspecies. Northwestern states south into Baja California, Nevada, Utah, and Colorado; into Canada. 4 described subspecies. Pleistocene of California.

Bufo canorus Camp.   Yosemite toad. Body length to 75 mm. Interorbital space narrow, with cranial crests absent. Parotoid glands large, rounded, frequently poorly defined and merging with adjacent warts. Strong sexual dichromism, with males lacking the conspicuous black dorsal spots or reticulations of females; throat not discolored in males. Call a trill. Enlarged tibial warts sometimes present. Central Sierra Nevada of California, 6500 to over 10,000 ft.

Bufo alvarius Girard.   Colorado River toad. Body length to 175 mm. Skin smooth, above green to brown to gray; no vertebral line; below immaculate or throat and pectoral area mottled; throat not pigmented in males. Glandular swellings on dorsal surface of thigh and shank; 1 or more white warts at angle of mouth. Cranial crests curved around eyes. Call of males very weak. Southeastern California and southwestern and south central Arizona; into Mexico.

Bufo speciosus Girard.   Texas toad. Body length to 90 mm. 2 dark sharp-edged metatarsal tubercles. Cranial crests absent or poorly developed. Dorsum greenish gray with no vertebral line; unspotted below; in males the discolored area of throat is confined to posterior part. Call a rapid trill, short and continuously repeated. Western Oklahoma to southeastern New Mexico; western two-thirds of Texas.

Bufo cognatus Say.   Great Plains toad. Body length to 100 mm. Supraorbital cranial crests united anteriorly in a horny boss. Parotoid glands oval, small. Dorsum usually with long, green, diagonal spots or stripes; venter immaculate; throat of male discolored posteriorly. Call a harsh, rapid trill, up to 50 seconds in length. Western Texas to Utah, southeastern California, and Nevada; central Oklahoma to western Minnesota; central Montana to eastern Colorado; into Mexico. Pliocene of Kansas and Pleistocene of Kansas and Oklahoma.

Bufo debilis Girard.   Green toad. Body length to 52 mm. Dorsum greenish; throat discolored in males. Cranial crests absent or faint. Parotoid glands large, dorsolateral in position. Head and body depressed. Call a rapid trill (buzz). Southwestern Kansas and southeastern Colorado to central and southern Texas and southeastern Arizona; into Mexico. 3 described subspecies.

Bufo retiformis Sanders and Smith.   Sonora toad. Body length to 57 mm. Dorsal pattern of black reticulum on greenish (in life) background; throat discolored in males. Cranial crests poorly developed or absent. Parotoid glands large. Head and body depressed. Call a rapid trill. Pima County, Arizona; into Mexico.

Bufo punctatus **Baird and Girard.** Red-spotted toad. Body length to 75 mm. Parotoid glands small, usually circular. Cranial crests absent or poorly defined. Head depressed, with wide interorbital space. Dorsum gray or reddish, lacking a vertebral line; below immaculate or with a few small spots; male with discolored throat. Call a trill. South-eastern California to southern Nevada and south-western Kansas; south central Oklahoma to Edwards Plateau of Texas; into Mexico. Pleistocene of Nevada.

Bufo quercicus **Holbrook.** Oak toad. Body length to 35 mm. A prominent vertebral line from tip of snout to anus; posterior part of throat discolored in males. Cranial crests weak. Parotoid glands oval, sometimes rounded. Call a peep. In Coastal Plain, North Carolina to eastern Louisiana; peninsular Florida. Pleistocene of Florida.

Bufo valliceps **Wiegmann.** Gulf Coast toad. Body length to 125 mm. Cranial crests very strongly developed. Parotoid glands small, subtriangular. Usually a light vertebral line; males with discolored throat. Call a trill. Hybridizes with *B. fowleri* and *B. woodhousei*. Southern Arkansas, Louisiana east along coast to Bay St. Louis, Mississippi; east, central, and south Texas, west to Pecos River; into Mexico. Pleistocene of Yucatán.

Bufo marinus **(Linnaeus).** Giant toad. Body length to 175 mm or more. Parotoid glands subtriangular, pitted, long as head. Dorsal color variable. Call a low, slow trill. Range extends south from extreme southern Texas; introduced and thriving in Miami area of Florida. Pleistocene of Yucatán.

Family Leptodactylidae. Vomerine and maxillary teeth present or absent; pectoral girdle arciferous; vertebrae procoelous; sacral diapophyses cylindrical or nearly so; terminal phalanges simple or T-shaped. Tips of digits expanded or simple. Many species with terrestrial development of eggs. South and Central America (25 genera), Africa (1 genus), Australia and New Guinea region (17 genera). 1 genus known from the early Eocene of India, 3 from the lower Oligocene of Argentina, and 1 each from the South American Miocene and Pleistocene; Miocene of Florida; Pleistocene of Texas.

KEY TO GENERA OF LEPTODACTYLIDAE

1. Tips of digits not expanded   *Leptodactylus*
   Tips of digits expanded   **2**
2. Vomerine teeth present (Fig. 3-43)
       Eleutherodactylus
   Vomerine teeth absent   *Syrrhophus*

Fig. 3-43. Roof of mouth of *Eleutherodactylus* showing vomerine teeth.

Genus *Leptodactylus* **Fitzinger.** Maxillary teeth present; vomerine teeth present, in elongate transverse series; mesosternum ossified. Tips of digits without distinct discs. About 50 species, Argentina to southern Texas.

Leptodactylus labialis **(Cope).** Mexican white-lipped frog. Body length to 50 mm. Abdominal disc present. A light labial line; a dark stripe from snout through eye and tympanum; below unspotted. Eggs laid in froth on land near water, with tadpole developing in water. Lower Rio Grande Valley of Texas, into Mexico and Central America.

Genus *Eleutherodactylus* **Dumeril and Bibron.** Maxillary teeth present; vomerine teeth usually present, not in strongly transverse series; mesosternum cartilaginous; terminal phalanges T-shaped. Tips of digits with distinct discs. About 200 species, central South America to southern United States. Pleistocene of Texas.

KEY TO SPECIES OF *ELEUTHERODACTYLUS*

1. A large ventral abdominal disc (Fig. 3-44); Texas and westward   *E. augusti*
   No abdominal disc; Florida   *E. ricordi*

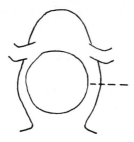

Fig. 3-44. Venter of *Eleutherodactylus augusti* showing abdominal disc.

Fig. 3-45. *Eleutherodactylus augusti.*
Eighteen miles west of Medina, Bandera
County, Tex., Apr. 22, 1954. (Coll.
W. Frank Blair, David Pettus, and J. S. Mecham.
Photo by Isabelle Hunt Conant.)

Eleutherodactylus augusti (Dugès). Barking
frog (Fig. 3-45). Body length to 94 mm. A large
prominent ventral disc. Head and body flat and
broad. Subarticular tubercles very conspicuous;
tips of digits expanded. Eggs laid on land. Edwards
Plateau in central Texas to southeastern Arizona;
into Mexico. 4 described subspecies, 2 of these in
the United States. Pleistocene of Texas.

Eleutherodactylus ricordi **Dumeril and Bibron.**
Greenhouse frog. Body length to 30 mm. Tips of
digits expanded. Dorsal pattern of stripes or mottling. Eggs laid on land. In isolated colonies in
peninsular Florida; introduced. 4 described subspecies, 1 in the United States.

Genus *Syrrhophus* Cope. Vomerine teeth absent; mesosternum cartilaginous; terminal phalanges T-shaped. About 22 species, Peru to Texas.
Pleistocene of Texas.

KEY TO SPECIES OF *SYRRHOPHUS*

1. Dorsal markings (poorly defined spots) few; extreme
southern tip of Texas *S. campi*
Dorsal markings (irregular lines and blotches) more
numerous; central to western Texas *S. marnocki*

Syrrhophus marnocki **Cope.** Cliff frog. Body
length to 35 mm. Tips of digits expanded. Tympanum distinct. Above greenish olive with darker
speckling or reticulation; throat gray; venter immaculate. Eggs laid on land. Edwards Plateau to
trans-Pecos Texas. Pleistocene of Texas.

Syrrhophus campi **Stejneger.** Rio Grande frog.
Body length to 27 mm. Tips of digits expanded.
Tympanum distinct. A dark stripe from snout
through eye and tympanum; above brown to olive
gray with dark markings; below unspotted. Eggs
laid on land. Lower Rio Grande Valley of Texas;
into Mexico.

Family Hylidae. Treefrogs. Pectoral girdle arciferous; vertebrae procoelous; urostyle attached
to 2 condyles; teeth in upper jaw; intercalary cartilages between ultimate and penultimate phalanges. Japan, northern Asia, Europe, North Africa,
and North America; best developed in South America, but well developed in Australia-New Guinea
region. Oligocene and Miocene of Europe and
Miocene and Pleistocene of the United States. 33
genera, 32 in the New World.

KEY TO GENERA OF HYLIDAE

1. Body length of adults not over 18 mm; southeastern
United States *Limnaoedus*
Body length of adults over 18 mm 2
2. Digital discs greatly reduced, but little wider than
digits 3
Digital discs not reduced, distinctly wider than digits
4

3. Webs on feet reduced   *Pseudacris*
   Webs on feet not reduced, extending nearly to tips of toes   *Acris*
4. Cranial skin coossified with skull; secondary bony growth forming a low ridge along edge of upper jaw   *Pternohyla*
   Cranial skin not coossified with skull, or if so then secondary bony growth not forming a low ridge along edge of upper jaw   *Hyla*

Genus *Hyla* Laurenti. Toe discs at least $\frac{1}{2}$ diameter of tympanum. Terminal phalanges claw-shaped; sacral diapophyses dilated; sphenethmoid not extending forward between vomers; palatines not well separated from vomers. Pupil horizontal. Tympanum present. Range as for family. Miocene of Florida and Miocene-Pliocene contact of Nebraska; Pleistocene of California; doubtfully from European Miocene. About 350 species, of which about 180 are New World.

KEY TO SPECIES OF *HYLA*

1. Back usually with 2 or 3 longitudinal stripes; Pacific states, Nevada, Idaho, and western Montana   *H. regilla*
   Back without 2 or 3 longitudinal stripes   2
2. Tympanum approximately as large as eye; extreme southern Texas   *H. baudini*
   Tympanum distinctly smaller than eye   3
3. In California, chiefly west of the Mojave Desert   *H. californiae*
   Not as above   4
4. A black-bordered light spot below posterior end of eye   5
   No black-bordered light spot below posterior end of eye   7
5. Rear of thigh with green in life   *H. avivoca*
   Rear of thigh with yellow or orange in life   6
6. Call a slow trill, about 25 pulses per second   *H. versicolor*
   Call a fast trill, about 50 pulses per second   *H. chrysoscelis*
7. With X-shaped mark on back   *H. crucifer*
   No X-shaped mark on back   8
8. No dark dorsal markings   9
   With dark dorsal markings   11
9. Finger discs approximately as large as tympanum; extreme southern Florida and Keys   *H. septentrionalis*
   Finger discs distinctly smaller than tympanum   10
10. A thin white line extending backward from top of tympanum   *H. andersoni*
    No lateral line, or a broad lateral line extending backward below tympanum   *H. cinerea*
11. Back with many small dark spots   12
    Back without many small dark spots   13
12. Tympanum small, hardly more than $\frac{1}{2}$ diameter of eye; southwestern United States   *H. arenicolor*
    Tympanum large, about $\frac{3}{4}$ diameter of eye; southeastern United States   *H. gratiosa*
13. A pair of longitudinal dark bars posteriorly on dorsum; Arizona and New Mexico   *H. wrightorum*
    No pair of longitudinal dark bars posteriorly on dorsum; southeastern United States   14
14. Rear of femur not spotted; usually a white line on upper jaw and along side of body   *H. squirella*
    Rear of femur spotted; no white line on upper jaw and along side of body   *H. femoralis*

Hyla versicolor Le Conte. Gray treefrog (Fig. 3-46). Body length to 60 mm. Dorsal color variable, but a large irregular dark area usually present; a light spot below eye; groin, axilla, and concealed surfaces of hind legs orange or yellow. Arboreal. Eastern half of the United States west to the Great Plains; Edwards Plateau of Texas; absent from most of peninsular Florida; into Canada. 2 described subspecies. Hybridizes with *H. femoralis* and *H. avivoca*. Pleistocene of Texas.

Fig. 3-46. Hyla versicolor. Taunton Lakes, Burlington County, N.J., July 4, 1950. (Coll. R. Conant. Photo by Isabelle Hunt Conant.)

Hyla chrysoscelis **Cope.** Southern gray tree-frog. The range of this sibling species of *H. versicolor* is poorly known but it may include most of the eastern half of the United States.

Hyla avivoca **Viosca.** Bird-voiced treefrog. Body length to 50 mm. Concealed color of groin and legs yellowish green to greenish white; a light spot below eye. Arboreal. South Carolina to south-eastern Mississippi and southeastern Oklahoma, up Mississippi River valley to southern Illinois. 2 described subspecies. Hybridizes with *H. versicolor*.

Hyla femoralis **Latreille.** Pine woods treefrog. Body length to 40 mm. Back with irregular blotches; dorsal color variable, commonly reddish brown; hind surface of thigh with yellow markings. Arboreal. Eastern Louisiana to Maryland in Coastal Plain; not in southern Florida. Hybridizes with *H. versicolor*. Pleistocene of Florida.

Hyla arenicolor **Cope.** Canyon treefrog. Body length to 54 mm. No well-defined eye stripe; dorsal ground color gray to brown to olive, with many small dark spots; below white or cream with orange or yellow in groin and on concealed surfaces of limbs. Terrestrial, usually near water. Southern Utah and Colorado, trans-Pecos Texas, New Mexico, and eastern two-thirds of Arizona; into Mexico.

Hyla californiae **Gorman.** California canyon treefrog. Body length to 44 mm. Dorsum finely warty; tympanum obscure, less than one-half diameter of eye. No dark stripe from jaw through eye as in *H. regilla;* irregular dark spots and blotches on dorsum; a subocular white spot may be present. Along canyon streams chiefly west of the Mojave-Sonora Desert from vicinity of Santa Margarita, California, south into Baja California; several desert populations in the Joshua Tree National Monument area. Reported to hybridize with *H. regilla* in central Los Angeles County, California.

Hyla baudini (**Dumeril and Bibron**). Mexican treefrog. Body length to 90 mm. A dark patch over arm insertion; a light greenish spot under eye; back variable in color, frequently with irregular dark markings. Vocal sacs of males lateral. Terrestrial and arboreal. Extreme southern Texas and southward.

Hyla cinerea (**Schneider**). Green treefrog. Body length to 63 mm. Body slender. Dorsum green, with or without a white or yellowish lateral stripe; below white or yellowish. In low vegetation of Coastal Plain, Maryland to southern Texas; peninsular Florida; up Mississippi River valley to southern Illinois. 2 described subspecies. Hybridizes with *H. gratiosa*. Pleistocene of Florida.

Hyla andersoni **Baird.** Pine barrens treefrog. Body length to 47 mm. Dorsum green in life with a light-bordered plum-colored band along the side of the body; orange in axilla and groin and on rear of thigh. Terrestrial and arboreal. Central New Jersey to South Carolina.

Hyla gratiosa **Le Conte.** Barking treefrog. Body length to 70 mm. Body stout. Dorsum variable in color, granular, usually with many dark spots; a light subtympanic stripe; groin and axilla yellow; below unspotted. Arboreal. In Coastal Plain, North Carolina to southeastern Louisiana; peninsular Florida; northern Georgia in piedmont and Cumberland Plateau; White County, Tennessee; Cape May County, New Jersey, probably introduced. Hybridizes with *H. cinerea*. Pleistocene of Florida.

Hyla squirella **Latreille.** Squirrel treefrog. Body length to 37 mm. Usually a light line along upper jaw; a transverse bar between eyes frequently present; an indistinct line along back; back with or without spots; metachrosis pronounced. Terrestrial and arboreal. Virginia to Texas in Coastal Plain; peninsular Florida. Pleistocene of Florida.

Hyla regilla **Baird and Girard.** Pacific treefrog. Body length to 47 mm. A black eye stripe; dorsal ground color very variable, capable of marked metachrosis; dorsal markings variable but often of lines or spots in longitudinal arrangement; venter unspotted, white, yellowish, or dusky; groin and concealed surfaces of limbs yellow to orange. Terrestrial or in low vegetation. Washington, Oregon, California except for southeastern part, Nevada, western Idaho, and Montana; sea level to 11,000 ft. 10 described subspecies.

Hyla wrightorum **Taylor.** Arizona treefrog. Body length to 48 mm. Dorsal markings usually a pair of stripes posteriorly and sometimes a pair of spots more anteriorly; venter immaculate; groin and rear of thigh greenish orange or gold. Arboreal and terrestrial. North central New Mexico to north central Arizona at 5000–7000 ft; Huachuca Mountains of southeastern Arizona; Yuma County, Arizona, at less than 250 ft. Considered a subspecies of *H. regilla* by Jameson, Mackey, and Richmond (1966).

Hyla crucifer **Wied.** Spring peeper. Body length to 33 mm. Back usually brown with an oblique dark cross; a dark stripe from snout through eye and tympanum; venter frequently spotted anteriorly. Terrestrial and arboreal. Eastern half of the United States as far west as plains, but not in lower half of peninsula of Florida; into Canada. 2 described subspecies.

Hyla septentrionalis **Boulenger.** Cuban treefrog. Body length to 100 mm. Skin on top of head united with skull. Above greenish to yellowish with indistinct spots; below yellowish, with yellow in groin and on thigh. Terrestrial and arboreal. Florida Keys and southern tip of peninsula, possibly introduced.

Genus *Pseudacris* Fitzinger. Sphenethmoid extending forward between vomers; palatines well separated from vomers; vomers with teeth; nasals relatively larger, more closely approximated, than in *Hyla*. Small terrestrial and fossorial hylids with toe webbing and discs reduced, probably a polyphyletic assemblage. North America. Miocene-Pliocene contact of Nebraska, Pleistocene of Kansas.

KEY TO SPECIES OF *PSEUDACRIS*

1. Dorsal pattern of 5 longitudinal stripes (sometimes broken into spots or even absent)   **2**
   Dorsal pattern not of 5 longitudinal stripes (*P. clarki* occasionally is striped)   **4**
2. Hind legs with longitudinal dorsal markings   ***P. brimleyi***
   Hind legs with transverse dorsal markings   **3**
3. Dark tibial bands broad with narrow light interspaces   ***P. nigrita***
   Dark tibial bands (or markings) not occupying more than ½ of dorsal surface of tibia   ***P. triseriata***
4. A distinct, uninterrupted dark line from eye to halfway between front- and hind-leg insertions; dorsal spots green in life   ***P. clarki***
   No distinct, uninterrupted dark line from eye to halfway between front- and hind-leg insertions; dorsal spots not green in life   **5**
5. With 2 broad dorsal stripes which approach or meet one another at middle of back   ***P. brachyphona***
   Not with 2 broad dorsal stripes which approach or meet one another   **6**
6. With 2 longitudinal dorsal stripes; southeastern United States   ***P. ornata***
   Back with irregular spots or vermiculations, or immaculate; central United States   ***P. streckeri***

Pseudacris nigrita (**Le Conte**). Southern chorus frog. Body length to 32 mm. Snout more pointed than in *P. triseriata*. Call more slowly pulsed than in *P. triseriata*. Coastal Plain, North Carolina to southern Mississippi; peninsular Florida. 2 described subspecies.

Pseudacris triseriata (**Wied**). Northern chorus frog. Body length to 37 mm. Snout less pointed than in *P. nigrita*. A dark triangular mark may be present between eyes. Call more rapidly pulsed than in *P. nigrita*. Eastern half of the United States (excluding New England, most of New York, much of Pennsylvania and West Virginia, and most of the range of *P. nigrita*) into eastern part of Great Plains; across northern Great Plains into Montana and southern Idaho, thence south to New Mexico and Arizona at higher elevations (to 11,000 ft in Uintah Mountains of Utah); into Canada. 4 described subspecies. Hybridizes with *P. brachyphona* and *P. clarki*.

Pseudacris clarki (**Baird**). Spotted chorus frog. Body length to 31 mm. Dorsal pattern of irregular greenish (in life) spots, occasionally in rows. In the plains, central Kansas to extreme southern Texas. Pleistocene of Texas. Hybridizes with *P. triseriata*.

Pseudacris brimleyi **Brandt and Walker.** Brimley's chorus frog. Body length to 33 mm. Sharply defined black lateral stripes and weak triseriate dorsal pattern; leg markings more or less longitudinal; venter yellowish with dark spots. Coastal Plain, southern Virginia to southeastern Georgia.

Pseudacris brachyphona (**Cope**). Mountain chorus frog. Body length to 35 mm. Dorsal markings frequently of 2 broad, irregular arcs approaching one another or fusing at center of back. Toe discs small but conspicuous. Southwestern Pennsylvania to northern Alabama and Georgia and northeastern Mississippi. Hybridizes with *P. triseriata*.

Pseudacris ornata (**Holbrook**). Ornate chorus frog. Body length to 36 mm. Pronounced metachrosis, with commonest dorsal color reddish brown; a broad dark stripe from tip of snout through eye, interrupted at shoulder. In Coastal Plain, North Carolina to southeastern Louisiana. Pleistocene of Florida.

Pseudacris streckeri **Wright and Wright.** Strecker's chorus frog. Body length to 46 mm. Body stout with broad head. Dorsal color variable; dorsal markings varying from heavy spots or irregular stripes to none. Oklahoma to central south Texas; east in Arkansas River valley into Arkansas; in Coastal Plain at edge of interior highland, southwestern Arkansas to extreme southern Illinois; an apparently isolated population in sand areas of west-central Illinois. 2 described subspecies.

Genus *Acris* Dumeril and Bibron. Toe discs reduced. Toes well webbed. Small diurnal and nocturnal hylids, more aquatic than any other hylids of the United States, found along the shores of lakes, ponds, and streams. Eastern North America. Miocene-Pliocene contact of Nebraska; Pliocene of Kansas; Pleistocene of Kansas, Texas, and Florida.

KEY TO SPECIES OF *ACRIS*

1. 1½ or fewer terminal phalanges of 4th toe of hind foot free of web *A. crepitans*
   2 or more terminal phalanges of 4th toe of hind foot free of web *A. gryllus*

Acris crepitans **Baird.** Northern cricket frog. Body length to 35 mm. 1½ or fewer terminal phalanges of 4th toe of hind foot usually free of web. Uppermost stripe on rear of femur irregular. Body stout with rounded snout; toes less slender than in *A. gryllus*. Dorsum warty, uniformly gray or with green, yellow, or reddish markings. Eastern two-thirds of United States west to Rocky Mountains but not in northeastern or northwestern part of this area; range more or less complementary to that of *A. gryllus*, with which it presumedly hybridizes in the overlap zone. 3 described subspecies. Pleistocene of Texas.

Acris gryllus **(Le Conte).** Southern cricket frog. Body length to 31 mm. 2 or more terminal phalanges of 4th toe of hind foot usually free of web. Uppermost stripe on rear of femur narrow and sharply defined. Body slender with pointed snout; toes slender. Dorsum less warty than in *A. crepitans*, dark brown or black with color markings as in *A. crepitans*. Coastal Plain, southeastern Virginia to southeastern Louisiana; peninsular Florida. 2 described subspecies.

Genus *Limnaoedus* Mittleman and List. Vomers lacking teeth; parasphenoid fitting in a posterior notch of the sphenethmoid; sphenethmoid incompletely fused dorsally; omosternum and xiphisternum not calcified; sacral diapophyses slender. Maximum body length about 18 mm. A single species in southeastern United States.

Limnaoedus ocularis **(Holbrook).** Little grass frog. Body length to 18 mm. Dorsal color variable; a dark stripe of variable length running from eye backward. In low vegetation of Coastal Plain, North Carolina to and including Florida except extreme western tip of panhandle.

Genus *Pternohyla* Boulenger. Skin of top of head fused to skull; a secondary bony growth forming a low ridge along edge of upper jaw; no teeth on mandibles, palatines, or parasphenoid. 2 Mexican species, 1 reaching southern Arizona.

Pternohyla fodiens **Boulenger.** Mexican burrowing frog. Body length to 60 mm. A plump, fossorial hylid with prominent elongate, dorsal spots. Pima County, Arizona, south into Mexico.

Family Ranidae. Frogs. Diapophyses of sacral vertebrae cylindrical to slightly dilated; maxillary teeth present, vomerine teeth present or absent. Tongue bicornuate posteriorly. Tips of digits simple or expanded. Toes webbed, fingers free. A large family found in every continent, with Africa presumedly the center of differentiation. Miocene of Europe (3 genera), Pliocene of the United States (2 genera), and Pleistocene of the United States.

Genus *Rana* Linnaeus. Skin relatively smooth. Pupil of eye horizontal. Vomerine teeth present. Tympanum present, enlarged in males of some species. Base of first finger seasonally swollen in males. Eggs in large masses in water. About 400 species, of these about 27 in New World; the only genus of the family in the New World. Miocene of Europe; Miocene-Pliocene contact of Nebraska; Pliocene of Kansas, Nevada, and California; Pleistocene of California, Kansas, Texas, and Florida.

KEY TO SPECIES OF *RANA*

1. Dark stripe through eye and tympanum (sometimes fragmentary or obscure); no large, regular, dark dorsal spots 2
   No dark stripe through eye and tympanum 6
2. Underparts red or yellow in life; range in the United States mostly west of central Montana 3

Underparts not red or yellow in life; range in the United States mostly east of eastern Idaho
R. sylvatica

3. Groin yellowish or pea green in life; back with intense black spots **R. cascadae**
Groin with red in life **4**

4. In California **R. aurora** (part)
In states north and northeast of California **5**

5. Heel of hind leg when extended forward falls short of external nares **R. pretiosa**
Heel of hind leg when extended forward reaches to or beyond external nares **R. aurora** (part)

6. With distinct dorsolateral folds (Fig. 3-47) **7**
Dorsolateral folds indistinct or absent **12**

Fig. 3-47. Dorsal view of ranid showing dorsolateral folds.

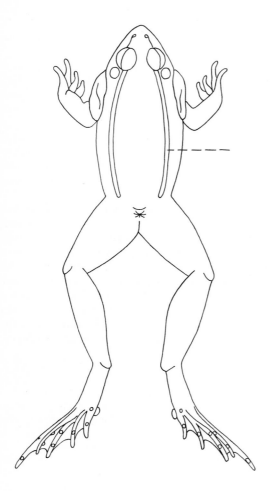

7. Dorsolateral folds extending not more than ⅔ length of body **R. clamitans**
Dorsolateral folds extending full length of body **8**

8. With rectangular spots in 2 rows between dorsolateral folds; bright orange in groin and underneath legs in life **R. palustris**
Spots, if present, not rectangular; no bright orange in groin and underneath legs **9**

9. A light-colored line on upper jaw (obscure in some far western populations) **R. pipiens**
No light-colored line on upper jaw **10**

10. Dorsal spots with distinct light borders **R. areolata**
Dorsal spots without distinct light borders **11**

11. Dorsal spots distinct against pale background
R. capito
Dorsal spots poorly differentiated from background
R. sevosa

12. Tympanum smaller than eye; western United States **13**
Tympanum as large as, or larger than, eye **15**

13. No outer metatarsal tubercle; southeastern Arizona **R. tarahumarae**
Outer metatarsal tubercle present; California, western Oregon, and extreme western Nevada **14**

14. A light stripe across snout; toe tips brown
R. boylei
No light stripe across snout; toe tips black
R. muscosa

15. Small frogs, to 75 mm body length **16**
Large frogs, seldom breeding below 80 mm body length **17**

16. Venter spotted or reticulated; Coastal Plain, southern New Jersey to Georgia **R. virgatipes**
Venter immaculate; northern New England and northern New York to Minnesota
R. septentrionalis

Fig. 3-48. Hind foot of Rana grylio.

17. Web of hind foot very full (Fig. 3-48)   *R. grylio*
    Web less full (Fig. 3-49)   **18**
18. Throat light, with irregular dark blotches
                            R. catesbeiana
    Throat slate-colored or smoky, with irregular white
      spots or blotches   *R. heckscheri*

Fig. 3-49. Hind foot of *Rana catesbeiana* or *R. heckscheri.*

Rana pipiens **Schreber.** Leopard frog (Fig. 3-50). Body length to 105 mm. Prominent pale dorsolateral folds. Color above variable, frequently brown or green, usually spotted, but may have elongate markings or none at all; below immaculate. Males with lateral vocal pouches. Eastern edge of Washington and Oregon, northeastern and southeastern California, and all of the United States to east of this; into Canada and Mexico. 6 described subspecies. Pleistocene of Kansas, Texas, and Florida.

Rana palustris **Le Conte.** Pickerel frog. Body length to 80 mm. Concealed surfaces of flanks and thighs yellow to bright orange in life; dorsal spots more or less rectangular, without light border. Small lateral vocal pouches in males. Eastern United States west to plains, but distribution erratic; not in Florida or southern halves of Alabama, Georgia, or South Carolina; into Canada.

Rana areolata **Baird and Girard.** Crawfish frog. Body length to 115 mm. Many large dorsal spots outlined with light; venter immaculate or with a few spots in the pectoral area or anterior to it. Males with large lateral vocal pouches. In burrows of other animals, especially crayfish, eastern Texas

to Ohio to southern Iowa and eastern Kansas. 2 described subspecies.

Rana sevosa **Goin and Netting.** Dusky gopher frog. Body length to 105 mm. Back with numerous prominent warts. Belly more spotted than in *R. capito*. Males with lateral vocal pouches. In burrows of other animals, especially the gopher tortoise, southeastern Louisiana, southern Mississippi, and southwestern Alabama.

Rana capito **Le Conte.** Florida gopher frog. Body length to 110 mm. Head broad. Dorsolateral folds very wide and low. Brown or black vermiculations on chin and throat and sometimes the anterior half of the venter. Males with lateral vocal pouches. In burrows of other animals, especially the gopher tortoise, in the Coastal Plain, North Carolina through the Florida peninsula. 2 described subspecies.

Rana boylei **Baird.** Foothill yellow-legged frog. Body length to 80 mm. A light stripe across snout. Skin papillate. Foothills of the Sierra Nevada;. Santa Barbara County, California, north to Marion County, Oregon; east in north central California to foothills of Sierra Nevada.

Rana muscosa **Camp.** Mountain yellow-legged frog. Body length to 90 mm. No light stripe across snout. Skin less papillate than in *R. boylei*. South-

Fig. 3-50. Rana pipiens. **Everglades National Park, Fla., Nov. 24, 1951. (Coll. R. and I. H. Conant. Photo by Isabelle Hunt Conant.)**

ern half of the Sierra Nevada of California; southern California south of Ventura County, as far south as San Diego County. A species of higher altitudes, or faster streams, than *R. boylei.* 2 described subspecies.

Rana tarahumarae **Boulenger.** Tarahumara frog. Body length to 115 mm. Olive tan above with more or less distinct black markings; limbs with blotches or crossbands; below light-colored or dusky. Tympanum indistinct. Southeastern Arizona and perhaps southwestern New Mexico; into Mexico.

Rana pretiosa **Baird and Girard.** Spotted frog. Body length to 95 mm. Dorsum with large black indistinctly bordered spots on yellowish, reddish, or brown background; dorsal spots often with light tubercles in center; venter and undersurface of limbs orange-yellow to salmon in life; melanophores scattered over venter; iris bright golden in life. Eye directed upward. Dorsolateral folds present. Tympanum small and inconspicuous. Heel of hind leg when extended forward falls short of external nares. Northwestern states, south into northern California, Nevada, and Utah; east into central Montana and western Wyoming; into Canada. 2 described subspecies.

Rana aurora **Baird and Girard.** Red-legged frog. Body length to 135 mm. Dorsum usually with large irregular spots with many small spots scattered between; venter and undersurfaces of limbs reddish in life, with heavy black mottling in groin area; melanophores scattered over venter; iris bronze-colored in life. Eye not directed upward as in *R. pretiosa.* Dorsolateral folds present. Heel of hind leg when extended forward reaches to or beyond external nares (in northern subspecies, *R. a. aurora,* only). In coastal areas of Washington, Oregon, and California; in Sierra Nevada of California but not in Central Valley; into Canada. Pleistocene of California. 2 described subspecies.

Rana cascadae **Slater.** Cascades frog. Body length to 75 mm. Dorsal spots distinct, inky black, with no smaller spots scattered between them; no melanophores in median abdominal area; concealed parts of limbs and lower abdomen light yellow in life; iris yellow in life. Eye directed as in *R. aurora.* Dorsolateral folds present. Heel of hind leg when extended forward reaches to or beyond external nares. A high-altitude species of the Cascade

Mountains of Washington and Oregon, Olympic Mountains of Washington, and northern California.

Rana sylvatica **Le Conte.** Wood frog. Body length to 80 mm. A dark "mask" from tip of snout through eye to beyond tympanum; back reddish brown to yellowish, sometimes with a light middorsal stripe; belly immaculate or speckled anteriorly. Prominent dorsolateral folds. In moist woodland, northeastern United States south to northwestern South Carolina and southwest to northwestern Arkansas; northwest to northern North Dakota; presumedly isolated populations in Lyon County, Kansas, in north central Colorado (Jackson and Grand Counties), and in Boundary County, Idaho; into Canada. 3 described subspecies.

Rana catesbeiana **Shaw.** Bullfrog. Body length to 185 mm. Back green to dark brown or black; venter immaculate to heavily reticulated. Eastern half of the United States, west across plains along streams; not in southern half of Florida; introduced in many places in western United States; into Mexico and Canada. Pleistocene of Kansas and Florida.

Rana clamitans **Latreille.** Green frog. Body length to 100 mm. Back olive to olive brown, sometimes with indistinct black markings; throat yellow in males in life. Tympanum very large in males. Eastern United States west to plains, but not in southern half of Florida; introduced into Montana, Washington, and possibly other western states; into Canada. 2 described subspecies.

Rana heckscheri **Wright.** River frog. Body length to 150 mm. Dorsum greenish black; venter with extensive gray marking. In Coastal Plain, southern Mississippi to North Carolina; not in lower two-thirds of peninsula of Florida.

Rana grylio **Stejneger.** Pig frog. Body length to 160 mm. Brown to blackish brown above, with some black spotting; below heavily reticulated. Body form slim. Sometimes with a suggestion of dorsolateral folds. Highly aquatic. In the Coastal Plain, southeastern Texas to southern South Carolina; peninsular Florida. Pleistocene of Florida.

Rana septentrionalis **Baird.** Mink frog. Body length to 75 mm. Dorsum brown-olive with dark spots or reticulations; venter immaculate back of pectoral area. Sometimes with indistinct dorsolateral folds. Northern New England and northern New York to Minnesota; into Canada.

Rana virgatipes **Cope.** Carpenter frog. Body length to 65 mm. Brown above with a pale stripe running backward from each eye; venter heavily reticulated. Males with lateral vocal pouches. Southern New Jersey to Georgia in Coastal Plain.

Identification of Larval Anurans

The identification of larval anurans involves considerable difficulty, and it is suggested that the beginning student attempt to rear unknown tadpoles through metamorphosis. This will not only facilitate identification but will acquaint the student with some of the details of life history of the species in question. The author has found shallow white enamel pans approximately $9 \times 12 \times 2$ in. excellent for bufonid and hylid tadpoles. Boiled lettuce and canned spinach are good basic food items for ranids, bufonids, hylids, and pelobatids (the author has found no satisfactory food for microhylid tadpoles); living *Elodea*, *Myriophyllum*, and filamentous algae in the pans will provide additional plant food. The diet of bufonids and pelobatids may be supplemented with uncooked bits of beef, pork, lamb, earthworm, crayfish, or other animal flesh. Overcrowding and the accumulation of uneaten food must be carefully avoided. At metamorphosis opportunity for getting out of the water must be provided. Vestigial-winged and curly-winged *Drosophila* have proved excellent food for newly metamorphosed hylids and bufonids.

Grace Orton has kindly allowed me to reprint her key to the genera of tadpoles of the United States and Canada (*Rhinophrynus*, recently found in extreme southern Texas, is not included; it can be recognized by the lipless mouth provided with long fringelike barbels). In the tooth formulas the numerator denotes the number of rows of teeth on the upper lip and the denominator the number of rows of teeth on the lower lip.

KEY TO GENERA OF TADPOLES

1. Mouth parts simple: no disclike lip differentiated around mouth, no labial teeth; margins of jaws soft, no horny beaks; sides of upper jaw with a pair of wide soft flaps that overhang lower jaw (Fig. 3-64). Spiracle median, opens just in front of median anus (Figs. 3-53, 3-58)   **2**

Mouth parts complex: mouth surrounded by a disc-like lip bearing transverse rows of horny labial teeth and more or less extensively edged with papillae; margins of jaws bear hard black beaks (Figs. 3-65–3-71)   **3**

2. Inner edge of each upper-jaw flap with several dis-

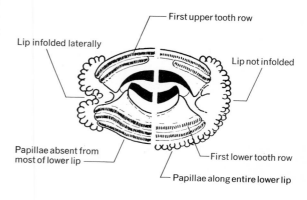

Fig. 3-51. Composite diagram illustrating characters of mouth parts. Left half of drawing based on *Bufo*, right half based on *Acris*.

First upper tooth row

Lip infolded laterally

Lip not infolded

Papillae absent from most of lower lip

First lower tooth row

Papillae along entire lower lip

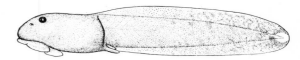

Fig. 3-52. *Ascaphus truei.*

Fig. 3-53. *Gastrophryne carolinensis.*

Fig. 3-54. *Bufo fowleri.*

tinct papillae; spiracular and anal openings more distinctly separated **Hypopachus**
Inner edges of upper-jaw flaps typically smooth, papillae absent or poorly defined (Fig. 3-64); spiracular and anal openings very closely adjacent **Gastrophryne**
3. Spiracle median, opens on chest region just posterior to lower lip; lips very greatly enlarged, suckerlike;

Fig. 3-55. Hyla cinerea.

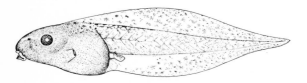

Fig. 3-56. Acris crepitans.

Fig. 3-57. Rana pipiens.

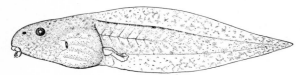

Fig. 3-58. Gastrophryne carolinensis (ventral view).

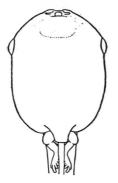

3 tooth rows on upper lip, 10–12 on lower lip; some tooth rows multiple, consisting of 2 or more lines of teeth on same tooth ridge; upper beak flat and platelike with white edge, lower beak reduced to a small vestige (Fig. 3-60) **Ascaphus**
Spiracle on left side of body (Figs. 3-54–3-57, 3-59). Lips small to moderately large; tooth rows various, but no more than 6 rows on lower lip; each row consists of a single line of teeth per tooth ridge; beaks variously modified, but lower beak is of normal size **4**
4. Anus median. Eyes dorsal, tend to be rather together and relatively small. Marginal papillae present or absent along lower lip; no well-defined submarginal row of papillae on lower lip **5**
Anus opens on right side of lower tail fin (Fig. 3-59). Eyes dorsal or lateral, of moderate to large size. Marginal papillae present along lower lip; a ventrolateral submarginal row of papillae present (Fig. 3-70) or absent (Fig. 3-68) on lower lip **7**

Fig. 3-59. Rana pipiens (ventral view).

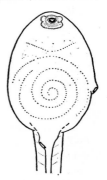

Fig. 3-60. Ascaphus truei (ventral view).

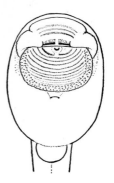

5. Lips infolded laterally; marginal papillae present only around sides of lips; tooth rows $\frac{2}{3}$, 1st upper row not reduced in length (Fig. 3-66). Nostrils often much enlarged .... *Bufo*

Lips not infolded laterally; marginal papillae present along full length of lower lip (Figs. 3-65, 3-67). Nostrils small, inconspicuous 6

6. Papillae border entire lip or absent only from very short median gap on edge of upper lip; 3–6 upper tooth rows, about 4–6 lower rows; first upper (outermost) row very short (Fig. 3-65). Beaks some-

Fig. 3-61. *Gastrophryne carolinensis* (dorsal view).

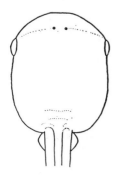

Fig. 3-62. *Hyla cinerea* (dorsal view).

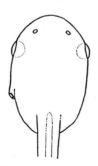

Fig. 3-63. *Bufo fowleri* (dorsal view).

Fig. 3-64. *Gastrophryne carolinensis* (mouth parts).

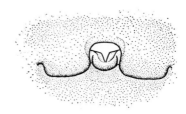

Fig. 3-65. *Scaphiopus holbrooki* (mouth parts).

Fig. 3-66. *Bufo fowleri* (mouth parts).

Fig. 3-67. *Leptodactylus albilabris* (mouth parts).

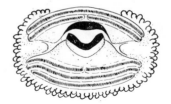

times much enlarged and modified with conspicuous median cusps **Scaphiopus**

Nearly entire length of upper lip free of marginal papillae. Tooth rows $\frac{2}{3}$ 1st upper tooth row not reduced in length (Fig. 3-67) **Leptodactylus**

7. Lips infolded laterally and typically have well-defined oblique row of submarginal papillae between ends of lower tooth rows and papillae bordering lip (Figs. 3-70, 3-71). Moderate to large tadpoles, attaining total length of 45–150 mm, depending on species **Rana**

Lips not infolded laterally and not having well-defined row of submarginal papillae (Figs. 3-68, 3-69). Maximum total length usually less than 50 mm **8**

8. Tooth rows $\frac{2}{3}$ ($\frac{2}{2}$ in some *H. crucifer*), labial teeth numerous and closely crowded. Outline of closed lips subtriangular with apex forward (Fig. 3-69). Eyes lateral, snout short and blunt (Fig. 3-62). Nostrils small and inconspicuous. Maximum total length of most species 25–40 mm; some reach 50–60 mm **Hyla, Pseudacris, and Limnaoedus**

Tooth rows $\frac{2}{2}$, labial teeth relatively fewer and usually rather widely spaced; lip outline not markedly triangular when closed (Fig. 3-68). Head narrower; eyes tend to be slightly dorsal, not always visible from ventral view. Nostrils sometimes much enlarged. Tail tip frequently jet black. Maximum total length about 45 mm, but usually considerably smaller **Acris**

Fig. 3-68. Acris crepitans (mouth parts).

Fig. 3-69. Hyla cinerea (mouth parts). Left side of drawing with lips open, right side with lips relaxed.

Fig. 3-70. Rana catesbeiana (mouth parts).

Fig. 3-71. Rana pipiens (mouth parts). *R. pipiens* sometimes has a third upper row as in *R. catesbeiana.*

REFERENCES

General

Bishop, S. C. 1943. *Handbook of Salamanders,* Comstock Publishing Associates, Inc., Ithaca, N.Y.

Blair, W. F. 1958. Mating Call in the Speciation of Anuran Amphibians, *Am. Naturalist,* **92**:27.

———. 1958 (printed in 1959). Call Structure and Species Groups in U.S. Treefrogs (*Hyla*), *Southwestern Naturalist,* **3**:77.

———. 1965. Amphibian Speciation, p. 543 in *The Quaternary of the United States,* H. E. Wright, Jr., and D. G. Frey (eds.), Princeton University Press, Princeton, N.J.

Bogert, C. M. 1960. The Influence of Sound on the Behavior of Amphibians and Reptiles, Animal Sounds and Communication, *A.I.B.S.,* **7**:137.

Cochran, D. M. 1961. *Living Amphibians of the World,* Hanover House, New York.

Cope, E. D. 1889. The Batrachia of North America, *Bull. U.S. Natl. Museum,* 34.

Dickerson, M. C. 1906. *The Frog Book,* Doubleday & Company, Inc., Garden City, N.Y.

Dunn, E. R. 1926. *The Salamanders of the Family Plethodontidae,* Smith College Anniversary Ser. 7.

Eaton, T. H., Jr. 1959. The Ancestry of Modern Amphibia: A Review of the Evidence, *Univ. Kans. Publ.,* **12:**155.

Estes, R. 1965. Fossil Salamanders and Salamander Origins, *Am. Zoologist,* **5:**319.

Goin, C. J., and O. B. Goin. 1962. *Introduction to Herpetology,* W. H. Freeman and Company, San Francisco.

Hecht, M. K. 1957. A Case of Parallel Evolution in Salamanders, *Proc. Zool. Soc.,* Calcutta, Mookerjee Memorial Volume, p. 283.

————. 1958. A Synopsis of the Mud Puppies of Eastern North America, *Proc. Staten Island Inst. Arts Sci.,* **21:**5.

————. 1962. A Reevaluation of the Early History of the Frogs, Part I, *Systematic Zool.,* **11:**39.

————. 1963. A Reevaluation of the Early History of the Frogs, Part II, *Systematic Zool.,* **12:**20.

Highton, R. 1962. Revision of North American Salamanders of the Genus *Plethodon, Bull. Fla. State Museum, Biol. Sci.,* **6:**235.

Holmes, S. J. 1927. *The Biology of the Frog,* 4th ed., The Macmillan Company, New York.

Livezey, R. L., and A. H. Wright. 1947. A Synoptic Key to the Salientian Eggs of the United States, *Am. Midland Naturalist,* **37:**179.

Mertens, R. 1960. *The World of Amphibians and Reptiles,* translated by H. W. Parker, McGraw-Hill Book Company, New York.

Noble, G. K. 1931. *The Biology of the Amphibia,* McGraw-Hill Book Company, New York.

Orton, G. L. 1957. Larval Evolution and Frog Classification, *Systematic Zool.,* **6:**79.

Regal, P. J. 1966. Feeding Specializations and the Classification. of Terrestrial Salamanders, *Evolution,* **20:**392.

Rugh, R. 1951. *The Frog: Its Reproduction and Development,* McGraw-Hill Book Company, New York.

Schmidt, K. P. 1953. *A Check List of North American Amphibians and Reptiles,* 6th ed., University of Chicago Press, Chicago.

Tihen, J. A. 1965. Evolutionary Trends in Frogs, *Am. Zoologist,* **5:**309.

Wright, A. H. 1929. Synopsis and Description of North American Tadpoles, *Proc. U.S. Natl. Museum,* vol. 74, no. 11.

———— and A. A. Wright. 1949. *Handbook of Frogs and Toads of the United States and Canada,* 3d ed., Comstock Publishing Associates, Inc., Ithaca, N.Y.

Zweifel, R. G. 1956. A Survey of the Frogs of the *Augusti* Group, Genus *Eleutherodactylus, Am. Museum Novitates,* 1813.

Regional

Bishop, S. C. 1941. The Salamanders of New York, *Bull. N.Y. State Museum,* 324.

Bragg, A. N., et al. 1950. *Researches on the Amphibia of Oklahoma,* University of Oklahoma Press, Norman.

Breckinridge, W. J. 1944. *Reptiles and Amphibians of Minnesota,* University of Minnesota Press, Minneapolis.

Brimley, C. S. 1939–1941. The Amphibians and Reptiles of North Carolina, *Carolina Tips,* vol. 2, nos. 1–7; vol. 3, nos. 1–7; vol. 4, no. 1.

Brown, B. C. 1950. *An Annotated Check List of the Reptiles and Amphibians of Texas,* Baylor University Press, Waco, Tex.

Carr, A., and C. J. Goin. 1955. *Guide to the Reptiles, Amphibians, and Fresh-water Fishes of Florida,* University of Florida Press, Gainesville.

Chermock, R. L. 1952. A Key to the Amphibians and Reptiles of Alabama, *Geol. Surv. Ala. Museum Paper* 33.

Conant, R. 1957. *Reptiles and Amphibians of the Northeastern States,* 3d ed., Zoological Society of Philadelphia, Philadelphia, Pa.

————. 1958. *A Field Guide to Reptiles and Amphibians,* Houghton Mifflin Company, Boston.

Crenshaw, J. W., Jr., and W. F. Blair. 1959. Relationships in the *Pseudacris nigrita* Complex in Southwestern Georgia, *Copeia,* **1959**(3):215.

Dowling, H. G. 1957. Amphibians and Reptiles in Arkansas, *Occasional Papers Univ. Ark. Museum,* no. 3.

Gordon, K. 1939. The Amphibia and Reptilia of Oregon, *Ore. State Monographs,* no. 1.

Grobman, A. B. 1944. The Distribution of the Salamanders of the Genus *Plethodon* in Eastern United States and Canada, *Ann. N.Y. Acad. Sci.,* **45:**261.

Gunter, G., and W. E. Brode. 1964. *Necturus* in the State of Mississippi, with Notes on Adjacent Areas, *Herpetologica,* **20:**114.

Hurter, J., Sr. 1911. Herpetology of Missouri, *Trans. Acad. Sci. St. Louis,* **20:**59.

Karlstrom, E. L. 1962. The Toad Genus *Bufo* in the Sierra Nevada of California, *Univ. Calif. Publ. Zool.,* vol. 62.

King, W. 1939. A Survey of the Herpetology of Great Smoky Mountains National Park, *Am. Midland Naturalist,* **21:**531.

Linsdale, J. M. 1940. Amphibians and Reptiles in Nevada, *Proc. Am. Acad. Arts Sci.,* **73:**197.

Martof, B. S. 1956. *Amphibians and Reptiles of Georgia,* University of Georgia Press, Athens.

Pickwell, G. 1947. *Amphibians and Reptiles of the Pacific States,* Stanford University Press, Stanford, Calif.

Ruthven, A. G., C. Thompson, and H. T. Gaige. 1928. The Herpetology of Michigan, *Mich. Handbook Ser. Univ. Mich.,* 3.

Slevin, J. R. 1928. The Amphibians of Western North America, *Occasional Papers Calif. Acad. Sci.,* no. 16.

Smith, H. M. 1956. Handbook of Amphibians and Reptiles of Kansas, 2d ed., *Univ. Kans. Museum Nat. Hist. Misc. Publ.* 9.

Smith, P. W. 1961. The Amphibians and Reptiles of Illinois, *Ill. Nat. Hist. Surv. Bull.* 28.

Stebbins, R. C. 1951. *Amphibians of Western North America,* University of California Press, Berkeley.

————. 1954. *Amphibians and Reptiles of Western North America,* McGraw-Hill Book Company, New York.

Storer, T. I. 1925. A Synopsis of the Amphibia of California, *Univ. Calif. Publ. Zool.,* 27.

Valentine, B. D. 1963. The Salamander Genus *Desmognathus* in Mississippi, *Copeia,* 1963(1):130.

Walker, C. F. 1946. The Amphibians of Ohio: I. The Frogs and Toads, *Ohio State Museum Sci. Bull.,* vol. 1, no. 3, Ohio State Archaeological and Historical Society, Columbus.

Wheeler, G. C. 1947. The Amphibians and Reptiles of North Dakota, *Am. Midland Naturalist,* 38:162.

Wright, A. H. 1932. *Life-histories of the Frogs of the Okefinokee Swamp, Georgia,* The Macmillan Company, New York.

Specific

Adler, K. K., and D. M. Dennis. 1962. *Plethodon longicrus,* a New Salamander (Amphibia: Plethodontidae) from North Carolina, *Ohio Herpetological Soc., Spec. Publ.* 4.

Baker, J. K. 1957. *Eurycea troglodytes:* A New Blind Cave Salamander from Texas, *Tex. J. Sci.,* 9:328.

Brandon, R. A. 1965. A New Race of the Neotenic Salamander *Gyrinophilus palleucus, Copeia,* 1965(3):346.

Burns, D. M. 1962. The Taxonomic Status of the Salamander *Plethodon vandykei larselli, Copeia,* 1962(1):177.

Chrapliwy, P. S., and K. L. Williams. 1957. A Species of Frog New to the Fauna of the United States: *Pternohyla fodiens* Boulenger, *Nat. Hist. Miscellanea, Chicago Acad. Sci.*

Gorman, J. 1960. Treetoad Studies, 1. *Hyla californiae,* New Species, *Herpetologica,* 16:214.

Grobman, A. B. 1959. The Anterior Cranial Elements of the Salamanders, *Pseudotriton* and *Gyrinophilus, Copeia,* 1959(1):60.

Highton, R. 1961. A New Genus of Lungless Salamanders from the Coastal Plain of Alabama, *Copeia,* 1961(1):65.

———— and A. H. Brame, Jr. 1965. *Plethodon stormi* species nov., *Pilot Register of Zoology,* Card No. 20.

James, P. 1966. The Mexican Burrowing Toad, *Rhinophrynus dorsalis,* an Addition to the Vertebrate Fauna of the United States, *Tex. J. Sci.,* 18:272.

Jameson, D. L., J. P. Mackey, and R. C. Richmond. 1966. The Systematics of the Pacific Tree Frog, *Hyla regilla, Proc. Calif. Acad. Sci.,* 33(19):551.

Kezer, J., T. Seto, and C. M. Pomerat. 1965. Cytological Evidence against Parallel Evolution of *Necturus* and *Proteus, Am. Naturalist,* 99:153.

Martof, B. S., and H. C. Gerhardt. 1965. Observations on the Geographic Variation in *Ambystoma cingulatum, Copeia,* 1965(3):342.

———— and F. L. Rose. 1962. The Comparative Osteology of the Anterior Cranial Elements of the Salamanders *Gyrinophilus* and *Pseudotriton, Copeia,* 1962(4):727.

———— and ————. 1963. Geographic Variation in Southern Populations of *Desmognathus ochrophaeus, Am. Midland Naturalist,* 69:376.

Mitchell, R. W., and J. R. Reddell. 1965. *Eurycea tridentifera,* a New Species of Troglobitic Salamander from Texas and a Reclassification of *Typhlomolge rathbuni, Tex. J. Sci.,* 17:12.

Neill, W. T. 1963. Notes on the Alabama Waterdog, *Necturus alabamensis* Viosca, *Herpetologica,* 19:166.

————. 1964. A New Species of Salamander, Genus *Amphiuma,* from Florida, *Herpetologica,* 20:62.

Pylka, J. M., and R. D. Warren. 1958. A Population of *Haideotriton* in Florida. *Copeia,* 1958(4):334.

Rose, F. L., and F. M. Bush. 1963. A New Species of *Eurycea* (Amphibia: Caudata) from the Southeastern United States, *Tulane Stud. Zool.,* 10:121.

Uzzell, T. M., Jr. 1964. Relation of the Diploid and Triploid Species of the *Ambystoma jeffersonianum* Complex (Amphibia, Caudata), *Copeia,* 1964(2):257.

Part Four

REPTILES

F. R. Cagle

THE modern reptiles are characterized by morphological and physiological traits suitable for land-dwelling animals living under warm conditions. They do not have the ability to maintain a constant body temperature, but through behavior may exercise some control. Their lives are dominated by changes in their environment, especially changes of light and temperature.

The skin is dry with few or no glands but with an outer layer of horny scales that is shed at intervals. Bony plates (osteoderms) are often present under the skin. Many reptiles have the ability to modify their color through control of melanophores. This ability is of value in temperature regulation.

Most reptiles have the capacity of moving rapidly and obtain their food through active pursuit. This activity is reflected in the skeletal structure. The bones are generally well ossified, and the head is carried high off the ground and rotated freely. The first 2 cervical vertebrae form an atlas and axis, and the skull has a single condyle. The usually procoelous trunk vertebrae may have supplementary articulating surfaces. The pectoral and pelvic girdles are comparatively larger than those of amphibians. The ventral elements are especially well formed and better ossified to provide attachment for the musculature of the legs. The skull has fossae in the temporal region to permit freedom of movement of the massive jaw muscles.

All reptiles seize their food with the teeth, horny jaws, or tongue. Although they have no true salivary glands, the labial glands may be modified to form poison glands. Typically the stomach is capable of great distension, permitting a large volume of food to be taken. The intestine empties into a chamber, the cloaca, which also receives the products of the kidneys and reproductive organs.

The lungs of snakes and lizards are expanded by movements of the ribs, but the turtle must enlarge the body cavity through rotation of the limb girdles or movement of abdominal muscles. If this does not move air into the lungs, the turtle must swallow air.

The land-dwelling reptiles have developed elaborate behavior patterns, and the nervous system has become accordingly modified. The cerebral hemispheres are larger than in fishes and amphibians as a result of the expansion of the ventral portions. The eyes have eyelids or protective covers as well as a nictitating membrane. Lacrimal and Harderian glands furnish secretions to keep the surface of the eye moist. A ring of bony plates supports the eye in many forms.

The sense of hearing is inferior. Lizards, snakes, and crocodiles are probably insensitive to most sounds, but turtles are responsive to sound in a range of approximately 110 cycles per second. This sensitivity is probably correlated with the presence of a well-developed columella auris.

The modern reptiles live successfully only in tropical and temperate areas. They find their optimum conditions in the tropics.

All reptiles have internal fertilization and either lay eggs or retain the developing young within the female. The reptile egg with a large yolk, an enveloping sac (amnion), and a firm, protecting shell

TABLE 4-1. *Classification of Reptiles*
Class Reptilia

	Families	
	Extinct	Living
Subclass Anapsida		
Order Cotylosauria	13	0
Order Chelonia	12	9
Subclass Lepidosauria		
Order Eosuchia	7	0
Order Rhynchocephalia	5	1
Order Squamata	9	32
Subclass Archosauria		
Order Thecodontia	8	0
Order Crocodilia	11	1
Order Saurischia	15	0
Order Ornithischia	10	0
Order Pterosauria	4	0
Subclass Uncertain		
Order Mesosauria	1	0
Subclass Ichthyopterygia		
Order Ichthyosauria	5	0
Subclass Euryapsida		
Order Protosauria	5	0
Order Sauropterygia	15	0
Subclass Synapsida		
Order Pelycosauria	8	0
Order Therapsida	49	0

was one of the most important developments in evolution, for its appearance freed the animal from a dependence on water. The young reptile is independent of its parent when it emerges from the egg or is born. The hatchling usually has an egg tooth (lizards or snakes) or a horny caruncle (turtles) which may be used to cut the enclosing membranes and is then lost.

The existing reptiles represent only 4 of 15 or 16 main lines of evolution from the amphibians of early Carboniferous time, about 275 million years ago. The snakes and lizards (order Squamata) are the most abundant reptiles in the world of today. The turtles (order Chelonia) have retained many of the characteristics of ancient Permian forms and must be considered highly successful in spite of their apparently clumsy shells. The crocodiles and alligators (order Crocodilia) are mere remnants of a once-abundant group. The tuatara, *Sphenodon*, is a relic of Triassic time without any close living relatives. As most of the orders of reptiles are extinct, our knowledge of them is determined primarily from fossils. Their classification is necessarily based on osteological features. Characteristics of the skull are especially useful.

The reptiles were the dominant group in most Mesozoic habitats. There are more than 6000 living kinds.

KEY TO THE ORDERS OF REPTILES IN NORTH AMERICA

1. Animal with a bony or leathery shell   **Order Chelonia**
   Animal without a bony or leathery shell   **2**
2. Cloacal opening a longitudinal slit   **Order Crocodilia**
   Cloacal opening a transverse slit
   Order Squamata   3
3. 2 pairs of legs usually present; eyelids and external ear openings usually present; ventral scales usually in multiple rows   **Suborder Lacertilia**
   Legs absent; no eyelids or external ear opening; ventral scales usually in single row
   Suborder Serpentes

Order Chelonia.   Turtles. Terrestrial or aquatic; both fresh-water and marine. Body enclosed in an upper shell, the carapace, and a lower shell, the plastron, composed of bony plates covered with horny shields; no sternum or sternal ribs. Skull anapsid; quadrate fixed; no teeth (except in embryonic *Trionyx* and fossil forms from the Triassic); jaws with a horny sheath. Eyelids present; pupils round. Tympanum and columella auris present. Vertebrae procoelous, amphicoelous, or opisthocoelous; 10 trunk vertebrae, 8 cervical vertebrae; caudals reduced. Ribs usually united with the carapace; ribs with capitular portions only; pectoral girdle internal to ribs. Neck retractible in most forms. Pentadactyl limbs; phalangeal formula usually 2,3,3,3,3 or 2,2,2,2,2; limbs elephantine in some terrestrial forms, webbed in fresh-water forms, and modified as flippers in marine turtles. Copulatory organ retracted into sheath in ventral floor of cloaca anterior to cloacal opening.

Breathing achieved by movements of head and limbs, movements of abdominal muscles, or by pumping movement of the hyoid. Aquatic forms may also respire by circulating water through the pharynx and the cloacal sacs. Vision probably the best-developed sense; olfactory organs well developed; hearing not acute, but ears responsive to a narrow range of sound.

Males of many species have much longer nails on forefeet than females and a longer preanal region. Oviparous. Eggs round or ovoid, soft- or hard-shelled; laid in nest excavated in the soil or sand. Eggs not attended by the female or male. Hatchling with a horny caruncle on upper jaw for slitting

Fig. 4-1. Shields of the carapace.

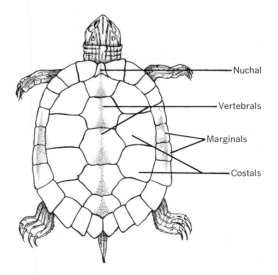

Nuchal

Vertebrals

Marginals

Costals

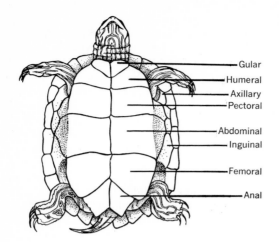

Gular
Humeral
Axillary
Pectoral

Abdominal
Inguinal

Femoral

Anal

Fig. 4-2. Shields of the plastron.

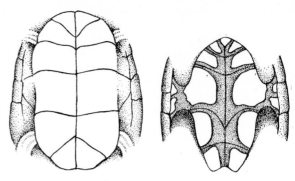

Fig. 4-3. (left). Plastron of *Chrysemys*. Fig. 4-4 (right). Plastron of *Sternothaerus*.

egg shell. Carnivorous, herbivorous, and omnivorous forms.

A very ancient group that has no close relationship to other extant orders. 9 families, 66 genera, and 211 species of living turtles. Permian.

IDENTIFICATION. The identification of turtles depends primarily on the configuration of the head and shell, the arrangement of the plastral and carapacial shields, the color pattern of the head and legs. The identifier must know the names of the shields (Figs. 4-1, 4-2).

The size of the turtle is expressed as the maximum length of the carapace measured in a straight line.

KEY TO THE FAMILIES

1. Digits elongated, flattened, and bound together; limbs paddlelike   **2**
   Digits not elongated, flattened, and bound together (some may have webs between the digits)   **3**
2. Feet scaleless; no claws; carapace covered with skin; no epidermal shields   **Dermochelyidae**
   Feet covered with scales; one or more claws; carapace with epidermal horny shields   **Cheloniidae**
3. Claws 3–3; carapace covered with skin; edge of carapace flexible   **Trionychidae**
   Claws 5–4 or 5–3; carapace covered with horny shields, or, if no horny shields are evident, cara-

pace skin with many projections, giving it a roughened appearance   **4**
4. Plastron with 12 shields, pectoral shield forming front part of bridge (Figs. 4-3, 4-14)   **Testudinidae**
   Plastron with 11 or fewer shields; pectoral shield not forming a part of bridge (Figs. 4-4, 4-5, 4-12)   **Chelydridae**

Family Chelydridae. Snapping turtles; mud turtles; musk turtles. Nuchal plate with long processes underlying the marginals; alveolar surface of maxilla without ridges; no ribs on 10th dorsal vertebra; centrum of 8th vertebra usually doubly concave anteriorly; one biconvex centrum in neck vertebrae.

KEY TO SUBFAMILIES OF CHELYDRIDAE

1. Tail generally long with a crest of large horny tubercles; rear margin of carapace strongly serrate   **Chelydrinae**
   Tail generally short with fleshy papillae; rear of carapace nearly smooth   **Kinosterninae**

Subfamily Chelydrinae. Snapping turtles. Large aquatic predaceous turtles. Carapace rough, ridged. Plastron small, cruciform, with 9 shields; joined to carapace by cartilage (Fig. 4-5); entoplastron present. Skull with narrow roof formed by parietals and postfrontals. Eggs spherical, hardshelled. Lower Cretaceous. North and South America; 2 genera, both in the United States.

KEY TO GENERA OF CHELYDRINAE

1. A single row of marginal shields. Tail with 2 rows of large scales underneath; 1 series of tubercles above
Chelydra
   5th to 8th marginal shields doubled. Tail with many small scales underneath; 3 series of tubercles above
Macroclemys

Genus *Chelydra* Schweigger. Carapace with 3 keels prominent in young but scarcely evident in large individuals; tail long. Oligocene. 2 species; 4 subspecies, 2 of them in the United States.

KEY TO SPECIES OF *CHELYDRA*

1. Temporal region and back of head with flat juxtaposed plates; dorsal surface of neck with rounded, wartlike tubercles; width of 3d vertebral (measure along anterior edge) much less than height of 2d costal
C. serpentina
   Temporal region and back of head with granular scales and scattered tubercles; dorsal surface of neck with long pointed tubercles; width of 3d vertebral the same as or greater than height of 2d lateral
C. osceola

Chelydra serpentina **Linnaeus.** Snapping turtle. Color brown to black. Maximum weight in excess of 60 lb; maximum carapace length 16 in. Carapace usually flattened and sloping gradually to the posterior marginals. Dorsal surface of neck has rounded projections. Eastern and central North America from Nova Scotia and Quebec westward through Ontario, Manitoba, Saskatchewan, and Alberta; southward, east of the Rocky Mountains to the Gulf except peninsular Florida (Fig. 4-6).

Chelydra osceola **Stejneger.** Florida snapping turtle. Color tan to black. Carapace high in region of 3d and 4th vertebrals and sloping abruptly to posterior marginals. Dorsal surface of neck with long tubercles, giving the neck a shaggy appearance. Peninsular Florida.

Genus *Macroclemys* Gray. Alligator snapping turtle. Largest fresh-water turtle of America; maximum weight exceeds 100 lb. Only American turtle that uses a tongue appendage to attract food. A single species. Miocene.

Macroclemys temmincki **(Troost).** Alligator snapping turtle. Carapace with 3 prominent keels; both upper and lower jaws with hooklike tips. General color brown to black. The Gulf Coast, Texas to Florida and southern Georgia; Mississippi Valley to Illinois; lower Missouri and Ohio Rivers.

Subfamily Kinosterninae Agassiz. Mud turtles and musk turtles. Small aquatic or semiterrestrial turtles. Carapace elongate, smooth, highly arched; 11 marginal shields on each side; plastron cruciform or with movable anterior and posterior lobes united to carapace by suture or ligament; plastron with 9–11 shields, 8 or 9 bony plates; no entoplastron (Figs. 4-11, 4-12). Omnivorous. Eggs ovoid,

Fig. 4-5. Plastron of *Macroclemys temmincki.*

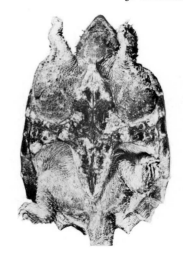

Fig. 4-6. Head of *Chelydra serpentina*. (Photograph by Cyrus Crites.)

hard-shelled. 4 genera, 2 in the United States. Pliocene. North and South America.

KEY TO GENERA OF KINOSTERNINAE

1. Plastron with 2 transverse hinges (not apparent in individuals less than 2 in. in length). Length of interfemoral suture much less than length of interhumeral suture; pectoral shield 3-sided (Fig. 4-7). No areas of uncornified skin along the median, longitudinal suture **Kinosternon**

Fig. 4-7 (left). Plastron of *Kinosternon. Fig. 4-8* (right). Plastron of *Sternothaerus*.

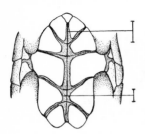
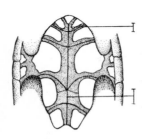

Plastron without transverse hinges. Length of interfemoral suture about equal to or greater than, length of interhumeral suture; pectoral shields 4-sided (Fig. 4-8). Areas of uncornified skin adjacent to median, longitudinal suture **Sternothaerus**

Genus *Kinosternon* Spix. Mud turtles. Plastron hinged (not evident in juveniles), large enough to close carapace opening (Fig. 4-11). Tail with a horny, clawlike tip. Pliocene. 18 species, 5 in United States, others in Central and South America.

KEY TO SPECIES OF *KINOSTERNON*

1. Carapace with 3 longitudinal yellow lines **K. bauri**
Carapace without 3 longitudinal yellow lines **2**
2. Both 9th and 10th marginals extended upward (Fig. 4-9) **K. flavescens**
9th marginal not extended upward **3**
3. 10th marginal not much higher than 9th **K. subrubrum**
10th marginal much higher than 9th (Fig. 4-10) **4**
4. Head and neck dark brown to black, with fine retic-

ulations above, coarser ones on side; an ill-defined light line from angle of jaw to ear region **K. hirtipes**
Head and neck mottled; no light line from angle of jaw to ear region **K. sonoriense**

Fig. 4-9 (left). Marginal shields of *Kinosternon flavescens. Fig. 4-10.* Marginal shields of *Kinosternon sonoriense.*

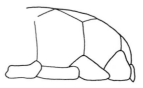
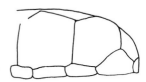

***Kinosternon bauri* Garman.** Striped mud turtle. Carapace brown to olive with 3 often poorly defined longitudinal yellow stripes; 2 yellow stripes on side of the head. 2 subspecies. Southern Georgia and Florida.

***Kinosternon flavescens* (Agassiz).** Yellow mud turtle. Carapace yellow to brown, depressed. No head stripes; head uniform yellow or rarely with irregular dark mottlings. Length 150 mm. 3 subspecies. Arizona and Texas northward to Colorado and Utah; through Oklahoma and Kansas to Iowa and Illinois.

***Kinosternon sonoriense* Le Conte.** Sonora mud turtle. Carapace brown. No head stripes; head conspicuously flecked or mottled with dark color. Length to 175 mm. Southeastern California to southwestern New Mexico; adjacent states of Mexico.

***Kinosternon subrubrum* (Lacépède).** Mud turtle. Carapace dark brown, unmarked. Head usually with 2 poorly defined lateral light stripes. 3 subspecies in the United States. Eastern United States. Connecticut southward to Florida and westward to Oklahoma and Texas (Fig. 4-11).

***Kinosternon hirtipes* Wagler.** Mexican mud turtle. Western Texas and southern Arizona. Mexican plateau, northward to western Texas.

Genus *Sternothaerus* Bell. Musk turtles. Plastron cruciform; smaller than carapace opening (Fig. 4-12). Common name from ability to produce musklike odor from glands.

KEY TO SPECIES OF *STERNOTHAERUS*

1. 2 light lines usually present on sides of head (if absent, head almost black); barbels on throat and chin. 3 dorsal keels present or absent in juveniles. Shields of carapace not overlapping   **S. odoratus**
Light stripes, if present, alternating with dark stripes;

Fig. 4-11. Plastron of *Kinosternon subrubrum.* (**Photograph by Josephine Cagle.**)

Fig. 4-12. Plastron of *Sternothaerus odoratus.* (**Photograph by Josephine Cagle.**)

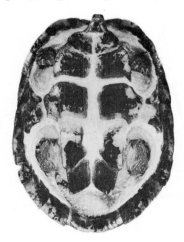

barbels restricted to chin region. Number of keels variable. Shields of carapace overlapping   **2**
2. Gular absent. Carapace with a sharp median keel sloping steeply to marginals; no lateral keels. 3 or more vertebrals generally longer than wide (except in juveniles)   **S. carinatus**
Gular present. Number of keels variable, but carapace sides not steeply sloping. 3 or more vertebrals wider than long   **3**
3. Head with network of narrow lines on a light background. Carapace low and flattened; no sharp middorsal keel; juveniles without lateral keels
   S. depressus
Head with spots, with spots and stripes and network on posterior part of head and neck. Carapace high and rounded; sharp middorsal keel or 3 keels
   S. minor

Sternothaerus carinatus (Gray). Keel-backed musk turtle. Carapace often steep-sided in the form of an inverted V from front view. Small individuals brightly colored with irregular mottling of amber and black dots; head often tinted with light orange. Large individuals blackish brown; heads much broadened. Length to 150 mm. Arkansas, Oklahoma, eastern Texas; eastward into Mississippi.
Sternothaerus odoratus (Latreille). Stinkpot. Carapace of juveniles with 3 keels which are lost in older individuals. Color brownish or black; carapace shields may be marked with black dashes in juveniles. Side of head usually with 2 yellow lines extending back from orbit. Length to 100 mm. Eastern and southern United States.
Sternothaerus minor (Agassiz). Loggerhead musk turtle. Carapace tan to brown; juveniles with 3 keels which are lost in older individuals. Head spotted or blotched. Length to 100 mm. 2 subspecies. Eastern Louisiana to northern Florida; northward to eastern Tennessee.
Sternothaerus depressus Tinkle and Webb. Flattened musk turtle. Carapace much flattened in juveniles; flat-topped with steeply arched sides in adults. Color tan to black. Length to 90 mm. Black Warrior River system of Alabama.
Family Testudinidae. Aquatic, semiterrestrial, and terrestrial turtles. Carapace oval, arched or depressed; nuchal plate without lateral processes; neck vertebrae with 2 biconvex centra; centrum of eighth vertebra typically doubly convex anteriorly; alveolar surface of maxilla with at least 1

ridge. 11 or 12 marginals on each side. Plastron with 12 shields; bony union with carapace.

KEY TO SUBFAMILIES OF TESTUDINIDAE

1. Top of head covered with shields; digits short, without rudiment of web   **Testudininae**
Top of head covered anteriorly with undivided skin; digits usually elongate, nearly always with some web between toes   **Emydinae**

Subfamily Emydinae.   Largest group of living turtles; includes both terrestrial and aquatic turtles. Carapace oval; arched or flattened. Toes usually with some web between; toes usually elongated, median digits with 3 phalanges (except *Terrapene*, which has 2 in median digits of forefeet).

KEY TO GENERA OF EMYDINAE

1. Plastron with 1 transverse hinge (has ability to close completely the carapace opening)   *Terrapene*
Plastron without or with 2 transverse hinges   **2**
2. 5th toe of rear foot absent (Fig. 4-13) *Terrapene*
5th toe of rear foot present (but clawless) and connected by web to 4th   **3**
3. Axillaries and inguinals absent; or if present, small and rudimentary; shortest distance between shields much greater than length of interabdominal suture   **4**
Axillaries and inguinals large, well developed (Fig. 4-14)   **6**
4. Front of upper jaw, seen from side, curved upward **5**
Front of upper jaw, seen from side, curved downward   *Clemmys*

Fig. 4-13. Rear foot of *Terrapene.*

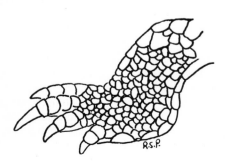

Fig. 4-14. Plastron of *Pseudemys scripta.* (Photograph by Josephine Cagle.)

Fig. 4-15 (top). Measurement of symphysis length on lower jaw. *Fig. 4-16* (left). Anterior marginal shields of *Chrysemys*. *Fig. 4-17* (right). Anterior marginal shields of *Pseudemys.*

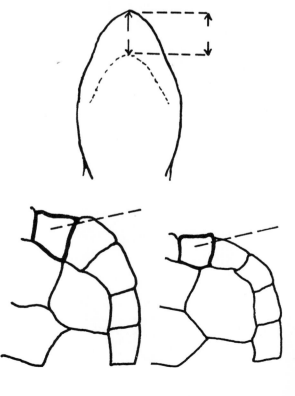

5. Lower jaw and chin immaculate yellow *Emydoidea*
Lower jaw and chin not immaculate yellow
Malaclemys
6. Each of the posterior marginals with a ∧-shaped
notch, or, if not, lower jaw spoon-shaped with long
symphysis (length of symphysis equal to or greater
than ½ length of intergular suture) (Fig. 4-15)
Graptemys
Each of the posterior marginals smooth; no ∧-
shaped notch   7
7. 1st marginal not extending beyond suture between
1st costal and 1st vertebral (Fig. 4-16); crushing
surface of upper jaw narrow in all individuals   8
1st marginal extending beyond suture between 1st
costal and 1st vertebral (Fig. 4-17)   *Pseudemys*
8. Carapace with a network of yellow lines (may be
absent in individuals longer than 8 in.). Neck
elongate; length from tip of nose to anterior edge
of carapace equal to ½ carapace length
Deirochelys
Carapace without a network of yellow lines; no
distinct, oval, black spot or blotch on bridge. Neck
short; length from tip of nose to anterior edge of
carapace equal to only ⅓ or less of carapace length
Chrysemys

Genus *Clemmys* Ritgen.   Pond turtles. Carapace
oval, depressed (carapace of *C. insculpta* often not
depressed). Eocene; Eurasia, North Africa, North
America. 13 species, 4 in the United States.

KEY TO SPECIES OF *CLEMMYS*

1. Found in eastern and central states; carapace with
or without a median keel   2
Found only on Pacific slope; carapace without a
median keel   *C. marmorata*
2. Carapace without yellow spots and often with a
median keel   3
Each shield of carapace with 1 or several yellow or
orange spots; carapace without a median keel
C. guttata
3. Skin salmon red; no orange blotch on temple
C. insculpta
Skin not salmon red; conspicuous orange blotch on
temple   *C. muhlenbergi*

Clemmys guttata (Schneider).   Spotted turtle.
Carapace brown with small round orange or yellow
spots; plastron black with yellow or orange blotches.
Carapace smooth; not keeled. Southern Maine to
northern Illinois; southward on east coast to

Georgia. Quebec to northern Florida; Pennsylvania
and New York westward to Wisconsin and Illinois.
Clemmys insculpta (Le Conte).   Wood turtle.
Carapace gray or brown, roughened by concentric
grooves and ridges on each plate; plastron yellow
with a black blotch on each plate. Carapace keeled.
Nova Scotia to northern Virginia; New York westward into Iowa.
Clemmys marmorata (Baird and Girard).
Western pond turtle. Carapace olive or brown to
black; each shield with many dark dots or dashes
often forming radiate or reticulate pattern; plastron
black and yellow. Length to 175 mm. 2 subspecies.
Pacific Coast from British Columbia to Baja California.
Clemmys muhlenbergi (Schoepff).   Bog turtle.
Carapace dark brown to black with yellow or red
markings. Head black with an orange or yellow
spot on each side. New York to western North
Carolina.
Genus *Emydoidea* Gray.   Semi-box turtles. Carapace high, arched; lobes of plastron connected by
hinge; plastron joined to carapace by cartilage.
Carapace opening cannot be closed. Paleocene.
North America, Eurasia.
Emydoidea blandingi (Holbrook).   Blanding's
turtle. Carapace black with numerous yellow spots;
plastron yellow with irregular black blotches laterally. Southern Vermont and New Hampshire to
central Pennsylvania; westward to Minnesota and
Nebraska.
Genus *Terrapene* Merrem.   Box turtles. Carapace high, arched; lobes of plastron connected by
a hinge; plastron joined to carapace by cartilage.
Carapace opening can be closed.

KEY TO SPECIES OF *TERRAPENE*

1. No trace of a keel on 2d vertebral (Fig. 4-18). Interfemoral suture more than half as long as interabdominal suture (Fig. 4-19)   *T. ornata*
Keel present on 2d vertebral. Interfemoral suture
less than half as long as interabdominal suture
(Fig. 4-20)   *T. carolina*

Terrapene ornata (Agassiz).   Western box turtle. Generally similar to *T. carolina* but without a
keel and having a long interabdominal suture.
Color of carapace light brown to black; each shield

Fig. 4-18 (left). *Terrapene ornata.* (From Stebbins, *Amphibians and Reptiles of Western North America,* McGraw-Hill Book Company.) *Fig. 4-19* (right). Plastron of *Terrapene ornata.*

Fig. 4-22. Chrysemys picta (From Stebbins, *Amphibians and Reptiles of Western North America,* McGraw-Hill Book Company.)

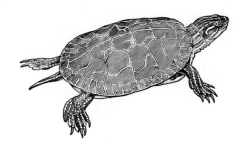

with pattern of radiating yellow lines. Occasional individuals and populations have the carapace uniformly yellow. Length to 150 mm. 2 subspecies. Wyoming, South Dakota, Iowa, Wisconsin, Illinois, and Indiana to Texas, New Mexico, Arizona, and northern Mexico.

Terrapene carolina (**Linnaeus**). Box turtle (Fig. 4-21). Carapace oval, highly arched; a keel present on at least 1 of the vertebral shields. Interabdominal suture short (Fig. 4-20). 4 subspecies. Eastern United States; Maine through Florida westward to eastern Kansas, Oklahoma, and Texas.

Genus *Malaclemys* Gray. Diamondback terrapins. Carapace keeled, depressed; each shield deeply marked with concentric rings. Carapace of

juveniles with conspicuous, enlarged knobs on each vertebral. 1 species; 6 subspecies.

Malaclemys terrapin (**Schoepff**). Diamondback terrapin. Brackish water. Once of much value as food, now eaten only locally. 6 subspecies. Coastal marshes and inshore waters from New England to Texas.

Genus *Chrysemys* Gray. Painted turtles. Carapace smooth, flattened; no keel. Plastron joined to carapace by a bony bridge. Feet moderately webbed. Eocene, North America. 1 species; 4 subspecies.

Chrysemys picta (**Schneider**). Painted turtle. One of the most abundant turtles of the middle western and eastern states. Generally brightly marked on carapace and head with red, orange, or yellow (Fig. 4-22). Length to 225 mm. Northern United States; extends southward into Mexico from Colorado; southward in the Mississippi Valley to the Gulf, southward along the Atlantic Coast to Florida.

Genus *Graptemys* Agassiz. Map turtles and sawbacks. Carapace smooth, flattened or steep-sided, and keeled; often with black-tipped spines on the vertebral shields. Oligocene. North America. 9 species, 3 subspecies.

Fig. 4-20 (left). **Plastron of *Terrapene carolina*.** *Fig. 4-21* (right). **Head of *Terrapene carolina*.** (Photograph by Cyrus Crites.)

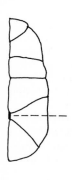

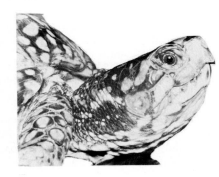

KEY TO SPECIES OF *GRAPTEMYS*

1. Length of lower jaw symphysis ¼ longer than shortest distance between orbits; a small, triangular yellow spot back of eye separated from eye by 2–3 diagonal, yellow lines (Fig. 4-23)

G. geographica

Fig. 4-23 (left). Head of *Graptemys geographica*.
Fig. 4-24 (right). Head of *Graptemys oculifera*.

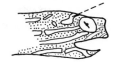

Fig. 4-25 (left). Lower jaw of *Graptemys barbouri*.
Fig. 4-26 (right). Lower jaw of *Graptemys pulchra*.

Fig. 4-27 (left). Head of *Graptemys kohni*.
Fig. 4-28 (right). Head of *Graptemys versa*.

Fig. 4-29. Head of *Graptemys nigrinoda*.

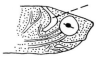

Fig. 4-30. Variation in head patterns of *Graptemys pseudogeographica*.

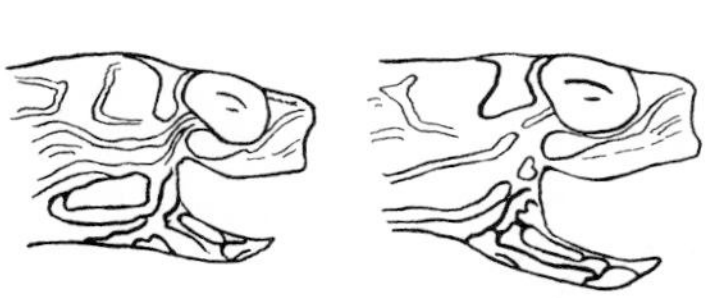

Length of lower jaw symphysis equal to or less than minimum distance between orbit; or, if symphysis is long, no triangular spot behind eye **2**
2. Each costal shield with a central, solid blotch of yellow or orange **G. flavimaculata**
No central, solid blotch of yellow or orange on each costal shield **3**
3. Each costal shield with a complete circle of yellow or orange; a postocular yellow mark with a diameter equal to half or less that of orbit (Fig. 4-24) **G. oculifera**
No such combination of markings as described **4**
4. An enlarged yellow or greenish blotch (greater than area of orbit) covering much of postocular area (Fig. 4-31) **5**
No yellow or greenish blotch covering much of postocular area **6**
5. An irregular bordered yellow bar extending across ventral surface of lower jaw (Fig. 4-25) **G. barbouri**
No bar as described (an elongated blotch may be present) (Fig. 4-26) **G. pulchra**
6. A postorbital line extending anteriorly under the eye (Fig. 4-27) **G. kohni**
No postorbital line extending anteriorly under the eye **7**
7. A postorbital line or spot terminating above eye and extending posteriorly as a neck stripe. Juvenile with a vertical, orange postorbital bar (Fig. 4-28) **G. versa**
No postorbital line or spot terminating above eye and extending posteriorly **8**
8. A postorbital, vertical, posteriorly curved, yellow line joining a diagonal line on upper surface of head (Fig. 4-29); spines of vertebral shields much broadened, knoblike **G. nigrinoda**
Postorbital spot oval, rectangular, or comma-shaped (Fig. 4-30); spines of vertebral shields not broadened, knoblike **G. pseudogeographica**

Graptemys flavimaculata Cagle. Yellowblotched sawback. Head with broad yellow or orange-tinted longitudinal lines. Each costal shield with a central yellow or orange blotch. Length to 200 mm. Head narrow. Pascagoula River, Mississippi.

Graptemys oculifera Baur. Ringed sawback. Head with broad yellow lines on a black background. Each costal shield with a complete yellow circle; width of circular stripe equal to width of broadest head stripe. Length to 220 mm. Pearl River in Louisiana and Mississippi.

Graptemys nigrinoda **Cagle.** Black-knobbed sawback. Head with bright yellow lines on a black background; a curved, yellow postorbital bar. Each costal shield with a circle or yellow mark; width of yellow line forming mark much less than width of evident head stripe. Head narrow. Length to 220 mm. Black Warrior and Alabama Rivers, Alabama.

Graptemys barbouri **Carr and Marchand.** Barbour's map turtle. Head with a large, irregular, greenish postorbital blotch. Lateral stripe on lower jaw. Head much broadened in females. Length to 330 mm. Gulf Coast streams from the Escambia River eastward to Flint River.

Graptemys pulchra **Baur.** Alabama map turtle. Head with large, irregular, greenish or yellowish postorbital blotch (Fig. 4-31). Longitudinal stripe at symphysis of lower jaw. Head much broadened in females. Length to 250 mm. Gulf Coast streams from Escambia River westward to Pearl River.

Graptemys pseudogeographica **Gray.** False map turtle. Head with a comma-shaped postorbital mark; several narrow, longitudinal yellow lines entering the orbit. Costals with a network of yellow lines. 3 subspecies. Length to 320 mm. Eastern Texas northward to Minnesota and Wisconsin; east to Mississippi in the South and Ohio in the North.

Graptemys geographica **(Lesueur).** Map turtle. Head with thin yellow lines; a triangular yellow postorbital mark. Carapace green with few or no irregularly distributed yellow lines; carapace flattened in adults; low spines on vertebrals. Length to 300 mm. Louisiana north through eastern Oklahoma and Kansas to northern Minnesota; eastward through Missouri, Illinois, Kentucky, and Tennessee to Vermont.

Graptemys versa **Stejneger.** Texas map turtle. Postorbital mark a longitudinal yellow line. Carapace shields wrinkled in those larger than 5 cm; carapace flattened; no spines on vertebrals. Length to 200 mm. Colorado River system, Texas.

Graptemys kohni **Baur.** Mississippi map turtle. A vertical yellow postorbital line continuing under orbit. Carapace not depressed; distinct spines on vertebrals. Length to 300 mm. Eastern Texas, Oklahoma; Kansas and Nebraska east to Mississippi River and southern Indiana.

Genus *Pseudemys* Gray. Sliders and cooters. Carapace shape variable: low, flattened in some groups, high and arched in others. Carapace smooth, no spines. Plastron joined to carapace by a bony bridge. Pliocene. North America. 5 species in the United States.

KEY TO SPECIES OF *PSEUDEMYS*

1. A yellow or red line or spot behind orbit and as wide as ½ diameter of orbit, or head black with but faint indications of a pattern (Fig. 4-32). Ridge on alveolar surface of upper jaw without many teethlike serrations but often with tubercles

P. scripta

No markings as described, but yellow lines or spots as wide as ⅓ orbit diameter sometimes present. Ridge on alveolar surface of upper jaw with many conspicuous, teethlike serrations **2**

Fig. 4-31. Head of *Graptemys pulchra.* (Photograph by Josephine Cagle.)

Fig. 4-32. **Variation in head markings of *Pseudemys scripta*.**

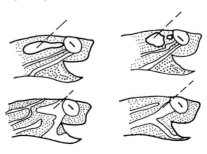

2. Upper jaw smooth or with a slight notch at apex and slightly serrate at tip (or, if notched, lines on head numerous, some broken to form spots and bars) (Figs. 4-33, 4-35)   **3**

    Upper jaw with a distinct notch bounded by cusps; lines on side of head not numerous   **4**

3. Second costal with a light-colored C-shaped mark; pattern of dark lines on plastron   ***P. concinna***

    Second costal with a light-colored vertical or inverted Y-shaped mark; usually no dark markings on plastron   ***P. floridana***

4. Markings on lower marginals enclosing light area. Markings on bridge oblong or a concentric pattern with light centers   ***P. rubriventris***

    Markings on lower marginals solid, circular or oval, or with light centers. No markings on bridge or a few large spots or bars   **5**

5. Paramedian head stripe ending posterior to eye (Fig. 4-33)   ***P. nelsoni***

    Paramedian head stripes continuing between the eyes and onto snout   ***P. alabamensis***

Fig. 4-33. **Dorsal side of head,** *Pseudemys nelsoni.*

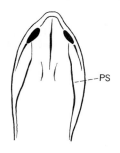

Fig. 4-34. Pseudemys scripta. (**From Stebbins,** *Amphibians and Reptiles of Western North America,* **McGraw-Hill Book Company.**)

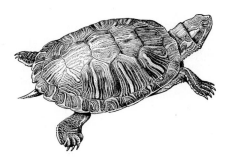

Fig. 4-35. **Head of** *Pseudemys concinna.* (**Photograph by Josephine Cagle.**)

Pseudemys scripta (**Schoepff**). Pond slider. Carapace generally low, flattened; length rarely greater than 23 cm (Fig. 4-34). 4 subspecies. Atlantic Coast from Virginia south through north Florida and west along Gulf Coast to southern Tamaulipas and western Texas. North through Mississippi Valley to Wisconsin and Ohio. Introduced into Michigan.

Pseudemys concinna (**Le Conte**). River cooter. Carapace generally low, flattened; carapace length to 40 cm; dark markings on submarginals large and in contact with dark markings on bridge. Typically a river turtle. 5 subspecies. Virginia into Florida westward through coastal states through Texas and into Mexico; northward in Mississippi Valley to southern Illinois, Missouri, and Kansas.

Pseudemys floridana (**Le Conte**). Cooter. Carapace generally high, arched; length to 37 cm; dark markings on submarginals few and reduced and not in contact with dark markings on bridge. Typically a pond turtle. 2 subspecies. Atlantic Coast from Maryland south through Florida and west along Gulf Coast to east Texas. North through

Mississippi Valley to southern Illinois, Missouri, and Kansas.

***Pseudemys nelsoni* Carr.** Florida red-bellied turtle. Carapace high, arched; highest point anterior to the central part. Peninsular Florida.

***Pseudemys alabamensis* Baur.** Alabama red-bellied turtle. Similar to *P. nelsoni* except for a larger number of head stripes and the anterior extension of the paramedian head stripe. Florida panhandle to vicinity of Mobile, Alabama.

***Pseudemys rubriventris* (Le Conte).** Red-bellied turtle. Carapace high, arched; highest point near central section. 2 subspecies. Atlantic Coast from southern New Jersey to North Carolina; isolated colony in Massachusetts.

Genus *Deirochelys*. Carapace high, arched; smooth or finely wrinkled. Neck longer than that of any other North American emydid. 1 species, 3 subspecies.

***Deirochelys reticularia* (Latreille).** Chicken turtle. Carapace marked with a network of yellow lines; distinct, elongate black blotch usually present on the bridge. North Carolina south through peninsular Florida and west through southern Mississippi and Alabama to eastern Texas; north in Mississippi Valley to Missouri.

Subfamily Testudininae. Land tortoises. Terrestrial turtles. Carapace high and arched but sometimes flattened dorsally. Plastron with 12 shields; joined firmly with carapace by a bony bridge. Legs elephantlike; toes short with thick-ened claws and without web; not more than 2 phalanges in any digit. 1 genus in the United States.

Genus *Gopherus*. Gopher tortoises. Carapace steep-sided; flattened dorsally. Forelegs with enlarged, thickened scales which provide a protective surface when legs are folded into opening of carapace (Fig. 4-36). 4 species, 3 in United States.

KEY TO SPECIES OF *GOPHERUS*

1. Carapace length less than twice maximum height; head wedge-shaped in front   ***G. berlandieri***
   Carapace length twice maximum height; head rounded in front   **2**
2. Plastron not bent upward in front; distance from base of 1st claw to base of 4th claw of forefoot equals distance from base of 1st claw to base of 4th claw of rear foot   ***G. agassizi***
   Plastron bent upward in front; distance from base of 1st claw to base of 4th claw of forefoot equal to distance from base of 1st claw to base of 3d claw of rear foot   ***G. polyphemus***

***Gopherus berlandieri* (Agassiz).** Texas tortoise. Carapace brown, often with yellow centers on carapace plates; plastron yellow. Head broad; jaw more pointed than in other forms. Hind foot is proportionally narrower than that of *G. agassizi* (width of head is 57–89% of foot width). Southern Texas and northeastern Mexico.

***Gopherus agassizi* (Cooper).** Desert tortoise. Carapace light brown or yellow; plastron yellow, often tinted with brown about the margins. Head narrower than that of other members of the genus. Hind foot large (width of head is 85–115% of foot width). Length to 350 mm. Southern Nevada and southwestern Utah to southern California, southwestern Arizona, and northern Mexico.

***Gopherus polyphemus* (Daudin).** Gopher tortoise. Carapace brown or brownish yellow; young turtles with distinct light centers to carapace plates. Head broad; jaws short and broadly angled. Hind foot narrow (width of head is 53–78% of width of hind foot). South Carolina through Florida; westward to Louisiana and north to Arkansas.

Family Cheloniidae. Sea turtles. Carapace covered with epidermal shields. Legs modified to form paddles; feet covered with scales. Skull roof incomplete.

Fig. 4-36. Gopherus agassizi. (From Stebbins, *Amphibians and Reptiles of Western North America,* McGraw-Hill Book Company.)

KEY TO THE GENERA OF CHELONIIDAE

1. Costals four on each side; one pair of prefrontals between the eyes (Fig. 4-37*B*); horny edge of lower jaw strongly dentate; bony alveolar surface of upper jaw with a low but regularly raised ridge *Chelonia*
   Costals 4–9 on each side; 2 pairs of prefrontals between the eyes (Fig. 4-37*C*); horny edge of lower jaw smooth or weakly dentate; bony alveolar surface of upper jaw smooth or with a sharp-crested ridge **2**
2. Costals 4 on each side, usually imbricate; nuchal not in contact with first costals; snout elongate, narrow; premaxilla deeply excavated and not terminally toothed; bony alveolar surface of upper jaw with a single sharp-crested ridge *Eretmochelys*
   Costals 5 or more on each side, not imbricate; nuchal in contact with first costals; snout relatively short and broad; premaxilla toothed; bony alveolar surface of upper jaw smooth or with a rounded ridge **3**
3. Costals 5–9 on each side; 4 enlarged inframarginals on bridge; maxillae not in contact, separated by vomer. Color gray to olive-green *Lepidochelys*
   Costals 5 on each side (Fig. 4-37*C*); 3 enlarged inframarginals on bridge; maxillae in contact. Color brown or reddish brown *Caretta*

Genus *Caretta*. Loggerhead turtle. General color reddish brown. A keel present in small individuals is inconspicuous in large adults. Carapace length to 215 cm. 1 species.

Caretta caretta (**Linnaeus**). Loggerhead (Fig. 4-37*C*). 2 subspecies. Atlantic Ocean, from Newfoundland south on east coast. Pacific Ocean, coast of southern California.

Genus *Chelonia*. Green turtle. General color brown; carapace sometimes irregularly marked with olive or dark-brown blotches. The common name is based on the green fat, not the color of the turtle. In demand for its excellent meat, this turtle is being rapidly reduced by commercial exploitation. Carapace length to 155 cm. 1 species.

Chelonia mydas (**Linnaeus**). Green turtle (Fig. 4-37*B*). 2 subspecies. Atlantic Ocean, Massachusetts south on east coast and into Gulf of Mexico. Pacific Ocean, coast of southern California.

Genus *Eretmochelys*. Hawksbill turtle. General color brown. Carapace keeled; shields overlapping. This turtle, the source of commercial tortoise shell, is still sought for its carapace shields. Carapace length to 90 cm. 1 species.

Eretmochelys imbricata (**Linnaeus**). Hawksbill. 2 subspecies. Atlantic Ocean, Massachusetts south on east coast and into Gulf of Mexico. Pacific Ocean, coast of Baja California southward to Peru.

Genus *Lepidocheyls* Fitzinger. Ridley turtles. General color gray. Carapace keeled. Carapace length to 70 cm. 1 species.

Lepidochelys olivacea (**Eschscholtz**). Ridley. 2 subspecies. Along coasts of the United States, Gulf of Mexico; north along east coast to Nova Scotia.

Family *Dermochelyidae*. Leatherback turtle. Carapace without shields. Seven ridges extend length of carapace. General color black or brown. Largest of living turtles (Fig. 4-37*A*). Carapace length to 240 cm. 1 species.

Dermochelys coriacea (**Linnaeus**). Leatherback. Along coasts of the United States.

Family Trionychidae. Softshell turtles. Aquatic. Carapace covered with a leathery skin; edges of carapace flexible cartilage (ossified only in very large individuals). Head narrow; a proboscislike snout. Feet webbed.

Genus *Trionyx* Geoffrey St. Hilaire. Softshell turtles. 14 species, 8 subspecies. 3 species, 8 subspecies in the United States.

KEY TO SPECIES OF *TRIONYX*

1. No tubercles on anterior margin of carapace; nasal septum without a lateral ridge projecting into nostril *T. muticus*
   Tubercles present on anterior margin of carapace (very low and rounded in hatchlings); nasal septum with a lateral ridge projecting into nostril **2**
2. Plastron uniform dark slate or blackish; soft parts of body blackish and with large pale marks dorsally; carapace with large black blotches, often fused along margin, on pale background. Carapace with many well-defined longitudinal ridges *T. ferox*
   Plastron whitish; blackish flecks or blotches sometimes present. Combination of characters not as above **3**
3. Carapace with pattern of white dots, or black ocelli and/or spots. Carapace sometimes gritty resembling sandpaper ***T. spiniferus***

Carapace uniform pale brownish or grayish, or having mottled and blotched pattern; white dots, black ocelli and/or spots may be present. Carapace not gritty   **4**

4. Marginal ridge present   *T. ferox*
     Marginal ridge absent   *T. spiniferus*

***Trionyx muticus* Lesueur.** Smooth softshell. Individuals greater than 200 mm in length with carapace uniform brown and head brown above but light on sides. Smaller individuals reddish brown with many dark flecks; a dark line from eye onto neck. Length to 300 mm. 2 subspecies. Pennsylvania to South Dakota; southward to the Gulf Coast.

***Trionyx ferox* Schneider.** Florida softshell. Carapace of large individuals dark gray to black. Carapace with rough appearance resulting from presence of many rounded bumps. Carapace of small individuals dark gray and marked with distinct, large dark spots. Southern South Carolina, southern Georgia, and Florida.

***Trionyx spiniferus* Lesueur.** Spiny softshell. Carapace of large individuals with olive to brown and marked by darker, vaguely defined blotches. Carapace of smaller individuals tan to brown and marked with blackish ocelli or spots. 6 subspecies. Vermont south to northern Florida; eastward to Montana and southwest into Mexico. Introduced to Colorado River system of California, Nevada, Arizona, and New Mexico.

Order Crocodilia. Aquatic or amphibious; fresh water but entering salt water. Body elongate; neck short. Bony plates (osteoderms) usually present beneath the horny scales. Tail elongated; laterally compressed. Limbs short; digits 5–4. Skull without pineal foramen; temporal region with an upper and a lower arch; teeth simple, conical, pleurodont; quadrate fixed. External tympanum covered with a skin fold; maxillae, palatines, pterygoids form a secondary palate. Vertebral column differentiated into cervical, thoracolumbar, sacral, and caudal regions; ribs bicipital; vertebrae procoelous. Abdominal ribs present. Cloacal opening a longitudinal slit. Lungs in a separate thoracic cavity; heart 4-chambered; copulatory organ single.

Oviparous. Eggs oval, hard-shelled; laid in nests constructed and guarded by the female. Carnivorous. Triassic. Modern forms are survivors of a once-

Fig. 4-37. Marine turtles. *A, Dermochelys coriacea; B, Chelonia mydas; C, Caretta caretta.* (From Stebbins, *Amphibians and Reptiles of Western North America,* McGraw-Hill Book Company.)

abundant group but have changed little since Triassic time.

Family Crocodilidae.

KEY TO GENERA OF FAMILY CROCODILIDAE

1. Snout broad; a tooth not visible when mouth is closed   *Alligator*
     Snout pointed; a tooth exposed when mouth is closed   *Crocodylus*

***Crocodylus* Laurenti.** 1 species in the United States.

***Crocodylus acutus* Cuvier.** American crocodile. Southern Florida and the Florida Keys.

Genus *Alligator* Cuvier. 1 species in the United States.

Alligator mississipiensis (**Daudin**). American alligator. Atlantic and Gulf Coast Plains; North Carolina to the Texas coast.

Order Squamata. Lizards and snakes. Terrestrial, fossorial, arboreal, or aquatic. **Modern remnants of the once-abundant diapsid reptiles. Bony border of lower temporal opening lost; quadratojugal absent; squamosal small; quadrate movable. Jurassic.

Suborder Lacertilia. Terrestrial, fossorial, arboreal, or aquatic. Body elongate, usually with 2 pairs of appendages (absent in *Ophisaurus, Anniella,* and *Rhineura* of the United States). Shape of the pentadactyl limbs often reflects the locomotion habits; short and weak in burrowing forms, elongated and slender in cursorial forms; short and strong in some climbing forms. Scales of digits variously modified; fringelike adhesive pads or ridged scales. Scale types exceedingly varied; scales may be flexible and granular as in the Gekkonidae, spiny or keeled as in the Iguanidae, smooth and shiny as in the Scincidae, beadlike and supported by osteoderms as in the Helodermatidae. Pits (femoral pores) often present on ventral surface of the rear legs are usually larger in males.

Skull diapsid; 2 temporal arches (upper sometimes absent). Mandibular rami usually suturally joined. Teeth acrodont or pleurodont. Tongue variable in shape; short and thickened or extensible, sometimes forked; primary use associated with olfaction. Eyes usually well developed; degenerate in some burrowing forms; eyelids generally present (absent as a movable structure in most Gekkonidae and Xantusidae); pupils round or vertical. External ear openings concealed in some groups (*Anniella, Holbrookia*).

Vertebral column differentiated into cervical, thoracolumbar, sacral, and caudal regions; sternum generally present. Vertebrae of tail may be broken by muscular contractions so as to shed the tail. Copulatory organs paired, retracted into postcloacal sheaths. Male with a broader tail base than female. Cloacal opening transverse. Breathing achieved by thoracic movements. Vision usually the best-developed sense; olfactory organs moderately developed; hearing poor.

Generally no conspicuous differences between the sexes. Males may be more brightly colored, have enlarged postanal scales, enlarged gular folds, broader heads, or better-developed femoral pores. Oviparous or viviparous; eggs flexible or hard-shelled, generally ovoid. Eggs attended by female in some species. Carnivorous, herbivorous. About 3000 species, 20 families.

The arrangement of the head and body scales, the color pattern, and the general configuration are the bases for identification of lizards. A knowledge of the general pattern of the head scales is essential (Fig. 4-38). The illustrations with each generic key must be consulted. The size is expressed as head-body length (tip of nose to cloacal opening).

KEY TO THE FAMILIES OF LACERTILIA

1. 2 pairs of legs	**2**
Legs absent	**9**
2. Eyelids present; pupils not vertical	**3**
Eyelids present or absent; if present, pupil is vertical	**8**
3. All scales around body about equal in size; scales smooth, shiny	**Scincidae**

Fig. 4-38. Head scales of lizards. (From Stebbins, *Amphibians and Reptiles of Western North America*, McGraw-Hill Book Company.)

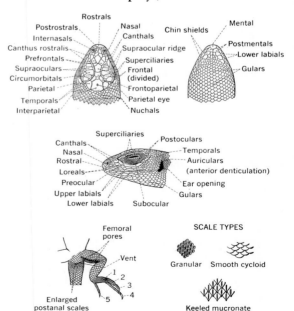

Fig. 4-39 (left). **Head scales uniform.** *Fig. 4-40* (right). **Head scales unequal in size.**

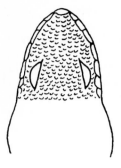
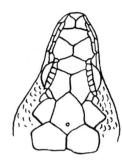

Fig. 4-41 (left). **Ventral scales, Helodermatidae.**
Fig. 4-42 (right). **Ventral scales, Gekkonidae.**

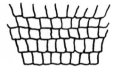

All scales around body not about equal in size or not smooth and shiny   **4**

4. A skin fold extending from the head to the rear leg; scales of fold much smaller than those above and below it (small scales of the fold sometimes concealed by edges of the ventral plates)   **Anguidae**

   No skin fold extending from head to rear leg; no lateral rows of scales smaller than dorsal or ventral scales   **5**

5. Head covered with very small scales of about equal size (Fig. 4-39)   **6**

   Head covered with scales unequal in size (Fig. 4-40)   **7**

6. Ventral scales arranged in transverse rows (Fig. 4-41)   **Helodermatidae**

   Ventral scales not in transverse rows (Fig. 4-42)   **Gekkonidae**

7. More than 10 scale rows across belly   **Iguanidae**

   Less than 10 scale rows across belly   **Teiidae**

8. Ventral plates rectangular; head plates large, 1 or 2 between eyes   **Xantusidae**

   Ventral plates rounded; head plates small, more than 10 between eyes   **Gekkonidae**

9. Ear opening present   **Anguidae**

   Ear opening absent   **10**

10. Eye opening absent; no eye evident   **Amphisbaenidae**

   Eye opening present   **Anniellidae**

Family Gekkonidae. Vertebrae procoelous or amphicoelous; tongue usually fleshy, nonextensible; soft skin with small, evenly distributed scales. Skull arches absent; clavicles broadened ventrally; parietals separate or united; eyelid immobile in most forms, transparent; extremely large eyes with a vertical pupil. Most of the geckos are arboreal and nocturnal. Some have a strong voice. Oviparous or viviparous. Cosmopolitan.

KEY TO GENERA OF GEKKONIDAE

1. Movable eyelids present; digits not widened or lobed   *Coleonyx*

   No movable eyelids; digits widened or not   **2**

2. Digits not widened   *Gonatodes*

   Digits widened (but may be so only at tips)   **3**

3. A single, round scale at tips of toes; scales uniform in size on back   *Sphaerodactylus*

   Not a single, round scale at tips of toes; scales not uniform in size on back   **4**

4. 2 groups of broadened scales at tips of toes; a claw between the 2 groups   *Phyllodactylus*

   Most of the scales on ventral surface of digit enlarged; last joint of digit narrowed, clawlike   *Hemidactylus*

Genus *Gonatodes* Fitzinger. Padless geckos. Digits not widened; no preanal or femoral pores. Most diurnal of the geckos. 15 species, 1 in the United States.

***Gonatodes fuscus* (Hallowell).** Yellow-headed gecko. Color generally gray tinged with brown. Markings very variable; may be in form of spots or bars, or color may be uniform. Head and neck yellow. Length about 38 mm. Introduced to Key West, Florida. Caribbean coasts, West Indies generally.

Genus *Phyllodactylus* Gray. Leaf-toed geckos. 2 scales at tip of each digit much enlarged; claw between scales. 40 species, 1 in the United States.

***Phyllodactylus tuberculosus* Wiegmann.** Leaf-toed gecko. Gray or brown with slate-colored, irregular markings tending to form crossbands or spots (Fig. 4-43). Length to 63 mm. Extreme southern counties of California, Baja California, and western Mexico.

Genus *Hemidactylus* **Oken.** House geckos. Digits much widened along most of length; terminal portion of each digit clawlike. Preanal or femoral pores present in males. 35 species, 1 species introduced into Florida and Louisiana.

Hemidactylus turcicus **(Linnaeus).** Mediterranean gecko. Color white to light brown with darker spots irregularly distributed over dorsal surface. Several conspicuous enlarged longitudinal rows of keeled tubercles on back. Introduced to Florida, Louisiana, and Texas. Mediterranean coasts and islands southward on coasts of the Red Sea; in eastern Africa to northern Kenya Colony and eastward to northwestern India.

Genus *Sphaerodactylus* **Wagler.** Dwarf geckos. A small, rounded pad on the tip of each digit; claws very small. No femoral or preanal pores. 26 species, 2 species introduced to Florida.

KEY TO SPECIES OF *SPHAERODACTYLUS*

1. Dorsal scales rather large, strongly keeled, overlapping; dorsal scales larger than ventral scales
   S. notatus

   Dorsal scales small, granular; smaller than ventral scales **S. cinereus**

Sphaerodactylus cinereus **Wagler.** Ashy gecko. Color reddish brown to gray with many small light-yellow spots over the back and legs. Young animals with red crossbands. Length about 35 mm. Introduced in Key West and Key Largo, Florida. Haiti, Cuba, and adjacent islands.

Sphaerodactylus notatus **Baird.** Reef gecko. Many small dark spots on a brownish-yellow background or with 3 broad, longitudinal dark stripes on the head and neck. Length about 28 mm. Introduced in southern Florida. Bahama Islands and Cuba.

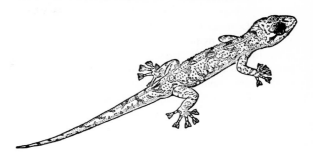

Fig. 4-43. Phyllodactylus tuberculosus. (**From Stebbins, *Amphibians and Reptiles of Western North America*, McGraw-Hill Book Company.**)

Genus *Coleonyx.* Banded geckos. Movable eyelids. Digits not widened; claws small and hidden between 2 lateral plates. Limbs small; skin with fine scales. 5 species, 2 in the United States.

KEY TO SPECIES OF *COLEONYX*

1. Scales bearing preanal pores separated by 1 or more scales in midventral line; usually 4 preanal pores
   C. brevis

   Scales bearing preanal pores continuous across midventral line; usually more than 4 preanal pores
   C. variegatus

Coleonyx brevis **Stejneger.** Texas banded gecko. Color pattern similar to that of the western banded gecko. Southwestern Texas, central New Mexico into Mexico.

Coleonyx variegatus **(Baird).** Banded gecko. Body of specimens smaller than 50 mm with 5 broad, brown crossbands. These bands break up in larger specimens to form spots (Fig. 4-44). Length to 75 mm. 5 subspecies. Desert areas of southern

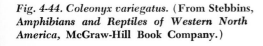

Fig. 4-44. Coleonyx variegatus. (From Stebbins, *Amphibians and Reptiles of Western North America*, McGraw-Hill Book Company.)

California, Utah, Nevada, Arizona, New Mexico, and adjacent parts of Mexico.

Family Iguanidae. Iguanas. Terrestrial, semiaquatic, arboreal. Teeth pleurodont; tongue thick and nonprotractile; femoral pores usually present. Vertebrae procoelous; clavicles not broadened; parietals united. Very varied body forms reflecting adaptations to many different conditions. Large, herbivorous, semimarine iguanas; small, arboreal anoles; large cursorial, crested lizards; grotesque horned lizards. Insectivorous and herbivorous. Largest family in the United States. About 44 genera in the world, 12 in the United States.

KEY TO GENERA OF IGUANIDAE

1. Scales on penultimate phalanx of digits widened (Fig. 4-45); no femoral pores   *Anolis*
   Scales on penultimate phalanx of digits not widened; femoral pores usually present, conspicuous in males   **2**
2. Head with bony spines or with a projecting ridge (Fig. 4-46)   *Phrynosoma*
   Head without bony spines or projecting ridge   **3**
3. Superciliary scales not overlapping; rostral divided, with a median seam reaching lip (Fig. 4-47)   *Sauromalus*
   Superciliary scales overlapping; rostral not divided, no median seam reaching lip   **4**
4. Seams between supralabials diagonal; median postmental present (Fig. 4-48)   **5**
   Seams between supralabials vertical; no median postmental present   **7**
5. No ear opening   *Holbrookia*
   A distinct ear opening   **6**
6. Toes with a lateral fringe of spinelike scales separated by tiny scales from ventral scales; ear

Fig. 4-45 (left). Digits of *Anolis*. (After Ditmars.) *Fig. 4-46* (right). Head of *Phrynosoma*. (From Stebbins, *Amphibians and Reptiles of Western North America*, McGraw-Hill Book Company.)

Fig. 4-47 (left). Front view of head scales of *Sauromalus*. (From Stebbins, *Amphibians and Reptiles of Western North America*, McGraw-Hill Book Company.) *Fig. 4-48* (center). Supralabials of *Holbrookia*. (From Stebbins, *Amphibians and Reptiles of Western North America*, McGraw-Hill Book Company.) *Fig. 4-49* (right). Head scales of *Uta*. (From Stebbins, *Amphibians and Reptiles of Western North America*, McGraw-Hill Book Company.)

Fig. 4-50 (left). Nasal region of *Urosaurus*. *Fig. 4-51* (right). Nasal region of *Uta*.

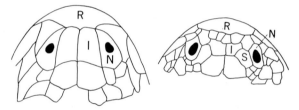

opening larger than interparietal scale; scales projecting over ear opening   *Uma*
Toe spines smaller and not separated from ventral scales by lines of smaller scales; ear opening smaller than or about same size as interparietal scale; no scale projecting over ear opening   *Callisaurus*

7. No transverse gular fold   **8**
   Transverse gular fold present   **9**
8. Central, dorsal, longitudinal row of large keeled scales; nasal in contact with rostral   *Leiocephalus*
   No central, dorsal, longitudinal row of large keeled scales; nasal separated from rostral   *Sceloporus*
9. A central, dorsal, longitudinal row of enlarged scales   *Dipsosaurus*
   No central, dorsal, longitudinal row of enlarged scales   **10**

10. Interparietal much larger than surrounding scales, larger than ear opening (Fig. 4-49)   **11**

    Interparietal as small as surrounding scales, smaller than ear opening   *Crotaphytus*

11. No scales between nasal and internasal (Fig. 4-50)   *Urosaurus*

    Scales (supranasals) separating nasals from internasals (Fig. 4-51)   *Uta*

Genus *Anolis* Daudin. Anoles. Generally small, slender-bodied; differentiated from all other lizards of the United States by the presence of a loose fold of skin at the throat which may be expanded to form a fanlike structure. Scales on underside of toes expanded to form a pad. Arboreal or terrestrial. Insectivorous; oviparous. More than 300 species and subspecies; 2 species in the United States.

KEY TO SPECIES OF *ANOLIS*

1. Tail without a dorsal ridge; ventral scales equal to or somewhat larger than dorsal scales   *A. carolinensis*

    Tail with a dorsal ridge; ventral scales much larger than dorsal scales   *A. sagrei*

Anolis carolinensis **Voigt.** Green anole. Individuals capable of rapid color change; color generally either green or brown. Gular fan pink when extended. Oviparous; many soft-shelled eggs laid from March to August. Atlantic and Gulf Coastal Plains; North Carolina to Florida and westward to central Texas and Oklahoma; northward in Mississippi Valley to Arkansas and Tennessee.

Anolis sagrei **Cocteau.** Brown anole. Individuals may change color rapidly; color green to dark brown. Oviparous. 3 subspecies. Island of Cuba to Central American coast, Bahamas, and Florida Keys. Introduced to Florida mainland.

Genus *Dipsosaurus* Hallowell. Desert iguanas. Large, rather stout lizards with an exceptionally long tail and strong legs. A middorsal row of enlarged scales distinguishes them from other lizards. Herbivorous. Oviparous. 5 species and subspecies; 1 species in the United States.

Dipsosaurus dorsalis **(Baird and Girard).** Desert iguana. Color grayish brown or with an irregular pattern of reddish brown. Laterally the reddish marks may tend to form a longitudinal pattern; the back may be marked with light, dark-bordered spots. Sand-dwelling. Length to 130 mm. Deserts of California, Nevada, Utah, Arizona, and adjacent Mexico.

Genus *Crotaphytus* Holbrook. Collared and leopard lizards. Large long-tailed lizards with a gular sac. Brightly marked with white, yellow, or orange. Often run on hind legs with tail erect. Oviparous. Insectivorous. 3 species, 5 subspecies.

KEY TO SPECIES OF *CROTAPHYTUS*

1. 1 or 2 black bands across neck (Fig. 4-52)   *C. collaris*

    No black bands across neck (Fig. 4-53)   **2**

2. Head width equal to or greater than distance between ear and nostril   *C. reticulatus*

    Head width less than distance between ear and nostril   *C. wislizeni*

Crotaphytus collaris **(Say).** Collared lizard. Conspicuous in its brilliant coloration and collar of 2 black bands. Length 115 mm; tail often twice the length of body (Fig. 4-52). 3 subspecies in the United States. Missouri westward to southern California, northward in the Great Basin to Idaho and Washington; northern Mexico.

Crotaphytus reticulatus **Baird.** Reticulate collared lizard. Form and size similar to *C. collaris* but differs in having a reticulated pattern on the back. Southern Texas and adjacent Mexico.

Fig. 4-52. Crotaphytus collaris. (From Stebbins, Amphibians and Reptiles of Western North America, McGraw-Hill Book Company.)

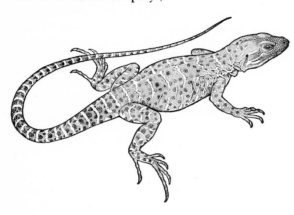

***Crotaphytus wislizeni* Baird and Girard.** Leopard lizard. Color light with many rounded or elongated dark-brown or black spots or color dark with light crossbars. Length to 120 mm; tail long (Fig. 4-53). 3 subspecies in the United States. Great Basin and adjacent arid areas, east to trans-Pecos Texas, southward into Mexico, and San Joaquin Valley in California.

Genus *Sauromalus* Dumeril. Chuckwallas. Large blunt-tailed lizards with very small scales; claws short and thickened; granular gular fold. Abundant, edible animals. 9 species, 1 in the United States. Southwestern California, southern Nevada, western Arizona southward into Mexico.

***Sauromalus obesus* Baird.** Chuckwalla. One of the largest lizards of the United States; length to about 210 mm. Body thick, broad. Color usually gray to black; sometimes marked with red splotches. Tail dark, light-banded, or uniformly light. Adult males usually with light tails. Herbivorous; oviparous. 3 subspecies in the United States. Colorado Desert region, southeastern California to Arizona, southern Nevada, and southwestern Utah.

Genus *Holbrookia* Girard. Earless lizards. Small slender flattened forms with small uniform scales; supralabials overlap, and seams are diagonal. Color generally light with distinct black slanting marks on sides (often absent or faint in females), ventral surface, or tail. No external ear opening. Chiefly insectivorous. Oviparous.

KEY TO SPECIES OF *HOLBROOKIA*

1. Tail flat; with broad, ventral, black bands
     H. texana
   Tail rounded; without black, ventral bands (but many have small black spots)   2
2. Dorsal central body scales distinctly keeled; maximum snout-vent length 61 mm   *H. propinqua*
   Dorsal central body scales not keeled; maximum snout-vent length often greater than 61 mm   3
3. No ventral black spots on tail   *H. maculata*
   Ventral black spots on tail   *H. lacerata*

***Holbrookia maculata* Girard.** Lesser earless lizard. Color light gray to brownish; middorsal light line; 2 short black diagonal bars on sides of belly in male; underside of tail unmarked. Length about 60 mm. 5 subspecies. South Dakota to Texas, Arizona, and into Mexico.

***Holbrookia lacerata* Cope.** Spot-tailed earless lizard. Body with many spots; dark spots surrounded by light color; usually several dark oval streaks at edge of abdomen. 2 subspecies. Central and southern Texas into Mexico.

***Holbrookia texana* (Troschel).** Greater earless lizard. Largest of the earless lizards. General color variable, usually gray to reddish; bluish tinge ventrolaterally; reddish spots irregularly distributed dorsally. Several broad black bands on ventral surface of tail. Length about 70 mm. 2 subspecies. Central and southeastern Arizona, southern New Mexico, and central Texas; southward into Mexico (Fig. 4-54).

***Holbrookia propinqua* Baird and Girard.** Keeled earless lizard. Color gray to brown, no blue color

Fig. 4-54. Holbrookia texana. (From Stebbins, Amphibians and Reptiles of Western North America, McGraw-Hill Book Company.)

above the diagonal black lines on belly sides in male. Length about 60 mm. Texas south of San Antonio; adjacent coastal plain in Mexico.

Genus *Uma* Baird. Fringe-toed lizards. Moderate-sized, flattened lizards. Adapted to living in sandy areas; snout wedgelike, lower jaw countersunk; ear flap with pointed scales, eyelids thick; body scales fine granular; fringe of long pointed scales on margins of toes. Chiefly insectivorous. Oviparous. 1 species; 3 subspecies in the United States.

***Uma notata* Baird.** Fringe-toed lizard. Color gray or white; irregular rows of dark lines sometimes forming a reticulum. Length to 120 mm. Colorado and Mojave Desert regions and the Coachella Valley, California.

Genus *Callisaurus* Blainville. Zebra-tailed lizards. Generally similar to *Holbrookia* and *Uma.* Ear opening present but without the fringed toes characteristic of *Uma.* Moderate-sized, long-tailed. Sand-dwelling. 1 species; 3 subspecies in the United States.

***Callisaurus draconoides* Blainville.** Zebra-tailed lizard. General color gray to brown with many white or yellow dots; 2 black chevronlike markings on sides of belly. The broad, transverse dark bands on tail readily identify the lizard in the field. Males with bright blue on belly sides. Length to 93 mm. West central Nevada southward through Colorado Desert region and Baja California.

Genus *Sceloporus* Wiegmann. Spiny lizards. Small to moderate-sized lizards; body generally short and broad. Scales usually strongly keeled, sometimes sharply pointed and bristling. Oviparous or viviparous. Largest genus in the United States; 15 species; 22 subspecies.

KEY TO SPECIES OF *SCELOPORUS*

1. Postfemoral pocket present (Fig. 4-55); males with pink belly patches   *S. variabilis*
   No postfemoral pocket; males without pink belly patches   **2**
2. Lateral scales small, granular, not overlapping
                                                     S. merriami
   Lateral scales not granular, overlapping   **3**
3. Lateral scales arranged in series parallel to body axis; 2 postrostrals   *S. scalaris*
   Lateral scales in oblique series ascending posteriorly (Fig. 4-56); 3–6 postrostrals   **4**

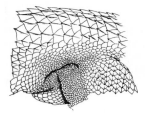
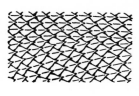

Fig. 4-55 (left). Postfemoral pocket of *Sceloporus variabilis.* (Smith, *Handbook of Lizards,* Comstock Publishing Associates.) *Fig. 4-56* (right). Lateral scales in oblique series. (From Stebbins, *Amphibians and Reptiles of Western North America,* McGraw-Hill Book Company.)

4. Supraoculars large, in a single row, usually 5 in number (Fig. 4-57); last 1 or 2 supraoculars not separated from the median head scales by a row of small scales as are the other supraoculars **5**
   Supraoculars smaller, in a single or double row; all supraoculars separated from median head scales by at least 1 complete row of small scales   **7**
5. No black shoulder patch or but poorly defined dark marks; dorsal scales with a shallow notch on either side of central spine   *S. orcutti*
   Black shoulder patch distinct and extending onto neck; dorsal scales deeply notched on either side of central spine   **6**
6. Ear scales generally 3 or 4; upper one longest (Fig. 4-58); forelegs distinctly crossbarred; no greatly enlarged occipital scale behind each parietal
                                                     S. clarki
   Ear scales generally 5–7, all elongate and pointed, median ones longest; forelegs not crossbarred; a

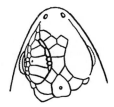

Fig. 4-57 (left). Supraoculars large, in a single row. (From Stebbins, *Amphibians and Reptiles of Western North America,* McGraw-Hill Book Company.) *Fig. 4-58* (right). Ear scales of *Sceloporus clarki.* (From Stebbins, *Amphibians and Reptiles of Western North America,* McGraw-Hill Book Company.)

much enlarged occipital scale posterior to each parietal **S. magister**

7. A broad black collar across back of neck, bordered by light lines **8**

   No distinct light-bordered black collar **10**

8. Supraoculars in a single series, undivided; adult males black above with a light spot on each scale **S. jarrovi**

   Supraoculars in 2 rows; adult males not black with a light spot on each scale **9**

9. Tail distinctly banded toward tip; supraoculars generally in 2 regular rows; median head scales (especially the frontal) usually divided irregularly **S. poinsetti**

   Tail indistinctly banded; supraoculars large, irregular; median head scales not usually divided **S. cyanogenys**

10. Scales on rear surface of thigh very small, granular, not overlapping **11**

    Scales on rear surface of thigh larger, overlapping, keeled **12**

11. Lateral neck scales much smaller than dorsal neck scales; lateral row of dorsal neck scales enlarged, strongly keeled, spinous; gular region bluish in males, not reticulated **S. grammicus**

    Lateral neck scales not much smaller than dorsal neck scales; lateral row of dorsal neck scales not enlarged, keeled, spinous; gular region often with bluish areas (spots) but usually reticulated **S. graciosus**

12. Posterior surface of thigh nearly immaculate; distinct dorsolateral light lines; dorsal bars nearly or quite absent in adult males **S. olivaceus**

    Posterior surface of thigh with a broad longitudinal dark line or irregular black marks; no dorsolateral light lines but crossbars on back visible in adult males **13**

13. Scales on posterior surface of thigh abruptly differentiated from dorsal scales of same member, median posterior scales not distinctly larger than adjacent scales **S. occidentalis**

    Scales on posterior surface of thigh gradually merging with larger dorsal scales of same member, at least median posterior scales distinctly larger than adjacent scales **14**

14. Lateral dark stripe distinct; dorsal pattern indistinct **S. woodi**

    No distinct lateral dark stripe or 2 dark lateral stripes **S. undulatus**

Sceloporus variabilis Wiegmann. Rose-bellied lizard. Color gray to brown; longitudinal lateral light line; faint middorsal line; dark bars between lateral lines and middorsal line. Oviparous.

Length to 53 mm. 9 subspecies, 1 in the United States. Southern Texas; southward through Mexico.

Sceloporus merriami Stejneger. Canyon lizard. Color gray; 10–14 dark spots on each side of middorsal line. Oviparous. Length to 58 mm. 2 subspecies. Trans-Pecos Texas, eastward to Devils River and adjacent Mexico.

Sceloporus scalaris Wiegmann. Bunch grass lizard. Color uniform brownish or a pattern of semicircular blotches on a background of brown; 2 narrow longitudinal lateral light lines usually present (Fig. 4-59). Probably viviparous. Length to 62 mm. 1 subspecies in the United States. Southern Arizona; northern and central Mexico.

Sceloporus grammicus Wiegmann. Mesquite lizard. Color gray; 4–6 narrow, dark, curved transverse lines on back. Large individuals may be almost uniform in color. Viviparous. Mexico northward to southern Texas.

Sceloporus jarrovi Cope. Yarrow's spiny lizard. Color black with a light spot on each scale; black collar bordered with light color. Belly blue, black, or gray (Fig. 4-60). Viviparous. Carnivorous. Length to 90 mm. Southern Arizona and New Mexico; southward to central Veracruz.

Sceloporus poinsetti Baird and Girard. Crevice spiny lizard. Color gray to reddish; wide black col-

Fig. 4-59. Sceloporus scalaris. (From Stebbins, Amphibians and Reptiles of Western North America, McGraw-Hill Book Company.)

A

B

lar on neck bordered by light zones. Tail banded with black. Throat and sides of belly of males blue (Fig. 4-61). Viviparous. Carnivorous. Length to 125 mm. Southern Mexico, central and western Texas through western Nuevo León and Coahuila to Durango in Mexico.

Sceloporus cyanogenys Cope. Blue spring lizard. Largest *Sceloporus*, length to 141 mm. Color bluish; males with more intense color than females; distinct black neck collar. Throats of males light blue; females gray. Viviparous. Southern Texas southward in Mexico to central Tamaulipas and Nuevo León (Fig. 4-62).

Sceloporus olivaceous Smith. Texas spiny lizard. Color gray to brown; a longitudinal light stripe on each side; several dark transverse bars. Males with blue belly patches. Oviparous. Length to 100 mm. Southern Oklahoma through Texas prairie region to the lower Rio Grande; in Mexico to southeastern Coahuila and southern Tamaulipas.

Sceloporus clarki Baird and Girard. Clark's spiny lizard. Color gray to brown; narrow dark dorsal crossbands; legs with dark bands. Large males may retain juvenile pattern or become uniform in color; blue throats and blue belly patches present (Fig. 4-63). Insectivorous. Oviparous. Length to 130 mm. Southern Arizona and southwestern New Mexico into northwestern Mexico.

Sceloporus orcutti Stejneger. Granite spiny lizard. Brightly colored; somewhat iridescent. Color dusky to black; males with blue color on each dorsal scale, light-yellow spot on many of head scales, and blue ventral surface; light males have a broad purple stripe down middle of back. Females banded, ventral surfaces unmarked but sometimes blue-tinged. Length to 110 mm. Southern California; southward in Baja California.

Fig. 4-60. Sceloporus jarrovi. (From Stebbins, Amphibians and Reptiles of Western North America, McGraw-Hill Book Company.)

Fig. 4-62. Sceloporus cyanogenys. (Photograph by Isabelle Hunt Conant.)

Fig. 4-61. Sceloporus poinsetti. (From Stebbins, Amphibians and Reptiles of Western North America, McGraw-Hill Book Company.)

Fig. 4-63. Sceloporus clarki. (From Stebbins, Amphibians and Reptiles of Western North America, McGraw-Hill Book Company.)

Fig. 4-64. *Sceloporus magister.* (From Stebbins, *Amphibians and Reptiles of Western North America,* McGraw-Hill Book Company.)

Fig. 4-65. *Sceloporus undulatus.* (From Stebbins, *Amphibians and Reptiles of Western North America,* McGraw-Hill Book Company.)

Fig. 4-66. *Sceloporus graciosus.* (From Stebbins, *Amphibians and Reptiles of Western North America,* McGraw-Hill Book Company.)

Sceloporus magister Hallowell. Desert spiny lizard. Color brown with dark crossbands; sides darker than back. Narrow black collar in the male. Belly sides and throat may be dark blue in male; white or cream in the female. Adult males with varied patterns; longitudinal stripes, longitudinal series of square or rectangular blotches, 6–7 dark crossbars, or with uniform dorsal color. 4 subspecies in the United States. Length to 140 mm. Southern Nevada, Utah, and California eastward to trans-Pecos Texas southward into northwestern Mexico (Fig. 4-64).

Sceloporus undulatus (Bosc, in Latreille). Eastern fence lizard. Color gray to brown with curved dark transverse marks or longitudinal dark stripes on the back. Male with a blue throat and sides of belly (Fig. 4-65). Oviparous. 7 subspecies in the United States. Length to 84 mm. Eastern United States, west to the Great Basin, southward into northern Mexico.

Sceloporus occidentalis Baird and Girard. Western fence lizard. Color grayish, brownish or greenish; dark transverse bands on back. Male with a blue throat and sides of belly. 6 subspecies in the United States. Oviparous. About 95 mm in length. Pacific coastal region and the Great Basin.

Sceloporus woodi Stejneger. Florida scrub lizard. Color gray to brown with a dark lateral band from head to tail. Back uniform or with 2 series of crossbars. Belly light with a light-blue lateral area. Peninsular Florida.

Sceloporus graciosus Baird and Girard. Sagebrush lizard. Color gray-brown or olive; yellow, orange, or reddish sides; a black spot is present just anterior to each foreleg. Throat light blue with white flecks; belly with light-blue lateral areas (Fig. 4-66). Oviparous. Length to 65 mm. 3 subspecies in the United States. Great Basin and adjacent mountain areas, southwestward into southern California.

Genus Urosaurus Hallowell. Tree lizards. External features similar to members of genus *Uta* except for absence of supranasals. Sternal plate long and relatively narrow; posterior margin tapering almost to a point. Arboreal. Oviparous. 1 species in the United States.

Urosaurus ornatus Baird and Girard. Tree lizard. Color gray, bluish to brown; vaguely defined, transverse bands of dark color on back (Fig. 4-67).

Length to 65 mm. 7 subspecies in the United States. Southwestern Wyoming to Mexico and from the lower Colorado River eastward to central Texas.

Genus *Uta* Baird and Girard. Side-blotched lizards. External features similar to members of the genus *Urosaurus* except for presence of supranasals. Sternal plate relatively short and broad; posterior margin truncate. Arboreal and terrestrial. Oviparous. 4 species in the United States.

KEY TO SPECIES OF *UTA*

1. A dark collar bordered with white. Dorsal scales granular and not much larger than lateral scales
   U. mearnsi
   No dark collar. Dorsal scales keeled and larger than lateral scales   **2**
2. A dorsolateral line or fold; no small blue blotch behind axilla   **3**
   No dorsolateral line or fold; usually a small blue or black blotch behind axilla (light in females)
   U. stansburiana
3. Tail at least twice the length of head and body combined; dorsals becoming abruptly smaller on either side of central band of enlarged scales   *U. graciosa*
   Tail not so long; dorsals becoming abruptly smaller or not; all dorsal scales small   *U. microscutata*

Uta stansburiana **Baird and Girard.** Side-blotched lizard. Small rock-dwelling lizards. Color brown with longitudinal stripes or many small spots. Distinguished by a large blue-black spot behind the axilla. Brightly colored; chevrons of black, brown, and white. 3 subspecies in the United States. Great Basin and Colorado Desert regions, westward into southern California, eastward through Texas, and southward through Baja California.

Uta microscutata **Van Denburgh.** Small-scaled lizard. Uniform dark gray or with several black, transverse bands on either side. Gular region light blue in the male, white in the female. Length to 50 mm. San Diego County, California, southward into central Baja California.

Uta graciosa **(Hallowell).** Long-tailed brush lizard. Color pale gray to brown; several narrow dark crossbands. Narrow dark lines on head. Tail as much as $2\frac{1}{2}$ times length of head and body combined. Length to 55 mm. Southern California,

Fig. 4-67. Urosaurus ornatus. (From Stebbins, *Amphibians and Reptiles of Western North America*, McGraw-Hill Book Company.)

Arizona, and extreme southern Nevada; adjacent Baja California and Sonora.

Uta mearnsi **Stejneger.** Banded rock lizard. Color olive gray; several dark curved crossbands on the back. Distinct dark tail bands. Many small light dots on back. Length to 90 mm. San Diego County, California, south to central Baja California.

Genus *Phrynosoma* Wiegmann. Horned lizards. Unique, flattened horned lizards totally unlike any other reptiles (except general similarity to genus *Moloch* of Australia). Tail very short; head usually with elongate sharp horns. Deserts of Southwest and Mexico. Insectivorous.

KEY TO SPECIES OF *PHRYNOSOMA*

1. 4 large spines on posterior part of head continuous with temporal spines to form a continuous crown (Fig. 4-68)   *P. solare*
   2 occipital spines, or occipital spines absent or reduced and not continuous with temporal spines   **2**

Fig. 4-68 (left). Head of *Phrynosoma solare*. (From Stebbins, *Amphibians and Reptiles of Western North America*, McGraw-Hill Book Company.) *Fig. 4-69* (right). Head of *Phrynosoma cornutum*.

1 2 3 4

2. More than 2 longitudinal series of enlarged scales on each side of throat, between the 2 series of large chin shields **P. coronatum**

2, 1, or no longitudinal series of enlarged scales on each side of throat between chin shields **3**

3. No enlarged scales at side of abdomen, chin shields in contact with infralabials throughout **P. modestum**

At least 1 row of enlarged scales at side of abdomen; chin shields separated from infralabials by at least 1 row of scales throughout most of length of series **4**

4. Posterior chin shields much larger than infralabials **5**

Posterior chin shields smaller than posterior infralabials or scales continuous with them and decreasing in size posteriorly **P. douglassi**

5. Belly white, unspotted; dark streak down back; horns very long **P. m'calli**

Belly usually spotted; no dark streak down back; horns short **6**

6. Ventral scales smooth; femoral pore series nearly meeting midventrally; horns nearly horizontal to head axis; 1 row of large scales at edge of abdomen **P. platyrhinos**

Ventral scales keeled; femoral pore series widely separated midventrally; horns at 45° angle to head axis (Fig. 4-69); 2 rows of enlarged scales at edge of abdomen **P. cornutum**

Phrynosoma coronatum Blainville. Coast horned lizard. Color yellow, reddish brown, or gray; dark patch on each side of neck; series of dark spots on each side of middorsal line. Belly light with dark mottling. Length to 100 mm. 2 subspecies in the United States. Western California and peninsula of Baja California.

Phrynosoma cornutum (Harlan). Texas horned lizard. Color tan to reddish brown; series of large oval light-bordered dark spots on each side of a light middorsal line. Length to 110 mm. Western Arkansas and Missouri, Kansas, southeastern Colorado through Oklahoma and Texas to southeastern Arizona; adjacent Mexico.

Phrynosoma douglassi (Bell). Short-horned lizard. Color gray or brown; series of dark blotches along each side of middorsal line. Tail banded above. Length to 93 mm. 5 subspecies in the United States. Western United States from the Canadian border into Mexico.

Phrynosoma m'calli (Hallowell). Flat-tailed horned lizard. Color white to tan; a narrow middorsal line; a brown blotch on each shoulder. 2

horns at least twice length of other horns. Length to 80 mm. Colorado Desert in southeastern California, southwestern Arizona, and adjacent Baja California and Sonora.

Phrynosoma modestum Girard. Round-tailed horned lizard. Large black blotch on each side of the neck and above groin. The absence of lateral abdominal fringe scales and the rounded, slender tail abruptly broadened at the base distinguish this species from others in the United States. Length to 45 mm. Southeastern Arizona to western Texas; northern New Mexico into Mexico.

Phrynosoma platyrhinos Girard. Desert horned lizard. Color gray, brown, or blackish; pattern ordinarily conspicuous. Large dark blotch on each side of neck; a light middorsal line sometimes present; 4 or more dark blotches on back. Tail banded with black or brown. Belly light with small dark spots. Length to 94 mm. 2 subspecies in the United States. Great Basin and Sonoran Desert regions.

Phrynosoma solare Gray. Regal horned lizard. Only *Phrynosoma* with 4 occipital horns. Length 100 mm. Arizona, Baja California, and Sonora.

Genus Leiocephalus. Curl-tailed lizards. General resemblance to *Sceloporus*. 1 species in the United States.

Leiocephalus carinatus Gray. Bahama curl-tailed lizard. Color gray to brown; dark streaks and spots on the back. Tail banded with dark. Length 90 mm. Cuba and Bahama Islands; introduced in Miami area, Florida.

Family Anguidae. Lateral fold lizards. Terrestrial, arboreal, burrowing; some elongate, snakelike. Tongue extensible, bifid. Femoral pores absent. Upper temporal fossa of skull roofed with bone. A lateral fold extends from the base of the neck to the insertion of the hind limb. Osteoderms often present. Oviparous or viviparous. 11 genera in the world, 2 in the United States.

KEY TO GENERA OF ANGUIDAE

1. Forelegs and hind legs present **Gerrhonotus**

No legs **Ophisaurus**

Genus Ophisaurus Daudin. Glass lizards. Snakelike but retain internal vestiges of the girdles or short modified legs. 10 species, 3 in the United States.

KEY TO SPECIES OF *OPHISAURUS*

1. 1 or 2 upper labials of each side in contact with orbit; frontonasal usually double; scales along lateral fold are 97 or less   *O. compressus*
Upper labials separated from orbit by scales; frontonasal single; scales along lateral fold are 94 or more   **2**
2. White markings (absent only in very young) occurring on posterior corners of scales, never primarily in middle of scales; no distinct middorsal stripe present; no dark stripes present on belly
    O. ventralis
White markings occurring in middle of scales, often forming smooth stripes; a distinct middorsal stripe generally present in adults, always in young; dark stripes usually present on scale rows 1 and 2 of belly   *O. attenuatus*

Ophisaurus ventralis (**Linnaeus**). Eastern glass lizard. Virginia to Florida and westward on the Coastal Plain to eastern Louisiana; northward to Oklahoma and southern Illinois.
Ophisaurus compressus **Cope.** Island glass lizard. Coastal areas and islands off South Carolina, Georgia, and Florida; peninsular Florida.
Ophisaurus attenuatus **Baird.** Slender glass lizard. 2 subspecies in the United States. Mississippi Basin, Atlantic and Gulf Coastal Plains.
Genus *Gerrhonotus* **Wiegmann.** Alligator lizards. Elongate body, legs feeble. Keeled scales form ridges extending length of the dorsal surface. Tail about twice body length (frequently lost). Unusually pugnacious lizards. Oviparous or viviparous.

KEY TO SPECIES OF *GERRHONOTUS*

1. A median scale immediately posterior to rostral; nasal separated from rostral by a single scale (Fig. 4-70); subocular not in contact with lowest temporal. No difference in color of back and sides
    G. liocephalus
No median scale immediately posterior to rostral; nasal scale not usually separated from rostral; subocular in contact with lowest temporal. Color of back lighter than that of sides   **2**

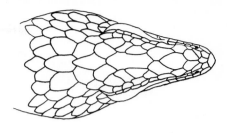

Fig. 4-70. Head of *Gerrhonotus liocephalus.*

2. Dorsal longitudinal scale rows 16–17; ventral scales of tail keeled in adults; no crossbands in adult
    G. coeruleus
Dorsal longitudinal scale rows 14; ventral scales of tail not keeled in adults; crossbands in adults   **3**
3. Body crossbands (behind occiput to anterior border of thighs) 7–10; dorsal transverse scale rows 41–44
    G. panamintinus
Body crossbands 8–13; dorsal transverse scale rows 44–50   **4**
4. Body crossbands 8–11; transverse dorsal scale rows 60–48; 6–8 dorsal scales keeled; young with crossbands   *G. kingi*
Body crossbands 9–13; transverse dorsal scale rows 50–39; all dorsal scales keeled; young with longitudinal stripes   *G. multicarinatus*

Gerrhonotus coeruleus **Wiegmann.** Northern alligator lizard. Color greenish, central group of brown blotches which may form crossbands; dark bands on legs and tail (Fig. 4-71). Length to 110 mm. 5 subspecies in the United States. British Columbia and Alberta to California and Montana.
Gerrhonotus kingi (**Gray**). Arizona alligator lizard. Color gray to brown, 9–12 crossbands of brown on head and body. Dark scales may have whitish spots. Length to 100 mm. Central and southeastern Arizona and southwestern New Mexico, adjacent part of Sonora and Chihuahua.
Gerrhonotus multicarinatus (**Blainville**). Southern alligator lizard. Color gray, reddish to greenish gray; several brown crossbands are much darker laterally than on the back. Each crossband sometimes has a broad red anterior border. Length to 137 mm. 3 subspecies in the United States. Southern Washington southward into northern Baja California.
Gerrhonotus liocephalus **Wiegmann.** Texas alligator lizard. Color yellow to brown; darker cross-

Fig. 4-71. Gerrhonotus coeruleus. (From Stebbins, *Amphibians and Reptiles of Western North America,* McGraw-Hill Book Company.)

Fig. 4-72. Anniella pulchra. (From Stebbins, *Amphibians and Reptiles of Western North America,* McGraw-Hill Book Company.)

bands irregular but distinct. Tail banded, but legs and head uniform in color. Length to 200 mm. Central and western Texas to southern Mexico.

Gerrhonotus panamintinus Stebbins. Panamint alligator lizard. Length to 122 mm. Panamint, Nelson, and Inyo Mountains, California.

Family Helodermatidae. Beaded lizards. Terrestrial, desert forms. Teeth grooved, venom glands present. Tongue extensible, bifid. Body short, stout; tail short, blunt. Osteoderms present, scales tuberclelike. Only poisonous lizards in the United States. Length to 400 mm. 1 genus, 2 species.

Genus *Heloderma* Wiegmann. Beaded lizards. 1 species in the United States.

Heloderma suspectum Cope. Gila monster. Color orange, pink, or yellow with a reticulate pattern of brown or black. Legs and sides of head often black. Carnivorous. Utah and Nevada to southwestern New Mexico and adjacent Mexico.

Family Anniellidae. Shovel-snouted legless lizard. Fossorial. Scales smooth, shiny. Body elongate, snakelike; no legs but with a vestigial pelvic girdle. Short, blunt tail. No ear opening; teeth relatively large but few; eyes almost hidden. Osteoderms present. 1 genus.

Genus *Anniella* Gray. Legless lizards.

Anniella pulchra Gray. California legless lizard. Color generally light cream to black. A longitudinal, middorsal black line, 2 or 3 similar lines laterally (Fig. 4-72). Length to 150 mm. Combination of no ear opening and the presence of movable eyelids will always identify this lizard within its geographic range. 2 subspecies. Coastal regions from San Francisco Bay area to northwestern Baja California; foothills of southern Sierra.

Family Xantusidae. Night lizards. Terrestrial, nocturnal. Small lizards with vertical pupils; large protruding eyes; no movable eyelids. Tongue short, broad, with many diagonal folds. Scales granular, sometimes tuberclelike. Ventral scales rectangular, platelike. Length to 60 mm. Most species are viviparous. Insectivorous. 4 genera, 2 in the United States.

KEY TO GENERA OF XANTUSIDAE

1. Ventral scales in 12–14 longitudinal rows   *Xantusia*
   Ventral scales in 16 longitudinal rows   *Klauberina*

Genus *Xantusia* Baird. Night lizards. Secretive, nocturnal animals found usually under branches of yuccas or in crevices of rocks.

KEY TO SPECIES OF *XANTUSIA*

1. Ventral scales usually in 12 rows; back marked with small dark flecks or spots   **2**
   Ventral scales usually in 14 rows; back marked with large blackish blotches   *X. henshawi*

Fig. 4-73. Xantusia arizonae. (From Stebbins, *Amphibians and Reptiles of Western North America,* McGraw-Hill Book Company.)

2. Granular scales across back at mid-body 43–50; lamellae under 4th toe 25–28   *X. arizonae*
   Granular scales across back at mid-body 33–40; lamellae under 4th toe 18–21   *X. vigilis*

Xantusia arizonae **Klauber.** Arizona night lizard. Marked with small black dots or bars forming 7–8 longitudinal dark rows on a light background (Fig. 4-73). Belly immaculate. Length to 95 mm. Arizona.

Xantusia henshawi **Stejneger.** Granite night lizard. Many large ovoid black or brown blotches on back and sides; background color tan or yellow. Belly immaculate. Length to 70 mm. Riverside County through San Diego County into the San Pedro Mártir in Baja California.

Xantusia vigilis **Baird.** Desert night lizard. Uniform cream or brown or with a pattern of many small brown or black spots and bars; these may form a reticulum or series of longitudinal lines. Belly immaculate. Length to 43 mm. 4 subspecies, 2 in the United States. San Bonito and Inyo Counties, California, south throughout Baja California; southern Nevada, Utah, and northwestern Arizona.

Genus *Klauberina* Savage. Island night lizard.

Klauberina riversiana **Cope.** Island night lizard. Dorsal pattern of brown blotches or a reticulum of dark brown or black on a gray background. Belly gray. Length to 70 mm. San Nicolas, Santa Barbara, and San Clemente Islands off the coast of California.

Family Teidae. Terrestrial, fossorial, and arboreal forms. Teeth varied in form; modified acrodont and pleurodont; eyelids present; femoral pores present. Many body forms reflecting habits; some with feeble limbs and elongate bodies, others with short stout bodies and legs. Head shields usually large. 40 genera, 1 in the United States.

Genus *Cnemidophorus* Wagler. Whiptails and racerunners. Elongate, stout-legged lizards. Granular dorsal scales and large transverse belly scales in 8–12 longitudinal rows; tail length twice or more snout-vent length and with keeled scales in rings. About 19 species in North and South America. Insectivorous. Oviparous.

KEY TO SPECIES OF *CNEMIDOPHORUS*

1. 1 frontoparietal scale   *C. hyperthrus*
   2 frontoparietal scales   **2**
2. Numerous dark bars on back between longitudinal stripes which may be poorly defined; arrangement of bars sometimes furnishing appearance of cross bands

No black bars on back; usually 6 or 7 longitudinal light dorsal stripes   **4**
3. Scales in front of gular fold relatively small; usually grading into granular scales of fold   *C. tigris*
   Scales in front of gular fold conspicuously enlarged and not grading into granular scales of fold
        C. tessellatus
4. Circumorbital scales usually separating 3d and often 2d supraocular from frontal (Fig. 4-74)
        C. perplexus
   Circumorbital scales not separating supraocular from frontals   **5**
5. Scales on posterior surface of forearm enlarged (Fig. 4-75); light spots usually present in dark field between light stripes; adult size to about 130 mm
        6

Fig. 4-74 (left). Head scalation of *Cnemidophorus perplexus*. (From Stebbins, *Amphibians and Reptiles of Western North America*, McGraw-Hill Book Company.) *Fig. 4-75* (right). Forearm of *Cnemidophorus septemvittatus*. (From Stebbins, *Amphibians and Reptiles of Western North America*, McGraw-Hill Book Company.)

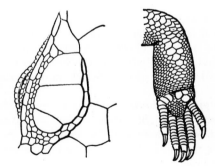

Scales on posterior surface of forearm not or very slightly enlarged; no light spots in dark fields between stripes; adult size to about 85 mm   **9**
6. Dorsal granular scales (counted around mid-body) 85–101; longitudinal light stripes often absent in adults   *C. burti*
   Dorsal granular scales 55–98; longitudinal light stripes present in adults   **7**
7. Dorsal granular scales 62–86; scales between paravertebral stripes 2–8; chin of male white
        C. exsanguis
   Dorsal granular scales 84–98; scales between paravertebral stripes 4–18; chin of male rarely white
        8

8. Dorsal granular scales 84–96; scales between paravertebral stripes 8–18; chin of male pink or red
    C. gularis

Dorsal granular scales 77–98; scales between paravertebral stripes 4–17; chin of male pale blue
with black spots, rarely white **C. septemvittatus**

9. Dorsal granular scales 68–110; scales between para-
vertebrals 8–22; chin of male white
    C. sexlineatus

Dorsal granular scales 55–85; scales between para-
vertebrals 5–11; chin of male pale blue   **10**

10. Dorsal granular scales 55–78; scales between para-
vertebrals 7–11; maximum size 70 mm
    C. inornatus

Dorsal granular scales 63–85; scales between para-
vertebrals 5–10; maximum size 85 mm *C. velox*

Cnemidophorus tessellatus (Say). Checkered
whiptail. Back marked with many dark bars
grouped so that intervening light areas form several
distinct longitudinal lines. Pattern very variable.
Length to 100 mm. Western Texas, New Mexico
northward to Colorado; probably southward to
Coahuila.

Cnemidophorus tigris Baird and Girard. Western whiptail. Color brownish; many black crossbars
or spots arranged in transverse rows; 4 or more
longitudinal stripes. Ventral surface white or gray
with black spots or checks. Length to 100 mm.
6 subspecies in the United States. Southern Washington and Idaho southeastward into northern
Mexico.

Cnemidophorus burti Taylor. Arizona whiptail. Pattern of 6 light stripes in juvenile usually
replaced in large individuals by light spots. Some
individuals retain evidences of stripes and develop
a reddish-brown or reddish-orange color on the
dorsal surface. Length to 130 mm. 2 subspecies in
the United States. Southern Arizona.

Cnemidophorus exsanguis Lowe. Whiptail.
Pattern of 6 light stripes persistent in large adults;
light spots in the dark fields commonly overlap the
light stripes. Length to 95 mm. Trans-Pecos, Texas,
westward through southern New Mexico to central
Arizona; southward into Sonora and Chihuahua.

Cnemidophorus gularis Baird and Girard.
Eastern spotted whiptail. Pattern of 6 light stripes
persistent in large adults; vertebral stripes usually
less well defined than others and may fuse. Light
spots in dark fields abundant. Males with pink or
red chin and blue-black chest and abdomen.

Length to 90 mm. Southern Oklahoma and Texas;
south to San Luís Potosí and Veracruz.

Cnemidophorus inornatus Baird. Little striped
whiptail. Pattern of 6 to 8 stripes persistent in some
adults; many adults without stripes; no spots or
other light marking in dark fields. Males with pale-
blue ventral surfaces; females whiter. Bright-blue
tails of hatchlings become gray-blue in large adults.
Length to 70 mm. West Texas through southern
New Mexico into southeastern Arizona; south into
Chihuahua Desert. 2 subspecies; 1 in the United
States.

Cnemidophorus perplexus Baird and Girard.
Devils River whiptail. Pattern of 7 light stripes
retained in adults; vertebral stripe irregular; light
spots in dark fields obscure. Bright-blue tail of
hatchling changes to blue-gray in adults. Length
to 86 mm. New Mexico and vicinity of El Paso,
Texas.

Cnemidophorus septemvittatus Cope. Big
Bend whiptail. Pattern of 6 or 7 stripes retained in
adults on anterior body; light spots in dark fields
abundant in adults. Ventral surface white or pale
blue; some individuals with black spots on chin
and chest. Length to 105 mm. Big Bend region of
Texas.

Cnemidophorus sexlineatus (Linnaeus). 6-
lined racerunner. Pattern of 6 distinct, longitudinal
lines on body; upper lateral line yellowish; lower
lines bluish or greenish; broad central brown stripe.
No spots in dark areas. Males with belly and throat
light blue (Fig. 4-76). Length to 77 mm. Maryland to Florida; westward in south to eastern Texas
and northward to Wisconsin and Minnesota, Indiana, and Illinois.

Cnemidophorus velox Springer. Plateau whiptail. Pattern of 6 or 7 stripes; 7th stripe, if present,
less distinct than others; no light spots in dark
fields. Ventral surface white or tinged with blue.
Length to 85 mm. Utah, Colorado, Arizona, New
Mexico.

Cnemidophorus hyperythrus Cope. Orange-
throated whiptail. Color gray to black; 6 longitudinal light lines; middorsal lines less well defined
than lateral lines; orange or yellow on ventral
surface. Length to 70 mm. Southern California
southward into Baja California.

Family Scincidae. Terrestrial, arboreal, fossorial. Extensible tongue with papillae and slightly

notched tip; eyelids usually well developed; femoral pores absent. Characterized by presence of flat, smooth, overlapping scales; osteoderms present. Oviparous or viviparous. 30 genera, 3 in the United States.

KEY TO GENERA OF SCINCIDAE

1. Legs very short with not more than 2 digits   *Neoseps*
   Legs longer, each with 5 digits   **2**
2. No supranasals; lower eyelid with a translucent disk
   Lygosoma
   Supranasals present; eyelids scaly   *Eumeces*

Genus *Lygosoma* Gray.   More than 150 species, only 1 in the United States.

***Lygosoma laterale* Say.**   Ground skink. Small, elongate, length 50 mm. Color chocolate brown with a wide middorsal light stripe; much yellow present in some individuals. Southeastern Kansas southwest to Pecos River in Texas, northeast to southern New Jersey, southward through all the eastern United States.

Genus *Eumeces*.   Striped skinks. Terrestrial, arboreal, fossorial. Elongate, shiny-scaled; scales near uniform in size. Teeth on ptrygoids and often on the palatines. Oviparous or viviparous. About 59 species, 15 in the United States.

KEY TO SPECIES OF *EUMECES*

1. Lateral scale rows at an angle to dorsal rows (Fig. 4-77); young black with white or orange spots on head; adults light with black border on each scale   *E. obsoletus*
   Lateral scale rows parallel to dorsal rows; young not black; adults not light with black border on each scale   **2**
2. Dorsolateral light stripes if present, including 2d scale row (count from middorsal line) on at least anterior part of body; no median light line on head or body   **3**
   Dorsolateral light stripes not present, or if present, not including 2d scale row at a point above forelimb; median light line present or not   **5**
3. Postnasals absent; 3 supraoculars; single large scale between last labial and parietal   *E. egregius*
   Postnasal present; 4 supraoculars; 2 large scales between last labial and parietal   **4**
4. Supralabials usually 8; usually 3 enlarged nuchals
   E. gilberti

Fig. 4-76. Cnemidophorus sexlineatus. (Photograph by Isabelle Hunt Conant.)

Supralabials usually 7; usually 4 enlarged nuchals
   E. skiltonianus
5. A dorsolateral light line extending length of body on 3d scale row only   *E. multivirgatus*
   Not as described   **6**
6. No light longitudinal lines on head or body   **7**
   Light longitudinal line or lines present on head or body   **10**
7. Postnasal present; median subcaudals sometimes widened   **8**
   Postnasal absent; median subcaudals not greatly widened   *E. tetragrammus*
8. Median subcaudals not or slightly widened
   E. inexpectatus
   Median subcaudals distinctly widened   **9**
9. No postlabials or 1 or 2 of small size; usually 8 upper labials, 5th below the eye; intercalary scales on 4th row of forefoot reaching on to next-to-last phalanx. Maximum snout-vent length 140 mm   *E. laticeps*
   2 postlabials; usually 7 upper labials, 4th below eye; lateral intercalary scales not extending onto last phalanx. Maximum snout-vent length 80 mm
   E. fasciatus
10. Postnasal present   **11**
    Postnasal absent   **12**

Fig. 4-77. Eumeces obsoletus. (From Stebbins, Amphibians and Reptiles of Western North America, McGraw-Hill Book Company.)

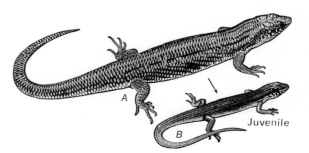

Juvenile

11. Subcaudal scales not wider than adjacent rows; light stripes, if present, 1 scale or less in width and never including 3d scale row

E. inexpectatus

Subcaudal scales distinctly wider than adjacent rows; light stripe, if present, more than 1 scale in width and often including 3d scale row   **9**

12. 2 light lines on top of head   **13**

No light lines on top of head   **15**

13. Parietals enclosing interparietal posteriorly; an elongate postlabial   *E. callicephalus*

Parietals not enclosing interparietal posteriorly; several postlabials   **14**

14. Lateral light lines extending length of body; lines on top of head fading posteriorly, not uniting

E. tetragrammus

Lateral light lines disappearing on posterior half of body; lines on top of head usually visible to point of union   *E. brevilineatus*

15. 1 postmental; limbs overlapping when adpressed (may not in large females)   *E. anthracinus*

2 postmentals; limbs of adults not overlapping when adpressed   *E. septentrionalis*

Eumeces fasciatus (**Linnaeus**).   Five-lined skink. Young animals with 5 longitudinal light lines on a blue-black background; tail blue. Older animals lose the stripes and become brown to reddish in color, and the tail loses the blue color. Length to 80 mm. Eastern United States westward into Texas, Oklahoma, Kansas, Nebraska, and South Dakota.

Eumeces laticeps **Schneider.**   Broad-headed skink. Young as in *E. fasciatus.* Large adult males are reddish brown with an orange or red head; large females have some evidence of the stripes (Fig. 4-78). Southeastern United States; southern Pennsylvania and Maryland into Kansas, Oklahoma, and Texas. Length to 130 mm.

Eumeces inexpectatus **Taylor.**   Southeastern five-lined skink. Color as in *E. laticeps* and *E. fasciatus,* but differentiated by the fact that the central row of subcaudals is not widened. Length to 90 mm. Virginia to Florida and along the Gulf Coast to Louisiana.

Eumeces brevilineatus **Cope.**   Short-lined skink. Greenish-gray to brown-colored; body without distinct markings, but a light line from jaw to shoulder; another light line passing through the tympanum. Length to 66 mm. Central and western Texas, Nuevo León.

Eumeces callicephalus **Bocourt.**   Mountain skink. Color gray, tan, or brown; a light dorsolateral stripe from head to base of tail on 4th scale row (from middorsal line); central back light in color; poorly defined light lines on head; brownish lateral band. Length to 65 mm. Southeastern Arizona, northern Mexico.

Eumeces tetragrammus (**Baird**).   Four-lined skink. Color gray dorsally, blue tinge on sides; lateral light lines from head to the base of the tail; another light line from jaws to the tail; no dark color below the lateral stripe; head reddish, longitudinal lines absent in large animals. Length to 71 mm. Southern Texas to northern Veracruz.

Eumeces obsoletus (**Baird and Girard**).   Great Plains skink. Color dark gray to brown; dark borders on most of dorsal scales. Young entirely shiny black except for a bright-blue tail and distinct white spots on each labial (Fig. 4-77). Length to 125 mm. Nebraska, Colorado, south through Texas, New Mexico, and Arizona; Chihuahua, Nuevo León, Tamaulipas.

Eumeces multivirgatus **Hallowell.**   Many-lined skink. Color and pattern variable. Usually numerous alternating light and dark longitudinal lines, these reduced in large individuals. Some individuals without lines. Young with only 4–5 lines. Length to 57 mm. 3 subspecies. Nebraska, eastern Colorado, Kansas, Oklahoma, Texas, and west through New Mexico into southern Arizona.

Eumeces anthracinus (**Baird**).   Coal skink. Gray to brown; a narrow light line from eye to base of tail; a broad dark band below the lateral stripe. Length to 65 mm. 2 subspecies in the United States. Eastern and central United States to the Gulf Coast.

Eumeces septentrionalis (**Baird**).   Prairie skink. Color brown or brownish gray; 3 brownish-gray longitudinal stripes separated by black lines; lower sides and belly tinged with blue. Young black with

Fig. 4-78. *Eumeces laticeps.* (Photograph by Isabelle Hunt Conant.)

7 light longitudinal lines. 3 subspecies. Length to 75 mm. Manitoba to Texas.

Eumeces skiltonianus **Baird and Girard.** Western skink. Color brown; 2 broad dorsal light lines from snout to tail; a broad lateral light line from labials to tail; area between the dorsal and lateral line much darker than area between dorsal lines. Tail blue in young but dark in older individuals. Length to 83 mm. British Columbia southward into Baja California. Eastern Nevada, western Utah, and northwestern Arizona.

Eumeces gilberti **Van Denburgh.** Gilbert's skink. Color and pattern similar to that of *E. skiltonianus* until a length greater than 60 mm is attained. The dorsal and lateral lines are lost in large adults. Tail blue, pink, reddish, purplish. Length to 113 mm. 5 subspecies. California, southern Nevada, Arizona.

Eumeces egregius **(Baird).** Florida skink. Color gray or brown; light longitudinal lines; tail tinted orange-red or blue. Length to 60 mm. 4 subspecies. Alabama and Georgia through Florida.

Genus *Neoseps* Stejneger. Sand skinks. Body extremely elongate, limbs reduced, feeble. Ear opening covered by scales, eyelid with transparent "windows." 1 species.

Neoseps reynoldsi **Stejneger.** Sand skink. Tan or brown, a series of dark spots from head to tail base. Dark streak from nostril through eye. Florida.

Family Amphisbaenidae. Ringed lizards. Fossorial. Elongate, usually limbless, bodies of nearly same diameter throughout length; tail short. Symmetrical rings of scales around body; no ear opening; eyes present but under a layer of skin. Oviparous or viviparous. 10 genera, about 100 species; 1 species in the United States.

Genus *Rhineura* Cope. Worm lizards. Differs from others of the family primarily in that the dorsal surface of the tail is flattened.

Rhineura floridana **Baird.** Florida worm lizard. Color pink to red in life. The only earless, legless, blind lizard of the United States. Often confused with earthworms. Length to 260 mm. Florida.

Suborder Serpentes. Body extremely elongated; without appendages. Sternum absent; no vestiges of forelimbs or pectoral girdle; vestigial pelvic girdle present in the boas and pythons. Skull diapsid; anterior part of braincase completely ossified; no temporal arch; teeth acrodont, specialized in some snakes for injection of poison; mandibular rami usually connected by an elastic ligament. Vertebrae numerous and differentiated into precaudal and caudal regions; ribs single-headed. No movable eyelids, external ear openings, or functional auditory apparatus. The extensible, forked tongue functions as a tactile and olfactory organ. Copulatory organ (hemipenis) bifid, retracted into sheath posterior to anal opening; cloacal opening transverse.

Oviparous or viviparous. Terrestrial or aquatic. The ovoid soft-shelled eggs are usually laid in a protected cavity and are not guarded by the female. Some snakes, as the python and the mud snake (*Farancia*), do remain with the eggs until they hatch. Generally carnivorous.

The snakes were specialized early in their evolution for burrowing and swallowing intact prey. Their ancestors were probably subterranean-dwelling lizards. Many of the anatomical features of snakes suggest that structures were lost and then developed again. Cretaceous. 13 families, 330 genera, and 3000 species of living snakes. Generally distributed in Tropical and Temperate Zones.

IDENTIFICATION. The pattern of scale arrangement and the configuration of scales provide the most useful basis for identification. The first step in learning to use the keys is to become acquainted with the typical scalation (Figs. 4-79 to 4-81).

Most of the snakes have the ventral scales (usually abbreviated "ventrals") transversely widened, a modification which makes them useful in locomotion. The cloacal opening is covered with a scale, the anal, which may be single or divided into 2 parts. The ventral scales posterior to the anal scales are named caudals. These are divided to form 2 rows in most snakes but are undivided in some forms. The number of ventrals is determined by counting from the first scale wider than long behind the chin to, but not including, the anal; the number of caudals is determined by counting from the first scale wide enough to meet its opposite to the tip of the tail.

The longitudinal rows of dorsal scales appear in lateral view as diagonal rows extending around the body. These scales are counted at mid-body when only 1 count is given, and about 5 scale lengths behind the head, mid-body, and 5 scale lengths in front of the anal when 3 scale counts are given.

Fig. 4-79. Lateral head scales of a snake. (From Stebbins, *Amphibians and Reptiles of Western North America*, McGraw-Hill Book Company.)

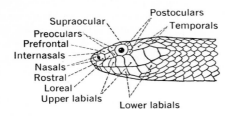

Fig. 4-80. Dorsal and ventral head scales of a snake. (From Stebbins, *Amphibians and Reptiles of Western North America*, McGraw-Hill Book Company.)

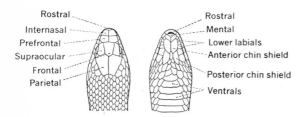

Fig. 4-81. Method of counting dorsal scales. (From Stebbins, *Amphibians and Reptiles of Western North America*, McGraw-Hill Book Company.)

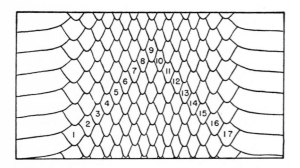

The difficulty of making counts results in frequent errors by the inexperienced; the number of scale rows (dorsals) should be carefully checked. Each of the dorsals may have a distinct longitudinal ridge, giving it a keeled appearance and making the entire scale appear rough rather than smooth.

The configuration and size of the snake are often of aid in identification. Such terms as long-headed or blunt-headed must be evaluated in relation to the figures. The size of the snake is reported here as total length (tip of nose to tip of tail). Juvenile snakes are often very different from adults in their color patterns.

KEY TO THE FAMILIES OF SERPENTES

1. No transversely elongated scales present on belly; eyes covered with scales; wormlike **Leptotyphlopidae**
   Transversely elongated scales (ventrals) present; eyes not covered with scales; not wormlike **2**
2. A pit between the eye and nostril (Fig. 4-82) **Viperidae (subfamily Crotalinae)**
   No pit between the eye and nostril **3**
3. No paired, elongate scales between lower labials of right and left sides (Fig. 4-83) **Boidae**
   At least 2 pairs of elongate scales between lower labials of right and left sides **4**
4. A pair of short permanently erect grooved fangs in anterior part of upper jaw (all forms in the United States with black, yellow, and red rings around body) **Elapidae**
   Without a pair of short, permanently erect grooved fangs in anterior part of upper jaw (some king snakes in the United States with black, yellow, and red rings around body) **Colubridae**

Fig. 4-82. Head of *Crotalus*.

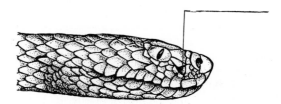

Fig. 4-83. Chin scales of *Charina*.

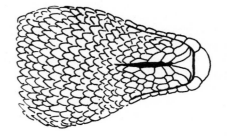

Family Leptotyphlopidae. Blind snakes. Fossorial. Small, wormlike snakes with the eyes covered by the head scales. Tail short and terminating in a sharp point; ventrals not transversely elongated, similar to scales of dorsal surface. Teeth only in the lower jaw. Pelvic girdle and vestiges of femur present. Living snakes may have a partially transparent skin and appear pinkish or purplish. Body cylindrical, not flattened ventrally. Often confused with earthworms. Oviparous. 1 genus occurs in North and South America and Africa; 2 species in the United States.

Genus *Leptotyphlops* Fitzinger. Slender blind snakes. Characters of the family.

KEY TO SPECIES OF *LEPTOTYPHLOPS*

1. Supraoculars present (Fig. 4-84)   ***L. dulcis***
No supraoculars   ***L. humilis***

Fig. 4-84. **Head scales of *Leptotyphlops dulcis*.**

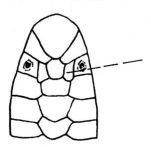

Leptotyphlops dulcis **(Baird and Girard).** Texas blind snake. Color light brown; labials and ventral surface of head lighter; no sharp demarcation between dorsal and ventral surfaces. 2 subspecies. Southwestern Kansas to southeastern Arizona and northeastern Mexico.

Leptotyphlops humilis **(Baird and Girard).** Western blind snake. Color light brown; labials and ventral surface lighter; usually a sharp demarcation between dorsal and ventral color. 4 subspecies. Southern California, Nevada, and southwestern Utah to southwestern Texas. Baja California; northern Coahuila, Mexico.

Family Boidae. Boas and pythons. Terrestrial,

arboreal. Moderate- to large-sized snakes. Both jaws with teeth. Vestiges of pelvic girdle and hind limb usually present; hind limb may appear externally as simple spurlike projections. Tropics of both hemispheres and into temperate regions of western United States. 25 genera, 2 in the United States.

KEY TO GENERA OF BOIDAE

1. 3 scales between eyes, central one larger   *Charina*
Several small scales of about equal size between eyes
Lichanura

Genus *Lichanura* Cope. Rosy boas. Heavy-bodied, short-tailed snakes with triangular heads; vestiges of the rear limbs externally apparent in some individuals. Southwestern United States and northwestern Mexico. 1 species in the United States.

***Lichanura roseofusca* Cope.** Rosy boa. Color gray to bluish; 3 broad longitudinal red or brownish stripes. Viviparous, 2–8 young about 12 in. long. Slow-moving, primarily nocturnal. 2 subspecies. Length to 900 mm. Southern California to southwestern Arizona into Baja California.

Genus *Charina* Gray. Rubber boas.

Charina bottae **(Blainville).** Rubber boa. Color uniform olive or brown; ventrals white or yellow; sometimes conspicuously spotted on edges of ventrals or lower scale rows. Young pink or tan. Viviparous, 1–8 young 6–9 in. long. Length to 600 mm. 3 subspecies. British Columbia and western Montana southwest to northern Utah, central Nevada, and southern California.

Family Colubridae. Fossorial, arboreal, terrestrial, aquatic. Small- to large-sized snakes; extremely variable in body form. Both jaws with teeth; some forms with rear teeth grooved to transport venom; no vestiges of girdles or limbs; coronoid and postfrontal bones absent; pupil of eye round, horizontal, or vertical. Viviparous or oviparous. Includes most of the snakes of the world and 75% of the species of the United States.

KEY TO GENERA OF COLUBRIDAE

1. Keels on some or all dorsal scales (Fig. 4-85)   **2**
All scales smooth (Fig. 4-86)   **19**

2. Anal plate divided (Fig. 4-87)   **3**
   Anal plate not divided   **15**
3. Rostral turned upward and keeled (Fig. 4-88)   *Heterodon*
   Rostral not turned upward and keeled   **4**
4. Loreal present (Fig. 4-89)   **5**
   No loreal (Fig. 4-90)   *Storeria*
5. 2 internasals   **8**
   1 internasal   **6**
6. 5 upper labials; 6 lower labials; 17 scale rows   *Virginia*
   7 or 8 upper labials; 8–10 lower labials; 19 or 21 scale rows   **7**
7. No preocular (loreal in contact with eye) (Fig. 4-91); ventrals more than 160   *Farancia*
   1 or 2 preoculars (Fig. 4-92); ventrals less than 140   *Liodytes*
8. 1 or 2 preoculars   **10**

No preoculars (loreal in contact with eye) (Fig. 4-93)   **9**
9. 5 or 6 upper labials (Fig. 4-93)   *Virginia*
   7 upper labials   *Abastor*
10. 17 scale rows   **11**
    More than 17 scale rows   **13**
11. More than 100 caudals   **12**
    Less than 50 caudals   *Seminatrix*
12. 7 upper labials; 7 or 8 lower labials; dorsal color grass green   *Opheodrys*
    9 upper labials; 10 or 11 lower labials; dorsal color dark with a light spot on each scale   *Drymobius*
        Natrix
13. 3 postoculars   **14**
    2 postoculars   *Natrix*
14. 19–23 scale rows; scales strongly keeled   *Natrix*
    25 or more scale rows; scales weakly keeled   *Elaphe*
15. 27 or more scale rows; usually 4 prefrontals   *Pituophis*
    Less than 27 scale rows; 2 prefrontals   **16**
16. Pupil of eye round; no suboculars; rostral normal   **17**
    Pupil elliptical; subocular present; rostral very large with free lateral edges   *Phyllorhynchus*
17. 8 or more lower labials   *Thamnophis*
    Less than 8 lower labials   **18**
18. A double row of black belly spots   *Tropidoclonion*
    Usually no belly spots   *Virginia*
19. Anal plate divided   **27**
    Anal plate not divided   **20**
20. All or most caudals divided   **21**
    Most caudals entire   *Rhinocheilus*
21. Pupil of eye round   **23**
    Pupil elliptical   **22**

Fig. 4-85 (left). **Keeled scales.** *Fig. 4-86* (right). **Smooth scales.**

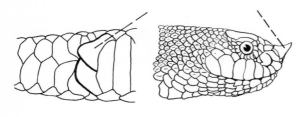

Fig. 4-87. (left). **Divided anal.** *Fig. 4-88* (right). **Head of *Heterodon nasicus*.**

Fig. 4-91 (left). **Head of *Farancia*. *Fig. 4-92* (right).** **2 preoculars.**

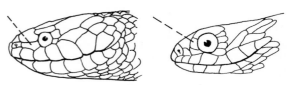

Fig. 4-89 (left). **Head of *Natrix*; loreal present.**
Fig. 4-90 (right). **Loreal absent.**

Fig. 4-93 (left). **No preoculars (loreal in contact with eye).** *Fig. 4-94* (right). **No loreal present.**

22. Suboculars present   *Phyllorhynchus*
    No suboculars (upper labials in contact with eye)
                       Trimorphodon
23. Loreal present   24
    No loreal (Fig. 4-94)   *Stilosoma*
24. Ventral surface light without markings   25
    Ventral surface with at least some dark markings
                                   26
25. 6 or 7 upper labials; 8 lower labials   *Cemophora*
    8 upper labials; 12–15 lower labials   *Arizona*
26. 17 scale rows   *Drymarchon*
    More than 17 scale rows   *Lampropeltis*
27. Less than 19 scale rows   28
    19 or more scale rows   45
28. Loreal present   29
    No loreal   41
29. 1 or more preoculars   31
    No preoculars; loreal and prefrontal in contact with
    eye (Fig. 4-93)   30
30. 13 scale rows; nasal plate entire; 5 upper labials
                        Carphophis
    More than 13 scale rows; nasal plate divided; 6
    upper labials   *Virginia*
31. 2 or 3 preoculars   32
    1 preocular   36
32. Rostral normal   33
    Rostral much enlarged with free lateral edges
                        Salvadora
33. 2 or 3 anterior temporals (usually); lower preocu-
    lar very small, wedged between adjacent upper
    labials (Fig. 4-95)   34
    1 anterior temporal; lower preocular not wedged
    between upper labials   35
34. Scale rows at posterior end of body 15   *Coluber*
    Scale rows at posterior end of body 13, 12, or 11
                       Masticophis
35. Nasal plate divided; a ring on neck or black spots
    on belly or both   *Diadophis*
    Nasal plate entire; no ring on neck; no black spots
    on belly   *Opheodrys*
36. 17 scale rows   40
    Less than 17 scale rows   37

37. Posterior chin shields about as long as anterior chin
    shields; caudals more than 65   *Opheodrys*
    Posterior chin shields much shorter than anterior
    chin shields; caudals less than 65   38
38. Ventral surface uniform light color, or if any dark
    markings, they are black rings that encircle body;
    no lateral light stripe on 4th or 5th row of scales
                                39
    Each ventral with a dark anterior border; usually a
    light stripe on 4th or 5th row of scales   *Contia*
39. Snout normal; no nasal valve   *Sonora*
    Snout flattened; shovellike; nasal valve present
                         Chionactis
40. More than 60 caudals; a dark line from rostral
    through eye to last upper labial: 7 upper labials
                        Rhadinaea
    Less than 60 caudals; no dark line from rostral to
    last upper labial; usually 8 upper labials
                        Seminatrix
41. 13 scale rows   *Chilomeniscus*
    More than 13 scale rows   42
42. 15 scale rows   43
    17 scale rows   44
43. Snout flattened, shovellike   *Chionactis*
    Snout normal   *Tantilla*
44. Rostral acute and turned up at tip; tail short and
    thick   *Ficimia*
    Head long and nose sharp; tail long and slender
    (more than half body)   *Oxybelis*
45. No preocular (loreal in contact with eye) (Fig.
    4-91)   46
    1 or more preoculars   47
46. Longitudinal stripes present   *Abastor*
    No stripes   *Farancia*
47. Pupil of eye round   48
    Pupil elliptical   49
48. 25 or more scale rows   *Elaphe*
    19 scale rows   *Coniophanes*
49. 1 loreal; 2 postoculars; 1 anterior temporal (Fig.
    4-96); 7 or 8 upper labials   50
    2 or more loreals; 3 or 4 postoculars; 2 or 3 anterior
    temporals (Fig. 4-97); usually 9 or more upper
    labials   *Trimorphodon*

Fig. 4-95 (left). **Lower preocular small.**
Fig. 4-96 (right). **Head of** *Hypsiglena.*

Fig. 4-97. **Head of** *Trimorphodon.*

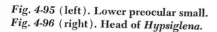
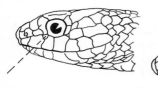
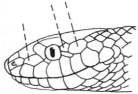
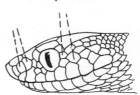

50. Dorsal pattern consists of small blotches or 1 or 2
    series of smaller alternating spots on sides
                                                Hypsiglena

    Dorsal pattern consists of 22–26 large blotches with
    no alternating lateral spots   *Leptodeira*

Genus *Natrix*. Water snakes. Moderate- to large-sized, heavy-bodied snakes with strongly keeled scales. Anal scent glands well developed. Viviparous. About 100 forms; North America, Eurasia, North Africa, and Australia.

KEY TO SPECIES OF *NATRIX*

1. Scale rows 19   **2**
   Scale rows more than 19   **5**
2. Lower labials 7; 1 preocular   *N. kirtlandi*
   Lower labials 9–11; usually 2 preoculars   **3**
3. 1 long dark median stripe on belly, or no markings
   except on end of the ventrals   *N. grahami*
   2 long dark median stripes near middle of belly, at
   least anteriorly   **4**
4. Light stripes present at sides of belly
                            N. septemvittata
   No light stripes present at sides of belly *N. rigida*
5. Scale rows 27–33 (if 25, a pattern of alternating dorsal and lateral spots present); lower labials usually
   11–13   **6**
   Scale rows 21–25 (rarely 27 in *erythrogaster*); lower
   labials usually 10   **7**
6. No subocular   *N. taxispilota*
   1 or more suboculars   *N. cyclopion*
7. Dorsal surface unicolor   *N. erythrogaster*
   Dorsal surface with crossbands or blotches **8**
8. Ventral surface immaculate or with dark anterolateral
   margins of ventrals   **9**
   Ventral surface not as above, but with conspicuous,
   dark, crescentic or vermiculate markings or blotches or with row of light spots down center of belly   *N. sipedon*
9. A maximum of 40 dark dorsal spots or blotches
                            N. erythrogaster
   A minimum of 55 dark dorsal spots or blotches
                            N. harteri

Natrix grahami (**Baird and Girard**). Graham's water snake. Slender, striped; color gray to brown with light-colored longitudinal stripes; lateral light stripe on the 3 lower scale rows. Belly cream or white with a central row of dark spots posteriorly (Fig. 4-98). Length to 900 mm. Illinois westward through eastern Kansas and southward to Louisiana and eastern Texas.

Natrix rigida (**Say**). Glossy water snake. Color gray to brown; 2 parallel brown stripes down back. Belly cream or yellow with 2 rows of black spots. Length to 610 mm. Virginia and South Carolina westward on the Coastal Plain to eastern Texas and southeastern Oklahoma. 3 subspecies.

Natrix septemvittata (**Say**). Queen snake. Color light to dark brown; a yellow longitudinal stripe on the 1st and 2d scale rows. Belly cream with 2 parallel rows of dark spots down middle. 2 subspecies. Length to 650 mm. New York southward to northern Florida, eastern Mississippi, northward to Wisconsin and Illinois. An isolated population in Arkansas and Missouri.

Natrix cyclopion (**Dumeril, Bibron, and Dumeril**). Green water snake. Large, heavy-bodied snake; green to brown, about 50 dark crossbars in smaller individuals. Belly yellow. 2 subspecies. Length to 1800 mm. South Carolina to Florida, westward to eastern Texas; in Mississippi Valley to southern Indiana.

Natrix erythrogaster (**Forster**). Plain-bellied water snake. Large, heavy-bodied snake. Larger individuals uniform gray, brown, or black; smaller ones with distinct crossbars. Belly yellow to red. 4 subspecies. Length to 1500 mm. Southeastern and central United States.

Natrix harteri **Trapido**. Brazos water snake. Color brown; 58–65 dark spots in 2 longitudinal rows, alternating with lateral spots. Belly pinkish. 2 subspecies. Length to 900 mm. Palo Pinto, Throckmorton, Coke, Tom Green, Concho, McCulloch, Brown, and Runnels Counties, Texas.

Natrix kirtlandi (**Kennicott**). Kirtland's water

Fig. 4-98. Natrix grahami. (**Photograph by Isabelle Hunt Conant.**)

snake. Marked with 4 rows of 50 black blotches on a background of gray to brown. Belly reddish; row of black spots down each side of belly. Length to 460 mm. Southern Wisconsin and Michigan southward to northern Kentucky; through Ohio to western Pennsylvania.

Natrix taxispilota (**Holbrook**). Brown water snake. Heavy-bodied snake. A pattern of square or diamond-shaped markings on back (Fig. 4-99). Belly yellow with crescent-shaped dark spots or brownish. 2 subspecies. Length to 1700 mm. Southeastern United States: Virginia to Florida westward into Missouri and southward into Mexico.

Natrix sipedon (**Linnaeus**). Common water snake. Stout-bodied snakes. Color and pattern extremely variable: may be uniform gray to brown; or with distinct reddish-brown or black crossbands; or with longitudinal lines. Length to 1500 mm. 10 subspecies in the United States. Minnesota eastward to Maine, southward to Gulf Coast.

Genus *Seminatrix* **Cope.** Black swamp snakes. Smooth-scaled, stout-bodied snakes living near the water. 1 genus, 1 species.

Seminatrix pygaea **Cope.** Black swamp snake. Uniform black above; belly red with narrow black bars on each ventral. Length to 400 mm. 3 subspecies. South Carolina through peninsular Florida.

Genus *Storeria* **Baird and Girard.** Brown snakes. Small terrestrial snakes with much shortened heads. Viviparous. 8 species, 2 in the United States.

KEY TO SPECIES OF *STORERIA*

1. Scales in 15 rows   *S. occipitomaculata*
  Scales in 17 rows   **2**
2. Dark markings on posterior upper labials *S. dekayi*
  No dark markings on posterior upper labials
              S. tropica

Storeria dekayi (**Holbrook**). Brown snake. Color gray to brown; yellowish, tan, or light-gray middorsal stripe bordered by black spots. Belly white to brown, sometimes pinkish, with black dots on the ends of the ventrals. Black vertical markings or blotches on the anterior temporal. Length to about 500 mm. 4 subspecies in the United States. Southern Quebec and Ontario southward to the Gulf Coast.

Storeria tropica **Cope.** Tropical brown snake.

Fig. 4-99. Natrix taxispilota. (**Photograph by Isabelle Hunt Conant.**)

Color as in *S. dekayi* but with a horizontal dark line through the long axis of the anterior temporal. 4 subspecies, 1 in the United States. Coastal area of Louisiana and Texas, extending through eastern Mexico into Honduras.

Storeria occipitomaculata (**Storer**). Red-bellied snake. Color gray to brown with a light middorsal stripe; light spot below and behind the eyes; 3 light spots at back of head. Belly red with the color of the sides extending onto ventrals. Length to 400 mm. 3 subspecies in the United States. Eastern North America westward to Manitoba and Wyoming; southward through eastern Kansas and Oklahoma to Gulf Coast, except southern Florida.

Genus *Thamnophis* **Fitzinger.** Garter snakes. Small- to moderate-sized snakes with strongly keeled dorsal scales. Close relatives of the water snakes but usually more terrestrial. Viviparous.

KEY TO SPECIES OF *THAMNOPHIS*

1. Lateral stripe anteriorly involving 4th scale row **2**
  Lateral stripe anteriorly not involving 4th scale row,
  or absent   **7**
2. Tail usually more than 27% of total length **3**
  Tail usually less than 27% of total length **4**
3. Parietal spots always present, fused, bright, and
  usually fairly large; brown pigment usually not
  extending onto the ventral scutes to form a dark
  ventrolateral stripe on each side, or, if present,
  covering less than ⅔ of the area of each scute
             T. proximus
  Parietal spots often absent, when present small and

rarely fused or bright; brown pigment always extending onto ventral scutes and usually covering ⅔ or more of the area of each scute **T. sauritus**

4. Scale rows 21–19–17; upper labials 8 (occasionally 9) **T. eques**
   Scale rows less than 21–19–17; upper labials less than 8 **5**

5. Lateral stripe anteriorly on 3d and 4th rows of scales; maximum scale rows 21 **T. radix**
   Lateral stripe anteriorly on 2d, 3d, and 4th rows of scales; maximum scale rows usually 17 or 19 **6**

6. Scale rows usually 19–19–17; upper labials usually 7 **T. butleri**
   Scale rows usually 17–17–17; upper labials usually 6 **T. brachystoma**

7. Lateral stripe anteriorly on 2d and 3d rows of scales, or absent **8**
   Lateral stripe anteriorly on 3d row of scales only **T. marcianus**

8. 7 upper labials **9**
   8 upper labials **10**

9. Lower labials 8 or 9; scale rows usually 17–15; ventrals usually less than 153; dorsal stripe yellow or red; pale yellowish flecks between scales in area between dorsal and lateral stripes; posterior chin shields but little longer than anterior **T. ordinoides**
   Lower labials usually 10; scale rows usually 19–17; ventrals usually more than 153; dorsal stripe bright yellow; no pale dorsolateral flecks; posterior chin shields markedly longer than anterior **T. sirtalis**

10. Maximum scale rows 19; ground color of side of body brown; on each side 2 series of conspicuous dark spots **T. cyrtopsis**
    Maximum scale rows 21; if less, ground color of side of body not brown and not 2 series of conspicuous dark spots on each side **11**

11. Usually no lateral stripe; dorsal spots numerous and prominent **T. rufipunctatus**
    Lateral stripe on 2d and 3d scale rows; dorsal spots not prominent **T. elegans**

Thamnophis rufipunctatus (Cope). Narrow-headed garter snake. Head long, narrow; poorly defined longitudinal stripes; many distinct dark dorsal spots. Southeastern Arizona and southwestern New Mexico southward into Chihuahua and Durango.

Thamnophis eques (Reuss). Mexican garter snake. Pattern of 3 distinct, longitudinal yellow and orange stripes; lateral stripes on the 2d and 3d scale rows. Crescent-shaped white or yellow markings behind the mouth. Length to 500 mm. 2 subspecies, 1 in the United States. Western Texas to southern Arizona, southward throughout the Mexican plateau.

Thamnophis cyrtopsis (Kennicott). Black-necked garter snake. Lateral stripe, when present, on the 2d or 3d scale rows; usually 3 stripes, a middorsal and 2 lateral stripes, these sometimes absent or poorly defined. A pattern of 2 alternating rows of spots between the stripes. 4 subspecies, 2 in the United States. Southern Utah and Colorado to Central America.

Thamnophis marcianus (Baird and Girard). Checkered garter snake. Color brownish yellow; lateral stripe restricted to the 3d scale row anteriorly, the 2d and 3d posteriorly; 2 rows of squarish black spots between the longitudinal stripes; a yellow crescent-shaped mark behind angle of mouth; black vertical mark on upper labials. Belly with dark spots on the ends of the ventrals. Length to 750 mm. 2 subspecies in the United States. Kansas, Oklahoma, and Texas westward to southeastern California; northeastern Mexico.

Thamnophis ordinoides (Baird and Girard). Northwestern garter snake. Middorsal stripe red, yellow, or orange; color between stripes bluish to black. Belly marked with red, often with black. Southern Vancouver Island and mainland of southwestern British Columbia; southward through Washington and Oregon chiefly west of the Cascade Mountains, and western Del Norte County, California.

Thamnophis elegans (Baird and Girard). Western garter snake. Large group of subspecies with much variation in their markings: some uniform in color, others with checkered pattern or longitudinal stripes. 11 subspecies. Western North America; Montana southeastward to western Oklahoma, westward to Pacific Coast.

Thamnophis radix (Baird and Girard). Plains garter snake. Color greenish gray to brownish; lateral yellow stripes on the 3d and 4th scale rows anteriorly; a distinct middorsal yellow or orange stripe; labials barred with black. Belly whitish or bluish green; black spots on ends of ventrals. 2 subspecies in the United States. Length to 900 mm. Great Plains, eastward through Illinois into Ohio.

Thamnophis butleri (**Cope**). Butler's garter snake. Background color brown; yellow or orange lateral stripes on the 2d, 3d, and 4th scale rows anteriorly; dark blotches may be present between lateral stripes. Belly greenish with dark marks. Head somewhat wider than neck. Length to 650 mm. Indiana, Ohio, southern Michigan, southeastern Wisconsin.

Thamnophis brachystoma (**Cope**). Short-headed garter snake. Lateral stripes as in *T. butleri*. Background color dark brown; dark blotches between lateral stripes faint or absent. Head not wider than neck. Length to 550 mm. Southwestern New York and northwestern Pennsylvania.

Thamnophis sauritus (**Linnaeus**). Eastern ribbon snake. 3 conspicuous, longitudinal, yellow or orange stripes. Background color dark; double row of poorly defined black spots between the stripes. 4 subspecies. Southern Ontario and southern Maine southward east of the Mississippi River to the Florida Keys and to Louisiana (Fig. 4-100).

Thamnophis proximus (**Say**). Western ribbon snake. Pattern of 3 longitudinal stripes; middorsal stripe yellow, orange, red, or brown. Ground color brown, black, or olive. Southern Wisconsin, Indiana, and the Mississippi Valley west through the Great Plains to southeastern Colorado and eastern New Mexico. 4 subspecies.

Thamnophis sirtalis (**Linnaeus**). Common garter snake. Pattern of 3 distinct light stripes; lateral stripes on 2d and 3d rows of scales anteriorly; color between stripes brown, black, or red with bars or blotches of contrasting color. Belly greenish or yellow; black spots sometimes present on tips of ventrals. 9 subspecies in the United States. Length to 1239 mm. North America, southern Canada southward.

Genus *Tropidoclonion* **Cope.** Lined snakes. 1 species.

Tropidoclonion lineatum (**Hallowell**). Lined snake. Color gray to brown; a white or yellow middorsal stripe and a lateral stripe on the 2d and 3d scale rows. Belly yellow or white with 2 rows of black spots. Viviparous, 7–8 young about 8 in. in length. 4 subspecies. Length to 550 mm. Central Illinois, Missouri, Iowa, South Dakota southward through eastern Nebraska, Kansas, Oklahoma, and Texas. An isolated population in Colorado and New Mexico.

Genus *Virginia* **Baird and Girard.** Earth snakes. Small, secretive, uniformly colored snakes. Viviparous. 2 species.

KEY TO SPECIES OF *VIRGINIA*

1. Scales strongly keeled, 5 upper labials; 1 postocular **V. striatula**

   Scales not strongly keeled; 6 upper labials; 2 postoculars **V. valeriae**

Virginia striatula **Linnaeus.** Rough earth snake. Uniform brown or reddish; belly yellow or pink. Length to 300 mm. Atlantic and Gulf states from Virginia to northern Florida, westward to central Texas; through eastern Oklahoma to southeastern Kansas, and eastward in Missouri, Arkansas, Tennessee, and Kentucky.

Virginia valeriae **Baird and Girard.** Smooth earth snake. Uniform gray or brown or with 4 rows of small black dots. Belly white. Length to 300 mm. 3 subspecies. Pennsylvania and New Jersey southward to central Florida and westward into Oklahoma and Texas.

Genus *Liodytes* **Cope.** Swamp snakes. 1 species.

Liodytes alleni (**Garman**). Striped swamp snake. A relatively stout-bodied aquatic snake. Brown middorsal stripe 5 or 6 scale rows in width; lateral brown stripe about 2 scale rows in width; color between stripes greenish to yellow. Belly uniform yellow or with a midventral row of dark spots. Viviparous. 2 subspecies. Extreme southern Georgia and peninsular Florida.

Fig. 4-100. *Thamnophis sauritus.* (**Photograph by Isabelle Hunt Conant.**)

Genus *Heterodon* Latreille. Hognose snakes. Stout-bodied, moderate-sized snakes with upturned shovellike snouts. Terrestrial, oviparous. All have ability to flatten the neck and head when alarmed and the habit of feigning death.

KEY TO SPECIES OF *HETERODON*

1. Prefrontals in contact   ***H. platyrhinos***
   Prefrontals separated by small scales   2
2. Belly immaculate or mottled with gray-brown
                              H. simus
   Belly black with small yellow patches   ***H. nasicus***

Heterodon platyrhinos Latreille. Eastern hognose snake. Color brown, black, yellow, or reddish; series of dark dorsal blotches alternating with a row of dark lateral blotches. Belly yellow marked with black. Melanistic (entirely black) individuals common. Scale rows usually 25–25–19. Length to 1150 mm. Eastern United States north to southern Vermont, New Hampshire, New York, and Ontario: west to South Dakota, eastern and central Nebraska, Kansas, Oklahoma, and eastern Texas.

Heterodon nasicus Baird and Girard. Western hognose snake. Color gray or brown; 23–52 dark dorsal blotches alternating with lateral blotches. Belly mostly black. Scale rows usually 23–23–19. 3 subspecies. Length to 790 mm. Illinois to Alberta; southward to southern Arizona and Texas.

Heterodon simus (Linnaeus). Southern hognose snake. Color similar to that of *H. platyrhinos* but paler; ventral surface of tail is not conspicuously lighter than the belly. Scale row usually 25–25–21. Length to 600 mm. North Carolina to central Florida, westward to southern Mississippi.

Genus *Rhadinea* Cope. Slender snakes.

Rhadinea flavilata (Cope). Yellow-lipped snake. Slender-bodied; uniformly colored tan to reddish brown; dark band from eye to the mouth; upper lip yellowish. Belly yellow. Oviparous. Length to 400 mm. Eastern North Carolina southward and westward on the Atlantic and Gulf Coastal Plains to eastern Louisiana.

Genus *Diadophis* Baird and Girard. Ringneck snakes. Usually small snakes with a yellow or orange ring around the neck.

KEY TO SPECIES OF *DIADOPHIS*

1. Ventral color usually extending on anterior portion of body, onto 1 or more of lowermost rows of dorsal scales   ***D. amabilis***
   Ventral color not extending onto lowermost rows of dorsal scales (except rarely at extreme anterior end)   ***D. punctatus***

Diadophis amabilis Baird and Girard. Western ringneck snake. Small (length to 600 mm) secretive snakes. Color uniform gray, greenish or bluish, with a conspicuous orange or yellow neck ring. Belly yellow, orange, or red; colors extending onto sides of body. 6 subspecies. Western United States; west of trans-Pecos Texas in the south; west of Utah and Colorado in the north.

Diadophis punctatus (Linnaeus). Eastern ringneck snake. Color gray to black. Belly yellow to reddish but the color not extending onto sides. Light neck ring distinct. Length to 570 mm. 6 subspecies. Nova Scotia to Minnesota and westward to Colorado; southward to Florida in the east and New Mexico in the west.

Genus *Carphophis* Gervais. Worm snakes.

Carphophis amoenus (Say). Worm snake. Small, wormlike, secretive snakes. Color uniform gray, brown, or black. Pink color of belly extending onto sides. Oviparous. Length to 375 mm. 3 subspecies. Eastern and central United States: Connecticut southward into Georgia, westward into Nebraska, Iowa, Kansas, Oklahoma, and eastern Texas.

Genus *Abastor* Gray. Rainbow snakes.

Abastor erythrogrammus (Latreille). Rainbow snake. Color bluish black, longitudinal red or yellow stripes. Belly red with 2 rows of dark spots. A sharp scale at end of tail. Secretive, burrowing; oviparous. Length to 1500 mm. Maryland through the lower Atlantic Coastal Plain to central Florida; westward to eastern Louisiana.

Genus *Farancia*. Mud snakes.

Farancia abacura (Holbrook). Mud snake. Shiny black and red; back black, belly red; red of belly extending upward onto sides of body. Burrowing, secretive, aquatic; oviparous. **Tail terminating in a sharp spinelike scale commonly and erroneously believed to be a "stinger." This snake often the source of "hoop snake" and "stinging** snake" stories. Length to 2100 mm. 2 subspecies. Southeastern United States.

Genus *Coniophanes* Hallowell. Black-striped snakes. 1 species 1 subspecies in the United States.

***Coniophanes imperialis* (Baird).** Black-striped snake. Color brown with a darker middorsal and lateral stripe; a yellow line from nostrils to rear of head. Belly reddish. Grooved teeth in back of upper jaw; venom present but not dangerous to man. Length to 500 mm. Southern Texas to northern Central America.

Genus *Coluber* Linnaeus. Racers. Large slender-bodied long-tailed nonvenomous terrestrial snakes. Scales smooth; very active. Oviparous.

***Coluber constrictor* Linnaeus.** Racer. A group of 8 subspecies in the United States intergrading in broad geographic zones; many intermediates are found. Color in adults uniform gray, greenish, tan, blue, or black, or with white flecks on a blue-gray background (Fig. 4-101). Juveniles (less than 600 mm in length) with a pattern of 50–85 dorsal blotches or crossbands. Length to 1800 mm. Eastern North America, from Maine to Florida; westward to the Pacific Coast, exclusive of the higher parts of the Rocky Mountains.

Genus *Masticophis* Baird and Girard. Whipsnakes. Large, slender-bodied, long-tailed, terrestrial snakes (tail longer than in the racers). Color not uniform except in very large adults. Active, carnivorous, oviparous.

KEY TO SPECIES OF *MASTICOPHIS*

1. 15 scale rows — *M. taeniatus*
   17 scale rows — 2
2. No well-defined longitudinal stripes — *M. flagellum*
   Well-defined longitudinal stripes — 3
3. 1 light lateral stripe continued onto tail — *M. lateralis*
   2 or 3 light lateral stripes, these not continued onto tail — *M. bilineatus*

***Masticophis flagellum* (Shaw).** Coachwhip. Color tan, brown, red, gray to black; uniform or with black crossbands anteriorly. Length to 2100 mm. 6 subspecies. Southern United States and Mexico.

***Masticophis lateralis* (Hallowell).** Striped racer. Color brown or black with usually a yellow to white lateral stripe. Upper and lower labials

Fig. 4-101. Coluber constrictor. (Photograph by Isabelle Hunt Conant.)

light. Belly usually cream to yellow anteriorly, pinkish posteriorly or orange. Length to 1500 mm. 2 subspecies. California and northwestern Baja California.

***Masticophis bilineatus* Jan.** Sonora whipsnake. Alternating longitudinal stripes of light yellow and brown. Belly cream. Length to 1200 mm. Southern Arizona and southwestern New Mexico; southward in Mexico to Oaxaca.

***Masticophis taeniatus* (Hallowell).** Striped whipsnake. Longitudinal dark stripes on a cream, gray, or reddish-brown background. Belly cream or tan; underside of tail may be pink. Length to 1500 mm. 4 subspecies in the United States. The Great Basin to central Texas and southward into Mexico.

Genus *Opheodrys* Fitzinger. Green snakes. Small, slender-bodied, uniform green snakes. Terrestrial, arboreal, oviparous.

KEY TO SPECIES OF *OPHEODRYS*

1. Scales keeled — *O. aestivus*
   Scales not keeled — *O. vernalis*

***Opheodrys aestivus* (Linnaeus).** Rough green snake. Color green, belly whitish or yellowish. Length to 1000 mm. 2 subspecies. Connecticut southward to Florida, westward to the Pecos River in Texas, and northward to southern Iowa, Illinois, Indiana, and Ohio.

***Opheodrys vernalis* (Harlan).** Smooth green snake. Uniform green dorsally; belly lighter green or yellowish. Length to 650 mm. 2 subspecies.

Northern United States and southern Canada eastward from North Dakota and Kansas; North Dakota southwestward to Utah and southern New Mexico.

Genus *Drymobius* Fitzinger. Speckled racers. 1 species, 1 subspecies in the United States.

***Drymobius margaritiferus* (Schlegel).** Speckled racer. Moderate-sized; color uniform dark green, brown, or black; a green or yellow spot on each scale; belly yellowish. Length to 900 mm. The Brownsville region of Texas; southward to northern Veracruz.

Genus *Drymarchon* Fitzinger. Indigo snakes. 1 species, 8 subspecies; 2 subspecies in the United States.

***Drymarchon corais* (Daudin).** Indigo snake. Among the largest nonvenomous snakes of the United States; length to 2500 mm. Color uniform bluish, brown, or black. 2 subspecies. South Carolina to Florida and southern Alabama. Central southern Texas into Mexico.

Genus *Salvadora* Baird and Girard. Patch-nosed snakes. Rostral scale very large and flattened with projecting free edges. Terrestrial, fossorial. Slender-bodied, rapidly moving. Oviparous. Southern United States to Guatemala. 3 species, 3 subspecies in the United States.

KEY TO SPECIES OF *SALVADORA*

1. Rear pair of enlarged scales under lower jaw in contact or separated by 1 scale; 8 upper labials **2**
   Rear pair of enlarged scales under lower jaw separated by 2 or 3 scales; 9 upper labials
   S. hexalepis
2. No narrow dark line on 3d or 4th row of scales
   S. grahamiae
   A narrow dark line on 3d scale row *S. lineata*

***Salvadora grahamiae* Baird and Girard.** Mountain patch-nosed snake. General color yellowish; yellow, middorsal, longitudinal stripe bordered by a darker lateral stripe; belly yellow. Length to 900 mm. Southeastern Arizona, southwestern New Mexico, and trans-Pecos Texas.

***Salvadora hexalepis* (Cope).** Western patch-nosed snake. Pattern as in *S. grahamiae* with an additional dark stripe on the 3d and 4th scale rows. Length to 1200 mm. 4 subspecies. Nevada, southward through California, Arizona, and southwestern New Mexico into Mexico.

***Salvadora lineata* Schmidt.** Texas patch-nosed snake. Pattern as in *S. grahamiae* with a narrow, dark stripe only on the 3d scale row. Central and southern Texas and adjacent Mexico.

Genus *Phyllorhynchus* Stejneger. Leaf-nosed snakes. Small nocturnal desert snakes with conspicuous dark dorsal blotches. General color white to yellow. Rostral scale much enlarged. Oviparous. 2 species.

KEY TO SPECIES OF *PHYLLORHYNCHUS*

1. Dorsal blotches (not including tail spots) less than 16 *P. browni*
   Dorsal blotches (not including tail spots) 17 or more
   P. decurtatus

***Phyllorhynchus browni* Stejneger.** Saddled leaf-nosed snake. About 9–15 dorsal brown blotches as large as or larger than interspaces; general color white or yellow. Interspaces punctuated with many dark spots. 2 subspecies. Length to 400 mm. South central Arizona, Sonora, and Sinaloa, Mexico.

***Phyllorhynchus decurtatus* (Cope).** Spotted leaf-nosed snake. About 17–59 dark blotches on a white or yellowish background; blotches equal or less than size of interspaces. 1 or 2 rows of lateral spots. Length to 500 mm. 2 subspecies. Colorado Desert region, southward in Sonora and Baja California.

Genus *Elaphe* Fitzinger. Rat snakes. Large, elongate, richly colored, terrestrial and arboreal snakes. All kill their food by constricting it. Oviparous. About 60 forms in North America, Eurasia, Malay Archipelago. 5 species, 10 subspecies in the United States.

KEY TO SPECIES OF *ELAPHE*

1. Suboculars present *E. subocularis*
   Suboculars absent **2**
2. Upper surface grayish or greenish *E. triaspis*
   Upper surface black or with blotches or stripes **3**
3. Neck bands crossing parietals and uniting on frontal
   E. guttata
   No neck bands crossing parietals and uniting on frontal **4**

4. More than 220 ventrals   *E. obsoleta*
   Less than 220 ventrals   *E. vulpina*

Elaphe guttata (**Linnaeus**). Corn snake. Color orange, reddish, or gray with about 40 reddish or brownish dorsal blotches bordered with black. Belly yellowish with square black marks. Length to 1900 mm. 3 subspecies. Southern New Jersey through Florida, westward to central Mexico, northward to Kentucky and Illinois; in the west to Colorado and Utah.

Elaphe vulpina (**Baird and Girard**). Fox snake. Color yellow to light brown with 28–51 brown blotches (Fig. 4-102). Belly yellowish, marked with black. Length to 1700 mm. 2 subspecies. Eastern Nebraska to southern Ontario and western Ohio, and northern Wisconsin and Michigan southward to the Wabash River.

Elaphe obsoleta (**Say**). Rat snake. 8 subspecies, each quite different but many intermediates occur where their ranges overlap. Young snakes with a pattern of dorsal blotches alternating with lateral blotches; older snakes uniform black or yellow, gray or brown, with dark longitudinal stripes or similar to young. Length to 2500 mm. Eastern North America: New England through Florida; westward into Minnesota; southward through Texas into Mexico.

Elaphe triaspis **Cope**. Green rat snake. Color uniform gray or green; belly white, without patterns. Young marked with dark blotches. 3 subspecies; 1 in the United States. Southeastern Arizona to Costa Rica.

Elaphe subocularis (**Brown**). Trans-Pecos rat snake. Color yellow-tinted or gray with about 24 dark square blotches. The presence of subocular scales separates this species from other *Elaphe*. Trans-Pecos Texas and southern New Mexico; Chihuahua, Coahuila, and Durango, Mexico.

Genus *Arizona* Kennicott. Glossy snakes. Moderate-sized, smooth, shiny scales. Rostral plate enlarged. Ventrals 185–241; anal undivided; caudals 39–63. Dorsals 25–31. Nocturnal. Oviparous. 1 species; 7 subspecies in the United States.

Arizona elegans **Kennicott**. Glossy snake. Color light brown or yellow; numerous brown dorsal blotches; supplementary series of smaller blotches laterally. Belly white. Length to 1380 mm. 7 subspecies in the United States. Southwestern United States and northern Mexico.

Fig. 4-102. Elaphe vulpina. (**Photograph by Isabelle Hunt Conant.**)

Genus *Pituophis* Holbrook. Bullsnakes. Large, harmless, but pugnacious snakes well known over much of the United States. Terrestrial and arboreal. Oviparous. Have ability to produce a loud hiss by forcing air over a membrane in front of the glottis. Kill prey by constricting it. 1 species.

Pituophis melanoleucus (**Daudin**). Pine snake. Color very variable: white, yellow, or gray with dorsal blotches alternating with lateral blotches; light brown with brown spots on back and tail; reddish brown with blotches blending with ground color anteriorly; black with gray or black below. 11 subspecies. Eastern United States; New York southwestward to eastern Texas; western North America from Wisconsin and Indiana westward to the Pacific Coast.

Genus *Lampropeltis* Fitzinger. Kingsnakes. Nonvenomous, smooth-scaled, terrestrial snakes. Moderate-sized. Oviparous. Constrictors.

KEY TO SPECIES OF *LAMPROPELTIS*

1. Pattern of black-bordered, orange-red blotches, bands, or rings alternating with white-bordered medium to dark-gray interspace markings
           L. mexicana
   Pattern not as above   **2**
2. Pattern with red or with dorsal blotches of brown, gray, or red, with black borders   **3**
   Pattern without red, without dorsal blotches of brown, gray, or red, with black borders   ***L. getulus***
3. Pattern of black-edged dorsal blotches of brownish or dark red, only narrowly in contact with 5th row of scales or extending no lower than the 6th or 7th rows (blotches obscure in large individuals)
           L. calligaster

Pattern in rings; or if in blotches or saddles of brown, gray, or red, these broadly in contact with 5th or a lower set of scales **4**

4. Whitish crossbands more than 40; top of head black; snout uniformly white **L. pyromelana**

Whitish crossbands on body and tail less than 40; or, if more than 40, snout not uniformly whitish **5**

5. Whitish rings usually more than 30 **L. zonata**

Whitish rings less than 30 **L. doliata**

Lampropeltis mexicana (**Garman**). Mexican kingsnake. Pattern of black-bordered red or orange blotches alternating with white-bordered gray areas. Head wide; distinct neck region. Mottled gray snout; dark postocular stripe or spot; red-centered blotch on frontal parietal region. 5 subspecies; 2 in the United States. Southwestern Texas.

Lampropeltis calligaster (**Harlan**). Prairie kingsnake. Color light brown or gray with pattern of squarish, brown, green, or red blotches bordered with black; 2 alternating lateral rows of dark blotches. Length to 1300 mm. 2 subspecies. Southeastern and central United States: Maryland southward into Florida; westward to central Texas, Arkansas, Missouri, and Iowa.

Lampropeltis getulus **Linnaeus**. Common kingsnake. This species includes 9 subspecies. Intermediates between the subspecies occur over broad geographic zones. Color varies from dark blue or black to yellow; pattern of light dots or flecks, yellow crossbands or a light middorsal stripe. Southern United States.

Lampropeltis doliata (**Linnaeus**). Milk snake. This species includes 7 subspecies generally similar in that they have bright patterns formed by combi-

nations of red or brown, yellow, and black rings. The common name is associated with the myth that these snakes suck cows. East of the Rocky Mountains.

Lampropeltis zonata (**Lockington**). California mountain kingsnake. Pattern of 24–48 black rings bordered on either side by white; most of the black rings may enclose red areas laterally and dorsally. 5 subspecies. Southern Washington through California into northern Baja California.

Lampropeltis pyromelana (**Cope**). Sonora mountain kingsnake. Pattern of 35–61 light rings separated by black rings that often have red centers. rings continue across belly. 4 subspecies, 3 in United States. Utah, Arizona, southwestern New Mexico, east central Nevada; Sonora and Chihuahua.

Genus *Stilosoma* **Brown.** Short-tailed snakes.

Stilosoma extenuatum **Brown.** Short-tailed snake. Small burrowing snake; silvery gray with 60–70 brown, black-bordered spots on the body; interspaces yellow to red. Belly marked with brown or black. Length to about 500 mm. 3 subspecies. Central peninsular Florida.

Genus *Cemophora* **Cope.** Scarlet snakes.

Cemophora coccinea (**Blumenbach**). Scarlet snake. Small, burrowing snake with red, black, and yellow rings that do not extend onto the belly (Fig. 4-103). Length to about 630 mm. 2 subspecies. Oviparous. New Jersey southward through the Atlantic and Gulf States, westward to Texas and Oklahoma; separate colonies in Illinois, Indiana, and Missouri.

Genus *Rhinocheilus* **Baird and Girard.** Long-nosed snakes. Medium-sized. Smooth scales; undivided subcaudals; undivided anal. Burrowing; nocturnal.

Rhinocheilus lecontei **Baird and Girard.** Long-nosed snake. Pattern of 26–46 dark brown or black dorsal blotches alternating with bands of red, yellow, or white. Sides spotted with yellow below the black blotches. Belly light with some dark spots. Length to about 950 mm. 2 subspecies in the United States. Oviparous. Southwestern United States and northern Mexico.

Genus *Contia* **Baird and Girard.** Sharp-tailed snakes. A horny, sharp-pointed scale at tip of tail. Smooth scales; divided anal.

Contia tenuis **Baird and Girard.** Sharp-tailed

Fig. 4-103. Cemophora coccinea. (Photograph by Isabelle Hunt Conant.)

snake. Color tan or dark brown with few spots or vague reticulation of darker color; light lateral stripe sometimes present. Belly gray or yellowish white with transverse black lines. Length to about 490 mm. Vancouver Island and Puget Sound south to south central California.

Genus *Sonora* Baird and Girard. Ground snakes. Small, secretive, burrowing snakes with a pattern of bands. Smooth scales; divided anal; divided subcaudals. Rear teeth of maxillary elongated and grooved. Body cylindrical. Length to 650 mm. Oviparous. 12 species and subspecies; 3 species, 7 subspecies in the United States.

KEY TO SPECIES OF *SONORA*

1. Scale rows 15, rarely 14 or 16; ventrals 134 or more in males, 140 or more in females   2
   Scale rows 13–13, rarely 14–13 or 14–14; ventrals 126–142 in males, 136–151 in females   *S. taylori*
2. Caudals in males 39–52, in females 31–44
            S. episcopa
   Caudals in males 53–59, in females 46–51
            S. semiannulata

Sonora episcopa **Kennicott.** Ground snake. Color very variable; gray, brown, or red; uniform in color or with dark crossbands or with light edges to each scale (Fig. 4-104). 2 subspecies. Southeastern Colorado, Kansas, and western Missouri southward to eastern New Mexico, central Texas; into Nuevo León, Mexico.

Sonora semiannulata **Baird and Girard.** Western ground snake. Color variable; uniform gray to reddish or marked with distinct dark crossbands, or with a wide middorsal line of red. 5 subspecies. Western Texas to California, north to Idaho and south into Chihuahua and Baja California, Mexico.

Sonora taylori **(Boulenger).** South Texas ground snake. Color uniform brownish, unbanded. Nuevo León, Mexico, and southern Texas.

Genus *Chionactis* Cope. Shovel-nosed snakes. Upper jaw longer than lower; lower jaw countersunk within margin of upper jaw; snout long and flattened. Nasal opening with valvelike structure. Ventral surface flattened (rounded in *Sonora*). Adapted for burrowing in sand.

KEY TO SPECIES OF *CHIONACTIS*

1. Snout convex above; dark bands of body usually fewer than 21   *C. palarostris*
   Snout flattened above; dark bands of body usually 21 or more   *C. occipitalis*

Chionactis occipitalis **(Hallowell).** Western shovel-nosed snake. Color white or yellowish with a pattern of brown or black rings. An orange or red saddle alternating with the black rings which may be continuous across the belly. Snout often darker than body color. 4 subspecies in the United States. Southwestern Arizona and southeastern California; adjacent Nevada and Sonora.

Chionactis palarostris **Klauber.** Sonora shovel-nosed snake. Color yellowish with pattern of alternating black and red blotches; black blotches extend across ventrals, but red blotches do not (Fig. 4-105). Snout cream-colored. 1 subspecies in the United States. Western Pima County, Arizona, and Sonora.

Fig. 4-104 (left). *Sonora episcopa.* (From Stebbins, *Amphibians and Reptiles of Western North America,* McGraw-Hill Book Company.) *Fig. 4-105* (right). *Chionactis palarostris.* (From Stebbins, *Amphibians and Reptiles of Western North America,* McGraw-Hill Book Company.)

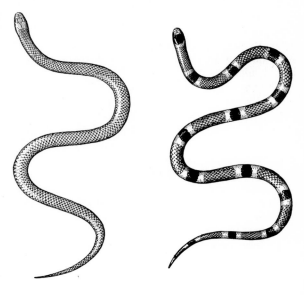

Genus *Ficimia* **Gray.** Hook-nosed snakes. Small, burrowing snakes with the rostral enlarged. Lengths to about 300 mm. 3 species.

KEY TO SPECIES OF *FICIMIA*

1. Rostral separating the small internasals and in contact with prefrontals **2**
   Rostral separating the prefrontals and in contact with frontal; no internasals **F. olivacea**
2. Black-edged brown markings on head and body; head markings not forming a continuous blotch or mark **F. cana**
   Uniformly colored black markings on head and body; a large black mark on head **F. quadrangularis**

Ficimia cana **(Cope).** Western hook-nosed snake. Color grayish brown with 30 or more dark-banded crossbands on the body. Dorsal scales 19–17. Western Texas to Arizona; southward into Mexico.

Ficimia olivacea **Gray.** Mexican hook-nosed snake. Color gray with 46 or more dark crossbands on the body. Dorsal scales 17–17. Southern Texas to Central America.

Ficimia quadrangularis **Günther.** Color light brown, cream, or red with 25–32 dorsal black blotches. A black Y-shaped mark covering head; base of Y extending posteriorly to connect with a black neck band 4 to 5 scales in length. Southeastern Arizona.

Genus *Chilomeniscus* **Cope.** Sand snake. Small snakes adapted to burrowing in sand. Snout shovel-like. 1 species.

Chilomeniscus cinctus **Cope.** Banded sand snake. Scales very smooth. Color yellow or red; 18–28 dark crossbands on the body. Belly white to yellow; rings may extend across ventral side posteriorly. Length to 250 mm. Southwestern Arizona; adjacent Sonora and Baja California.

Genus *Hypsiglena* **Cope.** Night snakes. Small, nocturnal, venomous snakes with enlarged teeth in the rear of the upper jaw; a vertical pupil. Length to 400 mm. 1 species in the United States.

Hypsiglena torquata **Günther.** Night snake. Color gray or yellow with a pattern of alternating series of dorsal and lateral spots. 6 subspecies in the United States. Western Washington southward to California and Arizona; east to central Oklahoma and Texas; south through Central America and northern South America.

Genus *Leptodeira* **Fitzinger.** Cat-eyed snakes. Slender, nocturnal snakes with enlarged grooved teeth in rear of the upper jaw. Oviparous. 1 species in the United States.

Leptodeira annulata **(Linnaeus).** Cat-eyed snake. Color gray to greenish with brown or black dorsal blotches extending almost to ventrals. Belly yellowish. Southern Texas to South America.

Genus *Trimorphodon* **Cope.** Lyre snakes. Slender-bodied snakes with broad heads, vertical pupils, and grooved fangs in rear of upper jaw. "Lyre-shaped" markings on the head of some forms. Oviparous. Occur primarily in rocky sections of desert and semidesert. 14 species and subspecies. Southern United States to Costa Rica. 3 species in the United States.

KEY TO SPECIES OF *TRIMORPHODON*

1. Blotches or bands on body usually less than 23; interspaces wider than blotches **T. vilkinsoni**
   Blotches on body usually more than 21; interspaces equal to or less than width of blotches **2**
2. Anal plate entire **T. vandenburghi**
   Anal plate divided **T. lambda**

Trimorphodon lambda **Cope.** Sonora lyre snake. Background grayish; a series of 21–34 brownish dorsal blotches each split by a transverse light bar. Juvenile with a series of dark-brown diamonds on a light-gray background. Length to 1030 mm. Southeastern California, southern Nevada, and southwestern Utah; southward in Sonora.

Trimorphodon vandenburghi **Klauber.** California lyre snake. Background light gray or light brown; a series of 28–43 brownish dorsal blotches split by a transverse light bar. Coastal and desert southern California, from Los Angeles County and the Argus Mountains, Inyo County, southward into northwestern Baja California.

Trimorphodon vilkinsoni **Cope.** Texas lyre snake. Background color blue-gray; a series of 17–22 brown crossbands; usually a series of small blotches between the larger crossbands. Length to 730 mm. Southeastern New Mexico and trans-Pecos Texas; adjacent Chihuahua.

Genus *Oxybelis* Wagler. Vine snakes. Body extremely slender; head much elongated. Arboreal. Habits largely unknown. 6 species. Southwestern United States south to northern Argentina. 1 species in the United States.

***Oxybelis aeneus* (Wagler).** Vine snake. Color brown to gray with dark dots irregularly spaced on the body. Belly reddish or grayish; color may be whitish to light yellow. Arizona southward to tropical South America.

Genus *Tantilla* Baird and Girard. Black-headed snakes. Small, slender, secretive, rear-fanged snakes. Head blunt; snout projecting beyond lower jaw. Color uniform pale brown; some species with a black head or a dark neck ring. Oviparous; probably 2–6 eggs. Length to about 400 mm. About 45 species and subspecies. Southern United States to South America. 6 species in the United States.

KEY TO SPECIES OF *TANTILLA*

1. A light neck band crossing tips of parietals (Fig. 4-106)   **2**
   No light neck band crossing tips of parietals   **4**
2. A dark band forming posterior margin of light band; dark band ½ to 1½ scale lengths in width; eye large, its diameter more than half its distance from snout   ***T. wilcoxi***
   A dark band forming posterior margin of light band, but band 2–4 scale rows in width; eye small, its diameter less than half distance to snout   **3**
3. Light neck band interrupted by black area extending posteriorly from parietals and 1st vertebral   ***T. diabola***
   Light neck band not interrupted by black area extending posteriorly from parietals and 1st vertebral   ***T. coronata***
4. Upper labials usually 7 (6th approximately as long as 5th); postoculars usually 2, rarely 1; head black or dark brown above, contrasting with general coloration   **5**
   Upper labials usually 6 (when 7, 6th usually much shorter than 5th); usually 1 postocular; head but little darker than dorsal body color   ***T. gracilis***
5. Black of head extending 3–5 scale lengths behind parietals, not bordered behind with a narrow white band; mental plate usually separated from chin shields by 1st lower labials   ***T. nigriceps***
   Black of head usually extending only 1 or 2 scale lengths behind parietals, usually bordered behind with a narrow white band; mental plate usually in contact with chin shields (Fig. 4-107)   ***T. planiceps***

***Tantilla coronata* Baird and Girard.** Crowned snake. Length to 320 mm. 3 subspecies. Southeastern United States; from Virginia and southern Indiana southward; west to Louisiana.

***Tantilla diabola* Foquette and Potter, Jr.** Val Verde black-headed snake. Length to 175 mm. General color tannish gray; ventral surface white. Black patch covering top of head; labials black with light areas; whitish collar. Val Verde County, Texas.

***Tantilla gracilis* Baird and Girard.** Flat-headed snake. Only species without the black head. Length to 230 mm. 2 subspecies. Central Missouri and eastern Kansas through Oklahoma and Arkansas to eastern Texas.

***Tantilla nigriceps* Kennicott.** Great Plains black-headed snake. Color pale brown or yellowish gray; whitish collar; labials gray or white. Caudals bright coral red. 3 subspecies. Great Plains region.

***Tantilla wilcoxi* Stejneger.** Huachuca black-headed snake. Color pale brownish gray; top and sides of head dark brown or gray, labials light in color; whitish collar. Belly coral red. Southeastern Arizona.

***Tantilla planiceps* Blainville.** Western black-headed snake. General color gray to brown; a fine, dark line on the middorsal scale row. 7 subspecies;

Fig. 4-106 (left). **Head of *Tantilla wilcoxi*. (From Stebbins, *Amphibians and Reptiles of Western North America*, McGraw-Hill Book Company.)**
Fig. 4-107 (right). **Head of *Tantilla planiceps*. (From Stebbins, *Amphibians and Reptiles of Western North America*, McGraw-Hill Book Company.)**

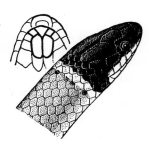

5 in the United States. Western Colorado to California and southeastward to western Texas.

Family Elapidae. Coral snakes. Terrestrial and arboreal; small to large snakes. Both jaws with teeth, anterior pair of teeth in upper jaw grooved or perforated for transporting of venom; fangs permanently erect; venom primarily neurotoxic. No vestiges of girdles or limbs; coronoid bone absent. Most abundant in Asia, Africa, and Australia; more than 30 genera. Only 2 genera in the United States. Some of the best-known poisonous snakes of the world are in this family, the mambas, cobras, kraits, and many of the poisonous snakes of Australia.

KEY TO GENERA OF ELAPIDAE

1. Ring on neck red   *Micruroides*
   Ring on neck black   *Micrurus*

Genus *Micrurus* Wagler. Eastern coral snakes. Brightly colored snakes usually ringed with black, red, and yellow. The only tooth on the maxillary bone modified as a fang. 68 species and subspecies. Southern United States to northern Argentina, Bolivia, and Peru. 1 species, 3 subspecies in the United States.

Micrurus fulvius (Linnaeus). Eastern coral snake. Pattern of wide black rings bordered by narrow yellow rings and separated from the next yellow-bordered black ring by a red wing. Length to 1300 mm. Southeastern United States, westward to Pecos River in Texas; also in adjacent Tamaulipas.

Genus *Micruroides* Schmidt. Western coral snake. A solid tooth on the rear of the maxillary in addition to the fang on the front. 1 species.

Micruroides euryxanthus (Kennicott). Arizona coral snake. Pattern of black, red and yellow rings; first ring behind the yellow on the head is red (black in *M. fulvius*). Southern Arizona and New Mexico to Sonora and Chihuahua.

Family Viperidae, Subfamily Crotalinae. Pit vipers. Terrestrial, arboreal. Both jaws with teeth; pair of teeth on anterior part of maxillary bone modified for venom transport; these teeth may be folded against roof of mouth. Venom primarily hematoxic. No vestiges of girdles or limbs; coronoid bone absent. Pupil of eye vertical. A heat-sensitive pit between eye and nostril. Viviparous. 5 genera, 3 in the United States. Central and South America, Asia.

KEY TO GENERA OF CROTALINAE

1. A rattle on end of tail   2
   No rattle on end of tail   *Agkistrodon*
2. Top of head covered with large scales (Fig. 4-108)
                                                     Sistrurus

   Top of head covered with small scales (if large scales are present, they do not occur back of eyes) (Fig. 4-109)   *Crotalus*

Fig. 4-108 (left). Head of *Sistrurus*. *Fig. 4-109* (right). Head of *Crotalus*.

 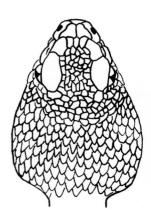

Genus *Agkistrodon* Beauvois. Copperheads and Cottonmouths. Heavy-bodied snakes with strongly keeled scales and short tails. United States to Nicaragua; Asia. 2 species in the United States.

KEY TO SPECIES OF *AGKISTRODON*

1. 23 scale rows; a loreal scale present   *A. contortrix*
   25 scale rows; no loreal scale   *A. piscivorus*

Agkistrodon contortrix (*Linnaeus*). Copperhead. Color brown to copper red with 10–20 darker crossbands on body; crossbands narrower at middle of body than on sides. Narrow black line from eye to angle of jaws. Belly light with dark blotches. Young with a yellow- or green-tipped tail. Terrestrial. 4 subspecies. Length to 1450 mm. Eastern and southeastern United States.

Agkistrodon piscivorus (Lacépède). Cottonmouth. Color olive brown to black with 10–16 darker crossbands. Belly light with dark blotches. Large adults uniform black. Young reddish brown with yellow- or green-tipped tails. Semiaquatic. Lengths to 1900 mm. 2 subspecies. Southeastern United States.

Genus *Sistrurus* Rafinesque. Pygmy rattlers and massasaugas. Small snakes with a rattle comparatively smaller than in the true rattlesnakes. 3 species; 2 in the United States.

KEY TO SPECIES OF *SISTRURUS*

1. Upper preocular not in contact with postnasal; tail long and slender **S. miliarius**
   Upper preocular in contact with postnasal; tail short **S. catenatus**

Sistrurus miliarius (Linnaeus). Pygmy rattlesnake. Gray to brownish black in color with 30–40 black dorsal blotches alternating with smaller dark lateral marks. Belly white or yellow with black blotches. Length to about 800 mm. 3 subspecies. North Carolina to Florida, westward to southeastern Missouri, Arkansas, Oklahoma, and central Texas.

Sistrurus catenatus (Rafinesque). Massasauga. Color gray to brown with 20–50 dark-red-brown or black dorsal blotches. 2 alternating lateral rows of similar markings. Belly dark and blotched or light with many irregular dark blotches. Length to about 950 mm. 3 subspecies. The Great Plains and prairie regions southward from Nebraska into northern Mexico; eastward to Pennsylvania and New York.

Genus *Crotalus* Linnaeus. Rattlesnakes. Head triangular; neck narrow. Pupil vertical. Body short and heavy. Anal plate and subcaudals not divided. The common name suggests the outstanding character of the genus. The rattle is a unique structure formed of horny, interlocking segments; present only as a horn button in newly born young. A new ring is added to the rattle each time the snake sheds its skin. 42 species and subspecies in North and South America; 13 species in the United States.

KEY TO SPECIES OF *CROTALUS*

1. A pair of hornlike processes above eyes **C. cerastes**
   No hornlike process above eyes **2**

2. Tip of snout and canthus rostralis raised into a sharp ridge; rostral and mental marked vertically by a narrow light line on a red-brown ground **C. willardi**
   Tip of snout and canthus rostralis not raised into a ridge; no light line on rostral and mental **3**
3. Prenasal curved under postnasal; usually a dorsal pattern of widely separated crossbars **C. lepidus**
   Prenasal not curved under postnasal; pattern not widely separated crossbars **4**
4. Prenasals usually separated from rostral by small scales or granules; upper preoculars often divided, horizontally, vertically, or both **C. mitchelli**
   Prenasals in contact with rostral; upper preocular not divided, or if divided, loreal longer than high **5**
5. Tail of alternating black and light-ash-gray rings, both colors in contrast with posterior belly color which may be gray, dark gray, cream, pink, red, red-brown, or olive-brown **6**
   Tail not of alternating black and light-ash-gray rings in strong color contrast to body color immediately anterior to tail **8**
6. Dark and light tail rings of approximately equal width; postocular light stripe, if present, intersects upper labials from 1 to 3 scales anterior to angle of mouth; proximal rattle black **7**
   Dark tail rings narrower than light; postocular light stripe, if present, passes backward above angle of mouth; lower half of proximal rattle light in color **C. scutulatus**
7. First lower labials usually not divided transversely; general color buff gray or gray-brown (or pink or red); dark punctuations conspicuous in marking **C. atrox**
   First lower labials usually divided transversely; general color pink, red, brick red, or red-brown; dark punctuations faint or absent from markings **C. ruber**
8. 2 internasals **9**
   More than 2 internasals **C. viridis**
9. A vertical light line on the posterior edge of the prenasals and first upper labials **C. adamanteus**
   No vertical light line on the posterior edge of the prenasals and first upper labials **10**
10. Supraoculars pitted, sutured, or with outer edges broken **C. mitchelli**
    Supraoculars not pitted, sutured, nor with broken outer edges **11**
11. Length of head contained in total length (adults) less than 25 times; width of proximal rattle contained in length of head more than 2½ times **12**
    Length of head contained in total length (adults)

25 times or more; width of proximal rattle contained in length of head less than 2½ times
C. tigris

12. Usually a definite division between the scales of frontal and prefrontal areas; scales in anterior part of frontal are larger than those behind; anterior body pattern not in chevron-shaped bands or not all black   13

No definite division between scales of frontal and prefrontal areas; scales in anterior part of frontal area not conspicuously larger than those behind; normal pattern a series of chevron-shaped crossbands, sometimes broken, or with the body all black   **C. horridus**

13. Dorsal pattern of 2 parallel rows of small brown blotches   **C. pricei**

Dorsal pattern of a single row of large blotches   14

14. Tail rings sharply contrasting in color; usually a single loreal   **C. scutulatus**

Tail often black or with rings faintly in evidence against dark backgrounds   **C. molossus**

Crotalus horridus **Linnaeus.** Timber rattlesnake. Color yellowish brown, gray to black with 18–33 black or dark-brown V-shaped crossbands on body. Tail black in larger individuals. Large individuals sometimes uniform black. Length to 1850 mm. 2 subspecies. Eastern North America.

Crotalus molossus **Baird and Girard.** Blacktailed rattlesnake. Color yellow or greenish with dark rhomboid blotches which may connect with lateral spots to make crossbands. Tail black. Central Texas to Arizona; southward into Mexico.

Crotalus pricei **Van Denburgh.** Twin-spotted rattlesnake. A dorsal pattern of 2 or 3 rows of small squarish spots on a gray-to-brown background. Southeastern Arizona; adjacent Sonora and Chihuahua, and southward into Mexico.

Crotalus lepidus **(Kennicott).** Rock rattlesnake. Color olive to greenish gray with narrow black widely spaced bands. Belly light and not marked. 2 subspecies in the United States. Central Texas to southern Arizona, southward on the Mexican plateau.

Crotalus scutulatus **Kennicott.** Mojave rattlesnake. Pattern of diamond-shaped marks bordered by light scales. Distinct black bands around the tail. Southern Nevada and California southeastward to Texas and into Mexico.

Crotalus adamanteus **Beauvois.** Eastern diamond rattlesnake. Color olive to brown with a pattern of distinct brown-black diamond-shaped marks with light centers and borders of yellow. Length to 2600 mm. Coastal North Carolina, South Carolina, and Georgia; Gulf Coastal Plain to Louisiana.

Crotalus atrox **Baird and Girard.** Western diamondback rattlesnake. Color tan, gray, or brown with 25–45 diamond-shaped or rhomboid darker blotches with light borders on body. Tail with distinct broad black rings separated by white. A light diagonal stripe back of eye. Length 2200 mm. Central Arkansas, western Oklahoma, and Texas, southwestward to southeastern California and adjacent states of Mexico.

Crotalus ruber **Cope.** Red diamond rattlesnake. Color reddish with a pattern of darker blotches as described for *C. atrox*. Most placid of the genus. Length to 1800 mm. Southwestern California southward to tip of Baja California.

Crotalus viridis **(Rafinesque).** Western rattlesnake. Color greenish gray, yellow, red, or black. Pattern of 35–55 round or square dorsal blotches of a darker shade. Side of head with 2 diagonal white lines. Length to 1500 mm. 8 subspecies in the United States. Western North America; western North Dakota south through western Texas; westward to Pacific Coast.

Crotalus mitchelli **(Cope).** Speckled rattlesnake. Color very variable; may be reddish, gray, or yellow with darker specks or may have a pattern of distinct dark blotches or bands on body and tail. Distinguished from all other rattlesnakes of the United States by the small scales between the rostral and nasal scales. 2 subspecies. Southwestern United States, Baja California, and Sonora.

Crotalus tigris **Kennicott.** Tiger rattlesnake. Color gray or tan, sometimes pink-tinted. Pattern of numerous narrow dark crossbands. 6–8 dark rings on tail. Southern and central Arizona; northeastern and central Sonora.

Crotalus willardi **Meek.** Ridge-nosed rattlesnake. Color brownish with a pattern of narrow, widely separated light crossbands. Tail with longitudinal stripes. Southeastern Arizona and southwestern New Mexico to the Mexican states of Durango and Zacatecas.

Crotalus cerastes **Hallowell.** Sidewinder. The pair of hornlike processes over the eyes distinguishes this species. Color light yellowish, pink, to gray with a dorsal series of light-brownish blotches. Nocturnal. 3 subspecies. Mojave and Colorado Deserts.

REFERENCES

General

Barbour, T. 1926. *Reptiles and Amphibians: Their Habits and Adaptations,* rev. ed., Houghton Mifflin Company, Boston.

Carr, A. 1952. *Handbook of Turtles,* Comstock Publishing Associates, Inc., Ithaca, N.Y.

Conant, R. 1958. *A Field Guide to Reptiles and Amphibians,* Houghton Mifflin Company, Boston.

——— and W. Bridges. 1939. *What Snake Is That?* Appleton-Century-Crofts, Inc., New York.

Cope, E. D. 1900. The Crocodilians, Lizards and Snakes of North America, *Rept. U.S. Natl. Museum* for 1898.

Curran, C. H., and C. Kauffeld. 1937. *Snakes and Their Ways,* Harper & Brothers, New York.

Ditmars, R. L. 1936. *The Reptiles of North America,* Doubleday & Company, New York.

———. 1939. *A Field Book of North American Snakes,* Doubleday & Company, New York.

Gadow, H. 1923. *Amphibia and Reptiles,* St. Martin's Press, Inc., New York.

Goin, J., and O. B. Goin. 1962. *Introduction to Herpetology,* W. H. Freeman and Company, San Francisco.

Pope, C. H. 1937. *Snakes Alive and How They Live,* The Viking Press, Inc., New York.

———. 1939. *Turtles of the United States and Canada,* Alfred A. Knopf, Inc., New York.

Romer, A. S. 1956. *Osteology of the Reptiles,* University of Chicago Press, Chicago.

Schmidt, K. P. 1953. *A Check List of North American Amphibians and Reptiles,* University of Chicago Press, Chicago.

——— and D. D. Davis. 1941. *Field Book of Snakes,* G. P. Putnam's Sons, New York.

Smith, H. M. 1946. *Handbook of Lizards,* Comstock Publishing Associates, Ithaca, N.Y.

Wright, A. H., and Anna A. Wright. 1957. *Handbook of Snakes of the United States and Canada,* Comstock Publishing Associates, Ithaca, New York, 2 vols.

Regional

Babcock, H. L. 1919. Turtles of New England, *Mem. Boston Soc. Nat. Hist.,* 8:325.

Blanchard, F. N. 1922. The Amphibians and Reptiles of Western Tennessee, *Occasional Papers Museum Zool. Univ. Mich.,* no. 117, 18 pp.

Breckenridge, W. J. 1944. *Reptiles and Amphibians of Minnesota,* University of Minnesota Press, Minneapolis.

Brown, B. C. 1950. *An Annotated Checklist of the Reptiles and Amphibians of Texas,* Baylor University Press, Waco, Tex.

Cagle, F. R. 1941. A Key to the Reptiles and Amphibians of Illinois, *Museum Nat. Soc. Sci. Contr.* 5.

———. 1952. *A Key to the Amphibians and Reptiles of Louisiana,* Tulane Book Store, New Orleans, La.

Cahn, A. R. 1937. The Turtles of Illinois, *Ill. Biol. Monograph* 16, 218 pp.

Carr, A. F., Jr. 1940. A Contribution to the Herpetology of Florida, *Univ. Fla. Publ. Biol. Sci.* 3, 118 pp.

Chermock, R. L. 1952. A Key to the Amphibians and Reptiles of Alabama, *Geol. Surv. Ala. Museum Paper* 33, 88 pp.

Conant, R. 1951. *The Reptiles of Ohio,* University of Notre Dame Press, Notre Dame, Ind.

Cook, F. A. 1942. Alligators and Lizards of Mississippi, *Bull. Miss. State Game Fish Comm.,* 20 pp.

———. 1954. Snakes of Mississippi, *Bull. Miss. State Game Fish Comm.,* 40 pp.

Dellinger, S. C., and J. D. Black. 1938. Herpetology of Arkansas. 1. The Reptiles, *Occasional Papers Univ. Ark. Museum,* no. 1, 47 pp.

Fowler, H. W. 1907. The Amphibians and Reptiles of New Jersey, *Ann. Rept. N.J. State Museum* for 1904.

Fowlie, J. A. 1965. *The Snakes of Arizona,* Azul Quinta Press, Fallbrook, Calif.

Gordon, K. 1939. The Amphibia and Reptilia of Oregon, *Ore. State Monographs Zool.*

Guthrie, J. E. 1926. The Snakes of Iowa, *Bull. Iowa State Agr. Expt. Sta.,* 239.

Hudson, G. E. 1942. The Amphibians and Reptiles of Nebraska, *Nebr. Conservation Bull.* 24, 146 pp.

Hurter, J. 1911. Herpetology of Missouri, *Trans. Acad. Sci. St. Louis,* 20, p. 59.

Jones-Burdick, W. H. 1939. Guide to the Snakes of Colorado, *Univ. Colo. Museum Leaflet* 1, 11 pp.

Kelly, H. A., A. W. Davis, and H. C. Robertson. 1936. Snakes of Maryland, *Nat. Hist. Soc. Md. Spec. Publ.*

Lamson, G. H. 1935. The Reptiles of Connecticut, *Conn. Geol. Nat. Hist. Surv. Bull.* 54, 35 pp.

McCauley, R. H. 1945. *The Reptiles of Maryland and the District of Columbia,* privately printed, Hagerstown, Md.

Maslin, T. Paul. 1959. An Annotated Checklist of the Amphibians and Reptiles of Colorado, *Univ. Colo. Stud., Ser. Biol.,* no. 6, 97 pp.

Netting, M. G. 1939. Hand List of the Amphibians and

Reptiles of Pennsylvania, *Bienn. Rept. Penn. Fish Comm.* for 1936–1938, 26 pp.

Oliver, J. A., and J. R. Bailey. 1939. Amphibians and Reptiles of New Hampshire, Exclusive of Marine Forms, *Biol. Surv. Conn. Watershed Rept.* 4, p. 195.

Ortenburger, A. I. 1929. A Key to the Lizards and Snakes of Oklahoma, *Publ. Univ. Okla. Biol. Surv.* 2, p. 209.

Over, W. H. 1923. Amphibians and Reptiles of South Dakota, *S.D. Geol. Nat. Hist. Surv. Bull.* 12, 34 pp.

Owen, R. P. 1940. A List of the Reptiles of Washington, *Copeia,* **1940:**169.

Perkins, C. B. 1942. A Key to the Snakes of the United States, *Bull. Zool. Soc. San Diego,* 24, 79 pp.

Pickwell, G. 1947. *Amphibians and Reptiles of the Pacific States,* Stanford University Press, Stanford, Calif.

Pope, T. E. B., and W. E. Dickinson. 1928. The Amphibians and Reptiles of Wisconsin, *Bull. Public Museum Milwaukee,* 8, 138 pp.

Ruthven, A. G., C. Thompson, and H. T. Gaige. 1928. The Herpetology of Michigan, *Mich. Handbook Ser. Univ. Mich.* 3.

Schwart, H. H. 1938. Reptiles of Arkansas, *Univ. Ark. Agr. Expt. Sta., Bull.* 357, 47 pp.

Smith, Hobart M. 1950. Handbook of Reptiles and Amphibians of Kansas, *Museum Zool. Misc. Publ. no.* 2, Lawrence, Kansas.

Smith, Philip W. 1961. The Amphibians and Reptiles of Illinois, *Bull. Ill. Nat. Hist. Surv.,* 28, 298 pp.

Stebbins, R. C. 1954. *Amphibians and Reptiles of Western North America,* McGraw-Hill Book Company, New York.

Surface, H. A. 1906. The Serpents of Pennsylvania, *Zool. Bull. Div. Zool. Penn. State Dept. Agr.* 4, p. 113.

———. 1907. The Lizards of Pennsylvania, *ibid.* 5, p. 233.

Tanner, W. W. 1941. The Reptiles and Amphibians of Idaho, *Great Basin Naturalist,* **2:**87.

Van Denburgh, J. 1922. The Reptiles of Western North America. 1. Lizards. 2. Snakes and Turtles, *Occasional Papers Calif. Acad. Sci.,* no. 10, 1028 pp.

———. 1942. Notes on the Herpetology of New Mexico with a List of Species Known from That State, *Proc. Calif. Acad. Sci.,* **13:**189.

Woodbury, A. M. 1931. A Descriptive Catalogue of the Reptiles of Utah, *Bull. Univ. Utah,* 21, 129 pp.

Wright, A. H., and S. C. A. Bishop. 1915. A Biological Reconnaissance of the Okefinokee Swamp in Georgia. The Reptiles, *Proc. Acad. Nat. Sci. Phila.,* 1915, 85 pp.

BIRDS arose from diapsid reptiles during late Jurassic time, and they still retain many traces of their reptilian ancestry. The skull is essentially reptilian, with a single occipital condyle, only 1 middle-ear bone, and a lower jaw composed of several bones. The coracoid remains as a distinct bone in the pectoral girdle, and the foot retains the reptilian formula of 2–3–4–5 phalanges for the first 4 digits. There are scales on the feet and in some cases separate shields covering the jaws. Reptilian features of the eye are the bony sclerotic ring, pecten, and striated ciliary muscles. The shelled telolecithal egg with its 4 extraembryonic membranes is also a legacy from the reptiles.

Avian Characters

Evolutionary changes for flight have modified this basically reptilian pattern. Such modifications are associated with an increase in metabolic rate, the formation of a flight surface, reduction in weight, and rigidity of the skeleton. Many adaptations for flying are retained even in those groups, like the ostriches and penguins, which have secondarily lost the power of flight.

Changes in several systems have given increased metabolic efficiency. Body temperature is higher than in other vertebrates, ranging between 100–110°F. Feathers form an insulating cover which helps retain body heat. The 4-chambered heart, shared with the mammals, is an efficient organ for the circulation of blood, but the presence of the right aortic arch in birds and the left one in mammals indicates their separate histories. The lungs of birds are molded to the ribs, instead of hanging free. Aeration is complete at each breath, and since there is no diaphragm, it is accomplished by rib action alone. A number of the ribs have uncinate processes, which overlap the following ribs.

A resistant yet flexible flight surface is supplied by the large quills, the remiges, and rectrices. The interlocking structure of these feathers affords a lightweight, elastic surface for support in the air.

Weight reduction is accomplished by reduction of muscles, by the lightness of the skeleton, and by the disappearance of unnecessary parts. The wing muscles and those of the lower leg are much reduced in volume, with very long tendons going out to the insertions, so that there is practically no flesh on the distal portions of wings and legs. The motive power for the wings is largely supplied by the superficial pectoral muscle on the downstroke and by the supracoracoideus muscle on the upstroke. These muscles make up much of the bird's weight and are associated with great development of the sternum, which is provided with a keel (carina) in flying species.

Much of the skeleton is hollow, reducing its weight in some cases to less than half that of the feathers. Air sacs, already present in a rudimentary state in some reptiles, extend from the lungs into many of the bones.

Further reduction of weight is accomplished by the loss or decrease in size or number of various organs. Teeth have been gone in all birds since the Cretaceous. The typical pentadactyl hand of vertebrates has been reduced to 3 fingers, each consisting of only 1 or 2 phalanges. The metacarpals are correspondingly reduced to 3, and the free carpals to only 2. The fibula is degenerate, and the fifth toe is always gone, except in rare mutations. The caudal vertebrae have decreased in number. There is no urinary bladder, and in most forms no penis. Usually only the left ovary persists, permitting the development of but a single egg at a time.

Rigidity of the skeleton is obtained through the fusion of bones. In the skull the fusion has progressed so far that most of the sutures are no longer visible. In the palate the previously paired prevomers have united to form a small median bone. In the vertebral column there is a tendency for some of the thoracic vertebrae to fuse and the most distal caudal vertebrae unite to form a pygostyle, to which the large tail feathers attach. The greatest fusion of vertebrae occurs to form the synsacrum, composed of the posterior thoracics, lumbars, sacrals, and anterior caudals. Furthermore, the right and left innominate bones tend to fuse to the synsacrum. In the pectoral girdle the 2 clavicles grow together to form a furculum, which in turn is sometimes ankylosed to the sternum. In the wing the second and third fingers are bound together. The metacarpals and some of the carpals unite into a single element, the carpometacarpus. In the leg the more proximal tarsals fuse to the tibia to form a tibiotarsus, and the distal tarsals unite with the second, third, and fourth metatarsals to form a tarsometatarsus. Since bird bones lack epiphyses, these compound leg bones function during growth in much the same manner as the epiphyses of mammalian bones. Since the ankle joint occurs between tarsal elements, it is useful in absorbing the shock of alighting.

Birds retain a large number (13–25) of cervical vertebrae. Their centra are unique in being saddle-shaped or heterocoelous, an adaptation which permits great flexibility and allows the bird to reach all parts of its body with the bill or to fold the neck backward in flight.

The voice of birds is produced in a special structure known as the syrinx, situated near the juncture of the bronchi, so that the voice originates near the base of the trachea, rather than at its upper end as in other vertebrates.

Ordinal and Family Characters

The major taxonomic units are based partly on the pterylography but in large part on internal structure. Skeletal characters which have been found useful in defining the orders and families include the structure of the palate, the number of cervical vertebrae and the shape of their processes, the structure of the sternum, the position of the coracoids and the development of the procoracoid process, the presence or absence of a bony bridge at the lower end of the tibiotarsus, and the number of canals in the hypotarsus.

The palatal types (Figs. 5-1–5-3) are based on variations in the maxillopalatine processes and the vomer. The maxillopalatines are fused together at the midline in the desmognathous palate, separate in the other types. The vomer is small or lacking in the desmognathous type, small and pointed in the schizognathous palate, and broad and anteriorly truncate or 2-pronged in the aegithognathous palate.

In the muscular system most emphasis is placed in the leg (Figs. 5-4, 5-5) and syringeal musculature. The presence or absence of certain thigh muscles is indicated by symbols. Thus the thigh-muscle formula $ABXY+$ (as in the pigeon) indicates the presence of the caudofemoralis muscle (A), its accessory part the iliofemoralis (B), the flexor cruris lateralis or semitendinosus (X), its

accessory part (Y), and the ambiens ($+$). The thigh formula A (as in the pelican) indicates the presence of the caudofemoralis only. Often a minus sign is used to denote the absence of the ambiens; so that in the case of the pelican the designation $A-$ is optional.

The 2 deepest plantar tendons, flexor digitorum longus and flexor hallucis longus, show important taxonomic variations. 8 or more variations are designated by numbers, but they fall into 3 main groups (Figs. 5-6–5-8).

In a schizopelmous condition (type 7) the 2 tendons cross but are entirely free from each other and thus control their respective toes only.

Fig. 5-1. Desmognathous palate of black vulture (*Coragyps atratus*). *BP*, basipterygoid process; *MXP*, maxillopalatine process; *P*, palatine; *PT*, pterygoid; *Q*, quadrate; *R*, basisphenoidal rostrum.

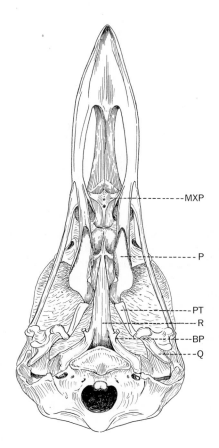

Fig. 5-2. **Schizognathous palate of chicken (***Gallus gallus***).** *MXP,* maxillopalatine process; *P,* palatine; *PT,* pterygoid; *Q,* quadrate; *R,* basisphenoidal rostrum; *V,* vomer.

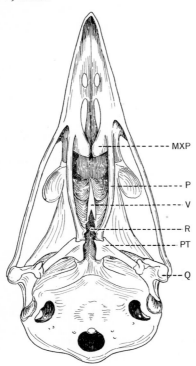

toes. In type 5 the tendons unite without crossing and then send more or less equal branches to all 4 toes. In type 6 flexor digitorum inserts only on the third toe; the first, second, and fourth toes receive branches from flexor hallucis, which also controls the third toe through a vinculum.

The number of pairs of intrinsic muscles of the syrinx and their place of insertion form useful characters in defining taxonomic groups. Likewise useful is the position of the syrinx itself, whether located on the trachea, on the bronchi, or at the juncture of the trachea and bronchi.

In the digestive system the presence or absence of a functional crop or caeca is of some importance, but the convolutions of the small intestine are of greater taxonomic value. The variations in the in-

Fig. 5-3. **Aegithognathous palate of fish crow (***Corvus ossifragus***).** *MXP,* maxillopalatine process; *P,* palatine; *PT,* pterygoid; *Q,* quadrate; *R,* basisphenoidal rostrum; *V,* vomer.

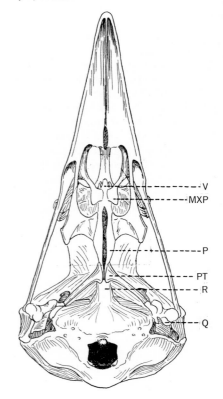

In the heteropelmous condition (type 8) flexor digitorum goes to the first 2 toes, which are directed backward; it also controls the anterior toes through a vinculum. Flexor hallucis controls the anterior toes only.

In a desmopelmous foot the 2 tendons are more or less united. 6 variations occur. In type 1 the tendons cross and unite, so that both control all 4 toes. In type 2 the hallux is supplied only above the union of the 2 tendons, so that flexor hallucis governs all 4 toes, but flexor digitorum controls the anterior toes only. In type 3 the crossing of the tendons is so distal that flexor hallucis controls only the first 2 toes, while flexor digitorum controls the 3 anterior toes. In type 4 the hallux has been lost, but both muscles are equally strong, with the common tendon dividing to insert on the anterior

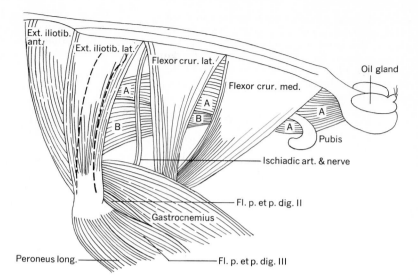

Fig. 5-4. Thigh muscles of whistling swan (*Olor columbianus*). Lateral view. *A*, caudofemoralis; *B*, iliofemoralis.

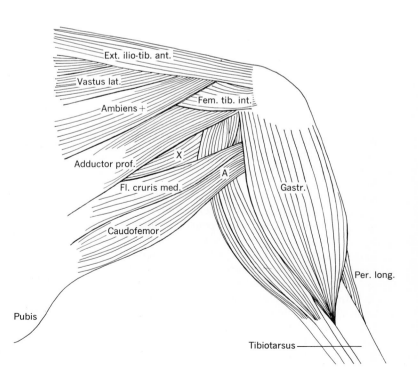

Fig. 5-5. Thigh muscles of whistling swan (*Olor columbianus*). Medial view. *X*, flexor cruris lateralis.

Fig. 5-6 (left). Desmopelmous foot of rock dove (*Columba livia*). *Fig. 5-7* (center). Schizopelmous foot of fish crow (*Corvus ossifragus*). *Fig. 5-8* (right). Heteropelmous foot of citreoline trogon (*Trogon citreolus*). *Fl. d.*, flexor digitorum longus; *Fl. h.*, flexor hallucis longus; I-IV, branches leading to respective toes.

Fig. 5-9 (left). Type 1 periorthocoelous intestine of common tern (*Sterna hirundo*). *Fig. 5-10* (right). Type 2 orthocoelous intestine of Leach's petrel (*Oceanodroma leucorhoa*).

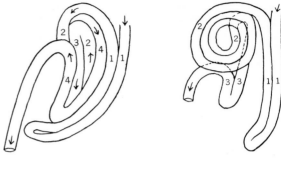

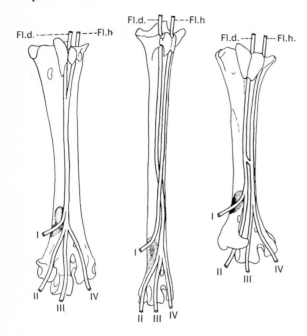

Fig. 5-11 (left). Type 3 telocoelous intestine of Hawaiian goose (*Nesochen sandvicensis*). *C*, caecum. (After Gadow.) *Fig. 5-12* (right). Type 4 telogyric intestine of European sparrow hawk (*Accipiter nisus*).

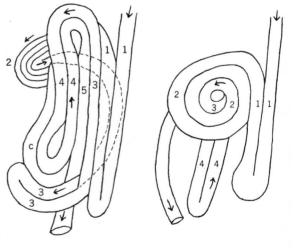

testine depend upon the number of loops, whether the loops are right- or left-handed, whether they are straight or spiral, and whether the descending and ascending branches of a loop are connected by a mesentery (closed) or are free and enclose another loop (open).

There are 8 principal variations in intestinal convolution, but unfortunately the numbering of the types does not follow a phylogenetic sequence. In the orthocoelous intestine (type 2, Fig. 5-10) the loops are all straight, alternating from right to left, and closed. The antorthocoelous condition (type 7, Fig. 5-15) is similar, but the fourth loop is open. In the telocoelous condition (type 3, Fig. 5-11) in the Anseriformes, the first 4 loops alternate, but the following ones are all left-handed; the second loop forms a spiral, and the second and third are open. In the telogyric intestine (type 4,

Fig. 5-12) the loops are all closed and alternating, but the second and third form a spiral. In the mesogyric intestine (type 8, Fig. 5-16) the loops alternate; the second and third form a spiral, and the fourth is open. In the plagiocoelous intestine (type 5, Fig. 5-13) the first 2 loops are right-handed and closed; the third and fourth are left-handed and may be either open or closed. In the isorthocoelous type (type 6, Fig. 5-14) the first and fourth loops are right-handed, the second and third left-handed; they may all be closed, or some may be partly or entirely open. In the periorthocoelous intestine (type 1, Fig. 5-9) the first and fourth loops are right-handed, the second and third left-handed, and the fifth and sixth when present are also left-handed; the second and third loops tend to form a spiral; the second and sometimes others are open.

Generic and Specific Characters

Although external features are often useful in helping to define the higher taxonomic categories, they are of most value in the definition and identification of genera, species, and subspecies. Differences in shape and structure of the bill, feet, wings, and tail are helpful in the definition of genera and, to a lesser extent, species. Specific characters, however, are based mostly on variations in size, color, or markings. Most birds attain their full growth within a few months of hatching, and the subsequent individual variation in size often amounts to less than 3% of the mean linear measurement among members of the same sex and subspecies.

Measurements in this manual are in millimeters and were taken in the following manner.

WING. Measured from the bend of the wing to the tip of the longest primary, with the wing pressed flat against the ruler. This results in a slightly larger but more accurate measurement than one obtained with calipers.

TAIL. Measured with one point of the dividers inserted between the bases of the 2 middle rectrices, and the other point at the tip of the longest rectrix.

CULMEN. Measured with dividers, from the juncture of the bill and the forehead (often concealed by feathers) to the tip of the bill. The measurement exposed culmen, occasionally used, is from the feathering over the dorsal ridge of the bill.

TARSUS. With one point of the dividers at the posterior side of the heel joint, the other at the articulation of the middle toe on the anterior side.

MIDDLE TOE. From the last-named point to the

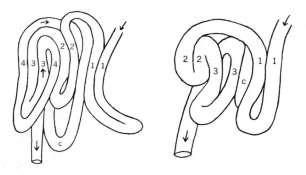

Fig. 5-13 (left). Type 5 plagiocoelous intestine of European quail (*Coturnix coturnix*). C, caecum. (After Gadow.) *Fig. 5-14* (right). Type 6 isorthocoelous intestine of European goatsucker (*Caprimulgus europaeus*).

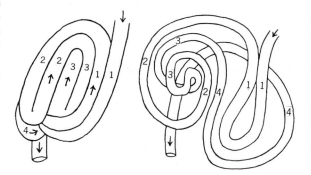

Fig. 5-15 (left). Type 7 antorthocoelous intestine of flicker (*Colaptes auratus*). *Fig. 5-16* (right). Type 8 mesogyric intestine of European blackbird (*Turdus merula*). C, caecum (After Gadow.)

base of the claw, along the dorsal side. The claw is never included unless middle toe with claw is specified.

Figure 5-17 shows many of the external parts important in the description of species of birds.

Classification

The classification adopted here is a modification of that of Wetmore (1960). The world orders and higher taxa are listed in Table 5-1. The counts of paleogenera and paleospecies are original; those of

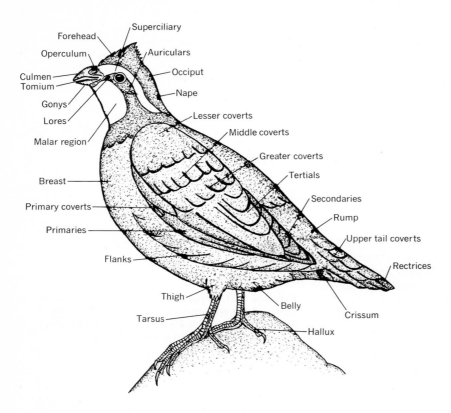

Fig. 5-17. Topography of bobwhite (*Colinus virginianus*).

neogenera and neospecies are modified from Fisher and Peterson (1964). The major divisions of the class and the 5 extinct and 7 extralimital living orders are briefly characterized below.

The subclass Sauriurae was transitional between reptiles and birds. Its only known order is the Archaeopterygiformes, of the uppermost Jurassic of Bavaria. The jaws were toothed, and the metacarpals and metatarsals unfused. The long, bony tail extended far beyond the body, and each pair of rectrices attached to a free caudal vertebra.

The subclass Odontoholcae includes the order Hesperornithiformes, aberrant divers of the Upper Cretaceous of North America. They retained the reptilian features of toothed jaws and a single-headed quadrate, but the tail was shortened, with the last 6 or 7 vertebrae forming an incipient pygostyle. Fused foot and ankle bones formed a

tarsometatarsus. *Hesperornis* had lost the power of flight, as shown by the unkeeled sternum and degenerate wings.

The subclass Ornithurae includes all living birds and certain extinct groups. The jaws are without teeth. The bones of the hand fuse to form a carpometacarpus, and those of the foot a tarsometatarsus. The tail vertebrae are few and do not extend beyond the body. This subclass may be divided into three infraclasses, Dromaeognathae, Ratitae, and Carinatae.

The infraclass Dromaeognathae has the single order Tinamiformes, the partridgelike tinamous of Central and South America. Combining some characters of both Ratitae and Carinatae, they may be close to the ancestral stock of those two groups, although their incompletely known fossil record extends only from the Pliocene. They agree with

TABLE 5-1. *Orders of Birds*

	Families		Genera		Species	
	Fossil	*Recent*	*Fossil*	*Recent*	*Fossil*	*Recent*
Subclass † Sauriurae °						
† Archaeopterygiformes °	1	0	1	0	2	0
Subclass † Odontoholcae						
† Hesperornithiformes	1	0	2	0	4	0
Subclass Ornithurae						
Infraclass Dromaeognathae °						
Tinamiformes °	0	1	2	9	4	42
Infraclass Ratitae °						
Struthioniformes °	1	1	1	1	7	1
Rheiformes °	1	1	1	2	4	2
Casuariiformes °	1	2	2	2	7	4
† Aepyornithiformes °	1	0	4	0	9	0
† Dinornithiformes °	2	0	7	0	27	0
Apterygiformes °	0	1	1	1	1	3
Infraclass Carinatae						
Gaviiformes	2	1	5	1	12	4
Podicipediformes	1	1	3	3	9	17
Procellariiformes	0	4	4	18	19	82
Sphenisciformes °	0	1	21	6	33	15
Pelecaniformes	6	6	19	8	67	55
Ardeiformes	6	6	32	50	75	112
Anseriformes	0	2	23	49	89	149
Accipitriformes	1	5	37	77	92	275
Galliformes	0	5	30	86	81	251
Ralliformes	8	12	77	80	153	197
† Ichthyornithiformes	2	0	3	0	9	0
Charadriiformes	1	12	26	112	74	301
Columbiformes	0	3	3	50	14	291
Psittaciformes	0	1	3	72	7	326
Strigiformes	1	2	5	24	34	133
Caprimulgiformes	0	5	0	23	0	92
Cuculiformes	0	2	3	39	6	144
Coliiformes °	0	1	0	1	0	6
Trogoniformes	0	1	2	8	4	35
Apodiformes	1	3	2	130	7	390
Coraciiformes	0	9	7	49	8	193
Piciformes	0	9	2	73	4	377
Passeriformes	2	52	15	1253	41	5159
Total	39	146	343	2227	902	8656

° Extralimital.
† Extinct.

the Ratitae in having a large vomer, single-headed quadrate, an ilioischiatic notch, and no pygostyle. They agree with the Carinatae in having 15 dorsal vertebrae, a keeled sternum, a furculum, a separate coracoid and scapula, functional wings, 2 free carpals, and divergent ilia.

The infraclass Ratitae includes gigantic to moderately large terrestrial birds that have secondarily lost the ability to fly. Sometimes thought to be polyphyletic, they share many structural characters, including a large vomer, single-headed quadrate (except in *Apteryx*), 20 dorsal vertebrae, unkeeled sternum, unfused clavicles, ankylosed and obtusely angled coracoid-scapula, degenerate wings, 3 free carpals, parallel ilia, an ilioischiatic notch, no oil gland (except in *Apteryx*), and no pygostyle. The 6 orders are characterized below.

The order Struthioniformes has a record reaching back to the Middle Eocene of Europe, although represented today only by the ostrich, of Africa and Asia Minor. It is characterized by a broad, short bill with basal nostrils, no aftershafts, no syrinx, long caeca, a pubic symphysis in the adult, unbridged tibiotarsus, 2 anterior toes, and no hallux.

The Rheiformes, the South American rheas, are similar but have a syrinx (the only case among Ratitae), an ischial rather than pubic symphysis, and 3 anterior toes. Their record extends from the Lower Eocene.

The Casuariiformes, the cassowaries and emus of Australia and New Guinea, differ in having large aftershafts, small caeca, no symphysis of the innominates, 3 anterior toes, and no hallux.

The Aepyornithiformes, the elephant-birds, occurred in North Africa in the Eocene and Oligocene and survived on Madagascar until about 1649. The tibiotarsus was unbridged. The foot had 3 anterior toes, with a hallux present in 1 genus. Some species reached a height of 9–10 feet and a weight approaching 1000 lb.

The Dinornithiformes, the moas of New Zealand, ranged from the Miocene or Pliocene to the seventeenth century. They had large aftershafts, a bridged tibiotarsus, 3 anterior toes, and a hallux in some forms. Of more slender build than the elephant-birds, some species attained a height of 13 ft, the tallest known birds.

In contrast with the other Ratitae, the Apterygiformes, the kiwis of New Zealand, are the size of chickens. The long, slender bill has terminal nostrils, used in detecting earthworms. Kiwis have no aftershafts, large caeca, a bridged tibiotarsus, 3 anterior toes, and a hind toe. The oil gland and double-headed quadrate are carinate rather than ratite features. The adult lacks the pecten of the eye, the only such case among birds. The record of the kiwis is unknown earlier than the Pleistocene.

The infraclass Carinatae has untoothed jaws, double-headed quadrate and small or subverted vomer (*Ichthyornis* is an exception), 15 dorsal vertebrae, keeled sternum, clavicles that nearly always fuse into a furculum, separate coracoid and scapula (in rare cases they ankylose but form less than an obtuse angle), divergent ilia, an ilioischiatic fenestra (an open notch was still present in the Ichthyornithiformes), an oil gland (lost in a few), and a pygostyle formed through fusion of several caudal vertebrae. The 1 extinct and 2 extralimital orders are characterized below.

The Sphenisciformes or penguins are secondarily flightless divers of cold southern oceans. They share many characteristics with the Procellariiformes, but have nontubular nostrils and wings modified as paddles. Their record goes back to the Lower Eocene.

The Ichthyornithiformes, of the Upper Cretaceous of North America, may have been ancestral to some of the living water birds. In appearance they were gull-like, but the quadrate was single-headed and the vomer large. The humerus had an enormous deltoid crest, though unspecialized in lacking the pneumatic fossae and pointed ectepicondylar process of gulls. A recent study (Gregory, 1952) concludes that the toothed jaws attributed to *Ichthyornis* are reptilian.

The Coliiformes, the colies or mousebirds, are an African group superficially resembling titmice, but in anatomical characters very close to the Cuculiformes. The head is crested and the tail long, with only 10 rectrices. All the toes may be held forward, although the first and fourth are reversible.

Nomenclature

The nomenclature is basically that of the American Ornithologists' Union *Check-list of North American Birds* (fifth edition, 1957). I have, however, adopted certain changes indicated by recent studies, giving the A.O.U. name in a footnote.

Some hold that such innovations are out of place in a work of this nature, but in my opinion it is more important for students to evaluate revisionary works than blindly to follow static authority. The interval between successive editions of the Check-list has increased to the extent that 26 years now separate the last two editions. At times, moreover, the Check-list committee has perhaps been ultra-conservative; it was not until the fourth edition (1931) that they could agree to adopt the current classification of higher groups, based on the work of Fürbringer (1888) and Gadow (1892).

Coverage

This manual includes descriptions of all native and established exotic species of birds of regular occurrence within the 48 metropolitan United States. The inclusion of Alaskan and Hawaiian species and those of accidental occurrence, unfortunately, would have overextended this volume.

The statements of geographic ranges are summary outlines of normal distribution. For more detailed ranges and accidental and casual occurrences the reader should refer to the works listed under References at the end of this Part.

KEY TO LIVING ORDERS OF CARINATE BIRDS

1. Hind toe connected to anterior toes by a web   **2**
   Hind toe free or absent   **3**
2. Wings paddlelike, without remiges
        (**Sphenisciformes, extralimital**)
   Wings normal, with primaries and secondaries
        Pelecaniformes

Fig. 5-18. Bill of Leach's petrel (*Oceanodroma leucorhoa*), showing tubular nostril of order Procellariiformes. *T*, nasal tube.

3. Nostrils in raised tubes (Fig. 5-18)   **4**
   Nostrils not in raised tubes   **5**
4. Anterior toes webbed   **Procellariiformes**
   Anterior toes free   **Caprimulgiformes**
5. Distance from bend of wing to tip of longest secondary less than from latter point to tip of wing (Fig. 5-19)   **Apodiformes**
   Distance from bend of wing to tip of longest secondary more than from latter point to tip of wing   **6**
6. First toe subject to being turned forward, giving 4 anterior toes   (**Coliiformes, extralimital**)
   With only 2 or 3 anterior toes   **7**
7. With only 2 anterior toes   **8**
   With 3 anterior toes   **12**
8. Outer hind toe slightly shorter than inner hind toe   **Trogoniformes**
   Outer hind toe decidedly longer than inner hind toe   **9**

Fig. 5-19. Wing of chimney swift (*Chaetura pelagica*), showing long primaries and very short secondaries of order Apodiformes.

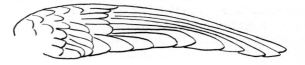

Fig. 5-20 (left). Bill of grass parakeet (*Melopsittacus undulatus*), showing cere surrounding nostrils in order Psittaciformes. *Fig. 5-21.* (right). Foot of horned grebe (*Podiceps auritus*), showing lobed toes of order Podicipediformes.

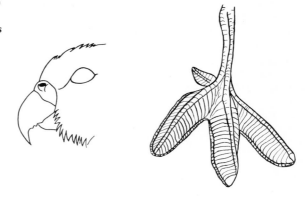

9. Bill with a cere (Fig. 5-20)   **10**
    No cere   **11**
10. Tarsus feathered   **Strigiformes**
    Tarsus unfeathered   **Psittaciformes**
11. Tail length more than wing length; culmen decurved   **Cuculiformes**
    Tail length nearly always less than wing length (except in some members of the tropical American Galbulidae, which have straight, sharp bill)   **Piciformes**
12. Tarsus much compressed laterally, its greatest depth at least twice its greatest width   **13**
    Greatest depth of tarsus less than twice its greatest width   **14**

Fig. 5-22. Bill of shoveler (*Spatula clypeata*), showing fringelike lamellae along edge of bill in order Anseriformes. *L*, lamella; *R*, rostral nail.

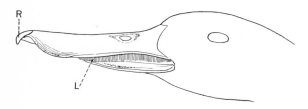

Fig. 5-23. Foot of belted kingfisher (*Megaceryle alcyon*), showing syndactyl foot of order Coraciiformes, with toes III and IV grown together.

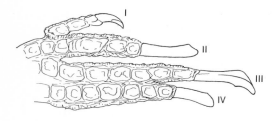

Fig. 5-24. Head of horned lark (*Eremophila alpestris*), showing rictal bristles at base of bill in order Passeriformes.

13. Anterior toes connected by full webs   **Gaviiformes**
    Anterior toes lobed, each with a separate web (Fig. 5-21)   **Podicipediformes**
14. Bill with a cere and hooked   **Accipitriformes**
    No cere; bill usually not hooked   **15**
15. Bill lamellate (or entirely serrate) and not abruptly bent down in middle (bill not lamellate in South American Anhimidae, which have a large double spur on wrist) (Fig. 5-22)   **Anseriformes**
    Bill usually not lamellate (lamellate in Phoenicopteridae, which have bill abruptly bent down in middle); wing spur occasionally present, but never double   **16**
16. Bill longer than head; legs very long; lower part of tibia bare; hind toe incumbent or slightly elevated, its claw shorter than basal segment but extending beyond first 2 segments of outer toe   **Ardeiformes**
    Without the above combination of characters   **17**
17. Tarsi short; feet usually syndactyl (Fig. 5-23) with outer and middle toes grown together at base and with a common sole (toes not syndactyl in the following extralimital families: Leptosomatidae of Madagascar, with powder downs, erect tuft on lores, stubby bill, and reversible outer toe; Coraciidae of Africa, Eurasia, and Australia, with stout, curved bill, slitlike nostrils, and rictal bristles; Upupidae of Eurasia and Africa, with awllike bill and long, black-tipped, sandy crest; Phoeniculidae of Africa, with long, slender bill, long, graduated tail, and metallic black plumage)   **Coraciiformes**
    Tarsal length variable; feet not syndactyl   **18**
18. Rictal bristles often present (Fig. 5-24); tarsus longer than longest front toe and claw; hind toe and claw as long as outer toe and claw; no frontal shield, no wing spur, no webbing of toes   **Passeriformes**
    Without the above combination of characters   **19**
19. Bill shorter than head; nostrils thin, imperforate slits, overhung by fleshy or horny operculum; hind toe incumbent or absent   **Columbiformes**
    Without the above combination of characters   **20**
20. Wing pointed, with outer primary longest (if 2d primary longest, then wing spurred, or tarsus reticulate and head crested; if 4th primary longest, the outer 3 primaries only ⅓ width of the others); secondaries distinctly shorter than primaries; toes sometimes webbed   **Charadriiformes**
    Wing rounded, with outer primary shorter than 2d; secondaries long, often concealing primaries; toes never fully webbed   **21**
21. Bill usually long, pointed (short in some Rallidae and in Old World Otididae); nostrils perforate

(imperforate in Rhynochetidae of New Caledonia); lower part of tibia bare (feathered in pantropical Heliornithidae) **Ralliformes**
Bill short, with culmen curved; nostrils imperforate; tibia entirely feathered **Galliformes**

Order Gaviiformes. Loons. Bill long, straight, pointed; nostrils perforate, linear, with a lobe on upper edge. Feathering entirely covering lores and extending into nostrils. Functional primaries 11, none emarginate; secondaries, aquintocubital; tertials shorter than primaries. Rectrices short but stiff. Aftershafts present; adult downs present on apteria and pterylae; oil gland tufted. Tarsus laterally compressed, reticulate, unserrated behind. Anterior toes webbed; claws narrow; hallux only slightly elevated (Fig. 5-25).

Palate schizognathous, without basipterygoid processes. Nares holorhinal. Mandibular process short. Cervical vertebrae 14. Sternum without internal spine; external spine small, notched; xiphial area unnotched, longer than lateral processes. Femur short, bowed, set far back on body. Patella reduced. Cnemial crest of tibiotarsus well developed. Supraorbital glands large (Fig. 5-26). Tongue small; intestine type 2; caeca long. Syrinx tracheo-bronchial. Both right and left carotid arteries present. Thigh-muscle formula $ABX+$; toe tendons type 2.

Diving birds; plantigrade on land, from which flight is not possible. Food fish. Nest of vegetation, placed on shore. Eggs 2, olive mottled with brown. Young downy and precocial. Lower Cretaceous onward. Holarctic.

Family Gaviidae. Tarsometatarsus with tendinal shelf below distal foramen; hypotarsus with

Fig. 5-25. Foot of common loon (Gavia immer).

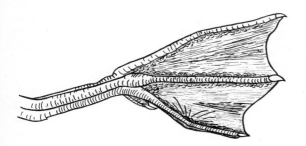

1 large canal; middle trochlea longest. Late Paleocene onward. Holarctic. A single living genus.

Genus *Gavia* Forster. 4 species, of which *G. adamsii* (Gray) is only accidental south of Vancouver Island.

KEY TO SPECIES OF *GAVIA*

1. Culmen 75–90 mm   *G. immer*
   Culmen 47–67 mm   2
2. Rectrices 20; tarsus about equal to middle toe *without* claw   *G. stellata*
   Rectrices 16–18; tarsus longer than middle toe *with* claw   *G. arctica*

Gavia immer (**Brünnich**). Common loon. Black above; back and wings spotted and barred with white; neck black, with a patch of black and white streaks on center of throat and 2 similar half rings on sides of lower neck; breast and belly white; crissum black and white. WINTER: Blackish above, the feathers with grayish margins; underparts white. Wing 325–393, culmen 75–90, tarsus 81–99 mm. Arctic south to Scandinavia, New York, Great Lakes, North Dakota, Manitoba, and northern California. Winters mainly on salt water, from Maine to Florida, off the coast of California, and in Europe to the Mediterranean.

Gavia arctica (**Linnaeus**). Arctic loon. Similar to *G. immer* but smaller; in breeding plumage with crown bluish gray; patch of black and white streaks on sides of neck lengthened, not forming half rings as in *G. immer*. Wing 284–342, culmen 51–67, tarsus 64–83 mm. Northern Eurasia and arctic America; winters on Pacific Coast from southern Alaska to Baja California; casual in the East.

Gavia stellata (**Pontoppidan**). Red-throated loon. Head and neck bluish gray; upper parts dusky, hind neck streaked with white; throat chestnut; remaining underparts white. In winter similar to the 2 preceding species, but feathers on back with V-shaped white edges. Wing 254–305, culmen 47–57, tarsus 68–77.5 mm. Circumpolar, breeding south nearly to the Canadian border; winters mainly on salt water, south to the Mediterranean, China, Baja California, and Florida.

Order Podicipediformes. Grebes. Bill usually long and straight, the tip sometimes decurved;

Fig. 5-26. Skull of red-throated loon (*Gavia stellata*). *HN*, holorhinal nostril; *PN*, premaxillary-nasal suture; *SG*, groove for supraorbital gland, used in secretion of salt.

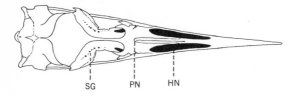

nostrils perforate, linear or oval, not lobed. Feathering not reaching nostrils; a bare strip from eye to bill. Functional primaries 11, the outer 3–4 emarginate; secondaries aquintocubital; tertials longer than primaries. Tail rudimentary, soft. Aftershafts present; adult downs on apteria and feather tracts; oil gland tufted. Tarsus laterally compressed, scutellate, serrated behind. Toes lobed, nails broad and flat, hallux raised.

Palate schizognathous, without basipterygoid processes. Nares holorhinal. Mandibular process short. Cervicals 17–21. Sternum without internal spine, external spine somewhat forked; xiphial area notched and shorter than lateral processes. Femur short, bowed, set far back on body. Patella large but not fused to the well-developed cnemial process of the tibiotarsus. Hypotarsus with 2–3 canals. Supraorbital glands small. Tongue small; intestine type 2; caeca large. Syrinx tracheobronchial. Only the left carotid artery present. Thigh-muscle formula *BX*–; toe tendons type 2.

Excellent divers; plantigrade on land, from which flight is impossible. Food mainly fish and aquatic arthropods; the stomach is often crammed with the bird's own feathers. Nest of vegetation, floating on water. Eggs 3–9, plain whitish or buffy. Young downy and precocial. Late Cretaceous onward. Cosmopolitan, except for Hawaii and some other oceanic islands. The northern species are migratory.

Family Podicipedidae. Wings short but normal. Patella imperforate (ambiens muscle lost). Tarsometatarsus with distal foramen. Miocene onward. 3 living genera, 17 species; in the United States 3 genera, 6 species.

KEY TO GENERA OF PODICIPEDIDAE

1. Length of culmen about twice depth of bill
   Podilymbus
   Length of culmen more than twice depth of bill   2
2. Length of culmen about 3 times depth of bill
   Podiceps
   Length of culmen about 5 times depth of bill.
   Aechmophorus

Genus *Podilymbus* Lesson. Bill stout, its length less than twice its depth; neck much shorter than body. Americas. 1 species.

***Podilymbus podiceps* (Linnaeus).** Pied-billed grebe. Bill whitish with a black band across middle; chin and throat black; upper parts and lower throat brownish; remaining underparts silvery white, spotted with dusky. WINTER: Bill brownish without black band, throat dull white, underparts without spots. Wing 121.5–138.5, culmen 17–23.5, depth of bill at anterior end of nostril 9.5–11.5, tarsus 36–43 mm. Southern Canada, United States, Antilles, Middle and South America.

Genus *Podiceps* Latham. Bill slender, its length more than twice and less than 4 times its depth; neck much shorter than body. Nearly cosmopolitan (absent from Hawaii, etc.). 15 species, 4 in the United States.

KEY TO SPECIES OF *PODICEPS*

1. Wing less than 100 mm   *P. dominicus*
   Wing more than 120 mm   2
2. Wing 155–206 mm   *P. grisegena*
   Wing 120–150 mm   3
3. Bill deeper than wide   *P. auritus*
   Bill wider than deep   *P. caspicus*

***Podiceps dominicus* (Linnaeus).** Least grebe. Upper parts blackish; throat black, neck dark gray; breast and belly silvery, mottled with gray. WINTER: Chin white, throat brownish gray, breast more silvery. Wing 89–97, culmen 18–24, depth of bill at anterior end of nostrils 6–7, tarsus 27.5–35 mm. Southeastern Texas and southern Baja California south to Argentina; Greater Antilles.

***Podiceps auritus* (Linnaeus).** Horned grebe. Bill deeper than wide. Sides of head with large crest, golden above and black below; upper parts blackish; throat and breast rufous; remaining underparts white. WINTER: Uncrested; above blackish; lower throat dull gray; remaining underparts and patch on upper sides of neck silvery white. Wing 135–150, culmen 21–24, depth of bill 7–7.5, width of bill 5.5–6.5, tarsus 44–50 mm. Northern Europe, Asia, and North America south to Maine, the Great Lakes, Nebraska, and British Columbia; winters from southern portion of breeding range south to Florida, Gulf states, and southern California; also to southern Europe, North Africa, China, and Japan.

Podiceps caspicus (**Hablizl**). Eared grebe. Head, neck, and breast black; a tuft of long yellowish feathers behind eyes; upper parts blackish; flanks chestnut; remaining underparts silvery white. WINTER: Identical in coloration with the corresponding plumage of *C. auritus*, but bill wider than deep. Wing 120–140, culmen 19–24, depth of bill 5.5–6, width of bill 6.5, tarsus 39–45 mm. Europe, Asia, and South Africa; also North America, from western Canada, south to Iowa, Arizona, and California. Winters from California and Louisiana to Guatemala; also in the Old World range.

Podiceps grisegena (**Boddaert**). Red-necked grebe. Pileum black; large cheek patches ashy with a white border; neck rufous; back grayish brown; belly white. WINTER: Pileum gray; neck ashy. Wing 155–206, culmen 35–61, depth of bill 14, tarsus 53–64 mm. Europe, Asia, and northern North America, south to Washington and Minnesota; winters mainly on salt water south to Mediterranean, China, California, and Florida.

Genus *Aechmophorus* Coues. Culmen more than 4 times as long as depth of bill; neck about as long as body. Western North America. 1 species.

Aechmophorus occidentalis (**Lawrence**). West-

ern grebe. Upper parts slaty black; lower parts, cheeks, and sides of neck silvery white. Wing 170–216, culmen 53–78, depth of bill 10–12.5, tarsus 69–76 mm. Southwestern Canada south to North Dakota, Utah, and northern California; winters from British Columbia to central Mexico.

Order Procellariiformes. Tube-nosed swimmers. Bill hooked; rhamphotheca compound; nostrils in raised tubes, imperforate. Wing pointed; functional primaries 10; secondaries aquintocubital. Tail short to medium in length. Aftershafts present but reduced; adult downs all over body; oil gland tufted. Anterior toes palmate; hallux absent or reduced to 1 phalanx.

Schizognathous. Holorhinal. Mandibular process truncate. Cervicals 15. Sternum without internal spine, external spine slightly forked. Humerus rotated distally, with large ectepicondylar process. Cnemial process of tibiotarsus well developed. Hypotarsus complex. Supraorbital glands large. Tongue small; intestine type 2; caeca variable. Syrinx tracheobronchial. Both carotids present. Thigh muscles $A(B)X(Y)\pm$; toe tendons type 4.

Marine or pelagic birds, coming to land to nest. Posture on land more or less plantigrade. Partly nocturnal. Food entirely animal. Nest on ground or in burrows. A single egg. Young downy but altricial. Eocene onward. Cosmomarine. In the United States 3 of 4 families.

KEY TO FAMILIES OF PROCELLARIIFORMES

1. Nasal tube single, divided only internally (Fig. 5-27)
Oceanitidae
   2 nasal tubes, opening forward (Fig. 5-28)   **2**

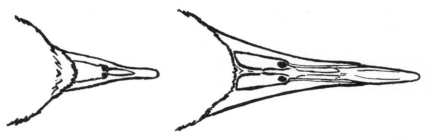

Fig. 5-27 (left). **Top of bill of Leach's petrel** (*Oceanodroma leucorhoa*).
Fig. 5-28 (right). **Top of bill of greater shearwater** (*Puffinus gravis*).

Fig. 5-29. Top of bill of short-tailed albatross (*Diomedea albatrus*).

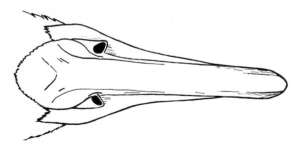

2. Nasal tubes lateral, separated by culmen (Fig. 5-29)
   Diomedeidae
   Nasal tubes together on culmen   **Procellariidae**

Family Diomedeidae. Albatrosses. Nasal tubes lateral, separated by culmen. Hallux absent or reduced. No basipterygoid processes; metasternum with 1 pair of notches. Caeca present. Size large, wingspread up to 3.5 meters, the largest of any bird. Food squid, blubber, and refuse. Nest on ground. Egg single, whitish with small reddish-brown specks. Eocene onward. All oceans except North Atlantic, where, however, present in the Pliocene. 1 genus, 12 species; in the United States 2 species.

Genus *Diomedea* Linnaeus. Tail short, 40% of the wing length or less, not pointed or cuneate. Lateral plates of mandibular rami not divided by a longitudinal sulcus.

KEY TO SPECIES OF *DIOMEDEA*

1. Culmen more than 135 mm   *D. albatrus*
   Culmen less than 115 mm   *D. nigripes*

Diomedea nigripes **Audubon.** Black-footed albatross. Dusky, becoming paler and grayer below, on head, and at base of tail; bill dusky; feet black. Wing 470–521, tail 165, culmen 102–108, tarsus 89–94 mm. North Pacific Ocean; on American side from Bering Strait to Baja California.

Diomedea albatrus **Pallas.** Short-tailed albatross. Adult white, the head and neck washed with yellowish; wings and tail blackish, with white patches on primaries and wing coverts; shafts of primaries yellow; bill yellow; feet flesh-colored. Young dusky, but primary shafts yellow as in adult; bill and feet pale brown. Wing 558–584, tail 140–152, culmen 140–142, tarsus 97–102 mm. North Pacific Ocean; on American side from Bering Sea to Baja California.

Family Procellariidae. Shearwaters, fulmars, and petrels. Nasal tubes located on culmen, opening forward. Hallux rudimentary. Outer primary equal to or longer than the next. Basipterygoid processes and caeca present. Sternum 4-notched. Size medium (gull-like). Food jellyfish, crustaceans, and offal. Nest on ground (fulmar) or in burrows. Egg single, usually plain white, sometimes with fine dots of reddish brown. Oligocene onward. All oceans. 10 genera, 51 species; in the United States 2 genera, 9 species.

KEY TO GENERA OF PROCELLARIIDAE

1. Gonys ascending toward tip   *Fulmarus*
   Gonys hooked downward toward tip   *Puffinus*

Genus *Puffinus* Brisson. Shearwaters. Bill slender, nearly as long as head and more than half length of middle toe, compressed distally, higher than broad at base; tips of both upper and under mandibles decurved, with gonys concave; nasal tubes short, ¼ length of culmen, broad and depressed, obliquely truncate, with thick partition; nostrils oval. Feet large, tarsus compressed, middle and outer toe subequal and as long as tarsus, inner toe short, hallux knoblike. Wing very long, outer primary longest. Tail rounded or wedge-shaped, rectrices 12. All oceans. 20 species, 8 in the United States.

KEY TO SPECIES OF *PUFFINUS*

1. Tail wedge-shaped, the difference between lateral and middle rectrices about 50 mm   *P. bulleri*
   Tail rounded, the difference between lateral and middle rectrices about 30 mm or less   2
2. Coloration sooty above and below   3
   Underparts whitish   4
3. Culmen 39 mm or more   *P. griseus*
   Culmen 33 mm or less   *P. tenuirostris*
4. Wing 235 mm or less   5
   Wing 292 mm or more   6

5. Wing 215 mm or less; crissum mostly white
   P. lherminieri
   Wing 225 mm or more; crissum black   *P. puffinus*
6. Crissum white, sometimes with ashy tips
   P. diomedea
   Crissum sooty or with only tips white   7
7. Upper tail coverts largely white   *P. gravis*
   Upper tail coverts sooty   *P. creatopus*

Puffinus tenuirostris (**Temminck**).   Slender-billed shearwater. Dark sooty gray above and on crissum, becoming black on remiges and tail; smoky gray below, paler on throat; under wing coverts ash gray. Iris brown; bill and feet blackish. Some specimens are much darker than others. Wing 263–277 tail 77–86 culmen 30–35 tarsus 49–53 mm. Breeds in vicinity of Tasmania; ranges north to Bering Sea occurring all along our Pacific Coast.

Puffinus griseus (**Gmelin**).   Sooty shearwater. Similar to the preceding but larger and under wing coverts white with gray tips. Wing 280–309, tail 84–99, culmen 38–47, tarsus 52–60 mm. Breeds in New Zealand and southern South America; ranges northward along both our coasts to the Arctic Circle.

Puffinus puffinus (**Brünnich**).   Manx shearwater. Above sooty slate, paler on head and neck; crissum sooty black; remaining underparts white. Bill dusky; feet pink. Wing 228–231, tail 83–96, culmen 33–36, tarsus 44–46 mm. Breeds on islands off west coast of Mexico and in eastern North Atlantic; ranges northward as far as Washington and to South Atlantic.

Puffinus lherminieri **Lesson**.   Audubon's shearwater. Like the preceding but smaller; eyelids and lining of wing white; crissum white except for the longest feathers. Iris brown; bill black; feet flesh, black on outside of tarsus and toes. Wing 190–213, tail 89–108, culmen 30–32, tarsus 38–42 mm. Breeds in Bermuda, West Indies, Cape Verde Islands, and South Pacific and Indian Oceans; occurs off Atlantic Coast from Florida to New Jersey or even to Long Island.

Puffinus creatopus **Coues**.   Pink-footed shearwater. Brown above; upper tail coverts sooty; below plain white, but often more or less barred with gray. Iris brown; bill flesh, blackish along culmen and tomia; feet flesh. Wing 321–342, tail 107–118, culmen 40–46, tarsus 51–56 mm. Breeds off Chile;

migrates to coast of California, Washington, and Alaska.

Puffinus diomedea (**Scopoli**).   Cory's shearwater. Above brownish gray with pale tips; base of primaries white; below white, the sides of head and neck with wavy bars of gray. Bill yellowish, tipped with black; feet reddish. Wing 330–368, culmen 45–59, tarsus 47–57 mm. Mediterranean Sea and Atlantic Ocean; off North America from Newfoundland to North Carolina.

Puffinus gravis (**O'Reilly**).   Greater shearwater. Above brown, darkest on head, and with paler tips on body; upper tail coverts white-tipped; cheeks and underparts white, with a patch of dark grayish brown on abdomen and crissum; axillars white, tipped with brown. Bill yellowish green; feet brownish, yellowish medially. Wing 290–337, tail 105–116, culmen 42–48, tarsus 53–62 mm. Breeds on Tristan da Cunha; ranges over the entire Atlantic, from the Arctic Circle to the Falklands.

Puffinus bulleri **Salvin**.   New Zealand shearwater; gray-backed shearwater. Back gray, with lighter edges; crown darker and uniform; below white; crissum mottled gray and white. Bill blue, tipped with black; feet flesh, blackish externally. Tail wedge-shaped. Wing 275–300, tail 114–132, culmen 39–44, tarsus 49–53 mm. Breeds in New Zealand; ranges northward to California.

Genus *Fulmarus* Stephens.   Bill stout, broad, not compressed; gonydeal angle prominent, gonys ascending toward tip of bill; nasal tubes more than $\frac{1}{3}$ length of culmen, divided by a thin partition; feet strong and rather short; rectrices 14. Monotypic.

Fulmarus glacialis (**Linnaeus**).   Fulmar. Head, neck, and underparts white; back and tail bluish gray; wings grayish brown, darker toward tips; a small black mark before and above eye. Bill yellow; legs whitish. In dark phase entirely smoky gray. Wing 300–349, culmen 33–43, tarsus 50–55 mm. North Atlantic and North Pacific Oceans, breeding in the Arctic; south in winter to British Isles, New Jersey, Baja California, and Japan.

Family Oceanitidae.[1]   Storm petrels. Bill shorter than head, slender and weak; nasal tube single, located on culmen, divided internally, nearly half length of bill. Legs long and slender; hallux spurlike. Wings long and narrow, with 2d primary long-

[1] Hydrobatidae of A.O.U., preoccupied and antedated.

est; secondaries 10–14. Basipterygoid processes absent; sternum solid or fenestrate. Caeca and accessory semitendinosus (Y) lacking. Size small, length 200 mm or less. Miocene onward. All oceans. 7 genera, 19 species; in the United States 3 genera, 5 species.

KEY TO GENERA OF OCEANITIDAE

1. Tarsus booted　　　　　　　　　　*Oceanites*
   Tarsus reticulate　　　　　　　　*Oceanodroma*

Genus *Oceanodroma* Reichenbach. Tarsus reticulate, about equal to middle toe with claw; basal phalanx of middle toe about equal to the next 2 phalanges together. Secondaries 14. Tail decidedly emarginate. Sternum solid. All oceans. 10 species, 4 in the United States.

KEY TO UNITED STATES SPECIES OF *OCEANODROMA*

1. Tarsus longer than middle toe with claw　*O. melania*
   Tarsus about equal to middle toe with claw　2
2. Upper tail coverts with white　　*O. leucorhoa*
   No white on upper tail coverts　　　　3
3. General color ashy; outer web of outer rectrix white
   　　　　　　　　　　　　　O. furcata
   General color sooty brown, without any white
   　　　　　　　　　　　　O. homochroa

***Oceanodroma furcata* (Gmelin).** Fork-tailed petrel. Bluish gray fading to whitish on throat and crissum; wing coverts dusky, edged with white. Wing 142–165, tail 75–100, culmen 14–16, tarsus 24–28 mm. Bering Sea to San Diego, California.

***Oceanodroma leucorhoa* (Vieillot).** Leach's petrel. Sooty brown, with a paler band on wing coverts; upper tail coverts white, at least laterally. Bill and feet black. Wing 153–164, tail 76–85, culmen 14–17, tarsus 23–26 mm. Both sides of the North Atlantic, from Iceland and the Hebrides to West Africa and from Greenland to Massachusetts; American side of the North Pacific, from Alaska to western Mexico.

***Oceanodroma homochroa* (Coues).** Ashy petrel. Sooty gray; wing coverts, rump, and upper tail coverts ashy gray. Wing 135–137, tail 84–89, tarsus 20–23 mm. Point Reyes, California, south to Baja California.

***Oceanodroma melania* (Bonaparte).**[1] Black petrel. Sooty black, browner below, and with a paler band on wing coverts. Wing 157–177, tail 73–87, culmen 15–17, tarsus 31–34 mm. Breeds off Baja California; ranges from Marin County, California, to central Peru.

Genus *Oceanites* Keyserling and Blasius. Tarsus booted, longer than middle toe with claw; basal phalanx of middle toe longer than the next 2 phalanges combined. Secondaries 10. Tail slightly emarginate. Sternum fenestrate. 1 species.

***Oceanites oceanicus* (Kuhl).** Wilson's petrel. Sooty black; band on wing coverts ashy; upper tail coverts uniform white; large patches of white on flanks, thighs, and crissum; lining of wing sooty. Bill and feet black; base of webs between toes yellow. Wing 130–158, tail 55–73, culmen 10–13, tarsus 31–37 mm. Breeds in Antarctica and southern South America; ranges northward to California, the Gulf of Mexico, Labrador, Britain, and the Mediterranean.

Order Pelecaniformes. Totipalmate swimmers. Bill long, with complex rhamphotheca; nostrils tiny or even without external openings; gular pouch present (vestigial in *Phaethon*). Wings large; functional primaries 11; secondaries 16–25, aquinto-cubital; aftershafts absent; mature downs nearly all over body; oil gland tufted. Legs short; tarsi reticulate; feet totipalmate (webs much reduced in *Fregata*).

Palate desmognathous, without basipterygoid processes. Holorhinal. Base of mandible blunt. Cervicals 15–20. Sternum without internal spine, external spine small; keel produced forward. Hypotarsus complex, with several canals. No supraorbital glands. Tongue rudimentary; proventriculus large; intestine type 2; caeca small. Syrinx tracheobronchial. Thigh muscles $A(XY)\pm$; toe tendons type 2; 1 pair of tracheosternal muscles.

Aquatic birds. Food fish. Nest of sticks, seaweed, or merely a hollow in the sand. Eggs chalky, plain white, white speckled with brown, or plain bluish white, 1–6. Both sexes incubate. Young altricial, hatched blind and often naked, but soon covered with thick down; fed by regurgitation. Late Cretaceous onward. Cosmopolitan. 6 living families, all occurring in this country.

---

[1] *Loomelania melania* of A.O.U.

KEY TO FAMILIES OF PELECANIFORMES

1. Bill hooked at tip 2
   Bill not hooked 4
2. Culmen about as long as tarsus **Phalacrocoracidae**
   Culmen about 3 times as long as tarsus 3
3. Tail long, deeply forked **Fregatidae**
   Tail short, not forked **Pelecanidae**
4. Chin feathered **Phaëthontidae**
   Chin naked 5
5. Tail rounded **Anhingidae**
   Tail pointed or wedge-shaped **Sulidae**

Family Phaëthontidae. Tropic-birds. Bill about as long as head; culmen curved, tomia minutely serrate; nostrils small, linear, better developed than in most of the order; gular pouch very small. Wings well developed. Middle rectrices of adult very long. Tarsus very short. Thigh-muscle formula AXY+. Cervicals 15. Furculum not ankylosed to sternum. Sternum double-notched. Caeca functionless. Nest on sand or in crevice; egg single, white, speckled with brown; young downy from time of hatching. Eocene onward. Tropicopolitan. 1 genus, 3 species (2 in the United States).

Genus *Phaëthon* Linnaeus. Characters as for the family.

KEY TO SPECIES OF *PHAETHON*

1. Rectrices 14; culmen 59–69 mm *P. aethereus*
   Rectrices 12; culmen 44–54 mm *P. lepturus*

Phaëthon aethereus **Linnaeus.** Red-billed tropic-bird. Mostly white; a black band through each eye, more or less continuous behind nape; upper parts transversely barred with black or gray; greater wing coverts and 5 outer primaries gray; shafts of outer rectrices dark; middle rectrices white, shafts dark basally; iris brown; bill red; feet yellowish, plantar surface gray. Wing 283–326, tail 94–119, central rectrices 414–790, culmen 59–69, tarsus 26–30 mm. Indian Ocean; in the Atlantic from Lesser Antilles to Brazil, Ascension Island, and Cape Verde Islands; in the Pacific from Gulf of California to Chile; occasional off southern California.

Phaëthon lepturus **Daudin.** White-tailed tropic-bird. White, without crossbarring, often suffused with salmon; a broad black stripe through eyes, more or less continuous around nape; a broad black band on secondaries, wing coverts, and scapulars; base of outer primaries black; tail white; shafts of primaries and rectrices white below, black basally above; flanks striped with dark gray; iris brown; bill orange-red; tarsi blue; toes and webs black. YOUNG: No long median tail plumes; bill yellow; upper parts with coarse wavy black bars. Wing 251–281, tail 87–113, middle rectrices 403–461, culmen 44–54, tarsus 21.5–23 mm. Tropical oceans, breeding north to Bermuda and the Bahamas; in the western Atlantic ranges north to latitude 40° occasionally along the coast of Florida, South Carolina, New York, and even Nova Scotia.

Family Pelecanidae. Pelicans. Bill long, straight, hooked at tip; culmen flat, with terminal nail; lower mandible very flexible; gonys very short; gular pouch large. Tail short, square, or slightly rounded, of 22–24 rectrices. Cervicals 17. Furculum fused to sternum. Sternum shallowly notched. Procoracoid process small. Olecranal fossa shallow. Thigh formula A—; carotids 1–2, syringeal muscles lacking. Caeca present. Air sacs remarkably developed. Feed by scooping up fish or plunging from air. Nest in colonies, of sticks, usually on ground, sometimes in trees; eggs 1–4, chalky white; young hatched naked, skin pink. Miocene onward. Nearly cosmopolitan, being absent from the far north, New Zealand, and Polynesia. 1 genus.

Genus *Pelecanus* Linnaeus. 6 species, 2 in the United States.

KEY TO SPECIES OF *PELECANUS*

1. Feathering extending onto mandibular ramus beyond level of eye *P. erythrorhynchos*
   Mandibular ramus unfeathered *P. occidentalis*

Pelecanus erythrorhynchos **Gmelin.** White pelican. Rectrices 24. White, with black remiges; occipital crest, lesser wing coverts, and lengthened plumes on breast yellow; middle of culmen with a horny crest; bill, pouch, and feet red. NONBREEDING PLUMAGE: Yellow plumes and knob on bill lacking; soft parts yellow. YOUNG: Lesser wing coverts and pileum mixed with gray. Wing 508–641, tail 152–167, culmen 280–381, tarsus 111–126 mm. Breeds from Great Slave Lake south to southern California, Nevada, and Utah; also on coast of Texas; winters from Florida, Gulf states, and California, south to Panama.

Pelecanus occidentalis Linnaeus. Brown pelican. Rectrices 22. Head and streak along pouch white, often tinged with yellow; neck dark reddish brown; a patch of yellow on lower foreneck; upper parts silver, striped with dark brown, the light color predominating on wing coverts; primaries black, their shafts white basally; secondaries and tail ashy; below smoky brown; sides with white shaft streaks; iris yellow; skin around eye blue; bill gray, spotted with red; pouch brown; legs black. NONBREEDING PLUMAGE: Entire head and neck white. YOUNG: Head and neck gray, upper parts dull brown tipped with paler; underparts white, becoming grayer on flanks. Wing 447–625, tail 114–200, culmen 251–425, tarsus 58–112 mm. Both coasts of America, from British Columbia to Chile and Galapagos, and from North Carolina to Texas, along coast of Mexico, Central America, and West Indies, south to Brazil.

Family Sulidae. Boobies and gannets. Bill longer than head, unhooked, conical, tomia serrate, maxilla notched near base; a long nasal groove, but no external nostrils. Wings long and pointed, outer primary longest. Tail wedge-shaped, rectrices 12–18. Feet large; outer toe about as long as middle toe. Cervicals 18. Furculum ligamentously attached to sternum; sternum shallowly notched. Coracoids overlap; procoracoid process large; hyposternal process well developed. Olecranal fossa deep. Thigh muscles AX+; syringeal muscles lacking. Carotids double or single. Caeca very small. Gall bladder large. Skin very pneumatic. Fish by plunging into water from air. Nest in colonies on rocks or sand, rarely in trees; eggs 1–2, white, calcareous; young with some down on hatching, skin black. Oligocene onward. Cosmomarine. 2 genera; 9 species, 5 in the United States.

KEY TO GENERA OF SULIDAE

1. Gular pouch naked **Sula**
Pouch feathered on sides, naked along strip in middle **Morus**

Genus Sula Brisson. Pouch entirely bare. Rectrices 12–18. Warm oceans. 6 species, 4 in the United States.

KEY TO SPECIES OF SULA

1. Bill dark (blue or greenish blue, sometimes red at base) 2
Bill light (yellow, orange, or red) 3
2. Feet blue; tarsus 47–57 mm **S. nebouxii**
Feet red or dull yellow; tarsus 33–40 mm **S. sula**
3. Feet yellow or pea green; rectrices 12–14; tarsus 42–50 mm **S. leucogaster**
Feet grayish or bluish; rectrices 16–18; tarsus 52–61 mm **S. dactylatra**

Sula dactylatra Lesson. Blue-faced booby; masked booby. Rectrices 16, rarely 18. White; remiges, some wing coverts, tips of long scapulars, and tail brown; iris yellow; skin of face black; bill orange-yellow or light red; feet grayish. YOUNG: Head, neck, and upper parts brown, mixed with white; iris pale green; feet purplish blue. Wing 406–486, tail 151–196, culmen 92–115, tarsus 52–61 mm. Warm waters of Indian, Pacific, and Atlantic Oceans; occurs on Dry Tortugas and rarely Gulf Coast.

Sula nebouxii Milne-Edwards. Blue-footed booby. Rectrices 16. Above cinnamon brown, more or less mottled with white; forehead, patch on base of hind neck, rump, and upper tail coverts mostly white; wings brown; tail brown, whitish at base; foreneck mottled cinnamon brown; rest of underparts white; iris yellow; skin of face and gular sac slaty; bill greenish blue; legs ultramarine blue. YOUNG: Iris brown. Wing 387–448, tail 165–243, culmen 94–115, tarsus 47–57 mm. Warm waters of American side of Pacific, occasional north of Baja California.

Sula leucogaster (Boddaert). Brown booby. Rectrices 12–14. Head, neck all around, wings, and tail brown; posterior underparts abruptly white; iris gray or whitish; eyelid bright blue; lores gray; bill, pouch, and feet yellow or pea green. YOUNG: Brown, paler below. Wing 360–415, tail 162–198, culmen 87–102, tarsus 42–50 mm. Warm waters of Indian, Pacific, and Atlantic Oceans; occurs on Dry Tortugas and rarely to Gulf Coast and California.

Sula sula (Linnaeus). Red-footed booby. Rectrices 14–16, narrower than in S. leucogaster. White; primaries and tips of secondaries blackish brown; iris gray or whitish (brown or yellow in female); eyelid blue; face green; bill bluish, red at base; pouch black (gray in female); legs red. YOUNG: Brownish, paler below; pouch pinkish; legs dull yellowish. Wing 362–405, tail 189–231, culmen 76–92, tarsus 33–40 mm. Warm waters of Indian, Pacific, and Atlantic Oceans, rarely to Gulf Coast.

Genus *Morus* Vieillot. Pouch feathered on sides, naked along strip in middle. Rectrices 12. North Atlantic, South Africa, Australia, and New Zealand. 3 species, 1 in the United States.

Morus bassanus (**Linnaeus**). Gannet. White, head washed with yellow; primaries and their coverts black; iris white; bill gray; feet black. YOUNG: Dark brown, spotted with white; breast and belly white, the feathers edged with gray; remiges and tail dusky; iris green; lores blue, bill brown; feet dusky. Wing 432–533, tail 229–254, culmen 90–112, tarsus 51–57 mm. Breeds on islands in Gulf of St. Lawrence, Newfoundland, Iceland, and Britain; winters from the coast of Virginia south to Cuba and Veracruz, also on North African coast, the Canaries, and Azores.

Family Phalacrocoracidae. Cormorants. Bill about as long as head, hooked, tomia smooth; a long nasal groove but no external nostrils; gular pouch small. Wings rounded, outer primary shorter than 2d and 3d. Tail rather long, rounded, of 12–14 rectrices. Tarsus equal to culmen in length; toes and webs well developed; outer toe longest. Occiput with styloid bone. Cervical vertebrae 19–20, 6th with overlapping postzygapophyses forming a slight kink; no bridge of Dönitz; 20th–21st vertebrae sometimes fused (*P. auritus*); free dorsals 3 or 4, opisthocoelous. Furculum ligamentously attached to sternum: sternum shallowly notched. Coracoids overlap; procoracoid and hyposternal processes small. Olecranal fossa deep. Patella large and free. Cnemial process somewhat developed. Thigh muscles $AX+$. Both carotids present. Pneumaticity not great. Fish by diving and swimming under water. Colonial; nest bulky, on cliffs, on ground, or in trees; eggs 2–6, bluish or greenish, chalky; young hatched naked, skin blackish. Paleocene onward. Nearly cosmopolitan. 2 genera (1 in the United States), 30 recent species.

Genus *Phalacrocorax* Brisson. Capable of flight. 2d and 3d primaries from outside longest. Plumage not hairlike. Nearly cosmopolitan (absent from Hawaii). 29 species, 5 in the United States.

KEY TO SPECIES OF *PHALACROCORAX*

1. Feathering on lower mandible extending forward beyond level of eyes   ***P. pelagicus***
   Lower mandible largely or entirely unfeathered   **2**
2. Feathering of posterior border of gular sac extending farther forward laterally than on midline   ***P. auritus***
   Feathering of posterior border of gular sac heart-shaped, extending forward on midline   **3**
3. Length of rostral nail $\frac{1}{4}$ or less length of culmen   ***P. penicillatus***
   Rostral nail decidedly more than $\frac{1}{4}$ length of culmen   **4**
4. Wing 328 mm or more; rectrices 14   ***P. carbo***
   Wing 305 mm or less; rectrices usually 12   ***P. olivaceus***

Phalacrocorax carbo (**Linnaeus**). Great cormorant. Rectrices 14. Posterior margin of gular sac heart-shaped. Head, neck, and underparts black glossed with green or purple; margin of gular pouch and patch on flanks white; threadlike white plumes scattered on head and neck; occiput crested; back and wing coverts bronzy gray edged with black; remiges and tail dark gray; iris green; bill black, whitish along tomia; circumorbital skin green, orange under eye; pouch yellow. WINTER: Lacks crest and white feathers. YOUNG: Pileum and hind neck brownish black; back and wing coverts brownish gray with dark margins; throat pale brown; undersurface of wings and belly dusky; underparts otherwise whitish; bare skin of head yellow. Wing 328–370, tail 184–197, culmen 58–72, tarsus 63–70 mm. Greenland, Iceland, Labrador, Nova Scotia, Europe, Asia, the African mainland, Australia, and New Zealand; winters south to Lake Ontario, Long Island, and occasionally to Maryland and South Carolina.

Phalacrocorax auritus (**Lesson**). Double-crested cormorant (Fig. 5-30). Rectrices 12. Gular sac convex behind. Outer primary equal to 4th from outside. Greenish black; scapulars and wing coverts ashy margined with black; a black or white crest behind each eye; sometimes with a few threadlike white feathers on head; iris green; eyelids blue; lores and pouch orange; feet black. NON-BREEDING PLUMAGE: Lacks the crests and blue eyelids; pouch and bill yellow, brown along culmen. YOUNG: Pale brown or whitish below. Wing 284–355, tail 133–178, culmen 48–65, tarsus 61–63 mm. Alaska, Saskatchewan, Ontario, and New-

Fig. 5-30. Head of double-crested cormorant (*Phalacrocorax auritus*).

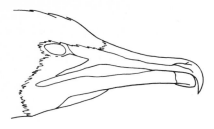

Fig. 5-31. Head of American anhinga (*Anhinga anhinga*).

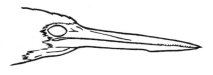

foundland, south to Florida, Louisiana, Baja California, Bahamas, and Isle of Pines.

Phalacrocorax olivaceus (**Humboldt**). Olivaceous cormorant. Rectrices usually 12, sometimes 13 or 14. Posterior border of pouch slightly heart-shaped. Outer primary shorter than 4th. Breeding plumage black with a slight purplish gloss; a border of white feathers along posterior margin of gular sac; no crests, but white filamentous feathers scattered on head, neck, and underparts. Otherwise similar to *P. auritus,* but the black-bordered feathers of scapulars and back are usually more lanceolate. Wing 240–305, tail 133–195, culmen 38–61, tarsus 45–59 mm. Southern Louisiana, southeastern Texas, and Sonora, south through Central and South America.

Phalacrocorax penicillatus (**Brandt**). Brandt's cormorant. Rectrices 12. Gular sac heart-shaped. Greenish black with violet or blue reflections on head and neck; in breeding plumage with a row of straw-colored long (50 mm) plumes along each side of neck, with even longer ones on scapulars; feathers bordering pouch brown; black margins of scapulars and wing coverts narrower than in other species; eye green; pouch dark blue; bill dusky. YOUNG: Brown below. Wing 267–298, tail 140–

165, culmen 66–75, tarsus 63 mm. Pacific Coast, from Vancouver Island to Baja California.

Phalacrocorax pelagicus **Pallas.** Pelagic cormorant. Rectrices 12. Gular sac heart-shaped, the feathering along midline of pouch and on sides of lower mandible extending far forward of level of eyes. Head and neck purplish black, the gloss becoming bluer on lower neck and dark green on lower parts and back; 2 black median crests, one on pileum, the other on nape; neck and rump with white threadlike plumes; a large patch of white on flanks; gular sac and skin of lores dull red. Nonbreeding birds lack the crests, the white plumes, and white flank patches. YOUNG: Brownish, darker above, with a slight greenish gloss. Wing 236–289 tail 152–178, culmen 42–53, tarsus 42–51 mm. Both coasts of the Pacific, from Bering Sea to Central Mexico, China, and Japan.

Family Anhingidae. Darters. Head very slender; bill twice length of head, straight, slender, sharp; tomia serrate; nostrils closed in adult; gular sac small. 3d primary longest, 2d subequal. Tail long, rounded; rectrices 12, the middle ones with numerous cross ribs. Tarsi short, but feet large; outer and middle toes subequal. Sexes unlike.

Styloid bone little developed or cartilaginous. Vertebrae with many ligaments; cervicals 20, 8th and 9th with overlapping postzygapophyses forming a permanent kink in neck, 9th with fibrous or ossified bridge of Dönitz forming loop for tendon; free dorsals 4, opisthocoelous. Furculum ligamentously attached to sternum. Sternum with a pair of rather deep rounded notches. Procoracoid process small. Olecranal fossa rather shallow. Only the left carotid artery. Pyloric lobe of stomach with hairlike processes; caeca small. Thigh muscles AX+. Fish by diving and swimming under water; can sink like a grebe and often swim partly submerged. Colonial; nest in bushes; eggs 3–5, blue, chalky; young hatched naked, skin buffy. Eocene onward. Tropics and subtropics of world. 1 genus.

Genus *Anhinga* **Brisson.** Fresh waters of Africa and Madagascar, tropical Asia, New Guinea, and Australia, warmer parts of America. 2 species, 1 in the United States.

Anhinga anhinga (**Linnaeus**). Anhinga (Fig. 5-31). Black, glossed with green; a mane of long black decomposed feathers on hind neck, this becoming grayish brown on crown and sides of neck;

upper back, lesser and middle coverts spotted with silvery; exposed portion of greater coverts silvery; scapulars and tertials with long silvery streaks; tail tipped with fawn; iris red; bill yellowish horn, dusky on culmen; gular sac black; lores yellowish green; eyelids black, bordered by blue and green; feet dull yellowish. FEMALE: Head, neck, and upper breast fawn color, darker on crown and hind neck, narrowly chestnut at breast; mane shorter than in male, mixed white and brown. In nonbreeding condition the mane is absent in both sexes. YOUNG: Resembles nonbreeding female but lacks ribbing and pale tip to tail; coloration somewhat paler and grayer. Wing 295–355, tail 218–265, culmen 67–97 mm. Texas, Arkansas, southern Illinois, and North Carolina, southward through tropical mainland to northern Argentina; Cuba and Isle of Pines.

Family Fregatidae. Frigate-birds. Bill longer than head, cylindrical, stout, both mandibles hooked down; tomia smooth; nostrils small, linear in a long groove; gular sac small but distensible. Wings very long, pointed; 1st primary longest. Tail deeply forked; rectrices 12. Tarsus minute, $\frac{1}{5}$ length of culmen, feathered; toes long, middle toe longest, its nail pectinate, all nails long; webbing reduced to basal portion.

Lacrimopalatine bone present. Cervicals 15. Sternum with posterior margin entire; whole pectoral girdle ankylosed to sternum; procoracoid process small. Ectepicondyle large. Thigh muscles A+. Caeca very small. Sexes different. Catch fish on wing and also rob boobies and terns. Colonial; nest in low bushes; 1 white egg; young naked at hatching. No fossil history. Warmer seas of world. 1 genus.

Genus *Fregata* Lacépède. Warmer seas, north to southern United States. 5 species, 1 in the United States.

Fregata magnificens **Mathews.** Magnificent frigate-bird. Plumage of male completely black with violet iridescence; iris brown; bill dull blue, chalky; gular sac inflatable, red or orange; feet black. FEMALE: Black above with some violet gloss; a light-brown bar on lesser coverts; throat and belly brown; breast white; gular sac not inflatable, dusky purple; orbital skin dark blue; feet reddish. YOUNG: Similar to female, but head, neck, and breast white. Wing 611–703, tail 339–543, culmen

105–130, tarsus 21–25 mm. Warmer parts of the Atlantic and Pacific, from Florida and Gulf states, through West Indies and Mexico to Brazil, from California to Ecuador and the Galapagos; Cape Verde Islands and West Africa.

Order Ardeiformes.[1] Heronlike birds and flamingos. Bill longer than head; nostrils usually perforate. Wings broad, rounded; functional primaries 10–11; secondaries aquintocubital. Tail short; rectrices 10–14, usually 12. Oil gland tufted. Legs very long; lower part of tibia bare.

Desmognathous, without basipterygoid processes. Sternum without internal spine, with 1–2 pairs of notches. Thigh formula $(AB)XY\pm$; toe tendons types 1, 4, or 7; 1 pair of sternotracheal muscles. Both carotids present, but sometimes joined. Caeca present; intestine types 2 or 4. Wading birds. Food aquatic animals. Nest on ground or in trees; eggs 2–6; young at least partly downy. Early Cretaceous onward. Cosmopolitan. 4 of 6 living families in United States.

KEY TO FAMILIES OF ARDEIFORMES

1. Anterior toes fully webbed   **Phoenicopteridae**
   Anterior toes not fully webbed   **2**
2. Middle claw pectinate (Fig. 5-32)   **Ardeidae**
   Middle claw not pectinate   **3**
3. With a long groove from nostril to tip of bill   **Plataleidae**
   Sides of maxilla without grooves   **Ciconiidae**

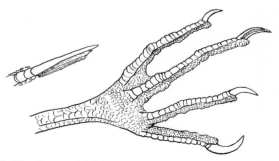

Fig. 5-32. Left foot of black-crowned night heron (*Nycticorax nycticorax*), showing pectinate middle claw.

[1] Ciconiiformes of A.O.U.

Family Ardeidae. Herons. Bill about as long as tarsus, straight, acute; sides of maxilla with a long groove; nostrils linear, slightly perforate. Functional primaries 10; rectrices 10–12. Front of tarsus scutellate, at least proximally; toes long, unwebbed; hallux incumbent, its claw the longest; middle claw pectinate. Head feathered except for naked lores; sides of neck with apteria; aftershafts present; adult downs confined to feather tracts; 2–3 pairs of powder-down tracts on rump and flanks.

Holorhinal. Mandibular process blunt. Cervicals 19–20. Sternum with 1 pair of notches. Furculum with hypocleidium. Hypotarsus complex. Thigh muscles $(B)XY—$; toe tendons types 1 or 7; syringeal muscles present. Supraorbital glands absent. Syrinx tracheobronchial. Tongue pointed; intestine type 2; caeca single. Nest usually a platform of sticks in trees, sometimes in cattails (*Ixobrychus*) or on ground (*Botaurus*); tree-nesting forms are more or less colonial; eggs 2–6, usually plain bluish or greenish (whitish in *Ixobrychus*, buffy in *Botaurus*); down of nestling sparse; altricial. Eocene onward. Cosmopolitan. 15 genera, 57 species; in the United States 9 genera, 12 species.

KEY TO GENERA OF ARDEIDAE

1. Rectrices 10; outer toe decidedly shorter than inner toe   **2**
   Rectrices 12; outer toe equal to or longer than inner toe   **3**
2. Tarsus longer than culmen   *Botaurus*
   Tarsus shorter than culmen   *Ixobrychus*
3. Lower half of front of tarsus reticulate   **4**
   Not more than lower third of acrotarsium reticulate, usually completely scutellate   **5**
4. Bare part of tibia shorter than hind toe and claw   *Nycticorax*
   Bare part of tibia longer than hind toe and claw   *Nyctanassa*
5. Tarsus decidedly shorter than culmen and shorter than middle toe with claw   *Butorides*
   Tarsus about equal to or longer than culmen, longer than middle toe with claw   **6**
6. Middle toe and claw almost as long as tarsus   *Ardeola*
   Middle toe and claw much shorter than tarsus   **7**
7. Claw of hind toe falling far short of end of 2d phalanx of middle toe   *Ardea*
   Claw of hind toe extending about to distal end of 2d phalanx of middle toe   **8**

8. Toes and claws short: middle toe and claw only about half length of tarsus   *Dichromanassa*
   Toes and claws long: middle toe and claw much more than half length of tarsus   *Egretta*

Genus *Ardea* Linnaeus. Bill straight, culmen shorter than tarsus, longer than middle toe with claw, nearly 5 times depth of bill; feathering on rami extending about as far forward as that on culmen, feathering on chin extending well beyond level of anterior end of nostrils. Tarsus longer than middle toe with claw, upper two-thirds scutellate in front; toes long, middle toe about 60% length of tarsus; outer toe longer than inner toe without claw; claw of hind toe falling far short of distal end of 2d phalanx of middle toe. 3d primary longest; longest secondaries conceal longest primaries. Rectrices 12. 3 pairs of powder-down patches. Cosmopolitan, except for some oceanic islands. 10 species, 1 in the United States.

***Ardea herodias* Linnaeus.** Great blue heron. Head white; sides of crown and occiput black, including long occipital plumes; neck fawn; midline of lower throat white streaked with black; lengthened plumes on sides of breast streaked white, gray, and black; median underparts black and white; sides gray; crissum white; thighs rusty; mantle and tail gray, with lengthened pale-gray scapular plumes; edge of wing deep rusty; primaries hoary black; bill yellowish, horn color on culmen; lores green; legs dark brown or blackish, soles olive yellow. YOUNG: No lengthened plumes; whole pileum blackish; black of underparts replaced by ashy; upper parts mixed with rusty; bill darker. Wing 430–533, tail 159–209, culmen 111–178, tarsus 146–232 mm. Southern Alaska and southern Canada south to Greater Antilles, Panama, and the Galapagos.[1]

Genus *Ardeola* Boie. Legs, toes and claws long; middle toe and claw longer than bill and almost as long as tarsus; bare part of tibia shorter than inner toe without claw. Africa and southern Eurasia. 2 species, 1 a recent invader of the Americas.

***Ardeola ibis* (Linnaeus).**[2] Cattle egret. White; in breeding plumage with decomposed buff plumes

---

[1] *Ardea occidentalis* of A.O.U., a pure white phase with blue lores and yellow eye, bill, and legs, occurs in southern Florida, Cuba, Jamaica, and Yucatan. Intermediates are like normal birds, but the crown is streaked.
[2] *Bubulcus ibis* of A.O.U.

on occiput, neck, breast, and scapulars; eye, bill, and legs yellow (legs dark in young). Wing 233–265, tail 88–98, culmen 52–60, tarsus 72–87 mm. Spain, Africa, southern Asia; has recently invaded northern South America, West Indies, and North America from Florida and Texas to Massachusetts and Ontario; introduced in Hawaii.

Genus *Egretta* **Forster.** Toes and claws long, with middle toe and claw much longer than half length of tarsus, but shorter than bill; bare part of tibia longer than inner toe. Practically cosmopolitan. 11 species, 4 in the United States.

KEY TO SPECIES OF *EGRETTA*

1. Large: wing 356–465 mm; tarsus 140–215 mm
   E. alba
   Smaller: wing under 300 mm; tarsus under 115 mm 2
2. Plumage entirely white   *E. thula*
   With at least tips of some primaries dark   3
3. Culmen distinctly shorter than tarsus   *E. caerulea*
   Culmen about as long as tarsus   *E. tricolor*

Egretta alba (**Linnaeus**).[1] Greater egret; common egret. White; with long, decomposed plumes on scapulars when breeding; eye, bill, and lores yellow; legs and feet black. Wing 356–465, tail 140–165, culmen 107–135, tarsus 140–215 mm. Eurasia east to New Zealand, Africa, Americas; breeds north to Oregon, Idaho, Oklahoma, Minnesota, Wisconsin, Indiana, Ohio, and New Jersey; in fall wanders north to southern Canada.

Egretta thula (**Molina**).[2] Snowy egret. White; long decomposed plumes on occiput, scapulars, and jugulum when breeding; eye, toes, and back of tarsi yellow; legs elsewhere black. Wing 230–290, tail 86–95, culmen 73–99, tarsus 74–114 mm. Breeds from Chile and Argentina north to California, Idaho, Oklahoma, Gulf states, and New Jersey, with some fall movement north.

Egretta caerulea (**Linnaeus**).[3] Little blue heron. Slate blue; head and neck tinged with maroon; occipital, scapular, and jugular plumes long but not decomposed; bill and lores blackish blue, base of mandible greenish; legs and feet blackish. YOUNG: White, but with tips of longer primaries

dusky; legs and feet greenish. Birds molting into adult plumage are pied. Wing 228–269, tail 84–108, culmen 64–86, tarsus 80–102 mm. Breeds from Uruguay and Peru north through Mexico and West Indies to Oklahoma, Missouri, Tennessee, and Massachusetts; in fall wanders north to Great Lakes and southern Canada.

Egretta tricolor (**Müller**).[4] Tricolored heron; Louisiana heron. Slaty blue above; occipital plumes lanceolate, not decomposed, maroon, the longer ones white; scapular plumes partly decomposed, pale purplish drab; jugular plumes lanceolate, slaty maroon; rump, breast, belly, and tail coverts white; throat white, passing down center of neck as a white line, where mixed with slate and rusty; iris red; tip of bill black; base of bill and lores blue; legs slate. WINTER: Bill black and yellow; lores yellow; legs yellowish green. YOUNG: No plumes; upper parts and neck duller, extensively mixed with rusty. Wing 204–281, tail 65–94, culmen 78–108, tarsus 81–106 mm. Brazil and Ecuador north to Mexico, Oklahoma, Missouri, Tennessee, and New Jersey; wanders north to California, Ohio Valley, and Massachusetts.

Genus *Dichromanassa* **Ridgway.** Bill stout; tarsus longer than culmen; toes short, about half length of tarsus. Gulf states, Central America, and Greater Antilles. 1 species.

Dichromanassa rufescens (**Gmelin**). Reddish egret. Head and neck cinnamon, with feathers lanceolate and lengthened during breeding; rest of plumage slaty, the scapular plumes long and somewhat decomposed; bill black, pinkish at base and on lores; iris white; legs dusky blue. YOUNG: Gray, mixed with rusty; bill and lores dark. A pure-white phase occurs, and dark birds are often blotched with white. Wing 294–342, tail 107–119, culmen 90–106, tarsus 124–145 mm. Gulf Coast of the United States and both coasts of Mexico, south to El Salvador; Bahamas, Cuba, Jamaica, and Hispaniola.

Genus *Butorides* **Blyth.** Tarsus much shorter than culmen, somewhat shorter than middle toe and claw; lateral toes equal in length, their claws nearly reaching base of middle claw; feathering on forehead extending farther forward than that on sides of mandible; 2d and 3d primaries longest. All continents except Europe. 1 species.

---

[1] *Casmerodius albus* of A.O.U.
[2] *Leucophoyx thula* of A.O.U.
[3] *Florida caerulea* of A.O.U.
[4] *Hydranassa tricolor* of A.O.U.

Butorides virescens (**Linnaeus**). Green heron Crown and lengthened occipital feathers greenish black; a white stripe down center of neck, streaked with black proximally; rest of head and neck cinnamon or maroon; upper parts glossy green, wing coverts edged with buffy, the lanceolate scapulars sometimes tinged with hoary; remiges slaty, inner ones narrowly tipped with white; edge of wing whitish; underparts ashy tinged with buff; iris yellow; bill dusky, yellowish below; legs greenish yellow to orange. YOUNG: No scapular plumes; underparts streaked white, buffy, and slate. Wing 156–202, tail 53–79, culmen 51–69, tarsus 43–58 mm. Washington, North Dakota, and southeastern Canada, south to Argentina; also Africa, Asia, and Australia.

Genus *Nyctanassa* Stejneger. Bill stout, culmen 3 times depth of bill, shorter than tarsus but about equal to middle toe with claw; tarsus longer than middle toe with claw, lower third of tarsus reticulate; hind toe and claw extending to distal end of second phalanx of middle toe, shorter than bare portion of tibia; primaries extending beyond secondaries. United States, West Indies, Central and northern South America. 1 species.

Nyctanassa violacea (**Linnaeus**). Yellow-crowned night heron. Crown, long occipital plumes, and auriculars yellowish white, the first tinged with rusty; rest of head and line down nape black; body bluish gray, the back feathers with black centers; iris orange; bill black; lores green; legs dusky green. YOUNG: Without occipital plumes; blackish brown, with white or buffy shaft streaks on head and neck, and with tips of back and wing feathers with a round buffy spot; breast and belly streaked with dark brown and buff; iris yellow. Wing 255–308, tail 97–118, culmen 62–81, tarsus 76–106 mm. Mexico, Kansas, the Ohio Valley, and South Carolina, south through West Indies and Central America to Peru and Brazil; Galapagos.

Genus *Nycticorax* Forster. Differs from *Nyctanassa* in having tarsus about equal to middle toe and claw; bare part of tibia shorter than hind toe and claw. Nearly cosmopolitan. 2 species, 1 in the United States.

Nycticorax nycticorax (**Linnaeus**). Black-crowned night heron. Forehead, long narrow occipital plumes, and throat white; crown and back greenish black; remaining upper parts deep gray;

underparts ashy; iris red; bill black, green at base and on lores; legs greenish yellow. YOUNG: No occipital plumes; grayish brown, with white shaft streaks on head, neck, wings, and underparts; remiges white-tipped. Wing 259–345, tail 97–114, culmen 63–86, tarsus 65–90 mm. Eurasia, Africa, and the Americas, north to Oregon, Wyoming, and southeastern Canada.

Genus *Botaurus* Stephens. Culmen shorter than tarsus; rectrices 10, very soft; 2 pairs of powder-down patches; primaries concealed by secondaries; bare portion of tibia shorter than middle toe without claw; tarsus scutellate, about equal to middle toe with claw; outer toe decidedly shorter than inner; claws long, slightly curved. Nearly cosmopolitan. 4 species, 1 in the United States.

Botaurus lentiginosus (**Rackett**). American bittern. Crown black mixed with rust; cheeks golden, continuing down sides of neck; malar stripe rusty, followed by a black stripe down sides of neck; throat white, with a buffy and dusky line down center; upper parts freckled with rust, buff, and brown; remiges plumbeous, the inner one freckled with rust at tips; underparts pale buff, the lower neck, breast, and sides with long streaks of brown, freckled and bordered with black; iris yellow; bill yellow, dusky along culmen; lores brown; legs yellowish green. Wing 238–296, tail 71–96, culmen 63–82, tarsus 78–103 mm. Middle Canada through the United States; winters south to Panama and West Indies.

Genus *Ixobrychus* Billberg. Differs from *Botaurus* in having culmen longer than tarsus; tips of primaries exposed. Cosmopolitan. 7 species, 1 in the United States.

Ixobrychus exilis (**Gmelin**). Least bittern. Crown, back, and tail black glossed with green; superciliaries, hind neck, and most of wing coverts chestnut; large patch in middle of coverts rich buff; narrow stripe on edge of scapulars buff; sides of neck rich buff; underparts white, with wide streaks of rich buff on throat, and tinged with buff on flanks; iris yellow; bill yellow, blackish on culmen; lores and tarsi green; toes yellow. FEMALE: Crown and back chestnut. YOUNG: Dorsal feathers buff-tipped. Wing 106–131, tail 37–47, culmen 41–52, tarsus 37–44 mm. Oregon, North Dakota, and southeastern Canada, south to West Indies and southern Mexico; reappears in most of South America; on migration in Central America.

Family Ciconiidae. Storks. Bill long, maxilla ungrooved; nostrils perforate. Head often naked, but no apteria on neck; functional primaries 10–11; adult downs on both pterylae and apteria; no powder downs; aftershafts present or absent. Tarsus reticulate; hallux slightly raised; claws unpectinate.

Holorhinal; mandibular process blunt. Cervicals 17. Sternum with 1 pair of notches. Furculum without hypocleidium. Hypotarsus simple. Syrinx without tracheobronchial muscles; thigh $AXY \pm$; toe tendons type 1. Supraorbital glands absent. Tongue rudimentary; intestine telogyric (type 4); caeca 2, rudimentary.

Nest large, of sticks, in trees or on chimneys; eggs 2–4, white, chalky; young altricial, downy.

Oligocene onward. Nearly cosmopolitan (absent from Oceania, New Zealand, and most of Australia). 10 genera; 17 species, 1 in the United States.

Genus *Mycteria* Linnaeus. Bill decurved; lower half of tibia bare; toes long, middle toe more than half length of tarsus; outer toe longer than inner toe; base of toes webbed; trachea flattened proximally, not convoluted. Head and neck of adult unfeathered, scaly (feathered in young). 1 species.

Mycteria americana **Linnaeus.** Wood ibis. White; remiges, primary coverts, alula, and tail purplish black; iris brown; bill yellowish; head plumbeous; legs bluish; toes black, yellowish at base. YOUNG: Head and neck with gray semiplumaceous feathers. Wing 447–495, tail 146–156, culmen 155–225, tarsus 174–216 mm. South Carolina and Gulf states south through West Indies and Mexico to Peru and Argentina.

Family Plataleidae.[1] Ibises and spoonbills. Bill grooved from nostril to tip, not acute; nostrils imperforate. Functional primaries 10; rectrices 12; no apteria on sides of neck, adult downs on both pterylae and apteria; no powder downs; aftershafts present. Toes with slight basal webs; middle claw slightly scalloped; hallux slightly raised.

Schizorhinal; with occipital fontanels; mandibular process hooked. Cervicals 17. Sternum with 2 pairs of notches. Coracoids crossed. Furculum without hypocleidium. Humerus with small, rounded ectepicondylar process. Hypotarsus simple. No syringeal muscles; thigh muscles complete, $ABXY+$; toe tendons type 1. Tongue reduced; intestine type 4; 2 rudimentary caeca. Supraorbital glands small.

[1] Threskiornithidae of A.O.U.

Nest of sticks or reeds, in bushes or cattails; eggs 3–5, greenish, sometimes spotted with brown; young altricial, downy.

Upper Eocene onward. Cosmopolitan, except Oceania. 20 genera, 30 species; in the United States 3 genera, 4 species.

KEY TO GENERA OF PLATALEIDAE

1. Bill spatulate   *Ajaia*
   Bill sickle-shaped   **2**
2. Claws almost straight; forehead feathered   *Plegadis*
   Claws curved; forehead of adult naked   *Eudocimus*

Genus *Plegadis* Kaup. Bill much longer than tarsus, decurved; sides of both upper and lower mandibles with a longitudinal groove along whole length; terminal portion of culmen with a median longitudinal groove; tarsus longer than middle toe with claw, scutellate above, reticulate near toes and behind; outer toe longer than inner; hallux slightly elevated, its claw not reaching beyond middle of second phalanx of middle toe; claws only slightly curved, the middle one shallowly scalloped; secondaries conceal primaries; lores and interramal area bare. Practically cosmopolitan. 3 species, 2 in the United States.

KEY TO UNITED STATES SPECIES OF *PLEGADIS*

1. Feathers around base of bill white   *P. chihi*
   Feathers around base of bill blackish (naked skin here often white)   *P. falcinellus*

Plegadis falcinellus (**Linnaeus**). Glossy ibis. Head, neck, lesser wing coverts, and underparts chestnut; remaining upper parts purplish green; crissum dusky green; iris brown; bill blackish; bare skin of head greenish or slate blue, with a posterior border of white skin (seasonal?); legs dark gray. WINTER AND YOUNG: Head, neck, and underparts grayish brown, the two former streaked with white. Wing 259–306, tail 102, culmen 100–145, tarsus 74–109 mm. Eurasia east to Australia; Africa and Madagascar; Puerto Rico, Haiti, Cuba, and Florida, and recently north to Long Island.

Fig. 5-33. Head of white ibis (*Eudocimus albus*).

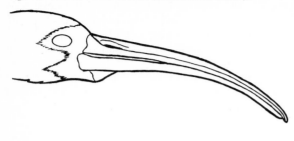

Plegadis chihi (Vieillot). White-faced ibis. Differs from the preceding in having white feathers around bare head skin in breeding plumage; gloss on plumage purplish rather than green; lores red. Wing 236–274, tail 95–108, culmen 95–152, tarsus 76–111 mm. Oregon and Utah south to Texas, Louisiana (?), and southern Mexico; reappears in South America from Peru and Brazil to Chile and Argentina.

Genus *Eudocimus* Wagler. Differs from *Plegadis* in having front of tarsus more extensively scutellate; claws curved; lateral groove on lower mandible confined to proximal half; dorsal groove on culmen lacking; forehead, anterior sides of head, and chin naked in adult. Tropical and subtropical America. 2 species, 1 in the United States.

Eudocimus albus (Linnaeus). White ibis (Fig. 5-33). White; tips of 4 outer primaries black; iris blue; tip of bill dusky; rest of bill, face, and legs red. YOUNG: Head and neck streaked brown and white; mantle, wings, and tip of tail brown; otherwise white. In older birds head and neck streaked; bastard wing and outer primaries black; rest of plumage white. Wing 262–300, tail 92–116, culmen 102–172, tarsus 78–112 mm. South Carolina and Gulf states, south through West Indies and Mexico to Venezuela and western Peru.

Genus *Ajaia* Reichenbach. Bill straight, only the extreme tip decurved, expanding into a horizontal disc on distal half, grooved along margins of upper mandible and on gonys; tarsus reticulate; claws curved; head of adult largely naked; trachea not convoluted. Tropical and subtropical America. 1 species.

Ajaia ajaja (Linnaeus). Roseate spoonbill. Neck white; rest of plumage pink, mixed with red

on lesser wing coverts, upper tail coverts, and jugulum; iris and legs red; bill and bare skin of head varied with green, yellow, orange, and black. YOUNG: Head mostly feathered; plumage white, tinged with pink on wings, tail, and belly; border of wings brown. Wing 353–388, tail 99–127, culmen 157–181, tarsus 95–118 mm. Southern Florida Gulf states, Greater Antilles, and central Mexico, south to Argentina and Chile.

Family Phoenicopteridae. Flamingos. Bill stout, lamellate, abruptly bent down at middle, distal portion of mandible broader than maxilla; nostrils perforate. Tibia largely bare; legs scutellate before and behind; anterior toes short, webbed; hallux absent or small and elevated. Functional primaries 11; secondaries extending beyond primaries; rectrices 14; aftershafts present; no lateral cervical apteria; adult downs on both pterylae and apteria.

Holorhinal; angular process long and hooked; supraorbital glands present. Cervicals 18–19. Sternum with 1 pair of notches; external spine a shallow Y. Coracoids cross. Furculum U-shaped, without hypocleidium. Hypotarsus simple. Thigh muscles $BXY+$; toe tendons type 4. Tongue very large; intestine type 4; 2 long caeca; crop present.

Colonial. Nest of mud, cone-shaped; eggs 2, chalky blue; young downy, precocial. Feeds by scooping up mud. Voice gooselike. Eocene onward. Eurasia, Africa, and Madagascar, South America, and the Caribbean area. 3 genera; 6 species, 1 in the United States.

Genus *Phoenicopterus* Linnaeus. Zygoma overlaps base of mandible; hallux present. Eurasia, Africa, America. 3 species, 1 in the United States.

Phoenicopterus ruber Linnaeus. American flamingo. Pink, more reddish on wings and flanks; remiges black; iris yellow; bill black at tip, yellow at base; legs red. YOUNG: Grayish white, more dusky on wings. Wing 388–419, tail 152, culmen 126–132, tarsus 302–368 mm. Breeds on Bahamas, Cuba, Hispaniola, Yucatan, Guiana, and Galapagos; occasional elsewhere in the Caribbean area, including Florida.

Order Anseriformes. Waterfowl and screamers. Bill with premaxillary division of sheath, neither pointed nor raptorial; nostrils perforate. Feet at least partly webbed. Functional primaries 10; secondaries aquintocubital; aftershafts rudimentary or absent; sides of neck feathered; adult downs on both apteria and pterylae; oil gland tufted.

Desmognathous, with vomer and basipterygoid processes. Holorhinal. Angular process long and hooked. Cervicals 16–25. Sternum with solid posterior border or 1 pair of notches or fenestrae; internal spine absent. Furculum U-shaped, without hypocleidium. Coracoids separate, with procoracoid process. Tongue well developed; intestine orthocoelous, type 3; 2 long caeca. Thigh muscles $ABX(Y)+$; toe flexors types 2 or 4; 2 pairs of sternotracheal muscles. Syrinx tracheobronchial. Both carotids present.

Aquatic or semiaquatic, all capable of swimming. Food animal or vegetable. Nest lined with down or feathers, usually on ground, sometimes in trees; eggs 3–15, plain glossy white, buffy, or greenish; young downy, precocial. Eocene onward. Cosmopolitan. 2 families; the peculiar Anhimidae, represented by 2 genera and 3 species, are confined to South America.

Family Anatidae. Ducks, geese, and swans. Bill lamellate, straight, with rostral and mental nails. Secondaries 15–24; rectrices 12–24; apteria present. Toes slender, palmate; hallux small, raised. Skeleton with normal pneumaticity. Basipterygoids articulate with anterior end of pterygoids. Sternum with 1 pair of notches or fenestrae. Ribs with uncinate processes. Hypotarsus complex. Thigh muscles $ABX+$; toe flexors type 2. Tongue fleshy, its edges fringed. Supraorbital glands and penis present. 47 genera, 149 species; in the United States 20 genera, 44 species.

KEY TO GENERA OF ANATIDAE

1. Neck as long as or longer than body; lores bare   **2**
   Neck shorter than body; lores feathered   **3**
2. Tail rounded   *Olor*
   Tail wedge-shaped   *Cygnus*
3. Tarsus longer than middle toe without claw   **4**
   Tarsus shorter than middle toe without claw   **6**
4. Serrations not visible from sides of closed bill   *Branta*
   Serrations visible from sides of closed bill   **5**
5. Width of bill greater than depth at posterior end of nostrils   *Philacte*

Width of bill less than depth at posterior end of nostrils   *Anser*
6. Sides of lower mandible serrated   **7**
   Sides of lower mandible lamellate   **8**
7. Serrations of lower mandible vertical   *Lophodytes*
   Serrations bent backward   *Mergus*
8. Lower part of acrotarsium reticulate   *Dendrocygna*
   Lower part of acrotarsium scutellate   **9**
9. Hind toe unlobed   **10**
   Hind toe with a broad membranous lobe   **13**
10. Rectrices broad, with rounded tips   *Aix*
    Rectrices narrower, at least middle pair with pointed tips   **11**
11. Width of bill near tip about twice width at base   *Spatula*
    Bill not much broader near tip than at base   **12**
12. Rectrices 14; culmen shorter than tarsus   *Mareca*
    Rectrices 16; culmen longer than tarsus   *Anas*
13. Rectrices 18   *Oxyura*
    Rectrices 14–16   **14**
14. Feathering on forehead or lores extending as far forward as level of posterior end on nostril   *Somateria*
    Feathering on forehead and lores not reaching forward to level of nostrils   **15**
15. Graduation of tail less than length of bill from nostril   *Aythya*
    Graduation of tail much more than length of bill from nostril   **16**
16. Graduation of tail less than distance from tip of bill to loral feathering   **17**
    Graduation of tail more than distance from tip of bill to loral feathering   **18**
17. Nostrils in middle of bill or beyond   *Melanitta*
    Nostrils in basal third of bill   *Camptorhynchus*
18. Rostral nail narrow; rectrices 16   *Bucephala*
    Rostral nail wide, occupying whole tip of bill; rectrices 14   **19**
19. Dorsal aspect of loral feathering forming a sharp angle; a fleshy lobe at base of commissure in male   *Histrionicus*
    Dorsal aspect of loral feathering gently rounded; no lobe at base of commissure   *Clangula*

Genus *Olor* Wagler. Culmen shorter than tarsus; lores bare; tail rounded, longer than middle toe with claw, rectrices 20–24; tarsus shorter than middle toe without claw, reticulate; hallux elevated, unlobed. Cervicals 22–24. Trachea looped within sternum, and symphysis of furculum correspondingly bent. Holarctic. 4 species, 2 in the United States.

KEY TO SPECIES OF *OLOR*

1. Distance from anterior corner of eye to posterior end of nostril much greater than from posterior end of nostril to tip of bill; rectrices 20   *O. columbianus*
Distance from anterior corner of eye to posterior end of nostril not greater than from posterior end of nostril to tip of bill; rectrices 24   *O. buccinator*

Olor columbianus (Ord). Whistling swan. Front of nostril nearer tip than base of bill. White; bill black, often with a yellow spot on lores; iris brown; feet black. YOUNG: Ashy, washed with reddish brown on head and neck; bill flesh-colored, dusky at tip; lores without yellow spot; feet flesh-colored. Wing 520–575, tail 153–203, culmen 97–106, tarsus 99–123 mm. Breeds in Arctic Canada, Alaska, and adjacent Siberia; winters on Atlantic Coast from Massachusetts to the Gulf states, and on Pacific Coast from southern Alaska to northern Mexico.

Olor buccinator Richardson. Trumpeter swan. Front of nostril in middle of bill. White; bill black with a reddish streak along cutting edges; lores entirely black. YOUNG: Bill blackish, middle of culmen and edge of lower mandible flesh-colored; feet yellowish olive brown. Wing 545–652, tail 173–204, culmen 109–119, tarsus 103–127 mm. Formerly bred from Alaska and Canada south to Wyoming, Nebraska, Iowa, Missouri, and Indiana; wintered south to California and Gulf of Mexico; now confined to Alaska (1,300 birds), western Canada, and Yellowstone region (500 birds).

Genus *Cygnus* Bechstein. Differs from *Olor* in having tail cuneate; trachea straight, not entering sternum; furculum simple. Palearctic and southern South America. 2 species, 1 introduced to the United States.

Cygnus olor (Gmelin). Mute swan. White; maxilla orange, its base, nostrils, knob at base of culmen, and mandible black; feet gray. YOUNG: Brownish gray; bill black; frontal knob smaller. Wing 535–622, culmen 70–90, tarsus 98–120 mm. Northern Europe and Asia; winters in southern Europe, North Africa, Asia Minor, and central Asia; feral in eastern New York, wandering to Massachusetts and New Jersey.

Genus *Branta* Scopoli. Bill shorter than head, higher than wide at nostrils; culmen nearly straight, about equal to shortest (inner) anterior toe and claw; tomia nearly straight, the serrations scarcely visible in closed bill; nostril in middle of nasal fossa; rectrices 14–20; tarsus longer than middle toe with claw. Cervicals 16–18; supraorbital glands scarcely impressed on frontals. Holarctic, 4 species; 2 in the United States.

KEY TO SPECIES OF *BRANTA*

1. Cheeks white   *B. canadensis*
   Whole head black   *B. bernicla*

Branta canadensis Linnaeus. Canada goose. Head and neck black; a white patch on each cheek, sometimes meeting on throat; some races with a white half collar at front of lower neck; body brownish gray, edged with white, paler on upper tail coverts and underparts; belly and crissum white; tail black; iris brown; bill and feet black. Wing 345–533, tail 127–152, culmen 26–59, tarsus 61–95 mm. North America, breeding from Arctic south to Gulf of St. Lawrence, Ohio Valley (formerly), South Dakota, Colorado, Utah, and California; in winter south to southern tier of states and to Mexico.

Branta bernicla (Linnaeus).[1] Brant. Head, neck, breast, and upper back black; a patch of white streaks on sides of neck; back grayish brown; sides of rump and upper tail coverts white; remiges and tail blackish; underparts whitish becoming grayer anteriorly, but sharply contrasting with black of upper breast; iris brown; bill and feet black. Wing 300–345, tail 91–114, culmen 30–41, tarsus 53–63 mm. Arctic Europe, Asia, and North America; winters on coast south to North Carolina, Baja California, China, Japan, and Mediterranean.

Genus *Philacte* Bannister. Resembles *Branta* but bill wider than high at nostrils; nostrils in lower anterior portion of nasal fossae; lamellae partly visible on closed tomia; supraorbital glands impressed on frontals. Siberia and Alaska, southward in winter. 1 species.

Philacte canagica (Sevastianoff). Emperor goose. Head and neck white, often tinged with rust; chin and foreneck black; body bluish gray, barred with black and white; remiges gray, blackish at tip; tail slate basally, white distally; iris

[1] Includes black brant, *Branta nigricans* of A.O.U.

brown; bill flesh, with white nail; feet orange. YOUNG: Head and neck spotted with dusky; black barring of upper parts replaced by brown. Wing 363–400, tail 127–152, culmen 35–42, tarsus 66–72 mm. Siberia and Alaska, in winter occasionally southward along Pacific Coast to California.

Genus *Anser* Brisson. Differs from *Branta* in having lamellae visible along whole length of tomia; culmen concave; frontals rising abruptly from bill. Holarctic; 5 species, 3 in the United States.

KEY TO SPECIES OF *ANSER*

1. Feet yellow   *A. albifrons*
   Feet pink   2
2. Base of bill warty; tomia pink   *A. rossii*
   Base of bill smooth; tomia black   *A. caerulescens*

Anser albifrons (Scopoli). White-fronted goose. Band on forehead white; upper parts brownish gray, tipped with paler; sides of rump, upper and under tail coverts white; tail tipped with white; greater coverts ashy, edged with white; remiges dusky, edged and tipped with white; below whitish blotched with black; iris brown; bill pink, nail white; feet orange. Wing 362–450, tail 113–139, culmen 42–57, tarsus 57–79 mm. Holarctic, breeding in the far north; winters south to North Africa, southern Asia, Mexico, Texas, and Louisiana; rare east of Mississippi Valley.

Anser caerulescens (Linnaeus).[1] Blue goose. Head and upper neck white, stained with rusty; upper parts gray, tipped with brown or ashy; rump white or blue-gray; primaries black; secondaries and tail dusky, edged with white; breast and belly like back or white; flanks white; iris brown; feet and bill pink, rostral nail whitish, tomia blackish. YOUNG: Head and neck slaty; chin white; bill and feet dusky. SNOW GOOSE: White, often stained with rust; primaries black, their bases and coverts gray; iris brown; bill pink, tomia black, rostral nail whitish; feet red. YOUNG: Gray mottled with dusky on head and neck; rump, tail, and its coverts white; primaries slaty, darker at tips; secondaries dusky; bill dusky pink. Wing 368–444, tail 140–165, culmen 49–68; tarsus 71–89 mm. Breeds in Arctic;

[1] *Chen caerulescens* of A.O.U.; includes snow goose, *Chen hyperborea* of A.O.U.

winters in Japan, interior California, and coast of Maryland, Virginia, North Carolina, Louisiana, and Texas; migrates through Mississippi Valley.

Anser rossii Cassin.[2] Ross's goose. Plumage like snow goose. Bill of adult covered with warts basally and without black tomia; weight about half. Wing 349–393, tail 127, culmen 38–43, tarsus 58–76 mm. Mackenzie and Keewatin; migrates through Great Basin to interior California.

Genus *Dendrocygna* Swainson. Tree-ducks. Bill longer than head; lamellae not visible from side; nostrils in basal half of bill; wings rounded, outer primary shorter than 4th; tail rounded; legs long, extending beyond tip of tail; tarsus reticulate, shorter than middle toe; hallux unlobed, more than $\frac{1}{3}$ length of tarsus. Warmer parts of world; 8 species, 2 in the United States.

KEY TO SPECIES OF *DENDROCYGNA*

1. Belly black   *D. autumnalis*
   Belly light brown, like breast   *D. bicolor*

Dendrocygna autumnalis (Linnaeus). Black bellied tree-duck. Pileum cinnamon brown; sides of head ashy gray, becoming whiter around eye and on throat; line down hind neck black; lower neck and breast cinnamon brown; back darker; rump, upper tail coverts, tail, lining of wing, and belly black, the last mixed with white; crissum mottled black and white; lesser wing coverts golden olive; middle coverts ashy; greater coverts whitish; remiges black; outer webs of middle primaries whitish near base; iris brown; bill raspberry, orange above nostrils; rostral nail pale blue; feet flesh. Wing 233–246, tail 61–70, culmen 45–54, tarsus 56–66 mm. Texas south to Brazil.

Dendrocygna bicolor (Vieillot). Fulvous tree-duck. Head yellowish brown, darker on pileum, paler on throat; stripe down hind neck black; foreneck with half collar of black and white streaks; back dark brown; rump and tail black; upper and under tail coverts white; breast buffy brown; belly paler; flanks with cream and dusky streaks; lesser coverts chocolate; rest of wing black; iris brown; bill blue-black with black nail; feet slate. Wing 196–242, tail 41–57, culmen 42–52, tarsus 50–60

[2] *Chen rossii* of A.O.U.

mm. California, Nevada, Arizona, Texas, and Louisiana, south to Argentina; also East Africa, Madagascar, and India; some winter in Florida.

Genus *Anas* Linnaeus. Culmen shorter than head but longer than tarsus; sides of bill nearly straight or slightly flaring; frontals merging with line of culmen; about 16–36 lamellae visible on side of closed maxilla; wing pointed, 1st or 2d primaries longest; tail cuneate; rectrices 16, with tips more or less pointed; legs short, not reaching tip of tail; lower part of acrotarsium scutellate; tarsus shorter than middle toe; hallux unlobed, less than one-third length of middle toe. External sternal spine long and pointed, forming an angle of 90° or more with keel. Cosmopolitan. 43 species, 9 in the United States.

KEY TO SPECIES OF *ANAS*

1. More than 30 lamellae visible on side of closed maxilla   *A. strepera*
About 16-25 lamellae visible on sides of closed maxilla   **2**
2. Axillars with prominent brownish markings   *A. acuta*
Axillars plain white   **3**
3. Width of maxilla near tip less than ⅛ length of commissure   *A. crecca*
Width of maxilla near tip more than ⅛ length of commissure   **4**
4. Lesser wing coverts bluish   **5**
Lesser wing coverts not bluish   **6**
5. Male with white crescent before eye; female with chin and upper throat unstreaked   *A. discors*
Male with head chestnut; female with only chin region unstreaked   *A. cyanoptera*
6. Tertials and lesser wing coverts without pale edges   *A. platyrhynchos*
Tertials and lesser coverts edged with buffy   **7**
7. Chin streaked with dusky   *A. rubripes*
Chin buffy, unstreaked   **8**
8. Speculum bordered with black only   *A. fulvigula*
Speculum bordered with black and white   *A. diazi*

Anas platyrhynchos Linnaeus. Mallard. Head and neck green; ring around lower neck white; back grayish brown; rump, upper and under tail coverts black glossed with green, the 4 longer upper coverts curled; rectrices gray, white externally; scapulars vermiculated with dusky, the outer ones with a brown patch; wing coverts grayish brown; speculum purplish blue with a complete black border and a white bar before and behind; remiges dark brown; upper breast deep chestnut; remaining underparts whitish, finely vermiculated with gray; lining of wing white; iris brown; bill yellow or olive green, its nail black; feet orange. FEMALE: Head and neck buff streaked with brown; throat plain buffy; a dark streak through eye; body streaked and mottled brown and buffy, paler below; crissum whitish streaked with brown; tail grayish brown, whitish laterally; wings as in male; bill as in male but spotted with dusky, nail dusky. Wing 240–304, tail 76–101, culmen 51–61, tarsus 38–49 mm. The Holarctic, North Africa, and India; in America breeds south to Mexican boundary, Kansas, Missouri, Ohio Valley, and Virginia; winters south to West Indies and Panama. This and the next 3 forms are probably conspecific

Anas diazi Ridgway. Mexican duck. Both sexes resemble female *A. platyrhynchos* but darker and more heavily streaked; wing coverts edged with buff; anterior white speculum bar sometimes absent; bill greenish yellow or orange, dusky on culmen in female, nail black; feet orange. Wing 226–279, tail 89–94, culmen 47–57, tarsus 41–48 mm. Upper Rio Grande, from El Paso to Douglas, Arizona; also Mexican highlands.

Anas rubripes Brewster. Black duck. Crown, hind neck, and upper parts dark brown, indistinctly streaked and edged with buffy; sides of head and neck, chin and throat buff, heavily streaked with brown; underparts dark brown broadly edged with buff; speculum purple bordered all around with black only; lining of wing white; iris brown; bill yellow or olive, often with dusky spots; nail black; feet red to brownish olive. Wing 267–292, culmen 51–60, tarsus 43–48 mm. Eastern Canada, breeding south to Great Lakes and Delaware; winters from Great Lakes and New England south to Gulf states.

Anas fulvigula Ridgway. Mottled duck. Paler than *A. rubripes:* chin and foreneck plain buff; streaks on sides of neck finer; edges of body feathers paler and broader; speculum blue, bordered with black; bill bright yellow or dull orange, sometimes with black spots; nail black; feet orange. Wing 256–258, tail 88–90, culmen 50–52, tarsus 43–47 mm. Peninsular Florida; coast of Louisiana and Texas.

Anas acuta **Linnaeus.** Pintail. Head and neck brown, glossed on occiput with purplish bronze; upper hind neck black, separated from brown of occiput by a white V-shaped stripe down sides of neck, the stripes confluent with white of lower foreneck and underparts; lower hind neck, back, and rump irregularly barred with black and white; tertials and longer scapulars black, edged with buff; wing coverts brownish gray; speculum purplish bronze green, bordered in front by a cinnamon bar, above by black, behind by a narrow black bar followed by a white bar; remiges brownish; upper tail coverts vermiculated and streaked black and white; rectrices brownish edged with white, the middle pair tapering to twice the length of the next; crissum black; wing lining and axillars mottled dusky brown and white; iris brown; bill grayish blue, black on culmen, base, and nail; feet grayish blue. FEMALE: Head and neck buffy streaked with brown; throat paler and less heavily streaked; upper parts, including wing coverts and tail, brown, barred or edged with pale buff or white; middle rectrices little longer than the rest; below buffy white, more or less streaked with brown; speculum much paler than in male. Wing 243–284, tail 184–241, (♀ 114–127), culmen 44–54, tarsus 38–47 mm. Holarctic, North Africa, and India; in America breeds south to California, Utah, Colorado, Nebraska, Iowa, and Great Lakes; winters south to West Indies and Panama.

Anas discors **Linnaeus.** Blue-winged teal. A large white crescent in front of eye; pileum, lores, chin, and borders of crescent black; rest of head and upper neck dull lead blue; upper back brown, irregularly barred with buffy; rump and upper tail coverts blackish; tail dark brown; long scapulars and tertials blackish, with long buffy shaft stripe, the 2 middle ones with large bright-blue tip; lesser coverts pale blue; speculum bright green, bordered in front by a white stripe; remiges dusky; below pale chestnut, spotted and barred with black; crissum black; axillars white; iris brown; bill blackish; feet dull yellowish. FEMALE: Above brown, irregularly barred with buffy; lesser coverts blue; speculum bronze, bordered before, and often behind, with white; underparts whitish or buffy, heavily streaked and spotted with dusky; throat immaculate; axillars white. Wing 169–190, tail 61–69, culmen 35–43, tarsus 29–33 mm. Breeds from Can-

ada south to California, New Mexico, Texas, Louisiana, Tennessee, and North Carolina; winters from California and South Carolina to West Indies, Central and South America.

Anas cyanoptera **Vieillot.** Cinnamon teal. Head, neck, upper back, scapulars, and underparts chestnut; forehead and center of crown black; upper back, scapulars, and flanks spotted and barred with black; rump and upper tail coverts olive brown, with broad chestnut bars; tail dark brown, edged with buffy; crissum blackish, mixed with chestnut; remiges blackish; lesser coverts blue; speculum bright green, bordered in front and behind with white, above and below with black; axillars white; iris orange; bill black; feet dull orange-yellow. FEMALE: Very similar to female *A. discors*, but throat streaked, only the chin being immaculate; underparts usually tinged with cinnamon; bill averaging larger and somewhat constricted at base. Wing 167–220, tail 62–95, culmen 36–49, tarsus 29–36 mm. Western North America from British Columbia, Montana, Wyoming, Kansas, and Texas to California and central Mexico; also in South America.

Anas crecca **Linnaeus.**[1] Green-winged teal. Head and neck chestnut; area around eye black; behind this a bright-green patch, forming a crest with the lengthened chestnut neck feathers; tip of crest black; chin blackish; lower neck and back vermiculated black and white; upper tail coverts black, edged with buff or with black and white vermiculations; tail and remiges brownish; lesser coverts olive gray; speculum bright green, bordered in front by pale cinnamon, behind by white, and above and below by black; upper breast pale cinnamon spotted with black; flanks like back, but with a white bar before wing; belly white; crissum black centrally, white or buffy laterally; axillars white; iris brown; bill black; feet olive gray. FEMALE: Above, including lesser coverts, brown variegated with buffy; speculum as in male; throat whitish, finely streaked with brown; breast brown and buffy; rest of underparts mostly whitish, faintly marked with brown; axillars white. Wing 159–204, tail 58–76, culmen 29–40, tarsus 27–35 mm. North America south to California, New Mexico, Nebraska, Great Lakes, and New York; winters south

---

[1] Includes *Anas carolinensis* Gmelin of A.O.U.

to West Indies and Central America. Also Europe and Asia.

Anas strepera Linnaeus. Gadwall. Lamellae very numerous, about 32–36 visible along side of closed maxilla. Crown and nape brown; upper back, breast, and flanks vermiculated black and white; rump dark brown; upper and under tail coverts black; throat and sides of head buffy white, streaked with dusky; belly whitish (spotted with brown in immature); axillars white; lesser coverts gray, middle lesser coverts and median coverts forming a large chestnut patch; speculum white, bordered (except behind) with black; remiges and tail brownish; iris reddish brown; bill bluish black; feet orange; webs blackish. FEMALE: Head and neck buffy, streaked with brown, darker above; back, breast, upper and under tail coverts brown, margined and barred with buff; wing coverts brownish gray, edged with ash; rump similar but darker; belly whitish; speculum as in male; bill orange spotted with black; feet dull yellow. Wing 245–280, tail 82–90, culmen 36–47, tarsus 36–44 mm. Holarctic, North Africa, and India; in America breeds south to California, Arizona, New Mexico, Texas, Iowa, Wisconsin, Pennsylvania, and North Carolina; winters south to Gulf states, Mexico, and Africa.

Genus Mareca Stephens. Differs from *Anas* in having bill tapering, shorter than tarsus, with few lamellae, only about 11–16 visible along side of closed maxilla; frontals slightly swollen; supraorbital glands slightly more impressed; rectrices 14. Holarctic, India, and South America. 3 species, 2 in the United States.

KEY TO SPECIES OF MARECA

1. Axillars white, or only slightly vermiculated near tip
   M. americana
   Axillars closely vermiculated with grayish; shafts dark
   M. penelope

Mareca americana (Gmelin). Baldpate. Crown white (spotted with slate in young); postocular stripe green; rest of head and neck buffy white, heavily streaked with dusky; lower hind neck, back, and scapulars vermiculated dusky and vinaceous; rump vermiculated dusky and white (plain brown in young); upper tail coverts black laterally, buffy gray medially; rectrices gray, the middle ones black and lengthened; breast and sides vinaceous; belly white; axillars white, sometimes lightly freckled with slate; bend of wing gray vermiculated with slate; speculum green, bordered all around with black; remaining coverts white; remiges gray; tertials black, edged outwardly with white; iris brown; bill blue-gray, black at tip and below; feet blue-gray. FEMALE: Head and neck buffy, streaked with brown; upper parts, including lesser and middle coverts, brown, edged and barred with paler; speculum black, tinged with green; tertials edged gray and white; upper breast vinaceous buff, barred with brown; sides reddish brown; belly white; crissum irregularly barred, dusky brown and white; middle rectrices not lengthened; axillars white. Wing 253–279, tail 94–133, culmen 33–38, tarsus 37–42 mm. North America, breeding south to California, Utah, Colorado, Nebraska, and Indiana; winters south to Panama.

Mareca penelope (Linnaeus). European widgeon. Male with crown cream buff; head and upper neck cinnamon. Both sexes with the white axillars heavily freckled with slate, the shafts dark. Otherwise resembles *M. americana*. Wing 235–290, tail 77–109, culmen 32–36, tarsus 36–41 mm. Europe and Asia, south in winter to North Africa and India; occurs with some regularity on both our coasts and in the Mississippi Valley.

Genus Spatula Boie. Differs from *Anas* in having culmen from base longer than head or tarsus; sides of bill much flared, twice width at base; lamellae very fine and numerous, about 105 visible on side of closed maxilla; rectrices 14. Holarctic, Africa, Australia, New Zealand, South America. 4 species, 1 in the United States.

Spatula clypeata (Linnaeus). Shoveler (Fig. 5-34). Head and neck green; upper back brown, barred and streaked with white; rump black, with a large white patch on each side; upper tail coverts black, glossed with green; middle rectrices dusky, edged with white; outer rectrices pale buff mottled with brown; breast white, often spotted with black; belly chestnut, more or less barred with slate, especially on flanks; crissum white basally, greenish black distally; scapulars white, the longer ones edged with greenish black or pale blue; tertials greenish black with white shaft stripe; lesser and median coverts pale blue; greater coverts white; speculum green; remiges brown, axillars white; iris yellow; bill black; feet orange. FEMALE: Head and neck buffy, streaked with slate; upper parts, including scapulars, brown, edged and barred with buff; underparts buff, spotted with brown; wing duller than in male; bill olive brown. Wing 216–253, tail 77–81, culmen 58–73, tarsus 34–39 mm. Europe, Asia, Africa; in North America breeds south to California, Arizona, New Mexico, Kansas, Iowa, and Illinois; winters south to West Indies, Central America, and Colombia.

Genus *Aix* Boie. Head crested; culmen shorter than head, about equal to tarsus; sides of bill tapering; no lamellae extending beyond edge of maxilla; bare frontal angle prominent; cranium swollen, with frontals rising abruptly from culmen; 2d primary longest; tail rounded, rectrices 14, broad at tip; tail coverts long; legs short, lower part of acrotarsium scutellate, upper part smooth; tarsus shorter than middle toe; hallux unlobed; sternum without external spine and with notches tardily open; ulna much shorter than tibia. Nests in hollow trees. North America, 1 species. Asia, 1 species.

***Aix sponsa* (Linnaeus).** Wood duck. Pileum green; superciliary white; line through eye purple; line from bill to below eye green; lower malar region and patch on sides of neck purple; throat white, extending upward behind eye to green stripe and also up sides of hind neck to green stripe; back brown; rump, upper and under tail coverts, and tail dusky green; sides of rump reddish purple; lower neck purplish chestnut, with fine white spots posteriorly; upper breast white squamate with brown; belly white; flanks buffy brown vermiculated with black, the last flank feathers boldly barred at tip with black and white; axillars white barred with brown; scapulars bronze, externally edged with black glossed with bluish violet; wing coverts grayish brown; proximal median and greater coverts bluish violet tipped with white; remiges brown, edged with silver and tipped on inner webs with blue; eye and lids red; bill black at tip, on culmen, and below, pink in middle, red bordered with yellow at base; feet dull yellow, webs dusky. FEMALE: Pileum grayish brown; eye ring white; back, scapulars, and rump bronzy brown; tail and upper coverts brown glossed with

Fig. 5-34. Bill of shoveler (*Spatula clypeata*).

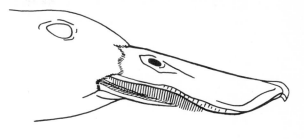

green; throat white; lower neck ashy; breast and flanks olive brown, mottled with buff; belly and crissum white, mottled with brown posteriorly; wings duller than in male; eyes brown; lids yellow; bill blue-gray, dusky on culmen, nail, and below, and with a white spot below nostril; feet dull yellow, webs dusky. Wing 214–241, tail 100, culmen 35, tarsus 32–39 mm. Southern Canada and entire United States; winters south to Jamaica and central Mexico.

Genus *Aythya* Boie. Culmen longer than tarsus; sides of bill straight or slightly flaring; lamellae coarse, scarcely extending beyond maxilla; wing pointed, first or second primaries longest; tail short, rounded or double-rounded; rectrices 14, tips pointed; legs short; acrotarsium partly scutellate; tarsus shorter than middle toe; middle and outer toes about equal; hallux lobed. External sternal spine short, forked; ulna and tibia about equal. All continents and New Zealand. 13 species, 5 in the United States.

KEY TO SPECIES OF *AYTHYA*

1. Speculum white   2
   Speculum gray   3
2. Width of rostral nail 6 mm or less   *A. affinis*
   Width of rostral nail 7 mm or more   *A. marila*
3. Culmen as long as head, 3 times width of bill
        A. valisineria
   Culmen shorter than head, less than 3 times width of bill   4
4. Gray speculum with narrow white tip and dusky subterminal band   *A. collaris*
   Gray speculum with narrow white tip but with no dusky subterminal band   *A. americana*

Aythya affinis (Eyton). Lesser scaup. Head, neck, back, rump, upper and under tail coverts, and breast black, glossed with purple; scapulars, interscapulars, flanks, and lower belly white, vermiculated with black; upper belly and axillars plain white; wing coverts, remiges, and rectrices dusky; speculum white, tipped with dusky; iris yellow; bill blue, nail black; feet and webs dusky. FEMALE: Head, neck, upper parts, upper breast, flanks, and crissum brown; a large white patch on lores; lower breast and axillars white; belly whitish, barred with brown; speculum as in male. Wing 187–210, tail 52–54, exposed culmen 37–40, width of rostral nail 5.5–6, tarsus 34–35.5 mm. North America, breeding south to Colorado and Idaho; winters throughout the United States and south to Panama and the West Indies.

Aythya marila (Linnaeus). Greater scaup. Differs from *A. affinis* in larger size, especially of bill; head of male glossed with green as well as purple; white of speculum extending onto 6–7 inner primaries; flanks of male unbarred. Wing 194–237, exposed culmen 42–47, width of nail 6.5–7, tarsus 34–41 mm. Holarctic; in America breeds south to Canadian border; winters mainly on coast, south to Florida, Louisiana, Texas, and Baja California.

Aythya collaris (Donovan). Ring-necked duck. Head, neck, upper parts, upper breast, and crissum black glossed with purple; small patch on chin white; narrow ring at base of neck chestnut; speculum gray with narrow white tip and dusky subterminal band; tertials black glossed with green; breast and belly white; flanks and lower belly whitish vermiculated with dusky; axillars white; iris yellow; bill blue, tipped with black, ringed with light blue near tip and at base; feet blue-gray, webs dusky. FEMALE: Upper parts brown; speculum as in male; lores and chin creamy; face and throat whitish mottled with brown; foreneck pale brown; upper breast and flanks reddish brown; belly white; lower belly and crissum grayish brown, the latter tipped with white; axillars white; iris brown. Wing 193–203, tail 56–61, exposed culmen 44–48, width of rostral nail 7, tarsus 33–36 mm. Breeds from southern Canada south to Arizona, Utah, Nebraska, Iowa, Wisconsin, and Michigan; winters in southern states, south to Greater Antilles and Panama.

Aythya americana (Eyton). Redhead. Head and upper neck reddish chestnut; lower neck, upper back, rump, upper and under tail coverts, and upper breast black; back, scapulars, and flanks white vermiculated with black; belly white; speculum gray with narrow white tip; axillars white; iris yellow; bill pale blue, tipped with black and ringed with white near tip; feet blue-gray, webs dusky. FEMALE: Head and neck yellowish brown; lores, chin, and area around eye whitish; upper parts brown, grayer on wings; speculum gray with narrow white tip; upper breast, flanks, and belly brownish; lower breast whitish (brownish in young); crissum brownish, white posteriorly; axillars white; iris brown. Wing 216–235, tail 52, exposed culmen 46–57, rostral nail 8, tarsus 42 mm. Middle Canada south to California, Arizona, New Mexico, Colorado, Nebraska, Iowa, and Wisconsin; winters in southern states, south to Mexico and West Indies.

Aythya valisineria (Wilson). Canvasback. Head and neck reddish chestnut; crown dusky; upper back, rump, upper and under tail coverts, and upper breast black; back, wing coverts, flanks, and belly white vermiculated with slate; lower breast plain white; speculum pale gray, tipped with white; axillars white; iris red or yellow; bill blackish; feet blue-gray, webs dusky. FEMALE: Head, neck, chest, upper back, rump, and upper tail coverts brown; scapulars, wing coverts, and flanks whitish vermiculated with brown; chin, throat, and lower breast whitish; belly grayish brown; crissum whitish mottled with brown; speculum gray tipped with white; axillars white; iris brown. Wing 222–235, culmen 53–63, tarsus 44 mm. North America, breeding south to Great Basin, Nebraska, and Wisconsin; winters through the United States to Mexico.

Genus *Bucephala* Baird. Head with puffy crest; culmen from base longer than tarsus, exposed culmen shorter than tarsus; sides of bill tapering toward tip, base constricted; lamellae extending slightly beyond edge of maxilla; wing pointed; tail short, rounded, rectrices 16; legs short but set far back; acrotarsium scutellate; hind toe lobed. Frontals swollen; sternum without external spine, posterior border fenestrate. Nest in hollow trees. Holarctic. 3 species, but only 1 occurs regularly on the Old World continent.

KEY TO SPECIES OF *BUCEPHALA*

1. Nostrils in basal half of bill; wing under 180 mm
   B. albeola
   Nostrils in terminal half of bill; wing over 190 mm   2
2. Male with oval white patch on lores; female with width of rostral nail 5 mm or less   *B. clangula*
   Male with crescent-shaped white patch on lores; female with width of rostral nail 6 mm or more
   B. islandica

Bucephala clangula (**Linnaeus**). Common goldeneye. Head and neck green; lores with an oval white patch; neck white; back and rump black; outer scapulars striped black and white; upper tail coverts black, tipped with white; tail, primaries, and outer secondaries dusky; edge of wing black; rest of coverts and most of secondaries white; underparts white; flanks and crissum edged with slate; axillars dusky; iris yellow; bill black; feet orange, webs dusky. FEMALE: Head, neck, and upper parts brown, edged with paler on back and tail coverts; inner secondaries white; inner greater (and sometimes middle and part of lesser) coverts white, often tipped with slate; lower foreneck grayish; underparts white; axillars brown; bill often with yellow tip. Wing 185–236, exposed culmen 28–38, width of rostral nail 4.5–6.5, tarsus 28–42 mm. Holarctic; in America breeds south to Montana, North Dakota, Minnesota, Michigan, New York, New Hampshire, and Vermont; winters south to South Carolina, Great Lakes, Iowa, Colorado, and Baja California, and sometimes to Gulf states.

Bucephala islandica (**Gmelin**). Barrow's goldeneye. Differs from *B. clangula* in having bill smaller and more pointed but with wider nail; frontals more swollen. MALE: Head glossed with bluish purple; white loral patch crescent-shaped; scapulars bifurcate, with white tear-shaped shaft spots; trachea with smaller secondary expansion than in *B. clangula*. FEMALE: Head darker than in *B. clangula*; gray of throat encroaching on breast; wing coverts often with less white; bill often all yellow. Wing 209–248, exposed culmen 30–36, width of nail 6–9, tarsus 33–40 mm. Breeds in Iceland, Greenland, Labrador, and in Rockies and Sierra Nevada from Alberta to Colorado and California; winters in breeding area and along coast from Gulf of St. Lawrence to Long Island, and from southern Alaska to San Francisco.

Bucephala albeola (**Linnaeus**). Bufflehead. Head and upper neck glossed with green, bronze, and purple; patch from behind eye to crown white; upper parts black; inner secondaries and most of wing coverts white; underparts white; flanks edged with black; axillars mottled dusky and white; iris brown; bill bluish flesh, dusky at tip and base; feet flesh. FEMALE: Grayish brown, ashy below; auricular patch and speculum white; axillars mottled; bill and feet dusky gray. Wing 150–175, culmen 23–29, tarsus 30–33 mm. North America, breeding south to Canadian border, California, and Montana; winters south to Gulf states and Mexico.

Genus *Clangula* Leach. Feathering at base of maxilla, viewed from above, gently rounded; culmen shorter than tarsus; sides of bill tapering; rostral nail wide, occupying whole tip of bill; lamellae extending slightly beyond edge of basal half of maxilla; nostrils basal; wings pointed; tail cuneate; rectrices 14, middle pair elongate in male, equal to length of body; legs set far back; tarsus shorter than anterior toes; lower acrotarsium scutellate; outer toe longest; hallux lobed. Holarctic. 1 species.

Clangula hyemalis (**Linnaeus**). Old-squaw. Head and neck mostly black; lores and malar region gray; postocular patch white; occiput and hind neck mixed black and white; upper back and scapulars black, broadly edged with clay color; back, rump, middle upper tail coverts, 4 central rectrices, wings, throat, breast, and axillars blackish brown; belly, crissum, flanks, lateral rectrices, and upper tail coverts white; iris reddish, brown, yellow, or white; tip of bill pink base and nail black; feet blue-gray webs dusky. WINTER MALE: Head neck and belly white; neck patches upper parts and breast dark brown; lores and sides of head gray; scapulars pale gray. SUMMER FEMALE: Upper parts, auriculars, and axillars brown; lores and occiput pale brown; postocular stripe, malar stripe, and underparts white. WINTER FEMALE: Crown, auriculars, hind neck, upper parts, and axillars dusky brown; scapulars edged with ash or cinnamon; below white, breast washed with brown or gray. Wing 198–233, tail 61–221, culmen 25–29, tarsus 33–38 mm. Holarctic, breeding on tundra; in America winters south to North Carolina, Great Lakes, and Washington, less frequently to Florida and California.

Genus _Histrionicus_ Lesson. Differs from _Clangula_ in having feathering at base of maxilla, viewed from above, forming 3 sharp angles; nostrils less basal in position; lamellae not extending beyond edge of maxilla; a fleshy wattle at base of maxilla of male; middle rectrices not elongate. Holarctic. 1 species.

Histrionicus histrionicus (**Linnaeus**). Harlequin duck. Plumage mostly slate blue; sides of crown and flanks rufous; tail coverts and center of crown black; black-bordered white areas on lores, auriculars, stripe on sides of neck, interrupted collar at base of neck, bar across sides of breast, shaft area of outer scapulars, tips of inner greater and central median coverts, and lateral spot at base of crissum; speculum violet; axillars dusky brown; iris brown; bill and feet blue-gray, rostral nail yellowish. FEMALE: Grayish brown, mottled grayish white below; a white spot on sides of forehead, lores, and auriculars; bill dusky. Wing 188–204, culmen 24–29, tarsus 36–38 mm. Eastern Siberia and Alaska to mountains of California and Colorado; Iceland, Greenland, and Labrador, extending in winter to Long Island.

Genus _Camptorhynchus_ Bonaparte. Bill wide and flaring, longer than tarsus; nail about $\frac{1}{3}$ width of bill; lamellae not visible beyond edge of maxilla; nostrils high and basal; cheek feathers stiff and bristly; tail short, rectrices 14; hind toe lobed. Atlantic seaboard. 1 species.

Camptorhynchus labradorius (**Gmelin**). Labrador duck. White on head, neck, upper breast, scapulars, wing coverts, secondaries, and axillars; stiff cheek feathers pale brown; rest of plumage black, including crown stripe and ring around lower neck; iris brown; bill black, orange at base and on tomia, culmen bluish; feet blue-gray, webs dusky. FEMALE: Brownish gray, more ashy on wing coverts, tertials, and secondaries; speculum and axillars white. Wing 216–226, tail 89, culmen 40–44, tarsus 38–40 mm. Labrador to New Jersey (extinct since 1878).

Genus _Melanitta_ Boie. Bill flaring, then tapering to nail; nostrils in distal half of bill; lamellae scarcely extending beyond edge of maxilla; chin feathers extending to about opposite nostrils; bill of male swollen above nostrils; tail pointed, longer than bill, rectrices 14–16; tarsus longer than exposed culmen, but shorter than culmen from base; hind toe lobed. Holarctic, breeding mainly north of the United States. 3 species.

KEY TO SPECIES OF _MELANITTA_

1. Commissure shorter than inner toe; rectrices 16
M. nigra
Commissure longer than inner toe; rectrices 14 **2**
2. Frontal feathering extending farthest forward on lores; a white speculum _M. fusca_
Frontal feathering extending farthest forward on forehead; no white speculum _M. perspicillata_

Melanitta fusca (**Linnaeus**).[1] White-winged scoter. Black; speculum and small postocular spot white; iris white, bill orange or red, black at base and along tomia, culmen white; feet orange, pink externally, webs dusky. FEMALE: Sooty brown; speculum white; iris brown; bill black, mixed with white or pink; feet reddish brown. YOUNG: Like female, but with a white spot on lores and another on auriculars. Wing 270–290, culmen 35–43, tarsus 45–53 mm. Holarctic, south in winter to Spain, California, Great Lakes, and North Carolina.

Melanitta perspicillata (**Linnaeus**). Surf scoter. Black; forehead and nape white; iris white; bill red, tip yellow, a round black patch near base; feet red, orange internally, webs dusky. FEMALE: Like juvenile _M. deglandi_ but lacks white speculum. Wing 223–256, tail 80, exposed culmen 33–40, tarsus 39–47 mm. Arctic America; winters south to California, Great Lakes, and South Carolina.

Melanitta nigra (**Linnaeus**).[2] Common scoter; black scoter. Black; iris brown; bill black, knob orange; feet olive, webs dusky; outer primary emarginate. FEMALE: Sooty brown, paler on belly and sides of head. Wing 222–241, culmen 42–46, tarsus 42–51 mm. Arctic America, Europe, and Asia; winters south to Mediterranean, China, California, Great Lakes, and North Carolina.

Genus _Somateria_ Leach. Bill about as long as head, culmen more than twice width of bill; sides of bill tapering to nail, base swollen above nostrils; lamellae not extending beyond edges of maxilla; nostrils near middle of bill; bare frontal angles prominent; scapulars and inner secondaries re-

---

[1] Includes _Melanitta deglandi_ (Bonaparte) of A.O.U.
[2] _Oidemia nigra_ of A.O.U.

curved; rectrices 14–16; hind toe lobed. Holarctic. 3 species, 2 in the United States.

KEY TO SPECIES OF *SOMATERIA*

1. Frontal feathering extending farthest forward on lores
   S. mollissima

   Frontal feathering extending farthest forward on forehead
   S. spectabilis

Somateria mollissima (**Linnaeus**). Common eider. Crown, rump, tail and coverts, lower breast, belly, and most of wing black; cheeks and nape pale green, separated by a white line; rest of head, neck, upper breast, patch on flanks, scapulars, inner secondaries, their coverts, most of lesser coverts, and axillars white; iris brown; bill orange, yellow, or green, nail white; feet yellow or green, webs dusky. FEMALE: Brown; head and neck finely streaked with black; upper parts barred with rust and buff; upper breast and flanks barred with black and buff; greater coverts and secondaries tipped with white; bill green, nail yellowish; feet dull yellowish, webs dusky. Wing 255–315, exposed culmen 44–62, tarsus 44–53 mm. Breeds on tundra of America east of 100th meridian and in European Arctic; winters in the Atlantic south to France, Massachusetts, and occasionally Virginia.

Somateria spectabilis (**Linnaeus**). King eider. Crown and nape pearl gray; sides of face pale green; neck, upper back, breast, spot on flanks, and axillars white, the breast tinged with creamy brown; rest of body, wings, and tail black, including a black V on throat and black border to bare frontal angles; eye brown, lids blue, subocular spot black; bill reddish orange; feet dull orange, webs dusky. FEMALE: Resembles same sex of *S. mollissima* but more tawny brown, with U-shaped dark markings rather than bars; feathering on forehead extends farther forward than on lores. Wing 245–292, culmen 30–36, tarsus 43–47 mm. Circumpolar; winters south to Britain, the Baltic, New York, and rarely to the Great Lakes and Virginia.

Genus *Lophodytes* Reichenbach. Head with fan-shaped crest; bill narrow, its length equal to head, longer than tarsus, with vertical serrations along entire tomia (Fig. 5-35); nostril linear, in basal third of bill; rostral nail deflected, occupying whole tip of bill; wing pointed; tail rounded, about twice length of bill; rectrices 18; tarsus reticulate, more than half length of outer toe; hallux lobed. Sternal spines reduced, metasternum fenestrate. Nests in hollow trees. North America. 1 species.

Lophodytes cucullatus (**Linnaeus**). Hooded merganser. Head, neck, upper back, and scapulars black; crest white, bordered with black; lower back, rump, upper tail coverts, and tail sooty brown; breast and belly white; 2 black bars extend forward from back to sides of breast; flanks reddish brown vermiculated with dusk; lesser and middle coverts ashy; greater coverts black, the inner ones tipped with white; tertials and inner secondaries black striped with white; primaries dusky; axillars white; crissum brownish gray speckled with white; iris yellow; bill black; feet olive brown, webs dusky. FEMALE: Head and neck brownish gray, crest browner; upper breast gray; chin, underparts, and axillars white; upper parts and sides sooty brown; speculum white, with a median black bar; iris brown; bill dusky, yellowish below. Wing 184–200, tail 82–89, culmen 35–42, tarsus 30–33 mm. Breeds in wooded areas from central Canada to

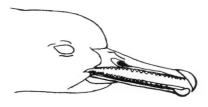

Fig. 5-35. Bill of hooded merganser (*Lophodytes cucullatus*).

Fig. 5-36. Bill of red-breasted merganser (*Mergus serrator*). S, serration.

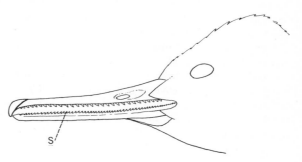

Oregon, Wyoming, Nebraska, Louisiana, and northern Florida; winters south to Cuba and Mexico.

Genus *Mergus* Linnaeus. Differs from *Lophodytes* in having crest smaller; serrations on tomia pointing backward (Fig. 5–36); tail about equal to bill. Holarctic and southern South America. 5 species, 2 in the United States.

KEY TO SPECIES OF *MERGUS*

1. Nostrils in middle of bill; feathering extending farther forward on forehead and on side of mandible
   M. merganser
   Nostrils in basal third of bill; feathering extending farther forward on lores   *M. serrator*

Mergus merganser **Linnaeus.** Common merganser. Head and neck green, scarcely crested; lower neck and underparts white tinged with pink, flanks vermiculated with gray; back black; rump, upper tail coverts, and tail ashy; edge of wing black; remaining lesser coverts white; outer greater coverts and outer secondaries black; inner greater coverts white, with a black bar across their base; middle secondaries white; inner secondaries and tertials white, edged with black; primaries dusky; axillars white; iris brown; bill red, tip and culmen dusky; feet red. FEMALE AND YOUNG: Head and neck reddish brown; an indistinct paler line below eye; throat white; lower hind neck and upper parts ashy gray; lesser and median wing coverts ashy gray; outer secondaries and their coverts and outer tertials black; inner greater coverts black tipped with white; speculum white; lower foreneck ashy, abruptly contrasting with the reddish-brown upper neck; underparts white. Wing 243–290, culmen 45–63, tarsus 42–53 mm. Holarctic; in America breeds south to California, Arizona, New Mexico, South Dakota, Minnesota, Wisconsin, Michigan, New York, New Hampshire, and Vermont; winters south to the Gulf states and northern Mexico.

Mergus serrator **Linnaeus.** Red-breasted merganser. Head and neck green, with long thin crest; foreneck with wide white half collar; back black; rump and upper tail coverts gray barred with black; tail ashy; upper breast reddish brown mottled with black; sides of upper breast black, with several rows of large white spots, the feathers lengthened; flanks white vermiculated with black; underparts white tinged with pale salmon; axillars white; edge of wing gray; remaining lesser and middle coverts white; inner tertials and outer secondaries and their coverts black; remaining greater coverts black, broadly tipped with white; middle secondaries white; inner secondaries and outer tertials white edged with black; iris red; bill red, dusky on culmen; feet red. FEMALE AND YOUNG: Differs from female *M. merganser* in having feathering on side of maxilla extending farther forward than on either the side of mandible or forehead; nostrils in basal part of bill; crest more straggly; reddish brown of head and neck paler and gradually merging into the color of the adjacent regions. Wing 214–257, tail 60–72, culmen 52–63, tarsus 40–48 mm. Holarctic; in America breeds south to Maine, New York, Michigan, Wisconsin, Minnesota, and southern Canada; winters south to Baja California and Gulf states.

Genus *Oxyura* Bonaparte. Bill flaring, depressed in middle, shorter than head, longer than tarsus, lamellae extending beyond middle of sides of maxilla; nail minute, hooked downward; wings pointed; tail longer than bill, rectrices 18, stiff, lanceolate, with enlarged shafts grooved underneath; tail coverts short; legs set far back; tarsus only about half length of middle and outer toes with claw; hallux webbed. External sternal spine forked, more or less fused to internal spine; metasternum notched. All continents. 7 species, 1 in the United States.

Oxyura jamaicensis (**Gmelin**). Ruddy duck. Crown and nape black; chin and cheeks white; neck, upper parts, upper breast, and flanks reddish chestnut; below silvery, more or less tipped with darker; wings and tail dark brown; crissum white; axillars gray, white at tip; iris brown; bill bright blue; feet blue-gray, webs dusky. FEMALE: Upper parts dark brown, vermiculated with light brown on crown, with gray on body, wings, and upper tail coverts; cheeks whitish, mottled with brown on malar region; throat whitish foreneck gray; underparts silvery, mottled with brown, upper breast mainly golden; bill dusky. Wing 136–152, tail 67–74, exposed culmen 38–40, tarsus 30–34 mm. Central Canada, breeding south to Baja California, Arizona, Nebraska, Iowa, and Lake Michigan area; also in Mexico, Guatemala, and West Indies; winters or migrates throughout the United States.

Order Accipitriformes.[1] Diurnal birds of prey. Bill hooked, with cere, nostrils located in middle of cere. Functional primaries 10; secondaries aquintocubital. Adult downs on apteria and pterylae. Anisodactyl.

Desmognathous (schizognathous in *Elanus*); holorhinal; no supraorbital grooves; mandibular process short. Cervical vertebrae 13–17. Furculum U-shaped, without enlargement of hypocleidium. Hypotarsus simple (with canal in *Pandion*).

Tongue normal, not fleshy or fringed; intestine type 4; caeca usually rudimentary or absent. Syrinx tracheobronchial or absent; 1 pair of sternotracheal muscles and not more than 2 pairs of intrinsic syringeal muscles; trachea without enlargements. Ambiens present; other thigh muscles variable; toe flexors type 3 or 5. Both carotid arteries present.

Land birds, feeding on carrion or prey. Nest usually of sticks; eggs 1–7, often marked with reddish brown; young downy and altricial. Paleocene onward. Cosmopolitan. 4 of the 5 living families occur in the United States.

KEY TO FAMILIES OF ACCIPITRIFORMES

1. Nostrils perforate (Fig. 5-37)   *Vulturidae*
   Nostrils imperforate   **2**
2. Claws all about same length, rounded beneath   **Pandionidae**
   Outer claw smaller than hind claw; claws usually grooved underneath   **3**
3. Nostrils either circular with a central bony tubercle, or if narrow slits, with hind end uppermost   **Falconidae**

Fig. 5-37. Head of turkey vulture (*Cathartes aura*). N, perforate nostril.

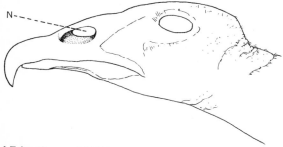

Nostrils either oval without a central tubercle, or if narrow slits, then with anterior end uppermost   **Accipitridae**

Family Vulturidae.[1] New World vultures. Head naked or with sparse down; nostrils large, oval, perforate; bill weakly raptorial; 3d or 4th primaries longest; 2d finger with a claw; middle toe nearly as long as tarsus, connected to 2d toe by a basal web; hallux slightly elevated, small, not adapted for grasping; rectrices 12–14; no aftershafts; oil gland naked.

Indirectly desmognathous, without vomer, maxillopalatine processes thin and scroll-like basipterygoid processes present, articulating with middle of pterygoids; olfactory grooves large; lacrimals fused to frontals, without superciliary plates; cervicals 15–17; sternal spines absent, metasternum with 2 pairs of notches or with 1 pair of notches and 1 pair of fenestrae; procoracoid process rudimentary; a posterior ilioischiatic notch; hypotarsus with 2 shallow grooves. Crop well developed; gall bladder present; caeca absent. Syrinx and intrinsic muscles absent. Thigh-muscle formula $(A)XY+$; toe flexors type 5.

Food carrion. Voice a hiss. No nest, eggs laid on ground or in cavity; eggs 1–2, elongate, whitish or spotted, with inner greenish or yellowish translucence; young at first naked, later downy, fed by regurgitation.

Paleocene onward. North and South America (Europe in Tertiary). 5 genera, 6 species, 3 species of as many genera in the United States.

KEY TO GENERA OF VULTURIDAE

1. Whole neck naked; breast feathers lanceolate   *Gymnogyps*
   Lower neck feathered; breast feathers broad   **2**
2. Tail rounded   *Cathartes*
   Tail square   *Coragyps*

Genus *Gymnogyps* Lesson. Head and whole neck bare, except for patch of bristly feathers from forehead to rictus; bill weak; commissure reaching only to point opposite nostrils; cere longer than rhinotheca, elevated posteriorly above forehead;

[1] Falconiformes of A.O.U.

[1] Cathartidae of A.O.U.

nostrils small, ovate; feathers of lower parts stiff and lanceolate; 4th or 5th primaries longest, extending well beyond secondaries and reaching tip of tail, the 7 outer primaries sinuate; tail truncate, less than half length of wing, rectrices 12, or sometimes 14; tarsus reticulate, shorter than head and bill combined, longer than middle toe. Cervicals 15, dorsal vertebrae 4, synsacral vertebrae 13; free caudals 5; metasternum with 2 pairs of shallow notches. Eggs plain gray. Western North America. 1 living species, now reduced to about 40 birds.

Gymnogyps californianus (Shaw). California condor. Black; remiges and rectrices grayish; tips of greater coverts, edges of secondaries, axillars, and under wing coverts white; iris red; bill whitish; cere and bare skin of head orange; throat red, hind neck pale bluish; feet pink. YOUNG: Head skin black, covered with brownish down; wing coverts without white; upper parts edged with brown. Wing 806–914, tail 330–380, culmen from cere 36.5–44, tarsus 109–124 mm. Formerly from Washington to Baja California; now confined to Coast Ranges of central and southern California.

Genus *Cathartes* Illiger. Head and upper neck bare; bill short but commissure long, extending to point opposite eye; cere elevated above nostrils; nostrils large, oval; 3d or 4th primary longest, much longer than secondaries but not reaching end of tail, outer 5 primaries sinuate on inner web; tail half length of wing, rounded, rectrices 12; tarsus reticulate, about equal to middle toe. Cervical and synsacral vertebrae both 13, both with 2 pairs of ribs; metasternum with 1 pair of notches and a lateral pair of fenestrae. Eggs whitish spotted with brown. North and South America. 2 species, 1 in the United States.

Cathartes aura (Linnaeus). Turkey vulture. Blackish brown, glossed with purple; upper wing coverts edged with brown; shafts of remiges and rectrices whitish below; iris brown; bill white; cere and bare head red; feet fleshy. YOUNG: Head and neck with sparse black down, skin blackish or yellow. Wing 458–559, tail 225–298, culmen from cere 21.5–26, tarsus 58–75 mm. Argentina and Chile north to Connecticut, New York, and southern Canada.

Genus *Coragyps* Le Maout. Head and upper throat bare, hind neck feathered; bill long; commissure extending nearly to below eye; cere straight, but forehead humped and then depressed beyond cere; nostrils linear; 4th or 5th primary longest, little longer than secondaries and not reaching tip of tail, outer 5 primaries sinuate on inner web; tail square, less than half length of wing, rectrices 12; tarsus reticulate; feet extending beyond tip of tail; middle toe about equal to tarsus. Cervical and sacral vertebrae both 14, each group with 3 pairs of ribs; sternum 4-notched. Eggs whitish spotted with brown. Continental North and South America. 1 species.

Coragyps atratus (Bechstein). Black vulture. Black; shafts of primaries white; base of outer primaries whitish below; iris and feet brown; bill and head black. Wing 381–454, tail 153–216, culmen from cere 34–40, tarsus 74–83 mm. Continental America from Chile and Argentina north to Mexico, Kansas, southern Illinois, southern Indiana, and Maryland.

Family Accipitridae. Hawks, eagles, Old World vultures. Bill usually strongly raptorial; maxilla unnotched (except in some kites); nostrils imperforate but without a central bony tubercle, oval or slitlike with the anterior end uppermost; head feathered (except in Old World vultures); aftershafts present; oil gland tufted; tarsus usually scutellate (sometimes reticulate, booted, or feathered); hallux incumbent, not reversible, about as long as shortest front toe; soles usually smooth: claws graduated, that of hallux largest, all usually grooved beneath.

Tracheobronchial muscles present; thigh formula A+; toe flexors type 3. Indirectly desmognathous (schizognathous in *Elanus*), with vomer not expanded nor closely applied to maxillopalatine processes; nasals incompletely ossified; lacrimals with a double superciliary plate (absent in Old World vultures); no mandibular foramen; cervicals 13–14; thoracic vertebrae all free; sternum without internal spine, metasternum solid or with a single pair of notches or fenestrae; postacetabular ilium bowed ventrally. Caeca small, nonfunctional; food animal prey. Both ovaries present. Female usually larger than male.

Nest usually bulky, of sticks; eggs 2–7, usually spotted with brown, sometimes plain whitish; young downy, altricial. Eocene onward. Cosmopolitan. 58 genera, 208 species; in the United States 11 genera, 22 species.

KEY TO GENERA OF ACCIPITRIDAE

1. Claws rounded underneath *Elanus*
   Claws with a longitudinal groove underneath 2
2. Tail forked for about half its length *Elanoides*
   Tail not deeply forked 3
3. Maxillary tomium with several notches *Ictinia*
   Maxillary tomium smooth or sinuate 4
4. Middle claw with shelf on inner edge *Rostrhamus*
   Middle claw normal 5
5. Face with an indistinct ruff *Circus*
   Face without a ruff 6
6. Tail more than ⅔ length of wing 7
   Tail less than ⅔ length of wing 8
7. Lores covered with short feathers *Accipiter*
   Lores nearly naked *Parabuteo*
8. 5 or more outer primaries emarginate on inner webs 9
   Not more than 4 outer primaries emarginated 10
9. Tarsus feathered to base of toes *Aquila*
   Lower third of tarsus unfeathered *Haliaeetus*
10. Primaries exceeding secondaries by about length of tarsus *Buteo*
    Primaries exceeding secondaries by about half length of tarsus *Buteogallus*

Genus *Elanus* Savigny. Bill small, maxillary tomium nearly straight, opening to beneath eyes; nostrils oval, horizontal, in middle of cere; lores feathered; wings long and pointed, 2d or 3d primaries longest, exceeding secondaries by about 4 times length of tarsus, 2 outer primaries emarginate on inner web; tail double-rounded, more than half as long as wing; tarsus reticulate, equal to middle toe with claw; claws rounded underneath. A pair of inguinal powder-down patches. Palate schizognathus; lacrimal with double superciliary plate; metasternum fenestrate. Warmer parts of world. 4 species, 1 in the United States.

***Elanus leucurus* (Vieillot).** White-tailed kite. Above bluish gray becoming white on head and tail; region before and around eyes black; lesser wing coverts and edge of lining of wing black; underparts white; iris orange; bill black; cere and feet yellow; claws black. YOUNG: Tinged with rust; wings and wing coverts tipped with white; tail with dusky subterminal band. Wing 302–328, tail 174–186, culmen from cere 18–19, tarsus 36–39 mm. Florida, the lower Rio Grande, and California, south on the mainland to Chile and Argentina.

Genus *Elanoides* Vieillot. Bill small; maxillary tomium sinuate; nostril obliquely oval, anterior end highest; wing very long, 2d or 3d primaries longest, exceeding secondaries by about 7 times length of tarsus; outer 2 primaries emarginate on inner webs; tail more than ¾ length of wing, forked for more than ½ its length; tarsus about equal to middle toe with claw, reticulate; claws with a slight longitudinal groove beneath; lores feathered. Palate desmognathous; no superciliary shield; metasternum notched. Monotypic.

***Elanoides forficatus* (Linnaeus).** Swallow-tailed kite. Head, neck, underparts, and band across lower back white; secondaries white, broadly tipped with black; back, wings, and tail glossy black. IMMATURE: Head and neck streaked with dusky; remiges and rectrices narrowly tipped with white. Wing 390–447, tail 275–370, culmen from cere 19–21, tarsus 32–33 mm. South Dakota, southern Minnesota, Ohio Valley, and North Carolina south through Gulf states and eastern Mexico to Argentina; winters in the tropics.

Genus *Ictinia* Vieillot. Bill short but stout; maxillary tomium with 2 teeth; nostril small, nearly circular; lores sparsely feathered; a naked superciliary shield; wing pointed, 3d primary longest, exceeding secondaries by more than length of tail; 2 outer primaries emarginate on inner web; tail about half as long as wing, square or slightly emarginate; tarsus scutellate, about equal to middle toe; 2d toe with basal 2 phalanges ankylosed; claws longitudinally grooved beneath. Tropical and subtropical America. 2 species, 1 in the United States.

***Ictinia misisippiensis* (Wilson).** Mississippi kite. Gray, becoming whitish on forehead; wings and tail blackish, the 8 inner primaries with chestnut spots on inner web and a basal chestnut edge on outer web; iris, rictus, and feet red; bill and cere black. IMMATURE: Superciliary whitish; upper parts streaked with rusty; underparts whitish streaked with slate; outer primary with a white spot on inner web; iris brown. Wing 286–315, tail 149–172, culmen from cere 15–17, tarsus 35–40 mm. Kansas, Missouri, Iowa, and South Carolina south to Florida, the Gulf, Texas, and Oklahoma; winters from Florida and Texas to Guatemala.

Genus *Rostrhamus* Lesson. Maxilla very narrow and much hooked; tomia smooth; nostril elliptical; lores bare; wing long but broad, 4th or

5th primary longest, exceeding secondaries by about 4 times length of tarsus; 4 outer primaries emarginate on inner web; tail nearly half length of wing, slightly emarginate; tarsus about equal to middle toe and claw, scutellate; claws very long and slender, grooved beneath, middle claw with inner edge produced in a shelf, somewhat pectinate. 1 species.

Rostrhamus sociabilis (**Vieillot**). Everglade kite. Dark slaty, browner on wing coverts; tail coverts and base of tail white; iris, eyelids, lores, cere, gape, and base of mandible red; bill black; feet orange; claws black. FEMALE: Throat and breast streaked with buff; secondaries, wing coverts, and belly tipped with rust; thighs rusty barred with gray. IMMATURE: Still more heavily marked with buff, rust, and slate; iris brown; feet dull yellow. Wing 325–382, tail 164–202, culmen from cere 22–33, tarsus 47–57 mm. Southern Florida; Cuba; eastern Mexico to Argentina and western Ecuador.

Genus *Accipiter* Brisson. Bill short but deep; maxillary tomium strongly sinuate; nostril ovate or almost circular; lores covered with small feathers; wing short and rounded, 4th or 5th primaries longest, exceeding secondaries by only about length of tarsus; 5 (3 in some extralimital species) outer primaries notched on inner web; tail long, more than 80% length of wing, square or rounded; tarsus slender, much longer than middle toe with claw, scutellate or occasionally booted; toes long and slender; claws slender, with a longitudinal groove beneath. Metasternum fenestrate. Nearly cosmopolitan. 44 species, 3 in the United States.

KEY TO SPECIES OF *ACCIPITER*

1. Tail square; wing 141–210 mm *A. striatus*
   Tail rounded; wing more than 210 mm **2**
2. Inner toe with claw much shorter than middle toe with claw; wing 214–278 mm *A. cooperii*
   Inner toe with claw about as long as middle toe with claw; wing 280–380 mm *A. gentilis*

Accipiter gentilis (**Linnaeus**). Goshawk. Pileum blackish; superciliary with dusky and white streaks; upper parts slate gray; tail rounded, tipped with white and crossed by 4 dusky bars; below white vermiculated with slate and often with dusky shaft

streaks; iris red; bill black; cere and feet yellow; claws black. YOUNG: Grayish brown above, margined with tawny to white; rectrices brown, tipped with white and with 4–6 dusky crossbars; below buffy white with broad brown shaft stripes; iris pale yellow. Wing 280–380, tail 227–301, culmen from cere 20–26, tarsus 67–95 mm. Holarctic; in America breeds in coniferous forests of Alaska and Canada, south to Massachusetts, New York, northern Michigan (in mountains to Maryland and Tennessee), New Mexico, Arizona, California, and northern Mexico; winters at intervals south to Ohio Valley, Oklahoma, and Texas, occasionally farther.

Accipiter cooperii (**Bonaparte**). Cooper's hawk. Pileum blackish; lores whitish; sides of head buffy brown; upper parts slaty; tail gray, tipped with white and crossed on exposed portion by 3 dusky bands; below white; throat streaked with brown; breast and belly with dusky shaft streaks and reddish-brown crossbars; crissum plain white; thighs and axillars barred rufous and white; iris red; bill black; cere greenish; feet yellow. IMMATURE: Pileum brown streaked with buff; upper parts dusky brown edged with buff or cinnamon and with more or less concealed white spots on scapulars; tail grayish brown, crossed by 4–5 dusky-brown bands; underparts white with large dusky-brown shaft stripes; thighs and axillars barred brown and white; crissum plain buffy white; iris yellowish brown. Wing 214–278, tail 181–242, culmen from cere 15–21, tarsus 62–75 mm. Southern Canada south to Florida, Gulf states, California, and northern Mexico; winters south to Costa Rica.

Accipiter striatus **Vieillot.** Sharp-shinned hawk. Resembles *A. cooperii* but smaller and with a square tail. Wing 141–210, tail 120–180, culmen from cere 9.5–14, tarsus 45–59 mm. All North America, breeding south to Florida, Greater Antilles, and northern Mexico; winters south to Panama.

Genus *Parabuteo* Ridgway. Bill stout; maxillary tomium sinuate; nostrils oval; lores bare except for bristles; 4th primary longest, exceeding secondaries by more than length of tarsus; outer 4 primaries notched; tail rounded, about 75% length of wing; tarsus scutellate, the bare portion about as long as middle toe and claw; 3d and 4th toes connected by a basal web; claws with a longitudinal groove underneath. Monotypic.

Parabuteo   unicinctus   (**Temminck**).   Dusky hawk; Harris's hawk. Sooty brown; tail coverts, base, and tip of tail white; lesser wing coverts, wing lining, and thighs chestnut; iris brown; bill bluish; cere, lores, eyelids, and feet yellow; claws black. IMMATURE: Upper parts blackish, varied with rust and buff; underparts buffy, streaked with dusky; tail grayish brown, whitish at base, narrowly barred with dusky. Wing 318–388, tail 213–262, culmen from cere 24–29, tarsus 80–92 mm. Southern Texas, Arizona, and southeastern California south to Argentina.

Genus *Buteo* Lacépède. Bill strong, maxillary tomium sinuate, nostrils obliquely oval to round; lores more or less feathered but mixed with bristles; 3d or 4th primary longest, exceeding secondaries by at least length of tarsus, 3–4 outer primaries notched on inner web; tail slightly rounded, 44–67% length of wing; tarsus short and thick, about twice length of middle toe and claw, scutellate or feathered; claws stout, grooved underneath. Metasternum fenestrate, notched, or entire. Nearly cosmopolitan. 26 species, 10 in the United States.

KEY TO SPECIES OF *BUTEO*

1. Acrotarsium feathered to base of toes   2
   Lower part of tarsus bare   3
2. Width of gape, from corners of mouth, 33–38 mm
       B. lagopus
   Width of gape 42–48 mm   *B. regalis*
3. 4 outer primaries strongly notched on inner web   4
   Only 3 outer primaries strongly notched   7
4. Dorsal surface of outer web of primaries distinctly barred with blackish and white or buffy
       B. lineatus
   Dorsal surface of outer web of primaries plain or faintly barred with dusky and brownish gray   5
5. Wing under 290 mm   *B. nitidus*
   Wing over 330 mm   6
6. Tail banded with at least some white
       B. albonotatus
   Tail without white bands   *B. jamaicensis*
7. Tail less than half length of wing   *B. albicaudatus*
   Tail more than half length of wing   8
8. Wing more than 360 mm   *B. swainsoni*
   Wing less than 325 mm   9
9. Middle toe without claw shorter than bare part of acrotarsium   *B. platypterus*
   Middle toe without claw longer than bare part of acrotarsium   *B. brachyurus*

Buteo nitidus (**Latham**).   Gray hawk; Mexican goshawk. Outer 4 primaries notched. Upper parts gray; remiges tipped with white; upper tail coverts white, the central ones dusky; tail black, narrowly tipped with gray and crossed by 1–3 white bands; throat gray vermiculated with white; underparts and axillars narrowly barred gray and white; crissum white; iris dark brown; bill and claws blackish; cere and feet yellow. IMMATURE: Above dark brown edged with rusty or ochraceous; tail grayish brown, with about 6–7 narrow dusky bands; sides of head buffy white with dark-brown postocular and malar stripes; sides of neck streaked brown and pale rusty; throat buffy white with a median fuscous stripe, not reaching chin; breast and belly white with large dark-brown stripes; thighs and axillars buffy white barred with brown; crissum plain buffy white; iris grayish brown. Wing 231–289, tail 146–195, culmen from cere 20–25, tarsus 64–79 mm. Southern Arizona and Rio Grande Valley south in lowlands to Bolivia, Paraguay, and northern Argentina.

Buteo lineatus (**Gmelin**).   Red-shouldered hawk. Outer 4 primaries notched. Head, throat, hind neck, and back streaked dusky and tawny; upper tail coverts barred black and white; tail black tipped with white and crossed by 3–4 narrow white bands on exposed portion; shoulders rusty, streaked with dusky; remiges and remaining coverts barred black and white on both webs; underparts and axillars buffy heavily barred with rufous, and with dusky shaft streaks on anterior portion, the bars becoming obsolete on crissum; iris brown; bill dusky blue; cere, eyelids, and gape yellow; feet yellowish green. IMMATURE: Above fuscous brown tinged with ochraceous; upper tail coverts barred fuscous and buffy; tail fuscous tipped with whitish and crossed by 4–6 narrow gray bars; shoulders mixed dusky and rufous; wing coverts fuscous; primaries dusky, barred on both webs with buff or pale rufous; below pale buff heavily streaked with fuscous; thighs and usually crissum barred fuscous and buff; axillars streaked or barred with the same; iris pale yellow. Wing 278–353, tail 169–236, culmen from cere 18–25, tarsus 70–87 mm. Southern Quebec, Ontario, Manitoba, and eastern United States, west to Great Plains, south to Florida Keys, Gulf states, and Tamaulipas; California and northwestern Mexico.

Buteo jamaicensis (Gmelin).[1] Red-tailed hawk. Outer 4 primaries notched. Upper parts dark brown, more or less mixed with whitish or buffy; tail rufous narrowly tipped with white, usually with a subterminal black band, and sometimes a series of such bands; upper tail coverts sometimes barred white and rufous or brown; sides of neck reddish brown, making a broken collar of broad streaks across foreneck and sides of breast; underparts white, blotched with sepia across belly; thighs and axillars plain or barred with brown; iris brown; bill and claws blackish; cere and feet yellowish. VARIATIONS: Dark, light, and reddish phases exist. In dark phase the whole plumage is sooty except for the rufous tail. The palest birds are almost white on head and tail, with the back very pale and the underparts immaculate. In reddish phase the light areas, both above and below are suffused with rufous and buff. In Harlan's hawk, the tail is gray or white mottled with slate. IMMATURE: Tail gray crossed by 9 dusky bands; iris yellow. Wing 330–436, tail 189–254, culmen from cere 22–33, tarsus 74–96 mm. All North America, from Alaska to Panama and the West Indies.

Buteo albonotatus Kaup. Zone-tailed hawk. Outer 4 primaries notched. Black; bases of body feathers white; lores whitish; inner webs of remiges pale gray barred with black; tail with narrow white tip and 3 broad ashy-white bands; iris brown; bill and claws black; cere and feet yellow. IMMATURE: Black, spotted with white; back and wing coverts edged with brown; tail dark brown, crossed by about 7 narrow black bands. Wing 375–438, tail 187–234, culmen from cere 21–25, tarsus 67–79 mm. Southern Texas, New Mexico, and Arizona, south to Brazil and eastern Bolivia.

Buteo lagopus (Pontoppidan). Rough-legged hawk. Outer 4 primaries notched; tarsi feathered to toes. Head and neck whitish streaked with fuscous; upper parts varied brown, white, and rusty; upper tail coverts white, streaked or barred with fuscous; basal two-thirds of tail white; distal third of tail grayish brown, narrowly tipped with white, with a dusky subterminal band and often with 3 narrow white bands; chin, throat, and breast whitish or creamy, streaked with fuscous; thighs and belly whitish, more or less spotted with fuscous;

crissum whitish; axillars white streaked with fuscous; under wing coverts blotched white, fuscous, and tawny; base of remiges white on undersurface; iris brown; bill and claws black; cere yellowish green toes yellow. DARK PHASE: Blackish; tail white on basal third; distal two-thirds dark brown, narrowly tipped with white, with a broad black subterminal bar and 4 narrower black bars. There is much individual variation connecting the 2 phases. IMMATURE: A large fuscous patch on sides of abdomen; terminal portion of tail unbarred brown, tipped with white. Wing 395–470, tail 210–241, culmen from cere 20–26, tarsus 65–76 mm. Holarctic, breeding in far north; winters in central Europe and Asia, and from southern Canada south to Great Basin, Missouri Valley, Ohio Valley, and Maryland, irregularly farther.

Buteo regalis (Gray). Ferruginous hawk. Outer 4 primaries notched; tarsi feathered to toes. Crown and hind neck streaked cinnamon brown and white; upper parts ferruginous, broadly streaked with black; upper tail coverts ferruginous barred with brown, white basally; tail grayish white, mottled with ferruginous and deep gray; chin and throat white; breast white, with cinnamon shaft streaks; belly and flanks white with wavy crossbars of ferruginous and fuscous; thighs deep ferruginous barred with fuscous; crissum white; axillars and under wing coverts mixed white and ferruginous; bases of outer primaries white internally; iris, cere, and toes yellow; bill blackish. DARK PHASE: Deep sepia, more or less edged with cinnamon; tail as in light phase but washed with dark brown. IMMATURE: Upper parts less rusty than in adult; underparts almost immaculate white; tail gray, crossed by 4 indistinct broad fuscous bars. Wing 419–450, tail 231–252, culmen from cere 26–30.5, tarsus 79–97 mm. Great Basin and Great Plains, from southern Canada, eastern Washington, eastern Oregon, and northeastern California, east to the Dakotas, Kansas, western Oklahoma, and western Texas; winters south to Arizona, New Mexico, and northern Mexico.

Buteo brachyurus Vieillot. Short-tailed hawk. Outer 3 primaries notched; middle toe without claw longer than bare part of acrotarsium. Above blackish; concealed bases of occipital feathers white; sides of rump tinged with rufous; tail grayish brown, narrowly tipped with white, with dusky

---

[1] Includes Harlan's hawk, *Buteo harlani* (Audubon) of A.O.U.

subterminal band and 4 narrow broken dusky bars; a white spot on lores; sides of throat rufous brown with dusky shaft streaks; underparts white, sides of breast with dusky shaft streaks, thighs washed with buff; a large dusky patch on under primary coverts, rest of wing lining and axillars white; iris brown; bill black; cere and feet yellow. DARK PHASE: Sooty brown; concealed bases of occipital feathers white; tail brownish gray with 6–7 narrow black bands; undersurface of primaries and tail silvery, barred with gray. YOUNG: Upper parts blackish, mixed and edged with buff; tail with 7–9 dusky bands; sides of head and neck streaked white or dusky; underparts plain white or streaked with brown. Wing 290–323, tail 160–190, culmen from cere 18–21, tarsus 58–62 mm. Peninsular Florida and tropical America from Mexico to Argentina.

Buteo playtypterus (Vieillot). Broad-winged hawk. Outer 3 primaries notched; middle toe without claw shorter than bare part of acrotarsium. Upper parts dark brown; occiput and nape with white bases and buffy edges; upper tail coverts tipped with white and with a concealed white bar; tail narrowly tipped with white, crossed at middle by a broad whitish band and by a narrower whitish band subbasally; lores buffy; cheeks dark brown streaked with tawny; malar stripe fuscous; chin and throat whitish, sometimes streaked with brown; underparts whitish streaked and barred with brown; lower belly and crissum often immaculate; under wing coverts whitish with a few brown marks; inner webs of remiges white, barred with dusky; iris reddish brown; bill and claws black; cere and feet yellow. IMMATURE: Above dark brown mixed with white and edged with cinnamon; tail buffy brown tipped with white, with a dusky subterminal band and 5–6 narrow indistinct bars of slate; chin and throat whitish with a median dark-brown streak; underparts whitish with dark-brown tear-shaped spots; lower belly and crissum plain white; under wing coverts buffy sparsely spotted with brown; iris pale brown. DARK PHASE: Sooty brown; occipital feathers white basally; outer primaries white internally; tail as in normal phase. Wing 244–296, tail 145–185, culmen from cere 17–20.5, tarsus 53–66 mm. Breeds in eastern North America, from southern Canada south to Florida and Gulf states, west to edge of Great

Plains, from eastern North Dakota to Austin, Texas; winters from southern Mexico to Matto Grosso; resident in West Indies.

Buteo swainsoni Bonaparte. Swainson's hawk. Outer 3 primaries notched. Forehead buffy; upper parts grayish brown, edged with tawny; concealed bases of nape feathers white; lateral upper tail coverts whitish barred with fuscous; tail brownish gray, tipped with whitish, with a broad subterminal band and 9–10 black bars; lores, chin, and throat white; breast light russet with dusky shaft streaks; remaining underparts white; iris brown; bill black; cere and feet yellow. RUFOUS PHASE: Black; crissum barred with brown. Intermediates between all 3 phases occur. IMMATURE: Upper parts more extensively buffy; sides of head buffy streaked with black; malar stripe black; underparts pinkish buff, with tear-shaped fuscous spots on breast; iris pale reddish brown. Often the entire underparts are spotted with black or tawny, with crissum and thighs barred. Wing 362–427, tail 185–234, culmen from cere 20.5–26, tarsus 62–76 mm. North America west of the Mississippi, breeding from Alaska to northern Mexico; migrates through Central America and winters in Argentina; casual in eastern North America.

Buteo albicaudatus Vieillot. White-tailed hawk Outer 3 primaries notched; tail only 44–48% of wing length. Crown, hind neck, upper back, and wings slate; concealed bases of nape feathers white; lesser wing coverts and scapulars largely cinnamon with dusky shaft streaks; rump and upper tail coverts white with wavy dusky bars; tail pale gray, with a broad subterminal black band and about 7 narrow wavy dusky bars; underparts white, with wavy dusky bars on sides, flanks, and thighs; axillars and under wing coverts barred dusky and white; iris brown; bill dusky horn; cere green; feet yellow. 1 extralimital subspecies has chin and throat dark gray like crown. A dark phase is said to exist, with this color extending over the entire underparts, with the belly and tail coverts suffused with rufous, the thighs barred with rufous and white. IMMATURE: Upper parts and anterior underparts blackish; wing coverts edged with pale cinnamon; rump edged with white; upper tail coverts whitish barred with brown; tail gray narrowly barred with darker; flanks, belly, thighs, and crissum buff blotched with fuscous; under wing coverts fuscous edged with buff. Wing 404–450, tail 194–211, culmen from cere 24–28, tarsus 85–95 mm. Southern parts of Texas, New Mexico, and Arizona south to Argentina.

Genus *Buteogallus* Lesson. Bill fairly strong; maxillary tomium sinuate; nostrils round; lores sparsely feathered or bristled; 3d to 5th primary longest, exceeding secondaries by only about half length of tarsus; 4 outer primaries shallowly notched; tail square, about half length of wing; tarsus scutellate; toes weak; claws slender, grooved underneath. Metasternum fenestrate. Tropical America. 3 species, 1 in the United States.

Buteogallus anthracinus (**Deppe**). Crab hawk; Mexican black hawk. Tail 52–60% of wing length. Black; lores and area below eye white; concealed bases of occipital feathers white; upper and under wing coverts, belly, and thighs narrowly tipped with ochraceous or white; upper and under tail coverts tipped with white; tail black, narrowly tipped with whitish and with a broad white band across middle; axillars black with broken bars of white or ochraceous; outer primaries white at base; iris brown; tip of bill blackish; base of bill, cere, and feet yellow. IMMATURE: Above fuscous; crown and nape broadly streaked with buffy; scapulars, wing coverts, and upper tail coverts varied with tawny; rectrices buffy, with 5–7 oblique black bands; chin and throat buffy with fine dusky shaft streaks; underparts tawny with broad fuscous shaft streaks; thighs and crissum tawny barred with fuscous; under wing coverts buffy streaked with fuscous. Wing 328–400, tail 178–244, culmen from cere 24–33, tarsus 74–107 mm. Southern parts of Texas, New Mexico, and Arizona, through Central America to north coast of South America and adjacent islands.

Genus *Aquila* Brisson. Bill short and stout; maxillary tomium slightly sinuate; nostril obliquely oval; 4th or 5th primary longest, exceeding secondaries by more than length of tarsus; 5 outer primaries deeply notched; tail slightly rounded, 56–60% of wing length; entire tarsus feathered; toes stout; claws very strong, grooved beneath; feathers of hind neck lanceolate. Metasternum entire or with tiny fenestrae. All continents except South America and Australia. 8 species, 1 in the United States.

Aquila chrysaëtos (**Linnaeus**). Golden eagle.

Chocolate; legs, crissum, and edge of wing paler; nape golden; middle of tail mottled with buffy or gray; iris brown; bill bluish horn; cere and toes yellow; claws black. IMMATURE: Fuscous; head and body feathers white basally; tail ashy white with a broad subterminal fuscous band; remiges white basally; tarsi whitish. Several molts are required before full adult plumage is reached. Wing 555–666, tail 320–390, culmen from cere 37–47, tarsus 101–123 mm. Holarctic; in America mainly confined to the Rockies and westward, from Alaska to northern Mexico; in the east south in the mountains to Tennessee and North Carolina.

Genus *Haliaeetus* Savigny. Bill long and stout; maxillary tomium practically straight; nostrils obliquely oval; 3d to 5th primary longest, exceeding secondaries by much less than length of tarsus; 5–6 outer primaries notched; tail rounded, 45–53% of wing length, rectrices 12; lower third of tarsus unfeathered, scutellate; claws grooved underneath. Metasternum entire. All continents except South America. 8 species, 1 in the United States.

Haliaeetus leucocephalus (**Linnaeus**). Bald eagle. Wings and body dark brown, many of the feathers edged with pale brown; head, neck, tail and its coverts white; iris, bill, cere, and feet yellow; claws black. IMMATURE: Dark brown, including head; bases of feathers whitish, giving a blotched appearance; tail dark brown mottled with dirty white; iris grayish; bill and cere brownish horn. Several molts are required before full adult plumage is reached. Wing 515–685, tail 232–365, culmen from cere 47–58, tarsus 95–110 mm. North America from the Arctic south to southern Florida, the Gulf Coast, the Mexican boundary, and Baja California.

Genus *Circus* Lacépède. Bill small, maxillary tomium slightly sinuate; nostril oval, more or less covered with bristles; feathers at posterior border of face stiff, forming an indistinct ruff; wing long, 3d or 4th primary longest, exceeding secondaries by twice length of tarsus; 3–4 outer primaries notched; tail slightly rounded, 58–72% length of wing; tarsus long and slender, scutellate; toes slender; claws grooved beneath. All continents. 12 species, 1 in the United States.

Circus cyaneus (**Linnaeus**). Marsh hawk. Tail 65–72% of wing length. Head, neck, upper breast, and upper parts gray; lores and eye region paler; occiput streaked dusky and buffy white; upper tail coverts white; tail gray, with 5–7 fuscous bands; outer primaries blackish; posterior underparts and lining of wing white, either immaculate or with broken bars of tawny; iris, cere, and feet yellow; bill blackish, horn at base; claws black. FEMALE: Upper parts dark brown; lores and eye region buffy; ruff streaked dark brown and buff; upper tail coverts white; tail dusky brown with 5 broad bars, gray on middle rectrices, ochraceous on lateral ones; primaries dusky with broad gray bands, the latter becoming whitish beneath; chin dirty white; throat and upper breast ochraceous broadly streaked with dark brown; remaining underparts buffy, more or less striped with brown; iris grayish brown; cere light green. IMMATURE: Resembles female but head, nape, scapulars, and wing coverts edged with tawny brown; light markings of wings and tail more tinged with cinnamon; underparts cinnamon rufous, the throat, upper breast, and sides streaked with dark brown; axillars brown barred with cinnamon. Wing 328–405, tail 223–251, culmen from cere 15–19, tarsus 70–84 mm. Holarctic; in North America breeds from the Arctic south to Baja California, New Mexico, Arizona, Texas, Missouri, Ohio Valley, Maryland, and Virginia; winters south to Panama and Greater Antilles.

Family Pandionidae. Ospreys. Bill raptorial; nostrils imperforate, obliquely oval with anterior end uppermost, without bony tubercle. Plumage dense, without aftershafts except on crissum; oil gland tufted; head feathered. Tarsus reticulate, about as long as middle toe and claw; toes short, soles prickly, hallux reversible; claws equal in length, rounded beneath.

Desmognathous, without basipterygoid processes; nasal septum imperfectly ossified; lacrimals fused to skull, without superciliary processes; no mandibular foramen; cervicals 15; thoracic vertebrae all free; sternum without internal spine; metasternum with lateral corners longest, with 1–2 pairs of slight scallops; procoracoid small; furculum much depressed dorsoventrally; pelvis very broad, with postacetabular portion depressed; tibia bridged as in rest of order; acrotarsium with a bony ring at upper end as in owls; hypotarsus with 1 canal. Tracheobronchial muscle present; thigh formula $A+$; toe flexors type 9. Caeca large; food fish.

Nest bulky, of sticks; eggs 2–4, buffy blotched with brown; young downy, altricial.

Pleistocene onward. Cosmopolitan, except for New Zealand. 1 genus.

Genus *Pandion* Savigny. Maxillary tomium slightly notched or nearly straight; lores almost bare; wing long, 3d primary longest, exceeding secondaries by twice length of tarsus; 3 outer primaries notched; rectrices 12, tail double-rounded, 43–47% of wing length. 1 species.

Pandion haliaëtus (**Linnaeus**). Osprey. Upper parts dark brown; head and neck white, often streaked with dark brown on crown; stripe through eye and down side of neck dark brown; tail with about 5 grayish bars, more distinct below; underparts and axillars white, more or less streaked with brown on breast and legs; under wing coverts banded or blotched with dark brown and white; bases of primaries banded with white below; iris yellow; claws and tip of bill black; cere, base of bill, and feet blue. Wing 450–512, tail 199–240, culmen from cere 30–36, tarsus 58–68 mm. Nearly cosmopolitan; in North America south to Guatemala; winters south to West Indies and Argentina.

Family Falconidae. Falcons. Bill raptorial, with maxilla decurved at tip, its tomium usually with 1–2 notches, sometimes only sinuate; nostrils imperforate, usually round with central bony tubercle, sometimes oblique slits with anterior end lowest. Tarsus usually reticulate (scutellate in *Caracara*); hallux incumbent, not reversible; claws moderately graduated, with longitudinal groove beneath. Crown feathered; aftershafts present; oil gland tufted.

Directly desmognathous with vomer expanded to meet maxillopalatines, without basipterygoid processes; nasals ossified; lacrimal without superciliary plate; a mandibular foramen. Cervicals 13–14; thoracic vertebrae ankylosed (not in *Herpetotheres*). Spina interna sterni present (rudimentary in *Caracara*); metasternum solid or with 1 pair of notches or fenestrae. Procoracoid large. Postacetabular ilium bowed ventrally. Acrotarsium without bony ring; hypotarsus without canals. A gap between 1st and 2d bronchial rings. Tracheobronchial muscles present; thigh formula $A+$; toe flexors type 3. Caeca minute.

Nest usually of sticks; eggs 2–5, buffy, speckled or mottled with brown; young downy, altricial.

Miocene onward. Cosmopolitan. 12 genera, 59 species; in the United States 2 genera, 7 species.

KEY TO GENERA OF FALCONIDAE

1. Nostril circular   *Falco*
   Nostril an oblique slit (Fig. 5-38)   *Caracara*

Fig. 5-38. Head of Audubon's caracara (*Caracara cheriway*).

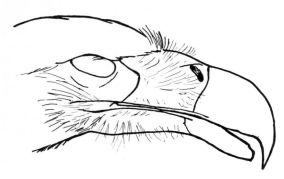

Genus *Caracara* Merrem. Bill long; nostril slit-like, obliquely vertical, with anterior end lowest; maxillary tomium with slight indication of falconine tooth; lores bare; 3d or 4th primary longest, exceeding secondaries by a little less than length of tarsus; 4 outer primaries notched. Tail 52–69% of wing length, slightly rounded. Tarsus much longer than middle toe and claw, scutellate in front; toes weak; claws slender, moderately curved. Food largely carrion. Cuba and southern United States, south on the mainland to southern South America. 3 species, 1 in the United States.

Caracara cheriway (Jacquin). Audubon's caracara. Crown, wings, lower back, sides, belly, thighs, and wing lining fuscous; hind neck, upper back, and breast barred fuscous and buffy white; throat, crissum, and upper tail coverts buffy white, the latter sometimes barred with fuscous; tail white, broadly tipped with fuscous and with 11–14 narrow fuscous bands; bases of 6 outer primaries barred with white on inner web; iris brown; bill whitish, bluish at base; cere and lores red; feet yellow; claws black. IMMATURE: Upper parts and

belly dark brown, streaked or tipped with white on interscapulars, lesser and middle wing coverts; sides of face, throat, and tail coverts buffy; breast and belly dark brown with buffy shaft stripes. Wing 355–418, tail 185–254, culmen from cere 30–36, tarsus 81–94 mm. Southern Florida, Cuba, coast of Louisiana, southern parts of Texas, New Mexico, and Arizona, south through Mexico and Central America to Brazil.

Genus *Falco* Linnaeus. Bill short, maxillary tomium with 2 notches and a tooth near tip (Fig. 5-39); nostril round with a central tubercle; lores feathered; 1st to 3d primary longest, exceeding secondaries by 2–3 times length of tarsus; 1–2 outer primaries notched on inner web; tail rounded or double-rounded, 46–70% of wing length; tarsus mostly reticulate, about equal to middle toe and claw; claws curved. Cosmopolitan, 37 species; 6 species in the United States.

KEY TO SPECIES OF *FALCO*

1. Outer primary longer than 3d   *F. peregrinus*
   1st primary shorter than 3d   2
2. Upper two-thirds of acrotarsium feathered
                             F. rusticolus
   Acrotarsium mostly unfeathered   3
3. Base of toes reticulate   4
   Toes entirely scutellate   5
4. Wing 289–302 mm   *F. mexicanus*
   Wing 182–228 mm   *F. columbarius*
5. Wing 248–302 mm   *F. femoralis*
   Wing 160–207 mm   *F. sparverius*

Falco peregrinus **Tunstall.** Peregrine falcon; duck hawk. Outer primary notched, longer than 3d. Less than upper half of tarsus feathered, lower part reticulate. Forehead, lores, and auriculars buffy; crown blackish; upper parts bluish gray barred with dusky; wing coverts tipped with paler gray; tail with 11–12 blackish bars; orbital and malar areas blackish; throat whitish; underparts creamy or buff, more or less spotted and barred with black; primaries blackish, inner webs whitish at base and barred with black; iris brown; bill bluish, blackish at tip; cere, orbital skin, and feet yellow. IMMATURE: Upper parts blackish, tipped with buffy; tail with 5–6 interrupted pale bars; underparts heavily streaked and barred with blackish; orbit, cere, and feet bluish. Wing 293–377, tail 135–196, culmen from cere 18–25, tarsus 46–62 mm. Cosmopolitan; in United States breeds south to California, Arizona, Texas, Kansas, Gulf states, Pennsylvania, and Connecticut; winters in United States, southward through West Indies and Central America to Chile and Argentina.

Falco rusticolus **Linnaeus.** Gyrfalcon. Outer primary notched, shorter than third. Upper two-thirds of tarsus feathered, lower part reticulate. Creamy white; crown, sides of head, and nape with black shaft stripes or spots; remiges tipped and barred with drab; tail plain creamy or with 9–11 dark bars, interrupted on lateral rectrices; underparts spotted or plain; iris brown; cere yellow; bill and feet yellowish gray. YOUNG: Back and wings margined with white; striped with grayish brown below. There is much variation in depth of coloring and extent of markings. Some birds are immaculate white on crown and below, with the ground color of the upper parts white; others are nearly plain dusky above and below, with the markings nearly obliterated. Wing 340–423, tail 203–266, culmen from cere 20.5–28, tarsus 53–72 mm. Circumpolar; in America south irregularly in winter to New England, New York, Pennsylvania, Great Lakes, Kansas, Montana, and Oregon.

Falco mexicanus **Schlegel.** Prairie falcon. Outer 2 primaries notched. Upper third of tarsus feathered, lower part reticulate. Forehead cream, streaked with black; upper parts dark brown tinged with cinereous; upper tail coverts tipped with cinnamon; tail drab, tipped buff, and with 8–9 buff bars; lores and superciliaries buffy, meeting on occiput; malar stripe sepia; below buffy white, the breast, sides, and thighs with brown tear-shaped spots; axillars brown with a few rusty spots near tip; wing lining white streaked with brown; iris brown; bill slate at tip, greenish at base; cere, bare orbit, and feet yellow; claws black. YOUNG: Back and tail barred with dark brown; cere, orbit, and feet pale blue. Wing 289–357, tail 159–201, culmen from cere 18–26, tarsus 50–64 mm. Southwestern Canada south to Baja California, Arizona, New Mexico, middle Texas, east to the Dakotas, Kansas, and Oklahoma.

Falco femoralis **Temminck.** Aplomado falcon. Outer 2 primaries notched. Acrotarsium partly, toes entirely scutellate. Above slate gray, darker on crown; rump and upper tail coverts barred with

Fig. 5-39. Bill of sparrow hawk (*Falco sparverius*). *C*, cere; *Md*, mandible; *Mx*, maxilla; *T*, nasal tubercle.

paler gray; tail dusky, tipped and with 8 bands of whitish; superciliaries and hind neck buff; malar stripe and auriculars dusky; chin, throat, and middle of breast buffy white; sides, flanks, and narrow band across belly blackish, narrowly tipped with white; thighs, abdomen, and crissum tawny buff; iris brown; tip of bill slate; base of bill, cere, and feet yellow; claws black. YOUNG: Breast with blackish streaks; tail with 10 white bands. Wing 230–302, tail 149–207, culmen from cere 15–20, tarsus 43–60 mm. Southern parts of Texas, New Mexico, and Arizona, south on the mainland to Argentina and Chile.

Falco columbarius **Linnaeus.** Pigeon hawk. Outer 2 primaries notched; middle of tarsus and basal portion of toes reticulate. Above bluish slate with dark shaft streaks; tip of tail whitish, with broad black subterminal band and broken dusky bars; primaries dusky, barred with white on inner webs; superciliaries and hind neck rusty, streaked with black; cheeks buffy streaked with brown; chin and throat pure white; breast, belly, and crissum deep buff streaked with brown; thighs plain cinnamon buff; axillars banded brown and white; iris brown; tip of bill dusky blue; base of bill, cere, and feet yellow; claws black. Wing 177–228, tail 114–149, culmen from cere 11–15, tarsus 35–45 mm. Holarctic, breeding in northern coniferous forest, south in the United States to New Hampshire, New York, Ohio, northern Michigan, northern Wisconsin, Iowa, North Dakota, Montana, and Oregon; winters from the southern tier of states through Mexico, Central America, and West Indies to Ecuador and Peru.

Falco sparverius **Linnaeus.** Sparrow hawk. Outer 2 primaries notched; toes scutellate. Crown and occiput slate, usually with a median patch of cinnamon rufous; nape with a black spot; back and scapulars cinnamon rufous, more or less barred with black; upper tail coverts cinnamon rufous; 2 middle pairs of rectrices cinnamon rufous, tipped with pale gray and with broad black subterminal band; outer rectrix gray on outer web, inner web with 4–5 black bands, gray tip and interspaces, rufous toward base; remaining rectrices with 2–3 black bands; wing coverts slate, more or less spotted with black; primaries black, banded with white on inner webs; malar stripe and auricular patch black; lores, chin, upper throat, and cheeks white; breast and upper abdomen cinnamon buff, more or less spotted with black; thighs, lower belly, and crissum pale buff; axillars banded black and white; iris brown; bill mainly bluish; base of bill, cere, and feet yellow; claws black. FEMALE: Head and neck as in male, but crown often with dark shaft streaks; back, wings, and tail rusty barred with black; breast whitish or buffy, streaked with brown or dusky; posterior underparts pale buff; axillars banded brown and white. Wing (U.S. races) 165–207, tail 110–142, culmen from cere 11–15, tarsus, 32–42 mm. North and South America, breeding north to the limit of trees, south throughout West Indies, Mexico, Central and South America, to Chile and Argentina.

Order Galliformes. Fowl. Bill short, culmen decurved, maxilla vaulted, rhamphotheca simple, nostrils imperforate. Feet large, hallux always present. Wings short, rounded, concave; primaries 10, stiff and curved; secondaries usually quintocubital (aquintocubital in some megapodes). Rectrices 8–32. Aftershafts present; adult downs scarce, on apteria only; oil gland tufted (naked in *Argus* and some megapodes).

Schizognathous, with basipterygoid processes (rudimentary or absent in *Opisthocomus*) and vomer; holorhinal; angular process long (except in *Opisthocomus*). Cervicals 16. Sternum with spina communis, 2 pairs of deep notches and very narrow xiphoid area (1 pair of notches and wide xiphoid area in *Opisthocomus*). Furculum Y-shaped. Coracoids cross. Ectepicondylar process small; second metacarpal with a downward projecting process (except in Cracidae and Numididae).

Thigh muscles $(A)BXY+$; toe tendons type 1. Tongue normal; intestine type 5; caeca long. Both carotids usually present (only the left in megapodes). No supraorbital glands.

Rasorial or arboreal birds. Food mainly vegetable. Nest usually a depression in ground, sometimes of sticks in trees; eggs 2–20, plain buffy, white, or mottled; young downy, precocial.

Eocene onward. Cosmopolitan (except Polynesia and Antarctica). In the United States 2 families, 20 species.

KEY TO FAMILIES OF GALLIFORMES

1. Hallux more than half length of lateral toes, incumbent   **Cracidae**
   Hallux less than half length of lateral toes, elevated   **Phasianidae**

Family Cracidae. Chachalacas, guans, curassows. Tomia smooth; nostrils usually exposed (feathered in *Oreophasis*). Wing rounded; secondaries quintocubital, as long as primaries. Tail rounded, about as long as wing. Tarsus entirely scutellate, unspurred; hallux incumbent, more than half length of lateral toes. Aftershafts reduced; oil gland tufted. Trachea coiled in male, but not entering sternum. Anterior lateral processes of sternum directed laterad; inner sternal notch less than half length of sternum. Both carotids present.

Largely arboreal. Nest of sticks, in trees; eggs 3, plain buffy white. Gregarious; call loud. Easily domesticated.

Eocene onward. Neotropical mainland, north to Rio Grande Valley of Texas (to Nebraska and Florida in Tertiary). 11 genera, 39 species; 1 species in the United States.

Genus *Ortalis* Merrem. Chachalacas. Bill depressed, nostrils longitudinal, exposed (Fig. 5-40). Crown slightly crested; orbits, lores, and sides of throat bare; a line of feathering down middle of throat. Texas to Argentina. 11 species, 1 in the United States.

Ortalis vetula (Wagler). Chachalaca. Olive brown, grayer on head; lower breast paler; thighs, belly, and crissum buffy brown; tail greenish olive, broadly tipped with buff or white; iris brown; bill and feet horn. Wing 172–219, tail 197–264, exposed culmen 19–28, tarsus 49–66 mm. Lower Rio Grande Valley in southern Texas, south through Caribbean part of Mexico and Central America to Honduras and Nicaragua. Introduced on Sapelo and Blackbeard Islands, Georgia.

Family Phasianidae. Quails, pheasants, grouse, turkeys. Tomia smooth or notched; nostrils exposed or feathered. Secondaries quintocubital, usually shorter than primaries. Tarsus scutellate or feathered, sometimes spurred; hallux elevated, less than half length of lateral toes. Aftershafts large; oil gland tufted (except in *Argus*). Trachea straight. Sternum with inner notch more than half length of bone; anterior lateral processes directed forward. Both carotids present.

Nest on ground; eggs 6–20, plain buff or speckled with brown. Oligocene onward. Worldwide, except Polynesia. 62 genera, 194 species, in the United States 16 genera, 19 species. 4 subfamilies.[1]

Subfamily Odontophorinae. American quails. Bill short, stout; mandibular (and often maxillary) tomium with 1 or more notches; nostrils exposed. Head and neck without bare areas. Secondaries shorter than primaries. Tail short; rectrices 10–14. Tarsus scutellate, unfeathered and unspurred, more than half length of tibia; toes neither fringed nor feathered. Oligocene onward. North and South America. 10 genera, 33 species; in United States 5 genera, 6 species.

Subfamily Phasianinae. Pheasants and partridges. Bill rather slender, unnotched; nostrils exposed. Face often with naked areas. Secondaries sometimes as long as primaries. Tail short to longer than wing; rectrices 8–18. Tarsus scutellate, unfeathered, often spurred, more than half length of tibia; toes neither fringed nor feathered. Oligocene onward. Old World. 39 genera, 142 species; 3 genera, 3 species feral in the United States.

Subfamily Tetraoninae. Grouse. Bill short, unnotched; nostrils concealed by feathers. Often with bare superciliaries or bare cervical sacs. Secondaries shorter than primaries. Tail about half length of wing to longer than wing; rectrices 16–20. Tarsus short, about half length of tibia, densely feathered for half to entire length; toes often feathered or laterally fringed. Miocene onward. Holarctic. 11 genera, 17 species; in United States 7 genera, 9 species.

Subfamily Meleagridinae. Turkeys. Bill rather long and narrow, unnotched; nostrils exposed. Head and upper neck naked, with caruncles; body feathers broad, nearly truncate; secondaries about as

[1] Grouse and turkeys treated as families Tetraonidae and Meleagrididae by A.O.U.

Fig. 5-40. **Head of chachalaca (*Ortalis vetula*).** *O*, operculum.

long as primaries in closed wing. Tail large, but shorter than wing, strongly rounded; rectrices 18, broad. Tarsus naked, spurred in male, more than half length of tibia; toes bare. Pliocene onward. North America. 2 genera; 2 species, 1 in the United States.

KEY TO GENERA OF PHASIANIDAE

1. Head naked   *Meleagris*
   Head largely feathered   **2**
2. Nostrils completely covered with feathers (Tetraoninae)   **3**
   Nostrils exposed   **9**
3. All rectrices pointed   *Centrocercus*
   Most rectrices wide, truncate or rounded   **4**
4. Lower half of tarsus naked   *Bonasa*
   Tarsus completely feathered   **5**
5. Middle rectrices abruptly longer than the rest   *Pedioecetes*
   Middle rectrices not abruptly lengthened   **6**
6. Rectrices 18; tail half length of wing or less   *Tympanuchus*
   Rectrices 16 or 20; tail more than half length of wing   **7**
7. Rectrices 20   *Dendragapus*
   Rectrices 16   **8**
8. Outer primary between 7th and 8th; outer 3 slightly emarginate   *Canachites*
   Outer primary between 6th and 7th; outer 4–5 slightly emarginate   *Lagopus*
9. Mandibular tomium smooth (Phasianinae)   **10**
   Mandibular tomium with 1 or more notches (Odontophorinae)   **12**
10. Tail longer than wing   *Phasianus*
    Tail shorter than wing   **11**
11. Rectrices 14   *Alectoris*
    Rectrices 16–18   *Perdix*

12. Outer and inner claws reaching to middle of middle
    claw   *Cyrtonyx*
    Inner claw not reaching base of middle claw   **13**
13. Rectrices 14 (crest bushy)   *Callipepla*
    Rectrices 12   **14**
14. Crest inconspicuous   *Colinus*
    Crest plumes long, narrow, clinging together   **15**
15. Crest plumes narrow throughout   *Oreortyx*
    Crest plumes broadening near tip   *Lophortyx*

Genus *Colinus* Goldfuss. Mandibular tomium with several serrations; crown feathers slightly lengthened; superciliaries feathered; outer primary shorter than 7th; tail rounded, less than half wing length; rectrices 12; tarsus scutellate, unspurred, slightly shorter than middle toe with claw; tips of outer and inner claws falling short of base of middle claw. North, Central, and northern South America. 4 species, 1 in United States.

Colinus virginianus (Linnaeus). Bobwhite. Superciliaries and throat white (black in 1 U.S. and several Mexican races); crown, neck, and subocular stripe, and upper breast russet mixed with black, the black predominating in some races; nape streaked with white; upper back and wing coverts russet; barred on edges of feathers with blackish and gray; tertials buffy gray, irregularly barred with black, russet, and whitish, and edged internally with buff; remiges grayish, the inner secondaries barred with pale cinnamon on outer webs; rump and upper tail coverts gray, vermiculated, streaked, and barred with black, white, and buffy; tail gray, speckled at tip with whitish; underparts buffy white barred with black; flanks similar but with long tawny shaft stripes; iris brown; bill black; feet gray. FEMALE: White areas of head and throat replaced by deep buff. Wing 99–119, tail 49–70, culmen 13–18, tarsus 27–34 mm. Maine, Ontario, Minnesota, South Dakota, and eastern Colorado, south to Gulf states and Cuba, west to New Mexico and Arizona, and south through most of Mexico to Guatemala.

Genus *Callipepla* Wagler. Crest bushy; tail strongly rounded, rectrices 14. Mexican plateau and southwestern United States. 1 species.

Callipepla squamata (Vigors). Scaled quail. Crown grayish brown; crest browner, tipped with white; nape gray, banded with brown; back and wing coverts grayish brown; tertials edged with buff; remiges drab; tail gray; chin and throat buff;

neck and breast bluish gray squamate with black; belly and crissum whitish or buffy, squamate with dusky, and often with a patch of chestnut in center of belly; flanks brown with white shaft streaks; iris brown; bill black; feet gray. Wing 113–121, tail 76–90, culmen 15–17, tarsus 31–35 mm. Western parts of Texas, Oklahoma, and Kansas, eastern Colorado, New Mexico, and southern Arizona, south to central Mexico.

Genus *Oreortyx* Baird. Crest of 2 long slender cohering plumes; rectrices 12. Pacific states and Baja California. 1 species.

Oreortyx picta (Douglas). Mountain quail. Crown slate; crest black; upper parts grayish olive brown; inner secondaries and coverts edged and tipped with whitish; tail fuscous speckled with olive brown; superciliary and chin white; cheeks and throat chestnut, becoming blackish posteriorly, and bordered behind by a white line; breast slate, tipped with chestnut, sides chestnut barred with white; flanks chestnut barred with black; thighs buff; middle of belly buffy white; crissum black streaked with russet; under wing coverts slaty; bill black; iris and feet brown. Wing 125–140, tail 69–92, culmen 14–18, tarsus 32–38 mm. Western Washington, Oregon, California, southwestern Idaho, western Nevada, northern Baja California.

Genus *Lophortyx* Bonaparte. Crest long, of about 6 recurved imbricated feathers, expanded at tip; outer primary shorter than 8th; rectrices 12; claws rather long and slender, outer claw about reaching base of middle claw, inner one shorter. Western United States and northern Mexico. 3 species, 2 in the United States.

KEY TO SPECIES OF *LOPHORTYX*

1. Sides chestnut with white shaft streaks   *L. gambelii*
   Sides olive with white shaft streaks   *L. californica*

Lophortyx californica (Shaw). California quail. Forehead white with fine black shaft streaks; a narrow white line on lores; a white line across center of crown, continuing as a white postocular stripe; a black line bordering this behind, and including crest; occiput brown; throat black, bordered behind by a white line; a narrow black line runs behind this from auriculars; sides of neck blue-gray, streaked with black and speckled with white; hind neck blue-gray squamate with dusky; scapulars olive gray streaked with white; upper parts olive brown; tertials edged with tawny; tail blue-gray; breast blue-gray; sides olive with large white shaft streaks, becoming buffy on flanks; lower breast buff or white, squamate with black; upper abdomen chestnut squamate with black; belly buffy white barred with olive; crissum buff streaked with olive; iris brown; bill black; feet gray. FEMALE: Throat and cheeks grayish white streaked with olive brown; breast grayish brown. Wing 105–119, tail 79–100, culmen 14–17, tarsus 28–35 mm. Oregon, California, western Nevada, Baja California. Introduced into Vancouver Island, Washington, Oregon, Colorado, Arizona, and New Mexico; also to Hawaii, Chile, and New Zealand.

Lophortyx gambelii Gambel. Gambel's quail. Male differs from *L. californica* in lacking white line on lores; forehead black, finely streaked with buffy; crown chestnut; nape and sides of neck blue-gray streaked with brown; sides chestnut with large white shaft streaks; breast uniform gray; middle of belly black; crissum buff with brown shaft streaks. FEMALE: Cheeks and throat white streaked with brown; belly whitish streaked with brown, without black patch. Wing 105–122, tail 83–107, culmen 12–16, tarsus 26–32 mm. Western Colorado, western Texas, New Mexico, Arizona, southern Utah, southern Nevada, southern California, and northwest Mexico. Introduced to Hawaii.

Genus *Cyrtonyx* Gould. Crest short, bushy; tail wedge-shaped, soft, little longer than coverts, rectrices 12; claws long and heavy, outer and inner claws reaching to about middle of middle claw, which is more than $\frac{1}{3}$ length of tarsus. Mountains of southwestern states to Nicaragua. 2 species, 1 in the United States.

Cyrtonyx montezumae (Vigors). Harlequin quail. Middle of forehead black, bordered by white; a black line from bill to above and behind eye; a blackish-slate line from bill along lower border of cheeks, where it bends forward toward lower throat; subocular area and superciliary stripe white; a black patch on auriculars, extending to below eye, bordered by white; middle of throat black, bordered by white; crown and nape brown, marked with black; back and tail brown, barred with black and with large white or buffy shaft streaks, the

latter absent on rump; inner secondaries olive gray, brokenly barred with black and with buff shaft stripes; primaries brown barred with white on outer webs; wing coverts olive gray, transversely spotted with black on greater, with white on lesser coverts; sides slate gray with large white spots, continuing in a narrow band across upper breast; lower breast deep chestnut; belly, thighs, and crissum black; iris brown; bill black, bluish below, feet blue. FEMALE: Crown buff, barred with black and with buffy shaft stripes; nape less heavily barred; back and tail similar but browner; rump more blackish and without shaft streaks; wing coverts brown, brokenly barred with sepia; cheeks brown flecked with blackish; circumorbital region and superciliaries whitish; throat white, bordered with dusky-flecked brown; breast, belly, and crissum brown with some whitish spots; middle of belly flecked and streaked with blackish. Wing 110–131, tail 48–63, culmen 14–16, tarsus 28–33 mm. Western Texas, New Mexico, Arizona, and Mexican tableland.

Genus *Phasianus* Linnaeus. Feathers of postocular region lengthened to form prominent tufts in male, less so in female; orbital area more or less bare; outer primary longer than 7th; tomia unnotched; tail longer than wing, graduated, rectrices 18; tarsus with spur in male, slight knob in female. Palearctic. Monotypic. Introduced to the United States, Britain, Hawaii, New Zealand.

Phasianus colchicus Linnaeus. Ring-necked pheasant. MALE: Superciliaries white; head and neck shiny blue-green or green; a white collar around base of neck; interscapulars bright buff, spotted with white, and streaked and barred with greenish black; scapulars and inner wing coverts glossy brown, striped with black and white; outer wing coverts light gray; remiges buffy brown, barred with whitish; rump and upper tail coverts greenish, streaked with dusky and barred with buff; tail olive buff, edged with brown, and with many blackish bars; breast hazel glossed with purple and barred with black; flanks light buff, streaked with glossy blue-black; center of belly greenish black; rest of belly and thighs brown; crissum purplish hazel; under wing coverts white; axillars barred with brown; bare face skin red; iris brown; bill yellow; feet brownish. FEMALE: Crown barred brown and buff; superciliaries and sides of face buffy, more vinaceous on neck; subocular white; interscapulars russet, with black and vinaceous squamations; scapulars and tertials brown, black medially, and edged with buff; wing coverts light buff barred with dusky; remiges brown barred with pale cinnamon; rump and upper tail coverts vinaceous, barred with buff and streaked with dusky; tail vinaceous, with irregular black and brown bars; throat buffy; lower neck vinaceous, squamate with black and cinnamon. Wing 194–245, tail 236–513, culmen 33–43, tarsus 61–75 mm. Established from southern Canada south to California, Idaho, Colorado, Kansas, Missouri, Kentucky, Ohio, Pennsylvania, and Maryland. Introduced from Europe and Asia.

Genus *Perdix* Brisson. A slight bushy crest; postocular area bare; tomia unnotched; outer primary longer than 8th; tail about half length of wing, slightly rounded, rectrices 16–18; tarsus unspurred, slightly longer than middle toe with claw. Palearctic. 3 species, 1 introduced to the United States.

Perdix perdix (**Linnaeus**). Hungarian partridge; gray partridge. Crown brown streaked with buffy; hind neck gray vermiculated with dusky, back and middle rectrices buffy brown, barred with chestnut; scapulars and wing coverts similar but with buffy-white shaft streaks; lateral rectrices chestnut; remiges brown barred with pinkish buff; head and throat pale cinnamon; breast gray vermiculated with dusky; flanks buffy gray barred with tawny; belly whitish, with a large chestnut inverted V on anterior portion; crissum buff; bare postocular red; iris brown; bill horn; feet gray. FEMALE: Nape more brownish gray; chestnut belly patch often obsolete. Wing 144–162, tail 73–84, culmen 12–16, tarsus 35–43 mm. Europe, western Siberia, and Persia. Established on interior plains of southern Canada, Great Basin, northern Great Plains, and around Great Lakes.

Genus *Alectoris* Kaup. Differs from *Perdix* in having only 14 rectrices; tarsus of male with rudimentary spur. Europe, North Africa, Asia. 4 species, 1 introduced to the United States.

Alectoris graeca (**Meisner**). Chukar partridge; rock partridge. Forehead gray; upper parts brownish gray tinged with pinkish; scapulars blue-gray, broadly edged with chestnut; lateral pairs of rectrices pale cinnamon; line through eye black; auriculars chestnut; throat white, bordered by a black line; breast gray; belly reddish buff; flanks slate, barred with black, white, and chestnut; eyelids, bill, and feet red. Wing 156–174, culmen 14–15, tarsus 43–47 mm. Asia and southern Europe; established in Washington, Idaho, Nevada, California, Arizona, and Colorado.

Genus *Dendragapus* Elliot. Superciliaries and small inflatable sac on side of neck of male bare; no lengthened tuft of feathers on side of neck; tail 62–81% length of wing, moderately rounded, rectrices 20; tarsus completely feathered, shorter than middle toe without claw. Western North America. 1 species.

Dendragapus obscurus (**Say**). Dusky grouse; sooty grouse; blue grouse. Grayish above, vermiculated with blackish; tail dark brown, often with a light-gray tip; below sooty gray; throat, sides, flanks, and thighs tipped and streaked with white; crissum sooty gray, tipped with white and banded with black; axillars mostly white; feathers bordering neck sac white, tipped with dusky; sac yellow (in Sierras) or purple (Rockies); iris brown; bill blackish; feet brownish. FEMALE: Upper parts blackish banded with tawny; wing coverts tawny mottled with black, the lesser and median ones banded with white; remiges brown tipped with tawny and mottled with tawny on outer webs of primaries; inner secondaries white-tipped; tail tawny, often tipped with gray and irregularly banded with black; cheeks mixed with white; throat white with blackish V-shaped marks; breast brown banded with cinnamon; posterior underparts as in male. Wing 178–248, tail 111–201, exposed culmen 16–24, tarsus 37–48 mm. Coniferous forests of southeastern Alaska and western Canada, south to California, northern Arizona, western New Mexico, Wyoming, and western South Dakota.

Genus *Canachites* Stejneger. Similar to *Dendragapus* but without inflatable neck sacs; tail 55–75% length of wing, rectrices 16. Northern coniferous forest of North America. 1 species.

Canachites canadensis (**Linnaeus**). Spruce grouse. Above gray barred with black; wing coverts browner; upper tail coverts like back but tipped with pale gray; tail blackish, vermiculated basally with brown or gray, often broadly tipped with tawny, and sometimes with extreme tip narrowly banded black and white; primaries edged with white; postocular stripe black-and-white streaked; nasal plumes, chin, and cheeks black; throat barred black and white; sides of neck and lower throat like back; band across upper breast and extending on midline of lower breast black; posterior underparts irregularly banded black and grayish white; tarsal feathers sooty gray; bare superciliary vermilion; iris brown; bill and feet dusky. FEMALE: Upper parts barred black, gray, and buff; throat and breast tawny buff, with dusky bars; abdomen and crissum white banded with dusky; tail as in male. Wing 159–194, tail 94–144, culmen 12–20, tarsus 32–39 mm. Spruce forests from Alaska and northern Canada, Oregon, Idaho, Montana, and Wyoming (Yellowstone), and south to northern parts of Minnesota, Wisconsin, Michigan, New York, New Hampshire, Vermont, and Maine.

Genus *Lagopus* Brisson. Superciliaries bare; neck without bare sacs or elongated tufts of feathers; tail short, 55–65% length of wing, rounded, rectrices 16; tarsus completely feathered, and in winter toes also. Undergoes 3 annual molts. Holarctic. 4 species, 2 reaching the United States.

KEY TO SPECIES OF *LAGOPUS*

1. Rectrices entirely white, except middle pair
   L. leucurus
   Tail black, tipped with white   *L. lagopus*

Lagopus lagopus (**Linnaeus**). Willow ptarmigan. SUMMER MALE: Chestnut, barred with black except on throat and breast; eye ring, remiges, and outer wing coverts white; middle rectrices like back, lateral ones black tipped with white; middle of belly and legs white; bare superciliary red; bill blackish. AUTUMN MALE: Chestnut replaced by hazel, and feathers of back tipped with white. WINTER MALE: White; middle rectrices and shafts of primaries black. FEMALE: Resembles male of corresponding season, but in summer chestnut replaced by tawny olive; in autumn grayer than male. Wing 171–216, tail 96–139, culmen from nostril 9–12 mm. Breeds on the Holarctic tundra, south in America to James Bay and Newfoundland; in winter south casually to Montana, North Dakota, Wisconsin, Michigan, New York, Maine, and Massachusetts.

Lagopus leucurus (**Swainson**). White-tailed ptarmigan. SUMMER: Above grayish buff, barred and vermiculated with black; breast, sides, and flanks whitish barred with black; remiges, wing coverts, rectrices (except middle pair), center of belly, and legs white; iris brown; bare superciliary red; bill black. FALL: Grayish buff of upper parts replaced by tawny buff; sides of head, throat, and breast white barred with brown; sides tawny buff mottled with black; remiges, wing coverts, outer rectrices, belly, crissum, and feet white. WINTER: Entirely white. Wing 155–194, tail 84–109, culmen 10–18, tarsus 30–35 mm. Mountains above timber line, from Alaska to Washington (south to Mount St. Helens), and in Rockies south into Montana, Wyoming, Colorado, and northern New Mexico.

Genus *Bonasa* Stephens. Side of neck without sac but with erectile tuft of feathers; tail rounded, 72–92% of wing length, rectrices 18–20; tarsus longer than middle toe, lower half unfeathered, scutellate. North America. Monotypic.

Bonasa umbellus (**Linnaeus**). Ruffed grouse. Above varied black, brown, buff, and gray; neck tufts black or rusty; scapulars, wing coverts, and rump streaked with pale gray; tail gray or rusty, with many irregular bars and a broad subterminal band of black; below buffy, barred with brown, except on throat. Wing 165–196, tail 119–181, culmen 21–31, tarsus 36–48 mm. Wooded North America, south to Virginia, mountains of Georgia and Alabama, Kentucky, Tennessee, Arkansas, Missouri Valley (formerly), South Dakota, Colorado, Utah, and northern California.

Genus *Tympanuchus* Gloger. Side of neck with erectile tuft of feathers (in male) and below this a bare inflatable sac; tail rounded, 40–53% of wing length, rectrices 18; tarsus feathered (more heavily in winter), about length of middle toe. North America, east of the Rockies. 2 nominal species, but *Pedioecetes* could be merged with this genus.

KEY TO SPECIES OF *TYMPANUCHUS*

1. Dark bars on lower back plain blackish, on flanks plain brown   *T. cupido*
   Dark bars on lower back and flanks bordered by 2 narrow black bars   *T. pallidicinctus*

Tympanuchus cupido (**Linnaeus**). Greater prairie chicken. Above barred with brown, buff, and black; tail grayish brown narrowly tipped with buff; primaries brown, spotted with white on outer webs; neck tufts blackish, striped with buff; throat buffy; sides of head buffy mottled with black; underparts whitish buff barred with grayish brown; bare throat sacs orange. Wing 195–241, tail 78–128, culmen 16–26, tarsus 41–52 mm. Prairie region from southern Canada south to Ohio, Indiana, and Illinois, west to eastern Colorado, southeastern Wyoming, Oklahoma, and northern Texas. Coast of Texas and Louisiana and formerly of Massachusetts to Washington, D.C. (heath hen, *T. c. cupido*, became extinct in 1932).

Tympanuchus pallidicinctus (**Ridgway**). Lesser prairie chicken. Differs in having the dark bars on back and flanks divided by a brown stripe; gular sacs yellow. Wing 195–220, tail 81–95, culmen 17–18, tarsus 42–47 mm. Plains of southwestern Kansas, southern Colorado, Oklahoma, northern Texas, and eastern New Mexico.

Genus *Pedioecetes* Baird. No tufts of feathers or bare sacs on sides of neck; tail 49–61% length of wing, graduated, with middle rectrices extending far beyond the rest, rectrices 18, their tips truncate; tarsus feathered, about equal to middle toe. Interior of western North America. Monotypic.

Pedioecetes phasianellus (**Linnaeus**). Sharp-tailed grouse. Above tawny brown, varied with black; wing coverts and scapulars spotted with white; lateral rectrices whitish terminally; throat buffy; underparts whitish, with dusky V-shaped marks on breast and sides. Wing 186–223, tail 92–135, culmen 10–13, tarsus 39–48 mm. Great Basin and Plains, from interior Alaska and western Canada south to northeastern California, Utah, New Mexico, Nebraska, Minnesota, Wisconsin, and Michigan; formerly to Oklahoma and Illinois.

Genus *Centrocercus* Swainson. Male with bare sac on sides of neck; feathers on sides of neck spiny; tail graduated or strongly rounded, rectrices 18, attenuate; tarsus longer than middle toe with claw, feathered. Sagebrush plains of interior North America. Monotypic.

Centrocercus urophasianus (**Bonaparte**). Sage hen. Pileum, nape, and interscapulars drab barred with dusky; posterior upper parts and sides grayish buff, tipped with ashy and vermiculated with black; wing coverts with white shaft streaks; sides of head brown; cheeks and throat whitish mottled with brown; a white V across throat to auriculars; bare neck sacs olive green, surrounded by white stiff feathers; breast white, tipped and streaked with black; abdomen black; crissum black, tipped with white; thighs drab, speckled with brown; under wing coverts white. Wing 251–323, tail 288–332, culmen 33–41, tarsus 44–59 mm. Saskatchewan Montana, and eastern parts of Washington, Oregon, and Colorado, to northern Nevada, Utah, New Mexico, Colorado, and the Dakotas; formerly to western Nebraska and western Kansas.

Genus *Meleagris* Linnaeus. Tomia smooth; nostrils exposed; head and upper neck bare or downy, with fleshy caruncles, and in male with an erectile caruncle on forehead; body feathers broad, nearly truncate; aftershafts large; wing rounded, outer primary about equal to 10th; tail large but shorter than wing, flat, rounded; rectrices 18, broad; tarsus scutellate, spurred in male, nearly twice length of middle toe; hallux elevated, less than half length of front toes. United States and Mexican plateau. 1 species.

Meleagris gallopavo Linnaeus. Wild turkey. Dark brown; body feathers glossed with coppery bronze, purple, and green, and tipped with black; upper tail coverts chestnut, with dusky bars, tipped with chestnut, buff, or white, and with a subterminal black band; tail similar but not iridescent and ground color paler; remiges and primary coverts dusky barred with white on both webs; beard greenish coppery black; belly and thighs nonmetallic brown tipped with grayish buff; bare head skin bluish mixed with red; iris brown; bill yellowish orange; feet reddish. FEMALE: Lacks spurs and "beard"; head with fewer caruncles and more feathers. Wing 354–550, tail 268–400, culmen 27–40, tarsus 124–182 mm. Eastern United States, west to South Dakota, Nebraska, Colorado, and Arizona, south through the Mexican plateau. Exterminated over large areas in the east and often mixed with domestic strains.

Order Ralliformes. Marsh birds. Bill usually longer than head; nostrils perforate (except in Rhynochetidae); rhamphotheca simple; lores feathered or bristled. Legs long; lower part of tibia bare; acrotarsium at least partly scutellate; toes not fully webbed; hallux elevated or absent. Wing rounded; primaries 10–11; secondaries long. Tail short, rectrices 10–18. Lateral neck apteria small; adult downs all over (only on apteria in Otidae), aftershafts present (except in Mesoenatidae and Heliornithidae).

Schizognathous (desmognathous in Cariamidae), with vomer and without basipterygoid processes; lacrimals loose; angular process blunt; cervicals 14–20; true ribs 5–7; sternum with external, without internal spine; coracoids not crossed; humerus without well-developed ectepicondylar process. Intestine periorthocoelous (type 1); no crop; caeca and gall bladder present. Thigh muscles $(AB)XY+$.

Swamp birds (waders or swimmers). Food mostly plants or mollusks. Nest on ground (arboreal in Eurypygidae); eggs buffy white, spotted or speckled with brown; young downy without stripes or blotches, mostly nidifugous.

Paleocene onward. Cosmopolitan. In the United States 3 families, 13 species.

KEY TO FAMILIES OF RALLIFORMES

1. Middle toe less than half length of tarsus   **Gruidae**
   Middle toe more than half length of tarsus   **2**
2. Outer primary half width of next, broader at tip, which is curved, and shorter than 7th primary
        Aramidae
   Outer primary more than half width of next
        Rallidae

Family Gruidae. Cranes. Bill straight, longer than head but shorter than tarsus; crown naked or with ornamental plumes; neck long. Primaries 11, shorter than tertials; secondaries aquintocubital. Tail short, slightly rounded, rectrices 12. Legs very long; toes short, about $\frac{1}{3}$ length of tarsus; hallux elevated, nonfunctional. Aftershafts very small; no powder downs; oil gland tufted.

Schizorhinal; with occipital fontanels and supraorbital glands; cervicals 19–20; external sternal spine expanded, metasternum entire; hypotarsus complex. Thigh muscles $(AB)XY+$; plantar tendons type 1. Caeca large, opposite.

Voice resonant. Performs mating dance; eggs 2, shells rough; downy young brownish, pale below.

Eocene onward. North America, Eurasia, Africa, Australia. 4 genera, 14 species; in the United States 1 genus, 2 species.

Genus *Grus* Pallas. Upper half of head naked, warty, with some short hairs; no pendant plumes on neck; trachea coiled and entering sternum; furculum fused to sternal keel; thigh muscles $ABXY+$. Eurasia, Africa (winter), Australia, North America. 10 species, 2 in North America.

KEY TO SPECIES OF *GRUS*

1. Plumage largely white   *G. americana*
   Plumage largely gray   *G. canadensis*

Grus americana (**Linnaeus**). Whooping crane. White; nape patch primaries, primary coverts, and alula slaty black; bare skin of head red; iris yellow; bill greenish yellow; legs black. YOUNG: White plumage blotched with buff; whole head feathered, with crown dusky. Wing 550–630, tail 194–245, culmen 129–148, tarsus 260–301 mm. Formerly bred in interior Canada south to Iowa and Nebraska; migrated practically throughout the United States and wintered on Gulf Coast and in central Mexico. Now (1966) reduced to 38 individuals which winter at Aransas, Texas, and breed in Wood Buffalo Park, northwestern Canada.

Grus canadensis (**Linnaeus**). Sandhill crane. Ashy gray, often tinged with rust; chin white; primaries and primary coverts blackish slate; bare head skin red; iris red; bill and legs olive. YOUNG: Crown feathered, buffy. Wing 425–590, tail 150–187, culmen 82–170, tarsus 186–266 mm. Arctic America and adjacent Siberia, breeding south to Michigan, Minnesota, Nebraska, Colorado, Arizona, and California; also breeds on Coastal Plain in Georgia, Florida, Alabama, and Louisiana and on Cuba and Isle of Pines; winters south to central Mexico.

Family Aramidae. Limpkins. Head feathered; bill somewhat decurved, nearly as long as tarsus; primaries 10, outer one narrow and bent; secondaries aquintocubital; rectrices 12, well developed; aftershafts small; oil gland tufted; anterior toes more than half length of tarsus; hallux slightly elevated but long and functional.

Schizorhinal; cervicals 17; true ribs 7; furculum free, without hypocleidium; hypotarsus complex.

Trachea straight, broader proximally but not entering sternum. Thigh muscles $BXY+$; plantar tendons type 1. Caeca lateral and close together.

Inhabits river swamp. Food snails. Cry raucous. Nest of vegetation, on wet ground or in bushes; eggs 4–7; downy young brownish, paler below.

Tropical America, 1 living species. Known from Oligocene to Pliocene of Nebraska and South Dakota.

Genus *Aramus* Vieillot. Monotypic.

Aramus guarauna (Linnaeus). Limpkin. Brown; glossed with purplish on remiges, rump, and tail; other parts with large white shaft stripes; chin whitish. Wing 290–339, tail 121–160, culmen 85–131, tarsus 102–142 mm. Southern Georgia, Florida, Greater Antilles, and lowlands of Mexico and Central America south to western Ecuador and Argentina.

Family Rallidae. Rails. Head feathered; primaries 10–11, longer than secondaries; aquintocubital; rectrices short, soft, 10–14; hallux long, nearly incumbent; oil gland tufted; no powder downs, aftershafts present.

Holorhinal; without occipital fontanels; cervicals 14–15; furculum free; metasternum with a deep pair of notches; hypotarsus with 2 grooves or 1 groove and 1 canal. Thigh muscles $ABXY+$; toe tendons type 1. Caeca long; gall bladder present.

Marsh dwellers (waders or swimmers). Flight weak, and several species have become flightless. Voice a cackle. Nest a platform of plant material; eggs speckled or spotted, 4–16; downy young usually black, sometimes with orange or white "hairs" about head; able to swim when hatched.

Late Cretaceous onward. Cosmopolitan; 51 genera, 131 species, several of which have become extinct. In the United States 5 genera, 10 species.

KEY TO GENERA OF RALLIDAE

1. A horny shield on forehead — 2
   No frontal shield — 4
2. Toes lobed — *Fulica*
   Toes unlobed — 3
3. Hind claw shorter than middle claw — *Gallinula*
   Hind claw longer than middle claw — *Porphyrula*
4. Bill as long as tarsus or longer — *Rallus*
   Bill not more than ⅔ length of tarsus — *Porzana*

Genus *Rallus* Linnaeus. Bill much longer than head, slightly longer than tarsus, slightly decurved; no frontal shield; outer primary shorter than secondaries, about equal to 7th or 8th primary; tail less than half length of wing, rectrices soft, scarcely exceeding under tail coverts; toes long, unlobed, middle toe and claw about as long as tarsus. Holarctic, Neotropical, Oriental, and Ethiopian regions. 16 species, 4 in the United States.

KEY TO SPECIES OF *RALLUS*

1. Wing 113 mm or less — *R. limicola*
   Wing 127 mm or more — 2
2. Side of head grayish — *R. longirostris*
   Side of head brownish — 3
3. Dark stripes on back more distinct, edged with buffy; breast cinnamon — *R. elegans*
   Dark stripes on back less distinct, edged with ashy gray; breast buffy — *R. obsoletus*

Rallus longirostris Boddaert. Clapper rail. Above dark olive brown, the feathers with darker centers; back edged with ashy gray; wings buffy brown; supraloral stripe buffy; lores grayish brown; cheeks slate gray; throat white; sides of neck and foreneck grayish buff or pale grayish cinnamon; breast pale cinnamon without grayish suffusion; middle of belly whitish; flanks and anterior crissum grayish brown barred with white; longer under tail coverts white with dusky spots. Wing 128–163, tail 49–72, culmen 46–69, tarsus 41–56 mm. Salt marshes of Atlantic Coast, from Connecticut south to the Gulf states, eastern Mexico, and British Honduras; also in West Indies and on both coasts of South America, south to western Peru and São Paulo.

Rallus obsoletus Ridgway.[1] California clapper rail. Pattern identical with that of *R. longirostris*. Cheeks brown, not gray; back browner, but edged with ashy as in *R. longirostris;* breast more buffy; size averages larger. Wing 138–170, tail 54–80, culmen 49–66, tarsus 45–63 mm. Salt marshes of the Pacific Coast, from Humboldt Bay, California, south to Nayarit; also in fresh-water marshes of Salton Sea, California, and lower Colorado River from Yuma to Laguna, Arizona; also in marshes of Valley of Mexico.

---

[1] Included in *Rallus longirostris* by A.O.U.

Rallus elegans Audubon. King rail. Pattern exactly as in *R. longirostris* but even larger and more brightly colored than *R. obsoletus.* Upper parts bright brown; back with dusky centers and edged with buff, giving a more striped effect; cheeks brownish cinnamon; breast bright cinnamon; belly tinged with cinnamon; flanks barred black and white. Wing 134–177, tail 52–72, culmen 46–65, tarsus 49–64 mm. Fresh-water marshes from Connecticut, New York, Ontario, Michigan, Wisconsin, and Minnesota, south to Florida and Gulf states, and west to Oklahoma, Kansas, and Nebraska; also Cuba and Isle of Pines.

Rallus limicola Vieillot. Virginia rail. Crown dusky edged with brown; nape and back black, edged with deep olive brown; wing coverts russet; supraloral stripe pinkish buff; lores black; cheeks lead gray; chin white; throat, breast, and sides deep dull cinnamon; belly paler cinnamon; crissum and flanks (not sides) black, narrowly barred with white. Wing 94–113, tail 38–54, culmen 36–44, tarsus 31–39 mm. Fresh-water marshes from southern Canada south to North Carolina, Alabama, Oklahoma, New Mexico, Arizona, and California and in parts of Mexico. Winters from southern border of breeding range south to Florida, Gulf states, Mexico, and Guatemala; occurs on both fresh-water and salt marshes in winter. Also breeds in South America, in the Andes and Patagonia.

Genus *Porzana* **Vieillot.** Bill compressed, shorter than head, 55–80% length of tarsus; length of outer primary between 5th and 8th; tail less than half length of wing, exceeding coverts; tarsus slightly shorter than middle toe without claw; hallux elevated, less than half length of middle toe. Cosmopolitan. 28 species, 3 in the United States.

KEY TO SPECIES OF *PORZANA*

1. Secondaries white   *P. noveboracensis*
   Secondaries without white   2
2. Wing 98 mm or more   *P. carolina*
   Wing 77 mm or less   *P. jamaicensis*

Porzana carolina (**Linnaeus**). Sora. Above olive brown; midline of crown black; hind neck streaked with black; back and tertials streaked

with black and white; wing coverts spotted with black and white; outer primary and alula edged with white; lores, chin, and line down throat black; superciliary, sides of throat, and breast gray; lower breast faintly barred with white; sides olive brown barred with black and white; belly and crissum white; vent buffy, indistinctly barred with black; axillars barred dusky and white; iris red; bill greenish yellow; feet green. YOUNG: Above more buffy brown; black markings of head reduced or absent; throat whitish; breast plain buff. Wing 98–116, tail 38–54, culmen 17–24, tarsus 27–36 mm. Breeds from Canada south to Baja California, Arizona, New Mexico, Colorado, Kansas, Missouri, Illinois, Indiana, Ohio, Pennsylvania, and Maryland. Winters from southern United States south through Mexico, Central America, West Indies, and northern South America.

Porzana jamaicensis (**Gmelin**).[1] Black rail. Head slate gray; hind neck dark brown; back dark brown speckled with white; wings and tail fuscous, brokenly barred with white; underparts slate gray; flanks, belly, thighs, crissum, and axillars barred with white; iris red; bill black; feet green. Wing 63–77, tail 25–35, culmen 12–16, tarsus 18–25 mm. Eastern Canada, Utah, and Oregon, south to Baja California, Kansas, Illinois, and Florida; also breeds in Greater Antilles; winters from Gulf states and California to Guatemala. Local and secretive.

Porzana noveboracensis (**Gmelin**).[2] Yellow rail. Crown blackish barred with white; back buff, barred with white and streaked with black; primaries olive brown; secondaries mostly white; tail black, barred with white and edged with brown; lores and cheeks tawny olive, the lower portion speckled with black and white; throat buffy; breast and sides of neck dark brown scalloped with white; crissum brown; axillars white; iris brown; bill yellow; feet pale olive. Wing 73–93, tail 28–39, culmen 11–15, tarsus 20–27 mm. Canada, south to California, North Dakota, Illinois, Ohio, and Massachusetts. Winters from North Carolina to Florida, Gulf states, Arizona, California, and Oregon. Local and secretive.

Genus *Gallinula* **Brisson.** Bill equal to length of head, shorter than tarsus; a small squarish frontal shield; outer primary longer than 7th; tail

[1] *Laterallus jamaicensis* of A.O.U.
[2] *Coturnicops noveboracensis* of A.O.U.

rounded, rectrices 12, firm; tarsus shorter than middle toe without claw, about equal to outer toe without claw; hallux large, about ⅓ length of middle toe, its claw shorter than middle claw. Cosmopolitan. 3 species, 1 in the United States.

Gallinula chloropus (Linnaeus). Common gallinule. Slate; back and upper tail coverts olive brown; edge of wing white; flanks striped with white; belly and under wing coverts mixed with white; crissum white, with a large black patch in center; iris reddish; frontal shield and bill scarlet; tip of bill light green; legs light green, with a band of scarlet on tibiae. Wing 152–181, tail 62–86, culmen 36–47, tarsus 45–57 mm. All continents except Australia. In North America fresh-water swamps, north to California, South Dakota, the Great Lakes, and southeastern Canada.

Genus *Porphyrula* Blyth. Bill as long as head, less than half length of tarsus; frontal shield round, covering most of crown; outer primary longer than 6th; tail stiff, rectrices 10; tarsus about as long as middle toe without claw, longer than lateral toes without claws; hallux large, more than one-third length of tarsus, its claw longer than middle claw. Warm parts of Africa and America. 3 species, 1 in the United States.

Porphyrula martinica (Linnaeus). Purple gallinule. Head and underparts purple; crissum white; exposed part of wings light blue; back and tail olive green; frontal shield light blue; base of bill scarlet; tip of bill and legs greenish yellow. YOUNG: Crown dark brown; back brownish olive; wing coverts tipped with buff; underparts buff or buffy brown; crissum white; bill entirely yellowish. Wing 161–184, tail 60–78, culmen 28–33, tarsus 52–65 mm. Fresh-water swamps; South Carolina and Gulf states, West Indies, Mexico, Central and tropical South America.

Genus *Fulica* Linnaeus. Bill slightly shorter than head, more than half length of tarsus; frontal shield small, pointed; outer primary longer than 6th, shorter than tertials; tail hidden by coverts; tarsus shorter than middle or outer toes, longer than inner toe without claw; toes strongly lobed; hallux nearly ⅓ length of middle toe without claw. Swimmers and divers. Cosmopolitan. 10 species, 1 in the United States.

Fulica americana Gmelin. American coot. Head and neck black; crissum white, black medially; rest

of plumage slate, paler below and mixed with white on belly; edge of wing and tip of secondaries white; shield rich brown; bill white, with a reddish-brown subterminal band; iris red; feet green. Wing 171–202, tail 41–61, culmen 30–34, tarsus 44–61 mm. North America, West Indies, Central and northern South America, on fresh water.

Order Charadriiformes. Shore-birds, gulls, and auks. Head feathered; nostrils usually perforate. Wings more or less pointed; primaries 11 (10 functional); secondaries 11, aquintocubital, shorter than primaries. Tail usually short; rectrices 10–28. Hallux small or absent (long in Jacanidae). Plumage dense; aftershafts present; oil gland tufted.

Schizognathous, with basipterygoid processes (except in Burhinidae, Dromadidae, and adult Alcidae), vomer present; supraorbital glands present, often large; schizorhinal (except Burhinidae, Glareolidae, Thinocoridae); lacrimals fused to skull. Cervicals 15–16; dorsal vertebrae opisthocoelous. Sternum with small external spine, no internal spine; metasternum with 1–2 pairs of notches or 1 pair of notches and 1 pair of fenestrae. Procoracoid and hyposternal processes prominent. Furculum U-shaped. Humerus with large bladelike ectepicondylar process (small in some Alcidae). Hypotarsus complex.

Thigh muscles $A(B)X(Y)\pm$; toe tendons type 1. Intestine type 1; caeca present. Both carotids present.

Water birds. Often colonial. Nest usually on ground; eggs 1–5, large, pyriform, nearly always buffy, blotched or spotted with brown; young downy, marked with distinctive pattern of stripes or spots; more or less nidifugous.

Late Cretaceous onward. Cosmopolitan. In the United States 10 families, 97 species.

KEY TO FAMILIES OF CHARADRIIFORMES

1. Mandible longer than maxilla; bill bladelike (Fig. 5-41)   **Rynchopidae**
   Mandible as short as or shorter than maxilla   **2**
2. Anterior toes fully webbed   **3**
   Anterior toes with only basal webs or none   **5**
3. Hallux absent; tail shorter than tarsus and middle toe together   **Alcidae**
   Hallux usually present; tail longer than tarsus and middle toe together   **4**

4. Sheath of maxilla in 3 parts   **Stercorariidae**
   Sheath of maxilla entire   **Laridae**
5. Middle toe with claw less than half length of tarsus and less than length of bare portion of tibia
          Recurvirostridae
   Middle toe with claw more than half length of tarsus and more than length of bare portion of tibia   **6**
6. Hind claw longer than its toe   **Jacanidae**
   Hind claw shorter than its toe   **7**
7. Toes lobed   **Phalaropodidae**
   Toes not lobed   **8**
8. Front of tarsus scutellate; hallux usually present
          Scolopacidae
   Front of tarsus reticulate; hallux usually absent   **9**
9. Bill much longer than tarsus   **Haematopodidae**
   Bill shorter than tarsus   **Charadriidae**

Fig. 5-41. **Bill of black skimmer (*Rynchops nigra*).**

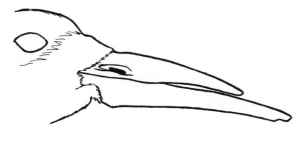

Suborder Charadrii. Shore-birds. Legs inserted near middle of body. Anterior toes not fully webbed. Adult downs on pterylae only. Middle of metasternum not rounded. Furculum without hypocleidium. Coracoid with furcular facet broad and much hollowed under brachial tuberosity, the procoracoid process bent upward. Ectepicondylar process of humerus large. Thumb present. Hypotarsus complex. Thigh muscles $A(B)XY+$. Intestine with less than 6 loops; caeca present, usually large. Nidifugous. 14 families, 6 in the United States.

Family Jacanidae. Jacanas. Bill ploverlike (swollen subterminally); wing slightly rounded, outer primary longer than 6th; a spur often present on wrist; legs long; toes very long, equal to tarsus; nails long and straight, that of hallux longer than toe; rectrices 10.

Without occipital fontanels or supraorbital grooves; cervicals 16; true ribs 5, free ribs 3; sternum with 1 pair of notches. Thigh muscles $ABXY+$; caeca nipplelike.

Swamp birds, walking around on lily pads. Food insects. The carpal spur is used for fighting. Nest a pile of damp vegetation; eggs 3–5, brown scrawled with black; young downy, precocial, running, swimming, and diving as soon as hatched.

Miocene onward. Tropics of world. 6 genera; 7 species; 1 species in the United States.

Genus *Jacana* Brisson. Bill longer than head; frontal lappet scalloped and free behind; metacarpal spur sharp; tail short. Tropical America. 1 species.

Jacana spinosa (**Linnaeus**). American jacana. Remiges pale greenish yellow, tipped and externally edged with dusky; head, neck, and breast black; rest of plumage maroon, darker below; iris brown; base of maxilla red; rest of bill and lappet yellow; legs greenish. YOUNG: Grayish brown above, buffy white below; remiges as in adult; lappet smaller. Wing 115–143, tail 41–48, culmen 26–34, tarsus 52–64 mm. Lower Rio Grande Valley of Texas, south through the mainland and West Indies to western Ecuador and Argentina.

Family Haematopodidae. Oystercatchers. Bill long, straight, much compressed laterally, deeper at middle than subbasally; outer primary longest; rectrices 12; tarsus stout, reticulate, much shorter than bill; toes with basal webs; hallux absent; middle toe little more than half length of tarsus; claws short.

Occipital fontanels and supraorbital grooves present; cervicals 15; true ribs 6, free ribs 2; sternum with 2 pairs of notches. Thigh muscles $ABXY+$. Caeca functional. Food mollusks.

Miocene onward. Seacoasts of world, except Polynesia. 1 living genus.

Genus *Haematopus* Linnaeus. Tibial bridge unossified. 4 species, 2 in the United States.

KEY TO SPECIES OF *HAEMATOPUS*

1. Plumage entirely sooty   *H. bachmani*
   Underparts largely white   *H. palliatus*

Haematopus palliatus **Temminck.** American oystercatcher. Head, neck, and upper breast black; back, scapulars, lesser and middle coverts grayish brown; tail white with a broad dusky tip; primaries dusky with a blaze of white; greater coverts, axillars, under wing coverts, and large part of secondaries white; breast, belly, upper and under tail coverts white, sometimes blotched with black; iris yellow; eyelids and bill red; legs pinkish. YOUNG: Crown speckled with pale brown; upper parts edged with buff; iris and bill brownish. Wing 241–275, tail 89–113, culmen 70–104, tarsus 48–60 mm. Both coasts of America, north to Virginia and (formerly) southern California; occasionally north to New Brunswick.

Haematopus bachmani Audubon. Black oystercatcher. Sooty; iris yellow; eyelids and bill vermilion; feet fleshy white. Wing 241–268, tail 92–107, culmen 64–83, tarsus 45–53 mm. Pacific Coast, from Alaska to Baja California.

Family Charadriidae.   Plovers. Bill hard, shorter than head, contracted at middle, swollen terminally; outer primary longest; tarsus usually reticulate, longer than bill; toes with no more than basal webs; hallux usually absent.

Occipital fontanels present; supraorbital grooves large, perforated; cervicals 15; true ribs 6, free ribs 2; sternum with 2 pairs of notches. Thigh muscles *ABXY* +. Caeca functional.

Some extralimital species have the head crested or with a fleshy lappet. Others have a metacarpal spur, and 1 species has the bill twisted to the right.

Oligocene onward. Cosmopolitan. 29 genera, 61 species; 3 genera, 8 species in the United States.

KEY TO GENERA OF CHARADRIIDAE

1. Hind toe minute but present; axillars black
   Squatarola
   Hind toe absent; axillars gray or white   **2**
2. Bare part of tibia longer than middle toe without claw; axillars gray   *Pluvialis*
   Bare part of tibia shorter than middle toe without claw; axillars white   *Charadrius*

Genus *Squatarola* Cuvier.   Head uncrested, without lappets; culmen longer than bare part of tibia, shorter than middle toe without claw; wing unspurred; tarsus reticulate, longer than middle toe with claw; hallux minute but with claw. Nearly cosmopolitan. 1 species.

Squatarola squatarola (Linnaeus).   Black-bellied plover. Forehead and long superciliary stripe white; upper parts irregularly barred grayish, black-

ish, and white; tail, upper and under tail coverts white barred with blackish; remiges dusky with a white blaze on each feather; under wing coverts and lower belly white; remaining underparts, including axillars, black; bill and feet black; iris dark brown. WINTER: Upper parts brownish gray margined with white; underparts white, with a few brownish-gray streaks on throat; tail, tail coverts, and axillars as in summer. Wing 178–199, tail 68–84, culmen 28–31, tarsus 42–51 mm. Breeds on Holarctic tundra; winters on coast from Carolinas, Gulf and Pacific states, south to South America, South Africa, India, and Australia. On migration occurs on inland bodies of water.

Genus *Pluvialis* Brisson.   Exposed culmen shorter than bare part of tibia; hallux absent. Cosmopolitan. 2 species, 1 in the United States.

Pluvialis dominica (Müller).   American golden plover. Forehead, long superciliaries continuing down sides of neck and breast white; crissum white with a few dusky streaks; wing lining and axillars grayish brown; flanks barred black and white; rest of underparts black; upper parts blackish, heavily spotted with golden; remiges dusky, without white blaze, but with shafts white subterminally; tail barred dusky and grayish; iris brown; bill and feet black. WINTER: Speckling of upper parts grayish, golden only on rump; underparts pale grayish brown, with some dusky mottling on breast. Wing 147–187, tail 58–75, culmen 21–25, tarsus 39–46 mm. Breeds on American and eastern Asiatic tundra. Winters in the Pacific, from Hawaii to China, Australia, and New Zealand; also winters on plains of southern South America, migrating southward in fall mainly over the ocean south of New England, and northward in spring through the Mississippi Valley; also occurs on migration on the coast of the Pacific states.

Genus *Charadrius* Linnaeus.   Resembles *Squatarola* but lacks hind toe; differs from *Pluvialis* in having culmen longer than bare part of tibia. Cosmopolitan. 24 species, 6 in the United States.

KEY TO SPECIES OF *CHARADRIUS*

1. Tarsus twice length of middle toe without claw
   C. montanus
   Tarsus less than twice length of middle toe without claw   **2**

2. Rump rufous; tail more than half length of wing
   C. vociferus
   Rump not rufous; tail less than half length of wing
   3
3. Culmen 19–22 mm; outstretched legs extending beyond tail *C. wilsonius*
   Culmen 11–16 mm; outstretched legs not reaching end of tail 4
4. A dark stripe on lores *C. semipalmatus*
   Lores entirely white 5
5. Legs orange *C. melodus*
   Legs gray *C. alexandrinus*

Charadrius montanus Townsend.[1] Mountain plover. Forehead and superciliary white; lores and front of crown black; upper parts pale grayish brown, margined with buff; tail with whitish tip and dusky subterminal band; greater coverts tipped with white; primaries dusky, white basally and on shafts; below white, washed with buff on breast; iris brown; bill black; feet brownish yellow. WINTER: Crown and lores like back. Wing 138–155, tail 57–69, culmen 19–22, tarsus 37–40 mm. Montana and western Nebraska to New Mexico and Texas. Winters from California and Texas to northern Mexico.

Charadrius vociferus Linnaeus. Killdeer. Tail more than half length of wing. Forehead white, continuing as a line to eye; forepart of crown black, reaching eyes; postocular stripe white, becoming reddish ochraceous posteriorly and bordered below by a black line; lores, auriculars, top of head, back, and lesser coverts grayish brown; rump and upper tail coverts reddish ochraceous; tail grayish brown with a subterminal black band, all but the middle pair of rectrices tipped with white or ochraceous and washed with ochraceous subbasally; wing band white, formed by tips of greater coverts and inner secondaries; remiges dusky externally, white internally; upper throat white, continuing as a white collar around hind neck; lower throat black, forming a ring around neck, broad ventrally; a black half collar on upper breast, separated from black neck ring by a patch of buff; remaining underparts, including axillars and crissum, white; iris brown; eye ring red; bill black; feet yellowish gray. Wing 147–170, tail 88–103, culmen 19–23, tarsus 32–37 mm. North America, West Indies, and northern South America.

[1] *Eupoda montana* of A.O.U.

Charadrius semipalmatus Bonaparte. Semipalmated plover. Forehead from eye to eye white, continuing as a narrow superciliary; band on forecrown black; throat white, continuing as a ring around hind neck; upper breast black, continuing as a ring around base of neck; remaining underparts white; lores, auriculars, crown, and upper parts grayish brown; narrow wing bar white; remiges dusky; outer rectrices white; rest of tail grayish brown, tipped with white, with a black subterminal bar; iris brown; bill black, yellow basally; feet yellow. WINTER: Black of head and breast replaced by grayish brown. Wing 114–138, tail 52–67, culmen 11–16, tarsus 21–24 mm. Holarctic; in America breeds through Canada and Alaska; winters from southern United States to South America; also to Africa and India.

Charadrius melodus Ord. Piping plover. Forehead, lores, nape, upper tail coverts, wing bar, and underparts white; a black bar across front of crown; a black bar on breast, usually interrupted in middle and with traces of black around base of hind neck; remiges dusky; remaining upper parts pale sandy; tail with subterminal black band and white tip; outer rectrices white; iris brown; bill black, orange basally; feet orange. WINTER: Lacks black crown band; breast ring brownish gray; bill entirely dark. Wing 112–124, tail 47–55, culmen 11–13, tarsus 21–23 mm. Breeds from Canada east of the Rockies south to Nebraska, Great Lakes, and on coast to Virginia (formerly North Carolina). Winters on Atlantic and Gulf Coasts from South Carolina to Mexico.

Charadrius alexandrinus Linnaeus. Snowy plover. Forehead, lores, anterior cheeks, wing bar, and underparts white; a black patch on each side of breast; front of crown and ear patch black; upper parts pale sandy; middle rectrices brownish gray, dusky at tip; outer rectrices white; iris brown; bill black; feet dusky. WINTER: Black head and breast markings replaced by grayish brown. Wing 98–108, tail 37–47, culmen 13–16, tarsus 22–25 mm. All continents. In America along Pacific Coast from Washington to Chile; salt plains of Utah, Kansas, and Oklahoma; Gulf Coast from Texas to Florida and West Indies.

Charadrius wilsonius Ord. Wilson's plover. Bill heavy, longer than toes; outstretched feet extending beyond tail. Forehead white; lores and band across front of crown black; postorbital stripe buffy white; auriculars and upper parts grayish brown, tinged with cinnamon on nape; wing bar white; tip of wings and tail dusky; lateral rectrices plain white; broad breast band black; rest of underparts white; iris brown; bill black; feet flesh. FEMALE: Crown and lores like back; breast band pale brown tinged with cinnamon. WINTER: Like female but breast band gray. Wing 106–121, tail 42–50, culmen 19–22, tarsus 27–31 mm. Atlantic and Gulf Coasts from Virginia to Texas, West Indies, and South America; Pacific Coast from Mexico to South America.

Family Scolopacidae. Sandpipers. Bill usually long, its tip often soft and pitted, and well supplied with nerves; outer primary usually longest; tail short, rectrices usually 12. Bare part of tibia shorter than middle toe with claw; tarsus less than twice length of middle toe with claw; toes with no more than basal webs; hallux usually present.

Occipital fontanels present; supraorbital grooves reduced or absent, imperforate; cervicals 15, true ribs 6, free ribs 2; sternum usually 4-notched (2-. Thigh muscles $ABXY+$ ($AXY+$ in *Arenaria*). Caeca functional.

Paleocene onward. Cosmopolitan. 26 genera, 77 species; in the United States 19 genera, 32 species.

KEY TO GENERA OF SCOLOPACIDAE

1. Tail emarginate; bill ploverlike   *Aphriza*
   Tail not emarginate; bill not ploverlike   2
2. Bill shorter than head, depressed in middle, with hard, pointed, ascending tip   *Arenaria*
   Bill about as long as head or longer, straight or decurved·   3
3. Hind toe absent   *Crocethia*
   Hind toe present   4
4. Thigh entirely feathered; 3 outer primaries narrow   *Philohela*
   Lower thigh naked; all primaries about same width   5
5. Ears located directly below eye   *Capella*
   Ears located below and behind eye   6
6. Tail about half length of wing   7
   Tail much less than half length of wing   8
7. Tarsus much longer than culmen   *Bartramia*
   Tarsus about equal to culmen or shorter   *Actitis*
8. Tarsus reticulate behind (bill curved down)   *Numenius*
   Tarsus scutellate before and behind   9

9. Distal third of both mandibles very rough and pitted   *Limnodromus*
   Tip of bill usually smooth; if rough, this area not occupying distal third of bill   10
10. Culmen much longer than tarsus   11
    Culmen about equal to tarsus   *Micropalama*
11. Tip of bill curved slightly upward; culmen 67–119 mm   *Limosa*
    Bill not recurved, shorter than 66 mm   12
12. Feathers of lores reaching nostrils   *Tryngites*
    Feathering not reaching nostrils   13
13. Outstretched feet not extending beyond tip of tail   *Calidris*
    Outstretched feet extending beyond tip of tail   14
14. Axillars black   *Catoptrophorus*
    Axillars white or barred   15
15. Tarsus decidedly more than half length of tail; tail barred black and white   16
    Tarsus about half length of tail or less; tail not barred   17
16. Tarsus yellow, much longer than exposed culmen   *Totanus*
    Tarsus green, about equal to exposed culmen   *Tringa*
17. Tail 71–80 mm, slightly rounded   *Heteroscelus*
    Tail under 67 mm, with middle rectrices projecting beyond the others   18
18. A small web between bases of anterior toes   *Ereunetes*
    Toes unwebbed   *Erolia*

Genus *Aphriza* Audubon.[1] Bill short, ploverlike; tail truncate; tarsus longer than culmen, scutellate in front, reticulate behind; toes fringed; hallux present. Pacific Coast of America. Monotypic.

Aphriza virgata (Gmelin). Surf-bird. Head, neck, and upper parts streaked black and white; back blotched with cinnamon rufous; wing coverts brownish gray; remiges gray, secondaries with broad white tips; upper tail coverts white; tail white, with a broad dusky subterminal band; underparts whitish, heavily marked with black crescents; axillars white; iris brown; bill black, orange basally; legs green. WINTER: Crown, hind neck, and upper parts brownish gray with dusky streaks; throat white with dusky flecks; breast and belly white, with dusky spots and streaks. Wing 164–183, tail 63–69, culmen 23–26, tarsus 29–31 mm. Breeds in mountains of Alaska. Winters on Pacific Coast, from southern Alaska to Straits of Magellan.

Genus *Arenaria* Brisson.[1] Bill short, cuneate,

[1] *Aphriza* and *Arenaria* included in Charadriidae by A.O.U.

compressed, with culmen concave and gonys ascending; tail slightly rounded; tarsus slightly longer than culmen, scutellate before and behind; toes fringed; hallux present. Cosmopolitan. 2 species.

KEY TO SPECIES OF *ARENARIA*

1. Chin and throat white   *A. interpres*
   Chin and throat sooty   *A. melanocephala*

Arenaria interpres (**Linnaeus**). Ruddy turnstone. Forehead white, with a more or less complete black line above lores; crown white or brownish gray streaked with black; mantle blackish, pied with rufous; rump white; shorter upper tail coverts black, distal ones white; tail white with broad black subterminal band, tinged with rufous; wing dusky with a broad white bar; lores, cheeks, and throat white; a large black patch below eye, confluent with black of breast and sides of neck; belly, crissum, and axillars white; iris brown; bill black; feet orange. WINTER: Without rufous; black areas mixed with white. Wing 140–160, tail 57–68, culmen 21–25, tarsus 23–26 mm. Cosmopolitan. Breeds in Arctic; migrates mainly along coast; winters from California and North Carolina southward.

Arenaria melanocephala (**Vigors**). Black turnstone. Head, neck, mantle, and breast sooty; lores white; forehead, superciliary, and auriculars sooty streaked with white; sides spotted with white; wing band, rump, longer upper tail coverts, and posterior underparts white; tail white, with broad black subterminal band. Wing 138–153, tail 58–66, culmen 21–24, tarsus 23–26 mm. Pacific Coast; breeds in Alaska and winters south to Baja California.

Genus *Totanus* Bechstein. Bill long, straight, unpitted; thighs largely bare; tarsus longer than culmen; tail barred. Cosmopolitan, breeding in Holarctic. 7 species, 2 in the United States.

KEY TO SPECIES OF *TOTANUS*

1. Wing 149–163 mm, culmen 30–39 mm   *T. flavipes*
   Wing 180–198 mm, culmen 52–61 mm

                    T. melanoleucus

Totanus melanoleucus (**Gmelin**). Greater yellowlegs. Above brownish gray; head and neck streaked with white; back and wings spotted or barred with white, and in summer blotched with black; upper tail coverts white, with a few dusky bars; middle rectrices gray barred with blackish, outer ones barred black and white; primaries black; underparts white; throat and breast with dusky streaks; axillars, sides, and sometimes crissum with dusky bars; iris brown; bill blackish; feet yellow. Wing 180–198, tail 71–83, culmen 52–61, tarsus 55–68 mm. Breeds in Alaska and Canada; winters from southern United States to Patagonia; frequents shallow ponds and mud flats on both fresh and salt water.

Totanus flavipes (**Gmelin**). Lesser yellowlegs. Plumage identical with *T. melanoleucus*. Wing 149–163, tail 55–67, culmen 30–39, tarsus 46–55 mm. Breeds in Alaska and Canada; winters mainly in southern South America; migrates mainly east of the Rockies.

Genus *Tringa* Linnaeus. Differs from *Totanus* in having tarsus and exposed culmen equal; metasternum 2-notched. Holarctic, wintering in tropics. 2 species, 1 in the United States.

Tringa solitaria **Wilson.** Solitary sandpiper. Entire upper parts, including middle rectrices, olive brown; crown streaked, back dotted with white; lateral rectrices barred black and white; primaries blackish, sometimes mottled with white on inner web; below white; breast and sides of neck with dusky streaks; axillars barred black and white; iris brown; bill black; feet green. Wing 122–142, tail 50–59, culmen 27–32, tarsus 27–33 mm. Alaska and Canada to Washington and Colorado; migrates throughout United States on fresh water; winters in tropical America.

Genus *Heteroscelus* Baird. Bill somewhat longer than tarsus, straight, unpitted; bare part of tibia less than half length of middle toe; tarsus partly reticulate; tail unbarred. Both shores of the Pacific. 2 species, 1 in the United States.

Heteroscelus incanus (**Gmelin**). Wandering tattler. Above brownish gray, darker on wings; superciliary whitish; cheeks grayish, with dusky streaks; below white; foreneck streaked and body barred with dusky; axillars barred dusky and white; iris brown; bill horn; feet yellow. WINTER: Back often spotted with white; underparts plain white, tinged with gray on throat. Wing 161–180, tail 71–80, culmen 34–42, tarsus 32–34 mm. Breeds in

Alaska; winters from Pacific Mexico to Galapagos, and in southwest Pacific.

Genus *Actitis* Illiger. Bill equal to tarsus, straight, unpitted; tail half length of wing, rounded; bare tibia about half length of middle toe; tarsus, scutellate, not much longer than toes. Practically cosmopolitan, breeding in Holarctic. 2 species; 1 in the United States.

Actitis macularia (**Linnaeus**). Spotted sandpiper. Above brownish olive; pileum streaked, back and middle rectrices with dusky bars; lateral rectrices barred black and white; superciliaries white; below white spotted with black; axillars white; iris brown; bill black above, yellow below; feet pale. WINTER: Black barring of upper parts confined to scapulars and wing coverts; below plain white. Wing 89–109, tail 45–53, culmen 22–25, tarsus 20–25 mm. North America, breeding south to South Carolina, Alabama, Louisiana, and the Mexican border states; winters south to Brazil.

Genus *Catoptrophorus* Bonaparte. Bill long, straight, unpitted, bare tibia nearly as long as middle toe; tarsus equal to or longer than culmen; toes semipalmate. North America (South America in winter). Monotypic.

Catoptrophorus semipalmatus (**Gmelin**). Willet. Above gray; crown and hind neck streaked with dusky; back barred with black; a large white wing patch, formed by secondaries and bases of primaries; primary coverts and ends of primaries black; upper tail coverts white; tail gray mottled with dusky; underparts white; foreneck and upper breast spotted and sides barred with dusky; axillars and wing lining black; iris brown; bill black; feet gray. WINTER: Lacks black barring and streaks; sides of neck pale gray; back mottled with buff in young. Wing 175–220, tail 67–88, culmen 53–65, tarsus 52–70 mm. Breeds on Atlantic and Gulf Coasts from New Jersey south to Bahamas and Texas; also in Great Plains and Basin, from Minnesota and Iowa to Oregon and California. Winters on salt water from California, Gulf states, and Florida south to Brazil.

Genus *Bartramia* Lesson. Bill about as long as head or middle toe, straight, unpitted; tail half length of wing, graduated; bare part of tibia nearly as long as middle toe; tarsus much longer than culmen, scutellate; toes short. North America (South America in winter). Monotypic.

Bartramia longicuada (**Bechstein**). Upland plover. Pileum and nape streaked black and buff; wing coverts, tertials, and upper back buff, barred with dusky and brown; lower back and rump black; longer upper tail coverts buff barred with black; middle rectrices gray, barred with black and edged with buff; outer rectrices cinnamon buff, barred with black and tipped with white; primaries and their coverts black, the 2 outer primaries barred black and white on inner webs; throat white; lores buffy; auriculars and lower throat buffy with dusky streaks; breast buffy with dusky squamations; sides, axillars, and under wing coverts barred black and white; abdomen whitish; crissum buff, sometimes with a few dusky bars; iris brown; bill blackish above, yellowish below; feet yellowish gray. Wing 157–181, tail 79–92, culmen 26–32, tarsus 44–50 mm. Dry grasslands of North America, south to Virginia, Ohio River, Missouri, Oklahoma, Colorado, Utah, and Oregon. Migrates east of Rockies, through West Indies, Mexico, and Central America. Winters on pampas of South America.

Genus *Numenius* Brisson. Bill longer than tarsus, decurved, unpitted; tail less than half length of wing, rounded; bare part of tibia about length of middle toe; tarsus scutellate in front, reticulate behind; toes short. Cosmopolitan, breeding in Holarctic. 8 species, 3 in the United States.

KEY TO SPECIES OF *NUMENIUS*

1. Axillars cinnamon, mostly unbarred; culmen over 100 mm .... *N. americanus*
   Axillars cinnamon, heavily barred with brown; culmen under 100 mm 2
2. Primaries barred .... *N. phaeopus*
   Primaries unbarred .... *N. borealis*

Numenius americanus Bechstein. Long-billed curlew. Crown, nape, sides of head, lower throat, and breast buffy streaked with dusky; back and wing coverts cinnamon buff, with dusky streaks and bars; remiges and rectrices cinnamon buff with dusky bars; primary coverts and outer webs of outer primaries plain dusky; upper throat and anterior part of superciliary white; posterior underparts cinnamon buff, with some dusky bars on sides; axillars and wing lining mostly plain cinnamon, with a few dusky bars; iris brown; bill dusky, flesh at base below; feet gray. Wing 252–308, tail 96–135, culmen 106–219, tarsus 70–93 mm. Breeds in Great Basin and and Plains from southwest Canada to Oregon, Idaho, Nevada, Utah, Wyoming, and South Dakota (formerly to Illinois); winters from California, Arizona, and New Mexico to Guatemala; rare east of Mississippi on migration.

Numenius phaeopus (**Linnaeus**). Hudsonian curlew; whimbrel. Crown sooty brown with a central stripe of buff; superciliary pale buff; loral stripe sooty brown; throat white; cheeks, neck all around, and breast buff streaked with brown; upper parts sooty brown mottled with grayish buff; tail and remiges grayish brown barred with dusky; sides, axillars, and wing lining pale cinnamon buff, heavily barred with brown; belly pale buff; crissum buff with a few dusky bars; iris brown; bill blackish, flesh basally below; feet gray. Wing 231–267, tail 88–102, culmen 77–95, tarsus 52–61 mm. Breeds in arctic Europe, Asia, and America; winters coastwise through Southern Hemisphere. In America migrates along both coasts, more rarely in interior in spring.

Numenius borealis (**Forster**). Eskimo curlew. General coloration as in *N. phaeopus;* primaries and secondaries unbarred but with narrow whitish margins; crown sooty brown streaked with buff, more heavily on midline; postocular stripe sooty brown. Wing 190–215, tail 76–83, culmen 47–60, tarsus 40–45 mm. Extinct or practically so; last specimen collected in 1925, with a few sight records since. Bred on tundra of Alaska and Mackenzie; migrated in fall along Atlantic Coast from Labrador to New Jersey, thence over water to South America; wintered on pampas of Argentina; northward spring migration passed through Mississippi Valley.

Genus *Limosa* Brisson. Bill long, slightly upturned, unpitted; legs long; bare part of tibia nearly as long as or longer than middle toe; tarsus scutellate, shorter than bill. Holarctic, migrating to Southern Hemisphere. 4 species, 2 in the United States.

KEY TO SPECIES OF *LIMOSA*

1. Axillars sooty brown; tail black with white base and tip   *L. haemastica*
   Axillars cinnamon; tail barred   *L. fedoa*

Limosa haemastica (**Linnaeus**). Hudsonian godwit. Head and neck buffy with dusky streaks; upper throat and supraloral stripe buffy; upper back and tertials sooty brown, spotted and barred with cinnamon buff; wing coverts grayish brown, greater coverts tipped with white; remiges dusky, inner primaries white at base; lower back, rump, and distal upper tail coverts black, broken by a white patch on proximal tail coverts; tail black tipped with white, and with white bases to lateral rectrices; below russet with dusky bars, crissum mixed russet and whitish, with dusky bars; axillars sooty brown; iris brown; bill dull yellowish; feet gray. WINTER: Upper back brownish gray, margined with buff; below whitish buff. Wing 196–222, tail 72–82, culmen 67–92, tarsus 54–62 mm. Breeds in Arctic and winters in southern South America; migrates mainly along Atlantic Coast in fall and through Mississippi Valley in spring.

Limosa fedoa (**Linnaeus**). Marbled godwit. Head and neck buffy with dusky streaks; throat buffy white; upper back and tertials dusky barred with buff; lesser coverts grayish brown edged with buff; middle and greater coverts and secondaries cinnamon buff marbled with black; primaries and coverts blackish flecked with cinnamon; rump, upper tail coverts, tail, axillars, and underparts cinnamon with dusky bars; iris brown; bill dusky, flesh at base; feet gray. Wing 212–234, tail 78–95, culmen 89–126, tarsus 67–76 mm. Breeds on plains of Canada and Dakotas (formerly to Utah and Wisconsin). Winters from California, Gulf states, and Georgia to northern South America.

Genus *Tryngites* Cabanis. Bill shorter than head or middle toe, grooved nearly to tip, unpitted; tail rounded with middle rectrices slightly projecting; bare tibia shorter than middle toe; tarsus much longer than culmen, scutellate; toes unwebbed. North America, wintering in South America. Monotypic.

Tryngites subruficollis (**Vieillot**). Buff-breasted sandpiper. Above grayish buff; pileum streaked with black; upper back, scapulars, and tertials blackish edged with buff; wing coverts grayish buff spotted with black; greater coverts edged with white; primaries brown tipped with white and at base of inner webs and with black subterminal spot; secondaries brown, white on most of inner webs where mottled with black and pale buff;

rump and upper tail coverts buff blotched with black; tail grayish brown, paler externally where lined with black, tip white; underparts and sides of head buffy tipped with white; axillars and wing lining white; iris brown; bill dusky; feet yellowish. Wing 122–136, tail 54–63, culmen 18–20, tarsus 29–37 mm. Breeds in arctic America; winters in Argentina; migrates through Mississippi Valley.

Genus *Micropalama* Baird. Bill about as long as tarsus, slightly decurved, grooved nearly to tip, tip slightly pitted; tail slightly double-emarginate, with middle rectrices not projecting; bare part of tibia about as long as toes; tarsus long, scutellate; toes with basal webs. North America, South America in winter. Monotypic.

Micropalama himantopus (**Bonaparte**). Stilt sandpiper. Head and neck streaked dusky and whitish; auriculars cinnamon rufous; superciliary and throat white; back mixed black, gray, and buffy; wing coverts brownish gray edged with pale gray; secondaries gray edged with white; primaries dusky; rump brownish gray mottled with pale gray and dusky; upper tail coverts white, with dusky bars or streaks; tail brownish gray, white on inner webs where streaked with gray; underparts barred dusky and white; axillars white with some gray marks; iris brown; bill dusky; feet olive. WINTER: Back plain brownish gray; below white, streaked with gray on neck, breast, and crissum. Wing 116–137, tail 44–58, culmen 36–44, tarsus 36–45 mm. Breeds in arctic America; winters on pampas; migrates mainly through Great Plains, occasionally on Atlantic Coast.

Genus *Calidris* Merrem. Bill slightly longer than tarsus, grooved nearly to tip, unpitted; tail square; bare part of tibia shorter than toes; tarsus scutellate; toes fringed, not extending beyond tip of tail. Holarctic, Southern Hemisphere in winter. 2 species, 1 in the United States.

Calidris canutus (**Linnaeus**). Knot. Upper parts gray, streaked and spotted with black and tinged with rufous; rump gray with dusky bars; upper tail coverts white barred with black; wing coverts gray, streaked and edged with white; 2 white wing bars, formed by tips of primary coverts and greater coverts; remiges dusky edged with white; tail gray, shafts white; superciliary and underparts rufous, becoming white on flanks, belly, and crissum; sides with dusky bars, crissum with

a few dusky streaks; axillars and wing lining white with V-shaped dusky bars; iris brown; bill and feet black. WINTER: Crown, hind neck, and back gray, with faint dusky shaft streaks (with dusky squamations in young); underparts white; throat and superciliary with dusky streaks; breast and sides squamate with dusky. Wing 152–176, tail 55–66, culmen 31–38, tarsus 30–33 mm. Circumboreal; winters on coasts of Southern Hemisphere; migrates almost entirely along seacoasts.

Genus *Crocethia* Billberg. Bill about equal to head or tarsus, grooved nearly to tip, unpitted; middle rectrices projecting and acuminate; bare part of tibia about half length of middle toe; tarsus scutellate except upper rear portion; toes fringed; hallux absent. Circumboreal, Southern Hemisphere in winter. Monotypic.

Crocethia alba (**Pallas**). Sanderling. Top and sides of head, hind neck, and back mixed rusty and pale gray, streaked or blotched with black; wing coverts dusky; tip of greater coverts forming a white band; remiges dusky, white basally; tertials, rump, medial upper tail coverts, and tail dusky margined with whitish; lateral tail coverts and bases of lateral rectrices white; throat and posterior underparts white; lower throat, breast, and sides white and rusty, with dusky spots or bars; axillars and wing lining white; iris brown; bill and feet black. WINTER: Crown, hind neck, and back sandy gray with faint dusky shaft streaks; below pure white. Wing 113–127, tail 45–55, culmen 23–28, tarsus 23–26 mm. Circumboreal; winters on sandy beaches from southern United States, southern Europe, and Japan, south through Southern Hemisphere. Migrates mainly along seashore, but also on Great Lakes.

Genus *Ereunetes* Illiger. Bill equal to or longer than tarsus, straight, grooved nearly to tip where much pitted; 4 middle rectrices projecting; bare part of tibia shorter than middle toe; tarsus scutellate; toes shorter than tarsus relatively thick, with basal webs. Arctic America, to South America in winter. 2 species.

KEY TO SPECIES OF *EREUNETES*

1. Exposed culmen 17–20 mm in male, 18–22 mm in female, about equal to or shorter than middle toe with claw *E. pusillus*

Exposed culmen 20.5–23.5 mm in male, 23–28 mm in female, longer than middle toe with claw
E. mauri

Ereunetes pusillus (**Linnaeus**). Semipalmated sandpiper. Upper parts brownish gray, streaked and spotted with black; scapulars tinged with rust; tips of greater coverts white; middle of rump, median upper tail coverts, and tail sooty; sides of rump and lateral upper tail coverts white; remiges dusky with white shafts; superciliary whitish; lores brownish; below white; breast with dusky streaks; iris brown; bill and feet black. WINTER: Above without rusty; breast only faintly streaked. Wing 88–101, tail 38–44, culmen 17–20 (♀ 18–22), tarsus 19–22 mm. Breeds from northeastern Siberia through Arctic to Labrador; winters from South Atlantic and Gulf states south along both coasts of South America; on migration inland mainly east of Rockies, on mud flats.

Ereunetes mauri **Cabanis**. Western sandpiper. Differs from *E. pusillus* in longer bill, more heavily streaked breast (even in winter), and with much more rust on upper parts in summer. Wing 90–100, tail 38–47, culmen 20.5–23.5 (♀ 23–28), tarsus 20–24 mm. Breeds in Alaska; winters on both coasts from Washington and North Carolina to northern South America; on migration occurs inland mainly west of Rockies.

Genus *Erolia* Vieillot. Bill equal to or slightly longer than tarsus, straight or somewhat decurved, grooved nearly to tip, where pitted; 2 middle rectrices projecting, acuminate; bare part of tibia shorter than middle toe; tarsus scutellate; toes almost as long as tarsus, thin, unwebbed. Northern Holarctic, to Southern Hemisphere in winter. 13 species, 7 in the United States.

KEY TO SPECIES OF *EROLIA*

1. Upper tail coverts mostly plain white   *E. fuscicollis*
   Upper tail coverts largely dark   2
2. Smaller; wing 82–91 mm   *E. minutilla*
   Larger; wing 103–146 mm   3
3. Bill and feet brownish or yellowish   4
   Bill and feet black   5
4. Tarsus 21–23, shorter than middle toe   *E. maritima*
   Tarsus 24–30, longer than middle toe   *E. melanotos*
5. Culmen about equal to tarsus (tarsus longer than middle toe with claw)   *E. bairdii*

Culmen much longer than tarsus   6
6. Tarsus longer than middle toe with claw; bill curved down   *E. alpina*
   Tarsus shorter than middle toe with claw; bill straight   *E. ptilocnemis*

Erolia alpina (**Linnaeus**). Red-backed sandpiper; dunlin. Bill decurved. Crown and hind neck brown, with dusky streaks; back brown, blotched with cinnamon and black; wing coverts brown, edged with gray; greater coverts tipped with white; primary coverts and remiges dusky, the inner ones white basally; rump grayish brown; tail and median tail coverts brown; lateral tail coverts white; upper throat, axillars, and wing lining white; lower throat and breast streaked dusky and white; belly black, tipped with white; flanks and crissum white, with a few dusky streaks; iris brown; bill and feet black. WINTER: Above brownish gray, indistinctly streaked with darker; superciliary, upper throat, belly, and crissum white; lower throat and breast gray, streaked with brown; sides often with a few brown streaks; scapulars (in young) margined with cinnamon. Wing 103–125, tail 42–56, culmen 26–42, tarsus 21–28 mm. Holarctic, breeding mainly on tundra; winters on salt water south to India, Mediterranean, and in America from New Jersey to Florida and Gulf states, and on Pacific Coast from southern Canada to northern Mexico.

Erolia ptilocnemis (**Coues**). Rock sandpiper. Tibia almost completely feathered. Crown and hind neck fulvous streaked with black; back and scapulars rusty, spotted with black and tipped with white; wing coverts sooty tipped with white, forming a bar on greater coverts; remiges slate, shafts white; rump, upper tail coverts, and middle rectrices sooty; lateral rectrices gray; superciliary whitish; lores slate; auriculars sooty brown; throat and breast buffy white spotted with slate, forming a patch on sides of breast; belly white; sides and crissum white with dusky streaks; iris brown; bill and feet blackish. Wing 108–134, tail 46–65, culmen 23–37, tarsus 21–24 mm. Kuriles, Commanders, Aleutians, and Pribilofs; winters south along Pacific Coast to Washington and Oregon.

Erolia maritima (**Brünnich**). Purple sandpiper. Crown and hind neck streaked dusky and buff; back, wing coverts, rump, upper tail coverts, and middle rectrices sooty, glossed with purple, edged with buff and white; lateral rectrices brownish gray, shafts white; greater coverts tipped with white, forming a band; remiges brownish gray, with white shafts, the inner ones largely white; superciliary whitish, streaked with dusky; sides of head dusky streaked with gray; underparts white with dusky streaks, becoming spotted on breast; iris brown; bill dusky at tip, yellowish at base; feet dull yellowish. WINTER: Crown and hind neck gray; back sooty, glossed with purple and edged with gray, and often with buff on scapulars; a white spot on lores and lower eyelid; chin white; throat ashy, streaked with white on upper part; breast brownish gray margined with white; belly and crissum white. Wing 115–130, tail 52–66, culmen 27–35, tarsus 21–23 mm. Eastern Canadian Arctic, east to Taimyr Peninsula in Siberia; winters on both coasts of Atlantic, south to Long Island, Britain, and the Baltic, occasionally farther.

Erolia melanotos (Vieillot). Pectoral sandpiper. Sides of head, neck all around, and breast buff streaked with black; crown, back, and wing coverts black, broadly edged with cinnamon and buff or ashy; rump and upper tail coverts sooty; lateral upper tail coverts white with black centers; middle rectrices sooty edged with buff, outer rectrices brown, narrowly tipped with white; superciliary white streaked with brown; loral stripe brown; chin, belly, and axillars white; flanks and crissum white, with a few brown streaks; iris brown; bill and feet dull greenish yellow. Wing 120–146, tail 51–65, culmen 24–29, tarsus 24–30 mm. Arctic America and east Asia; winters in southern South America; migrates through Mississippi Valley and along Atlantic Coast, rare on Pacific.

Erolia bairdii (Coues). Baird's sandpiper. Crown and hind neck grayish buff streaked with black; back grayish buff spotted with black; wing coverts grayish brown; greater coverts tipped with whitish; remiges grayish brown edged with ashy, shafts white subterminally; rump, upper tail coverts, and middle rectrices sooty, margined with whitish; lateral tail coverts white with dusky V's; lateral rectrices pale grayish brown edged with white; superciliary buffy white; sides of head, lower throat, and breast pale buff narrowly streaked with dusky; axillars, wing lining, and posterior under-

parts white; iris brown; bill and feet blackish. YOUNG: Scapulars tipped with white. Wing 114–126, tail 48–54, culmen 21–24, tarsus 20–23 mm. Arctic America; migrates mainly through Great Plains, rarer on coasts; winters in Argentina and Chile.

Erolia minutilla (Vieillot). Least sandpiper. Crown and hind neck buffy brown streaked with black; back, tertials, and middle coverts black margined with cinnamon; other coverts grayish brown, greater ones narrowly tipped with white; remiges dusky, shafts white; rump, upper tail coverts, and middle rectrices sooty; lateral tail coverts white streaked with brown; lateral rectrices grayish brown; superciliary and chin whitish; loral stripe brown; breast, throat, and sides of head buff streaked with brown; rest of underparts white; iris brown; bill dusky; legs olive. YOUNG: Cinnamon color above paler; scapulars tipped with white; wing coverts tipped with buff; throat and breast pale buff, only faintly streaked. WINTER ADULT: Anterior upper parts ashy gray, streaked and blotched with black; breast pale ashy, faintly streaked with brown. Wing 82–91, tail 34–41, culmen 16–20, tarsus 16–19 mm. Breeds in Alaska and Canada south of the tundra; migrates throughout the United States; winters from South Atlantic and Gulf states and California to Patagonia.

Erolia fuscicollis (Vieillot). White-rumped sandpiper. Crown, hind neck, and back rusty brown streaked with black; wing coverts grayish brown; greater coverts narrowly tipped with white; remiges dusky, secondaries edged with white; rump grayish brown edged with buffy; upper tail coverts white, outer ones with sagittate dusky marks; tail grayish brown, edged and laterally tipped with whitish; superciliary and underparts white; breast and sides streaked with dusky; iris brown; bill dusky, basally flesh; feet black. WINTER: Crown, hind neck, and back grayish brown, indistinctly streaked with dusky. Wing 117–124, tail 50–54, culmen 21–26, tarsus 22–24 mm. Arctic America; migrates mainly through Great Plains, less commonly on Atlantic Coast; winters in southern South America.

Genus *Limnodromus* Wied. Bill much longer than tarsus, straight, with distal third heavily pitted; tail nearly square; bare part of tibia shorter than front toes; tarsus scutellate; outer toe with a basal web; feet extending beyond tail. Holarctic; India and South America in winter. 3 species, 2 in the United States.

KEY TO SPECIES OF *LIMNODROMUS*

1. Crissum spotted   *L. griseus*
   Crissum barred   *L. scolopaceus*

Limnodromus griseus (Gmelin). Short-billed dowitcher. Crown and hind neck cinnamon buff streaked with black; back and tertials black, edged with cinnamon and buff; wing coverts grayish brown, edged with ashy; greater coverts tipped and edged with white; rump, upper tail coverts, and tail barred black and white; superciliary buff; lores brown; underparts pinkish cinnamon, often mixed with white on belly; throat, breast, and sides of breast with dusky spots; flanks often barred; axillars and wing lining white with dusky V's; iris brown; bill dusky brown; feet olive. WINTER: Crown, hind neck, back, and wing coverts gray; throat and breast gray flecked with white; belly white; sides and crissum white, with dusky spots, flanks often barred. YOUNG: Like winter plumage, but back and tertials mottled black and clay color; throat, breast, and sides dull buff, speckled with dusky. Wing 138–158, tail 47–60, culmen 54–68, tarsus 33–41.5 mm. Arctic America; migrates throughout United States; winters on mud flats from Gulf states and California to northern South America.

Limnodromus scolopaceus (Say). Long-billed dowitcher. Differs in having breast, sides, and crissum barred; chin with dusky flecks in spring. Wing 137–155, tail 47–60, culmen 57–78, tarsus 33–44 mm. Coast of east Siberia and northern Alaska; winters south to Guatemala, mainly on fresh water; occurs on both coasts of the United States on migration.

Genus *Capella* Frenzel. Bill more than twice length of tarsus, straight, heavily pitted for nearly distal half; ears directly below eyes; tail rounded, rectrices 14–16, outer pair narrow; lower fourth of tibia bare; tarsus scutellate, shorter than middle toe with claw; toes unwebbed, feet extending beyond tail. Cosmopolitan. 12 species, 1 in the United States.

Capella gallinago (**Linnaeus**). Common snipe. Crown sooty with a median buff stripe; hind neck streaked black and buff; back and tertials black barred with brown, with 4 long buff stripes; wing coverts dark brown, spotted with white or buff; tertials black, edged with white and barred with brown; remiges dusky; edge of outer primary and tips of primary coverts and secondaries white; upper rump black, narrowly barred with white; lower rump and upper tail coverts rich buff, irregularly barred with black; tail black, rufous terminally, where interrupted by a black bar; lateral pair of rectrices black-and-white-barred; next 2 pairs white beyond subterminal black bar; superciliary buff; lores brown; chin buffy white; throat and upper breast buff streaked or barred with brown; belly white; sides black-and-white-barred; crissum buff with dusky squamations; axillars and wing lining barred black and white; iris brown; tip of bill dusky; base of bill and feet greenish. Wing 118–135, tail 50–63, culmen 58–73, tarsus 28–33 mm. Wet meadows of Holarctic; in North America south to New Jersey, Great Lakes, Iowa, and in mountains to Colorado and California; winters from middle states to Brazil.

Genus *Philohela* Gray. Bill twice as long as tarsus, straight, with distal third moderately pitted; ears directly below eyes; outer 3 primaries very narrow and bent, 4th primary longest; tail rounded, rectrices 14; tibia entirely feathered; tarsus scutellate in front, reticulate behind, about equal to longest toes; toes unwebbed; feet extending beyond tail. Eastern North America. Monotypic.

Philohela minor (**Gmelin**). American woodcock. Forehead ashy gray, bordered laterally by a dusky loral streak, and halfway divided medially by a dusky stripe from culmen; occiput black, crossed transversely by 4 buff bands; orbital area and auriculars buffy, with a dusky postocular streak and auricular streak; sides of head and hind neck gray mixed with cinnamon buff; back, rump, and median tail coverts black, mottled and barred with buff, and with 4 wide gray stripes down back; lateral tail coverts cinnamon; tail black, notched with buff and tipped with dark gray (silvery below); wing coverts pale cinnamon, mixed with gray and barred with dusky; remiges dusky, the inner one mottled or barred with pale cinnamon; chin whitish; throat gray, mixed with cinnamon; breast, belly, axillars, and wing lining cinnamon; crissum cinnamon, streaked with dusky and with a few silver tips; iris brown; bill and feet pale brown. Wing 119–143, tail 54–65, culmen 60–74, tarsus 29–34 mm. Wet thickets in deciduous forests of southeast Canada and eastern United States, south to northern Florida and Louisiana, west to the plains; withdraws from the north in winter.

Family Recurvirostridae.   Avocets and stilts. Bill long, awllike, straight or recurved; outer primary longest; tail truncate, rectrices 12–14; legs long, extending far beyond tail; bare part of tibia much longer than middle toe, $\frac{1}{3}$ to more than $\frac{1}{2}$ length of tarsus; tarsus reticulate, longer than bill; toes with basal webs.

Waders on fresh- or salt-water marshes; able to swim and dive. Food mainly aquatic insects. Nest on ground; eggs 3–4, buffy, spotted or blotched with black; 2 birds sometimes lay in a single nest.

Eocene onward. Cosmopolitan. 4 genera, 7 species; in the United States 2 genera, 2 species.

KEY TO GENERA OF RECURVIROSTRIDAE

1. Bill nearly straight, shorter than bare part of tibia
    Himantopus
    Bill upturned, longer than bare part of tibia
    Recurvirostra

Genus *Recurvirostra* Linnaeus.   Bill upturned, hooked at tip, grooved for $\frac{1}{3}$ its length; folded primaries extending beyond tertials but not beyond tip of tail; legs stout; bare part of tibia shorter than bill; toes semipalmate; hallux minute. Feeds by swinging bill from side to side. Nearly cosmopolitan (absent from New Zealand and Oceania). 4 species, 1 in the United States.

Recurvirostra americana Gmelin.   American avocet. Feathers around base of bill whitish; head, neck, and upper breast pale cinnamon; a sooty patch on each side of back, formed by inner scapulars; outer scapulars, rest of back, rump, and tail coverts white; tail silvery gray; wing coverts sooty, becoming gray on inner tertials; primaries black; a broad white wing band, formed by most of secondaries and tips of greater coverts; belly, crissum, axillars, and most of wing lining white; iris brown; bill black; feet bluish. WINTER: Head, neck, and

upper breast dirty white. Wing 212–230, tail 79–90, culmen 75–93, tarsus 85–100 mm. Prairie provinces and western states, east to Iowa, Kansas, and Texas; winters from California and Texas to Guatemala; rare on Atlantic Coast.

Genus *Himantopus* Brisson.   Bill nearly straight (slightly upturned), grooved for half its length; folded primaries extending beyond tail and far beyond tertials; legs slender; bare part of tibia longer than bill; toes practically unwebbed; hallux absent. Warmer parts of world. 1 species.

Himantopus himantopus (Linnaeus).[1]   Black-necked stilt. Underparts, axillars, forehead, spot above and below eye, lower back, rump, and upper tail coverts white; rectrices ashy; crown, hind neck, upper back, wings, and wing lining black; iris red; bill black; feet pink. Wing 203–232, tail 66–76, culmen 60–70, tarsus 96–119 mm. Warmer parts of world; in America from Oregon, Idaho, Colorado, and Gulf and Atlantic Coasts from South Carolina southward through West Indies and Middle and South America.

Family Phalaropodidae.   Phalaropes. Bill straight, awllike, about equal to tarsus; wing pointed, tip extending far beyond tertials; legs short, not reaching tip of tail; lower part of tibia bare; tarsus scutellate, scarcely longer than toes, somewhat compressed; toes with basal webs and lateral lobes or membranes. Occipital fontanels variable; basipterygoid processes present; supraorbital grooves rudimentary; cervicals 15; true ribs 6; sternum 4-notched. Plumage dense. Male the smaller and duller sex, builds nest and incubates. Nest on ground, of grass; eggs 4, buff, spotted or scrawled with blackish; food aquatic animals. Unrecorded before the Pleistocene. Holarctic; Southern Hemisphere in winter. 3 genera, 3 species.

KEY TO GENERA OF PHALAROPODIDAE

1. Sides of bill expanded subterminally (tarsus shorter than middle toe and claw)   *Phalaropus*
    Sides of bill not expanded   2
2. Tarsus about equal to middle toe without claw
    Lobipes
    Tarsus much longer than middle toe without claw
    Steganopus

---

[1] *Himantopus mexicanus* (Müller) of A.O.U.

Genus *Phalaropus* Brisson. Bill broad, expanded subterminally; tail rounded; toes scalloped. Occipital fontanels present. Monotypic.

Phalaropus fulicarius (**Linnaeus**). Red phalarope. Pileum and chin slate; lores and cheeks white; underparts purplish cinnamon; axillars and wing lining white; hind neck mixed slate and cinnamon; back buff striped with black; rump and upper tail coverts black and chestnut; tail dusky, edged with buff or white; wing coverts slate; greater coverts with broad white tips, forming a band; primaries dusky; iris brown; bill dusky at tip, yellowish at base; feet bluish green, lobes and inner side of tarsus yellow. WINTER: Orbits and nape dusky; rest of head and neck and underparts white; upper parts pale blue-gray. YOUNG: Buffy white below; nape dusky; upper parts black edged with buff. Wing 119–142, tail 59–71, culmen 19–24, tarsus 20–24 mm. Circumarctic; migrates along coasts of Atlantic and Pacific, mainly offshore; winters mainly at sea off Chile, West Africa, and Arabia.

Genus *Lobipes* Cuvier. Bill needlelike; tail rounded; toes scalloped. Usually lacks occipital fontanels. Monotypic.

Lobipes lobatus (**Linnaeus**). Northern phalarope. Above blackish; back striped with buff; tips of greater coverts forming a white wing band; neck all around and upper breast rufous; underparts white; iris brown; bill black; feet bluish gray. WINTER: Forehead, superciliary, and underparts white; a black stripe passing below and behind eye; crown blackish; upper parts blackish streaked with white or buff. Wing 102–119, tail 46–52, culmen 20–24, tarsus 18–24 mm. Circumarctic and subarctic; migrates mainly offshore, less frequently along coast or through interior; winters mainly at sea in southern oceans.

Genus *Steganopus* Vieillot. Bill slender; tail double-emarginate; toes with unscalloped lateral membranes. Lacks occipital fontanels. 1 species.

Steganopus tricolor **Vieillot.** Wilson's phalarope. Crown pale bluish gray; occiput and nape whitish, becoming bluish gray on back and rump; supraloral stripe white; a black stripe through eye down each side of neck, changing to deep chestnut on lower neck and continuing along sides of back; upper tail coverts white; tail gray mottled with white; chin white; foreneck buffy cinnamon, fading to buff on breast; belly white; wings brownish gray, coverts and tertials with narrow white edge; iris brown; bill and feet black. WINTER: Upper parts light gray; upper tail coverts, superciliary, and underparts white. YOUNG: Upper parts dusky, margined with buff; upper tail coverts, superciliary, and underparts white; neck tinged with buff. Wing 116–137, tail 48–65, culmen 28–36, tarsus 29–33 mm. Breeds in marshes of northern Great Basin and Plains, south to California, Nevada, Utah, Nebraska, and Iowa, rarely east to tip of Lake Michigan; occurs on both Atlantic and Pacific Coasts on migration; winters inland in southern South America.

Suborder Lari. Gulls and allies. Bill compressed, gonys prominent; adult downs on both pterylae and apteria; rectrices 12; legs inserted near middle of body, short, not extending beyond tail; lower part of tibia naked; tarsus scutellate in front, reticulate behind; toes webbed; hallux small or absent.

No basipterygoid processes nor occipital fontanels; supraorbital grooves large, perforated; cervical vertebrae 15, dorsals 5; 6–7 true ribs, furculum with hypocleidium; coracoid with furcular facet broad and much hollowed under brachial tuberosity, procoracoid process bent upward; sternum with 1–2 pairs of notches or fenestrae; hypotarsus with 2 open grooves; humerus with large triangular ectepicondylar process; thumb present. Thigh muscle $A(B)XY\pm$. Intestine with less than 6 loops; caeca present.

Aquatic. Fish-eaters and scavengers. Semiprecocial. Cosmopolitan. 3 families.

Family Stercorariidae. Jaegers and skuas. Bill strongly hooked, shorter than middle toe without claw; rhamphotheca complex, with cere. Middle rectrices longest. Feet strong; claws relatively large and curved; hallux small. Sternum with 2 pairs of notches; coracoids separate. Caeca long.

Predacious, robbing gulls and boobies of their prey; also eats birds and their eggs and blubber. Nest on ground, of grass or seaweed; eggs 2–3 olive brown spotted with black.

Known from Pleistocene. Bipolar, following cold currents equatorwise or beyond. 2 genera, 4 species.

KEY TO GENERA OF STERCORARIIDAE

1. Tarsus shorter than middle toe with claw
   Catharacta
   Tarsus longer than middle toe with claw
   Stercorarius

Genus *Catharacta* Brünnich. Bill stout, its depth equal to or more than distance from nostril to tip; tail less than half wing, nearly square, but with middle rectrices slightly longest; tarsus shorter than middle toe with claw. Bipolar; to mid-latitudes and beyond in winter. 1 species.

***Catharacta skua* Brünnich.** Skua. Dark brown streaked with cinnamon; throat rufous streaked with buff; underparts grayish brown streaked with cinnamon; axillars and wing lining mostly dark brown with some cinnamon; remiges dusky with a white patch at base; tail dusky; iris brown; bill and feet black. Wing 365–420, tail 131–158, culmen 44–56, tarsus 58–71 mm. Breeds on Iceland, Faroes, Shetlands, and Orkneys; winters on both sides of Atlantic to Long Island and Gibraltar. Also breeds on coast of Chile; ranges north to Peru and less frequently to coastal California, Washington, and British Columbia. Also breeds in Antarctic and New Zealand.

Genus *Stercorarius* Brisson. Bill relatively slender, its depth less than distance from nostril to tip; middle rectrices of adults projecting far beyond the rest, more than half length of wing; tarsus longer than middle toe. Circumarctic; Southern Hemisphere in winter. 3 species.

KEY TO SPECIES OF *STERCORARIUS*

1. Bill deeper than wide; wing 349–374 mm
   S. pomarinus
   Bill wider than deep; wing 295–341 mm   2
2. Cere longer than dertrum; tarsi all black
   S. parasiticus
   Cere not longer than dertrum; tarsi at least partly
   light   *S. longicaudus*

***Stercorarius pomarinus* (Temminck).** Pomarine jaeger. Long middle rectrices of adult twisted. Hind neck and underparts white; auriculars and sides of neck tinged with yellow; anal region and crissum light brownish gray; pileum, lores, malar region, and upper parts sooty gray; iris brown; bill whitish, tip black; feet blackish, upper tarsus light blue. DARK PHASE: Sooty brown, tinged with yellow on auriculars. YOUNG: Head, neck, and underparts dull buff barred with dusky; back and wing coverts grayish brown, spotted with buff. Wing 349–374, middle rectrices 128–243, culmen 38–44, tarsus 48–55 mm. Arctic America and Eurasia; migrates off both coasts; winters in Gulf of Mexico and on southern oceans.

***Stercorarius parasiticus* (Linnaeus).** Parasitic jaeger. Upper parts, lores, and crissum grayish brown; face, neck all around, and underparts white; sides of neck tinged with yellow; shafts of primaries ivory; iris brown; bill horn, dusky at tip; feet black. DARK PHASE: Sooty brown, tinged with yellow on neck. YOUNG: Above dusky brown streaked with cinnamon or buff; below barred with brown and either white or cinnamon. Wing 301–341, middle rectrices 165–235, culmen 28–35, tarsus 43–48 mm. Circumarctic and subarctic; migrates along coasts; winters from southern U.S. waters to southern oceans.

***Stercorarius longicaudus* Vieillot.** Long-tailed jaeger. Crown, lores, orbital area, and nape sooty; hind neck and face straw yellow; upper parts brownish gray; shafts of only 2 outer primaries white on upper surface; upper breast white; posterior underparts gray; iris brown; bill horn, dusky at tip; tarsi wholly or partly gray; toes and webs black. Dark phase unknown. YOUNG: Above ashy brown barred with buff; below dull white barred with ashy brown. Wing 295–327, middle rectrices 238–350, culmen 27–31, tarsus 40–46 mm. Circumarctic and subarctic; migrates off coasts, rare off Pacific Coast of United States; winters mainly at sea in South Pacific and in Atlantic from latitudes 40° N to 50° S.

Family Laridae. Gulls and terns. Bill unhooked and without cere; claws weak; hallux small or absent. Sternum usually with 2 pairs of notches; coracoids in contact. Thigh muscles *ABXY+* (gulls) or *AXY+* (terns). Caeca moderate in length. Fish-eaters and scavengers; rob birds' nests. Nest on ground, sometimes of sticks in trees. Eggs 1–5, usually 2–3, white, buff, olive, or bluish, marked with dusky. Eocene onward. Cosmopolitan. 17 genera, 83 species; in the United States 10 genera, 32 species.

KEY TO GENERA OF LARIDAE

1. Depth of bill greater at gonydeal angle than at posterior end of nostril (subfamily Larinae, Fig. 5-42) **2**

    Depth of bill greater at posterior end of nostril than at gonydeal angle (subfamily Sterninae) **5**
2. Tail forked   *Xema*
    Tail square or rounded   **3**
3. Tarsus shorter than middle toe without claw   *Rissa*
    Tarsus longer than middle toe without claw   **4**
4. Planta tarsi serrate; bare part of tibia shorter than half middle toe without claw   *Pagophila*
    Planta tarsi not serrate; bare part of tibia longer than half middle toe without claw   *Larus*
5. Tail graduated and only slightly forked   *Anous*
    Tail forked   **6**
6. Head with an occipital crest   **7**
    Head uncrested   **8**
7. Inner webs of primaries bicolored; tarsus 25–36 mm   *Thalasseus*
    Inner webs of primaries gray; tarsus 40–46 mm   *Hydroprogne*
8. Gonys longer than middle toe without claw; tail forked for $\frac{1}{2}$ its length   *Sterna*
    Gonys shorter than middle toe without claw; tail forked for $\frac{1}{3}$ or less   **9**
9. Bill stout (depth about $\frac{1}{3}$ length of culmen)   *Gelochelidon*
    Bill slender (depth $\frac{1}{4}$ or less length of culmen)   *Chlidonias*

Genus *Larus* Linnaeus. Bill shorter than head (Fig. 5-42); tail square, less than half length of wing; tarsus longer than culmen or middle toe without claw; planta tarsi not rough; hallux free; front toes fully webbed. Cosmopolitan. 34 species, 14 in the United States.

KEY TO SPECIES OF *LARUS*

1. Wing 210–221 mm   *L. minutus*
    Wing over 240 mm   **2**
2. Exposed culmen not exceeding 35, tarsus not exceeding 43 mm   **3**
    Exposed culmen at least 38, tarsus at least 45 mm   **4**
3. Bill black; depth at gonys 6–7.5 mm   *L. philadelphia*
    Bill reddish; depth at gonys 8–9.5 mm   *L. pipixcan*
4. Shafts of outer primaries yellowish white above and below   **5**
    Shafts of outer primaries pale brownish to black above   **6**
5. Wing 424–474 mm   *L. hyperboreus*
    Wing 379–419 mm   *L. glaucoides*
6. Tail uniform black with or without narrow white or buffy tip   *L. heermanni*
    Tail either white or else mottled, banded, or speckled with brown or dusky   **7**
7. Adult with primaries gray subterminally; young with tail gray or with a few whitish markings (wing 411–452 mm)   *L. glaucescens*
    Adult with primaries black subterminally; young with terminal third of tail blackish or brownish   **8**
8. Tail 185–201 mm, tarsus 73–77 mm, middle toe without claw 69–74.5 mm   *L. marinus*
    Tail not exceeding 178 mm, tarsus 68 mm, middle toe 66 mm   **9**
9. Depth at gonys not more than $\frac{1}{3}$ exposed culmen   **10**

    Depth at gonys more than $\frac{1}{3}$ exposed culmen   **12**
10. Bill reddish black   *L. atricilla*
    Bill yellowish or whitish, at least for basal half   **11**
11. Bill yellow with black subterminal band   *L. delawarensis*
    Bill plain yellow, or in young blackish terminally, whitish basally   *L. canus*
12. Adult with broken black band on bill; young with abruptly black-tipped bill   *L. californicus*
    Adult with unbanded bill; young with black tip of bill fading toward lighter base   **13**
13. Adult with gray wedges on inner web of 2 outer primaries; young with basal part of tail extensively mottled   *L. argentatus*
    Adult without gray wedges on 2 outer primaries; young with little mottling on tail   *L. occidentalis*

***Larus hyperboreus* Gunnerus.** Glaucous gull. Mantle pale gray, fading into white on tips of remiges; rest of plumage white; iris yellow; bill yellow with a red spot on sides of mandible; feet pink. YOUNG: Grayish white; upper parts and tail streaked and mottled with pale brownish gray; tip of bill dusky. Wing 424–474, tail 173–207, culmen 49–67, tarsus 53–74 mm. Circumpolar; in America winters south to California, Great Lakes, and New York, casually farther.

***Larus glaucoides* Meyer.** Iceland gull. Colored like *L. hyperboreus*, but smaller. Wing 379–419, tail 155–170, culmen 40.5–48, tarsus 52–56 mm. Circumpolar; in America winters south to New Jersey and Great Lakes, occasionally farther.

***Larus glaucescens* Naumann.** Glaucous-winged gull. Mantle pale gray; remiges white-tipped, primaries deep gray subterminally; rest of plumage white; iris cream; bill yellow with a red spot on side of mandible; feet flesh. YOUNG: Deep gray; head and neck streaked, back and tail mottled with whitish; wings and tail with a glaucous cast; bill blackish, paler at base; feet dusky. Wing 411–452, tail 171–186, culmen 53.5–61.5, tarsus 62–70 mm. North Pacific Coast and islands, from Siberia to Alaska and Washington; winters south to California and Japan.

Larus marinus Linnaeus. Great black-backed gull. Mantle blackish gray; remiges tipped with white; 2d to 5th primaries with a black bar across white tip; rest of plumage white; iris whitish; bill yellow with a subterminal red spot on side of mandible; feet flesh. YOUNG: Mantle dusky margined with buffy; wing and tail dusky, narrowly tipped with whitish; head, neck, and underparts dirty white streaked and spotted with grayish brown; iris brown; bill dusky; feet dirty white. Wing 465–494, tail 185–201, culmen 59.5–68, tarsus 73–77 mm. Coasts of North Atlantic, south to Massachusetts, Britain, and the Baltic; winters south to Mediterranean and Caspian Seas, Great Lakes, Delaware Bay, and Florida.

Larus occidentalis Audubon. Western gull. Mantle deep gray; proximal remiges tipped with white; outer primary black, with white terminal area crossed by a black subterminal bar; 2d to 6th primaries narrowly tipped with white, black subterminally; iris brown; bill yellow with a red spot on side of mandible; feet yellow. YOUNG: Grayish brown mottled with dirty white; wings dusky; tail dusky with narrow white tip, base grayish brown; bill blackish, becoming brownish basally; feet flesh. Wing 368–447, tail 147–177, culmen 51–59, tarsus 58–68 mm. Pacific Coast from Washington to northern Mexico.

Larus argentatus Pontoppidan. Herring gull. Mantle pale gray; 6 outer primaries black subterminally, tipped with white; 1–2 outer primaries with a black bar across white tip; rest of plumage white; iris pale yellow; bill yellow with a red spot on rami; feet flesh. YOUNG: Grayish brown, streaked and mottled with whitish or buffy; remiges and exposed part of tail dark brown; base of tail mottled with dirty white; iris brown; bill blackish, fading into flesh basally; feet flesh. Wing 401–465, tail 159–177, culmen 45.5–63, tarsus 54–68 mm. Hol-

arctic, in America breeds south to North Dakota, Great Lakes, New York, and Maine; winters on Great Lakes and both coasts south to West Indies and Panama.

Larus californicus Lawrence. California gull. Mantle light gray; remiges tipped with white; 6 outer primaries black subterminally; outer 1–2 primaries with white tip broken by a black bar; rest of plumage white; iris brown; bill yellow crossed by an incomplete black subterminal ring; mandible red distally; feet greenish. YOUNG: Grayish brown spotted or streaked with buffy white; remiges dusky, some with narrow white tips; tail dusky mottled basally with whitish; bill dusky, darker at tip; feet flesh. Wing 371–428, tail 124–168, culmen 42–54.5, tarsus 47–63 mm. Inland lakes of western North America, from Canada to Great Salt Lake and to Mono Lake, California; winters on Pacific Coast from British Columbia to Mexico.

Larus delawarensis Ord. Ring-billed gull. Mantle pale gray; inner remiges tipped with white; first primary black terminally, with a white subterminal spot; 2d to 6th primaries narrowly tipped with white, black subterminally; 2d primary often with a 2d white subterminal spot; rest of plumage white; iris yellow; bill greenish yellow with a black subterminal band; feet greenish yellow. YOUNG: Above grayish brown margined with buffy; below white, more or less spotted with grayish brown; primaries dusky; distal third of tail dusky; base of tail abruptly pale gray mottled with brownish; iris brown; bill flesh with a broad black tip; feet flesh. Wing 334–389, tail 132–161, culmen 37–45.5, tarsus 47–61 mm. Inland lakes from Alaska and Canada south to Oregon, Idaho, Utah, Colorado, North Dakota, Minnesota, Wisconsin, and Michigan; winters on Great Lakes and both coasts, from British Columbia and Maine to Mexico.

Larus canus Linnaeus. Mew gull. Mantle light gray; secondaries white terminally; 2 outer primaries black terminally, interrupted by a white spot; 3d to 6th primaries black subterminally with narrow white tips; rest of plumage white; iris brown; bill and feet greenish yellow. YOUNG: Grayish brown margined with buff; primaries dusky; a dusky band on distal third of tail, base whitish with some dark mottling; bill flesh, tip dusky; feet flesh. Wing 328–380, tail 132–150, culmen 32–45, tarsus 45–56 mm. Northern Europe and Asia, coasts and lakes of Alaska and northwest Canada; in America winters from southern Alaska to California.

Larus atricilla Linnaeus. Laughing gull. Head and upper neck slate; eyelids white; mantle deep gray; proximal remiges tipped with white; 4 outer primaries mostly black, usually with small white tips; 5th and 6th primaries gray, with white tip crossed by a black bar; upper tail coverts, tail, and underparts white; iris brown; bill and feet red. WINTER: Head and neck white, mottled with gray on orbital region, auriculars, and occiput; bill and feet reddish black. YOUNG: Occiput, nape, and mantle grayish brown margined with buff; remiges blackish, the inner ones with white tips; distal third of tail blackish, base pale gray; upper tail coverts and underparts white; breast clouded with pale brown. Wing 311–348, tail 116–135, culmen 37–43, tarsus 46–53 mm. Both coasts, from Maine and California south to Middle America and West Indies; winters from South Carolina south to Peru and Brazil.

Larus pipixcan Wagler. Franklin's gull. Head and upper neck slate; eyelids white; mantle deep gray; inner remiges white-tipped; primaries pale gray tipped with white; outer 5–6 primaries with black bar across white tip; outer web of outer primary black; upper tail coverts, tail, and underparts white; iris brown; bill and feet red. WINTER: Head and neck white; orbits, auriculars, and occiput gray; bill dusky red tipped with bright red; feet dusky red. YOUNG: Occiput, posterior part of face, and mantle grayish brown; upper tail coverts, forehead, front of face, eyelids, and underparts white; remiges dusky; secondaries broadly, primaries narrowly, tipped with white; tail pale gray with broad blackish subterminal band; bill and feet dusky red. Wing 270–295, tail 94–109, culmen 30–34.5, tarsus 38–42.5 mm. Prairie lakes, from the prairie provinces south to Utah, South Dakota, and Minnesota; winters on Pacific Coast of South America, rarely on Gulf Coast of United States.

Larus philadelphia (**Ord**). Bonaparte's gull. Head and upper neck slate; eyelids white; mantle pale gray; tail and coverts, edge of wing, and underparts white; outer primaries white, tipped with black; outer web of 2–3 outer primaries black; 5th and 6th primaries pale gray, black subtermi-

Fig. 5-42. Bill of laughing gull (*Larus atricilla*).

nally and narrowly tipped with white; inner primaries pale gray with subterminal black spot; iris brown; bill black; feet orange. WINTER: Head and neck white; occiput tinged with gray; a slate spot on auriculars. YOUNG: Mantle mostly pale gray; band running along proximal wing coverts, scapulars, and tertials grayish brown tipped with buffy white; head, neck, upper tail coverts, and underparts white; occiput and auricular spot brownish; tail white with a dusky terminal band; remiges pale gray or white, black terminally, the inner ones with small white tips; feet yellowish flesh. Wing 247–274, tail 97–109, culmen 28–33, tarsus 32–36.5 mm. Nests in trees in forests of interior Alaska and western Canada; winters on coast from southern Alaska to Mexico, and from Massachusetts to Florida and Gulf states; Great Lakes on migration.

Larus minutus Pallas. Little gull. Head, including eyelids, and upper neck all around black; mantle pale gray; lower neck, rump, upper tail coverts, tail, and underparts white; remiges pale gray, broadly tipped with white; edge of outer primary black in middle; iris brown; bill and feet red. WINTER: Head and neck white; a blackish spot before eyes and behind ears. YOUNG: Like winter, but nape, wing coverts, outer webs of primaries, and distal third of tail sooty; bill blackish; legs flesh. Wing 210–221, tail 86–96, culmen 22–25, tarsus 23–26 mm. Europe and Asia; also breeds near Toronto; small numbers winter on Lake Erie, Lake Ontario, and coast from Maine to New Jersey or farther.

Larus heermanni Cassin. Heermann's gull. Head and upper neck white; mantle and underparts brownish gray; remiges dusky, proximal ones narrowly tipped with white; upper tail coverts light gray; tail black tipped with white; iris brown; bill red; feet black. WINTER: Head and neck grayish brown; bill red tipped with dusky. YOUNG: Sooty brown; mantle margined with whitish; bill flesh, tip dusky. Wing 338–376, tail 131–162, culmen 39.5–48, tarsus 47.5–54 mm. Breeds on Pacific Coast of Mexico; winters on coast from Vancouver Island to Guatemala.

Genus *Pagophila* Kaup. Differs from *Larus* in having scutella of tarsi and toes rough; webs of feet incised; tibia feathered almost to heel. Circumpolar. Monotypic.

Pagophila eburnea (**Phipps**). Ivory gull. White; iris brown; bill yellow, base slate, tip orange. YOUNG: Head, neck, and mantle flecked or smudged with dusky; narrow tips of remiges and narrow subterminal band on tail blackish; bill dusky. Wing 300–362, tail 129–156, culmen 31–36, tarsus 33–39 mm. Polar islands; winters on north coasts of Europe, Asia, and America, occasionally as far as New York.

Genus *Rissa* Stephens. Tarsus not rough, shorter than culmen or middle toe; hallux absent or rudimentary; webs full. Holarctic. 2 species, 1 in the United States.

Rissa tridactyla (**Linnaeus**). Kittiwake. Mantle light gray; secondaries tipped with white; primaries tipped with black; rest of plumage white; iris brown; bill greenish yellow; feet black. WINTER: Occiput, nape, and auriculars gray; tip of bill black. YOUNG: Mantle spotted with black; nape patch, tip of tail, and bill black. Wing 293–330, tail 115–142, culmen 34–43, tarsus 31–36 mm. Circumarctic and subarctic; winters south to Mediterranean, New Jersey, and California, sometimes farther.

Genus *Xema* Leach. Resembles *Larus* but tail forked. Circumpolar; South America in winter. Monotypic.

Xema sabini (**Sabine**). Sabine's gull. Head and upper neck slate, bordered by a black collar; mantle light gray; lower neck, underparts, tail coverts, tail, greater wing coverts, and inner remiges white; edge of wing black; 5 outer primaries mostly black, with inner wedges and narrow tips white; iris brown; bill black, tip yellow; feet gray. WINTER: Head and neck white; occiput, nape, and auriculars gray. YOUNG: Occiput, nape, line through eye, and mantle brownish gray margined with buffy; tail white, tip black; tail coverts, forehead, and underparts white; bill dusky. Wing 260–286, tail 111–139, culmen 25–28.5, tarsus 30–34.5 mm. Circumarctic coast, migrates off Pacific Coast of America; winters off Peru; casual on Atlantic Coast and in interior.

Genus *Gelochelidon* Brehm. Bill gull-like, stout, with deep gonys but shorter than rami; tail less than half length of wing, forked for $\frac{1}{3}$ its length; tarsus shorter than culmen. Nearly cosmopolitan. Monotypic.

Gelochelidon nilotica (**Gmelin**). Gull-billed tern. Crown and nape black; mantle, wings, and tail pale gray; underparts white; iris brown; bill black; feet reddish black. WINTER: Crown white; nape pale gray; a dusky spot in front of eye and on auriculars. Wing 275–338, tail 103–135, culmen 35–42, tarsus 27–36 mm. In America, Atlantic and Gulf Coasts from Virginia to Texas; Salton Sea, California; winters south to Panama and Ecuador.

Genus *Chlidonias* Rafinesque. Bill slender, with gonys about equal to ramus; tail less than half wing, slightly forked; tarsus much shorter than culmen, about equal to middle toe; webs deeply incised. Holarctic; Southern Hemisphere in winter. 3 species, 1 in the United States.

Chlidonias nigra (**Linnaeus**). Black tern. Head, neck, and underparts black; edge of wing and crissum white; mantle, wings, tail, and axillars deep gray; iris brown; bill black; feet purplish. WINTER: Head, neck, underparts, and axillars white; orbit, auriculars, and occiput dusky. Wing 191–215, tail 73–87, culmen 26–29, tarsus 14–17 mm. Holarctic; in America breeds south to California, Nevada, Colorado, Kansas, Missouri, Tennessee, Pennsylvania, and New York; winters in Southern Hemisphere.

Genus *Sterna* Linnaeus. Bill slender, gonys not prominent, usually about as long as rami; tail more than half wing and forked for more than half its length (somewhat less in *S. albifrons*); tarsus much shorter than culmen, about equal to middle toe; webs moderately incised. Cosmopolitan. 22 species, 6 in the United States.

KEY TO SPECIES OF *STERNA*

1. Mantle sooty or blackish 2
   Mantle silvery gray 3
2. Hind neck pale gray .... *S. anaethetus*
   Hind neck blackish like rest of upper parts
   .... *S. fuscata*

3. Wing 160–178 mm   *S. albifrons*
    Wing 218–300 mm   **4**
4. Both webs of outer rectrix entirely white
                      S. dougallii
    1 web of outer rectrix darker than the other   **5**
5. Outer web of lateral rectrix white, inner web gray
      terminally; tarsus 22–26, middle toe without claw
      19.5–22.5 mm   *S. forsteri*
    Outer web of outer rectrix blackish, inner web white;
      tarsus 14–21, middle toe 15–18 mm   **6**
6. Tarsus (18–22) longer than middle toe without claw;
      bill vermilion   *S. hirundo*
    Tarsus (14–17) shorter than middle toe; bill dark
      carmine   *S. paradisaea*

Sterna forsteri **Nuttall.** Forster's tern. Crown and nape black; upper parts silvery gray; upper tail coverts, outer web of outer rectrix, and underparts white; exposed part of wing silver; inner webs of outer primaries edged with dusky; inner web of outer rectrix dusky terminally; iris brown; bill orange, terminal third black; feet orange. WINTER: Pileum white; nape dusky; a broad black stripe through eye; bill mostly dusky, corners of mouth orange. YOUNG: Like winter plumage, but tip of outer rectrix clouded with gray on both webs. Wing 239–269, tail 140–218, culmen 37–42, tarsus 22–26 mm. Breeds on prairie lakes from southern Canada to California, Nevada, Utah, Colorado, Nebraska, Minnesota, and Illinois; also on Atlantic Coast from Virginia to Texas; winters from California and South Carolina to Guatemala.

Sterna hirundo **Linnaeus.** Common tern. Crown and nape black; mantle pale gray; upper tail coverts white; underparts white, clouded with pale gray on breast; inner webs of rectrices white; outer web of outer rectrix blackish, of others gray; 5 outer primaries dusky externally, inner wedges of shafts white; iris brown; bill vermilion, tip blackish; feet coral. WINTER: Forehead white; occiput and nape black; underparts pure white. Wing 235–273, tail 128–174, culmen 32–40.5, tarsus 18–22 mm, middle toe 16–18, nostril to gonydeal angle 8–9 mm. Holarctic, to Southern Hemisphere in winter; in United States breeds along Atlantic Coast, on Great Lakes, and on lakes of Minnesota and North Dakota; winters mainly on both coasts of Latin America.

Sterna paradisaea **Pontoppidan.** Arctic tern. Differs from *S. hirundo* in slightly darker under-parts; bill dark carmine with indistinct dark tip; feet dark carmine; tarsus usually shorter than middle toe. Wing 238–290, tail 135–213, culmen 28–45, tarsus 14–17, middle toe 14–17.5, nostril to gonydeal angle 5.5–7.5 mm. Circumarctic and subarctic, breeding south to British Columbia and coast of Massachusetts; winters in southern oceans, south to the Antarctic; migrates mainly offshore, but occurs coastwise south to California and New York.

Sterna dougallii **Montagu.** Roseate tern. Crown and nape black; mantle pearly gray; edge of outer primaries blackish tinged with silver; tail silvery; underparts white tinged with pink; iris brown; bill black, reddish basally; feet red. WINTER: Forehead white; underparts without pink. YOUNG: Similar to winter plumage but bill and feet dusky. Wing 219–240, tail 138–215, culmen 35–40, tarsus 18–21 mm. Breeds in isolated colonies in warmer oceans of world; in west Atlantic breeds from Nova Scotia to coast of Virginia and on Dry Tortugas; rare migrant along Gulf Coast; winters south to Brazil.

Sterna albifrons **Pallas.** Least tern. Fork of tail less than half its length. Forehead and underparts white; loral stripe, crown, and nape black; mantle and tail pearl gray; 2–3 outer primaries black externally; iris brown; bill yellow, often tipped with black; feet orange-yellow. WINTER: Head white; line from eyes to and across occiput and nape black; bill and feet duller. Wing 160–178, tail 66–97, culmen 23.5–31, tarsus 14–16 mm. Warmer waters of world. In the United States breeds on Atlantic and Gulf Coasts north to Massachusetts; Missouri-Mississippi River north to South Dakota; coast of California.

Sterna fuscata **Linnaeus.** Sooty tern. Gonys short. Forehead, sides of head, edge of wing, and underparts white; loral stripe, occiput, nape, and upper parts sooty; primary shafts whitish below; iris brown; bill and feet black. YOUNG: Feathers of mantle tipped with white; wing lining pearl gray; otherwise entirely sooty. Wing 271–300, tail 124–170, culmen 40–47, tarsus 21–25 mm. Warmer oceans; in United States breeds on Dry Tortugas and sporadically in Louisiana and Texas; wanders north rarely to Maine.

Sterna anaethetus **Scopoli.** Bridled tern. Forehead, superciliary, and underparts white; loral stripe and upper parts black; hind neck pale gray; primaries black, edged anteriorly with white and with white wedge on inner web; outer rectrices white, tipped with grayish; succeeding rectrices more extensively dark; middle 4 pairs blackish; bill and feet black. Wing 251–274, tail 147–178, culmen 38–44, tarsus 19–21 mm. Warm Atlantic, Indian, and Pacific waters; occurs storm-blown north to Alabama, Florida, and South Carolina.

Genus *Thalasseus* Boie. An occipital crest; gonydeal angle obsolete; tail about half wing, forked halfway; tarsus only about half length of culmen, but longer than middle toe; webs moderately incised. Warmer seas. 7 species, 3 in the United States.

KEY TO SPECIES OF *THALASSEUS*

1. Bill mostly black (tip often yellow)   *T. sandvicensis*
   Bill orange-red   **2**
2. Wing 292–320 mm; gonys longer than rami
                                    T. elegans
   Wing 357–393 mm; gonys shorter than rami
                                  T. maximus

Thalasseus maximus (Boddaert). Royal tern. Gonys shorter than rami; bill rather stout. Crown black; mantle and tail pearl gray; hind neck and underparts white; exposed part of primaries dusky, more or less silvered; iris brown; bill orange; feet black. WINTER: Forehead white; crown mixed black and white; crest black. Wing 357–393, tail 103–196, culmen 58–68, tarsus 29.5–35.5 mm. Atlantic Coast from Virginia to Texas, West Indies, both coasts of Mexico, and in winter to South America; West Africa.

Thalasseus elegans (Gambel). Elegant tern. Gonys longer than rami; size small, but bill long and slender. Colored like *T. maximus*, but underparts tinted with pink in breeding plumage. Wing 292–320, tail 115–165, culmen 58–65, tarsus 28–30.5 mm. Breeds in Baja California; migrates north to San Francisco; winters south to Chile.

Thalasseus sandvicensis (Latham). Sandwich tern; Cabot's tern. Plumage like that of *T. maximus* but much smaller. Bill long and slender, black with abrupt yellow tip (yellow obsolete in young). Wing 259–302, tail 99–130, culmen 50–54, tarsus 25–27 mm. Virginia to Texas, through the Carib-

bean area; Europe, wintering in Africa and India.

Genus *Hydroprogne* Kaup. Occipital crest short; gonys much shorter than rami; tail less than half wing, slightly forked; tarsus much shorter than bill, longer than middle toe; webs full. Cosmopolitan. Monotypic.

Hydroprogne caspia (Pallas). Caspian tern. Crown black; mantle and tail pearl gray; upper tail coverts and underparts white; iris brown; bill red; feet black. WINTER: Crown streaked with white; nape black. Wing 400–423, tail 130–150, culmen 64–75, tarsus 40–46 mm. Nearly cosmopolitan; in America breeds discontinuously on Atlantic Coast from Virginia to Texas, Gulf of St. Lawrence, Lakes Huron and Michigan, Great Salt Lake, Klamath Lake, lakes of northwest Canada, and Scammons Lagoon, Baja California; winters on both coasts from California (rare) and South Carolina to Mexico.

Genus *Anous* Stephens. Uncrested; tail more than half wing, graduated and slightly forked; tarsus much shorter than culmen or middle toe. Tropical oceans. 2 species.

KEY TO SPECIES OF *ANOUS*

1. Gonys shorter than mandibular rami   **A. stolidus**
   Gonys longer than mandibular rami   **A. tenuirostris**

Anous stolidus (Linnaeus). Brown noddy. Forehead white; crown gray; lores and area above eye black; rest of plumage sooty brown; bill black; feet reddish brown. Culmen curved; gonys shorter than rami; 4th rectrices longest. Wing 259–273, tail 138–148, culmen 40–44, tarsus 23–25 mm. Tropical islands and oceans; in the United States breeds only on Dry Tortugas; occurs storm-blown along coast from Louisiana to South Carolina.

Anous tenuirostris (Temminck). Black noddy. Differs in having crown and nape paler gray; rest of plumage and feet blacker. Bill thinner and straighter, with gonys longer than rami; 3d rectrices longest. Smaller: wing 211–241, tail 112–132, culmen 37–48, tarsus 18–21 mm. Tropical Pacific, Indian, and South Atlantic Oceans; western Caribbean and Dry Tortugas.

Family Rynchopidae. Skimmers. Bill bladelike, with mandible much longer than maxilla, rami very short and bent laterally; tail $\frac{1}{3}$ length of wing, moderately forked; tarsus half length of culmen, longer than middle toe; claws long and curved; webs moderately incised; pupil a vertical slit; tongue short. Sternum 4-notched. Ambiens absent; 2 tendons to ulnar side of arm.

Largely nocturnal. Fly with bill cutting water. Gregarious. Voice a grunt. Food fish, crustaceans. Eggs 3–5, blotched, laid on beach.

No fossil record. Coasts and rivers of North and South America, Africa, India. 1 genus, 3 species; 1 United States species.

Genus *Rynchops* Linnaeus.

Rynchops nigra **Linnaeus.** Black skimmer. Forehead, hind neck, and underparts white; a white wing band formed by tips of secondaries; tail mostly white, the central feathers dusky internally; crown and mantle black; iris brown; bill black, base and feet orange or red. YOUNG: Crown and wing coverts edged with white; bill shorter. Wing 338–430, tail 102–137, culmen 52–89, tarsus 25–37 mm. Coastwise from New Jersey (formerly Massachusetts) to Florida and Texas, West Indies, and both coasts of Mexico south to Argentina and Chile.

Suborder Alcae. Auks. A single living family.

Family Alcidae. Bill compressed, often with deciduous covering; wing pointed but short, not extending beyond tail; tail short, rectrices 12–18; legs set far back on body; tarsus usually shorter than middle toe, usually reticulate; outer toe without claw about equal to middle toe; hallux absent; toes webbed.

No basipterygoid processes; supraorbital grooves large, perforated (Fig. 5–43); fontanels present; vertebrae all free, cervicals 15, dorsals with large bifid hemapophysis; metasternum produced, sometimes indented at midline, notches none, 1 pair, or 1 pair with 1 pair of fenestrae; coracoid hooked dorsally with hyposternal process, procoracoid process at right angle to shaft, brachial tuberosity not hollowed; furculum U-shaped; pelvis compressed, preacetabular ilium longer than postacetabular, pubis very long, equal to ilium; humerus compressed and rotated distally, ectepicondylar process relatively small; thumb absent; hypotarsus with 3 grooves.

Thigh muscles $A(B)X$ —; 1 ulnar tendon. Caeca rudimentary; intestine with at least 6 loops; gall bladder present. Both carotid arteries present (except in *Plautus*).

Swimmers and divers; food fish and marine in-

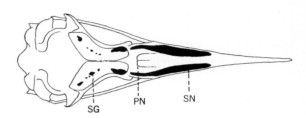

Fig. 5-43. **Skull of murre (*Uria aalge*). *PN*, premaxillary-nasal suture; *SG*, supraorbital groove for salt gland; *SN*, schizorhinal nostril.**

vertebrates. Nest in cavities among rocks or in burrows; eggs 1–2, pyriform or ovate, white, buff, or blue, plain or spotted with brown; young downy, tardily nidifugous.

Eocene onward. Holarctic coasts. 13 genera, 22 species; in the United States 12 genera, 17 species.

KEY TO GENERA OF ALCIDAE

1. Nostrils remote from feathers   2
   Feathering reaching nostrils   6
2. Front of tarsus scutellate   3
   Tarsus entirely reticulate   5
3. Depth of bill at base 3 times its width   *Fratercula*
   Depth of bill not more than twice its width   4
4. Inner claw largest and most curved   *Lunda*
   Inner claw not larger or more curved than the others   *Cerorhinca*
5. Width of bill at base greater than depth   *Ptychoramphus*
   Width of bill less than its depth   *Cyclorrhynchus*
6. Tarsus entirely reticulate   *Brachyramphus*
   Tarsus scutellate in front   7
7. Bill as wide as deep   *Plautus*
   Bill deeper than wide   8
8. Tarsus longer than middle toe without claw   9

Tarsus shorter than middle toe *Synthliboramphus*
9. Feathering on ramus forming an angle *Cepphus*
  Feathering on ramus forming an oblique line **10**
10. Exposed culmen twice length of tarsus *Pinguinus*
  Exposed culmen little longer than tarsus **11**
11. Depth of bill nearly equal to exposed culmen *Alca*
  Depth of bill less than ⅓ exposed culmen *Uria*

Genus *Plautus* Gunnerus. Bill quaillike, short, thick, and curved; feathering on ramus extending forward in an oblique line, feathering on culmen just reaching nostril; tail less than ⅓ wing, rounded or double-rounded, rectrices 12; tarsus about equal to shortest toe, longer than culmen, scutellate in front. Only 1 carotid artery. Atlantic quadrant of Arctic. Monotypic.

Plautus alle (**Linnaeus**). Dovekie. Upper parts glossy black; a white streak on upper eyelid; secondaries tipped and posterior scapulars streaked with white; throat and sides of head sooty brown; rest of underparts white; flanks striped with black; iris brown; bill black; feet flesh, webs dusky. WINTER: Throat and sides of head white; a faint dusky collar around base of neck. Wing 107–122, tail 31–41, culmen 12.5–15, tarsus 19–21 mm. Atlantic quadrant of Arctic from Baffin Land to Novaya Zemlya; winters south to coast of Europe and New Jersey, irregularly even to Florida and Cuba.

Genus *Alca* Linnaeus. Bill deep, compressed, and arched, sides with 2–3 curved vertical grooves; nostril feathered, feathering of ramus oblique; tail more than ⅓ wing, graduated, rectrices 12, tarsus about equal to culmen, shorter than middle toe, scutellate in front. North Atlantic. Monotypic.

Alca torda **Linnaeus.** Razor-billed auk. Head and neck all around dark brown, becoming black on rest of upper parts; narrow loral stripe white; secondaries white-tipped; underparts white; iris brown; bill black, grooves white; feet black. WINTER: Throat and sides of head white. YOUNG: As in winter, but bill without grooves or white. Wing 188–201, tail 72–88, culmen 32–35, tarsus 31–33 mm. North Atlantic south to Nova Scotia and Britain; winters south to Mediterranean and New York, occasionally farther.

Genus *Pinguinus* Bonnaterre. Bill deep, compressed, and arched, sides with 6–10 obliquely vertical grooves; nostril feathered, rami obliquely feathered; tail half length of wing, graduated, rectrices 14, tarsus half length of culmen or middle toe, scutellate. Flightless. North Atlantic (extinct). Monotypic.

Pinguinus impennis (**Linnaeus**). Great auk. Head and neck all around sooty brown; a large white oval on lores; upper parts black; secondaries white-tipped; underparts white; iris brown; bill black, grooves white; feet black. Throat probably white in winter. Wing 146, tail 76, culmen 80–89, tarsus 42 mm. Bred on islands from Newfoundland to Iceland and Orkneys; wintered south to Massachusetts, France, and Denmark, occasionally even to Florida and Spain. Extinct since 1844.

Genus *Uria* Brisson. Bill long, slender, compressed, ungrooved; nostrils feathered, rami obliquely feathered; tail less than ⅓ wing, rounded, rectrices 12; tarsus somewhat shorter than culmen or middle toe, scutellate on lower part of acrotarsium. Arctic and subarctic waters. 2 species.

KEY TO SPECIES OF *URIA*

1. Depth of bill at nostrils less than ⅓ exposed culmen
  U. aalge
  Depth of bill at nostrils more than ⅓ exposed culmen
  U. lomvia

Uria aalge (**Pontoppidan**). Common murre. Head and neck all around rich brown; upper parts dark grayish brown; secondaries white-tipped; underparts white; flanks streaked with grayish brown; iris brown; bill and feet black. WINTER: Throat and sides of head white. Wing 182–223, tail 39–53, culmen 40–58, tarsus 33–40 mm. Northern seas, breeding south to California, Nova Scotia, France, and Japan; winters south to Maine, occasionally farther.

Uria lomvia (**Linnaeus**). Thick-billed murre. Differs from *U. aalge* in having crown blackish, darker than the brown throat; base of maxillary tomium whitish; bill short and swollen at gonys. Wing 198–230, tail 42–57, culmen 31–46, tarsus 31–40 mm. Arctic seas, breeding south to Gulf of St. Lawrence; winters south to Long Island, North Sea, and Japan, irregularly farther south or inland.

Genus *Cepphus* Pallas. Bill long, slender, slightly deeper than wide, ungrooved; nostrils feathered; feathering on ramus making an angle; tail more than ⅓ wing, rounded, rectrices 12–14; tarsus about equal to exposed culmen, shorter than middle toe, scutellate. Arctic and subarctic coasts. 3 species, 2 in the United States.

KEY TO SPECIES OF *CEPPHUS*

1. Under wing coverts white   *C. grylle*
   Under wing coverts brownish gray   *C. columba*

Cepphus grylle (**Linnaeus**). Black guillemot. Rectrices 12. Black; a large white patch on upper wing coverts white; iris brown; bill black; feet red. WINTER: Upper parts varied black and white; underparts white. Wing 149–174, tail 44–56, culmen 28–34, tarsus 28–32 mm. Breeds from Arctic south to Maine and Scotland; winters south to Massachusetts, occasionally to New Jersey.

Cepphus columba **Pallas**. Pigeon guillemot. Differs in having 14 rectrices; white wing patch divided by a black bar; under wing coverts brownish gray; size larger. Wing 161–181, tail 45–52, culmen 30–34, tarsus 31–35 mm. Bering Sea south to Japan and California.

Genus *Brachyramphus* Brandt. Bill short, thin; feathering in contact with nostrils; feathering of ramus oblique; tail $\frac{1}{4}$ wing, rounded, rectrices 14; tarsus about equal to exposed culmen, much shorter than middle toe, reticulate. North Pacific. 4 species, 3 in the United States.

KEY TO SPECIES OF *BRACHYRAMPHUS*

1. Tarsus much shorter than middle toe without claw
              B. marmoratum
   Tarsus equal to or longer than middle toe without claw   **2**
2. Under wing coverts plain white   *B. hypoleucum*
   Under wing coverts mostly brownish gray
              B. craveri

Brachyramphus marmoratum (**Gmelin**). Marbled murrelet. Upper parts sooty brown; back barred with rust; underparts white barred with fuscous; axillars and wing lining fuscous; iris brown; bill black; feet flesh. WINTER: Crown fuscous; nuchal band white; upper parts fuscous tipped with gray; patch on scapulars white; underparts white; flanks striped with gray; axillars and wing lining fuscous. YOUNG: Resembles winter plumage but underparts mottled with dusky. Wing 112–

129, tail 28–35, culmen 14–16, tarsus 15–16 mm. North Pacific from Alaska to Japan and California.

Brachyramphus hypoleucum **Xantus**.[1] Xantus's murrelet. Upper parts slate; lower eyelid white; underparts, axillars, and under wing coverts white; flanks gray tipped with white; inner webs of primaries whitish; iris brown; bill black, pale blue at base of mandible; feet dusky, inner side of tarsus and upper side of toes and webs pale blue. Wing 113–127, tail 30–33, culmen 15–23, tarsus 22–25 mm. Coast of southern California and Baja California.

Brachyramphus craveri (**Salvadori**).[2] Craveri's murrelet. Differs in having upper parts browner; primaries without white; flanks without white tips; under wing coverts mostly brownish gray. Wing 117, tail 34–35, culmen 20, tarsus 22–23 mm. Breeds in Gulf of California; in fall north to Monterey Bay, California.

Genus *Synthliboramphus* Brandt. Frontal feathers semierect. Bill very short, thin; feathers reaching nostrils, on ramus slightly angled; tail about $\frac{1}{4}$ wing, slightly rounded, rectrices 14; tarsus slightly longer than middle toe without claw, twice as long as bill, scutellate. North Pacific. 2 species, 1 in the United States.

Synthliboramphus antiquum (**Gmelin**). Ancient murrelet. Crown black; face and throat dark brown; hind neck, sides of occiput, and above ears streaked with white; upper parts gray; wing and tail blackish; lower throat, underparts, and under wing coverts white; flanks sooty; iris brown; bill bluish white with a black stripe along culmen; feet bluish white. WINTER: Chin slate; throat white; no white streaks on head or neck; flanks striped gray and white. Wing 131–140, tail 33–39, culmen 13–14, tarsus 26–28 mm. North Pacific; south in winter to Japan and California; accidental inland.

Genus *Ptychoramphus* Brandt. Bill short, subconical; feathers remote from nostrils, on ramus angled; tail about $\frac{1}{4}$ wing, slightly rounded, rectrices 14; tarsus longer than culmen, shorter than middle toe, reticulate. American side of North Pacific. Monotypic.

Ptychoramphus aleuticum (**Pallas**). Cassin's auklet. Above blackish slate; eyelids white; throat, flanks, and wing lining paler slate; underparts

---

[1] *Endomychura hypoleuca* of A.O.U.
[2] *Endomychura craveri* of A.O.U.

white; iris white; bill black, yellowish at base of mandible; feet bluish. Wing 110–129, tail 25–34, culmen 19–20, tarsus 24–25 mm. Pacific Coast from Aleutians to Baja California.

Genus _Cyclorrhynchus_ Kaup. Bill short, deep, mandible convex, gonys much recurved; feathers remote from nostrils, angled on rami; tail about $\frac{1}{4}$ wing, slightly rounded, rectrices 14; tarsus longer than culmen, shorter than middle toe, reticulate. North Pacific. Monotypic.

Cyclorrynchus psittacula (**Pallas**). Parakeet auklet. Upper parts blackish slate; throat and flanks paler gray; wing lining pale gray with a few white streaks; long plumes behind eye white; bill orange-red, with deciduous nasal shield horn-colored and swellings at corners of mouth white; iris white; feet pale blue, blackish externally, on joints, and middle of webs. WINTER: Lacks ear plumes and swellings on bill; throat and flanks white. Wing 141–152, tail 39–43, culmen 14–16, tarsus 27–31 mm. Bering Sea; in winter south to Japan and California.

Genus _Cerorhinca_ Bonaparte. Bill deep and narrow, culmen curved; feathers remote from nostril, angled on ramus; tail about $\frac{1}{4}$ wing, rounded, rectrices 16–18; tarsus shorter than culmen, much shorter than middle toe, with lower half of acrotarsium scutellate. North Pacific. Monotypic.

Cerorhinca monocerata (**Pallas**). Rhinoceros auklet. Upper parts sooty black; sides of head and neck, axillars, under wing coverts, flanks, and crissum ashy gray; throat, breast, and belly white; 2 pairs of long white plumes, from behind eyes and corners of mouth; iris brown; feet ivory, planta tarsi and soles black; bill orange, the culmen and borders of hornlike appendage on cere black. WINTER: Plumage, including head plumes, as in summer; bill lacks horn. YOUNG: Lacks head plumes and horn. Wing 169–183, tail 43–60, culmen 32–39, tarsus 27–30 mm. North Pacific, south to Japan and California.

Genus _Fratercula_ Brisson. Bill short, compressed almost as deep as long, with 3–4 grooves on side; feathering remote from nostrils, perpendicular on ramus; tail less than $\frac{1}{3}$ wing, slightly rounded, rectrices 16; tarsus much shorter than middle toe, scarcely $\frac{1}{2}$ length of culmen, scutellate; inner claw largest and most curved. Nest in burrows. Arctic and subarctic. 2 species.

KEY TO SPECIES OF _FRATERCULA_

1. Black of throat not reaching bill; tail 42–53 mm
   F. arctica
   Black of throat reaching bill; tail 60–86 mm
   F. corniculata

Fratercula arctica (**Linnaeus**). Atlantic puffin. Bill grooves 4, broad, deep, oblique. Crown and nape grayish brown; upper parts blackish, continuing as a browner collar around upper throat; chin and sides of head ashy white, with a darker patch on malar region; wing lining ashy gray; underparts white; iris brown; eyelids red, with blue obtuse appendages; base of bill, tip terminal groove, and rictal rosette yellow; subbasal ridge blue; rest of bill red; feet red. WINTER: Appendages of bill and eyelids molted; bill grooved; lores and orbit blackish. YOUNG: No appendages or grooves. Wing 157–177, tail 42–53, culmen 44–55, tarsus 24–29 mm. North Atlantic and adjacent Arctic, south to Maine and Portugal; winters south to Mediterranean and Massachusetts or farther.

Fratercula corniculata (**Naumann**). Horned puffin. Bill more convex; grooves 3, narrow, shallow, vertical; appendage on upper eyelid long and slender, the lower one obtuse. Crown grayish brown; sides of head white; neck all around and upper parts black; chin brownish gray; underparts white; under wing coverts ashy gray; iris brownish gray; eyelids red, appendages blackish; feet red; bill mostly red, base yellow, rictal rosette orange. WINTER: Bill and eye appendages molted; base of bill dusky; eyelids brownish gray; sides of head gray; orbit and lores blackish. YOUNG: Without appendages or grooves; bill smaller and brownish. Wing 168–187, tail 60–68, culmen 46–55, tarsus 25–31 mm. North Pacific and adjacent Arctic, south to Japan and Queen Charlotte Islands, rarely to California.

Genus _Lunda_ Pallas. Bill compressed, arched, nearly as deep as long, maxilla with 3 vertical grooves, mandible smooth; tail about $\frac{1}{3}$ wing, double-rounded, rectrices 16; tarsus much shorter than bill or toes, scutellate; inner claw longest and most curved. Nest in burrows. North Pacific. Monotypic.

Lunda cirrhata (**Pallas**). Tufted puffin. Sooty black, becoming browner on face and underparts; area from eye to eye across forehead, lores, and suborbital area white; long postocular tufts cream;

bill red at base, green at tip, rictal rosette purplish; iris white; eyelids red; feet red. WINTER: Crests and bill sheaths molted. YOUNG: Bill ungrooved, brownish. Wing 180–206, tail 57–64, culmen 53–65, tarsus 29–34 mm. Bering Sea and North Pacific south to Japan and California.

Order Columbiformes. Pigeonlike birds. Plumage dense, easily detached; adult downs absent or on apteria only and scarce; aftershafts absent or very small; oil gland naked or absent; no lateral cervical apteria; skin thin. Bill small; culmen decurved; nostrils imperforate; rhamphotheca simple. Legs short; toes unwebbed. Primaries 11 (10 functional); secondaries 11–15, usually aquintocubital. Rectrices 12–20.

Schizognathous, with vomer rudimentary or absent; schizorhinal; supraorbital grooves absent. Vertebrae heterocoelous, cervicals 14–16 with hemapophyses forming half canals. Sternum with 1–2 pair of notches or 1 pair and 1 pair of foramina. Coracoids in contact, with procoracoid bent upward but imperforate. Furculum U-shaped. Humerus with 1 fossa, with prominent deltoid crest and minute but raised ectepicondylar process. Tibial bridge ossified. Hypotarsus complex, with 1–3 canals.

Thigh muscles $ABX(Y) \pm$; plantar tendons type 1. Tongue pointed; crop globular; gall bladder usually absent. Carotids double.

Food vegetable; imbibe like horse without raising head. Nest in tree, on ground, or on cliff; eggs 1–3, ellipsoidal; nestling with sparse or uniform down.

Oligocene onward. Cosmopolitan. In the United States 1 family.

Family Columbidae. Doves. Both adult downs and aftershafts absent. Bill slender, soft, and swollen at base with middle part constricted; nostrils slitlike, overhung by operculum (Fig. 5-44). Wings long, pointed. Tarsus scutellate in front; hallux functional.

Basipterygoid processes present, medial; cervicals 15; sternum with both internal and external spines (Fig. 5-45); furculum without hypocleidium. Caeca rudimentary; intestinal convolutions type 1; food seeds or fruit. Nest frail, of sticks, usually in tree; eggs 1–2, white. Nestling hatched blind with sparse down; altricial; fed at first by pigeon milk from crop lining of both parents.

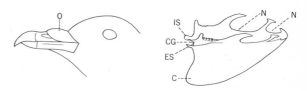

Fig. 5-44 (left). Bill of band-tailed pigeon (*Columba fasciata*). *O*, operculum. *Fig. 5-45* (right). Sternum of rock dove (*Columba livia*). *C*, carina; *CG*, coracoidal groove; *ES*, external spine; *IS*, internal spine; *N*, notch.

Miocene onward. Cosmopolitan, 50 genera, 291 species; in the United States 9 genera, 14 species.

KEY TO GENERA OF COLUMBIDAE

1. Tarsus shorter than outer toe with claw   2
   Tarsus longer than outer toe with claw   4
2. Middle rectrices pointed, tail graduated   *Ectopistes*
   Middle rectrices not pointed, tail square or round   3
3. Upper part of tarsus feathered   *Columba*
   Tarsus unfeathered   *Streptopelia*
4. Wing pointed, with outer primary about as long as longest   5
   Wing rounded, with outer primary decidedly shorter than longest   6
5. Tail graduated, rectrices pointed   *Zenaidura*
   Tail and rectrices rounded   *Zenaida*
6. Distal third of outer primary abruptly narrowed   *Leptotila*
   Outer primary normal   7
7. Tarsus longer than middle toe with claw   *Geotrygon*
   Tarsus shorter than middle toe with claw   8
8. Tail graduated, about as long as wing   *Scardafella*
   Tail rounded, shorter than wing   *Columbigallina*

Genus *Columba* Linnaeus. Wing long, 2d to 4th primaries longest; tail square or rounded, $\frac{1}{2}$ to $\frac{4}{5}$ length of wing; rectrices 12; tarsus shorter than outer toe with claw, feathered proximally. Nearly cosmopolitan, absent from New Zealand and Hawaii. 51 species, 4 in the United States.

KEY TO SPECIES OF *COLUMBA*

1. Rump white; throat metallic   *C. livia*
   Rump not white; throat not metallic   2

2. Pileum white or gray   *C. leucocephala*
    Pileum purplish or brownish   3
3. Bill tipped with black   *C. fasciata*
    No black on bill   *C. flavirostris*

Columba livia Gmelin. Rock dove; domestic pigeon. Tarsus longer than middle toe without claw; tail rounded. Bluish gray; rump white in most races; wing with 2 black bars; tail tipped with black; throat and breast metallic green or purple; iris orange-red; bill dusky, cere whitish; feet red. Wing 184–230, tail 118–136, culmen 14–22, tarsus 27–34 mm. Palearctic, India, and Africa. Feral birds widely established in the United States.

Columba leucocephala Linnaeus. White-crowned pigeon. Tarsus shorter than middle toe with claw; tail square. Crown white or smoky gray; hind neck of adult metallic green scaled with black; rest of plumage slate; iris white; bill dull red tipped with pale green; feet red. Wing 172–204, tail 112–145, culmen 15–20, tarsus 22–28 mm. West Indies to Florida Keys.

Columba fasciata Say. Band-tailed pigeon. Tarsus shorter than middle toe; tail rounded. Entire head, throat, and underparts purplish drab; crissum and middle of belly mostly white; sides and wing lining bluish slate; a white collar on nape; hind neck metallic bronze; back grayish brown glossed with bronze; wing coverts bluish slate; greater coverts edged with white; remiges dusky with narrow whitish edges; rump, upper tail coverts, and base of tail bluish slate; a black bar across middle of tail, tip brown; iris yellow with a pink ring externally; orbit red; bill yellow, tip black; feet yellow. YOUNG: Lacks white collar and metallic colors. Wing 191–230, tail 121–151, culmen 15.5–20, tarsus 24–29.5 mm. From British Columbia south, in mountains from Utah to Nicaragua.

Columba flavirostris Wagler. Red-billed pigeon. Tarsus shorter than middle toe; tail rounded. Head and neck all around, breast, and inner upper wing coverts purplish red; back and remiges brown; rest of wing coverts, rump, upper tail coverts, wing lining, belly, and crissum bluish slate; tail blackish slate; iris orange; eyelids and feet red; bill red, tip whitish (whole bill fades to yellow in skins). Wing 180–207, tail 105–127, culmen 13–16, tarsus 23–28 mm. Lowlands of Mexico and South Texas to Costa Rica.

Genus Streptopelia Bonaparte. Wing long, 2d to 3d primary longest; tail more than $\frac{2}{3}$ wing, rounded, rectrices 12; tarsus unfeathered, shorter than outer toe with claw. Palearctic, Oriental, and Ethiopian regions. 16 species, 2 established in the United States.

KEY TO SPECIES OF *STREPTOPELIA*

1. Black feathers of hind neck forked, each with 2 white apical spots   *S. chinensis*
    Black collar edged with whitish, feathers not forked   *S. risoria*

Streptopelia risoria (Linnaeus). Ringed turtle dove. Head, neck, and underparts pinkish buff; collar on hind neck black edged with whitish; upper parts sandy; tail drab, becoming whitish at tip and laterally; iris orange; eye ring and feet red; bill blackish. Wing 154–183, tail 112–115, culmen 13–17, tarsus 21–25 mm. Southeast Europe to Japan and India. A pale variety is widely domesticated and established in West Indies, Florida, and Los Angeles.

Streptopelia chinensis (Scopoli). Chinese spotted dove. Head and underparts pinkish gray, paler on throat and belly; crissum white; collar on hind neck black, each feather bifurcate with 2 white tips; upper parts earth brown; outer rectrices black with broad white tips; iris yellowish pink, orbit lead-colored; feet violet red. Wing 143–161, tail, 123–154, culmen 15–17 mm. China to India and Borneo; introduced to Hawaii and Los Angeles.

Genus Ectopistes Swainson. Wing long and pointed, outer 2 primaries longest; tail nearly twice as long as wing, graduated for half its length, rectrices 12, pointed; tarsus about equal to middle toe without claw, feathered at base. Eastern North America. Monotypic.

Ectopistes migratorius (Linnaeus). Passenger pigeon. Head, upper throat, hind neck, and rump bluish gray; sides of neck glossy purplish bronze; upper parts gray and grayish brown, blotched with black on scapulars and proximal wing coverts; remiges dusky, tawny externally and narrowly edged with whitish; middle rectrices brownish gray, dusky toward tip; lateral rectrices passing from pearl gray to white, with a black and chestnut spot subbasally on inner web; lower throat, breast, and sides purplish chestnut; belly and crissum white; wing lining pale bluish gray; iris, orbit, and feet red; bill black. FEMALE: Duller; head brownish gray; upper parts brownish, spotted with black; underparts drab gray; neck less glossy. Wing 175–214, tail 142–211, culmen 15–18.5, tarsus 25.5–29 mm. Eastern deciduous forest; bred south to Kansas, northern Mississippi, Kentucky, and Pennsylvania; migrated south to Florida and Louisiana. Extinct since 1914.

Genus *Zenaidura* **Bonaparte.** Wing pointed, 2d primary longest; tail ¾ wing length, sometimes exceeding wing, graduated, rectrices 14, pointed; tarsus bare, about equal to middle toe without claw. North and South America. 3 species, 1 in the United States.

Zenaidura macroura **(Linnaeus).** Mourning dove. Forehead and circumorbital area fawn; spot on auriculars black glossed with blue; sides of neck glossy bronze; crown and nape bluish slate; upper parts grayish brown; scapulars and tertials blotched with black; remiges slate, outer one narrowly edged with white; middle rectrices bluish slate; next 2 pairs slate crossed by a black medial bar; 4 outer pairs bluish slate, whitish at tip, crossed by black medial bar; outer web of outer rectrix white; chin whitish; throat and breast vinaceous; belly and crissum pinkish buff; sides and wing lining bluish gray; iris brown; orbital skin blue; bill black, rictus red, cere gray; feet red. FEMALE: Duller; upper parts more brownish; face dirty white; breast mixed gray and pink; belly pale buff; crissum whitish. YOUNG: Plumage with pale tips. Wing 130–156, tail 97–162, culmen 12–15, tarsus 18–22 mm. North America, breeding south to West Indies and Mexican plateau; winters from southern United States to Panama.

Genus *Zenaida* **Bonaparte.** 1st to 3d primaries longest; tail rounded, less than ¾ wing, rectrices 12, rounded; tarsus unfeathered, about equal to (*Z. asiatica*) or longer than (*Z. aurita*) middle toe without claw. Middle and South America. 2 species.

KEY TO SPECIES OF *ZENAIDA*

1. Wing coverts with a large white patch   **Z. *asiatica***
   Wing coverts spotted with black   **Z. *aurita***

Zenaida aurita **(Temminck).** Zenaida dove. Forehead fawn; crown brown; 2 blue-black spots on auriculars; sides of neck glossy purple; upper parts brown; tertials and wing coverts spotted with black; remiges dusky, inner ones white-tipped; primaries narrowly edged with white; tail gray with black subterminal bar, outer rectrices tipped with pale gray; chin buffy; underparts vinaceous; sides and wing lining bluish gray; iris brown; orbit blue; bill black; feet red. Wing 136–165, tail 77–111, culmen 12.5–17, tarsus 21–27 mm. West Indies, Yucatan, and Florida Keys (formerly).

Zenaida asiatica **(Linnaeus).** White-winged dove. Crown and hind neck purple, glossy on sides of neck; auricular patch blue-black; upper parts and middle rectrices brown; a large white patch on outer wing coverts; remiges dusky, secondaries broadly, primaries narrowly, edged with white; outer rectrices bluish gray, subterminal band black, tip white; throat and breast vinaceous; belly, crissum, sides, and wing lining bluish gray; iris red; orbit blue; bill black; feet red. Wing 142–166, tail 85–115, culmen 17.5–22.5, tarsus 22–26 mm. Southern parts of Arizona, New Mexico, and Texas, south to Chile; occurs on migration in Louisiana and Florida (breeds on Keys).

Genus *Scardafella* **Bonaparte.** Bill slender, weak; wings short, 2d to 4th primaries longest, outer primary somewhat bowed; tail equal to wing, rectrices 12, narrow, with outer pair much shorter than rest; tarsus longer than middle toe without claw, with base slightly feathered. Southwestern United States to Brazil. 2 species, 1 in the United States.

Scardafella inca **(Lesson).** Inca dove. Upper parts grayish brown tinged with pink, scaled with black; remiges rich chestnut, dusky at tips; middle pair of rectrices grayish brown; next 2 pairs grayish brown with black tips; outer rectrices black basally, broadly white distally; upper throat whitish; lower throat and breast pale violet scaled with dusky; belly and crissum buff barred with dusky; outer wing coverts chestnut, inner ones velvety black; axillars chestnut with narrow black tips; iris orange; bill black; feet pink. Wing 86–96, tail 85–103, culmen 11–14, tarsus 15–16.5 mm. Southern parts of Texas, New Mexico, and Arizona, south to Costa Rica.

Genus *Columbigallina* **Boie.** Bill short, weak;

2d to 3d primaries longest, slightly exceeding secondaries; tail less than ¾ wing, rounded; rectrices 12, broad; tarsus longer than middle toe without claw, base with some feathering. Southern United States to South America. 5 species, 1 in the United States.

Columbigallina passerina (**Linnaeus**). Ground dove. Forehead, sides of head, and underparts smoky pink; throat and breast squamate with dusky; crissum often edged with white; crown and hind neck bluish-gray squamate with dusky; back and upper tail coverts grayish brown; wing coverts smoky pink spotted with glossy purplish blue; remiges chestnut, tipped and edged with dusky; tail dull gray, broadly black terminally, and often edged with white at tip; wing lining chestnut; iris orange; bill red; feet pink. FEMALE: Forehead, lower throat, and breast pinkish gray scaled with dusky; wing coverts pinkish gray spotted with bluish purple; chin whitish; posterior underparts gray mixed with white on midline and crissum. Wing 83–91, tail 55–65, culmen 11–12, tarsus 15–17.5 mm. South Atlantic and Gulf states from South Carolina to Texas; southeast California and southwest Arizona south to Ecuador and Brazil; West Indies.

Genus *Leptotila* Swainson. Wing rounded, 3d or 4th primaries longest, outer primary short with terminal third very narrow, 3d to 6th primaries incised on outer web; tail ¾ wing, rounded; rectrices 12, broad; tarsus naked, longer than middle toe with claw. Middle and South America. 8 species, 1 in the United States.

Leptotila verreauxi (**Bonaparte**). White-fronted dove; white-tipped dove. Crown and hind neck grayish violet, paler on forehead; upper back glossed with bronze; upper parts and middle rectrices grayish brown; remiges dusky brown; lateral rectrices black, tipped with white; chin whitish; breast and sides of neck vinaceous; flanks buff; belly and crissum white; wing lining chestnut; iris yellow; orbit blue; bill black; feet red. Wing 135–160, tail 95–114, culmen 14–17.5, tarsus 27–33 mm. Vicinity of Brownsville, Texas, south to Argentina.

Genus *Geotrygon* Gosse. Wing rounded, bowed; 3d or 4th primary longest, 1st and 2d incised on inner web, 2 to 5th or 6th on inner web; tail less than ⅔ wing, rounded; rectrices 12, broad; tarsus

naked, longer than middle toe with claw. Middle and South America. 16 species, 1 in the United States.

Geotrygon chrysia **Bonaparte.** Key West quail-dove. Forehead and lores chestnut; crown and nape glossy green and purple; hind neck metallic bronze; upper parts chestnut, glossed with purple on back; a white stripe from bill to auriculars bordered below by a chestnut stripe; upper throat white; lower throat and breast pale vinaceous; belly and crissum buffy white; sides brownish; wing lining chestnut; iris orange; bill horn, red at base; feet pink. YOUNG: Forehead pale grayish brown; upper parts chestnut, tipped with cinnamon buff; white stripe from bill to auriculars; throat and breast grayish brown tipped with cinnamon; posterior underparts grayish buff. Wing 150–159, tail 90–103, culmen 14.5–19.5, tarsus 27–30.5 mm. Bahamas, Cuba, and Hispaniola; occasional on Florida Keys.

Order Psittaciformes. Parrots. Plumage harsh and dense; adult downs on both apteria and pterylae; aftershafts present; oil gland tufted; skin thick. Bill short, hooked, with cere; rostrum movable; nares imperforate; roof of mouth rough; tongue thick and blunt. Primaries 10; secondaries 8–14, aquintocubital. Rectrices 12 (14 in 1 genus). Tarsus shorter than longest toe, reticulate; feet zygodactyl with 4th toe reversible (Fig. 5-46).

Desmognathous without basipterygoid processes; holorhinal; supraorbital grooves absent. Dorsal vertebrae opisthocoelous; cervicals 13–14. Sternum without internal spine, external spine large and turned up; metasternum usually with 1 pair of fenestrae; keel usually deep. Procoracoid large, not perforated. Furculum U-shaped, without hypocleidium. Humerus with a perforated fossa and small ectepicondylar process. Tibial bridge ossified or not. Hypotarsus with 1 or more canals.

Thigh muscles *AXY* ±; plantar tendons galline (type 1). Crop present; intestine telogyrous (type 4); no caeca; gall bladder usually absent. Both carotids usually present (only the left in 1 genus).

Food seeds and fruits. Flight bulletlike. Always in pairs or flocks. Voice chattering. Nest in hollow trees; eggs white, subspherical, 2–6. Young usually hatched blind and naked, later downy; altricial.

Miocene onward. Tropicopolitan. 1 family.

Family Psittacidae. Parrots. 72 genera, 326

species; in the United States 3 genera, 3 species (1 introduced).

KEY TO GENERA OF PSITTACIDAE

1. Cere exposed   *Melopsittacus*
   Cere feathered   **2**
2. Exposed culmen shorter than longest toe with claw
            Conuropsis
   Exposed culmen longer than longest toe with claw
            Rhynchopsitta

Genus *Melopsittacus* Gould. Bill longer than deep; cere naked; outer primary thin, pointed, shorter than primary coverts; 2d or 3d primaries longest; tail ¾ wing or longer, strongly graduated, with middle rectrices acuminate and lengthened; tarsus equal to longest toe without claw. Australia; introduced in the United States.

Melopsittacus undulatus (**Shaw**). Budgerigar; grass parakeet. Forehead, line along side of bill, and large patch on lower cheeks yellow; crown, hind neck, subocular area, and auriculars yellowish green, barred with black; posterior malar feathers purple-tipped; back, scapulars, and lesser coverts dusky, barred with pale yellow; rump, upper tail coverts, and underparts green; remiges olive brown, the secondaries and inner primaries with concealed white wedge; rectrices dark green with broad, yellow subterminal bar; middle rectrices dark blue; bill horn; feet gray. White, blue, and yellow varieties exist in domestication. Wing 91–100, tail 75–106, culmen from cere 9–10, tarsus 12–13 mm. Australia; established around Tampa Bay, Florida.

Genus *Conuropsis* Salvadori. Bill deeper than long; cere feathered; wing long, pointed, 2d to 3d primaries longest; tail ¾ wing, graduated and wedge-shaped; lateral rectrices acuminate; tarsus longer than 2d toe, shorter than 3d or 4th. Eastern United States (extinct). Monotypic.

Conuropsis carolinensis (**Linnaeus**). Carolina parakeet. Green; forehead and front of face orange; head, neck, bend of wing, middle of belly, and lower part of thighs yellow; iris brown; orbit and feet pinkish white; bill cream. YOUNG: Forehead and lores orange; otherwise entirely green. Wing 175–203, tail 130–167, culmen 21.5–26, tarsus 15.5–18 mm. Florida and Gulf states north to Vir-

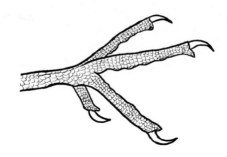

Fig. 5-46. Foot of grass parakeet (*Melopsittacus undulatus*), enlarged.

ginia and the Ohio and Missouri Valleys, casually farther. Last collected in 1901 in Florida, where more or less authentic sight records were made until about 1920.

Genus *Rhynchopsitta* Bonaparte. Bill as deep as long; cere feathered; wing long, pointed, 2d primary longest; tail ⅔ wing, graduated, tips of rectrices rounded; tarsus equal to 2d toe without claw, shorter than 3d or 4th. Northern part of Mexican plateau. 1 species.

Rhynchopsitta pachyrhyncha (**Swainson**). Thick-billed parrot. Green; forehead, lores, edge of wing, and thighs red; greater under wing coverts yellow; remiges and rectrices blackish underneath; iris reddish; bill and feet black. Wing 255–273, tail 165–195, culmen 37–41, tarsus 19–22 mm. Pine zone in mountains of northern Mexico, occasionally into southern Arizona.

Order Cuculiformes. Cuckoos and plantain-eaters. Plumage loose; adult downs scarce and only on apteria; primaries 10, secondaries quintocubital; tail longer than wing, usually graduated, rectrices 8–10. Bill more or less decurved, without cere; nares imperforate; tarsus scutellate; feet zygodactyl.

Desmognathous without basipterygoid processes; holorhinal; mandibular process hooked inward; vertebrate heterocoelous, cervicals 13–15; sternum fenestrate or notched; procoracoid bent upward, imperforate; humerus with 1 open fossa and small ectepicondylar process; pelvis broad posteriorly; tibial bridge ossified; hypotarsus with 2 canals. Thigh muscles $A(B)XY+$; plantar tendons type 1. Both carotids present. Nestling altricial.

Oligocene onward. Cosmopolitan. 1 family in the United States.

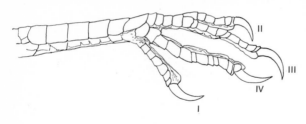

Fig. 5-47. Foot of ani (*Crotophaga ani*), enlarged.

Family Cuculidae. Cuckoos. Tomia smooth; 4th toe permanently reversed; planta tarsi as well as acrotarsium scutellate (Fig. 5-47); oil gland naked; no aftershafts. Vomer present but small; 14 cervicals; sternum with large common spine; coracoids separate; furculum with hypocleidium. Intestine with 4 loops; caeca present. Food insects. Some species build own nest, others are communal, and many are nest parasites. Oligocene onward. Cosmopolitan. 34 genera, 126 species; in the United States 3 genera, 6 species.

KEY TO GENERA OF CUCULIDAE

1. Culmen with a knifelike ridge; plumage black
   Crotophaga
   Culmen without a compressed ridge; plumage not black   **2**
2. Head crested   *Geococcyx*
   Head not crested   *Coccyzus*

Genus *Coccyzus* Vieillot. Culmen about equal to tarsus; wing rounded, outer primary not much longer than secondaries, 3d or 4th primary longest; tail slightly longer than wing, graduated; rectrices 10, rounded at tips; tarsus slightly longer than middle toe with claw, extreme base feathered. North and South America. 8 species, 3 in the United States.

KEY TO SPECIES OF *COCCYZUS*

1. Entire bill black   *C. erythropthalmus*
   Bill partly yellow   **2**
2. Auriculars black; under parts buff   *C. minor*
   Auriculars hardly darker than crown; underparts pure white   *C. americanus*

Coccyzus erythropthalmus (Wilson). Blackbilled cuckoo. Above brownish olive glossed with bronze; remiges slightly tinged with rufous along edge; tail with small whitish tips; lores and malar region gray; throat, breast, and crissum grayish buff; breast and belly white; wing lining buffy; iris brown; orbit red; bill black, bluish at base of mandible; feet bluish. Wing 133–147, tail 142–164, culmen 21.5–26, tarsus 21–25.5 mm. Eastern deciduous forest, breeding south to Georgia, Arkansas, and Kansas; winters in South America.

Coccyzus americanus (Linnaeus). Yellowbilled cuckoo. Above grayish bronze; head grayer; base of primaries rufous, especially on inner webs; lateral rectrices with broad white tips; auriculars slightly darker than crown; underparts white; throat and thighs tinged with pale gray; wing lining pale buff; iris brown; orbit and feet grayish; bill black, mandible mostly yellow. Wing 135–156, tail 134–156, culmen 24–30, tarsus 23–29 mm. Southern Canada south to West Indies and northern Mexico; winters in South America.

Coccyzus minor (Gmelin). Mangrove cuckoo. Crown gray; auriculars black; upper parts grayish brown; remiges pale cinnamon on inner webs and with a tinge of cinnamon on outer webs of primaries; lateral rectrices black, broadly tipped with white; underparts rich buff. Wing 125–148, tail 147–177, culmen 24.5–30, tarsus 25–31 mm. Coast of south Florida, West Indies, both coasts of Middle America, Caribbean coast of South America.

Genus *Crotophaga* Linnaeus. Culmen sharp, arched, shorter than tarsus; wing much rounded, outer primary shorter than secondaries, 4th to 5th longest; tail much longer than wing, graduated; rectrices 8, broadest terminally; tarsus longer than middle toe with claw. Tropical America. 3 species, 2 in the United States.

KEY TO SPECIES OF *CROTOPHAGA*

1. Bill with several oblique grooves   *C. sulcirostris*
   Bill smooth   *C. ani*

Crotophaga ani Linnaeus. Smooth-billed ani. Black; feathers of anterior half squamate with metallic purple; wings and tail glossed with purple and blue; iris brown; bill and feet black. Wing

141–161, tail 161–201, culmen 28–35, tarsus 34–40 mm. South Florida (breeds), West Indies, Panama, South America.

Crotophaga sulcirostris **Swainson.** Groove-billed ani. Maxilla with 3, mandible with 2 oblique grooves. Metallic squamations and gloss on wings and tail more bluish. Wing 128–157, tail 160–197, culmen 25–30, tarsus 31–36 mm. Lower Rio Grande Valley and Mexico south to Guiana and Peru.

Genus *Geococcyx* **Wagler.** Occiput crested; bill longer than head; wing much rounded, 4th to 6th primaries longest, secondaries about as long as primaries; tail more than half again as long as wing, graduated; rectrices 10, tapering; tarsus much longer than toes, about equal to culmen from base. Southwestern United States, Mexico, and Central America. 2 species, 1 in the United States.

Geococcyx californianus **(Lesson).** Roadrunner. Crown and nape glossy blue-black, streaked with rusty and buff; feathers of back olive glossed with bronze, with black, rusty, and whitish margins; rump grayish brown; upper tail coverts and middle rectrices bluish bronze with narrow white outer margins and broad white tips; remiges greenish bronze edged with white; chin white; auriculars streaked black and white; throat and upper breast buff streaked with black; posterior underparts buffy white; wing lining black; iris orange; orbit bright blue, becoming orange on bare skin of neck; bill dusky, mandible horn basally; feet pale blue, with patches of cream. Wing 162–196, tail 260–316, culmen 48–59, tarsus 55–67 mm. Arid plains from California and Nevada east to Arkansas and Louisiana, south to central Mexico.

Order Strigiformes. Owls. Plumage fluffy, covering base of bill, lores, legs, and often toes; aftershafts rudimentary or absent; adult downs scarce, on apteria only. Bill hooked, with cere; nostrils imperforate, usually located at anterior edge of cere; tomia unnotched. Feet zygodactyl, with 4th toe reversible; claws raptorial. Primaries 11 (10 functional), outer edge with barbs reversed; secondaries 11–18, acquintocubital. Rectrices usually 12 (10 in 1 genus), shorter than wing. Eyes directed forward; face surrounded by a ruff of feathers.

Schizognathous with desmognathous tendencies; vomer and basipterygoid processes present; maxillopalatine processes spongy; holorhinal. Cervicals 14. Sternum 2–4 notched; internal spine absent.

Coracoids cross, procoracoid bent upward and perforate. Fuculum weak, without hypocleidium. Humerus with 1 pneumatic fossa, ectepicondylar process small and distal. Tibial bridge unossified; fibula fused distally. Hypotarsus simple with 1 large groove.

Thigh formula A; plantar tendons type 1. No dilated crop; intestine type 6; caeca long; gall bladder usually present. 2 carotids. Syrinx bronchial.

Nocturnal birds of prey; flight silent. Voice a hoot or whistle. Nest in hollow tree, on ground, or an abandoned nest of hawk or crow; eggs white, obovate or subspherical, 2–10; incubation commences with first egg. Young altricial, covered with woolly down.

Eocene onward. Cosmopolitan. 2 living families.

KEY TO FAMILIES OF STRIGIFORMES

1. Middle claw pectinate on its inner edge **Tytonidae**
   Middle claw smooth **Strigidae**

Family Tytonidae. Barn owls. Primaries not notched; tail emarginate; legs long, extending beyond tail; feathers of planta tarsi reversed; middle claw pectinate on inner edge; oil gland with 2–3 filoplumes.

Skull narrow; palatines narrow and nearly straight; sternum with 1 pair of shallow notches, without manubrium; pelvis narrow; tarsus longer than femur, without bony ring. Eggs obovate, 4–6.

Miocene onward. Practically cosmopolitan (absent from New Zealand and Hawaii). 2 genera, 10 species. 1 species in the United States.

Genus *Tyto* **Billberg.** Facial disc complete; no ear tufts; outer primary about equal to 3d; toes bristled; tail about 40% wing length; inner toe as long as middle toe. Nearly cosmopolitan. 9 species, 1 in the United States.

Tyto alba **(Scopoli).** Barn owl. Above buff, everywhere vermiculated and spotted with gray and white; remiges and tail vermiculated but without spots and crossed by several dark bands; face white or buff; a brown spot before eye; facial ring brownish; underparts white or buff with dusky spots; iris brown; bill yellow; toes brownish. Wing 314–360, tail 126–157, culmen 21–24, tarsus 67–77 mm. All continents; in America from British

Columbia, North Dakota, Great Lakes, and Massachusetts to Tierra del Fuego.

Family Strigidae. Typical owls. Inner web of 1–6 primaries notched; tail rounded or truncate; legs usually not extending beyond tail; feathers of planta tarsi not reversed; middle claw not pectinate; oil gland naked.

Skull broad; palatines broad and bowed; sternum with external spine forming a manubrium; metasternum with 2 deep pairs of notches; pelvis broad; tarsus with a bony ring on upper anterior face, usually shorter than femur. Eggs subcylindrical.

Oligocene onward. Cosmopolitan. 22 genera, 123 species. In the United States 10 genera, 17 species.

KEY TO GENERA OF STRIGIDAE

1. Cere bulbous, with nostrils opening through it (Fig. 5-48)   2
   Cere flat, with nostrils opening at its anterior edge (Fig. 5-49)   4
2. Tarsus about twice middle toe without claw   *Speotyto*
   Tarsus about equal to middle toe with claw   3
3. Rectrices 10; tarsus bristled   *Micrathene*
   Rectrices 12; tarsus feathered   *Glaucidium*
4. Under tail coverts reaching tip of tail   *Nyctea*
   Under tail coverts falling short of tip of tail   5
5. With ear tufts   6
   Without ear tufts   8
6. Facial disc complete; eyes in middle of disc   *Asio*

Fig. 5-48 (left). Bill of burrowing owl (*Speotyto cunicularia*). *Fig. 5-49* (right). Bill of screech owl (*Otus asio*).

Facial disc incomplete above and below; eyes near top of disc   7
7. Tail more than half length of wing; wing over 300 mm   *Bubo*
   Tail half length of wing or less; wing under 200 mm   *Otus*
8. Graduation of tail much greater than middle toe with claw   *Surnia*
   Graduation of tail less than middle toe with claw   9
9. Outer primary shorter than secondaries; wing over 200 mm   *Strix*
   Outer primary equal to secondaries; wing under 200 mm   *Aegolius*

Genus *Otus* Pennant. With ear tufts; facial ring incomplete, with eyes near top; nostrils at edge of cere; cere flat, shorter than culmen; outer primary usually decidedly shorter than secondaries; 3d to 5th primaries longest; 2–6 outer primaries emarginate on inner web; tail half length of wing or less, nearly truncate; rectrices 12; tarsus usually entirely feathered, equal to or longer that middle toe with claw; toes bristled, feathered at extreme base. Europe, Asia to Moluccas, Africa and Madagascar, North and South America. 36 species, 3 in the United States.

KEY TO SPECIES OF *OTUS*

1. Toes entirely naked; outer primary longer than secondaries; culmen from cere 8.5–10 mm   *O. flammeolus*
   Base of toes feathered or bristled; outer primary shorter than secondaries; culmen from cere 10.5–17.5 mm   2
2. Extreme base of toes feathered; middle toe without claw free for 14.5 mm or more; 5–6 primaries emarginate on inner web   *O. asio*
   Base of toes bristled; middle toe without claw free for 14 mm or less; 4 outer primaries emarginate on inner web   *O. trichopsis*

Otus asio (**Linnaeus**). Screech owl. Above brownish gray, heavily streaked and finely vermiculated with black; inner edge of ear tufts spotted with white or buff; nape and upper back both often with an indistinct ring of white or buff spots; a broad white band along outer edge of scapulars; edge of wing white; another white band along edge of outer middle and greater coverts; primaries dusky, crossed by several buffy bands;

tail with 7–8 narrow buffy bands; face barred dusky and gray; a large brown spot above eye; superciliary, lores, and chin whitish; a broken black ring on sides of face; breast and flanks whitish, tinged with buff, heavily streaked with black and more lightly barred with gray; belly and crissum white; thighs buff; tarsi white, spotted with bay; wing lining pale buffy; iris yellow; bill and toes horn. RED PHASE (absent west of 100th meridian): Gray replaced by rufous. Wing 139–193, tail 63–101, culmen from cere 12–17.5 mm. North America from Alaska and southern Canada south to northern Mexico.

Otus trichopsis (**Wagler**). Spotted screech owl. Hardly specifically distinct from *O. asio,* of which it is the southern highland representative. Differs in having the facial bristles perhaps averaging longer; feathers not reaching toes, which are bristled; only 4 primaries (instead of 5–6) emarginate on inner web; buff or white spots on nape and upper back more distinct; black borders to sides of face smaller; size averaging less. Both red and gray phases occur. Wing 140–151, tail 64–75, culmen 10.5–13.5 mm. Huachuca and Catalina Mountains of Arizona south in mountains to Honduras.

Otus flammeolus (**Kaup**). Flammulated screech owl. Coloration as in *O. asio,* with red and gray phases. Differs in having ear tufts rudimentary; outer primary longer than secondaries; 4 outer primaries sinuate on inner webs; 3d or 4th primary longest; toes and lower part of tarsus completely naked; iris brown; size averaging smaller, especially the bill and feet. Wing 128–144, tail 58–67, culmen 8.5–10 mm. Mountains of West, from British Columbia south to Guatemala.

Genus Bubo Dumeril. With ear tufts; facial ring incomplete; outer primary about equal to secondaries; 3d to 4th primary longest; 2–4 primaries sinuate on inner web; tail more than half wing, rounded; tarsus about equal to middle toe and claw; feathering extending nearly to tip of toes. North and South American mainland, Palearctic and Oriental regions, continental Africa. 11 species, 1 in the United States.

Bubo virginianus (**Gmelin**). Great horned owl. Upper parts pale gray and ochraceous, heavily mottled and vermiculated with black; ear tufts mostly black; wings and tail barred with dusky; face whitish or buff, faintly barred with dusky,

and bordered laterally with black; chin with bold black streaks; a white bib on throat; base of throat with a collar of black blotches; below white or ochraceous regularly barred with blackish; feet and wing lining white or ochraceous, more or less spotted with dusky, becoming bars on axillars; iris yellow; bill and claws blackish. Wing 304–400, tail 175–252, culmen 21–33 mm. Whole mainland of North and South America.

Genus Nyctea Stephens. Similar to *Bubo,* but ear tufts rudimentary; bill nearly concealed by feathers; 4 primaries emarginate, outer primary equal to 5th; crissum reaching tip of tail; toes completely feathered, with even the claws partly hidden. Circumpolar. Monotypic.

Nyctea scandiaca (**Linnaeus**). Snowy owl. White, more or less barred with dusky; iris yellow; bill and claws black. Wing 394–465, tail 220–275, culmen 24.5–28 mm. Breeds on tundra; in winter south at intervals to central states or even farther.

Genus Surnia Dumeril. Related to *Nyctea* but without ear tufts; outer primary equal to 6th; tail graduated, ¾ wing; tarsus longer than middle toe; toes feathered nearly to tips. Boreal forests of Holarctic. Monotypic.

Surnia ulula (**Linnaeus**). Hawk owl. Crown and nape blackish spotted with white; upper parts dark brown; a stripe of white blotches along scapulars; rump and outer wing coverts spotted with white with black shafts; collar of foreneck dark brown, continuing as a black stripe bordering sides of face; a 2d black crescent behind and confluent with the 1st, separated by a white space; underparts, wing lining, and legs white barred with chestnut brown; iris and bill yellow; claws dusky. Wing 218–251, tail 160–191, culmen 17–20.5 mm. Northern coniferous forests of both hemispheres; migrates irregularly south to Canadian border or farther.

Genus Glaucidium Boie. Without ear tufts; facial ring incomplete, with eyes near top; cere bulbous, shorter than culmen, with nostrils opening through it; 4 primaries emarginate; 3d to 5th primaries longest, outer one much shorter than secondaries; tail nearly square, nearly ¾ wing, rectrices 12; tarsus about equal to middle toe without claw; toes bristled. Palearctic, Oriental, mainland Africa, North and South America. 13 species, 2 in the United States.

KEY TO SPECIES OF *GLAUCIDIUM*

1. Pileum heavily streaked   *G. brasilianum*
   Pileum with light dots or almost plain   *G. gnoma*

Glaucidium gnoma **Wagler.** Pygmy owl. Crown and subloral-auricular stripe brownish gray dotted with white or buff; base of hind neck with a black-and-white collar; back grayish brown with larger buff spots; tail dusky with 5–8 incomplete white bars; remiges dusky barred with white or buff; lores, chin, and malar area white; face bordered below with grayish brown, spotted with buff; sides brown with a few white dots; underparts white; lower breast broadly streaked with brown; under wing coverts whitish, the outer ones tipped with black; leg plumage grayish brown; iris, bill, and toes yellow; claws black. YOUNG: Crown gray with a few white streaks on forehead. Wing 82–105, tail 57–78, culmen 9.5–12 mm. Alaska south in mountains to Guatemala, east to the Rockies.

Glaucidium brasilianum **(Gmelin).** Ferruginous owl. Above rufous or hair brown; crown streaked with whitish; nape with a black-and-white collar; wing coverts and band along scapulars with white or buff spots; remiges and tail brownish barred with white or rufous; malar area white; face more or less bounded by brown; bib white; breast, flanks, and crissum white heavily streaked with brown; legs white or buff; iris, bill, and toes yellow; claws dark. Wing 86–102, tail 53–69, culmen 9.5–13 mm. Southern Texas and Arizona south in lowlands to Argentina.

Genus *Micrathene* Coues. Similar to *Glaucidium*, but tail less than half wing; rectrices 10; tarsus almost entirely bristled, equal to middle toe with claw. Southwestern United States to central Mexico. Monotypic.

Micrathene whitneyi **(Cooper).** Elf owl. Above brown or brownish gray speckled with tawny; a white collar on hind neck; scapular stripe white margined with black; middle and greater coverts spotted with white; edge of wing white; remiges with 5–6 buff bars; tail with 4–5 broken buff bars; superciliary white barred with black; face cinnamon; malar area white, ending in a black bar; throat cinnamon; underparts mixed buffy brown, gray, and white, narrowly barred with dusky; crissum and wing lining white, with some brown spots or dusky streaks; iris yellow; bill horn. Wing 99–115, tail 45–53, culmen 8–9.5 mm. Mexican border from southeast California to lower Rio Grande, south to central Mexico.

Genus *Speotyto* Gloger. Similar to *Glaucidium*, but 3 primaries emarginate, 3d longest, outer longer than 5th or secondaries; legs long and slender, extending beyond tail; tarsus twice middle toe without claw, the front part with bristlelike feathers. Nest in hole in ground. Plains of North and South America. Monotypic.

Speotyto cunicularia **(Molina).** Burrowing owl. Above brown spotted with white; auriculars brown streaked with buff; face white; throat collar black; small bib white; upper breast brown spotted with white; underparts white or buffy, barred with brown laterally; iris yellow; bill and feet horn. Wing 145–181, tail 65–86, culmen 38–48 mm. Western plains from Pacific to Iowa and Louisiana, south to Argentina; south Florida and West Indies; migratory in north.

Genus *Strix* Linnaeus. Without ear tufts; facial ring complete, with eyes near center; ear openings large crossed by ligamentous bridge, the right one larger; 4–5 primaries sinuate on inner web, 4th or 5th longest, outer one shorter than secondaries; tail rounded, $\frac{2}{3}$ wing; rectrices 12; tarsus as long as middle toe with claw or longer, completely feathered; toes completely or partially feathered. Palearctic and Oriental regions, North and South America. 12 species, 3 in the United States.

KEY TO SPECIES OF *STRIX*

1. Feathering of toes concealing base of claws; wing 410–465 mm   *S. nebulosa*
   Toe feathering not reaching claws; wing 310–355 mm
        2
2. Head and breast barred; flanks streaked   *S. varia*
   Head, breast, and flanks spotted   *S. occidentalis*

Strix varia **Barton.** Barred owl. Above brown barred with white or buff; face grayish barred with brown; a blackish spot before eye; facial ring dusky brown, barred with white except on sides of face and chin; breast and sides of neck barred brown and white; posterior underparts buffy; flanks and crissum streaked with brown; thighs mottled with brown; iris brown; bill yellow; toes yellowish gray;

claws blackish. Wing 315–355, tail 205–257, culmen 23–30 mm. Eastern deciduous forest and wooded parts of Great Plains, south in mountains to cloud forest of Honduras.

Strix occidentalis (**Xantus**). Spotted owl. Upper parts spotted with white; white bars on breast broken into spots; flanks spotted with brown. Wing 310–328, tail 195–225, culmen 19.5–23.5 mm. Pacific forests and mountains from British Columbia to southern California; Rockies from Colorado and western Texas to central Mexico.

Strix nebulosa **Forster.** Great gray owl. Above sooty brown barred and striped with grayish white; wing coverts spotted with whitish; wings and tail banded with pale brownish gray; face grayish white barred with brown; a dusky spot before eye; facial ring dark brown; chin blackish; underparts grayish white, striped anteriorly and barred posteriorly with sooty brown; iris and bill yellow. Wing 410–465, tail 300–347, culmen 23–29.5 mm. Taiga of both hemispheres, south in mountains to California, Idaho, and Montana; in winter south irregularly to Nebraska, Minnesota, Ohio, and New Jersey.

Genus *Asio* **Brisson.** Differs from *Strix* in having ear tufts; cere as long as culmen; 1–2 primaries emarginate; 2d to 4th primaries longest, outer longer than 6th; tail about half wing. Nearly cosmopolitan (absent from Australia, New Zealand, and much of Oceania). 6 species, 2 in the United States.

KEY TO SPECIES OF *ASIO*

1. Ear tufts long; upper parts vermiculated   *A. otus*
   Ear tufts rudimentary; upper parts streaked   *A. flammeus*

Asio otus (**Linnaeus**). Long-eared owl. Above blackish brown vermiculated with grayish; outer scapulars spotted with white; remiges buffy at base, rest grayish mottled with dusky and crossed by 7–10 dusky bars; tail barred with dusky; ear tufts black edged with buff and white; superciliary and lores grayish; eye ring black; face buffy; disc black; chin white; underparts buffy; breast and crissum streaked with sooty brown; sides and flanks streaked and barred with sooty brown; a large black patch on under primary coverts; iris

yellow; bill black. Wing 284–303, tail 122–160, culmen 15–18.5 mm. Holarctic forests south to Virginia, Arkansas, and California; in winter to the Gulf and central Mexico and to central Europe and China.

Asio flammeus (**Pontoppidan**). Short-eared owl. Above buff striped with brown; wings and tail banded with brown, basal portion plain whitish or ochraceous; eye ring black; front of face white; rest of face buffy with fine dusky shaft streaks; below buffy; breast, sides, and upper belly streaked with brown; a black blotch on under primary coverts; iris yellow; bill black. Wing 298–330, tail 137–161, culmen 15.5–20 mm. Practically the whole world, except for Australia, New Zealand, and much of Oceania; in the United States breeds south to New Jersey, Ohio, northern Indiana and Illinois, Missouri, Kansas, Colorado, and California; winters south to the Gulf and Guatemala.

Genus *Aegolius* **Kaup.** Differs from *Strix* in having skull asymmetrical; ear openings very large, right and left of equal size but asymmetrical; only 2 primaries sinuate, 3d and 4th longest, outermost about equal to secondaries. Holarctic, Central and South America. 3 species, 2 in the United States.

KEY TO SPECIES OF *AEGOLIUS*

1. Wing 163–182 mm; cere not tumid   *A. funereus*
   Wing 134–146 mm; cere tumid   *A. acadicus*

Aegolius funereus (**Linnaeus**). Boreal owl. Above chocolate; crown speckled with white; scapulars and wing coverts with white spots; remiges and tail with broken white bars; loral spot black; face white; facial rim dark brown speckled with white; below whitish; breast spotted with chocolate; flanks, crissum, and wing lining streaked with chocolate; iris and bill yellow; claws black. Wing 163–182, tail 96–107, culmen 13.5–16 mm. Taiga of both hemisphere; in winter south irregularly to northern states.

Aegolius acadicus (**Gmelin**). Saw-whet owl. Above reddish brown; crown with white shaft streaks; scapulars, wing coverts, and outer edge of primaries spotted with white; edge of wing white; tail with 2–3 broken white bars; eyelid mostly black; face white; facial rim chestnut streaked with white; bib and crissum white; below

Fig. 5-50. Foot of chuck-will's-widow (*Caprimulgus carolinensis*), enlarged.

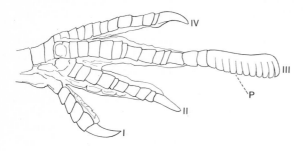

white striped with reddish brown; wing lining buff blotched with brown; iris yellow; bill black; toes yellowish; claws brown. YOUNG: Crown and entire upper parts plain chocolate; wings, tail, and face as in adult; lower rim of face reddish brown, streaked with white laterally; breast dull brown; belly, crissum, and legs rich buff. Wing 134–146, tail 65–73, culmen 11–14 mm. North America south to New England, mountains of Maryland, Great Lakes, Nebraska, and Oklahoma, and in mountains to California, Arizona, and New Mexico, and thence to Costa Rica; in winter irregularly to Georgia and Louisiana. Distribution spotty.

Order Caprimulgiformes. Goatsuckers and allies. Plumage fluffy, loose, aftershafts rudimentary; adult downs on apteria only; oil gland naked; skin thin. Bill short, somewhat hooked; nostrils imperforate. Feet small, weak, anisodactyl; tarsi partly feathered. Wing long; primaries 10, outermost longer than 5th; secondaries 12–15, aquintocubital. Tail long, rectrices 10.

Skull flat; vomer present; palatines wide; mandibular process very short. Cervicals 13–14. Sternum notched, without internal spine, external spine small or absent. Coracoids separate. Furculum U-shaped. Humerus with pneumatic fossa partially roofed, internal tuberosity produced, ectepicondylar process small. Hypotarsus complex.

No crop; intestine short, type 6; caeca usually long. Thigh muscles $(A)XY-$; plantar tendons synpelmous (type 5). Syrinx tracheal. 2 carotids.

Nocturnal or crepuscular. Food usually insects, caught mainly in flight. Eggs 2–4, white or speckled; young downy but altricial.

Nearly cosmopolitan (absent from New Zealand and Hawaii). Fossil history unknown before the Pleistocene. 1 family in the United States.

Family Caprimulgidae. Goatsuckers. Bill weak, flat; gape very wide; opening to below eyes; nostrils in raised tubes; rictal bristles usually long (small in *Chordeiles*). No powder downs. Tarsus twice length of hallux and claw; toe formula 2–3–4–4; outer toe shorter than inner toe; hallux turned sideways; middle claw pectinate (Fig. 5-50). Oil gland present.

Schizognathous (*Chordeiles desmognathous*), with basipterygoid processes; lacrimal well developed; sternum with 1 pair of notches; procoracoid rudimentary; tibia with distal end compressed, intercondylar pit deep, bridge ossified. Thigh muscles $AXY-$.

Food insects. No nest; eggs laid on ground, usually speckled, 2 in number. Geographic and geological range as for the order. 18 genera, 69 species; in the United States 4 genera, 7 species.

KEY TO GENERA OF CAPRIMULGIDAE

1. Tail emarginate; rictal bristles short   **Chordeiles**
   Tail rounded or square; rictal bristles extending beyond tip of bill   2
2. Tarsus longer than middle toe without claw   **Nyctidromus**
   Tarsus about as long as middle toe without claw   3
3. Tail square; tarsus nearly naked   **Phalaenoptilus**
   Tail rounded; upper half of tarsus feathered   **Caprimulgus**

Genus *Nyctidromus* Gould. Rictal bristles longer than bill; 3d primary longest; outer longer than 5th; tail 80% of wing or more, rounded; tarsus longer than middle toe without claw, feathered only at extreme base; outer toe slightly shorter than inner toe. Neotropical mainland. Monotypic.

Nyctidromus albicollis (Gmelin). Pauraque. Crown brownish or grayish vermiculated with dusky, with heavy black streaks; hind collar buff; middle of back brownish, vermiculated and streaked with blackish; upper tail coverts and 2 middle pairs of rectrices similar but barred rather than streaked; 3d and 4th pairs of rectrices white, edged with black distally, extreme base barred black and buff; outer pair almost entirely black;

scapulars and wing coverts brownish, vermiculated with dusky, blotched with black, and edged with golden buff; primaries dusky, with a white patch across 5–6 outer primaries; secondaries dusky barred with buff; eye ring and subocular area chestnut; chin and malar region buffy brown barred and streaked with black; a white rictal streak joining white throat; breast brown barred with dusky, bounded posteriorly by a buff band; belly, flanks, crissum, and wing lining buff barred with dusky; under primary coverts black; iris, bill, and feet brown. FEMALE AND YOUNG: Primary patch largely ochraceous; lateral pairs of rectrices mostly dusky barred with ochraceous, with white tips smaller; white throat patch more restricted; rictal stripe absent. Wing 141–187, tail 113–182, culmen 10.5–14, tarsus 22–28.5 mm. Neotropical mainland north to the Rio Grande.

Genus *Phalaenoptilus* Ridgway. Bristles long; 2d or 3d primary longest; tail square, less than $\frac{2}{3}$ wing; tarsus about equal to middle toe without claw. Western North America. Monotypic.

Phalaenoptilus nuttallii (**Audubon**). Poor-will. Above brownish gray with wavy dusky bars and often frosted with silver; scapulars, wing coverts, and inner secondaries with black bars larger, especially on middle of feather; remiges and primary coverts banded black and buff; lateral rectrices banded black and buff, with broad white tips; face sooty; chin and malar region brown freckled with dusky; rictal stripe and throat white; upper breast sooty bounded posteriorly with white; lower breast and sides buffy white barred with dusky; belly, flanks, crissum, and wing lining buffy; bill black; iris and feet brown. Wing 126–152, tail 78–95, culmen 10–13.5, tarsus 16–19 mm. British Columbia south to central Mexico, east to Dakotas, Iowa, Kansas, and Texas. Known to hibernate.

Genus *Caprimulgus* Linnaeus. Bristles long; 2d primary longest; tail rounded, $\frac{2}{3}$ to $\frac{3}{4}$ wing; upper half of tarsus feathered. North and South America, Eurasia to tropical Australia, Africa, and Madagascar. 38 species, 3 in the United States.

KEY TO SPECIES OF *CAPRIMULGUS*

1. With a distinct buff collar on hind neck
   C. ridgwayi
   Hind collar broken or obsolete   2

2. Wing 202–225 mm; rictal bristles with many lateral filaments   *C. carolinensis*
   Wing 147–178 mm; bristles with few if any lateral filaments   *C. vociferus*

Caprimulgus carolinensis **Gmelin.** Chuck-will's-widow. Above grayish brown streaked with black; scapulars and wing coverts blotched with black and rich buff; broken hind collar buffy; tail rich buff vermiculated and brokenly barred with black; 3 outer pairs of rectrices broadly white-tipped; remiges dusky with broken bars of rich buff; indistinct supraloral and rictal stripes white; face and throat tawny with dusky bars; forecollar white barred with black; breast sooty vermiculated with buff and gray, crossed by 2 rows of buff spots; belly, flanks, crissum, and wing lining tawny buff with dusky bars. FEMALE: Tips of 3 outer rectrices rich buff. Wing 202–225, tail 130–151, culmen 9–14.5, tarsus 17–19 mm. Southeastern United States to Texas, Kansas, Missouri, and southern parts of Illinois, Indiana, and Maryland; winters from Mexico and Greater Antilles to Colombia.

Caprimulgus vociferus **Wilson.** Whip-poor-will. Browns darker and buffs less warm; crown blotched and streaked with black; middle rectrices buffy gray vermiculated and barred with black; 3 lateral rectrices white-tipped (buff in female), basally blackish brokenly barred with buff; face and throat sooty brown lightly tipped with tawny. Wing 147–178, tail 105–134, culmen 10–16, tarsus 15.5–18.5 mm. Eastern Canada south to Kansas, Louisiana, Mississippi, Alabama, and Georgia; also in mountains of southern Arizona, New Mexico, and southwest Texas, south to Central America and Puerto Rico; winters from South Carolina and Gulf states to Florida and Central America.

Caprimulgus ridgwayi (**Nelson**). Ridgway's whip-poor-will. Coloration less dusky, more buffy, than in *C. vociferus;* hind collar unbroken from auriculars around hind neck, tawny buff; light tip of rectrices mainly on inner web. Wing 155–162, tail 115–121, culmen 14–15.5, tarsus 15–17 mm. Extreme southern Arizona and New Mexico south to Honduras.

Genus *Chordeiles* Swainson. Bristles shorter than bill; maxilla grooved near tip; 1st or 2d primaries longest; tail emarginate, less than $\frac{2}{3}$ wing; tarsus equal to middle toe without claw, about half feathered; palate desmognathous. North and South America. 4 species, 2 in the United States.

KEY TO SPECIES OF *CHORDEILES*

1. Basal part of primaries plain dusky, with a white
   patch on outer 5   *C. minor*
   Basal part of primaries spotted with rust; white
   patch confined to outer 4   *C. acutipennis*

Chordeiles minor (**Forster**). Common nighthawk. Upper parts black, freckled with buff and silver, lightest on inner wing coverts and tertials; a white patch on edge of wing; remiges dusky; a white patch across middle of 5 outer primaries, located proximal to tip of 7th primary; tail dusky, barred with dusky-freckled white, the last white bar immaculate and broader, but not extending to middle pair; upper throat white; face and upper breast black streaked with tawny; underparts barred black and white; iris brown; bill black; feet dusky brown. FEMALE: Lacks subterminal white tail bar; throat patch and underparts tinged with buff. Wing 159–213, tail 89–122, culmen 5–8, tarsus 12–16 mm. North America, south to Greater Antilles and Mexico; Panama; winters in South America.

Chordeiles acutipennis (**Hermann**). Lesser nighthawk. Outer primary shorter than 2d instead of longer; base of primaries spotted with tawny; white wing patch confined to 4 outer primaries, with 7th primary not extending beyond the patch; plumage more buffy. Wing 159–192, tail 90–119, culmen 5–7, tarsus 12–15 mm. Interior of California and southern parts of Nevada, Utah, New Mexico, and Texas, south to Brazil.

Order Apodiformes. Swifts and hummingbirds. Aftershafts present; oil gland naked. No rictal bristles; nostrils imperforate. Primaries 10, outermost usually, 2d or 3d sometimes, longest; secondaries 6–11, very short, usually quintocubital; no middle coverts; alula with not more than 3 feathers. Rectrices 10. Feet small, weak; tarsi with few if any scales; claws curved, sharp.

Holorhinal; basipterygoid processes and vomer present; mandible truncate basally with internal process developed. Cervicals 13–14. Sternum entire or fenestrate, with common spine, keel deep. Furculum U-shaped with small hypocleidium. Coracoids separate; procoracoid process perforate. Humerus very short and broad; internal tuberosity curved anconally; deltoid crest spinelike, curved palmarly; ectepicondylar process spinelike, raised to near deltoid crest; entepicondyle prominent. Carpometacarpus longer than humerus or ulna; phalanges long. Tibia bridged. Upper anterior face of tarsometatarsus with a bony bridge to proximal ligamental attachment. Gadow, and following him Ridgway, erred in defining osteological characters in this order, namely, for the vomer, basipterygoid processes, manubrium, tibial bridge, and hypotarsus.

Thigh muscles A—; plantar flexors synpelmous (type 5). Usually only the left carotid. Intestine type 6; caeca absent.

Food insects or nectar, feeding in flight. Nest of sticks or lichens; eggs 1–5, white, elliptical. Young altricial, usually hatched naked.

Oligocene onward. Nearly cosmopolitan (absent from New Zealand and Hawaii). In the United States 2 families.

KEY TO FAMILIES OF APODIFORMES

1. Gape large; bill tiny, flat, triangular   **Apodidae**
   Gape small; bill long, needlelike   **Trochilidae**

Family Apodidae. Swifts. Bill short, flat, triangular; gape large; nostrils close together, nonoperculate, opening vertically; tongue not extensile. Aftershafts large; adult downs on apteria only. Wing bladelike, with remiges clinging together; secondaries 8–11; alular feathers 2–3. Tail short, not reaching tip of wing. Tarsus longer than hallux; toes cleft to base; hallux and 2d toe reversible.

Aegithognathous; basipterygoid processes rudimentary; vomer expanded anteriorly to reach maxillopalatine processes; maxillopalatines unciform; palatines notched exteriorly. Metasternum convex or entire (foraminate in 1 genus); 6–7 pairs of ribs. Tibial bridge with canal open. Hypotarsus simple. Syrinx tracheal. No crop.

Food flying insects. Nest of sticks, cemented with saliva; eggs 2–5.

Oligocene onward. Cosmopolitan, except New Zealand and Hawaii. 8 genera, 67 species; in the United States 3 genera, 4 species.

KEY TO GENERA OF APODIDAE

1. Tarsus and base of toes feathered   *Aëronautes*
    Tarsus and toes naked   **2**
2. Tips of rectrices bare and spiny   *Chaetura*
    Tail without bare spiny shafts   *Cypseloides*

Genus *Chaetura* Stephens. Tail $\frac{1}{3}$ wing or less, square, shaft tips bare and spiny (Fig. 5-51); tarsus naked and unscaled. Americas, African mainland, Oriental region. 17 species, 2 in the United States.

KEY TO SPECIES OF *CHAETURA*

1. Wing 123–133 mm, rump about same shade as tail, only slightly paler than back   *C. pelagica*
    Wing 91–119 mm; rump much paler than back and tail   *C. vauxi*

Chaetura pelagica (**Linnaeus**). Chimney swift. Sooty; throat grayer. Wing 123–133, tail 40–45, culmen 5–6, tarsus 11–12.5 mm. Eastern North America, west to Montana, South Dakota, and Texas; migrates to West Indies, Mexico, and Central America.

Chaetura vauxi (**Townsend**). Vaux's swift. Above sooty; rump and upper tail coverts light grayish brown, paler on throat, darker on crissum. Wing 105–119, tail 34–39, culmen 4.5–5.5, tarsus 10–11.5 mm. Western North America, south to California; rare east of Cascades and Sierra Nevada to Montana and Nevada; winters in Louisiana and Central America.

Genus *Cypseloides* Streubel. Tail not spiny, emarginate (square in female), about $\frac{1}{3}$ wing. Americas 3 species, 1 in the United States.

Cypseloides niger (**Gmelin**). Black swift. Sooty; crown (and in female belly and crissum) white-tipped. Wing 149–175, tail 47–66, culmen 6–7.5, tarsus 12–13.5 mm. British Columbia, south through mountains to Costa Rica, east to Colorado and New Mexico; West Indies.

Genus *Aëronautes* Hartert. Tail forked, more than $\frac{1}{3}$ wing; tarsus and toes feathered. Mountains of western North and South America. 2 species, 1 in the United States.

Aëronautes saxatalis (**Woodhouse**). White-throated swift. Above sooty; sides of rump white;

Fig. 5-51. Tail of chimney swift (*Chaetura pelagica*).

lores whitish; spot before eye black; throat and middle of breast and belly white; sides, flanks, and crissum sooty; iris brown; bill black; feet pink. Wing 131–149, tail 53–63, culmen 5–6.5, tarsus 9.5–11 mm. Mountains of West, east to Black Hills, south to El Salvador.

Family Trochilidae. Hummingbirds. Bill long, needlelike; gape small; mandible with lateral groove; nostrils lateral, operculate; tongue long, extensile, split into 2 parallel tubes. Aftershafts small; no adult downs. Secondaries 6–7; alular feathers 0–1. Tail variable, sometimes long. Tarsus shorter than hallux; toes not reversible.

Schizognathous; basipterygoid processes well developed; vomer narrow and pointed; maxillo-palatine processes expanded terminally; palatines unnotched. Metasternum convex and entire; 8 pairs of ribs. Tibial bridge present, but canal plugged with bone. Hypotarsus complex, with 1 canal. Syrinx tracheobronchial. Crop present.

Food insects and nectar; feeds by hovering before flowers. Nest cuplike, of plant material, not cemented, placed on limb; eggs 1–2.

Fossil record fragmentary, only Pleistocene. North and South America. 121 genera, 320 species; in the United States 9 genera, 16 species.

KEY TO GENERA OF TROCHILIDAE

1. Nasal operculum largely exposed   **2**
    Nasal operculum concealed by feathers   **5**
2. Nasal operculum wholly exposed   **3**
    Feathers overlapping top of nasal operculum   **4**
3. Bill more than half length of tail; culmen 18.5–23.5 mm   *Cynanthus*
    Bill not more than half tail; culmen 14.5–18.5 mm   *Hylocharis*

4. Bill more than ⅓ wing; wing 51–59 mm   *Amazilia*
   Bill less than ⅓ wing; wing 69–79 mm   *Lampornis*
5. Bill distinctly decurved   *Calothorax*
   Bill nearly straight   6
6. Bill ⅓ wing; culmen 10.5–13 mm   *Atthis*
   Bill more than ⅓ wing; culmen 13.5–22 mm   7
7. Middle rectrices longest (tail rounded) *Selasphorus*
   Tail emarginate or double-rounded   8
8. Bill thin, deeper than broad; wing 33–51 mm
                                              Archilochus
   Bill wide, broader than deep; wing 67–76 mm
                                                 Eugenes

Genus *Cynanthus* Swainson. Bill more than ⅓ wing, slightly decurved; nasal operculum exposed; tail forked, less than ⅔ wing; rectrices broad. Mexican tableland. 2 species, 1 in the United States.

***Cynanthus latirostris* Swainson.** Broad-billed hummingbird. Crown and body green; tail blue; 4 middle rectrices tipped with gray; remiges dusky; edge of outer primary whitish; throat blue; crissum gray; tuft on sides of rump white; bill red, tip black; feet black. FEMALE: Middle rectrices and base of the others green; underparts gray; postocular spot white, bordered ventrally with dusky. Wing 49–57, tail 28–36, culmen 18.5–23.5 mm. Mountains of southern Arizona and New Mexico south over the Mexican tableland.

Genus *Hylocharis* Boie. Differs from *Cynanthus* in having bill less than ⅓ wing, nearly straight; tail square. South America north to Arizona. 7 species, 1 in the United States.

***Hylocharis leucotis* (Vieillot).** White-eared hummingbird. Forehead and face blue; long postocular stripe white; auricular stripe black; upper parts and middle rectrices bronze green; outer rectrices blackish, narrowly tipped with gray; remiges purplish dusky; lower throat bright green; underparts bronze green, mixed with whitish on midline; bill red, tip broadly black. FEMALE: Pileum and lores rusty bronze; postocular, auriculars, upper parts, and tail as in male; underparts pale grayish flecked with brownish or bronze. Wing 45–59, tail 30–35, culmen 14.5–18.5 mm. Mountains of southern Arizona south through highlands to Nicaragua.

Genus *Amazilia* Lesson. Bill nearly straight, more than ⅓ wing; feathers overlapping top of operculum; tail less than ⅔ wing, square, double-rounded, or emarginate. Texas to Argentina. 29 species, 2 in the United States.

KEY TO SPECIES OF *AMAZILIA*

1. Underparts white   *A. violiceps*
   Underparts green anteriorly, cinnamon posteriorly
                                            A. yucatanensis

***Amazilia violiceps* (Gould).**[1] Violet-crowned hummingbird; Salvin's hummingbird. Crown violet; back green; remiges dusky, with faint violet gloss; tail olive bronze; underparts white; bill bright pink, tip dusky; feet dusky. Wing 55–60, tail 31–35, culmen 20.5–23.5 mm. Southern Arizona to southern Mexico.

***Amazilia yucatanensis* (Cabot).** Buff-bellied hummingbird; fawn-breasted hummingbird. Above green; tail chestnut margined with bronze; throat and breast bright green; axillars and posterior underparts pale cinnamon; femoral tufts white; bill reddish. Wing 51–59, tail 31–37, culmen 19–22.5 mm. Gulf lowlands from Rio Grande to Yucatan.

Genus *Lampornis* Swainson. Bill nearly straight, ⅓ wing or less; top of operculum overlapped by feathers; tail nearly square, less than ⅔ wing. Southwest United States in mountains to Panama. 6 species, 1 in the United States.

***Lampornis clemenciae* (Lesson).** Blue-throated hummingbird. Above bronze green; tail steel blue broadly tipped with white; postocular stripe white; subocular and auriculars dusky bronze; throat blue; underparts gray, tinged with bronze on sides; crissum tipped with whitish; bill black. FEMALE: Throat gray. Wing 69–79, tail 41–50, culmen 21.5–27.5 mm. Mountains of Arizona, New Mexico, and western Texas, south to Oaxaca.

Genus *Eugenes* Gould. Bill broad, nearly straight, more than ⅓ wing; operculum mostly hidden by feathers; tail emarginate (sometimes double-rounded in female), less than ⅔ wing; rectrices broad. Mountains from Arizona to Panama. Monotypic.

***Eugenes fulgens* (Swainson).** Rivoli's hummingbird; magnificent hummingbird. Crown purple; anterior part of forehead and hind neck blackish bronze; back and tail bronze green; remiges purplish dusky; postocular and rictus narrowly

[1] *Amazilia verticalis* (Deppe) of A.O.U.

white; throat emerald green; breast blackish bronze; belly grayish; crissum grayish edged with buff; femoral tufts white; bill black. FEMALE: Upper parts and middle rectrices bronze green; distal half of lateral rectrices black tipped with gray; remiges purplish black, postocular and rictus narrowly whitish; auriculars dusky; underparts gray; throat flecked with brown; crissum bronze edged with buff; anal tufts white. Wing 67–76, tail 38–48, culmen 26–31 mm. Mountains from Arizona and New Mexico to Panama.

Genus *Calothorax* Gray. Bill slightly curved, slender, about half length of wing; operculum covered by feathers; tail ⅔ wing or more, forked with narrow rectrices in male, double-rounded with broad rectrices in female; throat feathers of male elongate. Mexican tableland. 2 species, 1 in the United States.

Calothorax lucifer (**Swainson**). Lucifer hummingbird. Upper parts and 4 middle rectrices bronze green; remiges and outer rectrices purplish dusky; postocular spot and rictus narrowly white; throat purple; breast white; belly and crissum grayish; femoral tufts white; bill black. FEMALE: Outer rectrices cinnamon basally, purplish black medially, broadly tipped with white; postocular cinnamon buff; auriculars brownish; underparts cinnamon buff, whitish posteriorly. Wing 36–44, tail 23–31, culmen 19.5–22.5 mm. Mountains of southern Arizona and western Texas to central Mexico.

Genus *Atthis* Reichenbach. Bill straight, less than ⅓ wing; operculum hidden; tail less than ⅔ wing, rounded; rectrices broad; throat feathers of male elongate. Mountains from Arizona to Honduras. Monotypic.

Atthis heloisa (**Lesson and Delattre**). Heloise's hummingbird: bumblebee hummingbird. Above bronze green; remiges purplish dusky; tail cinnamon basally, bronze green medially, black subterminally, white at tip; malar and postocular areas whitish; throat purple; underparts white; sides cinnamon mixed with bronze; bill black. FEMALE: Throat white spotted with bronze. Wing 33–38, tail 19–23, culmen 10.5–13 mm. Southern Arizona (rare), south in mountains to Honduras.

Genus *Archilochus* Reichenbach. Bill straight, narrower than deep, more than ⅓ wing; operculum hidden; tail ½ to ⅔ wing, emarginate or double-rounded; throat feathers of male elongate. North and Middle America; Cuba. 6 species, 5 in the United States.

KEY TO SPECIES OF *ARCHILOCHUS*

1. Outer rectrices edges with rufous basally; base of bill light   *A. calliope*
    Rectrices without rufous; bill entirely black   2
2. Tail without white tips (males)   3
    Tail white-tipped (females)   6
3. Crown greenish   4
    Crown violet or red   5
4. Throat red   *A. colubris* ♂
    Throat black followed by purple   *A. alexandri* ♂
5. Head red   *A. anna* ♂
    Head violet   *A. costae* ♂
6. Middle pair of rectrices shorter than outer pair   *A. colubris* ♀
    Middle pair of rectrices longer than outer pair   7
7. Middle rectrices 8 mm in width   *A. anna* ♀
    Middle rectrices 7 mm or less in width   8
8. Tail 26–27.5 mm, culmen 19.5–22 mm   *A. alexandri* ♀
    Tail 21.5–25.5 mm, culmen 17–20 mm   *A. costae* ♀

Archilochus colubris (**Linnaeus**). Ruby-throated hummingbird (Fig. 5-52). Male with inner primaries notched at tip; rectrices pointed. Above bronze green, darker on crown; remiges and rectrices purplish dusky; chin, lores, and auriculars

Fig. 5-52. Female ruby-throated hummingbird (*Archilochus colubris*), from above.

black; small postocular spot white; throat red; underparts buffy gray; sides bronze; femoral tufts white; bill black. FEMALE: Tail tipped with white; middle rectrices shorter than outer pair; throat and underparts dirty white, tinged with bronze on flanks. Wing 37–45, tail 25–28, width of outer rectrix 3–3.5 (♀ 4.5–5), culmen 15–19.5 mm. Eastern North America, west to Dakotas, Nebraska, Kansas, and eastern Texas; winters mainly in Mexico and Central America.

Archilochus alexandri (**Bourcier and Mulsant**). Black-chinned hummingbird. Inner primaries of male notched at tip; face (below eyes) and throat black; lower throat purple; otherwise as in *A. colubris*. FEMALE: Like *A. colubris*, but middle pair of rectrices longer than outer pair. Wing 42–48, tail 24–28, culmen 18–22 mm. Western North America, east to Montana, Colorado, and western Texas; winters in Mexico.

Archilochus anna (**Lesson**).[1] Anna's hummingbird. Primaries unnotched; rectrices rounded at tips. Male differs from *A. colubris* in having whole head rose red, except for small white postocular spot. FEMALE: As in *A. colubris*, but throat speckled with bronze or red; middle rectrices longer than outer ones. Wing 48–51, tail 25–32, width of outer rectrix 2.5–3 (♀ 5), culmen 17–22 mm. California, mainly west of Sierra Nevada, and adjacent Mexico.

Archilochus costae (**Bourcier**).[2] Costa's hummingbird. Outer rectrices narrower than in *A. anna*. Head of male violet. Female with paler gray tips to tail. Wing 43–46, tail 22–25, width of outer rectrix 1.5 (♀ 3.5), culmen 16–20 mm. Southern California, southwest Utah, Arizona, southern New Mexico, and adjacent parts of Mexico.

Archilochus calliope (**Gould**).[3] Calliope hummingbird. Rectrices subspatulate. Above bronze green; long throat feathers white tipped with purple; underparts grayish, tinged with cinnamon on sides; tail edged with cinnamon basally; bill black, flesh at base. FEMALE: Differs from others of the genus in having posterior underparts and edge of rectrices basally cinnamon; base of bill flesh; throat flecked with bronze. Wing 37–44, tail 19.5–22.5, culmen 13.5–16 mm. Western North America, east

to Rockies in Montana, Idaho, Colorado, and western Texas.

Genus *Selasphorus* Swainson. Differs from *Archilochus* in having tail rounded or graduated. Western North America, south in mountains to Panama. 8 species, 3 in the United States.

KEY TO SPECIES OF *SELASPHORUS*

1. Middle rectrices entirely green or only edged with rufous   *S. platycercus*
   Middle rectrices extensively rufous, at least basally   **2**
2. Lateral rectrix broader (over 2 mm in ♂, over 3 in ♀)   *S. rufus*
   Lateral rectrix narrow (under 2 mm in ♂, under 2.5 in ♀)   *S. sasin*

Selasphorus platycercus (**Swainson**). Broadtailed hummingbird. Upper parts and middle pair of rectrices bronze green; rest of tail purplish black, the submedian pairs slightly edged with rufous; remiges purplish black; throat reddish purple, crossed by concealed white bars; underparts grayish, tinged with pale cinnamon on flanks and crissum; femoral tufts white; bill black. FEMALE: Throat whitish flecked with bronze; 3 outer pairs of rectrices white-tipped, rufous basally. Wing 46–52, tail 27–35, culmen 16–20 mm. Western United States, east to Nebraska and western Texas, south to Guatemala.

Selasphorus rufus (**Gmelin**). Rufous hummingbird. Male with submedian pair of rectrices notched on inner web. Crown bronze green; upper parts rufous; tail rufous tipped with purplish dusky; remiges purplish dusky; throat scarlet; breast white; posterior underparts cinnamon; femoral tufts white; bill black. FEMALE: Above bronze green; middle rectrices bronze green, rufous basally; outer rectrices rufous, purplish black subterminally, tips white; throat whitish freckled with orange-red; posterior underparts cinnamon mixed with white. Wing 38–45, tail 25–29, culmen 15–19 mm. Western North America, from Alaska to California and Idaho; winters in southern Louisiana and Mexico; migrates east to Wyoming and Colorado.

Selasphorus sasin (**Lesson**). Allen's hummingbird. Rectrices unnotched, narrower than in *S. rufus*. Male with upper parts bronze green, only the upper tail coverts and tail rufous. Female

---

[1] *Calypte anna* (Lesson) of A.O.U.
[2] *Calypte costae* (Bourcier) of A.O.U.
[3] *Stellula calliope* (Gould) of A.O.U.

colored like *S. rufus*. Wing 37–42, tail 23–26, culmen 15–18.5 mm. Coast of California; on migration to Arizona and northwestern Mexico.

Order Trogoniformes. Trogons. Bill stout, shorter than head, base bristled, culmen curved; nostrils imperforate, feathered. Plumage soft, easily detached, colors metallic; skin thin; aftershafts large; adult downs lacking; oil gland naked. Wing short, rounded, concave; outer primary shortest; secondaries short, quintocubital. Tail longer than wing, graduated; rectrices 12, wide; upper tail coverts well developed. Legs short; tarsus feathered, shorter than longest toes; toes heterodactyl, anterior ones grown together at base (Fig. 5-53).

Schizognathous with basipterygoid processes; vomer pointed, ankylosed to palatines; holorhinal; postarticular process present but small. Cervicals 15. Sternum with large unforked external spine, without internal spine; metasternum with 4 deep notches; 4–5 pairs of sternal ribs. Coracoids touching; procoracoid process moderate. Humerus pneumatic with 1 fossa, without ectepicondylar process. Tibiotarsus bridged. Tarsometatarsus complex with 2–3 canals, distal foramen tending to close.

Thigh muscles *AX*; plantar tendons heteropelmous. Left carotid only. Syrinx tracheobronchial. Intestinal convolutions type 6; caeca small but functional.

Forest birds. Food fruit or insects, taken on wing. Voice a hoot. Nest in hollow trees; eggs 3–4, white or pale blue; nestling altricial, gymnopedic.

Tropical America, Indo-Malayan region, and Africa. Oligocene onward. 1 family.

Family Trogonidae. 8 genera; 35 species, 1 in the United States.

Genus *Trogon* Brisson. Tomia serrate; nostrils round, nonoperculate; auriculars, wing coverts, and upper tail coverts normal, not excessively developed; eyelids naked; rectrices truncate or subtruncate; anterior toes united for more than basal phalanx. Mexican border to Argentina. 15 species, 1 in the United States.

***Trogon elegans* Gould.** Coppery-tailed trogon. Forehead, face, and upper throat black; lower throat, crown, nape, back, and upper tail coverts metallic green; wing coverts and edge of secondaries vermiculated black and white; primaries dusky edged with white; middle rectrices coppery green with broad black tips; next 2 pairs dusky

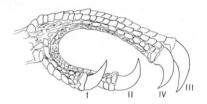

Fig. 5-53. **Foot of coppery-tailed trogon (*Trogon elegans*).**

tipped with black and edged with coppery green; 3 outer pairs vermiculated black and white, black basally, tips white; a white collar on upper chest; posterior underparts red; thighs black; iris brown; eyelids red; bill yellow; feet brown. FEMALE: Dull brown slate on forehead, wing coverts and edge of secondaries brown vermiculated with dusky, primaries brown edged with white; middle rectrices brown tipped with black; next 2 pairs dusky edged with brown; 3 outer pairs barred black and white, dusky basally, tips white; face and throat slaty brown; a white crescent from behind eye to auriculars, bordered behind by black; white chest collar inconspicuous; middle of breast white; sides of breast slaty brown barred with white; belly and crissum red; thighs blackish. YOUNG MALE: Above like male, below like female. Wing 124–137, tail 154–185, culmen 17–19, tarsus 14.5–16 mm. Southern Arizona and Texas to central Mexico, reappearing again from Guatemala to Costa Rica.

Order Coraciiformes. Rollerlike birds. Bill long; nostrils imperforate. Aftershafts small or absent; adult downs absent (except in kingfishers). Primaries 10–11, secondaries quintocubital (except in some kingfishers). Legs moderate or short; feet syndactyl (except in hoopoes and rollers).

Desmognathous, with basipterygoid processes and vomer absent or reduced; holorhinal. Cervicals 14–15. Sternum notched or fenestrate, with external spine grown to keel, often with common spine. Humerus with 1 pneumatic fossa; ectepicondylar process small or absent. Hypotarsus complex, with 1 canal.

Thigh muscles *AX(Y)* —; plantar tendons type 5. Syrinx tracheobronchial. Intestine type 6 or 7.

Nest in cavity; eggs 2–8, white or blue; nestling altricial, gymnopedic (except in hoopoes and woodhoopoes). Food animal matter.

Cosmopolitan except for Hawaii; Eocene onward. 1 family in the United States.

Family Halcyonidae.[1] Kingfishers. Bill longer than head, straight, acute, unserrate. Adult downs thick on apteria, thin on pterylae; aftershafts present; oil gland tufted. Primaries 11 (10 functional), secondaries 11–14; rectrices 12, rarely 10. Tarsus shorter than middle toe; feet syndactyl, soles flat.

Without basipterygoid processes or vomer; cervicals 15; sternum without internal spine, 4-notched; 3–4 pairs of sternal ribs; coracoids separate, procoracoid process large; pelvis broad; tibial bridge ossified. Thigh muscles AX. 2 carotids. Intestine type 6; caeca absent.

Nest in hole in bank, rarely in tree cavity; eggs white. Food fish, insects, or reptiles.

Cosmopolitan, except for Hawaii. Oligocene onward. 15 genera, 87 species; in the United States 2 genera, 2 species.

KEY TO GENERA OF HALCYONIDAE

1. Exposed culmen less than half length of wing
   Megaceryle
   Exposed culmen more than half length of wing
   Chloroceryle

Genus *Megaceryle* Kaup. Head crested; bill compressed, less than half wing; tail rounded; tarsus as long as inner toe without claw. Asia, Africa, North and South America, West Indies. 4 species, 1 in the United States.

Megaceryle alcyon (**Linnaeus**). Belted kingfisher. Above blue-gray; a white spot before eye; crown streaked with black; a white collar on hind neck; wing coverts and secondaries with a few white bars; middle rectrices with a black shaft stripe and a few white bars; other rectrices barred black and white and edged with blue-gray; primaries black with a large white patch across middle; sides of head, malar streak, and collar on upper breast blue-gray; rest of underparts white; flanks mixed with blue-gray; iris brown; bill black; feet slate. FEMALE: Flanks, axillars, and band across lower breast cinnamon rufous. Wing 145–169, tail 81–101, culmen 53–71, tarsus 10–12.5 mm. Alaska,

Canada, and all United States; winters from southern Alaskan coast and the Missouri and Ohio Rivers south to Central America.

Genus *Chloroceryle* Kaup. Crown plumage full but not forming a crest; bill more than half wing. Neotropical mainland. 4 species, 1 in the United States.

Chloroceryle americana (**Gmelin**). Green kingfisher. Sides of head and upper parts green; lower eyelid white; wing coverts, and sometimes forehead, speckled with white; remiges barred with white; outer rectrices white basally, green distally, where barred with white on inner webs; throat and sides of neck white; subrictal streak green; breast band chestnut; underparts white; sides, axillars, and crissum spotted with green; iris brown; bill and feet black. FEMALE: Lacks chestnut; breast crossed by 2 bands of green spots. Wing 79–90, tail 54–62, culmen 40–90, tarsus 8–10 mm. Southern Texas south in lowlands to Argentina.

Order Piciformes. Woodpeckers and allies. Nostrils imperforate. Adult downs absent; aftershafts present. Functional primaries 9–10, secondaries quintocubital; rectrices 10–12. Legs short; feet zygodactyl.

Basipterygoid processes absent. Cervicals 14. Sternum with large external spine, without internal spine; metasternum with 4 notches or 4 foramina. Coracoids separate; procoracoid process small. Humerus with 1 pneumatic fossa; ectepicondylar process small. Tibial bridge ossified. Tarsometatarsus complex, 2–5 canals.

Thigh muscles AX(Y) —; plantar tendons type 6. Syrinx tracheobronchial. Intestinal convolutions type 7; no crop; gall bladder very long.

Nest in holes in tree or in ground; eggs round, shining white; nestling nidicolous, gymnopedic, blind on hatching.

Holarctic, Neotropical, Oriental, African mainland. Miocene onward. 1 family in the United States.

Family Picidae. Woodpeckers. Bill strong, acute, straight or nearly so; nostrils usually feathered; tongue long, extensile. Aftershafts rudimentary; oil gland tufted. Wing rounded; 10 functional primaries. Tail shorter than wing, usually stiff and used as a prop; rectrices 12, but outer pair rudimentary. Tarsus scutellate, shorter than longest toe and claw; claws sharp, curved.

---

[1] Alcedinidae of A.O.U.

Skull very hard; palate schizognathous; vomer pointed; palatines and pterygoids both with forward-running processes; both hyoids very long and flexible, passing in a groove up over cranium to right nostril. Sternum 4-notched; external spine forked. Furculum U-shaped, without hypocleidium. Tarsometatarsus with a canal on proximal anterior face.

Thigh muscles *AX*. No caeca. Only the left carotid.

Food mainly grubs or ants. Flight undulating. Nest or dormitory drilled in trees.

North and South America, Europe, Asia, continental Africa. Miocene onward. 38 genera, 209 species; in the United States 8 genera, 20 species.

KEY TO GENERA OF PICIDAE

1. With only 3 toes   *Picoides*
   With 4 toes   **2**
2. Head with a long crest; wing 210 mm or more   **3**
   Without conspicuous crest; wing under 178 mm   **4**
3. Outer hind toe longest   *Campephilus*
   Outer front toe longest   *Dryocopus*
4. Outer hind toe longest   **5**
   Outer front toe longest   **6**
5. Distance from nostril to end of nasal groove less than from end of groove to tip of bill (Fig. 5-54)   *Sphyrapicus*
   Distance from nostril to end of nasal groove more than from end of groove to tip of bill (Fig. 5-55)   *Dendrocopos*
6. Without a distinct groove in front of nostrils; underparts spotted with black   *Colaptes*
   A distinct groove in front of nostrils; underparts unspotted   **7**
7. Plumage of underparts hairlike; rump and upper tail coverts blackish   *Asyndesmus*

Fig. 5-54 (left). Bill of yellow-bellied sapsucker (*Sphyrapicus varius*). *Fig. 5-55* (right). Bill of red-cockaded woodpecker (*Dendrocopos borealis*). *NG*, nasal groove.

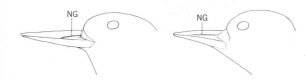

Feathers of underparts not hairy; rump and upper tail coverts white or barred   *Melanerpes*

Genus *Colaptes* Vigors. Bill slightly decurved, tip pointed, nasal groove obsolete, culmen ridged, nostrils concealed; 3d primary longest, outer shorter than 6th or about equal to 7th; tail ⅔ wing, slightly graduated, tips of rectrices abruptly pointed; outer front toe longest. North and South America. 4 species, 1 in the United States.

Colaptes auratus (**Linnaeus**).[1] Flicker. EASTERN RACES (yellow-shafted flicker): Crown and hind neck gray; nuchal collar red; back, wing coverts, and secondaries brownish barred with black; rump white, barred with black laterally; upper tail coverts white with black bars or U's; primaries and tail blackish dorsally, the shafts yellow basally; throat and sides of face vinaceous, malar stripe black (in male only); a black crescent on upper breast; underparts white spotted with black, tinged with fawn on sides; under wing coverts buffy yellow; undersurface of remiges and rectrices largely yellow, dusky at tips; iris reddish brown; bill and feet blackish. WESTERN RACES (red-shafted flicker): Pileum brown or gray; no nuchal collar; supraorbital patch cinnamon; malar stripe of male red; back grayer; remiges and tail with red instead of yellow; under wing coverts pink. SOUTHWESTERN RACES (gilded flicker): Like eastern birds, but without red nuchal crescent; malar stripe of male red; tail shafts almost entirely black dorsally. Wing 135–177, tail 84–124, culmen 28–43, tarsus 25–32 mm. North America south to Cuba and in mountains to Nicaragua.

Genus *Dryocopus* Boie. Head crested; bill straight, culmen ridged, tip chisellike, nasal grooves long and distinct, nostrils concealed; outer primary shorter than 7th, 3d longest; tail ⅔ wing; outer front toe longest. Palearctic and Oriental regions, North and South American mainland. 7 species, 1 in the United Sattes.

Dryocopus pileatus (**Linnaeus**). Pileated woodpecker. Black; crown and malar stripe red; a white line from nostrils down sides of neck; upper throat, bend of wing, patch on primaries, another patch on secondaries, and under wing coverts white; iris yellow; maxilla black; mandible mostly bluish

[1] Includes red-shafted flicker, *Colaptes cafer* (Gmelin), and gilded flicker, *C. chrysoides* (Malherbe), of A.O.U.

white; feet black. FEMALE: With red only on hind half of crown. Wing 210–253, tail 136–174, culmen 41.5–60, tarsus 31–36 mm. Forests of the United States and Canada.

Genus *Asyndesmus* Coues. Plumage of hind neck and underparts coarse and hairlike; bill nearly straight, tip pointed, culmen unridged, nostrils concealed, nasal groove distinct; wing long, 4th primary longest, outer one shorter than 6th; tail less than $\frac{2}{3}$ wing; outer front toe longest. Western North America. 1 species.

Asyndesmus lewis **(Gray).** Lewis's woodpecker. Forehead, face, and upper throat crimson; crown, upper parts, flanks, and crissum glossy greenish black; lower throat dull black tipped with whitish; collar on breast and hind neck silvery gray; underparts pinkish red, with white streaks; ventral surface of wings and tail black; iris brown; bill and feet dusky. YOUNG: Face black; no collar; underparts mostly pale gray with little red. Wing 162–180, tail 86–102, culmen 25.5–33, tarsus 23–26.5 mm. Western pine forests, east to Black Hills, western Kansas, and western Texas.

Genus *Melanerpes* Swainson. Bill stout, nearly straight, tip chisellike, culmen ridged, nostrils concealed, nasal ridge distinct but rather short; outer primary shorter than 5th, 2d to 4th longest; tail less than $\frac{2}{3}$ wing; outer front toe longest. North and South America. 18 species, 5 in the United States.

KEY TO SPECIES OF *MELANERPES*

1. Back plain black or at most tipped with gray   **2**
   Back barred black and white   **3**
2. A white band across forehead   *M. formicivorus*
   Without a white band on forehead
      M. erythrocephalus
3. Nasal plumes and middle of belly reddish
      M. carolinus
   Nasal plumes and middle of belly yellow or orange
   (in U.S. races)   **4**
4. Hind neck yellow, orange, or red; upper tail coverts
   unbarred   *M. aurifrons*
   Pileum and hind neck grayish brown; upper tail
   coverts barred   *M. hypopolius*

Melanerpes erythrocephalus **(Linnaeus).** Redheaded woodpecker. Head and neck red; back,

wing coverts, primaries, and tail black; outer rectrices tipped with white; secondaries, rump, upper tail coverts, and underparts white; belly tinged with reddish or yellowish; iris brown; bill bluish horn; feet gray. YOUNG: Head and neck dusky; throat streaked with whitish; back tipped with gray; secondaries banded black and white; breast and flanks brownish gray streaked with dusky. Wing 125–150, tail 63–85, culmen 25–31.5, tarsus 19–24.5 mm. Eastern North America, west to southeastern British Columbia, and eastern parts of Montana, Wyoming, Colorado, and New Mexico.

Melanerpes formicivorus **(Swainson).** Acorn woodpecker. Nasal tufts and chin black; band across forehead white, continuing in front of eye to throat, where pale yellow; crown red; circumorbital area, sides of face, and upper parts greenish black; rump, upper tail coverts, and base of primaries white; distal secondaries barred black and white on base of inner web; breast black streaked with white; a small red patch at upper border of breast; rest of underparts white; sides, crissum, and axillars streaked with black; iris whitish; bill black; feet dusky. FEMALE: Crown black. Wing 130–151, tail 65–88, culmen 22–23, tarsus 19–25 mm. Pine or oak woods from Oregon, California, Arizona, New Mexico, and western Texas, south in mountains to Colombia.

Melanerpes hypopolius **(Wagler).**[1] Gila woodpecker; gray-breasted woodpecker. Head, neck, and underparts drab; male with a red patch on middle of crown; orbit sometimes black; entire upper parts barred black and white; rectrices black, 2 outer pairs and inner web of middle pair barred black and white; primaries black, barred with white basally; belly yellow or white; flanks and crissum whitish barred with black; iris brown; bill black; feet grayish. Wing 120–140, tail 62–88, culmen 23.5–35, tarsus 20.5–25 mm. Southeastern California, southern Arizona, and New Mexico, south over Mexican plateau. Possibly conspecific with *M. aurifrons*.

Melanerpes aurifrons **(Wagler).**[2] Goldenfronted woodpecker. Nasal tufts, occiput, and belly yellow (red in some extralimital races); upper tail coverts, and often rump, plain white; otherwise as in *M. hypopolius*. Wing 127–143, tail 71–85, cul-

[1] *Centurus uropygialis* Baird of A.O.U.
[2] *Centurus aurifrons* (Wagler) of A.O.U.

men 27–35, tarsus 21–25 mm. Central Texas south through lowlands to Costa Rica. Possibly conspecific with *M. carolinus*.

Melanerpes carolinus (Linnaeus).[3] Red-bellied woodpecker. Nasal tufts, crown, and nape red; upper parts, including rump, usually upper tail coverts, secondaries, and base of primaries barred black and white; rectrices black; middle pair with a white basal stripe on outer web, inner web barred with white; outer pair barred black and white; next pair whitish at tip; sides of face and underparts gray; belly (sometimes face and throat) tinged with red; crissum barred black and white; iris reddish; bill black; feet gray. FEMALE: Crown gray; nasal tufts and occiput red. Wing 122–139, tail 68–85, culmen 25.5–33, tarsus 20–23 mm. Eastern United States, north to Delaware, Pennsylvania, southern Great Lakes, southern Minnesota, and southern South Dakota, west to eastern Nebraska, Kansas, Oklahoma, and central Texas.

Genus Sphyrapicus Baird. Bill straight, culmen ridged, nostrils concealed, nasal groove distinct and reaching tomium; outer primary usually longer than 5th, 3d longest; tail less than $\frac{2}{3}$ wing; outer hind toe slightly longest. North America. 2 species.

KEY TO SPECIES OF *SPHYRAPICUS*

1. Rump plain white; wing 132–143 mm *S. thyroideus*
   Rump mixed black and white, at least laterally; wing 118–133 mm *S. varius*

Sphyrapicus varius (Linnaeus). Yellow-bellied sapsucker. EASTERN AND ROCKY MOUNTAIN RACES: Crown red with black border; a whitish stripe from eye to eye around nape; a black stripe from auriculars down sides of neck; a whitish stripe from nasal tufts, below eye and down sides of neck; throat red; a black stripe from rictus down sides of throat to large black breast patch; back black blotched with white or buffy; rump and upper tail coverts white splotched with black laterally; rectrices black; middle pair barred black and white on inner web; outer 2 pairs with some white distally; wing black; a large white patch across distal coverts; remiges with broken white bars; sides of breast patch and median underparts yellowish; sides

brown; flanks and crissum whitish, both barred and streaked with dusky; iris and bill brown; feet gray. FEMALE: Chin and throat white. YOUNG: Head and neck brown flecked with whitish; breast brown squamate with dusky. PACIFIC RACES: Head, neck, and breast red; nasal tufts whitish; a short black preorbital and malar streak. Wing 118–132, tail 67–85, culmen 20–27, tarsus 18.5–22 mm. Northern coniferous forests, south to Massachusetts, mountains of North Carolina, Great Lakes, and in mountains to western Texas, New Mexico, Arizona, and California; winters south to West Indies and Central America.

Sphyrapicus thyroideus (Cassin). Williamson's sapsucker. Head, neck, back, and breast blue-black; a white stripe from eye to side of nape; another from nasal tufts to below auriculars; a red stripe on midline of throat; rump and upper tail coverts white; tail black; wings black with a large white patch across distal coverts; 2d to 5th primaries spotted with white; belly and middle of breast bright yellow; flanks and crissum white, striped or barred with black; iris brown; bill black; feet gray. YOUNG MALE: Throat stripe white. FEMALE: Head and neck drab; back and wings barred black and drab; no white wing patch; rump and upper tail coverts white; rectrices black, middle and lateral ones barred with white; throat sometimes with a red median stripe; breast and flanks barred black and buffy brown, sometimes forming a greenish black patch on breast; belly yellow; crissum white squamate with black. Wing 132–143, tail 71–89, culmen 21.5–28.5, tarsus 20–22.5 mm. Coniferous forests of southeastern British Columbia, east to Wyoming, south in mountains to California, Arizona, and New Mexico; Texas and northern Mexico in winter.

Genus Dendrocopos Koch. Bill rather short and straight, culmen ridged, nostrils concealed, nasal grooves long and reaching tomium; outer primary shorter than 5th, 3d longest; tail $\frac{1}{2}$ to $\frac{2}{3}$ wing; outer hind toe longest. Europe, Asia, Africa, North and South America. 32 species, 7 in the United States.

KEY TO SPECIES OF *DENDROCOPOS*

1. Pileum white *D. albolarvatus*
   Pileum blackish or brownish, sometimes with white marks 2

---

[3] *Centurus carolinus* (Linnaeus) of A.O.U.

2. Back plain brown   ***D. arizonae***
    Back with some white   3
3. A large white stripe down middle of back   4
    Back transversely barred with black   5
4. Lateral rectrices plain whitish; culmen 21–37.5 mm
                            D. villosus
    Lateral rectrices barred; culmen 14–19 mm
                           D. pubescens
5. Auriculars entirely white; lores black   ***D. borealis***
    Auriculars partly dark; lores dusky   6
6. Nasal tufts whitish; lateral rectrices with only 2–3
    bars   ***D. nuttallii***
    Nasal tufts brown; lateral rectrices with 4 or more
    bars   ***D. scalaris***

Dendrocopos villosus (**Linnaeus**). Hairy woodpecker. Nuchal band red in male; upper parts black; broad stripe down middle of back white; wing coverts more or less spotted and remiges barred with white; outer 3 pairs of rectrices largely white, without distinct bars; nasal tufts buffy; superciliary and subocular stripes white; auricular and malar stripes black; underparts white (brownish in Central American and some western races); iris brown; bill and feet blackish. YOUNG: No red nuchal patch; crown spotted with white (and red in male). Wing 98–138, tail 49–90, culmen 21–37.5, tarsus 17–25 mm. North America, south to Bahamas and Panama.

Dendrocopos pubescens (**Linnaeus**). Downy woodpecker. Plumage as in *D. villosus*, but outer pairs of rectrices with several black bars; crissum usually barred. Wing 86–105, tail 48–72, culmen 14–19, tarsus 14.5–18 mm. Alaska and Canada south through the United States.

Dendrocopos borealis (**Vieillot**). Red-cockaded woodpecker. Upper parts black; back and remiges regularly barred with white; wing coverts spotted with white; a small red spot on each side of nape in male; outer 3 pairs of rectrices white barred with black; nasal tufts and large auricular patch white; a black stripe from malar region down sides of neck; underparts white; sides and crissum spotted with black. YOUNG: No red nape spots; forehead streaked with white; male with large red patch in center of crown. Wing 95–126, tail 70–81, culmen 19–23, tarsus 19–21 mm. Oklahoma, Missouri, Kentucky, and North Carolina south to Florida and Gulf states.

Dendrocopos nuttallii (**Gambel**). Nuttall's woodpecker. Crown black streaked with white; occiput and nape red, mixed with black and white; back, lesser wing coverts, and 2 middle pairs of rectrices black; scapulars, rump, and remiges barred black and white; middle and greater coverts black spotted with white; outer 2 pairs of rectrices white, with 2–3 black bars, often incomplete; nasal tufts buffy white; auricular and malar stripes black; rictal and supraauricular stripes white; underparts buffy white; sides spotted with black; crissum and flanks barred with black; iris brown; bill horn; feet gray. FEMALE: Occiput and nape black streaked with white like crown. Wing 98–107, tail 59–67, culmen 18–22, tarsus 17.5–19 mm. Western California and northern Baja California.

Dendrocopos scalaris (**Wagler**). Ladder-backed woodpecker. Differs from *D. nuttallii* in having nasal tufts brownish; outer 2 pairs of rectrices with 4–5 black bars. Wing 84–111, tail 44–72, culmen 16.5–28.5, tarsus 15–21 mm. Colorado, Oklahoma, Texas, New Mexico, and southeastern California, south to Honduras.

Dendrocopos arizonae (**Hargitt**). Arizona woodpecker. Upper parts and 2 middle pairs of rectrices sooty brown; occipital band of male red; remiges with broken white bars; outer rectrices white barred with sooty; postocular and rictal stripes white; malar stripe dark brown; underparts whitish; breast spotted, flanks and crissum barred with dusky. Wing 103–121, tail 56–70, culmen 20–28, tarsus 17–21 mm. Upper Sonoran zone in southeastern Arizona, southwestern New Mexico, and northwestern Mexico.

Dendrocopos albolarvatus (**Cassin**). White-headed woodpecker. Black; head, neck, upper breast, and base of remiges white; occipital band of male red. Wing 122–131, tail 74–90, culmen 24–32, tarsus 20–23.5 mm. Mountains of western Idaho, Washington, Oregon, California, and western Nevada.

Genus *Picoides* Lacépède. Bill wide; culmen ridged; nasal grooves distinct, reaching tomium; tail less than $\frac{2}{3}$ wing; hallux absent; hind (4th) toe longest. Holarctic boreal forests. 2 species.

KEY TO SPECIES OF *PICOIDES*

1. Back plain black   ***P. arcticus***
    Back black barred with white   ***P. tridactylus***

Picoides arcticus (Swainson). Black-backed 3-toed woodpecker. Crown of male yellow, of female blue-black; upper parts, auriculars, and 2 middle pairs of rectrices glossy blue-black; 3 lateral pairs of rectrices whitish; outer webs of primaries spotted with white; a white stripe from lores, below auriculars, down sides of neck; a black stripe from malar region down sides of neck; underparts white; sides barred with black; iris chestnut; bill and feet slate. Wing 123–134, tail 74–85, culmen 28.5–35, tarsus 11–13.5 mm. Northern coniferous forests, south to New England, New York, Michigan, Minnesota, Black Hills, Montana, and in mountains to California. In winter south irregularly to Nebraska, Illinois, Indiana, Ohio, Pennsylvania, and New Jersey.

Picoides tridactylus (Linnaeus). Northern 3-toed woodpecker. Differs in having back dull black, barred with white along midline; forehead and occiput more or less spotted with white. Wing 109–129, tail 68–82, culmen 22–30.5, tarsus 18.5–22.5 mm. Holarctic coniferous forests; in America south to New England, New York, Michigan, Minnesota, and in mountains to New Mexico, Arizona, Idaho, and Oregon.

Genus *Campephilus* Gray. Head with recurved crest; bill wide; culmen strongly ridged; nasal groove distinct, reaching tomium; tail less than ⅔ wing; outer hind toe longest. Southeastern United States, Cuba, northern Mexico, Patagonia. 3 species, 1 in the United States.

Campephilus principalis (Linnaeus). Ivory-billed woodpecker. Glossy blue-black; nasal tufts white; a white stripe from below auriculars down sides of neck and along edge of scapulars; secondaries, tips of primaries, and wing lining white; crest of male red, of female blue-black; iris yellow; bill white; feet gray. Wing 240–263, tail 147–166, culmen 61–72.5, tarsus 40.5–46 mm. Formerly bottomland forests and adjacent pinelands of southeastern states, north to North Carolina, Indiana, southern Illinois, and Missouri, west to Oklahoma and Texas; Cuba. Now practically extinct; still existed in Florida in 1953, South Carolina probably until 1940, and Louisiana until about 1942; in Cuba in 1956.

Order Passeriformes. Perching birds. Adult downs usually absent or scarce, on apteria only; aftershafts usually present; oil gland naked. Primaries 10–11 (9–10 functional); secondaries 9, quintocubital; wing coverts in 3 regular series. Rectrices 10–16, usually 12. Nostrils imperforate. Feet anisodactyl; hallux incumbent, about as long as lateral toes or longer.

Aegithognathous (rarely schizognathous), with vomer but without basipterygoid processes; holorhinal (rarely schizorhinal); no supraorbital depressions; postarticular process small. Cervicals 14–15. Sternum usually with 1 pair of notches or fenestrate (rarely 2 pairs). Furculum U-shaped. Coracoids separate, procoracoid process small. Humerus with 1–2 pneumatic fossae; ectepicondylar process pres-

TABLE 5-2. *Suborders of Passeriformes*

	Eurylaimi	Tyranni	Menurae	Passeres
Hallux	Weak	Strongest toe	Strongest toe	Strongest toe
Cervical vertebrae	15	14	14	14
Spina externa	Simple	Forked	Forked	Forked
Syrinx	Anisomyodous	Anisomyodous	Diacromyodous	Diacromyodous
Syringeal muscles	1 pair	0–2 pairs	2–3 pairs	5–7 pairs
Intestine	Mesogyrous	Mesogyrous	Anticoelous	Mesogyrous or anticoelous
Plantar tendons	Desmopelmous	Schizopelmous	Schizopelmous	Schizopelmous
Families	1	10	2	39
Genera	8	314	2	929
Species	14	1078	4	4063
Distribution	Oriental, Ethiopian	Tropicopolitan, Nearctic, New Zealand	Australia	Cosmopolitan

ent. Tibial bridge ossified. Hypotarsus complex; metatarsals in a plane.

Thigh muscles $A(B)XY$ —, Y absent in 1 family; plantar tendons sometimes desmopelmous (type 1), usually schizopelmous (type 7). Left carotid only. Intestinal convolutions type 7 or 8; caeca rudimentary.

Cosmopolitan. Eocene onward. 4 suborders, 52 families, 1253 genera, 5159 species; in the United States 2 suborders, 26 families. The characters of the suborders are given in Table 52. Of the passeriform families included in this manual only the Tyrannidae and Cotingidae belong in the suborder Tyranni, all others being members of the Passeres.

KEY TO FAMILIES OF PASSERIFORMES

1. Tarsus cylindrical, rounded behind   2
   Tarsus compressed into a sharp ridge behind   4
2. Maxillary tomium without a subterminal notch   **Alaudidae**
   Maxillary tomium with a subterminal notch   3
3. Inner toe free; outer face of tarsus longitudinally undivided (Fig. 5-56)   **Tyrannidae**
   Basal phalanx of inner toe fused to middle toe; outer or posterior face of tarsus divided longitudinally (Fig. 5-57)   **Cotingidae**

Fig. 5-56 (left). **Outer face of right foot of eastern kingbird (*Tyrannus tyrannus*), showing exaspidean tarsus of Tyrannidae. *Fig. 5-57* (right). Outer face of right foot of rose-throated becard (*Pachyramphus aglaiae*), showing longitudinally divided, taxaspidean tarsus of Cotingidae.**

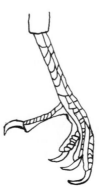

4. Longest primary more than twice length of secondaries   **Hirundinidae**
   Longest primary less than twice length of secondaries   5
5. Tail about half length of wing   6
   Tail at least $\frac{2}{3}$ length of wing   9
6. Length of gonys only half width across mandibular rami   **Bombycillidae**
   Length of gonys greater than width across rami   7
7. Inner toe definitely shorter than outer toe   **Sittidae**
   Inner toe nearly as long as outer toe   8
8. Tail rounded; tarsus booted   **Cinclidae**
   Tail emarginate; tarsus scutellate   **Sturnidae**
9. Bill hooked at tip   10
   Bill unhooked   11
10. Nostrils covered by bristles   **Laniidae**
    Nostrils exposed   **Vireonidae**
11. Gonys shorter than width of rami at feathers   **Ptilogonatidae**
    Gonys longer than width of mandible   12
12. Outer 2 primaries much shorter than secondaries   13
    Second primary little if any shorter than secondaries   14
13. Tarsus slightly longer than middle toe with claw   **Pycnonotidae**
    Tarsus much longer than middle toe with claw   **Timaliidae**
14. Primaries 10, outermost less than half length of longest [1]   15
    Primaries 9, outermost more than half length of longest   21
15. Hallux with claw longer than middle toe with claw   **Certhiidae**
    Hallux with claw shorter than middle toe with claw   16
16. Hallux with claw shorter than lateral toes with claw   17
    Hallux with claw longer than lateral toes with claw   18
17. Tarsus booted   **Turdidae**
    Tarsus scutellate   **Mimidae**
18. Most of basal phalanx of middle toe free from inner toe   19
    Half or more of basal phalanx of middle toe united to inner toe   20
19. Large (wing over 100 mm); bill and feet stout   **Corvidae**
    Small (wing under 95 mm); bill and feet slender   **Sylviidae**

---

[1] *Peucedramus*, in the otherwise 10-primaried family Sylviidae, has only 9 primaries, but differs from other 9-primaried birds in the following combination of characters: culmen with a flat area in advance of nostrils; rictal bristles few; nasal plumes falling short of nostril or operculum; nostril a horizontal slit; operculum swollen basally, its side folded into nostril; tail emarginate; tarsus almost booted.

20. Nostrils covered by feathers (rarely incompletely)
                                                Paridae
    Nostrils exposed   **Troglodytidae**
21. Nostrils minutely perforate   **Motacillidae**
    Nostrils imperforate   **22**
22. Length of mandibular ramus not more than half length of gonys; tarsus usually shorter than middle toe with claw   **Fringillidae**
    Length of mandibular ramus more than half length of gonys; tarsus longer than middle toe with claw   **23**
23. Bill slender, tenuirostral   **Parulidae**
    Bill stout, conirostral   **24**
24. Nostrils round   **Tanagridae**
    Nostrils oval or slits   **25**
25. With rictal bristles   **Emberizidae**
    Without rictal bristles   **Icteridae**

Family Cotingidae. Cotingas. Bill stout, bristled at base; maxilla with tip hooked and tomium notched subterminally; nostril round or elongate oval, nonoperculate, usually partly concealed by feathers. Wing rounded; primaries 10, with outer primary often longer than secondaries. Rectrices 12. Tarsus rounded behind, with outer face longitudinally divided (i.e., pycnaspidean, holaspidean, or taxaspidean; not exaspidean); first 2 phalanges of outer toe bound to middle toe; inner and outer toes about equal. Cranium hard; nares sometimes amphirhinal; humerus with 1 fossa. Syrinx tracheobronchial; vocal muscles inserted on dorsal end of semirings. Main thigh artery the femoral. Neotropical region, north to the Mexican boundary. No fossil record. 33 genera; 91 species, 1 in the United States.

Genus *Pachyramphus* Gray. Crown feathers erectile, forming a short bushy crest; bill stout, wide; nostril round, partly concealed; 2d primary of male sharp-tipped and much shorter than other remiges (Fig. 5-58); outer primary longer than secondaries; tail slightly rounded or double-rounded; tarsus taxaspidean. Tropical America, north to Arizona, Texas, and Jamaica. 16 species, 1 in United States.

***Pachyramphus aglaiae* Lafresnaye.**[1] Rose-throated becard. Above slate; crown black; lower throat pink in most races; rest of underparts gray; iris brown; bill and feet gray. FEMALE: Crown blackish or gray; back olive-cinnamon; tail and wings tawny; distal part of primaries dusky; under-

[1] *Platypsaris aglaiae* of A.O.U.

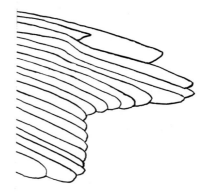

Fig. 5-58. Wing of rose-throated becard (*Pachyramphus aglaiae*), showing short second primary of male.

parts rich buff. Wing 83–98, tail 59–76, culmen 15–18, tarsus 20–23 mm. Costa Rica north to southern Arizona and southern Texas.

Family Tyrannidae. Tyrant flycatchers. Tip of maxilla hooked; bill usually flat, bristled at base; nostril round, usually nonoperculate, partly concealed by bristles. Feet relatively weak; tarsus exaspidean, rather rounded behind; middle toe adherent to lateral toes basally. Primaries 10; outer one longer than secondaries. Rectrices 12. Cranium spongy; humerus with 1 fossa. Main thigh artery sciatic. Syrinx tracheobronchial; intrinsic muscles inserted on ventral ends of bronchial semirings. North and South America; fossils unknown before Pleistocene. 118 genera, 375 species; in the United States 10 genera, 30 species.

KEY TO GENERA OF TYRANNIDAE

1. Bill about as deep as wide; bristles minute
                                        Camptostoma
    Bill wider than deep; bristles well developed   **2**
2. Tips of outer primaries abruptly attenuate   **3**
    Primaries normal   **4**
3. Tail shorter than wing   ***Tyrannus***
    Tail longer than wing   ***Muscivora***
4. Crissum red or orange   ***Pyrocephalus***
    Crissum without red or orange   **5**
5. With chestnut in tail   **6**
    Tail without chestnut   **8**
6. No yellow crown patch; tail 85–97% wing
                                           Myiarchus

A concealed yellow crown patch; tail 72–78% wing 7
7. Streaked below; outer primary longer than 6th
Myiodynastes
Plain yellow below; outer primary shorter than 6th
Pitangus
8. Tarsus less than ⅛ wing length *Contopus*
Tarsus more than ⅛ wing length 9
9. Lower mandible pale *Empidonax*
Bill entirely dark *Sayornis*

Genus *Tyrannus* Lacépède. Bill stout; head with concealed orange patch in adult; tips of outer primaries attenuate; outer primary longer than 7th; tail shorter than wing; tarsus shorter than middle toe with claw. Young birds lack the crown patch and attenuation of primaries. North and South America. 10 species, 5 in the United States.

KEY TO SPECIES OF *TYRANNUS*

1. Belly white 2
Belly yellow 3
2. Tip of tail white *T. tyrannus*
Tail without white tip *T. dominicensis*
3. Exposed culmen equal to or longer than tarsus
T. melancholicus
Exposed culmen shorter than tarsus 4
4. Outer primary shorter than 6th; outer web of lateral rectrix brownish with narrow gray edging
T. vociferans
Outer primary longer than 6th; outer web of lateral rectrix white to shaft *T. verticalis*

Tyrannus tyrannus (**Linnaeus**). Eastern kingbird. Above black; crown patch orange; wings edged with ashy; tail rounded, with white tip; below white, shaded with dusky across breast; axillars gray edged with white. Wing 110–125, tail 72–90, culmen 16.5–19.5, tarsus 16.5–19.5 mm. Eastern North America, west to Washington (Seattle), eastern Oregon, northeastern California, Utah, and northeastern Nevada and New Mexico; winters in tropical America.

Tyrannus dominicensis (**Gmelin**). Gray kingbird. Tail emarginate. Above gray, darker on auriculars; crown patch reddish orange; wings and outer rectrix edged with ashy; underparts whitish, shaded with ashy across breast; axillars yellow. Wing 106–127, tail 80–100, culmen 25.5–35, tarsus 15.5–21 mm. Coasts of South Carolina, Georgia, Alabama, and Florida, south through West Indies; winters in West Indies and northern South America.

Tyrannus melancholicus **Vieillot.** Tropical kingbird. Tail over 80% wing, emarginate. Crown gray, auriculars darker; crown patch reddish orange; back yellowish green; wings brown edged with ashy; tail brown; throat whitish; band across breast olive; rest of underparts bright yellow. Wing 104–131, tail 84–108, culmen 19.5–27, tarsus 15.5–21 mm. Lower Rio Grande Valley south through mainland of tropical America.

Tyrannus verticalis **Say.** Western kingbird. Tail less than 75% wing, emarginate. Crown dull gray; crown patch reddish orange; auriculars dusky; back dull olive; wings brown, edged with pale brown; tail black, outer web of lateral rectrix abruptly white; throat and upper breast light gray, paling anteriorly; rest of underparts dull yellow. Wing 119–135, tail 82–97, culmen 16–21, tarsus 16.5–20 mm. Western North America, east to Minnesota, Iowa, Kansas, and western Texas; winters in Mexico and Central America; irregular on migration east to Atlantic Coast.

Tyrannus vociferans **Swainson.** Cassin's kingbird. Tail less than 75% wing, emarginate. Differs from *T. verticalis* in having back gray with little or no olive wash; chin white in abrupt contrast to dark-gray throat and breast; outer web of lateral rectrix but little paler than rest of tail. Wing 121–137, tail 86–96, culmen 18–22, tarsus 18–20 mm. From Oregon and Montana east to Colorado, New Mexico, and western Texas, south to central Mexico; winters from California to Central America.

Genus *Muscivora* Lacépède. Like *Tyrannus*, but outer primary longer than 5th; tail longer than wing, deeply forked. Southern Great Plains to Argentina. 2 species, 1 in the United States.

Muscivora forficata (**Gmelin**). Scissor-tailed flycatcher. Crown and hind neck pearly gray; crown patch orange-red; upper back gray tinged with salmon; rump dusky; upper tail coverts and 3 middle pairs of rectrices black; 3 outer pairs pinkish white, black terminally; wings brown; anterior underparts white; flanks, crissum, and under wing coverts salmon; axillars and edge of scapulars orange-red. FEMALE: Crown patch obsolete or wanting; flanks, crissum, and wing lining buffy; axillar patch smaller and duller. Wing 112–129, tail 126–

256, culmen 16–19, tarsus 17.5–19 mm. Southern Great Plains from southern Nebraska and Kansas to western Louisiana and southern Texas; winters from southern Mexico to Panama.

Genus *Pyrocephalus* Gould. Head with a bushy crest; primaries not attenuate, outer one longer than 6th; tail less than $\frac{3}{4}$ wing, nearly even; tarsus longer than culmen or middle toe with claw. Southwestern United States to Argentina. Monotypic.

Pyrocephalus rubinus (**Boddaert**). Vermilion flycatcher. Crown and underparts vermilion; upper parts, wings, tail, and loral-auricular stripe sooty. FEMALE: Above brown; edge of lateral rectrix ashy; below whitish, streaked with brown on breast and sides; flanks and crissum vermilion, pink, or yellow. YOUNG: Like female but upper parts edged with whitish; flanks and crissum whitish streaked with brown like breast. Wing 66–86, tail 47–62, culmen 12–14, tarsus 14.5–17 mm. Southeastern California, southern Nevada, southern Utah, southern New Mexico, and southern Texas, south to Honduras; also in most of South America.

Genus *Sayornis* Gray. Head uncrested and without concealed patch; wing less than 6 times length of tarsus; primary tips not attenuate; outer primary longer than 7th; tail more than $\frac{3}{4}$ wing; tarsus longer than culmen or middle toe with claw; bill relatively narrow and entirely dark. North and South America. 3 species.

KEY TO SPECIES OF *SAYORNIS*

1. Belly cinnamon   *S. sayus*
   Belly white or buffy   2
2. Upper parts, throat, and breast black   *S. nigricans*
   Back olive; underparts buffy   *S. phoebe*

Sayornis phoebe (**Latham**). Eastern phoebe. Crown dusky; upper parts olive; wings and tail edged with whitish; underparts buffy, shaded with olive across breast. Wing 77–92, tail 63–78, culmen 13.5–16, tarsus 16.5–19.5 mm. Southern Canada and eastern United States, west to Black Hills and eastern Colorado, south to New Mexico, central Texas, northern Mississippi, and mountains of Georgia; winters from Ohio River south to Florida and central Mexico.

Sayornis nigricans (**Swainson**). Black phoebe. Upper parts, throat, breast, and sides dull black;

wings and tail edged with whitish; belly white; crissum white with dusky shaft streaks. Wing 81–96, tail 68–84, culmen 14–16.5, tarsus 15.5–19.5 mm. Southern Oregon, California, southern Arizona, New Mexico, and western Texas, south to Argentina.

Sayornis sayus (**Bonaparte**).[1] Say's phoebe. Ashy brown above, on throat and breast; tail dusky; belly and crissum cinnamon. Wing 96–111, tail 76–87, culmen 13.5–18, tarsus 18.5–22 mm. Western North America from Alaska to New Mexico, east to Dakotas and Kansas; winters from states along Mexican border to central Mexico.

Genus *Empidonax* Cabanis. Differs from *Sayornis* in having feet more delicate and lower mandible paler; size smaller. North and South America. 11 species, 8 in the United States.

KEY TO SPECIES OF *EMPIDONAX*

1. Breast, sides, and auriculars fulvous buff
                E. fulvifrons
   Breast, sides, and auriculars not fulvous buff   2
2. Wing tip (tip of secondaries to tip of primaries) longer than tarsus   3
   Wing tip shorter than tarsus   4
3. Mandible dusky   *E. hammondii*
   Mandible light   *E. virescens*
4. Eye ring and throat yellowish   5
   Eye ring and throat whitish   6
5. Outer primary longer than sixth   *E. flaviventris*
   Outer primary shorter than sixth   *E. difficilis*
6. Tail rounded or square; width of bill 5–6.5 mm
                E. traillii
   Tail slightly emarginate; width of bill 4–5.5 mm   7
7. Wing 58–67 mm, tail 47–58 mm; tail/wing ratio under 87%   *E. minimus*
   Wing 62–76 mm, tail 56–67 mm; tail/wing ratio over 88%   *E. affinis*

Empidonax virescens (**Vieillot**). Acadian flycatcher. Outer primary longer than 5th; wing tip longer than tarsus; tail even or slightly rounded. Above dull olive green; wings dusky edged with cream buff; 2 cream-buff bands on coverts; eye ring pale yellowish; below whitish, shaded with olive across breast; sides, belly, crissum, and wing lining pale yellow; maxilla dark; mandible ivory. Wing 66–80.5, tail 50.5–63, culmen 14–18, tarsus

[1] *Sayornis saya* of A.O.U.

14–17.5, width of bill 5.5–7 mm. Eastern deciduous forest, from eastern South Dakota, Iowa, southern parts of Wisconsin, Michigan, and Ontario, New York, Vermont, Connecticut, and Massachusetts, south to eastern Texas, Gulf states, and northern Florida. Winters in northern South America.

Empidonax traillii (**Audubon**). Traill's flycatcher; alder flycatcher. Outer primary shorter than 5th; wing tip about equal to tarsus; tail rounded or double-rounded. Above brownish or olive; wing bands olive buff; below whitish, shaded with brownish olive across breast; sides and crissum tinged with buffy yellow; maxilla black; mandible cream buff. Wing 64–76, tail 52.5–64, culmen 13–17.5, width of bill 5–6.5, tarsus 15.5–18 mm. North America, south of the tundra, south to Baja California, Sonora, New Mexico, central Texas, Arkansas, Illinois, Indiana, Ohio, Pennsylvania, northern New Jersey, and in mountains to North Carolina. Migrates through Mississippi Valley and westward, and winters from Mexico to South America.

Empidonax minimus (**Baird and Baird**). Least flycatcher. Outer primary usually shorter than 5th; wing tip about equal to middle toe with claw; tail emarginate. In color like *E. traillii*, but mandible dark olive buff or drab. Wing 58–67, tail 47–58, culmen 12–14, width of bill 4–5.5, tarsus 14.5–17.5 mm. Forests of southern Canada, south to Montana, eastern Wyoming, Oklahoma, Missouri, Illinois, Indiana, Ohio, Pennsylvania, New Jersey, and in mountains to northern Georgia. Migrates through Mississippi Valley; winters from Mexico to Panama.

Empidonax hammondii (**Xantus**). Hammond's flycatcher. Outer primary shorter than 5th, usually longer than 6th; wing tip longer than tarsus; tail emarginate; bill narrow. Above grayish olive; wing bands olive buff; eye ring and throat dirty white; breast shaded with olive; posterior underparts tinged with yellow; maxilla blackish; mandible hair brown. Wing 64–75, tail 52–61.5, culmen 12–14, width of bill 3.5–4.5, tarsus 14.5–17 mm. Spruce-fir forests from Alaska and northwest Canada south to California and Colorado; on migration to western Texas; winters from California and Arizona south to Honduras.

Empidonax affinis (**Swainson**).[1] Dusky fly-

catcher; gray flycatcher. Outer primary shorter than 6th, usually shorter than 7th; wing tip shorter than tarsus; tail more or less emarginate. Upper parts vary from grayish drab to deep olive or buffy olive (Mexican races); throat grayish white; breast light grayish olive; belly dull white, more or less tinged with yellow (entire underparts suffused with buffy yellow in Mexican races); mandible pale yellowish horn. Wing 62–79.5, tail 56–69.5, culmen 12.5–16.5, width of bill 4–4.5, tarsus 16–20 mm. Southwest Canada, western United States east to Montana, Black Hills, western Colorado, and New Mexico, and south in mountains to Guatemala; on migration to western Texas; winters in Mexico and Guatemala.

Empidonax flaviventris (**Baird and Baird**). Yellow-bellied flycatcher. Outer primary shorter than 5th but longer than 6th; wing tip shorter than tarsus; tail nearly even. Above greenish olive; wing bars and eye ring yellow or buff; underparts yellow, shaded with olive across breast; mandible ivory yellow. Wing 62.5–71.5, tail 48–56.5, culmen 12.5–15, width of bill 4.5–6, tarsus 14–17.5 mm. Coniferous forests of eastern Canada, south to North Dakota, Minnesota, northern Michigan, Pennsylvania, and Massachusetts; migrates through Mississippi Valley, more rarely on Atlantic Coast; winters from Mexico to Panama.

Empidonax difficilis **Baird**. Western flycatcher. Outer primary shorter than 6th; wing tip somewhat shorter than tarsus; tail nearly even. Above brownish olive (greenish in extralimital races); wing bands, eye ring, and throat olive buff; belly bright yellow; mandible ivory yellow. Wing 59–74, tail 51.5–64, culmen 12–16, width of bill 4.5–6.5, tarsus 14.5–19 mm. Western North America from Alaska to mountains of Panama, east to Black Hills, western Colorado, and western Texas; winters south of the United States.

Empidonax fulvifrons (**Giraud**). Buff-breasted flycatcher. Outer primary shorter than 6th; wing tip almost as long as tarsus; tail emarginate. Above light buffy brown, darker on crown, paler on nape and rump; wing bands, eye ring, and underparts buff or tawny; mandible cream buff. Wing 55–65, tail 44–53, culmen 11–12.5, width of bill 4–5, tarsus 13.5–14.5 mm. Mountains of Arizona and New Mexico south to Guatemala.

Genus *Contopus* Cabanis. Differs from *Empi-*

[1] Includes *Empidonax wrightii* Baird and *E. oberholseri* Phillips of A.O.U.

donax in having wing at least 6 times length of tarsus, its tip much longer than tarsus; tail emarginate. North and South America. 8 species, 4 in the United States.

KEY TO SPECIES OF *CONTOPUS*

1. Outer primary longer than 4th; wing 98–115 mm
                                          C. borealis
   Outer primary shorter than 4th   **2**
2. Wing 95–114 mm; outer primary shorter than 5th
                                        C. pertinax
   Wing 75–93.5 mm; outer primary longer than 5th   **3**
3. Mandible entirely dark; back not tinged with olive
                                      C. sordidulus
   Mandible mostly whitish; back with slight olive tinge   **C. virens**

Contopus virens (**Linnaeus**). Eastern wood pewee. Above dusky, tinged with olive on back; 2 wing bars and edges of secondaries ashy; eye ring obsolete; throat whitish; breast and sides shaded with olive gray; belly yellowish white; crissum similar but with dusky shaft streaks; maxilla and feet black; mandible whitish, often tipped with brown. Wing 77–90.5, tail 57.5–70, culmen 12.5–14.5, tarsus 12.5–14.5 mm. Forests of eastern North America, west to eastern Colorado and eastern Texas, south to Gulf states and central Florida; winters in Central and South America.

Contopus sordidulus **Sclater**. Western wood pewee. Differs in having back without olive tinge; belly and crissum paler, less yellowish; mandible wholly brown. Wing 75–93.5, tail 54.5–71.5, culmen 11.5–15, tarsus 12–14 mm. Forests of West, from Alaska to Panama, east to western North Dakota and western Texas; winters in Central and South America.

Contopus pertinax **Cabanis and Heine**. Greater pewee; Coues's flycatcher. Above grayish; wing bars indistinct; breast and sides pale grayish olive; throat, middle of belly, and crissum paler and tinged with yellow; maxilla brown; mandible ivory. Wing 95–114, tail 75.5–92, culmen 16–20.5, tarsus 15.5–18 mm. Pine forests of southern Arizona south in mountains to Nicaragua.

Contopus borealis (**Swainson**).[1] Olive-sided flycatcher. Above olive gray, darker on crown,

[1] *Nuttallornis borealis* of A.O.U.

wings, and tail; wing bands indistinct; a patch of fluffy white feathers on each side of rump; breast and sides brownish olive streaked with darker; throat and belly pale yellowish; crissum similar but squamate with brown; maxilla and feet black; mandible ivory. Wing 98–115, tail 63–77.5, culmen 15–18.5, tarsus 14.5–16 mm. Northern coniferous forest, south to Michigan, New York, and New Jersey, and in mountains to Baja California, Arizona, New Mexico, western Texas, and North Carolina; winters in South America.

Genus *Pitangus* Swainson. Head with concealed yellow patch; bill nearly as deep as wide; primaries normal, outer one shorter than 6th; tarsus nearly equal to culmen, longer than middle toe with claw; tail square, with cinnamon. Mexican border to South America. 2 species, 1 in the United States.

Pitangus sulphuratus (**Linnaeus**). Derby flycatcher; kiskadee flycatcher. Crown black, bordered all around by white and with a large yellow median patch; auricular and malar regions black; back olive brown; tail brown, rectrices broadly cinnamon medially; remiges brown edged with cinnamon; throat white; posterior underparts yellow; bill and feet black. Wing 111–130, tail 79.5–101, culmen 25–30.5, tarsus 23.5–28.5 mm. Rio Grande Valley south to Argentina.

Genus *Myiodynastes* Bonaparte. Differs from *Pitangus* in having outer primary longer than 6th; culmen rounded instead of ridged. Arizona to Argentina. 5 species, 1 in the United States.

Myiodynastes luteiventris **Sclater**. Sulphur-bellied flycatcher. Upper parts broadly streaked with brown and light brown; crown patch yellow; wings brown with 2 ashy bars; crissum and tail cinnamon with broad brown shaft streaks; middle of throat white; underparts yellow, boldly streaked with brown on neck, breast, and sides; bill and feet blackish. Wing 105.5–121.5, tail 77–88.5, culmen 19.5–23.5, tarsus 17.5–20 mm. Southern Arizona to Bolivia.

Genus *Myiarchus* Cabanis. Head without yellow patch; bill decidedly wider than deep; primaries normal, outer one shorter than 6th; tail more than 80% wing; tarsus longer than middle toe with claw; tail and often wings with cinnamon. North and South America. 17 species, 4 in the United States.

KEY TO SPECIES OF *MYIARCHUS*

1. Inner webs of rectrices mostly dusky   *M. tuberculifer*
Inner webs of rectrices mostly chestnut   **2**
2. Dusky shaft lines on inner web of lateral rectrix over
  3 mm wide   *M. tyrannulus*
Dusky shaft line on inner web of lateral rectrix under
  2 mm wide   **3**
3. Outer primary longer than 7th   *M. crinitus*
Outer primary shorter than 7th   *M. cinerascens*

Myiarchus crinitus (**Linnaeus**). Great crested flycatcher. Outer primary longer than 7th. Upper parts dull olive; wings brown with 2 ashy bars; primaries edged with chestnut; tail brown, inner webs largely chestnut; inner web of outer rectrix with a narrow dusky shaft line or none; throat and breast dark gray; belly and crissum sulfur yellow; bill and feet black; base of mandible brownish. Wing 95–109.5, tail 81.5–96, culmen 18–23, tarsus 19.5–21.5 mm. Southeastern Canada to Florida and Gulf states; winters from Mexico to Colombia.

Myiarchus cinerascens (**Lawrence**). Ash-throated flycatcher. Outer primary shorter than 7th. Back browner; throat and breast pale gray; belly pale yellow; base of mandible darker brown. Wing 88.5–104, tail 81.5–97, culmen 17–21, tarsus 19.5–24 mm. Washington, southern Idaho, Colorado, and Texas south to Costa Rica.

Myiarchus tyrannulus (**Müller**). Wied's crested flycatcher; Mexican crested flycatcher. Outer primary shorter than 7th. Colored like *M. cinerascens*, but dark shaft line on inner web of outer rectrix over 3 mm wide; bill entirely black. Wing 96–114, tail 85–105, culmen 20–26.5, tarsus 21.5–26 mm. Southern Arizona and New Mexico and lower Rio Grande south to Argentina; Lesser Antilles.

Myiarchus tuberculifer (**Lafresnaye and d'Orbigny**). Olivaceous flycatcher. Outer primary

Fig. 5-59. Foot of horned lark (*Eremophila alpestris*).

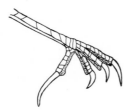

shorter than 7th. Crown dusky; back dark olive; wings and tail dusky, narrowly edged with cinnamon; throat and breast dark gray; belly and crissum yellow; bill and feet black. Wing 73–89.5, tail 64.5–87, culmen 16–19.5, tarsus 16–21 mm. Southern Arizona to Argentina.

Genus *Camptostoma* Sclater. Bill short, compressed, culmen curved; rictal bristles obsolete; outer primary shorter than 6th; tail nearly even; tarsus much longer than culmen or middle toe with claw. Southwestern United States to South America. 2 species, 1 in the United States.

Camptostoma imberbe **Sclater.** Beardless flycatcher. Above grayish olive, darker on crown, wings, and tail; 2 grayish wing bars (clay color in young); throat and breast pale grayish; belly and crissum yellowish white. Wing 46.5–58.5, tail 33–48.5, culmen 7.5–9.5, tarsus 12.5–15.5 mm. Southern Arizona and lower Rio Grande to Costa Rica.

Family Alaudidae. Larks. Bill conical, acute, culmen curved, maxilla unnotched; wing pointed, 9–10 obvious primaries, tertials well developed; tail short, rectrices 12; tarsus longer than bill, planta tarsi scutellate, rounded; middle toe shorter than tarsus, anterior claws slender and slightly curved; hind claw nearly straight, longer than its toe (Fig. 5-59). Humerus with 1 fossa and well-developed deltoid crest. Pliocene onward. Old World, North America, and Colombia. 15 genera, 75 species, 1 in the United States.[1]

Genus *Eremophila* Boie. A tuft of feathers on each side of occiput; primaries 9; tail nearly even, as long as from bend of wing to end of secondaries; middle toe about equal to culmen; hallux about equal to lateral toes. 2 species, 1 in the United States.

Eremophila alpestris (**Linnaeus**). Horned lark. Forehead and superciliaries white or yellowish; pileum and ear tufts black; occiput and underparts vinaceous sandy, usually streaked with brown; lateral rectrices black edged with white; stripe from lores to below eye and malar region black; throat yellow; lower throat patch black; posterior underparts white. FEMALE: Duller and more streaked above; ear tufts smaller; yellow paler. YOUNG: Speckled and spotted with whitish above and on breast. Wing 87–115.5, tail 52.5–75, cul-

[1] The skylark, *Alauda arvensis* (Linnaeus), was introduced on Long Island but disappeared about 1913.

men 8.5–13, tarsus 17.5–25 mm. Europe, Asia, North Africa, North America, Colombia.

Family Hirundinidae. Swallows. Bill short, flat, triangular; tip of maxilla uncinate; mouth wide; rictal bristles obsolete. Wing long, pointed; primaries 9; secondaries less than half length of primaries. Tail more or less emarginate; rectrices 12. Feet short, weak; tarsus scutellate or feathered, shorter than middle toe with claw. Known from Pleistocene. Cosmopolitan. 20 genera, 75 species; in the United States 6 genera, 9 species.

KEY TO GENERA OF HIRUNDINIDAE

1. Nostrils round, opening upward (Fig. 5-60) **2**
   Nostrils slitlike, overhung by operculum, opening laterally (Fig. 5-61) **4**

Fig. 5-60 (left). **Bill of purple martin** (*Progne subis*).
Fig. 5-61 (right). **Bill of barn swallow** (*Hirundo rustica*).

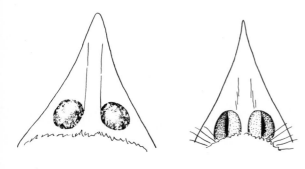

2. Bill stout, curved (wing 128–153 mm) *Progne*
   Bill weak, straight (wing 95–118 mm) **3**
3. Outer edge of primary often serrate; tail emarginate for more than 5 mm *Stelgidopteryx*
   Outer primary normal; tail emarginate for less than 4 mm *Petrochelidon*
4. A tiny tuft of feathers on inner side of acrotarsium above hallux *Riparia*
   Tarsus bare below **5**
5. Nostrils exposed; tail forked for over 15 mm *Hirundo*
   Nasal operculum at least partly feathered; tail emarginate for less than 15 mm *Tachycineta*

Genus *Progne* Boie. Bill stout, culmen curved; nostrils round, opening upward, without opercu-

lum; tail half wing, forked for $\frac{1}{4}$ its length; tarsus longer than culmen, shorter than middle toe without claw. North and South America. 5 species, 2 in the United States.

KEY TO SPECIES OF *PROGNE*

1. Male blue; female with a light area on forehead and sides of neck, the throat and breast gray-tipped
   P. subis
   Forehead and sides of neck not light; throat and breast not gray-tipped **P. chalybea**

Progne subis (**Linnaeus**). Purple martin. Iridescent blue; wings, tail, bill, and feet black. FEMALE: Above blue or sooty brown; a gray, often interrupted collar extending up around sides of neck; forehead sooty gray; throat and breast sooty gray edged with paler; belly and crissum whitish, often streaked with sooty gray. Wing 135–153, tail 65–79, tail fork 14–23, culmen 10–13, tarsus 14–16.5 mm. Southern Canada, south to Florida, Gulf coast, and northwestern Mexico (absent from much of Great Basin); winters in South America.

Progne chalybea (**Gmelin**). Gray-breasted martin. Upper parts blue; forehead sooty; throat and breast grayish brown; belly and crissum pure white. FEMALE: Above sooty brown, more or less glossed with blue; posterior underparts often gray. Wing 128–141, tail 54–74, tail fork 9–12, culmen 9–12, tarsus 12–15 mm. Lower Rio Grande to Argentina.

Genus *Petrochelidon* Cabanis. Bill weak; nostrils round; tail less than half wing, slightly emarginate. North and South America, Africa. 10 species, 2 in the United States.

KEY TO SPECIES OF *PETROCHELIDON*

1. Forehead buffy **P. pyrrhonota**
   Forehead chestnut **P. fulva**

Petrochelidon pyrrhonota (**Vieillot**). Cliff swallow. Forehead buff; crown blue-black; hind neck grayish brown; back blue-black streaked with whitish; rump cinnamon; wings and tail brown; lores black; throat and sides of face chestnut; a black patch on middle of lower throat and upper breast; sides of breast cinnamon; belly white; cris-

Fig. 5-62. Wing of tree swallow (*Tachycineta bicolor*).

sum pale cinnamon. YOUNG: Duller; forehead usually dusky; lower back and tertials margined terminally with pale cinnamon; throat and face dull cinnamon mixed with dusky and white. Wing 100–115, tail 44–52, culmen 6–8, tarsus 11–14 mm. Alaska and Canada, south to western North Carolina, northern Alabama, Tennessee, Missouri, and Mexico; winters in South America.

Petrochelidon fulva (**Vieillot**). Cave swallow. Forehead and face bright chestnut; no black patch on throat. Wing 101–110, tail 42–49.5, culmen 6.5–7, tarsus 11–12.5 mm. South Texas, New Mexico, and West Indies, south to Peru.

Genus *Stelgidopteryx* Baird. Bill weak; nostrils round, without operculum; tail emarginate; tarsus longer than middle toe without claw; male with outer web of outer primary with recurved barbs. North and South America. 1 species.

Stelgidopteryx ruficollis (**Vieillot**). Roughwinged swallow. Above grayish brown; throat, breast, and sides paler; belly white; crissum white, sometimes splotched with dusky. YOUNG: Wing coverts and tertials margined with cinnamon. Wing 95–119.5, tail 43–56.5, culmen 5–8, tarsus 10–12 mm. Throughout the United States (except northern New England), south to Argentina.

Genus *Riparia* Forster. Nostrils slitlike, opening laterally, overhung by partly feathered operculum; tail half wing, emarginate; tarsus longer than middle toe without claw, with a tuft of feathers above hallux. Palearctic, Africa, North and South America. 4 species, 1 in the United States.

Riparia riparia (**Linnaeus**). Bank swallow. Above grayish brown; tertials and lower back margined with ashy; below white; a broad grayish-brown band across breast. Wing 95–111, tail 44–58, culmen 5.5–7, tarsus 10–12 mm. Holarctic and Africa; in the United States breeds south to California, Arizona, Texas, Alabama, and Virginia; winters in South America, Africa, and India.

Genus *Tachycineta* Cabanis. Nostrils overhung by partly feathered or naked operculum; tail emarginate. North and South America. 6 species, 2 in the United States.

KEY TO SPECIES OF *TACHYCINETA*

1. Back shining; lateral toe and claw over 10 mm
   T. bicolor
   Back velvety; lateral toe and claw under 10 mm
   T. thalassina

Tachycineta bicolor (**Vieillot**).[1] Tree swallow (Fig. 5-62). Above glossy greenish blue; wings and tail dusky; tertials often white-tipped; lores black; below white. YOUNG: Above dark gray; tertials white-tipped. Wing 109–125, tail 51–60, tail fork 7–11.5, culmen 5.5–7, tarsus 10.5–12.5 mm. North America, south to California, Nevada, Utah, Colorado, Kansas, Arkansas, Mississippi, and Virginia; winters from California, Texas, Gulf states, and South Carolina, south to Honduras and Cuba.

Tachycineta thalassina (**Swainson**). Violet-green swallow. Above velvety green; often with purple on back and bluish on rump; wings and tail dusky; tertials often white-tipped; underparts and patch on each side of rump white. FEMALE: Duller; crown grayish brown. YOUNG: Above sooty brown; wings and tail dusky with slight gloss; tertials white-tipped; rump patches and underparts white, tinged with brownish on breast. Wing 99–125, tail 40–52, tail fork 3–7, culmen 4.5–6, tarsus 10–12 mm. Alaska to Mexico, east to Rockies; on migration to South Dakota and Texas; winters from Mexico to Costa Rica.

Genus *Hirundo* Linnaeus. Nostrils overhung by unfeathered operculum; tail $\frac{2}{3}$ wing, deeply forked. 13 species, 1 in the United States.

Hirundo rustica Linnaeus. Barn swallow. Above glossy steel blue, extending down sides of chest; inner webs of rectrices with a patch of white or pale cinnamon; forehead, throat, and chest cinnamon; posterior underparts pale cinnamon. YOUNG: Duller; forehead dull brown; crown sooty with slight blue gloss. Wing 110–124.5, tail 71–117,

[1] *Iridoprocne bicolor* (Vieillot) of A.O.U.

culmen 7–9, tarsus 11–13 mm. All continents except Australia; breeds through the United States except peninsular Florida; winters from Mexico to South America, Africa, and southern Asia.

Family Corvidae. Crows and jays. Bill stout, conical, nearly straight but slightly notched or bent down at tip; rictus not angled; nostrils usually hidden by nasal tufts (exposed in *Gymnorhinus*). Wing much rounded; primaries 10. Tail rounded or graduated; rectrices 12. Tarsus longer than middle toe with claw, scutellate; lateral toes much shorter than middle toe; hallux with claw shorter than middle toe with claw. Humerus with 1 fossa. Miocene onward. All regions except Polynesia and Madagascar. 35 genera, 119 species. In the United States 8 genera, 15 species.

KEY TO GENERA OF CORVIDAE

1. Tail graduated for half its length   *Pica*
   Tail not graduated for half its length   2
2. Nostrils exposed   *Gymnorhinus*
   Nostrils at least partly covered by nasal tufts   3
3. Tail less than ⅔ wing   4
   Tail more than ⅔ wing   5
4. Culmen curved; wing over 250 mm   *Corvus*
   Culmen nearly straight; wing under 200 mm   *Nucifraga*
5. Head with a pointed crest   *Cyanocitta*
   Head without a distinct crest   6
6. Secondaries nearly as long as primaries; plumage largely green   *Cyanocorax*
   Secondaries decidedly shorter than primaries; no green   7
7. 2d primary shorter than secondaries; plumage with blue   *Aphelocoma*
   2d primary equal to secondaries; plumage without blue   *Perisoreus*

Genus *Corvus* Linnaeus. Nasal bristles extending forward beyond nostrils; 3d or 4th primary longest, outer 4 sinuate on inner web; tail less than ⅔ wing, rounded or slightly graduated. Holarctic, Africa, Australia, southwest Pacific. 35 species, 5 in the United States.

KEY TO SPECIES OF *CORVUS*

1. Throat feathers lanceolate; outer primary longer than secondaries   2

Throat feathers normal; outer primary shorter than secondaries   3
2. Concealed bases of neck and breast feathers pure white   *C. cryptoleucus*
   Concealed bases of neck and breast feathers gray   *C. corax*
3. Middle toe without claw more than 70% tarsus (tarsus 44–50 mm)   *C. ossifragus*
   Middle toe without claw less than 70% tarsus   4
4. Tarsus 45–53 mm; culmen over 90% tarsus   *C. caurinus*
   Tarsus 53–66.5 mm; culmen under 85% tarsus   *C. brachyrhynchos*

***Corvus corax* Linnaeus.** Common raven. Black glossed with purplish. Throat feathers long and lanceolate; outer primary longer than secondaries. Wing 380–464, tail 208–260, culmen 64–92, tarsus 62–74 mm. Holarctic, south to Honduras; uncommon in eastern United States.

***Corvus cryptoleucus* Couch.** White-necked raven. Differs from *C. corax* in having white concealed bases to feathers of upper parts, throat, and breast. Wing 328–379, tail 182–214, culmen 49.5–59, tarsus 55.5–68.5 mm. Southern Arizona, New Mexico, western Texas, Oklahoma, and western Kansas, south to northern Mexico.

***Corvus ossifragus* Wilson.** Fish crow. Throat feathers normal, blended; outer primary shorter than secondaries; feet slender, planta tarsi scutellate. Black glossed with bluish. Wing 264–300, tail 137.5–176.5, culmen 39–45, tarsus 45–53, middle toe 28–35 mm. Culmen/tarsus ratio 92%, middle toe/tarsus ratio 73–80%. Atlantic Coastal Plain from Rhode Island to Florida and west along Gulf to eastern Texas; north to Memphis.

***Corvus caurinus* Baird.** Northwestern crow. Proportions intermediate between *C. ossifragus* and *C. brachyrhynchos;* color as in *C. brachyrhynchos*, of which it is often considered a subspecies. Wing 256.5–292.5, tail 144.5–170.5, culmen 41.5–49, tarsus 45–53, middle toe 28–35 mm. Culmen/tarsus ratio 92%, middle toe/tarsus ratio 62–66%. Pacific Coast from Alaska to Washington.

***Corvus brachyrhynchos* Brehm.** Common crow. Feet stout. Back black glossed with purple except at tips of feathers, giving a scaled appearance; below duller, planta tarsi extensively booted. Wing 278–337, tail 153–198, culmen 43–55.5, tarsus 53–66.5, middle toe 30.5–40 mm. Culmen/tarsus ratio 81–85%, middle toe/tarsus ratio 55–60%. Canada and the United States.

Genus *Nucifraga* Brisson. Bill long, compressed; outer primary much shorter than secondaries; 4th to 5th primaries longest, 5 outer ones sinuate on inner web; tail rounded, less than ⅔ wing. Holarctic. 2 species, 1 in the United States.

Nucifraga columbiana (**Wilson**). Clark's nutcracker. Smoky gray, darker on upper tail coverts; forehead and face white; wings and middle rectrices black glossed with purple; crissum and outer rectrices white; iris brown; bill and feet black. Wing 187–199, tail 111–119, culmen 37–45.5, tarsus 32.5–38 mm. Mountains of West from Alaska to northern Mexico.

Genus *Gymnorhinus* Wied. Bill long, slender, flat at tip; nostrils wholly exposed; outer primary shorter than secondaries, 3d to 4th longest, 4–5 sinuate; tail nearly even, less than ¾ wing. Western United States. 1 species.

Gymnorhinus cyanocephalus **Wied.** Piñon jay. Grayish blue, darker on crown, brighter on cheeks; throat streaked with grayish white; bill and feet black; iris brown. Wing 138–158, tail 97–116.5, culmen 29.5–42, tarsus 35–39 mm. Piñon and juniper woods from eastern Washington, Montana, and western South Dakota, south to Oklahoma, New Mexico, Arizona, and northern Baja California.

Genus *Pica* Brisson. Bill shorter than head, stout; nostrils concealed; 5th primary longest, outermost very short, narrow, and falcate; 7 primaries sinuate; tail much longer than wing, graduated. Holarctic. 2 species.

KEY TO SPECIES OF *PICA*

1. Bill black   *P. pica*
   Bill yellow   *P. nuttalli*

Pica pica (**Linnaeus**). Black-billed magpie. Head and neck all around, breast, back, lesser wing coverts, wing lining, upper and under tail coverts black; rump mixed black and white; tail black, glossed above with green and at tip with purple; scapulars white; secondaries, middle, and greater coverts glossy blue; primaries largely white internally, black glossed with green externally; lower breast and belly white; iris brown; bill and feet black. Wing 182–212, tail 232–302, culmen 31–

39.5, tarsus 43–50 mm. Palearctic and western North America in Great Basin and Plains, east to Dakotas, Nebraska, Kansas, and Oklahoma, south to eastern California, Arizona, and New Mexico; occasional east of the Mississippi.

Pica nuttalli (**Audubon**). Yellow-billed magpie. Perhaps only a race of *P. pica*, from which it differs in having bill, bare skin around eyes, and claws yellow. Wing 182–196, tail 229–254, culmen 30–32.5, tarsus 43.5–49.5 mm. California west of Sierra Nevada, from Shasta County south to Ventura and Kern Counties.

Genus *Cyanocorax* Boie. 4th primary longest, outer 5 sinuate; secondaries nearly concealing primaries; tail longer than wing, graduated for less than half its length. Tropical and subtropical America. 11 species, 1 in the United States.

Cyanocorax yncas (**Boddaert**). Green jay. Nasal tufts, crown, spot on upper eyelid, and triangular patch from malar region to posterior border of lower eyelid bright blue; narrow line across forehead whitish; a black mask running from lores up over and behind eye to auriculars, chin, and throat; back and wings yellowish green; 4 middle rectrices bluish green, 4 lateral pairs yellow; breast, belly, and crissum pale yellowish green; bill black; feet brown. Wing 104–134, tail 118–151, culmen 23–30, tarsus 34–40 mm. Lower Rio Grande to Honduras; Colombia to Bolivia.

Genus *Aphelocoma* Cabanis. 4th, 5th, and 6th primaries longest; secondaries decidedly shorter than longest primary, but longer than 2d primary; tail graduated for about ¼ its length. North and Central America. 3 species, 2 in the United States.

KEY TO SPECIES OF *APHELOCOMA*

1. Tail longer than wing   *A. coerulescens*
   Tail shorter than wing   *A. ultramarina*

Aphelocoma coerulescens (**Bosc**). Scrub jay. Crown, nape, and sides of neck blue; forehead and narrow superciliary usually whitish; sides of face blue or black; back grayish brown or bluish gray; wings and tail blue; throat and upper breast pale gray streaked with dusky; a dull bluish pectoral collar usually present; belly pale gray, often obscurely streaked with darker; crissum blue, gray, or white; bill and feet black. YOUNG: Crown and sides of head brown. Wing 100–151, tail 113–164, culmen 23–30, tarsus 33.5–48 mm. Scrub from Oregon, southern Idaho, southern Wyoming, Colorado, western Kansas, southwestern Oklahoma, and western Texas, south to central Mexico; peninsular Florida.

Aphelocoma ultramarina (**Bonaparte**). Mexican jay. Top and sides of head and neck dull blue; lores black; back dull bluish gray; throat and breast brownish gray; belly and crissum white. Wing 138–187, tail 113–182, culmen 24–32, tarsus 36–46 mm. Southern Arizona, New Mexico, and western Texas to central Mexico.

Genus *Cyanocitta* **Strickland**. Differs from *Aphelocoma* in having a long crest; tail somewhat shorter than wing. North and Central America. 2 species.

KEY TO SPECIES OF *CYANOCITTA*

1. Secondaries and tail tipped with white   *C. cristata*
   No white in wings or tail   *C. stelleri*

Cyanocitta cristata (**Linnaeus**). Blue jay. Above mainly purplish blue; a black line extending across base of forehead to lores, thence through eyes to meet a complete black collar; greater coverts and secondaries bright blue, barred with black and broadly tipped with white; primaries dusky edged with blue; tail broadly tipped with white, the upper surface blue barred with black, the lower surface black; throat and breast grayish; belly and crissum white; bill and feet black. Wing 118–148, tail 108–147.5, culmen 22–28, tarsus 31.5–37mm. Eastern North America, west to North Dakota, eastern Wyoming and Colorado, Oklahoma, and Texas.

Cyanocitta stelleri (**Gmelin**). Steller's jay. Head and neck blackish; forehead (and sometimes crest) streaked with blue; a whitish spot often present above eye; back dark brown; rump, upper tail coverts, breast, belly, and crissum blue; wings and tail blue on exposed portion, black below; secondaries and often tail barred with black. Wing 133–165, tail 122–160, culmen 23–32, tarsus 38–50.5 mm. Coniferous forests of West, east to Montana, Wyoming, Colorado, and western Texas, south to Nicaragua.

Genus *Perisoreus* **Bonaparte**. Bill short, weak, gonys convex; 5th and 6th primaries longest, 2d equal to secondaries; secondaries short; tail slightly shorter than wing, graduated for less than $\frac{1}{5}$; plumage fluffy; feathers of pileum long but not forming a crest. Holarctic. 3 species, 1 in the United States.

Perisoreus canadensis (**Linnaeus**). Canada jay. Head all around white; nape whitish, followed by an indistinct whitish half collar on upper back; body slaty gray, lighter (sometimes whitish) below; narrow tips of remiges and rectrices whitish; back often with pale shaft streaks; bill and feet black. YOUNG: Head and neck dusky. Wing 130–158, tail 122–151.5, culmen 17.5–24.5, tarsus 32–38 mm. Northern coniferous forest, south to New Hampshire, Vermont, northern New York, northern Michigan and Wisconsin, and in mountains to Black Hills, northern New Mexico, Arizona, and California.

Family Paridae. Titmice. Bill short, conical, unnotched; rictus straight; nostrils covered (rarely incompletely) by nasal tufts. Wings rounded; primaries 10; secondaries much shorter than longest primaries. Tail about as long as wing, rounded or graduated. Tarsus scutellate, longer than middle toe with claw; hallux with claw longer than lateral toes with claw. Plumage soft, fluffy. Humerus with 2 fossae. Recorded from Eocene. Holarctic, African mainland, Asia, Australia, New Zealand. 11 genera, 62 species; in the United States 4 genera, 11 species.

KEY TO GENERA OF PARIDAE

1. Nostrils partly exposed   *Auriparus*
   Nostrils concealed   **2**
2. Tail graduated   *Psaltriparus*
   Tail rounded   **3**
3. Crested   *Baeolophus*
   No crest   *Parus*

Genus *Baeolophus* **Cabanis**. Head crested; bill stout, culmen and gonys convex, nostrils concealed; rictal bristles present; 4th primary longest, 1st and 2d shorter than secondaries; tail slightly rounded, 30–90% wing length; tarsus about twice length of exposed culmen. North America. 3 species.

KEY TO SPECIES OF *BAEOLOPHUS*

1. Flanks rusty   *B. bicolor*
   Flanks grayish   2
2. Head and neck marked with black   *B. wollweberi*
   Head and neck without black   *B. inornatus*

Baeolophus bicolor (**Linnaeus**).[1] Tufted titmouse; black-crested titmouse. Forehead black, white, or chestnut; crest gray or black; lores white; above slate gray; below whitish gray; flanks rusty; iris brown; bill black; feet gray. Wing 65.5–79.5, tail 55.5–75, culmen 9–13, tarsus 18–22 mm. Deciduous forest of East, north to Connecticut, New York, Michigan, Wisconsin, and Iowa, west to Kansas, Oklahoma, and southwestern Texas, and into northeastern Mexico.

Baeolophus wollweberi (**Bonaparte**).[2] Bridled titmouse. Crown and crest black, gray medially; a black stripe through eye meets another across auriculars; superciliary and sides of face white; throat black; upper parts olive gray; underparts pale gray; bill black. Wing 60–67.5, tail 53–59, culmen 8–9, tarsus 15.5–17.5 mm. Oak woods of southern Arizona and New Mexico south to Mexican plateau.

Baeolophus inornatus (**Gambel**).[3] Plain titmouse. Above grayish brown; below grayish white, auriculars mottled with darker, sides and crissum more buffy; bill gray. Wing 63–75.5, tail 50–63, culmen 10.5–14, tarsus 19.5–22.5 mm. Oak woods from southern Oregon, southern Idaho, southern Wyoming, Colorado, western Oklahoma, and western Texas, south to Baja California, Arizona, and New Mexico.

Genus *Parus* Linnaeus. No crest; bill slender, gonys straight; 2d primary about as long as secondaries; tail sometimes equal to wing. Holarctic and Africa. 22 species, 6 in the United States.

KEY TO SPECIES OF *PARUS*

1. Pileum brown   2
   Pileum black   3
2. Back grayish brown; tail 54.5–70 mm
        P. hudsonicus
   Back chestnut; tail 43.5–52.5 mm   *P. rufescens*
3. A white superciliary   *P. gambeli*
   No superciliary   4

[1] *Parus bicolor* Linnaeus and *P. atricristatus* Cassin of A.O.U.
[2] *Parus wollweberi* (Bonaparte) of A.O.U.
[3] *Parus inornatus* Gambel of A.O.U.

4. Flanks dark gray resembling back   *P. sclateri*
   Flanks buffy or whitish   5
5. Secondaries and greater coverts edged with whitish; tail more than 90% wing   *P. atricapillus*
   Secondaries and greater coverts edged with gray; tail less than 90% wing   *P. carolinensis*

Parus atricapillus Linnaeus. Black-capped chickadee. Crown and hind neck black; back olive gray; wings and tail dusky, edged with whitish; throat black, the posterior border broken with white; sides of face and neck, breast, and belly white; sides and crissum buff; iris brown; bill black; feet gray. Wing 58–73, tail 53–71, culmen 8–11, tarsus 15–18.5 mm. Tail/wing ratio 91–97%. Northern Hemisphere south in America to northern California, Arizona, New Mexico, Kansas, Missouri, central Illinois, Tennessee, and mountains of North Carolina.

Parus carolinensis Audubon. Carolina chickadee. Differs from *P. atricapillus* in smaller size; back ashy gray; little if any white on wings and tail; black throat patch extending farther back and with posterior border scarcely mixed with white; tail relatively shorter. Wing 52.5–66.5, tail 44.5–58, culmen 7.5–9, tarsus 13.5–17 mm. Tail/wing ratio 82–88%. Southeastern United States, north to New Jersey, Pennsylvania, Ohio, central Indiana, central Illinois, central Missouri, southeastern Kansas, and west to Oklahoma and central Texas.

Parus sclateri Kleinschmidt. Mexican chickadee. Like *P. carolinensis* but larger; back dark gray; sides, flanks, and crissum almost as dark gray as back and without buff. Wing 64.5–71.5, tail 54.5–61.5, culmen 8–10, tarsus 17–19.5 mm. Pines of Mexican highlands north to southern Arizona and New Mexico.

Parus gambeli Ridgway. Mountain chickadee. Like *P. sclateri* but back and sides paler; superciliary white. Wing 65–72.5, tail 54.5–65.5, culmen 9.5–11.5, tarsus 16.5–19 mm. Mountains of West, south to Baja California, Arizona, New Mexico, and western Texas, east to Rockies.

Parus hudsonicus Forster. Boreal chickadee. Above grayish brown; sides of face white, not extending to neck; throat blackish; below whitish; sides and crissum rusty brown. Wing 58.5–70, tail 54.5–70, culmen 8–9.5, tarsus 15.5–17.5 mm. Northern coniferous forest, south to Montana, northern Minnesota, northern Wisconsin, northern

Michigan, and New York, occasionally farther in winter.

Parus rufescens Townsend. Chestnut-backed chickadee. Crown and nape dark brown; back chestnut; wings and tail brownish gray edged with paler; sides of face and neck white; throat sooty brown; breast and belly white; flanks chestnut; crissum brownish. Wing 53.5–64, tail 43.5–52.5, culmen 8–10, tarsus 16–17 mm. Coniferous forest of Montana and Pacific Coast from Alaska to California.

Genus *Psaltriparus* Bonaparte. No crest; bill and feet slender; tail longer than wing, strongly rounded or graduated; tarsus more than twice length of exposed culmen. Western North America. 1 species.

Psaltriparus minimus (Townsend).[1] Bush-tit. Pileum gray or brown; back gray or olive; wings and tail dusky edged with paler; below buffy or pinkish white; sides of head often black, continuing around nape as a narrow collar; iris brown. FEMALE: Lacks black on head, but black nuchal collar may be present; iris cream. Wing 44.5–53.5, tail 46.5–62, culmen 6–7.5, tarsus 14.5–17 mm. Southern British Columbia south to Guatemala, east to eastern Oregon, Colorado, western Oklahoma, and western Texas.

Genus *Auriparus* Baird. No crest; culmen as well as gonys straight, nostrils partly exposed, rictal bristles absent; outer primary narrow, 2d primary longer than outer one, 3d longest; tail shorter than wing, rounded. Southwestern United States and adjacent part of Mexico. 1 species.

Auriparus flaviceps (Sundevall). Verdin. Head and throat yellow; upper parts brownish gray; lesser wing coverts reddish chestnut; below grayish white; iris brown; bill and feet gray. YOUNG: Lacks yellow and chestnut. Wing 47–55, tail 39–50, culmen 8–9.5, tarsus 14.5–16 mm. Southeastern California, southern Nevada, southern Utah, southern New Mexico, and southern Texas to northern Mexico.

Family Sittidae. Nuthatches. Bill nearly as long as head, slender, straight; nostrils partly concealed by nasal tufts; rictal bristles short. Primaries 10, outer one rudimentary, 2d primary much longer than secondaries. Tail even, shorter than wing.

Tarsus about equal to culmen, shorter than middle toe with claw; hallux longer than lateral toes, inner toe much the shortest. Humerus with 2 large fossae. Eggs spotted. Pliocene onward. Holarctic, Oriental, and Australian regions; Madagascar. 8 genera, 35 species; in the United States 1 genus, 4 species.

Genus *Sitta* Linnaeus. Bill unhooked; rictal bristles present but small; nostrils without operculum; tail only about half length of wing; tarsus scutellate; middle toe decidedly the longest. Holarctic. 10 species, 4 in the United States.

KEY TO SPECIES OF *SITTA*

1. Superciliaries white   2
   No superciliaries   3
2. Breast white   *S. carolinensis*
   Breast ochraceous   *S. canadensis*
3. Middle rectrices largely white; crown grayish olive
      S. pygmaea
   Middle rectrices without white or only a narrow shaft stripe; crown brown   *S. pusilla*

Sitta carolinensis Latham. White-breasted nuthatch. Crown and hind neck glossy black; back and middle rectrices bluish gray; wings dusky, edged with bluish gray; rest of tail black with broad white subterminal band; lores, superciliary region, sides of face, underparts, and axillars dull white; under wing coverts black; crissum mixed chestnut and white; iris brown; bill black, gray at base of mandible; feet gray. FEMALE: Crown more or less gray. Wing 79.5–98.5, tail 39.5–53, culmen 16–23, tarsus 16–20 mm. Southern Canada through the United States to central Mexico.

Sitta canadensis Linnaeus. Red-breasted nuthatch. Crown and nape black; long superciliary white, bordered below by a long black stripe; back and middle rectrices bluish gray; lateral rectrices black, tipped with gray and with a white subterminal spot; sides of face and neck white; underparts buff or tawny. FEMALE: Crown lead gray. Wing 63–70, tail 33–39, culmen 12.5–17.5, tarsus 14.5–17 mm. Northern coniferous forest south to Massachusetts, New York, northern Michigan, northern Wisconsin, and Minnesota and in mountains to North Carolina, Tennessee, Colorado, Arizona, and California. Winters irregularly south to Gulf states.

---

[1] Includes *Psaltriparus melanotis* (Hartlaub) of A.O.U.

Fig. 5-63. Head and right foot of red-whiskered bulbul (*Pycnonotus jocosus*).

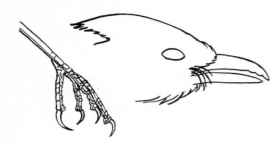

Fig. 5-64. Wing of red-whiskered bulbul (*Pycnonotus jocosus*).

Sitta pusilla Latham.   Brown-headed nuthatch. Crown and hind neck brown mixed with paler; a dark-brown stripe through eye; nuchal spot white; back bluish gray; middle rectrices bluish gray, sometimes with a narrow white shaft stripe; lateral rectrices black, tipped with gray, white subterminally; suborbital area and sides of neck white; underparts grayish white, more or less tinged with buff. YOUNG: Crown gray; stripe through eye dark brown; underparts washed with tawny. Wing 58.5–69, tail 28–34.5, culmen 13–16, tarsus 13–16 mm. Pine woods of southeastern United States, north to Delaware, Maryland, Virginia, northern Mississippi, and southeastern Missouri, west to eastern Texas and Arkansas; Bahamas.

Sitta pygmaea Vigors.   Pygmy nuthatch. Similar to *S. pusilla*, but crown grayish olive; streak through eye dusky; white nuchal spot smaller, often obsolete; middle rectrices largely white basally. Wing 60–70, tail 31.5–38.5, culmen 12.5–17.5, tarsus 13.5–16.5 mm. Coniferous forests of West, from British Columbia south to mountains of central Mexico, east to Montana, Wyoming, Colorado, Oklahoma, and western Texas.

Family Certhiidae.   Creepers. Bill long, very slender; nostrils exposed, operculate; rictal bristles obsolete. Wing rounded; primaries 10, outer one less than half length of 2d. Tarsus longer than middle toe, scutellate; hallux longer than lateral toes, inner toe shortest; claws long and curved. Eggs spotted. Known from Pleistocene. Holarctic, Oriental, and Australia. 1 genus; 5 species, 1 in the United States.

Genus *Certhia* Linnaeus.   Bill much curved; tail graduated, about as long as wing, rectrices stiff and pointed; hallux with claw longer than middle toe without claw. Holarctic. 5 species, 1 in the United States.

Certhia familiaris **Linnaeus.**   Brown creeper. Above dark brown streaked with black and white; rump tawny with concealed white spots; tail grayish brown; wing coverts brown spotted with buffy; remiges dark brown tipped and edged with gray or buffy, all but the 3 outer ones crossed by a buffy band; short superciliary whitish or buffy; underparts dirty white or pale buffy; flanks and crissum more strongly buff; iris brown; bill black above and at tip, base of mandible yellowish; feet brown. Wing 58–69.5, tail 52.5–70, culmen 11–17, tarsus 14–16 mm. Holarctic coniferous forests; in America south to Massachusetts, mountains of North Carolina, northern Michigan, Minnesota, Nebraska, Wyoming, and in high mountains to California, Arizona, and New Mexico, and thence at high altitudes through Mexico to Nicaragua. Winters south to Gulf Coast.

Family Pycnonotidae.   Bulbuls. Head (Fig. 5-63) often crested; nape with filoplumes. Bill short, straight, about as deep as wide; culmen curved; nostrils slit, exposed, operculate, often slightly perforate; maxillary tomium notched subterminally, its anterior half with slight denticulations; rictal bristles large. Wing short, rounded; primaries 10, outer 2 shorter than the other remiges (Fig. 5-64). Tail about as long as wing, rounded or slightly emarginate; rectrices 12, not soft. Legs and feet weak; tarsus slightly scutellate to booted, only a little longer than middle toe with claw; outer and inner toes subequal; about half of basal phalanx of middle toe fused to lateral toes; hallux strong, its claw reaching claw of middle toe. Africa and southern Asia. Recorded from Pleistocene. 15 genera; 119 species, 1 introduced in the United States.

Genus *Pycnonotus* Boie. Head usually crested; anterior half of maxillary tomium denticulate; tarsus scutellate; tail rounded. Africa and southern Asia. 28 species, 1 introduced to Florida.

Pycnonotus jocosus (**Linnaeus**). Red-whiskered bulbul (Fig. 5-63). Crown and long crest sooty; upper parts sooty brown; all but 2 middle pairs of rectrices with white tips; auriculars white, with a black line below; long subauricular tuft scarlet; underparts white; crissum orange-red; iris brown; bill and feet black. Wing 76–92, tail 78–91, culmen 15–17.5, tarsus 22.5 mm. Southern China to India and Siam; introduced in Australia and Dade County, Florida.

Family Timaliidae.[1] Babblers. Head crested or not; lores bristly. Bill straight, deeper than wide, short to longer than head, usually strong; culmen curved; nostrils exposed slits, operculate; maxillary tomium with slight notch or none, indenticulate; rictal bristles long. Wing short, rounded; primaries 10, outer 2 shorter than rest of remiges. Tail short to longer than wing, rounded; rectrices 12. Tarsus much longer than middle toe with claw, usually strong and heavily scutellate (slender and nearly booted in *Chamaea*); basal phalanx of outer toe fused to middle toe; inner toe nearly free; outer and inner toes subequal; hallux strong, its claw stout and reaching to about base of middle claw. Africa, southern Asia, Australia; Pacific Coast of North America. Known from Pleistocene. 60 genera; 277 species, 1 in the United States.

Genus *Chamaea* Gambel. Plumage very lax; pileum subcrested. Bill very short; maxillary tomium unnotched (Fig. 5-65); tail much longer than wing, strongly graduated; tarsus weakly scutellate in front, nearly booted on sides. Pacific Coast of North America. 1 species.

Chamaea fasciata (**Gambel**). Wren-tit (Fig. 5-66). Above olive brown, grayer on crown and sides of face; below grayish buff to buffy cinnamon, indistinctly streaked with dusky across breasts; iris white; bill and feet horn. Wing 54–63, tail 72–87, culmen 10–12, tarsus 22–27 mm. West of Sierra

[1] Includes family Chamaeidae of A.O.U.

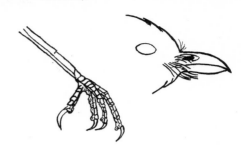

Fig. 5-65. Head and right foot of wren-tit (*Chamaea fasciata*).

Fig. 5-66. Wing of wren-tit (*Chamaea fasciata*).

Nevada, in chaparral of Oregon, California, and Baja California.

Family Cinclidae. Dippers. Contour feathers very dense, underlaid with down. Bill shorter than head or middle toe, straight, slender, compressed; no rictal bristles; nostrils operculate, with feathers reaching posterior end. Wing short, concave, pointed; primaries 10, first spurious and falcate; secondaries short. Tail somewhat more than half wing, nearly square; rectrices 12, broad; coverts reaching nearly to tip of tail. Tarsus booted except near toes, longer than middle toe with claw, about $\frac{1}{3}$ wing; lateral toes and hallux all equal; claw of hallux shorter than its digit; inner edge of middle claw slightly pectinate (Fig. 5-67). Wades in and under mountain streams. Eggs white. Recorded from Pleistocene. Palearctic south to Indochina; western North America to Argentina. 1 genus; 4 species, 1 in the United States.

Genus *Cinclus* Borkhausen.

Cinclus mexicanus Swainson. Dipper. Head and neck brownish; body slate, often tipped with whitish below; wings and tail blackish, slightly

Fig. 5-67. Head and right foot of water ouzel (*Cinclus mexicanus*).

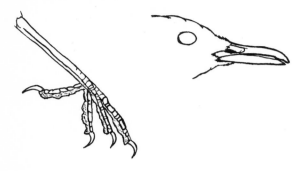

tipped with whitish; a white mark on both eyelids; iris brown; bill horn; legs yellowish. Wing 79–100, tail 41–56, culmen 16–19, tarsus 25–31 mm. Mountains of West, east to Black Hills, south to Guatemala.

Family Troglodytidae. Wrens. Bill about equal to middle toe or longer, curved, slender; rictal bristles small or absent; nostrils exposed, usually slitlike and operculate, usually slightly perforate. Wing short, curved, rounded; primaries 10, first at least half length of 2d, broad, shorter than secondaries. Tail usually shorter than wing, rounded; rectrices 12 (10 in 1 genus), broad, soft. Tarsus longer than culmen, scutellate in front; basal phalanx of middle toe fused to lateral toes; claw of hallux shorter than digit. Maxillopalatine processes long and thin. Eggs white or brown, usually speckled. Recorded from Pleistocene. Neotropical and Holarctic regions. 14 genera, 59 species; in the United States 6 genera, 9 species.

KEY TO GENERA OF TROGLODYTIDAE

1. Inner toe without claw short, not reaching last phalanx of middle toe   *Salpinctes*
   Inner and outer toes subequal, both reaching last phalanx of middle toe   2
2. Inner toe more closely grown to hallux than to middle toe   *Cistothorus*
   Inner toe more closely grown to middle toe than to hallux   3
3. Tarsus scutellate behind as well as in front
                                                 Campylorhynchus
   Planta tarsi laminiplantar   4

4. Tail about equal to wing   *Thryomanes*
   Tail shorter than wing   5
5. Maxillary tomium slightly notched; a few rictal bristles present   *Thryothorus*
   Maxilla unnotched; no rictal bristles   *Troglodytes*

Genus *Cistothorus* Cabanis. Bill unnotched, shorter to slightly longer than middle toe; culmen curved; nostrils linear, operculate; no rictal bristles; 3d or 4th primary longest, outermost narrow, about half length of 2d; 2d primary about equal to secondaries; tail more than 80% wing, graduated for more than $\frac{1}{3}$; tarsus longer than middle toe with claw, scutellate before, laminiplantar behind; lateral toes subequal, longer than hallux; inner toe more closely grown to hallux than to middle toe. North and South America. 4 species, 2 in the United States.

KEY TO SPECIES OF *CISTOTHORUS*

1. Crown with buff or white shaft streaks *C. platensis*
   Crown plain dusky, often with a brown median line
                                                 C. palustris

Cistothorus platensis (Latham). Short-billed marsh wren; sedge wren. Above brown streaked with dusky and buffy or white; tail, upper tail coverts, and wing coverts brown barred with dusky; remiges dusky, barred with buff on outer webs; superciliary and sides of head and neck pale buff; throat and lower breast white; upper breast, sides, flanks, belly, and crissum buff; flanks sometimes barred with brown; iris brown; maxilla dusky; mandible and feet flesh. Wing 41–47, tail 35–41.5, culmen 9.5–11, tarsus 15–17.5 mm. Sedgy swales of eastern Canada south to Delaware, Indiana, Missouri, and Kansas; also from eastern Mexico south to Argentina and Chile; winters south to Florida and Gulf Coast.

Cistothorus palustris (Wilson).[1] Long-billed marsh wren. Crown dusky, usually brown along midline; upper back black striped with white; wing coverts and lower back plain brown; tail brown barred with black; tertials blackish, brokenly barred with brown; remiges dusky, barred with brown on outer webs; upper tail coverts brown,

[1] *Telmatodytes palustris* (Wilson) of A.O.U.

plain or with dusky bars; superciliary whitish; sides of face buff or grayish, with dusky faint bars; below white; breast tinged with buff and sometimes with dusky speckles; flanks brown, buff, or gray; crissum whitish or buff, either plain or with dusky bars; iris brown; bill dusky above, brown below; feet brown. Wing 40–57, tail 33–50.5, culmen 12–15.5, tarsus 17.5–22 mm. Fresh and salt-water marshes of the United States and southern Canada; winters from southern United States to central Mexico.

Genus *Campylorhynchus* Spix. With 2 or 3 rictal bristles; outer primary more than half length of longest; tail graduated for only $\frac{1}{5}$ its length; tarsus scutellate on planta tarsi (sometimes fused medially) as well as on acrotarsium; inner toe more closely grown to middle toe than to hallux. Southwestern United States to South America. 11 species, 1 in the United States.

***Campylorhynchus brunneicapillum* (Lafresnaye).** Cactus wren. Crown and hind neck dark brown; back grayish brown speckled with black and white, becoming bars on wing coverts; upper tail coverts and middle rectrices dusky, barred with brownish; lateral rectrices black, tipped and brokenly barred with buffy white; remiges dusky, banded with pale buff; superciliary white; auriculars brown; throat and breast white, spotted with black; belly buff; flanks and crissum buff spotted with black; iris brown; bill dusky, brownish below; feet brown. Wing 76–92, tail 70–86, culmen 18.5–26, tarsus 25–30 mm. Lower Sonoran deserts from southern parts of California, Nevada, Utah, New Mexico, and Texas, south to central Mexico.

Genus *Thryothorus* Vieillot. Differs from *Campylorhynchus* in having the planta tarsi at least largely laminiplantar; maxillary tomium slightly notched. North and South America. 21 species, 1 in the United States.

***Thryothorus ludovicianus* (Latham).** Carolina wren. Above brown; rump with concealed white spots; remiges, rectrices, and upper tail coverts barred with dusky; long superciliary white or buff; postocular brown; sides of face white flecked with dusky; chin white; underparts buff or chestnut; crissum buff barred with black; wing lining black and white; iris brown; bill dusky above, horn below; feet brownish. Wing 52–66.5, tail 43–55.5, culmen 14.5–19.5, tarsus 20–23.5 mm. Southeast-

ern states north to Massachusetts, southern Ontario, southern Michigan, Illinois, southern Iowa, and southeastern Nebraska, west to Kansas, Oklahoma, and central Texas, and northeastern Mexico.

Genus *Thryomanes* Sclater. Tail about as long as wing; rictal bristles obsolete; tomium slightly notched; planta tarsi smooth. Southern United States and Mexico. 2 species, 1 in the United States.

***Thryomanes bewickii* (Audubon).** Bewick's wren. Above brown; remiges barred with black; rump with concealed white spots; tail tipped with gray; middle rectrices grayish brown barred with black; intermediate rectrices black, with gray tips broader; lateral rectrices similar but barred externally near tip with black and white; superciliary white; postocular brown; sides of face white flecked with dusky; underparts dull white; crissum barred black and white; iris brown; bill dusky above, horn below; feet brownish. Wing 46.5–62, tail 46–64, culmen 12–16, tarsus 17–21 mm. Southern United States north to Pennsylvania, southern Michigan, southern Wisconsin, southern Iowa, Nebraska, southwestern Wyoming, southern Utah, southern Nevada, and along Pacific Coast to southern British Columbia, south into central Mexico.

Genus *Troglodytes* Vieillot. Tail shorter than wing; tomium unnotched; no rictal bristles; planta tarsi smooth. Holarctic and Neotropical regions. 5 species, 2 in the United States.

KEY TO SPECIES OF *TROGLODYTES*

1. Tail 25–35, not reaching outstretched toes
   T. troglodytes
   Tail 38–50, extending beyond toes   *T. aëdon*

***Troglodytes troglodytes* (Linnaeus).** Winter wren. Above brown; back, wings, and tail barred with dusky; rump, outer primaries, and outer rectrix also barred with whitish; sides of face mixed brown and buffy; narrow superciliary, throat, and breast pale cinnamon; posterior underparts barred dusky, cinnamon, and whitish; iris and maxilla brown; mandible and feet light brown. Wing 40–55, tail 25–35, culmen 10–15, tarsus 16.5–21 mm. Holarctic boreal forests, south to Mediterranean and Asia Minor, and in America to Massachusetts, mountains of Georgia, northern Michigan and Minnesota, Montana, Idaho, and mountains of California. In winter extends south to Florida and Gulf Coast.

***Troglodytes aëdon* Vieillot.**[1] House wren. Above brown; rump with concealed black and white spots; wings and tail barred with dusky; faint superciliary, throat, and breast very pale brown; lower breast dull white; flanks, belly, and crissum barred dusky and buff or dull white; iris and maxilla dark brown; mandible and feet pale brown. Wing 47–55.5, tail 38–50, culmen 11–14, tarsus 15.5–18.5 mm. Canada and the United States (except southeastern Coastal Plain), south through Middle and South America; Lesser Antilles. Withdraws from Canada and northern states in winter.

Genus *Salpinctes* Cabanis. Bill longer than middle toe with claw; maxillary tomium smooth or slightly notched; nostrils linear, operculate; rictal bristles obsolete; 3d primary longest; tail about ¾ wing, rounded; tarsus longer than culmen, planta tarsi indistinctly scutellate; anterior toes fused basally; inner toe short, with base of claw not reaching end of 2d phalanx of middle toe. Western North America and Central America. 2 species.

KEY TO SPECIES OF *SALPINCTES*

1. Bill slightly notched, shorter than tarsus   *S. obsoletus*
   Bill unnotched, longer than tarsus   *S. mexicanus*

***Salpinctes obsoletus* (Say).** Rock wren. Above sandy gray, more or less speckled with dusky and whitish; rump tinged with cinnamon; wings and middle rectrices grayish brown narrowly barred with dusky; outer rectrix cinnamon buff with 2–4 black bars; remaining rectrices grayish brown, tipped with gray-mottled cinnamon buff, and with broad black subterminal band; superciliary and underparts dull white, tinged with buff posteriorly; crissum barred with black; iris brown; bill horn; feet black. YOUNG: Upper parts barred with dusky instead of spotted. Wing 63.5–75, tail 45.5–62, culmen 15.5–22.5, tarsus 18.5–23 mm. Southwestern Canada to Costa Rica, east to Dakotas, Nebraska, Kansas, western Oklahoma, and central Texas.

[1] Includes *Troglodytes brunneicollis* Sclater of A.O.U.

***Salpinctes mexicanus* (Swainson).**[1] Canyon wren. Above brown, spotted with black and white on back; wings barred with dusky; tail chestnut with 5–7 black bars; throat and breast white; belly, sides, and crissum chestnut brown, speckled with black and white; iris brown; bill horn, yellowish below at base; feet brown. Wing 55.5–72, tail 46–60.5, culmen 17–26, tarsus 16.5–21 mm. Southern British Columbia to Chiapas, east to Black Hills, Colorado, Oklahoma, and central Texas.

Family Mimidae. Mockingbirds. Bill slender, slightly to decidedly curved; rictal bristles well developed; nostrils oval, exposed, operculate, usually slightly perforate. Wing short, curved, rounded; primaries 10. Tail long, rounded; rectrices 12. Tarsus longer than culmen; acrotarsium scutellate; inner toe free at its base, slightly shorter than outer toe, base of its claw extending to 3d phalanx of middle toe. Maxillopalatine processes swollen and perforate. Eggs blue or greenish, plain or spotted. Young usually spotted. Known from Pleistocene. North and South America. 13 genera, 31 species; 4 genera, 10 species in the United States.

KEY TO GENERA OF MIMIDAE

1. Outer primary less than half 2d   *Oreoscoptes*
   Outer primary more than half 2d   2
2. Maxilla unnotched   *Toxostoma*
   Maxilla with a slight subterminal notch   3
3. Nostrils slightly perforate   *Mimus*
   Nostrils imperforate   *Dumetella*

Genus *Mimus* Boie. Bill shorter than head or middle toe, slightly notched; nostrils slightly perforate; outer primary more than half 2d; 2d primary shorter than 7th; tail equal to or longer than wing, strongly rounded. North and South America. 9 species, 1 in the United States.

***Mimus polyglottos* (Linnaeus).** Mockingbird. Above gray; wings dusky, edged with ashy; middle and greater coverts white-tipped; primary coverts white with black subterminal spot; base of primaries white; outer rectrix white; next 2 pairs dusky, with inner web largely white distally; rest of tail dusky; below white, shaded with grayish across breast and with buff on crissum; iris yellow; bill and feet black. YOUNG: Breast and flanks

[1] *Catherpes mexicanus* of A.O.U.

spotted with dusky. Wing 94–122.5, tail 94–134, culmen 15.5–20, tarsus 27.5–34.5 mm. Mexico, West Indies, and the United States, north to Oregon, Nevada, Utah, Wyoming, Nebraska, Iowa, Illinois, Indiana, Ohio, Pennsylvania, and New Jersey, occasionally farther.

Genus *Dumetella* S. D. W. Differs from *Mimus* in having culmen less arched; nostrils imperforate. North America. 1 species.

Dumetella carolinensis (**Linnaeus**). Catbird. Slate, paler below; cap black; remiges and rectrices blackish; crissum chestnut; iris brown; bill and feet black. Wing 84–96, tail 82–103, culmen 15–18, tarsus 27–29 mm. Southern Canada south to Oregon, Utah, New Mexico, Gulf states, northern Florida, and Bermuda; winters in southeastern United States, Bahamas, Cuba, Mexico, and Central America.

Genus *Toxostoma* Wagler. Bill unnotched, curved, equal to or longer than middle toe; tail longer than wing. North America. 10 species, 7 in the United States.

KEY TO SPECIES OF *TOXOSTOMA*

1. Underparts immaculate 2
   Underparts spotted or streaked 4
2. Crissum chestnut .... *T. dorsale*
   Crissum rich buff 3
3. Tarsus 28.5–32 mm; above pale brown ...... *T. lecontei*
   Tarsus 35–40.5 mm; above dark brown
   .... *T. redivivum*
4. Above reddish brown 5
   Above grayish brown 6
5. Above rufous brown; crissum usually unspotted
   .... *T. rufum*
   Above chestnut brown; crissum spotted
   .... *T. longirostre*
6. Culmen 21.5–25.5 mm; breast spots darker than back
   .... *T. bendirei*
   Culmen 27–34.5 mm; breast spots same shade as back
   .... *T. curvirostre*

Toxostoma rufum (**Linnaeus**). Brown thrasher. Above tawny rufous; middle and greater coverts and tertials tipped with white or buff, black subterminally; outer pairs of rectrices indistinctly white-tipped; sides of head grayish brown; below buff, becoming nearly white on middle of throat and belly; malar region, breast, and flanks with elongate dark-brown spots, with sometimes a few streaks on crissum; iris yellow; bill blackish, flesh at base of mandible; feet light brown. Wing 95.5–115, tail 109–141, culmen 22–29, tarsus 32.5–35.5 mm. Eastern North America, west to Alberta, and eastern parts of Montana, Wyoming, Colorado, and Texas; winters in southeastern United States.

Toxostoma longirostre (**Lafresnaye**). Long-billed thrasher. Like *T. rufum*, but bill more curved and darker at base; above darker chestnut; rectrices scarcely paler at tips; sides of head brownish gray; below whiter, with spots blacker; crissum spotted. Wing 93–103, tail 120–132, culmen 25–32, tarsus 33–37 mm. Southern Texas to central Mexico.

Toxostoma curvirostre (**Swainson**). Curve-billed thrasher. Above brownish gray; middle and greater coverts white-tipped; tips of 4 lateral pairs of rectrices white; below buffy white, spotted with brownish gray on all but middle of throat; iris golden; bill black; feet brown. Wing 98.5–116.5, tail 99–125.5, culmen 27–34.5, tarsus 29.5–35.5 mm. Southern parts of Arizona, New Mexico, and Texas, south to Oaxaca.

Toxostoma bendirei (**Coues**). Bendire's thrasher. Resembles *T. curvirostre*, but browner above; breast spots triangular, darker than color of back; bill shorter; base of mandible pale. Wing 97.5–108.5, tail 101.5–115.5, culmen 21.5–25.5, tarsus 31.5–34.5 mm. Deserts of California, Arizona, New Mexico, Chihuahua, Sonora, and Sinaloa; occasional in Utah, Nevada, and Colorado.

Toxostoma redivivum (**Gambel**). California thrasher. Above dark grayish brown; wing coverts margined with paler; cheeks and auriculars dusky brown streaked with whitish; throat buffy white; malar region flecked with dusky; breast pale grayish brown; belly buff; crissum cinnamon; iris brown; bill blackish, paler at base below; feet brown. Wing 96–106.5, tail 120–138, culmen 32–39.5, tarsus 35–40.5 mm. California and Baja California.

Toxostoma lecontei **Lawrence**. Leconte's thrasher. Like *T. redivivum*, but above pale grayish brown; tail dark, tipped with paler; submalar streak dusky; below pale buff, deeper on flanks and crissum; iris reddish brown. Wing 93–101, tail 114–127, culmen 30–35.5, tarsus 28.5–32 mm. Lower Sonoran deserts of eastern California, southern

Nevada, Utah, western Arizona, Sonora, and Baja California.

***Toxostoma dorsale* Henry.** Crissal thrasher. Above grayish brown; tail darker, tipped with buffy brown; submalar streak blackish; throat white; below pale grayish brown; flanks and crissum chestnut; iris brown; bill and feet blackish. Wing 93.5–105, tail 123–150.5, culmen 32–38.5, tarsus 30.5–34.5 mm. Southern Nevada and Utah, Arizona, New Mexico, western Texas, and northern Mexico.

Genus *Oreoscoptes* Baird. Outer primary less than half 2d, narrow; 2d primary longer than 6th; tail shorter than wing. Great Basin and Plains. 1 species.

***Oreoscoptes montanus* (Townsend).** Sage thrasher. Above brownish gray, indistinctly streaked with darker; wings and tail with paler edges; middle and greater coverts tipped with white; rectrices broadly tipped with white; indistinct superciliary white; sides of face grayish streaked with white; below buffy white streaked with dusky; iris yellow; bill dark, paler at base below; feet brown. Wing 94–103, tail 85–95, culmen 14.5–17.5, tarsus 28.5–32 mm. British Columbia and Pacific states east of Cascades and Coast Ranges, east to Saskatchewan, Montana, Wyoming, western Nebraska, Colorado, and Oklahoma, south to New Mexico, Arizona, and California.

Family Turdidae. Thrushes. Bill shorter than head or middle toe; maxilla notched; nostrils oval, operculate, more or less exposed, imperforate; rictal bristles well developed. Primaries 10; outer primary much less than half 2d. Tail shorter than (sometimes equal to) wing; rectrices 12. Tarsus longer than culmen, equal to or longer than middle toe with claw, booted; hallux not longer than lateral toes; inner toe free; middle and outer toes united at extreme base. Maxillopalatine processes swollen and perforate; humerus with 2 fossae. Eggs bluish or greenish, spotted or plain; young spotted. Known from Pleistocene. Cosmopolitan. 49 genera, 308 species; in the United States 5 genera, 11 species.

KEY TO GENERA OF TURDIDAE

1. Width of bill at nasal feathering equal to or greater than length of gonys   ***Myadestes***
   Width of bill at nasal feathering much less than length of gonys   **2**

2. Wing tip much longer than tarsus   ***Sialia***
   Wing tip about equal to tarsus   **3**
3. Tarsus about equal to middle toe with claw; maxillary notch obsolete   ***Ixoreus***
   Tarsus much longer than middle toe with claw; maxilla distinctly notched   **4**
4. Tarsus short, 29–34% length of tail   ***Turdus***
   Tarsus long, 39–45% length of tail   ***Hylocichla***

Genus *Turdus* Linnaeus. Bill compressed, straight, notched; exposed culmen longer than inner toe with claw; frontal feathering reaching base of nostrils; outer primary usually shorter than primary coverts; 3d, 4th, and 5th primaries longest; 2d and 6th primaries next longest and subequal; tail even or slightly rounded; tarsus much longer than middle toe with claw. Neotropical, Holarctic, Africa, Polynesia. 62 species, 1 in United States.

***Turdus migratorius* Linnaeus.** Robin. Top and sides of head black; eyelids and interrupted superciliary white; upper parts gray, darker on wings and tail; tips of lateral rectrices often white; upper throat white, streaked with black; lower throat, breast, upper belly, sides, and wing lining tawny rufous (margined with white in winter); middle of belly white; crissum white with gray base; iris brown; bill yellow, tip dusky; feet brown. FEMALE: Duller; black head plumage broadly edged with gray; breast margined with white. YOUNG: Back gray, barred with black and streaked with buffy white; upper throat plain white, bordered with black streaks on malar region; breast and upper belly pale tawny spotted with black. Wing 117.5–145, tail 87.5–111.5, culmen 18.5–21, tarsus 29.5–35.5 mm. North America, south to South Carolina, mountains of Georgia, northern Alabama and Mississippi, Arkansas, Texas, and south through western states to Oaxaca. Winters from middle United States to Florida, Gulf Coast, and Mexico.

Genus *Hylocichla* Baird. Near *Turdus*, but tail short, tarsus and toes long (tarsus/tail ratio 39–45%); outer primary always shorter than primary coverts; 2d primary always longer than 6th. North America. 5 species.

KEY TO SPECIES OF *HYLOCICHLA*

1. Flanks heavily spotted with black   *H. mustelina*
   Flanks unspotted   **2**

2. Tail more rufous than back; 2d primary shorter than
    5th   **H. guttata**
   Tail and back concolor; 2d primary longer than 5th
                                           3
3. Eye ring buffy   **H. ustulata**
   No buffy eye ring   4
4. Above olive brown; chest spots black on pale back-
    ground   **H. minima**
   Above tawny brown; chest spots dusky on rich buff
    background   **H. fuscescens**

Hylocichla mustelina (**Gmelin**). Wood thrush (Fig. 5-68). Above tawny brown, brightest on crown, becoming olive brown posteriorly; sides of face streaked black and white; underparts white; breast, sides, and flanks heavily spotted with black; wing lining mixed brown and white; iris brown; bill horn, yellowish at base below; feet yellow. Wing 103–113, tail 63–77.5, culmen 16–19, tarsus 28.5–33 mm. Eastern deciduous forest, west to eastern parts of South Dakota, Nebraska, Kansas, Oklahoma, and Teaxs, south to Louisiana, Alabama, and northern edge of Florida. Winters mainly in Central America.

Hylocichla guttata (**Pallas**). Hermit thrush. Above brown; upper tail coverts and tail rufous brown; eye ring buffy or white; sides of face brown flecked with buff; throat and breast buff; sides of throat streaked with black; breast spotted with black; flanks olive brown; belly white; crissum buff; wing lining mainly buff, flecked with brown; inner webs of primaries edged with cinnamon basally on lower surface; feet yellowish. Wing 81–106, tail 58.5–79.5, culmen 11.5–16, tarsus 26.5–32 mm. Northern coniferous forest, south to New Jersey, mountains of Maryland and Virginia, northern parts of Michigan, Wisconsin, and Minnesota, and in mountains to New Mexico, Arizona, California, and Baja California. Winters along Pacific Coast and southern United States to Guatemala.

Hylocichla minima (**Lafresnaye**). Gray-cheeked thrush. Above brown; sides of head gray streaked with white; eye ring white, indistinct; sides of throat and breast pale buff spotted with black; flanks pale olive gray; rest of underparts white; bill dusky; feet brown. Wing 85–109, tail 60.5–78.5, culmen 12.5–15, tarsus 26–32.5 mm. Northern coniferous forest, south to Massachusetts, Catskills, and Adirondacks, west to northeastern Siberia. Migrates through eastern United States

Fig. 5-68. Head and right foot of wood thrush (*Hylocichla mustelina*).

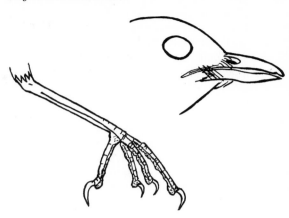

and winters in Hispaniola and northern South America.

Hylocichla ustulata (**Nuttall**). Olive-backed thrush. Like *H. minima,* but eye ring buff; auriculars mixed olive brown and buff; throat and breast buffier; base of mandible flesh; feet pale. Wing 88.5–105, tail 61.5–79.5, culmen 11.5–15, tarsus 25.5–31 mm. Northern coniferous forest, south to New England, New York, Pennsylvania, West Virginia, mountains of North Carolina, northern Michigan, northern Wisconsin, northern Minnesota, Colorado, Utah, Nevada, and California. Winters in South America, east of the Andes.

Hylocichla fuscescens (**Stephens**). Veery; willow thrush. Above tawny brown; no distinct eye ring; auriculars streaked brown and whitish; throat pale buff, bordered laterally by dusky streaks; chest rich buff, lightly spotted with brown; flanks gray; belly and crissum white; bill dusky, flesh at base below; feet brown. Wing 89–105, tail 65.5–79, culmen 13–15.5, tarsus 27.5–32 mm. Moist mixed forest, south to mountains of Georgia, Kentucky, Ohio, northern Indiana, northern Illinois, northern Iowa, North Dakota, Wyoming, Colorado, Utah, Nevada, and Oregon. Winters in eastern South America.

Genus *Ixoreus* Bonaparte. Similar to *Turdus,* but maxillary notch obsolete; nostrils nearly concealed by feathers; tarsus about equal to middle toe with claw. Western North America. 1 species.

Ixoreus naevius (**Gmelin**). Varied thrush. Above slate; wings dusky edged with gray and buff; middle and greater coverts tipped with tawny spots; lateral rectrices white-tipped; superciliary tawny; sides of face black; below ochraceous; a broad black band across chest; belly white; crissum slate, tipped with white and ochraceous; axillars slate, white basally; a subbasal band of whitish across inner surface of remiges; iris brown; bill dusky, pale at base of mandible; feet pale brown. FEMALE: Duller; above brownish; chest band brownish or obsolete. YOUNG: Like female, but breast squamate with grayish brown. Wing 119–132.5, tail 78.5–93.5, culmen 18–23, tarsus 29.5–33 mm. Northwestern coniferous forest from Alaska to northern California, east to Montana and northern Idaho. In winter extends to southern California and western Nevada.

Genus *Sialia* Swainson. Exposed culmen shorter than inner toe with claw; nostrils nearly covered by feathers; wing pointed; wing tip much longer than tarsus; outer primary much shorter than primary coverts; 2d or 3d primaries longest; tail short, emarginate; tarsus little longer than middle toe with claw. North America. 3 species.

KEY TO SPECIES OF *SIALIA*

1. Without rufous below; young streaked with grayish brown below **S. currucoides**
   With rufous below; young streaked with sooty brown below **2**
2. Throat of male blue, of female and young gray **S. mexicana**
   Throat of male and female chestnut, of young whitish **S. sialis**

Sialia sialis (**Linnaeus**). Eastern bluebird. Upper parts and sides of face blue; back sometimes narrowly edged with chestnut; remiges tipped with dusky; chin and submalar line grayish; throat, breast, and flanks chestnut; belly and crissum white; wing lining pale blue or gray; iris brown; bill and feet black. FEMALE: Above bluish gray tinged with brown; wings and tail blue; outer primary and rectrix edged with white; throat, breast, and flanks brownish cinnamon; belly and crissum white; wing lining gray and white. YOUNG: Above sooty brown; back with elongate whitish spots; wings and tail as in female; eye ring white; below dirty white; throat and breast heavily squamate with sooty brown. Wing 93–110.5, tail 56.5–72.5, culmen 11–14, tarsus 18–21.5 mm. Eastern North America, west to eastern parts of Montana, Wyoming, Colorado, Oklahoma, and Texas; southern Arizona through mountains of Mexico to El Salvador; Bermuda. Withdraws south of Ohio and Missouri Rivers in winter.

Sialia mexicana Swainson. Western bluebird. Upper parts, throat, middle of breast, crissum, and axillars blue; back often with a chestnut patch; sides of breast chestnut, sometimes extending across chest; belly and flanks pale gray; iris brown; bill and feet black. FEMALE: Crown, hind neck, and back brown, often tinged with blue; rump, tail, lesser wing coverts, and remiges blue; throat and breast brownish gray, sometimes tinged with blue; sides of breast dull chestnut; belly and flanks pale brownish gray; crissum pale bluish gray, streaked with dusky; axillars dull blue gray. YOUNG: As in young *S. sialis*, but throat darker gray. Wing 100–122, tail 58–77, culmen 11–13.5, tarsus 18.5–22.5 mm. Western North America from British Columbia to central Mexico, east to western Montana, western Wyoming, western Colorado, New Mexico, and western Texas.

Sialia currucoides (**Bechstein**). Mountain bluebird. Turquoise blue; wings tipped with dusky; throat, breast, and wing lining paler blue; belly and crissum bluish white; iris brown; bill and feet black. FEMALE: Head, neck, and back smoke gray; rump, tail, and wings blue; eye ring white; throat and breast pale brownish gray, becoming white on belly and crissum. YOUNG: Like female, but back, throat, and breast streaked with white; rump and upper tail coverts ashy; wings and tail margined with white. Wing 107–121.5, tail 64.5–76, culmen 12–14.5, tarsus 21.5–24 mm. Mountains of West, south to California and Arizona, east to western part of Dakotas, Nebraska, Colorado, Oklahoma, and New Mexico. Winters south to northern Mexico; on migration to western Texas and Kansas.

Genus *Myadestes* Swainson. Bill notched, about equal to inner toe with claw, much wider than deep; nostrils exposed, slightly perforate; wing long, rounded; wing tip about equal to tarsus; outer primary slightly longer than coverts; 4th primary longest; tail about equal to wing, double-rounded; rectrices tapering; tarsus a little longer than middle toe with claw. North and South America, West Indies. 6 species, 1 in the United States.

Myadestes townsendi (**Audubon**). Townsend's solitaire. Gray, paler below; lores dusky; eye ring white; wings edged with paler; inner remiges buff basally, blackish subbasally; tips of tertials and outer rectrix margined with whitish; 2 outer pairs of rectrices with broad white tips; iris brown; bill and feet black. Wing 110.5–123, tail 95–109.5, culmen 11–13, tarsus 19.5–22 mm. Canadian zone of West, east to Montana, Black Hills, western Nebraska, western Colorado, and New Mexico, south to northern Mexico. On migration to Kansas and western Texas.

Family Sylviidae. Old World warblers. Bill slender, straight; culmen ridged between nostrils and often with a flat area in advance of nostrils; nostril horizontally slit, operculate, usually exposed; operculum usually swollen basally, usually with edge turned in on nostril; maxillary tomium slightly notched; rictal bristles few (Fig. 5-69). Primaries 10, outer one (2 in *Ramphocaenus*) shorter than secondaries (only 9 primaries in *Peucedramus*). Tail rounded or emarginate, usually shorter than wing (about equal to wing in *Polioptila*); rectrices 12. Tarsus booted to scutellate, much longer than middle toe with claw; hallux and lateral toes subequal. Maxillopalatines thin and flat; humerus with 2 fossae. Eggs spotted; young unspotted. Recorded from Pleistocene. Cosmopolitan, but mostly Old World. 87 genera, 416 species; in the United States 3 genera, 5 species.

KEY TO GENERA OF SYLVIIDAE

1. Primaries 9, outer one long   *Peucedramus*
     Primaries 10, outer one less than half length of 2d   **2**
2. Tail emarginate, shorter than wing   *Regulus*
     Tail rounded, about as long as wing   *Polioptila*

Genus *Peucedramus* Coues.[1] Bill shorter than head; culmen with a flat area in advance of nostrils; nasal plumes falling short of nostril and operculum; operculum swollen basally, its side folded into nostril; tail emarginate, shorter than wing; pri-

[1] Referred to family Parulidae by the A.O.U.

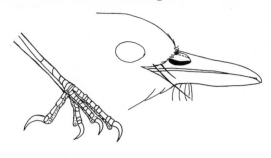

Fig. 5-69. **Head and foot of olive warbler** (*Peucedramus taeniatus*), **enlarged.**

Fig. 5-70. **Wing and tail of olive warbler** (*Peucedramus taeniatus*).

maries 9, outermost nearly as long as longest; tarsus almost booted. North America. 1 species.

Peucedramus taeniatus (**DuBus**). Olive warbler (Figs. 5-69, 5-70). Head, neck, and breast tawny; mask from bill through eye to auriculars black; back gray; wings black edged with olive; 2 wing bands and patch at base of primaries white;

Fig. 5-71. Head and foot of ruby-crowned kinglet (*Regulus calendula*), enlarged.

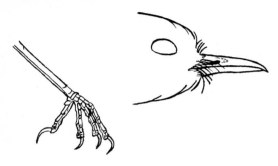

Fig. 5-72. Head and foot of blue-gray gnatcatcher (*Polioptila caerulea*), enlarged.

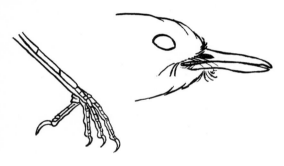

tail black, 2 outer pairs with large white subterminal patch; belly and crissum white; iris brown; bill dusky, paler at base below; feet dusky. FEMALE AND YOUNG: Crown olive green; mask dull and mixed with whitish; throat yellow, continuing behind and above mask. In still younger birds crown and back dull olive gray; throat and sides of face yellowish buff; no mask, but a black spot on auriculars. Wing 67–78, tail 47.5–56, culmen 9–12, tarsus 17–20 mm. Pines at high altitudes from Arizona and New Mexico to Nicaragua.

Genus *Regulus* Cuvier. Bill very thin, shorter than middle toe with claw; culmen with flat area in advance of nostrils; nostril partly or entirely covered by nasal plumes; operculum swollen basally and folded laterally into nostril (Fig. 5-71); primaries 10, outer one longer than primary coverts, 4th–5th longest; tail emarginate, ¾ wing; tarsus booted. Holarctic. 4 species, 2 in the United States.

KEY TO SPECIES OF *REGULUS*

1. Sides of crown bordered with black   *R. satrapa*
   Crown without black   *R. calendula*

Regulus satrapa **Lichtenstein.** Golden-crowned kinglet. Crest orange, bordered by yellow and externally by black; superciliaries grayish white, meeting on forehead; sides of face, nape, and upper back gray; rest of upper parts olive green; wings and tail dusky edged with yellowish olive; an interrupted black band across wing, formed by primary coverts and subbasal spot on secondaries; below dirty whitish olive; iris brown; bill black; feet brown. FEMALE: Crest yellow, bordered by black. Wing 52.5–60, tail 38–46.5, culmen 6–9, tarsus 15.5–18 mm. Northern coniferous forest, south to Massachusetts, Catskills, northern Pennsylvania, northern Michigan, northern Wisconsin, and northern Minnesota, and in mountains of West south through Mexico to Guatemala. Winters all over the United States except southern Florida.

Regulus calendula **(Linnaeus).** Ruby-crowned kinglet (Fig. 5-71). Above olive, greener posteriorly; crown patch red; wings and tail dusky edged with greenish olive; 2 white wing bands; tertials edged with white; a black band across base of secondaries; eye ring white; below buffy olive; bill black; iris and feet brown. FEMALE: Without red. Wing 50–61.5, tail 39–45, culmen 7–10.5, tarsus 17.5–20 mm. Nearctic coniferous forest, south to Massachusetts, northern New York, northern Michigan, and in mountains to New Mexico, Arizona, and California; Guadalupe Island. Winters in southern United States and Mexico.

Genus *Polioptila* Sclater. Exposed culmen about as long as middle toe with claw, without flat area in advance of nostrils; nostrils exposed (Fig. 5-72); primaries 10, outermost much longer than primary coverts, 3d–5th longest; tail about equal to wing, strongly rounded; tarsus scutellate, twice as long as middle toe. North and South America, West Indies. 9 species, 2 in the United States.

KEY TO SPECIES OF *POLIOPTILA*

1. Exposed part of outer rectrix entirely white
      P. caerulea
   Exposed part of outer rectrix black for basal half
      P. melanura

Polioptila caerulea (**Linnaeus**). Blue-gray gnatcatcher (Fig. 5-72). Above blue-gray; a black line before eye, meeting on forehead; eye ring white; wings dusky; tertials edged with ash; tail black; 3 lateral rectrices with broad white tips, the outer one black only at concealed base; below white, shaded with pale blue-gray across breast; iris brown; bill black, pale at base below; feet black. FEMALE AND WINTER: No black U on forehead. Wing 42.5–54.5, tail 41–55.5, culmen 9–11, tarsus 15–18.5 mm. Bahamas and southern Mexico, north to California, Nevada, Idaho, Wyoming, Nebraska, Iowa, southern Wisconsin, southern Michigan, Pennsylvania, and New York. Winters from South Carolina, Gulf states, Arizona, and California, south to Mexico, Guatemala, Cuba, and Bahamas.

Polioptila melanura **Lawrence**. Black-tailed gnatcatcher. Pileum glossy black; back blue-gray; wings dusky edged with ash; tail and upper tail coverts black; outer 2 rectrices broadly white-tipped, but with black bases larger than tip; below white, shaded with pale gray across breast. FEMALE AND YOUNG: Pileum gray like back. Wing 44–48.5, tail 47–53.5, culmen 8.5–10, tarsus 16–18.5 mm. Northern Mexico, north to southern parts of California, Nevada, New Mexico, and Texas.

Family Motacillidae. Wagtails. Bill slender, shorter than head or middle toe, notched; rictal bristles present; nostrils exposed, slitlike, operculate, minutely perforate. Wing pointed; primaries 9; secondaries short; tertials long, sometimes longer than primaries. Tail $\frac{2}{3}$ wing or longer; rectrices 12. Tarsus slender, acrotarsium scutellate; lateral and hind toes subequal; inner toe nearly free; claw of hallux equal to or longer than digit. Miocene onward. Cosmopolitan. 5 genera, 53 species; in the United States 1 genus, 2 species.

Genus *Anthus* Bechstein. Tail shorter than wing, double-rounded; middle claw slightly pectinate. Cosmopolitan. 34 species, 2 in the United States.

KEY TO SPECIES OF *ANTHUS*

1. Hind toe and claw shorter than tarsus   *A. spinoletta*
   Hind toe and claw longer than tarsus   *A. spragueii*

Anthus spinoletta (**Linnaeus**). Water pipit (Fig. 5-73). 4th primary much longer than 5th;

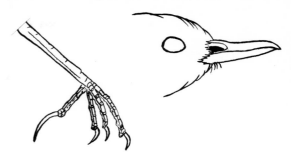

Fig. 5-73. Head and foot of water pipit (*Anthus spinoletta*), enlarged.

outstretched toes not reaching tip of tail; hind toe and claw shorter than tarsus. Above brownish gray obscurely streaked with dusky; wings and tail edged with ashy olive; 2 ashy wing bands; outer 2 rectrices largely white externally and at tip; superciliary and underparts pinkish buff; breast, sides, and flanks more or less streaked with dusky; iris, bill, and feet brown; base of mandible paler. WINTER: Above olive brown, streaked with dusky; superciliary and underparts cream buff; streaks below more prominent. Wing 78–92, tail 57.5–70, culmen 11–13, tarsus 20.5–23, hind claw 7–11 mm. Holarctic, breeding on tundra and alpine zone of mountains, south to Maine, New Mexico, and California. Winters along Pacific Coast and from central and southern states to northern Central America.

Anthus spragueii (**Audubon**). Sprague's pipit. 4th and 5th primaries nearly equal; outstretched toes extending beyond tail; hind toe and claw longer than tarsus. Above grayish brown heavily streaked with dusky; wings and tail dusky margined with grayish buff; 2 grayish-buff wing bands; 2 outer rectrices largely white externally and at tip; superciliary and underparts buffy white; chest and sides streaked with black; axillars white. Wing 77–84, tail 52–59, culmen 11.5–13, tarsus 22–24, hind claw 11.5–15.5 mm. Northern Great Plains, from Alberta, Saskatchewan, and Manitoba to Montana and Dakotas. Winters from Gulf states to northern Mexico.

Family Bombycillidae. Waxwings. Head crested; plumage soft; secondaries often with waxy tips. Bill short, broad, triangular, thick; culmen and gonys curved; maxilla notched; nostrils concealed by velvety feathers; gape wide; rictal bristles absent. Wing long, pointed; primaries 9; wing tip about equal to tarsus and middle toe together. Tail less than ⅔ wing, nearly even; coverts long. Legs short; tarsus scutellate, about equal to middle toe and claw. Maxillopalatines swollen and hollow. Eggs spotted; young streaked below. Known from Pleistocene. Holarctic. 1 genus, 3 species; 2 species in the United States.

Genus *Bombycilla* Brisson.

KEY TO SPECIES OF *BOMBYCILLA*

1. Crissum chestnut; secondaries white-tipped
   B. garrulus
   Crissum white; secondaries without white
   B. cedrorum

Bombycilla garrulus (**Linnaeus**). Bohemian waxwing. Grayish brown, grayer on upper tail coverts; forehead and malar streak cinnamon; ante-

Fig. 5-74. Head (enlarged) and wing of cedar waxwing (*Bombycilla cedrorum*).

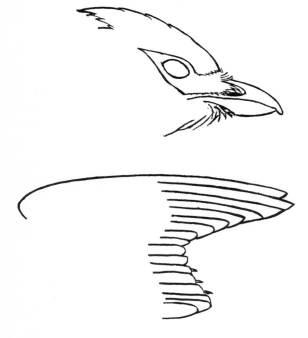

rior end of malar streak white, extending to below eye; chin and line from nostrils through eyes black; wings dusky; primary coverts and secondaries tipped with white; several secondaries often with red waxy tips; primaries edged at tip with yellow, or yellow and white; tail tipped with yellow, black subterminally; middle of belly whitish; crissum chestnut; iris brown; bill and feet black. Wing 110–121, tail 59–70, culmen 10–12, tarsus 19–21.5 mm. Holarctic coniferous forests, south to Washington, northern Idaho, and Montana. Winters southward irregularly to California, Arizona, Colorado, Arkansas, Illinois, Indiana, Ohio, Pennsylvania, and New England, sometimes farther.

Bombycilla cedrorum **Vieillot.** Cedar waxwing (Fig. 5-74). Chin and stripe across forehead through eyes black; short malar stripe white; posterior border of forehead and lower side of postocular stripe narrowly white; head, neck, and breast reddish brown; back brown; rump and tail grayish; tail tipped with yellow, black subterminally; wings dusky, edged with gray and ashy; secondaries often with red waxy tips; belly yellowish; crissum whitish; iris brown; bill and feet black. YOUNG: Black of face duller; back, breast, and sides streaked with white; postocular streak white. Wing 91–99, tail 51–61, culmen 9–11, tarsus 16–18 mm. Forests of North America, south to Georgia, Tennessee, Kentucky, Indiana, Illinois, Missouri, Nebraska, Colorado, Utah, and northern California. Winters from central states to Cuba, Mexico, and Central America.

Family Ptilogonatidae. Silky flycatchers. Related to Bombycillidae, but plumage without waxy tips; nostrils exposed, oval, operculate; rictal bristles well developed. Wing rounded; primaries 10, outer primary about half length of 2d; wing tip shorter than tarsus. Tail rounded, about equal to or even longer than wing. No fossil record. Mexico and Central America. 3 genera, 3 species; 1 in the United States.

Genus *Phainopepla* Sclater. Crested; bill relatively narrow; wing and tail about equal; tarsus shorter than middle toe with claw; acrotarsium and part of planta tarsi scutellate. Southwestern United States and Mexico. 1 species.

Phainopepla nitens (**Swainson**). Phainopepla (Fig. 5-75). Shining black; primaries with a large white patch; iris red; bill and feet black. FEMALE:

Gray; crest blackish; primary patch paler gray. Wing 86–99, tail 83–105, culmen 10–12, tarsus 16–19.5 mm. Southern California, southern Nevada, southern Utah, New Mexico, and southwest Texas, south over Mexican tableland.

Family Laniidae. Shrikes. Bill shorter than head, strong, compressed, strongly hooked, and notched; rictal bristles well developed; nostril covered with bristles but slightly perforate. Wing short, rounded; primaries 10. Tail about as long as wing; rectrices 12. Tarsus stout, longer than bill, scutellate; middle toe with claw shorter than tarsus; outer and middle toes coherent basally; inner toe nearly free. Maxillopalatines flat; external nares with a bony tubercle; humerus with 1 pneumatic fossa.

Predaceous; prey impaled on thorns. Eggs speckled; young barred. Miocene onward. Africa, Asia to New Guinea, Europe, North America. 12 genera, 74 species; in the United States, 1 genus, 2 species.

Genus *Lanius* Linnaeus. Outer primary about half 2d; tail graduated. Holarctic, Africa, Indian region. 23 species, 2 in the United States.

KEY TO SPECIES OF *LANIUS*

1. Lores gray, often with a black spot; wing 110–121 mm *L. excubitor*
   Lores entirely black; wing 88–106 mm
   L. ludovicianus

Lanius excubitor **Linnaeus.** Northern shrike. Above bluish gray; forehead, superciliary, rump, upper tail coverts, and scapulars white; lores gray, but often with a black spot near nostrils and another before eye; lower eyelid and auriculars black; wings black; a white spot at base of primaries; secondaries tipped with white; tail black, tipped with white, most broadly laterally; below white; breast with wavy dusky bars; iris brown; bill and feet black. YOUNG: Lores entirely gray; upper parts tinged with brownish, barred posteriorly with brown. Wing 110–121, tail 104.5–118, culmen 17–19, tarsus 25–28.5 mm. Northern Holarctic; winters south irregularly to Maryland, West Virginia, Ohio, Indiana, Illinois, Missouri, Kansas, Colorado, Utah, Nevada, and northern California, rarely farther south.

Fig. 5-75. Head (enlarged) and wing of phainopepla (*Phainopepla nitens*).

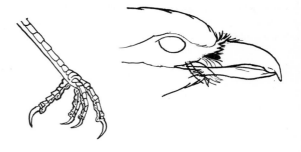

Fig. 5-76. Head and foot of loggerhead shrike (*Lanius ludovicianus*).

Lanius ludovicianus **Linnaeus.** Loggerhead shrike (Fig. 5-76). Differs in having middle toe with claw longer than exposed culmen, instead of shorter; lores entirely black; underparts of adult unbarred. Wing 88–106, tail 86–107, culmen 13–17, tarsus 25–29 mm. Southern Canada and the United States to southern Mexico; winters from central states southward.

Family Sturnidae. Starlings. Bill longer than head, straight, wider than deep; nostrils exposed,

Fig. 5-77. Head and foot of common myna (*Acridotheres tristis*), family Sturnidae.

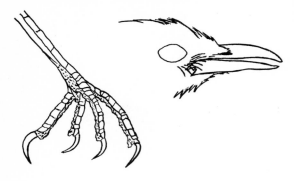

operculate, imperforate, situated nearer rictus than to culmen; rictal bristles obsolete. Wing pointed; primaries 10. Rectrices 12. Tarsus short, stout, scutellate, scarcely longer than middle toe with claw; toes free at base (Fig. 5-77). Africa, Asia, Polynesia, Europe. 25 genera, 111 species. 1 species introduced to the United States.

Genus *Sturnus* Linnaeus. Plumage glossy; anterior feathers lanceolate; head without crest or wattles; outer primary minute; tail only half wing, emarginate. Recorded from Eocene. Europe, Asia, Africa. 16 species; 1 introduced to the United States.

***Sturnus vulgaris* Linnaeus.** Starling. Black, glossed with green and purple; back speckled with clay color; wings, tail, and crissum edged with clay color; iris brown; bill yellow, gray on rami; feet reddish. WINTER: Crown speckled with clay color; underparts spotted with white; bill dusky. YOUNG: Grayish brown, paler below; wings and tail margined with clay color; underparts streaked with white; bill dusky. Wing 120–142, tail 56–65, culmen 23–26.5, tarsus 28–30 mm. Palearctic; introduced in New York City in 1890 and has since spread throughout the country, though still less common in the West.

Family Vireonidae. Vireos. Bill straight, uncinate, and notched, shorter than head; culmen and gonys curved; nostrils at least partly exposed, operculate, imperforate; rictal bristles present. Wing rounded; primaries 9–10. Tail shorter than wing; rectrices 12. Tarsus longer than middle toe with claw, scutellate on acrotarsium; inner toe shortest; anterior toes coherent at base. Maxillopalatines hooked and perforate. Known from Pleistocene. Americas. 8 genera, 42 species; in the United States 1 genus, 12 species.

Genus *Vireo* Vieillot. Bill about as wide as deep; tail practically even; inner toe without claw shorter than hallux without claw. Americas. 20 species, 12 in the United States.

KEY TO SPECIES OF *VIREO*

1.	Primaries 10; outer primary spurious	2
	Primaries 9; outer primary long	8
2.	Lores and eye ring bright yellow	*V. griseus*
	Lores and eye ring white or grayish white	3
3.	Wing 61 mm or more	4
	Wing 59 mm or less	6
4.	No trace of wing bands	*V. gilvus*
	Wing with 1 or 2 whitish wing bands	5
5.	1 indistinct whitish wing band; wing 61–67 mm	
		V. vicinior
	2 distinct white wing bands; wing 67–87 mm	
		V. solitarius
6.	Lores and eye ring white; axillars and under wing coverts yellow	*V. atricapillus*
	Lores and eye ring not distinctly white; wing lining whitish, sometimes tinged with yellow	7
7.	A dusky mark on lores; wing 52–59 mm	*V. bellii*
	No dusky mark on lores; wing 59–72 mm	
		V. huttoni
8.	Wing coverts with 2 white bands	*V. flavifrons*
	Wing without bands	9
9.	Entire underparts olive yellow; wing 62–69 mm	
		V. philadelphicus
	Median underparts white; wing 73–86 mm	10
10.	A dusky submalar streak	*V. altiloquus*
	No dusky submalar streak	11
11.	Crissum and axillars yellow	*V. flavoviridis*
	Crissum and axillars buffy white	*V. olivaceus*

***Vireo atricapillus* Woodhouse.** Black-capped vireo. Above olive green; crown and nape black or gray; lores and eye ring white; wings and tail dusky edged with yellowish green; 2 yellow wing bands; underparts white; flanks greenish yellow; axillars yellow; iris reddish brown; maxilla black; mandible and feet bluish. YOUNG: Crown and nape brownish gray. Wing 54–57, tail 40–46, culmen 9, tarsus 18.5–19 mm. Southern Kansas, Oklahoma, and western Texas; winters in Mexico.

Vireo griseus (Boddaert). White-eyed vireo. Above olive green; crown grayer; lores and eye ring yellow; 2 yellowish wing bands; below whitish; throat and breast shaded with brownish; sides, flanks, and crissum yellow; iris white; maxilla black; mandible and feet bluish. Wing 54–65, tail 44–53, culmen 9–12, tarsus 18–22 mm. Southern parts of New England, New York, Ontario, Michigan, Wisconsin, and Iowa, through southeastern United States to Bermuda, and northern Mexico, west to central Texas, and eastern parts of Oklahoma, Kansas, and Nebraska. Winters from southern states to Mexico and Guatemala.

Vireo huttoni Cassin. Hutton's vireo. Above grayish olive; wings and tail edged with pale greenish; 2 dull-white wing bars; lores, eye ring, and underparts pale yellowish olive; iris brown; maxilla black; mandible and feet grayish. Wing 59–72, tail 47–55, culmen 8–11, tarsus 18–20 mm. Oak woods of central Mexico; north along Pacific Coast to Vancouver Island and in interior to California, Arizona, New Mexico, and Chisos Mountains, Texas.

Vireo vicinior Coues. Gray vireo. Above gray; rump tinged with olive; wings and tail dusky edged with whitish; 1 indistinct grayish wing band; lores, eye ring, and underparts pale grayish; flanks tinged with olive; abdomen and axillars tinged with pale yellow; iris brown; maxilla black; mandible and feet bluish gray. Wing 61–67, tail 55–61, culmen 9–11, tarsus 18.5–20 mm. Southern California, southern Nevada, southern Utah, New Mexico, and western Texas south to northern Mexico. Winters in western Mexico.

Vireo bellii Audubon. Bell's vireo. Above olive green; crown and nape grayish brown; wings and tail brownish edged with olive green; 2 dull-white wing bars; eye ring dull white, interrupted by a dusky spot in front of eye; below buffy white, yellower on breast, sides, crissum, and axillars; iris brown; maxilla brown; mandible and feet grayish. Wing 52–58.5, tail 41.5–54, culmen 9–10, tarsus 18–20 mm. Southern parts of California, Nevada, Utah, and New Mexico, eastern Colorado, Nebraska, southeastern South Dakota, southern Minnesota, southern Wisconsin, east to Illinois, western Tennessee, Arkansas, and Texas, south to central Mexico. Winters in Mexico and Central America.

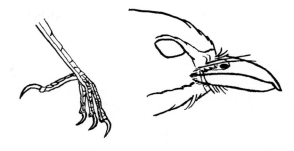

Fig. 5-78. Head (enlarged) and foot of solitary vireo (*Vireo solitarius*).

Vireo solitarius (Wilson). Solitary vireo (Fig. 5-78). Crown and sides of face slaty; lores and eye ring white; a black spot before eye; back olive green or grayish; wings and tail dusky edged with olive; 2 yellowish or whitish wing bars; below white, shaded with buffy across breast; sides, flanks, and axillars yellowish; iris brown; maxilla black; mandible and feet bluish. Wing 67–87, tail 47–61, culmen 9.5–12, tarsus 17.5–20 mm. Southern Canada, south to Georgia, Ohio, Michigan, Wisconsin, Minnesota, North Dakota, and in mountains of West, through Mexico to northern Central America. Winters from southern United States southward.

Vireo flavifrons Vieillot. Yellow-throated vireo. Crown, hind neck, and upper back yellowish olive; rump, wing coverts, and upper tail coverts gray; wings and tail black; 2 white wing bands; tertials and tail edged with white; remiges edged with olive; spot before eye dusky; lores, eye ring, throat, breast, and axillars yellow; belly and crissum white; iris brown; bill black, bluish below; feet bluish. Wing 72–80, tail 47–52, culmen 10.5–12, tarsus 18–20 mm. Eastern deciduous forest from Canada to Florida and Gulf states, west to eastern parts of Dakotas, Nebraska, Kansas, Oklahoma, and Texas. Winters from Mexico to Colombia.

Vireo olivaceus (Linnaeus). Red-eyed vireo. Crown gray; upper parts olive green; long superciliary whitish, bordered above and below by a dusky line; below white; axillars and crissum pale yellow; iris reddish; maxilla dusky; mandible and feet bluish. YOUNG: Iris dark brown. Wing 76–85, tail 47–60, culmen 11–12, tarsus 17–19 mm. Canada south to central Florida, Gulf states, and northern Mexico, west to Oregon, Utah, Montana, Black Hills, Kansas, Oklahoma, and Texas. Winters in South America.

Vireo flavoviridis (Cassin). Yellow-green vireo. Near *V. olivaceus*, of which perhaps a subspecies. Sides, flanks, and crissum bright yellow or greenish yellow. Wing 74–82, tail 49–60, culmen 13–15, tarsus 17–19 mm. Cameron County, Texas, south through Mexico and northern Central America. Winters in South America.

Vireo altiloquus (Vieillot). Black-whiskered vireo. Resembles *V. olivaceus*, but crown brownish olive without dusky margin; back duller; a dusky malar stripe; below suffused with buff; crissum yellow. Wing 72–85, tail 48–61, culmen 14–18, tarsus 17–20.5 mm. West Indies; Florida Gulf Coast and Keys.

Vireo philadelphicus (Cassin). Philadelphia vireo. Crown gray; back olive green; wings and tail brownish edged with olive; lesser and middle coverts olive gray; superciliary whitish; streak through eye dusky; below dull yellow, whiter on chin and belly; iris brown; maxilla horn; mandible and feet bluish. YOUNG: Crown olive. Wing 62–69, tail 43–48, culmen 10, tarsus 16–18 mm. Eastern Canada south to Maine, New Hampshire, New York, northern Michigan, and North Dakota. Winters in Central America.

Vireo gilvus (Vieillot). Warbling vireo. Crown gray; back grayish olive; wings and tail dusky edged with olive; superciliary whitish; sides of face buff; below dull white, tinged with buff on breast and yellow on flanks and crissum; iris brown; maxilla horn; mandible and feet bluish. YOUNG: Crown grayish buff. Wing 62–75, tail 45–55, culmen 9–12, tarsus 16–19 mm. Southern Canada, south to northern Alabama, Mississippi, Louisiana, Texas, New Mexico, California, and northern Mexico. Winters in Mexico and Central America.

Family Parulidae. Wood warblers. Bill straight, unhooked, shorter than head, usually slender (sometimes wide and flat); nostrils exposed, operculate, imperforate. Primaries 9. Tail shorter than wing; rectrices 12. Tarsus longer than middle toe with claw; acrotarsium weakly scutellate; lateral toes and hallux subequal; anterior toes coherent basally. Maxillopalatines hooked and perforate; humerus with a large divided fossa. Recorded from Miocene. Americas. 27 genera, 118 species; in the United States 15 genera, 53 species.

KEY TO GENERA OF PARULIDAE

1. Rictal bristles long, extending at least half length of bill   **2**
   Rictal bristles short or obsolete   **4**
2. Bill at nostril deeper than wide (rump white)   *Cardellina*
   Bill at nostril wider than deep (rump not white)   **3**
3. Tail fan-shaped, strongly rounded; tarsus/wing ratio 23–28%   *Setophaga*
   Tail normal, slightly rounded; tarsus/wing ratio 29–34%   *Wilsonia*
4. Outer primary decidedly shorter than 5th   **5**
   Outer primary equal to or longer than 5th   **7**
5. Culmen and rictus nearly straight   *Geothlypis*
   Culmen and rictus strongly arched   **6**
6. Maxilla notched   *Chamaethlypis*
   Maxilla unnotched   *Icteria*
7. Hind toe and claw as long as bare part of acrotarsium   *Mniotilta*
   Hind toe and claw much shorter than bare part of acrotarsium   **8**
8. Only a few small rictal bristles   **9**
   Rictal bristles well developed   **10**
9. Exposed culmen about as long as middle toe with claw   *Limnothlypis*
   Exposed culmen shorter than middle toe with claw   *Vermivora*
10. Exposed culmen longer than wing tip   *Protonotaria*
    Exposed culmen shorter than wing tip   **11**
11. Culmen swollen and arched   *Helmitheros*
    Culmen neither swollen nor arched   **12**
12. Tail rounded   *Oporornis*
    Tail not rounded   **13**
13. Inner toe nearly free from middle toe   *Seiurus*
    Anterior toes fused basally   **14**
14. Bill thin and sharp, culmen and gonys straight   *Parula*
    Bill not sharp, culmen curved   *Dendroica*

Genus *Mniotilta* Vieillot. Bill slender, slightly notched, about as long as middle toe without claw or hallux with claw; rictal bristles few, short; wing pointed, outer 3 primaries longest; wing tip longer than tarsus; tail double-rounded (nearly even), shorter than from bend of wing to tip of secondaries; tarsus about $\frac{1}{4}$ wing; middle toe with claw as long as tarsus; hallux and claw about as long as bare part of acrotarsium; inner toe shortest;

outer toe bound to base of middle toe. America. 1 species.

Mniotilta varia (**Linnaeus**).   Black-and-white warbler. Above black; central crown stripe white; back striped with white; edges of tertials and upper tail coverts white; 2 white wing bars; remiges and rectrices edged with gray; superciliary white; throat and sides of face black; malar stripe white; below white, streaked with black; iris brown; bill black, paler at base below; feet brown. FEMALE: Throat white. YOUNG: Throat white; below tinged with buff and with a few streaks. Wing 65–71, tail 43–51, culmen 10.5–13, tarsus 16.5–17.5 mm. Eastern North America, west to Montana, Nebraska, Kansas, and Texas, south to Louisiana, Mississippi, Alabama, and central Georgia. Winters from southern states to West Indies and northern South America.

Genus *Protonotaria* Baird.   Bill wedge-shaped; culmen ridged; only about 2 minute rictal bristles; wing tip about equal to tarsus; outer 3 primaries longest; tail slightly rounded; all toes with claw much shorter than tarsus; inner and outer toes equal and longer than hallux without claw. America. 1 species.

Protonotaria citrea (**Boddaert**).   Prothonotary warbler. Head and underparts yellow; upper back yellowish olive; rump, upper tail coverts, wing coverts, and tertials gray; remiges black; tail black, inner web of all but middle pair white, except at tip; crissum white; iris brown; bill black (pale brown with dusky tip in winter); feet blackish. FEMALE: Crown like back; belly whitish. Wing 64–74, tail 40–50, culmen 13–15, tarsus 18.5–20 mm. Wooded swamps of Southeast, from central Florida and Gulf states north to New Jersey, West Virginia, western New York, Ohio, southern Michigan, southern Wisconsin, and southern Minnesota, west to Iowa, Kansas, eastern Oklahoma, and eastern Texas. Winters in Central and South America.

Genus *Helmitheros* Rafinesque.   Bill stouter and even more wedge-shaped than in *Protonotaria*; outer 4 primaries longest. America. 1 species.

Helmitheros vermivorus (**Gmelin**).   Worm-eating warbler. Head, throat, and breast buff; a black stripe along each side of crown; stripe through eye black; above olive green; belly whitish; flanks pale brownish olive; crissum buffy; iris

brown; maxilla brown; mandible and feet flesh. Wing 65–73, tail 45–51, culmen 13–14.5, tarsus 17.5–19.5 mm. Georgia, northwestern Florida, Tennessee, Arkansas, and Texas, north to Connecticut, New York, Ohio, Indiana, Illinois, Iowa, Kansas, and Nebraska. Winters in West Indies, Mexico, and Central America.

Genus *Limnothlypis* Stone.   Bill stout, sharp, about equal to middle toe with claw, ridged and elevated between nostrils; rictal bristles minute; outer primary longer than 4th, the 2d and 3d longest; wing tip shorter than tarsus; tail even. America. 1 species.

Limnothlypis swainsonii (**Audubon**).   Swainson's warbler (Fig. 5-79). Above brown; superciliary buffy white; line through eye brown; below dirty white tinged with yellow; iris and bill brown; feet flesh. Wing 67.5–72, tail 46–50.5, culmen 14.5–16, tarsus 17.5–19 mm. Northern Florida and Louisiana north to Maryland, West Virginia, Kentucky, and southern Illinois, west to Missouri, Arkansas, and Oklahoma. Winters in Cuba, Jamaica, and eastern Mexico.

Genus *Vermivora* Swainson.   Bill thin, sharp, shorter than middle toe with claw; culmen and gonys straight; a few small rictal bristles; outer 4 primaries longest; tail nearly even; all toes much shorter than tarsus; lateral and hind toes equal; anterior toes fused basally. America. 11 species, 9 in the United States.

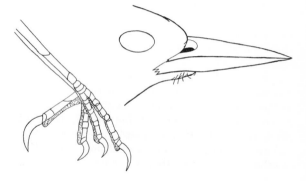

Fig. 5-79. Head and foot of Swainson's warbler (*Limnothlypis swainsonii*), enlarged.

KEY TO SPECIES OF *VERMIVORA*

1. Tail with a white patch on each side   2
   Tail without white patch   4
2. Wings unbanded   *V. bachmanii*
   Wing coverts banded   3
3. Wing bands yellow   *V. chrysoptera*
   Wing bands white   *V. pinus*
4. Crissum white   5
   Crissum yellow or chestnut   6
5. Upper tail coverts olive green like back   *V. peregrina*
   Upper tail coverts chestnut   *V. luciae*
6. Crissum chestnut   *V. crissalis*
   Crissum yellow   7
7. Eye ring yellow or obsolete   *V. celata*
   Eye ring white   8
8. Back gray in contrast to olive-green upper tail coverts   *V. virginiae*
   Back and upper tail coverts olive green   *V. ruficapilla*

Vermivora chrysoptera (**Linnaeus**). Golden-winged warbler. Crown yellow; superciliary white; back and middle rectrices gray; a large yellow patch on middle and greater coverts; remiges dusky edged with gray or olive green; 3 outer rectrices dusky with large white terminal patch on inner web; throat and sides of face black; malar stripe and underparts white; iris and feet brown; bill black. FEMALE: Yellow of crown largely replaced by olive green; throat and sides of face gray; back tinged with olive yellow; wing patch smaller, forming 2 yellow bands; belly often tinged with yellow. Wing 57.5–65, tail 43–48, culmen 10.5–11.5, tarsus 17–18 mm. Eastern states south to mountains of Georgia, Tennessee, Ohio, Indiana, Illinois, and southern Minnesota. Winters in Central and South America.

Vermivora pinus (**Linnaeus**). Blue-winged warbler. Crown and underparts yellow; lores and postocular spot black; back olive green; wing coverts blue-gray with 2 white bands; remiges and tail dusky, edged with blue-gray; outer 3–5 rectrices with large white subterminal patch on inner web; iris and feet brown; bill black (brown in winter). FEMALE: Crown more or less masked with olive green; lores and postocular spot gray; wing coverts tinged with olive green. Wing 56–63, tail 43.5–48.5, culmen 10.5–11.5, tarsus 16.5–18 mm. Massachusetts, New York, southern Pennsylvania, Ohio, southern Michigan, Illinois, southern Wisconsin, and southern Minnesota, south to northern Georgia, northern Alabama, Tennessee, and northern Arkansas. Winters from Mexico to Colombia. Hybrids with *V. chrysoptera* have been described as *V. leucobronchialis* (Brewster) and *V. lawrencii* (Herrick).

Vermivora bachmanii (**Audubon**). Bachman's warbler. Forehead, eye ring, cheeks, and chin yellow; anterior crown black margined with gray; posterior crown and nape gray; back olive green; edge of wing yellow; remiges and rectrices dusky edged with olive; 3–4 outer rectrices with large white subterminal patch on inner web; throat and breast black; crissum mixed white and yellow; iris brown; maxilla black; mandible and feet horn. FEMALE: Lacks black crown and throat patch. Wing 56.5–61.5, tail 42.5–46.5, culmen 11–12, tarsus 17–17.5 mm. Wooded swamps of southern Missouri, southern Indiana, Kentucky, Arkansas, Louisiana, Alabama, Georgia, and the Carolinas. Migrates through Florida; winters in Cuba. Rare.

Vermivora peregrina (**Wilson**). Tennessee warbler. Crown and hind neck gray; back olive green; remiges and rectrices dusky edged with olive green; inner edge of rectrices narrowly white; narrow superciliary white; postocular spot dusky; below white; iris brown; bill and feet horn. WINTER: Tinged with yellow below. YOUNG: Above entirely olive green; superciliary, throat, and breast pale olive yellow. Wing 58.5–68, tail 39–46, culmen 9.5–10, tarsus 15.5–17.5 mm. Northern coniferous forest, south to Massachusetts, New York, northern Michigan, northern Wisconsin, and northern Minnesota. Winters from southern Mexico to northern South America.

Vermivora celata (**Say**). Orange-crowned warbler. Above olive green; concealed crown patch tawny; faint superciliary, eye ring, and underparts olive yellow; an indistinct dusky spot before and behind eye; iris brown; bill and feet horn. YOUNG: Lacks crown patch. Wing 56–63, tail 45–52.5, culmen 9.5–12, tarsus 17.5–18.5 mm. North America, south in mountains of West to California, Arizona, New Mexico, and western Texas. Winters from South Carolina, Nevada, and Washington to Florida, Gulf states, Mexico, and Guatemala.

Vermivora ruficapilla (**Wilson**). Nashville warbler. Crown, hind neck, and sides of face gray; center of crown chestnut; back, wings, and tail olive green; lores and eye ring whitish; below bright yellow; iris brown; bill and feet horn. Wing 54–62.5, tail 39–48, culmen 8.5–10, tarsus 16.5–18 mm. North America south to New Jersey, West Virginia, Pennsylvania, Ohio, Michigan, Wisconsin, Minnesota, South Dakota, Idaho, Nevada, and California. Winters in Mexico and Guatemala.

Vermivora virginiae (**Baird**). Virginia's warbler. Above gray; center of crown chestnut; eye ring white; rump and upper tail coverts olive green; chin yellowish gray; underparts dirty white; upper breast and crissum yellow; iris brown; bill and feet horn. Wing 57.5–61.5, tail 45.5–47.5, culmen 9–10, tarsus 16–18 mm. Southern Rockies and Great Basin, in eastern California, Nevada, Utah, southern Idaho, Colorado, New Mexico, and Arizona. Winters in western Mexico; on migration in western Texas.

Vermivora crissalis (**Salvin and Godman**). Colima warbler. Above olive brown; forehead gray; center of crown tawny; eye ring white; rump yellowish; below gray; belly white; crissum rich yellow; wing lining white; bill and feet horn. Wing 60–68.5, tail 50.5–61, tarsus 19 mm. Deciduous woods (6000–7500 ft) of Chisos Mountains, Texas, and adjacent part of Mexico. Winters in western Mexico.

Vermivora luciae (**Cooper**). Lucy's warbler. Above gray; center of crown and upper tail coverts chestnut; lores, eye ring, and underparts white; breast shaded with buff; iris brown; bill and feet horn. Wing 51.5–58, tail 37.5–44, culmen 7.5–9, tarsus 15.5–16.5 mm. Mesquite in southern Utah, Colorado, New Mexico, Arizona, California, and Sonora. Winters in western Mexico.

Genus *Parula* Bonaparte. Similar to *Vermivora* but with rictal bristles well developed. Americas. 2 species.

KEY TO SPECIES OF *PARULA*

1. A white spot on eyelids   *P. americana*
   Eyelids without white   *P. pitiayumi*

Parula americana (**Linnaeus**). Parula warbler (Fig. 5-80). Sides of face and upper parts blue-gray; a large yellowish olive patch on middle of back; 2 white wing bars; 3 outer rectrices with a large white subterminal spot on inner web; lores

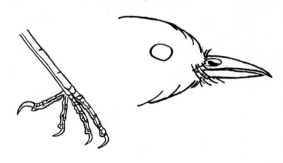

Fig. 5-80. Head and foot of parula warbler (*Parula americana*), enlarged.

blackish; a white spot on lower eyelid and a smaller one on upper lid; throat and breast yellow, interrupted by a tawny orange patch which is usually preceded by a black patch; belly and crissum white; anterior part of sides chestnut; iris brown; maxilla black; mandible yellow; feet brown. FEMALE: Without black; upper parts duller and tinged with olive green, especially in winter. Wing 52–63, tail 37.5–45, culmen 8–11.5, tarsus 14.5–18 mm. Eastern Canada and United States, west to Minnesota, Iowa, eastern Kansas, eastern Oklahoma, and eastern Texas. Winters in West Indies, southern Mexico, and Central America.

Parula pitiayumi (**Vieillot**). Olive-backed warbler. No white on eyelids; wing bands sometimes absent; only 2 outer pairs of rectrices with white patches; yellow of underparts more orange, extending back to cover most of abdomen; no chestnut patch on breast or sides; no black breast patch. Wing 45.5–59, tail 33–52, culmen 9–10.5, tarsus 14.5–19.5 mm. Southern Texas to Argentina.

Genus *Dendroica* Gray. Bill strong; culmen and gonys curved; rictal bristles well developed; 4th primary usually not longest; tail about $\frac{3}{4}$ wing, not deeply if at all emarginate; legs and feet weak. Americas. 27 species, 22 in the United States.

KEY TO SPECIES OF *DENDROICA*

1. Inner web of rectrices partly yellow   *D. petechia*
   Inner web of rectrices partly white   2
2. A white patch at base of primaries   *D. caerulescens*
   No white patch at base of primaries   3
3. Rump yellow, in contrast with back   4

Rump not yellow   7
4. A yellow crown patch   5
   Crown gray or black without yellow   6
5. Chin and throat yellow   *D. auduboni*
   Chin and throat white   *D. coronata*
6. Only the middle pair of rectrices without white
                                         D. magnolia
   2 middle pairs of rectrices without white
                                       D. tigrina
7. Crissum yellow   *D. palmarum*
   Crissum white or buffy, sometimes tinged with yellow   8
8. Wing bands indistinct or yellow   9
   2 white wing bands   10
9. Crown and back olive green; wing 51–59 mm
                                         D. discolor
   Crown bluish gray; wing 64–72 mm   *D. kirtlandii*
10. Back gray (sometimes streaked with black)   11
    Back bluish, olive, or brownish (sometimes streaked with black)   14
11. Center of throat yellow   12
    Center of throat black or whitish   13
12. A white patch on side of neck, extending to white superciliary   *D. dominica*
    No white on neck; superciliary yellow, short
                                       D. graciae
13. Sides of head yellow   *D. occidentalis*
    Sides of head black, or grayish and white
                                       D. nigrescens
14. Crown blue or greenish blue   *D. cerulea*
    Crown not blue or greenish blue   15
15. Eye ring, throat, breast, and belly plain white
                                       D. pensylvanica
    No white eye ring; throat, breast, and belly not plain white   16
16. Superciliary yellow   17
    No superciliary   20
17. An orange, yellow, or whitish spot on center of crown   *D. fusca*
    No crown spot   18
18. Auriculars entirely black or olive   *D. townsendi*
    Auriculars at least partly yellow   19
19. Outer 3 rectrices with white   *D. virens*
    Outer 4 pairs of rectrices with white
                                       D. chrysoparia
20. Back plain   *D. pinus*
    Back streaked   21
21. Crissum white   *D. striata*
    Crissum buff   *D. castanea*

Dendroica petechia (**Linnaeus**). Yellow warbler. Above greenish yellow, purer yellow on forehead (entire head chestnut in some extralimital races); back sometimes streaked with chestnut; wings and tail dusky, edged with yellow; inner webs of rectrices yellow, tipped with dusky; face and underparts yellow; breast, sides, and flanks streaked with chestnut; iris brown; bill black; feet brown. FEMALE: Duller; underparts often unstreaked. YOUNG: Still duller; above dull olive or olive gray; forehead, rump, and upper tail coverts tinged with yellow; wings and tail edged with olive; below pale olive yellow or buffy white; crissum yellow; bill horn. Wing 53–71, tail 39–56, culmen 9–13, tarsus 17–22 mm. Alaska and Canada south to northern Georgia, northern Alabama, and northern Mississippi, and to central Mexico; in mangroves south to Galapagos, Peru, and Colombia; all West Indies, north to Bahamas and Florida Keys. Winters from Mexico to South America.

Dendroica magnolia (**Wilson**). Magnolia warbler. Crown slate gray; eyelids white, the upper one continuous with white stripe bordering posterior crown; mask black, continuing on sides of neck to black back; lower back mixed with greenish olive; rump yellow; upper tail coverts black, edged with gray; all but median pair of rectrices with a large white band across middle; wing coverts with 2 prominent white bands, more or less confluent; throat, breast, and flanks yellow; breast and flanks with broad black stripes; crissum and middle of belly white; iris and feet brown; bill black. FEMALE: Back gray, more or less blotched with black; mask duller; wing bands not confluent; streaks below duller and fewer. YOUNG: Like female, but sides of face and band across breast ashy; maxilla horn; mandible yellowish. Wing 54.5–67, tail 46–53.5, culmen 8.5–10, tarsus 17–21 mm. Northern coniferous forest, south to Massachusetts, mountains of North Carolina, Pennsylvania, Ohio, northern Michigan, northern Wisconsin, and northern Minnesota. Winters from Mexico to Panama.

Dendroica tigrina (**Gmelin**). Cape May warbler. Crown black; center of crown sometimes with a cinnamon spot; back olive green blotched with black; rump yellow; wings and tail blackish edged with olive; 3–4 outer rectrices with a white subterminal patch on inner web; middle wing coverts white; superciliary yellow, becoming cinnamon posteriorly; streak through eye black; cheeks cinnamon; throat and breast yellow, becoming white on belly and crissum; throat spotted with cinnamon;

breast and flanks streaked with black; iris and feet brown; bill black. MALE IN WINTER: Cinnamon replaced by yellow; black of crown and back largely concealed by gray tips. FEMALE: Above olive, yellower on rump; crown often spotted with black; an indistinct whitish band on middle coverts; short superciliary yellow; below whitish, tinged with yellow; breast and flanks streaked with dusky. Wing 61–70, tail 43.5–49.5, culmen 9.5–10.5, tarsus 16.5–19 mm. Northern coniferous forest, south to Maine, New Hampshire, Vermont, and northern parts of New York, Michigan, Wisconsin, and Minnesota. Winters in West Indies.

Dendroica caerulescens (**Gmelin**). Black-throated blue warbler. Above dark blue; back often spotted with black; wings and tail black, edged with blue; patch at base of primaries white; 3 outer rectrices with white subterminal patch on inner web; sides of face, throat, and flanks black; breast, belly, and crissum white; iris and feet brown; bill black. FEMALE: Above olive; patch at base of primaries white; outer 2 rectrices with faint whitish subterminal spot on inner web; superciliary dirty white; below pale buffy olive. Wing 58.5–68, tail 43.5–54, culmen 8.5–10, tarsus 17.5–19.5 mm. Northern coniferous forest, south to mountains of Georgia, Kentucky, and northern parts of Ohio, Michigan, Wisconsin, and Minnesota. Winters in West Indies.

Dendroica coronata (**Linnaeus**). Myrtle warbler. Above bluish slate streaked with black; center of crown and rump yellow; wings and tail black edged with gray; 2 white wing bands; 3–4 outer rectrices with subterminal white patch on inner web; narrow superciliary and lower eyelid white; sides of face black; below white; breast and flanks heavily streaked with black; a yellow patch on sides of breast; iris and feet brown; bill black. FEMALE: Duller; upper parts and flanks tinged with brown. YOUNG: Above brown streaked with dusky; yellow crown patch concealed; rump yellow; superciliary indistinct; sides of face brownish; below brownish white; breast and flanks streaked with dusky. Wing 67–78, tail 50–60, culmen 8.5–11, tarsus 18–21 mm. Northern coniferous forest, south to Massachusetts, New York, Pennsylvania, Michigan, Wisconsin, and Minnesota. Winters from New Jersey, Ohio Valley, Arkansas, Oklahoma, Texas, Arizona, and Oregon south to Mexico and Central America.

Dendroica auduboni (**Townsend**). Audubon's warbler. Near *D. coronata*, but throat yellow; no superciliary; auriculars often blue gray; white wing bands of male confluent; 4–5 rectrices with white subterminal patch. Wing 73–84, tail 53–64.5, culmen 9.5–11, tarsus 18–22 mm. Western coniferous forests from coast east to Black Hills, western Nebraska, Rockies and western Texas, and south at high altitudes in Mexico and Guatemala. Winters from British Columbia, Arizona, and Texas to Guatemala.

Dendroica nigrescens (**Townsend**). Black-throated gray warbler. Head, neck, and breast black; postorbital and malar stripes white; spot above lores yellow; back slate gray, streaked with black; wings and tail edged with gray; 2 white wing bands; 4 outer rectrices tipped with white; posterior underparts white streaked with black; iris and feet brown; bill black. FEMALE: Crown gray streaked with black; middle of throat white. Wing 54–66.5, tail 47–55, culmen 8.5–9.5, tarsus 16.5–19 mm. British Columbia to northern Mexico, east to Colorado and New Mexico. Winters in southern Arizona and Mexico.

Dendroica townsendi (**Townsend**). Townsend's warbler. Head and neck black; superciliary and malar stripe yellow, confluent behind auriculars; lower eyelid yellow; back olive green with black wedges; wings and tail black, edged with gray; 2 large white wing bands; 3 lateral rectrices white, tipped with black externally; breast yellow, streaked with black laterally; belly white; crissum and flanks white streaked with black; soft parts brown. MALE IN WINTER: Black of crown more or less obscured by olive green, of throat by yellow; auriculars olive green. FEMALE: Like winter male but with little or no black above; streaks obscured below. Wing 63–69, tail 48–51, culmen 8–10, tarsus 18–19 mm. Northwest coniferous forest, south to Washington, northern Idaho, Montana, and northwestern Wyoming. Winters in California, Mexico, and Central America.

Dendroica virens (**Gmelin**). Black-throated green warbler. Above olive green; back sometimes with black wedges; wings and tail black, edged with gray; 2 white wing bands; 3 outer rectrices largely white subterminally; forehead and sides of face yellow; auriculars mixed yellow and green; throat and breast black; belly, crissum, and flanks white;

flanks heavily streaked with black; bill black; iris and feet brown. FEMALE: Black of throat and breast largely replaced or concealed by whitish. Wing 58–64, tail 45–49, culmen 9–10, tarsus 16–19 mm. Northern coniferous forest, south to mountains of Georgia, Alabama, and Kentucky, and to northern Michigan, Wisconsin, and Minnesota. Winters from Texas to Panama.

Dendroica chrysoparia Sclater and Salvin. Golden-cheeked warbler. Above black; rump and scapulars often mixed with olive green; often with a yellow spot in center of forehead; superciliary and sides of face yellow; line through eye black; wings and tail edged with gray; 2 white wing bands; 4 outer rectrices white subterminally; throat and breast black; belly, crissum and flanks pure white; flanks heavily streaked with black. FEMALE AND YOUNG: Above olive green, often streaked with black; chin and throat yellow; lower throat blotched with black; flank streaking obscure. Wing 58–65.5, tail 47.5–54.5, culmen 9–10.5, tarsus 17.5–19 mm. Cedar woods of central Texas, from Kerr and Travis Counties to Bexar and Medina Counties. Winters in Mexico and Central America.

Dendroica occidentalis (**Townsend**). Hermit warbler. Crown and sides of face yellow; above black, feathers edged with olive anteriorly, with gray posteriorly and on wings and tail; 2 white wing bands; 3 outer rectrices largely white subterminally; throat and upper breast black; rest of underparts white; soft parts brown. FEMALE AND YOUNG: Crown suffused with olive green and spotted with black; back grayish olive with streaks obscured or absent, throat patch mostly obscured by buffy tips. Wing 62–69, tail 46.5–52, culmen 9–11, tarsus 16.5–21 mm. Pacific conifer forest, in Washington, Oregon, and California. Winters in Mexico and Central America.

Dendroica cerulea (**Wilson**). Cerulean warbler. Above and sides of face light blue, brighter on crown; occiput and back streaked with black; wings and tail black, edged with bluish gray; 2 white wing bands; inner webs of rectrices with white subterminal patch; faint superciliary white; postocular streak dusky; below white; interrupted breast band blue and dusky; flanks streaked with dusky; bill black; iris and feet brown. FEMALE: Above and auriculars glaucous olive; wing coverts and upper tail coverts bluish olive; 2 white wing

bands; tail as in male; postocular stripe dusky; superciliary and underparts yellowish white; flanks streaked with olive. Wing 58–67.5, tail 41–47.5, culmen 9.5–10.5, tarsus 15.5–17 mm. Deciduous forest, from southern parts of New York, Ontario, Michigan, Wisconsin, and Minnesota, south to northern parts of Georgia, Alabama, Louisiana, and Texas, west to Nebraska, Kansas, and Oklahoma. Winters in South America.

Dendroica fusca (**Müller**). Blackburnian warbler. Above black; center of crown yellow; a yellowish white stripe along scapulars; middle and greater coverts white; remiges black, edged with gray; middle rectrices black; other rectrices white tipped with black; lores and auriculars black; superciliary, sides of neck, and throat orange; breast and belly yellowish white; sides and flanks streaked with black; crissum white. FEMALE: Duller; back grayish olive streaked with dusky; orange replaced by yellow; 2 white wing bars. YOUNG: Still duller, with yellow areas largely replaced by buff or white; scapular stripe absent; streaks below obscure. Wing 63–69.5, tail, 46–49, culmen 9.5–10.5, tarsus 17–18 mm. Northern coniferous forest, south to New Jersey, mountains of Georgia and Tennessee, Pennsylvania, Ohio, Michigan, Wisconsin, and Minnesota. Winters in South America.

Dendroica dominica (**Linnaeus**). Yellow-throated warbler. Above slate gray; forehead black; wings and tail black, edged with gray; 2 white wing bands; inner web of 3 outer rectrices tipped with white; superciliary white, usually yellow before eye; lower eyelid white; patch on side of neck white; throat and upper breast yellow, bordered laterally with black; auriculars black; underparts white; sides and flanks streaked with black. Wing 63–69.5, tail 46–53.5, culmen 11–15, tarsus 16–18 mm. Southeastern United States to Delaware, Maryland, West Virginia, Ohio, Indiana, Illinois, Kansas, Oklahoma, and eastern Texas. Winters from Louisiana and South Carolina to Florida, West Indies, Mexico, and Central America.

Dendroica graciae **Baird**. Grace's warbler. Near *D. dominica*, but superciliary yellow and not extending behind eye; no white patch on neck; auriculars and sides of face gray; yellow of breast only bordered with black; crown and back often streaked with black. Wing 53–69, tail 44–54.5, culmen 11.5–13.5, tarsus 17–18 mm. Pines of southern Utah, southern Colorado, New Mexico, Arizona, and western Texas, south to Mexico and Central America.

Dendroica pensylvanica (**Linnaeus**). Chestnut-sided warbler. Crown yellowish olive, whitish on forehead; a yellowish spot on occiput; above streaked black and olive or whitish; wings and tail black, edged with olive; 2 broad yellowish wing bands; inner webs of 3 outer rectrices broadly white-tipped; stripe through eye and malar region black; auriculars and underparts white; a chestnut stripe on sides. YOUNG: Above olive green; back and upper tail coverts obscurely streaked with dusky; eye ring, sides of head, and underparts dirty white, becoming purer posteriorly; chestnut flank stripe sometimes indicated. Wing 58–62, tail 45–52.5, culmen 9–10, tarsus 17–18 mm. Second growth, south to Maryland, mountains of Georgia, Kentucky, Illinois, Iowa, and North Dakota. Winters in Central America.

Dendroica castanea (**Wilson**). Bay-breasted warbler. Crown chestnut, bordered all around by black; sides of face black; back buffy gray, streaked with black; wings and tail black, edged with gray; 2 white wing bands; inner webs of 2–3 outer rectrices white-tipped; chin black or whitish; throat, breast, sides, and flanks chestnut; belly and crissum buff; iris, bill, and feet brown. FEMALE: Chestnut restricted or almost absent; crown olive streaked with black, sometimes tinged with chestnut. YOUNG: Above olive green; back often obscurely streaked with black; wing bands tinged with yellowish; sides of head yellowish olive; streak through eye dusky; eyelids yellowish white; below pale buff; crissum sometimes tinged with chestnut; feet dark brown. Wing 68–76, tail 48–56.5, culmen 9.5–11, tarsus 17.5–20 mm. Northern coniferous forest, south to Maine, New Hampshire, northern New York, and northern Michigan. Winters in Panama, Colombia, and Venezuela.

Dendroica striata (**Forster**). Black-poll warbler. Crown black; back olive gray streaked with black; wings and tail black, edged with gray; 2 white wing bands; inner webs of 3 outer rectrices white-tipped; malar region and auriculars white; below white; streaked with black on middle of throat, sides of neck, sides, and flanks; feet pale brown. FEMALE: Crown and upper parts greenish olive with dusky streaks; white below tinged with yellowish; streaks fewer and duller; crissum pure white. ADULT IN WINTER AND YOUNG: Above greenish olive, with a few streaks on middle of back; below pale whitish, tinged with yellow, scarcely streaked; crissum pure white; feet pale brown. In this plumage resembles young *D. castanea*, but the white crissum and pale legs are diagnostic. Wing 69–77.5, tail 45–54, culmen 9–11, tarsus 18–20.5 mm. Spruce forest of Alaska and Canada, south to Massachusetts and northern New York. Winters in South America.

Dendroica pinus (**Wilson**). Pine warbler. Above olive green; wings and tail dusky, edged with grayish; 2 white wing bands; 2 (sometimes 3) outer rectrices with a white patch at tip of inner web; faint superciliary and eyelids yellow; sides of face olive green; throat, breast, and sides greenish yellow; breast and sides streaked with olive; belly and crissum whitish; iris and feet brown; bill black. FEMALE: Above grayish olive; wings and tail as in male; lower eyelid white; throat and breast pale yellow; base of mandible pale. YOUNG: Like female, but above brownish, with slight olive tinge; little or no yellow below, where largely pale buff or dirty white. Wing 62–76, tail 50.5–58, culmen 10–13.5, tarsus 17–20 mm. Pines of southern Canada and eastern United States, west to Minnesota, Kansas, Oklahoma, and Texas; resident races in Bahamas and Hispaniola. Winters in southeastern United States, north to Virginia and Arkansas.

Dendroica kirtlandii (**Baird**). Kirtland's warbler. Above bluish gray streaked with black; wings and tail dusky, edged with gray; 2 dull-white wing bands; 2 outer rectrices with a white terminal patch on inner web; forehead and sides of face black; eyelids white; below lemon yellow; a band of black streaks and speckles across breast; sides and flanks gray, streaked with dusky; crissum white; iris, bill, and feet brown. FEMALE AND WINTER: Above brownish, streaked with black. Wing 64–72, tail 53–65, culmen 11–13, tarsus 21–23 mm. Jack pines of northern part of lower Michigan, from Montmorency and Alpena Counties, south to Wexford, Clare, Roscommon, Ogemaw, and Iosco Counties. Winters in Bahamas.

Dendroica discolor (**Vieillot**). Prairie warbler. Above yellowish olive green; scapulars usually blotched with chestnut or black; wings and tail dusky, edged with olive green; 1–2 indistinct yellowish wing bands; 3 outer rectrices white terminally on inner web; superciliary, lower eyelid, and underparts yellow; stripe through eye, patch below eye, and patch on side of neck black; sides and flanks streaked with black; iris and feet brown; bill blackish. FEMALE AND YOUNG: Scapular patch obsolete; black on sides of face replaced by gray; wing bands and streaks below obsolete; bill brown. Wing 51–59, tail 41–50, culmen 9–10, tarsus 17.5–19.5 mm. Southeastern United States, north to New Hampshire, New York, Ontario, Michigan, Illinois, Iowa, Kansas, Oklahoma, and Louisiana. Winters in Florida, Bahamas, and West Indies.

Dendroica palmarum (**Gmelin**). Palm warbler. Crown chestnut, streaked with dusky; back brown, with a few dusky streaks; rump and upper tail coverts yellowish olive; wings and tail dusky, edged with brownish olive; 2 indistinct pale-brown wing bands; 2 outer rectrices tipped with white on inner web; superciliary yellow; lores and postocular spot dusky; auriculars brown; throat and crissum yellow; breast and belly whitish or yellow; sides of throat, band across breast, and flanks streaked with brown; iris and feet brown; bill black. WINTER AND YOUNG: Crown brown, sometimes mixed with chestnut; back brown, rump and upper tail coverts yellowish olive; superciliary and underparts dirty white or buff, often tinged with yellow; sides of throat, breast band, and flanks streaked with brown; crissum yellow; bill brown. Wing 62–70.5, tail 51.5–57, culmen 9.5–10, tarsus 19.5–20.5 mm. Muskegs of Canada, south to Maine, northern Michigan, northern Wisconsin, and northern Minnesota. Winters from Louisiana, Alabama, and South Carolina to Florida and West Indies.

Genus *Oporornis* Baird. Near *Dendroica*, but rictal bristles less developed; tail rounded, without white; rectrices narrower terminally. America. 4 species.

KEY TO SPECIES OF *OPORORNIS*

1. Superciliary yellow		***O. formosus***
No superciliary		2
2. A complete white eye ring		***O. agilis***
Eye ring absent or incomplete		3
3. A white spot on eyelids		***O. tolmiei***
Eyelids entirely dark		***O. philadelphia***

Oporornis formosus (**Wilson**). Kentucky warbler. Above greenish olive; crown black, tipped with gray; mask from bill, below eye, to auriculars and sides of neck black; superciliary, lower eyelid, edge of wing, and underparts yellow; edge of outer primary whitish; iris and bill brown; feet flesh. FEMALE: Only forehead dusky, tipped with gray; mask dusky, not reaching neck; auriculars olive. YOUNG: Crown, forehead, and sides of face and neck olive; lores dusky; short superciliary, lower eyelid, and underparts yellow. Wing 62.5–74.5, tail 45.5–52.5, culmen 11.5–13, tarsus 20.5–23.5 mm. Broad-leafed forest of southeast, to southern New York, Pennsylvania, Ohio, Indiana, Illinois, southern Wisconsin, Iowa, and eastern parts of Nebraska, Kansas, Oklahoma, and Texas. Winters from Mexico to northern South America.

Oporornis agilis (**Wilson**). Connecticut warbler. Head, neck, and breast slate gray; eye ring white; back, wings, and tail olive green; outer primary edged with white; posterior underparts pale yellow; iris and bill brown; feet flesh. FEMALE AND YOUNG: Head grayish olive or brownish olive; eye ring whitish; throat pale buffy. Wing 67.5–75.5, tail 46.5–53, culmen 11.5–12.5, tarsus 19–23 mm. Canadian zone, south to northern Michigan, Wisconsin, and Minnesota. Winters in South America. In spring crosses West Indies to Florida, thence across mountains to Mississippi Valley; in fall migrates through Great Lakes states to New England and south along Atlantic seaboard.

Oporornis philadelphia (**Wilson**). Mourning warbler. Head and neck slate gray; lores dusky; throat and breast black, more or less concealed by gray tips anteriorly; above olive green; posterior underparts and edge of wing yellow; iris and bill brown; feet flesh. FEMALE AND YOUNG: Crown brownish gray; throat and breast pale brownish gray. Wing 54.5–65, tail 42.5–52.5, culmen 10.5–12, tarsus 20.5–22 mm. Canadian zone, south to New England, New York, mountains of Maryland and West Virginia, Ohio, northern Illinois, Wisconsin, Minnesota, and North Dakota. Winters in Central and South America.

Oporornis tolmiei (**Townsend**). MacGillivray's warbler. Like *O. philadelphia*, but upper and lower eyelids white; tail longer; male with black from lores to auriculars. Wing 54.5–65, tail 48.5–63 culmen 10.5–12, tarsus 20–23 mm. Canadian zone of West, south to Black Hills, New Mexico, Arizona, and California. Winters from Mexico to Colombia.

Genus *Seiurus* Swainson. Bill stout; tip of culmen deflected, gonys ascending; 2–3 rictal bristles well developed; outer 3 primaries longest; tail short, $\frac{2}{3}$ wing, even or slightly emarginate; legs and feet stout; inner toe nearly free. Americas. 3 species.

KEY TO SPECIES OF *SEIURUS*

1. Midline of crown orange, bordered by black
   S. aurocapillus
   Crown plain olive or brown   **2**
2. Crissum buff; outer primary shorter than 3d
   S. motacilla
   Crissum pale yellowish; outer primary longer than 3d
   S. noveboracensis

Seiurus aurocapillus (**Linnaeus**). Ovenbird. Crown orange, black laterally and in front; superciliary and upper parts olive green; eye ring white; auriculars buffy olive; below white; malar stripe black; breast, sides, and flanks streaked with black; wing lining pale yellow; iris and maxilla brown; mandible and feet flesh. Wing 70–79, tail 50–58.5, culmen 11.5–12.5, tarsus 20.5–23 mm. Canada south to northern Georgia and Alabama, Tennessee, Arkansas, Colorado, Black Hills, and Montana. Winters in southern Georgia, Florida, West Indies, Mexico, Central America, and Colombia.

Seiurus motacilla (**Vieillot**). Louisiana waterthrush. Upper parts and sides of face brown; superciliary and lower eyelid white; below white; breast, sides, and flanks streaked with brown; crissum and vicinity rich buff; iris brown; bill brown, paler below; feet flesh. Wing 75–84, tail 49.5–55.5, culmen 12.5–14, tarsus 21.5–23.5 mm. Damp woods, from New Hampshire, Vermont, New York, southern Ontario, and Michigan, Wisconsin, and southern Minnesota, south to Georgia, Alabama, Mississippi, Louisiana, and eastern Texas, west to Oklahoma, Kansas, and Nebraska. Winters in West Indies and from Mexico to Colombia.

Seiurus noveboracensis (**Gmelin**). Northern waterthrush. Upper parts and sides of face sooty olive; superciliary and lower eyelid buff; below pale yellow; chin with a few dusky speckles; malar stripe sooty; lower throat, breast, sides, and flanks streaked with sooty; iris brown; bill brown, paler below; feet pale brown. Wing 68.5–81, tail 45–58, culmen 11–16, tarsus 20–22.5 mm. Alaska and Canada, south to North Carolina, Ohio, northern Michigan, Wisconsin, northern Minnesota, North Dakota, Montana, and northern Idaho. Winters from West Indies and Mexico to South America.

Genus *Geothlypis* Cabanis. Bill rather stout; culmen curved, gonys straight; maxillary notch and rictal bristles obsolete; wing rounded, outer primary shorter than 5th; wing tip shorter than culmen; tail strongly rounded, about as long as wing; feet strong, reaching nearly to end of tail; hind toe and claw extending to claw of middle toe. Americas. 8 species, 1 in the United States.

Geothlypis trichas (**Linnaeus**). Yellowthroat (Fig. 5-81). Above olive green or olive brown; a black mask across forehead and extending from superciliary over side of face to side of neck; posterior border of mask ashy or white; throat and breast yellow; belly yellow or whitish; crissum yellow; sides and flanks brown; bend of wing yellowish; iris brown; bill black (pale brown below in winter); feet pale brown. FEMALE AND YOUNG: Black mask and gray border absent, replaced by brownish; crown tinged with cinnamon; throat and breast pale yellowish or buffy; crissum and bend of wing pale yellow; belly buffy white; flanks brownish; mandible pale brown. Wing 46–61.5, tail 42–61.5, culmen 9–13, tarsus 18.5–22 mm. North America south to central Mexico. Winters

Fig. 5-81. Head and foot of yellowthroat (*Geothlypis trichas*), enlarged.

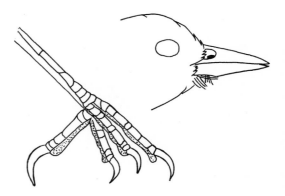

from southern United States to Greater Antilles and Panama.

Genus *Chamaethlypis* Ridgway. Near *Geothlypis*, but bill stouter; culmen and commissure strongly curved; maxillary notch distinct; 2–3 rictal bristles well developed; wing more strongly rounded, with outer primary shortest; tail graduated, longer than wing. Central America. 1 species.

Chamaethlypis poliocephala (**Baird**). Ground-chat (Fig. 5-82). Above greenish olive; crown gray; forehead and mask to eyes black; often with a spot of yellow or white on eyelids; bend of wing and underparts yellow, more buffy on belly; flanks buffy olive; iris brown; bill dusky, flesh below; feet brownish flesh. FEMALE: Olive of back encroaching and sometimes covering crown; no mask but forehead and lores dusky. Wing 51–63, tail 54–68.5, culmen 10–14, tarsus 19–25 mm. Bunch grass from vicinity of Brownsville, Texas, to Panama.

Genus *Icteria* Vieillot. Near *Chamaethlypis* but size large; no maxillary notch or rictal bristles; tail strongly rounded rather than graduated; tarsus less than $\frac{1}{3}$ wing, instead of nearly $\frac{1}{2}$ wing. North America. 1 species.

Icteria virens (**Linnaeus**). Yellow-breasted chat. Above olive green; eyelids and stripe before eye white; malar stripe white, squamate with dusky; lores and area between lower eyelid and malar stripe blackish; auriculars grayish olive; bend of wing, throat, and breast yellow; belly white; crissum buff; flanks buffy gray; iris brown; bill dusky (flesh below in winter and female); feet dusky. Wing 72–84, tail 69–86, culmen 13–15, tarsus 25–28 mm. Most of Mexico and the United States, north to New Hampshire, Vermont, New York, southern parts of Ontario, Michigan, and Minnesota, North Dakota, and western Canada; absent from Gulf Coast and peninsular Florida. Winters from Mexico to Panama.

Genus *Cardellina* DuBus. Bill shorter than distance from eye to nostril, as wide as deep; maxilla notched; rictal bristles extending half length of bill; wing rounded; 3d primary longest; wing tip shorter than tarsus; tail more than $\frac{3}{4}$ wing, slightly rounded; tarsus about $\frac{1}{4}$ wing, nearly booted; middle toe with claw much shorter than tarsus; lateral toes equal and longer than hallux. Guatemala, Mexico, and southwestern United States. 1 species.

Cardellina rubrifrons (**Giraud**). Red-faced warbler (Fig. 5-83). Forehead, lores, eyelids, malar region, throat, and postauricular region red; crown and auriculars black; posterior border of crown whitish; back, wings, and tail gray; a whitish wing bar on middle coverts; rump white; breast, belly, and crissum whitish, tinged with pink anteriorly; iris, bill, and feet brown. Wing 63–71, tail 55.5–61, culmen 7.5–9, tarsus 17–18.5 mm. Pines at high altitudes from southern Arizona and New Mexico to Guatemala.

Fig. 5-82. Head and foot of ground-chat (*Chamaethlypis poliocephala*), enlarged.

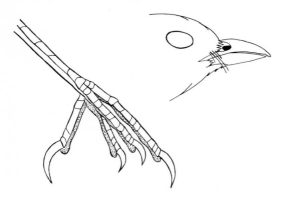

Fig. 5-83. Bill of red-faced warbler (*Cardellina rubrifrons*), enlarged.

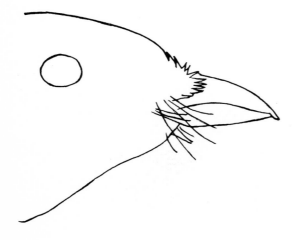

Genus *Wilsonia* **Bonaparte.** Near *Cardellina*, but bill wider than deep; nostrils exposed; tarsus more than ¼ wing. America. 3 species.

KEY TO SPECIES OF *WILSONIA*

1. Crissum white   *W. canadensis*
   Crissum yellow   **2**
2. Tail without white   *W. pusilla*
   Tail with large white patch   *W. citrina*

Wilsonia citrina (**Boddaert**). Hooded warbler. Mask yellow, covering forehead, lores, malar region, and auriculars; rest of head and neck black; back, wings, and tail olive green; inner web of 3 outer rectrices mostly white terminally; posterior underparts yellow; bend of wing and axillars yellowish; iris brown; bill black (pale brown in winter); feet brown (flesh in winter). YOUNG: Entire upper parts olive green, more yellowish on forehead; throat yellow; bill and feet flesh. FEMALE: Variously intermediate. Wing 60–69, tail 53–60, culmen 10–11, tarsus 18–20 mm. Eastern deciduous forest, from northern Florida, Gulf Coast, and eastern Texas, north to Iowa. Winters in Mexico and Central America.

Wilsonia pusilla (**Wilson**). Wilson's warbler. Above olive green; crown glossy black; forehead, lores, superciliary, and underparts yellow; iris and feet brown; bill brown, paler below. FEMALE: Crown largely (sometimes entirely) olive green. Wing 50–60, tail 45–52, culmen 7–9, tarsus 17–20 mm. Northern coniferous forest, south to New Hampshire, Vermont, northern Michigan and Minnesota, and in mountains to western Texas, New Mexico, and California. Winters in Mexico and Central America.

Wilsonia canadensis (**Linnaeus**). Canada warbler. Above gray; crown feathers with black centers; lores yellow, bordered above by a black line; a dusky spot before eye; below yellow; a necklace of black spots from below eye, down sides of neck, and across breast; crissum and axillars white; iris brown; bill horn, paler below; feet flesh. YOUNG: Above olive gray without black; lores and eye ring pale yellow; below yellow; breast streaked with olive; crissum white. FEMALE: Intermediate. Wing 60.5–67, tail 51–57.5, culmen 10–11.5, tarsus 18–19.5 mm. Damp forests from Canada south to New

Jersey, mountains of Georgia and Tennessee, Ohio, Michigan, Wisconsin, and Minnesota. Winters in South America, migrating through Mexico and Central America.

Genus *Setophaga* **Swainson.** Bill wide and flat; nasal plumes not covering nostrils, although numerous bristles extend over nostrils and in some cases to tip of bill; outer primary shorter than 3d to shorter than 7th; tail fan-shaped; otherwise like *Cardellina* and *Wilsonia*. Americas. 11 species, 2 in the United States.

KEY TO SPECIES OF *SETOPHAGA*

1. Outer rectrix white   *S. picta*
   Outer rectrix black on terminal third, orange or yellow basally   *S. ruticilla*

Setophaga ruticilla (**Linnaeus**). American redstart (Fig. 5-84). Upper parts, throat, and breast glossy black; patch at base of remiges orange; 4 outer rectrices orange on basal ⅔; sides, flanks, and wing lining orange; middle of belly whitish; crissum white, tipped with dusky; iris, bill, and feet dark brown. FEMALE: Crown gray; back grayish olive; wings and tail dusky, with yellow patches replacing the orange; below grayish or buffy white; axillars and sides of breast pale yellow. YOUNG MALE: Intermediate. Wing 58–67, tail 49–58, culmen 7–9, tarsus 15–19 mm. Canada, south to Georgia, Alabama, Mississippi, Louisiana, Oklahoma, Colorado, Utah, and Oregon. Winters in West Indies, Mexico, Central and South America.

Fig. 5-84. Bill of American redstart (*Setophaga ruticilla*), enlarged.

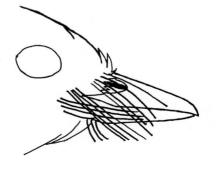

Setophaga picta **Swainson.** Painted redstart. Upper parts, throat, breast, sides, and flanks glossy black; middle and greater coverts forming a large white patch; outer 2 rectrices largely white; 3d rectrix with white subterminal spot; middle of breast and upper belly red; wing lining and lower belly white; crissum black, tipped with white; iris brown; bill and feet black. YOUNG: Above sooty black; wings and tail as in adult, but wing patch tinged with buff; throat and breast sooty gray, streaked with darker on breast; middle of belly whitish. Wing 66–75, tail 60–68, culmen 8–9, tarsus 16–17.5 mm. Pine-oak woods, from Arizona, New Mexico, and western Texas to Honduras.

Family Icteridae. Troupials. Bill about equal to head, stout; gonys and depth of bill at base less than distance from nostril to tip of maxilla; culmen more or less swollen, its tip sometimes deflected; nostrils exposed, operculate; tomium unnotched; rictus with angle and bristles obsolete. Primaries 9; outer webs of several sinuate. Tail more than half wing, rounded or graduated; rectrices 12. Tarsus slightly longer than middle toe with claw; acrotarsium scutellate; anterior toes slightly coherent at base. Maxillopalatines weak and curved. Eggs bluish or white, usually scrawled or spotted. Known from Pleistocene. America. 35 genera, 88 species. In the United States 10 genera, 19 species.

KEY TO GENERA OF ICTERIDAE

1.	Outstretched feet reaching beyond tail	***Sturnella***
	Feet not reaching end of tail	2
2.	Rectrices sharp-tipped	***Dolichonyx***
	Rectrices normal	3
3.	A ruff of elongate feathers around base of neck	***Tangavius***
	No neck ruff	4
4.	Exposed culmen shorter than middle toe without claw	5
	Exposed culmen longer than middle toe without claw	8
5.	Tail short, only ⅔ wing	***Molothrus***
	Tail ¾ wing or longer	6
6.	Culmen straight, slightly depressed in middle	***Xanthocephalus***
	Culmen curved	7
7.	Exposed culmen less than ⅔ tarsus	***Euphagus***
	Exposed culmen more than ⅔ tarsus	***Icterus***
8.	Culmen straight, slightly depressed in middle	***Agelaius***
	Culmen curved	9
9.	Tail graduated for ¼ its length or less	***Quiscalus***
	Tail graduated for ¼ its length	***Cassidix***

Genus *Tangavius* Lesson. Bill (Fig. 5-85) stout, conical, longer than middle toe without claw, ¾ tarsus, depth more than half its length; culmen curved, flattened between nostrils; side of maxilla with a groove parallel with culmen; gonys straight; commissure sinuate, deflected below nostril; nasal feathers reaching nostril; nostril round; neck feathers lengthened, forming a ruff; outer primary shorter than 3d, 2d longest, 2d–5th sinuate on outer web, 1st–3d notched on inner web; tail ⅔ wing, rounded. Nidification parasitic. America. 1 species.

Tangavius aeneus **(Wagler).** Bronzed cowbird. Black; glossed with blue or purple on wings, tail, and crissum; elsewhere glossed with bronze; iris red; bill and feet black. FEMALE AND YOUNG: Head, neck, and upper back sooty brown; wings, tail, and posterior part of body paler brown; iris orange. Wing 97.5–124, tail 64–98, culmen 20–24, tarsus 26–32 mm. Southern Arizona and Texas south to Colombia.

Genus *Molothrus* Swainson. Near *Tangavius*, but bill shorter than middle toe without claw (Fig. 5-86), about ⅔ tarsus; neck without ruff; primaries usually unnotched on inner web. Nidification parasitic. America. 4 species, 1 in the United States.

Molothrus ater **(Boddaert).** Brown-headed cowbird. Head and neck chocolate; rest of plumage black, glossed with purple on breast and upper back, with green elsewhere; iris brown; bill and feet black. FEMALE: Grayish brown above, paler below; chin brownish white. YOUNG: Like female, but squamate with buffy white above, streaked with sooty below. Wing 85.5–116, tail 58–80, culmen 14–19.5, tarsus 22.5–28 mm. Southern Canada, south to Mexico, Texas, Louisiana, Tennessee, and Georgia. Winters from Potomac and Ohio Valleys south to Florida, Gulf Coast, and Mexico.

Genus *Xanthocephalus* Bonaparte. Bill deep, shorter than middle toe without claw, about ⅔ tarsus; culmen slightly depressed in middle, flat between nostrils, narrow at tip; gonys and rictus straight; nostrils horizontally oval, operculum swollen; nasal plumes reaching nostril; outer 3 primaries longest; 2d–4th sinuate on outer web;

tail nearly $\frac{3}{4}$ wing, nearly even. Western North America. 1 species.

Xanthocephalus xanthocephalus (Bonaparte). Yellow-headed blackbird. Black; head, neck, and breast orange yellow; lores to area around eye and base of chin black; a white patch on primary coverts and greater coverts; anus yellow; iris brown; bill and feet black. FEMALE: Brown; superciliary, malar region, throat, breast, and anus pale yellow; lower breast indistinctly streaked with whitish; no white on wing. Wing 110–145.5, tail 79–108.5, culmen 19.5–25, tarsus 30–37 mm. Fresh-water marshes of West, east to Wisconsin, Illinois, Indiana, Kansas, and Arizona, south to Mexico. Winters from Louisiana, Texas, Arizona, and California to Mexico.

Genus *Agelaius* Vieillot. Near *Xanthocephalus*, but culmen more than $\frac{2}{3}$ tarsus, slightly longer than middle toe without claw; outer primary shorter than 2d and 3d, 2d–5th sinuate on outer web. America. 8 species, 2 in the United States.

KEY TO SPECIES OF *AGELAIUS*

1. Strongly glossed with bluish green; belly of ♀ unstreaked **A. tricolor**
   Little or no green gloss; ♀ entirely streaked below **A. phoeniceus**

Agelaius phoeniceus (Linnaeus). Red-winged blackbird. Black; lesser coverts bright red; middle coverts buff or whitish; iris brown; bill and feet black. YOUNG MALE: Feathers of upper parts edged with buff and brown, of underparts with buff; lesser coverts orange red, streaked with dusky. FEMALE: Brown; crown streaked with black and with a buff line down center; upper back streaked with buff and black; rump and upper tail coverts squamate with buff; wings and tail edged with paler; lesser coverts often mixed with russet or orange; middle and greater coverts tipped with buff or whitish; superciliary buff, sometimes tinged with salmon; below heavily streaked with whitish or buff; throat often tinged with salmon and often unstreaked. Wing 87.5–144, tail 63–105.5, culmen 17–26.5, tarsus 24.5–34 mm. Fresh-water marshes of Canada, the United States, Bahamas, Cuba, and Mexico south to Costa Rica. Withdraws from North in winter.

Fig. 5-85. Bill of bronzed cowbird (*Tangavius aeneus*).

Fig. 5-86. Bill of brown-headed cowbird (*Molothrus ater*).

Agelaius tricolor (Audubon). Tricolored redwing. Both sexes glossed with blue or green. Male with lesser coverts brownish red; middle coverts white. FEMALE: Crown with numerous buff streaks, without median line; posterior underparts unstreaked. Wing 104.5–123.5, tail 74–95, culmen 20–24, tarsus 25.5–30.5 mm. Oregon, California, and northern Baja California, west of Cascades and Sierra Nevada.

Genus *Euphagus* Cassin. Bill slender, depth less than half length; culmen gently curved, shorter than middle toe without claw, less than $\frac{2}{3}$ tarsus; gonys straight; outer primary equal to 4th, 2d longest, 2d–4th sinuate on outer web; tail rounded, about $\frac{3}{4}$ wing. North America. 2 species.

KEY TO SPECIES OF *EUPHAGUS*

1. Wing 114–117 mm (♀ 103–112); wing tip under 30 mm **E. carolinus**
   Wing 120–134 mm (♀ 116–120); wing tip over 30 mm **E. cyanocephalus**

Euphagus carolinus (Müller). Rusty blackbird. Black, faintly glossed with blue; iris pale yellow; bill and feet black; in winter upper parts overlaid with rusty brown, underparts with cinnamon buff. FEMALE: Slate gray, somewhat glossy above; iris pale yellow; in winter largely rusty brown above, pale brown below; superciliary buffy. Wing 103–117, tail 74–92.5, culmen 17.5–21.5, tarsus 29.5–32 mm. Alaska and Canada, south to New Hampshire, Vermont, and New York. Winters from Delaware and Ohio Rivers south to Gulf Coast.

Euphagus cyanocephalus (Wagler). Brewer's blackbird. Black; head and neck glossed with purple, rest with green; iris pale yellow; bill and feet black. FEMALE: Brownish gray, glossed as in male but more faintly; iris brown; in winter with pale-brown superciliary. Wing 116–134, tail 87–107, culmen 19–23.5, tarsus 28.5–33 mm. Western Canada and the United States, east to Wisconsin, Illinois, Kansas, Oklahoma, and western Texas. Winters mainly in Mexico, more rarely east to Florida.

Genus *Quiscalus* Vieillot. Bill strong, depth about half its length; culmen curved, flattened between nostrils, longer than middle toe without claw, about $\frac{2}{3}$ tarsus; maxillary tomium concave in middle, deflected below nostril; gonys nearly straight; outer primary shorter than 4th 2d–4th longest, 2d–5th emarginate on outer web; tail much more than $\frac{3}{4}$ wing, graduated for $\frac{1}{4}$ its length, plicate. Usually nests in colonies in trees. Eastern North America. 1 species.

Quiscalus quiscula (**Linnaeus**). Common

Fig. 5-87. Head and foot of boat-tailed grackle (*Cassidix mexicanus*).

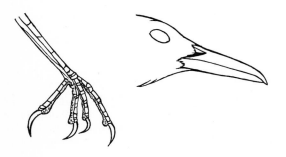

grackle. Blackish; head and neck glossed with purple or greenish blue; wings and tail glossed with purple; body glossed with plain bronze or squamate with bronze green and purplish; iris pale yellow; bill and feet black. FEMALE: With little gloss on body, which is mainly sooty. YOUNG: Plain sooty; wings and tail with slight gloss. Wing 116.5–153, tail 101–139.5, culmen 28–36, tarsus 31.5–38.5 mm. Eastern North America, west to base of Rockies, in Montana, Wyoming, Colorado, and New Mexico. Winters from Delaware and Ohio Valleys southward.

Genus *Cassidix* Lesson. Bill longer than head (Fig. 5-87), depth only $\frac{1}{3}$ its length; culmen curved, rounded between nostrils; maxillary tomium straight to behind nostril; outer primary longer than 4th, 2d–3d longest; tail often longer than wing, graduated for $\frac{1}{3}$ its length, plicate as in *Quiscalus*. American mainland. 3 species, 1 in the United States.

Cassidix mexicanus (**Gmelin**). Boat-tailed grackle. Black; head and neck glossed with purple, rest with blue; iris brown or yellow; bill and feet black. FEMALE: Postocular stripe, upper parts, and crissum dark brown; crown usually more cinnamon brown; back, wings, and tail with faint-blue gloss; superciliary and underparts buffy brown, palest anteriorly. YOUNG: Like female; but without gloss; underparts streaked with dark brown. Wing 131.5–204, tail 118–235, culmen 30.5–48.5, tarsus 37–54.5 mm. Northern South America, Central America, and Mexico to southern Arizona, New Mexico, and Texas, thence along Gulf and Atlantic Coastal Plains to Virginia.

Genus *Icterus* Brisson. Bill about as long as head, about $\frac{2}{3}$ tarsus, shorter than middle toe without claw, its depth less than half length; culmen and gonys usually curved (rarely straight), rounded between nostrils; side of maxilla usually smooth (with prenasal groove in *I. gularis*); nostril oval or crescentic, the feathers usually reaching its posterior end; commissure smooth; outer primary shorter than 2d to shorter than 7th, 2d or 3d longest, 2d–4th, 5th, or 6th sinuate on outer web; tail $\frac{3}{4}$ wing to longer than wing, strongly rounded or graduated; legs short, middle toe with claw almost as long as tarsus. Arboreal, with hanging woven nest. North and South America; West Indies. 30 species, 7 in the United States.

KEY TO SPECIES OF *ICTERUS*

1. Sides of breast spotted with black   *I. pectoralis*
   Sides of breast unspotted   2
2. Tail equal to or longer than wing   3
   Tail shorter than wing   4
3. Depth of bill at base 10–11.5 mm   *I. graduacauda*
   Depth of bill at base 7–8.5 mm   *I. cucullatus*
4. Wing 68.5–82.5 mm   *I. spurius*
   Wing 85 mm or more   5
5. Exposed culmen longer than middle toe with claw   *I. parisorum*
   Exposed culmen shorter than middle toe with claw   6
6. Superciliary stripe yellow   *I. bullockii*
   No superciliary   *I. galbula*

Icterus spurius (**Linnaeus**). Orchard oriole. Entire head and neck, back, wings, and tail black; rump, upper tail coverts, 2 wing bands, and underparts chestnut; remiges edged with whitish; iris brown; bill black, bluish at base below; feet bluish. FEMALE: Upper parts and tail olive green; wings dusky, edged with whitish; 2 whitish wing bands; below greenish yellow. YOUNG MALE: Like female, but lores, malar region, chin, and midline of throat black. Wing 68.5–82.5, tail 63.5–75, culmen 15–17.5, tarsus 20.5–23 mm. Eastern United States, from Gulf states and northern Florida north to Connecticut, southern parts of New York, Ontario, Michigan, Wisconsin, Minnesota, and southern North Dakota, west to Nebraska, eastern Colorado, and Texas. Winters from southern Mexico to Colombia.

Icterus cucullatus **Swainson**. Hooded oriole. Forehead, back, wings, tail, sides of face, throat, and upper breast black; 2 white wing bands; remiges and tips of lateral rectrices edged with whitish; elsewhere orange yellow; iris brown; bill black, bluish at base below; feet bluish. FEMALE: Sides of face and upper parts yellowish olive, browner on back; wings dusky; 2 white wing bands; tail yellowish olive; below yellow. YOUNG MALE: Like female, but throat black. Wing 74–90, tail 75–99, culmen 17–22; tarsus 20–24 mm. Mexico north to southern California, Arizona, New Mexico, and southern Texas.

Icterus graduacauda **Lesson**. Black-headed oriole. Head and throat black; back and upper tail coverts yellow, tinged with olive green; scapulars partly black; wings and tail black; remiges edged

with ashy; greater coverts tipped with ashy; lesser coverts and underparts lemon yellow; iris brown; bill black, blue at base below; feet bluish. FEMALE: Back and upper tail coverts greenish olive. YOUNG: Black of head and neck replaced by olive green dorsally, by lemon yellow ventrally; wings and tail dusky, edged with olive. Wing 89.5–102.5, tail 89–106, culmen 22–28, tarsus 25–28 mm. Mexico and lower Rio Grande Valley.

Icterus parisorum **Bonaparte**. Scott's oriole. Entire head and neck, upper breast, and back black; lesser and middle coverts, rump, upper tail coverts, and underparts lemon yellow; remiges and greater coverts black, edged with white; tail lemon yellow, broadly tipped with black, margined with whitish terminally. FEMALE: Above olive, streaked with black; wings dusky, edged with white; 2 white wing bands; tail yellowish olive, tipped with darker; below yellow, more olive laterally. Wing 94.5–105.5, tail 79–92, culmen 20.5–24.5, tarsus 23–25.5 mm. Piñon-juniper formation of Mexico, north to western Texas, New Mexico, southern Utah, and Nevada, and interior California.

Icterus galbula (**Linnaeus**). Baltimore oriole. Head, neck, and back black; rump, upper tail coverts, lesser wing coverts, and underparts orange; wings black, edged with white; a white band on greater coverts; tail orange, with a broad black band across middle. FEMALE AND YOUNG: Above grayish olive, tinged with yellow on crown, rump, and upper tail coverts; wings dusky, edged with whitish; 2 white wing bands; tail yellowish olive; below dull yellow, often pale gray on chin, breast, belly, and flanks; chin sometimes black. Wing 85–102, tail 66–80, culmen 16–20, tarsus 22.5–25.5 mm. Eastern North America, south to Georgia, northern Alabama and Louisiana, and eastern Texas, west to base of Rockies in Montana, Wyoming, and Colorado. Winters from Mexico to Colombia.

Icterus bullockii (**Swainson**). Bullock's oriole. Like *I. galbula*, but superciliary and sides of face orange; line through eye black; lesser coverts mostly black; edge of wing yellow; middle and greater coverts largely white, forming a patch; tail orange, tipped with black. FEMALE: Like female *I. galbula*, but superciliary and auriculars orange yellow. Wing 89.5–102.5, tail 69.5–92, culmen 16.5–20.5, tarsus 23.5–25.5 mm. Western North

America, east to South Dakota, Kansas, Oklahoma, and Texas, south into Mexico. Winters in Mexico. Hybridizes with *I. galbula.*

Icterus pectoralis (Wagler). Spotted-breasted oriole (Fig. 5-88). Head orange; rump, upper tail coverts, lesser and middle coverts, breast, belly, and crissum yellow; lores, eye ring, throat, spots on sides of breast, wings, and tail black; a small white spot at base of 2d–5th primaries; exposed base of tertials broadly edged with white; bill and feet dark gray. YOUNG: Face and throat without black; remiges and rectrices olive. Wing 92–114, tail 90–109, culmen 21–24.5, tarsus 25.5–28 mm. Southern Mexico to Costa Rica; introduced in southeastern Florida.

Fig. 5-88. Head and foot of spotted-breasted oriole (*Icterus pectoralis*).

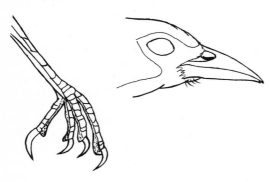

Fig. 5-89. Head and foot of eastern meadowlark (*Sturnella magna*).

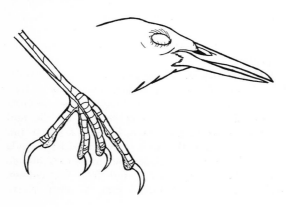

Genus *Sturnella* Vieillot. Bill about as long as head, straight, acute; culmen flattened; nostrils linear, with prominent operculum; wing short but pointed; outer primary shorter than 4th; tertials extending beyond secondaries, often equal to primaries; tail less than ¾ wing, rounded; rectrices rather stiff and pointed; legs strong, extending beyond tail; middle toe shorter than culmen. America. 2 species, which rarely hybridize.

KEY TO SPECIES OF *STURNELLA*

1. Dark bars on middle rectrices separate   *S. neglecta*
   Dark bars on middle rectrices confluent along shaft
   S. magna

Sturnella magna (Linnaeus). Eastern meadowlark (Fig. 5-89). Crown dusky, with a median buff stripe; hind neck, rump, and upper tail coverts streaked dusky and buff; back blackish, margined with buff and brown; lesser coverts dusky edged with gray; middle coverts dusky margined with grayish brown; greater coverts, terials, and secondaries blackish, with interrupted brown bars and margins; primaries dusky, edged with ashy; 4 outer rectrices white, margined terminally with a dusky brown streak in fresh plumage; 2 middle pairs blackish, edged and brokenly barred with brown; superciliary yellow before eye, buff behind; stripe through eye dusky; malar region, auriculars, and sides of neck pale buffy gray; throat, breast, belly, and edge of wing yellow; pectoral collar black; sides, flanks, and crissum buff, streaked with brown; iris brown; bill bluish horn; feet pale pinkish. Wing 89–129, tail 53–86.5, culmen 26.5–36.5, tarsus 35–46 mm. Eastern North America, west to Minnesota, Nebraska, Kansas, and Arizona, and thence south to South America; Cuba.

Sturnella neglecta Audubon. Western meadowlark. Paler throughout; dusky on secondaries, tertials, rump, upper tail coverts, and rectrices forming complete bars and much narrower than the light interspaces; yellow of throat encroaching somewhat on malar region; streaks on flanks and crissum tend to make brown and dusky bars. Wing 104.5–129, tail 50.5–82.5, culmen 27.5–36.5, tarsus 33.5–41.5 mm. Western North America, east to Wisconsin, Michigan, Illinois, Missouri, Oklahoma, and Texas. Where the ranges overlap, the present species is in the higher and drier fields.

Genus *Dolichonyx* Swainson. Bill much shorter than head, stout, conical; culmen nearly straight; nostril oval; wing long and pointed; outer primary longest; 2 primaries sinuate on outer web, often toothed on inner web; tertials longer than secondaries; tail ⅔ wing, rounded, stiff, with acuminate tips; legs long, reaching practically to end of tail; middle toe with claw longer than tarsus; claws slender, long, that of hallux equal to its toe. America. 1 species.

Dolichonyx oryzivorus (Linnaeus). Bobolink. Black; hind neck golden buff or whitish; upper rump gray; lower rump, upper tail coverts, and scapulars white; tertials margined and primaries edged with buff; extreme tips of rectrices gray; iris brown; bill black; feet brown. FEMALE AND WINTER: Crown dusky, with buff median stripe; upper parts buffy olive, streaked with dusky; interscapulars blackish, margined with pale buff and buffy olive; wings and tail dusky, edged with pale buff; postocular stripe dusky; superciliary, sides of face and neck, and underparts pale buffy, becoming brighter on sides of face and neck and across breast; sides, flanks, and crissum streaked with dusky; maxilla brown; mandible flesh. Wing 85–102, tail 59–70, culmen 14.5–17.5, tarsus 25–28.5 mm. North America, south to northeastern California, northern Nevada, Utah, Colorado, Missouri, Illinois, Indiana, Ohio, West Virginia, Pennsylvania, and New Jersey. Winters on pampas of southern South America; migrates through eastern states and West Indies.

Family Tanagridae.[1] Tanagers. Bill conirostral; culmen curved with tip deflected; nostril usually round, with operculum reduced or absent (elongate, with operculum in some extralimital species); commissure sometimes angled or toothed; mandibular ramus from sheath more than half length of gonys; rictal bristles present, not bushy. Primaries 9, several sinuate; wing tip shorter than tarsus. Tail usually shorter than wing, rounded or nearly even; rectrices 12. Tarsus usually longer than middle toe with claw; acrotarsium scutellate; anterior toes slightly coherent; hind toe with claw usually shorter than middle toe without claw. Maxillopalatines terminally expanded and perforate; palatomaxillaries well developed and free posteriorly. Humerus with 2 deep fossae, one of which is pneumatic and usually with trabeculae. Known from Pleistocene. Americas. 103 genera, 329 species; in the United States 6 genera, 14 species.

KEY TO GENERA OF TANAGRIDAE

1. Commissure gently curved   2
   Commissure with an angle near base   3
2. Exposed culmen longer than middle toe without claw   *Piranga*
   Exposed culmen much shorter than middle toe without claw   *Thraupis*
3. Head crested   *Pyrrhuloxia*
   Head uncrested   4
4. Maxillary angle anterior to nostril   *Guiraca*
   Maxillary angle under nostril   5
5. Basal width of mandible equal to depth of bill   *Passerina*
   Basal width of mandible less than depth of bill   *Pheucticus*

Genus *Piranga* Vieillot. Bill stout, swollen, longer than wide or deep, more than ⅔ tarsus, equal to or exceeding middle toe without claw; commissure curved, sometimes toothed; outer primary shorter than 4th; tail ¾ wing, nearly even; tertials short. America. 9 species, 4 in the United States.

KEY TO SPECIES OF *PIRANGA*

1. 2 light wing bands   *P. ludoviciana*
   No wing bands   2
2. Maxillary tomium without pronounced tooth   *P. rubra*
   Maxillary tomium with distinct tooth   3
3. Under wing coverts white, abruptly black at edge of wing   *P. olivacea*
   Under wing coverts salmon or yellow, without dark edge   *P. flava*

Piranga rubra (Linnaeus). Summer tanager (Fig. 5-90). Above dull red; wings brownish edged with dull red; below bright red; iris brown; bill horn, pale on tomia; feet brown. FEMALE: Above olive green; crown and rump often tinged with salmon; wings brown, edged with olive or gray; tail

---

[1] The A.O.U. uses the junior family name Thraupidae and refers all United States genera except *Piranga* (and *Thraupis*) to the Fringillidae.

Fig. 5-90. Head and foot of summer tanager (*Piranga rubra*).

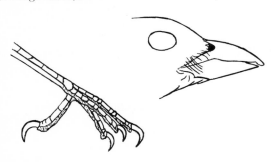

brown tinged with salmon; below yellowish buff, often tinged with salmon buff; crissum salmon buff. YOUNG MALE: Intermediate. Wing 88–106, tail 63–86, culmen 17–20, tarsus 18.5–21 mm. Southern United States, north to Delaware, Maryland, southern Pennsylvania, Ohio, southern parts of Indiana, Illinois, Iowa, and Nebraska, Texas, New Mexico, Arizona, southerrn Nevada, and southern California, south to northern Mexico. Winters from Mexico to South America; Cuba on migration.

Piranga flava (Vieillot). Hepatic tanager. Crown and underparts orange-vermilion; auriculars and back dull grayish pink; lores dusky; iris brown; bill horn; feet brown. FEMALE: Above olive green; lores, malar region, and auriculars grayish, barred with dusky; below yellow. YOUNG MALE: Intermediate. Wing 91–105, tail 73–85.5, culmen 16.5–18.5, tarsus 20.5–24 mm. Pine woods of western Texas, New Mexico, and Arizona, south to southern South America.

Piranga olivacea (Gmelin). Scarlet tanager. Scarlet; wings and tail black; under wing coverts white, abruptly black along edge of wing; iris brown; bill gray, tinged with blue at base, with yellowish green at tip; feet gray. WINTER MALE: Above olive green; wings and tail black; below yellow. FEMALE: Like winter male, but wings and tail grayish brown edged with olive green; under wing coverts whitish, with olive edge. Wing 87.5–99, tail 64–71.5, culmen 14.5–15.5, tarsus 18–21 mm. Eastern Canada south to South Carolina, northern Georgia, northern Alabama, northern Arkansas, and Kansas. Winters in South America; migrates through Caribbean region.

Piranga ludoviciana (Wilson). Western tanager. Head red or orange; neck, rump, upper tail coverts, 2 wing bands, underparts, wing lining, and edge of wing yellow; wings and tail black; secondaries, tertials, and rectrices tipped with whitish; iris brown; bill yellowish; feet gray. WINTER MALE: Head yellow; back margined with olive. FEMALE AND YOUNG: Above olive green; wings and tail dusky, edged with olive green; band on median coverts yellow, band on greater coverts whitish; below yellow; often tinged with salmon around bill. Wing 90–99, tail 67–75.5, culmen 13.5–16, tarsus 19.5–21 mm. Western North America, east to Black Hills, Rockies, and western Texas. Winters in Mexico and Central America.

Genus *Thraupis* Boie. Similar to *Piranga*, but exposed culmen much shorter than middle toe without claw; commissure untoothed. Mexico, Central and South America. 8 species, 1 introduced in Florida.

Thraupis virens (Linnaeus). Blue-gray tanager. Pale grayish blue; wings and tail greenish blue; lesser and middle coverts purplish blue; tips of primaries blackish; iris brown; maxilla black; mandible and legs bluish gray. Wing 83–92, tail 58–67, culmen 11–13, tarsus 19–21 mm. Southern Mexico, south to Peru and Brazil; introduced in southern Florida (Hollywood, breeding; St. Petersburg).

Genus *Pyrrhuloxia* Bonaparte. Head crested; bill short, stout, deeper than wide; commissure abruptly angled; mandibular tomium toothed; gonys straight; outer primary shortest; tail rounded, longer than wing; tarsus more than $\frac{1}{4}$ wing. North America and northern South America. 3 species, 2 in the United States.

KEY TO SPECIES OF *PYRRHULOXIA*

1. Chin reddish or buff; bill nearly as deep as long
   P. sinuata
   Chin black or grayish; bill less deep   *P. cardinalis*

Pyrrhuloxia sinuata Bonaparte. Pyrrhuloxia. Above brownish gray; crest, wings, and tail dull red; forehead, lores, throat, median underparts, thighs, and wing lining bright red; sides pale buffy gray; iris brown; bill yellow; feet brown. FEMALE: Face and underparts buff; crest, wings, and tail red. Wing 82–99, tail 84–108, culmen 15–17.5, tarsus 22.5–27 mm. Northern Mexico to southern Texas, New Mexico, and Arizona.

Pyrrhuloxia cardinalis (**Linnaeus**).[1] Cardinal (Fig. 5-91). Forehead, lores, and throat black; rest of head and neck, underparts, and wing lining bright red; wings and tail dull red; back dull red, edged in fresh plumage with gray; iris brown; bill orange-red; feet brown. FEMALE: Crest dull red; wings and tail dull red edged with gray; back brownish gray; forehead, lores, and chin grayish or dusky; underparts buff, becoming whitish on middle of belly; wing lining bright red. Wing 78–104, tail 82–127, culmen 15.5–22, tarsus 22–29 mm. British Honduras and Mexico north to Arizona, New Mexico, South Dakota, Iowa, and southern parts of Wisconsin, Michigan, Ontario, Pennsylvania, and New York.

Genus *Guiraca* Swainson. Uncrested; bill about 80% tarsus, narrow but very deep, width about half depth; maxillary angle anterior to nostril, with large rounded tooth; outer primary longer than 6th, often longer than 4th; wing tip shorter than middle toe without claw; tail slightly rounded, about $\frac{3}{4}$ wing; tarsus less than $\frac{1}{4}$ wing. North America. 1 species.

Guiraca caerulea (**Linnaeus**). Blue grosbeak (Fig. 5-92). Ultramarine blue; back mixed with dusky; feathers around base of bill black; wings and tail black, edged with blue; middle coverts and tips of greater coverts chestnut; tertials edged with cinnamon; crissum and rectrices tipped with white; iris brown; maxilla black; mandible bluish; feet black. WINTER: Blue more or less concealed by light brown above and by buff below; bill horn. FEMALE: Blue replaced by brown, becoming buffy on throat and belly. Wing 77–97, tail 59–79, culmen 14–21, tarsus 15.5–23 mm. Maryland, Kentucky, southern Illinois, Missouri, Nebraska, Colorado, southern Nevada, and California, south to northern Florida, Mexico, and Central America. Winters in Mexico and Central America; on migration to Cuba and Florida.

Genus *Passerina* Vieillot. Near *Guiraca*, but bill less stout, nearly as wide as deep; maxillary angle under or behind nostril, with tooth absent or obsolete; tarsus more than $\frac{1}{4}$ wing. North America. 6 species, 4 in the United States.

[1] *Richmondena cardinalis* of A.O.U.

Fig. 5-91. **Bill and foot of cardinal (*Pyrrhuloxia cardinalis*), enlarged.**

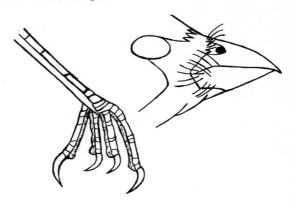

Fig. 5-92. **Head and foot of blue grosbeak (*Guiraca caerulea*), enlarged.**

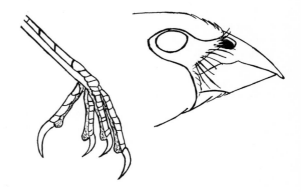

KEY TO SPECIES OF *PASSERINA*

1. Belly red or yellow — *P. ciris*
   Belly neither red nor yellow — 2
2. 2 whitish wing bands on coverts — *P. amoena*
   Wing bands brown or absent — 3
3. Commissure straight from tip to angle, where abruptly deflected — *P. cyanea*
   Commissure gently curved throughout — *P. versicolor*

Passerina cyanea (**Linnaeus**). Indigo bunting. Blue; deeper on head; lores and base of wing coverts black; remiges and rectrices black edged with blue; iris brown; maxilla black; mandible blue-gray; feet brown. FEMALE: Above brown; wings and tail edged with glaucous; 2 brown wing bands; tertials edged with brown; below olive buff, becoming whitish on throat and belly; breast, sides, and flanks streaked with brownish; bill horn. YOUNG MALE: Intermediate. Wing 63–71, tail 47–54, culmen 10–11, tarsus 16–18 mm. Eastern North America, south to northern Florida, west to Dakotas, Nebraska, Kansas, Oklahoma, and eastern Texas. Winters in Mexico, Central America, and Greater Antilles.

Passerina amoena (Say). Lazuli bunting. Head, neck, and upper parts blue, darker on wing coverts and back; lores black; 2 white wing bands; wings and tail blackish, edged with blue; breast tawny; belly and crissum white; iris brown; maxilla black; mandible blue-gray. FEMALE: Above brown; rump and upper tail coverts tinged with greenish blue; wings and tail dusky, edged with greenish blue; 2 buffy white wing bands; throat and breast buff; belly and crissum white. Wing 66–77, tail 51–59, culmen 9–10.5, tarsus 16–18 mm. Western North America, east to Dakotas, Nebraska, Kansas, Oklahoma, and western Texas. Winters in Mexico. Hybridizes with *P. cyanea*.

Passerina ciris (Linnaeus). Painted bunting. Eye ring, throat, and underparts red; rest of head and neck and lesser coverts purplish blue; scapulars and greater coverts yellowish green; middle coverts, rump, and upper tail coverts reddish purple; wings and tail dusky; primaries and rectrices edged with reddish purple; secondaries edged with green and reddish; iris brown; maxilla black; mandible blue-gray; feet brown. FEMALE: Above green; throat and breast olive yellow; belly and crissum yellow. YOUNG: Above olive brown; below olive buff. Wing 64–74, tail 48–57, culmen 10–11, tarsus 18–20 mm. North Carolina, Tennessee, Missouri, Kansas, and southeastern New Mexico, south to northern Mexico, Gulf Coast, and northern Florida. Winters in south Florida, Bahamas, Cuba, Mexico, and Central America.

Passerina versicolor (Bonaparte). Varied bunting. Feathers around bill black; crown, sides of face, lesser wing coverts, rump, upper tail coverts, and wing lining bluish purple; eyelids, occiput, nape, and throat dull red; rest of plumage dusky reddish purple; iris brown; bill and feet dusky. FEMALE: Above brown; rump and upper tail coverts bluish gray; wings and tail edged with bluish; 2 indistinct brownish wing bars; throat, belly, and

crissum whitish; breast grayish brown. YOUNG: Browner above and below. Wing 58–71, tail 46–57, culmen 9.5–11, tarsus 16.5–19 mm. Guatemala and Mexico to southern California, Arizona, and lower Rio Grande.

Genus *Pheucticus* Reichenbach. Like *Guiraca*, but bill more swollen; maxillary angle below nostril, without tooth in our species; culmen less than 80% tarsus; wing tip longer than middle toe without claw. Americas. 5 species, 2 in the United States.

KEY TO SPECIES OF *PHEUCTICUS*

1. Middle of lower breast yellow   *P. melanocephalus*
   Middle of lower breast pink or white
                                   P. ludovicianus

Pheucticus ludovicianus (Linnaeus). Rose-breasted grosbeak (Fig. 5-93). Head, neck, scapulars, and lesser coverts black; rump and middle coverts white; primaries black, with large white basal patch; secondaries, greater coverts, and upper tail coverts black, tipped with white; rectrices black, outer 3 with terminal $\frac{1}{2}$ of inner web white; middle of throat and breast and wing lining pink; belly, crissum, sides, and flanks white; iris brown; maxilla and feet horn; mandible gray. MALE IN WINTER: Upper parts veiled with brown; median crown stripe, superciliary, and malar stripe buffy; pink of throat veiled with buff and streaked with dusky. FEMALE: Upper parts, remiges, and rectrices brown; median crown stripe, lores, superciliary, lower eyelid, and malar stripe white; back streaked with buff; middle and greater coverts, tertials, and upper tail coverts white-tipped; below white; breast washed with buff; sides of neck, breast, flanks, and crissum streaked with brown; wing lining yellow. YOUNG MALE: Like female, but wing lining pink. Wing 95–105, tail 70–78, culmen 15–17.5, tarsus 21–24 mm. Eastern North America, south to New Jersey, mountains of Georgia, Pennsylvania, Ohio, Indiana, Illinois, Iowa, and eastern Kansas. Winters in West Indies, Mexico, Central and South America.

Pheucticus melanocephalus (Swainson). Black-headed grosbeak. Head and chin black; postocular streak often tawny; scapulars streaked black and tawny; rump and collar tawny; wings and tail black; middle and greater coverts, secondaries, and upper tail coverts tipped with white; primaries with large white basal patch and white subterminal edge; outer $\frac{2}{3}$ of rectrices white on terminal half of inner web; throat, breast, sides, and flanks tawny; wing lining and middle of lower breast yellow; middle of belly whitish; crissum pale buff; iris brown; bill horn, paler below; feet brown. FEMALE: Upper parts, wings, and tail dusky brown; median crown stripe, lores, superciliary, malar stripe, and collar buffy white; back streaked with buffy; 2 wing bands and tips of secondaries white; white primary patch reduced; below pale tawny, becoming buffy white posteriorly; breast, sides, and flanks streaked with brown; wing lining and midline of breast yellow. YOUNG MALE: Intermediate. Wing 93–102, tail 69–86, culmen 15–20, tarsus 21–25.5 mm. Mexico and western North America, east to Dakotas, Nebraska, Kansas, and western Texas. Winters in Mexico. Hybridizes with *P. ludovicianus*.

Family Fringillidae.[1] Finches. Bill short, stout, conical, about equal to middle toe without claw; nostril round, concealed or exposed; nasal fossa and operculum absent or obsolete; commissure with little or no basal angle; unfeathered mandibular ramus very short, much less than $\frac{1}{2}$ gonys. Primaries 9 (all our species) or 10 with outermost rudimentary. Tail emarginate, nearly even, or long and graduated; rectrices 12. Tarsus shorter or but little longer than middle toe with claw; acrotarsium scutellate. Palatomaxillaries absent; prepalatine bar with lateral projection; mediopalatine processes fused; inferior process of nasal short and approaching the vertical; nasal foramen roundish. Mandible nearly straight; symphysis long, about half length of ramus. Humerus with medial bar short, directed pálmarly or proximally, to reach floor of subtrochanteric fossa near external tuberosity; floor of the 2 fossae on the same plane. Femur shorter than humerus. Tibiotarsus with internal condyle often nearly as wide as external condyle. Tarsometatarsus short, about $\frac{2}{3}$ tibiotarsus. Social. Reported from Miocene. Nearly cosmopolitan (except New Zealand and Polynesia), but mostly Old World. 68 genera, 374 species; in the United States 9 genera, 18 species.

[1] Includes weaver finches, family Ploceidae of A.O.U.

Fig. 5-93. Head and foot of rose-breasted grosbeak (*Pheucticus ludovicianus*), enlarged.

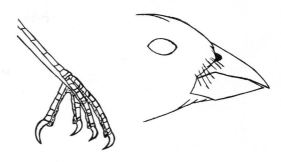

KEY TO GENERA OF FRINGILLIDAE

1. Bill laterally crossed near tip   *Loxia*
   Bill not laterally crossed   **2**
2. Bill depth more than hind toe and claw length
           Coccothraustes
   Bill depth less than length of hind toe with claw   **3**
3. Side of mandible with an oblique ridge   *Leucosticte*
   Side of mandible smooth   **4**
4. Rictal bristles bushy   **5**
   Rictal bristles normal, bare shafts   **8**
5. Width of lower mandible at feathers much less than length of gonys   **6**
   Width of mandible equal to or more than length of gonys   **7**
6. Culmen longer, 71–88% tarsus   *Carduelis*
   Culmen shorter, 51–62% tarsus   *Acanthis*
7. Outer primary shorter than 4th   *Pinicola*
   Outer primary longer than 4th   *Carpodacus*
8. Nostrils at least partly concealed   *Passer*
   Nostrils exposed   *Spiza*

Genus *Spiza* Bonaparte. Bill stout, conical, depth more than $\frac{3}{4}$ length; commissure sinuate in middle; gonys about equal to width of lower mandible at feathers, longer than bare part of ramus; nostril round, exposed, without operculum; rictal bristles normal, not bushy; outer 3 primaries longest; wing tip about equal to tarsus; tail slightly less than $\frac{3}{4}$ wing, nearly even to slightly double-rounded; tips of unworn rectrices pointed; coverts covering more than basal half of tail; tarsus more than $\frac{1}{4}$ wing; middle toe with claw about equal to tarsus, exceeding lateral claws by length of its own claw. North America. 1 species.

Spiza americana **Gmelin.** Dickcissel (Fig. 5-94). Crown olive gray, faintly streaked with

Fig. 5-94. Head and foot of dickcissel (*Spiza americana*), enlarged.

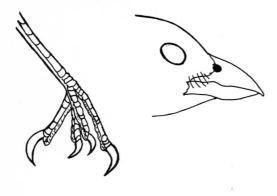

Fig. 5-95. Head and foot of European tree sparrow (*Passer montanus*), enlarged.

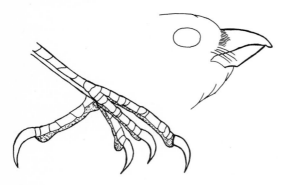

dusky; nape and sides of face and neck gray; superciliary yellow; malar spot white, followed by yellow; scapulars grayish brown streaked with dusky; rump and upper tail coverts grayish brown; lesser and middle coverts chestnut; bend of wing yellow; remiges and rectrices dusky, edged with clay color; chin and lower malar area white; lower throat black; breast yellow; sides, flanks, and crissum whitish; iris and feet brown; bill horn. FEMALE AND YOUNG: Duller; whole throat whitish; necklace, breast, and flanks streaked with dusky; wing coverts often without chestnut. Wing 74–86, tail 50–62, culmen 12.5–15.5, tarsus 21–24 mm. Interior plains, from Manitoba and Ontario south to Texas, Louisiana, and central parts of Mississippi,

Alabama, and Georgia; eastern breeding limits fluctuate and formerly included Coastal Plain from Massachusetts to South Carolina; winters from Mexico to northern South America.

Genus *Passer* Brisson. Bill stout, conical, depth $\frac{3}{4}$ length or more; commissure sinuate in middle; gonys longer than width of lower mandible, longer than bare ramus; nostril round, at least partly concealed, without operculum; rictal bristles normal, not bushy; outer 3 primaries longest; wing tip longer than tarsus; tail about $\frac{3}{4}$ wing, nearly even to slightly double-rounded; tips of rectrices rounded; coverts extending over more than half of tail; tarsus $\frac{1}{4}$ wing; middle toe with claw about equal to tarsus, exceeding lateral claws by length of its own claw. Eurasia, Africa. 10 species, 2 introduced in the United States.

KEY TO SPECIES OF *PASSER*

1. 2 white wing bands (on greater and middle coverts)
   P. montanus
   1 white wing band (on middle coverts only)
   P. domesticus

Passer domesticus (**Linnaeus**). House sparrow; English sparrow. Crown and nape gray; supraloral stripe and spot behind eye white; lores and subocular area black; cheeks and auriculars white, tinged with gray posteriorly; broad postocular stripe chestnut, extending down sides of neck; scapulars brown streaked with black; rump and upper tail coverts gray; wing coverts chestnut; middle coverts tipped with white, forming a band; wings and tail dusky, edged with brown; breast and middle of throat black; belly white; crissum buffy white; sides and flanks buffy gray; iris brown; bill black (mandible horn in winter); feet brown. FEMALE: Above grayish brown; scapulars streaked with black and buffy; middle coverts black subterminally, tipped with white; superciliary buffy white; cheeks buffy brown, darker above and posteriorly; below pale buffy gray; maxilla horn; mandible yellowish. YOUNG MALE: Like female but with indistinct dusky throat patch. Wing 71–82, tail 52–58, culmen 10–13.5, tarsus 18–20 mm. Europe, western Asia, North Africa; introduced to Australia, New Zealand, southern South America, Cuba, Bahamas, Bermuda; introduced to eastern United States in

1850; now occurs through Canada, the United States, and Mexico.

***Passer montanus* (Linnaeus).** European tree sparrow (Fig. 5-95). Crown and nape vinaceous brown; back, rump, and upper tail coverts buffy brown; scapulars streaked with black and bright brown; wings and tail dusky, edged with buffy brown; wing coverts light chocolate; greater and middle coverts tipped with white, black subterminally; a black area extends from throat to lores, below eye, and above auriculars; cheeks white, with a large black spot in middle; below grayish white; sides and flanks buffy brown; iris brown; bill black, paler at base; feet brown. Wing 64–74, tail 51–53, culmen 8–11, tarsus 17–18 mm. Europe and all Asia; introduced to Philippines and Australia; also to St. Louis, Missouri, and has spread to adjacent portion of Illinois.

Genus *Carduelis* Brisson. Bill conical, tip compressed, acute; culmen 71–88% tarsus, curved, straight, or even slightly concave; gonys straight or concave; nasal plumes covering nostrils; rictal bristles short and bushy; commissure nearly straight, but basal half of mandibular tomium a strong crest (concealed in closed bill); tail $\frac{2}{3}$ (60–72%) wing, emarginate; tarsus short, $\frac{1}{5}$ wing, middle toe with claw about equal to tarsus. All continents except Australia. 30 species, 5 in the United States.

KEY TO SPECIES OF *CARDUELIS*

1. Crown and throat red   *C. carduelis*
   No red in plumage   2
2. Tail with yellow at base   *C. pinus*
   Tail with white patches at base   3
3. Crissum yellow   *C. psaltria*
   Crissum white   4
4. Remiges and wing coverts edged with yellow
        C. lawrencei
   Not more than lesser converts edged with yellow
        C. tristis

***Carduelis carduelis* (Linnaeus).** European goldfinch. Crown, cheeks, and throat red to behind eyes; lores, back of crown, and sides of neck black; nuchal spot whitish; back, sides of breast, and flanks brown; upper tail coverts black, broadly edged and tipped with whitish; belly and crissum white; wings black, remiges broadly yellow basally

Fig. 5-96. Head and foot of American goldfinch (*Carduelis tristis*), enlarged.

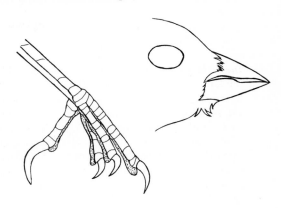

and narrowly tipped with white; tail black, large subterminal patch and narrow tips white; eye brown; bill whitish; feet flesh. Wing 73–81, tail 46–52, culmen 11–14, tarsus 14–16 mm. Europe, Asia Minor, North Africa; widely introduced in America but established only on Long Island and Bermuda.

***Carduelis pinus* (Wilson).**[1] Pine siskin. Above brown streaked with darker and often tinged with yellowish; wings and tail dusky; middle and greater coverts tipped with buffy white; tertials edged with buffy white; edges and base of remiges and rectrices yellow; below dirty or buffy white, streaked with dusky; iris brown; bill horn; feet brown. Wing 67–78, tail 40–48, culmen 9.5–12, tarsus 13–15 mm. Northern coniferous forest, south to New York, Pennsylvania, mountains of North Carolina, northern Michigan, Minnesota, Nebraska, and through mountains of West to Mexico. Winters south to Gulf states.

***Carduelis tristis* (Linnaeus).**[1] American goldfinch (Fig. 5-96). Yellow; wings and tail black; upper and under tail coverts white; secondaries, greater and middle coverts white-tipped; inner webs of rectrices with white patch at tip; iris brown; bill yellow, tip black; feet pale brown. FEMALE AND WINTER MALE: Above olive brown; wings and tail dusky, marked as in male; upper tail coverts ashy; face and chin dingy yellow; below dirty cinnamon buff; belly and crissum whitish;

[1] These species separated by the A.O.U. in a genus *Spinus* Koch, based on coloration alone.

bill horn. Wing 66–78, tail 40–52, culmen 9.5–11, tarsus 12.5–14.5 mm. North America, south to central Georgia and Alabama, Arkansas, Oklahoma, Colorado, Nevada, and Baja California. Winters south to Gulf Coast, Florida, and northern Mexico.

***Carduelis lawrencei* Cassin.**[1] Lawrence's goldfinch. Head and throat black; above brownish gray; rump tinged with olive; wing coverts, remiges, and rectrices edged with yellow; inner web of rectrices with white subterminal patch; breast yellow; belly and crissum white; sides and flanks pale brownish gray. FEMALE AND YOUNG: No black on head and neck; yellow duller. Wing 63–70, tail 42–51, culmen 8–9, tarsus 12.5–13.5 mm. California and Baja California, west of Sierra Nevada. Winters east to Sonora, Arizona, and New Mexico.

Carduelis psaltria (**Say**).[1] Lesser goldfinch. Crown glossy black; auriculars and upper parts glossy black, olive green, or mixed; wings and tail black; tertials white-tipped; primaries and inner webs of rectrices white basally; below yellow; iris and feet brown; bill horn. FEMALE: Above olive green; wings and tail dusky, with white patches smaller or obsolete; below olive yellow. Wing 53–68.5, tail 33.5–44.5, culmen 8.5–10.5, tarsus 11.5–13 mm. Southern Oregon, Utah, Colorado, New Mexico, and southern Texas, south to Peru.

Genus *Acanthis* Bechstein. Barely separable from *Carduelis* by its shorter bill, 51–62% tarsus.

Fig. 5-97. Head and foot of rosy finch (*Leucosticte arctoa*), enlarged.

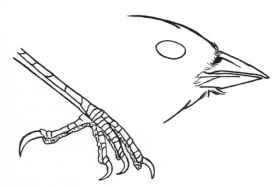

American species have the bill much compressed and the tail/wing ratio higher (70–78%), but these characters are not shared in extralimital species. Holarctic. 4 species, 2 in the United States.

KEY TO SPECIES OF *ACANTHIS*

1. Rump and sides streaked — ***A. flammea***
   Rump and sides unstreaked — ***A. hornemanni***

Acanthis flammea (**Linnaeus**). Common redpoll. Crown red; above brown, streaked with dusky and whitish; rump pinkish white, streaked with dusky; wings and tail brown, edged with whitish; 2 white wing bands; lores and chin dusky; cheeks, throat, and breast pinkish; rest of underparts white; sides, flanks, and crissum streaked with dusky; iris and feet brown; bill horn, tip dusky. FEMALE: Lacks pink on underparts. Wing 69–84, tail 48.5–65.5, culmen 7.5–10.5, tarsus 13.5–17.5 mm. Holarctic, south to northern Canada and Alps. Winters south irregularly to South Carolina, Alabama, Ohio, Indiana, Illinois, Kansas, Colorado, Idaho, and northern California.

Acanthis hornemanni (**Holboell**). Hoary redpoll. Paler; rump plain white or pale pink; below with few if any streaks. Wing 70–91, tail 53.5–68.5, culmen 7–10.5, tarsus 13.5–17 mm. Circumpolar. Winters irregularly south to Japan, northern Europe, New York, Michigan, Illinois, Minnesota, and Montana.

Genus *Leucosticte* Swainson. Bill stout; culmen nearly straight, shorter than middle toe without claw; nasal tufts scarcely extending beyond nostrils; a ridge extending forward from each nostril; a ridge along mandibular tomium; an oblique ridge from ramus to tomium; wing pointed; wing tip longer than tarsus and middle toe combined; tail $\frac{2}{3}$ wing, emarginate; tarsus $\frac{1}{5}$ wing, longer than middle toe with claw. Eastern Asia and western North America. 3 species, 1 in the United States.

Leucosticte arctoa (**Pallas**).[2] Rosy finch (Fig. 5-97). Nasal tufts whitish; crown black; occiput and sides of crown often ashy; sides of head, neck, back, and underparts cinnamon, chocolate, or blackish; wing coverts, rump, upper tail coverts, flanks, and crissum broadly pink-tipped; remiges

---

[1] These species separated by the A.O.U. in a genus *Spinus* Koch, based on coloration alone.

[2] Includes *Leucosticte tephrocotis* Swainson, *L. atrata* Ridgway, and *L. australis* Ridgway of the A.O.U.

and rectrices dusky, edged with pink; iris brown; bill black (yellow with dusky tip in winter); feet black. Wing 96–124, tail 60–85, culmen 10–15, tarsus 18–25 mm. Breeds above timber line from Siberia and Alaska south to California, Utah, and Colorado. Winters lower, spreading to Korea, Japan, Arizona, New Mexico, and Nebraska.

Genus *Pinicola* Vieillot. Bill short, deep, wide, about as deep as long; width of mandible greater than length of gonys; culmen strongly curved, with tip overhanging; nasal tufts well developed; outer primary shorter than 4th; wing tip shorter than tarsus and middle toe together; tail more than $\frac{3}{4}$ wing, emarginate; tarsus $\frac{1}{5}$ wing. Holarctic. 1 species.

Pinicola enucleator (**Linnaeus**). Pine grosbeak (Fig. 5-98). Pinkish red, the feathers gray basally; scapulars, flanks, belly, and crissum ashy gray; back mottled with dusky; wings and tail dusky; 2 white wing bands; tertials edged with white; remiges and rectrices edged with ashy; nasal tufts and lores dusky; iris brown; bill horn, paler below; feet black. FEMALE AND YOUNG: Smoky gray, paler below; red areas replaced by golden-olive tips. Wing 104–127, tail 79.5–107, culmen 13–17, tarsus 21–24 mm. Boreal coniferous forests of Asia, Europe, and North America, south to Maine, New Hampshire, and in mountains to New Mexico and California. Winters south irregularly to New Jersey, Pennsylvania, Indiana, Illinois, Iowa, Kansas, and mountain states.

Genus *Carpodacus* Kaup. Resembles *Pinicola*, but culmen more gently curved; depth of bill more than $\frac{3}{4}$ its length; outer primary longer than 4th; tail sometimes even. Holarctic. 20 species, 3 in the United States.

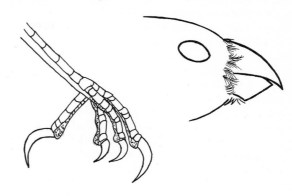

Fig. 5-98. Head and foot of pine grosbeak (*Pinicola enucleator*), enlarged.

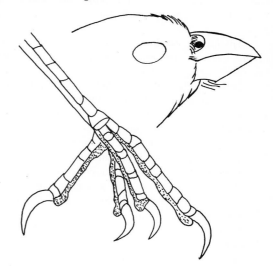

Fig. 5-99. Head and foot of house finch (*Carpodacus mexicanus*), enlarged.

KEY TO SPECIES OF *CARPODACUS*

1. Crissum plain or with only a few streaks
   C. purpureus
   Crissum heavily streaked 2
2. Tail deeply emarginate, less than $\frac{3}{4}$ wing   *C. cassinii*
   Tail slightly, if at all, emarginate, more than $\frac{3}{4}$ wing
   C. mexicanus

Carpodacus mexicanus (**Müller**). House finch (Fig. 5-99). Nasal tufts and lores whitish; forehead, crown laterally (sometimes entirely), rump, malar region, throat, and breast red or pink; back brown, streaked with darker, sometimes tinged with red; wings and tail dusky, edged with ashy; 2 ashy or buff wing bands, often indistinct; belly, sides, flanks, and crissum whitish or buffy, streaked with brown; iris and feet brown; bill horn. FEMALE AND YOUNG: Without red. Wing 68.5–84, tail 51–68.5, culmen 9.5–14, tarsus 16–21 mm. Southern Mexico north to western Texas, Oklahoma, western Kansas, Wyoming, Idaho, and Oregon. Introduced to Hawaii and vicinity of New York City, wintering south to North Carolina.

Carpodacus purpureus (**Gmelin**).   Purple finch. Nasal tufts and lores whitish; crown, rump, throat, and breast purplish red; sides of face brownish red; back reddish brown, streaked with darker; wings and tail dusky, edged with reddish brown; 2 reddish wing bands; belly and crissum white; flanks pinkish, streaked with brown; iris and feet brown; bill horn, paler below. FEMALE AND YOUNG: Above olive, streaked with dusky and whitish; wings and tail dusky, edged with gray; postocular stripe olive; supraauricular and malar stripes whitish streaked with olive; below white or buffy; throat, breast, and sides streaked with olive. Wing 75–86.5, tail 53.5–61.5, culmen 10.5–12.5, tarsus 15.5–19 mm. Coniferous forest, south to New Jersey, mountains of Maryland, Pennsylvania, northern Michigan and Illinois, Minnesota, North Dakota, and California. In winter extends to Gulf states, Texas, and Arizona.

Carpodacus cassinii **Baird.**   Cassin's purple finch. Crown crimson; back pink, mixed with brownish gray, streaked with dusky; wings and tail dusky, edged with reddish; rump, postocular stripe, malar region, throat, and breast pink, belly white; crissum white, streaked with dusky; iris and feet brown; bill horn, paler below. FEMALE AND YOUNG: Above olive gray, streaked with dusky; below white, streaked with dusky except on belly. Wing 87–96.5, tail 60–69, culmen 12–13, tarsus 17–20.5 mm. Coniferous belt of western mountains, from British Columbia, Montana, and Wyoming, south to Baja California, Arizona, and New Mexico. In winter extends to central Mexico.

Genus *Loxia* **Linnaeus.**   Bill long, $\frac{1}{5}$ wing, crossed at tip, indistinctly grooved; depth of bill less than $\frac{2}{3}$ length; width of mandible less than gonys; culmen curved, tip overhanging; outer primary longer than 4th; wing tip nearly twice tarsus; tail less than $\frac{2}{3}$ wing, emarginate; tarsus less than $\frac{1}{5}$ wing. Holarctic. 3 species, 2 in the United States.

KEY TO SPECIES OF *LOXIA*

1. 2 white wing bands on coverts   *L. leucoptera*
   Wings unbanded   *L. curvirostra*

Loxia curvirostra **Linnaeus.**   Red crossbill. Red or orange-red; nape and scapulars with dusky centers; line through eye and broken malar stripe brown; belly gray; crissum brownish, edged with ash; wings and tail brown, tinged with reddish; iris and feet brown; bill horn, paler below. FEMALE: Grayish olive, paler below; crown, rump, and breast tipped with dull yellow; crissum brown, edged with white; wings and tail faintly edged with olive. YOUNG: Like female, but heavily streaked with dusky and without yellow. Wing 79–104.5, tail 38–59, culmen 13–22, tarsus 15–19 mm. Holarctic coniferous forest, south to Maine, New Hampshire, Adirondacks, Pennsylvania, mountains of Tennessee and North Carolina, northern Michigan, northern Wisconsin, Minnesota, Black Hills, and through mountains of West and Mexico to Nicaragua. Occurs sporadically throughout the United States.

Loxia leucoptera **Gmelin.**   White-winged crossbill. Pinkish red; line through eye dusky; scapulars black; wings and tail black, edged with white; 2 white wing bands; belly and flanks gray; crissum black, edged with white; iris brown; bill horn, pale at base; feet dusky. FEMALE: Above dusky margined with olive yellow; rump light yellow; below olive gray; wings and tail as in male. YOUNG: Dull whitish streaked with dusky; wing bands tinged with buff. Wing 83–91.5, tail 49–59.5, culmen 15–17.5, tarsus 15–17 mm. Holarctic coniferous forest, south to Maine, New Hampshire, Adirondacks, and northern Michigan; Hispaniola; northern Europe and Asia. In winter south irregularly to North Carolina, Ohio, Illinois, Kansas, Colorado, Nevada, and Oregon.

Genus *Coccothraustes* **Brisson.**   Bill stout, swollen, less than $\frac{1}{5}$ wing, its depth more than $\frac{3}{4}$ length; culmen curved; width of mandible about equal to gonys; outer primary longer than 4th; wing tip shorter than tarsus and middle toe combined; tail less than $\frac{2}{3}$ wing, emarginate; tarsus less than $\frac{1}{5}$ wing, little longer than culmen. Holarctic. 3 species, 1 in the United States.

Coccothraustes vespertina (**Cooper**).[1]   Evening grosbeak. Forehead, superciliary, back, and rump yellow; crown, upper tail coverts, tail, and wings black; exposed part of tertials and tertial coverts pale gray; nape and face olive; below yellowish olive, becoming yellow posteriorly; wing lining yellow, black at edge; iris brown; bill green; feet

[1] *Hesperiphona vespertina* of A.O.U.

light brown. FEMALE: Above gray, head darker, rump paler; wings black; 6 inner primaries white basally; tertials and coverts gray; upper tail coverts and tail black; inner webs of 3 lateral rectrices white-tipped; submalar streak dusky; throat, belly, and crissum white; breast and sides yellowish gray; wing lining yellow, dusky at edge. YOUNG: Like female but tinged with brownish above, with buffy below; bill horn. Wing 104–117, tail 61–73, culmen 16–22.5, tarsus 19–23 mm. Canadian forest, south to Massachusetts, northern parts of New York, Michigan, and Minnesota, and Sierra Nevada and Rockies to Mexico. Winters sporadically south and east to Arkansas, Alabama, and Georgia.

Family Emberizidae.[1] Buntings. Bill short, stout, conical, much shorter than middle toe without claw; culmen usually straight or slightly concave, rarely curved; nostril exposed (except in *Plectrophenax*), usually elongate; nasal fossa and operculum well developed; commissure with basal angle; bare part of ramus more than half gonys. Primaries 9. Tail about $\frac{2}{3}$ wing to slightly longer than wing, rounded or double-rounded, rarely emarginate; rectrices 12. Tarsus longer than middle toe with claw; acrotarsium scutellate. Palatomaxillaries fused to prepalatine bar; prepalatine bar with lateral projection; mediopalatine processes unfused; inferior process of nasal long, inclined sharply forward; nasal foramen elongate. Mandible strongly deflected; symphysis short, only $\frac{1}{3}$ length of ramus. Humerus with medial bar long, directed toward shaft; upper subtrochanteric fossa on a higher plane than lower fossa. Femur little if any shorter than humerus. Tibiotarsus with internal condyle narrow. Tarsometatarsus long, 70% tibiotarsus or more. Pliocene onward. Mostly in Americas; a few genera in Eurasia and Africa. 59 genera, 213 species; in the United States 20 genera, 45 species.

KEY TO GENERA OF EMBERIZIDAE

1. Wing tip as long as combined tarsus and middle toe
   with claw   *Plectrophenax*
   Wing tip shorter than combined tarsus and middle
   toe with claw   **2**
2. Wing tip about equal to combined tarsus and middle toe without claw   **3**

[1] Family Fringillidae of A.O.U., in part.

Wing tip shorter than combined tarsus and middle
   toe without claw   **4**
3. Bill short and deep, $\frac{2}{3}$ its length   *Rhynchophanes*
   Depth of bill about half its length   *Calcarius*
4. Wing tip much longer than tarsus   *Chondestes*
   Wing tip shorter than tarsus   **5**
5. Tips of rectrices sharp and narrow   **6**
   At least some of the rectrices with normal, rounded
   tips   **7**
6. Tail graduated   *Ammospiza*
   Tail double-rounded   *Ammodramus*
7. Wing strongly rounded, outer primary shortest   **8**
   Outer primary not shortest   **9**
8. Tail long, more than 3 times tarsus   *Pipilo*
   Tail less than 3 times tarsus   *Arremonops*
9. 4 outer primaries about equally long   *Pooecetes*
   Outer primary distinctly shorter than 2d and 3d   **10**
10. Outer claw reaching middle of middle claw
           Passerella
    Outer claw not reaching beyond base of middle
    claw   **11**
11. Bill about as deep as long; culmen much curved
           Sporophila
    Bill longer than deep; culmen nearly straight   **12**
12. Bill depth much more than length of gonys
           Calamospiza
    Bill depth less than length of gonys   **13**
13. Wing less than 3 times tarsus   *Melospiza*
    Wing more than 3 times tarsus   **14**
14. Outer primary longer than 5th   **15**
    Outer primary shorter than 5th   **16**
15. Middle rectrices hardly longer than outer pair
           Passerculus
    Middle rectrices much shorter than outer pair
           Spizella
16. Mandibular tomium strongly sinuate   *Chlorura*
    Mandibular tomium straight or only slightly sinuate
           17
17. Outer rectrices longer than middle pair   *Junco*
    Outer rectrices shorter than middle pair   **18**
18. Graduation of tail more than length of lateral toes
    without claw   *Aimophila*
    Graduation of tail not more than half length of lateral toes without claw   **19**
19. Folded wing extending beyond upper tail coverts
           Amphispiza
    Folded wing falling short of end of tail coverts
           Zonotrichia

Genus *Plectrophenax* Stejneger. Bill short, 10% wing; culmen slightly indented; nostrils round, practically concealed; maxilla shallower than mandible; gonys short, about equal to length of ramus; wing long, pointed; wing tip nearly twice tarsus;

Fig. 5-100. Head and foot of snow bunting (*Plectrophenax nivalis*), enlarged.

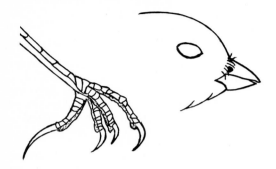

tail less than $\frac{2}{3}$ wing, emarginate; tarsus $\frac{1}{5}$ wing; lateral toes subequal; claws long, slender; lateral claws reaching base of middle claw. Holarctic. 1 species.

Plectrophenax nivalis (**Linnaeus**). Snow bunting (Fig. 5-100). White; back, alula, greater coverts, inner secondaries, primaries, and 2–3 middle pairs of rectrices black; iris brown; bill and feet black. FEMALE: Above black, margined with buffy gray; nape white, streaked with dusky; 2 white wing bands; secondaries and 3 outer rectrices white, edged with black at tip; below white. WINTER: White overlaid with rusty brown, especially on crown, auriculars, upper tail coverts, and breast; black of back more or less concealed by buffy margins; bill yellow. Wing 99–120, tail 60.5–74, culmen 9.5–13.5, tarsus 20.5–24 mm. Circumpolar tundra. Winters south to Mediterranean, China, Japan, and northern United States to North Carolina, Ohio, Indiana, Kansas, Colorado, and Oregon, sometimes farther.

Genus *Calcarius* Bechstein. Near *Plectrophenax,* but nostrils exposed; maxilla as deep as mandible; outer primary shorter than 2d; wing tip $1\frac{1}{2}$ times tarsus; tail emarginate or double-rounded, 63–72% of wing; tarsus more than $\frac{1}{5}$ wing. Holarctic. 3 species.

KEY TO SPECIES OF *CALCARIUS*

1. 2 lateral pairs of rectrices entirely white at base
   C. ornatus
   Inner web of 2 lateral pairs of rectrices dusky at base
   2

2. Belly white   *C. lapponicus*
   Belly rich buff   *C. pictus*

Calcarius lapponicus (**Linnaeus**). Lapland longspur. Head, throat, and upper breast black; postocular stripe buff; sides of neck and breast white; nape chestnut; back streaked black and buffy; wings and tail dusky, edged with grayish buff; 2 white wing bands; tertials and greater coverts edged with rich brown; a white terminal or subterminal patch on 2–3 outer rectrices; below white; sides and flanks tinged with buff and streaked with dusky; iris brown; bill yellow, tip black; feet blackish. FEMALE AND WINTER: Black and chestnut of head and neck largely concealed by buffy or whitish tips, especially on chin. Wing 86–100.5, tail 55.5–66.5, culmen 10–12.5, tarsus 20.5–22.5 mm. Circumpolar tundra. Winters south irregularly to central Europe and Asia, and to Virginia, Kentucky, Missouri, Oklahoma, Colorado, Utah, Nevada, and Oregon, or farther.

Calcarius pictus (**Swainson**). Smith's longspur. Crown and sides of head black; postocular, auricular, and malar stripes white; nape rich buff, streaked with dusky; back streaked black and buffy; lesser coverts black, white near body; below rich buff; wing lining white; rectrices dusky, 2–3 outer ones largely white terminally; iris brown; bill brownish, tip dark; feet flesh. FEMALE AND WINTER: Head brownish, streaked with dusky; throat and breast streaked with dusky; middle and greater coverts with 2 white wing bands. Wing 86.5–96, tail 55.5–69, culmen 10–11.5, tarsus 20–20.5 mm. Tundra, from Hudson Bay to Yukon. Winters on plains of Kansas, Oklahoma, and Texas, east to Illinois and Indiana.

Calcarius ornatus (**Townsend**). Chestnut-collared longspur. Crown, postocular stripe, and auricular spot black; nape chestnut; back blackish, edged with grayish brown; lesser coverts black, with a white patch; wings dark brown, edged with paler; 2–3 outer rectrices white; middle rectrices brown; intermediate rectrices white basally, brown terminally; superciliary, throat, lower belly, crissum, and wing lining white; breast and upper belly black; iris and feet brown; bill blackish. FEMALE: Above buffy brown, streaked with dusky; nape tinged with rufous; tail as in male; below grayish buff, paler on crissum; breast and belly sometimes streaked with dusky. Wing 75.5–90, tail 48–61, culmen 9.5–11, tarsus 18–21 mm. Northern Great Plains, south to western Minnesota, Kansas, Wyoming, and Montana. Winters from Nebraska, Colorado, and Arizona, south to Mexico, occasionally eastward.

Genus *Rhynchophanes* Baird. Differs from *Calcarius* in having depth of bill more than $\frac{2}{3}$ its length; tail emarginate, 57–60% wing. Great Plains. 1 species.

Rhynchophanes mccownii (**Lawrence**). McCown's longspur. Crown black; upper parts gray, streaked with dusky; middle coverts chestnut; wings gray, edged with ashy or white; greater coverts and secondaries tipped with white; tail white, with broad black tip; middle rectrices grayish; auriculars gray; superciliary, cheeks, and underparts white; rictal stripe and breast patch black; soft parts brown, tip of bill blackish. FEMALE: Crown grayish, streaked with dusky; middle coverts tipped with buffy; superciliary and anterior underparts buffy; rictal stripe and breast patch gray. Wing 80–94, tail 45.5–56, culmen 10.5–13, tarsus 18–21 mm. Northern Great Plains, south to western Minnesota, North Dakota, Wyoming, Montana, and Colorado. Winters from Kansas, Colorado, and Arizona, south to Mexico.

Genus *Chondestes* Swainson. Depth of bill less than $\frac{3}{4}$ length; gonys twice as long as bare part of ramus; nostril oval, partly concealed; wing tip longer than tarsus; tail more than $\frac{3}{4}$ wing, strongly rounded; rectrices broad with rounded tips; tarsus $\frac{1}{4}$ wing or less. North America. 1 species.

Chondestes grammacus (**Say**). Lark sparrow. Crown chestnut, becoming blackish near nostrils, and with a buff or whitish median stripe; upper parts brownish; back streaked with blackish; wings dusky, edged with brown; edge of wing white; 2 white or buff wing bands; 2d to 5th primaries with a whitish or buff basal patch; tail dusky, with large white tip; superciliary white or buff; stripe through eye black; suborbital area white; rictal streak black; auriculars chestnut; malar stripe white, continuing below and behind auriculars; submalar streak black; throat white; upper breast, sides, and flanks buffy gray; a black pectoral spot; lower breast and belly white; crissum dusky, broadly tipped with white or buffy; iris brown; bill horn above, paler below; feet yellowish. Wing

79–94, tail 61–78, culmen 10.5–14.5, tarsus 19–22 mm. Plains of Mexico and western North America, east to Great Lakes, West Virginia, and Alabama. Winters in Guatemala and Mexico, north to California and Gulf states.

Genus *Pooecetes* Baird. Wing tip shorter than tarsus; tail about $\frac{3}{4}$ wing, double-rounded; rectrices rather narrow and pointed; tarsus more than $\frac{1}{4}$ wing; lateral claws reaching about to base of middle claw. North America. 1 species.

Pooecetes gramineus (**Gmelin**). Vesper sparrow. Above grayish brown, streaked with dusky; lesser wing coverts cinnamon, streaked with dusky; edge of wing white; 2 buffy wing bands; wings and tail dusky, edged with buffy or ashy; outer rectrices white terminally and laterally; superciliary grayish white; auriculars brown; rictal and postocular stripes dark brown; malar stripe whitish; below whitish; sides of neck, breast, sides, and flanks tinged with buff and streaked with dusky; iris brown; maxilla brown; mandible pinkish; feet pale brown. Wing 73.5–86.5, tail 53–68.5, culmen 10–12.5, tarsus 20–22.5 mm. North America, south to North Carolina, Kentucky, Missouri, Kansas, western Texas, New Mexico, Arizona, and California. Winters from Ohio Valley, Texas, and California, to Florida, Gulf Coast, and Mexico.

Genus *Calamospiza* Bonaparte. Bill stout, depth more than $\frac{3}{4}$ length; outer primary shorter than 4th; wing tip shorter than tarsus; lateral claws falling short of base of middle claw. Great Plains. 1 species.

Calamospiza melanocorys Stejneger. Lark bunting. Black; middle and greater coverts white; tertials, upper and under tail coverts, and outer rectrices edged with white; iris and feet brown; bill horn, paler below. FEMALE AND YOUNG: Above grayish brown, streaked with dusky; middle and greater coverts whitish; tail blackish, tipped with white laterally; below white; sides of neck, breast, and sides streaked with dusky. MALE IN WINTER: Resembles female; chin black. Wing 82–92, tail 60.5–71, culmen 12.5–14.5, tarsus 22.5–26 mm. Great Plains, east to western Minnesota, Nebraska, Kansas, western Oklahoma, and western Texas, west to Montana, Wyoming, eastern Colorado, and eastern New Mexico. Winters from Texas and Arizona to Mexico.

Genus *Arremonops* Ridgway. Bill more than $\frac{1}{2}$ tarsus, depth less than $\frac{2}{3}$ length; culmen slightly curved; wing short, reaching little beyond base of tail, rounded; outer primary shorter than secondaries; wing tip very short; tail about equal to wing, graduated; rectrices broad; tarsus more than ⅓ tail; lateral claws falling short of middle claw. Texas to South America. 4 species, 1 in the United States.

Arremonops rufivirgatus (**Lawrence**). Olive sparrow (Fig. 5-101). Above olive; bend of wing yellow; sides of crown and stripe through eye brown; median crown stripe olive or gray; superciliary, cheeks, throat, breast, sides, and flanks

Fig. 5-101. Head and foot of olive sparrow (*Arremonops rufivirgatus*), enlarged.

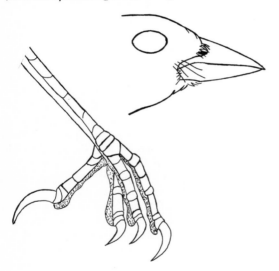

Fig. 5-102. Head and foot of rufous-sided towhee (*Pipilo erythrophthalmus*), enlarged.

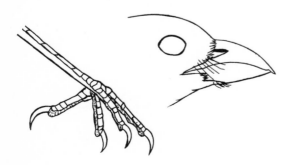

grayish or buff; belly white; crissum buffy; iris brown; bill horn, pale below; feet pale brown. Wing 59–67.5, tail 54.5–70, culmen 12–14, tarsus 23–25.5 mm. Mexico to southeastern Texas.

Genus *Chlorura* Sclater. Near *Arremonops*, but depth of bill about ⅔ length; culmen straight; outer primary about equal to 6th; tail longer than wing; tarsus less than ⅓ tail; lateral claws reaching beyond base of middle claw. Western North America. 1 species.

Chlorura chlorura (**Audubon**). Green-tailed towhee. Forehead black; crown chestnut; upper parts grayish olive; wings and tail olive green; bend of wing yellow; anterior part of lores, short malar stripe, and throat white; cheeks, sides of neck, breast, and sides gray; lower breast and belly white; crissum buff; iris reddish brown; bill black, bluish below; tarsi pale brown; toes dark brown. YOUNG: Crown grayish olive; back, breast, and sides streaked with dusky; 2 buffy wing bands. Wing 71–83.5, tail 74.5–87, culmen 11.5–13, tarsus 22–25.5 mm. Transition zone, from Washington, Idaho, and Montana, south to California, New Mexico, and western Texas. Winters from California and Texas to Mexico.

Genus *Pipilo* Vieillot. Near *Chlorura*, but culmen curved, straight, or indented; outer primary shortest; lateral claws scarcely reaching base of middle claw. North America to Guatemala. 6 species, 3 in the United States.

KEY TO SPECIES OF *PIPILO*

1. Rectrices with broad white tips
                               P. erythrophthalmus
   Rectrices with narrow buff tips or none   **2**
2. A necklace of dusky streaks   *P. fuscus*
   No necklace, but lores and chin dusky   *P. aberti*

Pipilo erythrophthalmus (**Linnaeus**). Rufous-sided towhee (Fig. 5-102). Head and neck black; back plain black, or olive streaked with black and white; rump and upper tail coverts black or olive; wings black; 2d to 7th primaries with a white basal patch; edge of 2d to 5th primaries partly white; tertials edged with white; middle and greater coverts sometimes white-tipped, forming 2 bands; edge and lining of wing white; tail black; 3–4 outer rectrices with large white tips; breast and belly white; sides and flanks chestnut; crissum rich buff; thighs mixed black and white; iris red, orange, or pale yellow; bill blackish; feet brown. FEMALE: Black usually replaced by brown; bill horn, paler below. YOUNG: Above brown; crown, nape and scapulars streaked with buff; wings and tail as in adult, but white marking small and tinged with buff; only 2 outer pairs of rectrices white-tipped; below buffy white; breast, sides, and flanks streaked with brown; thighs and crissum buff, speckled with brown; iris brown. Wing 71–92.5, tail 80.5–111.5, culmen 12–15.5, tarsus 25–31 mm. Southern Canada and the United States, south to high mountains of Mexico and Guatemala. Winters from Potomac and Ohio Valleys, Nebraska, Colorado, Nevada, and Pacific states southward.

Pipilo fuscus **Swainson.** Brown towhee. Above grayish brown; throat buff, bordered laterally and behind by dusky spots; breast, sides, and flanks buffy gray; belly white; crissum rich buff; iris brown; bill dusky, brownish below; feet brown. Wing 83–105, tail 85–119.5, culmen 13.5–17.5, tarsus 23–29 mm. Mexico to California, southwestern Oregon, Nevada, Arizona, New Mexico, southern Colorado, western Texas, and western Oklahoma.

Pipilo aberti **Baird.** Abert's towhee. Grayish brown; below paler and more pinkish buff; crissum pinkish cinnamon; lores dusky; chin streaked with dusky; bill and tarsi pale brown; toes darker. Wing 82–98, tail 102–121, culmen 15–16, tarsus 26–30 mm. Deserts of Sonora, southeastern California, southern Nevada, southwestern Utah, southern Arizona, and southwestern New Mexico.

Genus *Sporophila* Cabanis. Bill short, swollen, nearly as deep as long; culmen strongly curved; maxilla grooved; wing short; outer primary about equal to 6th; wing tip short; tail more than ¾ wing, rounded; tarsus about ⅓ tail; lateral claws not extending beyond base of middle claw. Tropical America. 32 species, 1 in the United States.

Sporophila torqueola (**Bonaparte**). White-collared seedeater (Fig. 5-103). Above black; rump white; wing coverts white-tipped, forming 2–3 bands; base of inner primaries white; sides of neck and underparts white; pectoral collar black; iris brown; bill black; feet dusky. Extralimital races may have rump and underparts cinnamon, or pectoral collar extending nearly to chin, or lack white

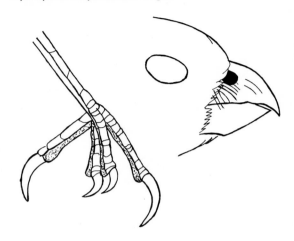

Fig. 5-103. Head and foot of white-collared seedeater (*Sporophila torqueola*), enlarged.

tips to wing coverts. FEMALE: Above olive brown; 2 buffy wing bands; below buff; bill horn; feet brown. Wing 48–57, tail 38.5–49, culmen 8–9.5, tarsus 14–16 mm. Lower Rio Grande south to Costa Rica, in grass.

Genus *Passerculus* Bonaparte. Bill conical, depth less than ⅔ length; culmen slightly indented; outer primary longer than 5th; wing tip somewhat shorter than tarsus; tertials long; tail hardly ¾ wing, double-rounded (Fig. 5-105); rectrices rather acuminate, middle pair about as long as outer pair; tarsus more than ⅓ tail; hallux as long as inner toe; lateral claws not reaching base of middle claw. North America. 1 species.

Passerculus sandwichensis (**Gmelin**).[1] Savannah sparrow (Fig. 5–104). Above brown, streaked with black, and on back with ashy; usually with a pale median crown stripe; wings and tail dusky, edged with brown or ashy; superciliary whitish or yellow; postocular, submalar, and rictal streaks dusky; auriculars brownish; malar stripe white or buffy; below white; sides of neck, breast, sides, and flanks streaked with dusky brown; edge of wing white or yellowish; iris brown; bill horn, paler below; feet flesh. Wing 61–84, tail 42.5–65, culmen 9.5–13, tarsus 18–24 mm. North America, in open country, south to New Jersey, West Vir-

[1] Includes *Passerculus princeps* Maynard of A.O.U.

Fig. 5-104. Head and foot of Savannah sparrow (*Passerculus sandwichensis*), enlarged.

Fig. 5-105 (left). Tail of Savannah sparrow (*Passerculus sandwichensis*). *Fig. 5-106* (right). Tail of seaside sparrow (*Ammospiza maritima*).

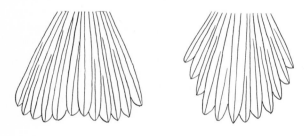

ginia, Ohio, Indiana, Illinois, Iowa, South Dakota, New Mexico, Utah, Nevada, and California, and in parts of Mexico and Guatemala. Winters from New York, Indiana, Arkansas, and Oklahoma south to Gulf Coast and Mexico, and from British Columbia south along Pacific Coast.

Genus Ammodramus Swainson. Bill stout, depth ⅔ length; culmen faintly indented; nostril oval; maxillary tomium sinuate before nostril; wing short; outer 4 primaries longest, scarcely exceeding tertials; wing tip about equal to middle toe without claw; tail short, double-rounded; outer rectrices shorter than middle pair; tips of rectrices sharp,

narrow; hallux longer than lateral toes. America. 2 species.

KEY TO SPECIES OF *AMMODRAMUS*

1. Breast and flanks streaked with dusky   **A. bairdii**
   Breast and flanks immaculate or faintly streaked
       A. savannarum

Ammodramus savannarum (**Gmelin**). Grasshopper sparrow (Fig. 5-107). Crown dusky, with a buff median stripe; nape gray, streaked with brown; back gray, blotched with black, chestnut, and buff; rump similar but without buff; upper tail coverts and rectrices blackish, broadly edged with brown; edge of wing yellow; lesser coverts olive; middle and greater coverts dusky, tipped with buff; remiges dusky, edged with buffy; superciliary buffy, yellow in front of eye; postocular stripe brown; auricular spot blackish; belly and wing lining white; rest of underparts buff; iris brown; bill horn, pale below; feet flesh. Wing 53.5–64, tail 37.5–51.5, culmen 10–12.5, tarsus 18–21 mm. Open fields, from Maine, New Hampshire, southern parts of Ontario, Michigan, Wisconsin, and Minnesota, North Dakota, Montana, and British Columbia, south to northern Georgia, and Alabama, Louisiana, Texas, and California; also in central Florida, West Indies, southern Mexico, Central America, and northern South America. Winters from North Carolina, Illinois, Texas, and California south.

Ammodramus bairdii (**Audubon**). Baird's sparrow. Crown buff, streaked laterally with black; upper parts brown, blotched with black and buffy; edge of wing white; sides of head buffy; postocular, rictal, and submalar stripes black; below pale buff; breast, sides, and flanks streaked with black; iris brown; bill brown, flesh below; feet flesh. Wing 66–72.5, tail 48–53.5, culmen 10–11, tarsus 19.5–21 mm. Northern Great Plains, south to Minnesota, North Dakota, and Montana. Winters from Texas to northwestern Mexico; Arizona and New Mexico on migration.

Genus Ammospiza Oberholser. Near *Ammodramus*, but depth of bill ½ to ⅔ length; tail graduated, more than 80% wing (Fig. 5-106); hallux shorter than inner toe. North America. 4 species.

KEY TO SPECIES OF *AMMOSPIZA*

1. Center of crown gray   2
   Center of crown buff or greenish   3
2. Superciliary buff   **A. caudacuta**
   Superciliary olive, yellow in front of eye
                                 A. maritima
3. Nape chestnut   **A. leconteii**
   Nape greenish   **A. henslowii**

Ammospiza leconteii (**Audubon**).[1] Leconte's sparrow. Crown blackish laterally, median stripe buff; nape chestnut, edged with gray; upper parts black, edged with buff and chestnut; superciliary and malar stripes buff; postocular streak black; auriculars slate; throat, breast, sides, and flanks buff; sides and flanks with a few black streaks; belly and edge of wing white; iris dark brown; bill horn, pale below; feet flesh. Wing 49–54, tail 46–56, culmen 8.5–10, tarsus 17.5–19 mm. Swales of northern Great Plains, south to Minnesota and North Dakota. Winters from Kansas, Missouri, Tennessee, and South Carolina, south to Florida and Texas.

Ammospiza henslowii (**Audubon**).[2] Henslow's sparrow. Sides of crown black; mid-crown stripe and superciliary olive green; nape olive green, streaked with black; back chestnut, blotched with black and margined with ash; rump and upper tail coverts deep buff, streaked with black; tail and wings brown; bend of wing with a slight green tinge; edge of wing yellowish white; wing coverts and tertials chestnut, streaked with black; rictal and submalar streaks black; sides of head pale buff; throat and belly white; breast, sides, and flanks buff, streaked with black; crissum buff; iris brown; bill brown above, yellow below; feet flesh. Wing 49–56.5, tail 44.5–53, culmen 10–14, tarsus 15–18.5 mm. Swales, from southern parts of New Hampshire, Vermont, New York, Ontario, Michigan, and Wisconsin, Minnesota, and South Dakota, south to northern Virginia, North Carolina, Ohio, Indiana, Illinois, Missouri, and eastern Kansas. Winters from South Carolina to Florida, thence west to Texas.

Ammospiza caudacuta (**Gmelin**). Sharp-tailed sparrow. Sides of crown brown, streaked with black; median stripe gray; upper parts grayish

[1] *Passerherbulus caudacutus* (Latham) of A.O.U.
[2] *Passerherbulus henslowii* (Audubon) of A.O.U.

Fig. 5-107. Head and foot of grasshopper sparrow (*Ammodramus savannarum*), enlarged.

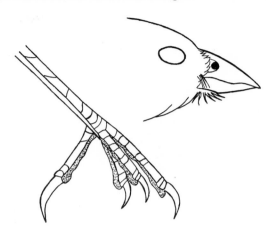

olive, usually streaked with black and ash; tail and wings brown, edged with olive or buff; edge of wing yellow; superciliary and malar stripes buff; postocular streak brown; auricular gray; submalar streak dusky; belly and wing lining white; throat often whitish; rest of underparts buff; breast, sides, and flanks usually streaked with black; iris brown; bill horn, pale below; feet light brown. Wing 52–61, tail 42.5–53, culmen 10–12.5, tarsus 18–23 mm. Salt marshes from Quebec to Virginia; freshwater marshes of northern Great Plains, south to Minnesota and South Dakota. Winters in salt marshes from New Jersey to Florida, west to Texas.

Ammospiza maritima (**Wilson**).[3] Seaside sparrow (Fig. 5-108). Crown brownish, streaked with black; faint median stripe gray; upper parts olive gray, streaked with dusky, and sometimes edged with ashy; tail and wings dusky, edged with olive or brown; edge of wing yellow; superciliary olive, yellow in front of eye; auriculars gray; postauricular spot dusky; malar stripe and chin white; submalar stripe dusky; breast, sides, flanks, and crissum gray, streaked with dusky, and often tinged with buff; thighs brown; iris brown; bill black, horn below; feet dark brown. YOUNG: Above brown, streaked with dusky; lores, malar stripe, and un-

[3] Includes *Ammospiza nigrescens* (Ridgway) and *A. mirabilis* (Howell) of A.O.U.

Fig. 5-108. Head and foot of seaside sparrow (*Ammospiza maritima*), enlarged.

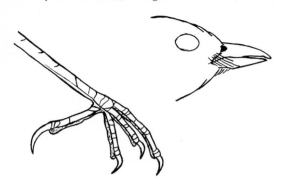

derparts buffy white; submalar stripe dusky; breast, sides, flanks, and crissum streaked with dusky; edge of wing whitish. Wing 54.5–65, tail 48.5–59, culmen 12.5–15, tarsus 21–24 mm. Salt marsh, from Massachusetts to Florida, west to Texas.

Genus Aimophila Swainson. Nostril longitudinal; wing and wing tip short; outer primary shorter than 6th; tail equal to or longer than wing, graduated; rectrices rather narrow, but tips rounded; tarsus less than ⅓ wing, slightly longer than middle toe with claw; claws weak; lateral claws falling short of base of middle claw. America. 13 species, 4 in the United States.

KEY TO SPECIES OF *AIMOPHILA*

1. Bend of wing yellow   2
    Bend of wing white   3
2. Flanks streaked with dark brown   *A. cassinii*
    Flanks unstreaked   *A. aestivalis*
3. Lesser coverts chestnut   *A. carpalis*
    Lesser coverts not chestnut   *A. ruficeps*

Aimophila aestivalis (**Lichtenstein**).[1] Bachman's sparrow; Botteri's sparrow. Above rusty brown, streaked with gray and usually with black; tail dusky, edged with brown or gray and indistinctly tipped with paler; wings dusky, edged with brown or gray; bend of wing yellow; lesser coverts rusty, washed with olive yellow; superciliary and sides of face buff or buffy gray; postocular stripe

[1] Includes *Aimophila botterii* (Sclater) of A.O.U.

brown; indistinct submalar streak dusky; breast, sides, and crissum buff; throat paler; belly white; iris brown; bill horn, pale below; feet straw. YOUNG: Breast, sides, and flanks spotted or streaked with black. Wing 56.5–70, tail 56.5–70.5, culmen 11–13, tarsus 18–23.5 mm. Pine or oak woods, from Virginia, Maryland, southern Pennsylvania, southern Indiana, southern Illinois, and Iowa, south to Florida, Gulf Coast, and eastern Texas, through Mexican tableland; formerly southern Arizona. Winters from Gulf states southward.

Aimophila cassinii (**Woodhouse**). Cassin's sparrow. Near *A. aestivalis,* but black markings on back tending to form spots or bars; dusky shaft stripe on middle pair of rectrices tending to form confluent bars; flanks streaked with dusky; below paler. Wing 59.5–67.5, tail 61–72, culmen 10–12, tarsus 18–20.5 mm. Western Kansas, southern Colorado, and southern Nevada, south to Rio Grande and northern Mexico. Winters in Arizona and northern Mexico.

Aimophila ruficeps (**Cassin**). Rufous-crowned sparrow. Forehead narrowly blackish, with an indistinct whitish midline; crown chestnut; upper parts brown or grayish, streaked with chestnut or dusky; tail rusty brown; superciliary, sides of head, and underparts dirty white, darker across breast; postocular streak chestnut; submalar streak black. YOUNG: Crown brown, streaked with darker; breast streaked with dusky. Wing 55.5–70.5, tail 57–74, culmen 10.5–14, tarsus 19–22 mm. California, Arizona, New Mexico, southern Colorado, Oklahoma, and western Texas, south over Mexican plateau.

Aimophila carpalis (**Coues**). Rufous-winged sparrow. Crown brown, streaked with black, and with a gray median stripe; back brownish gray, streaked with black; rump and tail brownish gray; outer rectrices white-tipped; lesser wing coverts cinnamon; wing brown, edged with ashy; superciliary and sides of head light gray; postocular stripe rusty; rictal and submalar streaks black; below grayish white; bill dusky, pale below. YOUNG: Crown and lesser coverts brown, streaked with black; breast streaked with black. Wing 59.5–66, tail 62–68, culmen 9.5–10.5, tarsus 18–20 mm. Southern Arizona, Sonora, and Sinaloa.

Genus Melospiza Baird. Near *Aimophila,* but tail slightly shorter than wing; tarsus more than ⅓ wing. North America. 3 species.

KEY TO SPECIES OF *MELOSPIZA*

1. Breast unstreaked   *M. georgiana*
   Breast heavily streaked   **2**
2. A black pectoral spot; outer primary shorter than 6th
                              M. melodia
   No pectoral spot; outer primary longer than 6th
                              M. lincolnii

Melospiza melodia (Wilson). Song sparrow. Crown brown, striped with black, gray on midline; upper parts buffy brown, streaked with dark brown; back also streaked with black and ash; tail brown; lesser coverts brown; middle and greater coverts and tertials brown, dusky subterminally, whitish at tip, forming 2 indistinct bands; remiges dusky, edged with brown; edge of wing white; superciliary, lores, and auriculars pale buffy gray; postocular, rictal, and submalar stripes dark brown; malar stripe buffy white; below white; breast, sides, and flanks tinged with buff and heavily streaked with dark brown; breast streaks coalescing to form a midpectoral spot; iris brown; bill horn, pale below; feet brown. Wing 53.5–87.5, tail 50–86.5, culmen 10–18, tarsus 20–29 mm. North America, south to Virginia, northern Georgia and Alabama, Kentucky, Illinois, Iowa, western Kansas, Colorado, northern New Mexico, Utah, Nevada, southwestern Arizona, California, and Mexican plateau. Winters from Massachusetts and Great Lakes south to Florida and Gulf states, and throughout the West.

Melospiza lincolnii (Audubon). Lincoln's sparrow. Sides of crown brown, streaked with black, median stripe olive gray; upper parts buffy olive, streaked with black; tail and wings dusky, edged with brown; superciliary gray; postocular and rictal stripes dark brown; malar stripe buff; auriculars olive brown; chin and belly white; sides of neck, breast, sides, flanks, and crissum buff, streaked with black; iris brown; bill horn, pale below; feet light brown. Wing 54.5–66.5, tail 51–62, culmen 9.5–12, tarsus 19–22 mm. Northern coniferous forest, south to New York, northern Minnesota, northern Michigan, and in mountains to New Mexico and California. Winters from Mississippi, Oklahoma, and California to Mexico and Guatemala. Rare on migration in South Atlantic states.

Melospiza georgiana (Latham). Swamp sparrow. Forehead black with a gray midline; crown chestnut; back rusty brown, streaked with black, buff, and gray; rump buffy brown, sometimes streaked with black; upper tail coverts rusty brown, streaked with black; tail and wings dusky, edged with rusty brown; tertials also edged with buffy white at tip; edge of wing white; superciliary and sides of neck gray; postocular, rictal, and submalar stripes dusky; auriculars brown; chin and malar stripe pale gray; breast darker and obscurely mottled with whitish; belly white; sides and flanks buffy brown; crissum buff; iris brown; bill horn, pale below; feet light brown. IMMATURE: Sides of crown brown, streaked with black; midline of crown gray; lores and malar stripe tinged with yellowish. YOUNG: Crown and upper parts grayish buff, streaked with dusky; 2 buffy-white wing bands; breast and sides shaded with buff and streaked with dusky; rest of underparts white. Wing 57.5–65.5, tail 52.5–64, culmen 10.5–12, tarsus 20.5–22 mm. Eastern North America, south to New Jersey, Pennsylvania, West Virginia, Indiana, Illinois, Missouri, and Nebraska. Winters from New Jersey, Ohio Valley, and Nebraska, south to Florida, Gulf Coast, Texas, and northern Mexico.

Genus *Zonotrichia* Swainson. Near *Melospiza* and *Aimophila*, but tail rounded or only slightly graduated; tarsus $\frac{1}{3}$ wing or less. North America. 4 species.

KEY TO SPECIES OF *ZONOTRICHIA*

1. Entire crown black   *Z. querula*
   Crown with a pale median stripe   **2**
2. A yellow spot in front of eye   *Z. albicollis*
   No yellow in front of eye   **3**
3. Median crown stripe yellow   *Z. atricapilla*
   Median crown stripe white or buffy   *Z. leucophrys*

Zonotrichia querula (Nuttall). Harris's sparrow. Crown black; upper parts brown; back streaked with black; 2 whitish wing bands; sides of head buffy brown; postauricular spot black; throat black, bordered with whitish; middle of breast blotched with black; belly white; sides and flanks buffy brown, streaked with darker; crissum pale buff; iris brown; bill horn; feet light brown. YOUNG: Crown black, squamate with grayish buff; chin black; throat white, bordered with black; middle of breast blotched with black. Wing 80–

91.5, tail 77–86, culmen 12–13, tarsus 23–25 mm. Breeds in interior Canada; winters in Nebraska, Missouri, Kansas, Oklahoma, and Texas, irregularly farther eastward and westward.

Zonotrichia atricapilla (**Gmelin**). Golden-crowned sparrow. Sides of crown black, midline yellow; nape gray; above grayish brown; back streaked with dusky and chestnut brown; greater coverts and tertials chestnut brown externally; 2 white wing bands; edge of wing yellow; sides of head, throat, and breast gray; belly whitish; sides and flanks buffy brown; crissum paler; iris brown; bill horn, pale below; feet brown. YOUNG: Front of crown yellowish olive, flecked with dusky; posterior crown olive brown, streaked with dusky. Wing 73.5–83.5, tail 69–83.5, culmen 11–13, tarsus 23.5–25.5 mm. Alaska and British Columbia; winters west of Sierra, from Oregon to Baja California.

Zonotrichia leucophrys (**Forster**). White-crowned sparrow. Sides of crown black, median stripe white; nape grayish; back grayish brown, streaked with chestnut brown; rump, upper tail coverts, tail, and wings brown; tertials dusky, edged with chestnut brown and terminally with whitish; 2 white wing bands; edge of wing white or yellow; lores black or white; superciliary white; long postocular stripe black; auriculars, sides of neck, throat, and breast gray; belly white; flanks buffy brown; crissum pale buff; iris brown; bill pink or yellow, tip dusky; feet light brown. YOUNG: Sides of crown chestnut, median stripe pale cinnamon. Wing 67.5–83.5, tail 63.5–82, culmen 10–12, tarsus 21.5–24.5 mm. Northern coniferous forest, south to mountains of New Mexico, Arizona, and California, and coastal forest to Santa Barbara County, California. Winters from Potomac and Ohio Valleys, Kansas, Utah, and British Columbia, south to Gulf states and Mexico.

Zonotrichia albicollis (**Gmelin**). White-throated sparrow. Sides of crown black mixed with brown; median stripe whitish; back rusty brown, streaked with black and gray; rump, upper tail coverts, and tail brown; wings dusky, edged with rusty brown; primaries edged with pale buff; 2 whitish wing bands; edge of wing pale yellow; superciliary white, yellow in front of eye; postocular stripe brown; auriculars, sides of neck, and breast gray; throat and malar stripe white; rictal and indistinct submalar stripes black; belly white;

sides and flanks buffy brown, indistinctly streaked with darker; crissum pale buff; axillars whitish; iris brown; bill horn, paler below; feet light brown. YOUNG: Duller and tinged with buffy; breast with obscure dusky streaks. Wing 68–77, tail 67–76, culmen 10.5–12, tarsus 22.5–24.5 mm. Northern coniferous forest, south to Massachusetts, New York, Pennsylvania, Michigan, Wisconsin, Minnesota, and Montana. Winters from Massachusetts, Ohio Valley, Missouri, and Oklahoma, south to Florida, Gulf Coast, and northeastern Mexico.

Genus *Spizella* Bonaparte. Bill short, depth equal to width and $\frac{3}{4}$ length; nostril longitudinal, partly concealed; maxillary tomium slightly sinuate; gonys less than width of bare rami; outer primary longer than 7th to longer than 5th; tail 90–105% wing, double-rounded; middle rectrices shorter than outer pair; tail coverts extending beyond end of closed wing; legs slender; inner toe shorter than outer. North America. 7 species.

KEY TO SPECIES OF *SPIZELLA*

1. A black pectoral spot	*S. arborea*
No pectoral spot	2
2. A blackish streak through eye	*S. passerina*
No blackish streak through eye	3
3. Crown plain gray	*S. atrogularis*
Crown not plain gray	4
4. Crown rusty	5
Crown heavily streaked with black	6
5. Postocular stripe rusty	*S. pusilla*
No postocular stripe	*S. wortheni*
6. Crown paler on midline	*S. pallida*
Crown not paler on midline, evenly streaked	
	S. breweri

Spizella arborea (**Wilson**). Tree sparrow. Crown chestnut; back buffy brown, streaked with black and rusty; rump and upper tail coverts buffy brown; tail dusky edged with buffy or whitish; wings dusky, edged with rusty or buff; 2 whitish wing bands; edge of wing white; superciliary pale gray; postocular stripe, spot before eye, short rictal stripe, and patch on sides of breast chestnut; auriculars, sides of neck, throat, and breast pale gray; pectoral spot dusky; belly and crissum pale buff; sides and flanks deeper; iris brown; maxilla black; mandible yellow, tip black; feet dark brown. Wing 70–82, tail 64–73, culmen 9–10.5, tarsus 20–21.5

mm. Breeds in Alaska and northern Canada. Winters south to South Carolina, Tennessee, Arkansas, Texas, New Mexico, Utah, Nevada, and Oregon.

Spizella passerina (Bechstein). Chipping sparrow. Forehead black, divided by a gray line; crown chestnut, often streaked with black; nape gray, streaked with black; back brown, streaked with black and rusty; rump and upper tail coverts gray, obscurely streaked with dusky; tail dusky, edged with gray; wings dusky, edged with brown; 2 whitish wing bands; edge of wing and axillars white; superciliary whitish; stripe through eye black; auriculars and sides of neck gray; below white; breast and flanks shaded with pale gray; iris brown; bill black; feet pale. WINTER: Chestnut of crown largely concealed by buffy brown, streaked with black; wing bands, superciliary, auriculars, breast, and flanks tinged with buff; an indistinct dusky rictal stripe; bill brown, pale below. YOUNG: Crown, back, rump, and upper tail coverts buffy brown streaked with black; nape, sides of face, and underparts whitish, heavily streaked with black; wing bands buffy. Wing 63–76, tail 51–65, culmen 8.5–10.5, tarsus 15.5–18 mm. North America, south to Georgia, northwestern Florida, Mississippi, Louisiana, Texas, and in pines through West to Mexico and northern Central America. Winters from southeastern United States, Texas, and California southward.

Spizella pusilla (Wilson). Field sparrow. Crown rusty, with obscure gray midline; back rusty, streaked with black and buff; rump and upper tail coverts buffy brown; tail dusky, edged with gray; wings dusky, edged with brown; 2 whitish wing bands; edge of wing white; superciliary and sides of neck gray; postocular stripe rusty; auriculars brownish; below pale grayish buff; belly whitish; a rusty patch on sides of breast; iris brown; bill pink; feet flesh. YOUNG: Crown, breast, and sides obscurely streaked with dusky. Wing 59.5–71, tail 54.5–72, culmen 8.5–10, tarsus 17–20 mm. Southeastern Canada, Maine, Michigan, Minnesota, South Dakota, and eastern Montana, south to Georgia, Florida panhandle, and northern parts of Alabama, Mississippi, Louisiana, and Texas. Winters from New Jersey, Ohio Valley, Missouri, and Oklahoma, through southeastern United States, and into northeastern Mexico.

Spizella wortheni Ridgway. Worthen's sparrow. Near *S. pusilla*, but forehead gray; back without rusty; wing bands indistinct; auriculars and sides of head gray; eye ring white; no rusty postocular stripe or patch on sides of breast; feet dark brown. Wing 65–70, tail 57.5–64, culmen 9–10, tarsus 17–18.5 mm. Northeastern Mexico to Silver City, New Mexico.

Spizella atrogularis (Cabanis). Black-chinned sparrow. Crown and nape gray; back rusty, streaked with black; rump, upper tail coverts, and lesser wing coverts gray; tail and remiges dusky, edged with gray; tertials, middle and greater coverts dusky, edged with brown; lores and throat black; sides of head, sides of neck, and underparts gray; belly whitish; bill pink; feet dark brown. YOUNG: Chin and throat gray. Wing 60–70, tail 60–74, culmen 8.5–10, tarsus 17.5–20.5 mm. Mexican plateau, north to southern California, Arizona, and New Mexico; western Texas on migration.

Spizella pallida (Swainson). Clay-colored sparrow. Sides of crown light brown streaked with black, midline buffy gray; nape gray; upper parts buffy brown, streaked with black; rump and upper tail coverts often unstreaked; tail brown, edged with gray; lesser coverts buffy brown; wings dusky, edged with buffy brown; 2 buffy wing bands; superciliary buffy white; auriculars buffy, bordered with dark brown; malar stripe white; indistinct submalar stripe brown; below whitish; breast and sides shaded with buff; bill light brown, tip dusky; feet pale. YOUNG: Breast streaked with dusky. Wing 58–64, tail 53–62, culmen 8.5–10, tarsus 17–18 mm. Great Plains, south to northern Michigan, northern Illinois, Iowa, Nebraska, Colorado, and Montana. Winters in Mexico, north to Texas and New Mexico, irregularly eastward.

Spizella breweri Cassin. Brewer's sparrow. Near *S. pallida*, but crown pale brown, uniformly streaked with black, and without gray median stripe; auriculars grayer. Wing 56–66, tail 57.5–63.5, culmen 8.5–9, tarsus 16–18 mm. Sagebrush of western Canada and western United States, east to Montana, Wyoming, western Nebraska, Colorado, New Mexico, and western Texas. Winters in Mexico, north to California and Texas.

Genus *Amphispiza* Coues. Near *Spizella*, but bill deeper than wide; gonys longer than width across bare rami; tail always slightly shorter than wing, rounded; upper tail coverts falling short of end of closed wing. Western North America. 2 species.

KEY TO SPECIES OF *AMPHISPIZA*

1. Superciliary stripe white   *A. bilineata*
   Only supraloral spot and eye ring white   *A. belli*

Amphispiza bilineata (**Cassin**).   Black-throated sparrow. Above brownish gray; outer rectrix edged and tipped with white; intermediate rectrices white-tipped on inner web; superciliary white, bordered above by a black line; sides of head gray; malar stripe white; lores, throat, and middle of breast black; lower breast and belly white; sides, flanks, and crissum buffy; iris brown; maxilla black; mandible grayish blue, tipped with black; feet dusky. YOUNG: Throat white; breast white, streaked with gray; greater coverts and tertials edged with buffy. Wing 60.5–70.5, tail 55.5–68, culmen 9–10.5, tarsus 17.5–20 mm. Desert scrub and sagebrush from Mexico north to California, Nevada, Utah, western Colorado, New Mexico, and western Texas.

Amphispiza belli (**Cassin**).   Bell's sparrow; sage sparrow. Above brownish gray, usually streaked with black; sometimes a narrow white median crown stripe; tail blackish, edged with light brown; wings blackish, edged with gray; 2 buffy-brown wing bands; edge of wing pale yellow; lores black; supraloral spot and eye ring white; auriculars brownish gray; malar stripe and underparts white; midpectoral spot and stripe bordering throat black; sides and flanks buffy, streaked with dusky; iris brown; maxilla black; mandible bluish gray; feet dusky. YOUNG: Tinged with buff; black border of throat broken; breast streaked with black. Wing 59–81, tail 58.5–78.5, culmen 8–10.5, tarsus 19–22.5 mm. Desert scrub, sagebrush, and chaparral of Mexico, north to eastern Washington, Idaho, southwestern Montana, western Colorado, and western New Mexico, and on coast to central California. In winter spreads to western Texas.

Genus *Junco* Wagler.   Bill slender, deeper than wide, depth less than $\frac{2}{3}$ culmen; tomium before angle nearly straight; gonys much longer than width across bare rami; outer primary longer than 7th; tail double-rounded, more than $\frac{3}{4}$ **wing; tail**

coverts extending beyond closed wing; lateral toes short, subequal. North America. 4 species, 3 in the United States.

KEY TO SPECIES OF *JUNCO*

1. Iris yellow; mandible yellow; maxilla black
                                                     J. phaeonotus
   Iris brown; mandible flesh; maxilla usually flesh, rarely black   **2**
2. 2 white bands on wing coverts   *J. aikeni*
   No wing bands   *J. hyemalis*

Junco aikeni **Ridgway.**   White-winged junco. Slate gray; middle and greater coverts white-tipped; 3 outer rectrices entirely, 4th partly white; belly and crissum white; iris brown; bill and feet flesh. Wing 81–93, tail 71–79, culmen 11.5–13, tarsus 20–22 mm. Black Hills of South Dakota and Wyoming and adjacent parts of Montana and Nebraska; winters south to Arizona, New Mexico, western Texas, Oklahoma, and western Kansas.

Junco hyemalis (**Linnaeus**).[1]   Slate-colored junco; brown-eyed junco. Iris dark brown; mandible and feet flesh; maxilla usually flesh (black in *J. h. dorsalis*). EASTERN RACES: Colored like *J. aikeni*, but without white wing bands; only 2 outer pairs of rectrices largely white, 3d and 4th with at most only a white streak. WESTERN RACES: Head and neck black or gray; middorsum brown or reddish brown; sides and flanks brownish or pinkish; 2 (rarely 3) outer pairs of rectrices largely white. SOUTHWESTERN RACES: Back and sometimes edges of wing coverts and tertials reddish brown; rump gray; lores black; head, neck, and sides gray; 3 outer rectrices usually largely white. Wing 66.5–86.5, tail 58.5–77, culmen 10–12, tarsus 18.5–23 mm. Coniferous forest, south to Massachusetts, New York, in mountains to Georgia, northern parts of Michigan, Wisconsin, and Minnesota, and mountains of West (except Black Hills) south to central New Mexico, central Arizona, California, and Baja California; in winter extends to Gulf states.

Junco phaeonotus **Wagler.**   Mexican junco; yellow-eyed junco (Fig. 5-109). Crown and auriculars gray; back and edges of wing coverts and tertials reddish brown; rump and upper tail coverts gray-

---

[1] Includes Oregon and gray-headed juncos, *J. oreganus* (Townsend) and *J. caniceps* (Woodhouse) of A.O.U.

ish brown; rectrices dusky, 2 outer pairs extensively white; lores black; throat, breast, sides, and flanks pale gray; belly and crissum almost white; iris, mandible, and feet yellow; maxilla black. Wing 70.5–84, tail 62.5–76.5, culmen 10.5–12.5, tarsus 20.5–22.5 mm. Pines at high altitudes in Guatemala and Mexico, north to southwestern New Mexico and southeastern Arizona.

Genus *Passerella* Swainson. Bill short, stout, depth about ¾ culmen; maxillary tomium nearly straight before angle; gonys longer than width across bare rami; outer primary longer than 6th; tail slightly double-rounded ¾ to as long as wing; upper tail coverts short; legs stout; lateral toes subequal. North America. 1 species.

Passerella iliaca (Merrem). Fox sparrow. Above gray or rusty brown; back sometimes streaked with chestnut; rump and upper tail coverts cinnamon rufous; tail and wings dusky brown, edged with reddish brown; 2 white or rusty wing bands; below white, breast and sides of neck spotted and sides and flanks streaked with chestnut rufous; crissum white or buffy, plain or streaked with brown; iris and feet brown; bill dusky, yellowish below. Wing 74–91.5, tail 66–91, culmen 10.5–16.5, tarsus 21.5–26.5 mm. Northern coniferous forest, south to Maine, Wyoming, Idaho, Nevada, and southern California. Winters from Potomac and Ohio Valleys, Arkansas, and Oklahoma to Gulf states, and from British Columbia to Baja California, Arizona, and New Mexico.

REFERENCES

General

Allen, A. A. 1947. *Ornithology Laboratory Notebook,* 5th ed., Comstock Publishing Company, Ithaca, N.Y.

——. 1961. *The Book of Bird Life,* 2d ed., D. Van Nostrand Company, Inc., Princeton, N.J.

Allen, G. M. 1925. *Birds and Their Attributes.* Marshall-Jones Company, Boston (reprinted 1962, Dover Publications, Inc., New York).

American Ornithologists' Union. 1957. *Check-list of North American Birds,* 5th ed., Lord Baltimore Press, Baltimore.

Austin, O. L., Jr. 1961. *Birds of the World,* Golden Press, New York.

Bent, A. C. 1919–1958. *Life Histories of North Amer-*

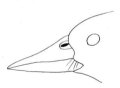

Fig. 5-109. **Head of yellow-eyed junco (*Junco phaeonotus*), enlarged.**

ican Birds, 20 vols. to date, U.S. National Museum, Washington (early numbers reprinted by Dover Publications, Inc., New York).

Berger, A. J. 1961. *Bird Study,* John Wiley & Sons, Inc., New York.

Brodkorb, Pierce. 1963–1967. *Catalogue of Fossil Birds,* 3 parts to date, Florida State Museum, Gainesville.

Cory, C. B., C. E. Hellmayr, and H. B. Conover. 1918–1948. *Catalogue of Birds of the Americas,* 15 parts, Field Museum of Natural History, Chicago.

Fisher, James, and R. T. Peterson. 1964. *The World of Birds,* Doubleday & Company, Inc., Garden City, N.Y.

Grassé, P. P. (ed.). 1950. *Oiseaux. Traité de zoologie,* vol. 15, Masson et Cie, Paris.

Gregory, Joseph T. 1952. The Jaws of the Cretaceous Toothed Birds, Ichthyornis and Hesperornis. *Condor,* **54:**73–88, figs. 1–9.

Grube, G. E. 1964. *Introductory Ornithology,* Wm. C. Brown Company, Dubuque, Iowa.

Lanyon, W. E. 1963. *Biology of Birds,* Natural History Press, Garden City, N.Y.

Marshall, A. J. (ed.). 1960–1961. *Biology and Comparative Physiology of Birds,* 2 vols., Academic Press, Inc., New York.

Palmer, R. S. (ed.). 1962. *Handbook of North American Birds,* 1 vol. to date, Yale University Press, New Haven, Conn.

Peters, J. L., and others. 1931–1964. *Check-list of Birds of the World,* 10 vols. to date, Harvard University Press, Cambridge, Mass.

Peterson, R. T. 1947. *A Field Guide to the Birds,* 2d ed., Houghton Mifflin Company, Boston.

——. 1961. *A Field Guide to Western Birds,* 2d ed., Houghton Mifflin Company, Boston.

——. 1963. *The Birds,* Time, Inc., New York.

Pettingill, O. S., Jr. 1956. *A Laboratory and Field Manual of Ornithology,* 3d ed., Burgess Publishing Company, Minneapolis.

Ridgway, Robert, and Herbert Friedmann. 1901–1950. *The Birds of North and Middle America,* 11 parts to date, U.S. National Museum, Washington.

Stresemann, Erwin. 1927–1934. Aves, in W. Kükenthal

and T. Krumbach, *Handbuch der Zoologie,* vol. 7, part 2, Walter de Gruyter & Co., Berlin.

Strong, R. M. 1939–1959. *A Bibliography of Birds,* 4 parts, Field Museum of Natural History, Chicago.

Thompson, Sir A. L. (ed.). 1964. *A New Dictionary of Birds.* McGraw-Hill Book Company, New York.

Van Tyne, Josselyn, and A. J. Berger. 1959. *Fundamentals of Ornithology,* John Wiley & Sons, Inc., New York.

Wallace, G. J. 1963. *An Introduction to Ornithology,* 2d ed., The Macmillan Company, New York.

Welty, J. C. 1962. *The Life of Birds,* W. B. Saunders Company, Philadelphia.

Wetmore, Alexander. 1960. *A Classification for the Birds of the World,* Smithsonian Institution, Washington.

Wing, L. W. 1956. *Natural History of Birds,* The Ronald Press Company, New York.

Zimmer, J. T. 1926. *Catalogue of the Edward E. Ayer Ornithological Library,* 2 parts, Field Museum of Natural History, Chicago.

Regional

Allen, G. M. 1903. *A List of the Birds of New Hampshire,* Manchester Institute of Arts and Sciences, Manchester. ·

Arvey, M. D. 1947. *A Check-list of the Birds of Idaho,* University of Kansas Museum of Natural History, Lawrence. Additions published in *Condor,* **52**:275 (1950).

Baerg, W. J. 1951. *Birds of Arkansas,* rev. ed., University of Arkansas Agricultural Experiment Station, Fayetteville.

Bailey, A. M., and R. J. Niedrach. 1965. *Birds of Colorado,* 2 vols., Denver Museum of Natural History, Denver.

Bennitt, Rudolf. 1932. *Check-list of the Birds of Missouri,* University of Missouri Studies, Columbia.

Borror, D. J. 1950. *A Check List of the Birds of Ohio* [etc.], Ohio Journal of Science, Columbus.

Brooks, Maurice. 1944. *A Check-list of West Virginia Birds,* West Virginia University Agricultural Experiment Station, Morgantown.

Bull, John. 1964. *Birds of the New York Area,* Harper & Row, Publishers, Incorporated, New York.

Burleigh, T. D. 1958. *Georgia Birds,* University of Oklahoma Press, Norman.

Butler, A. W. 1898. *The Birds of Indiana,* Department of Geology and Natural Resources, Indianapolis.

Coffey, B. B., Jr. 1936. *A Preliminary List of the Birds of Mississippi,* privately mimeographed, Memphis, Tenn.

DuMont, P. A. 1933. *A Revised List of the Birds of Iowa,* University of Iowa Studies in Natural History, Iowa City.

Eaton, E. H. 1923. *Birds of New York,* 2d ed., 2 vols., New York State Museum, Albany.

Fortner, H. C., W. P. Smith, and E. J. Dole. 1933. *A List of Vermont Birds,* Vermont Department of Agriculture, Montpelier.

Gabrielson, I. N., and S. G. Jewett. 1940. *Birds of Oregon,* Oregon State College, Corvallis.

———— and F. C. Lincoln. 1959. *Birds of Alaska,* Wildlife Management Institute, Washington.

Ganier, A. F. 1933. *A Distributional List of the Birds of Tennessee,* Tennessee Ornithological Society, Nashville.

Griscom, Ludlow, and D. E. Snyder. 1955. *The Birds of Massachusetts: An Annotated and Revised Check List,* Peabody Museum, Salem.

Gromme, O. J. 1963. *Birds of Wisconsin,* University of Wisconsin Press, Madison.

Haecker, F. W., R. A. Moser, and J. B. Swenk. 1945. *Check-list of the Birds of Nebraska,* Nebraska Bird Review, Lincoln.

Howe, R. H., Jr., and Edward Sturtevant. 1903. *A Supplement to the Birds of Rhode Island,* privately printed, Middletown.

Imhof, T. A. 1962. *Alabama Birds,* University of Alabama Press, Tuscaloosa.

Jewett, S. G., et al. 1953. *Birds of Washington State,* University of Washington Press, Seattle.

Ligon, J. S. 1961. *New Mexico Birds* [etc.], University of New Mexico Press, Albuquerque.

Linsdale, J. M. 1951. *A List of the Birds of Nevada,* Condor, Berkeley, Calif.

Lowery, G. H., Jr. 1955. *Louisiana Birds,* Louisiana State University Press, Baton Rouge.

McCreary, Otto. 1939. *Wyoming Bird Life,* Burgess Publishing Company, Minneapolis.

Mengel, R. M. 1965. *Birds of Kentucky,* American Ornithologists' Union, Lawrence, Kans.

Miller, A. H. 1951. *An Analysis of the Distribution of the Birds of California,* University of California Press, Berkeley.

Munro, G. C. 1960. *Birds of Hawaii,* rev. ed., Tuttle, Rutland, Vermont.

Murray, J. J. 1952. *A Check-list of the Birds of Virginia,* Virginia Society of Ornithology, Lexington.

Over, W. H., and C. S. Thoms. 1946. *Birds of South Dakota,* rev. ed., University of South Dakota, Vermillion.

Palmer, R. S. 1949. *Maine Birds,* 1949. Museum of Comparative Zoology, Cambridge, Mass.

Pearson, T. G., C. S. Brimley, and H. H. Brimley. 1959. *Birds of North Carolina* (revised by D. L. Wray and

H. T. Davis), North Carolina State Museum, Raleigh.

Peterson, R. T. 1960. *A Field Guide to the Birds of Texas,* Houghton Mifflin Company, Boston.

Pettingill, O. S., Jr., and N. R. Whitney. 1965. *Birds of the Black Hills,* Cornell Laboratory of Ornithology, Ithaca, N.Y.

Phillips, Allan, Joe Marshall, and Gale Monson. 1964. *The Birds of Arizona,* University of Arizona Press, Tucson.

Poole, E. L. 1964. *Pennsylvania Birds. An Annotated List,* Livingston Publishing Co., Narberth, Pa.

Rhoads, S. N., and C. J. Pennock. 1905. Birds of Delaware: A Preliminary List, *The Auk,* 22:194–205.

Roberts, T. S. 1932. *The Birds of Minnesota,* 2d rev. ed., 2 vols., University of Minnesota Press, Minneapolis.

Sage, J. H., L. B. Bishop, and W. P. Bliss. 1913. *The Birds of Connecticut,* Connecticut Geological and Natural History Survey, Hartford.

Saunders, A. A. 1921. *A Distributional List of the Birds of Montana,* Pacific Coast Avifauna, Berkeley, Calif.

Smith, H. R., and P. W. Parmalee. 1955. *A Distributional Check List of the Birds of Illinois,* Illinois State Museum, Springfield.

Sprunt, Alexander, Jr. 1954. *Florida Bird Life,* Coward-McCann, Inc., New York.

—— and E. B. Chamberlain. 1949. *South Carolina Bird Life,* University of South Carolina Press, Columbia.

Stewart, R. E., and C. S. Robbins. 1958. *Birds of Maryland and the District of Columbia,* North American Fauna, Washington.

Stone, Witmer. 1909. *The Birds of New Jersey,* New Jersey State Museum, Trenton.

Sutton, G. M. 1967. *Oklahoma Birds* [etc.], University of Oklahoma Press, Norman.

Todd, W. E. C. 1940. *Birds of Western Pennsylvania,* University of Pittsburgh Press, Pittsburgh.

Tordoff, H. B. 1956. *Check-list of the Birds of Kansas,* University of Kansas Museum of Natural History, Lawrence.

Wood, N. A. 1923. *A Preliminary Survey of the Bird Life of North Dakota,* University of Michigan Museum of Zoology, Ann Arbor.

——. 1951. *The Birds of Michigan,* University of Michigan Mueseum of Zoology, Ann Arbor.

Woodbury, A. M., Clarence Cottam, and J. W. Sugden. 1949. *Annotated Check-list of the Birds of Utah,* University of Utah, Salt Lake City.

Living mammals are sharply set off from other living vertebrates. Each side of the jaw consists of a single bone, the dentary, of dermal origin, which articulates with the squamosal, also of dermal origin. The skull articulates with the atlas of the vertebral column by 2 laterally placed occipital condyles. A tympanic bone supports the tympanic membrane. The middle ear has 3 auditory ossicles, which are remnants of the ancestral gill skeleton. At least some of the bones of the body possess epiphyses at their extremities. Teeth are restricted to the margins of the jaw. The dentition is thecodont and usually diphyodont and heterodont. The jaw teeth (molars and premolars) often have 2 or more roots. There is a single bony nasal opening. Mammals are homoiothermal animals with hair (in some stage of development) usually serving an insulative function. Mammary glands are present and secrete milk for nourishment of the young. An amnion, chorion, and allantois are present during embryologic development. All living mammals except the monotremes are viviparous. In the higher mammals a placenta permits exchange of materials between the blood stream of the embryo and that of the mother. The heart is 4-chambered with complete division into left and right halves. The left element of the fourth aortic arch persists. The brain is large principally because of enlargement of the cerebral hemispheres and development of a neopallium in their roof.

The mammals apparently evolved in the late Triassic from reptiles (subclass Synapsida) in which there was a single lateral temporal fenestra. Synapsid reptiles of the order Therapsida are known from Middle Permian to Middle Triassic. Some of these showed modification in the mammalian direction in (1) presence of 2 occipital condyles instead of 1, (2) enlargement of the dentary and corresponding reduction of other jaw elements, (3) heterodont dentition restricted to the margin of the jaw, (4) presence of a well-developed secondary palate, (5) loss of the pineal eye, (6) expansion of the temporal fenestra and loss of the postorbital bar, and (7) various other features of the skeleton. Definitely mammalian fossils are known from the middle of the Jurassic. Fragmentary remains from the Upper Triassic, classed as the order Ictidosauria, have been variously regarded as reptilian or mammalian, and the transition from therapsid reptiles to primitive mammals is incompletely known.

Mammals apparently remained relatively small in size and few in kind until after the extinction of the giant Mesozoic reptiles. Then, aided by the advantages of homoiothermy, viviparity, and development of the brain, mammals spread over most of the land surface of the earth and reinvaded the aquatic environment. Adaptive radiation of the mammals has brought the exploitation by this group of the many opportunities available on land and many of those in the seas.

A classification of extinct and living mammals is shown in Table 6-1. The Prototheria (platypus and echidnas) of Australia and New Guinea are remarkable in being the only known oviparous mammals. The pectoral girdle is reptilian in character, for there is an interclavicle, a large coracoid, and a precoracoid, and the scapula lacks a spine. These primitive mammals possess hair, homoiothermy, and mammary glands, and the lower jaw has a single bony element. Fossil Prototheria are known from no earlier than Pleistocene. They are generally regarded as having originated from reptiles independently of other living mammals.

The Multituberculata were probably the earliest herbivorous mammals and apparently represented another independent origin from reptilian

Fig. 6-1. Ringtail (*Bassariscus astutus*). (From Bailey, Biological Survey of Texas, *North Am. Fauna* 25.)

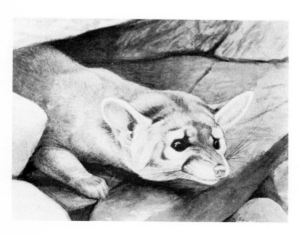

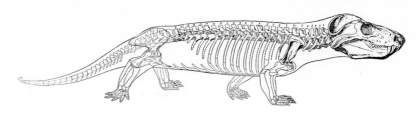

ancestors. This group appeared in the Upper Jurassic and persisted to late Paleocene. The Triconodonta were Jurassic mammals of uncertain relationship to other groups. They were probably carnivorous; the molars typically had the 3 sharp, conical cusps arranged in a row along the long axis of the tooth (Fig. 6-3).

The subclass Theria, with its 3 infraclasses, includes most of the extinct mammals and all living ones except the Monotremata. The Pantotheria and Symmetrodonta were Jurassic mammals. The former group was possibly ancestral to higher mammals. The molars were 3-cusped, with the cusps arranged in an asymmetrical triangle. The Symmetrodonta had the cusps arranged in a symmetrical triangle, with the base of the triangle external in the upper jaw and internal in the lower.

The infraclasses Metatheria and Eutheria comprise 2 distinct lines of evolution with a long history of coexistence. The Metatheria, with their less efficient mode of embryologic development, have been supplanted by the Eutheria (placental mammals) over most of the earth. Only in the Australian region, under virtual isolation from eutherian competition, has a wide variety of marsupial mammals persisted to the present. 6 of the 8 living families of Metatheria are restricted to this region, where the group shows adaptive radiation remarkably paralleling the radiation of eutherians in other parts of the world.

The infraclass Eutheria contains 16 living and 10 extinct orders, which have been arranged by G. G. Simpson in 4 natural groupings or cohorts. The cohort Unguiculata includes 2 extinct orders and the orders Pholidota and Dermoptera in addition to the 5 orders treated in this book. The Pholidota contains the single living genus *Manis* (pangolin) of Asia and Africa. The body is covered with horny, imbricated scales. The rostrum and tongue are elongated, and teeth are lacking

as adaptations to the ant-eating habit. The Tillodontia were upper Paleocene to middle Eocene omnivores or herbivores with large, gnawing incisors (Fig. 6-4). The Taeniodonta were lower Paleocene to upper Eocene, probably herbivorous offshoots of primitive insectivores. In some, the teeth were peglike and rootless, with high crowns and with enamel restricted to the sides of the teeth (Fig. 6-5).

The cohort Glires contains the orders Lagomorpha and Rodentia. The rabbits and rodents were placed in the same order in the older classifications, but it is now generally believed that they represent independent lines of evolution, known

Fig. 6-3. Theoretical sequence of development of tooth cusps from a single conical tooth. (After Osborn, from Weichert, *Anatomy of the Chordates*, McGraw-Hill Book Company.)

Fig. 6-4. An Eocene tillodont, *Tillotherium*. (After Marsh, from Romer, *Vertebrate Paleontology*, University of Chicago Press.)

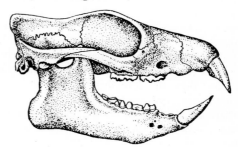

Fig. 6-5. A Paleocene taeniodont, *Wortmania.* (After Matthew, from Gregory, *Evolution Emerging,* The Macmillan Company.)

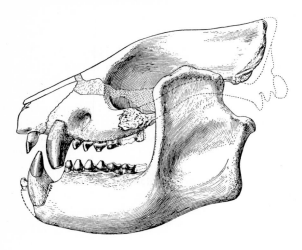

from as far back as Paleocene. The cohort Mutica includes the single order Cetacea. The relationships of the whales to other mammals are obscure. It has been suggested that they were derived from primitive carnivores (creodonts).

The cohort Ferungulata includes 8 extinct orders and the orders Tubulidentata, Proboscidea, and Hyracoidea, in addition to the 4 orders treated in this book. The order Tubulidentata is represented by the single genus *Orycteropus* (aard-vark) which is known from the Pliocene of Europe and is living today in Africa. The aard-vark is an ant-eater with a produced snout and a few peglike teeth which lack enamel. The teeth are peculiar in having numerous small canals in the dentine instead of a pulp cavity. The order Proboscidea was once widely distributed over Eurasia and the Americas, but only 2 genera and species remain today, of which one inhabits Asia and the other Africa. The group disappeared from the Americas in the Pleistocene. The huge size, graviportal limb skeleton, large hypsodont and lophodont molars, and enormously developed second incisors (tusks) are a few of the peculiarities of this group. The order Hyracoidea (hyraxes) includes 3 living genera restricted to Africa and several extinct genera known from as early as lower Oligocene. The jaw teeth are sepa-

rated from the incisors by a diastema. The incisors are enlarged and rootless, with persistent pulps. They are sharp-pointed and triangular in cross section, with enamel on the anterointernal and anteroexternal surfaces.

The Condylarthra were primitive Paleocene and Eocene ungulates of the Americas and Europe, with brachyodont and bunodont dentition and with claws or hooves. The Litopterna were a Paleocene to Pleistocene, South American offshoot of the Condylarthra. Hooves were present, and mesaxonic toe reduction to 3-toed or even 1-toed forms occurred. The tooth row usually lacked a diastema, and the jaw teeth were usually brachyodont and lophodont. The Notoungulata were a diverse group of Paleocene to Pleistocene ungulates which apparently originated in Asia and reached its greatest development in South America. The feet were mesaxonic with the number of digits tending to be reduced to 3, with hooves or claws. The upper molars were lophodont, and the lower molars typically developed 2 crescentic ridges. The Astrapotheria were a small group of ungulates which lived in South America from Eocene to Miocene. The posterior molars tended to be greatly enlarged. The Pantodonta were large, ungainly ungulates of Eurasia and North America which lived from middle Paleocene to late Eocene. The canines reached large size in some of these, and some had claws rather than hooves. The Dinocerata were heavily built ungulates of the upper Paleocene to upper Eocene of Eurasia and North America. Most had pairs of hornlike, bony swellings on the anterior part of the skull. The Pyrotheria were a small group of large South American ungulates of the Eocene and Oligocene. Members of this group showed numerous parallelisms to the proboscideans. The Embrithopoda includes a single genus known from the lower Oligocene of Egypt. This ungulate was notable for huge horns on the nasals and a smaller pair on the frontals. The tooth row was without diastema, and the molars were hypsodont.

Identification

External measurements, color, pattern, and character of the pelage, and skeletal characters, principally of the skull, are used in the description and

TABLE 6-1. *Classification of Mammals [Modified from Simpson (1945)]*
Class Mammalia

	Families		Genera	
	Extinct	Living	Extinct	Living
Subclass Prototheria				
Order Monotremata	0	2	0	3
†Subclass Allotheria				
†Order Multituberculata	5	0	35	0
Subclass uncertain				
†Order Triconodonta	1	0	8	0
Subclass Theria				
†Infraclass Pantotheria				
†Order Pantotheria	4	0	22	0
†Order Symmetrodonta	2	0	5	0
Infraclass Metatheria				
Order Marsupialia	5	8	81	57
Infraclass Eutheria				
Cohort Unguiculata				
Order Insectivora	12	8	88	71
Order Dermoptera	1	1	2	1
Order Chiroptera	2	17	16	118
Order Primates	7	11	99	59
†Order Tillodontia	1	0	4	0
†Order Taeniodonta	1	0	7	0
Order Edentata	7	3	113	19
Order Pholidota	0	1	3	1
Cohort Glires				
Order Lagomorpha	1	2	23	9
Order Rodentia	10	30	275	344
Cohort Mutica				
Order Cetacea	9	9	137	35
Cohort Ferungulata				
Order Carnivora	6	9	261	114
†Order Condylartha	6	0	42	0
†Order Litopterna	2	0	41	0
†Order Notoungulata	14	0	105	0
†Order Astrapotheria	2	0	9	0
Order Tubulidentata	0	1	1	1
†Order Pantodonta	3	0	9	0
†Order Dinocerata	1	0	8	0
†Order Pyrotheria	1	0	6	0
Order Proboscidea	5	1	22	2
†Order Embrithopoda	1	0	1	0
Order Hyracoidea	2	1	10	3
Order Sirenia	3	2	14	2
Order Perissodactyla	9	3	152	6
Order Artiodactyla	16	9	333	86

† Fossil forms.

identification of mammals. The standard external measurements are:

TOTAL LENGTH. Measured from tip of nose to tip of tail sheath.

TAIL LENGTH. Measured from proximal to distal end of tail sheath when tail is held perpendicular to long axis of body.

HIND-FOOT LENGTH. Measured from heel to tip of longest claw.

EAR LENGTH. Measured from deepest part of notch on external border to tip of pinna.

Total length is a measurement without much meaning except that body length may be determined by subtracting tail length from total length. As used in this book the term body length therefore refers to the length of the head and body and is exclusive of the tail. Where body length is given in averages and ranges (e.g., "body length averaging 135–155 mm"), the extremes are averages of geographic races as reported in the literature. Total length and tail length are ordinarily measured to whole millimeters, and foot and ear length are measured to tenths of millimeters. The measurement of ear length has been generally omitted from the present work because this measurement on fresh material is missing from much of the literature, and some workers have measured the ear from the crown instead of the notch. The length of the forearm and the length and shape of the tragus are important external characters in bat classification. The number, position, and shape of plantar pads are useful in some groups, particularly in rodents. The presence or absence and position of special integumental glands are useful, particularly in some rodents, ungulates, and carnivores. The number and position of mammae in females are useful characters for classification in some groups.

The external appearance of a mammal is greatly affected by the quality and color of its pelage. The pelage usually consists of 2 types of hairs. Underhairs are generally thick and soft and lie next to the skin. Longer, coarser, guard hairs project beyond the underhairs and protect them from wear. The relative abundance of the 2 types, the length of the hairs, and their relative coarseness or fineness vary greatly among mammals. The quills of porcupines are specialized hairs. In some heteromyid rodents, stiff, grooved hairs (spines) are scattered among the unspecialized hairs on the rump and back.

Color is one of the most difficult attributes of a mammal to describe objectively. Uniform areas of color are rare because underhairs and guard hairs are usually of different color and because the color of the hairs is in bands. The basal band is usually gray, followed by the band which gives most of the color to the pelage. The guard hairs are often tipped with a color different from that of the underhairs. In formal descriptions, an attempt is usually made to describe colors in terms of some standard reference such as Ridgway's *Color Standards and Color Nomenclature*. The color terms used in the present book are commonly used ones which do not have reference to any standard color scheme. The only objective method of measuring mammal colors is to use one of the photoelectric devices which are adaptable for this purpose.

Color is also generally one of the least reliable characters by which to identify species of mammals. Colors often vary geographically because they tend to match the color of the soils on which the animals live. Most mammals are countershaded in the coloration pattern. The underparts are palest, while the upper parts are darkest, and the darkest dorsal coloration is often along the middorsal line.

Characters of the skull and teeth are generally used in identification of species and higher categories of mammals. The most commonly used skull measurements are:

BASILAR LENGTH (OF HENSEL). From anteriormost border of foramen magnum to posterior border of alveolus of first upper incisor.

CONDYLOBASAL LENGTH. From anterior median point between bases of first upper incisors to most posterior point of occipital condyle.

GREATEST LENGTH. From tip of nasals to most posterior part of braincase.

PALATAL LENGTH. Greatest anteroposterior measurement of palate in median line.

INTERORBITAL BREADTH. Least measurement across skull between orbits.

CRANIAL BREADTH. Greatest distance across braincase.

MASTOID BREADTH. Greatest distance of skull across mastoid processes.

MAXILLARY TOOTH ROW. Length of upper mo-

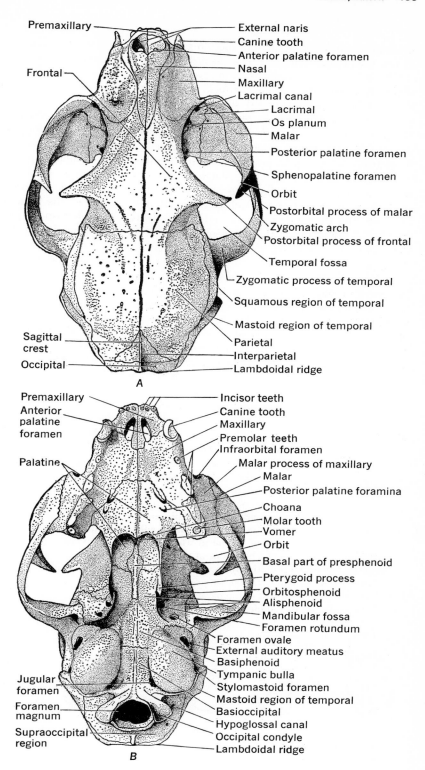

Fig. 6-6. Skull of the domestic cat *Felis catus* showing various structures used in mammal classification: *A,* dorsal view; *B,* ventral view. (From Weichert, *Elements of Chordate Anatomy,* McGraw-Hill Book Company.)

lariform tooth row from anterior margin of alveolus of first to posterior margin of alveolus of last.

Characters of the teeth are extensively used in mammal classification. The number of teeth of different kinds are expressed in the dental formula. The basic formula in placental mammals is the primitive one of $\frac{3}{3}, \frac{1}{1}, \frac{4}{4}, \frac{3}{3}$, meaning that there are 3 incisors, 1 canine, 4 premolars, and 3 molars in each side of the upper jaw and a similar number in each side of the lower jaw. Starting from this basic formula, one can show the number of teeth of each type present in a kind of mammal. For example, the dental formula for cricetid rodents is $\frac{1}{1}, \frac{0}{0}, \frac{0}{0}, \frac{3}{3}$, which means that the dentition has been reduced to 1 incisor and 3 molars in each half of the upper and lower jaws.

The teeth of mammals are thecodont, being set in alveoli or sockets in the jaw bones. The bases of the teeth may become constricted and often divided into separate roots, each in a separate pocket of the alveolus. Such teeth cease to grow after they reach mature size; they are called "rooted" teeth. Teeth which are subjected to much wear, as in the incisors of rodents, may have bases which remain widely open through life and are capable of continuous elongation in compensation for the wearing away of their crowns. Such teeth are called "rootless" teeth.

The jaw teeth of mammals are primitively low-crowned or brachyodont. Herbivores have tended to develop high-crowned (hypsodont) jaw teeth through great lengthening of their cusps and the growth of cement over the entire tooth. The grinding surface is a complex one, made up of layers of cement, enamel, and dentine.

The pattern of the molar crown varies greatly in mammals. Primitively, there are 3 major cusps in a triangular pattern (Fig. 6-7). This basic pattern is little changed in carnivorous and insectivorous mammals generally. In omnivores such as man the cusps are low and rounded (bunodont). In herbivores the pattern generally becomes complicated as the grinding surface develops. Expansion of each cusp into a longitudinal crescent results in the selenodont tooth of artiodactyls. The cusps may fuse into ridges to produce the lophodont teeth of rodents and perissodactyls. The nomenclature of the cusps and folds of the molar teeth of a representative rodent is shown in Fig. 6-8.

Other parts of the skeleton than the skull and teeth are used in classification of some groups of mammals. Characters of the baculum are useful key characters in rodents and other groups; preservation of this bone is helpful for use of several of the keys to rodent genera.

In the following accounts, the earliest occurrence of the taxon in the geological record is given, followed by the Recent geographical distribution.

Subclass Theria. Viviparous, with teats. Heart with completely membranous and complete right auriculoventricular valve. Brain usually with corpus callosum. Coracoid vestigial and not reaching

Fig. 6-7. Diagrams of molar-tooth patterns of placental mammals: A, right upper molar of a primitive form (outer edge of tooth above, front edge to right); B, the same of a type in which the tooth has been "squared up" by addition of a hypocone at the back inner corner; C, left lower molar of a primitive form with 5 cusps (outer edge of tooth above, front end to right); D, the same of a type in which the tooth has been "squared up" by the loss of the paraconid. (From Romer, *The Vertebrate Body,* W. B. Saunders Company.)

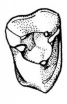

A   B   C   D

sternum; scapula with a spine; no interclavicle. Ribs typically bicipital. Middle Jurassic; 3 infraclasses, including extinct Pantotheria and 2 living.

KEY TO INFRACLASSES OF THERIA

1. Epipubic bones present; fur-lined belly pouch in females; incisors $\frac{5}{4}$ in U.S. forms **Metatheria**
   Epipubic bones lacking; no belly pouch; incisors fewer than $\frac{5}{4}$ **Eutheria**

Infraclass Metatheria. Terrestrial, arboreal, fossorial, or rarely aquatic. Skull with jugal reaching to glenoid cavity; angle of mandible usually inflected. Epipubic bones present, supporting marsupium or pouch. 4th toe usually most highly developed; claws always present. Molars usually $\frac{4}{4}$. Corpus callosum of brain very small. No well-developed auditory bullae; middle ear protected by a process of alisphenoid. Usually no well-developed placenta; young born small and very immature,

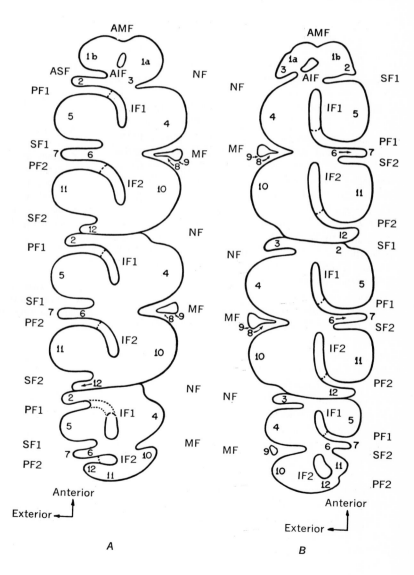

Fig. 6-8. Stylized diagram of a complex enamel pattern of *A* upper molar and *B* lower molar teeth of a rodent (*Reithrodontomys*). Names of cusps: *A*, upper molars; 1, anterocone (1*a*, anterolingual conule, 1*b*, anterolabial conule); 2, anterior cingulum; 3, anteroloph; 4, protocone; 5, paracone; 6, mesoloph; 7, mesostyle; 8, enteroloph; 9, enterostyle; 10, hypocone; 11, metacone; 12, posterior cingulum. *B*, lower molars: 1, anteroconid (1*a*, anterolabial conulid, 1*b*, anterolingual conulid); 2, anterior cingulum; 3, anterolophid; 4, protoconid; 5, metaconid; 6, mesolophid; 7, mesostylid; 8, ectolophid; 9, ectostylid; 10, hypoconid; 11, entoconid; 12, posterior cingulum. Names of folds (all molars): *MF*, major; *NF*, minor; *PF*, primary (1, first, 2, second); *SF*, secondary (1, first, 2, second); *AMF*, anteromedian; *ASF*, anterosecondary; *IF*, internal (1, first internal, 2, second internal); *AIF*, anterointernal. [From Hooper, *A Systematic Review of the Harvest Mice (Genus Reithrodontomys) of Latin America*, University of Michigan Press.]

becoming attached to teats in marsupium, where development continues. Upper Cretaceous. 1 order.

Order Marsupialia. Characters of infraclass. 5 extinct and 8 living families, of which 6 are restricted to Australian region. 2 families in South America, 1 ranging into North America.

Family Didelphidae. Opossums. Omnivorous, mostly arboreal, usually with long, partially naked, prehensile tail. Pouch often absent. Limbs pentadactyl with short, compressed, sharp, curved claws; hallux opposable and without claw. Incisors small; canines large; premolars with compressed, pointed crowns; molars with numerous sharp cusps. Dentition: $\frac{5}{4}, \frac{1}{1}, \frac{3}{3}, \frac{4}{4}$. Upper Cretaceous; South and North America. About 22 extinct and 11 living genera, 1 reaching the United States.

Genus *Didelphis* Linnaeus. Relatively large, with long, scaly, prehensile tail. The woolly fur overlaid by numerous long, coarse guard hairs. Pouch present and complete. Skull with sagittal and occipital crests well developed, particularly in old age. Incisors small; canines large. Pliocene; South and North America. 1 species reaching the United States.

Didelphis marsupialis **Linnaeus.** Common opossum. Body length about 380–500 mm; hind foot about 55–80 mm. Tail about 60–95% of body length. Color a mixture of blacks and whites, variable, from mostly black to nearly all white. Feet blackish, with more or less white on toes; ears blackish, usually whitish at tips. Tail black for variable distance from base. From Brazil in South America north through Mexico and eastern half of the United States to southeastern South Dakota, central Wisconsin, central Michigan, and central New York. Established by introduction in California, Oregon, and Washington.

Infraclass Eutheria. Highly diversified group. A well-developed allantoic placenta present. Angle of mandible usually not inflected. No pouch or epipubic bones. Brain with well-developed corpus callosum. Teeth usually diphyodont, with complete replacement of all except the molars. Dentition usually some modification of the basic formula $\frac{3}{3}, \frac{1}{1}, \frac{4}{4}, \frac{3}{3}$, and more than that number of teeth rarely present. Upper Cretaceous. 10 extinct and 16 living orders, 11 in the United States.

KEY TO ORDERS OF EUTHERIA

1. Teeth homodont; often vestigial or absent (Fig. 6-9*B*)   **2**
   Teeth heterodont (Fig. 6-9*A*)   **3**
2. Forelimbs modified into flippers; hind limbs absent externally   **Cetacea**
   Forelimbs not modified into flippers; hind limbs present   **Edentata**
3. Forelimbs modified for flight; fingers longer than forearm, supporting membranous wing   **Chiroptera**
   Forelimbs not modified for flight; fingers shorter than forearm and without flight membrane   **4**
4. Hind limbs absent   **Sirenia**
   Hind limbs present   **5**
5. Feet unguligrade; hooves present   **6**
   Feet not unguligrade; digits with claws or nails   **7**
6. Axis of foot passing between almost equally developed 3d and 4th digits (Fig. 6-10*A*, *B*)   **Artiodactyla**
   Axis of foot passing through 3d digit (Fig. 6-10*C*, *D*)   **Perissodactyla**
7. Canines absent; anterior and posterior teeth separated by a wide diastema (Fig. 6-11)   **8**
   Canines present; anterior and posterior teeth not separated by a wide diastema   **9**
8. Incisors $\frac{2}{1}$, the 2d small and set immediately behind the 1st   **Lagomorpha**
   Incisors $\frac{1}{1}$ **Rodentia**

Fig. 6-9. Representative mammals, showing *A*, heterodont dentition of *Bassariscus astutus*; *B*, homodont dentition of *Dasypus novemcinctus*.

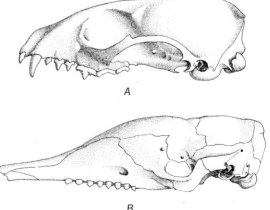

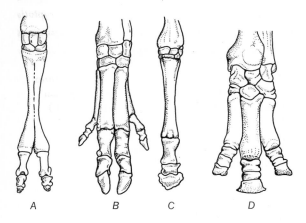

Fig. 6-10. Ungulate feet: *A*, paraxonic foot of camel with 2 main metapodials fused to form cannon bone; *B*, paraxonic foot of pig without cannon bone; *C*, mesaxonic foot of horse in which first and fifth toes are gone and second and fourth remain as vestigial splints; *D*, mesaxonic foot of rhinoceros. (After Flower, from Romer, *The Vertebrate Body*, W. B. Saunders Company.)

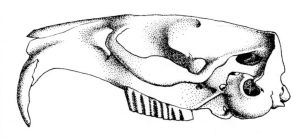

Fig. 6-11. Skull of *Ondatra zibethicus*. Note diastema between incisors and jaw teeth.

9. Orbit enclosed by a bony ring formed by junction of postorbital process of frontal with zygomatic arch **Primates**
Orbit not enclosed by complete bony ring **10**
10. Canines small, no larger than incisors **Insectivora**
Canines large, much larger than incisors **Carnivora**

Order Insectivora. Generally small, terrestrial, plantigrade placentals. Digits unguiculate and generally pentadactyl; pollex and hallux not opposable to other digits. Teeth rooted, molars with sharp cusps. Usually no fewer than 2 incisors in either side of mandible. Canines usually weak; incisors, canines, and anterior premolars usually not clearly differentiated from one another. Cranial cavity usually relatively small; braincase little elevated above facial line. Rostrum generally much produced anteriorly. Zygomatic arches generally slender or absent (in most). Cerebral hemispheres smooth, not extending backward over cerebellum. Body covered with fur or protected by spines (hedgehogs). No scrotum; uterus bicornuate. Mostly insect or other invertebrate feeders, except *Potomogale* which feeds on fish. Upper Cretaceous. 12 extinct and 8 living families, including alimiqui (Solenodontidae) of Cuba and Haiti, tenrecs (Tenrecidae) of Madagascar, jes (Potomogalidae) of West Africa, Cape golden moles (Chrysochloridae) of South Africa, hedgehogs (Erinaceidae) of Europe, Asia, and Africa, elephant shrews (Macroscelididae) of Africa, and the 2 families represented in North America.

KEY TO FAMILIES OF INSECTIVORA

1. Forefeet slender, held horizontally; zygomata and auditory bullae absent; humerus long and slender

Fig. 6-12. Skulls of *A, Blarina brevicauda; B, Scalopus aquaticus.* (A from Merriam, Revision of the Shrews of the American Genera *Blarina* and *Notiosorex*, North Am. Fauna 10.)

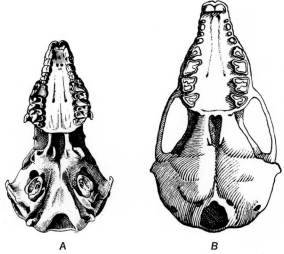

A   B

(length more than twice the width) (Fig. 6-12*A*)
                                    Soricidae

Forefeet greatly enlarged, held vertically; zygomata
and auditory bullae present; humerus short and
heavy (length less than twice the width) (Fig.
6-12*B*)   **Talpidae**

Family Soricidae. Shrews. Small; with minute
eyes, sharp-pointed snouts, and small ears. 21 living genera, of which 13 occur in Asia and 7 are
limited there; others in Africa, Europe, North
America, and northern South America. Late Oligocene. 5 genera in the United States.

Fig. 6-13. Teeth of *Sorex bendirei* showing principal
cusps: *A*, left upper row, *B*, left lower row: *Me*,
metacone; *Ms*, mesostyle; *Mts*, metastyle; *Pa*, paracone;
Ps, parastyle; *Hy*, hypocone; *Pr*, protocone; *End*,
entoconid; *Med*, metaconid; *Pad*, paraconid; *Hyd*,
hypoconid; *Prd*, protoconid. (From Jackson, A
Taxonomic Review of the North American Long-tailed
Shrews, *North Am. Fauna* 51.)

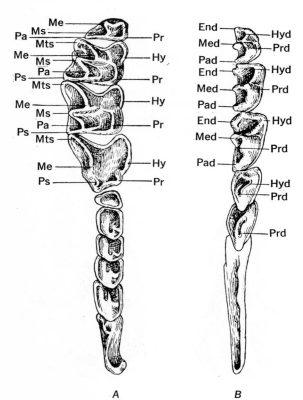

A   B

KEY TO GENERA OF SORICIDAE

1. Lower incisors 1		**2**
Lower incisors 2		**3**
2. 3d and 5th upper unicuspids scarcely, if at all, visible in lateral view (Fig. 6-14*A*)		***Microsorex***
All 5 upper unicuspids visible in lateral view (Fig. 6-14*B*)		***Sorex***
3. Incisors $\frac{4}{2}$ (total teeth 32)		***Blarina***
Incisors $\frac{3}{2}$ (total teeth 30 or fewer)		**4**
4. Premolars $\frac{2}{1}$ (total teeth 30)		***Cryptotis***
Premolars $\frac{1}{1}$ (total teeth 28)		***Notiosorex***

Fig. 6-14. Tooth rows of *A*, *Microsorex hoyi*; *B*, *Sorex
fumeus*. (From Merriam, Synopsis of the American
Shrews of the Genus Sorex, *North Am. Fauna* 10.)

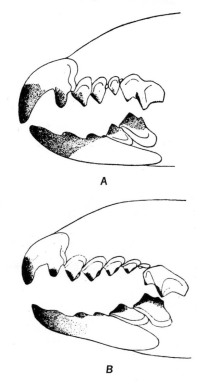

A

B

Genus *Sorex* Linnaeus. Long-tailed shrews.
Size small. Pelage soft and velvetlike. Tail long
(half to equal to body length), more or less completely haired. Ears small, moderately haired, nearly

concealed by fur. Eyes minute. Snout pointed. Soles of hind feet naked, normally with 6 plantar pads. Mammae 6. Dentition: $\frac{3}{1}, \frac{1}{1}, \frac{3}{1}, \frac{3}{3}$. Miocene; Europe, Asia, North America. 20 species currently recognized in the United States.

KEY TO SPECIES OF *SOREX*

1. Size large; hind foot 18 mm or more; color grayish, never distinctly brown **2**
   Size small; hind foot less than 18 mm; if hind foot over 16 mm, color distinctly brown **3**
2. Rostrum relatively short and little down-curved; anterior end of premaxilla scarcely narrower dorsoventrally than middle portion; hind foot distinctly fringed with hair **S. palustris**
   Rostrum relatively long and distinctly down-curved; anterior end of premaxilla much narrower dorsoventrally than middle portion; hind foot slightly fringed **S. bendirei**
3. 3d unicuspid not smaller than 4th **4**
   3d unicuspid smaller than 4th **10**
4. Infraorbital foramen with posterior border lying caudad to plane of interspace between 1st and 2d upper molars **S. dispar**
   Posterior border of infraorbital foramen even with or anterior to plane of interspace between 1st and 2d upper molars **5**
5. Maxillary breadth less than 4.6 mm **6**
   Maxillary breadth more than 4.6 mm **8**
6. Condylobasal length less than 15 mm **S. preblei**
   Condylobasal length 15 mm or more **7**
7. Interorbital breadth 3.3 mm or more **S. lyelli**
   Interorbital breadth 3.2 mm or less **S. cinereus**
8. Condylobasal length less than 17.5 mm; cranial breadth usually less than 8.5 mm; maxillary tooth row usually less than 6.1 mm **S. merriami**
   Condylobasal length more than 17.5 mm; cranial breadth 8.5 mm or more; maxillary tooth row 6.1 mm or more **9**
9. Color of back sharply darker than sides; tail less than 45 mm **S. arcticus**
   Back of same color as sides; tail 45 mm or more **S. fumeus**
10. Geographic range east of the central grasslands **S. longirostris**
    Geographic range west of the central grasslands **11**
11. Tail sharply bicolor; underparts of body scarcely, if any, paler than upper parts. Ridge extending from apex of unicuspid toward interior edge of cingulum separated from cingulum by longitudinal groove **S. trowbridgei**

Tail not sharply bicolor; underparts of body distinctly paler than upper parts. Ridge extending from apex of unicuspid toward interior edge of cingulum not separated from cingulum by longitudinal groove **12**

12. Foramen magnum placed relatively ventrad, encroaching less into supraoccipital and more into basioccipital (Fig. 6-15*B*) **13**
    Foramen magnum placed relatively dorsad, encroaching more into supraoccipital and less into basioccipital (Fig. 6-15*A*) **15**
13. Size larger; total length more than 130 mm; condylobasal length 20.0 mm or more; length of maxillary tooth row more than 7.0 mm **S. pacificus**
    Size smaller; total length less than 130 mm; condylobasal length less than 20.0 mm; length of maxillary tooth row less than 7.0 mm **14**
14. Metaconid of 1st upper molar comparatively high; upper border of foramen magnum less acute **S. obscurus**
    Metaconid of 1st upper molar comparatively low; upper border of foramen magnum more acute **S. vagrans**
15. Restricted to Santa Catalina Island, California **S. willetti**
    Not on Santa Catalina Island, California **16**
16. Condylobasal length more than 16.3 mm; maxillary breadth usually more than 4.7 mm **17**
    Condylobasal length less than 16.3 mm; maxillary breadth usually less than 4.7 mm **18**
17. Color blackish; palatal length less than 6.7 mm **S. sinuosus**
    Color never blackish; palatal length 6.7 mm or more **S. ornatus**
18. Tail less than 40 mm; condylobasal length less than 15.4 mm; cranial breadth less than 7.2 mm **19**
    Tail more than 40 mm; condylobasal length more than 15.4 mm; cranial breadth more than 7.2 mm **20**
19. Hind foot less than 11 mm; condylobasal length

Fig. 6-15. Ventral view of foramen magnum: **A,** *Sorex ornatus;* **B,** *Sorex obscurus.* (From Jackson, A Taxonomic Review of the North American Long-tailed Shrews, *North Am. Fauna* 51.)

A   *B*

less than 14.8 mm; Colorado and northern Arizona **S. nanus**

Hind foot more than 11 mm; condylobasal length more than 14.8 mm; California and Nevada **S. tenellus**

20. Condylobasal length more than 15.8 mm; palatal length more than 6 mm; California **S. ornatus**

Condylobasal length less than 15.8 mm; palatal length less than 6 mm; Oregon **S. trigonirostris**

Sorex cinereus **Kerr.** Masked shrew. Size small; body length about 55–61 mm; hind foot about 10–14 mm. Tail about 50–70% of body length. Upper parts grayish brown; underparts pale gray; tail indistinctly bicolor. Skull relatively weak; rostrum narrow, molariform teeth relatively narrow; 4th unicuspid generally smaller than 3d, rarely about equal. From Alaska across Canada, south in the United States to northern Nebraska, southern Illinois, and Maryland; south through Rocky Mountains to southern New Mexico; south through Appalachian chain to western North Carolina.

Sorex lyelli **Merriam.** Mount Lyell shrew. Size small; body length about 62–63 mm; hind foot about 11–12 mm. Tail about 65% of body length. Upper parts brownish, sides paler; underparts pale gray, faintly buffy. Skull flatter than in *S. cinereus* and relatively broader interorbitally. High elevations (6900 ft and above) in central Sierra Nevada of California. Closely related to *S. cinereus*, from which its range is widely separated.

Sorex preblei **Jackson.** Malheur shrew. Size small; body length 50–59 mm; hind foot about 11 mm. Tail about 60% of body length. Upper parts brown, sides scarcely paler; underparts pale gray, faintly buffy. Skull rather flat, with relatively broad rostrum and short tooth row. Eastern Oregon, at elevations of about 4500 ft. Range allopatric to that of the closely related *S. cinereus*.

Sorex fumeus **Miller.** Smoky shrew. Size medium; body length about 66–81 mm; hind foot about 13–15 mm. Tail about 55–70% of body length. Upper parts dull brownish gray; underparts only slightly paler gray than upper parts; feet pale buffy, the outer edge dusky. Skull relatively broad and short, with broad interorbital region; infraorbital foramen large; molariform teeth rather deeply emarginate posteriorly. Northeastern United States from New England states south in Appala-

chian chain to western North Carolina; west to central Ohio. A disjunct population in western Kentucky and another in southeastern Wisconsin.

Sorex arcticus **Kerr.** Arctic shrew. Size medium; body length 70–78 mm; hind foot about 13–14 mm. Tail about 50–55% of body length. Color pattern tricolor; back distinctly darker than sides, which in turn are distinctly darker than underside. Skull moderate; dentition moderately heavy; 4th unicuspid smaller than 3d. Eastern and central Canada, into the United States from northwestern North Dakota to northeastern Nebraska and central Wisconsin.

Sorex merriami **Dobson.** Merriam shrew. Body length about 55–69 mm; hind foot about 12–13 mm. Tail about 55–65% of body length. Upper parts brownish; underparts distinctly whitish to only slightly paler than back; tail indistinctly bicolor. Skull relatively short and broad, flattened through braincase, relatively high and swollen interorbitally, with a short, broad rostrum, which is abruptly truncate anteriorly. 4th unicuspid usually smaller than 3d. In apparently isolated montane populations over an area from Arizona and Nevada north to southeastern Washington and western North Dakota.

Sorex longirostris **Bachman.** Long-nosed shrew. Size small; body length 46–69 mm; hind foot about 10–12 mm. Tail relatively short, about 55% of body length. Upper parts brown; underparts grayish; tail indistinctly bicolor. Skull with short rostrum and crowded unicuspid tooth row; 1st and 2d unicuspids about equal in size, the 3d and 4th decidedly smaller than 1st and 2d, the 3d somewhat smaller than the 4th; 5th unicuspid much smaller than 4th, almost minute. Southeastern United States from northern Florida and central Alabama along Coastal Plain to Maryland; an apparently disjunct population in Illinois.

Sorex dispar **Batchelder.** Gray shrew. Size medium; body length about 65–70 mm; hind foot about 14–15 mm. Tail long, about 80–85% of body length. Upper parts dull grayish; underparts scarcely if any paler than back. Tail slightly if at all bicolor. Skull smooth, nonangular, long and narrow, moderately flattened; rostrum long and narrow, depressed; infraorbital foramen with posterior border lying behind plane of interspace between 1st and 2d upper molars; 3d unicuspid about equal to the 4th in size. Mountains in northeastern United States from northern New York and Western Massachusetts south to western Virginia.

Sorex trowbridgei **Baird.** Trowbridge shrew. Size medium; body length 57–71 mm; hind foot about 13–15 mm. Tail relatively long, about 83–95% of body length. Upper parts dark gray to dark brownish gray; underparts scarcely, if any, paler than back; tail sharply bicolor, dark above, nearly white below. Skull moderately depressed; 3d unicuspid smaller than 4th; ridge extending from apex of unicuspid toward interior edge of cingulum only slightly pigmented and rarely pigmented to cingulum, separated from cingulum by anteroposterior groove and never ending in distinct cusplet. Upland and lowland forests; southwestern British Columbia south through Washington and Oregon, mostly west of Cascades, to central California.

Sorex vagrans **Baird.** Vagrant shrew. Size small; body length 57–68 mm; hind foot about 11–13 mm. Tail about 60–70% of body length. Upper parts blackish; sides and flanks usually paler than back; underparts pale gray, usually tinged with pale pinkish buff, to brownish; tail more or less bicolor. Skull moderately flattened; rostrum comparatively short and broad. 3d unicuspid distinctly smaller than 4th, the ridge extending internally from apex of unicuspid to border of cingulum well developed, usually heavily pigmented and tending apically to form a distinct cusplet. Difficult to distinguish from the sympatric *S. obscurus,* from which it differs in smaller size, narrower skull, shorter palate, and more acute upper border of foramen magnum. From southwestern British Columbia south to central California and Nevada and through the Rocky Mountain chain into Mexico. Regarded by some as conspecific with *S. obscurus* and *S. pacificus;* if so, this is an overlapped polytypic species with broad overlap from northern California to British Columbia.

Sorex obscurus **Merriam.** Dusky shrew. Body length about 60–78 mm; hind foot about 13–15 mm. Tail about 60–80% of body length. Upper parts brownish; sides usually somewhat paler than back; underparts pale gray to buffy brown. Skull generally similar to that of *S. vagrans,* but broader, particularly interorbitally and through rostrum, palate longer and upper border of foramen magnum

less acute. From Alaska south through Cascade–Sierra Nevada chain to southern California and through Rocky Mountain chain to southern New Mexico.

Sorex pacificus **Coues.** Pacific shrew. Size relatively large; body length about 74–86 mm; hind foot about 16–18 mm. Tail about 70–80% of body length. Upper parts dark brown to blackish, changing gradually on sides to slightly paler of underparts; tail essentially unicolor. Skull large and broad, with heavy rostrum; zygomatic ridge of squamosal well developed, forming a distinct shelflike process extending posteriorly nearly to mastoid region; dentition heavy, the unicuspids relatively broad and swollen. West of Cascade–Sierra Nevada chain from middle Oregon coast south to San Francisco Bay, California. Includes the nominal species *S. yaquinae.*

Sorex ornatus **Merriam.** Ornate shrew. Size small; body length about 57–68 mm; hind foot about 12–13 mm. Tail relatively short, about 56–66% of body length. Upper parts grayish brown, gradually changing to pale gray of underparts; tail indistinctly bicolor. Skull rather flattened through braincase, depressed interorbitally; foramen magnum placed dorsad, encroaching more into supraoccipital and less into basioccipital. Southern two-thirds of California, south into Baja California. The range of this species is allopatric to that of the populations referred to as *S. trigonirostris, S. tenellus,* and *S. nanus* and the insular populations known as *S. sinuosus* and *S. willetti.*

Sorex trigonirostris **Jackson.** Ashland shrew. Body length about 61 mm; hind foot about 12 mm. Tail about 55% of body length. Upper parts dark brown; underparts pale gray. Mastoid region of skull more angular and prominent than in *S. ornatus.* Known only from the vicinity of Ashland, Oregon. This apparently disjunct population is closely related to and possibly a subspecies of *S. ornatus.*

Sorex willetti **Von Bloeker.** Santa Catalina shrew. Body length about 66 mm. Tail about 58% of body length. A member of *S. ornatus* group. Santa Catalina Island, California.

Sorex sinuosus **Grinnell.** Suisun shrew. Body length about 55–64 mm; hind foot about 12 mm. Tail about 60% of body length. Upper parts dark, almost black; underparts brown. Skull generally similar to that of S. *ornatus*. Restricted to Grizzly Island, near Suisun, Solano County, California.

Sorex tenellus **Merriam.** Inyo shrew. Body length about 57–61 mm; hind foot about 12–13 mm. Tail about 70% of body length. Upper parts grayish brown; underparts pale gray. Skull smaller and relatively narrower and teeth smaller than in S. *ornatus*. Known only from scattered mountains at 5000–10,000 ft elevation in southeastern California and southern Nevada.

Sorex nanus **Merriam.** Dwarf shrew. Size small; body length about 63 mm; hind foot about 10 mm. Tail about 67% of body length. Upper parts grayish brown; underparts buffy gray; tail indistinctly bicolor. Skull generally similar to that of S. *ornatus* but smaller. Known only from mountains of northern Arizona and central Colorado.

Sorex palustris **Richardson.** Water shrew. Size large; body length about 74–88 mm; hind foot about 18–21 mm. Hind foot with conspicuous fringe of stiff hairs. Tail long, about 70–105% of body length. Upper parts blackish; underparts pale to dark gray; tail distinctly to indistinctly bicolor. Rostrum comparatively short, scarcely curved ventrally at anterior end; anterior end of premaxilla scarcely narrower dorsoventrally than middle part; unicuspid row relatively short. Across Canada and south along Appalachians to West Virginia; south into northern Michigan and Wisconsin, central Minnesota, and northeastern South Dakota; in the West, follows Rocky Mountain chain south to northern New Mexico and southwestern Utah, and follows Cascade–Sierra Nevada chain south to central California. A disjunct population in southeastern Arizona.

Sorex bendirei **(Merriam).** Marsh shrew. Size large; body length about 83–94 mm; hind foot about 18–21 mm. Hind foot with moderate fringe of stiff hairs. Upper parts dark brown to blackish, scantily flecked with whitish hair tips; underparts slightly paler than back to whitish; tail unicolor to distinctly bicolor. Rostrum comparatively long, distinctly curved ventrally at anterior end; anterior end of premaxilla decidedly narrower dorsoventrally than middle part. From southwestern British Columbia south through western Washington and Oregon, mostly west of Cascades, and along California coast nearly to San Francisco Bay.

Genus *Microsorex* **Coues.** Pygmy shrews. Small, externally similar to *Sorex*, but with shorter tail than usual in that genus. Skull relatively flat and narrow, with short, broad rostrum; infraorbital foramina relatively small; mandible short and heavy. All 5 unicuspids evident in lateral view, but the 5th sometimes minute and indistinct; 3d unicuspid not disclike, not anteroposteriorly flattened; primary (anterior) lobe of 1st upper incisor relatively broad, the length less than twice the width and usually less than twice the length of the secondary lobe. Dentition: $\frac{3}{1}, \frac{1}{1}, \frac{3}{1}, \frac{3}{3}$ Pleistocene; North America. 1 living species.

Microsorex hoyi **(Baird).** Pygmy shrew. Size tiny; body length 50–67 mm; hind foot about 9–12 mm. Tail about 50–60% of body length. Upper parts grayish brown; underparts gray. Tail indistinctly bicolor. From Alaska across Canada; enters the United States in northeastern Washington, also from North Dakota to northeastern Iowa and northern Michigan; east of Great Lakes south to eastern Ohio and south in Appalachians to western North Carolina. Relictual in Rocky Mountains of northern Colorado.

Genus *Blarina* **Gray.** Short-tailed shrew. Relatively large, robust, with very short tails. Pelage short, dense, and velvety. No apparent external ears. All teeth heavily tipped with dark chestnut, which usually reaches far down on the crowns. Basal lobe of middle incisor elongated anteroposteriorly. Unicuspids 5, the anterior 4 in 2 pairs; 1st and 2d largest and subequal; 3d and 4th abruptly much smaller and subequal; 5th minute. Dentition: $\frac{4}{2}, \frac{1}{0}, \frac{2}{1}, \frac{3}{3}$. Upper Pliocene; North America. 1 living species.

Blarina brevicauda **(Say).** Short-tailed shrew. Body length averaging 74–100 mm; hind foot about 12–18 mm. Tail about 24–33% of body length. Upper parts leaden gray; underparts paler gray. Forests and grasslands of eastern half of the United States and adjacent Canada west to a line from Corpus Christi Bay, Texas, to north central North Dakota.

Genus *Cryptotis* **Pomel.** Least shrews. Small, relatively slender, with very short tails. No apparent external ears. Teeth heavily tipped with chestnut. Unicuspids 4, never in 2 pairs; 4th always smallest and usually minute. Dentition: $\frac{3}{2}, \frac{1}{0}, \frac{2}{1}, \frac{3}{3}$. Recent; North America, northern South America. 1 species in the United States.

Cryptotis parva (Say). Least shrew. Body length averaging about 59–67 mm; hind foot about 11–12 mm. Tail short, averaging about 25–30% of body length. Upper parts brownish; underparts grayish. Grassy habitats in eastern United States south of a line from central New York to northeastern Iowa and northwestern Kansas and west through Kansas, Oklahoma, and all but trans-Pecos Texas.

Genus *Notiosorex* Baird. Desert shrews. Small, moderately slender, with moderate tails and conspicuous external ears. Braincase flat and broadly rounded. Anterior teeth lightly tipped with orange; molars white. Unicuspids 3, forming a uniform series, the 3d more than half as large as the 2d, never minute. Unicuspids narrow at base, without trace of secondary cusplet on inner side. Dentition: $\frac{3}{2}, \frac{1}{0}, \frac{1}{1}, \frac{3}{3}$. Pleistocene; North America. 1 living species.

Notiosorex crawfordi (Baird). Desert shrew. Body length about 59–61 mm; hind foot about 10–11 mm. Tail about 45–52% of body length. Upper parts medium to dark gray; underparts paler gray; tail indistinctly bicolor. Arid southwest from southern California and southern Nevada east to northwestern Arkansas and middle Texas coast, south into western Mexico.

Family Talpidae. Moles. Fossorial insectivores. Pelage fine and silky. Body stout, eyes vestigial, ear conch lacking, snout proboscislike. Distribution limited to sandy, loamy, or marshy soils. 17 living genera: 12 in Eurasia, 5 in North America. Middle Oligocene. 5 extinct genera from Europe. 2 from North America.

KEY TO GENERA OF TALPIDAE

1. Body length less than 95 mm   *Neürotrichus*
   Body length more than 100 mm   **2**
2. End of snout with a fringe of fleshy processes (Fig. 6-16B)   *Condylura*
   End of snout without a fringe of fleshy processes (Fig. 6-16A)   **3**
3. Tail densely haired; nostrils lateral   *Parascalops*
   Tail naked or scantily haired; nostrils superior   **4**
4. Foretoes webbed; lower incisors 2; east of Rocky Mountains   *Scalopus*
   Foretoes not webbed; lower incisors 3; west of Rocky Mountains   *Scapanus*

Fig. 6-16. Ventral view of snouts of A, *Parascalops breweri*; B, *Condylura cristata*. (From Jackson, A Review of the American Moles, *North Am. Fauna* 38.)

A   B

Genus *Neürotrichus* Günther. Size small; body little depressed. Tail long, constricted at base, distinctly annulated, sparsely haired. Nostrils lateral. Forefeet longer than broad. 6 tubercles on sole of each hind foot. Mammae 8. Auditory bullae incomplete. Dentition: $\frac{3}{3}, \frac{1}{1}, \frac{2}{2}, \frac{3}{3}$. Recent; North America. 1 species.

Neürotrichus gibbsi (Baird). Gibb's mole. Body length about 76–87 mm; hind foot about 16–17 mm. Tail about 44–50% of body length. Color dark gray. Moist environments in vicinity of swamps, marshes, or streams or in moist, dense woods. Less subterranean than most American members of family. Tunnels often open to surface. West of Cascades and Sierra Nevada from middle California north to southern British Columbia.

Genus *Parascalops* True. Body robust, scarcely depressed. Tail densely covered with hair. Head conoidal, depressed. Nostrils lateral, crescentic. Sole of each hind foot with 2 tubercles and a distinct heel pad. Rostrum slender. Mammae 8. Pleistocene; North America. 1 living species. Auditory bullae incomplete. Dentition: $\frac{3}{3}, \frac{1}{1}, \frac{4}{4}, \frac{3}{3}$.

Parascalops breweri (Bachman). Hairy-tailed mole. Body length about 116–126 mm; hind foot 18–20 mm. Tail about 25% of body length. Color blackish. Hairs on nose and tail sometimes white in old age. Northeastern United States and southeastern Canada, from Maine and New Brunswick southwest to northeastern Ohio, south in Appalachian chain to western North Carolina.

Genus *Scapanus* Pomel. Body robust. Tail thick and fleshy, tapering apically and slightly constricted proximally, scantily haired. Forefeet as broad as long. Soles of hind feet with 1–3 distinct tubercles. Fur dense, soft; hairs nearly equal in length. Mammae 8. Males larger than females. Auditory bullae complete. Dentition: $\frac{3}{3}, \frac{1}{1}, \frac{4}{4}, \frac{3}{3}$. Pleistocene; North America. 3 living species.

KEY TO SPECIES OF *SCAPANUS*

1. Unicuspid teeth usually crowded and not evenly spaced   **S. *latimanus***
   Unicuspid teeth evenly spaced, not crowded   **2**
2. Total length more than 200 mm; greatest length of skull more than 40 mm   **S. *townsendi***
   Total length less than 200 mm; greatest length of skull less than 40 mm   **S. *orarius***

Scapanus latimanus (Bachman). California mole. Body length 117–150 mm; hind foot 18–24 mm. Tail about 23–29% of body length. Color brown or gray. Skull wide and heavy; rostrum short and broad. Sublacrimal-maxillary ridge heavy. Mountains of California north to southern Oregon and northwestern Nevada.

Scapanus orarius True. Coast mole. Body length about 135 mm; hind foot about 20–23 mm. Tail about 25% of body length. Color almost black. Feet and claws relatively small. Rostrum narrow; sublacrimal-maxillary ridge indistinct. Mandible weak. Humid coast region of northern California, Oregon, and Washington, east through Cascades in north central Oregon.

Scapanus townsendi (Bachman). Townsend's mole. Body length about 150–183 mm; hind foot about 24–28 mm. Tail 22–30% of body length. Color almost black. Rostrum long and relatively narrow; sublacrimal-maxillary ridge well developed. Extreme northwestern California, Oregon, and Washington west of Cascades.

Genus *Scalopus* Geoffroy. Body robust; tail short, nearly naked. Head conoidal, depressed; nose elongated, apical portion naked back to nasals. Nostrils superior, crescentic. Fur dense, soft, silky, the hairs nearly equal in length. Forefeet broader than long. Mammae 6. Auditory bullae complete. Dentition: $\frac{3}{2}, \frac{1}{0}, \frac{3}{3}, \frac{3}{3}$. Pliocene; North America. 1 living species.

Scalopus aquaticus (Linnaeus). Eastern mole. Body length about 113–185 mm; hind foot 15–24 mm. Tail 15–20% of body length. Males larger than females. Sandy or moist, loamy soils in eastern United States south of a line from Massachusetts to

central Minnesota and east of a line from western Nebraska to south central Texas. Pelage color changing from darker in the East to pale silvery in the West.

Genus *Condylura* Illiger. Body fairly robust, little depressed. Tail long, distinctly annulated, covered with coarse, blackish hairs. Head narrow, little depressed; snout terminating in a naked disc, surrounded on its margin by 22 fleshy processes. Nostrils nearly circular, on anterior surface of nasal disc. 1st to 4th toes of forefoot each with 3 flat, triangular processes on outer, inferior edge. 1 large and 5 small tubercles on sole of hind foot. Mammae 8. Auditory bullae incomplete. Dentition: $\frac{3}{3}, \frac{1}{1}, \frac{4}{4}, \frac{3}{3}$. Recent; North America. 1 species.

Condylura cristata (Linnaeus). Star-nosed mole. Body length about 118–125 mm; hind foot about 26–30 mm. Tail about 55–63% of body length. Color blackish brown to nearly black. Wet meadows or marshes. Users of surface runways as well as subterranean tunnels. Southeastern Canada and northeastern United States from southern Labrador, central Quebec, and Ontario and southeastern Manitoba, south to northeastern Illinois and northern Indiana and Ohio, south in Appalachian chain to western North Carolina. In Atlantic Coast region, south to Virginia; a disjunct population in Georgia.

Order Chiroptera. Bats. Flying mammals with forelimbs modified as wings. Long bones greatly elongated and slender. Pollex excluded from wing and with claw. Metacarpals of other 4 digits greatly elongated and, with the slender phalanges, supporting the wing membrane. Pectoral girdle greatly developed; sternum usually keeled. Hind limbs rotated outward by wing membranes so that knees are directed backward. Hind feet with short tarsus, slender, laterally compressed toes, and much-curved claws. A calcar spur extending posteriorly from heel and supporting the interfemoral membrane. Ear conch with a prominent central lobe, the tragus. Zygomata slender; absent in some. Cerebral hemispheres smooth and not extending backward over cerebellum. Penis pendent, testes abdominal or inguinal. Mammae thoracic. Mating in most North American species in fall, the sperm stored over the winter in female genital tract. Ovulation and fertilization in spring. Cold-sensitive because of their great heat-radiating surfaces, and the majority tropical or subtropical in distribution.

Temperate Zone species avoid cold by hibernating or migrating to warmer regions. All U.S. forms insectivorous except *Choeronycteris* and *Leptonycteris* which are nectar feeders, in part at least. Others frugivorous, piscivorous (*Noctilio, Pizonyx*), carnivorous (*Phyllostomus, Megaderma*), or sanguivorous (*Desmodus*). Eocene; cosmopolitan. 17 living families, 3 in the United States.

KEY TO FAMILIES OF CHIROPTERA

1. With a distinct cutaneous flap (nose leaf) projecting dorsally from nasal region (Fig. 6-17A); if nose leaf absent there are cutaneous folds across chin and tail is above interfemoral membrane; 3d phalanx of middle finger bony   **Phyllostomatidae**
   Nasal region without a dorsal leaflike structure (Fig. 6-17B); 3d phalanx of middle finger cartilaginous except at extreme base   **2**
2. Tail reaching to back edge of interfemoral membrane or barely beyond (Fig. 6-18A); fibula very slender or rudimentary   **Vespertilionidae**
   Tail reaching well beyond back edge of interfemoral membrane (Fig. 6-18B); fibula robust, its diameter usually about half that of tibia   **Molossidae**

Fig. 6-17. Bat faces: *A, Macrotus waterhousei; B, Lasiurus cinereus.* (From Allen, **A Monograph of the Bats of North America, Bull. U.S. Natl. Museum 43.**)

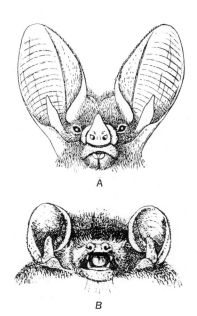

A

B

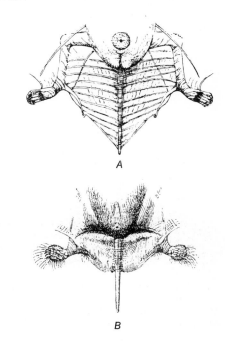

Fig. 6-18. Interfemoral membranes: *A, Antrozous pallidus; B, Tadarida brasiliensis.* (From Allen, **A Monograph of the Bats of North America, Bull. U.S. Natl. Museum 43.**)

A

B

Family Phyllostomatidae. Leaf-nosed bats. A tropical and subtropical group. Cutaneous nasal outgrowths present in most. 3 complete bony phalanges in 3d finger. Premaxillae fused with one another and with maxillae. Fibula slender and incomplete. Molar teeth well developed. Recent; North and South America. 35 genera; 4 in the United States.

KEY TO GENERA OF PHYLLOSTOMATIDAE

1. Total teeth 34; snout short (Fig. 6-19A)   **2**
   Total teeth 30; snout elongated (Fig. 6-19B)   **3**
2. Prominent nose leaf present   *Macrotus*
   Nose leaf lacking   *Mormoops*

Fig. 6-19. Bat heads: *A, Mormoops megalophylla; B, Choeronycteris mexicana.* (From Goodwin, **Mammals of Costa Rica,** *Bull. Am. Museum Nat. Hist.,* vol. 87, art. 5.)

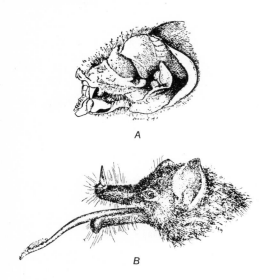

3. Incisors $\frac{2}{0}$; molars $\frac{3}{3}$; tail extending halfway to edge of interfemoral membrane   ***Choeronycteris***
   Incisors $\frac{2}{2}$; molars $\frac{2}{2}$; tail absent   ***Leptonycteris***

Genus *Mormoops* **Leach.** Leafchin bats. Tropical group in which lower lip has platelike outgrowths. Humerus without secondary articulation with scapula. Both rostrum and braincase broader than long; basioccipital region conspicuously elevated; lower rim of foramen magnum above level of rostrum. Dentition: $\frac{2}{2}, \frac{1}{1}, \frac{2}{3}, \frac{3}{3}$. Recent; tropical America. 2 species, one of which reaches southern United States.

Mormoops megalophylla **Peters.** Leafchin bat. Size large; forearm about 51–57 mm. Ears low, well furred on posterior surface. Tail projecting free dorsally about halfway of broad interfemoral membrane. Pelage brown. Wing membranes blackish, contrasting with the thin, gray, naked interfemoral membrane. Northern South America, north into southern Texas as far as southern edge of Edwards Plateau and into southern Arizona. Caves and mine tunnels.

Genus *Macrotus* **Gray.** Leaf-nosed bats. Ears large, joined across forehead; tail long, projecting beyond broad interfemoral membrane. Muzzle with nose leaf; lower lip without platelike outgrowths. Humerus with definite secondary articulation with scapula. Braincase rising gradually in front, forming only a slight angle with rostrum; rostrum distinctly flattened above; auditory bullae large. Dentition: $\frac{2}{2}, \frac{1}{1}, \frac{2}{3}, \frac{3}{3}$. Recent; southern North America, West Indies. About 4 species, one of which reaches southwestern United States.

Macrotus waterhousei **Gray.** Leaf-nosed bat. Size large; forearm about 50 mm. Ear large, with prominent, pointed tragus. Nose leaf simple, entire, rounded at muzzle and fixed to upper lip, while free at sides. Tail projecting about $\frac{1}{6}$ of its length beyond posterior border of interfemoral membrane. Bases of hairs white above and below; tips brownish gray; face colored like underparts. From western Mexico north through southeastern Arizona and southern California to southern Nevada. Caves and mine tunnels.

Genus *Choeronycteris* **Tschudi.** Hog-nosed bats. Small, with elongated muzzles, highly extensible tongues, well-developed nose leaves and small, separate ears. Tail extending less than halfway to edge of very wide interfemoral membrane. Weak but distinct calcar present. Skull with rostrum very greatly elongated. Zygomata incomplete. Dentition: $\frac{2}{0}, \frac{1}{1}, \frac{2}{3}, \frac{3}{3}$ Recent; tropical America. 1 species.

Choeronycteris mexicana **Tschudi.** Hog-nosed bat. Forearm about 44 mm. Rostrum long and slender, with small nose leaf. Ears small. Tail short, extending less than halfway to border of wide interfemoral membrane. Color dark brown. From Central America through western Mexico and Baja California into southern United States from southwestern New Mexico to California. Occurring in buildings.

Genus *Leptonycteris* **Lydekker.** Long-nosed bats. With elongated muzzles, highly extensible tongues, well-developed nose leaves, and small separate ears. Tail absent; interfemoral membrane very narrow; calcar small but distinct. Skull with rostrum greatly elongated; zygomata slender but complete. Dentition: $\frac{2}{2}, \frac{1}{1}, \frac{2}{3}, \frac{2}{2}$. Recent; tropical America. 1 species.

Leptonycteris nivalis (**Saussure**). Long-nosed bat. Forearm about 55–59 mm. Muzzle long, with a diamond-shaped nose leaf. Ears short and broad; tragus large. Membranes thick and leathery. General color brown; bases of hairs buffy white; underparts and shoulders paler than back. Central America north through Mexico; barely reaching the United States in trans-Pecos Texas and southeastern Arizona.

Family Vespertilionidae. Vespertilionid bats. Large, nearly cosmopolitan group. Muzzle and lips simple; without accessory cutaneous processes. Ears usually separate; tragus well developed, straight or slightly curved. Tail long, extending to edge of wide interfemoral membrane, but never much beyond. Only 2 bony phalanges in 3d finger, the 3d phalanx cartilaginous except at extreme base. A well-developed double articulation between scapula and humerus; ulna rudimentary; bony palate conspicuously emarginate anteriorly. Late Oligocene; worldwide. At least 26 living genera, 9 in the United States.

KEY TO GENERA OF VESPERTILIONIDAE

1. Total teeth 34 or more   **2**
   Total teeth less than 34   **6**
2. Total teeth 34   **3**
   Total teeth more than 34   **4**
3. Ears very large; forearm more than 40 mm; 3 large white spots on dorsal surface (Fig. 6-20)
                                       Euderma

   Ears small; forearm less than 40 mm; dorsal surface unspotted   *Pipistrellus*
4. Total teeth 38   *Myotis*
   Total teeth 36   **5**
5. Ears very large (more than 30 mm from notch) and joined across head; interfemoral membrane naked
                                       Plecotus

6. Ears small (less than 20 mm from notch) and separate; interfemoral membrane haired above on basal half   *Lasionycteris*
6. Total teeth 32   **7**
   Total teeth less than 32   **8**
7. Incisors $\frac{2}{3}$; premolars $\frac{1}{2}$; color brown   *Eptesicus*
   Incisors $\frac{1}{3}$; premolars $\frac{2}{2}$; colors reddish or mahogany, never brown   *Lasiurus*
8. Total teeth 28; ears very large, more than 25 mm from notch   *Antrozous*
   Total teeth 30; ears small, less than 20 mm from notch   **9**
9. Skull short and deep; depth of braincase, including auditory bullae, about half greatest length (Fig. 6-21A); metacarpal of 3d, 4th, and 5th fingers successively much shortened; pelage color yellowish   *Lasiurus*
   Skull not short and deep; depth of braincase, including auditory bullae, considerably less than half of greatest length (Fig. 6-21B); metacarpal of 3d, 4th, and 5th fingers about equal in length; color never yellowish   *Nycticeius*

Fig. 6-21. Bat skulls: *A, Lasiurus intermedius; B, Nycticeius humeralis.* (From Miller, Revision of the North American Bats of the Family Vespertilionidae, *North Am. Fauna* 13.)

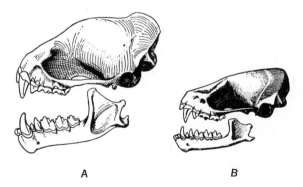

A   B

Fig. 6-20. *Euderma maculata.* (From Hall, *Mammals of Nevada,* University of California Press.)

Genus *Myotis* Kaup. Brown bats. Relatively unspecialized group of small, widely distributed bats. Ear slender; tragus slender and nearly or quite straight. Interfemoral membrane large, furred at extreme base above. Skull slender and light; rostrum nearly as long as braincase; sagittal crest low but usually distinct; palate deeply emarginate in front. Dentition: $\frac{2}{3}, \frac{1}{1}, \frac{3}{3}, \frac{3}{3}$. Middle Oligocene; virtually worldwide. Numerous species; 13 in the United States. Many are cave inhabitants.

KEY TO SPECIES OF *MYOTIS*

1. Underside of wing furred to level of elbow; skull with rostrum shortened and occiput unusually elevated (Fig. 6-22A)   **M. volans**

    Underside of wing not furred to level of elbow; skull with normal rostrum and occiput (Fig. 6-22B)   **2**

2. Foot small, about 40–46% of tibial length   **3**

    Foot large, usually about 48–60% of tibial length   **4**

3. Hairs of back with long, shiny tips; 3d metacarpal not so long as forearm; skull with flattened braincase and gradually rising profile (Fig. 6-23A)   **M. subulatus**

    Hairs of back dull-tipped; 3d metacarpal usually as long as forearm; skull with rounded braincase and abruptly rising profile (Fig. 6-23B)   **M. californicus**

4. Wing membrane attached to tarsus; fur of back without obviously darker basal area; foot usually about 60% of tibial length   **M. grisescens**

    Wing membrane attached to side of foot; fur of back with obviously darkened basal area; foot usually less than 57% of tibial length   **5**

5. Fur of back with an obvious tricolor pattern; calcar usually with a small but evident keel   **M. sodalis**

    Fur of back without an obvious tricolor pattern; calcar usually with no trace of keel   **6**

6. Ear when laid forward extending noticeably beyond tip of muzzle   **7**

    Ear when laid forward not extending noticeably beyond tip of muzzle   **9**

7. Free border of interfemoral membrane with inconspicuous, scattered stiff hairs   **M. keeni**

    Free border of interfemoral membrane usually with a noticeable fringe of stiff hairs   **8**

8. Size larger; forearm usually 41–46 mm; ear smaller, usually less than 19.0 mm from notch; tail fringe conspicuous   **M. thysanodes**

    Size smaller; forearm usually less than 40 mm; ear larger, usually more than 19.0 mm from notch; tail fringe not conspicuous   **M. evotis**

9. Cheek teeth heavy; breadth of maxillary molars relatively great compared with that of intervening palate (Fig. 6-24A); length of maxillary tooth row 5.8 mm or (usually) more   **10**

    Cheek teeth normal; their breadth less when compared with intervening palate (Fig. 6-24B); length of maxillary tooth row 5.9 mm or (usually) less   **11**

10. Pelage glossy; braincase flattened   **M. occultus**

    Pelage dull; braincase highly arched   **M. velifer**

11. Dorsal fur dense, woolly; a low but evident sagittal crest always present in adults   **M. austroriparius**

    Dorsal fur normal, silky; a sagittal crest rarely present   **12**

Fig. 6-22. Bat skulls: **A,** *Myotis volans;* **B,** *Myotis thysanodes.* (From Hall, *Mammals of Nevada,* University of California Press.)

A   B

Fig. 6-23. Bat skulls: **A,** *Myotis subulatus;* **B,** *Myotis californicus.* (From Hall, *Mammals of Nevada,* University of California Press.)

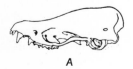

A   B

Fig. 6-24. Palatal view of bat skulls: **A,** *Myotis velifer;* **B,** *Myotis austroriparius.*

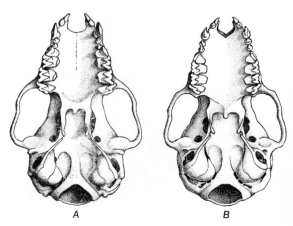

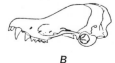

A   B

12. Forearm 32–37 mm; greatest length of skull 13.2–14.2 mm; dorsal hairs without conspicuous, burnished tips   *M. yumanensis*

  Forearm 36–40 mm; greatest length of skull 14.3–15.3 mm; dorsal hairs with conspicuous, burnished tips   *M. lucifugus*

Myotis lucifugus (**Le Conte**). Little brown bat. Size medium; forearm about 36–41 mm. Tail about 76% of body length. Ear reaching to nostril when laid forward; tragus about 8 mm, with nearly straight inner margin and narrowly rounded tip. Interfemoral membrane with a few minute hairs along free edge; calcar about 17 mm long and unkeeled. Pelage brown; hairs of back with long glossy tips which give pelage a metallic sheen. Skull with gradually rising forehead and broad braincase; usually no distinct sagittal crest. From Alaska across Canada and the United States as far south as southern California, northeastern New Mexico, southeastern Oklahoma, and southeastern Georgia. Colonial in caves and buildings.

Myotis yumanensis (**Allen**). Yuma brown bat. Size small; forearm about 32–38 mm. Tail about 81% of body length. Ear reaching to nostril when laid forward; tragus about 7 mm, anterior edge nearly straight, tip bluntly rounded. Interfemoral membrane furred to about line joining knees and with numerous short, stiff hairs in several rows along free edge, but not forming a fringe. Calcar long, extending ¾ way from heel to tail and ending in a minute lobule, unkeeled. Pelage dull brown. Skull with abruptly rising forehead. From southern Mexico north into southwestern British Columbia, west of a line from the Pecos River in Texas to eastern Montana.

Myotis austroriparius (**Rhoads**). Florida brown bat. Size medium; forearm about 36–41 mm. Tail about 66–74% of body length. Hind foot about 8.4–10.0 mm; calcar unkeeled. Pelage thick, woolly, dull yellowish brown, with little contrast in color between tips and bases of hairs. Skull with relatively narrow interorbital constriction and low, but distinct, sagittal crest. Peninsular Florida west along Gulf Coast to western Louisiana and southeastern Oklahoma and north in Mississippi embayment to southern Indiana.

Myotis grisescens **Howell**. Tennessee brown bat. Size large; forearm about 41–46 mm. Tail about 76% of body length. Ear reaching to or slightly beyond nostril when laid forward. Unique among American species in that wing membrane is inserted at tarsus instead of side of foot. Pelage grayish brown; hairs without dark bases. Skull with an obvious sagittal crest in adults. Kentucky, Tennessee, and northern Alabama, west to northeastern Oklahoma and northern Missouri, south to northern Florida. Colonial in caves.

Myotis velifer (**Allen**). Mexican brown bat. Size large; forearm about 40–47 mm. Tail about 80% of body length. Ear reaching to or slightly beyond nostril when laid forward. Wing membrane arising from base of toes. Hairs moderately long, with dark bases; general color dull brown. Skull with well-developed sagittal crest in adults; cheek teeth large, very broad in relation to palatal width. From Guatemala and Mexico north into the United States to southeastern California, northern Utah, and south central Kansas. Colonial in caves.

Myotis occultus **Hollister**. Arizona brown bat. Forearm about 33–40 mm. Tail about 82% of body length. General color yellowish; bases of hairs blackish; tips of hairs burnished, giving glossy appearance. Ears and membranes brownish to blackish. Skull with low, flattened braincase, enlarged rostrum, and distinct sagittal crest; cheek teeth very large, as in *M. velifer*. Arizona, western New Mexico, southeastern California, and southward into western Mexico.

Myotis keeni (**Merriam**). Northern brown bat. Forearm about 35–39 mm. Tail about 85–90% of body length. Ear relatively long and extending slightly beyond tip of nose when laid forward. 3d, 4th, and 5th metacarpals approximately equal in length. Calcar long, extending about halfway to end of tail. Pelage fine and full; color somewhat glossy, dark brown. Membranes brown to black. Skull relatively slender; sagittal crest present or absent. Southeastern Canada and northeastern United States, south to northern Florida and southern Arkansas and west to northeastern Oklahoma and western North Dakota. An apparently disjunct population in western Washington and western British Columbia.

Myotis evotis (**Allen**). Long-eared brown bat. Forearm about 36–41 mm. Tail about 90–97% of body length. Ear long, extending 5–7 mm beyond tip of nose when laid forward. 3d, 4th, and 5th metacarpals usually about equal in length. Free border of interfemoral membrane with a thin and inconspicuous fringe of minute hairs. Calcar long, sometimes with a rudimentary keel. Pelage full and soft; general color light brown; ears and membranes usually brownish black. Auditory bullae slightly enlarged; a small and inconspicuous sagittal crest usually present in adults. From southern Mexico north to southwestern Canada, west of a line from Pecos River in Texas to northwestern North Dakota. Buildings.

Myotis thysanodes **Miller.** Fringed brown bat. Forearm about 40–46 mm. Tail about 75–81% of body length. Ears long, reaching about 3–5 mm beyond nose when laid forward. Free edge of interfemoral membrane with conspicuous fringe of short stiff hairs in clumps of about 15 to a tuft. Calcar without distinct keel, but skin along its free margin thickened and compressed to an evident edge. Pelage full and rather long; hairs dark at base except on sides of abdomen, with slightly buff tips above and whitish tips below. Skull with well-developed sagittal crest. From southern Mexico north through western United States to northwestern Oregon and southeastern British Columbia; east as far as trans-Pecos Texas and southeastern South Dakota. Buildings and caves.

Myotis sodalis **Miller and Allen.** Kentucky brown bat. Forearm about 36–41 mm. Tail about 81% of body length. Ear reaching nostril when laid forward; tragus rather short (about 6 mm) and blunt, curving slightly forward. Metacarpals regularly graduated, 3d longest, 4th and 5th successively a little shorter. Calcar long (about 16.5 mm), nearly equal to free border of interfemoral membrane, usually with a low keel. Hairs of upper surface tricolor; basal two-thirds black, followed by a narrow, grayish band and tipped with cinnamon brown. Ventral hairs gray at base, with grayish-white tips. Skull with a relatively narrow braincase and a slight but definite sagittal crest present in adults. Eastern United States from western Vermont south along Appalachian chain to western Florida, west to southwestern Wisconsin and northeastern Oklahoma.

Myotis volans (**Allen**). Western brown bat. Forearm about 35–41 mm. Tail about 93% of body length. Ear short, bluntly rounded, barely reaching nostril when laid forward. Calcar about as long as free edge of interfemoral membrane, with a low,

elongate keel arising about length of tarsus from ankle. Pelage full and long; interfemoral membrane furred above over an area approximately the length of femur; wing membrane furred below as far out as a line joining elbow and knee. Upper parts buffy to reddish brown; underparts grayish buffy to pale buffy; bases of hairs blackish. Skull with rostrum shortened and profile of braincase abruptly elevated; temporal ridges bowed outward, uniting anterior to occiput to form low sagittal crest. Southern Mexico through western United States to western Canada; east to eastern New Mexico and western South Dakota.

Myotis californicus (**Audubon and Bachman**). California brown bat. Size small; forearm about 29–36 mm. Tail about 91–98% of body length. Ear comparatively long, exceeding muzzle when laid forward. Feet small, slender. Calcar less than length of free border of interfemoral membrane, usually with a distinct keel arising abruptly at about length of metatarsus from heel and gradually tapering off. Pelage long, full, and fine-textured; tips of hairs usually yellow or brown, bases dark gray. Skull with relatively long, tapering rostrum; profile rising sharply to forehead and flat-topped braincase; sagittal crest usually inconspicuous or absent. Southern Mexico through western United States to western British Columbia; east to trans-Pecos Texas, north central Colorado and eastern Idaho.

Myotis subulatus (**Say**). Masked brown bat. Size small; forearm about 31–36 mm. Tail about 94% of body length. Ear relatively long, reaching tip of snout or slightly exceeding it when laid forward. Foot small, slender. Calcar keeled, long, and slender, equaling or exceeding length of free border of interfemoral membrane. Pelage full and silky; general color yellowish; face and ears black. Membranes blackish. Braincase relatively broad and flat; profile sloping gradually from rostrum to forehead. A low but sharply defined sagittal crest sometimes present in adults. Transcontinental through approximately middle third of United States, reaching east coast from Virginia northward; from trans-Pecos Texas westward, ranging from Mexican border northward to Canada.

Genus *Lasionycteris* Peters. Silver-haired bats. Moderate-sized, with short, broad ears and broad, flattened skull. Dentition: $\frac{2}{3}, \frac{1}{1}, \frac{2}{3}, \frac{3}{3}$. Recent; North America. 1 species.

Lasionycteris noctivagans (**Le Conte**). Silver-haired bat. Forearm about 39–43 mm. Tail about 67–76% of body length. Ear barely reaching nostril when laid forward; tragus short, straight, and bluntly rounded at tip. Top of interfemoral membrane furred on basal half. Fur deep, blackish chocolate brown throughout; many hairs on back, underparts and interfemoral membrane tipped with silvery white. Skull flattened; rostrum relatively very broad; dorsal profile of skull nearly straight. No sagittal crest. A migrant form that may be taken rarely anywhere in the United States. Ranges northward to southern Alaska and across southern Canada. Apparently breeds only in northern part of range.

Genus *Pipistrellus* Kaup. Pipistrelles. Small, with ears distinctly longer than broad and tapering to a narrowly rounded tip. Tragus straight or slightly curved forward. Interfemoral membrane sprinkled dorsally with hairs on basal third. Skull small and light; braincase relatively inflated; rostrum relatively broad. Dentition: $\frac{2}{3}, \frac{1}{1}, \frac{2}{2}, \frac{3}{3}$. Pleistocene; North America and most of Eastern Hemisphere. Numerous species; 2 in the New World and the United States.

KEY TO SPECIES OF *PIPISTRELLUS*

1. Tragus blunt, with tip bent forward (Fig. 6-25A); western North America   ***P. hesperus***
   Tragus tapering and straight (Fig. 6-25B); eastern North America   ***P. subflavus***

Fig. 6-25. Bat heads: A, Pipistrellus hesperus; B, Pipistrellus subflavus. (From Allen, A Monograph of the Bats of North America, Bull. U.S. Natl. Museum 43.)

     A   **B**

Pipistrellus hesperus (**Allen**). Western pipistrelle. Size very small; forearm about 28–33 mm. Tail about 67–75% of body length. Ear small, barely reaching to nostril when laid forward. Calcar about as long as tibia, with indistinct keel on posterior edge. General color light yellowish gray to whitish gray. Ears, muzzle, face, and membranes black. Rostrum broad and flattened; dorsal profile nearly straight, with only a slight angle between rostrum and forehead. Western United States from central Washington and northern California to southern Mexico; west of a line from southeastern Washington to eastern trans-Pecos Texas.

Pipistrellus subflavus (**Cuvier**). Eastern pipistrelle. Size small; forearm about 34–36 mm. Tail about 93% of body length. Ear reaching slightly beyond nostril when laid forward. Calcar longer than tibia, indistinctly keeled on posterior edge. Membranes thin and delicate. General color light yellowish brown, uniform below but mixed above with darker brown. Rostrum narrow and arched; profile rising in a distinct angle to forehead. Eastern United States west to central Minnesota, western Oklahoma, and southwestern Texas; south along east coast of Mexico.

Genus *Eptesicus* Rafinesque. Big brown bats. Large, with heavily built skull, relatively broad rostrum, and nearly straight dorsal profile of skull. Ears short; tragus straight, short, directed slightly forward. Top of interfemoral membrane naked except for a sprinkling of hairs on basal fourth. Dentition: $\frac{2}{3}, \frac{1}{1}, \frac{1}{2}, \frac{3}{3}$. Pleistocene; Asia, Africa, Australia, South and North America. Numerous species, 1 in the United States.

Eptesicus fuscus (**Beauvois**). Big brown bat. Forearm about 42–52 mm. Calcar slightly longer than foot, keeled. Color brown, paler below; ears and membranes blackish. From Central America throughout Mexico and the United States into Canada.

Genus *Nycticeius* Rafinesque. Evening bats. Small, with small ears and thick, leathery ears and membranes and essentially naked interfemoral membrane. Dentition: $\frac{1}{3}, \frac{1}{1}, \frac{1}{2}, \frac{3}{3}$. Recent; Asia, Africa, Australia, North America. Several species, 1 in North America.

Nycticeius humeralis Rafinesque. Evening bat. Forearm about 34–39 mm. Tail about 65% of body length. Ear small, naked except at extreme base above; tragus short, broad, and blunt, bent slightly forward. Pelage dull brownish, slightly paler below. Skull short, broad, and low; dorsal profile nearly straight, slightly convex over front of braincase. Eastern United States north to Virginia and southern Michigan, west to eastern Nebraska, southeastern Oklahoma, and through central and southern Texas into northeastern Mexico.

Genus *Lasiurus* **Gray.** Red bats and yellow bats. Moderate-sized, with dorsal surface of interfemoral membrane thickly furred. Coloration bright. Skull broad, short, and deep. Dentition: $\frac{1}{3}, \frac{1}{1}, \frac{1}{2}$ or $\frac{2}{2}, \frac{3}{3}$. Recent; North and South America, West Indies, Hawaii. 6 species in the United States.

KEY TO SPECIES OF *LASIURUS*

1. Each upper jaw usually with **7** teeth (tiny first premolar sometimes missing)   **2**
   Each upper jaw with 6 teeth   **4**
2. Forearm more than 45 mm   *L. cinereus*
   Forearm less than 45 mm   **3**
3. Color mahogany brown   *L. seminolus*
   Color red, orange, or yellow   *L. borealis*
4. Length of upper tooth row less than 6.2 mm   *L. ega*
   Length of upper tooth row more than 6.2 mm   **5**
5. Size larger; forearm usually more than 50 mm; southern Texas   *L. intermedius*
   Size smaller; forearm usually less than 50 mm; southeastern United States   *L. floridanus*

Lasiurus borealis (**Miller**). Red bat. Forearm about 37–44 mm. Tail about 82–114% of body length. Ear short, when laid forward reaching slightly over halfway from angle of mouth to nostril; outer side densely furred throughout basal two-thirds. Color variable, ranging from yellowish red to yellowish gray; a whitish area in front of shoulder. Skull with broad rostrum and flaring zygomata; dorsal profile nearly straight. From Central America north through eastern Mexico and eastern and central United States to southern Canada, west to Pecos River, Texas, eastern Colorado and western North Dakota, absent from peninsular Florida; also north through western Mexico and Baja California through Arizona and southern and western California to southwestern British Columbia. Migratory.

Lasiurus seminolus (**Rhoads**). Seminole bat. Similar to *L. borealis* in all characters except color. General color rich, mahogany brown throughout. Eastern Mexico around Gulf Coastal Plain to southern New York.

Lasiurus cinereus (**Beauvois**). Hoary bat. Size large; forearm about 46–55 mm. Tail about 67–74% of body length. Ear relatively broad, outer side densely furred to a little above middle, inner side with conspicuous patch of yellowish hairs above and in front of middle and a border of similar hairs along lower part of anterior edge. Margin of ear membrane dark brown or blackish. Forearm with a distinct patch of fur near base. General color a mixture of yellowish brown, deep brown, and white; hairs on dorsal surface mostly tipped with silvery white. Skull with short, broad rostrum and flaring zygomata. Migratory species, occurring from southern Mexico throughout the United States and into Canada. Apparently breeds only in northern part of range and on higher mountains farther south.

Lasiurus ega Gervais. Southern yellow bat. Forearm about 43–52 mm. Tail about 77–78% of body length. Color light yellowish brown. Pelage full and soft. Membranes thick and leathery. From Brazil northward through western Mexico, reaching the United States from southern California to New Mexico.

Lasiurus intermedius **Allen.** Northern yellow bat. Relatively large; forearm averaging 55 mm. Tail averaging about 83% of body length. Color and general appearance about as in preceding species. Gulf Coastal Plain from Houston, Texas, to Yucatan.

Lasiurus floridanus (**Miller**). Florida yellow bat. Forearm about 45–50 mm. Tail about 82–88% of body length. Color and external characters about as in *L. intermedius,* from which it differs principally in smaller size. Forested Coastal Plain from eastern Texas east to Virginia and south through peninsular Florida.

Genus *Plecotus* **Geoffroy Saint-Hilaire.** Big-eared bats. Very large-eared. Muzzle with dorsolateral glandular masses, which in some project as prominent lumps. Braincase relatively short. Dentition: $\frac{2}{3}, \frac{1}{1}, \frac{2}{3}, \frac{3}{3}$. Pleistocene; temperate North America, Eurasia, northern Africa. 3 species in the United States.

KEY TO SPECIES OF *PLECOTUS*

1. Nostril unspecialized; muzzle glands not enlarged; calcar keeled   ***P. phyllotis***

Nostril with posterior elongation; muzzle glands greatly enlarged; calcar unkeeled **2**

2. Tips of ventral hairs white or whitish, sharply contrasting with blackish bases *P. rafinesquei*

Tips of ventral hairs not contrasting sharply with their bases *P. townsendi*

Plecotus phyllotis (Allen). Allen's big-eared bat. Forearm about 46 mm. Tail about 76% of body length. Color brown, paler ventrally. Ears very large (about 40 mm from notch) with accessory anterior basal lobe developed into projecting lappet. Skull with supraorbital region sharply ridged; median postpalatal process absent. Southeastern Arizona, southward into Mexico.

Plecotus rafinesquei Lesson. Eastern big-eared bat. Forearm about 40–46 mm. Tail about 86–106% of body length. Ears very large (about 32–36 mm from notch), joined across forehead. Color brown, paler ventrally. Skull with supraorbital region smoothly rounded or faintly ridged. Median postpalatal process a prominent spine. Southeastern United States north to southern Ohio and central Indiana; west to southeastern Oklahoma.

Plecotus townsendi Cooper. Western big-eared bat. Forearm about 39–48 mm. Tail about 79–112% of body length. Color brown, paler ventrally. Ears very large (about 31–38 mm from notch), joined across forehead. Skull with supraorbital region smooth or faintly ridged. Median postpalatal process a prominent spine. Western United States from southern British Columbia and northwestern South Dakota south through Mexico. Disjunct populations eastward in Ozarks and central Appalachians.

Genus *Euderma* Allen. Spotted bats. Very large-eared and without glandular swellings on the face. Skull with elevated braincase; rostrum narrow and pointed; auditory bullae greatly expanded. Dentition: $\frac{2}{3}, \frac{1}{1}, \frac{2}{2}, \frac{3}{3}$. Recent; North America. 1 species.

Euderma maculatum (Allen). Spotted bat. Forearm about 50 mm. Tail about 83% of body length. Ears very long (about 34 mm from meatus) and joined across forehead. Dorsal color blackish, with a prominent white patch at base of tail and one on each shoulder; underparts white. Ears and membranes light brown. Southwestern United States from southern and central California east

to central New Mexico and north to southern Montana and southwestern Idaho.

Genus *Antrozous* Allen. Pallid bats. Very large-eared, with anterior bases of ears close together but not joined. Nostrils surrounded by a horseshoe-shaped ridge. Skull with broad braincase, palate, and rostrum. Dentition: $\frac{1}{2}, \frac{1}{1}, \frac{1}{2}, \frac{3}{3}$. Pleistocene; North America. 1 species in the United States (a peripheral, allopatric population in Kansas has been regarded as a distinct species by some authors).

Antrozous pallidus (Le Conte). Pallid bat. Forearm about 48–60 mm. Tail about 66–72% of body length. Ear large (28–30 mm from meatus), reaching about 20 mm beyond tip of nose when laid forward. Pelage sparse and short. Upper parts gray; underparts paler. Membranes thick and leathery. From western Mexico north through United States to southern British Columbia, east to central Texas, southwestern Kansas, northeastern Utah, and western Idaho.

Family Molossidae. Freetail bats. Narrow-winged, with thick leathery membranes and much-shortened 5th finger. Tail projecting conspicuously beyond the short interfemoral membrane. Ears variable, sometimes joined; tragus small. Nostrils usually opening on thickened pad, the upper surface of which is often set with fine, horny excrescences. A well-developed double articulation between scapula and humerus; 3d phalanx of 3d finger cartilaginous except at extreme base. The slender ulna about half as long as radius; fibula complete and bowed outward from tibia, its diameter about half that of tibia. Lower Oligocene; worldwide in warmer regions. Several genera, 2 in the United States.

KEY TO GENERA OF MOLOSSIDAE

1. Bony palate with conspicuous median emargination extending back of roots of incisors (Fig. 6-26A) *Tadarida*

Bony palate without conspicuous median emargination (Fig. 6-26B) *Eumops*

Genus *Tadarida* Rafinesque. Freetail bats. Large, varied, and widely distributed group. Ears large and rounded, arising from same point on forehead; tragus small, flattened, squarely truncate

Fig. 6-26. Palatal view of bat skulls: *A, Tadarida molossa; B, Eumops perotis.* (From Miller, The Families and Genera of Bats, *Bull. U.S. Natl. Museum* 57.)

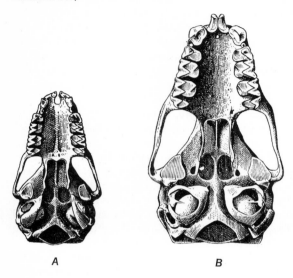

A   B

above. Muzzle pad sharply outlined, its upper margin thickly set with horny points. Upper lips full and wrinkled. Skull with rounded or somewhat flattened braincase; sagittal crest weak or lacking; anterior palatal emargination about as large as base of canine, wider posteriorly than between incisors. Dentition: $\frac{1}{2}$ or $\frac{1}{3}$, $\frac{1}{1}$, $\frac{2}{2}$, $\frac{3}{3}$. Upper Oligocene; worldwide in warmer regions. Numerous species, 3 in the United States.

KEY TO SPECIES OF *TADARIDA*

1. 3 incisors usually present in each lower jaw, total teeth usually 32; 2d phalanx of 4th finger more than 5 mm; ears extending to or short of rostrum when laid forward   *T. brasiliensis*

   2 incisors in each lower jaw, total teeth 30; 2d phalanx of 4th finger less than 5 mm; ears extending well beyond tip of rostrum when laid forward   2

2. Size smaller; forearm less than 53 mm
                            T. femorosacca

   Size larger; forearm more than 53 mm   *T. molossa*

Tadarida brasiliensis (Geoffroy Saint-Hilaire). Brazilian freetail bat. Forearm about 41–45 mm.

Tail about 45–60% of body length. Ear about 16–19 mm from notch. Color brown, slightly paler below; hairs dark to bases; ears and membranes dark brown to blackish. Skull with narrow rostrum and well-inflated braincase. Southern Brazil northward into United States to South Carolina, southeastern Nebraska, and southwestern Oregon.

Tadarida femorosacca (**Merriam**). Pocketed freetail bat. Forearm about 44–51 mm. Tail about 68% of body length. Ear about 19–23 mm from notch. Color dull brown. A fold of membrane from inner third of femur to middle of tibia, forming deep pocket between it and interfemoral membrane. 2 incisors in each lower jaw. From western Mexico barely reaches the United States from southern New Mexico to southern California, where it is rare.

Tadarida molossa (Pallas). Big freetail bat. Size large; forearm about 58–64 mm. Tail averaging about 61% of body length. Ear about 26–29 mm from notch. Color brown, slightly paler below; bases of hairs whitish; membranes, ears, and muzzle nearly black. Skull large; rostrum long and narrow. 2 incisors in each lower jaw. From South America north through Mexico and western United States to southwestern British Columbia, central Idaho, northern Nebraska, and central Iowa. Caves and buildings, rare.

Genus *Eumops* Miller. Mastiff bats. Ears large, rounded or squarish in outline, joined across forehead; tragus small, flat, its upper edge squarish or rounded. Muzzle pad well developed, deeply emarginate above, its lower edge and median ridge with minute, horny processes. Skull strong, rather slender, with sagittal crest absent or slightly indicated; rostrum always more than half as long as braincase; dorsal profile of skull from nares to occiput without strongly contrasting elevations and depressions. Bony palate not emarginate anteriorly. Dentition: $\frac{1}{2}$, $\frac{1}{1}$, $\frac{2}{2}$, $\frac{3}{3}$ in U.S. species (1 upper premolar lost in Central American species). Recent; tropical America. About 6 species, 2 in the United States.

KEY TO SPECIES OF *EUMOPS*

1. Size smaller; forearm less than 65 mm; Florida
                               E. glaucinus

   Size larger; forearm more than 65 mm   2

2. Ears smaller (keel averaging about 21 mm); interorbital region distinctly hourglass-shaped
E. underwoodi
Ears larger (keel averaging about 30 mm); interorbital region nearly cylindrical *E. perotis*

Eumops perotis (Schinz). Western mastiff bat. Size large; forearm about 73–78 mm. General color sooty brown, paler below; bases of hairs pale drab gray. Skull large, its greatest length about 30.3–32.0 mm. Brazil northward, reaching the United States along the Mexican border from western Texas to California and north to central California.

Eumops underwoodi Goodwin. Sonoran mastiff bat. Forearm about 65–70 mm. General color brown; bases of hairs dirty white. Central America north to Pima County, Arizona.

Eumops glaucinus (Wagner). Eastern mastiff bat. Forearm about 58–61 mm. General color brown; the hairs whitish at bases. Skull small, its greatest length about 22.4–24.3 mm. From Colombia and Ecuador in South America and Cuba and Jamaica in the West Indies reaching the United States rarely in southern Florida.

Order Primates. Generalized, principally arboreal, with well-developed clavicles. Tail usually long, sometimes prehensile. Limbs usually pentadactyl, or pollex rudimentary or absent. Hallux usually opposable; pollex usually more or less opposable when present. Terminal phalanges usually flattened and usually with nails rather than claws. Orbit usually large, always surrounded by a bony ring. Braincase relatively large. Cerebral hemispheres well developed and tending to overgrow cerebellum. Incisors usually $\frac{2}{2}$ (3 lower in family Tupaiidae); molars usually $\frac{3}{3}$, more or less bunodont, usually with 4 major tubercles. Paleocene; 2 suborders.

Suborder Anthropoidea. Eyes large and directed forward. Orbits completely separated from temporal fossa by vertical plate of bone. Snout short. Mandibles fused. Braincase usually much expanded; foramen magnum tending to be under rather than at back of skull. Cerebral hemispheres completely or nearly overgrowing cerebellum and much convoluted. Lower Oligocene; 1 extinct and 5 living families.

Family Hominidae. Bipedal, with short toes; hallux not opposable. Forelimbs relatively short.

Tooth row short; canines moderate to small. Braincase relatively large. Neopallium of cerebrum greatly developed. Pleistocene; 1 extinct and 1 living genus.

Genus *Homo* Linnaeus. Modern man. Braincase large, averaging about 1500 cc capacity. By virtue of relatively high intelligence capable of modifying and to some extent controlling environment. Dentition: $\frac{2}{2}, \frac{1}{1}, \frac{2}{2}, \frac{3}{3}$. 1 living species.

Homo sapiens **Linnaeus.** Modern man. Worldwide.

Order Edentata. A variable group of New World vegetation and insect eaters. Posterior thoracic and lumbar vertebrae with accessory zygapophyses. Clavicle present, although sometimes rudimentary. Teeth homodont (or absent); lacking enamel and usually monophyodont; never rooted, but with persistent pulp. Testes abdominal. 3 living families: Anteaters (Myrmecophagidae) of South and Central America, tree sloths (Bradypodidae) of South and Central America, and armadillos (Dasypodidae) of South and North America. Paleocene.

Family Dasypodidae. Armadillos. Fossorial, insect feeders. Body with a more or less rigid covering of bony plates embedded in skin and overlain by horny epidermal tissue. Hairs sparse and located in apertures between bony scutes. Forefeet with strongly developed, curved claws; hind feet plantigrade, pentadactyl. Tongue long, pointed, and protrusible. Teeth numerous, simple, usually monophyodont. Zygomatic arch of skull complete. Cervical vertebrae short, broad, depressed; atlas free, 2d, 3d, and often other cervicals fused. Paleocene; South and North America. About 26 extinct and 9 living genera; 1 in the United States.

Genus *Dasypus* Linnaeus. Body elongated and narrow; carapace with 7–9 movable bands. Tail moderate to long, tapering, the dermal scutes forming complete rings for most of length. Forefeet with 4 visible toes and concealed, clawless, vestigial 5th. A pair of inguinal mammae present in addition to pectoral pair. Ears large, ovate. Head narrow, with long, narrow subcylindrical, obliquely truncated snout; pterygoids meeting in midline below nasal passage. Teeth small, subcylindrical, diphyodont, about $\frac{7}{7}$ to $\frac{8}{8}$. Pleistocene; centering in South America, but into southern North America. 1 species in the United States.

Dasypus novemcinctus **Linnaeus.** Nine-banded armadillo. With 9 movable bands in carapace. Body length about 425 mm; hind foot about 100 mm. Tail long, about 85% of body length. Very sparsely haired. Color pale to dark brown. Young are always identical quadruplets. From South America through eastern Mexico to southern United States; all of Texas except trans-Pecos and panhandle, north to southeastern Kansas and east to southwestern Mississippi; well established by introduction in Florida. Range expanded eastward and northward in recent years.

Order Lagomorpha. Terrestrial, short-tailed herbivores. Teeth diphyodont. No canine teeth; incisors and molariform teeth separated by a wide diastema. 2 pairs of upper incisors, the 2d pair small and set directly behind the 1st. Teeth with persistent pulps and continuous growth. Incisive foramina large and confluent posteriorly. Bony palate narrow anteroposteriorly. Facial portion of maxillary incomplete. Fibula fused to tibia. Testes permanently external. Paleocene; Eurasia, Africa, North and South America. 2 living families.

KEY TO FAMILIES OF LAGOMORPHA

1. 5 teeth in upper molar row; hind legs little longer than forelegs; ears short, about as wide as high
Ochotonidae

6 teeth in upper molar row; hind legs distinctly longer than forelegs; ears higher than wide   *Leporidae*

Family Ochotonidae. Pikas. Small, short-eared, short-legged, with no visible tail. Skull flattened; no postorbital process of frontal; rostrum slender, nasals widest anteriorly; jugal long and projecting posteriorly to zygomatic process of squamosal. Dentition: $\frac{2}{1}, \frac{0}{0}, \frac{2}{2}, \frac{3}{3}$. Upper Oligocene; Eurasia, North America. 1 living genus.

Genus *Ochotona* Link. Characters and range of the family. 2 species in North America, 1 in the United States.

Ochotona princeps **(Richardson).** Pika. Body length 155–215 mm; hind foot 25–35 mm. Grayish to cinnamon buff above; washed with buff below. On talus slopes at high elevations in mountains from northern New Mexico and southern Sierra Nevada of California northward to central British Columbia.

Family Leporidae. Rabbits and hares. Hind legs elongated, longer than forelegs; hind feet large; ears long, longer than wide. Fur long and soft; tail densely furred, short. Skull elongated; rostrum broad; frontal with a supraorbital process which always projects posteriorly and sometimes anteriorly. Females generally larger than males. Dentition: $\frac{2}{1}, \frac{0}{0}, \frac{3}{2}, \frac{3}{3}$ (except 1 Japanese form, M $\frac{2}{3}$). Upper Eocene; Eurasia, Africa, North and South America. 8 living genera, 3 in North America, 2 in the United States.

KEY TO GENERA OF LEPORIDAE

1. Interparietal fused with parietals (Fig. 6-27A)
Lepus

Interparietal distinct, not fused with parietals (Fig. 6-27B)   *Sylvilagus*

Fig. 6-27. **Skulls of leporids:** A, *Lepus californicus;* B, *Sylvilagus auduboni.*

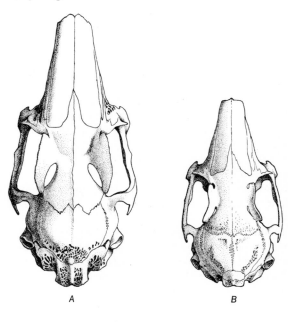

A   B

Genus *Lepus* Linnaeus. Hares. Moderate to large, short-tailed, moderate to long-eared. Upper parts grayish, brownish, or black, changing in some to white in winter. Interparietal not distinguishable in adult because of fusion with surrounding bones; supraorbital usually more or less broadly winglike and subtriangular in outline; cervical vertebrae long, 2d and 3d longer than wide; 3d to 5th ribs broad and flattened; ulna much slenderer and more tapering than radius. Pleistocene; Africa, Eurasia, North America. 10 native and 1 introduced species in North America; 6 in the United States.

KEY TO SPECIES OF *LEPUS*

1. All pelage white (except tips of ears sometimes dusky or black)   **2**
   Dorsal pelage brownish or grayish   **3**
2. Ear from notch more than 87 mm; least interorbital breadth more than 26 mm   *L. townsendi*
   Ear from notch less than 87 mm; least interorbital breadth less than 26 mm   *L. americanus*
3. Tail blackish or brownish all around   *L. americanus*
   Tail partly or wholly white   **4**
4. Tail all white or with faint buffy or dusky median line that does not extend onto rump   *L. townsendi*
   Tail white below with dorsal black or blackish line   **5**
5. Upper side of hind foot without trace of white; upper parts tawny   *L. europaeus*
   Upper side of hind foot white or whitish; upper parts grayish or brownish   **6**
6. Ears with terminal black patch on outside   *L. californicus*
   Ears without terminal black patch   **7**
7. Ear from notch more than 137 mm   *L. alleni*
   Ear from notch less than 137 mm   *L. callotis*

Lepus americanus **Erxleben.** Varying hare. Body length averaging 375–469 mm; hind foot about 112–150 mm. Ear from notch about 65–75 mm. Summer pelage: upper parts brownish or dusky grayish; underparts white, more or less encroached on by dorsal color. Hind foot brownish or white. Winter pelage: white at tips of hairs except for blackish tips of ears; or in California, Oregon, and Washington may be similar to summer pelage. Basilar length of skull less than 67 mm. Upper incisors inscribing arc of a circle with radius of less than 9.6 mm. Forests and swamps; Alaska across all but extreme north of Canada; south into the United States from northern Montana to north-eastern Ohio; south in Appalachian chain to eastern Tennessee; south through Rocky Mountains to northern New Mexico; south through Cascade–Sierra Nevada chain to northeastern California.

Lepus townsendi **Bachman.** White-tailed jackrabbit. Body length averaging 496–523 mm; hind foot about 145–172 mm. Ear from notch about 100–120 mm. Summer pelage: upper parts grayish brown; underparts white except neck, which is buffy. Tail white above and below or with a dusky or buffy median dorsal stripe which does not extend onto back. Winter pelage: white in northern part of range (Montana, Wyoming, the Dakotas, Minnesota); paler buffy than summer pelage in southern part of range. On plains and in mountains; southwestern Kansas and north central New Mexico west to eastern California; north into southern Canada; east through Iowa and to western Minnesota.

Lepus californicus **Gray.** Black-tailed jackrabbit. Body length averaging 454–521 mm; hind foot about 112–145 mm. Ear from notch about 105–135 mm. Winter pelage never all white. Upper parts gray to blackish; underparts dull buff to white. Tail with black median dorsal stripe which extends onto back. Outside of ears black at tip. Grasslands; from southern South Dakota, eastern Nebraska, western Missouri, and Arkansas and eastern Texas west to Pacific Coast and north to southern Washington; south into Mexico.

Lepus callotis **Wagler.** White-sided jackrabbit. Body length averaging 391–456 mm; hind foot about 124–133 mm. Ear from notch about 115 mm. Upper parts dark buff, heavily washed with black; flanks white, rump gray, nape buff. Underparts white except underside of head and neck, which are dull buff. Inner side of ears dull buff; front half of outer side dull buff, back half white and without black tip. Grasslands of southwestern New Mexico and south into Mexico. Range allopatric to the range of *L. alleni*.

Lepus alleni **Mearns.** Antelope jackrabbit. Body length averaging 543 mm; hind foot about 127–150 mm. Ear from notch about 138–173 mm. Middle of back and middle and side of head creamy buff, washed with black; middle of nape dingy buff; tail white except for median dorsal line of black extending onto rump; sides of shoulders, flanks, sides of abdomen, rump, and outside of legs gray; underside of head, chest, and middle of abdomen back to include sides of base of tail, front of hind legs, and tops of hind feet pure white. Front half of ears buffy gray; back half paler, more whitish; tip of ears pale buff or buffy white. Deserts of south central Arizona, south along west coast of Mexico.

Lepus europaeus Pallas.   European hare. Body length about 640–680 mm; hind foot about 130–150 mm. Ear from notch about 85–105 mm. Upper parts tawny, mixed with blackish hairs on back; underparts white, including underside of tail. Upper side of tail black; outside of ears with black terminal patch; upper side of feet tawny. Introduced into the United States and Canada from Connecticut west through Great Lakes region.

Genus Sylvilagus Gray.   Rabbits. Small to moderate-sized; with short to moderate-sized ears. Upper parts grayish to dark brownish. Interparietal distinct in adults; supraorbital process narrower and more strap-shaped or tapering to a more pointed tip posteriorly than in *Lepus*, the posterior notch or foramen usually much narrower or even absent because of fusion of postorbital process to braincase along its entire length; 2d to 4th cervical vertebrae broader than long; anterior ribs of nearly uniform width throughout their length, rodlike in form; radius and ulna about equal in size. Pleistocene; South America, North America. 8 species in the United States.

KEY TO SPECIES OF *SYLVILAGUS*

1. Anterior extension of supraorbital process more than half length of posterior extension (Fig. 6-28A); 1st upper molariform tooth with only 1 reentrant angle on anterior face   **S. idahoensis**
   Anterior extension of supraorbital process less than half length of posterior extension (Fig. 6-28B); 1st upper molariform tooth with more than 1 (usually 3) reentrant angles on anterior face   **2**
2. Anterior extenison of supraorbital process absent (or if a point is barely indicated, then ⅝ to all of posterior process fused to braincase) (Fig. 6-29A)   **3**
   Anterior extension of supraoribtal process present; posterior extension free of braincase or leaving a slit between process and braincase (Fig. 6-29B)   **5**
3. Size large; basilar length of skull more than 63 mm   **S. aquaticus**

Size moderate to small; basilar length of skull less than 63 mm   **4**
4. Underside of tail white; posterior extension of supraorbital process tapering to a slender point, this point free of braincase or barely touching it and leaving a silt or long foramen   **S. transitionalis**
   Underside of tail brown or gray; posterior extension of supraorbital process always fused to skull, usually for entire length   **S. palustris**
5. Auditory bullae large, greatly inflated (Fig. 6-30A)   **S. auduboni**
   Auditory bullae small (Fig. 6-30B)   **6**
6. Hind foot less than 81 mm; Pacific coastal strip from Columbia River south through California west of Cascade–Sierra Nevada chain   **S. bachmani**
   Hind foot usually more than 81 mm; east of Pacific coastal strip   **7**
7. In Arizona, New Mexico, and southern Colorado, posterior extension of supraorbital process free of braincase and supraoccipital shield, posteriorly pointed; from central Colorado north into Canada, diameter of external auditory meatus more than crown length of last 3 molariform teeth   **S. nuttalli**
   In Arizona, New Mexico, and southern Colorado, posterior extension of supraorbital process with its tip

Fig. 6-28. Rabbit skulls: A, *Sylvilagus idahoensis*; B, *Sylvilagus bachmani*. (From Hall, A Synopsis of the North American Lagomorpha, *Univ. Kans. Publ. Museum Nat. Hist.*, vol. 5, no. 10.)

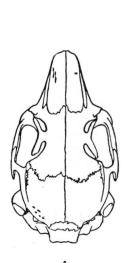
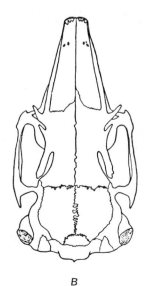

A   B

Fig. 6-29. Rabbit skulls: *A, Sylvilagus aquaticus; B, Sylvilagus nuttalli.* (From Hall, A Synopsis of North American Lagomorpha, *Univ. Kans. Publ. Museum Nat. Hist.,* vol. 5, no. 10.)

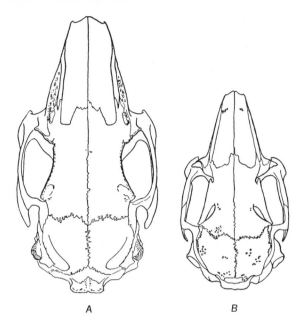

A　　　　　　B

Fig. 6-30. Ventral view of rabbit skulls: *A, Sylvilagus auduboni; B, Sylvilagus floridanus.*

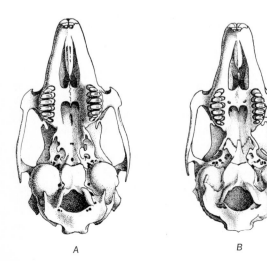

A　　　　　　B

against or fused to braincase and supraoccipital shield posteriorly truncate or notched; from central Colorado north into Canada, diameter of external auditory meatus less than crown length of last 3 molariform teeth　　　　**S. floridanus**

Sylvilagus idahoensis (**Merriam**). Pygmy rabbit. Size small; body length averaging 272 mm; hind foot about 65–72 mm. Ear short (about 40–55 mm), relatively broad and woolly. Tail very short (20–30 mm) and nearly unicolor, gray. Upper parts blackish to dark grayish; underparts white, except neck, which is buffy. Skull short, very broad posteriorly; auditory bullae very large; rostrum short, pointed. Infolded enamel dividing each molar tooth into 2 parts not crenate. Sagebrush plains from southwestern Montana to southwestern Utah and west to eastern and northeastern California and central Oregon. A disjunct population in southeastern Washington.

Sylvilagus bachmani (**Waterhouse**). Brush rabbit. Size small; body length averaging 278–323 mm; hind foot about 64–81 mm. Ear about 55–70 mm. Tail short, about 20–43 mm. Legs relatively short. Upper parts uniformly dark brown or brownish gray, nape rusty red; underparts dull white with the gray underfur showing through, underside of neck dull buffy. Braincase broad and rostrum rapidly tapering; ridge of enamel separating each molariform tooth into anterior and posterior sections only slightly crenulated. Chaparral west of crest of Cascade–Sierra Nevada chain from Columbia River south through Baja California.

Sylvilagus palustris (**Bachman**). Marsh rabbit. Size moderate; body length averaging 387–403 mm; hind foot about 88–91 mm. Tail short (about 33–39 mm), brownish or dingy gray below. Ear short, about 50–60 mm from notch. Upper parts blackish brown or reddish brown; nape reddish; abdomen buffy or with median white stripe. Braincase comparatively short, broad, and rounded; auditory bullae small; posterior and anterior extensions of supraorbital process joined to braincase along most (or all) of their length. Marshes and lowland thickets; Florida and Coastal Plain from Mobile Bay to southeastern Virginia.

Sylvilagus floridanus (**Allen**). Eastern cottontail. Size moderate; body length averaging 349–394 mm; hind foot about 87–104 mm. Ear from notch about 55–75 mm. Tail 39–65 mm, dark above, white below. Upper parts brownish or grayish, varying from darker in East to paler in West. Skull with smooth, moderate-sized auditory bullae; posterior extension of supraorbital process transversely thick. From Costa Rica through Mexico into Arizona and New Mexico and throughout the United States east of Rocky Mountain chain. Meets but does not overlap the range of S. *nuttalli* from Arizona, New Mexico, and Colorado north to northwestern North Dakota. Mostly in upland forests in eastern United States; along streams in plains; in brush and forests of mountains in Southwest.

Sylvilagus nuttalli (Bachman). Mountain cottontail. Size moderate; body length averaging 308–339 mm; hind foot about 88–110 mm. Ear from notch about 60–70 mm. Tail about 44–50 mm, dark above, white below. Pelage thick and heavy; feet heavily furred. Upper parts yellowish gray. Auditory bullae moderate, compactly rounded, smooth, as in closely related S. *floridanus*, to which its range is closely complementary. Differs from S. *floridanus* along eastern border of range in more slender rostrum and larger external auditory meatus; in New Mexico and Arizona differs in posteriorly pointed and unnotched supraoccipital shield and in posterior extension of supraorbital process, the tip of which projects free from the braincase or merely lies free (not fused) against the braincase. Eastern Arizona and northern New Mexico north to Canada, west to the Cascade–Sierra Nevada chain. Usually in sagebrush in North; usually in montane forests in South.

Sylvilagus transitionalis (Bangs). Allegheny cottontail. Size moderate; body length averaging 349 mm; hind foot about 95 mm. Ear relatively short, rounded, about 55 mm from notch. Upper parts almost uniform pinkish buffy, back overlaid with a blackish wash, giving a finely streaked rather than grizzled effect; top of head with a narrow black patch between ears. Auditory bullae very small; supraorbital process decreasing in width anteriorly and ending in a point against skull with no anterior process or notch. Forests, Atlantic Coast from southern Maine to southern New Jersey and west to eastern New York and central Pennsylvania; south in Appalachian chain to northern Alabama.

Sylvilagus auduboni (Baird). Desert cotton

tail. Size moderate to fairly small; body length averaging 298–354 mm; hind foot about 75–100 mm, with short hairs. Ears relatively long (about 60–75 mm from notch) and sparsely haired. Upper parts varying from pale, buffy gray in arid environments to dark, buffy brown in more humid ones. Skull generally with a straight, narrow, and rather pointed rostrum; posterior extension of supraorbital process usually broad, terminal end of the blunt, posterior point touching braincase; auditory bullae large, roughly rounded. From southern Mexico north into the United States west of a line from southern Texas to southwestern South Dakota; in West ranges north to northern California, central Nevada, northwestern Utah, and north central Montana. Deserts and up to 8500 ft in mountains.

Sylvilagus aquaticus (Bachman). Swamp rabbit. Size large; body length averaging about 465–468 mm; hind foot about 105–110 mm. Tail about 67–71 mm. Ear about 68–72 mm from notch. Upper parts blackish brown or reddish brown; underparts with some white; underside of tail white. Skull large; posterior extension of supraorbital process joined for entire length with side of braincase or with a small foramen between the braincase and the base of the posterior extension of the supraorbital process. From eastern Texas and Oklahoma and southeastern Kansas east to northwestern South Carolina; north in Mississippi embayment to southern Illinois and Indiana. Meets but does not overlap range of S. *palustris* in southern Alabama and Georgia. Usually in marshes and swamps; in flood-plain forests in western edge of range.

Order Rodentia. Worldwide group of mostly small, unguiculate herbivores, with dentition adapted for gnawing and grinding. Incisors, $\frac{1}{1}$; with heavy coat of enamel on anterior surface and lacking from lateral and posterior surfaces. Wear on the comparatively soft dentine maintains a sharp, chisellike cutting edge on incisors, which grow continuously from persistent pulps. No canines; molariform teeth separated from incisors by a wide diastema. Molars rooted or rootless. Paleocene; nearly worldwide. 3 suborders.[1]

---

[1] Regarded as artificial grouping by many; for attempted phylogenetic classifications see A. E. Wood, 1955, A Revised Classification of the Rodents, *J. Mammology,* **36**:165–187.

KEY TO SUBORDERS OF RODENTIA

1. Infraorbital opening larger than foramen magnum (Fig. 6-31*A*) **Hystricomorpha**
   Infraorbital opening smaller than foramen magnum **2**

2. Infraorbital opening vertically elongate (Fig. 6-31*B*); mandibular molariform teeth, 3 **Myomorpha**
   Infraorbital opening minute (Fig. 6-31*C*); mandibular molariform teeth, 4 **Sciuromorpha**

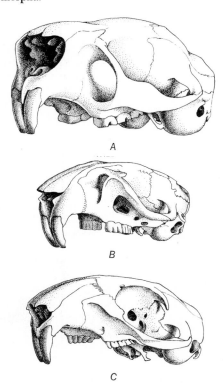

Fig. 6-31. Skulls of rodents, showing differences in infraorbital openings: *A, Erethizon dorsatum,* suborder Hystricomorpha; *B, Neotoma micropus,* suborder Myomorpha; *C, Citellus variegatus,* suborder Sciuromorpha.

Suborder Sciuromorpha. Rodents in which infraorbital opening is not enlarged for passage of medial masseter (except African Anomaluridae and Pedetidae, included here by some authors). Zygomatic arch slender, chiefly formed by jugal, which is not supported by long maxillary process; angle of mandible arising from lower surface of incisive alveolus. 1 lower and 1 or 2 upper premolars. Clavicles well developed; tibia and fibula separate. Paleocene. 5 generally recognized families, all of which occur in North America.

KEY TO FAMILIES OF SCIUROMORPHA

1. Prominent postorbital process of frontal present (Fig. 6-32*A*) **Sciuridae**
   No postorbital process (Fig. 6-32*B*) **2**
2. Fur-lined cheek pouches present and opening outside mouth **3**
   Cheek pouches never fur-lined, not opening outside mouth **4**
3. Left and right infraorbital canals connected through nasal septum (Fig. 6-33*A*); forelimbs smaller than hind, claws small **Heteromyidae**
   Nasal septum not perforated (Fig. 6-33*B*); forelimbs larger than hind and with much larger claws **Geomyidae**
4. Infraorbital canal round and opening in anterior face of zygomatic arch; skull very flat; tail less than $\frac{1}{10}$ of body length **Aplodontidae**
   Infraorbital canal minute and opening on side of rostrum anterior to zygomatic plate; tail flattened dorsoventrally, about $\frac{1}{3}$ of body length **Castoridae**

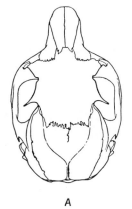
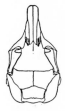

Fig. 6-32. Skulls of sciuromorph rodents: *A, Tamiasciurus douglasi,* family Sciuridae; *B, Perognathus formosus,* family Heteromyidae. (From Hall, *Mammals of Nevada,* University of California Press.)

Fig. 6-33. Skulls of sciuromorph rodents: *A, Perognathus longimembris,* family Heteromyidae; *B, Thomomys talpoides,* family Geomyidae. (From Hall, *Mammals of Nevada,* University of California Press.)

A   B

Fig. 6-34. Skull of *Marmota flaviventris.* (From Hall, *Mammals of Nevada,* University of California Press.)

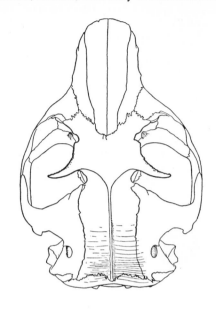

Family Aplodontidae. Sewellel. Primitive rodents in which the masseter originates principally from the lower edge of the zygomatic arch and in which the brain is small, braincase not swollen. **Dentition:** $\frac{1}{1}, \frac{0}{0}, \frac{2}{1}, \frac{3}{3}$. Upper Eocene. 1 extinct genus from Asia, 5 from North America. 1 living genus and species, restricted to North America.

Genus *Aplodontia* Richardson. Stout, heavy, short-limbed rodents with pentadactyl, plantigrade feet, small eyes, and very short tails.

Aplodontia rufa (Rafinesque). Sewellel. Body length about 285–295 mm; hind foot about 55 mm. Tail about 10% of body length. Colonial burrowers in dense, wet forests. Sierra Nevada of California through western Oregon and Washington.

Family Sciuridae. Squirrels. Mostly unspecialized rodents, with the lateral masseter arising from a conspicuous channel on the rostrum anterior to the zygomatic arch. A prominent postorbital process of the frontal present on each side. Teeth low-crowned, tuberculate, and capped with enamel. Premolars $\frac{2}{1}$, the first one very small or deciduous. Miocene; cosmopolitan, except Australia. 8 of the 42 generally recognized genera occur in North America.

KEY TO GENERA OF SCIURIDAE

1. Infraorbital opening a foramen piercing the zygomatic plate of the maxillary   **2**
   Infraorbital opening a canal passing between the zygomatic plate and rostrum   **3**

Fig. 6-35. Skulls of sciurid rodents: *A, Cynomys ludovicianus; B, Sciurus niger.*

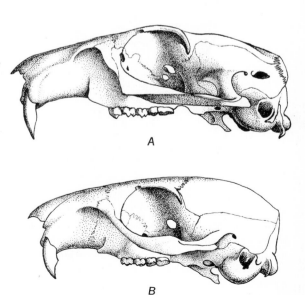

A

B

2. Premolars $\frac{1}{1}$ *Tamias*

    Premolars $\frac{2}{1}$ *Eutamias*

3. Postorbital processes very broad, projecting at nearly right angles to long axis of skull at about middle of orbit; skull flat or concave between postorbital processes (Fig. 6-34)   *Marmota*

    Postorbital processes projecting backward and downward (Fig. 6-31C); skull convex between postorbital processes   **4**

4. Zygomata converging anteriorly, angular portion twisted toward a horizontal plane (Fig. 6-35A)   **5**

    Zygomata nearly parallel, not twisted (Fig. 6-35B)   **6**

5. Upper molar rows strongly convergent posteriorly   *Cynomys*

    Upper molar rows nearly parallel   *Citellus*

6. Body skin extending laterally to form planing membrane between fore and hind legs   *Glaucomys*

    No planing membranes   **7**

7. Anterior border of orbit ventrally opposite last premolar; baculum vestigial   *Tamiasciurus*

    Anterior border of orbit ventrally opposite 1st molar; baculum well developed   *Sciurus*

Genus *Sciurus* Linnaeus. Tree squirrels. Skull characterized by broad interorbital region and deep braincase; jugal with an angular process on its upper surface. Slender-bodied, arboreal, with long bushy tails. Dentition: $\frac{1}{1}, \frac{0}{0}, \frac{2}{1}$ or $\frac{1}{1}, \frac{3}{3}$. Miocene; cosmopolitan (except Australia), numerous living species, about 6 in North America north of Mexico.

KEY TO SPECIES OF *SCIURUS*

1. 5 molariform teeth in upper jaw   **2**

    4 molariform teeth in upper jaw   **4**

2. Ears tufted with long black or blackish hairs   *S. aberti*

    Ears not tufted   **3**

3. California, Oregon, and Washington   *S. griseus*

    Eastern half of the United States   *S. carolinensis*

4. Dorsal surface and tail gray, belly and tail fringe white (Arizona)   *S. arizonensis*

    Dorsal surface and tail reddish or yellowish, belly slightly to very yellowish   **5**

5. Eastern half of the United States   *S. niger*

    Extreme southeastern Arizona and southwestern New Mexico   *S. apache*

***Sciurus griseus* Ord.** Western gray squirrel. Body length about 300 mm; hind foot about 74–80 mm. Tail about 90% of body length. Upper parts bright silvery gray; underparts white. Tail gray above and below. Pacific Coast and ranges from central Washington south to northern Baja California.

***Sciurus carolinensis* Gmelin.** Eastern gray squirrel. Body length about 220–275 mm; hind foot about 55–75 mm. Tail about 75–90% of body length. Upper parts gray, often with rusty wash, or sometimes blackish; underparts white to gray. Tail hairs gray, with more or less conspicuous white tips. Forests of eastern North America, extending west to western border of deciduous forest in eastern Texas and north to eastern North Dakota and southern Manitoba.

***Sciurus aberti* Woodhouse.** Tassel-eared squirrel. Body length about 260–290 mm; hind foot about 65–80 mm. Tail about 80–85% of body length. Tail all white or white below. Body white or black below, gray on sides and reddish above. Sometimes melanistic. Ear tufts prominent. Rocky Mountain yellow pine forests in Arizona, New Mexico, and Colorado. Including *S. kaibabensis* of some authors.

***Sciurus niger* Linnaeus.** Eastern fox squirrel. Body length about 200–300 mm; hind foot about 51–82 mm. Tail long, bushy, about 75–105% of body length. Usually reddish gray above, rusty below. Tips of tail hairs buffy. Tending to melanism and to white nose in Southeast; buffy tones very pale (underparts nearly white) in Southwest; steel gray throughout mid-Atlantic coast. Upland forests of eastern North America, and in stream-valley forests west nearly across Texas, Oklahoma, Kansas, Nebraska, and North Dakota. North into upper Minnesota and Wisconsin, and Michigan and to southwestern New York and southern Pennsylvania.

***Sciurus apache* Allen.** Apache fox squirrel. Body length about 320 mm; hind foot about 73–80 mm. Tail about 85% of body length. Upper parts reddish brown, with broad dorsal stripe in winter; underparts rusty. Mixed oak and pine forests, entering the United States only in mountains of southeastern Arizona and southwestern New Mexico.

***Sciurus arizonensis* Coues.** Arizona gray squirrel. Body length about 270 mm; hind foot about 73 mm. Tail about 105% of body length. Upper parts dark gray; underparts white. Tail dark gray with silvery margins. Ears untufted. Pine and oak forests, entering the United States only in mountains of southeastern Arizona and southwestern New Mexico.

Genus *Tamiasciurus* Trouessart.   Red squirrels; chickarees. Small, more or less reddish tree squirrels, with a conspicuous, black side stripe in summer pelage. Ear tufts in winter. Dorsal surface of skull comparatively flat; braincase shallow; postorbital processes short. Anterior upper premolar absent or small and nonfunctional, being covered by crown of last premolar. Baculum vestigial. Dentition: $\frac{1}{1}, \frac{0}{0}, \frac{2}{1}$ or $\frac{1}{1}, \frac{3}{3}$. Pleistocene; North America. 2 currently recognized, allopatric species, which may constitute a single species.

KEY TO SPECIES OF *TAMIASCIURUS*

1. Belly reddish or orange; range from coastal, southwestern British Columbia through Washington and Oregon west of east base of Cascades through mountains of California   ***T. douglasi***
   Belly white or whitish; range east of California or north or east of western Oregon, Washington, and southwestern British Columbia   ***T. hudsonicus***

Tamiasciurus hudsonicus **(Erxleben).**   Red squirrel. Body length about 185–215 mm; hind foot about 35–57 mm. Tail about 60–75% of body length. Upper parts reddish gray, reddest along dorsal line; underparts white. Throughout Canada and Alaska, north to limits of tree growth. In the United States east of central grasslands from North Dakota and southern Iowa, east to Virginia, follows Appalachian chain south to western North Carolina. In the West, Rocky Mountains south to southern Arizona and New Mexico.

Tamiasciurus douglasi **(Bachman).**   Douglas squirrel; chickaree. Body length 185–200 mm; hind foot about 45–55 mm. Tail about 65–75% of body length. Upper parts dusky olive; underparts orange. Pacific Coast forests from sea level to high in Cascades. Ranges south in Sierra Nevada of California. A disjunct population in mountains of northern Baja California. The Douglas squirrel has been considered by some authors as a subspecies of the red squirrel, the range of which it approaches in Oregon, Washington, and British Columbia.

Genus *Marmota* Frisch.   Marmots. Largest sciurids. Heavy-bodied, short-tailed, short-legged burrowers. Underfur soft and dense, with long and coarse overhairs. Braincase broad; zygomata widespreading, their anterior portions thickened; postorbital processes broad and nearly at right angles to long axis of skull. Dentition: $\frac{1}{1}, \frac{0}{0}, \frac{2}{1}, \frac{3}{3}$. Middle Pliocene. About 12 living species in Asia, Europe, North America; 5 in North America, 4 in the United States.

KEY TO SPECIES OF *MARMOTA*

1. Posterior edge of postorbital process forming approximately a right angle with long axis of skull; upperparts without conspicuous patches of buffy, black, or white   ***M. monax***
   Posterior edge of postorbital processes forming a distinctly acute angle with long axis of skull; upper parts with patches of buff, white, or black   **2**
2. Upper parts mainly black and white   ***M. caligata***
   Upper parts mainly brownish, yellowish, drab, or buffy   **3**
3. Upper parts of soild colors (not grizzled)
   M. olympus
   Upper parts of mixed colors (grizzled)
   M. flaviventris

Marmota monax **(Linnaeus).**   Woodchuck. Body length about 354–512 mm; hind foot about 68–89 mm. Tail about 30–37% of body length. Grizzled, with black or dark-brown feet and without white markings except around nose. Forelegs overlaid with deep reddish-colored hairs. Posterior pad on sole of hind foot oval and located near middle of sole. Mammae 8. From central Alabama and southeastern Oklahoma, northward east of the central grasslands into Canada, and westward nearly across Canada, north of the grasslands.

Marmota flaviventris **(Audubon and Bachman).** Yellow-bellied marmot. Body length about 339–520 mm; hind foot about 68–92 mm. Tail about 35–50% of body length. Yellowish brown, with yellow belly and light-buff to dark-brown hind feet. Plantar pads as in *M. monax*. Sides of neck with conspicuous buffy patches; head usually with white markings between eyes. Mammae 10. Mountains, foothills, and rocky canyons from southern British Columbia in Sierra Nevada to California;

east to Black Hills; south in Rocky Mountains to northern New Mexico.

Marmota caligata (**Eschscholtz**). Hoary marmot. Body length 450–527 mm; hind foot about 90–113 mm. Tail about 39–49% of body length. Conspicuously marked, with head and foreparts of mixed black and white in varying proportions. Belly whitish; feet black or blackish brown. Posterior pad on sole of hind foot nearly circular and near edge of sole. Mammae 10. High mountains at and above timber line in northwestern Montana, northeastern Idaho, and in Cascade Range of Washington north through Alaska.

Marmota olympus (**Merriam**). Olympic marmot. Body length about 490–521 mm; hind foot about 90–112 mm. Tail about 36–42% of body length. Brownish drab with intermixed white hair; closely related to *M. caligata.* Broad white patch in front of eyes; sides of nose, lips, and chin white. Above timber line in Olympic Mountains, Washington.

Genus *Cynomys* Rafinesque. Prairie dogs. Large, stout, fossorial, and colonial squirrels. Tail short and flat. Ears small. Molariform teeth very heavy, much broader than long. Anterior premolar large, more than half the size of the last premolar. Dentition: $\frac{1}{1}, \frac{0}{0}, \frac{2}{1}, \frac{3}{3}$. Pleistocene; North America. 2 species in the United States, 1 in Mexico.

KEY TO SPECIES OF *CYNOMYS*

1. Tail tipped with black   *C. ludovicianus*
   Tail tipped with white   *C. gunnisoni*

Cynomys ludovicianus (**Ord**). Black-tailed prairie dog. Body length 260–320 mm; hind foot 55–67 mm. Tail relatively long, 24–44% of body length. Upper parts reddish; underparts buffy; tail tipped with black. Mammae 8. Jugal heavy, thickened; outer surface at angle of ascending branch very broad, triangular. Teeth large, expanding laterally. Principally in the central grasslands, from southeastern Arizona, east to north central Texas, north to western North Dakota, and west to eastern front of Rocky Mountains. A related species, *C. mexicanus,* in southeastern Coahuila and northern San Luis Potosi, Mexico.

Cynomys gunnisoni (**Baird**). White-tailed prairie dog. Body length 253–322 mm; hind foot

52–67 mm. Tail relatively short, 10–26% of body length. Upper parts reddish; underparts buffy; tail tipped with white. Mammae 10 (rarely 12). Jugal weak, thin, and flat; outer surface at angle of ascending branch only slightly thickened, the margin rounded. Teeth smaller and less expanded laterally than in *C. ludovicianus.* Discontinuously distributed in high plains, broad valleys, and mountain parks, from northwestern New Mexico and northern Arizona, north through mountainous parts of Colorado, Utah, and Wyoming to southern Montana. As here recognized, includes allopatric forms usually referred to as species *C. leucurus* and *C. parvidens.*

Genus *Citellus* Oken. Ground squirrels. Generalized, ground-living and burrowing squirrels. Mostly smaller than tree squirrels, with moderate to short tails and small ears. Frontal bone of skull with prominent postorbital processes projected backward and downward. Braincase moderately shallow; zygomata convergent anteriorly and twisted in jugal region from vertical to horizontal plane. Infraorbital opening oval or nearly triangular in shape. Dentition: $\frac{1}{1}, \frac{0}{0}, \frac{2}{1}, \frac{3}{3}$. Pliocene; North America and Eurasia. About 50 living species; about 19 in the United States.

KEY TO SPECIES OF *CITELLUS*

1. Molars relatively hypsodont; parastyle ridge on 1st and 2d upper molars joining the protocone with an abrupt change of direction   **2**
   Molars relatively brachyodont; parastyle ridge on 1st and 2d upper molars rising evenly to join the protocone without abrupt change of direction   **11**
2. Metaloph on last upper premolar continuous   **3**
   Metaloph on last upper premolar not continuous   **9**
3. Upper parts unspotted   **4**
   Upper parts spotted or mottled   **7**
4. Hind foot less than 39 mm   *C. townsendi*
   Hind foot more than 39 mm   **5**
5. Underside of tail grayish   *C. armatus*
   Underside of tail buffy or reddish   **6**
6. Underside of tail buffy   *C. richardsoni*
   Underside of tail reddish   *C. beldingi*
7. Hind foot more than 40 mm   *C. columbianus*
   Hind foot less than 40 mm   **8**
8. Upper parts grayish   *C. washingtoni*
   Upper parts brownish   *C. brunneus*
9. Upper parts striped   *C. tridecemlineatus*
   Upper parts spotted   **10**

10. Dorsal spots in rows   ***C. mexicanus***
    Dorsal spots not in rows   ***C. spilosoma***

11. Anterior upper premolar with 2 cusps and a functional cutting edge, more than ¼ size of last premolar   ***C. franklini***
    Anterior upper premolar simple, less than ¼ size of last premolar   **12**

12. Upper incisors relatively slender, not distinctly recurved   **13**
    Upper incisors relatively stout and distinctly recurved   **15**

13. Upper parts with black and white stripes; postorbital processes long and slender   ***C. lateralis***
    Upper parts plain; postorbital processes short and stout   **14**

14. Tail white below   ***C. mohavensis***
    Tail not white below   ***C. tereticaudus***

15. Upper parts unstriped or indistinctly striped; braincase rounded on upper surface   **16**
    Upper parts distinctly striped; braincase flattened on upper surface   **17**

16. Side of neck and shoulders whitish; tail less than 44% of total length   ***C. beecheyi***
    Side of neck and shoulders not whitish; tail more than 44% of total length   ***C. variegatus***

17. Tail gray below   ***C. harrisi***
    Tail white below   **18**

18. Upper parts buffy   ***C. nelsoni***
    Upper parts grayish   ***C. leucurus***

Citellus townsendi (**Bachman**). Townsend ground squirrel. Small; body length about 142–182 mm; hind foot about 29–38 mm. Tail about 24–31% of body length. Color plain gray to indistinctly dappled, shaded with buff. Skull relatively short and broad; zygomata heavy and widely expanded; rostrum stout, with nearly parallel sides; postorbital processes long, slender, decurved; auditory bullae moderately inflated. Densely colonial in dry, sandy sagebrush valleys and on juniper-covered ridges. Southern Nevada, through eastern California and western Utah, north to southern Washington and southern Idaho.

Citellus washingtoni **Howell.** Washington ground squirrel. Body averaging about 163–179 mm; hind foot about 30–38 mm. Tail averaging about 24–28% of body length. Color gray, spotted with whitish. Skull similar to that of *C. townsendi*, but relatively longer and narrower. Colonial in dry prairies of Columbia Basin, Washington and Oregon, east of Columbia River.

Citellus brunneus **Howell.** Idaho ground squirrel. Body length averaging about 175 mm; hind foot 33–37 mm. Tail averaging about 33% of body length. Generally similar to *C. washingtoni* but distinguished by larger ears, longer and darker tail, smaller dorsal spots, and larger, relatively broader skull with longer nasals. Restricted to Payette and Weiser Valleys in western Idaho.

Citellus richardsoni (**Sabine**). Richardson ground squirrel. Body length averaging about 189–211 mm; hind foot about 39–49 mm. Tail averaging about 35–43% of body length. Upper parts gray, washed with buff; underparts pale buff. Skull convex dorsally, the highest point being between postorbital processes; braincase narrow, deep, much constricted anteriorly; postorbital processes long, slender, decurved; zygomata heavy and broad, widely expanded posteriorly, narrowing anteriorly; nasals ending nearly on plane of posterior ends of premaxillae, or shorter. Open plains and prairies, from northern Colorado and Nevada, north through Idaho into Canada and east through the Dakotas into western Minnesota.

Citellus armatus (**Kennicott**). Uinta ground squirrel. Body length averaging about 221 mm; hind foot about 42–46 mm. Tail about 33% of body length. Generally similar to *C. richardsoni*, but distinguished by grayish instead of reddish tail and slightly longer skull with broader cranium, interorbital region, and rostrum. Open ground in foothills and mountains; nearly to timber line in mountain meadows. Central Utah, north through western Wyoming and eastern Idaho to southwestern Montana.

Citellus beldingi (**Merriam**). Belding ground squirrel. Body length averaging about 208–212 mm; hind foot about 40–47 mm. Tail averaging about 32% of body length. Generally similar to *C. richardsoni*. Distinguished by a well-defined, reddish-brown band along middorsal surface and by skull differences. Skull relatively longer and narrower across zygomata, but broader interorbitally; zygomata slenderer; auditory bullae smaller. Colonial on open ground in valleys and in mountain meadows nearly to timber line. From Sierra Nevada of California, north through eastern Oregon, east through southwestern Idaho and northern Nevada.

Citellus columbianus (**Ord**). Columbian ground squirrel. Body length averaging about 249–

269 mm; hind foot about 48–58 mm. Tail averaging about 37–41% of body length. Bushy-tailed; mottled grayish above. Dorsal outline of skull relatively flat, the highest point behind postorbital processes; interorbital region relatively broad and flat. Colonial on open prairies, sparsely forested hillsides, and in mountain parks. Enters the United States from eastern Washington and Oregon east to western Montana, south in Idaho nearly to Snake River. A related species, *C. undulatus*, ranging widely in Canada, Alaska, and eastern Siberia.

Citellus tridecemlineatus (**Mitchill**). 13-lined ground squirrel. Body length averaging 118–180 mm; hind foot 27–41 mm. Tail averaging 49–60% of body length. Dorsal surface with pattern of 5 or more narrow, longitudinal dark and about 6 narrow, light stripes. Skull relatively long, narrow, weak; braincase usually longer than broad; interorbital region relatively long; rostrum long and tapering gradually; zygomata stout but not widely expanded. Noncolonial in grasslands of central North America, from southern Texas and eastern Arizona north through Montana; limited on East by the deciduous forest, but following prairies and cleared land east into Ohio and lower Michigan.

Citellus mexicanus (**Erxleben**). Mexican ground squirrel. Body length averaging about 183 mm; hind foot about 38–51 mm. Tail averaging about 64% of body length. Upper parts with squarish white spots arranged in usually 9 longitudinal rows on brownish background; underparts white or whitish. Skull generally similar to that of *C. tridecemlineatus*, but larger, with less elongate, more nearly square braincase, more widely expanded zygomata, and larger, more smoothly rounded auditory bullae. Noncolonial in mesquite brushlands. Southeastern New Mexico and western two-thirds of Texas, north to base of panhandle, entering the United States from Mexico.

Citellus spilosoma (**Bennett**). Spotted ground squirrel. Body length averaging about 135–165 mm; hind foot about 28–38 mm. Tail averaging about 39–50% of body length. Upper parts with many squarish, whitish spots (often small and indistinct); 2 color phases, with grayish cinnamon or reddish background dorsally. Skull generally similar to that of *C. tridecemlineatus*, but relatively shorter and broader, particularly rostrum and interorbital region; auditory bullae much larger. Non-

colonial, mostly on sandy soils; south from southern South Dakota and western half of Nebraska through western half of Texas into Mexico; west across eastern two-thirds of Arizona.

Citellus franklini (**Sabine**). Franklin ground squirrel. Body length averaging about 238–244 mm; hind foot about 51–58 mm. Tail averaging about 59–62% of body length. Upper parts plain gray; underparts only slightly paler than back. Skull long, narrow, with flattened dorsal outline. Colonial in prairies and fields; from northern half of Kansas and northern half of Missouri north through eastern two-thirds of North Dakota and most of Minnesota into Canada, east into northwestern Indiana.

Citellus variegatus (**Erxleben**). Rock squirrel. Body length averaging 258–282 mm; hind foot about 53–65 mm. Tail averaging 73–82% of body length. Upper parts gray, frequently with black head or black hood and sometimes completely black. Skull with flat dorsal profile and relatively broad, shallow braincase; rostrum relatively broad, tapering gradually; postorbital processes stout, decurved. Infraorbital foramen narrowly oval. Upper incisors stout, recurved; molariform teeth low-crowned. Usually in rocky situations. Edwards Plateau and trans-Pecos Texas, Arizona, and New Mexico, northward through Utah and the Colorado Rockies. Widely distributed in Mexico.

Citellus beecheyi (**Richardson**). California ground squirrel. Body length averaging 235–279 mm; hind foot about 50–64 mm. Tail averaging 62–77% of body length. Upper parts brownish; flecked with white; sides of neck and shoulders white or whitish, separated by a dark triangle. Skull generally similar to that of *C. variegatus* but smaller. Colonial in many situations except dense woods and brush. Pacific slope of Oregon, south through all of California except southeastern desert, into Baja California.

Citellus harrisi (**Audubon and Bachman**). Gray-tailed antelope squirrel. Body length averaging about 150 mm; hind foot about 38–42 mm. Tail averaging about 55–58% of body length. Upper parts cinnamon (gray in winter); tail with mixed black and white, both above and below. A horizontal white stripe on each side of back. Skull with zygomata only slightly expanded posteriorly; postorbital processes small, slender; antorbital foramen narrowly oval. Upper incisors stout, recurved; molariform teeth low-crowned. Noncolonial in deserts of southern and western Arizona and southwestern New Mexico, southward in Sonora, Mexico.

Citellus leucurus (Merriam). White-tailed antelope squirrel. Body length averaging 142–158 mm; hind foot about 36–43 mm. Tail averaging about 44–49% of body length. Similar to *C. harrisi* in color and pattern except that the tail is white below. Skull essentially similar to that of *C. harrisi.* Noncolonial, usually in rocky and brushy situations. From trans-Pecos Texas, through northwestern New Mexico, and extreme western Colorado to southwestern Idaho and southeastern Oregon, south through eastern California into Baja California. Does not overlap range of *C. harrisi,* which it partially encircles. Includes *C. interpres* of some authors.

Citellus nelsoni (Merriam). San Joaquin antelope squirrel. Body length averaging about 161 mm; hind foot about 40–43 mm. Tail averaging about 43% of body length. Similar to *C. leucurus* from which it differs in more buffy coloration in both summer and winter pelage and larger skull with heavier, more spreading zygomata and larger auditory bullae. Noncolonial, in San Joaquin Valley, California.

Citellus mohavensis (Merriam). Mohave ground squirrel. Body length averaging about 157 mm; hind foot about 32–40 mm. Tail averaging about 42% of body length. Upper parts plain cinnamon gray; tail whitish below. Skull short, broad; postorbital processes short, stout. Upper incisors moderately stout and slightly recurved; molars low-crowned. Noncolonial in sparse sagebrush on sandy or gravelly soils of Mohave Desert, California.

Citellus tereticaudus (Baird). Round-tailed ground squirrel. Body length averaging 150–158 mm; hind foot about 32–40 mm. Tail averaging 46–62% of body length. Upper parts plain, grayish cinnamon. External ears reduced to a mere rim. Tail pencillike, not bushy. Skull similar to that of *C. mohavensis.* In isolated colonies in hot, sandy deserts. From southern tip of Nevada south through southeastern California and southwestern Arizona into Mexico.

Citellus lateralis (Say). Mantled ground squir-rel. Body length averaging 168–194 mm; hind foot 35–49 mm. Tail averaging 43–60% of body length. Each side of back with a longitudinal white stripe, bordered on each side by a black stripe. Inner stripe reduced or absent in some races. In summer pelage a more or less distinct mantle of buffy or reddish over head and shoulders. Skull with upper incisors relatively slender and not distinctly recurved; postorbital processes long and slender. Open, forested mountain slopes and foothills. From southern California, central Arizona, and northern New Mexico, north to Canada, generally in and west of Rocky Mountains. Including *C. saturatus* of Washington and British Columbia, which has been considered a distinct species. A related species, *C. madrensis,* in Sierra Madre of southern Chihuahua.

Genus *Tamias* Illiger. Eastern chipmunk. Small, fairly stout-bodied, with moderately long tail (tail about one-third of total length). Tail well haired but not bushy. Ears prominent, rounded. Skull fairly long and narrow; zygomata rather weak, evenly curved, and not widely expanded; notch in posterior edge of zygomatic plate of maxillary opposite last premolar or anterior edge of 1st molar. Postorbital processes of frontal broad at base and rather short. Infraorbital openings large, rounded, piercing zygomatic plate. Auditory bullae small. Dentition: $\frac{1}{1}, \frac{0}{0}, \frac{1}{1}, \frac{3}{3}$. Pleistocene; North America. 1 species.

Tamias striatus (Linnaeus). Eastern chipmunk. Body length averaging 139–167 mm; hind foot 32–38 mm. Tail about 59–62% of body length. Dorsal pattern of 5 blackish and 2 whitish longitudinal stripes. Side of face striped; a narrow buffy stripe from nose, above eye, nearly to ear, and a wider buffy stripe from beneath eye to ear. Forests of eastern North America, west to eastern Louisiana, eastern Oklahoma, eastern Kansas, eastern Iowa and eastern North Dakota. Throughout eastern United States north of a line from eastern Louisiana to central North Carolina, but south to western Florida; north into southern Canada.

Genus *Eutamias* Trouessart. Western chipmunks. Closely related to *Tamias.* The distinctive color pattern consists of 5 blackish and 4 whitish longitudinal stripes, all of approximately equal width. Rostrum shorter and more abruptly constricted at base than in *Tamias,* palate relatively shorter, terminating on the plane of last molars or a little posterior to it; zygomatic plate of the maxillary usually opposite middle or posterior part of the last premolar. Bullae relatively large. Dentition: $\frac{1}{1}, \frac{0}{0}, \frac{2}{1}, \frac{3}{3}$. Pliocene; Asia, North America. About 15, possibly fewer, species in the United States.

KEY TO SPECIES OF *EUTAMIAS*

1. Dorsal stripes (except median) more or less indistinct   **2**
   Dorsal stripes all distinctly marked   **3**
2. Restricted to Charleston Mountains, Nevada; baculum with distal 50% or more of shaft markedly compressed laterally, base markedly widened   *E. palmeri*
   Not in Charleston Mountains, Nevada; baculum with less than distal 50% slightly compressed laterally, base not markedly widened   *E. dorsalis*
3. Size large, skull usually 37 mm or more in greatest length   **4**
   Size smaller, skull usually less than 37 mm in greatest length   **8**
4. Southeastern New Mexico and northern trans-Pecos Texas   *E. canipes*
   Pacific states, British Columbia to Baja California   **5**
5. Tips of nasals not separated by median notch; backs of ears distinctly bicolor in all pelages   **6**
   Tips of nasals separated by median notch; backs of ears unicolor or nearly so in summer pelage (bicolor in winter)   **7**
6. Submalar dark stripe expanding to conspicuous black area below ear; ear very long and pointed; shaft of baculum thick, more than 0.25 mm in diameter at widest point   *E. quadrimaculatus*
   Submalar dark stripe not black below ear; ears relatively short, not pointed; shaft of baculum thin, less than 0.20 mm at widest point   *E. townsendi*
7. Color comparatively reddish; range north of San Francisco Bay, California; height of keel on tip of baculum 10% of length of upturned tip   *E. sonomae*
   Color comparatively grayish; range south of San Francisco Bay, California; height of keel on tip of baculum at least 14% of length of upturned tip   *E. merriami*
8. Size small; body length usually less than 110 mm   **9**
   Size larger; body length usually more than 110 mm   **10**
9. Interorbital distance greater; more than 24% of greatest length of skull; tip of baculum forming distinct angle with shaft   *E. alpinus*
   Interorbital distance narrower; less than 24% of greatest length of skull; tip of baculum forming indistinct angle with shaft   *E. minimus*
10. Yellow pine belt and above in Charleston Mountains, Nevada   *E. palmeri*
    Distribution otherwise than in yellow pine belt and above of Charleston Mountains, Nevada   **11**
11. Baculum with distal 50% or more of shaft markedly compressed laterally, base markedly widened   **12**
    Baculum with less than distal 50% slightly compressed laterally, base not markedly widened   **14**
12. Baculum with distal half of shaft laterally compressed and curved downward to base of tip   *E. umbrinus*
    Baculum with distal two-thirds of shaft laterally compressed and curved downward to base of tip   **13**
13. Skull flattened dorsally; rump principally gray; angle formed by tip of baculum and shaft more than 102°   *E. panamintinus*
    Skull rounded dorsally; rump with more or less buffy; angle formed by tip of baculum and shaft less than 100°   *E. speciosus*
14. Shaft of baculum thin, less than 0.20 mm in diameter at widest point   *E. amoenus*
    Shaft of baculum thick, more than 0.25 mm in diameter at widest point   **15**
15. Angle formed by tip and shaft of baculum more than 140°; central and southern Arizona and southern half of New Mexico   *E. cinereicollis*
    Angle formed by tip and shaft of baculum less than 135°; range otherwise than southern Arizona and southern half of New Mexico   **16**
16. Underside of tail dark rufous; shaft of baculum usually more than 3.65 mm in length, but when shorter, diameter is 0.60 mm or more at widest point   *E. ruficaudus*
    Underside of tail buffy or tawny; shaft of baculum less than 3.65 mm in length and 0.55 mm or less in diameter at widest point   *E. quadrivittatus*

Eutamias alpinus (Merriam). Alpine chipmunk. Small, body length averaging about 106 mm; hind foot 28–31 mm. Tail about 74% of body length. Color grayish, with dark stripes tawny, the median one darkest and usually mainly blackish; median light stripes smoke gray, outer pair broader, creamy white. Tip of tail fuscous black for about 20 mm. Higher Sierra Nevada (8000 ft to timber line) of eastern California from Olancha Peak northward to Mount Conness.

Eutamias minimus (Bachman). Least chipmunk. Widely distributed, showing much geographic variation in size and color. Body length averaging about 98–115 mm; hind foot about 26 mm in smallest to about 35 mm in largest races. Tail about 79–90% of body length. Dorsal color ranging from pale yellowish with pale fulvous dark stripes to dark grayish fulvous with black dark stripes; dark stripes reaching to base of tail. Widely distributed in Canada; ranging south through western half of North and South Dakota and through Rocky Mountains to Sacramento Mountains, New Mexico, and White Mountains, Arizona, and west to eastern California, Oregon, and Washington. Also entering the United States in northeastern Minnesota, northern half of Wisconsin, and Upper Peninsula of Michigan.

Eutamias amoenus (**Allen**). Yellow pine chipmunk. Body length averaging 113–126 mm; hind foot 29–35 mm. Tail averaging 73–85% of body length. Colors bright; dark dorsal stripes black or brownish black, often mixed with tan; inner light dorsal stripes smoke gray, usually more or less mixed with cinnamon; outer pair of light stripes clear, creamy white. Undersurface of tail pinkish buff to dark fulvous. Sierra Nevada of California from Mammoth Pass northward into Canada, east in northern Rockies to northwestern Utah, western Wyoming, and central Montana.

Eutamias ruficaudus **Howell.** Red-tailed chipmunk. Body length averaging 125 mm; hind foot 32–36 mm. Tail averaging about 85% of body length. Very similar to *E. amoenus,* within the range of which it occurs. Differs in whiter underparts, more reddish tail, more whitish dorsal stripes, and larger skull with longer rostrum. Eastern slope of Rocky Mountain divide in western Montana, west to northeastern Washington. In and above yellow pine belt of mountains.

Eutamias panamintinus (**Merriam**). Panamint chipmunk. Body length averaging 115 mm; hind foot 31–33 mm. Tail averaging 80% of body length. Closely related to *E. amoenus,* with which its range is allopatric. Differs from it in more grayish rump. Desert mountain ranges of southeastern California and southwestern Nevada and east slope of Sierra Nevada from Olancha Peak north to Bishop Creek. Inhabits piñon belt.

Eutamias quadrivittatus (**Say**). Colorado chipmunk. Body length averaging 122–124 mm; hind foot 26–35 mm. Tail averaging 81–82% of body

length. Head, rump, and thighs gray; underside of tail fulvous, bordered with white or light fulvous. Mountains of Colorado, northern Arizona, and New Mexico, and southern and eastern Utah.

Eutamias umbrinus (**Allen**). Uinta chipmunk. Body length averaging 116–126 mm; hind foot 30–35 mm. Tail averaging about 73–81% of body length. Head, rump, and thighs gray; underside of tail fulvous, bordered with buff. Mountains of northwestern Colorado; from northwestern Wyoming south through central Utah to northwestern Arizona and west through southeastern half of Nevada to summit of Sierra Nevada in east central California.

Eutamias speciosus (**Merriam**). Lodgepole pine chipmunk. Body length averaging about 122–127 mm; hind foot 30–36 mm. Tail averaging about 70% of body length. Dorsal stripes sharply contrasting; median stripe black, usually fading to brown over shoulder; inner light stripes washed with ochraceous; outer light stripes conspicuously broad and white. Crown of head brown, grizzled with gray; dark facial stripes black, bordered with brown; submalar stripe black centrally below eye. Lake Tahoe area of Nevada and Sierra Nevada of California from Mount Lassen south to San Bernardino and San Jacinto Mountains.

Eutamias palmeri **Merriam.** Palmer chipmunk. Body length averaging 125 mm; hind foot 31–35 mm. Tail averaging 76% of body length. Closely related to *E. quadrivittatus,* with which its range is allopatric. Dorsal stripes (except median) more or less indistinct. Tail blackish above. Charleston Mountains of southern Nevada. Yellow pine belt and higher.

Eutamias cinereicollis (**Allen**). Gray-collared chipmunk. Body length averaging 122–128 mm; hind foot 33–36 mm. Tail averaging 74–84% of body length. Distribution allopatric with that of *E. quadrivittatus.* Differs in being generally more gray; nape, shoulders, rump, and thighs smoke gray. Dark stripes black or brownish black; inner light stripes smoke gray, outer ones grayish white. Central Arizona, eastward through southern New Mexico west of Rio Grande. Occurs in yellow-pine belt and above in mountains.

Eutamias canipes (**Bailey**). Gray-footed chipmunk. Body length averaging 128–143 mm; hind foot 32–36 mm. Tail averaging about 71–78% of body length. Color similar to that of *E. cinereicollis*. Baculum distinctive; shaft 2.97–3.40 mm, base wide, 0.84–1.00 mm. Southern New Mexico east of Rio Grande, and Guadalupe Mountains, Texas.

Eutamias townsendi (**Bachman**). Townsend chipmunk. Size large; body length averaging 139–149 mm; hind foot 34–39 mm. Tail averaging 76–79% of body length. Dorsal color dull, dark ochraceous. Stripes wide and not sharply delimited. Dark stripes dark brown or brownish; inner light stripes pale buffy gray, lateral ones whitish. Tail edged with smoke gray or pale buff. Western half of Washington and Oregon, south in Sierra Nevada of California and along coast to Sonoma County, California. Inhabitant of dense, humid forests.

Eutamias sonomae **Grinnell.** Sonoma chipmunk. Body length averaging 133 mm; hind foot 32–39 mm. Tail averaging about 80–86% of body length. Closely related to *E. townsendi*, from which it differs in brighter summer pelage and more grayish head. Median dark stripe blackish, bordered by cinnamon, outer ones black or brownish black, more or less overlaid with ochraceous; light stripes more or less yellowish. Ear unicolor (gray) behind. Coast ranges of northwestern California from Marin County north into Siskiyou County and east to Lassen County. Altitudinal range 900–4500 ft. Inhabits open or brushy situations on hillsides. Said to occur with *E. townsendi* without interbreeding.

Eutamias quadrimaculatus (**Gray**). Long-eared chipmunk. Body length averaging 135 mm; hind foot 34–37 mm. Tail averaging about 77% of body length. Closely related to *E. townsendi*, with which it occurs without interbreeding. Distinguished by large postauricular white patch, which is larger than depressed ear. Tail edged with gray; dark stripe below ear black. Western slope of Sierra Nevada of California from Mariposa County north to northern Plumas County and east to Glenbrook, Nevada. Fir and redwood forests.

Eutamias merriami (**Allen**). Merriam chipmunk. Body length averaging 120–133 mm; hind foot 33–39 mm. Tail averaging 85–97% of body length. Postauricular spots indistinct, grayish, or absent. Line through eye brown. Outer light stripes of dorsum grayish. Mountains of southern half of California and Baja California, ranging from sea level to 9000 ft.

Eutamias dorsalis (**Baird**). Cliff chipmunk. Body length averaging 121–129 mm; hind foot 32–39 mm. Tail averaging about 83% of body length. Characterized by the dorsal stripes (except median) being indistinct or obsolete. Dorsal color mostly pale smoke gray or neutral gray. Ears pale smoke gray or grayish white. Mountains of southwestern New Mexico and southeastern Arizona, northwest to southwestern Nevada and north to southern Idaho and southwestern Wyoming. Piñon and lower yellow pine belts.

Genus *Glaucomys* **Thomas.** Flying squirrels. Small to medium-sized, with moderately long, dense, silky pelage. A planing membrane extending from wrists to ankles, supported anteriorly by slender, cartilaginous process from wrist. Tail broad, much flattened, densely haired. Skull with zygomata vertical (as in *Sciurus*); interorbital and postorbital regions extremely constricted; postorbital processes slender. Incisors slender, not recurved as in most tree squirrels. Dentition: $\frac{1}{1}, \frac{0}{0}, \frac{2}{1}, \frac{3}{3}$. Pleistocene; North America. Several related genera in Eurasia. 2 species.

KEY TO SPECIES OF *GLAUCOMYS*

1. Size small, hind foot less than 33 mm   **G. volans**
   Size larger, hind foot more than 33 mm   **G. sabrinus**

Glaucomys volans (**Linnaeus**). Southern flying squirrel. Body length averaging 125–135 mm; hind foot 28–33 mm. Tail averaging about 70–83% of body length. Upper parts brownish, bases of hairs dark gray; underparts creamy white. Forests, eastern United States, north to southern Vermont, west to border of forest from eastern Texas to Minnesota. Disjunct populations in central and southern Mexico.

Glaucomys sabrinus (**Shaw**). Northern flying squirrel. Body length averaging 150–194 mm; hind foot 34–46 mm. Tail averaging 73–89% of body length. Upper parts brownish; underparts creamy white, the hairs usually gray at bases. Forests of interior Alaska, completely across Canada; enters eastern United States in northern part of border states from eastern North Dakota to Michigan, and in New England states, south in Appalachian chain to western North Carolina. In western mountains, south to southern Sierra Nevada of California and in southern Utah. A disjunct population in Black Hills of South Dakota.

Family Geomyidae. Pocket gophers. Small to medium-sized, highly specialized for fossorial life. Body robust; head short, broad; cervical region short; legs short; forefeet large, with long claws, hind feet smaller; tail short, scantily haired, tactile in function. Eyes small; external ears greatly reduced, frequently only a fleshy rim; skin loosely attached to body. External, fur-lined cheek pouches which open outside of mouth. Females smaller than males. Skull wide and flat, with heavy, widely spreading zygomata. Mandibles massive. Upper and lower incisors long, heavy. All teeth rootless and continual-growing. Dentition: $\frac{1}{1}, \frac{0}{0}, \frac{1}{1}, \frac{3}{3}$. Middle Oligocene; North America. About 8 living genera, 3 in the United States, the others in Mexico and Central America.

KEY TO GENERA OF GEOMYIDAE

1. Upper incisors without conspicuous longitudinal grooves on anterior face (Fig. 6-36A)
   Thomomys

   Upper incisors with 1 or 2 deep longitudinal grooves on anterior face **2**
2. Upper incisors with 1 longitudinal groove (Fig. 6-36B) **Cratogeomys**

   Upper incisors with 2 longitudinal grooves (Fig. 6-36C) **Geomys**

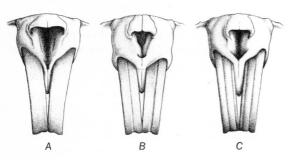

Fig. 6-36. Incisors of pocket gophers: A, *Thomomys*; B, *Cratogeomys*; C, *Geomys*.

Genus Thomomys Maximilian. Western pocket gophers. Small to medium-sized, with forefeet weaker and claws smaller than in other geomyids.

Upper incisors superficially smooth, but with fine longitudinal groove on anterior surface of each; groove invisible to naked eye in some. An anterior and a posterior vertical enamel plate present on all upper and all lower molars and premolars. Upper Miocene. Present range approximately the western half of North America between about 18–50° north latitude. A taxonomically difficult group because of the enormous differentiation that has occurred under extreme conditions of local isolation and local environmental selection. Numerous allopatric "species" of some authors are included under a single name in the following key.

KEY TO SPECIES GROUPS OF *THOMOMYS*

1. Rostrum slender, abruptly arched in front of upper molars (Fig. 6-37A); sphenoidal fissure absent **2**
   Rostrum deep and evenly sloping in front of upper molars (Fig. 6-37B); sphenoidal fissure present **3**
2. Ears rounded; mammae 10–12 **T. talpoides**
   Ears pointed; mammae usually 8 **T. monticola**
3. Size large; hind foot of male usually 40 mm or more; claws relatively small; pterygoids concave on inner surface, convex on outer **T. bulbivorus**
   Size smaller; hind foot of male usually less than 40 mm; claws relatively heavy; pterygoids flat and straight **4**
4. Size larger; hind foot more than 35 mm in male and more than 30 mm in female; color gray or black

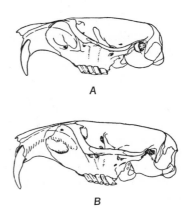

Fig. 6-37. Skulls of *Thomomys* showing rostral profile: A, *Thomomys talpoides*; B, *Thomomys bottae*. (From Hall, *Mammals of Nevada*, University of California Press.)

(dichromatic) often with white patches on chin, feet, throat, or lining of pouches; distribution from southeastern Lassen County, California, and valleys of northwestern Nevada north into Snake River valley **T. townsendi**

Size smaller; hind foot usually less than 35 mm in male and less than 30 mm in female; color variable but usually with more or less buffy or brownish; distribution west, south, or east of southwestern Lassen County, California, and northwestern Nevada **5**

Fig. 6-38. Dorsal view of anterior end of zygomatic arch in *A, Thomomys bottae; B, Thomomys umbrinus.* [From Davis and Buechner, Pocket Gophers (Thomomys) of the Davis Mountains, Texas, *J. Mammalogy,* vol. 27, no. 3.]

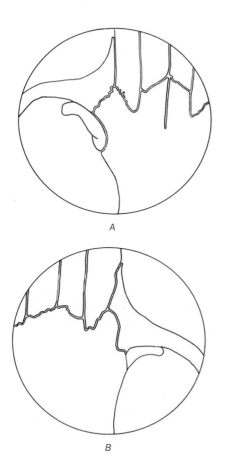

A

B

5. Margin of anterior base of zygomata at junction with frontal definitely concave or emarginate in outline (Fig. 6-38*B*); mammae usually 6 **T. umbrinus**

Margin of anterior base of zygomata at junction with frontal convex (Fig. 6-38*A*); mammae usually 8 **T. bottae**

Thomomys bulbivorus (**Richardson**). Camas pocket gopher. Largest of the genus; body length averaging about 210 mm in males; hind foot of adult males 40–43 mm. Ear a thickened rim; tail usually naked. Winter pelage long and furry; summer coat short and harsh. Color dark brown or sooty. Mammae 8. Skull short and wide; pterygoids convexly inflated; incisors slender and greatly protruding. Groove on upper incisors very obscure. Willamette Valley, Oregon, in rich valley soils.

Thomomys townsendi (**Bachman**). Townsend pocket gopher. Size large, body length of males averaging 181–205 mm; hind foot of males about 34–41 mm. Ears small but pointed. Mammae 8. Dichromatic, occurring in a dark gray and black phase. Skull wide, with spreading zygomata; sphenoidal fissure present. Dentition heavy; incisors long and slightly projecting, groove obscure. Southeastern Lassen County, California, and valleys of northwestern Nevada north into Snake River valley of southern Idaho. Inhabits deep alluvial soils. Distribution allopatric to that of *T. bottae* to the south.

Thomomys bottae (**Eydoux and Gervais**). Valley pocket gopher. A widely distributed and extremely variable species in size and color; as treated here may represent a complex of several species as claimed by some authors. Body length of males averaging 125–185 mm; hind foot of males 26–34 mm. Color usually more or less brownish, but varying geographically from nearly white to very dark. Mammae normally 8. Skull rather heavy, with broad rostrum and heavy zygomata. Margin of anterior base of zygomata at junction with frontal convex. Skull with sphenoidal fissure. Upper incisors broad and heavy, grooves obscure. All of California except northeast; southern half of Nevada, southern Utah and Colorado; south through trans-Pecos Texas, New Mexico, and Arizona. Over 150 named subspecies, reflecting differentiation in small, more or less isolated populations. The nominal species *T. baileyi, T. fulvus, T. suboles, T. harquahalae, T. pectoralis,* and *T. lachuguilla* are considered distinct by some authors. Regarded by some as conspecific with *T. umbrinus.*

Thomomys umbrinus (**Richardson**). Pygmy pocket gopher. Size small; body length of males averaging 135–150 mm; hind foot of males about 26–31 mm. Dorsal color brown with median area usually blackish. Black, usually large, postauricular spots present, these sometimes confluent with black of back. Mammae usually 6. Skull slender, with braincase smoothly rounded. Anterior base of zygomata emarginate or concave in outline where they join the frontal. Center of distribution in central Chihuahua; enters United States and overlaps range of *T. bottae* in extreme southeastern Arizona and western Texas.

Thomomys talpoides (**Richardson**). Northern pocket gopher. Size about as in *T. bottae;* general coloration duller and with larger black postauricular spots. Body length averaging 135–170 mm in males; hind foot 24–31 in males. Mammae 8–12, usually 10. Distinguished by slender rostrum, which is abruptly arched in front of upper molars. Zygomata slender, depressed anteriorly. Anterior opening of infraorbital canal posterior to anterior palatine foramina. No sphenoidal fissure. Upper incisors narrow and thin. From western Canada, ranges south through Dakotas and Rocky Mountains to northwestern New Mexico and northwestern Arizona; through northern half of Nevada, all of Oregon, and in California east of Sierra Nevada south nearly to Inyo County. Interdigitates the range of *T. bottae.* In southern part of range limited to thin soils of high elevations.

Thomomys monticola Allen. Sierra pocket gopher. Size small; body length of males averaging 134–154 mm; hind foot of males about 25–31 mm. Feet and claws very slender; ears relatively large, thin, and pointed. Mammae usually 8. Skull long, slender, and low; zygomata very slender. Rostrum slender, abruptly arched in front of upper molars. Sphenoidal fissure absent. Upper incisors distinctly grooved. From Fresno County, California, north through Sierra Nevada, through southeastern, central, and northwestern Oregon. Generally in higher mountains, altitudinal range 3600–10,350 ft.

Genus Geomys Rafinesque. Eastern pocket gophers. Small to large pocket gophers, with relatively large forefeet and heavy claws. Each upper incisor with a major, medial groove and a smaller,

inner groove. Upper premolar with 3 vertical enamel plates (the posterior absent); 1st and 2d upper molars each with an anterior and a posterior plate; each lower molar with a single (posterior) plate. Pliocene; North America. Gulf Coastal Plain and central grasslands; east of Rocky Mountains. Mostly restricted to sandy soils. The 4 nominal species in the following key are all allopatric in distribution.

KEY TO SPECIES OF *GEOMYS*

1. Width of rostrum equal to or less than greatest length of basioccipital **2**
   Width of rostrum greater than greatest length of basioccipital **3**
2. Squamosal arm of zygomata ending in prominent knob over middle of jugal **G. arenarius**
   Squamosal arm of zygomata lacking a prominent knob over middle of jugal **G. personatus**
3. Nasals distinctly hourglass-shaped, strongly constricted near the middle **G. pinetis**
   Nasals not hourglass-shaped, slightly, if at all, constricted near the middle **G. bursarius**

Geomys pinetis Rafinesque. Piney woods gopher. Size moderate; body of males about 170–195 mm; hind foot of males about 33–37 mm. Dorsal color varying from cinnamon brown to dark plumbeous. Skull with a weak sagittal crest or with temporal ridges failing to unite into a crest. Nasals hourglass-shaped, constricted near the middle. Gulf Coastal Plain from Mobile Bay east through southern half of Alabama and Georgia, south nearly through peninsula of Florida. The nominal species *G. colonus, G. fontanelus,* and *G. cumberlandius* are regarded by some as distinct.

Geomys bursarius (**Shaw**). Prairie pocket gopher. Size small to large; body length of adult males about 155–205 mm; hind foot of males about 27–43 mm. Dorsal color varying from dark chestnut brown in East to pale buffy ochraceous in the West. Rostrum broad, exceeding greatest length of basioccipital. Nasals little or not constricted near the middle. Northwestern Minnesota to Gulf of Mexico; east of eastern front of Rocky Mountains; mostly west of Mississippi River to western Wisconsin and across Illinois into northwestern Indiana.

Geomys arenarius **Merriam.** Desert pocket gopher. Size medium; body length of adult males about 175–180 mm; hind foot of males about 30–34 mm. Color pale brown above; more or less white below. Jugal long, its dorsal exposure greater than width of rostrum below infraorbital openings. Zygomata narrow and nearly parallel; squamosal arm of zygomata ending in a prominent knob over middle of jugal. Rio Grande Valley in western tip of Texas and adjacent New Mexico and in Tularosa Basin, New Mexico.

Geomys personatus **True.** Tamaulipan pocket gopher. Size moderate to large; body length of adult males about 175–215 mm; hind foot of males about 30–42 mm. Upper parts brownish drab. Skull heavy, with a high sagittal crest; zygomata project at right angles to long axis of skull. Dorsal exposure of jugal longer than width of rostrum below infraorbital openings. Width of rostrum equal to or less than greatest length of basioccipital. Brushlands of southern Texas south of Balcones escarpment and east to and slightly across lower Nueces River; south along Mexican coast.

Genus *Cratogeomys* **Merriam.** Mexican gophers. Medium to large pocket gophers, with relatively large forefeet and heavy claws. Each upper incisor with a single deep median groove. Upper premolar with 3 vertical enamel plates (posterior absent); 1st and 2d upper molars each with 1 (anterior) enamel plate; each lower molar with a single (posterior) plate. Upper Pliocene; North America. Center of distribution of the several poorly known species in Mexico. Generally associated with plains and basins, but ecologically tolerant; has altitudinal range from below 1500 to over 12,000 ft on mountains in Mexico. 1 species enters United States.

Cratogeomys castanops (**Baird**). Plains pocket gopher. Body length of males about 190–205 mm; hind foot 31–42 mm. Tail about 42–51% of body length. Color generally yellowish brown. Principally on high plains and mountain basins from central Mexico through trans-Pecos Texas, north between Rio Grande in New Mexico and escarpment of High Plains in Texas to southeastern Colorado.

Family Heteromyidae. Pocket mice and kangaroo rats. Very small, small, or medium-sized rodents. More or less saltatorial. Body fairly slender; tail typically long; hind feet proportionally large, hind limbs long. Forelimbs and forefeet relatively small. Ear moderate to small; eyes moderate to large. External, fur-lined cheek pouches present, opening outside of mouth. Skull light; rostrum slender. Squamosal much expanded, and jugal extending forward to lacrimal. Mastoids more or less expanded (enormously in some), tending to show in dorsal aspect. Incisors weak; molars rooted or rootless, with transverse laminae. Middle Oligocene; North America, northern South America. 5 living genera, 4 in the United States.

KEY TO GENERA OF HETEROMYIDAE

1. Mastoid region of skull greatly expanded; width of interparietal less than ¼ of greatest width of skull (Fig. 6-39A)   **2**

   Mastoid moderately or little expanded; width of interparietal more than ¼ of greatest width of skull (Fig. 6-39B)   **3**

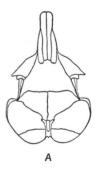

Fig. 6-39. Heteromyid skulls: *A, Dipodomys merriami; B, Perognathus penicillatus.* (From Hall, *Mammals of Nevada,* University of California Press.)

A   B

Fig. 6-40. Anterior lateral view of heteromyid skulls showing zygomatic process of maxillary: *A, Microdipodops megacephalus; B, Dipodomys ordi.* (From Hall, *Mammals of Nevada,* University of California Press.)

A   B

Fig. 6-41. Heteromyid skulls: A, *Perognathus hispidus;* B, *Liomys irroratus.*

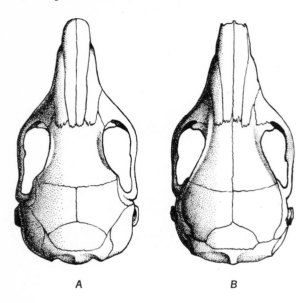

A   B

2. Size small, hind foot less than 32 mm; tip of tail not tufted; zygomatic processes of maxillaries not expanded dorsally (Fig. 6-40A)   *Microdipodops*
Size larger, hind foot more than 32 mm; tip of tail tufted; zygomatic processes of maxillaries broadly expanded dorsally (Fig. 6-40B)   *Dipodomys*
3. Upper incisors grooved; mastoids large, visible dorsally (Fig. 6-41A); pelage normal or with scattered spines, mostly on rump   *Perognathus*
Upper incisors not grooved; mastoids small, not visible dorsally (Fig. 6-41B); pelage principally of stiff, flattened, grooved spines   *Liomys*

Genus *Perognathus* Maximilian. Pocket mice. Very small to large-mouse-sized, with small ears and moderate to long (longer than body) tail. Hind feet and hind limbs moderately enlarged. 4 toes on forefeet, 5 on hind. A fur-lined cheek pouch present and opening externally on each side of mouth. Skull light; mastoids expanded and visible in dorsal aspect. Auditory bullae inflated, more or less triangular in outline, anteriorly apposed to pterygoids. Jugals light and threadlike; rostrum slender, pointed. Infraorbital foramen a small lateral opening in maxilla. Upper incisors grooved. Molars rooted and tuberculate. Dentition: $\frac{1}{1}, \frac{0}{0}, \frac{1}{1}, \frac{3}{3}$. Miocene; North America. Center of distribution arid grasslands of southwestern United States and northern Mexico. 18 species in the United States.

KEY TO SPECIES OF *PEROGNATHUS*

1. Size smaller; pelage soft, without spines; mastoids large, projecting beyond plane of occiput (Fig. 6-42A); interparietal width usually less than interorbital width; auditory bullae meeting or nearly meeting anteriorly   2
Size larger; pelage harsh, often with rump spines; mastoids smaller, not projecting beyond plane of occiput (Fig. 6-42B); interparietal width equal to or greater than interorbital width; bullae separated by nearly full width of basisphenoid   12
2. Antitragus lobed; hind foot more than 20 mm   3
Antitragus not lobed; hind foot 20 mm or less   5
3. Tail long and heavily crested; soles naked; shaft and base of baculum slender, basal width contained more than 15 times in length of baculum (Fig. 6-43A)   *P. formosus*
Tail moderate; soles of hind feet somewhat hairy; basal width of baculum contained less than 15 times in its length (Fig. 6-43B)   4
4. Ears white; dorsal side of tail faintly dusky at tip   *P. alticolus*

Fig. 6-42. Pocket-mouse skulls: A, *Perognathus flavus;* B, *Perognathus intermedius.*

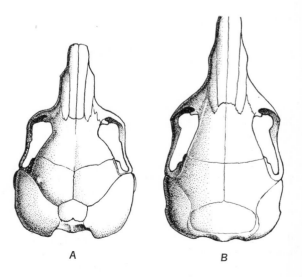

A   B

Fig. 6-43. Lateral (left) and ventral (right) views of bacula of *Perognathus*: A, *Perognathus formosus*; B, *Perognathus inornatus*. (From Burt, A Study of the Baculum in the Genera Perognathus and Dipodomys, *J. Mammalogy*, vol. 17, no. 2.)

Fig. 6-44. Lateral (left) and ventral (right) views of bacula of *Perognathus*: A, *Perognathus hispidus*, B, *Perognathus baileyi*; C, *Perognathus penicillatus*. (From Burt, A Study of the Baculum in the Genera Perognathus and Dipodomys, *J. Mammalogy*, vol. 17, no. 2.)

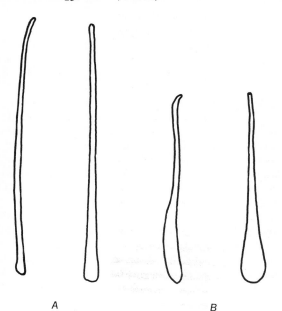

Ears buffy or dusky; dorsal side of tail dusky throughout **P. parvus**

5. Tail longer than body **6**
   Tail about as long as, or shorter than, body **8**
6. Mastoids very large; length of baculum more than 6.5 mm, width of base more 0.7 mm **P. amplus**
   Mastoids moderate; length of baculum less than 6.5 mm, width of base less than 0.7 mm **7**
7. Size larger, body length more than 65 mm; length of baculum more than 5.5 mm, width of base more 0.5 mm **P. inornatus**
   Size smaller, body length less than 65 mm; length of baculum less than 5.5 mm; width of base less than 0.5 mm **P. longimembris**
8. Interparietal relatively large, its width contained less than 4.0 times in basilar length **9**
   Interparietal medium or small, its width contained more than 4.0 times in basilar length **11**
9. Size larger, hind foot more than 18.0 mm **P. apache**
   Size smaller, hind foot less than 18.0 mm **10**
10. Color olive gray **P. fasciatus**
    Color yellowish buff **P. flavescens**

11. Tail usually less than 50 mm; interparietal narrower, contained more than 4.7 times in basilar length **P. flavus**
    Tail usually more than 50 mm; interparietal wider, contained less than 4.7 times in basilar length **P. merriami**
12. Rump with more or less distinct spines **13**
    Rump without spines **16**
13. Lateral line faint or indistinguishable; pelage coarse; spines strong, extending to sides **P. spinatus**
    Lateral line well marked; pelage not coarse; spines moderate, usually confined to rump **14**
14. Ears elongate, 10 mm or more **P. californicus**
    Ears rounded, less than 10 mm **15**
15. Color rich brown; range in southern California, west of Colorado River **P. fallax**
    Color paler, gray or brownish (black on lava beds); range east of Colorado River **P. intermedius**
16. Tail not crested, shorter than body; skull in adults with a supraorbital bead; tip of baculum trifid (Fig. 6-44A) **P. hispidus**
    Tail crested, longer than body; skull without supraorbital bead; tip of baculum simple, not trifid **17**
17. Size large; length of interparietal more than 3.7 mm; baculum slightly curved, not sigmoid in outline (Fig. 6-44B) **P. baileyi**

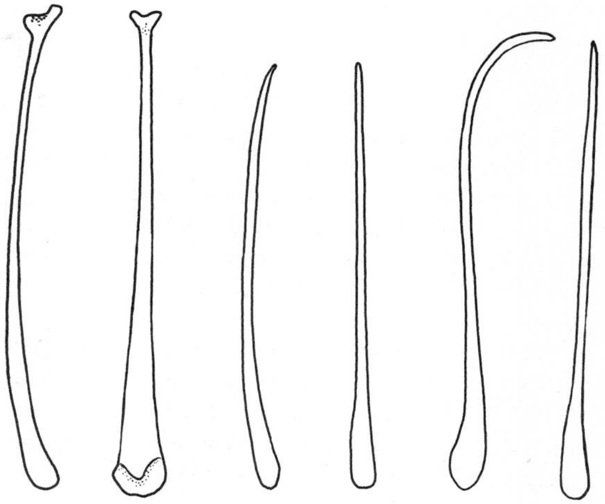

Size smaller; length of interparietal less than 3.7 mm; baculum sigmoid in outline (Fig. 6-44C)
P. penicillatus

Perognathus fasciatus **Maximilian.** Wyoming pocket mouse. Size small; body length averaging about 70 mm; hind foot about 16–18 mm. Tail averaging about 85–93% of body length. Proximal half of sole of hind foot hairy. Pelage soft, grayish olive above, white or buffy below. Skull small; mastoids slightly projecting posteriorly; auditory bullae scarcely meeting anteriorly. Interparietal pentagonal, large (width about 4.3–4.8 mm). Lower premolar about equal to or slightly smaller than last molar. Northern plains of the Dakotas, south in Nebraska to Platte River; eastern Montana and eastern half of Wyoming, south in Colorado to Larimer County.

Perognathus flavescens **(Merriam).** Plains pocket mouse. Proportions about as in *P. fasciatus,* but slightly smaller; pelage harsher. Body length averaging 68 mm; hind foot about 17 mm. Tail averaging about 90% of body length. Dorsal color buff, lined with black, never markedly olivaceous. Interparietal large, pentagonal, its width about 4.8 mm. Lower premolar smaller than last molar. Center of distribution in plains of western Nebraska; ranges north through southeastern North Dakota, east in Iowa to Mississippi River, south through western Kansas to Texas Panhandle, west to east base of Rocky Mountains. Usually on sandy soils.

Perognathus merriami **Allen.** Merriam pocket mouse. Size small; body length averaging about 60 mm; hind foot about 16 mm. Tail averaging about 97% of body length. Proximal half of sole of hind foot hairy. Pelage fairly soft, bright ochraceous above, more or less mixed with black. Interparietal pentagonal, fairly large, its width about 3.3–3.6 mm. Lower premolar usually about equal to last molar. Center of distribution in plains of western Texas; ranging throughout the western half of Texas, west to Pecos Valley in New Mexico, north to southwestern Kansas. Usually associated with mesquite; on many soil types.

Perognathus flavus **Baird.** Silky pocket mouse. Size small, body length averaging 62–65 mm; hind foot about 16–18 mm. Tail averaging 81–100% of body length. Pelage very soft; dorsal color buffy with more or less intermixed black (nearly black

on lava beds). Interparietal very small, its width about 2.8–3.1 mm. Lower premolar smaller than last molar. From southeastern Wyoming and northwestern Nebraska, south through western Texas and New Mexico; west into southeastern Arizona and northern Arizona and southeastern Utah.

Perognathus apache **Merriam.** Apache pocket mouse. Size moderate; body length about 66–77 mm; hind foot about 17–21 mm. Tail averaging about 86–103% of body length. Pelage soft; dorsal color buffy or yellowish with intermixed black; underparts white. On White Sands of New Mexico, dorsal color may be white except for indefinite median gray stripe. Mastoids large; auditory bullae apposed anteriorly; interparietal large, its width about 4.0–4.3 mm. Lower premolar smaller than last molar. Southeastern Utah, southwestern Colorado, northeastern Arizona, and southward through western New Mexico to Mexico.

Perognathus amplus **Osgood.** Arizona pocket mouse. Size moderate; body length about 75 mm; hind foot about 17–22 mm. Tail about 107% of body length, slightly penicillate. Antitragus lobed. Pelage soft, full, and long; dorsal color pinkish buff sprinkled with black; underparts white; a fairly wide, buff lateral line, extending onto forelegs nearly to wrist. Orbital area pale; a white spot present above and below base of ear. Mastoids greatly expanded, bulging in all directions; interparietal small, its width about 3.3 mm. Lower premolar about equal to or slightly larger than last molar. Western two-thirds of Arizona, in desert; south into Sonora.

Perognathus longimembris **(Coues).** Little pocket mouse. Size small; body length averaging 58–65 mm; hind foot 15–20 mm. Tail averaging 110–138% of body length. Pelage long and silky; dorsal color buff to grayish buff. Nasals long; maxillary branches of zygomata gradually narrowing anteriorly; interparietal width about 3.5–3.8 mm. Lower premolar larger than last molar. Western and northwestern Arizona and southern and western Utah, through Nevada to southeastern Oregon and northeastern California and central, southeastern, and extreme southern California; into Mexico. Sandy or gravelly soils.

Perognathus inornatus **Merriam.** San Joaquin pocket mouse. Size moderate; body length averaging about 71 mm; hind foot about 18–21 mm. Tail averaging about 104% of body length. Antitragus not lobed. Pelage soft; dorsal color buffy. Mastoids large; auditory bullae apposed anteriorly. Interparietal width about 3.8 mm. Lower premolar larger than last molar. San Joaquin Valley, California.

Perognathus parvus (Peale). Great Basin pocket mouse. Size large; body length averaging 80–91 mm; hind foot 19–24 mm. Tail about 110–122% of body length. Ears moderate; antitragus prominently lobed. Pelage soft; dorsal color blackish olive gray to buffy olive gray, with a buffy lateral line. Auditory bullae meeting anteriorly; interparietal large, its width about 5.4 mm. Lower premolar smaller than last molar. Nevada and adjacent parts of California and northwestern Arizona, north through eastern Washington and Oregon, western, central, and northern Utah, southwestern Wyoming, and Idaho. Sandy areas with sagebrush and other desert shrubs.

Perognathus alticolus Rhoads. White-eared pocket mouse. Size large; body length about 82 mm; hind foot 21–23 mm. Tail about equal to body length. Color and skull about as in *P. parvus*, but distinguished by having hair in, on, and below ears white. Tail faint buff above, terminal fourth slightly dusky. San Bernardino Mountains and southwestern Kern County, California. Closely related to the allopatric species *P. parvus*. Tree yucca and pine belts of mountains.

Perognathus formosus Merriam. Long-tailed pocket mouse. Size large; body length averaging 78–86 mm; hind foot about 22–26 mm. Tail 100–133% of body length and heavily crested. Ears relatively large, somewhat attenuate; antitragus prominently lobed. Soles of feet naked. Pelage soft. Interparietal large, its width about 5.8 mm. Lower premolar larger than last molar. Western and southern Utah, southern and western Nevada, and adjacent California, south into Baja California. Rocky slopes on low, hot desert.

Perognathus baileyi Merriam. Bailey pocket mouse. Size large; body length averaging about 94 mm; hind foot about 26–28 mm. Tail averaging about 129% of body length, distinctly crested. Pelage soft; dorsal color grayish. Interparietal large, its width about 6.8 mm. Lower premolar smaller than or about equal to last molar. From central Arizona south into Mexico, west into extreme southern California and south through Baja California. Low, hot deserts and low foothills.

Perognathus penicillatus Woodhouse. Desert pocket mouse. Size large; body length averaging 80–96 mm; hind foot about 22–26 mm. Tail averaging 104–122% of body length, distinctly crested. Pelage moderately soft, without spines. Dorsal color buffy, lightly mixed with black. No distinct lateral line. Interparietal rounded, its width about 6.9–7.6 mm. Auditory bullae widely separated anteriorly. Lower premolar larger than last molar. Entering from Mexico from trans-Pecos Texas to southern California and ranging north through western Arizona and eastern California to southern tip of Nevada. Usually restricted to sandy soils.

Perognathus intermedius Merriam. Rock pocket mouse. Size medium; body length averaging about 77–78 mm; hind foot about 19–24 mm. Tail averaging about 133% of body length, crested. Pelage harsh, rump spines present. Dorsal color brownish or grayish, with intermixed black. Some individuals may be blackish to black on lava beds. Interparietal large, its width about 7.2–8.0 mm. Lower premolar larger than last molar. Western and southern Arizona and through central Arizona north to southern Utah; southwestern and central New Mexico to trans-Pecos Texas and south into Mexico. Rocky situations in deserts. As here recognized, includes the nominal species *P. nelsoni*.

Perognathus fallax Merriam. San Diego pocket mouse. Size medium; body length about 79 mm; hind foot about 21–26 mm. Tail about 132% of body length, crested. Rump spines present. Dorsal color rich brown; middle of back and rump blackish. Lateral line pinkish buff. Tail bicolor. Interparietal wide (about 7.8 mm) with a slight anterior angle. Southwestern California and extending into Baja California. Deserts and foothills.

Perognathus californicus Merriam. California pocket mouse. Size medium; body length averaging 89–93 mm; hind foot about 24–29 mm. Tail averaging 116–126% of body length, crested. Rump with numerous spines. Dorsal color grayish brown. Tail bicolor. Mastoids very small; mastoid width greatly reduced. Interparietal wide (about 8.1–8.3 mm), rounded but with slight anterior angle. Lower premolar slightly larger than last molar. Coastal California from San Francisco Bay south into Baja California, and foothills of west slope of

Sierra Nevada from Placer County south into Baja California.

***Perognathus spinatus* Merriam.** Spiny pocket mouse. Size medium; body length averaging about 80 mm; hind foot about 20–28 mm. Tail averaging about 126% of body length, crested. Rump spines large and prominent; scattered spines on flanks and sides and often extending to shoulders. Dorsal color pale yellowish, often mottled in appearance. Lateral lines very faint or lacking. Mastoids small. Interparietal broad (width about 7.6–7.7 mm) with slight anterior angle. Lower premolar about equal to last molar. Deserts of southern California south through Baja California.

***Perognathus hispidus* Baird.** Plains pocket mouse. Size large; body length averaging 104–114 mm; hind foot about 25–28 mm. Tail averaging about 95% of body length, not crested. Pelage harsh, but without spines. Ears small; antitragus lobed. Skull heavy, with a distinct supraorbital bead. Mastoids relatively small. Interparietal large (width about 7.2–8.0 mm), rounded-pentagonal. Lower premolar about equal to last molar. Entering United States from Mexico from southern Texas to southeastern Arizona, ranging northward between eastern front of Rocky Mountains and eastern forest, through Nebraska to southern North Dakota. Inhabitant of level brushlands and grassy plains on various types of soil.

Genus *Microdipodops* Merriam. Kangaroo mice. Small, mouse-sized; hind legs long, hind foot proportionally large (about 23–27 mm); forelegs relatively short. Soles of hind feet densely furred. Tail about equal to or somewhat longer than body length; tail without dark dorsal and ventral stripe and white lateral stripes. Skull with mastoids enormously inflated, nearly meeting in midline, and projecting far posteriorly; supraoccipital, interparietal, and parietals much reduced. Rostrum and anterior part of skull relatively small. Shelf of palate ending posterior to plane of last molars. Angular process of mandible truncated and thickened instead of ending in a point. Dentition: $\frac{1}{1}, \frac{0}{0}, \frac{1}{1}, \frac{3}{3}$. Recent; North America. 2 species.

KEY TO SPECIES OF *MICRODIPODOPS*

1. Dorsal color blackish or brownish; postauricular patches (if present) buff   ***M. megacephalus***

     Dorsal color whitish or pale buff; postauricular patches white   ***M. pallidus***

***Microdipodops megacephalus* Merriam.** Dark kangaroo mouse. Body length averaging 69–80 mm; hind foot about 23–27 mm. Tail about 100–130% of body length. Upper parts brownish, blackish or grayish; underparts paler, the hairs usually gray at bases and white in distal half, sometimes white to bases and sometimes with buffy tips. Tail with a black tip above. Skull with premaxillae extending little back of nasals; incisive foramina widest posteriorly or at middle. Deserts in Nevada, adjacent California border, western Utah, and southeastern Oregon.

***Microdipodops pallidus* Merriam.** Pale kangaroo mouse. Body length averaging 71–77 mm; hind foot about 25–27 mm. Tail about 113–123% of body length. Color pale; upper parts pale buffy; underparts white, the white extending to bases of hairs. Postauricular patch prominent, white. Tail without black tip. Skull with premaxillae extending well back of nasals; sides of incisive foramina parallel; auditory bullae greatly inflated. Deserts of southwestern Nevada and adjacent California border. Prefers fine sand with some plant growth.

Genus *Dipodomys* Gray. Kangaroo rats. Highly saltatorial; hind feet large; hind limbs long. Forelimbs and feet reduced. Tail long, penicillate. Ears moderate to small. Soles of feet hairy; hind toes 5 or 4, sometimes variable within a species. Fur-lined cheek pouches present. Tail black or blackish above and below, white laterally. A conspicuous white hip mark usually present. Skull light, roughly conical in outline. Mastoids enormously expanded Interparietal small. Jugal slender; zygomatic process of maxilla broad. Incisors weak. Dentition: $\frac{1}{1}, \frac{0}{0}, \frac{1}{1}, \frac{3}{3}$. Upper Pliocene; North America. Center of distribution in arid southwestern United States. About 14 species in the United States.

KEY TO SPECIES OF *DIPODOMYS*

1. Hind foot with 4 toes   2
     Hind foot with 5 toes   7
2. Tip of tail dusky   3
     Tip of tail abruptly and distinctly white   4
3. Dark dorsal and ventral tail stripes usually narrower than lateral white tail stripes; sides of rostrum

markedly divergent posteriorly (Fig. 6-45A); nasals longer, 12.2 mm or more   *D. merriami*
Dark dorsal and ventral tail stripes usually broader than lateral white stripes; sides of rostrum nearly parallel posteriorly (Fig. 6-45B); nasals shorter, 12.6 mm or less   *D. nitratoides*

4. Distribution west and north of Sierra Nevada; base of baculum relatively large, greatest dorsoventral width of base contained less than 7 times in length of baculum (Fig. 6-46A)   *D. heermanni*
Distribution east of Sierra Nevada; base of baculum relatively small, greatest dorsoventral width of base contained more than 7 times in length of baculum (Fig. 6-46B)   5

5. Size smaller, body length less than 130 mm; restricted to north central Texas and south central Oklahoma   *D. elator*
Size larger, body length more than 130 mm; distributed west of 100th meridian   6

6. White tip of tail preceded by a black band completely encircling tail; distributed east of central Arizona; dark dorsal and ventral stripes present on tail   *D. spectabilis*
White tip of tail not preceded by encircling black band; no ventral dark stripe on tail   *D. deserti*

7. Width of maxillary arch at middle less than 3.9 mm; lower incisors chisel-shaped; basal part of baculum relatively small, greatest dorsoventral width contained more than 8.0 times in length of baculum (Fig. 6-47A)   *D. microps*
Width of maxillary arch at middle more than 3.9 mm; lower incisors awl-shaped; basal part of baculum relatively large, greatest dorsoventral width contained less than 8.0 times in length of baculum (Fig. 6-47B)   8

8. Size large, body length more than 130 mm   *D. ingens*

Fig. 6-45. Rostral region of skull of *Dipodomys*: A, *Dipodomys merriami*; B, *Dipodomys nitratoides*. (From Grinnell, A Geographical Study of the Kangaroo Rats of California, *Univ. Calif. Publ. Zool.*, vol. 24, no. 1.)

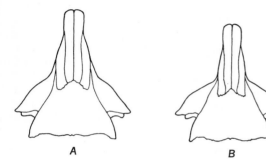

Fig. 6-46. Lateral (left) and ventral (right) views of bacula of *Dipodomys*: A, *Dipodomys heermanni*; B, *Dipodomys spectabilis*. (From Burt, A Study of the Baculum in the Genera Perognathus and Dipodomys, *J. Mammalogy*, vol. 17, no. 2.)

Fig. 6-47. Lateral (left) and ventral (right) views of bacula of *Dipodomys*: A, *Dipodomys microps*; B, *Dipodomys agilis*. (From Burt, A Study of the Baculum in the Genera Perognathus and Dipodomys, *J. Mammalogy*, vol. 17, no. 2.)

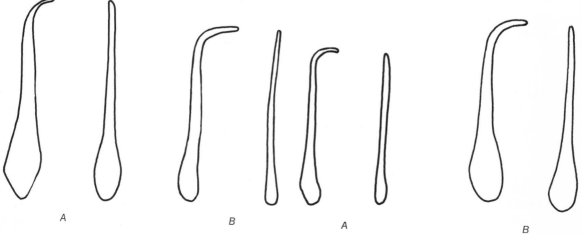

Size smaller, body length less than 130 mm　　9
9. Ear relatively large　　10
　Ear relatively medium or small　　12
10. Color dark, brownish; ear mostly blackish; dark ventral stripe midway of tail wider than lateral white stripe　　***D. venustus***
　Color paler, buffy; ear mostly brownish; dark ventral stripe midway of tail narrower than lateral white stripe　　11
11. Larger, total length usually more than 315 mm; ear larger; nasals flaring at ends　　***D. elephantinus***
　Smaller, total length usually less than 315 mm; ear smaller; nasals not flaring at ends　　***D. agilis***
12. Size smaller, total length usually less than 265 mm; greatest length of skull usually less than 38 mm; maxillary arch relatively narrow, width of arch at middle usually contained more than 8.4 times in greatest length of skull (Fig. 6-48A)　　***D. ordi***
　Size larger, total length usually more than 265 mm; greatest length of skull usually more than 38 mm; maxillary arch moderate to relatively broad, width of arch at middle usually contained 8.4 or less times in greatest length of skull (Fig. 6-48B)　　13
13. Maxillary arch relatively broad, width of arch at middle usually contained less than 7.3 times in greatest length of skull　　14
　Maxillary arch moderate, width of arch at middle usually contained more than 7.3 times in greatest length of skull　　15
14. Auditory bullae nearly globular in shape; range, San Jacinto Valley, California　　***D. stephensi***
　Auditory bullae not globular in shape; range not San Jacinto Valley, California　　***D. panamintinus***
15. Maxillary arch relatively broader, width of arch at middle usually contained less than 8.0 times in greatest length of skull; ear smaller　　***D. heermanni***
　Maxillary arch relatively narrower, width of arch at middle usually contained more than 8.0 times in greatest length of skull; ear larger　　***D. agilis***

Fig. 6-48. Outlines of maxillary arches of *Dipodomys*: A, *Dipodomys ordi*; B, *Dipodomys panamintinus*. (From Grinnell, A Geographical Study of the Kangaroo Rats of California, *Univ. Calif. Publ. Zool.*, vol. 24, no. 1.)

A　　　　B

Dipodomys ordi Woodhouse. Ord kangaroo rat. Size medium; body length about 95–130 mm; hind foot about 35–45 mm. 5 toes on hind foot. Tail relatively short; from about equal to body length to about 126% of body. Skull with relatively short rostrum, moderate to large auditory bullae and relatively wide interparietal. Maxillary arch moderately narrow, its width about 4.0–4.3 mm. Color variable, from pale to dark. From Mexico, ranging into the United States from southern Texas to western Arizona and extending northward between eastern California and central Oklahoma and Kansas to northern Montana and into Canada.

Dipodomys microps (Merriam). Great Basin kangaroo rat. Size small to moderate; body length averaging 106–124 mm; hind foot about 38–46 mm. 5 toes on hind foot. Ear small. Tail averaging about 128–143% of body length. The very narrow maxillary arch (width about 3.4 mm at middle), narrow nasals, and chisel-shaped (rather than awl-shaped) lower incisors are distinctive. Color variable, but frequently relatively dark; dark ventral tail stripe usually extending to tip; black dorsal and ventral tail stripes wider than white lateral ones; lining of cheek pouches dusky. Southeastern Oregon and northeastern California to central Utah, northern Arizona, southern Nevada, and adjacent California.

Dipodomys panamintinus (Merriam). Panamint kangaroo rat. Size medium; body length averaging 122–126 mm; hind foot about 42–47 mm. 5 toes on hind foot. Tail averaging about 135–142% of body length. Skull with broad (about 5.4 mm at middle) maxillary arch as in *D. heermanni*. Color moderately dark to pale; facial arietiform marking solidly blackish to obscure. Eastern California and adjacent Nevada from southeast of Mono Lake, south to Mohave Desert. Found at about 4300–8900 ft elevation.

Dipodomys stephensi (Merriam). Stephens kangaroo rat. Size medium; body length averaging 119 mm; hind foot about 41–43 mm. 5 toes on hind foot. Tail about 145% of body length. Ear relatively small. Skull with broad maxillary arch, its width at middle about 5.1–6.0 mm. Auditory bullae globular in shape, expanded posteroexternally. Dorsal color dusky cinnamon buff; facial arietiform marking solidly blackish. Restricted to San Jacinto Valley, California. Distribution allopatric to that of *D. panamintinus*, with which it may be conspecific.

Dipodomys heermanni **Le Conte.** Heermann kangaroo rat. Size medium, variable; body length averaging 112–122 mm; hind foot about 38–47 mm. Hind foot with 4 or 5 toes (usually 4). Tail long, about 147–167% of body length. Color variable, but dorsal color frequently dark, dusky cinnamon buff. Facial arietiform markings usually distinct. Tip of tail white to blackish. Skull characterized by broad maxillary arch, its width at middle about 4.8–5.3 mm. California, from Tehachapi Mountains north between Sierra Nevada and coastal redwood forest to southern Oregon.

Dipodomys agilis **Gambel.** Pacific kangaroo rat. Size medium; body length averaging 112–119 mm; hind foot about 40–46 mm. 5 toes on hind foot. Tail long, about 153–157% of body length. Skull with bullae moderately inflated, supraoccipital and interparietal fairly broad. Maxillary arch relatively fairly narrow, its width about 4.7–4.9 mm. Color may be dark or pale. Southern San Joaquin Valley and Pacific slope of southern California and into Baja California.

Dipodomys venustus **(Merriam).** Santa Cruz kangaroo rat. Size moderately large; body length averaging about 121–122 mm; hind foot about 44–47 mm. 5 toes on hind foot. Tail long, about 150–159% of body length. Ears relatively large. Skull generally as in *D. agilis,* with which its range is allopatric, but size larger and rostrum proportionally longer. Dorsal color dark, cinnamon brown. Facial arietiform figure blackish, complete. Range a narrow strip of coastal California from San Francisco Bay south nearly to Santa Barbara County.

Dipodomys elephantinus **(Grinnell).** Big-eared kangaroo rat. Size large; body length averaging about 127 mm; hind foot about 44–50 mm. 5 toes on hind foot. Ear relatively large. Tail long, about 155% of body length. Skull with large auditory and mastoid bullae; supraoccipital and interparietal narrow; nasals flaring at ends; maxillary arch relatively narrow, its width at middle about 5.2 mm. Dorsal color moderately dark. Limited to chaparral slopes of southern Gabilan Range, California. Range allopatric to that of *D. agilis* and *D. venustus,* and all may belong to same species.

Dipodomys ingens **(Merriam).** Giant kangaroo rat. Size large, body length averaging 144 mm; hind

foot 46–55 mm. 5 toes on hind foot. Tail relatively short, about 128% of body length. Ear relatively small. Skull massive, with widely spreading maxillary arches. Width of maxillary arch at middle about 5.6–6.2 mm. Dorsal color moderately pale; the dusky nose and whisker patches not ordinarily connected to form a continuous arietiform marking. Restricted to narrow strip of semiarid, more or less level territory along southwestern border of San Joaquin Valley and the nearby Carrizo Plain and Cuyama Valley, California.

Dipodomys spectabilis **Merriam.** Bannertail kangaroo rat. Size large; body length about 135–155 mm; hind foot about 47–53 mm. 4 toes on hind foot. Tail long, about 152% of body length. Skull with mastoids large, but not obscuring supraoccipital and interparietal. Interparietal elongate, tapering posteriorly. Maxillary arch moderately wide, its width at middle about 5.0–5.6 mm. Dorsal color moderately dark, dusky facial markings forming complete arietiform figure. Tail with a conspicuous white tip; lateral white tail stripes distinctly narrower than dorsal and ventral dark stripes, narrowing distally and becoming obsolete at varying distances from white tip. Both dorsal and ventral tail stripes blackish or black. Northwestern New Mexico, southeast through trans-Pecos Texas; southeastern and southern Arizona; south into Mexico.

Dipodomys deserti **Stephens.** Desert kangaroo rat. Size large; body length averaging about 135–143 mm; hind foot about 50–58 mm. 4 toes on hind foot. Tail about 143–145% of body length. Skull with greatly inflated mastoids; supraoccipitals and interparietal usually hidden by appressed mastoids. Maxillary arch narrow, its width at middle about 3.6–4.8 mm. Pelage exceptionally long and silky. Color very pale; tail without ventral dark stripe, but with conspicuous white tip. No dark facial markings. Deserts of southeastern California and southern and western Nevada, south through southwestern Arizona into Sonora, Mexico.

Dipodomys elator **Merriam.** Mesquite kangaroo rat. Size medium; body length about 122 mm; hind foot about 46 mm. 4 toes on hind foot. Tail long, about 150–156% of body length. Skull with mastoids less inflated than in *D. spectabilis;* interparietal rounded, and about as wide as long. Maxillary arch broad, its width at middle about 6.4 mm.

Color dark; dusky facial markings nearly or quite forming complete arietiform figure. Tail with a distinct white tip; lateral white tail stripes about as wide as dorsal and ventral dark tail stripes; lateral stripes narrowing toward tip of tail but not becoming completely obsolete to produce complete black subterminal band. Dark ventral tail stripe dusky blackish for its full length, slightly paler than dorsal one. Dorsal tail stripe densely blackish, becoming blackish or black anterior to white tip. Mesquite plains from Coryell County, Texas, to Chattanooga, Oklahoma, west to Vernon, Texas.

Dipodomys merriami **Mearns.** Merriam kangaroo rat. Size small, body length averaging 95–105 mm; hind foot about 36–41 mm. 4 toes on hind foot. Tail slender, about 131–153% of body length. Maxillary arch broad and sharply angled, its width at middle about 4.6–5.5 mm. Dorsal color varying from pale to dark, dull buffy. Facial arietiform markings dusky to blackish, usually distinct. From western and southern Nevada south through deserts of southeastern California, Arizona (except Colorado Plateau in northeast), southern New Mexico, and trans-Pecos Texas, south through much of Mexico.

Dipodomys nitratoides **Merriam.** Fresno kangaroo rat. Size small; body length averaging about 88–102 mm; hind foot about 33–37 mm. 4 toes on hind foot. Tail moderate to long, about 132–148% of body length. Skull generally similar to that of *D. merriami,* but rostrum shorter and angle formed by side of rostrum and anterior margin of maxillary arch less obtuse. Maxillary arch relatively broad, its width about 4.3–4.7 mm. Pelage coarser and dorsal coloration slightly duller than in *D. merriami.* Dark dorsal and ventral tail stripes wider than lateral white ones. San Joaquin Valley, California.

Genus *Liomys* **Merriam.** Spiny pocket mice. Large-mouse-sized, with tail equal to or longer than body. Ears small. Pelage harsh, principally of stiff, flattened spines. Tail moderately well haired. Sole of hind foot hairy posteriorly. Skull usually broad, with narrow rostrum; interpterygoid fossa U-shaped, broad, and rounded anteriorly. Mastoid little expanded, not visible in dorsal view. Supraorbital ridge moderately well developed. Last molars narrower than premolars; tubercle over root of lower incisor large. Dentition: $\frac{1}{0}, \frac{0}{0}, \frac{1}{1}, \frac{3}{3}$. Upper

Pliocene; North America. Center of distribution of some 10 species in Mexico; 1 species enters the United States.

Liomys irroratus **(Gray).** Spiny pocket mouse. Body length averaging about 115 mm; hind foot 25–35 mm. Tail averaging 106% of body length. Upper parts gray; underparts and feet white. Widely distributed in Mexico, from Oaxaca to southern Chihuahua; entering the United States in lower Rio Grande Valley of southern Texas. In Texas an inhabitant of thorny brushlands.

Family Castoridae. Beavers. Large, aquatically adapted rodents, with tail greatly broadened and flattened dorsoventrally. Hind feet webbed. Skull massive; lacking postorbital processes. Infraorbital canal inconspicuous, opening on side of rostrum anterior to zygomatic plate. Molariform teeth rootless, with complicated enamel folds. Angle of mandible rounded. Dentition: $\frac{1}{1}, \frac{0}{0}, \frac{1}{1}, \frac{3}{3}$. Lower Oligocene; Eurasia, North America. 1 living genus.

Genus *Castor* **Linnaeus.** Beavers. Characters and distribution of the family. 1 species in North America.

Castor canadensis **Kuhl.** Beaver. Body length about 550–860 mm; hind foot about 155–200 mm. Tail about 50–80% of body length. Nasals shorter and somewhat wider than in European beaver. Formerly ranged along streams and lakes over most of North America; extirpated over much of its former range but reintroduced in many places.

Suborder Myomorpha. Rodents in which the infraorbital canal is moderately enlarged for transmission of the medial masseter; opening usually V-shaped. Zygomatic plate broadened and tilted more or less upward; lateral masseter arising from channel in front of zygomatic arch. Zygomatic arch slender; jugal seldom extending far forward, usually supported by long zygomatic process of maxilla. No postorbital processes. Angle of mandible rising from lower surface of incisive alveolus. Clavicles well developed. Tibia fused with fibula. Oligocene; cosmopolitan. 9 generally recognized living families, 2 of which are native, 1 introduced, in North America and New World.

KEY TO FAMILIES OF MYOMORPHA

1. Infraorbital canal oval (Fig. 6-49); molars low-crowned and with a complex pattern of transverse enamel folds   **Zapodidae**

Fig. 6-49. Skull of *Zapus princeps.*

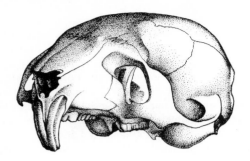

Infraorbital canal V-shaped (Fig. 6-31B); molars high-crowned, with prismatic pattern of enamel, or low-crowned with cusps   2

2. Upper molars with cusps in 3 longitudinal rows
                                                          Muridae
   Molars with cusps in 2 longitudinal rows or with prismatic pattern of dentine bordered by enamel
                                                          Cricetidae

Family Cricetidae. New World mice and voles. Small, mostly generalized or slightly fossorial-specialized. Molars rooted or rootless; with 2 longitudinal rows of cusps or 2 rows of triangular prisms placed alternately. Dentition: $\frac{1}{1}, \frac{0}{0}, \frac{0}{0}, \frac{3}{3}$. Oligocene; cosmopolitan. 5 subfamilies, 2 in North America and the New World.

Fig. 6-50. **Enamel pattern of left upper molar row: A,** *Pitymys pinetorum;* **B,** *Onychomys leucogaster.*

A   B

KEY TO SUBFAMILIES OF CRICETIDAE

1. Molars rooted, tuberculate or prismatic (Fig. 6-50B); if prismatic, skull without special ridges for attachment of masseter muscle   **Cricetinae**
   Molars usually rootless; crown with prismatic pattern (Fig. 6-50A); skull with ridges for attachment of masseter   **Microtinae**

Subfamily Cricetinae. New World rodents. A highly variable group of terrestrial, semiarboreal, or semiaquatic rodents. Skull little modified for attachment of jaw muscles; no squamosal crests; no median interorbital crest. Molars rooted, crowns cuspidate, laminate, or prismatic; when cuspidate, the cusps arranged in 2 longitudinal rows in both uppers and lowers. Lower Oligocene; Eurasia, Africa, South and North America. About 58 currently recognized genera, of which 51 are restricted to the New World. 8 genera in the United States.

KEY TO GENERA OF CRICETINAE

1. Upper molars flat-crowned, cusps not apparent at any stage; enamel in open or narrow folds (Fig. 6-51A)   2
   Upper molars with cusps apparent in at least some stage of wear, not flat-crowned (Fig. 6-51B)   3
2. Enamel folds in young adults widely open (resembling microtines); supraorbital ridges not prominent   *Neotoma*
   Enamel folds of upper molars deep and narrow; not widely open in young adults; molars appear compressed; supraorbital ridges prominent, extending onto parietals   *Sigmodon*

Fig. 6-51. **Left upper molar row: A,** *Neotoma micropus;* **B,** *Oryzomys palustris.*

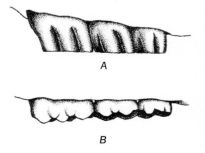

A

B

Fig. 6-52. Left mandible of A, *Peromyscus leucopus;* B, *Onychomys leucogaster.*

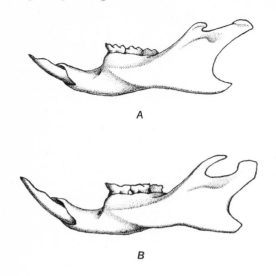

A

B

3. Upper incisors with a prominent groove
   Reithrodontomys
   Upper incisors ungrooved   **4**
4. Supraorbital ridges present; cusps of inner and outer rows on upper molars opposite one another
   Oryzomys
   Supraorbital ridges lacking; cusps of inner and outer rows more or less alternating   **5**
5. Coronoid process of mandible well developed (Fig. 6-52*B*)   **6**
   Coronoid process reduced (Fig. 6-52*A*)   **7**
6. Molars narrow, cusps elongated; plantar pads of hind feet 4   *Onychomys*
   Molars not distinctly narrowed, cusps low; plantar pads 6   *Baiomys*
7. Posterior palatine foramina about midway between interpterygoid fossa and anterior palatine foramina; baculum relatively long, its length contained less than 20 times in body length (Fig. 6-56*B*)   *Peromyscus*
   Posterior palatine foramina nearer to interpterygoid fossa than to anterior palatine foramina; baculum very short, its length contained more than 20 times in body length (Fig. 6-56*A*)   *Ochrotomys*

Genus *Oryzomys* Baird. Rice rats. A highly variable group of moderate-sized rodents; terrestrial, arboreal, or semiaquatic in habits. Tail long and typically scantily haired. Mammae 8. Skull variable; interorbital constriction moderate to considerable. Supraorbital ridges weak to well developed. Posterior border of palate behind molar row; lateral pits present. Bullae usually small or very small. Upper molars 3-rooted, lowers 2-rooted. Upper molars low-crowned with fairly low cusps. Pleistocene; South and North America. About 130 nominal species, many of which may represent subspecies. Center of distribution in South America; 1 species entering the United States.

Oryzomys palustris (Harlan). Rice rat. Size large for the genus; body length averaging 114–152 mm; hind foot about 27–40 mm. Tail long, from slightly shorter to longer than body length. Ears small, inconspicuous, and well haired. Toes of hind feet webbed near base. Pelage moderately coarse. Skull with short rostrum and high braincase; temporal ridges well developed. Dorsal color grayish to tawny; ventral color white to pale buffy. From Central America and Mexico along Gulf Coast into southeastern United States; following Mississippi embayment to southern Illinois. Reported from southeastern Kansas.

Genus *Reithrodontomys* Giglioli. Harvest mice. Small to very small; tail long, often exceeding body length, slender, moderately or scantily haired, with conspicuous annulations. Ears prominent, more or less haired. Hind feet long, narrow; plantar pads 6. Mammae 6. Skull with smoothly rounded, moderately inflated braincase; no supraorbital ridges. Bullae small to moderate. Palate square posteriorly, ending about even with last molars. Coronoid process of mandible low. Upper incisors with a conspicuous longitudinal groove. Molars low-crowned, cuspidate. Pleistocene; North America, northern South America. 17 living species with center of distribution in southern Mexico and Central America; 5 species in the United States.

KEY TO SPECIES OF *REITHRODONTOMYS*

1. Tail usually more than 110% of body length; in last lower molar, 1st primary enamel fold as long as or longer than 2d primary fold and extending more than halfway across tooth; major fold clearly visible (Fig. 6-53*C*)   *R. fulvescens*
   Tail usually less than 110% of body length; in last lower molar, 1st primary enamel fold distinctly

Fig. 6-53. Occusal view of left lower molar row of harvest mice; *A, Reithrodontomys humulis; B, Reithrodontomys megalotis; C, Reithrodontomys fulvescens.* [From Hooper, *A Systematic Review of Harvest Mice* (*Genus Reithrodontomys*) *of Latin America,* University of Michigan Press.]

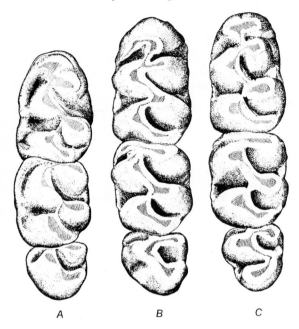

A   B   C

shorter than 2d fold and extending less than halfway across tooth; major fold indistinct or absent (Fig. 6-53*A, B*)   2

2. A distinct labial shelf or ridge, often with distinct cusplets, on 1st and 2d lower molars (Fig. 6-53*A*)   *R. humulis*

No distinct labial shelf or ridge on 1st and 2d lower molars (Fig. 6-53*B*)   3

3. Tail less than 90% of body length; braincase narrow, usually less than 9.6 mm; rostrum short and broad   *R. montanus*

Tail length about equal to, or longer than, body; braincase broader, usually more than 9.5 mm; rostrum longer and narrower   4

4. Ears blackish, dorsal fur long, dense, and dark-colored   *R. raviventris*

Ears buffy or brownish, when brownish, the lower inner and upper outer parts darker than remainder; dorsal fur shorter and less dense, varying from pale buff to reddish brown   *R. megalotis*

Reithrodontomys humulis (**Audubon and Bachman**). Eastern harvest mouse. Size small; body length averaging about 59–66 mm; hind foot about 15–17 mm. Tail short, about 84–91% of body length. Dorsal color mixed blackish and pinkish cinnamon; ventral color grayish white. Ears blackish brown; tail bicolor, fuscous above, grayish white below. Braincase narrow, highly arched; rostrum short and broad; nasals broad, ending nearly on a line with end of premaxillae. Deciduous forest region of southeastern United States from southeastern Texas and western Arkansas east to Florida, north to southeastern Ohio and Maryland.

Reithrodontomys montanus (**Baird**). Plains harvest mouse. Size small; body length averaging about 72–82 mm; hind foot about 15–17 mm; tail short, about 85% of body length. Ears small, about 13–16 mm. Dorsal color mixed black or blackish and pale buff, darkest along median line; ventral color white. Tail sharply bicolor, dark brown above, white or whitish below. Ears with a large blackish or brownish patch on outer surface. Central grasslands from southwestern South Dakota and eastern Wyoming south through Nebraska, Kansas, southwestern Missouri, and Oklahoma (west of deciduous forest) to Edwards Plateau in Texas; west through eastern half of Colorado, central New Mexico, and southeastern Arizona.

Reithrodontomys megalotis (**Baird**). Desert harvest mouse. Size medium; body length averaging about 67–77 mm; hind foot about 14–20 mm. Tail moderate, about 80–100% of body length. Color about as in *R. montanus.* In skull, zygomatic plate broad. The characters of this species parallel those of *R. montanus,* which occurs with it over a large area. Best distinguished from *R. montanus* by relatively longer tail, larger ear, and relatively broader braincase and narrower rostrum and by its preference for brushy rather than grassy habitats. From southern Mexico, north through western United States to southwestern Canada. In north, ranging east to southern Wisconsin and northeastern Arkansas. In south, ranging east to Oklahoma and Texas panhandles and into trans-Pecos Texas.

Reithrodontomys raviventris Dixon. Salt marsh harvest mouse. Size medium; body length averaging 66–74 mm; hind foot about 15–19 mm. Tail about equal to or longer than body. Dorsal pelage long, dense, and very dark; ventral pelage reddish; tail nearly unicolor. Skull generally similar to that of *R. megalotis*, but with relatively shorter rostrum, nasals, and palatal foramina; zygomata more widely expanded anteriorly. Restricted to salt marshes bordering San Francisco Bay, California.

Reithrodontomys fulvescens **Allen.** Fulvous harvest mouse. Body length averaging about 71–73 mm; hind foot about 16–22 mm. Tail markedly longer than body, usually about 130% or more of body length. Pelage coarse for the genus. Dorsal pelage reddish brown and ventral pelage buffy in humid areas to pale buff above and pale buff or whitish below in arid regions. Rostrum relatively heavy; braincase elongate. Length of incisive foramina slightly greater than width of rostrum. From northern Central America, north through Mexico to southern Arizona and western and southern Texas, north through central and eastern Texas and Oklahoma and central and western Arkansas to southern Missouri, east to southwestern Mississippi.

Genus *Peromyscus* **Gloger.** White-footed mice. Small, with moderate to long tail; the tail more than half of body length and frequently equal to or greater than body length. Hair moderately long and typically soft. Ears comparatively large. Skull with thin-walled braincase, little ridged. Supraorbital border smoothly rounded, sharply edged, or beaded; interparietal large, conspicuous; zygomata slender, depressed to level of palate. Posterior border of palate squared or rounded, without lateral pits, about even with plane of posterior roots of last molars. Auditory bullae more or less inflated. Nasals usually projecting forward over incisors. Coronoid process of mandible short and slightly developed. Baculum a simple, slender rod, typically with a broad, flattened base (at least in U.S. forms). Lower Pliocene; North America. Numerous species, with center of distribution in Mexico.

KEY TO SPECIES OF *PEROMYSCUS*

1. Plantar pads 5; baculum slender, with basal part subglobular, greatest width included more than 11 times in length (Fig. 6-54A)   *P. floridanus*
   Plantar pads usually 6; baculum heavier, basal part flattened, greatest width included less than 11 times in length (Fig. 6-54B)   **2**

2. 1st and 2d upper molars without accessory tubercles between outer primary tubercles or with rudimentary ones (Fig. 6-55A); mammae 4, inguinal   **3**
   1st and 2d upper molars with more or less well developed accessory tubercles between outer primary ones (Fig. 6-55B); mammae 6 (4 inguinal, 2 pectoral)   **6**
3. Size large; hind foot 25 mm or more   *P. californicus*
   Size smaller; hind foot 24 mm or less   **4**
4. Tail with long soft hair, terminating in a distinct tuft; zygomata compressed anteriorly; nasals attenuate, slightly or not at all exceeded by premaxillae   *P. crinitus*
   Tail with short hairs, with slight or no tuft at tip;

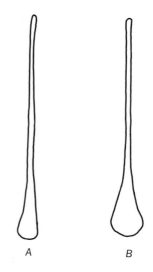

Fig. 6-54. Bacula of *Peromyscus: A, Peromyscus floridanus; B, Peromyscus maniculatus.* (From Blair, Systematic Relationships of Peromyscus and Several Related Genera as Shown by the Baculum, *J. Mammalogy,* vol. 23, no. 2.)

Fig. 6-55. Left upper molars of *A, Peromyscus eremicus; B, Peromyscus leucopus.* (From Osgood, Revision of the Mice of the American Genus Peromyscus, *North Am. Fauna* 28.)

zygomata less compressed anteriorly; nasals broader and flatter, definitely exceeded by premaxillae   5

5. Mastoid breadth usually less than 11.4 mm; shaft of baculum usually dorsally curved, base usually obovate   ***P. eremicus***

    Mastoid breadth usually more than 11.4 mm; shaft of baculum usually ventrally curved, base usually spatulate   ***P. merriami***

6. Tail usually shorter than body, slightly or not tufted; ears relatively small; no distinct ochraceous band between dark dorsal color and white underparts; baculum slightly curved dorsoventrally, small, less than 12 mm in length   7

    Tail usually longer than body, more or less tufted; ears relatively large; a more or less distinct ochraceous band between dark dorsal color and white underparts; baculum strongly curved dorsoventrally, large, more than 12 mm in length   12

7. Tail usually distinctly bicolor; baculum shorter (length less than 8 mm), base proportionally broader (contained less than 7.2 times in length)   8

    Tail usually indistinctly bicolor; baculum longer (length more than 8 mm), base proportionally narrower (contained more than 7.2 times in length)   10

8. Size small; hind foot usually less than 19 mm; tail usually less than 60 mm; Florida, southern Georgia, Alabama, and South Carolina   ***P. polionotus***

    Size larger; hind foot usually more than 19 mm; tail usually more than 60 mm; other than in southeastern United States   9

9. Hind foot usually more than 22 mm; Pacific Northwest from lower Columbia River into southwestern British Columbia   ***P. oreas***

    Hind foot usually less than 22 mm; over most of North America   ***P. maniculatus***

10. Distribution west of central Texas or north of a line from northeastern Oklahoma to southeastern Virginia   ***P. leucopus***

    Distribution east of central Texas and south of a line from northeastern Oklahoma to southeastern Virginia   11

11. Size larger; hind foot usually more than 22 mm; usually inhabiting lowlands   ***P. gossypinus***

    Size smaller; hind foot usually less than 22 mm; usually inhabiting uplands   ***P. leucopus***

12. Ears moderate, usually 18 mm or less (dry); tail sheath easily broken   13

    Ears larger, 18 mm or more (dry); tail sheath relatively tough   14

Fig. 6-56. Bacula: *A, Ochrotomys nuttalli; B, Peromyscus boylei.* (From Blair, Systematic Relationships of Peromyscus and Several Related Genera as Shown by the Baculum, *J. Mammalogy,* vol. 23, no. 2.)

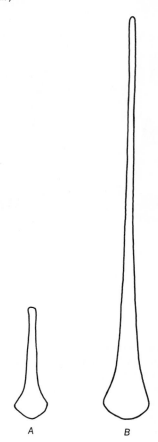

A   B

13. Dusky of hind leg reaching to or over tarsal joint; cartilage cap of baculum 0.15 mm or less in length   ***P. boylei***

    Dusky of hind leg not reaching to tarsal joint; cartilage cap of baculum 0.50 mm or more in length   ***P. pectoralis***

14. Restricted to canyons in Texas Panhandle   ***P. comanche***

    Mountains of southwestern North America   15

15. Ears larger; rostrum shorter (nasals about 10 mm); auditory bullae larger (Fig. 6-57A)   ***P. truei***

    Ears smaller; rostrum longer (nasals about 11 mm); auditory bullae smaller (Fig. 6-57B)   ***P. difficilis***

Fig. 6-57. Ventral view of skull of A, *Peromyscus truei;* B, *Peromyscus difficilis.*

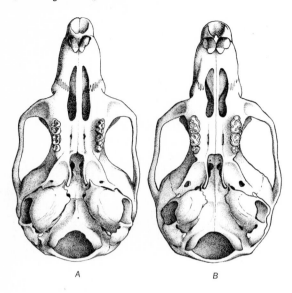

A   B

Peromyscus eremicus (**Baird**). Cactus mouse. Size medium; body length about 90–95 mm; hind foot about 19–22 mm. Tail long, about 106–114% of body length, finely annulated and covered with short hairs, giving a naked appearance. Ears large, thinly haired. Penis oval in cross section. Sole of hind foot naked to end of calcaneum, at least medially. Plantar pads 6. Mammae 4. Pelage soft and silky, generally pale buff above (blackish on lava beds). An ochraceous buff lateral line bordering white of ventral surface. Tail indistinctly bicolor, distinctly elastic. Skull medium; braincase high and elongate; maxillaries always ending posterior to nasals. Auditory bullae moderately inflated. Molar teeth simple; no accessory cusps between primary cusps of 1st and 2d upper molars. Deserts from central Mexico and Baja California into southwestern United States. Entering the United States from Laredo, Texas, through southern New Mexico to southwestern Utah, southern Nevada, and southern California.

Peromyscus merriami **Mearns.** Merriam mouse. Similar to *P. eremicus,* but larger. Body length averaging about 92 mm; hind foot about 20–24 mm. Tail long, about 109% of body length. Color generally similar to *P. eremicus* but usually with a large pectoral spot of cinnamon. Skull larger; mastoid process visible in dorsal profile. From western Mexico into southern Arizona.

Peromyscus californicus (**Gambel**). California mouse. Largest of U.S. species; body length averaging 99–110 mm; hind foot about 25–29 mm. Tail long, averaging about 121–135% of body length, well haired, but annulations not completely concealed. Ears large, thinly haired. Sole of hind foot naked to end of calcaneum. Plantar pads 6. Mammae 4. Pelage long and lax. Dorsal pelage dark; ventral pelage white. Tail bicolor. Skull large; auditory bullae large; molars heavy, without accessory cusps between primary cusps of 1st and 2d upper molars. From northern Baja California north through chaparral of western California to San Francisco Bay.

Peromyscus crinitus (**Merriam**). Canyon mouse. Size medium; body length averaging 76–85 mm; hind foot about 17–23 mm. Tail long, averaging about 109–124% of body length, densely covered with long, soft hairs, ending in a distinct tuft. Hind foot hairy on proximal fourth or naked in median part to calcaneum. Ears large, about 18–22 mm. Plantar pads 6. Mammae 4. Upper parts buffy, underparts white. Tail bicolor. In skull, rostrum elongate, depressed, and rounded; zygomata compressed anteriorly; premaxillae not exceeding posterior ends of nasals. Enamel pattern of 1st and 2d upper molars simple (as in *P. eremicus*) or with rudimentary accessory tubercles between primary cusps. From northern Baja California and northwestern Mexico, north through eastern California to northern Oregon, east through northern Arizona to northwestern New Mexico and western Colorado.

Peromyscus maniculatus (**Wagner**). Deer mouse. Small to medium-sized; body length 70–100 mm; hind foot about 15–22 mm; ears small to moderate; highly variable geographically in all body proportions. The most distinctive characters are the usually sharply bicolor tail and the rather short, compact skull and small auditory bullae. Plantar pads 6. Mammae 6. Forest inhabitants in the Appalachians, northern United States, Rocky Mountains, and Pacific Coast ranges and forests. Desert and grassland inhabitants in central and southwestern United States. Forest races tend to longer tails (approaching or exceeding body length), larger feet and ears, and generally darker color than grassland and desert races. Dichromatic in southwestern deserts, occurring in buff and gray phases. From southern Mexico north to northern Canada; throughout the United States except Gulf and Atlantic Coastal Plains. About 65 described subspecies.

Peromyscus oreas **Bangs.** Subalpine deer mouse. Size relatively large; body length about 77–100 mm; hind foot 21–25 mm, usually more than 22. Tail long, usually over 110 mm and averaging about 124% of body length. Ears relatively large. Color dark and rich. Eastern foot of Cascade Mountains and British Columbia Coast Range to Pacific; lower Columbia River north into southwestern British Columbia. Some overlap with *P. maniculatus,* with which it apparently hybridizes.

Peromyscus polionotus (**Wagner**). Beach mouse. Size smallest of U.S. species. Body length averaging 78–87 mm; hind foot about 15–19 mm. Tail short, about 55–65% of body length. Body proportions and skull generally similar to small, short-tailed representatives of *P. maniculatus.* Plantar pads 5 or 6. Mammae 6. Color and pattern highly variable in correlation with soil color. Inland populations dark above, white below, as in *P. maniculatus.* Coastal and insular populations with reduced intensity of dorsal color and with the white ventral pattern encroaching to a lesser or greater degree on the dorsum. Restricted to Florida, Alabama, Georgia, and South Carolina, where it prefers coastal dunes and inland fields. Range allopatric to that of *P. maniculatus.* Several described subspecies, of which the most striking is the nearly white beach mouse population of Santa Rosa Island, Florida.

Peromyscus leucopus (**Rafinesque**). Wood mouse. Size medium; body length averaging about 89–105 mm; hind foot about 19–24 mm; varying from smaller in the East and Southeast to larger in the Northwest and West. Tail moderate, about 70–90% of body length. Ears moderate. Plantar pads 6. Mammae 6. Color reddish brown or grayish; tail usually indistinctly bicolor. Difficult to distinguish from related species in some parts of range. Usually distinguishable from the related, sympatric *P. gossypinus* in Southeast by smaller size (hind foot usually less than 22 mm) and preference for upland habitats. Usually separable from grassland races of *P. maniculatus* in the central United States by its larger size, longer tail, larger ears, more slender body, its indistinctly bicolor tail, and its preference for forests. Usually separable from forest races of *P. maniculatus* in the northeastern United States by less densely haired and indistinctly bicolor tail and smaller ears. Forest and brush habitats of eastern United States north of a line from southwestern Mississippi to northeastern North Carolina; west to western Montana and central Arizona; south through eastern Mexico to Yucatán.

Peromyscus gossypinus (**Le Conte**). Cotton mouse. Size medium, body length averaging 98–112 mm; hind foot about 22–26 mm. Tail moderate, about 70–80% of body length. Body and skull proportions similar to those of *P. leucopus,* from which it differs principally in larger size, darker color, and in preference for lowland habitats. Hybrids between the 2 species have been reported. Plantar pads 6. Mammae 6. Forests throughout peninsular Florida and mostly on Coastal Plain from southeastern Virginia to southeastern Oklahoma and eastern Texas.

Peromyscus boylei (**Baird**). Brush mouse. Size moderate, body length averaging 92–102 mm; hind foot about 19–26 mm. Tail long, about as long as or slightly longer than body. Ears moderate, about 16–20 mm from notch. Tail moderately haired, bicolor, and with a distinct tuft. Dorsal color buffy gray, ventral color white, the two separated by a bright buff lateral line. Skull slender with moderately enlarged auditory bullae. Plantar pads 6. Mammae 6. Difficult to distinguish from other species that occur with it. From *P. pectoralis* usually distinguished by extension of dark color of hind leg down over tarsal joint, by the very short apical cartilage of the baculum (0.14–0.15 mm), and by a preference for higher elevation and more moist environment. From *P. truei, P. difficilis,* and *P. comanche,* usually distinguishable by smaller auditory bullae, smaller ears, and more easily broken tail sheath. Rocky and forested habitats from Central America north to southwestern Missouri and northern Arkansas, northern Utah, and northern California.

Peromyscus pectoralis **Osgood.** Encinal mouse. Size moderate, body length averaging about 90 mm; hind foot about 19–23 mm. Tail slightly longer than body, with a distinct tuft. Plantar pads 6. Mammae 6. Body proportions and skull very similar to those of *P. boylei,* but generally distinguishable by white tarsal joint and longer (0.56–1.27 mm) apical cartilage of baculum and by usual occurrence in lower, more arid situations. Rocky and forested habitats from central Mexico to western and north central Texas and southeastern Arizona.

Peromyscus truei (**Shufeldt**). Piñon mouse. Size moderate; body length averaging about 94–102 mm; hind foot about 21–27 mm. Tail moderate (about 85–115% of body length), bicolor, well haired, and with a terminal tuft. Pelage long and silky. Ears large. Plantar pads 6. Mammae 6. Dorsal color buffy gray with a buffy lateral stripe. Auditory bullae large and inflated. Difficult to distinguish from *P. difficilis,* from which it differs in larger auditory bullae, slightly smaller size, relatively shorter tail, less gray dorsal color, and larger ears (ear from notch longer than hind foot). Rocky, mountain slopes, usually in piñon belt and above, from southern Mexico north to northern Oregon, northwestern Colorado, and northwestern Oklahoma; west through mountains of California.

Peromyscus difficilis (**Allen**). Juniper mouse. Size moderate; body length averaging about 96–102 mm; hind foot about 22–28 mm. Tail long (usually slightly longer than body), bicolor, and with a terminal tuft. Plantar pads 6. Mammae 6. Dorsal color grayish (blackish on lava beds). Auditory bullae and ears usually larger than in *P. boylei* and smaller than in *P. truei.* Usually separable from *P. truei* by the longer rostrum, by the ear from notch being shorter than hind foot, by more gray dorsal color and relatively longer tail. Difficult to distinguish from *P. boylei,* from which it differs in slightly more inflated bullae and larger ears. Mountain slopes from southern Mexico into New Mexico, west into eastern Arizona, east to northwestern Oklahoma, and north to central Colorado.

Peromyscus comanche **Blair.** Palo Duro mouse. Size moderate; body length averaging about 88 mm; hind foot about 22–24 mm. Tail long (equal to body length to 115% of body length), bicolor, and with a terminal tuft. Ears large, about 21–23 mm from notch. Plantar pads 6. Mammae 6. Dorsal color buffy gray, with a buffy lateral stripe. Differs from *P. boylei* in larger ears and larger, more inflated auditory bullae. Differs from *P. difficilis* in more buffy coloration, slightly larger auditory bullae, and a triangular rather than elliptical interparietal. Restricted to cedar-forested canyons along the escarpment of the high plains in the Texas Panhandle. Distribution allopatric to that of *P. difficilis,* but specific distinctness indicated by behavioral isolating mechanisms and reduced genetic compatibility.

Peromyscus floridanus (**Chapman**). Gopher mouse. Size large; body length averaging about 113 mm; hind foot about 24–29 mm. Tail moderate, about 75% of body length. Ears large, 22–25 mm from notch. Dorsal color buffy gray; tail indistinctly bicolor. Plantar pads 5. Mammae 6. Molar teeth large and broad; accessory tubercles between outer salient angles small and inconspicuous. Baculum slender, relatively small, and with a rounded base. Restricted to peninsular Florida, where it prefers scrub vegetation on sand.

Genus *Ochrotomys* **Osgood.** Golden mouse. Generally resembling *Peromyscus,* but pelage dense and soft; ears of same color as upper parts. Molariform teeth relatively wide, with relatively thick dentine; dentine areas never confluent except in extremely old individuals. Baculum very small. Entepicondylar foramen present near head of humerus. Recent; North America. 1 species.

Ochrotomys nuttalli (**Harlan**). Golden mouse. Size medium; body length averaging about 90–96 mm; hind foot about 19–20 mm. Tail shorter than body, about 85–95% of body length. Dorsal surface and ears a rich, reddish buff; underparts creamy. Plantar pads 6, with a rudimentary 7th. Mammae 6. Forests in eastern Texas and Oklahoma and southern Missouri, east to the Atlantic and south into northern, peninsular Florida.

Genus *Baiomys* **True.** Pygmy mice. Very small, body length usually less than 75 mm; tail short, less than 75% of body length; ears small; plantar pads 6. Hair moderately coarse. Body compact; general appearance somewhat volelike. Skull generally similar to that of *Peromyscus,* but rostrum short, nasals not projecting over incisors; coronoid process of mandible large, broad, and strongly recurved. Baculum relatively short and otherwise distinctive; the base narrow with thin wings anterior to it projecting dorsolaterally to form channel. Baculum narrowest about ⅔ of distance from base. Upper Pliocene; North America. 2 species, with center of distribution in Mexico, 1 entering the United States.

Baiomys taylori (Thomas). Pygmy mouse. Very small; body averaging about 59–66 mm; hind foot 13–15 mm; tail averaging about 64–67% of body length. Color grayish. Inhabitant of thorny brushlands and dense prairies, where it makes well-beaten trails. Enters southern Texas and follows Coastal Plain north to northcentral Texas and east nearly to Louisiana. Also entering southeastern Arizona.

Genus Onychomys Baird. Grasshopper mice. Principally insectivorous, moderate- to small-sized mice, with short tails, small ears, and fairly long, very soft fur. Dorsal color more or less grayish; underparts white. Soles of feet furred. Plantar pads 4. Mammae 6. Tail tending to be spindle-shaped, with greatest diameter near the middle. Skull with pronounced interorbital constriction; no supraorbital ridges. Zygomata narrow; bullae small; palate ending slightly behind 3d molar. Cusps on molar teeth high, slightly alternating, but less so than in *Peromyscus*. Coronoid process of mandible well developed. Upper Pliocene; North America. 2 species, both in the United States.

KEY TO SPECIES OF *ONYCHOMYS*

1. Tail less than half of body length where the 2 species occur together; 1st upper molar usually less than half the length of molar row; length of upper molar row contained 1.2 or (usually) less in interorbital breadth; length of mandible usually more than 14 mm   *O. leucogaster*

   Tail more than half of body length; 1st upper molar usually more than half the length of molar row; length of upper molar row contained 1.2 or more in interorbital breadth; length of mandible usually less than 14 mm   *O. torridus*

Onychomys leucogaster (Maximilian). Grasshopper mouse. Body stout. Body length about 94–118 mm; hind foot about 17–25 mm. Tail 31–57% of body length. Tail less than half the body length

where this species occurs with *O. torridus*. Interorbital region of skull relatively narrow; 1st upper molar usually less than half the length of the molar row. Dorsal color gray or pinkish cinnamon; tip of tail usually white. Central grasslands from western Minnesota to northern Montana and adjacent Canada, south to Corpus Christi Bay, Texas, west to Sierra Nevada of California and Cascades of Washington and Oregon; south into eastern Mexico. Grasslands and deserts.

Onychomys torridus (Coues). Scorpion mouse. Body length about 90–104 mm. Body less stout and tail relatively longer than in *O. leucogaster* where the two occur together. Hind foot about 18–24 mm. Tail about 48–56% of body length. Skull smaller and less stout than in *O. leucogaster*, but with relatively wide interorbital region; teeth lower-crowned; 1st upper molar usually more than half the length of molar row. Color about as in *O. leucogaster*. Deserts from trans-Pecos Texas through southwestern Utah to central, western Nevada and southern half of California, south into Mexico.

Genus Sigmodon Say and Ord. Cotton rats. Compact, coarse-haired, moderately short-tailed (tail shorter than body) with short, rounded ears. Tail tapered to a slender tip. Plantar pads 6. Mammae 10. Skull with heavy rostrum and prominent supraorbital ridges which extend onto parietals. Zygomatic plate cut back very sharply above, with a forward-projecting process on upper border. Palate broad, ending behind last molar, with well-developed lateral pits. Prominent coronoid process of mandible. Upper molars heavy, flat-crowned, with long, narrow, enamel folds. Upper Pliocene; North and South America. Habits vole-like. Generally associated with dense grass, where they make conspicuous trails. About 20 nominal species, of which several tropical ones may be subspecies. Center of distribution in Mexico. 3 species entering the United States.

KEY TO SPECIES OF *SIGMODON*

1. Size larger; belly whitish; habitat usually lowlands   *S. hispidus*
   Size smaller; belly buff or gray; habitat mountains **2**
2. Belly buff; nose gray   *S. minimus*
   Belly gray or grayish; nose yellowish   *S. ochrognathus*

Sigmodon hispidus **Say and Ord.** Common cotton rat. Size large; body length about 130–160 mm; hind foot about 28–41 mm. Tail moderate, about 60–90% of body length. Ear relatively small and rounded, about 18–22 mm from notch and partially concealed in fur. Dorsal color buffy gray to blackish. Ventral color white to dirty white. From Central America throughout Mexico, north through southern Arizona and New Mexico, through Texas to northeastern Kansas and southeastern Colorado, east across the Coastal Plain to southern Virginia and all of Florida. A color cline in the United States from darker in the East to paler in the West.

Sigmodon minimus **Mearns.** Least cotton rat. Size relatively small; body length about 130–150 mm; hind foot about 28–31 mm. Tail moderate, about 75% of body length and indistinctly bicolor. Upper parts dark gray; underparts buffy. Skull relatively short and broad. Mountains and valleys of central New Mexico and southeastern Arizona, south into Mexico.

Sigmodon ochrognathus **Bailey.** Yellow-nosed cotton rat. Size relatively small; body length about 115–145 mm; hind foot about 25–29 mm. Tail moderate, about 80% of body length. Upper parts blackish gray; underparts gray, sometimes with a buffy tinge. Nose, orbital rings, and upper parts of forelegs bright ochraceous. Mountains of trans-Pecos Texas, southwestern New Mexico, and southeastern Arizona and adjacent Mexico.

Genus *Neotoma* Say and Ord. Packrats. Size large; tail long, well-haired, sometimes bushy. Ears large. Skull with marked interorbital constriction; supraorbital ridges moderately developed. Zygomatic plate slightly cut back above. Bullae moderate to large. Palate ending about even with front of 1st molar. Coronoid process of mandible well developed. Molars high- and flat-crowned, prismatic. Upper Pliocene; North America. About 31 nominal species (some probably subspecies) with center of distribution in southwestern United States and Mexico; 8 in the United States.

KEY TO SPECIES OF *NEOTOMA*

1. Tail flattened and bushy; sole of hind foot normally densely furred from heel to posterior plantar pad; base of baculum quadrate in cross section
   N. cinerea

Tail rounded, not flattened and bushy; sole of hind foot naked along outer side at least to tarsometatarsal joint; base of baculum not quadrate in cross section   **2**

2. Middle loop on last upper molar partially or completely divided by inner reentrant angle; upper molar row only slightly narrower posteriorly than anteriorly; base of baculum concave above and below, dumbbell shape in outline (Fig. 6-58).
   N. fuscipes

   Middle loop on last upper molar not divided by inner reentrant angle; upper molar row much narrower posteriorly than anteriorly; base of baculum not markedly concave on both dorsal and ventral sides, not dumbbell shape in outline   **3**

3. 1st upper molar with anterointernal reentrant angle deep, reaching more than halfway across anterior lobe   **4**

   1st upper molar with anterointernal reentrant angle

Fig. 6-58. End view of base of baculum in *Neotoma fuscipes*. [From Burt and Barkalow, A Comparative Study of the Bacula of Wood Rats (Subfamily Neotominae), *J. Mammalogy*, vol. 23, no. 3.]

Fig. 6-59. Bacula of *Neotoma:* **A,** end view of base in *Neotoma mexicana;* **B,** lateral view of *Neotoma lepida;* **C,** lateral view in *Neotoma albigula.* [From Burt and Barkalow, A Comparative Study of the Bacula of Wood Rats (Subfamily Neotominae), *J. Mammalogy*, vol. 23, no. 3.]

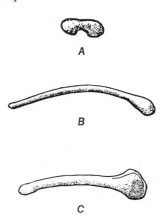

Fig. 6-60. Bacula of *Neotoma*: A, *Neotoma floridana*; B, *Neotoma micropus*. [From Burt and Barkalow, A Comparative Study of the Bacula of Wood Rats (Subfamily Neotominae), *J. Mammalogy*, vol. 23, no. 3.]

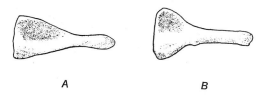

A   B

shallow, reaching less than halfway across anterior lobe   **5**

4. Hind foot less than 40 mm; base of baculum with lower surface inverted U-shaped in cross section (Fig. 6-59A); southwestern United States
N. mexicana

Hind foot more than 40 mm; base of baculum with lower surface very slightly convex or straight in cross section, Appalachian mountain chain of eastern United States   *N. floridana*

5. Baculum longer than 10 mm, slender, strongly curved dorsoventrally (Fig. 6-59B)   *N. lepida*

Baculum shorter than 10 mm, heavier, weakly curved or straight (Fig. 6-59C)   **6**

6. Tail semibushy; baculum 5 mm or less in length
N. stephensi

Tail not semibushy; baculum more than 5.5 mm in length   **7**

7. Interpterygoid fossa less than 3.5 mm wide
N. albigula

Interpterygoid fossa more than 3.5 mm wide   **8**

8. Palate concave or emarginate posteriorly; color never slaty gray in adults; expanded basal part of baculum as long as, or longer than, the shaft (Fig. 6-60A)   *N. floridana*

Palate with posterior median projection; color slaty gray above; expanded basal part of baculum shorter than the shaft (Fig. 6-60B)
N. micropus

Neotoma floridana (Ord). Florida packrat. Size large; body length averaging about 199–232 mm; hind foot about 35–46 mm. Tail length about 75–95% of body length. Ear medium. Skull long; ascending branches of premaxillae very long, reaching posteriorly far beyond nasals. Auditory bullae small, short, and rounded; palate without posterior median spine. 1st upper molar with anterointernal reentrant angle moderately to quite deep. Dorsal color brownish gray; ventral color white or grayish. Southern Florida to southeastern North Carolina and northward in Appalachian chain to Connecticut; west on Coastal Plain to Edwards Plateau in Texas; north to southern Indiana, northeastern Kansas, and southwestern South Dakota.

Neotoma micropus Baird. Plains packrat. Size large; body length averaging about 188–193 mm; hind foot about 36–41 mm. Body robust; tail thick, moderately haired, moderately long, its length about 70–85% of body length. Pelage short. Skull generally similar to that of *N. floridana* but more angular; rostrum heavier, nasals narrower posteriorly; posterior border of palate with more or less developed median spine. Dorsal color bluish gray; ventral color white. From eastern and northern Mexico, entering the United States from mouth of Rio Grande to southwestern New Mexico, north through western half of Texas and Oklahoma to southwestern Kansas and southeastern Colorado, west to east base of Rockies in New Mexico and in Rio Grande Valley. Range meeting and interdigitating that of *N. floridana*.

Neotoma albigula Hartley. White-throated packrat. Size medium; body length averaging about 176–214 mm; hind foot about 33–39 mm. Body moderately slender; tail about 85% of body length, sharply bicolor. Ears moderately large, about 30 mm from notch. Upper parts brownish or buffy gray (black or blackish on lava beds); underparts white (blackish on lava beds, except throat and chest, which are white to base of fur throughout the species). Skull medium; interpterygoid fossa narrow (about 3.2 mm or less); auditory bullae relatively large. Central Mexico north into the United States from western Texas to southeastern California, north to southeastern Utah and southeastern Colorado.

Neotoma lepida Thomas. Desert packrat. Size small; body length averaging about 150–170 mm; hind foot about 28–41 mm. Tail moderate, about 80–95% of body length. Ear moderate, about 25–30 mm from notch. Upper parts gray or buffy gray; underparts grayish, the hairs gray to their bases throughout. The long, dorsoventrally curved baculum is distinctive. Western Arizona through southern half of California, north in Utah and Nevada to southeastern Oregon and southwestern Idaho, south through Baja California.

Neotoma mexicana **Baird.** Mexican packrat. Size medium; body length averaging about 178–193 mm; hind foot about 31–41 mm. Tail moderate, bicolor, about 80–85% of body length. Ears moderate. Upper parts more or less brownish gray; underparts white or whitish, but with the hairs gray at the base. Melanistic above and below on lava beds. 1st upper molar with deep anterointernal reentrant angle; nasals constricted near middle. Last lower molar sometimes with a small accessory anterointernal reentrant angle. Mountains, usually in piñon or yellow pine belts, from Central America north through trans-Pecos Texas, New Mexico, and eastern Arizona and southwestern Utah to northern Colorado.

Neotoma stephensi **Goldman.** Stephen's packrat. Size medium; body length averaging 148–203 mm; hind foot about 28–34 mm. Tail semibushy, averaging about 71–76% of body length. Dorsal color yellowish to grayish buff; fur dark basally except on throat, pectoral, and inguinal regions and on inner sides of forelegs. Anterointernal reentrant angle of 1st upper molar shallow or absent. Rostrum relatively narrow; nasals generally truncate posteriorly. Baculum very small, wedge-shaped. Western New Mexico to west central Arizona and southern Utah.

Neotoma fuscipes **Baird.** Dusky-footed packrat. Size large; body length averaging about 187–229 mm; hind foot about 32–47 mm. Tail long, about 80–100% of body length. Ears large. Upper parts grayish brown; underparts grayish or whitish. Last upper molar large and with the middle enamel loop partially or completely divided by the deep reentrant angle. The short, heavy baculum, with a dumbbell-shaped base in cross section, is distinctive. Principally in chaparral of lower elevations but up to nearly 9000 ft, from northern Baja California to northwestern Oregon.

Neotoma cinerea **(Ord).** Bushy-tailed packrat. Size large; body length averaging 200–238 mm; hind foot about 35–52 mm. Tail about 75–90% of body length. Ears large, about 30–35 mm from notch. Soles of hind feet usually densely furred from heel to posterior plantar pad. The bushy tail is distinctive. Skull large and angular; temporal ridges prominent; frontal region narrow; auditory bullae large; interpterygoid fossa narrow. Upper parts pale buffy gray to blackish; underparts white or whitish. Usually in yellow pine belt and above in mountains, from northern New Mexico and Arizona and eastern and northern California into western Canada; east to northwestern Nebraska and central North Dakota.

Subfamily Microtinae. Voles and lemmings. Stout-bodied, short-legged, short-muzzled, terrestrial, burrowing forms. Ears short; tail shorter than body. Skull with prominent ridges for attachment of jaw muscles; zygomatic plate broad and strongly tilted upward; infraorbital foramen small and narrowed. Squamosal usually with a postorbital crest; supraorbital ridges tending to fuse into a median interorbital crest. Deep pits for muscle attachment usually present in mandible between molars and outer side of jaw. Palate usually terminating anterior to back of last molars. Zygomatic plate never cut back anteriorly. Zygomata robust. Upper Miocene; Eurasia, North Africa, North America. 31 generally recognized genera; 10 in North America, of which 8 occur in the United States.

KEY TO GENERA OF MICROTINAE

1. Lower incisors short, ending posteriorly opposite or in front of alveolus of last molar (Fig. 6-61A) *Synaptomys*
   Lower incisors long, extending behind the molar row and ending in or near the condylar process (Fig. 6-61B, C) 2
2. Molars rooted in adults 3
   Molars rootless in adults 5
3. Size large, hind foot more than 60 mm; tail laterally compressed *Ondatra*
   Size small, hind foot less than 60 mm; tail not compressed 4
4. Palate terminating posteriorly as a simple transverse shelf (Fig. 6-62A) *Clethrionomys*
   Palate terminating posteriorly as a sloping median ridge and 2 lateral pits; median spinous process present and forming a sloping septum between the shallow lateral pits (Fig. 6-62B) *Phenacomys*
5. Size large, hind foot more than 35 mm; fur very soft; plantar pads 5 *Neofiber*
   Size small, hind foot less than 35 mm; fur various; plantar pads 5 or 6 6
6. Ears much reduced, hidden in fur; bullae and mastoids visible dorsally, reaching plane of occipital condyles; sole densely haired *Lagurus*
   Ears visible above fur; bullae and mastoids not

Fig. 6-61. Relation of incisor root to molar row in microtine rodents: *A, Synaptomys; B, Phenacomys; C, Microtus.* (From Miller, The Genera and Subgenera of Voles and Lemmings, *North Am. Fauna* 12.)

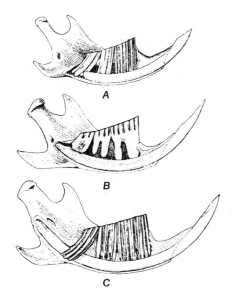

Fig. 6-63. Molar enamel patterns: *A, Microtus xanthognathus; B, Pitymys pinetorum.* (From Bailey, Revision of American Voles of the Genus Microtus, *North Am. Fauna* 17.)

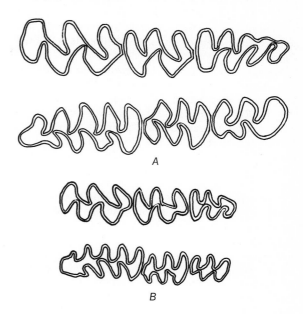

Fig. 6-62. Palatal view of microtine skulls: *A, Clethrionomys; B, Phenacomys.* (From Miller, The Genera and Subgenera of Voles and Lemmings, *North Am. Fauna* 12.)

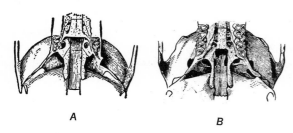

Genus *Synaptomys* Baird. Bog voles. Small, short-tailed, with long coarse hair. Ears small, nearly concealed by fur. Plantar pads 6. Mammae 6 or 8. Skull with supraorbital ridges fusing in adult to form median interorbital crest. Squamosal crests well developed; rostrum very thick and short; auditory bullae large. Upper molars with wide folds, partly filled with cement. Upper incisors very broad, longitudinally grooved. Upper Pliocene; North America. 2 species; Canada and the United States. Bogs and dense meadows.

KEY TO SPECIES OF *SYNAPTOMYS*

1. Mammae 6; 1st lower molar with 3 closed triangles; palate ending posteriorly in a broad, blunt median projection   *S. cooperi*
   Mammae 8; 1st lower molar without closed triangles; palate ending posteriorly in a sharply pointed median projection   *S. borealis*

Synaptomys cooperi **Baird.** Southern bog lemming. Body length averaging 100–113 mm; hind

greatly enlarged; sole of hind foot not densely haired   **7**

7. 1st lower molar with 5 closed triangles (Fig. 6-63A)   *Microtus*
   1st lower molar with 3 closed triangles (Fig. 6-63B)   **8**

8. Supraorbital ridges not united to form medial interorbital crest; mammae 4   *Pitymys*
   Supraorbital ridges united in adult; mammae 6   *Microtus*

foot about 16–24 mm. Tail about 15–20% of body length. Ear about 8–14 mm from notch. Upper parts bright cinnamon to brown, often blackish; underparts gray. Lower incisors heavy. From southeastern Canada west to western Minnesota, south to southwestern Kansas, northeastern Arkansas, southeastern Tennessee, and northeastern North Carolina.

Synaptomys borealis (**Richardson**). Northern bog lemming. Body averaging 97–117 mm; hind foot about 16–22 mm. Tail about 20–25% of body length. Ear about 12–13 mm from notch. Upper parts grizzled gray to brown, with some black hairs; underparts gray; tail bicolor; feet grayish to nearly black; a few hairs at base of ear brighter than remainder of pelage. Lower incisors relatively slender and sharply pointed. From Alaska across Canada; barely ranging into the United States from northern Washington to northwestern Montana, in northwestern Minnesota, and in Maine and New Hampshire.

Genus *Clethrionomys* **Tilesius.** Red-backed voles. Small, moderately short-tailed inhabitants of cool, damp forests. Ears medium. Plantar pads 6. Mammae 8. Upper parts usually red or reddish; lateral surfaces often gray. Skull weak, with slightly developed squamosal crests; supraorbital ridges weak and usually widely separated in adults, never fusing to form crest. Interorbital constriction moderate; braincase broad. Auditory bullae large. Palate ending posteriorly as a straight, transverse shelf, with no median septum; lateral pits present but free from median spinous process of palate when the latter is present. Molars rooted in adult. Pleistocene; Eurasia, North America. About 20 species, 5 in North America, 2 in the United States.

KEY TO SPECIES OF *CLETHRIONOMYS*

1. Dorsal surface usually with a reddish longitudinal stripe contrasting with the gray sides; postpalatal bridge usually truncate posteriorly   **C. gapperi**
   Dorsal surface darker and duller, reddish stripe inconspicuously set off from sides; postpalatal bridge with a median posteriorly directed spine (sometimes very small)   **C. occidentalis**

Clethrionomys gapperi (**Vigors**). Boreal red-backed vole. Body length about 90–112 mm; hind foot about 16–21 mm. Tail about 35–45% of body length. Ear about 12–16 mm from notch. Dorsal stripe reddish to yellowish brown (reddish stripe sometimes absent); sides gray to buffy gray; underparts silvery. Tail bicolor. Teeth relatively light. Terrestrial in forests. Transcontinental in Canada; following Appalachian chain south to western North Carolina. Also entering the United States from northern Michigan to northern Iowa and western North Dakota. In the West, following Rocky Mountains south to southwestern New Mexico and adjacent Arizona.

Clethrionomys occidentalis (**Merriam**). Western red-backed vole. Body length about 87–112 mm; hind foot about 18–21 mm. Tail about 40–50% of body length. Ear about 10–13 mm from notch. Dorsal stripe indistinct because of intermixed black hairs; sides light gray to dark buffy gray; underparts buffy over gray underfur. Tail sharply bicolor. Teeth relatively heavy. Terrestrial in humid coast forests from northern California north to Columbia River.

Genus *Phenacomys* **Merriam.** Phenacomys. Moderately small, terrestrial or arboreal microtines, with small ears mostly hidden by body hairs. Tail varying from short (25 mm) to long (85 mm). Pelage usually long and silky; short and coarse in some. Mammae usually 8. Skull characterized by posterior border not being shelflike, but terminating as a median, sloping ridge bounded on each side by a lateral pit. Molars rooted in adults. On lower molars, inner reentrant angles much deeper than outer. Upper Pliocene; North America. 6 species, 2 restricted to Canada and Alaska, 4 in the United States.

KEY TO SPECIES OF *PHENACOMYS*

1. Tail short; less than 50 mm   *P. intermedius*
   Tail long; more than 50 mm   2
2. Incisors not strongly decurved (Fig. 6-64A); tail slender, scantily haired; terrestrial   *P. albipes*
   Incisors strongly decurved (Fig. 6-64B); tail thick, well haired; arboreal   3
3. Braincase strongly ridged; postorbital processes slight and inconspicuous; coloration brownish, nose sooty   *P. silvicola*
   Braincase relatively unridged; prominent postorbital processes present; coloration reddish; nose not sooty   *P. longicaudus*

Fig. 6-64. Skulls of *Phenacomys*: A, *Phenacomys albipes*; B, *Phenacomys longicaudus*. (From Hall and Cockrum, A Synopsis of the North American Microtine Rodents, *Univ. Kans. Publ. Museum Nat. Hist.*, vol. 5, no. 2.)

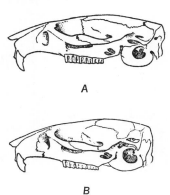

A

B

Phenacompys intermedius **Merriam.** Mountain phenacomys. Body length averaging about 111–115 mm; hind foot about 16–18 mm. Tail about 25–35% of body length. Upper parts gray to brownish, face sometimes yellowish; underparts whitish. Tail sharply bicolor. Ranging across Canada and extending southward at high elevations in Rocky Mountains to northern New Mexico and in Sierras to central California. Terrestrial in mountain meadows.

Phenacomys albipes **Merriam.** Pacific phenacomys. Body length averaging about 108 mm; hind foot about 19–20 mm. Tail about 60% of body length. Upper parts brown; underparts clear gray. Found near streams in Pacific Coast forest areas from northwestern California to northwestern Oregon.

Phenacomys silvicola **Howell.** Dusky tree phenacomys. Body length about 110 mm; hind foot about 20–22 mm. Tail about 70–80% of body length. Upper parts brownish; nose sooty; underparts whitish. Tail well haired. Restricted to a small area in northwestern Oregon.

Phenacomys longicaudus **True.** Red tree phenacomys. Body length averaging about 99–109 mm; hind foot about 19–22 mm. Tail about 65% of body length. Upper parts reddish brown; underparts whitish. Tail blackish, well haired. Arboreal

in Pacific Coast forest from San Francisco Bay, California, to northern Oregon.

Genus *Microtus* Schrank. Meadow voles. Small to moderate-sized, stout, short-legged, terrestrial, and semifossorial forms. Ears small and more or less hidden in fur. Tail usually short, usually less than half as long as body. Plantar pads usually 6, sometimes 5. Mammae 4–8. Skull angular; supraorbitals usually fusing into a median interorbital crest; squamosals with more or less well developed crests. Rostrum short; braincase broad. Bases of lower incisors extending far behind and on outer side of molars. Upper incisors not grooved. Molars rootless. Crown of molars with inner and outer reentrant angles approximately equal. Pleistocene; Eurasia, North America, North Africa. 11 species in the United States.

KEY TO SPECIES OF *MICROTUS*

1. 3d upper molar with 2 closed triangles (Fig. 6-65*B*); mammae 6   **2**
   3d upper molar with 3 closed triangles (except *M. breweri* and *M. nesophilus* with 2 usually confluent, *M. chrotorrhinus* with 5 closed) (Fig. 6-65*A*); mammae 8 or 4   **3**
2. Range in grasslands of central United States
      M. ochrogaster
   Range restricted to coastal prairie in southeastern Texas and southwestern Louisiana
      M. ludovicianus
3. Plantar pads 5   **4**
   Plantar pads 6   **5**
4. Size small, hind foot less than 22 mm; side glands obscure or lacking   ***M. oregoni***
   Size large; hind foot more than 22 mm; side glands conspicuous on flanks of adult males
      M. richardsoni
5. 2d upper molar with 4 closed angular sections and a rounded posterior loop   **6**
   2d upper molar with 4 closed sections and no posterior loop (except irregularly in *M. californicus*)   **9**
6. 3d upper molar with 2 of the 3 triangles usually confluent; insular species   **7**
   3d upper molar with 3 closed triangles   **8**
7. Interparietal approximately as wide as long; restricted to Muskeget Island, Massachusetts
      M. breweri
   Interparietal wider than long; restricted to Great Gull Island and Little Gull Island, New York; extinct   ***M. nesophilus***

Fig. 6-65. Molar enamel patterns: *A, Microtus pennsylvanicus; B, Microtus ochrogaster.* (From Bailey, Revision of American Voles of the Genus Microtus, *North Am. Fauna* 17.)

Fig. 6-66. Ventral view of skulls: *A, Microtus californicus; B, Microtus townsendi.* (From Bailey, Revision of American Voles of the Genus Microtus, *North Am. Fauna* 17.)

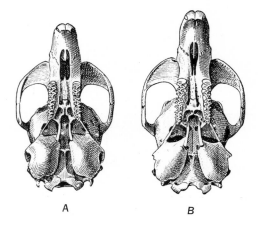

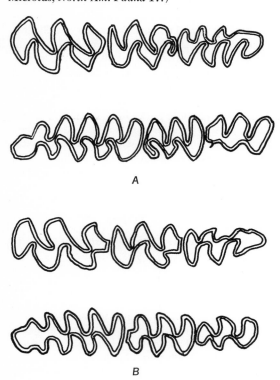

Tail shorter, contained in body length 2.0 or more times; skull prominently ridged   *M. californicus*
13. Dorsal pelage dark brown to blackish; hind foot more than 23.7 mm; west of crest of Cascade Mountains from extreme southwest British Columbia to 40° latitude in California
          M. townsendi
   Distribution mostly east of Pacific coastal area; where it occurs with *M. townsendi* upper parts yellowish or grayish and hind foot less than 23.7 mm   *M. montanus*

8. Nasals truncate posteriorly; restricted to Block Island, Rhode Island   *M. provectus*
   Nasals rounded posteriorly; mainland
          M. pennsylvanicus
9. Mammae 4   *M. mexicanus*
   Mammae 8   **10**
10. Nose yellowish; flanks of males with obscure glands or lacking glands   *M. chrotorrhinus*
   Nose not yellowish; a pair of glands on hips of males   **11**
11. Incisive foramina gradually tapered posteriorly, not abruptly constricted (Fig. 6-66A)   **12**
   Incisive foramina abruptly constricted and narrower posteriorly than anteriorly (Fig. 6-66B) **13**
12. Tail long, contained in body length less than 2.0 times; skull relatively smooth, not prominently ridged even in adults   *M. longicaudus*

Microtus pennsylvanicus (Ord). Meadow vole. Body length averaging 105–132 mm; hind foot about 18–24 mm; ear about 12–16 mm; tail variable, about 30–50% of body length. Upper parts dark brown to gray with a faint brown wash; underparts gray, usually washed with whitish or buffy. 2d upper molar with 4 closed angular sections and a rounded posterior loop. From Alaska throughout all but extreme north of Canada; south in eastern United States to northeastern Georgia, central Illinois, northern Missouri, and northeastern Nebraska; south through Rocky Mountains to central Utah and northern New Mexico. Relictual in southwestern New Mexico. Closely related to *M. agrestis* of Eurasia.

Microtus provectus Bangs. Block Island vole. Body length 123–133 mm; hind foot about 21–23 mm. Tail indistinctly bicolor, about 35–40% of body length. Differing from *M. pennsylvanicus* in having always gray underparts, in larger and narrower skulls of adults, in nasals truncate posteriorly rather than rounded, and in wider interparietal which extends farther forward. Insular form, restricted to Block Island, Rhode Island. Possibly conspecific with *M. pennsylvanicus*.

Microtus nesophilus **Bailey.** Gull Island vole. Body length about 144 mm; hind foot about 21–22 mm. Tail bicolor, about 28% of body length. Differing from closely related *M. pennsylvanicus* in darker pelage, shorter and wider braincase, more widely spreading zygomata, smaller auditory bullae, and deeper prezygomatic notch. Insular form, known only from Little Gull Island and Great Gull Island, New York; extinct.

Microtus breweri **(Baird).** Beach vole. Body length averaging 128 mm; hind foot about 22–25 mm. Tail bicolor, averaging about 42% of body length. Differing from closely related *M. pennsylvanicus* in paler pelage color, longer and coarser pelage, anteriorly wider nasals, and longer interparietal. Insular form, restricted to Muskeget Island, Massachusetts.

Microtus montanus **(Peale).** Mountain vole. Body length averaging 102–155 mm; hind foot about 18–25 mm. Tail about 35–45% of body length. Upper parts brownish; sides lighter and more buffy; underparts white to gray, sometimes with a buffy wash. Tail bicolor. Hip glands conspicuous in adult males. Incisive foramina narrow and constricted posteriorly; middle upper molar with 4 closed triangles. Marshes, meadows, and tule swamps, eastern Arizona, southeastern Wyoming, and central Montana west to eastern California and north through most of Oregon and central Washington continuing to British Columbia in Canada.

Microtus californicus **(Peale).** California vole. Body length averaging 108–143 mm; hind foot about 20–25 mm. Tail about 40–50% of body length. Upper parts brownish; sides paler; underparts gray, often with white-tipped hairs. Tail bicolor. Inconspicuous hip glands in adult males. Adult skull heavy, angled, and ridged; incisive foramina rounded at both ends and widest in middle; 2d upper molar with 4 closed triangles. Dry meadows and grassy uplands from northern Baja California through California to southwestern Oregon.

Microtus townsendi **(Bachman).** Townsend vole. Body length about 123–155 mm; hind foot about 22–26 mm. Ear about 15–17 mm, prominent above fur. Tail about 43–50% of body length. Fur thin and harsh. Upper parts brown, with many long, black hairs; sides buffy gray; underparts grayish. Tail indistinctly bicolor. Hip glands conspicuous in adult males. Skull long and little arched; incisive foramina long, narrow, and constricted posteriorly; 2d upper molar with 4 closed triangles. Open grasslands in low country west of Cascade Mountains from northern California into southwestern British Columbia.

Microtus longicaudus **(Merriam).** Long-tailed vole. Body length averaging 114–132 mm; hind foot about 20–25 mm. Ear moderate. Tail long, about 50–65% of body length, indistinctly to distinctly bicolor. Upper parts brownish gray to dark brown, with numerous black-tipped hairs; sides slightly paler; underparts gray with a wash of whitish or buffy. Skull relatively smooth, even in adults; incisive foramina gradually tapered posteriorly or as wide as anteriorly; middle upper molar with 4 closed triangles. Banks of mountain streams and in mountain meadows, from southern Alaska south in the United States, from Rocky Mountains westward, to southern New Mexico, eastern and northern Arizona, and central California. Disjunct populations in southern California, southeastern Arizona, and in the Black Hills.

Microtus mexicanus **(Saussure).** Mexican vole. Body length averaging about 105–109 mm; hind foot about 17–21 mm. Ear about 12–15 mm from notch. Tail short, about 25–30% of body length, indistinctly bicolor. Pelage coarse and lax. Upper parts cinnamon buff to brownish, with black hairs; sides paler; underparts washed with grayish buff, sometimes whitish. Mammae 4. Skull wide; incisive foramina short, wide, and posteriorly truncate; 1st lower molar usually with 6 inner salient angles. Open grassy places, principally in yellow pine belt, in mountains; southern central and western Mexico, north through southern and western New Mexico and eastern Arizona to southern Utah. A disjunct population in Hualpai Mountains of western Arizona.

Microtus chrotorrhinus **(Miller).** Yellow-cheeked vole. Body length averaging about 120–122 mm; hind foot about 19–22 mm. Ear moderate. Tail about 35–40% of body length, indistinctly bicolor. Upper parts grayish, with black-tipped hairs; face from eyes to nose yellowish to reddish orange; underparts gray, sometimes with faint whitish wash. Hip glands absent or vestigial. Skull light and smooth; auditory bullae large; incisive foramina short and wide; last upper molar with 5 closed triangles. Usually in rocky places near water in mountains; from southeastern Canada reaching the United States in northeastern Minnesota, and east of Great Lakes ranging southward through Appalachian chain to eastern Tennessee and western North Carolina.

Microtus richardsoni (DeKay). Water vole. Size large; body length averaging about 147–153 mm; hind foot about 25–30 mm. Ear about 15–20 mm from notch, mostly concealed in fur. Tail about 40–50% of body length, bicolor. Plantar pads 5. Side glands of males conspicuous. Pelage long and heavy. Upper parts grayish brown to dark reddish brown; underparts gray, with white or whitish wash. Skull large; zygomata widely spreading; auditory bullae small; incisors protruding far beyond premaxillae; molars with constricted and tightly closed sections; last upper molar with 3 closed triangles. Stream banks and wet meadows in Rocky Mountains south to central Utah and Cascades south to southern Oregon.

Microtus oregoni (Bachman). Creeping vole. Size small; body length about 97–99 mm; hind foot about 16–19 mm. Ear small (9–10 mm from notch) but visible above the short fur. Tail about 35–50% of body length, indistinctly bicolor. Eye very small and nearly hidden in fur. Plantar pads 5. Side glands absent or vestigial. Fur woolly in appearance. Upper parts reddish or grayish brown, ears blackish; underparts grayish white. Skull short and low; molars small; last upper molar with 2 or 3 closed triangles; 1st lower molar with 5 closed triangles; 2d lower molar with anterior pair of triangles usually confluent; 3d lower molar with 3 transverse loops. Usually in dry grassland, sea level to 6000 ft, southwestern British Columbia, south through western Washington and Oregon to northwestern California.

Microtus ochrogaster (Wagner). Prairie vole. Body length averaging 95–133 mm; hind foot about 17–22 mm. Ear about 11–15 mm from notch. Tail short, about 25–?5% of body length, sharply bicolor. Plantar pads 5. Mammae 6. Fur long and coarse. Upper parts gray, with a peppery appearance; sides paler; underparts gray or washed with whitish or pale cinnamon. Skull high and narrow; auditory bullae small and narrow; incisive foramina wide posteriorly; molars with wide reentrant angles; last upper molar with 2 closed triangles; 1st lower molar with 3 closed and 2 open triangles; 2d lower molar with anterior pair of triangles confluent. Prairies from central Oklahoma into southern Canada; west to central Colorado and Wyoming and southeastern Montana; east into Ohio and western West Virginia. This and the following species regarded by some as belonging to a separate genus, *Pedomys*.

Microtus ludovicianus Bailey. Louisiana vole. Body length averaging about 131 mm; hind foot about 18–19 mm. Tail and ear about as in *M. ochrogaster*, except tail indistinctly bicolor. Plantar pads 5. Mammae 6. Upper parts dark gray, with peppery appearance; underparts reddish to dark buffy. Skull generally similar to that of *M. ochrogaster*, to which this widely disjunct form appears closely related. Restricted to coastal prairies of extreme southeastern Texas and southwestern Louisiana.

Genus *Pitymys* McMurtrie. Pine voles. Short-tailed, small-eared, semifossorial forms with soft, dense pelage. Sole partially haired. Mammae 4. Plantar pads 5. Skull with relatively small squamosal crests; supraorbital ridges widely separated in adults in the interorbital region. Regarded by some as a subgenus of *Microtus*. Pleistocene; Eurasia, North America. 2 allopatric species in North America. 1 restricted to mountains of eastern Mexico.

Pitymys pinetorum (Le Conte). Pine vole. Common pine vole. Body length averaging 88–106 mm; hind foot about 13–20 mm. Tail averaging about 19–22% of body length. Ear about 8–12 mm, more or less hidden in fur. Tail indistinctly bicolor. Eastern United States west to eastern Texas, central Oklahoma, eastern Kansas, southeastern Nebraska, and southeastern Minnesota. Disjunct, relict populations on Edwards Plateau of Texas and in Wichita Mountains of Oklahoma.

Genus *Lagurus* Gloger. Sagebrush voles. Small, thick-bodied voles with short ears more or less hidden in fur. Tail short, about length of hind foot or slightly longer. Sole of hind foot densely haired, the 5 plantar pads concealed by hair. Mammae 8. Skull angular, with wide, flat braincase and prominent peglike squamosal crests. Supraorbital ridges not fused into interorbital crest. Bullae very large, with spongy bone internally; mastoids inflated and more or less visible in dorsal view. Posterior margin of palate a sloping median ridge, with 2 lateral pits (as in *Microtus*). Molars lacking cement in the widely open enamel folds. Upper Pliocene; Asia, North America. 1 species in North America.

Lagurus curtatus (**Cope**). Sagebrush vole. Body length averaging 95–115 mm; hind foot about 14–19 mm. Ear about 9–13 mm from notch. Tail short, about 21–24% of body length. Fur long and lax. Upper parts pale buffy gray to ashy gray; sides paler; underparts whitish to buffy. Usually found in sagebrush; from northern Colorado north through western North Dakota into Canada, west through most of Utah and Nevada to eastern California, and north through eastern Oregon to east central Washington.

Genus *Neofiber* True. Florida water rat. Large, semiaquatic microtines. Fur very soft; ears very small, mostly hidden in fur; tail rounded, partly clothed in long hairs, but scales apparent. Swimming fringes on tail and feet not greatly developed. Plantar pads 5. Mammae 6. Skull generally similar to that of *Microtus*. Molars rootless. 1st lower molar with 5 triangles and anterior and posterior loop; 3d lower molar with 1 outer fold and 2 salient angles. Pleistocene; North America. 1 living species.

Neofiber alleni **True.** Florida water rat. Size large; body length averaging about 200 mm; hind foot about 40–50 mm. Tail about 65% of body length. Upper parts brown to blackish brown; underparts grayish white, with a buffy wash. Marshes. Peninsular Florida west to Carrabelle and north to Okefinokee Swamp.

Genus *Ondatra* Link. Muskrats. Largest of the microtines; body length may exceed 300 mm. Adapted for aquatic life; fur dense and soft; ears short; tail long, laterally compressed, more or less naked, with well-developed swimming fringe below. Hind foot much larger than forefoot; with conspicuous swimming fringes; sole naked; plantar pads 5 or 4. Mammae 6. Skull generally similar to

that of *Microtus*. Molars rooted in adult. 1st lower molar with 6 triangles (the 1st not closed) between anterior loop and posterior loop. 3d lower molar with 3 outer salient angles. Upper Pliocene; North America. 2 nominal species, of which 1 is insular (Newfoundland).

Ondatra zibethicus (**Linnaeus**). Muskrat. Body length averaging about 240–346 mm; hind foot about 64–88 mm. Tail averaging about 73–93% of body length. Upper parts rusty red to almost black; underparts whitish or pale brown. Inhabitants of stream or lake shores or marshes. Distributed over most of North America north of Mexico, except Florida and a coastal strip in Georgia and South Carolina, and most of Texas (excluding panhandle, trans-Pecos, and extreme southeast) and most of California.

Family Muridae. Old World rats and mice. Mostly generalized rodents; molars laminate or with cusps; cusps in 3 longitudinal rows. Molars rooted. Skull without squamosal or interorbital crests. Dentition: $\frac{1}{1}, \frac{0}{0}, \frac{0}{0}, \frac{3}{3}$. Pliocene. Native in Europe, Asia, Australia; cosmopolitan as result of introduction. About 90 generally recognized genera, 2 introduced into North America.

KEY TO GENERA OF MURIDAE

1. 1st upper molar 5-rooted; its crown shorter than combined crowns of 2d and 3d molars   *Rattus*
   1st upper molar 3-rooted; its crown longer than combined crowns of 2d and 3d molars   *Mus*

Genus *Rattus* Fischer. Old World rats. Largest and most variable genus of mammals. Body length ranging from 100 mm (southeastern Asian species) to 290 mm (Australasian species). Plantar pads 6. Fur harsh; tail long, scantily haired in American species. Pleistocene; Old World. About 275 named species currently recognized, 2 introduced in the United States.

KEY TO SPECIES OF *RATTUS*

1. Tail shorter than body; no distinct notches on anterior row of cusps on 1st molar   *R. norvegicus*
   Tail longer than body; distinct outer notches on anterior row of cusps on 1st molar   *R. rattus*

Rattus norvegicus (Berkenhout). Norway rat. Body relatively robust. Body length averaging about 195 mm; hind foot about 35–44 mm. Tail always shorter than body (about 80% of body length), tapering, scantily haired, with scales prominent. Fur thin and coarse. Upper parts buffy brownish; underparts soiled whitish. Introduced throughout the United States, where it is usually associated with human habitations.

Rattus rattus (Linnaeus). Roof rat. Body relatively slender. Body length averaging about 175 mm; hind foot about 33–39 mm. Tail longer than body (about 110% of body length), tapering, scantily haired, with scales prominent. Upper parts brownish or black; underparts white or gray. Introduced species. Coastal areas of the United States and through much of Mexico. Usually around human habitations, but feral in southern Texas, southern Florida, and possibly other southern localities.

Genus *Mus* Linnaeus. Old World mice. Small, less than 100 mm in body length; hind foot generally narrow. Fur soft, harsh, or more or less spiny. Skull light, generally flat. Border of zygomatic plate cut back above, a small knob usually present on its lower border. Incisive foramina long, usually extending between anterior molars. 1st molar 3-rooted; crown elongated so that it is distinctly longer than 2d and 3d molars combined. Pleistocene; native Eurasia, cosmopolitan by introduction. About 44 species, 1 introduced into North America.

Mus musculus Linnaeus. House mouse. Size small; body length averaging about 80–85 mm; hind foot about 16–21 mm. Ear about 11–18 mm. Tail about equal to body length or longer, slender and tapering, scantily haired; scales prominent. Upper parts brownish gray; underparts paler. Occlusal surface of upper incisors notched. Introduced throughout the United States; usually associated with human habitations, but feral in many parts of the country.

Family Zapodidae. Jumping mice. Saltatorial; hind feet large, hind legs long; tail considerably longer than body. Mammae normally 8. Infraorbital canal large and oval; molars low-crowned and with enamel much folded. Upper incisors much curved, deeply grooved, and deep orange in color. Oligocene; Eurasia, North America. 3 genera, 2 in North America.

KEY TO GENERA OF ZAPODIDAE

1. 4 teeth in upper molar row   *Zapus*
   3 teeth in upper molar row   *Napaeozapus*

Genus *Zapus* Coues. A small, peglike premolar present in upper jaw. 2d upper molar similar in shape to first, but small, 3d reduced and nearly circular. Tail without white tip in American forms. Dentition: $\frac{1}{1}, \frac{0}{0}, \frac{1}{0}, \frac{3}{3}$. Pleistocene; Asia, North America. 3 species in North America.

KEY TO SPECIES OF *ZAPUS*

1. Premolars with crescentic fold on occlusal surface; tip of baculum spade-shaped   *Z. trinotatus*
   Premolars without crescentic fold on occlusal surface; tip of baculum not spade-shaped   **2**
2. Incisive foramina shorter than 4.6 mm; total length of baculum less than 5.1 mm   *Z. hudsonius*
   Incisive foramina longer than 4.7 mm; total length of baculum more than 5.1 mm   *Z. princeps*

Zapus trinotatus Rhoads. Pacific jumping mouse. Body length averaging about 90 mm; hind foot about 31–34 mm. Tail about 150–170% of body length. Upper parts ochraceous, sides paler; underparts white, usually with suffusion of ochraceous. Skull relatively broad and deep; pterygoid fossa broad. Southwestern British Columbia south through western Washington and Oregon and in humid coastal strip south in California to San Francisco Bay.

Zapus hudsonius (Zimmermann). Eastern jumping mouse. Size moderate; body length averaging 75–90 mm; hind foot about 28–32 mm. Ears small, with white or buffy edging. Tail long (about 120–155% of body length), tapering, scantily haired, bicolor, without white tip. Upper parts yellowish, with a broad, darker median stripe; underparts white or somewhat washed with yellowish. Skull relatively light; braincase narrow; incisive foramina and molars small. Principally in wet meadows and grasslands; from Alaska across Canada, ranging south in the United States to eastern Montana, northern Colorado, northeastern Oklahoma, and eastern Alabama.

Zapus princeps Allen. Western jumping mouse. Size relatively large; body length averaging about 88–100 mm; hind foot about 30–35 mm. Ears relatively small, with white or buffy edging.

Tail about 150–175% of body length, indistinctly to sharply bicolor, and without white tip. Color similar to that of *Z. hudsonius*. Skull relatively large and heavy, with rather heavy dentition; incisive foramina large. Mountain meadows and similar habitats; from western Canada south in Rocky Mountains to central New Mexico and eastern Arizona, south in Sierra Nevada to central California; east across North Dakota.

Genus *Napaeozapus* Preble. Woodland jumping mice. Premolars lacking; 2d molar equal to 1st in size. Tail tipped with white. Dentition: $\frac{1}{1}, \frac{0}{0}, \frac{0}{0}, \frac{3}{3}$. Pleistocene; North America. 1 currently recognized species.

Napaeozapus insignis (Miller). Woodland jumping mouse. Body length averaging about 90 mm; hind foot about 29–33 mm. Tail long (about 145–170% of body length) and tipped with white. Skull broad and stout; interorbital constriction broad. Upper parts buffy with a dark, median dorsal band; underparts white. Usually in climax forest; from eastern Canada, entering northeastern Minnesota, northern Wisconsin, and Michigan; from eastern Ohio east to Atlantic and south in Appalachian chain to northeastern Georgia.

Suborder Hystricomorpha. Rodents in which the infraorbital opening is greatly enlarged for passage of medial masseter, lateral masseter retaining its primitive position. Angular region of mandible typically with a prominent out-turned ridge for insertion of masseter. Zygomata stout; jugal not supported below by a continuation of the maxillary zygomatic process. Angular part of mandible arising from outer side of alveolus of the lower incisor. 1 premolar in each jaw. Tibia and fibula separate. Oligocene; Africa, Eurasia, Americas. At least 14 families, 2 in United States.

KEY TO FAMILIES OF HYSTRICOMORPHA

1. Hind foot webbed; no spines in fur **Capromyidae**
   Hind feet unwebbed; stout heavy spines in fur
   Erethizontidae

Family Erethizontidae. New World porcupines. Large, mostly arboreal, with many stout, sharp spines in the pelage. Body stout, legs short; tail prehensile in most. Soles naked; 4 toes on front feet; 5 on rear. Facial part of skull short and broad. Inferior border of angular process of mandible strongly inflected. Molars rooted, their crowns with internal and external enamel folds. Dentition: $\frac{1}{1}, \frac{0}{0}, \frac{1}{1}, \frac{3}{3}$. Oligocene; South and North America. 1 genus in the United States.

Genus *Erethizon* Cuvier. Porcupine. Robust, arboreal, with intermixed quills and fur. Tail short, nonprehensile. Upper Pliocene; North America. 1 species.

Erethizon dorsatum (Linnaeus). Porcupine. Size large; body length about 550–565 mm; hind foot about 86–124 mm. Tail about 33–55% of body length. Winter pelage blackish, with long yellow or yellow-tipped overhairs; the fur concealing the quills; in summer pelage sparsely haired, mainly with conspicuous, black-tipped, white quills and long overhairs. From Alaska across Canada, south through western United States east to central Texas, western Oklahoma, western Nebraska, and western Dakotas; east of plains ranging into the United States from northern Minnesota to central Michigan and northern West Virginia.

Family Capromyidae. Moderate to large, terrestrial, fossorial, arboreal, or semiaquatic. Clavicles complete. Skull with long incisive foramina extending into maxillae. Molars with internal and external enamel folds. Premolars $\frac{1}{1}$. Mammae high on sides of body. Miocene; South and Central America, West Indies. 7 extinct and 5 living genera, 1 introduced into the United States.

Genus *Myocaster* Kerr. Nutria. Semiaquatic, with long, round, sparsely haired tail. Hind feet large, webbed. Incisors very large, deep orange-red in color. Upper molars with 1 inner and 3 outer enamel folds; lower molars with 1 outer and 3 inner folds; last molar the largest. Pliocene; South America. 1 species, which has been introduced into the United States.

Myocaster coypus (Molina). Nutria. Size large; body length about 635 mm; hind foot about 140 mm. Tail about 70% of body length. General color dark yellowish brown or reddish brown. Native of temperate parts of South America. Introduced into various parts of the United States.

Order Cetacea. Whales. Body fusiform, without cervical constriction and tapering gradually toward tip of tail. Flukes present as lateral expansions of skin and dense, fibrous tissue in caudal region. Head very large, up to $\frac{1}{3}$ of total length;

gape wide. Forelimbs reduced to flattened, ovoid paddles without external segmentation; hind limbs absent, but vestiges of pelvic girdle remaining internally. Skin naked except for a few fine bristles present near mouth in some species in young or persisting through life. A thick layer of fat (blubber) beneath skin. A compressed, median dorsal fin of integumental origin usually present. Eye small, without nictitating membrane. External auditory meatus a minute opening behind eye, without pinna. Nostrils opening separately or by valvular aperture on top of head. Bones spongy, their cavities filled with oil. Cervical vertebrae often more or less fused into solid structure; no sacrum. Clavicles absent. Skull with short, broad, high braincase; supraoccipital bone reaching upward and forward from foramen magnum and meeting frontals at top, thus excluding parietals from upper part of braincase. Frontals expanded laterally to form roof of orbits. Teeth monophyodont and homodont, variable in number, lacking in some except in embryo. Middle Eocene; seas of world. 2 living suborders and 1 extinct (Archaeoceti).

KEY TO SUBORDERS OF CETACEA

1. Baleen present; no calcified teeth present after birth
      Mysticeti
  Baleen lacking; calcified teeth always present after birth **Odontoceti**

Suborder Odontoceti. Toothed whales. Calcified teeth always present after birth. No baleen. Upper surface of skull more or less asymmetrical. Nasal bones represented by nodules or flattened plates, applied closely to frontals and not forming part of roof of nasal passage, which is directed upward and backward. External respiratory aperture single, the 2 nostrils uniting before they reach surface. Manus always pentadactyl, but 1st and 5th digits usually little developed. Upper Eocene; all oceans. About 5 extinct and 6 living families, 5 off the United States.

KEY TO FAMILIES OF ODONTOCETI

1. No functional teeth in upper jaw; cranium with elevated crest behind nares (Fig. 6-67A); pterygoids thick, produced backward   **2**

Numerous teeth usually present in upper and lower jaws; cranium without elevated crest (Fig. 6-67B); pterygoids short, thin   **3**

2. Numerous teeth in lower jaw   **Physeteridae**
  Functional teeth in lower jaw 2 or less   **Ziphiidae**

3. Teeth with compressed, spade-shaped crowns, the crown separated from root by narrow neck (Fig. 6-68) or truncated, not conical   **Phocoenidae**

4. Cervical vertebrae free   **Monodontidae**
  At least 2 (usually more) anterior cervicals united   **Delphinidae**

Fig. 6-67. Skull of A, *Physeter catodon;* B, *Delphinus delphis.* (A from Flower and Lydekker, *Mammals Living and Extinct,* Adam and Charles Black Co., London; B from Norman and Fraser, *Great Fishes, Whales and Dolphins,* Putnam, London.)

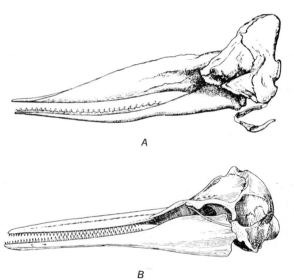

A

B

Fig. 6-68. Teeth of *Phocoena.* (From Flower and Lydekker, *Mammals Living and Extinct,* Adam and Charles Black Co., London.)

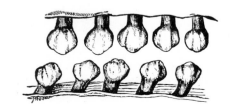

Family Ziphiidae. Beaked whales. Moderate-sized, up to about 915 cm in length, with elongated rostrum and falcate dorsal fin located far posteriorly. Throat grooved; blowhole median, single, crescentic in form. Skull with elevated cranial crest behind nares; pterygoids thick, produced backward; lacrimal bone distinct. Functional teeth limited to 1 (or rarely 2) in each lower jaw, large (particularly in males). Lower Miocene; all oceans. About 12 extinct and 5 living genera; 4 in waters off the United States.

KEY TO GENERA OF ZIPHIIDAE

1. Snout a distinct beak, abruptly set off from forehead (Fig. 6-69A, B)   2

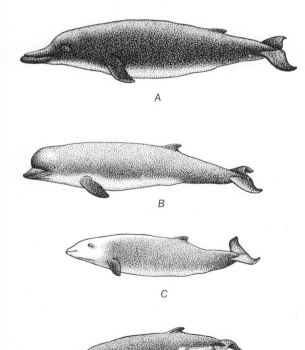

Fig. 6-69. Ziphiid whales: *A, Berardius bairdi; B, Hyperoodon ampullatus; C, Ziphius cavirostris; D, Mesoplodon mirus.* (From Burt, *A Field Guide to the Mammals,* Houghton Mifflin Company.)

Snout merging gradually into forehead (Fig. 6-69C, D)   3
2. 2 teeth in each mandible, the anterior one about 75 mm in length; snout longer than vertical distance from eye to plane of occiput (Fig. 6-69A)
Berardius
1 tooth in each mandible (usually concealed by gum) and less than 40 mm in length; snout shorter than vertical distance from eye to plane of occiput (Fig. 6-69B)   *Hyperoodon*
3. Mandibular tooth conical, located near symphysis; crest of skull formed by joined nasals   *Ziphius*
Mandibular tooth much compressed and pointed, usually located near middle of mandible; nasals sunk between upper ends of premaxillae
Mesoplodon

Genus *Mesoplodon* Gervais. "Cow-fish." Moderate-sized, slender, with muzzle produced into beak, and with falcate dorsal fin well back of middle of body. Rostrum of skull long, solid, and narrow throughout; mesethmoid ossified, coalescing with surrounding bones. Nasal bones narrow and sunk between upper ends of premaxillae. A single, large tooth in each lower jaw, none in upper. Anterior 2 or 3 cervicals united, remainder usually free. Upper Miocene; Atlantic, Pacific, Indian Oceans. About 9 species, 6 off the United States, all poorly known.

KEY TO SPECIES OF *MESOPLODON*

1. Teeth at tip of mandible and projecting obliquely forward (Fig. 6-70)   *M. mirus*
Teeth well back of tip of mandible (Fig. 6-71)   2
2. Teeth moderate, less than 35 mm in anteroposterior length in males (Fig. 6-71C, D)   3
Teeth massive, their anterioposterior length more than 70 mm in males (Fig. 6-71A, B)   4
3. Teeth in front of posterior edge of mandibular symphysis   *M. gervaisi*
Teeth near posterior edge of mandibular symphysis
M. bidens

Fig. 6-70. Mandible of *Mesoplodon mirus.* (From Norman and Fraser, *Great Fishes, Whales and Dolphins,* Putnam, London.)

Fig. 6-71. Mandibles of *Mesoplodon*: A, *Mesoplodon stejnegeri*; B, *Mesoplodon densirostris*; C, *Mesoplodon bidens*; D, *Mesoplodon gervaisi*. (From Norman and Fraser, *Great Fishes, Whales and Dolphins*, Putnam, London.)

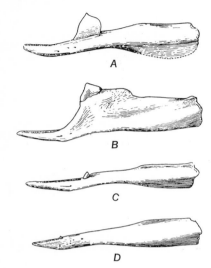

4. Mandible deepest at middle (Fig. 6-71*B*)
M. densirostris

Mandible normal, deepest posteriorly (Fig. 6-71*A*) 5
5. Teeth entirely behind mandibular symphysis
M. stejnegeri

Teeth overlapping mandibular symphysis
M. carlhubbsi

Mesoplodon bidens (Sowerby). Sowerby whale. Length up to about 490 cm. Upper parts blackish; underparts gray or white. Teeth flattened, triangular, located about $\frac{1}{3}$ of distance back from tip of mandible and near posterior edge of mandibular symphysis. North Atlantic.

Mesoplodon gervaisi (Deslongchamps). Gervais beaked whale. Length up to 670 cm. Teeth anterior to posterior edge of mandibular symphysis. Atlantic.

Mesoplodon densirostris (De Blainville). Heavy-jawed beaked whale. Length about 430 cm. Teeth massive, the mandible elevated about midway into crest from which crown of tooth projects.

Mesoplodon stejnegeri True. Stejneger's beaked whale. Length about 520 cm. Upper parts black-ish; underparts gray. Teeth massive, located entirely behind mandibular symphysis. Posterior edge of tooth rounded, anterior raised into acute point by projection of dentine as distinct, sharp cusp. North Pacific Coast.

Mesoplodon carlhubbsi Moore. Hubbs beaked whale. Length about 520 cm. Black above and below; beak whitish. Teeth massive, overlapping mandibular symphysis; similar in shape to those of *M. stejnegeri* except that bases are squared off. North Temperate Pacific.

Mesoplodon mirus True. True's beaked whale. Length about 430–520 cm. Upper parts blackish; underparts paler. Teeth laterally compressed, located at extreme tip of mandible and projecting obliquely forward. Atlantic.

Genus *Ziphius* Cuvier. Goosebeak whales. Moderate-sized, with falcate dorsal fin well back of middle of body and without prominent forehead; angle between beak and head very oblique. Skull with mesethmoid ossified and coalesced with surrounding bones of rostrum; nasals joined together and forming vertex of skull. Each side of lower jaw with a single conical tooth at anterior end and directed forward and upward. Vertebrae about 49. 3 anterior cervical vertebrae united, remainder free. Recent; all oceans. 1 species.

***Ziphius cavirostris* Cuvier.** Goosebeak whale. Length up to 855 cm. Color variable, generally blackish above, paler below; head and anterior part of body sometimes white. All oceans.

Genus *Berardius* Duvernoy. Moderate-sized, with tapering snout and well-defined forehead. Lower jaw projecting beyond tip of upper. Skull with long, narrow rostrum; mesethmoid partly ossified. Upper ends of premaxillae nearly symmetrical, moderately elevated, very slightly expanded and not curved forward over nares. Nasals broad, massive, and rounded, of nearly equal size, forming vertex of skull. 2 moderate-size, pointed teeth on each side of mandibular symphysis, the anterior one the larger. 3 anterior cervicals fused, remainder free. Recent; Pacific. 2 species, 1 off the United States.

***Berardius bairdi* Stejneger.** Baird beaked whale. Length up to 1280 cm. Head about $\frac{1}{8}$ of body length. Dorsal fin small, low, set far back. Upper parts gray to black; underparts paler to whitish. North Pacific.

Genus *Hyperoodon* Lacépède. Bottlenose whales. Moderate-sized, with distinct snout and prominent, bulging forehead. Skull with upper ends of premaxillae rising sharply behind nares to vertex and expanded laterally, their outer edges overhanging nares; the right larger than left. Nasals lying in hollow between upper extremities of premaxillae. A small, conical tooth at tip of each ramus of mandible, concealed by gum during life. All cervicals united. Recent; North Atlantic, Mediterranean, South Pacific, Antarctic. 2 species, 1 off the United States.

Hyperoodon ampullatus (Forster). Bottlenose whale. Length 610–915 cm. Dorsal fin small, falcate, set behind middle of body. Upper parts dark gray to black; underparts light gray to white. North Atlantic.

Family Physeteridae. Sperm whales. Size relatively small to very large. Cranium strongly asymmetrical in region of narial openings because of left opening greatly exceeding right. No distinct lacrimal bone. A spermaceti organ, a reservoir of clear oil, present in hollow above cranium and projecting over rostrum. Mandibular teeth numerous, set in a groove rather than in distinct alveoli and held in place by strong, fibrous gum; upper jaw without functional teeth. Lower Miocene; Atlantic, Pacific, Indian Oceans. About 17 extinct, 2 living genera.

KEY TO GENERA OF PHYSETERIDAE

1. Size large; body length 1220 cm or more; no dorsal fin   ***Physeter***
   Size small; body length less than 430 cm; dorsal fin present   ***Kogia***

Genus *Physeter* Linnaeus. Sperm whale. Very large, with enormous, truncated head (about $\frac{1}{3}$ of total length). Lower jaw relatively small, ending well short of end of snout. Atlas free, other cervical vertebrae fused into single mass. Mandibular symphysis more than half length of ramus. Teeth stout, conical, about 18–28 in each side of lower jaw. Upper Miocene; all oceans. 1 species.

Physeter catodon Linnaeus. Sperm whale. Length up to 1830 cm in males; about half of that in females. Dorsal fin absent, replaced by series of low ridges along posterior third of upper surface. Upper parts blackish, changing gradually to silvery gray or white below. All oceans. The whale of "Moby Dick."

Genus *Kogia* Gray. Pygmy sperm whale. Small, with short head (about $\frac{1}{6}$ of total length) and low, falcate dorsal fin, set behind middle of body. Lower jaw ending considerably short of end of snout. All cervical vertebrae united. Mandibular symphysis less than half length of ramus. Teeth slender, pointed, curved, about 9–14 on each side of lower jaw, these fitting into sockets in tissue of upper jaw when mouth closed. Recent: Atlantic, Pacific, Indian Oceans. 1, possibly 2, species.

Kogia breviceps (De Blainville). Pygmy sperm whale. Length about 275–395 cm. Upper parts black; underparts light gray to pinkish. Atlantic, Pacific, Indian Oceans.

Family Monodontidae. White whale and narwhal. Small whales with short rounded head without distinct beak. No dorsal fin. Cervical region relatively long, the vertebrae free or irregularly fused. Rostrum of skull about as long as cranial portion; pterygoids small, not meeting in midline. Dentition: $\frac{8}{8}$ to $\frac{10}{10}$, or in narwhal (*Monodon*) a single tooth (presumably an upper left premolar) functional in males and enormously produced in horizontal plane to form a spirally grooved tusk 213–244 cm in length. Pleistocene; cooler waters of Northern Hemisphere. 2 genera, 1 rarely off the United States.

Genus *Delphinapterus* Lacépède. White whale. Slender, with a poorly defined neck constriction behind head. Flippers broad, short, rounded. Skull narrow, elongated, depressed, with rostrum about equal to cranium in length. Cervical vertebrae free. Teeth small, conical, in anterior $\frac{3}{4}$ of jaw, about $\frac{8}{8}$ to $\frac{10}{10}$ Pleistocene; circumpolar in northern oceans. 1 species.

Delphinapterus leucas (Pallas). White whale. Length about 365–430 cm, rarely 550 cm. Young dark gray, changing successively through mottled and yellow to white in adult. Pacific and Atlantic, south as far as New Jersey in Atlantic.

Family Delphinidae. Dolphins. Variable group of small to moderate-sized cetaceans. Maxillae without large crests; lacrimal not distinct from jugal; pterygoids short, thin, often meeting in midline and enclosing an air space, open behind. Teeth conical, usually numerous in both jaws. Anterior ribs bicipital; posterior ones attached only to transverse processes. Lower Miocene; all seas. About 21 extinct and 13 living genera; 10 off the United States.

KEY TO GENERA OF DELPHINIDAE

1. No teeth in upper jaw; 3–7 in lower   *Grampidelphis*
   Upper and lower jaws each with 8 or more teeth   2
2. Dorsal fin absent   *Lissodelphis*
   Dorsal fin present   3
3. Dentition: $\frac{14}{14}$ or less   4
   Dentition: $\frac{21}{21}$ or more   6
4. Dorsal fin low, rounded   *Pseudorca*
   Dorsal fin high, falcate   5
5. Teeth along most of length of jaw; snout pointed (Fig. 6-72B)   *Grampus*
   Teeth restricted to anterior half of jaw; snout rounded (Fig. 6-72A)   *Globicephala*
6. Beak short, projecting no more than 75 mm beyond forehead or not set off from forehead (Fig. 6-73A)   7
   Beak long, projecting more than 75 mm beyond forehead from which it usually is prominently set off (Fig. 6-73B)   8
7. Teeth large; beak distinct   *Tursiops*
   Teeth small, beak indistinct   *Lagenorhynchus*

Fig. 6-72. Whales: A, *Globicephala melaena;* **B,** *Grampus orca.* **(From Burt,** *A Field Guide to the Mammals,* **Houghton Mifflin Company.)**

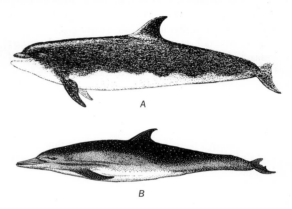

Fig. 6-73. Dolphins: *A, Tursiops truncatus; B, Delphinus delphis.* (*A* from Norman and Fraser, *Giant Fishes, Whales and Dolphins,* Putnam, London; *B* from Flower and Lydekker, *Mammals Living and Extinct,* Adam and Charles Black Co., London.)

8. Dentition: $\frac{27}{27}$ or less; crowns of teeth grooved   *Steno*
   Dentition: $\frac{30}{30}$ or more; crowns of teeth smooth   9
9. Palatal border of maxillae deeply grooved   *Delphinus*
   Palatal border of maxillae ungrooved   *Stenella*

Genus *Steno* Gray. Long-beaked dolphins. Beak long, but continuous with forehead, not set off by distinct step. Rostrum of skull long, narrow, compressed, very distinct from cranium. Mandibular symphysis long, equal to $\frac{1}{4}$ or more of length of ramus. Teeth moderate, with finely grooved crowns, about $\frac{20}{20}$ to $\frac{27}{27}$. Lower Pliocene; all oceans except polar. Several nominal species, 1 off the United States.

***Steno bredanensis* (Lesson).** Long-beaked dolphin. Length about 213–244 cm. Upper parts, flippers, and dorsal fin blackish; underparts pinkish white with gray spots; beak white. Warmer parts of Atlantic, Pacific, and Indian Oceans.

Genus *Stenella* Gray. Spotted dolphins. Head with narrow, elongated beak which is set off from forehead by deep V-shaped groove. Mandibular symphysis short, less than $\frac{1}{5}$ of length of ramus. Palate ungrooved. Teeth numerous, small (up to 3 mm in diameter), about $\frac{30}{30}$ to $\frac{52}{52}$ Recent; Atlantic, Pacific, and Indian Oceans. Several nominal species, 2 reported from U.S. coasts.

KEY TO SPECIES OF *STENELLA*

1. Teeth usually more than 44 in each half of upper and lower jaws; narrow black band from eye to anus and from eye to base of flipper   *S. styx*
   Teeth usually less than 40 in each half of upper and lower jaws; without narrow black band from eye to anus or from eye to base of flipper   2
2. Dorsal color purplish-gray; greatest breadth of skull more than 192 mm   *S. plagiodon*
   Color blackish; greatest breadth of skull less than 192 mm   *S. frontalis*

Stenella styx (Gray). Styx dolphin. Length about 245 cm. Upper parts blackish; underparts gray. Teeth about $\frac{44}{44}$ to $\frac{50}{50}$. Atlantic and Pacific; reported from coast of Oregon and Washington.

Stenella plagiodon (Cope). Spotted dolphin. Length about 152–213 cm. Upper parts and fins blackish, spotted with white or light gray; underparts pale gray, spotted with dark gray. Teeth about $\frac{34}{34}$ to $\frac{37}{37}$. Atlantic Coast north to North Carolina and Gulf of Mexico coast.

Stenella frontalis (Cuvier). Cuvier's dolphin. Length about 180 cm. Upper parts blackish, gray laterally with blackish mottling; underparts whitish. Teeth about $\frac{35}{35}$ to $\frac{44}{44}$. Atlantic, north to North Carolina.

Genus *Delphinus* Linnaeus. Common dolphins. Head with distinct, narrow beak about 127–152 mm long, which is sharply set off from low forehead by a deep V-shaped groove. Flippers narrow, pointed, somewhat falcate; 2d and 3d digits well developed, others rudimentary. Rostrum of skull about twice length of cranium. Pterygoids meeting in midline throughout their length. Palate with deep lateral grooves. Teeth small, conical, distributed through most of length of jaw, about $\frac{40}{40}$ to $\frac{60}{60}$. Lower Pliocene; all seas. 1 widely distributed species, possibly others.

Delphinus delphis Linnaeus. Common dolphin. Length about 200–245 cm. Upper parts, flippers, and flukes black; underparts white; sides with undulating bands or stripes of gray, yellow, and white. Eye with black ring from which dark streak extends toward snout. All seas. Some regard Pacific dolphin as a distinct species, *D. bairdi*.

Genus *Grampidelphis* Iredale and Troughton. Risso's dolphin. Head without beak; forehead rising almost perpendicularly from tip of upper jaw. Dorsal fin high, falcate, about midway of back. Teeth about $\frac{0}{0}$ to $\frac{0}{7}$, restricted to region of symphysis. Recent; seas, except polar. 1 species.

Grampidelphis griseus (Cuvier). Risso's dolphin. Length about 365–395 cm. Upper parts gray, becoming black on fins and tail; underparts paler to whitish, grading into color of upper parts. Seas, except polar.

Genus *Tursiops* Gervais. Bottlenose dolphins. Head with distinct beak. Rostrum of skull elongated, much longer than cranium, tapering moderately from base to apex. Palate ungrooved; symphysis of mandible short. Teeth large, about $\frac{20}{20}$ to $\frac{25}{25}$. Upper Pliocene; all seas, except polar. Several nominal species, 2 off the United States.

KEY TO SPECIES OF *TURSIOPS*

1. Portion of parietal bone forming border of temporal fossa broad throughout; Atlantic   *T. truncatus*
   Portion of parietal bone forming border of temporal fossa distinctly narrowed ventrally; Pacific   *T. gilli*

Tursiops truncatus (Montague). Atlantic bottlenose dolphin. Length about 335–365 cm. A prominent, falcate dorsal fin about midway of back. Upper parts, flukes, and flippers dark grayish brown; underparts whitish. Lower jaw and edge of upper lip white, with some mottling. Teeth large (about 9 mm in diameter) about $\frac{20}{20}$ to $\frac{22}{22}$. Atlantic.

Tursiops gilli Dall. Pacific bottlenose dolphin. Length about 305–365 cm. A prominent, falcate dorsal fin about midway of back. Upper parts, flukes, and flippers blackish; underparts paler except region from anus to tail which is blackish. Edge of upper lip white. Teeth about $\frac{24}{23}$. Pacific Coast of California.

Genus *Lagenorhynchus* Gray. Striped dolphins. Head with short, indistinct beak; high falcate dorsal fin located about middle of back. Prominent dorsal and ventral ridges behind dorsal fin and anus. Rostrum scarcely exceeding length of cranium; pterygoids meeting in midline. Teeth small, about $\frac{22}{22}$ to $\frac{46}{46}$. Vertebrae about 73–92. Ribs 15–16, of which 6 are bicipital. Recent, all seas except polar. Several species, 2 off the United States.

KEY TO SPECIES OF *LAGENORHYNCHUS*

1. Interorbital breadth less than 200 mm; total vertebrae less than 75   *L. obliquidens*
   Interorbital breadth more than 200 mm; total vertebrae more than 75   *L. acutus*

Lagenorhynchus acutus (**Gray**).   Atlantic
white-sided dolphin. Length about 213–275 cm.
Back black from snout to tip of flukes; belly white,
shading into gray anteriorly; side with a light longitudinal band extending from below dorsal posteriorly toward base of flukes, this band with an
upper white and a lower yellowish portion. Flippers black; a narrow black streak from flipper to
angle of mouth. Teeth about $\frac{30}{30}$ to $\frac{37}{37}$. North Atlantic to Cape Cod.

Lagenorhynchus obliquidens **Gill.**   Pacific
white-sided dolphin. Length about 213–245 cm.
Upper parts greenish black, with variable longitudinal stripes of dull black, gray and white along
sides; underparts white. Teeth about $\frac{28}{28}$ to $\frac{32}{32}$. North
Pacific, south to California.

Genus *Grampus* Gray.   Killer whales. Body
robust, streamlined, with high falcate dorsal fin
about middle of back. Flippers broad and rounded
in outline. Flippers, flukes, and dorsal fin becoming disproportionately large in old males. 1st and
2d, and sometimes 3d, vertebrae united, remainder
free. Teeth large, compressed anteroposteriorly,
$\frac{10}{10}$ to $\frac{12}{12}$ in number. Middle Pliocene; all seas.
1 species.

Grampus orca (**Linnaeus**).   Killer whale. Males
much larger (up to 915 cm) than females (up to
460 cm). Upper parts black, with a lens-shaped
patch of white above and back of eye and a saddle-
shaped patch of gray behind dorsal fin; underparts
white, the white extending upward and back into
black of side anterior to tail. All seas. The Pacific
killer whale is considered a distinct species, *G.
rectipinna* (Cope) by some.

Genus *Pseudorca* Reinhardt.   False killer whale.
Relatively slender, with small falcate dorsal fin
located at about middle of back. Head ending in
rounded snout which projects slightly beyond tip
of lower jaw. Flippers small; tapering, about 10%
of body length. 1st to 6th or 7th cervical vertebrae
unite. Teeth large, powerful, circular in cross

section, and about $\frac{8}{8}$ to $\frac{11}{11}$ in number. Upper
Pliocene; all seas. 1 species.

Pseudorca crassidens (**Owen**).   False killer
whale. Body length about 425–550 cm. Color entirely black. All seas.

Genus *Globicephala* Lesson.   Blackfish. Forehead bulging as a subglobular projection above
upper jaw, because of a cushion of fat on rostrum
in front of blowhole. Flippers very long and narrow; the 2d digit the longest, and the 1st, 4th, and
5th very short. Skull broad and depressed; rostrum
and cranium about equal in length. Teeth about
$\frac{8}{8}$ to $\frac{12}{12}$, confined to anterior half of jaw, small,
conical, curved, sometimes deciduous in old age.
Pleistocene; seas, except polar. Several nominal
species, 3 off the United States.

KEY TO SPECIES OF *GLOBICEPHALA*

1. Flippers relatively long, about 20% of body length;
   color black, with ventral white markings
                                        G. melaena
   Flippers less than 20% of body length; color black,
   with ventral gray markings   2
2. Premaxillae expanded anteriorly to extent of obscuring maxillae in dorsal view; Atlantic
                                        G. macrorhyncha
   Premaxillae not obscuring entire border of maxillae
   in dorsal view; Pacific   *G. scammoni*

Globicephala melaena (**Traill**).   Common blackfish. Length up to 855 cm. Dorsal fin long, low,
falcate, situated just back of head. Flipper about
20% of body length. Color almost entirely black.
Atlantic.

Globicephala scammoni **Cope.**   Pacific blackfish. Length about 460–490 cm. Dorsal fin long,
low, falcate, situated just back of head. Flipper
about 18% of body length. Color entirely black.
North Pacific.

Globicephala macrorhyncha **Gray.**   Short-finned
blackfish. Length about 460–610 cm. Dorsal fin
long, low, falcate, situated just back of head. Flipper about 17% of body length. Color entirely black.
Atlantic Coast of North America.

Genus *Lissodelphis* Gloger.   Right whale dolphins. Dorsal fin absent. Pterygoids separate. Teeth
numerous, about $\frac{44}{44}$. Recent; all oceans. 2 species;
1 off the United States.

Lissodelphis borealis (Peale). Right whale dolphin. Length about 120–245 cm. Form slender; snout slightly produced. Color black, with a white, lanceolate spot on breast extending in a narrow line to tail. Dentition about $\frac{44}{47}$. North Pacific.

Family Phocoenidae. Porpoises. Small cetaceans, 120–235 cm long, with low dorsal fin or none, and without beak. Rostrum of skull shorter than cranium, broad at base. Premaxillae raised into tuberosities anterior to nares. Pterygoids very small and widely separated. Symphysis of mandible very short. 1st to 6th (and sometimes 7th) cervical vertebrae coalesced. Teeth numerous, small, and distributed along nearly whole length of jaw; teeth with compressed, spade-shaped crowns, separated from root by a constricted neck. Miocene; oceans except polar. 3 extinct and 2 living genera, 1 in U.S. waters.

Genus *Phocoena* Cuvier. Porpoises. Dorsal fin present, usually with a row of tubercles along margin. Dentition: $\frac{16}{16}$ to $\frac{27}{27}$. Pterygoids not meeting. Recent; all seas except polar. 2 species along U.S. shores.

KEY TO SPECIES OF *PHOCOENA*

1. Sides uniformly blackish; vertebrae about 68
   P. phocoena
   Flanks sharply white; vertebrae about 98   **P. dalli**

Phocoena phocoena (Linnaeus). Harbor porpoise. Length about 120–180 cm. Form short and thick; head short, with blunt snout. Dorsal fin small, triangular, and about middle of back. Black above, paler gray or white below. Vertebrae about 68. Dentition variable, about $\frac{23}{23}$ to $\frac{27}{27}$. Atlantic and Pacific. In Pacific considered by some a separate species, *P. vomerina* Gill.

Phocoena dalli True. Dall porpoise. Length about 150–235 cm with pointed head and moderately stout body. Dorsal fin moderately high and falcate. Color shiny black, flanks white. Vertebrae about 98. Dentition about $\frac{23}{23}$ to $\frac{27}{27}$, teeth very small. Pacific Coast from Alaska to California. Regarded by some as comprising a separate genus, *Phocoenoides*.

Suborder Mysteceti. Whalebone whales. Mostly large whales, with large heads and large gape. No functional teeth. Roof of mouth with projecting plates of horny ectodermal material (whalebone or baleen) which form strainers for separating food from water taken into mouth. External respiratory opening double; skull symmetrical; rami of mandible arched outward and not forming true symphysis but connected instead by fibrous tissue. Sternum always a single bone. Ribs articulating only with transverse processes of vertebrae. Middle Oligocene; all oceans. 1 extinct and 3 living families.

KEY TO FAMILIES OF MYSTICETI

1. Throat ungrooved (Fig. 6-74A)   **Balaenidae**
   Throat with longitudinal grooves (Fig. 6-74B)   **2**
2. Dorsal fin present; throat grooves numerous
       Balaenopteridae
   Dorsal fin absent; throat grooves 2 (or rarely 4)
       Eschrichtidae

Fig. 6-74. Whalebone whales: *A, Eubalaena; B, Megaptera.* (After Scammon, from Bailey, The Mammals and Life Zones of Oregon, *North Am. Fauna* 55.)

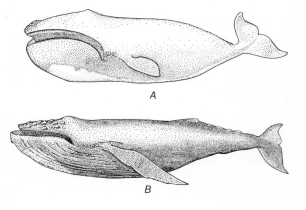

Family Eschrichtidae. Gray whales. Size large; body elongated; head relatively small. Dorsal fin absent; pectoral fin narrow. Throat with 2 longitudinal grooves. Anterior part of skull narrow; premaxillae pinched up in midline and visible from side. Cervical vertebrae free. Humerus relatively long, straight. Recent; North Pacific. 1 genus.

Genus *Eschrichtius* Gray. Gray whale. Characters of family. Recent; North Pacific. 1 species.

***Eschrichtius glaucus* (Cope).** Gray whale. Size moderate; body length about 1070–1370 cm. General color blotched grayish black. Baleen short, coarse, and thick, about 300 mm in length. North Pacific.

Family Balaenopteridae. Large whales with relatively small head; dorsal fin present; throat with numerous longitudinal grooves. Baleen plates relatively short. Cervical vertebrae usually free. Manus tetradactylous. Upper Pliocene; all oceans. 7 extinct and 3 living genera.

KEY TO GENERA OF BALAENOPTERIDAE

1. Flippers very long, narrow, about ¼ length of body; dorsal fin low, humplike   ***Megaptera***
   Flippers relatively shorter, less than ¼ length of body; dorsal fin not humplike   **2**
2. Size very large (up to 3050 cm in length); flippers about ⅐ of body length; dorsal fin relatively very small, with straight edges   ***Sibbaldus***
   Size smaller (up to 2135 cm in length); flippers less than ⅐ of body length; dorsal fin larger, more or less falcate   ***Balaenoptera***

Genus *Balaenoptera* Lacépède. Rorquals. Head relatively small, flat, pointed in front. Body long, slender. Dorsal fin small, falcate. Baleen short, coarse. Cervical vertebrae free. Scapula low and broad, with large acromion and coracoid process. Lower Pliocene; all oceans. 3 species.

KEY TO SPECIES OF *BALAENOPTERA*

1. Baleen yellowish white; body length less than 1070 cm   ***B. acutorostrata***
   Baleen not yellowish white; body length more than 1070 cm   **2**
2. Baleen more or less variegated with stripes of yellowish or purplish; more than 65 throat folds between flippers; body length more than 1830 cm   ***B. physalus***
   Baleen black, becoming white in fibers at free end of plates; less than 65 throat folds between flippers; body length less than 1830 cm   ***B. borealis***

***Balaenoptera physalus* (Linnaeus).** Finback whale. Size large; length 1980–2500 cm. A small,

falcate dorsal fin set far back. About 70–80 throat folds between flippers. Baleen about 610 mm in length, more or less variegated with stripes of yellow or purplish. Upper parts dull browinsh gray; underparts, including lower surface of tail flukes, white. All oceans.

***Balaenoptera acutorostrata* Lacépède.** Pike whale. Size small; length about 610–915 cm. Dorsal fin relatively high. Baleen short, yellowish white. Upper parts grayish black; underparts white, including ventral surface of flukes; inner side of flippers white and a broad white band across outer side. Atlantic and Pacific.

***Balaenoptera borealis* Lesson.** Rorqual. Length about 1525 cm up to 1735 cm. Dorsal fin larger than in *B. physalus* and located about ⅔ of distance from snout to fork of flukes. About 40–62 throat folds between flippers. Baleen black, becoming white in fibers at free end of plates. Color of upper parts variable, but some shade of gray; midventral region white back to about middle of body. All oceans.

Genus *Megaptera* Gray. Humpback whale. Head moderate-sized; dorsal fin low, humplike. Flippers long, slender, scalloped along margins, tetradactylous. Baleen plates short, broad. Cervical vertebrae free. Scapula with acromion and coracoid process absent or rudimentary. Lower Pliocene; all oceans. 1 species.

***Megaptera novaeangliae* (Borowski).** Humpback whale. Body stout; length about 1220–1525 cm. Throat folds 21–36 between flippers. Color variable; blackish above; gray below with whitish throat, often mottled or sometimes pure white under fins and flukes and on belly. Prominent nodules on head and flippers. Baleen plates gray to blackish, and the bristles white to grayish white. All oceans.

Genus *Sibbaldus* Gray. Blue whale. Size very large; dorsal fin small, triangular, and located far to rear. Baleen plates averaging about 325 per rank. Throat grooves extending far back (slightly beyond umbilicus) and averaging about 90 between flippers. Recent; all oceans. 1 species.

***Sibbaldus musculus* (Linnaeus).** Blue whale. Largest known mammal; length about 2285–3050 cm. Upper parts bluish black to gray; underparts yellowish or speckled with white. Atlantic and Pacific.

Family Balaenidae. Right whales. Mostly large, with large head equal to about ¼ or more of total length. Dorsal fin usually absent. Throat smooth, without longitudinal grooves. Baleen very long, slender, elastic. Cervical vertebrae united. Lower Miocene; all oceans. 4 extinct, 3 living genera. 1 off the United States.

Genus *Eubalaena* Gray. Right whale. Moderate to large (length 1070–2130 cm). Dorsal fin absent. Head very large; baleen very long. Manus pentadactylous. Pleistocene; oceans, except Arctic. About 3 species, 2 off the United States.

KEY TO SPECIES OF *EUBALAENA*

1. Atlantic Ocean; size smaller, length about 1070–1675 cm *E. glacialis*
   Pacific Ocean; size larger, length about 1830–2130 cm *E. sieboldi*

Eubalaena glacialis (**Borowski**). Atlantic right whale. Size moderate, body length 1070–1675 cm. Color blackish. Baleen black. North Atlantic and reported from Gulf of Mexico.

Eubalaena sieboldi (**Gray**). Pacific right whale. Size large, body length 1830–2130 cm. Color black or blackish. Baleen blackish, 215–260 cm long. North Pacific.

Order Carnivora. Flesh-eaters. Unguiculate, with claws usually sharp, rarely rudimentary or absent. Pollex and hallux not opposable. Diphyodont and heterodont; teeth rooted. Teeth modified for eating flesh; incisors usually 3 on each side; canine strong, conical, recurved; premolars sectorial; molars sectorial or tuberculate. Last upper premolar and 1st lower molar usually with opposed trenchant edges to form carnassial (flesh-cutting) mechanism. Uterus bicornuate. Mammae abdominal, variable in number. Clavicle often small or incomplete. Radius and ulna distinct. Lower Paleocene; nearly worldwide. 1 extinct (Creodonta) and 2 living suborders.

KEY TO SUBORDERS OF CARNIVORA

1. Feet modified into flippers; toes joined by thick membrane; hind limbs directed backward and used for swimming **Pinnipedia**

Feet normal, used in walking on land; hind limbs directed forward and used in walking **Fissipeda**

Suborder Fissipeda. Land carnivores. Land-living, with body form and limbs adapted for walking on land. Feet used in walking, although webbed in some. 1st digit of forefoot and 1st and 5th digits of hind foot never longer than others. Ears well developed. Middle Paleocene; 1 extinct and 7 living families, 5 in the New World and the United States.

KEY TO FAMILIES OF FISSIPEDA

1. 3 molars in each lower jaw; total teeth 42 (anterior premolars often lost in Ursidae) **2**
   1 or 2 molars in each lower jaw; total teeth less than 42 **3**
2. Tail vestigial; foot plantigrade; hind foot with 5 toes **Ursidae**
   Tail well developed; foot digitigrade; hind foot normally with 4 toes **Canidae**
3. 2 molars in each upper jaw; total teeth 40 **Procyonidae**
   1 molar in each upper jaw; total teeth less than 40 **4**
4. Foot plantigrade; 2 molars in each lower jaw; total teeth 32 or more ***Mustelidae***
   Foot digitigrade; 1 molar in each lower jaw; total teeth 30 or less **Felidae**

Family Canidae. Dogs and foxes. Medium-sized, long-legged, cursorial, principally carnivorous. Tail long and bushy; foot digitigrade; 4 toes on hind feet; usually 5 on forefeet; claws non-retractile. Skull with narrow, elongate rostrum; auditory bullae inflated, with closely appressed paraoccipital processes. Canines long and powerful; carnassial teeth well developed, with sharp cutting edges. Infraorbital foramen opening anteriorly above interspace between 3d and 4th upper premolars. Upper Eocene. About 56 extinct and 12 living genera, 3 in the United States.

KEY TO GENERA OF CANIDAE

1. Postorbital processes thickened, convex dorsally; basilar length of skull more than 147 mm ***Canis***
   Postorbital processes thin, concave dorsally; basilar length of skull less than 147 mm **2**

Fig. 6-75. Fox mandibles: *A, Vulpes fulva; B, Urocyon cinereoargenteus.* (From Burt, *The Mammals of Michigan,* University of Michigan Press.)

A   B

2. Adult skull with sagittal crest or with temporal ridges low; ventral margin of mandible without prominent step (Fig. 6-75A)   *Vulpes*
   Adult skull with high, lyrate temporal ridges; ventral margin of mandible with a distinct step (Fig. 6-75B)   *Urocyon*

Genus *Canis* Linnaeus. Dogs. Relatively large, with large, erect ears (except in some domestic types) and long, bushy tails. Pelage along dorsum from nape to behind shoulders elongated to form a mane which is erectile and conspicuous in anger. Skull with postorbital processes smoothly rounded above and evenly decurved; sagittal crest long, prominent. Dentition: $\frac{3}{3}, \frac{1}{1}, \frac{4}{4}, \frac{2}{3}$. Upper Pliocene; worldwide. 3 native species in the United States.

KEY TO SPECIES OF *CANIS*

1. Size large; greatest length of skull usually more than 250 mm   *C. lupus*
   Size smaller; greatest length of skull usually less than 250 mm   2
2. Body large, robust; feet large; zygomatic breadth usually more than 105 mm; rostral breadth posterior to canines usually more than 32 mm   *C. niger*
   Body smaller, more slender; feet smaller; zygomatic breadth usually less than 105 mm; rostral breadth posterior to canines usually less than 32 mm   *C. latrans*

Canis latrans Say. Coyote. Relatively small; body length about 810–940 mm; hind foot about 150–250 mm. Tail about 35–40% of body length. General color grayish or buffy grayish above, with varying amounts of black-tipped hairs; underparts whitish; feet, legs, and ears more or less reddish. Skull relatively long and slender, with slender ros-

trum and moderately spreading zygomata. Sometimes difficult to distinguish from *C. niger* in Texas and Oklahoma. From Central America through Mexico, western United States, and western Canada to Alaska; range in the United States expanding eastward with clearing of forests in recent times as far as New York.

Canis lupus Linnaeus. Gray wolf. Relatively large; body length about 1090–1220 mm; hind foot about 222–323 mm. Tail about 35% of body length. Color variable, frequently a mixture of gray, brown, and black above, whitish below; sometimes blackish above, paler below; nearly white in arctic populations. Skull large and heavy; 1st upper molar with indistinct cingulum on outer side. Original range circumpolar, widely distributed in Eurasia and throughout most of North America south to southern Mexico. A few remain in the United States today in parts of Rocky Mountain and Cascade–Sierra Nevada chains and in areas bordering Lake Superior.

Canis niger Bartram. Red wolf. Size large; body length about 1065–1245 mm; hind foot about 210–254 mm. Tail about 35% of body length. Color variable; upper parts usually a mixture of buff, gray, and black and more or less heavily overlaid with black, underparts whitish or buffy; upper parts sometimes black or brownish black and underside of neck and pectoral region similar, remainder of underparts mixed with white. Skull differing from that of *C. latrans* in larger size, higher cranium, deeper rostrum, more widely spreading zygomata and larger auditory bullae; differing from that of *C. lupus* in smaller, less massive size, more slender form, more deeply cleft crowns of molar teeth, and more prominent cingulum on outer edge of first upper molar. Formerly from peninsular Florida and Georgia west to central Texas and central Oklahoma and north through Mississippi embayment to northwestern Indiana. Now rare west of Mississippi River and extirpated east of it.

Genus *Vulpes* Oken. Red foxes. Small to moderate-sized, with large ears and long, bushy tails. Skull with postorbital processes thin and concave on dorsal surface; temporal ridges closely parallel or uniting to form sagittal crest; ventral margin of mandible without a distinct step. Dentition: $\frac{3}{3}, \frac{1}{1}, \frac{4}{4}, \frac{2}{3}$. Pleistocene; Eurasia, North Africa, North America. 3 nominal species in the United States.

KEY TO SPECIES OF *VULPES*

1. Larger; hind foot more than 135 mm; zygomatic breadth more than 63 mm; tip of tail usually white; feet black   **V. *fulva***
   Smaller; hind foot less than 135 mm; zygomatic breadth less than 63 mm; tip of tail black; feet not black   **2**
2. Ears larger; range west of Rocky Mountains east across southern New Mexico to trans-Pecos Texas   **V. *macrotis***
   Ears smaller; range east of Rocky Mountains, from Texas Panhandle to southern Alberta   **V. *velox***

Vulpes fulva (Desmarest). Red fox. Body length about 560–635 mm; hind foot about 150–172 mm in males, slightly smaller in females. Tail about 65% of body length. Color usually rusty reddish; back of ears and front of legs usually blackish; tail mixed buffy and black, abruptly and conspicuously tipped with white. Occurring also in "cross," "silver," and "black" phases, which are progressively blacker. Skull relatively large; ventral margin of mandible without a distinct step. From Alaska throughout all but extreme north of Canada; south in Cascade–Sierra Nevada chain to central California and in Rocky Mountain chain to southern New Mexico; south in plains to northern South Dakota; east of plains, south to eastern Texas, southern Alabama, and western South Carolina. Introductions as game have confused original distribution. It has been suggested that this is conspecific with *V. vulpes* of Eurasia.

Vulpes macrotis Merriam. Desert kit fox. Body length about 380–525 mm; hind foot about 122–140 mm in males and slightly smaller in females. Tail about 60–65% of body length. Ears large, tail round in cross section. General color grayish above; sides and chest buffy; underparts white; tip of tail black. Skull small; auditory bullae relatively large and close together; anterior palatine foramina extending posteriorly past alveoli of upper canines. Deserts of southwestern United States and adjacent Mexico from trans-Pecos Texas west to southern and central California, north to southeastern Oregon and northwestern Utah.

Vulpes velox (Say). Plains kit fox. Body length averaging about 565–610 mm in males. Hind foot averaging about 130 mm in males. Tail about 50% of body length. Rather similar to *V. macrotis*, with which it may be conspecific. Differs in much smaller ears, smaller auditory bullae, and relatively shorter skull. Range east of Rocky Mountain chain, from Texas Panhandle north through plains to southern Canada. Range allopatric to that of *V. macrotis*.

Genus *Urocyon* Baird. Gray foxes. Moderate-sized, with long legs and relatively short heads. A crest of hairs along top of tail making the brush triangular in cross section. Skull with relatively short rostrum, enlarged braincase, and with temporal ridges far apart, forming a lyre-shaped pattern and never uniting to form a sagittal crest; dorsal surfaces of postorbital processes with deep pits; upper incisors simple, unlobed; lower outline of mandible with a distinct step. Dentition: $\frac{3}{3}, \frac{1}{1}, \frac{4}{4}, \frac{2}{2}$. Pleistocene; North America. 1 species in the United States.

Urocyon cinereoargenteus (Schreber). Gray fox. Body length about 530–740 mm; hind foot about 120–150 mm in males and slightly smaller in females. Tail about 50–55% of body length. Upper parts blackish gray, darkest along midline; underparts reddish brown, becoming whitish on throat and middle of abdomen. Tail with sharply defined dorsal stripe which ends in a black tip. Feet, legs, back of ears, and sides of neck yellowish. From Central America through Mexico, throughout eastern half of the United States and in West north to northwestern Oregon, central Nevada, and northern Colorado. Populations of Santa Barbara Islands, California, have been treated as a separate species (*U. littoralis*) by some.

Family Ursidae. Bears. Large, generalized feeders, with 5-toed, plantigrade feet and very short tails. Lips loose and protrusible; digits clawed, longer on front feet; soles naked; eyes and ears small. Skull with auditory bullae depressed and little inflated; paraoccipital processes large, broad, and independent of bullae; postorbital processes well developed. Jaw teeth bunodont; carnassials not developed; the 3 premolars above and below rudimentary, often deciduous. Dentition: $\frac{3}{3}, \frac{1}{1}, \frac{4}{4}, \frac{2}{3}$. Middle Miocene; Eurasia, North and South America. About 8 extinct and 6 living genera, 1 in the United States.

Genus *Ursus* Linnaeus. Black and grizzly bears. Feet broad, completely plantigrade; each toe with long, compressed, moderately curved, nonretractile claw. Palms and soles naked. Ears moderate, erect, rounded, hairy. Fur usually long, soft, and shaggy. Skull more or less elongated; orbits small and incomplete posteriorly; palate extending considerably behind plane of last molars. The anterior premolars above and below single-rooted, rudimentary, and frequently absent, the 2d rarely present in adults; crowns of molars longer than broad. Dentition: $\frac{3}{3}, \frac{1}{1}, \frac{4}{4}, \frac{2}{3}$. Lower Pliocene; Eurasia, North America. 2 species in the United States.

KEY TO SPECIES OF *URSUS*

1. Claws on front feet more than 55 mm; mane present; last upper molar more than 1.5 times length of 1st upper molar ***U. horribilis***
   Claws on front feet less than 55 mm; mane absent; last upper molar not more than 1.5 times length of 1st upper molar ***U. americanus***

Ursus horribilis **Ord.** Grizzly bear. Size large, up to 215 cm in length and 850 lb in weight. General color brownish to blackish; dorsal hairs usually white-tipped, giving grizzled effect. A noticeable mane between shoulders. Skull long and narrow, with long rostrum and concave facial contour. Last upper molar large, 31 mm or more in length. Once ranged over most of western North America; now reduced to Alaska, western Canada, and Rocky Mountain chain in the United States south to northern New Mexico.

Ursus americanus **Pallas.** Black bear. Size moderate, up to about 183 cm in length and 400 lb in weight. Color black or brown; face brown. Skull relatively wide; rostrum short; facial contour flat. Last upper molar small, less than 31 mm in length. From Alaska across Canada; once ranged throughout most of the United States; now extirpated in many parts of its former range.

Family Procyonidae. Raccoons. Partially arboreal, with plantigrade or semiplantigrade feet, each with 5 toes with nonretractile or semiretractile claws. Tail moderate to long, more or less bushy, usually ringed with alternating light and dark colors. Skull with rounded and well-inflated braincase and short rostrum; paraoccipital processes separate from bullae; cheek teeth somewhat bunodont. Dentition: $\frac{3}{3}, \frac{1}{1}, \frac{4}{4}, \frac{2}{2}$ in American genera. Lower Miocene; North and South America, Asia.

About 11 extinct and 8 living genera, including panda (*Ailurus*) and giant panda (*Ailuropoda*) of Asia. 3 genera in the United States.

KEY TO GENERA OF PROCYONIDAE

1. Snout long, mobile, extending forward well beyond lower lip; tail markedly tapering ***Nasua***
   Snout short, normal, not projecting markedly beyond lower lip; tail not markedly tapering **2**
2. Dark rings on tail complete; posterior border of palate extending far behind posterior border of last molars (Fig. 6-76A) ***Procyon***
   Dark rings on tail incomplete below; posterior border of palate not extending markedly behind posterior border of last molars (Fig. 6-76B) ***Bassariscus***

Fig. 6-76. Ventral view of procyonid skulls: *A, Procyon lotor; B, Bassariscus astutus.* (From Hall, *Mammals of Nevada,* University of California Press.)

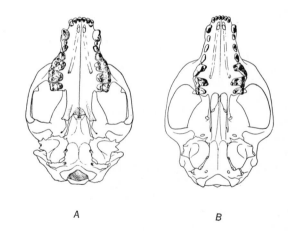

A   B

Genus *Bassariscus* Coues. Ringtails. Slender-bodied, very long tailed, with pointed nose, large eyes, and prominent ears. Claws semiretractile. Pelage soft. Skull with braincase flattened, but well expanded laterally; postorbital processes prominent; zygomata relatively slender; palate extending only slightly posterior to posterior border of molars; auditory bullae well inflated. Last upper premolar sectorial. Dentition: $\frac{3}{3}, \frac{1}{1}, \frac{4}{4}, \frac{2}{2}$. Upper Miocene; North America. 1 species in the United States.

Bassariscus astutus (**Lichtenstein**). Ringtail. Size moderate; body length about 370–400 mm; hind foot about 57–70 mm. Tail about 100–170% of body length. Upper parts buffy, with numerous dark-tipped hairs; eye usually ringed with blackish, with supraorbital, suborbital, and subauricular patches of white or buff; underparts white or tinged with pale buff. Tail normally with 7 light and 8 dark rings, the dark rings incomplete ventrally. Southern Mexico north into southwestern United States from eastern and central Texas to central New Mexico, southwestern Wyoming, and southwestern Oregon.

Genus *Procyon* **Storr.** Raccoons. Terrestrial and semiarboreal, generalized feeders, with robust bodies, rather short tails, short, pointed muzzles, and moderate ears. Feet plantigrade; soles naked, smooth; digits very long; claws nonretractile. Face with a black mask. Skull broad, with broad rostrum and broad braincase; auditory bullae large, inflated on inner side. Molar teeth with moderately high crowns and prominent cusps. Last upper premolar not carnassial, but subquadrate, about as long as broad, with 5 principal cusps. Dentition: $\frac{3}{3}, \frac{1}{1}, \frac{4}{4}, \frac{2}{2}$. Upper Pliocene; North and South America. 1 species in the United States.

Procyon lotor (**Linnaeus**). Raccoon. Body length about 430–615 mm in males and 385–500 in females. Hind foot about 96–138 mm in males and 83–123 mm in females. Tail about 42–52% of body length. Underfur soft, dense; many long coarse guard hairs. General color of upper parts varying geographically from pale gray to blackish, more or less suffused with buff and with varying numbers of black-tipped guard hairs; face with a sharply delimited black mask; sides of muzzle, lips, and chin white. Underparts with long grayish or buffy overhairs thinly overlying dense brownish underfur. Tail with 5–7 conspicuous black rings and a black tip alternating with broader grayish or buffy rings, the black rings less sharply defined below. Panama, northward throughout Mexico and the United States to southern Canada.

Genus *Nasua* **Storr.** Coatis. Mainly arboreal, with somewhat elongated bodies, stout, plantigrade limbs, long, nonprehensile, usually tapering tails and small ears. Snout long, mobile, extending well beyond lower lip. Skull with facial portion elongated and narrow; braincase elevated above rostrum; palate extending far beyond plane of last molars. Canines compressed laterally, with trenchant anterior and posterior edges; molars small. Dentition: $\frac{3}{3}, \frac{1}{1}, \frac{4}{4}, \frac{2}{2}$. Pleistocene; South America, southern North America. 1 species reaching the United States.

Nasua narica (**Linnaeus**). Coati. Body length about 505–635 mm. Hind foot about 115–135 mm in males and smaller in females. Tail about 90–110% of body length. General color grizzled brownish black above and brownish black mixed with buffy below. Muzzle and upper and lower border of eye white; ear edged with buffy white. Tail with indistinct blackish bands. Northern South America through Mexico, barely reaching the U.S. border from southern Texas to southern Arizona.

Family Mustelidae. Mustelids. Variable group of generalized, fossorial, or semiaquatic highly predaceous forms with 1 molar in upper jaw and 2 in lower. In upper molar, the inner, tubercular portion always longer anteroposteriorly than the sectorial, external portion. Palate generally much produced behind last molars. Postglenoid process of cranium usually considerably curved over glenoid fossa and holding tightly the condyle of mandible. Lower Oligocene; North and South America, Eurasia, Africa. About 47 extinct and 29 living genera, 9 in the United States.

KEY TO GENERA OF MUSTELIDAE

1. Total teeth 38   **2**
   Total teeth less than 38   **3**
2. Body slender; feet digitigrade   *Martes*
   Body robust; feet plantigrade   *Gulo*
3. Total teeth 36   *Lutra*
   Total teeth less than 36   **4**
4. 2 incisors in each lower jaw; feet fully webbed; marine   *Enhydra*
   3 incisors in each lower jaw; feet not fully webbed; terrestrial   **5**
5. Total teeth 32   *Conepatus*
   Total teeth 34   **6**
6. Palate nearly on a line with posterior border of last upper molars; colors a combination of black and white; tail hairs very long   **7**
   Palate behind upper molars; colors never a simple combination of black and white; tail hairs moderate or short   **8**
7. Infraorbital canal opening above anterior half of 4th

Fig. 6-77. Skulls of skunks: A, Spilogale gracilis; B, Mephitis mephitis. (From Hall, Mammals of Nevada, University of California Press.)

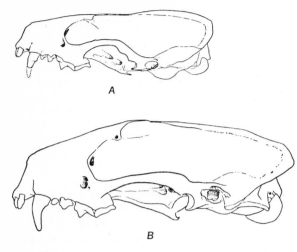

Fig. 6-78. Skulls of mustelids: A, Taxidea taxus; B, Mustela vison. (From Hall, Mammals of Nevada, University of California Press.)

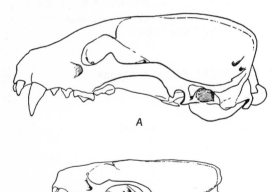

upper premolar (Fig. 6-77A); size smaller; **4** broken white stripes on back **Spilogale**
Infraorbital canal opening above posterior half of 4th upper premolar (Fig. 6-77B); size larger; dorsal pattern of 2 continuous white stripes on black, or solid black **Mephitis**

8. Body robust; size large; basilar length of skull more than 95 mm; facial angle of skull steep (Fig. 6-78A) **Taxidea**
  Body slender; size smaller; basilar length of skull less than 75 mm; facials angle of skull slight (Fig. 6-78B) **Mustela**

Genus *Mustela* **Linnaeus.** Weasels. Short-legged, long, slender-bodied, with short, rounded ears. Skull with slight facial angle; auditory bullae greatly inflated and with paraoccipital processes closely appressed to bullae; palate behind upper molars. Dentition: $\frac{3}{3}, \frac{1}{1}, \frac{3}{3}, \frac{1}{2}$ (premolars $\frac{2}{2}$ in a South American species). Some changing to white pelage in winter. Males much larger than females in some species. Upper Miocene; North and South America, Eurasia, Africa. Numerous species, 5 in North America and the United States.

KEY TO SPECIES OF *MUSTELA*

1. Length of upper tooth row less than 20 mm in males and 17.8 mm in females **2**
  Length of upper tooth row more than 20 mm in males and more than 17.8 mm in females **4**
2. Postglenoid length of skull less than 47% of condylobasal length **M. frenata**
  Postglenoid length of skull more than 47% of condylobasal length **3**
3. Tail without a black tip (at most a few black hairs at extreme tip); in females mastoid breadth usually greater than breadth of braincase **M. rixosa**
  Tail with a conspicuous black tip; in females mastoid breadth usually less than breadth of braincase **M. erminea**
4. Abdomen all white; face with blackish mask **M. nigripes**
  Abdomen dark brown; face uniformly brown, without mask **M. vison**

Mustela erminea **Linnaeus.** Short-tailed weasel. Size small to medium; body length about 150–230 mm in males and 125–190 mm in females; hind foot about 26–44 mm in males and 23–33 in females. Tail about 30–45% of body length and with a distinct black tip. Upper parts brown; underparts usually whitish from chin to inguinal region, but sometimes interrupted by brown of upper parts encircling body in abdominal region. Winter pelage white, with black-tipped tail, in northern part of range. Soles of feet densely haired in winter, and with only a relatively small area of pads exposed in summer. Skull with long braincase and short precranial portion. Circumpolar in Northern Hemisphere; in North America throughout Alaska and Canada and south in the United States both west and east of central grasslands; in West through and west of Rocky Mountain chain to northern New Mexico and central California; in east south to a line from northern Iowa to Maryland.

Mustela rixosa (**Bangs**). Least weasel. Size small; body length about 150–165 mm in males and 140–150 mm in females; hind foot about 21–31 mm in males and 18–26 mm in females. Tail 25% or less of body length, and without a black tip. Upper parts brown; underparts with at least some white in thoracic region; winter pelage all white in northern part of range and in some individuals in southern part. Skull with long braincase and short precranial portion; greatest breadth of braincase less than mastoid breadth. Circumpolar, from Norway and Switzerland across Siberia and across northern North America; south in the United States to northern Montana and to a line from southwestern Nebraska to southern Ohio; in Appalachian chain from Pennsylvania to western North Carolina.

Mustela frenata **Lichtenstein.** Long-tailed weasel. Size large; greatly variable geographically; body length about 230–270 mm in males and 200–230 mm in females; hind foot about 38–60 mm in males and 29–46 mm in females. Tail about 40–70% of body length, with distinct black tip. Upper parts brown; underparts pale, tinged with buffy or yellowish and continuous from chin to inguinal region. Face with white or yellowish markings in some populations. Changing in winter to all white except black tail tip in about northern half of the United States northward. Skull with long precranial portion. From southern Canada south through all of the United States, except Sonoran Desert, and southward to northern Bolivia in South America.

Mustela vison **Schreber.** Mink. Size large; body length about 330–510 mm in males and about $\frac{1}{4}$ smaller in females; hind foot about 56–72 mm in males and about 54–60 mm in females. Tail about 40–57% of body length, somewhat bushy. Color dark brown, paler below; white spots sometimes present on chin and throat. Underfur soft, dense; overhairs coarse. With strong anal scent glands. Skull with flattened auditory bullae; last upper molar dumbbell-shaped and smaller than preceding tooth. Alaska and most Canada, south throughout eastern half of the United States to Florida; south in the West to northwestern Texas, northern New Mexico, northern Nevada, and central California.

Mustela nigripes (**Audubon and Bachman**). Black-footed ferret. Size large; body length about 380–460 mm; hind foot about 60–70 mm in males and smaller in females. Tail about $\frac{1}{3}$ of body length. Ears relatively large. General color buffy, with indistinct patch of brownish over middle of back; feet and end of tail black and a black mask across face and eyes. Skull large and massive, very broad between orbits and deeply constricted behind postorbital processes. Central grasslands from western Texas and northern Arizona to southern Canada, mostly east of Rocky Mountain chain and west of a line from north central Texas to central North Dakota.

Genus *Martes* Pinel. Martens. Largely arboreal, with long, more or less slender bodies, short limbs, moderately long, more or less bushy tails. Muzzle pointed; eyes large; ears conspicuous, broad, furred inside and outside. Feet rounded, with 5 short toes and short, curved, sharp-pointed claws; soles densely furred between the naked pads. Outer fur long and glossy; underfur dense. Skull elongated and depressed; facial angle slight; palate behind last molars. Dentition: $\frac{3}{3}, \frac{1}{1}, \frac{4}{4}, \frac{1}{2}$. Lower Pliocene; Eurasia, North America. 2 species in the United States.

KEY TO SPECIES OF *MARTES*

1. Yellowish brown, with orange on throat and chest; size smaller; greatest length of skull less than 95 mm .... *M. americana*
   Dark brown, without orange on throat or chest; size larger; greatest length of skull more than 85 mm .... *M. pennanti*

Martes americana (**Turton**). Marten. Relatively slender, weasellike, with short legs and large feet. Body length about 405–430 mm in males and 355–380 mm in females. Tail about half of body length, bushy. Hind foot about 90–95 mm in males and about 75–83 mm in females. Upper parts yellowish brown; underparts paler; throat and chest yellowish orange. Skull with short rostrum and well-inflated auditory bullae. From Alaska across Canada, south into New England states, and in West follows Rocky Mountain chain south to northern New Mexico and Cascade–Sierra Nevada chain south to central California.

Martes pennanti (**Erxleben**). Fisher. Body relatively robust, weasellike, with short legs and large feet. Size large; body length about 505–635 mm; hind foot about 125–128 mm in males. Tail long and bushy, about 62% of body length. Upper parts dark brown to blackish, paler below; many hairs tipped with whitish. Skull massive, with well-inflated auditory bullae. Across Canada and south into New England states and New York; in West south through western Montana and south through Cascade–Sierra Nevada chain to central California.

Genus *Gulo* Pallas. Wolverines. Stout-bodied, with stout limbs and large, powerful, subplantigrade feet, with large, much curved, sharp-pointed claws. Soles of feet, except pads of toes, covered with thick, bristly hairs. Ears small; eyes small; tail short, thick, and bushy. Fur full and long. Skull with steep facial angle; palate behind upper molars. Crowns of teeth robust; upper molar much smaller than carnassial; lower carnassial large; 3d upper incisor large, caninelike. Dentition: $\frac{3}{3}, \frac{1}{1}, \frac{4}{4}, \frac{1}{2}$. Pleistocene; Eurasia, North America. 1 species in North America and the United States.

Gulo luscus (**Linnaeus**). Wolverine. Size large; body length about 735–815 mm; hind foot of males about 200 mm. Tail short and bushy, about 25% of body length. Pelage long and shaggy. Forelegs, back and end of tail dark blackish brown; forehead grayish; 2 broad yellowish-brown bands beginning on side of shoulders and passing low on sides of hips and joining across back and base of tail. From Alaska across Canada; in the United States today only in Cascade–Sierra Nevada chain of Washington and California and in Montana. Range has retreated northward in recent times.

Genus *Taxidea* Waterhouse. Badger. Fossorial, with robust, depressed body and short, strong, subplantigrade limbs, short tail, and short ears. Forefeet with long, heavy claws. Skull very wide in occipital region; occipital crest greatly developed; facial angle steep; palate behind upper molars. Upper molar about same size as carnassial, triangular, with apex turned backward, often with

cusps in transverse rows. Dentition: $\frac{3}{3}, \frac{1}{1}, \frac{3}{3}, \frac{1}{2}$. Upper Pliocene; North America. 1 living species.

Taxidea taxus (**Schreber**). Badger. Size large; body length about 455–560 mm; hind foot about 87–150 mm. Tail short, about 25% of body length. Upper parts grizzled yellowish brown; underparts buffy; legs, feet, top of head, ears, and small areas on cheeks blackish; a white stripe from nose backward between eyes onto shoulders. From northern Mexico through western United States to southern Canada; east to central Texas, eastern Kansas, northwestern Ohio, and Michigan.

Genus *Mephitis* Geoffroy and Cuvier. Striped skunks. Terrestrial, somewhat fossorial, with generalized food habits. Body moderately elongated; head small; ears short; limbs moderate, subplantigrade; feet with blunt, slightly curved claws. Tail long, with very long hairs. Anal glands highly developed. Color pattern of black and white. Skull highly arched and deepest in frontal region; posterior border of palate nearly on line with posterior border of last molars; mastoid bullae not inflated. Upper molar larger than carnassial, squarish, broader than long. Infraorbital canal opening above posterior half of 4th upper premolar. Dentition: $\frac{3}{3}, \frac{1}{1}, \frac{3}{3}, \frac{1}{1}$. Pleistocene; North America. 2 species; both in the United States.

KEY TO SPECIES OF *MEPHITIS*

1. Back usually with a white stripe, divided posteriorly; anterior palatine foramina usually small and narrow; auditory bullae not markedly inflated
   M. mephitis
   Back usually either wholly black or wholly white; anterior palatine foramina large and rounded; auditory bullae markedly inflated *M. macroura*

Mephitis mephitis (**Schreber**). Common striped skunk. Body length averaging 345–460 mm in males and 305–415 mm in females; hind foot about 60–90 mm in males and about 60–80 mm in females. Tail length variable geographically, about 44–107% of body length. Color pattern usually black, with an anterior, median white stripe dividing into 2 lateral stripes and a narrow median white stripe from nose to nape. Tail hairs white at bases. Color variable, from white stripes reduced or absent to white stripes greatly expanded. Skull long and relatively narrow interorbitally; interpterygoid fossa broad; palate ending squarely or with a median notch or spine; auditory bullae not greatly inflated; anterior palatine foramina usually small and narrow. From Mexico throughout the United States and into Canada.

Mephitis macroura **Lichtenstein.** Hooded skunk. Body length averaging about 312 mm in males and females. Hind foot about 65–73 mm in males. Tail long, averaging about 115% of body length. Hairs on nape elongated and directed laterally, forming a neck ruff or hood. Color variable, but generally upper parts chiefly white, under parts black, or upper parts nearly all black with narrow lateral stripes and undersurface of tail white. Skull short and broad; interpterygoid fossa narrow, U-shaped; palate without notch or spine; anterior palatine foramina large and rounded; auditory bullae markedly inflated. From Central America north through Mexico, barely reaching the United States from trans-Pecos Texas to southeastern Arizona.

Genus *Spilogale* Gray. Spotted skunks. Small, rather slender, terrestrial, semiarboreal, and somewhat fossorial, with generalized food habits. Similar in general features to *Mephitis.* Color pattern of 6 dorsal and lateral white stripes (variously interrupted by black) on black background. A white facial spot and a white spot anterior to and below each ear. Tip of tail usually with at least some all-white hairs; tail hairs unicolor (black or white) to bases. Skull small, flattened, with rostrum only slightly depressed below plane of braincase; auditory bullae inflated; mastoid and paraoccipital processes obsolete or very small; postorbital processes well developed; infraorbital canal opening above anterior half of 4th upper premolar; anteroposterior diameter of upper molar less than transverse diameter. Dentition: $\frac{3}{3}, \frac{1}{1}, \frac{3}{3}, \frac{1}{2}$. Upper Pliocene; North America. 2 nominal, allopatric species; their relationships poorly known and possibly both belonging to same species; relationships of these to named Mexican forms unknown at present.

KEY TO SPECIES OF *SPILOGALE*

1. Range west of a line from central Texas to north central Colorado .... **S. gracilis**
   Range east of a line from central Texas to north central Colorado .... **S. putorius**

Spilogale putorius (**Linnaeus**). Eastern spotted skunk. Body length averaging 240–308 mm in males and 230–303 mm in females; hind foot about 37–51 mm in males and 38–45 mm in females, variable geographically in size. Tail about 48–66% of body length. Where this species approaches range of western species in Texas and Colorado distinguished by larger size, very small facial and preauricular white patches, restricted white tail tip, and (in Texas at least) by smaller and less inflated mastoid bullae, which have flat ventral surfaces. Eastern United States from peninsular Florida north to Ohio River and southern Pennsylvania, and, west of Mississippi River, north to northern Minnesota and northern South Dakota; west to central Texas, eastern Colorado, and eastern Wyoming.

Spilogale gracilis **Merriam.** Western spotted skunk. Body length averaging 238–316 mm in males and 210–280 mm in females; hind foot about 40–54 mm in males and 40–47 mm in females. Tail about 48–65% of body length. Where this species approaches range of eastern species in Texas and Colorado distinguished by smaller size, larger, more prominent facial and preauricular white spots, much more extensive white tail tip, and (in Texas at least) by more inflated mastoid bullae, which are convex on their ventral surfaces. Western United States, west of a line from central Texas to southeastern Wyoming and north to central Idaho, southern Washington, and along the Washington coast to the southwestern part of British Columbia.

Genus *Conepatus* Gray. Rooter skunks. Robust, terrestrial, with generalized (principally insectivorous) food habits. Soles of hind feet naked; claws of forefeet long and stout. Snout long, projecting well beyond lower jaw and with thickened, naked pad on upper side. Anal glands well developed. Color a combination of blacks and whites. Skull large, with moderately inflated braincase and marked interorbital constriction. Zygomata relatively slender, widely spreading; palate ending well behind plane of last molars; interpterygoid space relatively narrow; auditory bullae moderate. Upper molar relatively large, subquadrate. Dentition: $\frac{3}{3}, \frac{1}{1}, \frac{2}{3}, \frac{1}{2}$. Pleistocene; North America, South America. 2 nominal, allopatric species in the United States.

KEY TO SPECIES OF *CONEPATUS*

1. White dorsal stripe narrower; size larger; body length usually more than 425 mm   *C. leuconotus*
White dorsal stripe wider; size smaller; body length usually less than 425 mm   *C. mesoleucus*

Conepatus mesoleucus (**Lichtenstein**). Western rooter skunk. Body length about 290–425 mm in males and 315–375 mm in females. Hind foot about 65–75 mm in males and 64–70 in females. Tail about 51–79% of body length. Color black, with a single wide dorsal stripe from top of head to base of tail, the stripe truncate anteriorly and approximately the same width throughout. Tail white throughout except for a few scattered black hairs below. From South America north through central and western Mexico into the United States from central Texas west to southern Arizona and north through New Mexico to southeastern Colorado.

Conepatus leuconotus (**Lichtenstein**). Tamaulipan rooter skunk. Body length about 440–510 mm in males and 420–465 mm in females. Hind foot about 70–90 mm in males and 65–80 mm in females. Tail about 46–80% of body length. Color black, with a single white dorsal stripe, which is wedge-shaped anteriorly and reduced in width or absent on rump. Dorsal side of tail white; ventral side black toward base and white toward tip. East coast of Mexico north through brushlands of southern Texas. Probably conspecific with *C. mesoleucus*.

Genus *Lutra* Brisson. River otters. Highly aquatic, with elongated body, short limbs, short, broad, webbed pentadactyl feet. Skull broad and depressed, facial portion very short; braincase large; palate behind upper molars. Upper molar squarish, cuspidate, broader than long. Dentition: $\frac{3}{3}, \frac{1}{1}, \frac{4}{3}, \frac{1}{2}$. Lower Pliocene; North and South America, Eurasia, Africa. 1 species in the United States.

Lutra canadensis (**Schreber**). River otter. Size large; body length about 63–76 cm; hind foot about 112–133 mm. Tail long, sleek, about 65% of body length, tapering from a thick base. Pelage short, dense, and soft. Upper parts rich, dark, chocolate brown; underparts slightly paler. From Alaska and Canada throughout the United States except arid regions from central Texas west to southern Arizona and of southern California and southern Nevada.

Genus *Enhydra* Fleming. Sea otter. Marine, with fully webbed toes. Forefeet small, with naked palms and short, compressed claws; hind feet very large, depressed, and finlike, with flattened phalanges. Skull broad and depressed. Upper molar large and quadrate, with inner tubercular portion much expanded anteroposteriorly, with rounded massive crown and blunt cusps. Dentition: $\frac{3}{2}, \frac{1}{1}, \frac{3}{3}, \frac{1}{2}$. Upper Pliocene; shores of northern Pacific.

Enhydra lutris (**Linnaeus**). Sea otter. Size large; adults reaching body length of about 102–112 cm. Hind foot about 220 mm. Tail moderately haired, about 27–32% of body length. Color brownish black; hairs white at tips; head and neck paler than back. Pacific Coast, southern California to Washington. Highly aquatic.

Family Felidae. Cats. Highly carnivorous, usually long-tailed. Feet digitigrade, with 5 toes in front and 4 behind. Dentition highly specialized for eating flesh. Carnassials highly developed, upper with 3-lobed blade, lower without talon or inner cusp; other jaw teeth reduced in size and number. Only 1 upper and 1 lower molar in living forms. Upper Eocene; worldwide. About 30 extinct genera. Living species have been lumped in as few as 3 genera by some; U.S. cats are customarily referred to 2 genera, which are, however, lumped by some.

KEY TO GENERA OF FELIDAE

1. Tail long, more than 30% of body length; premolars $\frac{3}{2}$; total teeth 30   *Felis*
Tail short, less than 30% of body length; premolars $\frac{2}{2}$; total teeth 28   *Lynx*

Genus *Felis* Linnaeus. Cats. Moderately long tailed, with sharp, completely retractile claws. Includes largest and smallest living members of family. Eyes large; ears moderate; tongue thickly covered with sharp-pointed, recurved, horny papillae. Skull typically short and rounded, the facial portion short and broad; zygomata wide and strong; auditory bullae large, rounded, and smooth. Dentition: $\frac{3}{3}, \frac{1}{1}, \frac{3}{2}, \frac{1}{1}$. Lower Pliocene; worldwide. 5 native species in the United States.

KEY TO SPECIES OF *FELIS*

1. Size large; greatest length of skull more than 170 mm   **2**
   Size moderate or small; greatest length of skull less than 170 mm   **3**
2. Color pattern plain, unspotted, in adults; dorsal profile of skull convex, without sagittal concavity   ***F. concolor***
   Color pattern spotted (rosettes); dorsal profile of skull with sagittal concavity   ***F. onca***
3. Color pattern unspotted (red or gray)   ***F. yagouaroundi***
   Color pattern spotted   **4**
4. Size larger; greatest length of skull more than 125 mm   ***F. pardalis***
   Size smaller; greatest length of skull less than 125 mm   ***F. wiedi***

Felis onca (**Linnaeus**). Jaguar. Largest American cat; body length 1115–1475 mm. Tail about 45% of body length. Dorsal color tawny, with a median chain of black spots bordered on each side by 5 longitudinal rows of black rosettes; crown and neck tawny, spotted with black; tail heavily spotted with black. Underparts buffy white, spotted with black; outside of legs tawny, spotted with black. Skull relatively elongate; rostrum short and broad, zygomatic arches wide and strong. From Patagonia in South America north through Mexico to the U.S. border; rarely in the United States today in Arizona, southwestern New Mexico, and southern Texas. Once ranged north at least to central Texas.

Felis pardalis **Linnaeus**. Ocelot. Moderate-sized; body about 685–970 mm; hind foot about 155–180 mm in males. Tail about 35–50% of body length. Upper parts buffy grayish, heavily spotted with black and with parallel black stripes on neck; dark spots on limbs and feet smaller than those on back; tail with irregular bands of black and buff. Underparts paler and heavily spotted with black. Skull with heavy zygomata, abruptly truncated rostrum, and large, well-inflated bullae. From Paraguay in South America north through Mexico, reaching the United States in southern Texas and southeastern Arizona.

Felis wiedi **Schinz**. Margay. Size small, body length about 510–750 mm; hind foot about 110–130 mm. Tail long, about 60–75% of body length. Upper parts buffy, paler on lower back and sides; nape with 3 narrow black stripes; a more or less unbroken middorsal line present, remainder of upper parts with heavy, black, sharply defined spots and linear marks. Underparts pale buffy, with large black spots and bands across throat and neck and blackish brown spots across abdomen and inner sides of limbs. Skull relatively weak and slender, with broad interorbital region and rounded braincase. From northern Argentina north to southern border of the United States in western Texas.

Felis concolor **Linnaeus**. Mountain lion. Large, lithe, with long tail and short, rounded ears. Body about 1075–1375 mm. Tail long, about 60–70% of body length. Upper parts grizzled gray to dark brown, darkest along middorsal area from top of head to base of tail, abruptly paler on shoulders and flanks. Underparts dull whitish, more or less buffy across abdomen. Sides of muzzle usually black; ears black externally; upper lips, chin, and throat white. Young spotted with black on buffy background. Skull short and rounded; sagittal crest convex in profile. Formerly ranged across southern Canada, throughout the United States, and south throughout South America. Now extirpated in much of its range in the United States and some parts of South America. Still common locally from Rocky Mountain chain westward, along Mexican border in Texas, and in southern Florida.

Felis yagouaroundi **Geoffroy**. Yaguaroundi. Small, slender-bodied, short-legged, with comparatively small head. Body about 560–760 mm; tail long, about 60–80% of body length. Color pattern plain, unspotted; occurring in 2 color phases, red and gray. Skull elongate; cranium compressed laterally; rostrum sharply elevated; bullae large and constricted laterally. From Argentina northward, barely reaching the United States in southern Texas and southeastern Arizona.

Genus *Lynx* Kerr. Lynxes. Long-legged, short-tailed cats with pointed, tufted ears; elongated hairs on cheeks forming "sideburns." Only 2 premolars in upper jaw. Dentition: $\frac{3}{3}, \frac{1}{1}, \frac{2}{2}, \frac{1}{1}$. Recent; Eurasia, Africa, North America. Several species; 2 in North America and the United States.

KEY TO SPECIES OF *LYNX*

1. Tip of tail black; hypoglossal foramen separate from foramen lacerum posterius (Fig. 6-79A)   ***L. canadensis***

Tip of tail black only on top; hypoglossal foramen confluent with foramen lacerum posterius (Fig. 6-79*B*)   *L. rufus*

Fig. 6-79. Posterior ventral view of A, *Lynx canadensis*; B, *Lynx rufus* skull. (From Burt, *The Mammals of Michigan*, University of Michigan Press.)

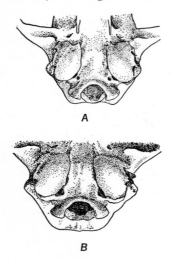

A

B

***Lynx canadensis* Kerr.**   Lynx. Size rather large; body length about 810–915 mm; hind foot about 205–245 mm. Tail about 12% of body length. Legs long; feet large; pads well furred. Upper parts a mixture of buffy, dark brown, and black, sprinkled with white; guard hairs black at tip and white subterminally, underfur brown. Top of head with many white-tipped hairs; eyelids white; ears buffy brown at base and on dorsal margin, with central white spot; ears white inside, terminating in a conspicuous, slender, black tuft. Sideburns conspicuous, consisting of white, black, and brown and white hairs. Sides and flanks buffy, sprinkled with white. Feet and legs buffy white; chin and throat white. Underparts buffy white, sparsely mottled with light brown. Skull with small postorbital processes; posterior palatine foramina near orbital rim of palate; hypoglossal canal separate from foramen lacerum posterius. From Alaska across Canada, south in Rocky Mountain chain to northwestern Colorado and in Cascades to central Oregon; also entering the United States in northern tip of New England states. Range in the United States apparently has shifted northward because of extirpation.

***Lynx rufus* (Schreber).**   Bobcat. Size moderate; body length about 635–765 mm; hind foot about 155–180 mm. Tail about 21–27% of body length. Ear tufts relatively short and inconspicuous; ears whitish on inside, blackish and grayish outside. Upper parts brownish to pale yellowish in general color, the pelage a mixture of tawny hairs tipped with black and white, giving dappled effect. Chin and underparts white or whitish. Sideburns with distinct black markings; top of tail with blackish bars. Skull robust; postorbital processes prominent; auditory bullae large, rounded. Once ranged from central Mexico, probably throughout the United States to southern Canada. Now extirpated in the densely settled parts of eastern United States.

Suborder Pinnipedia.   Seals. Adapted for aquatic life. Body streamlined; limbs highly specialized for swimming; the 2 proximal segments of limbs short and partially enclosed in body integument, and 3d segment elongated and expanded. Toes fully webbed, 5 on each limb; 1st and 5th digits of hind limb usually longer than others. Tail very short. Eyes large. Clavicle absent. Incisors always less than $\frac{3}{3}$; carnassials not developed; jaw teeth never more than 2-rooted, with conical, pointed crowns, never broad and tuberculate. Lower Miocene; practically all seas and coasts. 1 extinct and 3 living families; 2 in waters off the United States.

KEY TO FAMILIES OF PINNIPEDIA

1. External ear present; postorbital processes present in skull   **Otariidae**
   No external ear; no postorbital processes in skull   **Phocidae**

Family Otariidae.   Eared seals. Hind feet capable of being turned forward under trunk and supporting and moving body on land. Small external ears present. Palms and soles naked. Testes in a distinct, external scrotum. Skull with postorbital processes and alisphenoid canal. Angle of mandible inflected. 1st and 2d upper incisors small, with crowns divided by transverse groove into anterior and posterior cusp; 3d incisor large and caninelike. Lower Miocene; shores of Pacific and

South Atlantic. 7 extinct and 5 living genera, 4 off Pacific Coast of the United States.

KEY TO GENERA OF OTARIIDAE

1. Size large; body length more than 210 cm in males and 150 cm in females; hair short, coarse, without underfur   **2**
   Size smaller; body length less than 210 cm in males and 150 cm in females; soft, dense underfur present   **3**
2. Body length more than 245 cm in males and 200 cm in females; wide space in front of posterior premolars   *Eumetopias*
   Body length less than 245 cm in males and 200 cm in females; premolars evenly spaced   *Zalophus*
3. Facial portion of skull slender, narrow, and elongated; upper profile of skull sloping   *Arctocephalus*
   Facial portion of skull relatively broad, upper profile of skull nearly flat   *Callorhinus*

Genus *Arctocephalus* Geoffroy-St.-Hilaire and Cuvier. Southern fur seals. Hair short, densely filled with soft underfur. Skull with facial portion slender, narrow, and elongated, its upper profile sloping. Jaw teeth $\frac{6}{5}$, relatively large. Pliocene; Antarctic and southern Pacific and north to California coast. Several nominal species, 1 reaching the United States.

Arctocephalus townsendi **Merriam.** Guadalupe fur seal. Body length about 170 cm in males and 135 cm in females. General color dark brown; head and neck with whitish grizzling. Formerly common on coast and islands off southern California and Baja California. Now near extinction.

Genus *Callorhinus* Gray. Northern fur seal. Males much larger than females. Hair short, densely filled with soft underfur. Skull light, with slight crests and long, narrow interorbital ridge; upper profile of skull nearly flat. Jaw teeth $\frac{6}{5}$, small and sharp-pointed. Recent; North Pacific.

Callorhinus ursinus **(Linnaeus).** Northern fur seal. Body length about 215–245 cm in males and about 120 cm in females. Color dark brown in males and grayish brown in females and young. Breeding on Pribilof Islands, Alaska, and females migrating south along Pacific Coast to California.

Genus *Zalophus* Gill. California sea lion. Body cylindrical and streamlined; neck thin; head small;

eyes and ears small. Skull with high arched sagittal crest in old males, giving high forehead; rostrum narrow. Jaw teeth $\frac{5}{5}$, small, pointed, and evenly spaced. Pleistocene; Pacific from California to New Zealand, Australia, Japan. 1 species.

Zalophus californianus **(Lesson).** California sea lion. Body length about 215–245 cm in males and about 185 cm in females. Color dark brown, fading to light brown or yellowish brown; blackish when wet. Reaching the United States along Pacific Coast.

Genus *Eumetopias* Gill. Steller sea lion. Body heavy and stout, particularly in neck and chest. Skull with sagittal crest low in males, lacking in females, the forehead consequently low; rostrum broad. Jaw teeth $\frac{5}{5}$, small, pointed, with wide space in front of posterior premolars. Pleistocene; northern Pacific, south along coast of Asia to Japan and south along coast of North America from Bering Strait to central California.

Eumetopias jubata **(Schreber).** Steller sea lion. Size large; body length about 305–365 cm in males and 245–275 cm in females. Color uniformly brown or yellowish brown all over when dry, dark brown when wet. U.S. coast from Santa Rosa Island, California, northward.

Family Phocidae. Hair seals. Hind limbs extending posteriorly and incapable of being turned forward. Palms and soles hairy. Fur stiff, without woolly underfur. No external ear pinna. Testes abdominal. Skull without postorbital processes or alisphenoid canal; angle of mandible not inflected. Incisors with simple, pointed crowns; canines moderately developed; jaw teeth $\frac{5}{5}$. Middle Miocene; virtually all seas and coasts. About 11 extinct and 10 living genera; 2 on U.S. coasts.

KEY TO GENERA OF PHOCIDAE

1. Incisors $\frac{2}{2}$ *Monachus*
   Incisors either $\frac{3}{2}$ or $\frac{2}{2}$ **2**
2. Incisors $\frac{3}{2}$; 1st and 5th toes of hind foot not greatly exceeding others in length   *Phoca*
   Incisors $\frac{2}{1}$; 1st and 5th toes of hind foot greatly exceeding others in length   *Mirounga*

Genus *Phoca* Linnaeus. Common seals. Head round and short. Forefeet short, with 5 strong, sharp, slightly curved claws; hind feet with 5 narrower and less curved claws. Jaw teeth, except first premolar above and below, 2-rooted and with accessory cusps. Dentition: $\frac{3}{2}, \frac{1}{1}, \frac{4}{4}, \frac{1}{1}$. Middle Miocene; seas of Northern Hemisphere and Lake Baikal, Siberia. Several species, 1 on U.S. coasts.

Phoca vitulina Linnaeus. Harbor seal. Body length about 185 cm, the sexes about equal in size. Color pattern usually much spotted and mottled with brown and black on a grayish or yellowish background; paler below. Widely distributed in both Pacific and Atlantic; along Pacific Coast of the United States and south along Atlantic Coast to South Carolina.

Genus *Monachus* Fleming. Tropical seals. Large, with coarse pelage and naked palms and soles. Nails on both fore and hind feet small and rudimentary. Dentition: $\frac{2}{2}, \frac{1}{1}, \frac{4}{4}, \frac{1}{1}$. Recent; tropical and subtropical seas, West Indies, Mediterranean, Central Pacific. 1 species rarely reaches the United States.

Monachus tropicalis (Gray). West Indian seal. Body length about 230 cm. Upper parts dark brown; underparts paler and grayer. Central America and West Indies to southern coast of Florida and Texas.

Genus *Mirounga* Gray. Elephant seals. Hind foot without claws. Nose of adult males elongated into a short, tubular proboscis which is normally flaccid but capable of dilation and elongation under excitement. All teeth except canines relatively small; all jaw teeth single-rooted. Dentition: $\frac{2}{1}, \frac{1}{1}, \frac{4}{4}, \frac{1}{1}$. Recent: Antarctic, South Pacific, Indian Ocean, Pacific North America. 2 species, 1 reaching California coast.

Mirounga angustirostris (Gill). Northern elephant seal. Size large; body length about 455–550 cm in males and 245–275 in females. General color grayish to brown; underparts paler. Pacific Coast.

Order Sirenia. Manatees and dugongs. Wholly aquatic herbivores. Body robust; hind limbs absent; forelimbs paddle-shaped. Tail broad, flat, expanded transversely. Nostrils on upper surface of snout. Testes abdominal. Clavicle absent; pelvis vestigial. Skin tough, finely wrinkled, or very rugose, with sparsely scattered, fine hairs. Eyes very small; no ear pinnae. Dentition of incisors

and molariform teeth, the two separated by a wide diastema. Eocene; Atlantic Coast of North and South America and Africa and coasts of Red Sea, Indian Ocean, and western Pacific. 3 extinct and 2 living families, 1 in the United States.

Family Trichechidae. Manatees. Tail entire, rounded, or shovel-shaped. Rudimentary nails on forelimbs. Cervical vertebrae, only 6. Skull with relatively small braincase, short, narrow rostrum, small orbits, and massive zygomata. Nasal bones absent or vestigial. Dentition peculiar; incisors $\frac{2}{2}$, rudimentary, concealed beneath horny plates and lost before maturity; canines absent; jaw teeth $\frac{11}{11}$, but rarely more than $\frac{6}{6}$ present at one time, the anterior ones being lost before posterior ones come into use. Jaw teeth similar from front to back, with quadrate, enameled crowns, the grinding surfaces raised into tuberculate, transverse ridges. Upper teeth 3-rooted, lowers 2-rooted. Pleistocene; Atlantic Coast of southern North America, South America, and Africa. 1 living genus.

Genus *Trichechus* Linnaeus. Manatees. Characters and distribution of the family. Pleistocene. 3 living species, 1 in the United States.

Trichechus manatus Linnaeus. West Indian manatee. Size large; body length averaging about 215 cm. General color uniform dull grayish; skin virtually naked, but muzzle with stiff bristles. West Indies and Atlantic Coast from northern South America to southern United States; shallow waters of bays, estuaries, and large rivers in Florida, rarely north to North Carolina and on Texas coast in summer.

Order Perissodactyla. Odd-toed ungulates. Usually hooved herbivores, with mesaxonic foot. Upper end of astragalus with well-keeled surface for articulation of tibia, but lower end flattened and resting more on navicular than on cuboid. Femur with well-developed 3d trochanter. Premolars similar to molars in size and structure. Teeth lophodont in living forms. Facial part of skull usually long; nasals broad posteriorly and usually extending free well over nasal aperture. Lower Eocene; 2 suborders, 1 introduced into the United States.

Suborder Hippomorpha. Upper molars with prominent S-shaped ectoloph. Lower Eocene. 1 living and 3 extinct families, including the horned Brontotheriidae and clawed Chalicotheriidae.

Family Equidae. Horses. Hind toes 3 or less. Metapodials elongated, with hooves on terminal phalanges. Horns never present. Incisors $\frac{3}{3}$. Lower Eocene.

Genus *Equus* Linnaeus. Modern horses; zebra. Humerus and femur short; radius and tibia long, powerful. Ulna reduced and fused to radius; fibula present as an upper splint and lower vestige fused to tibia. Third digit well developed, with greatly elongated metapodial and short phalanges. Terminal phalanx expanded, semicircular, supporting hoof. Orbit surrounded by complete bony ring formed by junction of postorbital process of frontal with zygomatic arch. Incisors $\frac{3}{3}$, chisel-shaped and used in cropping; a wide diastema present, containing small, variable canine which is often lost in female; functional jaw teeth $\frac{6}{6}$, the first lower premolar being absent and first upper premolar small, deciduous. Upper Pliocene, world-wide by domestication. Several nominal species; 2 feral in the United States by introduction.

KEY TO SPECIES OF *EQUUS*

1. A small, bare callosity on inner side of hind leg below heel joint   ***E. caballus***
   No callosity on inner side of hind leg below heel joint   ***E. asinus***

***Equus caballus* Linnaeus.** Horse. Highly variable because of artificial selection under domestication. Tail with abundant long hairs, growing from base as well as from sides and tip. Mane long and flowing. A small bare callosity on inner side of hind limb below ankle joint and another on forelimb above carpus. Feral by introduction in parts of southwestern United States.

***Equus asinus* Linnaeus.** Ass. Tail without long hairs at base. Callosities on inner side of forelimbs above carpus, but none on hind limbs. Ears long; mane erect. Feral by introduction in parts of southwestern United States.

Order Artiodactyla. Even-toed ungulates. Hooved herbivores with paraxonic foot, i.e., the axis of the foot passing between the almost equally developed 3d and 4th digits. Astragalus with rolling surface above and equally developed lower pulley surface, resting equally on navicular and cuboid. Femur without 3d trochanter. Premolars and molars usually unlike; premolars single-lobed; molars 2-lobed, except last which is usually 3-lobed. Teeth bunodont or selenodont. Lower Eocene; nearly worldwide. 3 suborders, of which Tylopoda (camels, etc.) are limited to Asia and South America.

KEY TO SUBORDERS OF ARTIODACTYLA

1. Teeth bunodont (Fig. 6-80*A*)   **Suiformes**
   Teeth selenodont (Fig. 6-80*B*)   **Ruminantia**

Fig. 6-80. Left upper-jaw teeth: *A*, bunodont, of *Tayassu tajacu*; *B*, selenodont, of *Odocoileus hemionus*.

A

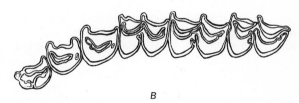

B

Suborder Suiformes. Pigs and hippos. 3d and 4th metapodials not completely fused to form "cannon" bone. Jaw teeth bunodont. Dental count high. Lower Eocene; Eurasia, Africa, North and South America. 10 extinct and 3 living families, 1 native and 1 introduced in the United States, the third (Hippopotamidae) restricted to Africa.

KEY TO FAMILIES OF SUIFORMES

1. Upper canines directed downward   **Tayassuidae**
   Upper canines directed outward or upward   **Suidae**

Family Suidae. Pigs. Feet narrow with 4 complete toes on each, but only the medial pair reaching ground. Head with an elongated, mobile snout, with nostrils opening in its nearly naked, flat, oval terminal surface. Stomach simple except for a more or less developed pouch near cardiac opening. 3d and 4th metapodials separate. Incisors rooted; upper canines curving more or less upward or outward. Lower Oligocene; Eurasia, Africa, worldwide by introduction. About 17 extinct and 5 living genera, 1 feral in the United States by introduction.

Genus *Sus* Linnaeus. Pigs. Skull with very high occipital crest, long, narrow nasals, and long, narrow palate which extends back beyond plane of last molars. Upper incisors decreasing in size from 1st to 3d; lower incisors long, narrow, close together, and almost horizontal in position. Canines prominent (particularly in males) with persistent roots and partially covered with enamel. Jaw teeth gradually larger and more complex from anterior to posterior of jaw. 3d molar very large, nearly as long as 1st and 2d combined. Dentition: $\frac{3}{3}, \frac{1}{1}, \frac{4}{4}, \frac{3}{3}$. Lower Pliocene; Eurasia and worldwide by introduction.

***Sus scrofa* Linnaeus.** Wild boar. Native of Europe; feral populations introduced into several parts of the United States, particularly mountains of Georgia, North Carolina, Tennessee, and on islands off California coast; throughout the United States in domestication.

Family Tayassuidae. Peccaries. Snout truncated as in pigs and nostrils opening in its terminal surface. Forefeet with 4 toes; hind feet with only 3 toes, the 5th being lacking. Stomach complex. 3d and 4th metapodials united at their upper ends. Incisors rooted; upper canines directed downward, with sharp, cutting posterior edges which form shearing mechanism with lower canines; jaw teeth in continuous series, gradually increasing in size; last premolar nearly as complex as molars. Dentition: $\frac{2}{3}, \frac{1}{1}, \frac{3}{3}, \frac{3}{3}$. Lower Oligocene; North and South America. 10 extinct and 1 living genera.

Genus *Tayassu* Fischer. Peccaries. Piglike, with short, vestigial tails, coarse, bristly hair, and small ears. A prominent musk gland on back. Pleistocene; North and South America. 2 living species, 1 in the United States. The generic name *Pecari* is also used.

***Tayassu tajacu* (Linnaeus).** Collared peccary. Head and body laterally compressed; body length about 900 mm; hind foot about 200 mm. General color grizzled, grayish and blackish, with a collar of lighter-colored shoulder stripes. From Patagonia in South America north to the United States from central Texas west to south central Arizona.

Suborder Ruminantia. Ruminants. 3d and 4th metapodials usually fused to form cannon bone. 2d and 5th digits usually rudimentary. Molar teeth selenodont. Upper incisors usually lacking, never more than 1 pair present. Horns or antlers usually present as outgrowths of frontal bones. Stomach complex, with at least 3 (usually 4) compartments. Upper Eocene; virtually worldwide. 5 extinct and 5 living families, including chevrotains (Tragulidae) of Asia and Africa and giraffes (Giraffidae) of Africa. 3 families in the United States.

KEY TO FAMILIES OF RUMINANTIA

1. Antlers of solid bone usually present, at least in males; 1st molar (and usually others) above and below brachyodont   **Cervidae**
   Bony, keratin-covered horns usually present; molars usually hypsodont   **2**
2. Horns unbranched, permanent   **Bovidae**
   Horns bifurcated, the keratin covering shed periodically   **Antilocapridae**

Family Cervidae. Deers. Antlers usually present in males or in both sexes. A large lacrimal vacuity present and excluding lacrimal bone from contact with nasal. Lateral digits usually present on both fore and hind feet. Lacrimal duct with 2 openings on or inside rim of orbit. Molars usually brachyodont. Upper canines usually present in both sexes. Gall bladder usually absent. Lower Oligocene; Eurasia, North and South America. 34 extinct and 17 living genera, 4 in North America and the United States.

KEY TO GENERA OF CERVIDAE

1. Antlers cylindrical, not palmate (Fig. 6-81*A*)   **2**
   Antlers more or less palmate (Fig. 6-81*B, C*)   **3**
2. Upper canines present   *Cervus*
   Upper canines absent   *Odocoileus*
3. Males only with antlers; lowest prongs not extending over face (Fig. 6-81*B*)   **Alces**
   Both sexes with antlers; lowest prongs extending over face (Fig. 6-18*C*)   *Rangifer*

Fig. 6-81. Skull of males: *A, Cervus canadensis; B, Alces alces; C, Rangifer caribou.* (From Seton, *Life Histories of Northern Animals,* Charles Scribner's Sons.)

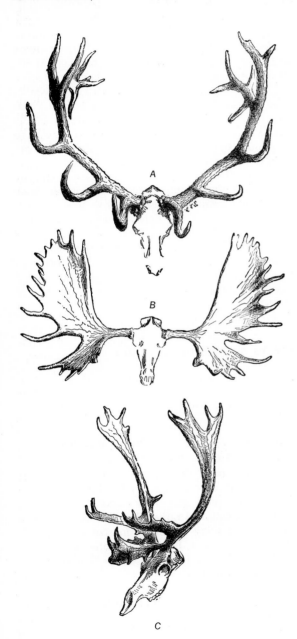

Genus *Cervus* Linnaeus. Old World deer. Large deer. Males with large antlers 2 or 3 times length of head. Skull without prominent frontal ridges; canines never large. Lacrimal vacuity and fossa moderate to large. Proximal portion of 2d and 5th metacarpals present. Dentition: $\frac{0}{3}, \frac{1}{1}, \frac{3}{3}, \frac{3}{3}$. Middle Pliocene; Eurasia, North America. Several species with center of distribution in Asia, 1 in North America.

***Cervus canadensis* Erxleben.** Elk, wapiti. Size large; body length about 215–275 cm in males and 200–230 cm in females; hind foot about 635 mm. Large antlers present in males, commonly with 5–7 points, as much as 1675 mm long measured along beam and following curves, with spread of up to 1525 mm. Mane present on neck. Sides and back grayish brown; head, neck, and legs dark brown; underparts blackish; rump and tail pale, buffy whitish. Originally ranged over much of the United States, now mainly in Rocky Mountain chain south to northern New Mexico and in a few scattered localities in western half of the United States.

Genus *Odocoileus* Rafinesque. American deer. Small to moderate-sized deer. Antlers moderate, never greatly exceeding length of head. Skull with vomer dividing posterior nares into 2 distinct chambers; premaxillae not reaching nasals. Lacrimal vacuity very large; lacrimal fossa small. Auditory bullae slightly inflated. Dentition: $\frac{0}{3}, \frac{0}{1}, \frac{3}{3}, \frac{3}{3}$. Pleistocene; North and South America. 2 species in the United States.

KEY TO SPECIES OF *ODOCOILEUS*

1. Antler with a main beam curving sharply out and forward and all subsidiary points except basal emerging from dorsal side of main beam (Fig. 6-82*A*)   *O. virginianus*
   Antler with more or less equal dichotomous branches (Fig. 6-82*B*)   *O. hemionus*

***Odocoileus virginianus* (Zimmermann).** White-tail deer. Size highly variable; body length about 1260–1605 mm; tail relatively long, about 245–270 mm; hind foot about 400–485 mm. Antler with branches from a main beam. Metatarsal gland small, about 25 mm in length. General color reddish in summer pelage, grayish in winter; paler

Fig. 6-82. Heads of *Odocoileus: A, Odocoileus virginianus; B, Odocoileus hemionus.* (Drawn by William F. Martin.)

below. Tail white below, similar to upper parts above and white-tipped. From northern South America, north through Mexico and the United States to southern Canada. Formerly through all of the United States except possibly the most arid parts of the Southwest; now locally extirpated in many parts of former range.

Odocoileus hemionus (**Rafinesque**). Mule deer. Body length about 1245–1600 mm; tail about 130–230 mm; hind foot about 400–585 mm. Ears longer than in *O. virginianus.* Antlers with more or less even dichotomous branches. Metatarsal gland large, about 100–150 mm in length. General color reddish in summer, grayish in winter. Tail black or black-tipped above, white or whitish below. Large white rump patches sometimes present at sides of tail. Formerly ranged from northwestern Mexico through western half of the United States into western Canada. Range has retreated westward in recent years; now mostly west of a line from Pecos River in Texas north to northeastern Colorado and northeastern North Dakota.

Genus *Alces* **Gray.** Moose. Large, with long legs and short neck and tail. Antlers (present in males) very large, palmate, and lacking the brow

and bez tines. A pendent structure (bell) 200–250 mm long on throat of both sexes. Upper lip long, thick, overhanging the lower. Nasal bones very short; narial opening very large. Dentition: $\frac{0}{3}, \frac{0}{1}, \frac{3}{3}, \frac{3}{3}$. Pleistocene; Eurasia, North America. 1 living species.

Alces alces (**Linnaeus**). Moose. Largest living deer; body length about 300 cm; tail very short, about 100 mm. Fur thick, coarse, longest on neck and throat. General color blackish or brownish; paler on legs. Northern Eurasia and northern North America. In North America from Alaska across Canada, south in Rocky Mountain chain to western Wyoming; in Maine in Northeast. Range has retreated northward in recent times.

Genus *Rangifer* **Smith.** Reindeers. Heavily built, with short limbs. Antlers present in both sexes, large, branching; bez and brow tines present, branched or palmated. Lateral hooves well developed; cleft between main hooves very deep. Skull with small lacrimal vacuity; posterior nares divided by vomer. Dentition: $\frac{0}{3}, \frac{0}{1}, \frac{3}{3}, \frac{3}{3}$. Pleistocene; Eurasia, North America. 2 native and 1 introduced species in North America; 2 native in the United States.

KEY TO SPECIES OF *RANGIFER*

1. Color paler; main beam of antlers nearly cylindrical
   R. arcticus
   Color darker; main beam of antlers flattened
   R. caribou

Rangifer caribou (**Gmelin**). Woodland caribou. Body length about 180 cm; tail about 100 mm. Upper parts dark brown; underparts, rump, and a band above each hoof white; in winter pelage upper parts gray, neck whitish. Antlers with main beam flattened rather than cylindrical. Lower incisors decreasing gradually in size from middle out. From eastern British Columbia east across southern Canada; formerly in the United States in northern Maine, Vermont, and New Hampshire.

Rangifer arcticus (**Richardson**). Barren grounds caribou. Body length about 240 cm; tail about 175 mm. Upper parts brownish; underparts whitish; in winter pelage dirty white in northern parts of range. Antlers with main beam nearly cylindrical, with long, backward, then forward curved beam, branching dichotomously at tip. Lower incisors decreasing in size from middle out by conspicuous steps. From northern Alaska throughout northern Canada; south through British Columbia to northeastern Washington, where it occurs rarely.

Family Antilocapridae. Pronghorns. Horns present in both sexes, branched, the bony core permanent, but horn shed annually by new growth from below pushing old horn off at tip. Vestigial hooves of 2d and 5th digits absent. Middle Miocene; North America. 13 extinct and 1 living genera.

Genus *Antilocapra* Ord. Pronghorns. Size small; body slender, graceful; neck maned. Horns with 1 flattened prong and a recurved tip; the bony core not branched. Hair coarse, brittle. Skull with large lacrimal vacuity. Teeth narrow. Dentition: $\frac{0}{3}, \frac{0}{1}, \frac{3}{3}, \frac{3}{3}$. Pleistocene; North America. 1 living species.

***Antilocapra americana* (Ord).** Pronghorn. Body length about 120 cm; tail about 110 mm; hind foot about 400 mm. Upper part buffy; underparts white; face and neck with strongly contrasted black and white markings; rump with large patch of long, white, erectile hairs. Formerly ranged over most of western half of the United States; now from northern Mexico and northern Baja California north to southern Canada; east to southern Texas, eastern Colorado, central Nebraska, and western North Dakota; west to central and northeastern California, southeastern Oregon, southern Idaho, and western Montana.

Family Bovidae. Cattle, etc. Horns, when present, hollow, unbranched, nondeciduous, the epidermal covering being renewed by growth from base. Lateral digits absent or often represented by hooves alone, sometimes supported by rudimentary skeleton of irregular nodules of bone. Distal ends of lateral metapodials always absent. Gall bladder usually present. Lacrimal canal usually with 1 opening, inside rim of orbit. Lacrimal bone usually articulating with nasal. Molars frequently hypsodont; canines absent in both sexes. Lower Miocene; Eurasia, Africa, North America. About 86 extinct and 54 living genera, 3 in the United States.

KEY TO GENERA OF BOVIDAE

1. Size very large; shoulder region much broader and higher than rump **Bison**
   Size smaller; shoulder region not much broader and higher than rump **2**
2. Horns massive, curving backward, then outward and forward (Fig. 6-83A); no beard on chin **Ovis**
   Horns small, slender, curving slightly backward (Fig. 6-83B); chin with a beard **Oreamnos**

Fig. 6-83. Heads of *A, Ovis canadensis; B, Oreamnos americanus.* (After Frick, from Gregory, *Evolution Emerging,* The Macmillan Company.)

Genus *Bison* Smith. Bison, wisent. Large, with massive skulls and unbranched horns. Higher and broader at shoulder than at rump. Body hair short, crisp, woolly; head and neck with abundant long hairs forming a mane which conceals eyes, ears, and bases of horns. A long beard beneath chin; a line of long hairs from head nearly to tail; tip of tail tufted. Skull with upper part of forehead transversely arched; space between horns elevated in middle; horns below plane of occiput. Nasals short, separated from premaxillae by a wide space.

Dentition: $\frac{0}{3}, \frac{0}{1}, \frac{3}{3}, \frac{3}{3}$. Pleistocene; Europe, North America. 2 living species, 1 in North America and the United States.

***Bison bison* (Linnaeus).** Bison. Size large; body length about 275 cm in males, 200 cm in females. Head, neck, legs, tail, and underparts blackish brown; upper parts paler. Formerly ranged through plains of central North America from Mexico north into Canada and east as far as New York and Florida. A few scattered herds remain under protection in the Plains states and in Canada.

Genus *Ovis* Linnaeus. Sheep. Body stout; legs and neck short. Horns massive in males, curving backward, outward, and forward in spiral; smaller and less curved in females; usually with more or less prominent tranverse ridges. Suborbital gland and lacrimal fossa usually present, but generally small. Foot glands on all feet. Chin not bearded. Basioccipital of skull wider anteriorly than posteriorly, with anterior pair of tubercles widely separated and much larger than posterior. Molars very hypsodont. Dentition: $\frac{0}{3}, \frac{0}{1}, \frac{3}{3}, \frac{3}{3}$. Upper Pliocene; Eurasia, North America, and worldwide by domestication. Several species; 2 native in North America, 1 in western United States.

***Ovis canadensis* Shaw.** Mountain sheep. Body length about 160 cm; tail about 125 mm. Horns massive in males, with prominent transverse ridges, brown in color. General color grayish brown. A conspicuous rump patch; back of limbs and end of muzzle white to creamy white. Rocky Mountain chain from central British Columbia south to trans-Pecos Texas, west to eastern California and eastern Oregon and south into Baja California. Range contracting in recent years.

Genus *Oreamnos* Rafinesque. Mountain goat. Stout-bodied, with large head, short legs, short, thick neck, and with a definite hump on shoulders. Horns slender, round, curving upward and slightly backward. Males with a beard on chin. Hair very long. Hooves relatively large. Cannon bones short. Tail short. Skull with shallow lacrimal fossa, but no fissure. Molars very hypsodont. Dentition: $\frac{0}{3}, \frac{0}{1}, \frac{3}{3}, \frac{3}{3}$. Pleistocene; western North America. 1 living species.

***Oreamnos americanus* (De Blainville).** Mountain goat. Body length about 160 mm. Length of horns about 250 mm. Color white; horns and hooves black. From southern Alaska and northwestern Canada south in Rocky Mountains to central Idaho and in Cascades to southern Washington. A high-mountain form.

REFERENCES

General

Allen, G. M. 1942. *Extinct and Vanishing Mammals of the Western Hemisphere with the Marine Species of All the Oceans,* Intelligencer Printing Co., Lancaster, Pa.

Anthony, H. E. 1928. *Field Book of North American Mammals,* G. P. Putnam's Sons, New York.

Beddard, F. E. 1902. *Mammalia,* Macmillan & Co., Ltd., London.

Burt, W. H. 1952. *A Field Guide to the Mammals,* Houghton Mifflin Company, Boston.

———. 1960. Bacula of North American Mammals, *Misc. Publ. Museum Zool. Univ. Mich.,* no. 113.

Cockrum, E. L. 1955. *Laboratory Manual of Mammalogy,* Burgess Publishing Company, Minneapolis.

———. 1962. *Introduction to Mammalogy,* The Ronald Press Company, New York.

Davis, D. E., and F. B. Golley. 1963. *Principles in Mammalogy,* Reinhold Publishing Corporation, New York.

Flower, W. H., and R. Lydekker. 1891. *An Introduction to the Study of Mammals Living and Extinct,* A. and C. Black, Ltd., London.

Glass, B. P. 1951. *A Key to the Skulls of North American Mammals,* Burgess Publishing Company, Minneapolis.

Hall, E. R., and K. R. Kelson. 1959. *The Mammals of North America,* 2 vols., The Ronald Press Company, New York.

Hamilton, W. J. 1939. *American Mammals,* McGraw-Hill Book Company, New York.

Miller, G. S., and R. Kellogg. 1955. List of North American Recent Mammals, *Bull. U.S. Natl. Museum,* 205.

Palmer, R. S. 1954. *The Mammal Guide,* Doubleday & Company, Inc., Garden City, N.Y.

Palmer, T. S. 1904. Index Generum Mammalium: A List of the Genera and Families of Mammals, *North Am. Fauna* 23.

Scott, W. B. 1913. *A History of Land Mammals in the Western Hemisphere,* The Macmillan Company, New York.

Simpson, G. G. 1945. The Principles of Classification and a Classification of Mammals, *Bull. Am. Museum Nat. Hist.,* 85.

Young, J. Z. 1959. *The Life of Mammals,* Oxford University Press, Fair Lawn, N.J.

Regional

Bailey, V. 1931. Mammals of New Mexico, *North Am. Fauna* 53.

———. 1936. The Mammals and Life Zones of Oregon, *North Am. Fauna* 56.

Burt, W. H. 1946. *The Mammals of Michigan,* University of Michigan Press, Ann Arbor, Mich.

———. 1957. *Mammals of the Great Lakes Region,* University of Michigan Press, Ann Arbor, Mich.

Cockrum, E. L. 1952. Mammals of Kansas, *Univ. Kans. Publ. Museum Nat. Hist.,* vol. 7, no. 1.

———. 1961. *The Recent Mammals of Arizona: Their Taxonomy and Distribution,* University of Arizona Press, Tucson, Ariz.

Dalquest, W. W. 1948. Mammals of Washington, *Univ. Kans. Publ. Museum Nat. Hist.,* vol. 2.

Davis, W. B. 1939. *The Recent Mammals of Idaho,* Caxton Printers, Ltd., Caldwell, Idaho.

Durrant, S. D. 1952. Mammals of Utah, Taxonomy and Distribution, *Univ. Kans. Publ. Museum Nat. Hist.,* vol. 6.

Grinnell, J., J. S. Dixon, and J. M. Linsdale. 1937. *Fur-bearing Mammals of California,* vols. 1 and 2, University of California Press, Berkeley, Calif.

Gunderson, H. L., and J. R. Beer. 1953. *The Mammals of Minnesota,* The University of Minnesota Press, Minneapolis.

Hall, E. R. 1946. *Mammals of Nevada,* University of California Press, Berkeley, Calif.

Hoffmeister, D. F., and C. O. Mohr. 1957. *Field Book of Illinois Mammals,* Illinois Natural History Survey, Urbana, Ill.

Jackson, H. H. T. 1961. *Mammals of Wisconsin,* The University of Wisconsin Press, Madison, Wis.

Schwartz, C. W., and E. R. Schwartz. 1959. *The Wild Mammals of Missouri,* University of Missouri Press, Columbia, Mo.

Insectivora

Jackson, H. H. T. 1915. A Review of the American Moles, *North Am. Fauna* 38.

———. 1928. A Taxonomic Review of the American Long-tailed Shrews (Genera Sorex and Microsorex), *North Am. Fauna* 51.

Merriam, C. H. 1895. Revision of the American Shrews of the Genera Blarina and Notiosorex, *North Am. Fauna* 10.

Chiroptera

Allen, H. 1893. A Monograph of the Bats of North America, *Bull. U.S. Natl. Museum,* 43.

Hall, E. R., and W. W. Dalquest. 1950. A Synopsis of the American Bats of the Genus Pipistrellus, *Univ. Kans. Publ. Museum Nat. Hist.,* 1(26):591–602.

Handley, C. O. 1959. A Revision of American Bats of the Genera Euderma and Plecotus, *Proc. U.S. Natl. Museum,* vol. 110, no. 3417.

Miller, G. S. 1897. Revision of the North American Bats of the Family Vespertilionidae, *North Am. Fauna* 13.

———. 1907. The Families and Genera of Bats, *Bull. U.S. Natl. Museum,* 57.

——— and G. M. Allen. 1928. The American Bats of the Genera Myotis and Pizonyx, *Bull. U.S. Natl. Museum,* 144.

Shamel, H. H. 1931. Notes on the American Bats of the Genus Tadarida, *Proc. U.S. Natl. Museum,* vol. 78, art. 19.

Lagomorpha

Hall, E. R. 1951. A Synopsis of North American Lagomorpha, *Univ. Kans. Publ. Museum Nat. Hist.,* 5(10): 119–202.

Howell, A. H. 1924. Revision of the American Pikas (Genus Ochotona), *North Am. Fauna* 47.

Lyon, M. W. 1904. Classification of the Hares and Their Allies, *Smithsonian Misc. Coll.* 45, pp. 321–447.

Nelson, E. W. 1909. The Rabbits of North America, *North Am. Fauna* 29.

Rodentia

Bailey, V. 1900. Revision of American Voles of the Genus Microtus, *North Am. Fauna* 17.

———. 1915. Revision of the Pocket Gophers of the Genus Thomomys, *North Am. Fauna* 39.

Davis, W. B. 1940. Distribution and Variation of Pocket Gophers (Genus Geomys) in the Southwestern United States, *Tex. Agr. Expt. Sta. Bull.* 590.

Ellerman, J. R. *The Families and Genera of Living Rodents.* Publ. Brit. Museum (Nat. Hist.), London, vol. 1 (rodents other than Muridae) 1940; vol. 2 (Family Muridae) 1941.

Goldman, E. A. 1910. Revision of the Wood Rats of the Genus Neotoma, *North Am. Fauna* 31.

———. 1911. Revision of the Spiny Pocket Mice (Genera Heteromys and Liomys), *North Am. Fauna* 34.

———. 1918. The Rice Rats of North America (Genus Oryzomys), *North Am. Fauna* 43.

Grinnell, J. 1922. A Geographical Study of the Kangaroo Rats of California, *Univ. Calif. Publ. Zool.,* vol. 24, no. 1.

Hall, E. R., and E. L. Cockrum. 1953. A Synopsis of North American Microtine Rodents, *Univ. Kans. Publ. Museum Nat. Hist.,* 5(27):373–498.

Hollister, N. 1911. A Systematic Synopsis of the Muskrats, *North Am. Fauna* 32.

———. 1914. A Systematic Account of the Grasshopper Mice, *Proc. U.S. Natl. Museum,* 47:427–489.

———. 1916. A Systematic Account of the Prairie-dogs, *North Am. Fauna* 40.

Hooper, E. T. 1952. *A Systematic Review of the Harvest Mice (Genus Reithrodontomys) of Latin America, Misc. Publ. Museum Zool. Univ. Mich.,* no. 77.

———. 1957. Dental Patterns in Mice of the Genus *Peromyscus, Misc. Publ. Museum Zool. Univ. Mich.,* no. 99.

———. 1958. The Male Phallus in Mice of the Genus *Peromyscus, Misc. Publ. Museum Zool. Univ. Mich.,* no. 105.

Howell, A. B. 1926. Voles of the Genus Phenacomys, *North Am. Fauna* 48.

———. 1927. Revision of the American Lemming Mice (Genus Synaptomys), *North Am. Fauna* 50.

Howell, A. H. 1914. Revision of the American Harvest Mice (Genus Reithrodontomys), *North Am. Fauna* 36.

———. 1915. Revision of the American Marmots, *North Am. Fauna* 37.

———. 1918. Revision of the American Flying Squirrels, *North Am. Fauna* 44.

———. 1929. Revision of the American Chipmunks (Genera Tamias and Eutamias), *North Am. Fauna* 52.

———. 1938. Revision of the North American Ground Squirrels, *North Am. Fauna* 56.

Johnson, D. H. 1943. Systematic Review of the Chipmunks (Genus Eutamias) of California, *Univ. Calif. Publ. Zool.,* 48(2):63–148.

Krutzsch, Philip H. 1954. North American Jumping Mice (Genus Zapus), *Univ. Kans. Publ. Museum Nat. Hist.,* 7(4):349–472.

Merriam, C. H. 1895. Monographic Revision of the Pocket Gophers, Family Geomyidae, *North Am. Fauna* 8.

Osgood, W. H. 1900. Revision of the Pocket Mice of the Genus Perognathus, *North Am. Fauna* 18.

———. 1909. Revision of the Mice of the American Genus Peromyscus, *North Am. Fauna* 28.

Preble, E. A. 1899. Revision of the Jumping Mice of the Genus Zapus, *North Am. Fauna* 15.

White, J. A. 1953. Taxonomy of the Chipmunks, *Eutamias quadrimaculatus* and *Eutamias umbrinus, Univ. Kans. Publ. Museum Nat. Hist.,* 5(33):563–582.

Cetacea

Beddard, F. E. 1900. *A Book of Whales,* G. P. Putnam's Sons, London.

Moore, J. C. 1963. Recognizing Certain Species of Beaked Whales of the Pacific Ocean, *Am. Midland Naturalist,* 70(2):396–428.

Norman, J. R., and F. C. Fraser. 1948. *Giant Fishes, Whales and Dolphins,* 2d ed., Putnam & Co., Ltd., London.

Scheffer, V. B., and Dale W. Rice. 1963. A List of the Marine Mammals of the World, *U.S. Fish and Wildlife Serv. Spec. Sci. Rep. Fisheries,* no. 431.

Slijper, E. J. 1962. *Whales,* Hutchinson & Co. (Publishers), Ltd., London.

Carnivora

Goldman, E. A. 1950. Raccoons of North and Middle America, *North Am. Fauna* 60.

Hall, E. R. 1951. American Weasels, *Univ. Kans. Publ. Museum Nat. Hist.,* no. 4.

Howell, A. H. 1901. Revision of the Skunks of the Genus Chincha, *North Am. Fauna* 20.

———. 1906. Revision of the Skunks of the Genus Spilogale, *North Am. Fauna* 26.

Scheffer, V. B. 1958. *Seals, Sea Lions and Walruses: A Review of the Pinnipedia,* Stanford University Press, Stanford, Calif.

Van Gelder, R. G. 1959. A Taxonomic Revision of the Spotted Skunks (Genus *Spilogale), Bull. Am. Museum Nat. Hist.,* vol. 117, art. 5.

Young, S. P., and E. A. Goldman. 1944. *The Wolves of North America,* Monumental Printing Co., Baltimore.

———. 1946. *The Puma, Mysterious American Cat,* Monumental Printing Co., Baltimore.

Acrodont Type of tooth attachment in which there are no sockets and the teeth are consolidated with the summit of the jaw; having acrodont teeth.

Acromion process The ventral projection of the spine of the scapula of mammals.

Acrotarsium The anterior part of the tarsal sheath in birds.

Adipose fin A fleshy fin, without rays, located behind the dorsal fin.

Adnate Congenitally grown together. In catfishes, applied to adipose fins that are joined in their full length to the back.

Aegithognathous A type of bird palate in which the vomer is truncate or V-shaped anteriorly and the maxillopalatines are free from each other.

Aftershaft A fluffy branch on the inner side of the feather, arising from the superior umbilicus; the hyporachis.

Air bladder A membranous gas-filled sac present or absent in the dorsal portion of the abdominal cavity of fishes. It may consist of 1–3 chambers.

Alisphenoid One of a pair of winglike bones of the braincase of mammals, derived from the epipterygoid of lower vertebrates.

Allopatric Occupying different geographic regions.

Altricial With the young bird hatched in a helpless condition.

Alveolus A pit; the socket of a tooth in mammals and crocodilians; an air cell of the lungs of higher vertebrates.

Ambiens A slender muscle on the medial side of the thigh, originating on the pectineal process of the ilium and inserting on the patellar tendon.

Amphicoelous Biconcave; used with reference to centra of vertebrae.

Amplexus The sexual embrace of amphibians.

Anal fin (A.) A median unpaired fin situated posterior to the anus (vent) and in front of the caudal peduncle.

Angular A bone of the lower jaw, primitively located between splenial and surangular bones;

it is absent in recent amphibians and mammals.

Anisodactyl With 3 front toes and 1 hind toe (the hallux).

Anisomyodous With the intrinsic syringeal muscles unequally inserted, i.e., either at the middle or at one end only (dorsal or ventral) of the bronchial semirings.

Annulation The circular arrangement of epidermal scales on the tail of a mammal.

Anteriad Toward the anterior, frontward.

Antitragus The ventral part of the ear pinna of mammals, adjacent to the tragus.

Antorbital Anterior to the orbit.

Anuran A tailless amphibian; a frog or toad.

Apical At the apex or end; the apical field of a fish scale is that portion opposite the basal field and on the posterior exposed end.

Apodan A legless amphibian; a member of the tropical order Apoda.

Appressed Pressed against, pressed together, in contact. In salamander taxonomy the front limb is extended backward along the axis of the body, whereas the hind limb is extended forward, the number of costal folds or grooves between the appressed limbs then being counted.

Apsidospondyl An amphibian in which the vertebral centra are ossified from cartilaginous arches.

Apterium (pl. apteria) An area of skin devoid of feathers, between the pterylae.

Aquintocubital Lacking the fifth secondary; diastataxic.

Arboreal Living in trees.

Arciferous Refers to nonrigid type of pectoral girdle in which the 2 epicoracoids overlap.

Arietiform figure The dark facial marking of kangaroo rats extending across the nose from one tuft of vibrissae to the other.

Articulation The point of juncture between 2 movable bones.

Astragalus One of the tarsal bones of higher vertebrates; it evidently represents a fusion of tibiale with intermedium and a central element.

Auditory bulla A bony capsule enclosing the inner and middle ear of most mammals.

Auriculars The loose-textured feathers covering the ear opening.

Axilla The armpit.

Axillars The elongated feathers growing from the armpit.

Baculum The penis bone or os priapi of mammals.

Balancer A transient lateral head appendage of rodlike form found in larvae of salamanders of quiet waters.

Baleen Horny plates of epidermal origin in the upper jaw of certain whales.

Barbel A fleshy protuberance in the form of a thread, flap, or conelike structure, usually rather small, but sometimes (as in catfishes) quite long.

Basal field The anterior end of a fish scale, opposite the apical field and usually covered by the scale in front.

Base (of fins) That line along which a fin is attached to the body.

Basibranchial tooth In fish, one of the teeth on the basibranchials, a series of median bones immediately behind the tongue.

Basilar length In mammals, the length of the skull from the anterior border of the foramen magnum to the posterior border of the alveolus of the first upper incisor.

Basioccipital The bone which forms the ventral margin of the foramen magnum; in mammals it fuses with the 2 exoccipitals and the supraoccipital to form the occipital bone.

Basipterygoid process An articular facet on either side of the basisphenoidal rostrum, for the pterygoids to abut against and glide over.

Basisphenoid A median ventral bone of the skull, lying anterior to the basi-occipital and (in mammals) between the auditory bullae.

Bastard wing The alula, a group of about 3 feathers growing from the thumb.

Bend of wing The point where the carpometacarpus joins the ulnare; the wrist.

Bez tine The first tine above the first or brow tine in artiodactyls.

Bicipital 2-headed; used with reference to ribs which have dual articulation with vertebrae.

Bicornuate Having 2 horns or extensions.

Bicuspid Having or ending in 2 points, as bicuspid teeth.

Bifid Divided into 2 equal lobes.

Booted With the tarsal sheath entire, not divided into scutellae; holothecal.

Boss A bump or raised area.

Branchiostegal ray In fishes, one of the long, curved, and often flattened bones supporting the branchiostegal membranes just below the gill cover.

Bridge In turtles, the narrow connection between plastron and carapace on each side of the body.

Bunodont Used with reference to low-crowned, enamel-capped teeth with low, rounded cusps.

Caecum (pl. caeca) A blind sac. In fishes the pyloric caeca are slender fingerlike structures arising from the junction between stomach and intestines; in tetrapods there are usually 1 or 2 colic caeca at the junction of small and large intestines.

Calcaneum The heel bone; the fibulare.

Calcar A process of the calcaneum which supports the interfemoral membrane in bats. A keel or widening is sometimes present on the outer edge.

Cancellate With many spaces as in 3-dimensional lattice work, as cancellate bone.

Canine A member of the dog family (Canidae); doglike; the primitively long, stout, cone-shaped, pointed tooth just behind the incisors in mammals, also similarly shaped teeth in fishes.

Canthus rostralis An angular ridge from the anterior border of the eye to the nostril in certain amphibians and reptiles.

Carapace The dorsal shell of a turtle.

Carinate Descriptive term applied to a sternum, claw, or other structure which is provided with a keel or ridge.

Carnassial Flesh-cutting.

Carnivorous Flesh-eating.

Carpus The wrist.

Caruncle A horny, spinelike projection on the upper jaw of hatchling turtles; in birds, a naked fleshy outgrowth, as the wattles and comb of certain birds.

Caudal Pertaining to the tail region.

Caudal fin (C.) The posteriormost unpaired fin

of fishes, borne on the distal extremity of the caudal peduncle.

Caudal peduncle The slender portion of the fish body behind the anal fin and bearing the caudal fin.

Centrum (*pl. centra*) The body of a vertebra.

Cere A membranous swollen covering of the base of the upper bill, through which the nostrils open.

Cerebellum A dorsal expansion of the anterior end of the hind brain; it is a center of coordination.

Cerebral hemisphere One of a pair of dorsal lobes of the forebrain.

Cervical Pertaining to the neck; pertaining to the cervix of an organ; a vertebra of the neck.

Chevron A V-shaped mark; a V-shaped bone (hemal arch) of the caudal appendage of a vertebrate.

Chromatophore A pigment cell.

Cingulum A ridge on a tooth around the base of the crown.

Circulus One of many concentric and continuous lines on scales and opercles of fishes.

Cirrus A slender, usually flexible, appendage. In some salamanders, one of 2 fleshy protuberances from the anterior edge of the upper lip.

Clavicle A dermal bone of the anterior margin of the pectoral girdle.

Cleithrum A dermal bone of the pectoral girdle of lower vertebrates.

Cloaca The common chamber into which discharge the digestive tract and the urogenital ducts in monotremes, birds, reptiles, amphibians, and some fishes.

Columella auris A slender bone connecting the tympanum with the internal ear in amphibians, reptiles, and birds. It is homologous with the hyomandibular bone of fishes and the stapes of mammals. Frequently referred to simply as the columella.

Commissure The line of closure of the bill of birds; a transverse fiber tract of the brain.

Compressed Flattened or narrowed laterally; higher than wide.

Condyle An articular protuberance on a bone.

Conoidal Cone-shaped.

Contour feathers The exposed body and flight feathers; those which make the contour or outline of the bird.

Coracoid A chondral bone in the pectoral girdle of pelycosaurs, therapsids, and monotremes, not homologous with the anterior coracoid of lower vertebrates.

Coronoid process A dorsal projection of the mandible of mammals anterior to the condylar process; it is a point of attachment of temporal muscles.

Corpus callosum A mass of nerve fibers connecting the cerebral hemispheres in higher mammals.

Costal Pertaining to a rib. In turtles, one of the shields of the carapace between the median shields (vertebrals) and the laterals (marginals).

Costal fold The area between 2 costal grooves.

Costal groove Vertical grooves in the sides of salamanders.

Covert One of the special small feathers covering the base of the flight feathers.

Crenate Notched or scalloped.

Crested Refers to tail of mammals with enlarged dorsal hairs anterior to terminal tuft.

Crissum The under tail coverts.

Crossopterygian One of the primitive bony fish ancestral to the amphibians.

Crotch In fish the angle enclosed by the branches of soft rays; the area is membranous and is sometimes heavily pigmented.

Cruciform Cross-shaped.

Ctenoid scale A fish scale bearing tiny spines on its posterior edge or field.

Cuboid One of the tarsal bones of higher vertebrates.

Culmen The dorsal ridge of the bill of birds.

Cuneate Wedge-shaped.

Cursorial Adapted for running or walking.

Cuspidate With a cusp or cusps; ending in a point or points.

Cusplet A small cusp of mammal tooth.

Cycloid scale A fish scale that is roughly circular, lacks spines, but bears circuli and usually radii.

Deciduous Shed during life.

Dentary One of a pair of dermal bones of the lower jaw. In fishes the dentaries are the anteriormost bones of the lower jaw; in mammals they remain as the only bones of the lower jaw.

Dentate Having teeth; toothlike.

Denticulation A small toothlike process; also, state of being finely toothed.

Depressed Flattened dorsoventrally; wider than high.

Depth That vertical measurement of a part (in fish usually taken by means of dividers), the greatest or least, as specified.

Dertrum The rostral nail; the anterior tip of the upper bill.

Desmognathous A type of bird palate in which the maxillopalatines are fused at the midline.

Desmopelmous Type of bird foot in which the plantar tendons are united so that the hind toe cannot be bent independently.

Diacromyodous With the intrinsic syringeal muscles inserted at both dorsal and ventral ends of the bronchial semirings.

Diapophysis A transverse process of a vertebra.

Diastema The gap separating the jaw teeth from anterior ones in various mammals.

Dichromism (dichromatism) Condition of having 2 color phases, as sexual dichromism in which the males are colored differently from the females.

Digit A finger or toe.

Digitigrade Condition in which only the toes contact the ground in walking.

Diphyodont With 2 sets of teeth, deciduous and permanent, during life.

Dorsad Toward the dorsal or upper side.

Dorsal fin (D.) A median unpaired and rayed fin inserted on the back, variously spinous or soft-rayed, sometimes partially or completely divided into 2 parts.

Dorsum The upper part of the body; the back.

Double-rounded With the median and lateral rectrices shortest, the intermediate ones longest.

Ectepicondylar process A lateral extension of the distal end of the humerus, above the external condyle.

Ectoderm The outer germ layer.

Ectoloph The ridge between the paracone and metacone in a lophodont tooth.

Ectopterygoid A paired bone of the mouth roof articulating with the palatine and the quadrate, found in fishes, amphibians, and primitive reptiles.

Edentulous Without teeth.

Emarginate With a notch; used to describe edges of structures that are indented or forked. The tail of a bird is emarginate if the lateral rectrices are longer than the middle ones; the wing of a bird is emarginate if the primaries are abruptly narrowed or cut out distally.

Entire Without indentations or interruptions; said of the sternum of a bird when neither notched nor fenestrate; said of the preopercle of fishes when neither serrate nor fimbriate.

Entopterygoid In fish a paired dermal bone of the roof of the mouth lying just below the orbit and beneath the suborbital series.

Epibranchial A gill cartilage, typically located between pharyngobranchial and ceratobranchial cartilages.

Epicnemial process A triangular projection at the proximal end of the tibiotarsus of certain birds, often united with the patella.

Epidermal Referring to the outer skin layer, of ectodermal origin; arising from the epidermis.

Epigean Living near the ground. Used to refer to fishes inhabiting surface waters in contrast to subterranean species.

Epipleural spine A spine borne on the upper side of a pleural rib.

Epipubic bone One of a pair of bones extending forward from the pubes in marsupials and monotremes; also called marsupial bones.

Exaspidean Used with reference to birds in which the tarsal envelope is continuous around the outer side of the tarsus, so that the 2 edges meet on the inner side.

External auditory meatus The external ear opening of the skull.

Falcate Curved or sickle-shaped.

Falciform Sickle-shaped. Used to describe the pharyngeal bones of some fishes.

Femoral Pertaining to the femur or thigh; also, one of a pair of shields of the plastron of turtles.

Femoral pores Integumental glands which appear as openings in scales on the undersurface of the thigh of most lizards.

Fibula The bone on the little-toe side of the shank of the hind leg.

Filoplume A hairlike feather with naked shaft and a tuft of barbs at the tip.

Fimbriate Fringed with slender elongate processes.

Firmisternal Term applied to rigid type of pectoral girdle of amphibians, as opposed to arciferous type.

Flight feathers The remiges and rectrices.

Flipper Paddlelike limb of an aquatic turtle or mammal.

Fluke A lateral expansion of the tail of a whale.

Focus The space enclosed by the smallest circulus of a fish scale. The focus is not usually in the center of a scale.

Fontanel An unossified area in a bony surface; in fish there may be a fontanel (anterior or posterior) between bones of the cranial roof.

Foramen lacerum posterius An opening between auditory bulla and basioccipital in mammals; through it pass cranial nerves IX, X, and XI and the internal jugular vein.

Foramen magnum The opening in the rear of the skull through which the spinal cord passes.

Fossa A depression, as the nasal fossa in which one of the nostrils lies.

Fossorial Adapted for digging or burrowing in the ground.

Frenum In fishes a bridge of tissue that connects upper lip and snout and prevents free movement of the upper lip.

Frontal A paired dermal bone of the dorsal part of the skull, located between nasal and parietal.

Frontal angle The angle formed by the forward extension of the feathering onto the culmen.

Frontal shield A fold of horny skin growing from the base of the culmen or forehead.

Frugivorous Fruit-eating.

Furculum (or furcula) The fused clavicles or wishbone of a bird.

Fusiform Tapering gradually at both ends; spindle-shaped.

Ganoid scale A type of fish scale covered with an enamellike substance. Ganoid scales seldom overlap one another but are arranged like the bricks in pavement.

Gape The opening of the mouth. In birds the gape includes both the cutting edges of the bill and the rictus.

Genital papilla A fleshy protuberance at the genital opening just anterior to the anal fin in some fishes.

Gill A structure covered by a thin epithelium and functioning in the exchange of respiratory gases, chiefly confined to the pharyngeal region in vertebrates. Gills may be internal or external.

Gill raker One of the slender rodlike to blunt knoblike projections from the anterior face of a gill arch. In fish taxonomy the gill rakers of the first arch only are counted; dissection is often necessary to obtain an accurate count which includes all rudimentary rakers.

Gill slit A paired opening from the pharynx to the outside.

Girdle Skeletal elements joining limbs to the body; the pectoral girdle is associated with the front limbs, the pelvic girdle with the hind limbs.

Gonopodium The modified anterior anal ray of male poeciliid fishes, used to inseminate the female.

Gonydeal angle A dorsoventral bend in the gonys.

Gonys The united ventral line of the right and left mandibles of a bird.

Graduated Increasing in length at regular intervals, from the sides toward the midline, as a graduated tail of a bird.

Granular With structure or texture suggesting small granules or grains.

Greater coverts The row of contour feathers covering the bases of the secondaries on the upper side of the wing.

Gular Pertaining to the throat.

Gular plate A large dermal bone on the throat and between the rami of the lower jaws, as in the bowfin (*Amia*).

Gular pouch A naked pouch of skin between the forks of the mandibular rami of certain birds.

Gymnopedic With the young bird hatched naked.

Hallux The first digit of the posterior limb; in birds usually directed backward.

Hemal spine The ventral projection from the ventral bony arch (hemal arch) of a caudal vertebra.

Heterocercal A type of fish tail in which the vertebrae bend upward to enter the upper caudal lobe, which is larger than the lower.

Heterocoelous Said of the saddle-shaped vertebral centrum of a bird in which the anterior face is concave from side to side and convex dorsoventrally, and the posterior face is convex from side to side and concave dorsoventrally.

Heterodactyl Type of bird foot in which there are 2 front toes and 2 hind toes, but differing from the zygodactyl condition in that the first and second toes are the hind ones.

Heterodont Having teeth differentiated for various functions.

Heteropelmous Type of bird foot in which the branches of flexor digitorum longus lead to the third and fourth toes, and the branches of flexor hallucis longus control the first and second toes.

Holorhinal With the nasal opening in the skull (bird) located forward from the frontal-premaxillary suture.

Holotype A single specimen designated by the author or authors of a species to represent that species.

Homocercal A type of fish tail in which the vertebrae are not markedly graduated in size and end at the base of the caudal fin, the lobes of which are about equal.

Homodont Having teeth essentially similar throughout the row.

Hyaline Glassy and translucent.

Hyoid A structure formed from gill-bar remnants, found in metamorphosed amphibians and higher forms and serving to support the tongue.

Hyoid tooth One of a number of teeth on the tongue of fishes.

Hypocleidium A median extension from the symphysis of the 2 halves of the furculum of birds.

Hypoglossal foramen An opening in the basioccipital bone through which the twelfth cranial nerve passes.

Hyposternal process A lateral projection of the lower end of the coracoid of birds.

Hypotarsus The posterior part of the proximal end of the tarsometatarsus of birds, corresponding to the calcaneum.

Hypural The modified terminal vertebra at the caudal base in fishes, often bearing thin bones in the shape of a fan.

Ilium (*pl. ilia*) A paired, dorsal bone of the pelvic girdle. It articulates with the sacrum.

Imbricate Overlapping.

Immaculate Spotless.

Imperforate Said of birds in which the nostrils are separated from each other by a partition.

Incisive foramen An opening, paired or single, in the anterior part of the palate behind the incisors.

Incumbent Said of a hind toe of a bird when it is inserted at the same level as the anterior toes, so that its whole length rests on the ground.

Inflected Bent inward.

Infraorbital canal A lateral line canal in the suborbital bones beneath and behind the orbit in fishes. The canal may be complete or interrupted.

Infraorbital foramen An opening from the anterior face of the orbit to the side of the rostrum in mammals; it is often enlarged into a canal.

Inguinal In the groin region.

Insectivorous Insect-eating.

Insertion The mode or place of attachment of a part. In fish that end of the base of a paired fin (pectoral or pelvic) closest to the most highly developed ray; the anteriormost or uppermost end of the fin base.

Intercalary Having reference to a part or structure which is inserted between the usual elements, as the intercalary cartilage between the ultimate and penultimate phalanges in hylid anurans.

Intercentrum The anterior element of the vertebral column of a given body segment in primitive vertebrates.

Interclavicle An unpaired dermal bone in the pectoral girdle.

Interfemoral membrane A flight membrane of bats located between the hind limbs and often involving the tail.

Intermuscular bone One of a number of bones lying between the myotomes of the dorsal (epaxial) musculature and the ventral (hypaxial) musculature.

Interparietal An unpaired bone in the roof of the skull located between parietals and supraoccipital.

Interpterygoid vacuity One of a pair of openings in the ventral surface of the skull of primitive fish and amphibians; it lies between the pterygoid and the parasphenoid.

Interradial membrane In fishes the membrane connecting adjacent fin rays.

Interramal area   The area between the mandibular rami.

Interscapular   One of the feathers of the dorsal plumage between the scapulars or shoulders.

Intertemporal (*postorbital*)   A paired dermal bone, posterior to the orbit and lateral to the parietal, of the skull of fish and primitive tetrapods.

Intromittent   Used in copulation, as the penis or analogous organs.

Isocercal   Type of tail in fishes in which the terminal vertebrae become progressively smaller and end in the median line of the caudal fin.

Isthmus   In fishes the narrow anterior portion of the breast lying between the lower jaws and to which the branchiostegal membranes may or may not be attached.

Jugal   A dermal bone located between maxilla and squamosal; the middle element of the zygomatic arch in mammals.

Jugular   Pertaining to the throat or neck. In fishes jugular pelvic fins are those located anterior to the pectoral insertion.

Keel   A ridge; an elevated line; a carina.

Labial   Pertaining to the lips.

Labyrinthodont   A primitive amphibian, one of the first land vertebrates, with complex folded teeth; they were extinct by the end of the Triassic.

Lacrimal   A dermal bone of the skull located in front of the orbit; in amniotes the tear duct opens through it.

Lamellate   Composed of thin plates, as the bill of a duck, with its transverse fringelike ridges or lamellae just within the cutting edges.

Lamina (*pl. laminae*)   A layer.

Laminiplantar   With the 2 rows of tarsal scutes meeting behind in a sharp ridge (birds).

Lanceolate   Lance-shaped.

Lateral field   The 2 sides of a fish scale, one dorsal and the other ventral when the scale is in place.

Lateral line (*L.l.*)   That part of a sensory system consisting of a series of tubes in lateral scales between the caudal fin and the upper corner of the gill cleft. Variously developed or absent in fishes; present also in larval amphibians.

Lepospondyl   An amphibian in which the vertebral centra are formed by deposition of bone around the embryonic notochord.

Lesser coverts   The several poorly defined rows of upper wing coverts proximal to the middle coverts.

Liebespiel   Courtship; sex play.

Lining of wing   The under wing coverts, including the axillars.

Lobed   Having flaps of skin along the sides of the individual toes.

Lore   The region between the eye and the bill of a bird, also the corresponding region in fishes and reptiles; by custom spoken of in the plural.

Loreal   Pertaining to the area between eye and bill in birds, or the corresponding region in reptiles or fish. Also, in reptiles, one of a number of paired scales between preocular and nasal or postnasal(s); if a single scale is present between eye and nasal, it should be considered a loreal if its length is greater than its height (width); if the height (width) of the scale is greater than its length, it should be considered a preocular.

Lyrate   Lyre-shaped.

Malar stripe   An area extending backward and downward from the corner of the mouth.

Mandible   The lower jaw or lower bill.

Mandibular ramus (*pl. rami*)   The right or left fork of the mandible.

Mantle   The upper surface of the folded wings plus the back, when uniformly colored as in a gull.

Manus   The hand.

Marginal   A shield of the lateral row of the carapace of turtles.

Marsupium   A pouch.

Masseter   A mammalian jaw muscle which moves the lower jaw forward and upward.

Mastoid region   Region of the exposed mastoid process of the temporal bone, behind the ear.

Maxilla   A bone of the upper jaw lying above (or behind) and parallel to the premaxilla; the upper bill of birds.

Maxillary arch   The bones of the upper jaw taken collectively. In kangaroo rat taxonomy the zygomatic process of the maxilla is called the maxillary arch.

Maxillopalatine A backward and inward pointing process of the maxillary bone.

Melanophore A chromatophore containing the brown-black pigment melanin.

Mental Pertaining to the chin; in reptiles, a single, central scale at the tip of the lower jaw.

Meristic Divided into segments or serial parts.

Mesaxonic foot Type of foot in which the axis passes through the middle digit which is larger than the others and symmetrical in itself.

Mesethmoid A chondral bone separating the right and left nasal passages in certain mammals; a bone of fish, apparently not homologous with the bone of the same name in mammals.

Mesic Having moderate moisture conditions.

Mesosternum Element of the sternum anterior to the xiphisternum, typically bony.

Metacarpal Referring to that region of the hand lying between the digits and the wrist; also, a bone of this region.

Metachrosis The changing of color.

Metaconid An internal, posterior cusp of the lower molar.

Metamerism Segmentation.

Metapodial Pertaining to the hand or foot.

Metapterygoid One of the paired bones of the maxillary arch articulating with the quadrate, symplectic, hyomandibular, and entopterygoid; in various fishes.

Metasternum The posterior part of the sternum; the xiphisternum.

Metatarsal tubercle A tubercle on the sole of the hind foot. In amphibians there are typically 2, a large inner tubercle and a smaller outer tubercle.

Middle coverts The row of upper wing coverts which overlaps the bases of the greater coverts.

Molariform In the shape of molar teeth.

Monophyodont Having a single set of teeth without replacement.

Nape The back of the neck. In fishes that region behind the occiput and before the dorsal fin.

Naris (*pl. nares*) An opening of the nasal cavity; nares may be internal nares (choanae) or external nares.

Nasal A paired dermal bone medial and posterior to the external nares.

Nasolabial groove A groove from the nostril to the upper lip, found in plethodontid salamanders.

Navicular Term used in mammals for the single remaining bone of the 4 centralia of the tarsus of primitive vertebrates.

Neopallium Association center in the dorsal part of the cerebrum of mammals.

Neoteny Condition of having the larval period indefinitely prolonged.

Neural spine The dorsal projection from the dorsal bony arch (neural arch) of a vertebra.

Nictitating membrane A transparent fold of skin which can be moved over the eye.

Nidicolous Reared in a nest (birds).

Notched Cut out along the edge, as the primaries or the tomia of the bill.

Notochord A longitudinal dorsal rod of tissue which in lampreys and the embryos of higher vertebrates gives support to the body. Persistent in some adult fishes and extending through the vertebral centra.

Nuchal Pertaining to the back of the neck; in turtles, the median anterior shield of the carapace; in lizards, enlarged scales immediately posterior to the head.

Nuptial tubercle A hardened process on the skin of a fish, usually a breeding male.

Obovate Inversely ovate.

Occipital Of or pertaining to the rear aspect of the head; the compound bone surrounding the foramen magnum in mammals; in lizards, one of a number of scales posterior to the parietals and interparietal.

Occipital crest A bony ridge across the skull near the posterior border of the parietals.

Occiput The posterodorsal extremity of the head; in fish the occiput is often the point between the naked head and the scaly nape.

Ocellus A simple eye, also an eyelike spot of color. A fish fin with an ocellus is said to be ocellated.

Olecranal fossa A depression on the anconal side of the distal end of the humerus, for the reception of the olecranon process of the ulna.

Operculum A covering flap, as the gill cover of fishes and larval amphibians or the nasal operculum of birds.

Opisthocoelous With the posterior end concave; used with reference to vertebral centra.

Opisthotic One of the bones of the otic capsule, ventral and posterior in location.

Orbit The bony pocket in which the eye is located.

Origin In fish the anteriormost end of the dorsal or anal fin base.

Osteoderm A bony dermal plate in the skin of crocodilians and certain lizards.

Otic notch A notch in the posterior end of the skull of primitive tetrapods, bounded above by the tabular and below by the squamosal; it represents the spiracular slit of fish.

Ovoid Egg-shaped.

Oviparity The egg-laying habit.

Palatine One of a pair of dermal bones of the anterior ventral surface of the skull, frequently just behind the internal naris.

Palmate Said with regard to bird feet in which the anterior toes are connected by webs.

Papilla A small fleshy protuberance; in fish, lips bearing papillae are said to be papillose.

Paraoccipital process A projection of the occipital bone in the region of the mastoid process.

Parasphenoid An unpaired dermal bone of the ventral surface of the skull of bony fishes and lower tetrapods.

Paraxonic foot Type of foot in which the axis passes between the third and fourth digits, which are almost equally developed.

Parietal Pertaining to the walls of a cavity; pertaining to the region of the parietal bones of the skull; a paired dermal bone of the roof of the skull, located between frontal and occipital; a paired scale of the posterior dorsal part of the head of reptiles.

Parotoid gland A glandular swelling behind the eye of various anurans and urodeles.

Parr mark One of a number of vertical dark bars on the sides of some young fishes, particularly the salmonoids.

Pecten A comblike membrane extending from the inner posterior wall of the eye into the vitreous chamber.

Pectinate Comblike; said of a claw with a series of toothlike projections.

Pectoral Pertaining to the chest; in turtles, one of a pair of shields of the plastron.

Pectoral fin (P_1) The uppermost or anteriormost of the paired fins in fishes. Homologous with the arm or foreleg in other vertebrates.

Pectoral girdle The girdle of the anterior limb.

Pelvic fin (P_2) One of a pair of fins in a ventral position well behind the pectoral fins (abdominal position) or below the pectoral fins (thoracic position). Rudimentary rays are included in the pelvic-ray count.

Penicillate Tufted at the tip.

Pentadactyl 5-toed.

Penultimate Next to the last.

Perforate Having an opening between right and left nostrils (birds and mammals).

Peritoneum The lining of the body cavity.

Phalanx (*pl. phalanges*) One of the bones of a digit.

Pharyngeal Pertaining to the pharynx; a toothed bone of the throat region of fishes.

Physoclistic Having no tube connecting the pharynx and the air bladder.

Pileum The top of the head or crown of a bird.

Pineal eye A simple, median, light-sensitive structure (retina, lens, and cornea may be developed) found in primitive fish and amphibians; an analogous organ, the parapineal or parietal eye, is found in *Sphenodon* and many lizards; in lampreys both parapineal and pineal organs form eyelike structures.

Placenta Tissues through which the embryo receives nourishment and respiratory gases from the mother.

Placoid scale A scale with a base of dentine covering a pulp cavity and a superficial layer of hard shiny material; such scales frequently have a single centrally located spine which gives the fish a rough surface; found in sharks, rays, and skates.

Plantar pad An integumental thickening on the sole of the foot.

Planta tarsi The posterior edge of the tarsus.

Plantigrade Type of locomotion in which the entire sole of the foot contacts the ground.

Plastron The ventral shell of a turtle.

Pleurocentrum One of the 2 posterior elements of the vertebral column of a given body segment in primitive vertebrates.

Pleurodont   With teeth attached to the side of the jaw.

Plica (*pl. plicae*)   A fold: lips of fishes bearing plicae are said to be plicate.

Pointed   Said of the wing when the outer primary is longest.

Pollex   The thumb or inner digit of the hand.

Polyphyletic   Having a number of evolutionary origins.

Postauricular patch   A distinctively colored spot behind the ear in various mammals.

Posteriad   In the posterior direction, backward, toward the tail.

Postglenoid   Behind the mandibular fossa of the squamosal bone.

Postmental   A paired or unpaired scale posterior to the mental, in reptiles.

Postorbital process   A projection from the frontal or jugal bone partly separating the orbit from the temporal fossa in mammals.

Postpalatal bridge   The posterior part of the bony palate in certain microtine rodents.

Powder down   A modified down feather which grows continuously, disintegrating at the tip.

Prearticular   A dermal bone of the inner surface of the lower jaw, anterior to the articular; it is present in many submammalian vertebrates.

Precocial   Term descriptive of birds able to run about when hatched.

Predorsal scale   A scale of the rows of scales which cross the midline between the dorsal origin and the occiput in fishes.

Prefrontal   A paired dermal bone of the roof of the skull, primitively at the upper anterior margin of the orbit and lateral and posterior to the nasal bone.

Prehensile   Capable of grasping, as the prehensile tail of an opossum.

Premaxilla   A paired dermal bone, the most anterior bone of the upper jaw.

Preopercle   An L-shaped bone in front of the opercle and with its lower segment directed forward. It may be serrated (with sharp points) or entire (lacking sharp points).

Preoperculomandibular pore   One of the pores opening into the lateral-line canal extending along the preoperculum and onto the mandible.

Preorbital scale   In fish one of a number of scales in front of the eye.

Primary   A large remex growing from the metacarpals or fingers.

Primary coverts   The row of feathers covering the dorsal side of the base of the primaries.

Procoelous   Concave anteriorly; used with reference to centra of vertebrae.

Procoracoid process   A medial extension of the upper part of the shaft of the coracoid.

Proximal   Nearest to the body or place of attachment.

Pseudobranch   A gill-like structure on the inner surface of the gill cover of fish near its upper edge.

Pterygoid   A paired dermal bone of the ventral surface of the skull; it is primitively large and more or less posterior to the palatine bones, but reduced in mammals and forms the walls of the interpterygoid fossa.

Pteryla (*pl. pterylae*)   An area of skin from which a group of feathers grows; a feather tract.

Pubis (*pl. pubes*)   Anterior ventral element (cartilaginous or bony) of the pelvic girdle.

Pygostyle   The fused terminal caudal vertebrae of birds, bearing the rectrices.

Quadrate   The skull bone which articulates with the lower jaw in bony fishes, amphibians, reptiles, and birds; in mammals it has become an ear ossicle, the incus.

Quintocubital   Possessing the fifth secondary; eutaxic.

Radius (*pl. radii*)   The bone on the thumb side of the forearm; one of a number of straight, but often incomplete, lines radiating outward from the focus of a fish scale.

Ramus (*pl. rami*)   A branch; a projecting part.

Raptorial   With the feet adapted for grasping prey, as in a hawk or owl.

Rasorial   With the feet adapted for scratching the ground, as in a chicken.

Rectrix (*pl. rectrices*)   A tail quill, exclusive of the tail coverts.

Reentrant angle   An inward directed angle in the side of a hypsodont tooth.

Remex (*pl. remiges*)   A flight feather of the wing, i.e., one of the primaries, secondaries, or tertials.

Reniform   Kidney-shaped.

Reticulation   A network. The tarsus of a bird is

said to be reticulate if there are small irregular scales not arranged in definite rows.

Reticulum A network; the second stomach of ruminants.

Retrorse Curved backward. In the spines of catfishes serrae are retrorse if they are directed toward the base of the spine.

Rhamphotheca The horny skin covering the bill of a bird.

Rhinotheca The horny skin covering the upper jaw of a bird.

Rictus That part of the gape of a bird from the proximal edge of the bill to the angle of the mouth.

Rostral A small paired dermal bone of the anterior end of the skull of fish and primitive tetrapods, located above or posterior to the premaxilla; also the scale, usually undivided, at the tip of the snout of reptiles.

Rostral fold In fish, a fold of skin just behind the upper lip and on top of the snout

Rostral nail The terminal plate of the rhamphotheca of the maxilla

Rostrum The preorbital part of the skull.

Rounded Descriptive of the tail of a bird with the middle rectrices longest and the others progressively slightly shorter.

Rugose Wrinkled.

Sacrum The vertebra or vertebrae which articulate with the pelvic girdle.

Sagittal crest A median, longitudinal ridge on the braincase, often formed by fusion of temporal ridges.

Saltatorial Adapted for jumping.

Sanguivorous Blood-eating.

Scapula A chondral bone in the pectoral girdle; shoulder blade.

Scapular One of the feathers, often elongated, growing from the shoulders.

Schizognathous A type of bird palate in which the maxillopalatines are free from each other.

Schizopelmous Type of bird foot in which the 2 flexor tendons are separate, with flexor hallucis going to the hind toe only.

Schizorhinal With the nasal openings in the skull of the bird extending proximal to the frontal-premaxillary suture.

Sclera The outer hardened layer of the eyeball.

Scute A horny shield or scale of a reptile; also a scale of some fishes.

Scutellate Said of the tarsus of a bird when large scales or scutella are arranged in definite series.

Secondary A large quill growing from the ulna; a cubital.

Sectorial Cutting.

Selenodont Type of tooth in which the enamel of the crown is in the form of longitudinal crescents.

Semipalmate Having the bases of the anterior toes joined by partial webs.

Septomaxilla A small dermal bone located at the back edge of the external naris or within the nasal cavity.

Serra (*pl. serrae*) A sawlike organ or structure, or one of the teeth of such a structure. A feature is serrate if it is toothed on the edge.

Sigmoid Roughly S-shaped.

Sinuate With a wavy edge.

Snout That part of a fish head from the anterior margin of the orbit to the tip of the upper lip.

Spatulate Shaped like a spatula (thin, flat).

Speculum A distinctively colored patch on the secondaries of ducks.

Spermaceti Whale oil.

Spermatophore A packet enclosing a number of spermatozoa; found in certain salamanders.

Sphenethmoid A chondral bone of lower vertebrates which presumably corresponds to the presphenoid and orbitosphenoids of mammals.

Sphenoidal fissure An opening in the alisphenoid bone of the mammal skull for passage of eye-muscle nerves.

Sphenopalatine foramen An opening on the lateral face of the palatine bone.

Spina sterni communis An anterior projection of the sternum, formed by the fused tips of the internal and external spines.

Spina sterni externa An anterior projection of the sternum, ventral to the coracoid groove.

Spina sterni interna An anterior projection of the sternum, dorsal to the coracoid groove.

Spine An unsegmented ray in a fish fin, usually stiff, but sometimes flexible. Also, any of several bony processes on the bones of the operculum.

Spiracle The excurrent opening of the gill chamber of a tadpole; blowhole of a whale; an opening on the upper back part of the head of an elasmobranch or some ganoids, representing the first postoral gill slit.

Squamation   Scale arrangement.

Squamosal   A dermal bone forming part of the posterior skull wall; in mammals the lower jaw articulates with it.

Square   Said of the tail of a bird when the rectrices are all equal in length.

Standard length   The distance from the snout tip to the end of the hypural plate in fishes.

Stereospondyl   A degenerate labyrinthodont (Triassic chiefly) characterized by vertebrae with reduced pleurocentra and enlarged intercentra.

Sternum   The breastbone.

Striate   Streaked or striped.

Subarticular tubercle   In amphibians, a small prominence on the ventral surface of a digit, located at the junction of 2 phalanges.

Sublacrimal-maxillary ridge   A ridge on the side of the rostrum above the maxillary tooth row in mammals.

Submalar stripe   Lowermost of the 3 dark facial stripes of chipmunks.

Subopercle   The lower posteriormost bone of the gill cover of fish.

Suborbital   One of a series of small bones lying just beneath the eye of fishes; below the eye.

Subquadrate   Roughly square in form.

Superciliary   Above the eye; a spot or marking above the eye; one of several scales above the eye of lizards.

Supramarginal   One of several small shields between the marginal and costal shields of the carapace of turtles.

Supramaxilla   A small movable bone lying near and adherent to the upper posterior edge of the maxilla in fishes.

Supraoccipital   The uppermost bone of the occipital complex, located above the foramen magnum.

Supraoccipital process   A flattened process extending upward and backward from the supraoccipital bone in fishes.

Supraorbital canal   One of a pair of lateral-line canals lying between the eyes and on the top of the head of fishes.

Supraorbital gland   A nasal lacrimal gland, which in many water birds is enlarged and lies in a depression on the dorsal side of the frontal above the eye.

Supratemporal canal   A lateral-line canal extending from the main lateral line across the occiput to the lateral line of the other side. Sometimes incomplete.

Sympatric   Living in the same region.

Symphysis   Junction between 2 bones in the median plane of the body, chiefly applied to the pubic symphysis and the mandibular symphysis.

Symplectic   A small bone lying at the junction of the hyomandibular, quadrate, and metapterygoid in certain fishes.

Syndactyl   Type of foot (birds, kangaroos) in which the anterior toes are grown together for most of their length and possess a common sole.

Synpelmous   Type of bird foot in which the flexor digitorum and flexor hallucis join before sending branches to the toes.

Syrinx   The vocal organ of birds, located at the lower end of the trachea, at the upper end of the bronchi, or at the junction of trachea and bronchi.

Tabular   A paired dermal bone, lateral to the postparietal, of the skull of fish and primitive tetrapods.

Tail coverts   The small feathers covering both dorsal and ventral bases of the rectrices.

Tapetum lucidum   A variously constructed mechanism in the eyes of animals making possible the return of light to visual cells which it has already traversed.

Tarsus (*pl. tarsi*)   The ankle; the shank (tarsometatarsus) of a bird's leg, above the toes.

Taxon   A category of classification.

Temporal ridge   One of a pair of ridges for attachment of the upper border of the temporal muscle; the 2 ridges sometimes fuse to form a sagittal crest.

Terete   Tapering, with circular cross section.

Terrestrial   Inhabiting the ground, as opposed to aquatic, arboreal, etc.

Tertial   One of the last 3 or more secondaries growing at the proximal end of the ulna of the bird; they are often somewhat longer than the other secondaries and distinctively colored.

Tessellated Marked with little checks or squares, like mosaic work.

Tetradactylous 4-toed.

Tetrapod A 4-footed animal.

Tibia The bone on the great-toe side of the shank of the hind leg.

Tibial bridge A bony bridge passing over the tendons on the anterior face of the distal end of the tibiotarsus of birds.

Tine A prong of an antler.

Tomium (*pl. tomia*) A cutting edge of the bill of a bird.

Toothed Said of a bird bill with 1 or 2 toothlike angles along the edge.

Totipalmate Said of a bird foot in which all 4 toes are joined by a web.

Tragus The central lobe anterior to the pinna of the ear.

Trans-Pecos That part of Texas lying west of the Pecos River.

Trenchant Cutting.

Tricuspid With 3 points; a tooth with 3 points.

Trifid 3-pronged.

Triseriate With 3 stripes.

Trochanter Any of several processes on the proximal part of the femur for muscle attachment.

Truncate Terminating abruptly as if the end were cut off.

Tuberculate With tubercles; with cusps in mammal teeth.

Tympanum The eardrum.

Ulna The bone on the little-finger side of the forearm.

Umbilicus The navel.

Uncinate Hooked or bent at the tip, as the bill of a hawk.

Uncinate process A backward projecting process of the vertebral ribs of birds and certain reptiles.

Unguiculate Provided with claws.

Unguligrade Type of locomotion in which only the tips of the digits contact the ground.

Unicuspid A single cusped tooth. In shrews applied to a row of 3–5 small teeth behind the enlarged pair of incisors.

Urodele A salamander.

Urostyle A rodlike bone, representing a number of fused vertebrae, making up the posterior part of the vertebral column in all anurans.

Vent The posterior opening of the digestive tract.

Venter The belly; the lower side of an animal.

Vermiculate With irregular wormlike markings.

Vermiform Having the shape of a worm.

Villiform Having the form of minute fingerlike processes. The teeth of fishes are said to be villiform when they are slender, equal in length, and close-set in bands.

Viviparous Producing living young.

Vomer A paired or unpaired dermal bone of the anterior end of the ventral side of the skull, just behind the premaxilla.

Weberian apparatus Modified anterior vertebrae joining the ear with the air bladder in suckers, minnows, catfishes, and characins.

Wing covert A wing feather other than a remex.

Wing tip The length by which the longest primary exceeds the outermost secondary, in the folded wing.

Xeric Characterized by minimal moisture conditions; dry.

Xiphisternum The most posterior element of the sternum, typically cartilaginous.

Ypsiloid cartilage A cartilage extending forward from the pelvic girdle in the ventral body wall of certain salamanders.

Zygapophysis One of a pair of processes for articulating a given end of a vertebra with the adjacent vertebra.

Zygodactyl Yoke-toed, i.e., with 2 front toes (the second and third) and 2 hind toes (the first and fourth).

Zygoma (*pl. zygomata*) The zygomatic arch, the bony bar lateral to the orbit and temporal fossa in mammals, formed by maxilla, jugal, and squamosal.

Zygomatic plate of maxillary That part of the maxillary bone entering into formation of the zygomatic arch.

Page references in **boldface** *type indicate illustrations.*